ENGINEERING
EVALUATION
OF ENERGY
SYSTEMS

Arthur P. Fraas
Consulting Engineer

ENGINEERING EVALUATION OF ENERGY SYSTEMS

McGraw-Hill Book Company

New York   St. Louis   San Francisco   Auckland   Bogotá
Hamburg   Johannesburg   London   Madrid   Mexico   Montreal
New Delhi   Panama   Paris   São Paulo   Singapore
Sydney   Tokyo   Toronto

Library of Congress Cataloging in Publication Data

Fraas, Arthur P.
    Engineering evaluation of energy systems.

    (McGraw-Hill series in energy, combustion, and environment)
    Includes bibliographical references and index.
    1. Power (Mechanics) 2. Renewable energy sources.
3. System analysis. I. Title. II. Series.
TJ163.9.F67   621.4   81-11783
ISBN 0-07-021758-0   AACR2

ENGINEERING EVALUATION OF ENERGY SYSTEMS

1 2 3 4 5 6 7 8 9 0 KPKP 8 9 8 7 6 5 4 3 2

ISBN 0-07-021758-0

This book was set in Times Roman by Allen Wayne Technical Corp.
The editors were Diane D. Heiberg and Susan Hazlett;
the designer was Elliot Epstein;
the production supervisor was John Mancia.
New drawings were done by Allen Wayne Technical Corp.
Kingsport Press, Inc., was printer and binder.

CONTENTS

PREFACE

The 1970s have seen an increasing realization that our supplies of cheap fossil fuels are dwindling inexorably. Spurred by major changes in the cost and availability of fuels, many programs for conserving energy and for increasing the efficiency with which the energy in fuel is converted into electric power have been initiated or proposed. One consequence of this has been a strong interest in advanced power plants as a means of obtaining increased efficiency, reduced capital cost, and the capability of burning coal without producing serious emissions. In addition, there has been a strong interest in unconventional energy sources—solar, wind, geothermal, and fusion, for example. To convert the energy from these various sources into electricity, many different advanced power conversion systems are being advocated. These include very high temperature gas turbines, advanced versions of combined gas turbine and steam cycles, bottoming cycles, special cycles employing unusual fluids, several varieties of magnetohydrodynamic systems, and direct conversion schemes such as fuel cells. In each case the advocates of a system paint a glowing picture of what their system will be able to do if they are given x millions of dollars for its development, but they usually do not delineate the vital development problems, particularly those associated with the limitations imposed by materials. Thus it is exceedingly difficult to discriminate between the claims made by the various advocates of competing systems. The interrelationships of thermodynamics, fluid flow, heat transfer, machine design, metallurgy, corrosion chemistry, erosion, combustion, fuel characteristics, instrumentation, control, operating life, environmental effects, and costs are so involved, and so subtle, that even the experts often become confused.

The author is fortunate to have worked on almost every type of power conversion system. This has led to many invitations to lecture on advanced power plants, including a graduate course in the subject in 1976. This experience showed that many graduate students and practicing engineers are interested in getting a broad perspective on the whole range of power conversion systems. It showed their need for a text that would not only relate the dry abstractions of thermodynamics, fluid mechanics, and heat transfer to the performance characteristics and costs of actual plants but would at the same time delineate the problems that have been encountered in the developmental efforts and thus give a basis for judging the feasibility of each concept.

The first ten chapters of the book, for which a basic background in thermodynamics, fluid mechanics, and heat transfer is assumed, describe the engineering principles employed in electric power plants. Each of the next eleven chapters treats a particular advanced energy conversion system or related set of systems. The last chapter compares the efficiencies, costs, operating characteristics, and developmental status of the various types of advanced energy conversion systems.

In preparing the text, it has been necessary to compromise between the brevity that some readers would prefer and the thorough, comprehensive treatment desired by others. Inasmuch as each of the chapters could be expanded into a full-sized book, and still would not do justice to all aspects of the problems, a line has been drawn at what seemed to be the key essentials to providing a clear picture of the problems and status of each type of power plant. Selected references have been included for those who wish to obtain more detailed information.

This book differs in a basic way from most engineering texts because the subject is basically different: the energy situation is studded with controversial environmental questions and political issues. In fact, one friend in the field stated that never in his life "except in the fields of religion and politics" had he ever seen "such divergent views held with such great conviction and so passionately defended." The problems involved cannot possibly be understood unless this situation is appreciated, and no text attempting to treat the problems can put them in good perspective without some mention of these subtleties and subjective aspects. In this vein it is pertinent to mention that by a deft selection of quotes from the author's papers and reports one could make a plausible case for the thesis that the author is personally prejudiced in favor of virtually any of these systems. It is hoped that this detailed familiarity with the full range of problems has made possible a reasonably good degree of objectivity. In any event, some personal notes and anecdotes have been included occasionally to convey at least a suggestion of the intensely human elements involved and their influence on what might otherwise appear to be purely technical problems.

ACKNOWLEDGMENTS

The author is deeply indebted to many people not only for help in the direct preparation of the manuscript for this book, but even more for their contributions to the 45 years of background on which the text has been based. In the earlier years the author was fortunate in learning about fluid flow problems through close associations with John Weske, Theodor Troller, Theodore Theodorsen, and George Wislicenus. Their patience and insights provided an invaluable background that led to work at the Oak Ridge National Laboratory. The work there engendered a special debt to Alvin Weinberg, who over the years not only asked the author to take a close look at an exceptionally wide variety of power plants, but also in each case posed incisive questions and created the staff and atmosphere that made the Oak Ridge National Laboratory an exceptionally fine place to work on such projects. A particular advantage was a close working relationship with other government laboratories, industry, and universities; thus the author is further indebted to many, many engineers, physicists, scientists, and economists, not only at ORNL but also throughout the United States and abroad, who have given generously of their time and expertise in work on problems of mutual interest.

The detailed preparation of this text has been helped enormously by friends and colleagues who have reviewed portions and contributed criticisms and suggestions and often provided additional material. David J. Rose reviewed the initial, very rough draft and gave invaluable advice on both the scope and presentation. Millard M. Myers contributed a wealth of material from his personal library along with his helpful reviews. The author further deeply appreciates the contributions of the other friends who have helped with reviews, notably Truman D. Anderson, Sidney J. Ball, Van D. Baxter, Lochlin W. Caffey, Ernest L. Daman, Stephen J. Ditto, Thomas Dolan, Robert H. Eustis, Judy H. Ferguson, Richard W. Foster-Pegg, Arthur G. Fraas, Lewis M. Fraas, Ronald L. Graves, William O. Haring, Robert S. Holcomb, David Hume, Ritchey Hume, Jr., William D. Jackson, John Keefer, Ralph Lane, Edward K. Levy, John W. Michel, George P. Palo, Richard Y. Pei, Alfred M. Perry, Garland Samuels, Eugene E. Stansbury, James T. Tanner, Joseph Turnage, H. Joe Wilkerson, and Hsuan Yeh.

Copy for the greater part of the illustrations has been provided by a wide variety of organizations; special thanks are due these organizations and the personnel who kindly took the time to supply the author with reproducible copy and permission for its publication.

Arthur P. Fraas

1

ENERGY PRODUCTION, UTILIZATION, AND RESOURCES

The prime objective of this book is to give a broad perspective on advanced energy conversion systems. Since the past provides a basis for understanding the present, it seems best to begin with a brief overview of the use of energy through history.

HISTORICAL BACKGROUND

Looking back into the history of the human race, one finds that probably the most important element in the transition from *Homo erectus* to *Homo sapiens* was the mastery of the use of fire some 300,000 years ago. Human beings began to employ the energy of domesticated animals around 7000 years ago and began to use fire to smelt metals around 5000 years ago. The first inanimate source of energy used to carry out mechanical work of which we have a written record is a waterwheel described in a Byzantine manuscript of the seventh century. (A model built from that description is in the Technische Museum in Vienna.) The crusaders found both waterwheels and windmills in fairly widespread use in the eastern Mediterranean area and brought the concepts utilized there back to Europe, where somewhat different types of these machines were already in use, apparently independently invented. By the twelfth century a fair number were employed for grinding grain and for operating both bellows and hammers in forge shops. They became widely used in the subsequent centuries; e.g., about 120 windmills were in use around Ypres in Flanders in the thirteenth century. In fact, the first machine in the history of the world to be produced in a standardized design was a windmill; over 1000 were built to pump water for the first major land drainage and reclamation project in the Netherlands in the fifteenth century.[1] These machines were built of wood with huge oak shafts for the rotors that weighed well over a ton; the marvelous art of the early millwrights in raising these into position awed the populace. Note the wide use of the root word "mill" in our language for machinery—e.g., sawmill, paper mill, mill building, etc.—derived from the term for those first machines that came into general use for grinding or milling grain.

Even when well greased, the wooden bearings for these mills presented wear problems, as did the cross-grain stresses imposed on the rotor hub by the arms of the windmill.[2] The development of water-powered blast furnaces and related facilities for making artillery provided foundries capable of making large iron castings and water-mill-driven lathes capable of turning and boring them. Thus by the latter part of the seventeenth century cast-iron hubs and bearings were in fairly widespread use in windmills. Concurrently, a shortage of trees for making charcoal to use in smelting iron led to the use of coal in England, where it became the principal source of energy for the metallurgical and ceramic industries.[3] The market for coal led to a need for pumps to remove water from the coal mines, and that in turn led to the development of the Newcomen steam engine almost 300 years ago. Watt's subsequent development of the condensing type of steam engine coupled with his adaptation of the flyball governor (which already had been developed for use in windmills[2]) made possible the spectacular industrial developments of the nineteenth century that were based on the reciprocating steam engine. This in turn led first to the development of the reciprocating internal combustion engine and then by 1880 to the production and use of electricity. The steam turbine began to come into use toward the end of the nineteenth century and the gas turbine in the 1930s. By the middle of the twentieth century, although waterpower was still an important element in our energy industry, windmills had largely faded out, as did the reciprocating steam engine, while the internal combustion engine, the steam turbine, and the gas turbine produced the bulk of our power. The availability of cheap coal, gas, and petroleum largely conditioned these developments.

It became evident by 1950 that our reserves of easily recovered petroleum and natural gas would be severely depleted by the end of the twentieth century. The discovery of nuclear fission during the previous decade and the development of fission reactors led those who examined the problems closely to conclude that nuclear fission offered a fortuitous solution to the

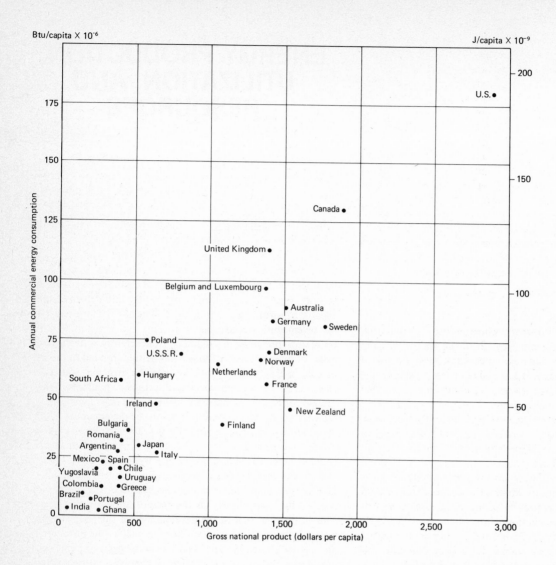

Figure 1.1 Relations between per capita energy consumption by industry and commerce and the per capita gross national product in typical countries in the early 1960s.[4]

energy problem. The subsequent 30 years of intensive development efforts have given us the pressurized and boiling-water fission reactors that by 1979 were providing over 12 percent of the electric power produced by U.S. central stations, and that 12 percent is greater than the total output of U.S. electric utilities in 1946. The cost to the consumer of this power has run 40 to 70 percent of that for fossil fuel plants. By 1985 over 25 percent of the power from U.S. central stations will come from fission reactors built or under construction in 1980. At the same time, shortages of natural gas and petroleum that developed in the 1970s have led to big programs to develop wind, solar, geothermal, and other ''natural'' or ''renewable'' energy sources on which there has been some experimental work since the beginning of the twentieth century but which have been little used to produce useful power.

ENERGY PRODUCTION AND CONSUMPTION

The use of inanimate sources of energy has made possible tremendous improvements in the standard of living. One indication of this is given by Fig. 1.1, which shows the relation between the per capita gross national product and the per capita energy consumption. There is surprisingly little scatter in the points plotted in Fig. 1.1 for different countries and in a similar plot for different time periods.[5] Much of the scatter, in fact, stems from differences in the degree to which a country such as Switzerland is forced by a lack of coal and oil to concentrate on the production of high-value merchandise, such as watches and tools requiring little energy input, and to trade these goods for raw steel, aluminum, and other commodities requiring high-energy inputs. Similarly, the purchase of steel scrap from the United States makes it possible for a country such as Denmark to produce steel with a lower nominal energy input per ton than is required for United States steel production because a much smaller fraction of the Danish production requires reduction from the ore. Agricultural production is also heavily dependent on the input of processed energy; Fig. 1.2 shows both the annual average yield of corn per acre in Illinois and the amount of nitrogen fertilizer employed as a function of year.[6] In countries such as India, the dependence of the ''green revolution'' on energy input is even greater, as indicated by the food shortages of 1979, when lower than normal rainfall in the summer monsoon in India not only reduced the amount of moisture in the soil but also reduced the amount of water stored in dams and hence the hydroelectric power available for driving pumps to draw water from deep wells for irrigation.

Energy Sources and Usage

Both the absolute and the relative amounts of energy obtained from various souces in the United States are indicated graphically in Fig. 1.3. The ordinate in this figure is quadrillion (10^{15}) British thermal units, a widely used parameter commonly abbreviated to *quads*, and easily converted to joules (1 Btu = 1055 J). Figure 1.3 shows that practically all the energy used in the United States has come from petroleum, natural gas, or coal—in that order.

Table 1.1 indicates the relative energy consumption by the principal segments of the U.S. economy.[8] Although not shown as a separate item in Table 1.1, agriculture consumes 2.6 percent of the energy used in the United States, ranking third in energy consumption for basic commodities; only steel and petroleum refining consume greater fractions. The components of the agricultural energy consumption are buried in the industrial and transportation sections of Table 1.1; hence Table 1.2 has been included to give an insight into the energy requirements of this vital sector of the economy. Note particularly in Table 1.1 that U.S. industrial production coupled with com-

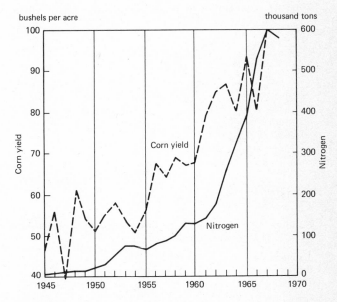

Figure 1.2 Increase in annual corn yield with increased nitrogen fertilizer usage in the 1945–1970 period for the state of Illinois. (*Source: J. H. Dawes, T. E. Larson, and R. H. Harmeson,* Proceedings of the Twenty-Fourth Annual Meeting, Soil Conservation Society of America, Fort Collins, Colo.: *SCSA, 1968.*)

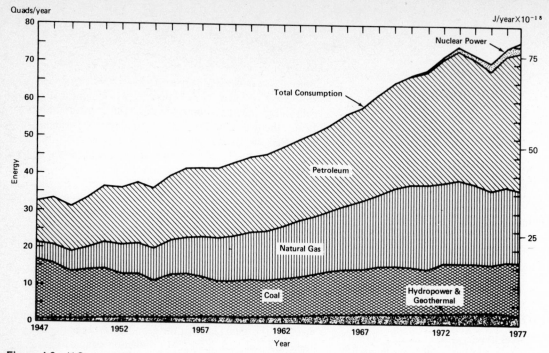

Figure 1.3 U.S. consumption of energy in the 1947–1977 period as derived from the principal sources.[7]

TABLE 1.1 Distribution of Gross Energy Usage in the U.S. in 1977 (Ref. 8)

INDUSTRIAL

	Quads	Percent of U.S. energy
Process steam	9.1	12.0
Direct heat	6.3	8.3
Electric drives/lighting	7.2	9.5
Feedstocks	4.7	6.2
Subtotal	27.3	36.0

TRANSPORTATION

	Quads	Percent of U.S. energy
Roads		
Automobiles	11.2	14.8
Buses	0.1	0.1
Trucks	4.9	6.5
Other		
Railroads	0.8	1.0
Air	1.7	2.3
Ships	1.1	1.5
Pipelines	0.8	1.0
Other	0.6	0.8
Subtotal	21.2	28.0

RESIDENTIAL

	Quads	Percent of U.S. energy
Space heating	8.6	11.3
Water heating	2.4	3.2
Large appliances	2.3	3.0
Lighting	1.8	2.4
Air conditioning	1.0	1.3
Subtotal	16.1	21.2

COMMERCIAL

	Quads	Percent of U.S. energy
Space heating	4.7	6.2
Water heating	0.6	0.8
Large appliances	0.8	1.1
Lighting	2.6	3.4
Air conditioning	1.7	2.2
Miscellaneous	0.8	1.1
Subtotal	11.2	14.8
Total for U.S.	75.8	100

modity transportation used almost half the energy consumed in the United States. At a national meeting on energy in 1972, a leading environmentalist proposed that we might save this energy for consumers by shutting down industry, but she became appalled at her own proposal when asked what might be done about the unemployment that would result or how consumers might be supplied with food and other goods.

The energy consumption situation becomes more complicated if one takes into consideration the form in which the energy is used, i.e., the form of heat or electricity (Fig. 1.4).[9] Many industrial plants, particularly in the paper and chemical industries, have their own power plants to generate both electric power and steam for process heat. If steam turbines are employed, they may discharge the steam from the turbines against a substantial back pressure determined by the temperature at which steam is required for the industrial processes. In other plants, gas turbines serve to generate electricity and the high-temperature exhaust gas is employed to generate steam for industrial processes. These plants are thus able to employ as much as 90 percent of the chemical energy ideally available from the fuel in meeting their electric and process heat requirements. Similarly, in a few cities the central stations employ back-pressure turbines and supply low-pressure steam to district heating systems for heating commercial buildings and apartment houses. Cogeneration of electricity and heat in this fashion is clearly desirable, and this subject is treated in Chap. 17. Unfortunately, district heating systems with cogeneration are not widely used because of side effects of political and legal barriers designed to limit the profits of electric utilities.

The listing for the transportation segment of Fig. 1.4 does not specify the source of the energy employed; actually, virtually all of it in the 1950–1980 period has entailed the use of liquid hydrocarbon fuels in gasoline and diesel engines or gas turbines. Much of the ship propulsion has involved steam turbines (some naval vessels use nuclear heat sources), and a few railroads have been electrified. Still, the great bulk of the U.S. transportation is dependent on petroleum at the time of writing.

Petroleum Usage

As can be seen in Fig. 1.3, petroleum provides a greater fraction of the energy used in the United States than any other source of fuel. The distribution of refinery products to the principal end uses is shown in Fig. 1.5. The distillate fractions are in particular demand, mainly for motor vehicles and heating systems for buildings. Steam boilers for ships, industrial plants, and utilities commonly employ residual fuel oil, the dregs of the refining processes. The availability of distillate fuels, particularly for transportation power plants, is complicated by the limitations inherent in refining processes. Such limitations make it progres-

TABLE 1.2   Energy Consumed in 1974 by Agriculture in the U.S., in Quads, Kilocalories, and as a Percentage of the Total U.S. Energy Budget (18,900 × 10¹² kcal/year)

	Quads	10¹² kcal	%
Farm production			
Crops			
Petroleum	0.75	189	1.0
Electricity	0.012	3	0.02
Livestock			
Petroleum	0.25	63	0.3
Electricity	0.036	9	0.05
Purchased inputs			
Fertilizer	0.55	140	0.7
Petroleum	0.20	49 ⎫	
Feeds and additives	0.10	25 ⎬	
Animal and marine oils	0.036	9 ⎬	0.5
Farm machinery	0.032	8 ⎬	
Pesticides	0.012	3 ⎭	
Total	1.98	498	2.6

Source: Economic Research Service, 1974.

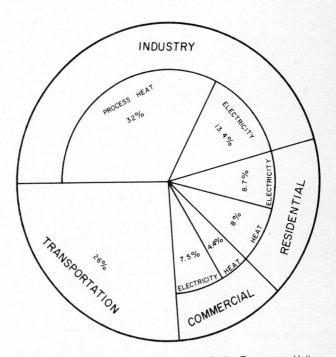

Figure 1.4   Distribution of energy usage in the Tennessee Valley area in 1974. (The energy fractions given for electricity are for the total energy in the fuel used to produce the electricity.) (*Courtesy Tennessee Valley Authority.*)

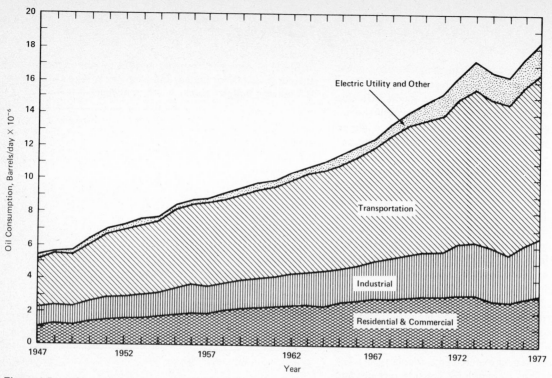

Figure 1.5 U.S. consumption of refined petroleum products by end-use sector in the 1947–1977 period.[7]

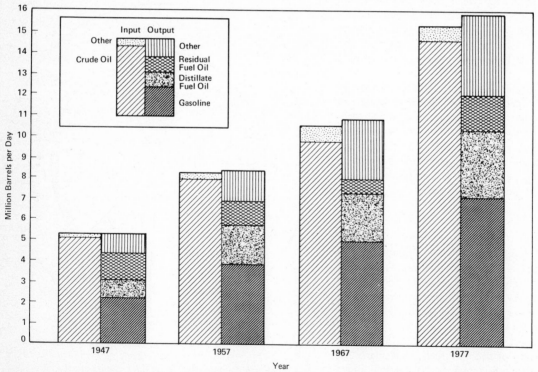

Figure 1.6 U.S. refinery input and output in the 1947–1977 period.[7]

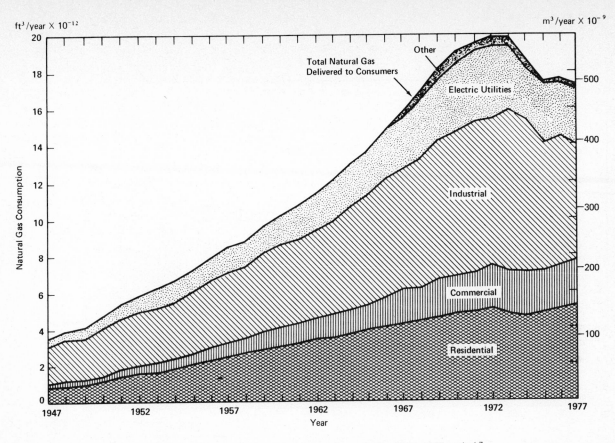

ft³/year × 10⁻¹²

m³/year × 10⁻⁹

Figure 1.7 U.S. consumption of natural gas by the principal types of user in the 1947–1977 period.[7]

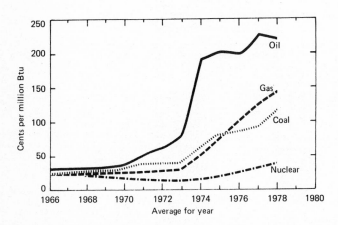

Source: FPC (FERC), EEI, AIF, BLS

Figure 1.8 Costs of fuel as burned in current year dollars, 1966 to 1978.[10]

sively more expensive in both energy and dollars to increase the yield of light hydrocarbons and reduce the amount of heavy residual fuel oil, though, as can be seen in Fig. 1.6, much progress has been made since 1947. Another factor also evident in Fig. 1.6 is the increasing demand for petroleum products for higher-value uses than fuel, e.g., plastics, paint, pesticides, and a host of other petrochemicals. In short, a complex of demands has drastically reduced the availability of distillate fuels for gas turbines in central station power plants.

Natural Gas Usage

Natural gas has been the favored fuel for all the major usage areas in which the equipment is stationary rather than mobile—a point that can be seen readily by comparing Fig. 1.7 with Fig. 1.5. It is clean, and, as shown in Fig. 1.8, it has been lower in cost than coal or oil in most areas until the latter 1970s. However, in the United States the fields from which it can be obtained at low cost have been depleted, and so the amount available began to drop about 1973, forcing cutbacks in its use. This was

Short tons $\times 10^{-6}$ t $\times 10^{-6}$

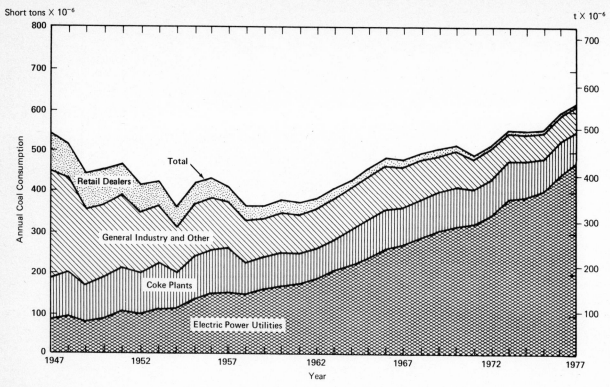

Figure 1.9 U.S. consumption of coal and lignite by end-use sector in the 1947–1977 period.[7]

TABLE 1.3 Sources of the Energy Consumed in the United States for Electric Power Production in the 1920–1978 Period in Percent of the Total and the Capacity of the Electric Power Plants Utilizing these Sources (Ref. 8)

	Source of energy, percent					Total usage, quads	Capacity of electric power plants, 1000 MWe				
	Coal	Oil	Gas	Hydro	Nuclear		Type of plant	1973	1974	1975	1976
1920	52.9	5.6	1.5	40.0	0	19.8	Steam fossil	321	338	353	369
1930	55.5	3.1	6.9	34.5	0	22.3	Steam nuclear	21	32	40	43
1940	54.7	4.2	7.8	33.3	0	23.9	Hydro	62	64	66	68
1950	47.0	10.2	13.6	29.2	0	34.0	Gas turbine	33	40	44	47
1960	53.6	6.1	21.0	19.3	0	44.6	Internal comb.	5	5	5	5
1970	46.1	11.9	24.3	16.2	1.4	66.9	Total	442	479	508	532
1975	44.6	15.1	15.6	15.8	9.0	70.6					
1976	46.4	15.9	14.4	13.9	9.4	74.4					
1977	46.3	17.1	14.4	10.4	11.8	75.8					
1978*	40.9	20.9	12.0	13.2	12.9						

*First quarter data from EIA.

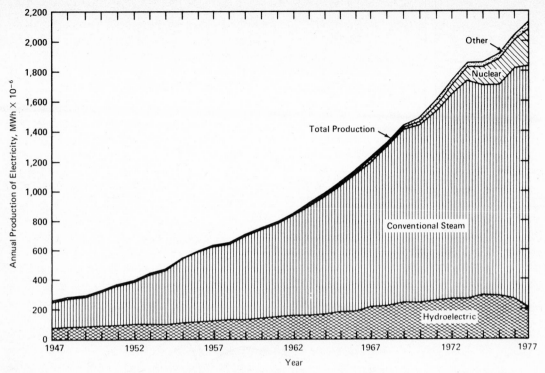

Figure 1.10 U.S. production of electricity from the principal types of power plant in the 1947–1977 period.[7]

particularly true for large utility and industrial boilers, because they could be converted to use coal more readily than any of the other users of natural gas.

Coal Usage

Figure 1.9 shows the marked changes in the use of coal by major sectors of the U.S. economy. Retail sales of coal for residential heating had faded to almost nothing by the 1970s, and consumption by industry had dropped to about one-third of that in 1947. On the other hand, coal consumption by the electric utilities rose rapidly from 1947 to the 1970s, and it is expected to continue to rise in the coming decades.

Electric-Power Production and Use

This text is concerned mainly with power plants for generating electricity; hence, following a review of the main sources of en-

ergy, a closer look at electric-power production is in order. Table 1.3 shows the energy sources employed for electricity generation in the United States through the 1920–1978 period, together with the total plant capacity through the 1973–1976 period for each type of electric-power conversion system. The total annual production of electricity by each type of plant over the 1947–1977 period is shown in Fig. 1.10. Note the major roles played by conventional and nuclear steam plants, the decreasing percentage of the contribution of hydro, and the minor role played by all other types of plant combined. (The low hydroelectric output in 1977 was the result of a severe drought.)

The increasing cost of fuels is leading to the greater use of co-generation systems to supply low-temperature heat, as well as electricity, and thus use of a much larger fraction of the energy in the fuel—a subject treated at some length in Chap. 17. At the same time, if one examines past trends in energy utilization, one finds that electric power represents a continually increasing fraction of the energy used by both industry and consumers.

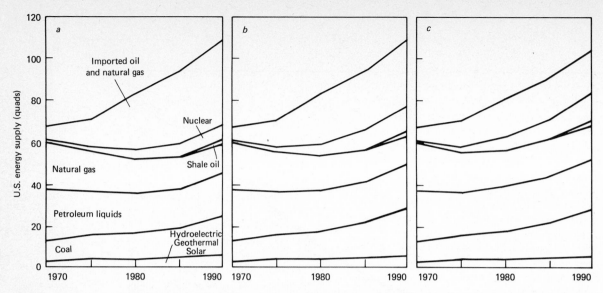

Figure 1.11 The three different projections for U.S. energy supply and consumption shown above are representative of the middle of the range covered by the Congressional Research Service study. The three sets of conditions shown are (*a*) medium demand, low supply, (*b*) medium demand, low oil and gas supply with high coal and nuclear supply, (*c*) low demand, high coal and nuclear supply.[12] (*Courtesy Scientific American.*)

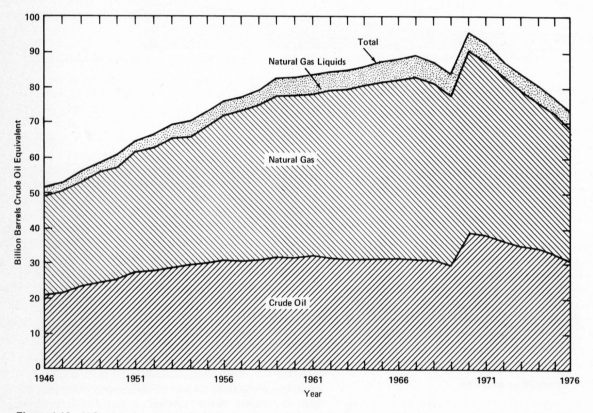

Figure 1.12 U.S. proved reserves of petroleum and natural gas as estimated at the end of each year in the 1946–1976 period.[7]

The use of heat pumps for building heating is increasing rapidly. One of the reasons is that a heat pump has a high *coefficient of performance* (cop) of ~ 2.5.* Therefore, the energy consumption for heating a house with a heat pump using electricity produced by coal at 35 percent efficiency is about the same as directly heating the house with oil or gas at a good residential furnace efficiency of ~ 70 percent, but the cost will be less. Similarly, electric furnaces for melting metals or glass can be built with no stack losses and good thermal insulation to give not only a lower overall energy consumption but also a cleaner product and better quality control. Factors such as these, coupled with consideration for the rapidly escalating cost of fuels, imply that power plants of the future should be designed to produce both heat and electricity and should yield increasing amounts of electricity per unit of heat required for industrial processes and building heating. This in turn implies that progressively higher thermal efficiencies must be obtained from the thermodynamic cycle if the overall energy utilization is to be as efficient as possible.

Projections of U.S. Energy Demand and Supply

Of the many projections of the future energy demand and supply situation that have been prepared in recent years, one of the best available at the time of writing was prepared for the period up to 1990 at the request of Congress and published in November 1977.[11,12] Various projections were made for the U.S. energy consumption using high, medium, and low assumptions for supplies of gas, oil, coal, and nuclear power. The most surprising thing about the different projections of the study is their similarity; it is clear that not only is it difficult to change the existing capital equipment structure, but also it is difficult to change trends in energy production and utilization. Three representative diagrams from this study which illustrate this point are shown in Fig. 1.11. Figures 1.11*a* and *b* are for a medium level of consumption; Fig. 1.11*a* is for a low domestic fossil fuel and nuclear power supply, while Fig. 1.11*b* is for a high domestic coal and nuclear power supply. Figure 1.11*c* is for a low demand and a high coal and nuclear fuel supply. As in all the cases considered, the difference between the assumed total U.S. energy consumption and domestic supplies would have to be made up by imported gas and oil. These projections were based on the assumption that there will be no major international upheaval that would drastically reduce imports and completely change the situation.

* The *cop* is the heat input to the house divided by the electric energy input to the heat pump.

ESTIMATED RESOURCES

Estimates of fuel resources are confusing in that some are for proven *reserves* whereas others take account of total *resources* that include what the estimator thinks will be found by further exploration. They differ even more widely in that some include the total amounts in the ground whereas others cover only those fuels recoverable at some specified price. Thin or low-grade coal seams at depths of over 300 m (1000 ft), for example, are hardly in the same class as thick seams near the surface. Similarly, uranium is a widely dispersed element usually found in concentrations too low to be recoverable. There is an enormous amount of uranium in seawater, for instance, but its concentration is only 3×10^{-3} grams per metric ton (g/t). This gives an amount of fission energy ideally available of 67 kWh/t of seawater, which might give an electric output of 20 kWh/t from a breeder reactor, but in a water reactor only 0.2 kWh/t of seawater. However, just the energy required to pump the seawater through a chemical processing plant, assuming a required pump head of 1.0 atm, would be about 1.5 kWh/t. This, of course, does not include the energy required for the rest of the processing.

Reserves and Resources of Petroleum and Natural Gas

The U.S. reserves of petroleum and natural gas as estimated from data for wells that have actually been drilled are given in Fig. 1.12 as a function of the year of the estimate. Two points are especially disturbing. First, the amount of U.S. reserves has been declining since 1969, and second, by 1976 our petroleum reserves were down to only about six times the annual rate of consumption. This raises questions as to the magnitude of total U.S. resources, together with the availability of oil from foreign sources and the magnitude of world petroleum resources.

Total World Resources

Table 1.4 summarizes estimates of both U.S. reserves and total resources of petroleum and natural gas. These values are broken down sufficiently to provide an insight into the uncertainties involved. Note particularly that two columns list "undiscovered" resources, one column based on an estimated 95 percent probability of discovery and the other on a 5 percent probability that such a large amount will be discovered. A less detailed set of estimates for the world as a whole is presented in Table 1.5. The Table 1.5 estimates do not include natural gas, which is much more difficult and expensive to ship, around 25 percent of the energy being lost in the liquefaction and shipping processes.

The relation between world petroleum consumption and

TABLE 1.4   Estimates of U.S. Reserves, Undiscovered Resources, and Total Resources of Petroleum and Natural Gas in 1975 (Ref. 8)

U.S. RECOVERABLE OIL, quads

Region	Reserves			Undiscovered		Total	
	Measured + indicated	Inferred	Total	(1)	(2)	(1)	(2)
Onshore							
Alaska	58	35	93	35—110		128—	203
Cont. 48	147	83	230	168—371		398—	601
Total	205	118	323	214—470		537—	793
Offshore							
Alaska	1	—	1	17—180		18—	181
Cont. 48	20	15	35	29—104		64—	139
Total	21	15	36	58—284		82—	320
Total	226	133	359	290—737		649—1,096	

U.S. RECOVERABLE NATURAL GAS, quads

Region	Reserves			Undiscovered		Total	
	Measured + indicated	Inferred	Total	(1)	(2)	(1)	(2)
Onshore							
Alaska	33	15	48	16— 59		64—	107
Cont. 48	175	123	298	254—467		552—	765
Total	208	138	346	272—521		618—	867
Offshore							
Alaska	—	—	—	8— 82		8—	82
Cont. 48	37	69	106	27—114		133—	220
Total	37	69	106	43—187		149—	293
Total	245	207	452	332—675		784—1,127	

(1) 19 in 20 chance undiscovered level will exist.
(2) 1 in 20 chance undiscovered level will exist.
Source: USGS-725 (1975)

TABLE 1.5   1975 Estimate of the World's Recoverable Reserves, Undiscovered Resources, and Total Resources of Petroleum in Quads (Ref. 8)

	Reserves	Undiscovered resources (1)	(2)	Total (1)	(2)
Socialist countries	744	404 —	4,060	1,148 —	4,804
United States	298	289 —	872	587 —	1,170
Canada	51	234 —	638	285 —	689
Middle East	2,521	334 —	1,624	2,855 —	4,145
Greater North Sea	128	115 —	463	243 —	591
Other Western Europe	13	38 —	98	51 —	111
North Africa	230	85 —	349	315 —	579
Gulf of Guinea	174	85 —	289	259 —	463
Other Africa	—	17 —	85	17 —	85
NW S. America	145	115 —	289	260 —	434
Other Latin America	81	132 —	553	213 —	634
Southeast Asia	132	106 —	289	238 —	421
Other Far East	43	115 —	697	158 —	740
Antarctica	—	30 —	289	30 —	289
TOTAL	4,560	3,482 —	8,120	8,042 —	12,680

(1) 19 in 20 chance level will exist.
(2) 1 in 20 chance undiscovered level will exist.

resources is shown in Fig. 1.13. Two curves are given, one for a high value for the total world resource and one for a more probable value. Note the similarity in shape between the two curves in Fig. 1.13, with the production rate peaking in 1969 for the United States and around 1990 for the world as a whole if the lower, more probable, estimate of the total resource is used. The implications—however unpleasant—are inescapable; the United States must shift as rapidly as possible from petroleum to other sources of energy.[14,15]

Secondary and Tertiary Recovery of Petroleum

Recovery of petroleum from a deposit normally occurs in stages. Initially, the pressure imposed by the overlying rock combined with the compressibility of the layer of sand or porous rock containing the petroleum may be sufficient to force the oil out at a high rate to give a classical "gusher." As oil is removed from the formation, the forced flow stops and it becomes necessary to pump the oil out of the well. The flow rate through the porous formation to the well slows down with pumping until the percolation rate is too low to justify the cost of operating the well. At this point, 20 to 35 percent of the oil in the original formation will have been recovered.

One factor affecting the rate of recovery from a field is the spacing of the wells—usually about one well to each 20 hectares (ha) (50 acres), depending on the porosity of the sand or rock. When the yield drops to a low level, the recovery rate can be increased by injecting water into wells at the perimeter of the field while drawing oil from wells toward the center. Addition of a detergent to the water can enhance the recovery. The recovery rate can also be increased by injecting high-temperature steam—an especially helpful if not essential step for heavy, viscous oils with their high tar content. The efficacy of these measures for enhanced recovery varies widely with the type of oil and the character of the sand or rock formation. At best, the overall recovery is generally limited to ~ 65 percent of the total oil in the field, the percentage depending on both the characteristics of the field and market prices, which determine the point at which further efforts become uneconomic.

The recovery of natural gas from a field can be greatly increased by injecting water through wells near the field perimeter as cited above for petroleum. The fraction of the gas recovered can be quite high if the gas-bearing stratum is dome-shaped and the water injection rate is kept low enough so that pockets of gas are not trapped by nonuniform water percolation rates. The recovery from such fields can be so high that they make excellent gas-storage reservoirs and, where available near centers of high gas consumption, are often used for that purpose. In such in-

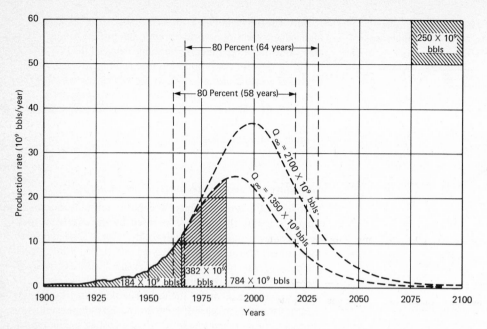

Figure 1.13 Complete cycle of world crude-oil production for two values of the estimated total original world resource.[13]

stances gas is pumped in from pipelines in the summer and withdrawn under peak load conditions in the winter.

Substantial deposits of hydrocarbons are present as very heavy oils or tars in tar sands such as those in California and Canada. The material is often so viscous that it must be recovered by digging it out and feeding it to a retort, and so the cost of the product is relatively high and a substantial amount of energy is consumed in the recovery process. Actual production of hydrocarbons from tar sands is beginning at the time of writing, but it is not yet clear how substantial a contribution to the fuel economy these will be.

Shale Oil Reserves and Resources

In the Colorado-Utah-Wyoming area the United States has immense amounts of oil shale containing hydrocarbons in the form of a waxy material called *kerogen*. A relatively small fraction of the deposits contain as much as 25 percent by weight kerogen, but most of the deposits have a much lower hydrocarbon content. The hydrocarbons can be recovered by destructive distillation, a process that inherently entails substantial heat losses. These constitute a progressively higher fraction of the energy recovered in the form of shale oil as the hydrocarbon fraction in the shale is reduced, and so only the richer deposits can be exploited economically. The shale-oil-recovery operations require large amounts of water—a crucial problem in the water-short areas in the west in which the richest deposits occur. They have the additional disadvantage that the crushed rock residue has a large void fraction and hence a specific volume roughly half again as great as that of the original rock strata, and this presents a major disposal problem. Underground gasification processes have been tried, but have yet to prove satisfactory. As with the underground gasification of coal, they depend on supplying a controlled amount of air to passages cut in the strata so that a portion of the combustible material is burned to raise the shale temperature to the desired level. It is inherently difficult to control the airflow distribution through the maze of passages so that the local shale temperatures run in the relatively narrow range desired; hence the process efficiency tends to fall far short of the value that would be obtainable with a uniform temperature distribution. Because of all these factors, estimates of the amount of shale oil recovery that will prove economically worthwhile vary widely and are commonly quite low, as shown in Table 1.6.

TABLE 1.6 Estimated Magnitudes of the World's Original Supply of Fossil Fuels Recoverable under 1975 Conditions and their Energy Contents (Ref. 13)

Fuel	Quantity	Energy content			%
		Quads	10^{21} thermal J	10^{15} thermal kWh	
Coal and lignite	2.35×10^{12} t	51,900	53.2	14.80	63.78
Petroleum liquids	2400×10^9 bbl	13,600	14.2	3.95	17.03
Natural gas	$12,000 \times 10^{12}$ ft³	12,500	13.1	3.64	15.71
Tar-sand oil	300×10^9 bbl	1,700	1.8	0.50	2.16
Shale oil	190×10^9 bbl	1,050	1.1	0.31	1.32
TOTALS		80,000	83.4	23.20	100.00

TABLE 1.7 Demonstrated Coal Reserve Base by Rank and Potential Method of Mining, Jan. 1, 1976 (Ref. 7)
(Billion Short Tons)

Mining method and area	Anthracite	Bituminous	Sub-bituminous	Lignite	Total*
Underground					
East of the Mississippi River	7	155	0	0	162
West of the Mississippi River	(†)	27	108	0	135
Total underground	7	182	108	0	297
Surface					
East of the Mississippi River	(†)	39	0	1	41
West of the Mississippi River	(†)	8	61	33	101
Total surface	(†)	47	61	34	141
Grand total*	(7)	229	168	34	438

*Sum of components may not equal total owing to independent rounding.

†Less than one-half billion tons.

Note: Includes measured and indicated categories as defined by the Bureau of Mines and U.S. Geological Survey and represents 100% of the coal in place. Recoverability varies between 40 and 90% for individual deposits. Fifty percent or more of the overall coal reserve base in the United States is estimated to be recoverable.

Source: Bureau of Mines.

Coal Reserves and Resources

U.S. coal reserves are among the greatest in the world. Table 1.7 summarizes U.S. coal reserves by type of coal, while Table 1.8 is similar but classifies the coal by sulfur content. Other considerations are the cost of mining and the ash content. Not surprisingly, the coal seams lowest in ash and sulfur and most easily mined have been largely mined out. Hence the average quality of the coal being mined has been dropping in the 1960s and 1970s, while the cost per ton has been rising. Both trends must be expected to continue.

Uranium and Thorium Reserves and Resources

As mentioned above, uranium is fairly abundant in the earth's crust, but it is widely distributed and mostly in low concentrations. Thus the magnitude of the resource depends heavily on how low a concentration can be recovered economically, and this depends on the market price. If the uranium is to be used for fueling reactors that do not breed, only about 0.5 percent of the uranium mined can be fissioned with no recycle (about 1 percent with recycle); the price that can be justified is rather low, and the ore reserves are small. As a consequence, with the

TABLE 1.8   Demonstrated Coal Reserve Base by Sulfur Content and Potential Method of Mining, Jan. 1, 1976 (Ref. 7)
(Billion Short Tons)

Mining method and area	Sulfur range				
	<1%	1.1–3%	>3%	Unknown	Total
Underground					
East of the Mississippi River	26	47	64	26	162
West of the Mississippi River	102	11	8	14	135
Total underground	126	58	73	39	297
Surface					
East of the Mississippi River	7	8	19	7	41
West of the Mississippi River	66	26	3	5	101
Total surface	76	34	20	11	141
Grand total	201	93	93	51	438

Note: As of Jan. 1, 1976, data may not add to totals shown because of rounding. Includes measured and indicated categories as defined by the Bureau of Mines and Geological Survey, and represents 100% of the coal in place. Recoverability varies between 40 and 90% for individual deposits. Fifty percent or more of the overall coal reserve base in the United States is estimated to be recoverable.

Source: Bureau of Mines.

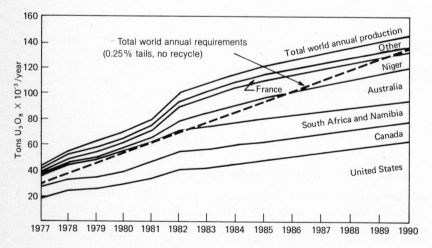

Figure 1.14   Estimated total world uranium requirements and the annual production capability of the principal producing nations.[16]

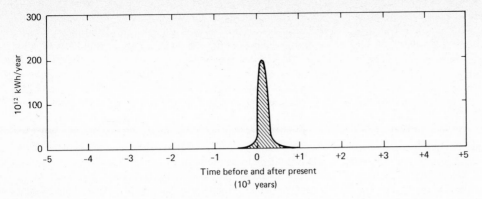

Figure 1.15 Epoch of fossil-fuel exploitation in perspective of human history from 5000 years in the past to 5000 years in the future.[13]

world's rapidly expanding use of fission power plants, it appears likely that a shortage of uranium will develop by the year 2000 unless breeder reactors are brought into use. This situation is shown graphically in Fig. 1.14, which gives estimates for the contributions to the total world production by the principal uranium-producing nations, together with the estimated requirements. The situation can be eased enormously through the use of breeder reactors, which will yield over 100 times as much energy per unit weight of uranium ore. This has the further effect of greatly increasing the price that can be justified for mining and processing the uranium and hence makes it economically attractive to extract uranium from very low grade deposits, thus vastly increasing the effective size of the resource base.

Although less well defined, thorium resources represent an energy source of the same order as uranium ore deposits. Thorium itself is not fissionable, but it can be converted to ^{233}U by absorption of a neutron. The characteristics of thorium are sufficiently different from those of uranium that different design compromises must be made in the reactor, and the reactor types that use thorium to breed fissionable ^{233}U have been little used up to the time of writing. If used in molten-salt or gas-cooled converter reactors, the thorium resources would roughly double the supply of fissionable material. Some perspective on the mining problem is given by typical ore contents. Relatively small deposits of high-grade uranium ores have uranium contents of 2000 g/t, about the same as the copper content of the lowest grade copper ores currently being mined. On the other hand there are huge deposits of Chattanooga shale having uranium contents of 20 to 60 g/t, about the same as the gold content of currently mined gold ore.

SUMMARY

In reviewing the above, a number of points become evident. First, energy is vital to our way of life, representing about 10 percent of our gross national product in 1981. Second, the world's supply of fossil fuels is being exhausted rapidly. In another century people will look back and be struck by the fact that the bulk of the world's fossil fuel, accumulated over hundreds of millions of years, will have been consumed in a scant century, a point shown graphically in Fig. 1.15.

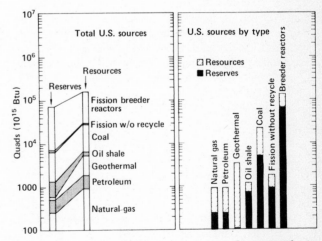

Figure 1.16 U.S. exhaustible energy sources. *Reserves* refer to an assured supply of raw materials that can be converted to energy, while *resources* refer to raw materials that may become recoverable at higher prices, with more exploration, or with new technology.[14]

The civilized world's need for energy is so great and the resources of the familiar fossil fuels are so limited that one must resort to a logarithmic scale to envision the problems. A reasonable set of estimates of both proven reserves and probable total resources are summarized on this basis in Fig. 1.16.[14] Implicit in Figs. 1.11 and 1.16 are some important conclusions. It should be remembered particularly that the United States appears to be much more fortunately endowed with energy resources than most other countries. If the value of U.S. energy consumption could be held at roughly 100 quads per year (as given by Fig. 1.11 for the 1980s), total dependence on coal would exhaust U.S. proven reserves in only 40 years. While the total coal resources might last as much as 200 years, the uranium resources, if burned in fission reactors without breeding, would last only 10 to 20 years. With breeding, the world's supply of recoverable uranium would serve the world economy for around 1000 years. This points up the long-range need for fusion; the reserves of lithium and deuterium would last 10^9 and 6×10^{11} years, respectively, i.e., over 10 times the estimated life of the sun.[15]

There is a widespread notion that solar energy can provide the earth with a virtually inexhaustible supply of energy. This is true for low-temperature heat for buildings, but, as will be shown in later chapters (particularly Chaps. 20, 21, and 22), the costs for solar electric power are inherently so high that it seems unlikely to be the solution. Thus the prudent course appears to be to base national planning on nuclear power sources as the mainstay of the U.S. economy. This is the course being followed in Europe.

PROBLEMS

1. If all the U.S. energy requirements were to be met by coal, and if half the total U.S. coal resources were to prove recoverable, how long would the U.S. coal supply last? Assume that the U.S. energy consumption levels out by the year 2000 at 100 quads per year (Fig. 1.16).

2. Over 25 percent by weight of moisture-free and ash-free coal consists of hydrocarbons. If these were removed by pyrolysis of the raw coal, they could be employed for motor fuel while the remaining char was burned in electric-utility steam plants. If all the fossil-fuel utility plants were fed coal that had been subjected to such treatment, what fraction of the U.S. motor fuel could be supplied from this source?

(See Table 1.1 and Figs. 1.5 and 1.9. Consider 1977 as a typical year.)

3. If a new lightweight and low-cost storage battery were developed for automotive use so that it replaced liquid hydrocarbons for the automotive sector, by what percentage would the capacity of the U.S. electric-utility system have to be increased?

4. For the same conditions as given in Prob. 1, estimate the number of years the U.S. uranium reserves would last if all the U.S. energy were obtained from fission without recycle or breeding. Repeat assuming recycle and breeding. (Use data in Fig. 1.16.)

REFERENCES

1. Spier, Peter: *Of Dikes and Windmills*, Doubleday & Company, Inc., Garden City, N.Y., 1969.

2. Freese, Stanley: *Windmills and Millwrighting*, David and Charles Limited, South Devon House, England, 1971.

3. Harris, J. R.: "The Rise of Coal Technology," *Scientific American*, vol. 230, no. 2, August 1974, pp. 92–97.

4. Cook, E.: "The Flow of Energy in an Industrial Society," *Scientific American*, vol. 224, no. 3, September 1971, pp. 135–160.

5. Linden, H. R.: *Testimony at Public Seminar, Energy Policy and Resources Development*, President's Energy Resources Council, U.S. Department of State, Washington, D.C., Dec. 10, 1974.

6. Heichel, G. H.: "Agricultural Production and Energy Resources," *American Scientist*, vol. 64, January–February 1976, pp. 63–72.

7. Energy Information Administration: *Annual Report to Congress*, vol. III, 1977, *Statistics and Trends of Energy Supply, Demand, and Prices*, DOE/EIA-0036/3.

8. *United States Energy Data Book, 1979*, General Electric Company, Energy Systems and Technology Division, Sunnyvale, Calif.

9. Rodgers, B. R.: Tennessee Valley Authority, personal communication, Mar. 1, 1977.

10. Brandfon, W. W.: "Comparative Costs for Central Station Electricity Generation," *Atomic Industrial Forum Conference on Energy for Central Station Electricity Generation*, Apr. 18, 1978.

11. *Project Interdependence: U.S. and World Energy Outlook through 1990*, Congressional Research Service, U.S. Government Printing Office, November 1977.

12. Rose, D. J., and R. K. Lester: "Nuclear Power, Nuclear Weapons, and International Stability," *Scientific American*, vol. 238, no. 4, April 1978.

13. Hubbert, M. K.: "Survey of World Energy Resources," *Proceedings of Conference on Energy and the Environment: Cost-Benefit Analysis*, School of Nuclear Engineering, Georgia Institute of Technology, June 23–27, 1975, pp. 3–38.

14. Maniscalco, J., et al.: "Civilian Applications of Laser Fusion," Lawrence Livermore Laboratory Report No. UCRL-52349, Nov. 17, 1977.

15. Fraas, A. P.: "Fusion Power—Likelihood and Promise," *Proceedings of Conference on Energy and the Environment: Cost-Benefit Analysis*, School of Nuclear Engineering, Georgia Institute of Technology, June 23–27, 1975, pp. 39–58.

16. "Fuel Resources: DOE Ups U Estimates," *Nuclear News*, vol. 21, no. 7, May 1978, pp. 46–47.

2

COMPARISON OF IDEAL AND ACTUAL THERMODYNAMIC CYCLES

In attempting to appraise the relative merits of the wide variety of energy systems and thermodynamic cycles that are under development or being offered, it is difficult to discriminate between the advantages inherent in a particular thermodynamic cycle and the advantages that may have been conferred on that cycle by the optimistic assumptions of its advocates. Because of this, it seems desirable to examine and compare the characteristics of the various ideal cycles likely to be of interest, together with the basic limitations on both the cycles and the machinery required to implement them. The discussion includes the effects of the principal parameters on the efficiency of the ideal cycles, the various losses associated with heat transfer, fluid flow, and mechanical friction, such vital considerations as the size, weight, and cost of the machines, and some mention of such major design details as runout, clearances, and bearing loads. It is assumed for the purposes of this discussion that the reader is familiar in a general way with the thermodynamics involved; the prime objective here is to highlight the principal relations, characteristics, and limitations on the various cycles to facilitate a critical appraisal of the performance to be expected from actual systems.

IDEAL CYCLES

Perhaps the first point to remember in dealing with heat-engine cycles is that all of them depend on compressing a working fluid, heating it, and then expanding the increased volume in an engine. The greater volume of the hot fluid in the expansion process gives a greater work output from the expansion than required for the compression and hence a net work output from the system. The thermodynamic efficiency is the ratio of the net energy output in the form of useful work to the heat energy input. A good understanding of the overall thermodynamic cycle efficiencies that one may hope to obtain in an actual power plant depends on, first, establishing the thermodynamic efficiency of the ideal cycle employed and the effects on it of the

major operating parameters and then, subsequently, establishing the losses in the actual machine that will detract from the efficiency of the idealized cycle. The first step in this chapter will be to review the ideal Carnot cycle and define its characteristics to provide a standard for subsequent comparison with the characteristics of the principal actual cycles.

Carnot Cycle

One of the greatest advances in all engineering history was made by Sadi Carnot in 1824 when in a flash of intuition he saw that the drop in temperature in the working fluid of a heat engine is analogous to the drop in head in a waterwheel.[1,2] Carnot, like his contemporary Benoît Fourneyron, who developed the theoretical basis for hydraulic turbines (Chap. 3), was inspired by a concern for efficiency—that basic concept which has been a major cornerstone of modern civilization (a concept generally absent in underdeveloped countries). Carnot saw that the flow of energy in a heat engine entailed a drop in temperature, and that the useful work output is proportional to this drop in temperature, just as the useful work output of the waterwheel is proportional to the drop in hydraulic head. From this he rationalized that the ideal efficiency of the heat engine is given by the difference between the thermal energy in the fluid entering the machine and that leaving it, divided by the total thermal energy entering the machine. He was able to use the then-new concept of absolute temperature to reduce his relation to the simple equation

$$\text{Carnot cycle efficiency} = \frac{T_2 - T_1}{T_2} \qquad (2.1)$$

Over 150 years of subsequent work on the thermodynamics of heat engines have demonstrated that this simple expression, involving only the peak and minimum temperatures of the cycle, defines the upper limit on the thermal efficiency of the most efficient cycles that can be devised. This provides an excellent criterion for comparing the potential of the various cycles for which practical machinery has been developed and a basis for estimating the effects of changes in the peak temperature in the cycle.

Carnot devised a thermodynamic cycle that would in principle yield this ideal thermal efficiency. The first step in the cycle is the isentropic compression of a gas, i.e., a compression in which no heat is added to or removed from the working fluid in the course of the work addition during the compression process. Heat is then added at constant temperature and the gas expands, doing some work in the process. This is followed by an isentropic expansion that yields additional external work with no heat interchange to or from the working fluid. The last phase of the cycle is a constant-temperature compression with heat rejection to a heat sink. The thermodynamic processes involved are commonly represented in the form of temperature-entropy (T-S) or pressure-volume (P-V) diagrams such as those of Fig. 2.1. The now-classical temperature-entropy diagram for the Carnot cycle is rectangular (which follows directly from Carnot's definition of the processes). The pressure-volume diagram, while geometrically more complex, is more helpful in some ways because the net work from either the ideal or the actual cycle is directly proportional to the net area of the diagram and the character of the losses is clearly indicated. For the temperature-entropy diagram, the area enclosed by the diagram for ideal cycles with isentropic processes is directly proportional to the useful work, but this is not the case for the actual processes, and hence the diagram cannot be used for quantitative analyses of losses.

Practical Limitations on the Carnot Cycle

Although the Carnot cycle is attractive from the standpoint of the ideal thermal efficiency, it has never been used in a practical engine because of losses that are inherent in any actual machine designed to work on that cycle. It is losses such as these that are the prime concern of this chapter; hence these losses in the Carnot cycle deserve mention. First, as can be seen from the P-V diagram of Fig. 2.1, the area under the compression (or work-input) phases of the cycle is almost as great as the area under the expansion (or work-output) phases. Thus the net work represented by the area of the thin region between them represents a small fraction of the other two areas. Fluid and mechanical frictional losses in any actual engine are proportional to the total work in each compression or expansion process, and their cumulative effect is so large as to reduce greatly the net work from the cycle. A second factor is that the power output from an engine of a given size is directly proportional to the flow rate of the working fluid through the engine, and to obtain a usefully

large output requires rather high fluid velocities, which give little time for heat transfer during compression or expansion processes in either piston or turbine engines. These temperature losses cause large temperature deviations from the ideal thermodynamic processes and cause severe losses in the cycle efficiency. Because of this, it is desirable to employ cycles in which the heat addition and removal processes occur outside the machine involved in putting work into or removing it from the working fluid, or else the heat should be added in a rapid combustion reaction within the working fluid so that no heat transfer between the fluid and a solid surface is required.

Rankine Cycle

The first thermodynamic cycle to be put to extensive practical use, the Rankine cycle, conforms with the precepts just cited and is one of the most important of all thermodynamic cycles. Originally invented by Watt, the thermodynamics of the cycle

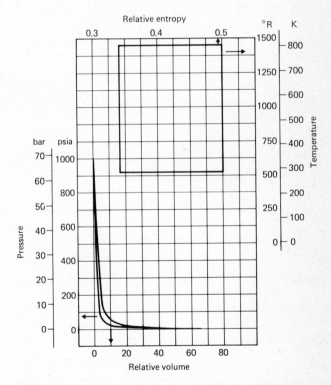

Figure 2.1 Temperature-entropy and pressure-volume diagrams for a Carnot cycle with peak temperatures and pressures of 538°C (1000°F) and 69 bar (1000 psia), respectively, and an initial temperature of 15°C (60°F).

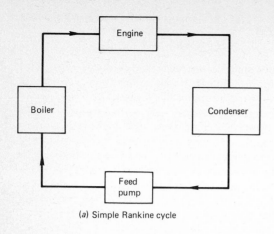

(a) Simple Rankine cycle

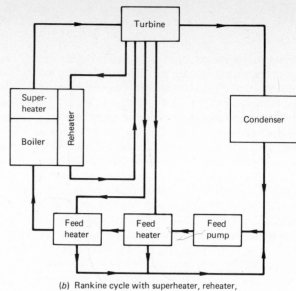

(b) Rankine cycle with superheater, reheater, and regenerative feed-water heaters

Figure 2.2 Schematic diagrams for two typical Rankine cycles, i.e., a simple Rankine cycle with saturated steam fed to the engine (or turbine) and a Rankine cycle with superheat, reheat, and regenerative feed-water heating.

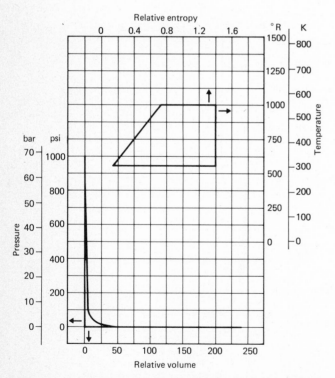

Figure 2.3 Temperature-entropy and pressure-volume diagrams for a simple ideal Rankine cycle with saturated steam supplied at 69 bar (1000 psia) and 285°C (545°F), and a condenser temperature of 100°F.

were subsequently formally delineated by Prof. W. J. M. Rankine of the University of Glasgow. A schematic diagram of the system for the Rankine cycle is shown in Fig. 2.2. The compression phase of the cycle is accomplished with the feed-water pump, heat addition to expand the working fluid takes place in the boiler at constant pressure, the large volume of steam produced in the boiler is expanded isentropically through the engine, and the working-fluid volume is reduced drastically in the condenser at constant pressure for return to the feed pump. Note that both the heat-addition and heat-removal processes for the cycle are carried out in components separated from the feed pump and engine which are responsible for the work-input and -output phases of the cycle. Thus the working fluid can and should flow through the engine rapidly with as little heat transfer as possible. In fact, the earliest steam engines—before Watt invented the condenser—lacked this vital feature; they depended on condensation of the steam in the cylinder itself to produce a vacuum and thus provide the pressure differential that drove the piston. Those early steam engines not only operated very slowly because of the substantial period of time required for heat transfer, they also were afflicted with terribly wasteful heat losses because the large mass of metal in the cylinder walls and piston had to be heated and then cooled in the course of each cycle. With Watt's system, the boiler and condenser each ran continuously at their proper operating temperature, and the engine cylinder was actually lagged with thermal insulation (in the form of wooden blocks) to minimize parasitic heat losses. This reduced the fuel consumption for a given output by a factor of 4.

Temperature-entropy and pressure-volume diagrams for the Rankine cycle are shown in Fig. 2.3. In comparing these diagrams with those of Fig. 2.1 for the Carnot cycle, it can be seen that the Rankine cycle has the major advantage that the work required for the compression process is very small because the volume of the working fluid is reduced by a factor of $\sim 10^4$ in the condenser. In fact, the feed-pump work is so small in relation to the work from the expansion process that the area in Fig. 2.3 corresponding to the feed-pump work is too thin a line to show in the diagram. Thus even if the mechanical and fluid friction losses in the feed pump and engine are quite large, there will still be a substantial net output from the cycle—a situation very different from that for the Carnot cycle.

The efficiency of the Rankine cycle for a given set of boiler and condenser pressures can be improved by superheating, regenerative feed heating, and reheating. The effects of these on the *T-S* diagram are shown in Fig. 2.4. Superheating increases the peak cycle temperature for a given boiler pressure, while

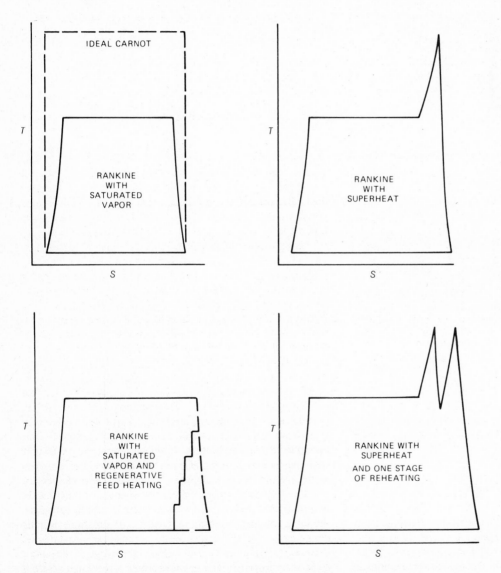

Figure 2.4   Temperature-entropy diagrams for a typical set of Rankine cycles.

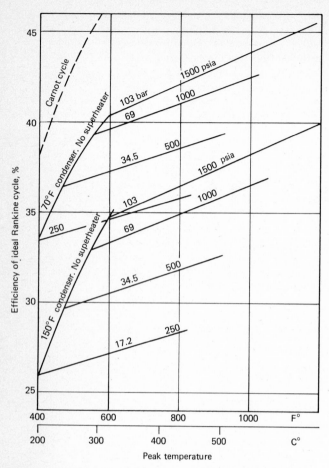

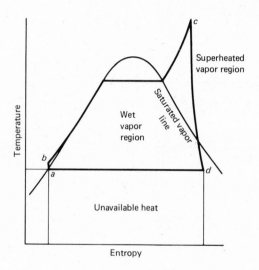

Figure 2.5   Effects of peak temperature on the efficiencies of the Carnot cycle and a series of steam Rankine cycles with various amounts of superheat for two condenser temperatures.[3]

Figure 2.6   Temperature-entropy diagram for a Rankine cycle with superheating showing the manner in which the expansion process (from *c* to *d*) extends into the wet vapor region.

regenerative feed heating and reheating serve to give a better approximation to the ideal rectangular *T-S* diagram of the Carnot cycle.

Effects of Peak Temperature on Rankine and Carnot Cycle Efficiencies

An interesting comparison between the effects of temperature on the efficiencies of the Carnot cycle and a series of idealized Rankine cycles is given by Fig. 2.5. For a given peak boiler pressure, superheating increases the thermal efficiency, but for a given peak temperature, the use of superheating in a Rankine cycle leads to a loss in cycle efficiency. This loss becomes progressively greater as the amount of superheat is increased. Thus the superheated-steam cycle would be less attractive than a saturated-steam cycle were it not for the fact that practical design and operating conditions not only limit the peak pressure but also introduce additional effects, such as turbine bucket erosion by droplets of moisture and losses in turbine efficiency caused by moisture churning. These considerations will be discussed further in Chaps. 3 and 11. It can be pointed out here, however, that a major consideration in the design of the equipment is the pressure that it must withstand. Hence, if a design is limited by the peak pressure in the system rather than the peak temperature, it may be advantageous to employ superheat in order to increase the efficiency of the cycle.

Moisture Limitations

The moisture problem in saturated-steam Rankine cycles stems from condensation of moisture from the lower stages of the turbine, because the shape of the saturated-vapor line on the *T-S* diagram is such that an isentropic expansion from saturated conditions proceeds into the wet region (Fig. 2.6.) This problem can be avoided through the use of a working fluid other than steam in the Rankine cycle. An organic working fluid such as Dowtherm A, for example, has thermodynamic characteristics such that the inclination of the saturation line on the *T-S* diagram leads to vapor expansion into the superheated region as shown in Fig. 2.7. Expansion in the dry region has the disadvantage that a regenerator must be used to avoid a loss in cycle efficiency that would be large for the diagram of Fig. 2.7. The problems associated with this effect will be discussed later in this chapter in connection with the losses experienced in actual cycles.

Effects of Condenser Temperature on Cycle Efficiency

The general effects of condenser temperature on a simple steam cycle can be inferred from the ideal Carnot cycle. An explicit indication of these effects is shown graphically in Fig. 2.8 not only for the ideal Carnot cycle and an ideal Rankine cycle but also for an actual Rankine cycle in which the turbine has an efficiency of 80 percent.[3] These curves show vividly the importance of reducing the condenser temperature in a cycle as much as practicable once the peak temperature and pressure are established by practical design considerations. The effects on power output and efficiency are particularly important if one must consider the design of a system for use in a hot desert environment, where allowances for temperature differences in the heat-transfer equipment will make it difficult to obtain condenser temperatures below perhaps 71°C (160°F), whereas if cooling water is

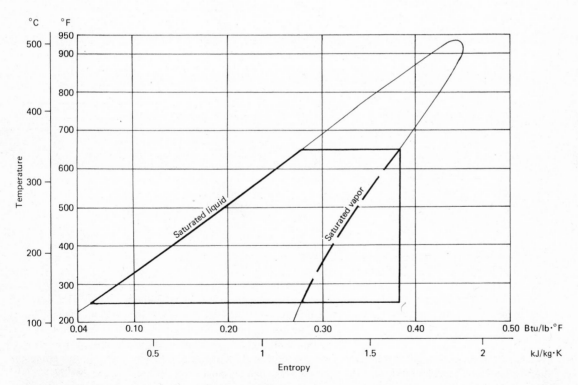

Figure 2.7   Temperature-entropy and pressure-volume diagrams for a simple ideal Rankine cycle with saturated Dowtherm A supplied at 343°C (650°F) and 3.1 bars (45.5 psia) and condensed at 212°C (250°F) and 0.013 bar (0.20 psia). (*Courtesy Oak Ridge National Laboratory.*)

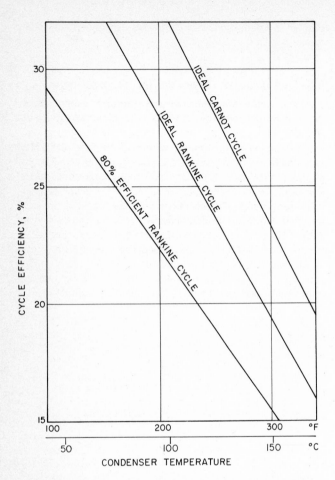

Figure 2.8 Effects of condenser temperature on the thermal efficiency of a series of Rankine cycles supplied with steam at 41.3 bar and 252°C (600 psia and 486°F).[3]

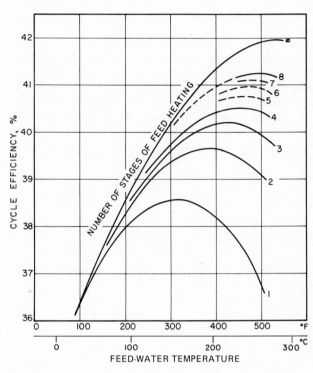

Figure 2.9 Effects of number of stages of feed-water heating and boiler inlet temperature on the thermal efficiency of a typical Rankine cycle with turbine inlet conditions of 69 bars and 482°C (1000 psia and 900°F).[4]

available at a temperature of 10°C (50°F), it is not at all difficult to get a condensing steam temperature of 32°C (90°F).

Effects of Regenerative Feed Heating on Cycle Efficiency

The improvement in the efficiency of a Rankine cycle through the use of multiple stages of regenerative feed-water heating is shown in Fig. 2.9. Steam bled off from appropriate points be-

tween stages in the turbine supplies the heat required to raise the temperature of the feed water from the condenser temperature to the boiling point, and it thus reduces the heat input to the boiler required per pound of steam. In the limit, the efficiency of the saturated-steam Rankine cycle with an infinite number of feed-water heating stages reaches the efficiency of the Carnot cycle. In practice, the size, weight, cost, and complexity of the heat exchangers and piping required for the feed-water heating system is such that six to eight stages of feed heating represent the upper limit of what it has been found practical to employ even in very large plants.

Effects of Reheat on Cycle Efficiency

The second means indicated in Fig. 2.4 for modifying Rankine cycles employing superheat so that they more closely approach the ideal Carnot cycle is the use of reheat.[4,5] Taking the steam out of the turbine and reheating it after about half of its expansion not only improves the efficiency of the cycle but also increases the output of the cycle for a given size of feed pump, boiler, condenser, and regenerative feed heater complex, thus reducing the capital costs per kilowatt in a well-proportioned system. The improvements in cycle efficiency obtainable in a steam system through the use of reheat are indicated in Fig. 2.10.

Brayton Cycle

The Brayton cycle is similar to a saturated-vapor Rankine cycle except that the working fluid is in gaseous form throughout the cycle and so a compressor is used rather than a feed-water pump.[6,7] A schematic diagram of a simple system is shown in Fig. 2.11a, and temperature-entropy and pressure-volume diagrams for the cycle (also known as the Joule cycle) are shown in Fig. 2.12. It can be seen by inspection of the temperature-entropy diagram that the thermal efficiency of the ideal cycle is much less than that of the Carnot cycle because the areas of both the upper left and the lower right corners of the area enclosed by the circumscribed rectangular Carnot diagram are not utilized. As a consequence, the ideal Brayton cycle inherently gives a poorer approximation to the ideal Carnot cycle than is given by the Rankine cycle operating with saturated vapor. On the other hand, practical design considerations permit operation with a

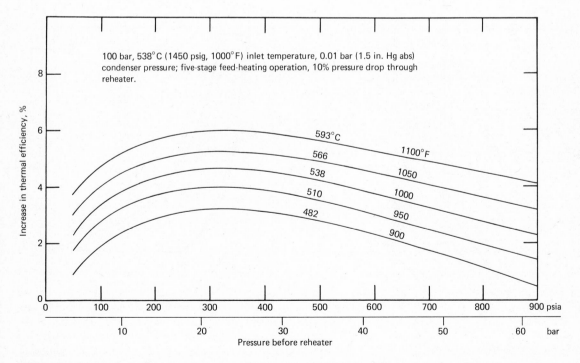

Figure 2.10 Increase in thermal efficiency of steam cycles obtainable through reheat:[5] 100 bar, 538 °C (1450 psig, 1000 °F) inlet temperature; 0.01 bar (1.5 in Hg abs) condenser pressure; five-stage feed-heating operation; 10 percent pressure drop through reheater.

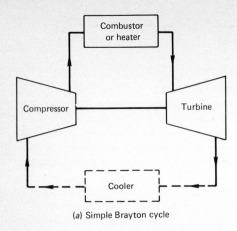

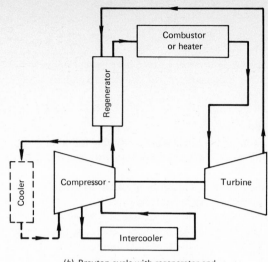

(a) Simple Brayton cycle

Figure 2.11 Schematic diagrams for typical Brayton cycles: (a) simple Brayton cycle and (b) Brayton cycle with regenerator and intercooler. Both cycles could be operated either as open cycles with the atmosphere serving as the heat sink or as closed cycles with a cooler included in the circuit as shown by the dotted lines.

(b) Brayton cycle with regenerator and intercooler

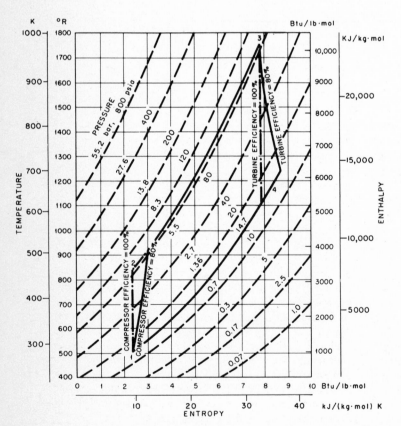

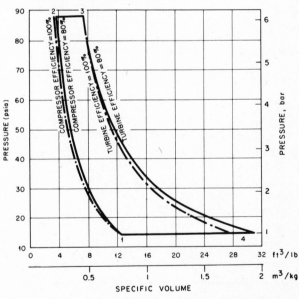

Figure 2.12 Temperature-entropy and pressure-volume diagrams for a simple ideal Brayton air cycle for both 100 and 80 percent efficiencies in the turbine and compressor.[7]

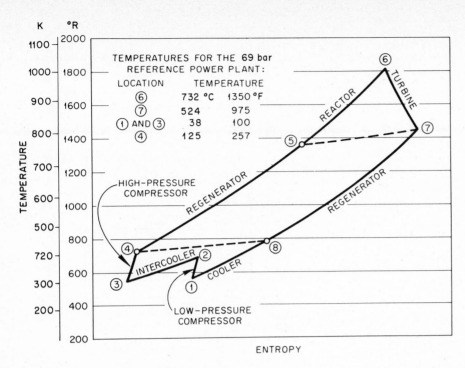

Figure 2.13 Temperature-entropy diagram for a Brayton cycle with a recuperator.[6] (*Courtesy Oak Ridge National Laboratory.*)

higher peak temperature than could be obtained in a steam Rankine cycle, and this helps to make the Brayton cycle attractive. The energy input required for the compressor is much greater than for the feed pump of a Rankine cycle, making the efficiency of the actual Brayton cycle strongly dependent on the efficiencies of the compressor and turbine—a point that will be discussed in a later section on losses.

Brayton Cycle with Regeneration and Intercooling

As in the Rankine cycle system, it is possible to obtain a marked improvement in thermodynamic cycle efficiency through the use of a regenerative heating system that employs a substantial portion of the heat in the exahust gas leaving the turbine to heat the gas leaving the compressor before it goes to the heat source. Figure 2.11*b* shows a schematic diagram for the system, Fig. 2.13 gives a temperature-entropy diagram for the process, and Fig. 2.14 shows the effect on the efficiency of the idealized cycle of the degree to which heat is recirculated in the cycle in terms of

the parameter "regenerator effectiveness." The regenerator effectiveness—a subject treated in Chap. 4—is the ratio of the temperature rise in the gas being heated to the temperature difference between the two gas streams entering the heat exchanger. In practice, the effectiveness is ordinarily 75 to 80 percent, but in special cases system optimization studies may favor a regenerator effectiveness as high as 95 percent. It may be mentioned that if a regenerator is employed in an actual cycle, the best efficiency is obtained with a relatively low pressure ratio, because this reduces the compressor outlet temperature and thus increases the amount of heat that can be recovered from the turbine exhaust and recirculated through the cycle. This is especially important for automotive gas turbines for which regenerators are designed to give an effectiveness of 90 percent (Chap. 12).

Figure 2.13 includes another modification to the cycle to improve its efficiency, i.e., intercooling between stages of the compressor to reduce the compressor power requirement. Intercooling is not often used because the extra fluid-pumping losses and heat-exchanger costs are not justified by the small improvement in cycle efficiency obtainable.

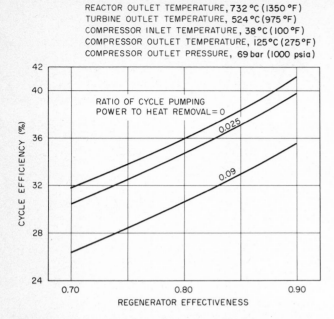

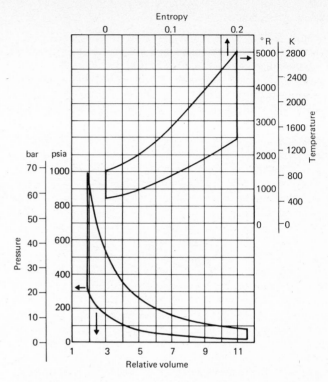

Figure 2.14 Effects of recuperator effectiveness on the efficiency of an idealized Brayton cycle. Reactor outlet temperature, 732°C (1350°F); turbine outlet temperature, 524°C (975°F); compressor inlet temperature, 38°C (100°F); compressor outlet temperature, 125°C (275°F); compressor outlet pressure, 69 bar (1000 psia).[6] (*Courtesy Oak Ridge National Laboratory.*)

Figure 2.15 Temperature-entropy and pressure-volume diagrams for an ideal Otto cycle.

Otto Cycle

This cycle is of interest for comparative purposes because it is used in virtually all modern high-speed gasoline and diesel engines. (Modern diesel engines actually follow an Otto cycle more closely than they do the constant-pressure-combustion classical ''diesel'' cycle employed in the large, old, low-speed diesel engines.[7]) The familiar pressure-volume diagram for an Otto cycle is presented in Fig. 2.15 along with the temperature-entropy diagram. A fresh charge of air is compressed isentropically in the cylinder as the piston rises to top dead center, the air is heated at constant volume by combustion of fuel, and the hot combustion gases expand isentropically, giving a larger work input to the piston than was required for compression. The hot combustion gas is exhausted to the atmosphere so that no cooler is required.

The overall thermal efficiency of the ideal Otto, or constant-volume, cycle making use of an ideal working fluid with constant specific heats is given by

$$\eta_{th} = 1 - \left(\frac{1}{V_1/V_2}\right)^{k-1} \tag{2.2}$$

Thus it is evident that the thermal efficiency of the Otto cycle is a function of the volumetric compression ratio; the peak temperatures for the expansion phase of the cycle are implicitly dependent on the compression ratio for a given inlet temperature and heat addition per unit weight of working fluid. This means that changing the power output by changing the amount of heat added at constant volume each cycle, although it changes the peak temperature, has relatively little effect on the thermal efficiency of the ideal cycle.

Stirling Cycle

Interest in the Stirling cycle has flared up every few decades since the cycle was first proposed in 1815,[8,9,10] and at the time of writing it is being considered for small solar power plants of the sort advocated by some environmentalists for dispersed electric-energy generating systems and for automotive engines. Many engines operating on the cycle have been built. Generally, they have been notable for their quiet operation, but, because they must operate at relatively low speeds to provide time for heat transfer, even with elegantly intricate heat-transfer matrices they have been heavy, bulky, and expensive in comparison with other types of engine of the same power output.

Temperature-entropy and pressure-volume diagrams for the cycle are shown in Fig. 2.16. The cycle normally employs two pistons which operate 90 degrees out of phase in the same or adjacent connected cylinders, one controlling the volume of a hot region and the other controlling the volume of a cold region (Fig. 2.17). The working fluid is sealed in the engine; heat is added or removed through heat-transfer surfaces built into the engine. These include a regenerative heat-transfer matrix through which the fluid flows in passing between the hot and cold regions. As can be seen in the diagrams of Figs. 2.16 and 2.17, the compression portion of the cycle begins with the bulk of the fluid in the cold region. The piston in that region moves to compress the gas at constant temperature (from points 1 to 2 of Fig. 2.16), with the cold heat-transfer surfaces acting to remove the heat of compression. This is followed by a constant-volume increase in pressure and temperature from points 2 to 3 as the working fluid is heated by being forced back through the regenerator heat-transfer matrix as it flows from the cold to the hot region of the engine. Addition of heat to the gas in the hot region continues during the subsequent expansion so that it is carried out isothermally from 3 to 4. Then the working fluid is cooled as it is forced from the hot region back to the cold region through the regenerative heat-transfer matrix at constant volume from points 4 to 1 so that the temperature is reduced to the initial value. Note that ideally these processes are reversible. Hence the thermal efficiency of the ideal Stirling cycle is equal to that of the Carnot cycle, and is independent of the volumetric compression ratio.

Supercritical CO₂ (Feher) Cycle

The supercritical CO_2 cycle was given serious consideration in the latter 1950s as one of the more promising thermodynamic cycles suited to the direct coupling of gas-cooled fission reactors to turbine-generators for central station applications,[11] and it subsequently was advocated for small power plants.[12] The cycle has an advantage over the Brayton cycle in that the compression

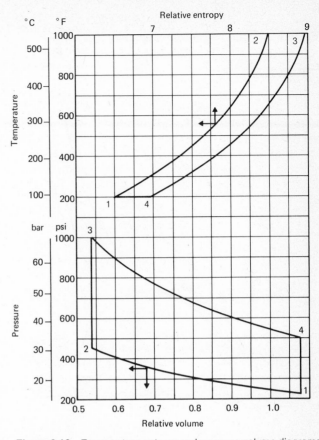

Figure 2.16 Temperature-entropy and pressure-volume diagrams for an ideal Stirling cycle.

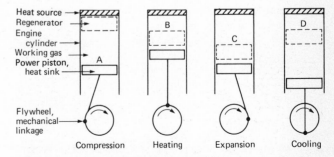

Figure 2.17 Diagrams showing the four phases in a Stirling cycle engine. The displacer piston-regenerator whose position is shown in dotted form may be operated by a small diameter piston rod that passes through a packing gland in the cylinder head at the top or by a more complex mechanism.

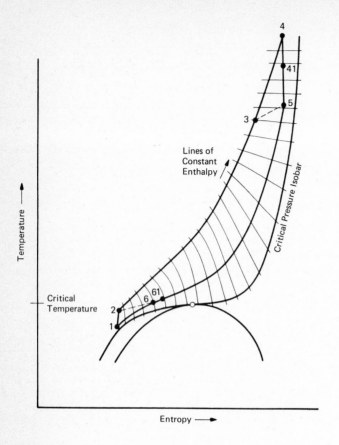

Figure 2.18 Temperature-entropy diagram for an ideal supercritical CO_2 cycle.[13]

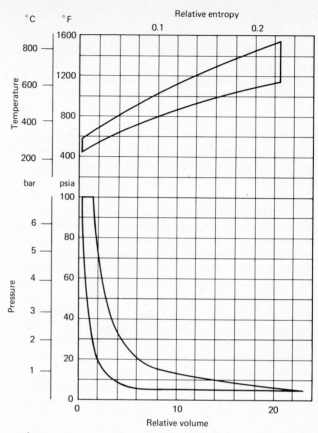

Figure 2.19 Temperature-entropy and pressure-volume diagrams for an ideal $AlCl_3$-Al_2Cl_6 cycle.[14]

operation is carried out in the region close to the triple point where the density of the gas is close to that of the liquid. Thus the compression work is much less than it would be for a Brayton cycle, and, in fact, falls roughly midway between the compression work for a Brayton cycle and that for a Rankine cycle. This effect can be deduced from Fig. 2.18, which shows a temperature-entropy diagram for the supercritical CO_2 cycle.[12,13] To take full advantage of operation close to the triple point, it is necessary to reject heat from the cycle at 38° C (100°F) or less, a condition that is easy to satisfy for some power plant sites but difficult to satisfy for a power plant that must operate under desert, tropical, or even U.S. summer conditions if cooling towers are used.

Dissociating Gas Cycles

A number of cycles have been proposed that are somewhat similar to the supercritical CO_2 cycle in that they fall somewhere between a Brayton cycle and a Rankine cycle. Some of these employ a working fluid that dissociates during the heating phase of the cycle to give a large increase in the volume of gas to be expanded through the engine, and then the molecular species reassociate in the cooler to reduce the volume of gas that must be compressed, thus increasing the net work from the cycle. This effect is shown in Fig. 2.19 for an aluminum chloride cycle in which the gas in the compressor is Al_2Cl_6 whereas that leaving the heater for the expansion engine is $AlCl_3$.[14] Somewhat similar cycles can be obtained using nitrogen oxides,[15] aluminum bromide, SO_2, etc. These cycles are treated further in the discussion of closed-cycle gas turbines in Chap. 14.

COMPARISON OF IDEAL AND ACTUAL CYCLES EMPLOYING RECIPROCATING ENGINES

An actual cycle differs from the corresponding idealized cycle as a consequence of many effects, mostly small, but cumulatively substantial. The major losses in reciprocating engines are those stemming from mechanical friction, heat losses, fluid-pumping-power requirements, and rounding of the corners of the pressure-volume diagram as a consequence of valving and other effects associated with going from one phase to another in the cycle.[3] While these losses are common to all thermodynamic cycles employing a reciprocating engine, their relative importance differs substantially from one cycle to another.

Mechanical Friction

Friction, particularly that between the piston and cylinder wall, represents a major loss, especially in high-speed engines.[7] To avoid large losses associated with gas blow-by past the piston, it is necessary to use piston rings which are forced firmly outward against the cylinder wall by the pressure of the gas acting behind the ring in the bottom of the ring groove. Only a limited amount of lubricant can be made available on the cylinder wall; otherwise, the amount of oil picked up and carried off by the cycle working fluid will be so large that the oil consumption will be excessive. Further, in a closed-cycle system (such as that of a Brayton or Stirling engine) deposition of oil on the hot surfaces of the heater will lead to carbon formation and serious fouling of these surfaces.

The effects of engine speed on the total frictional losses in a typical engine are shown in Fig. 2.20. Tests of many types of engine have shown that the principal parameter is not crankshaft rpm but the nominal piston speed calculated by multiplying the rpm by twice the stroke. The frictional losses increase as nearly the square of the piston speed, and they become excessive for piston speeds above about 15 m/s (3000 ft/min).

Limitations Imposed by Wear Rates

The life of a reciprocating engine is often limited by the rate at which the piston rings and cylinder wall wear in the course of operation. Under favorable conditions, reciprocating steam engines have been known to operate almost continuously for as much as 20 years with scarcely any detectable wear; under un-

favorable conditions in air-cooled aircraft engines operated with cylinder-wall temperatures above 150°C (300°F), the writer has observed severe wear to occur in only 150 h of operation. If there is no problem with either abrasion by dirt in the oil or corrosion (e.g., by moisture containing sulfurous acid from combustion products), the most important single parameter influencing wear is cylinder-wall temperature. Under otherwise favorable conditions the wear rates may be undetectable at cylinder-wall temperatures of 100°C (212°F), substantial at 150°C (300°F), and catastrophic at 163°C (325°F).

Rankine Cycle Engines

Perhaps the most important factor limiting the thermal efficiency of an actual Rankine cycle reciprocating engine is the practicable expansion ratio. If adequate allowances are made for clearances over the top of the piston and for recesses for valves and ports, it is difficult indeed to get a clearance volume less

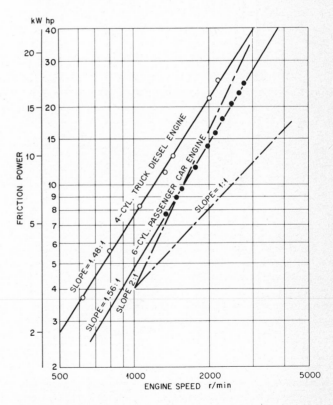

Figure 2.20 Effects of engine speed on frictional losses in reciprocating engines.[7]

than about 2.5 percent of the displacement volume of the cylinder; hence the expansion ratio available in a cylinder can hardly exceed about 40:1.

For a steam-engine cylinder this would correspond to an expansion from 50 psia at 280 °F down to 1 psia at 100 °F, or from 500 psia at 467 °F down to 10 psia at 193 °F. It is, of course, possible to make use of a double- or a triple-expansion engine with two or three cylinders in series. Many such engines have been built, but the blowdown losses between successive cylinders are large (Fig. 2.21), the difference in cylinder size and output are very large, and the upper limit on temperature imposed by cylinder-wall lubrication problems is such that it is hardly practicable to make use of a steam pressure higher than about 500 psia, which gives a saturation temperature of 467 °F. As discussed in the following paragraph, the high heat-transfer

coefficient for condensing steam will heat the cylinder wall to temperatures that are excessive from the lubrication standpoint if a steam pressure above about 500 psia is employed. Thus the thermal efficiency of a reciprocating steam engine is limited in practice to a modest value—around 12 percent. (The efficiency of steam locomotives was generally less than 8 percent because they operated noncondensing.)

Heat losses from the gas to the cylinder wall during the high-temperature portion of the cycle are a problem in any reciprocating engine, but they are particularly likely to be serious in a steam engine if the steam supply pressure corresponds to a saturation temperature higher than the temperature of the cylinder head and the upper portion of the cylinder wall.[3] This comes about because steam engines present a very different problem from gasoline or diesel engines in that the condensing heat-transfer coefficient for steam is extremely high—on the order of 5.7 W/(cm² · °C [10,000 Btu/(h · ft² · °F)] as compared to around 0.057 W/(cm² · °C) [100 Btu/(h · ft² · °F)) for combustion gases. As a consequence, steam will condense during the first portion of the cycle and then, when the pressure of the cylinder drops to a low value toward the latter part of the cycle, the film of condensed steam will reevaporate, thus chilling the inner surfaces of the cylinder head and cylinder wall. Consequent heat losses can easily run 15 percent of the heat in the steam being admitted to the cylinder.[3] To avoid these losses, the design of the engine should be such that the heat losses from the cylinder head and upper portion of the cylinder barrel are kept to a minimum and the saturation temperature associated with the steam admission pressure is no higher than the normal cylinder head operating temperature.[3]

Otto Cycle

In addition to the frictional losses indicated at the beginning of this section, Otto cycle engines are subject to substantial losses in the area of the pressure-volume diagram, as indicated in Fig. 2.22. The most important is the heat loss from the high-temperature combustion gases to the cylinder wall. The cylinder-wall temperature can be kept low enough to assure good lubrication for the piston rings with a total heat loss from the engine corresponding to about one-third of the total heat input into the cycle, but, since much of this heat loss is from the exhaust port, the consequent reduction in cycle efficiency is closer to 10 percent.

A substantial period of time is required to obtain smooth and complete combustion in Otto cycle engines. As a consequence, the left corners of the pressure-volume diagram are rounded and several percent of the available energy is lost (Fig. 2.22). The corners at the right of the pressure-volume diagram are

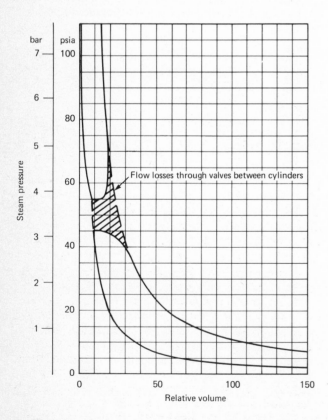

Figure 2.21 Pressure-volume diagram for a double-expansion steam engine showing the losses associated with blowdown from the high-pressure to the low-pressure cylinder for the cycle of Fig. 2.3.

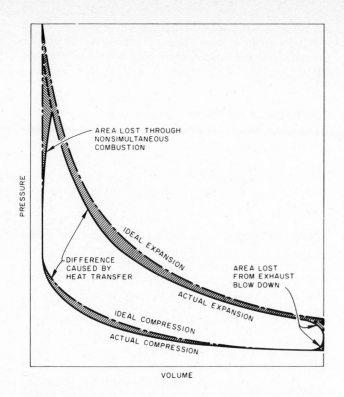

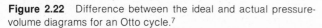

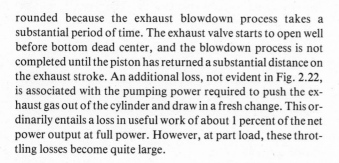

Figure 2.22 Difference between the ideal and actual pressure-volume diagrams for an Otto cycle.[7]

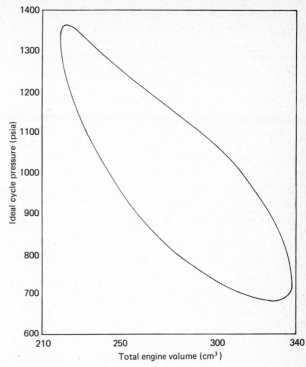

Figure 2.23 Pressure-volume diagram for an actual cycle in a Stirling engine operating at 3600 rpm with a mean working pressure of 70 atm.[8]

rounded because the exhaust blowdown process takes a substantial period of time. The exhaust valve starts to open well before bottom dead center, and the blowdown process is not completed until the piston has returned a substantial distance on the exhaust stroke. An additional loss, not evident in Fig. 2.22, is associated with the pumping power required to push the exhaust gas out of the cylinder and draw in a fresh change. This ordinarily entails a loss in useful work of about 1 percent of the net power output at full power. However, at part load, these throttling losses become quite large.

Stirling Cycle

The factor causing the greatest departure of the actual Stirling cycle from the ideal is that the piston motion given by the actual crank mechanism and the consequent phasing of the heat-transfer processes in the cooler, regenerator, and heater do not yield the ideal constant-temperature or constant-volume compression and expansion processes; instead, these processes tend to become mixed.[10] As a consequence, the actual pressure-volume diagram looks quite different from the ideal. These effects can be visualized by comparing Fig. 2.23 with Fig. 2.16.

As in other reciprocating engines, frictional losses are an important factor that detracts from the thermal efficiency of the actual cycle. The problems in the Stirling engine are aggravated by the fact that a seal around the piston rod for the displacer piston represents an appreciable frictional loss in addition to that of the piston rings or their equivalent on the main power piston. Further, the work per cycle is small unless the pressure in the cylinder is kept high, and this means high pressures throughout the cycle because the volumetric compression ratio is necessarily low. (A relatively large residual gas volume is inherently present in the heat-transfer matrices for heating, regenerating, and cooling the gas as it flows between the hot and cold regions of the cylinder.) To keep the pumping power losses within acceptable limits as the gas is forced back and forth through the heat-transfer matrices, the operating speed of the engine must be kept to about one-quarter of that for a conventional internal combustion engine. Inasmuch as the high cylinder pressures require a massive construction and the relatively low piston speeds reduce the number of cycles obtainable per minute, the Stirling engine generally tends to be larger and heavier for a given output than a diesel engine.

The peak cycle temperature in the Stirling engine is limited in part by materials considerations in the heater; in practice, a peak temperature of the order of 1300 °F appears to be the max-

imum practicable. Similarly, rather substantial heat losses to the cylinder walls are inherent, because, where the walls must be lubricated, the metal temperature must be kept below about 300 °F.

COMPARISON OF IDEAL AND ACTUAL CYCLES EMPLOYING TURBINES

The largest single set of losses in cycles employing turbines occurs in the turbines themselves. It happens that the deviations between ideal and actual cycles in turbine engines are relatively insensitive to the choice of working fluid for a given fluid Reynolds number. Thus consideration of these losses requires less attention to the type of cycle or working fluid than when reciprocating engines are concerned. A fairly detailed discussion of the losses involved is included in the next chapter. Suffice it to say here that the greatest single source of loss in turbine engines commonly stems from aerodynamic effects. With refined aerodynamic passages and well-designed turbine blades, the aerodynamic efficiency of an individual stage can be raised to 90 percent or even a little higher if the machine is large enough so that a blade Reynolds number of 10^6 or more can be obtained. However, in the smaller-size units this is not ordinarily possible, and the smaller the unit the lower the Reynolds number. For this reason it is advantageous in small machines to choose a working fluid with a high molecular weight, because this will usually give a higher Reynolds number in the turbine than would be obtained for the same power output with a lower-molecular-weight fluid. It is for this reason that mercury, organic vapors, and inert gases such as krypton and xenon have looked attractive as working fluids for small turbines. This factor is an important consideration if one considers the relative merits of large central stations and dispersed small power plants of the sort advocated by Amory Lovins and popular with many environmentalists.

Other smaller losses in turbines include leakage through shaft seals and mechanical friction in the bearings. These become less important as the design output is increased.

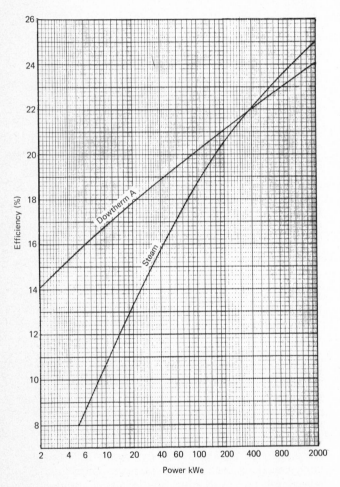

Figure 2.24 Effects of design power output on the overall thermal efficiency of a steam-turbine cycle designed for dry saturated steam at 44 bars (640 psia) and 256 °C (493 °F) turbine inlet with the condenser at 66 °C (150 °F) and 0.25 bar (3.7 psia) and a Dowtherm A system designed for 343 °C (650 °F) with a condenser pressure of 0.07 bar (1 psia).[16]

Rankine Cycle

The overall effects of size on the various losses and hence the normal efficiency of a steam-turbine cycle are shown in Fig. 2.24 for small turbines for a range of net outputs of 10 kWe to 1 MWe for a fixed set of turbine inlet and exhaust steam conditions.[16] These data were plotted from a parametric study in which the overall turbine and system design were reasonably well proportioned to give close to the best possible efficiency within the limits of practical design considerations. A similar curve for a Dowtherm vapor is also shown. These curves in-

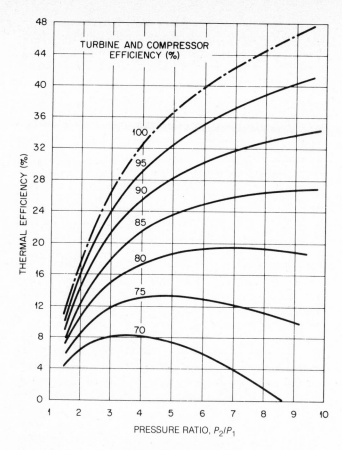

Figure 2.25 Effects of turbine and compressor aerodynamic efficiency on the efficiency of a simple Brayton cycle.[7]

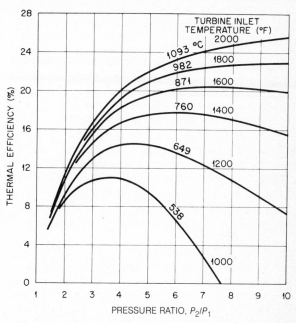

Figure 2.26 Effects of turbine inlet temperature on the efficiency of a simple Brayton cycle. As in Figs. 2.25 and 2.27 conditions for the base case were an efficiency of 80 percent in the compressor and turbine with no other allowance for pressure losses in the system and temperatures at the inlet to the compressor and turbine of 16°C (60°F) and 815°C (1500°F) respectively.[7]

dicate the pronounced effect of design output on the relative importance of the various losses, and they suggest the advantages associated with the use of a high-molecular-weight fluid which permits an increase in the Reynolds number and thus reduces the aerodynamic losses in the smaller-size units. In practical terms, the advantage of organic fluids over steam largely disappears for units designed to deliver 300 kWe or more, because the poorer heat-transfer properties of organic fluids give less favorable conditions in the heat exchangers required.

Brayton Cycle

The aerodynamic, bearing, and seal leakage losses are substantially the same in a Brayton cycle as in the Rankine cycle so far as the turbine is concerned. In addition, essentially the same types of fluid machine losses appear in the compression portion of the cycle. Here, however, their importance is very much greater than in the Rankine cycle because the compression work in a Brayton cycle ordinarily represents about two-thirds of the turbine work, whereas the feed-pump work in the Rankine cycle ordinarily represents only about 1 percent of the turbine work. As a consequence, the Brayton cycle is very sensitive to the aerodynamic design of the turbine and compressor, and it is very difficult to achieve high efficiency in these components in small sizes. However, through the use of highly refined design and fabrication techniques and high-molecular-weight gases, it is possible to achieve attractive overall thermal efficiencies in units as small as 10 kWe, provided that high gas temperatures can be used at the inlet to the turbine.

Some notion of the relative importance of the various losses in the Brayton cycle is indicated by Figs. 2.25, 2.26, and 2.27, which show, respectively, the effects of turbine and compressor efficiency, turbine inlet temperature, and compressor inlet temperature on the overall thermal efficiency of the cycle. The system pressure loss is also an important factor, because it increases the compressor work required.

LOSSES CHARACTERISTIC OF TYPICAL HEAT SOURCES

The previous discussion was concerned only with the cycle efficiency as based on the heat actually flowing into the thermodynamic cycle itself; it did not include losses characteristic of the heat source employed. In a fossil-fuel-fired steam plant, for example, the energy in the hot combustion gases discharged up the stack commonly runs about 10 percent of the energy in the fuel. As a consequence, if the actual steam cycle itself gives a thermal efficiency of 40 percent, the stack losses will act to reduce the plant efficiency to about 36 percent. These stack losses are not present in a nuclear plant—a factor that offsets in part the disadvantage of the relatively low steam temperatures obtainable with water reactors (about 600°F versus about 1000°F with fossil fuel plants).

Pumping losses for circulating the condenser cooling water and the combustion air through fossil fuel plants or the reactor coolant through nuclear plants commonly run about 1 or 2 percent of the net electric output. Other, similar but usually smaller

losses include the power for coal pulverizers, coal- and ash-handling machinery, and miscellaneous plant equipment. In addition, there may be large energy losses involved in sulfur-removal processes. All of these detract from the overall thermal efficiency of the plant.

OVERALL THERMAL EFFICIENCY OF ACTUAL CYCLES

It is interesting to consider the cumulative effects of the various losses cited above for some actual power plants. Figure 2.28 shows a set of curves for the Carnot cycle efficiency as a func-

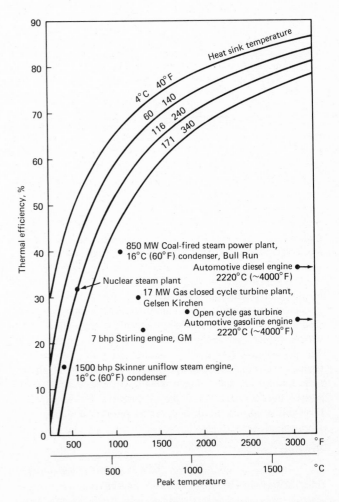

Figure 2.28 Comparison of the thermal efficiencies of large power plants with ideal Carnot cycle efficiencies.

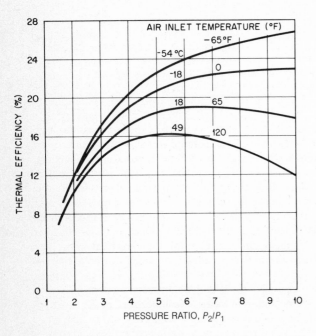

Figure 2.27 Effects of compressor inlet temperature on the efficiency of a simple Brayton cycle.[7]

TABLE 2.1 Summary of Calculations for a Closed-Cycle Gas Turbine

Working fluid = air
Compressor inlet air temperature = 60°F
Turbine inlet air temperature = 1500°F
Regenerator effectiveness = 0.80
Compressor efficiency = 0.86

Compressor inlet pressure = 4 atm
Compressor pressure ratio = 4:1
System pressure drop = 10%
Turbine pressure ratio = 3.6:1
Turbine efficiency = 0.88

Point in cycle	Temperature, °R	Pressure, atm	Enthalpy, Btu/lb	Relative pressure	Isentropic change in enthalpy, Btu/lb	Actual change in enthalpy, Btu/lb
Compressor inlet	520	4	124.3	1.2147		
Compressor outlet	771.5	16	184.9	4.859	60.6	
	812	16	194.8			70.5
Turbine inlet	1960	15.2	493.64	160.37		298.4
Turbine outlet	1414	4.22	346.57	44.55	147	
	1481	4.22	364.2			129.4

Net work = 58.9 Btu/lb
Regenerator inlet temperature difference
   = 669°R
Regenerator temperature rise = 535°R
Regenerator outlet temperature
   = 1347°R

Regenerator outlet enthalpy
   = 3291 Btu/lb
Heat added in regenerator
   = 134.3 Btu/lb
Heat added in combustor
   = 164.5 Btu/lb
Thermal efficiency of cycle
   = 58.9/164.5 = 35.8%

tion of peak cycle temperature for a series of heat-sink temperatures. Plotted on the same coordinates are points for a wide variety of actual power plants operating on different thermodynamic cycles with different heat sources. Note that, in general, increasing the peak cycle temperature leads to an increase in thermal efficiency, but that the deviation between the efficiency of the actual cycle from that of the ideal Carnot cycle increases as the peak cycle temperature is increased. This comes about as a consequence of the increasing importance of losses at the higher temperatures and the greater difficulty in making effective use of these higher temperatures in a practical machine subject to differential thermal expansion of parts, lubrication problems, and the like. Note that even with a turbine inlet temperature of 1800°F, a simple open-cycle gas turbine does not give as good an efficiency as a steam plant with a peak steam temperature of 1050°F. Note, too, that by eliminating stack losses, a nuclear steam system with a turbine inlet temperature of only 540°F yields almost as high an efficiency as a coal-fired steam plant with stack gas scrubbers operating at a turbine inlet temperature of 1050°F.

THERMODYNAMIC CYCLE CALCULATIONS

To do a proper job of thermodynamic cycle analysis for steam and gas turbine systems, one ought to use the complete steam tables[17,18] or the gas tables,[18,19] which take into account temperature variations of the specific heats, but some brief summaries of these data are included in the Appendix for rough calculations and for handling the problems included at the end of each chapter in this text. Brief sets of thermodynamic data are also included in the Appendix for helium, argon, Dowtherm, mercury, potassium, cesium, NH_3, and Freon 12 with a reference in each case to a source of much more complete data. Note that for the monatomic gases such as helium and argon the specific heats do not vary with temperature and hence the perfect gas laws can be used. In open-cycle gas turbines the combustion products are a sufficiently small fraction of the molecular species that the air tables give a good approximation. For gasoline and diesel engines, however, the higher temperatures and higher fuel-air ratios make it desirable to use thermodynamic charts designed for this work. An excellent set for fuel-air ratios of 80, 90, 100, 110, and 120 percent of the stoichiometric value are given in Ref. 20 (and included in Ref. 7). A small version of one of these is included in the Appendix as Fig. A2.7.

Typical Calculational Procedure

A typical calculation for a simple closed-cycle gas turbine is summarized in Table 2.1 to illustrate a convenient procedure. The first step is to fill in line 1 with the given data and the enthalpy and relative pressure from an air table (e.g., Table A2.3). The second step is to obtain the relative pressure at the compressor outlet for an isentropic compression by multiplying the relative pressure at the inlet by the pressure ratio and then filling in the enthalpy and temperature for that relative pressure as found in the air table to obtain the conditions at the end of an isentropic compression. The isentropic change in enthalpy during compression can then be divided by the compressor efficiency of 0.86 to obtain the actual change in enthalpy and hence the actual enthalpy and temperature at the compressor outlet. The temperature at the turbine inlet is specified; hence the enthalpy and relative pressure can be taken from the air table. The conditions at the turbine outlet can then be calculated by following a procedure similar to that used for the compressor outlet except that the isentropic drop in enthalpy is multiplied by the turbine efficiency of 0.88 to obtain the actual turbine enthalpy drop and thus the outlet temperature. Note that the turbine expansion ratio is less than the compressor pressure ratio because of the pressure drop in ducts and whatever combustor and/or heat exchangers are employed. The net work output is simply the turbine enthalpy drop minus the enthalpy rise in the compressor, and the thermal efficiency of the cycle is the net work divided by the enthalpy added in the combustion chamber between the compressor and turbine. Note that for combustion turbines this approach neglects the effects of the increased weight flow through the turbine from the fuel added in the combustion chamber. This is a small factor, however, because the weight flow rate of the fuel is only about 2 percent of the airflow and hence can be neglected in preliminary design analyses and calculations.

DIRECT CONVERSION

Several highly intriguing concepts for the direct conversion of heat or chemical energy into electricity have been under study since the early nineteenth century, and some of these have been developed into useful small power units for special applications,

notably for spacecraft. The principal concepts are fuel cells, thermoelectric junctions, and thermionic diodes. Although in principle these systems avoid both the isentropic processes and the losses associated with mechanical and fluid friction characteristic of the heat engines treated in the first portion of this chapter, their performance is also limited by basic thermodynamic principles,[21-24] and in practice their performance is seriously degraded by losses similar or analogous to those characteristic of heat engines. In view of the fact that these systems are favorites of some environmentalists for use in small dispersed power plants to replace our large centralized utility systems, some discussion of their characteristics and limitations is in order to permit comparisons with conventional power plants.

Fuel cells are the leading contender of the direct-conversion systems, so much so that an entire chapter of this text (Chap. 18) is devoted to their characteristics and problems; hence in view of their special character they will not be discussed further here. Thermoelectric and thermionic systems are not treated elsewhere in this text, and so their operating principles, construction, and characteristics are outlined in this section. The treatment is brief because there seems little likelihood that either system will make a significant contribution to the U.S. energy economy in the foreseeable future—a situation that stems from their formidable materials problems and their requirements for exceptionally highly refined technology and elegant construction, which lead to high costs and limited life. Further, their thermal efficiencies are relatively low.

Thermoelectric Generators

At first thought a thermoelectric power unit might be as simple as a bundle of thermocouples connected in the form of a thermopile to provide enough voltage to give a useful output. This is true if one is willing to accept an extremely low efficiency—less than 1 percent—but to obtain a higher efficiency it is necessary to employ special materials and a complex construction.

Principles of Operation

Figure 2.29 shows the elements of a thermoelectric generator utilizing semiconductor materials specially chosen and treated to give as high a voltage as possible for the temperature differential available between the heat source and the heat sink. In the N-type semiconductors, free electrons diffuse to the hot junction, making it positive in charge, while in P-type semiconductors the free electrons diffuse to the cold junction. Using these materials to form pairs of junctions arranged in series to form a chain as in Fig. 2.29 gives a thermoelectric generator.

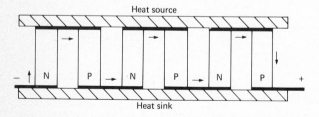

Figure 2.29 Arrangement of N and P junctions in a thermoelectric generator.

TABLE 2.2 Typical Values for the Controlling Physical Properties of Representative Thermoelectric Generator Materials

Property	Lead telluride	Silicon-germanium
Temperature range ΔT, °C	50–550	300–700
Seebeck coefficient α, μV/°C	218	230
Resistivity ϱ, $\Omega \cdot$ cm $\times 10^{-3}$	3.2	1.8
Thermal conductivity k, mW/(°C$\cdot$cm)	14	49
Performance parameter, $Z - \alpha^2/\varrho k$	1.06×10^{-3}	0.6×10^{-3}

The material most commonly used in such generators is lead telluride (PbTe); the N-type material is formed by adding a few parts in 10^4 of PbI, while the P-type material is formed by adding a few parts in 10^3 of sodium. The PbTe material is generally used up to ~ 600°C (1112°F), while germanium silicide (GeSi) suitably doped with trace amounts of other elements is used for hot-junction temperatures up to ~ 1000°C (1832°F). Other, more exotic semiconductor materials have also been used in research studies.[24]

The thermodynamic limitations on the output of thermoelectric generators are well defined.[22] The open-circuit voltage available from a junction is given by

$$E = \alpha \Delta T \qquad (2.3)$$

where E = voltage
α = average Seebeck coefficient over temperature range used
ΔT = temperature difference between hot and cold junctions

The electric-power output P is the product of the current I and the net voltage available, which is the open-circuit voltage minus the loss from the internal resistance R; that is,

$$P = I(\alpha \Delta T - IR) \qquad (2.4)$$

Most of the heat input to thermoelectric junctions is transmitted through them by thermal conduction. To this heat input must be added the energy leaving the hot junction as electrons, i.e., the Peltier heat effect ($I\alpha T_h$). The electrical resistance loss (I^2R), on the other hand, adds heat to the system, and half of this heat flows to each of the two junctions. Thus the thermal efficiency of the idealized thermoelectric generator may be approximated by the electric power output of Eq. (2.4) divided by the energy input to the hot junction; that is,

$$\text{Thermal efficiency} = \frac{I(\alpha \Delta T - IR)}{K\Delta T + I\alpha T_h - \frac{1}{2}I^2R} \qquad (2.5)$$

The maximum efficiency for a given ΔT and set of materials can be obtained by setting the derivative of Eq. (2.5) equal to zero, and this yields a value for the maximum thermal efficiency for the idealized system of

Maximum thermal efficiency =

$$\frac{T_h - T_c}{T_h} \left\{ \frac{\left[1 + \dfrac{\alpha^2}{\varrho k} \left(\dfrac{T_h + T_c}{2} \right) \right]^{1/2} - 1}{\left[1 + \dfrac{\alpha^2}{\varrho k} \left(\dfrac{T_h + T_c}{2} \right) \right]^{1/2} + \dfrac{T_c}{T_h}} \right\} \qquad (2.6)$$

where h and c signify hot and cold junctions and ϱ represents the electrical resistivity of thermoelectric material.

Note that the first term of Eq. (2.6) is the Carnot cycle efficiency. In the second term the value of the quantity $\alpha^2/\varrho k$ is commonly ~ 0.001, and so for $T_h = 900$ K and $T_c = 300$ K the second term would be ~ 0.17, giving a thermal efficiency of ~ 11 percent. (Table 2.2 gives the physical properties for PbTe and GeSi.)

Typical Construction

The losses in any actual thermoelectric generator are much greater than implied by Eq. (2.6). To gain some insight into these losses, together with an appreciation for other problems, it is necessary to consider a representative unit. The construction of a typical thermoelectric generator designed by the author for an undersea power unit is shown in Fig. 2.30.[25] This module employed 208 lead telluride couples in series. The couples were arranged around a central tubular heat source in 16 rows with 13 couples per row. The voltage output was about one-eighth of a volt per couple with a hot-junction temperature of 538°C (1000°F) and a cold-junction temperature of 38°C (100°F). The total output was 200 W at 24 V. The module was designed for coupling to an isotope heat source by a heat pipe in which potassium served as the heat transport medium by boiling at the

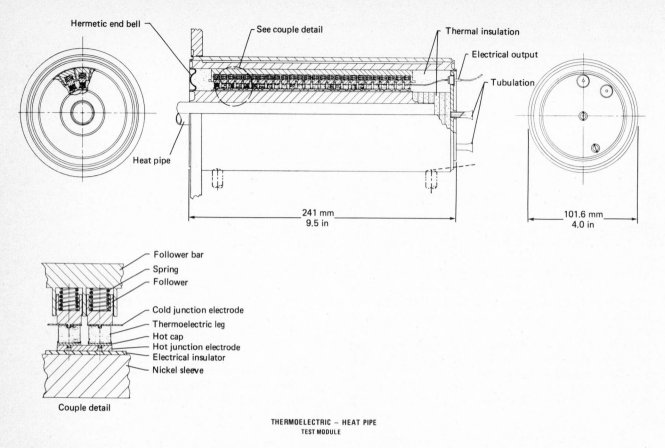

Follower bar
Spring
Follower
Cold junction electrode
Thermoelectric leg
Hot cap
Hot junction electrode
Electrical insulator
Nickel sleeve

Couple detail

THERMOELECTRIC — HEAT PIPE
TEST MODULE

Figure 2.30   Construction of a 200-W, 24-V, thermoelectric generator module employing 208 lead telluride junctions. The unit was designed for coupling through a potassium heat pipe to an isotope heat source operating at 538°C (1000°F)), with the outer shell cooled by seawater at 15°C.[25] (*Courtesy Oak Ridge National Laboratory.*)

bottom end, where it absorbed heat from a radioisotope, and condensing at the top inside the module. (This heat pipe proved highly effective; the temperature drop from one end of the heat pipe to the other ran less than 10°C.)

The generator module employed a widely used couple, a detailed drawing of which is shown in the lower left corner of Fig. 2.30. This drawing illustrates the usual method of coping with the differential thermal expansion problems inherent in the large temperature differences involved. There are sliding contacts that permit the hot central nickel sleeve to grow both axially and radially in relation to the cold outer cylindrical shell. The stack of components for each lead telluride thermoelectric leg of the junction is kept in close contact by a spring and follower shoe mounted at the cold end. Even with this axial spring load, however, the thermal and electric contacts are far from perfect,

and so there is a substantial accumulation of contact resistances across the nine interfaces in the electric-current path through each leg of the junction. As a consequence, the overall electric output and the thermal efficiency of a complete module are substantially less than for individual junctions tested under more nearly ideal conditions in the laboratory. For example, the overall thermal efficiency of the module of Fig. 2.30 was found to peak at about 6 percent, roughly half the ideal value indicated by Eq. (2.6).

Life

The temperature of the hot junction is sufficiently high that degradation of the semiconductor occurs because of both solid-state diffusion and volatilization effects. Degradation rates are of the order of 5 percent per year. In addition, thermal-strain cycling may cause failures and open circuits, galling or scuffing of rubbing surfaces acts to increase contact resistances, and one of the many thin layers of electric insulation may fail under the scuffing action and give a short circuit.

Costs

It can be seen from the complexity of construction shown in Fig. 2.30 that the costs of thermoelectric generators are inherently high: two of the units shown, purchased together, cost $91,000, or $455,000/kW, in 1970. The costs in production would be lower, but not as much lower as one might think, because the basic junction employed was a production item. In any case, the combination of high cost, limited life, and a thermal efficiency of only ~ 6 percent make thermoelectric generators unattractive for commercial electric power.

Thermionic Cells

Thermionic cells have been under active development, principally for spacecraft power units, since the latter 1950s.[23] They depend on the fact that electrons are emitted from a hot cathode and are absorbed in a cold anode. If this is done in a hard vacuum, the efficiency is poor unless the spacing between the emitter and collector is very small: ~ 10 to 20 μm,[26] an impractically small value when manufacturing tolerances and thermal distortion are considered. If cesium, the most easily ionized of the elements, is introduced as a very low pressure vapor (~ 0.001 atm), the conducting plasma formed permits a current density of as much as 10 A/cm² with a gap between electrodes of ~ 0.5 mm. The larger gap is sufficient so that a little thermal distortion in the electrodes can be tolerated, and hence virtually all the effort on thermionic cells has been directed toward the cesium plasma diode.

The relations involved in establishing the upper limit to the efficiency of thermionic cells are too involved for presentation here.[23] Basically, the situation is analogous to that for the thermoelectric junctions treated above: energy flows from the hot cathode to the cold anode in the form of both electrons and thermal radiation, and the efficiency of the cell is the ratio of the useful electric power obtained divided by the total energy emitted. Inasmuch as the bulk of the energy emitted is in the form of heat, the thermal efficiency is relatively low—of the order of 10 percent, although values as high as 18 percent have been reported for laboratory tests of single cells.[27] Voltage outputs are low—around 0.5 V; hence many cells must be connected in series to give a useful output, and I^2R losses in the interconnections tend to be rather large.

Emitter Temperature

The development and application of thermionic cells is completely dominated by the need for emitter temperatures in the range of 1700 to 2200 K (2600 to 3500 °F) in order to get a current density high enough to yield a useful output (Fig. 2.31). In the first place, it is exceedingly difficult to heat the emitter to this temperature range with any heat source other than fission, and hence most of the development effort has been devoted to thermionic cells designed to operate in fission reactors with the reactor fuel contained within the emitters of the thermionic

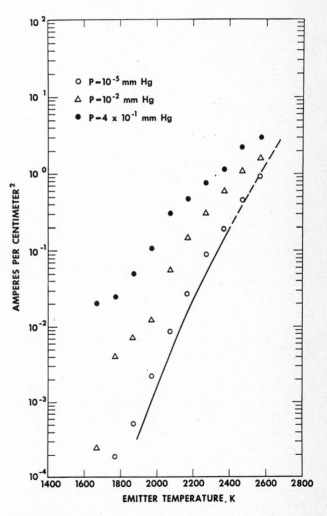

Figure 2.31 Effects of emitter temperature and cesium vapor pressure on the current density obtained with a thermionic diode. (*Courtesy Oak Ridge National Laboratory.*)

cells. Conceptual designs for power units employing combustion or solar heat sources have been prepared. However, the problems with corrosion and stresses in the materials of which the systems might be built, coupled with problems in maintaining the high degree of leak-tightness required throughout a long life, getting the assumed uniform heat input and temperature distribution in a large array of cells, etc., make it doubtful that such systems will be used for commercial power generation. Unfortunately, these vital engineering problems are rarely addressed in proposals made by those involved in thermionic cell research.

Typical Construction

In attempting to envision the characteristics, operation, and problems of thermionic cells, it is helpful to examine a typical cell design such as that of Fig. 2.32. The thermionic cells are arranged in series in a reactor fuel element much like dry cells in a

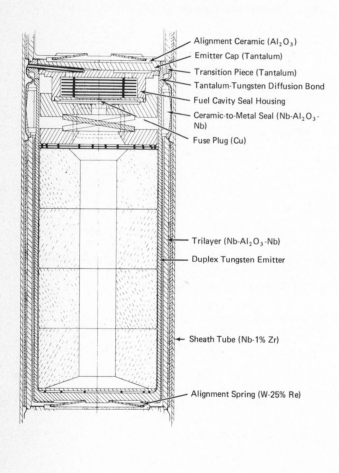

Figure 2.32 Typical construction of a thermionic cell for a fission reactor.[28] (*Courtesy General Atomic Co.*)

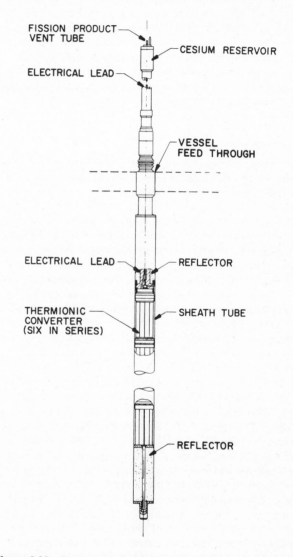

Figure 2.33 Typical construction of a fuel element for a thermionic cell reactor.[28] (*Courtesy General Atomic Co.*)

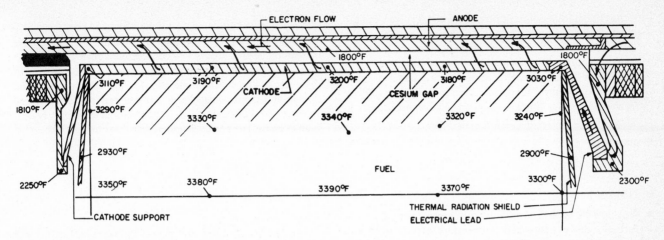

Figure 2.34 Temperature distribution in a typical thermionic converter at operating conditions; 1-MW nuclear thermionic space power plant. (*Courtesy NASA.*)

large flashlight (Fig. 2.33). Fission heat from the nuclear fuel in the center of the element flows radially out to the emitter surface. Electrons flow through the plasma in the ~0.5-mm gap between the emitter and collector, then axially through the collector to the next cell or, in the case of the last cell in the stack, out the end of the reactor to an electrode for a series connection to the next fuel element. Heat radiated from the emitter to the collector is conducted radially outward through electric insulation to an outer sheath, which is cooled by a circulating stream of liquid metal. The temperature distribution in a typical cell is shown in Fig. 2.34. The construction is complex, the dimensional tolerances must be held within close limits, the thermal and electric conductivities of interfaces between parts must be high where this is needed and low where they must be low, and a host of joints in dissimilar materials must be leaktight. The temperature differentials are large, coefficients of expansion vary from one material to another, and yet good leaktightness and good electric and thermal contacts must be maintained. In addition, though not shown clearly in Fig. 2.32 or Fig. 2.33, one set of passages must be provided to convey cesium vapor to each diode, and a second set is needed to vent fission product gases from the fuel, as otherwise these will cause swelling of the emitter shell and shorting to the collector.

Other types of cell construction have been proposed to avoid the problems implicit in Fig. 2.32, but they introduce other problems. A "button" type of construction is often used in experimental cells with the emitter and collector in the form of flat diaphragms, but their diameter must be kept small to prevent difficulty in maintaining the small gap between electrodes. Distortion or "oil-canning" of the diaphragms tends to occur

because of temperature differentials both across the thickness of the diaphragms and between the center and and the edges.

Developmental Problems

A variety of refractory materials including metals, cermets, and ceramics have been used in emitters, but all give a limited life at the high temperatures required. The current density obtainable is quite sensitive to the crystal structure in the emitter—even in pure tungsten the emission characteristics change from one face to another of the crystallites. Trace impurities, coupled with solid-state diffusion, tend to cause deterioration in the emission characteristics over a year's time. The vapor pressure of the emitter material, though low, is sufficient to give a little sublimation even of tungsten, and this forms a black deposit on the collector, greatly increasing its absorptivity for thermal radiation and thus degrading the efficiency of the cell.

Perhaps the most serious set of problems, and a set rarely faced by research personnel, is presented by cell reliability in operating arrays. The cells must be small in order to maintain the close gap required between the emitter and collector, and this means that the output of individual cells must be small—at most 500 W. This implies around 1000 cells in each of 100 arrays for a 100-MWe power output. The voltage output of individual cells is ~0.5 V; hence each array of ~1000 cells must be connected in a series-parallel arrangement. Even with a vastly higher reliability than has ever been obtained with single cells, failures of individual cells are certain to occur. If a failure leads to an electric short circuit, the current flow can be large and the damage is likely to be extensive. If a failure entails an open circuit, the temperature of the emitter of the cell involved will increase, and this is likely to lead to another failure that would cause a short circuit. There seems to be no reasonable way to provide diagnostic instrumentation that would detect such failures before serious damage developed. It was largely for this reason that the Central Electrical Generating Board (CEGB) research laboratory in England dropped its substantial research program on thermionic cells in the early 1960s.

Costs

Although the literature contains many claims that thermionic cells might be produced at low costs, either a review of the costs of the many units that have been built or a close examination of the costs of the special materials, fabrication processes, and tight quality control procedures required leads to the conclusion that the cost of thermionic cell systems would inherently be very high even for large-scale production.

PROBLEMS

1. A 200-MWe coal-fired steam plant operates with a steam-turbine inlet pressure and temperature of 100 bars and 538°C, giving an overall thermal efficiency of 32 percent when operating with no reheat, no regenerative feed heaters, a condenser temperature of 30°C on the steam side, and no stack gas scrubbers. Stating assumptions for component efficiencies, estimate the following:

   (a) The efficiency of the corresponding ideal Carnot cycle.
   (b) The efficiency of the corresponding ideal Rankine cycle.
   (c) The losses in the boiler (to the stack), the turbine, the generator, the feed pump, the auxiliaries, and miscellaneous. Express each loss as a percentage of the energy in the fuel.

2. A simple open-cycle 100-MWe oil-fired gas turbine operates with a pressure ratio of 10:1 and a 1000°C turbine-inlet temperature to give an overall thermal efficiency of 27 percent. Stating assumptions for component efficiencies, estimate the following:

   (a) The efficiency of the corresponding ideal Carnot cycle.
   (b) The compressor power input, the turbine output, and the efficiency of the corresponding ideal Brayton cycle.

   (c) The fuel-air ratio.
   (d) The losses in the compressor and turbine, the pressure loss in the combustion chamber and ducts between the compressor and turbine, and the energy loss in the hot gas discharged up the stack. Express each of these losses as a percentage of the energy available from the fuel supplied. Sum these losses to obtain the miscellaneous losses, i.e., those not specifically included.

3. Estimate the airflow, fuel flow, and maximum output of a 20-L-displacement four-stroke-cycle diesel engine at 1800 rpm if operated without supercharging. Approximately what would the output be if it were supercharged to 2 atm intake manifold pressure?

4. Suppose that you are approached by a man who claims to have a dope for gasoline that will permit an increase in compression ratio from 7:1 to 8.5:1 and further claims that such an increase will give 25 percent more power and 25 percent more gas mileage. Estimate the maximum improvement one might reasonably expect to realize, assuming that the doped fuel will permit the increase in compression ratio claimed.

REFERENCES

1. Carnot, N. L. S.: *Reflections on the Motive Power of Heat and on Machines Fitted to Develop That Power*, 1824, reprinted by John Wiley & Sons, Inc., New York, 1897.

2. *Steam, Its Generation and Use,* The Babcock & Wilcox Company, New York, 1972.

3. Fraas, A. P.: "Applications of Modern Heat Transfer and Fluid Flow Experience to the Design of Boilers for Automotive Steam Power Plants," paper presented at International Automotive Engineering Congress, Detroit, Jan. 13–17, 1969.

4. Salisberry, J. K.: *Steam Turbines and Their Cycles*, John Wiley & Sons, Inc., New York, 1950.

5. Reynolds, R. L.: "Reheating in Steam Turbines," *Trans. ASME*, vol. 71, 1949, p. 701.

6. Fraas, A. P., and M. N. Ozisik: "A Comparison of Gas-Turbine and Steam-Turbine Power Plants for Use with All-Ceramic Reactors," Oak Ridge National Laboratory Report No. ORNL-3204, July 1965.

7. Fraas, A.P.: *Combustion Engines*, McGraw-Hill Book Company, Inc., New York, 1948.

8. Meijer, R. J.: "The Philips Stirling Thermal Engine," thesis for Technische Hogeschool Delft, November 1960.

9. Kolin, I.: "The Stirling Cycle with Nuclear Fuel," *Nuclear Engineering International*, vol. 13, no. 151, December 1968, p.1028

10. Flynne, G., Jr., et al.: "GMR Stirling Thermal Engine, Part of the Stirling Engine Story — 1960 Chapter," *SAE Trans.*, vol. 68, 1960, p.665.

11. Bender, M., et al.: "Comparative Features of Gas-Cooled Fast Reactors," *Reactor Study,* Oak Ridge National Laboratory, 1958.

12. Feher, E. G., et al.: "System Analysis and Design Concept of a 150 kWe Supercritical Thermodynamic Cycle Power Conversion Module," Report No. DAC-60998-F, McDonnell Douglas Astronautics Company, January 1969.

13. Feher, E. G., et al.: "Investigation of Supercritical (Feher) Cycle," Air Force Aero Propulsion Laboratory Report No. AFAPL-TR-68-100, October 1968.

14. Blander, M., et al.: "Aluminum Chloride as a Thermodynamic Working Fluid and Heat Transfer Medium," USAEC Report ORNL-2677, Oak Ridge National Laboratory, Sept. 21, 1959.

15. Krasnin, A. K., and V. B. Nesterenko: "Dissociating Gases: A New Class of Coolants and Working Substances for Large Power Plants," IAEA, *Atomic Energy Review*, vol. 9, no. 1, 1971.

16. Fraas, A. P.: "Design Considerations and Proposed Designs for Steam Turbine-Generator Units for Terrestrial Power Plants of 30 kW(e) to 2 MW(e)," USAEC Report ORNL-TM-1837, Oak Ridge National Laboratory, June 1967.

17. *ASME Steam Tables*, American Society of Mechanical Engineers, 1967 (English units).

18. Irvine, T. F., and J. P. Hartnett: *Steam and Air Tables in SI Units: Including Data for Other Substances and a Separate Mollier Chart for Steam*, Hemisphere Publishing Corp., Washington, D.C., 1976.

19. Keenan, J. H., and J. Kaye: *Gas Tables*, John Wiley & Sons, Inc., New York, 1948.

20. W. J. McCann: *Thermodynamic Charts for Combustion Engine Fluids*, NACA RB3G28, 1943.

21. Somers, E.V.: "Energy Generation and Conversion, Conference on New Concepts in Physics for the Chemical Engineer," Franklin Institute Laboratories, March 30 and 31, 1965.

22. Hatsopoulos, G. N.: "Thermoelectric Effects and Irreversible Thermodynamics," *J. Applied Mech. Trans. ASME,* vol. 80, 1958, p. 428.

23. Hatsopoulos, B. N., and F. N. Huffman: "The Growth of Thermionic Energy Conversion," *Proceedings of the Tenth Intersociety Energy Conversion Engineering Conference,* 1975, pp. 342–350.

24. O'Riordan, P. A.: "Thermoelectricity: Executive Summary," *Proceedings of the Eleventh IECEC*, 1976, pp. 1525–1526.

25. Fraas, A. P., and M. E. LaVerne: "Reference Design for a Thermoelectric Isotope Power Unit Employing Heat Pipe Modules," Oak Ridge National Laboratory Report No. ORNL-TM-2959, November 1971.

26. Nottingham, W. B.: "Review of the Physics of Thermionics," *Energy Conversion for Space Power*, Academic Press, Inc., New York, 1961, pp. 125–136.

27. Morris, J. F.: "NASA Thermionic Conversion Program," *Proceedings of the Twelfth IECEC,* 1977, pp. 1540–1547.

28. *Thermionic Fuel Element Development Status*, Gulf General Atomic, 1970.

3

PERFORMANCE CHARACTERISTICS OF TURBINES, COMPRESSORS, AND PUMPS

The performance characteristics of turbines, compressors, and pumps play such a vital role in both the efficiency and the control of power plants that a chapter on the subject appeared to be needed to provide some background for those not familiar with the subject. The proper design of turbomachines involves complex thermodynamic, aerodynamic, and stress-analysis work,[1,2,3] but the basic relations needed for the systems analysis with which this text is concerned can be outlined and established by means of elementary thermodynamics and fluid mechanics in this chapter.

It happens that most of the basic concepts and relationships associated with the design and performance of turbines are also common to compressors and pumps. The similarity in appearance of axial-flow compressors and turbines is evident in Fig. 3.1, which is a cutaway view of a gas turbine. Not only is it convenient to develop the concepts and relationships for turbines and compressors together, but also the very points of similarity and difference serve to emphasize the essential elements of the theoretical background. Fortunately, the basic relationships are not nearly so formidable as they at first appear. The basic function of the rotor or stator blades in either a turbine or a compressor (or pump) is to change the velocity of the fluid stream by changing the flow direction, i.e., to turn the stream through some angle. This is accompanied by a change in momentum which produces a reaction force in the rotors. The distinction between the turbine and the compressor is simply that the energy flow in the former is from the fluid to the rotor whereas in the latter it is from the rotor to the fluid.

BASIC MECHANICS OF TURBOMACHINERY

The basic mechanics of turbines were recognized and worked out formally around 1830 by Benoit Fourneyron, an outstanding French engineer,[4] who set out to develop a machine that would give a higher efficiency and be suitable for use with higher heads than the waterwheels of his day. He succeeded remarkably well, obtaining an efficiency of 83 percent in a 6-hp impulse turbine operating with a head of 1.4 m. He designed and built over a hundred turbines. One developed 200 hp with a head of 5 m, and another developed 60 hp with the then incredibly high head of 114 m.

Impulse Turbines

The ideal impulse turbine represents a special type of turbomachine that offers an excellent means of introducing the basic principles. As shown in Fig. 3.2, the machine is quite simple: a stream of high-velocity fluid is directed through a nozzle into a set of scoop-shaped blades or buckets (a term used originally for hydraulic turbines), mounted on the periphery of the wheel. The fluid flow is directed into the wheel at an acute angle so that the flow is almost tangential to the wheel. The impact of the fluid jet on the buckets causes the wheel to rotate so that the machine can be made to do work. This type of blading was originally developed for hydraulic turbines and has been widely used in both steam and gas turbines. For relatively small fluid flow rates, only one or a few discrete nozzles may be used to direct the

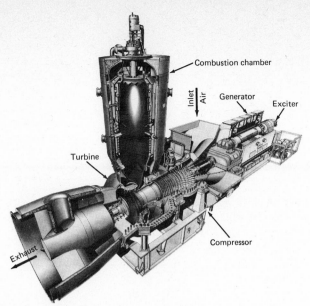

Figure 3.1 Cutaway view of a commercial gas turbine. (*Courtesy Brown Boveri and Co., Ltd.*)

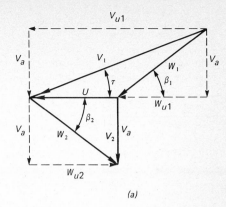

(a)

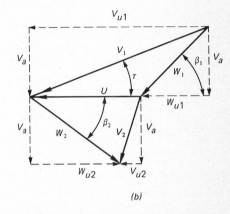

(b)

Figure 3.3 Velocity vector diagrams for an impulse turbine wheel under (a) design conditions and (b) a 20 percent overspeed condition.

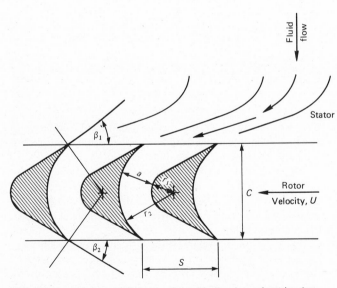

Figure 3.2 Schematic diagram of the blading in an impulse turbine wheel. Legend:

a: Dimension defining the minimum flow passage area in the rotor
C: Chord of the rotor blade
r_1: Radius of blade suction surface
r_2: Radius of blade pressure surface

S: Blade spacing, or pitch
U: Rotor velocity
β_1: Angle between blade and relative inlet velocity
β_2: Angle between blade and relative exit velocity

flow into the rotor, but for large fluid flow rates, the nozzles occupy the entire perimeter of the wheel, in which case the term *nozzle box* or *stator* is used, with *stator vanes* or *stator blades* serving as the nozzle walls that determine the direction of the flow entering the rotor.

Velocity Vector Diagrams

Velocity vector diagrams for an axial-flow impulse turbine are shown in Fig. 3.3 with the symbols traditionally employed. Two diagrams are included, one for operation at the design point where the absolute tangential velocity of the fluid leaving the wheel is zero and one for an overspeed condition where the tangential velocity of the fluid leaving the wheel is ~20 percent of the wheel velocity. Note that to simplify this presentation it is assumed that the fluid density and axial-flow-passage cross-sectional area are the same for the inlet and outlet of the rotor, and hence the axial velocity component of the fluid V_a is constant. (The subscript a denotes the axial component of the velocity.) The vector difference between the fluid inlet absolute velocity V_1 and the velocity of the blades U gives the relative velocity W_1 of the fluid with respect to the blades. A similar velocity triangle is given for the rotor outlet, the subscript 2 being used for the outlet velocity components. The subscript u is used to denote the tangential components of the fluid velocities (the components parallel to the velocity vector U for the rotor), that is, V_{u1}, W_{u1}, V_{u2}, W_{u2}.

Force on the Rotor

The force acting on the blades of an impulse turbine can be readily calculated from the change in momentum of the fluid stream. Thus the force F on the wheel is simply the product of the mass flow rate m and the change in the fluid velocity component in the plane of the wheel; that is,

$$F = m\,\Delta V = m\,\Delta V_u = m(V_{u1} - V_{u2}) \qquad (3.1)$$

Note that if V_{u1} is taken as positive, V_{u2} will be positive if in the same direction and negative if in the opposite direction.

Useful Power

The useful power developed by the machine then is

$$\text{Useful power} = F \times U = mU(V_{u1} - V_{u2}) \qquad (3.2)$$

The power input in the form of the kinetic energy in the fluid jets at the rotor inlet is given by

$$\text{Power input} = \frac{m}{2}V_1^2 \qquad (3.3)$$

Efficiency

The efficiency of the turbine, η_t, then becomes

$$\eta_t = \frac{\text{useful power}}{\text{power input}} = \frac{mU(V_{u1} - V_{u2})}{(m/2)V_1^2}$$

$$= \frac{2U(V_{u1} - V_{u2})}{V_1^2}$$

But $\qquad U = V_{u1} - W_{u1} = V_{u2} - W_{u2}$

and so $\qquad V_{u1} - V_{u2} = W_{u1} - W_{u2}$

For a pure impulse turbine operating at the design point, by definition the inlet and outlet axial velocities are equal and the angles β_1 and β_2 are equal for the fluid relative velocities W_1 and

W_2 entering and leaving the wheel. Hence $W_{u1} = -W_{u2}$, and so

$$V_{u1} - V_{u2} = 2W_{u1}$$

$$W_{u1} = V_{u1} - U$$

and $\qquad \eta_t = \dfrac{4U(V_{u1} - U)}{V_1^2} = \dfrac{4U(V_{u1} - U)}{V_a^2 + V_{u1}^2} \qquad (3.4)$

For any given values of V_a and V_{u1}, the efficiency will be maximum when the numerator of Eq. (3.4) is maximum, i.e., when its derivative with respect to U is zero. Thus, the condition for maximum efficiency when V_a and V_{u1} are given becomes

$$\frac{d}{dU}(4UV_{u1} - 4U^2) = 0$$

$$4V_{u1} - 8U = 0$$

$$U = \frac{V_{u1}}{2}$$

And the maximum efficiency becomes

$$\eta_{t\text{max}} = \frac{2V_{u1}^2 - V_{u1}^2}{V_a^2 + V_{u1}^2} = \frac{V_{u1}^2}{V_a^2 + V_{u1}^2} = \frac{V_{u1}^2}{V_1^2}$$

But, from Fig. 3.3, it is evident that $V_{u1}/V_1 = \cos\tau$. Hence

$$\eta_{t\text{max}} = \cos^2\tau \qquad (3.5)$$

By inspection of Eq. (3.5) it is evident that the maximum efficiency ideally obtainable from an impulse turbine approaches 100 percent as the angle τ approaches zero.

The implications of the above demonstration can be best understood from the vector diagrams in Fig. 3.3. Note that for maximum efficiency, V_2 is axial; i.e., the kinetic energy in the tangential component is equal to zero, and the kinetic energy at the outlet is a minimum.

This discussion of turbine efficiency points up a fundamental requirement for all types of turbines and compressors; namely, *for good efficiency there must be both a stator and a rotor* in order to have both a substantial tangential component for the absolute fluid velocity in the rotor and an essentially zero tangential component in the absolute fluid velocity leaving the machine. An appreciation for this fundamental point is absolutely essential to an understanding of the fluid mechanics of turbomachines. Only windmills, ventilation fans, and propellers can give good efficiency with only a rotor, because the absolute tangential velocity component of the fluid stream is small in relation to the axial velocity.

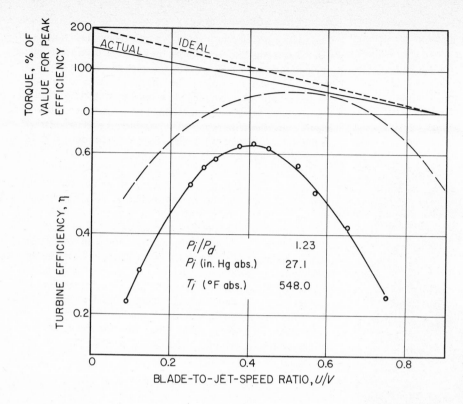

Figure 3.4 Efficiency and torque of a simple impulse turbine as a function of the velocity ratio U/V for a small experimental turbine.[5]

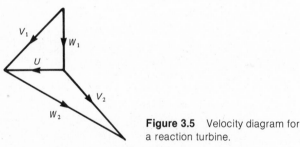

Figure 3.5 Velocity diagram for a reaction turbine.

Performance Characteristics of Impulse Turbines

The basic performance characteristics of an impulse turbine can be deduced from Eqs. (3.1) and (3.4). Figure 3.4 shows curves for both the torque acting on the rotor and the overall efficiency of the machine. Two sets of curves are given, one for the ideal case and the other for an actual machine.[5] Note that the torque is maximum for the condition with the rotor stalled, while the efficiency is a maximum for $U = V_{u1}/2$.

It should be pointed out that the dashed lines in Fig. 3.4 represent an ideal case in which the blade inlet and outlet angles are in effect adjusted for each condition to accommodate them to the fluid flow pattern for the ideal case. Such an adjustment is not practicable in an actual machine, and hence under off-design conditions, turbulence losses result. The difference between the two sets of curves is largely due to such turbulence losses. These effects will be discussed in more detail later.

Reaction Turbines

It was pointed out earlier that, by definition, a purely impulse turbine is a turbine in which the static pressure at the rotor outlet equals that at the rotor inlet, and hence all the energy passing from the fluid to the rotor must come from the kinetic energy in the fluid at the rotor inlet. This requires that the inlet and outlet velocities relative to the rotor, that is, W_1 and W_2, be of equal magnitude (but in opposite directions). In contrast, by definition the purely reaction turbine is one in which all the energy passing from the fluid to the rotor arises from an increase in velocity relative to the rotor developed by a pressure drop across the rotor, with the kinetic energy in the fluid stream at the rotor outlet being equal to that at the inlet. That is, the magnitude (though not the direction) of V_2 is equal to that of V_1. Figure 3.5 shows a typical velocity diagram.

In gas or vapor turbines, large increases in velocity within the rotor can be obtained from an isentropic expansion. Inasmuch as there is a continuum of possible vector configurations between those for purely impulse and those for purely reaction turbines as well as beyond them, the *reaction coefficient* has been devised to indicate the degree of reaction. This coefficient has been taken as the ratio of the static pressure drop to the total pressure drop across the rotor, and will be discussed further in a later section.

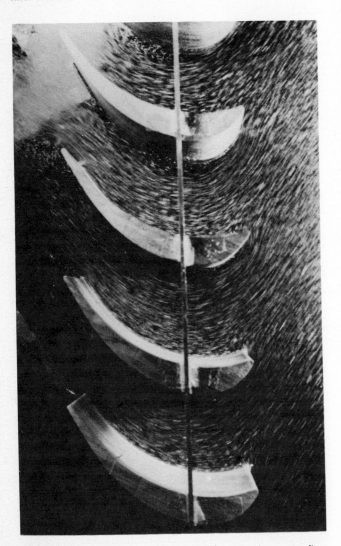

Figure 3.6 Cascade of turbine blades under test in a water flow channel. Flow separation regions are visible along the backs of the blades, together with the broad wakes that they induced. (*Courtesy Escher-Wyss, Ltd.*)

Axial-Flow Compressors

The basic principles outlined above for turbines also apply to axial-flow compressors or pumps, the key difference being that, in a compressor, the energy flow is from the wheel to the fluid rather than from the fluid to the wheel. A helpful insight into the interrelationship between flow velocity vectors, the static and dynamic pressures, and momentum interchange between the fluid and the wheel is gained by considering a rectangular set of blades, or *cascade*, of the sort employed in the development of airfoil sections for turbine and compressor blades (Fig. 3.6). To simplify the flow test work, the axes of the blades are parallel rather than on lines radiating from a central axis as they would be in a rotor. One can treat an element of such a cascade of compressor stator blades as if it were the duct of the diagram in Fig. 3.7. This case can be considered as one in which the fluid has been accelerated tangentially in the rotor and then changes direction in the stator so that it leaves the stator in a nearly axial direction.

The flow passage cross-sectional area between streamlines increases across the cascade, and hence the fluid velocity is reduced between the inlet and outlet. From Bernoulli's theorem it follows that an increase in static pressure must result, and, assuming a frictionless fluid of density ϱ, this static pressure increase Δp is

$$\Delta p = p_2 - p_1 = \frac{\varrho_1 W_1^2}{2} - \frac{\varrho_2 W_2^2}{2} \tag{3.6}$$

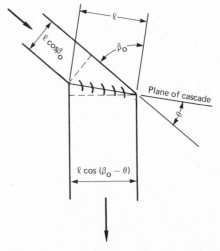

Figure 3.7 Diagram showing the principal geometric parameters for a cascade of blades turning the flow through an angle in a channel.

The basic variables can be correlated readily. As indicated in Fig. 3.7, the geometric variables are simply the length of the cascade l across the stream, the angle θ through which the fluid stream is turned, and the angle β_0 between the plane of the cascade and the fluid inlet velocity. Since the mass flow rate entering the cascade equals that leaving it, we may write

$$\varrho_1 W_1 l \cos \beta_0 = \varrho_2 W_2 l \cos (\beta_0 - \theta)$$

Neglecting the effects of changes in density, we may write

$$W_2 = W_1 \frac{\cos \beta_0}{\cos (\beta_0 - \theta)} \tag{3.7}$$

Substituting the value for W_2 given by Eq. (3.7) in Eq. (3.6), we obtain the static pressure increase Δp as

$$\Delta p = \frac{\varrho_1 W_1^2}{2} - \frac{\varrho_2 W_1^2}{2} \frac{\cos^2 \beta_0}{\cos^2 (\beta_0 - \theta)}$$

$$\Delta p = \frac{\varrho_1 W_1^2}{2} \left\{ 1 - \left[\frac{\cos^2 \beta_0}{\cos^2 (\beta_0 - \theta)} \right] \right\} \tag{3.8}$$

Equation (3.8) can also be derived for a moving cascade of blades such as a compressor rotor. The same identical expression will result from such a derivation if W_1 and W_2 are taken to be relative velocities of the fluid with respect to the cascade. Thus it follows that: *The static pressure change in a frictionless fluid stream passing through any cascade of blades is given by Eq. (3.8), where W_1, θ, and β_0 are the entrance velocity and entrance and exit angles of the fluid stream relative to the blades of the cascade.* The torque input to the rotor, the momentum change in the fluid passing through the rotor, and the energy exchange between the rotor and the fluid follow the same equations as derived above for the impulse turbine.

An excellent set of data from a comprehensive series of tests of static cascades of compressor blades is presented in Ref. 6.

Radial In-Flow Turbines

The Fourneyron turbines were impulse turbines of the radial inflow type, and a hydrodynamically refined version of this type developed in 1849 by J. B. Francis, an American engineer, is now the most widely used type of turbine in hydroelectric power plants. A view looking into the outlet end of a turbine wheel of this type is included in Fig. 3.8.

Figure 3.8 includes a view of another type of impulse turbine —a Pelton wheel—which is used for high-head installations (generally over 100 m). The water jet enters the wheel tangentially in the plane of rotation and is deflected to either side. The

PELTON
260 MW

FRANCIS
255 MW

KAPLAN
440 MW

Figure 3.8 Principal types of turbine rotor used in hydroelectric power plants. (*Courtesy Dominion Engineering Works, Ltd.*)

53

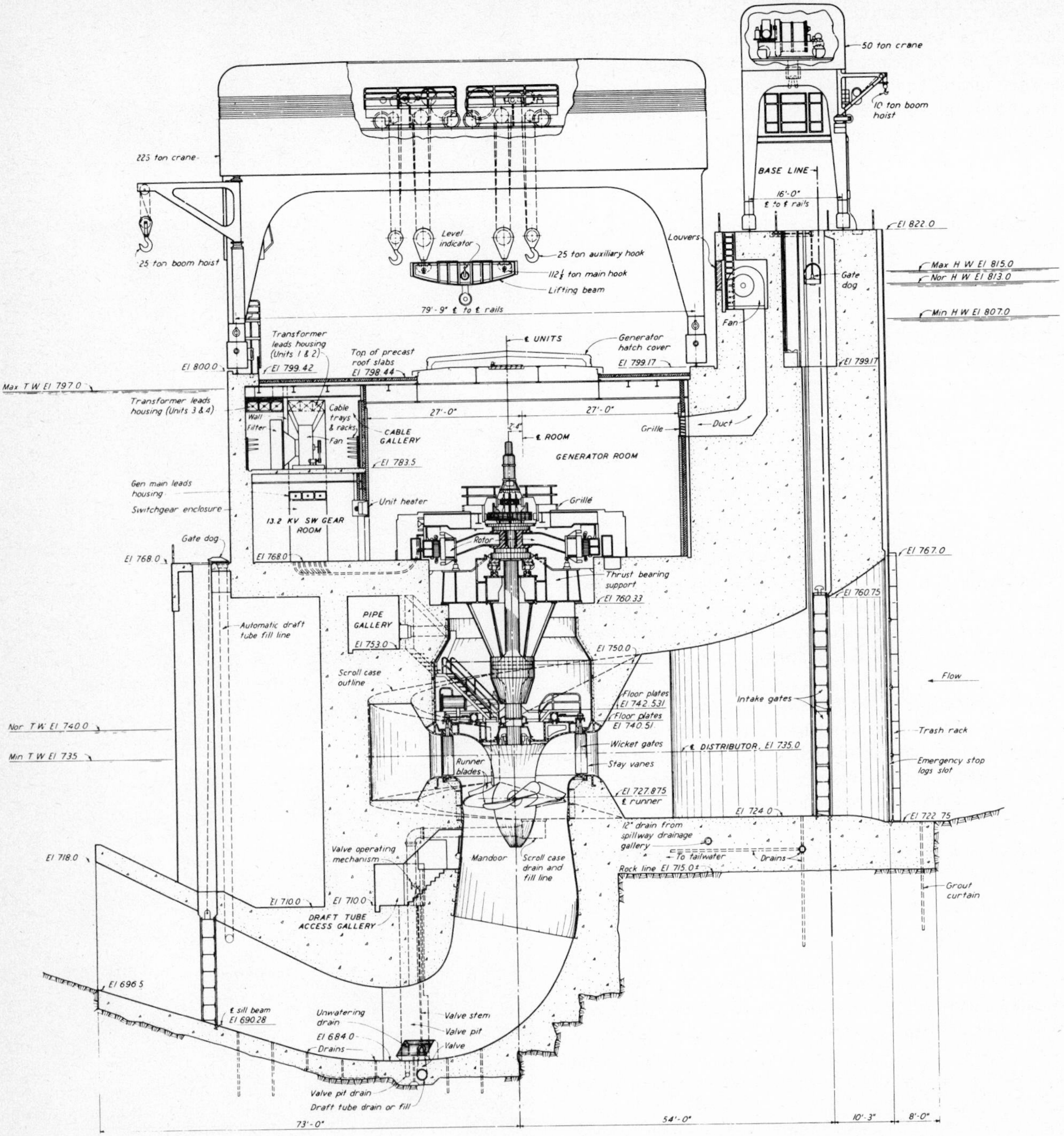

Figure 3.9 Section through a typical powerhouse showing a 22-m (66-ft) head, 35-MW Kaplan turbine installation in the Fort Loudon dam. (*Courtesy TVA.*)

NOTES:
Turbines rated: 44,000 hp, 65 f⁺ head, 105.8 rpm, Lest efficiency 70 ft head. Manufactured by Baldwin Southwark Division, Baldwin Locomotive Works, Philadelphia, Pennsylvania
Generators rated: 35,555 kva, 13,800 volts, 3 ph 60 cycle, 105.8 rpm, 60° rise. Manufactured by Allis-Chalmers Manufacturing Company, Milwaukee, Wisconsin

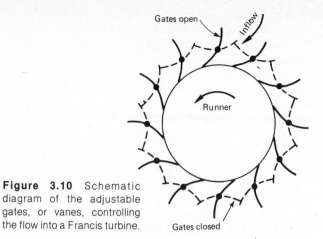

Figure 3.10 Schematic diagram of the adjustable gates, or vanes, controlling the flow into a Francis turbine.

Figure 3.11 Impeller for a centrifugal compressor. (*Courtesy Creare, Inc.*)

wheel is not submerged but operates above the water level of the tailrace. Hence some of the head ideally available is lost, but this is a small fraction of the total head and is partly offset by the lower drag on the wheel resulting from its rotation in air rather than water. The output is controlled by needle valves in the nozzles, giving good efficiency over a wide load range. Also included in Fig. 3.8 is a propeller-type turbine that is best suited for low-head installations, a matter that will be discussed in a later section.

Figure 3.9 shows a typical installation of a propeller-type turbine with a vertical shaft so that the generator can be mounted well above the water level in the tailrace. Essentially the same installation geometry is used for Francis turbines except that the water is directed tangentially into the turbine wheel (often called the *runner* in hydraulic turbines) through a set of guide vanes, or *gates*, whose angle can be adjusted as indicated in Fig. 3.10 to give a high efficiency over a wide range of loads. (See Fig. 19.8.) The basic velocity vector diagrams outlined above are also applicable here with appropriate allowances for the three-dimensional character of the flow.

Centrifugal Compressors and Pumps

Centrifugal compressors and pumps operate in essentially the same way as axial-flow machines except that they also take advantage of the centrifugal field to develop additional head, which amounts to about half the total increase in head produced. A typical rotor is shown in Fig. 3.11. Three basic types of rotor, or *impeller*, are used; i.e., the impeller vanes are backward-swept, radial, or forward-swept. The tangential velocity relative to the casing is less than the wheel tip speed for the backward-swept impeller vanes, roughly equal to the impeller tip

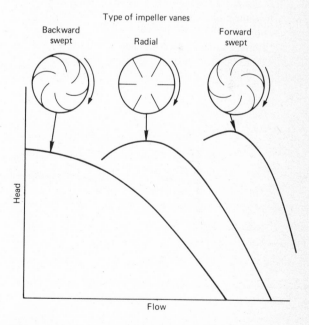

Figure 3.12 Types of centrifugal impeller and typical performance curves.

speed for the radial vanes, and higher than the impeller tip speed for the forward-swept vanes, giving the head-flow characteristics shown in Fig. 3.12. The radial-vaned, or nearly radial-vaned, impellers are generally used in centrifugal compressors, while impellers with backward-swept vanes are used mainly in pumps. The forward-swept vanes are rarely used except in low-head blowers for ventilating systems where noise (which is primarily a function of wheel tip speed) is a factor and it is desired that a maximum head be obtained with a given wheel tip speed.

The impeller in a centrifugal pump or blower may discharge directly into a snail shell–shaped casing (ideally an Archimedes spiral) commonly called a *volute* or *scroll*. Better recovery of the tangential velocity energy in the fluid leaving the wheel is obtained if a *vaned diffuser* is employed in the radial region between the impeller and the volute as shown in Fig. 3.13. Velocity vector diagrams for the flow through a centrifugal compressor are shown in Fig. 3.13.

For an ideal incompressible fluid flowing through a radial-vaned impeller in which the flow passage areas at the inlet and outlet are equal, the static-pressure rise in the impeller is equal to $\varrho V^2/2$, where ϱ is the fluid density and V is the change in the tangential velocity of the fluid in passing through the impeller. Since the tangential velocity of the fluid at the impeller inlet is normally zero, for practical purposes V can be taken as equal to the tip speed of the impeller. If the impeller vanes are not radial —as is sometimes the case—the reasoning becomes more complex but the same basic relations between changes in static pressure and changes in fluid velocity still hold.

An additional static-pressure rise can be obtained after the fluid leaves the impeller by converting its high tangential-velocity energy into a rise in pressure. A set of expanding passages called a *diffuser* can be arranged to slow down the high-velocity stream with a minimum of eddy losses. The pressure rise ideally attainable is the same as that from the acceleration of the fluid in the impeller, viz., $\varrho V^2/2$, where V may again be taken as equal to the tip speed of the impeller if the vanes are radial. If compressibility effects on the density are small, a first approximation to the total ideal pressure rise for the compressor is the sum of the two, or, algebraically,

$$\text{Ideal pressure rise} = \text{impeller } \Delta p + \text{diffuser } \Delta p$$

$$= \varrho \frac{V^2}{2} + \varrho \frac{V^2}{2} = \varrho V^2$$

Basic Differences between Turbines and Compressors in Limitations Imposed by Flow Separation

There is a basic difference between turbines and compressors that stems from a difference in the tendency for flow separation to occur and lead to gross deviations from the smooth streamline flow conditions represented by the idealized velocity vector diagrams. The phenomenon involved can be envisioned by examining the sketches of flow patterns in Fig. 3.14. A common form of flow separation is shown at the top for flow around an elbow; momentum carries the fluid toward the outer radius in going around the bend so that the streamlines do not follow the inner wall and a vortex develops giving a region of highly turbulent separated flow that extends some distance downstream of the elbow. The fluid following the streamlines accelerates in the elbow, where the streamlines draw more closely together, leading to a drop in the local static pressure. The bulk of the increase in velocity energy is dissipated in turbulence in the fluid

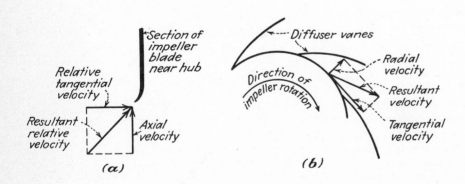

Figure 3.13   Components of the relative velocity for the airstream passing through a centrifugal compressor for (a) an impeller inlet guide vane and (b) the inlet to the diffuser vanes.

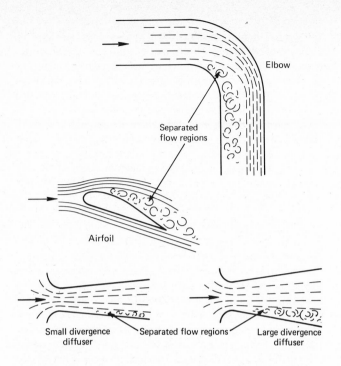

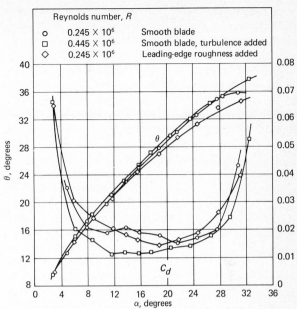

Figure 3.15 Effects of the airfoil angle of attack α on the turning angle θ and the drag coefficient C_d for a cascade of NACA 65-(12)10 airfoils inclined at an angle of 45 degrees relative to the inlet airstream.[6]

Figure 3.14 Typical examples of flow separation.

flow downstream of the elbow, and this is responsible for a net loss in static pressure equal to around one velocity head based on the fluid velocity entering the elbow.

Another example of flow separation is shown in Fig. 3.14 for flow over an airfoil. In this case the separated flow condition is known as *stall*, because in an airplane the transition from normal to separated flow is accompanied by a large increase in drag and a loss in lift that has caused many aircraft to crash in the course of taking off or landing.

The above extreme cases are included here because they help one to envision the more subtle flow separation problem in diffusers such as those shown at the bottom of Fig. 3.14. As the diffuser channel widens and the fluid slows down, the static pressure increases. This creates an unstable condition in the boundary layer close to the wall, where the velocity and momentum are nearly zero. At the wall there is an adverse pressure gradient; i.e., the increasing static pressure in the direction of flow tends to cause the fluid in the boundary layer to flow back up the diffuser in the opposite direction from the main stream. This may cause only a small degree of flow separation in a channel with a relatively small angle of divergence. The separation may become severe, however, if the angle of divergence is large, giv-

ing a steep adverse pressure gradient. The larger degree of flow separation and the greater turbulence give a large loss in the efficiency for the conversion of velocity energy into static pressure. It is for this reason that the rate of divergence in straight diffusers is kept to that of an equivalent cone having an included angle of 7 degrees, which happens to be small enough so that shear forces in the boundary layer are sufficient to prevent backflow and separation. Similarly, in a cascade of airfoils employed to produce an increase in static pressure in a compressor, the turning angle must be kept small enough so that flow separation does not occur. (Note that flow separation is beginning to develop in the cascade of Fig. 3.6.) The quantitative effects of increasing the turning angle on the performance of a typical cascade are shown in Fig. 3.15, taken from a comprehensive study of the effects of airfoil shape and flow-passage geometry on the characteristics of cascades designed for axial-flow compressors.[6] The curves in Fig. 3.15 were obtained by varying the nominal angle of attack of the airfoils with respect to the entering airstream. Note the initially gradual and progressively more rapid increase in the drag coefficient as the angle of attack is increased beyond 18 degrees, and note also the concurrent dropping off in the rate of increase in the turning angle effected by the cascade as a consequence of increasing flow separation. In a compressor this would mean a decreasing efficiency, and at still higher angles of attack than shown in Fig. 3.15 the flow separation would become so severe that the lift coefficient would drop rapidly and the blades would stall, giving an abrupt drop in the pressure rise.

PERFORMANCE CHARACTERISTICS

The wide variety of rotor and stator geometries surveyed in the previous section implies that a correspondingly wide variety of performance characteristics are available to the power plant system designer. Although confusing to the uninitiate, these performance characteristics can be related in a logical fashion to the velocity vector diagrams outlined in the previous section to give a consistent rational picture for the complete range of turbomachines.

Coefficients Used to Describe Turbomachine Performance

The performance characteristics of turbomachines are quite complex. An analysis of the performance of any particular machine is greatly facilitated by the use of a number of coefficients. Turbines or compressors are always designed to satisfy some particular set of conditions, called the *design condition*, which represents a particular geometric configuration of velocity vectors such as those in Fig. 3.3. The blade inlet and outlet angles are established on the basis of these vectors, and any deviation in fluid flow direction from that of the design condition should be expected to result in turbulence losses. Hence it seems reasonable to derive the commonly used performance coefficients from the generalized velocity vector diagram.

Head

Perhaps the most important parameter in turbomachine performance is the change in fluid pressure across the machine. This is commonly called the *head*, because the concept stemmed originally from work with hydraulic turbines and pumps. For hydraulic machines this pressure differential is usually expressed as the equivalent height of a column of the fluid in meters (or feet). For example, the difference in elevation between the water levels behind a dam and in the tailrace downstream of the turbine is the static head H_s. For gas and vapor turbines the terms *adiabatic* or *isentropic head* (H_{ad} or H_{is}) are often used, and they are defined as the change in enthalpy across the turbine divided by the heat equivalent of work. For example, Btu/lb ÷ Btu/ft·lb = ft, or J/kg ÷ J/kg·m = m.

Pressure Ratio

In machines handling compressible fluids the pressure rise in a compressor or the pressure drop in a turbine for a given velocity diagram is directly proportional to the inlet pressure. Thus the *pressure ratio* across the machine is a useful parameter. The pressure ratio obtained per stage depends on the tip speed, and the overall pressure ratio across the machine is the product of the number of stages, n, and the stage pressure ratio raised to the nth power. (This convenient relationship is a prime reason for using the pressure ratio rather than the pressure rise.)

In turbines, the pressure ratio is taken as the absolute inlet pressure divided by the absolute outlet pressure. Large pressure ratios per stage (of the order of 2) can be used without difficulty with flow separation and eddy losses from backflow in the boundary layer, because the pressure gradient is favorable and induces a flow in the boundary layer that is in the same direction as that of the main stream.

A steam turbine having an inlet pressure of 240 atm and a condenser pressure of 0.05 atm provides a good example of the use of the presure ratio. The overall pressure ratio is 4800, while the stage pressure ratio would be no more than 2. Thus the approximate number of stages, n, can be determined from the expression $2^n = 4800$, or $n = \sim 13$ stages. For a gas turbine, on the other hand, the overall pressure ratio might be ~ 16 making the number of stages ~ 4.

In axial-flow compressors, the pressure ratio is taken as the ratio of the outlet pressure to the inlet pressure. The static pressure increases in the direction of the main stream flow, and hence in axial-flow machines the adverse pressure gradient tends to cause backflow in the boundary layer. This backflow can be avoided only by keeping the flow passage aerodynamically clean, the surfaces smooth (to keep the boundary layer thin), and the adverse pressure gradient sufficiently small that it is overcome by the shear forces in the boundary-layer region without inducing instabilities and flow separation. As a consequence, the pressure ratio in axial-flow compressors was commonly kept below 1.14 up to 1970. It is for this reason that the number of stages in the axial-flow compressor of a gas turbine is much higher than in the turbine (Fig. 3.1). For a stage pressure ratio of 1.14 and an overall compressor pressure ratio of 16, for example, the number of compressor stages required would be ~ 21. Elegant computer programs developed in the 1970s have made it possible to obtain stage pressure ratios of 1.3, but engines with these high blade loadings are much more subject to stall in the blades, surging, and flow reversal. (See Fig. 3.24.)

It may be noted here that in centrifugal compressors the centrifugal field in the rotor acts to keep the boundary layer moving in the same direction as the main stream so that backflow in the boundary layer is less of a problem. This, coupled with the pressure rise obtained from the centrifugal field, makes possible pressure ratios of 2 or more per stage, and units having pressure ratios in this range are often employed. Centrifugal compressors are treated only briefly here because they are inherently restricted to relatively small airflow rates—for which they are indeed well suited.

Tip Speed

As will be discussed later, compressibility effects in gases and vapors or cavitation in liquid streams may limit the velocities that can be used and hence the pressure change or *head* per stage. The rotor tip speed is a convenient parameter to indicate the velocities. There is usually an incentive to obtain as high a head per stage as possible without excessive losses in efficiency in order to minimize the number of stages. In practice, for gases this commonly leads to compressor wheel tip speeds of the order of 305 m/s (1000 ft/s), i.e., approaching sonic velocity at the inlet. The higher sonic velocities at the higher temperatures near the compressor outlet and in the turbine would permit higher tip speeds, but mechanical design considerations normally militate against this.

RPM

Turbine-generator units and electric motor–driven pumps and compressors are constrained to operate at one of a number of discrete shaft speeds because of the basic characteristics of ac generators and motors. For most applications the electric system operates at 60 hertz (Hz), and this means that the highest shaft speed is 60 rps, or 3600 rpm, obtained with a two-pole machine with the rotor making one revolution per electric-current cycle. The next highest speeds are 1800, 1200, and 900 rpm for four-, six-, and eight-pole machines, respectively. Still larger numbers of poles are used for large hydroelectric turbine units; e.g., a 72-pole machine would operate at 100 rpm. Similar relations hold for motor-driven compressors except that slip between the rotating field and the rotor in induction motors causes a small loss in speed, e.g., 1750 rpm instead of 1800 rpm for a four-pole machine. In Europe most electric systems operate at 50 Hz, which gives a set of shaft speeds that are five-sixths of those for 60 Hz.

Rotor Diameter

The rotor tip speed desired and rpm define the rotor diameter. This presents an additional problem, because for favorable proportions of the velocity vector diagrams and good efficiency the fluid flow rate should be related properly to the rotor diameter. Generally, the axial velocity through a machine should be

roughly one-third of the wheel tip speed, but the situation varies somewhat with the type of machine. For example, for a given rotor diameter, the flow-passage area available in the impeller of a centrifugal compressor such as that of Fig. 3.11 is much smaller than that available in an axial-flow machine such as that of Fig. 3.1. Further, the pressure rise per stage is much higher for a given rotor tip speed. Thus the centrifugal compressor is best suited for low flows and high heads, whereas the axial compressor is best suited to high flow rates, and the number of stages can be increased to suit the head required.

Flow Coefficient

Probably the most important single feature of the velocity vector diagram is the ratio of the axial-flow velocity of the fluid to the tip speed of the blade. Swiss and German designers use the symbol ϕ for this ratio in the ideal machine and call it the *coefficient of flow*. Since it is desirable to put this in terms of more commonly used quantities such as the total volume flow rate Q, the rotor diameter D, and the rotational speed n, and since actual hub diameters differ from one machine to the other, they define their coefficient for an ideal machine with zero rotor hub diameter to make performance data for all machines directly comparable. Thus, by definition,

$$\phi = \frac{V_a}{U_{\text{ideal}}} = \frac{Q/(\pi D^2/4)}{\pi n D}$$

$$= \frac{Q}{(\pi^2/4)nD^3} \qquad (3.9)$$

British and U.S. designers more commonly use a similar quantity proportional to ϕ but a little bit more conveniently calculated from test results because the constant $\pi^2/4$ has been dropped. That is,

$$\text{Flow coefficient} = \frac{Q}{nD^3} \qquad (3.10)$$

Specific Speed

The relationships between power requirements, fluid flow rate, rotor diameter, and shaft speed are simplified in hydraulic turbines and pumps, because the liquid density is a constant. One of Fourneyron's major contributions was the recognition that the basic characteristics of a geometrically similar set of turbines or pumps are similar irrespective of size. Thus, once one such machine has been tested, it is possible to define its characteristics with a single parameter, the *specific speed*. The basis for this is the following set of fundamental relationships derived from the vector diagrams presented earlier. These key relationships between the head H_s, the flow Q, the power P, the rotor diameter D, and the rotor speed N, are as follows:

$$H_s \sim N^2 D^2 \quad \text{or} \quad D \sim \frac{H_s^{1/2}}{N}$$

$$Q \sim \text{(fluid velocity)(flow area)} \sim ND \times D^2 \sim ND^3$$

$$P \sim \text{(volumetric flow)(head)} \sim QH_s$$

For a given rotor geometry it is possible to derive from these expressions a parameter that is independent of the diameter, and hence a set of test data from a model defines the head-power-speed characteristics of that rotor geometry irrespective of size. That is,

$$P \sim QH_s \sim ND^3 H_s \sim N \left(\frac{H_s^{1/2}}{N}\right)^3 H_s \sim \frac{H_s^{5/2}}{N^2}$$

$$N \sim \frac{H_s^{5/4}}{P^{1/2}}$$

$$\frac{NP^{1/2}}{H_s^{5/4}} = \text{constant}$$

The parameter *specific speed* N_s is defined from this relation as

$$N_s = \frac{C_s NP^{1/2}}{H_s^{5/4}}$$

where C_s is a constant that depends on the units employed for H_s, N, and P.

The specific speed is an extremely valuable parameter in preliminary design work because, once the head, power, and rotor speed are defined for a proposed application, one can calculate the specific speed and then choose the rotor geometry to be employed from a set of pump or turbine designs for which models have been built and tested so that their specific speeds and related characteristics are well defined.

In working with the specific speed, one must be careful about the units employed, because values given in the literature vary widely as a consequence of the use of different systems of units. Most of the work in the United States has simply employed rpm, horsepower, and head in feet with the constant $C_s = 1$, and so the specific speed is sometimes defined as the speed at which a reduced-scale model would yield 1 hp with a head of 1 ft. The problem with units is further confused because it is sometimes more convenient to work with a given flow than a given power, in which case the expression $P \sim QH_s$ can be used to obtain a somewhat different expression for the specific speed. That is,

$$N_s \sim \frac{NP^{1/2}}{H_s^{5/4}} \sim \frac{N(QH_s)^{1/2}}{H_s^{5/4}} \sim \frac{NQ^{1/2}}{H_s^{3/4}}$$

The effect of the choice of units on the value of the specific speed can be seen from the following set of terms given by Csanady,[3] who uses the symbol Ω for the specific speed with N in

Practical quantity, x	Consistent quantity, Ω
$x_1 = \dfrac{\text{rpm } \sqrt{\text{cfs}}}{\text{ft}}$	$\Omega = \dfrac{x_1}{129}$
$x_2 = \dfrac{\text{rpm } \sqrt{\text{gpm*}}}{\text{ft}}$	$\Omega = \dfrac{x_2}{2730}$
$x_3 = \dfrac{\text{rpm } \sqrt{\text{hp}\dagger}}{\text{ft}}$	$\Omega = \dfrac{x_3}{42}$
$x_4 = \dfrac{\text{rpm } \sqrt{\text{metric hp}\dagger}}{\text{m}}$	$\Omega = \dfrac{x_4}{187}$
$x_5 = \dfrac{\text{rpm } \sqrt{\text{m}^3/\text{s}}}{\text{ft}}$	$\Omega = \dfrac{x_5}{53}$

*gpm = gallons per minute (U.S. gallons).

†Working fluid is water.

radians per second, Q in cubic feet per second, and the head in square feet per second squared (that is, gH_s).

One use of the specific-speed parameter is illustrated by Fig. 3.16, which shows the ranges of specific speed and head for which the three principal types of hydraulic turbine are suited. In this case the specific speed was calculated from the rpm, the horsepower, and the head in feet with $C_s = 1$. Referring to the three principal types of turbine shown in Fig. 3.8, note that Pelton wheels, which are used with high heads and low-inlet flow areas for a given rotor diameter, are characterized by low specific speeds of 3 to 7; Francis-type rotors, which are well suited to intermediate heads and fairly large flow passage areas, are characterized by moderate specific speeds of 15 to 100; and propeller-type turbines, which are well suited to low heads and have large flow-passage areas, are characterized by specific speeds of 90 to 180. Thus the specific speed may be thought of as an indication of the ratio of the minimum water-flow-passage area in the machine to the square of the rotor diameter.

Turbine Efficiency

The turbine efficiency η_t is normally defined as the ratio of the actual work output obtained to the work ideally available from the hydraulic head or an isentropic expansion. That is,

$$\eta_t = \frac{\text{actual work output}}{\text{ideal work output}} \tag{3.11}$$

Compressor Efficiency

The compressor efficiency η_c is normally defined in much the same manner as the turbine efficiency, except that to maintain the proper sense of efficiency it is the ratio of the isentropic work to the actual work input. That is,

$$\eta_c = \frac{\text{ideal work input}}{\text{actual work input}} \quad (3.12)$$

Note that the *stage efficiency* (i.e., the efficiency of a single stage) is inherently greater than the efficiency of a complete machine because of the cumulative effects of deviations from ideality in coupling a series of less-than-perfectly matched stages. This effect is greater in compressors than in turbines.

The Pressure Coefficient

The ratio of the total pressure change in the fluid passing through a rotor to a fictitious velocity pressure calculated from the rotor tip speed is widely used as a measure of the rate of energy exchange between the rotor and the fluid. German and Swiss practice has been to call this ψ, the *coefficient of pressure*. Specifically, they define this as

$$\psi = \frac{\text{change in total pressure}}{\varrho(U^2/2)} \quad (3.13)$$

British and U.S. practice has been to use a similar coefficient that was developed for centrifugal compressors having radial-vaned impellers. In that particular case, the ideal total pressure rise in the rotor is twice the fictitious velocity pressure based on rotor tip speed. Hence most British and U.S. engineers define the *pressure coefficient* as

$$\text{Pressure coefficient} = \frac{\text{change in total pressure}}{\varrho U^2} \quad (3.14)$$

Thus the pressure coefficient of a given machine has just half the numerical value of the quantity ψ.

Either of these coefficients may be related to the velocity vector diagram. Since the change in total pressure, ΔH (H is the total head), in passing through a turbine rotor is equal to the sum of the changes in the velocity and static pressures, we may write

$$\Delta H = \frac{\varrho}{2}(V_1^2 - V_2^2) + \frac{\varrho}{2}(W_2^2 - W_1^2) \quad (3.15)$$

The first term of Eq. (3.15) represents the drop in velocity pressure between the rotor inlet and outlet, while the second term represents the drop in static pressure. The basis for the latter was developed above. The expression for ΔH can be put into a somewhat different form by substituting the tangential and axial components for the fluid velocities; that is,

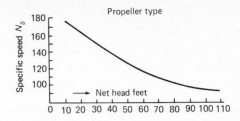

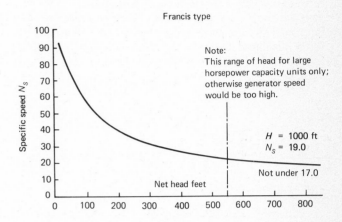

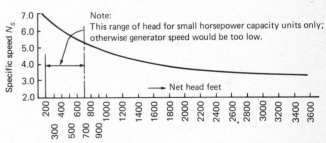

Figure 3.16 Recommended upper limit values for the specific speed based on one runner and one jet for impulse-type machines. (*Courtesy Allis Chalmers.*)

$$\Delta H = \frac{\varrho}{2}(V_{u1}^2 + V_a^2 - V_{u2}^2 - V_a^2) +$$

$$\frac{\varrho}{2}(W_{u2}^2 + V_a^2 - W_{u1}^2 - V_a^2)$$

$$= \frac{\varrho}{2}(V_{u1}^2 - V_{u2}^2 + W_{u2}^2 - W_{u1}^2) \quad (3.16)$$

Thus, by definition,

$$\psi = \frac{\Delta H}{\varrho(U^2/2)} = \frac{(\varrho/2)(V_{u1}^2 - V_{u2}^2 + W_{u2}^2 - W_{u1}^2)}{\varrho(U^2/2)}$$

$$= \frac{(V_{u1} - V_{u2})(V_{u1} + V_{u2}) + (W_{u2} - W_{u1})(W_{u2} + W_{u1})}{U^2}$$

$$(3.17)$$

Note that all the operations indicated above are *scalar*; this will be true throughout the balance of this derivation. However, there is a caveat, because the quantities are components of vectors, and the sense of those components must be kept in mind. A vector pointing to the left in Fig. 3.3 has been taken as positive, and one to the right as negative. By inspection of Fig. 3.3 it is also evident that

$$U = V_{u1} - W_{u1} = V_{u2} - W_{u2} \qquad (3.18)$$

Hence

$$2U = (V_{u1} - W_{u1}) + (V_{u2} - W_{u2})$$

$$= (V_{u1} + V_{u2} - W_{u2} - W_{u1}) \qquad (3.19)$$

Further, from Eq. (3.13),

$$V_{u1} - V_{u2} = W_{u1} - W_{u2} \qquad (3.20)$$

Substituting these values for U and $(W_{u2} - W_{u1})$ in Eq. (3.17), we obtain

$$\psi = \frac{2(V_{u1} - V_{u2})(V_{u1} + V_{u2}) - (V_{u1} - V_{u2})(W_{u2} + W_{u1})}{U(V_{u1} + V_{u2} - W_{u2} - W_{u1})}$$

$$= \frac{2(V_{u1} - V_{u2})}{U} = \frac{2\Delta V_u}{U} \qquad (3.21)$$

Thus, by definition, the pressure coefficient for the ideal case is equal to twice the tangential velocity increment in the rotor divided by the rotor tip speed.

The Reaction Coefficient

As mentioned above, the *reaction coefficient* is used to indicate the degree of reaction in turbines designed to fall in the region between pure impulse and pure reaction turbines, i.e., in the region in which both the fluid velocity and the static pressure drop as the fluid passes through the rotor. The reaction coefficient $\varkappa$ has been defined as the ratio of the static head H_s to the total head H for the stage; that is,

$$\varkappa = \frac{\Delta H_s}{\Delta H}$$

From Eq. (3.16),

$$\Delta H_s = \frac{\varrho}{2}(W_2^2 - W_1^2)$$

$$= \frac{\varrho}{2}(W_{u2} - W_{u1})(W_{u2} + W_{u1}) \qquad (3.22)$$

Hence, substituting for ΔH from Eq. (3.16) and ΔH_s from Eq. (3.22),

$$\varkappa = \frac{(W_{u2} - W_{u1})(W_{u2} + W_{u1})}{(V_{u1} - V_{u2})(V_{u1} + V_{u2}) + (W_{u2} - W_{u1})(W_{u2} + W_{u1})}$$

Substituting for $(V_{u1} - V_{u2})$ from Eq. (3.20),

$$\varkappa = \frac{(W_{u2} - W_{u1})(W_{u2} + W_{u1})}{(W_{u1} - W_{u2})(V_{u1} + V_{u2}) + (W_{u2} - W_{u1})(W_{u2} + W_{u1})}$$

$$= \frac{-(W_{u2} + W_{u1})}{(V_{u1} + V_{u2}) - (W_{u2} + W_{u1})} \qquad (3.23)$$

Substituting $2U$ from Eq. (3.19), we obtain

$$\varkappa = \frac{-(W_{u2} + W_{u1})}{2U} \qquad (3.24)$$

Since the sign of $(W_{u2} + W_{u1})$ is negative for reaction turbines the coefficient ψ becomes positive for these machines.

Use of Coefficients for Compressors

All the above coefficients except Reynolds number have been formulated by definition. Hence, although they were originally prepared for turbines, they can be extended to compressors simply by using similar definitions for these machines. The only change that occurs is that there is a static pressure rise instead of a drop in compressors and hence the coefficient ψ becomes negative for compressors. As a matter of convenience, the negative sign is often omitted in practice, however.

The pressure coefficient is limited in compressors by flow separation and "stalling" of the blade when the adverse pressure gradient becomes too great. This effect is shown in Fig. 3.17 for a typical stage in an axial-flow compressor. Note that curves are given both for the pressure coefficient as a function of the flow coefficient and for the pressure ratio as a function of the corrected airflow ratio. Compressors cannot be operated at flow coefficients smaller than that at which stall occurs, i.e., to the left of point 1 on these curves.

Reynolds Number

It is evident that the actual fluid flow pattern differs from the ideal in that there are low-velocity regions in the boundary lay-

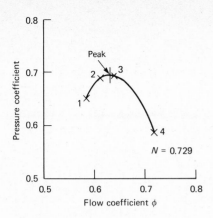

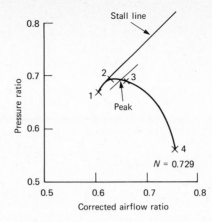

Figure 3.17 Determination and location of the peak pressure coefficient value and the stall line for a particular speed.[7]

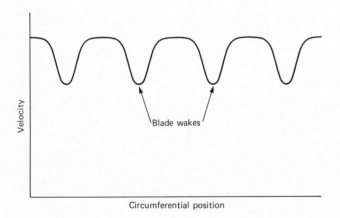

Figure 3.18 Velocity distribution one chord length downstream of a compressor stator plotted from test data given in Ref. 8.

ers. In addition to the frictional losses thus entailed, there are other losses resulting from the disparity between the velocity vector diagrams for the design condition and the low-velocity regions in the wakes of blades resulting from the boundary layers. This effect can be visualized with the help of Fig. 3.18.[8] While a theoretical demonstration of the effect of Reynolds number, Re, on these losses is very complex, it is not difficult to visualize that these losses diminish with a reduction in the ratio of the volume of fluid flowing in the boundary layer to the total fluid flow through the machine. In short, an increase in Re in effect gives a reduction in boundary-layer thickness and hence a reduction in losses arising from the presence of the boundary

layer. Tests run with geometrically similar machines with different sizes of blades, at different speeds, etc., have demonstrated the magnitude of these effects. Figure 3.19 shows the results of a set of German tests run in this way on a series of compressors. Note the serious loss in efficiency that accompanies a drop in Re below 100,000, the so-called critical Reynolds number for turbines. For 2.5-cm (1.0-in) chord blades, this means that the rotor tip speed should be at least 58.5 m/s (192 ft/s) in axial-flow compressors for gas turbines.

Mach Number

Shock losses increase very rapidly in turbomachines as fluid velocities approach sonic values. These losses are so important in axial-flow compressors that most designs are made on the basis of a maximum Mach number of about 0.9, and very often it is considered that for high efficiency a value of 0.8 should not be exceeded.

To allow for Mach number effects, compressor performance data are commonly reduced to standard atmospheric conditions by multiplying the flow rate Q by the factor $\theta = \sqrt{T_1/520}$, where T_1 is in degrees Rankine, and dividing the tip speed U by this same factor θ. This eliminates the effect of variations in sonic velocity with air temperature.

Velocity Vector Components and Performance Coefficients

The magnitude of the components of the principal vectors can be expressed in terms of the velocity U and the coefficients ϕ, ψ,

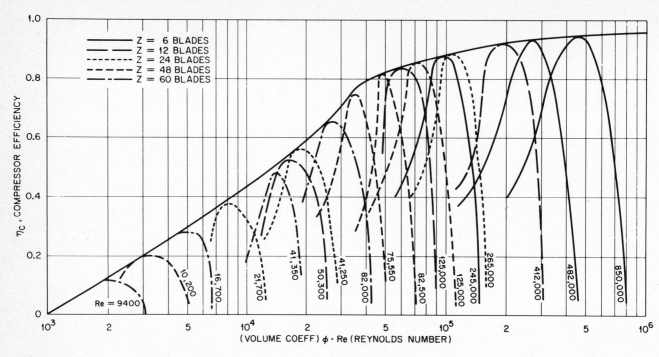

Figure 3.19 Effect of Reynolds number on the efficiency of a series of geometrically similar axial-flow compressors. All proportions except the number of blades and all basic dimensions except the blade chord were held constant.[8]

and $\varkappa$. Substituting $W_{u1} - W_{u2}$ from Eq. (3.20) for $V_{u1} - V_{u2}$ in Eq. (3.21), we obtain

$$\psi = \frac{2(W_{u1} - W_{u2})}{U}$$

From Eq. (3.24),

$$\varkappa = -\frac{W_{u2} + W_{u1}}{2U}$$

This may be rewritten to give

$$W_{u1} - W_{u2} = \frac{\psi U}{2}$$

$$W_{u1} + W_{u2} = -2U\varkappa$$

Adding, we find that

$$2W_{u1} = \frac{\psi U}{2} - 2U\varkappa$$

$$W_{u1} = U\left(\frac{\psi}{4} - \varkappa\right) \tag{3.25}$$

Subtracting,

$$-2W_{u2} = \frac{\psi U}{2} + 2U\varkappa$$

$$W_{u2} = -U\left(\frac{\psi}{4} + \varkappa\right) \tag{3.26}$$

Further, from Eqs. (3.19) and (3.21),

$$V_{u1} + V_{u2} = 2U + (W_{u2} + W_{u1}) = 2U - 2\varkappa u$$

$$V_{u1} - V_{u2} = \frac{\psi U}{2}$$

Adding,

$$2V_{u1} = 2U - 2\varkappa U + \frac{\psi U}{2}$$

$$V_{u1} = U\left(1 - \varkappa + \frac{\psi}{4}\right) \tag{3.27}$$

And subtracting,

$$2V_{u2} = 2U - 2\varkappa U - \frac{\psi U}{2}$$

$$V_{u2} = U\left(1 - \varkappa - \frac{\psi}{4}\right) \qquad (3.28)$$

Summarizing, it is evident that the velocity vector diagram can be constructed for any condition given θ, ψ, $\varkappa$, and U by using the following equations:

$$V_a = \phi U$$

$$V_{u1} = U\left(1 - \varkappa + \frac{\psi}{4}\right)$$

$$V_{u2} = U\left(1 - \varkappa - \frac{\psi}{4}\right)$$

$$W_{u1} = U\left(\frac{\psi}{4} - \varkappa\right)$$

$$W_{u2} = -U\left(\frac{\psi}{4} + \varkappa\right)$$

Velocity Vector Diagrams for Representative Conditions

A set of velocity vector diagrams covering most of the normal range of design conditions for axial-flow turbines and compressors is presented in Fig. 3.20. Schematic diagrams for the corresponding stator and rotor blading are also given. A high value of the coefficient ψ signifies a high turbine output per stage. A value of $\psi = 0$ signifies a condition intermediate between a turbine and a compressor, while a high negative value of ψ signifies a high pressure rise per stage in a compressor. These effects are

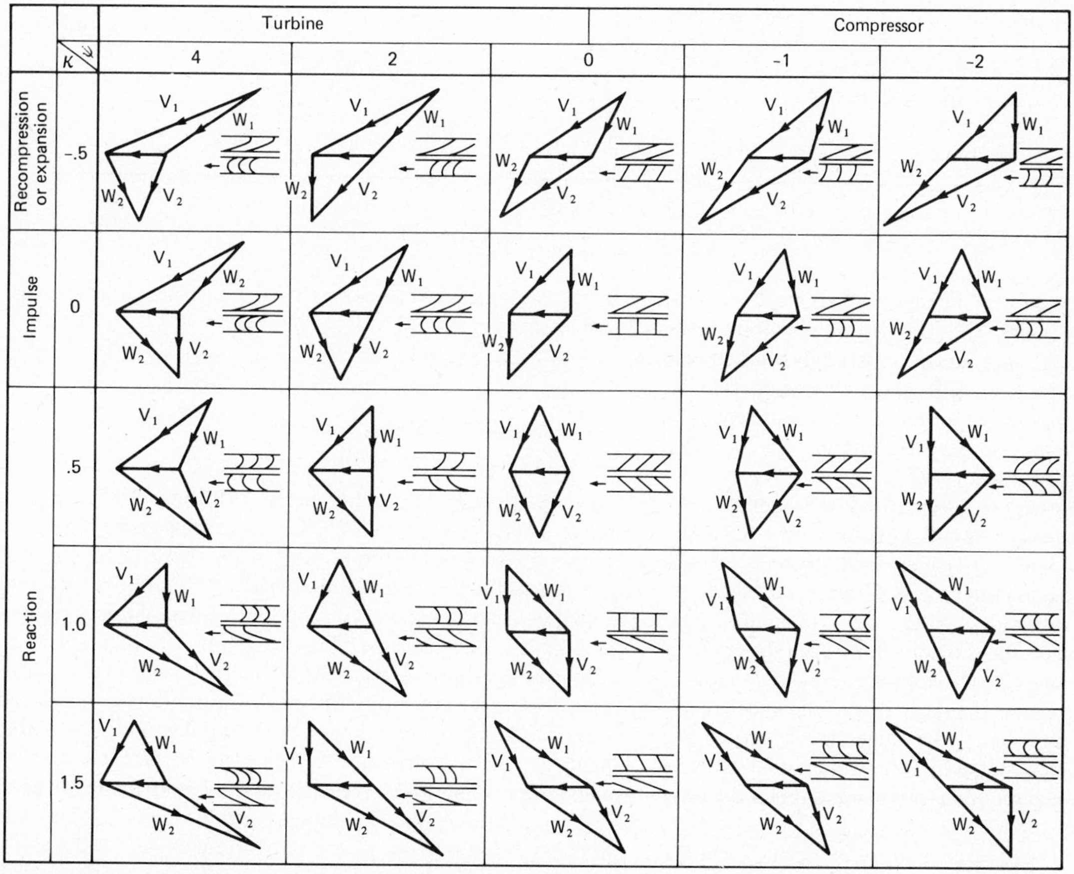

Figure 3.20 Velocity vector diagrams and rotor and stator blade forms for representative turbomachine design conditions for blade speed = fluid axial velocity = constant.[8]

TABLE 3.1   Approximate Fields of Operation of the More Common Turbomachines

	$\varkappa$	ψ	Turbine 4	Turbine 2	Turbine 0	Compressor 1	Compressor 2
Recompression or Reexpansion	−0.5		Impulse turbine operating below normal rotor speed				
Impulse	0		Normal condition for single-stage impulse turbine	Impulse turbine operating above normal speed	Impulse turbine at twice normal speed	Some types of blower	
Reaction	0.5		Normal conditions for 50% reaction turbine	Reaction turbines operating above normal rotor speed. Also conditions at blade tips where ψ at root is higher	Reaction turbines at run-away speed conditions	Turbojet engine compressors	Centrifugal compressor impeller inlets
	1.0		Normal conditions for 100% reaction turbine			Compressors for stationary gas turbine power plants	

most easily traced by looking at the row of diagrams for $\varkappa = 0$, that is, for impulse machines. The effects of the reaction coefficient $\varkappa$ are also very interesting, and are most easily traced by examining the diagrams for $\psi = 4$. By inspection of the relative velocity vectors, it is evident that an actual rise in static pressure occurs for $\varkappa = -0.5$, no change takes place for $\varkappa = 0$, and a drop in static pressure results for $\varkappa > 0$.

Many of the diagrams in Fig. 3.20 are only of academic interest, particularly those for $\psi = 0$ and those for $\varkappa = -0.5$ at the top of the chart. Others are very widely used. Table 3.1 gives a rough idea of the usage of each type of diagram.

Figure 3.20 was drawn with turbines used as the point of departure, and for turbines the stator must precede the rotor to impart a strong tangential velocity component to the flow entering the rotor. In compressors the stator often follows the rotor, because for maximum efficiency the tangential velocity of the fluid leaving the machine should be zero, which usually means that there must be a final set of stator blades to direct the flow into a purely axial direction.

As discussed earlier, large turning angles in the blading must be avoided in axial-flow compressors to prevent serious losses from flow separation (i.e., stalling of the blades), while much

larger turning angles can be employed in turbines without excessive losses.

Ideal Performance Characteristics of Impulse and Reaction Turbines

One of the most important factors affecting turbine performance characteristics is U/V_1, the ratio of the rotor tip speed to the fluid velocity at the rotor inlet. A set of curves for the efficiency and the torque as a function of U/V_1 for both an actual impulse turbine and the corresponding ideal case were presented in Fig. 3.4. The effects of varying degrees of reaction on the ideal case are shown in Fig. 3.21 for a series of turbines designed to operate at the same rotor tip speed. The only losses considered were the kinetic energy losses in the fluid leaving the rotor; hence any actual turbine would give poorer performance because of frictional and turbulence losses. The difference in peak efficiency between the ideal impulse turbine in Fig. 3.4 and that in Fig. 3.21 is due to the difference in nozzle angle. See Eq. (3.5).

The curves for efficiency and torque are substantially similar for all three cases in Fig. 3.21. While at first glance the range of efficient operation appears to increase with the degree of reaction, this is not actually the case. For a given fluid velocity at the rotor inlet, the rotor tip speed can be varied only about ± 25 percent if the resulting loss in efficiency is to be kept to less than 10 percent. This is true for the 100 percent and 50 percent reaction turbines, as well as for the impulse turbine.

It is interesting to note that the torque curves in Fig. 3.21 are straight lines for all three ideal cases. Actual turbines usually show the same characteristics.

The effect of the factor U/V_1 on the total and static pressure drops, as well as the reaction coefficient, are also shown in Fig. 3.21. Note that x varies widely with U/V_1 for turbines that are designed for either 0 percent or 100 percent reaction but stays constant for turbines designed for 50 percent reaction.

Losses

The curves given in Fig. 3.21 were drawn for the ideal case of the perfect frictionless fluid. The only loss considered was that of the kinetic energy in the fluid leaving the rotor. While these idealized curves given in Fig. 3.21 are important in that they represent the best performance conceivable, it is also necessary to consider the various types of frictional loss. These losses may be large or small, depending on the design and operating conditions.

Boundary-Layer Losses

The most important source of loss—and one that cannot be avoided—is caused by fluid friction in the boundary layer between the fluid stream and the surfaces of the blades, the rotor hub, and the housings. As pointed out earlier (Fig. 3.19), this loss diminishes with an increase in Reynolds number. For values of Re greater than 100,000, this loss is commonly less than 5 percent *if there is no flow separation*. If flow separation occurs either in the flow over the blades or in the boundary layer at the bounding surfaces of the hub or external housing, the frictional losses in the resulting eddies will be large. In addition, the entire flow pattern is adversely affected so that some portions of the blade airfoil section operate with an effective angle of attack greater than the design value while other portions operate at less than the design value. The magnitude of the resulting losses increases rapidly with the degree of flow separation, and these losses can easily amount to 10, 20, or 30 percent of the machine output. Thus turbines and compressors must be designed and built with great care to minimize all possibilities of flow separation. This in turn means small turning angles in compressor blading, gradual changes in section for diverging passages, smooth surfaces, etc.

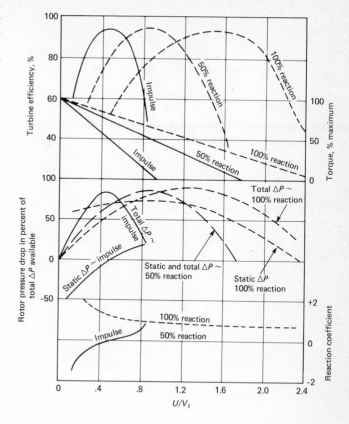

Figure 3.21 Ideal performance characteristics of a series of three similar turbines differing only in the degree of reaction for which they were designed.

The boundary-layer losses are greatly increased if deposits form on the blades in the course of service. Some notion of the extent to which the initial polished surface can become rough is given by Fig. 3.22. This problem is treated further in Chap. 12.

In addition to the direct losses from skin friction, the losses in the boundary layers produce low-velocity wakes in the flow leaving a rotor or stator as indicated in Fig. 3.18. The resulting deviations from the ideal uniform velocity distribution entering the next cascade causes deviations in the velocity vector diagrams in the downstream cascade, and these lead to losses in efficiency. Passage of a blade through these wakes also produces fluctuations in the aerodynamics forces acting on the blade, and this may induce serious blade vibration. It is for this reason that streamlined lacing wires are commonly used to tie the blades together and snub vibrations in the lower stages of large steam turbines and in the inlet stages of compressors, where the blades are relatively long and slender.

Figure 3.22 Ash deposits on the blades of a gas turbine rotor after operation on blast furnace gas. The turbine inlet temperature was 720°C (1328°F), the dust content was 3 to 5 mg/m³, and the operating time since the last cleaning was 7700 h. The total operating time on the turbine was 137,000 h. (*Courtesy Brown Boveri and Co., Ltd.*)

Tip Leakage Losses

A portion of the flow tends to bypass through the clearance between the blade tips and the casing. This leads to a loss not only because of the mass of fluid bypassed but also as a consequence of changes in the flow pattern at the tips of the blade. The latter effect can be reduced by shrouding the blade tips—i.e., enclosing the tips with a cylindrical shroud that rotates with the blades—and this is sometimes done. If it is not, the loss is approximately proportional to the ratio of twice the radial tip clearance divided by the blade height. For large blade heights this loss is likely to be only 1 or 2 percent, but for small-diameter turbines with small blade heights it may run as high as 10 or 15 percent.[10]

Seal Leakage

A small but nontrivial loss occurs as a consequence of flow bypassing the blade rows through the seals both between stages and at the ends of the shaft, particularly at the high-pressure end. This loss commonly runs at about 1 percent, but it may run substantially higher in a small turbine.

Partial Admission

In turbines designed for a small output, the flow may not be sufficient to permit a reasonable blade height with the flow uniformly distributed around the entire perimeter of the turbine wheel. When this is the case, it is usually advantageous to make use of a partial admission turbine, i.e., one in which the high-velocity gas stream from the nozzles enters only a fraction of the total perimeter of the wheel. This increases the blade height and thus reduces the tip losses for that portion of the wheel, but there is a loss associated with churning of the gas in the casing around the balance of the perimeter. The loss commonly runs on the order of 10 percent, but it of course varies widely with the proportions and the exhaust pressure.[10,11]

Moisture Churning

The expansion process in vapor turbines is often carried out well into the saturated region, where as much as 30 percent of the vapor admitted to the turbine will have condensed. This moisture will appear initially as droplets of the order of a few microns in diameter, but these tend to be centrifuged out and deposited on the surfaces of rotor and stator blades, from which they are thrown off as much larger droplets. These large droplets may cause severe erosion of the turbine buckets (depending on detail design effects), and they always cause a loss in efficiency as a consequence of churning of the moisture, i.e., the momentum interchange between the rotor and the liquid droplets when they impinge on or are thrown off from the rotor. This loss commonly amounts to about 1 percent per 1 percent of moisture (Fig. 3.23).[3]

Inlet and Exit Losses

Eddy losses at the inlet and outlet of the machine must be expected because of changes in section or flow direction. Such losses can usually be kept to 1 or 2 percent of the machine output.

Bearing Losses

Mechanical friction in the bearings on the rotor shaft causes a small loss. The force required to shear the air film in the clearances between the rotor and the housings also causes a small loss which is difficult to separate from that in the bearings. The two losses together, however, are commonly less than 1 percent of the machine output.[1,10]

Design Charts

A detailed study of the losses in a wide range of compressors and turbines has led Dr. O. E. Balje to devise a basis for estimation of the performance obtainable from aerodynamically clean machines on the basis of the key parameters.[10,11] Space does not

permit a summary of this work, but typical design charts for turbines are presented in the Appendix.

Performance Maps

In choosing machines for particular applications, it is important to have performance maps showing the effects of the operating speed and flow rate on the pressure ratio and efficiency. Such maps are essential for analyzing the stability and control characteristics of the integrated power plant. Figure 3.24 shows such a map for a typical axial-flow compressor.[12]

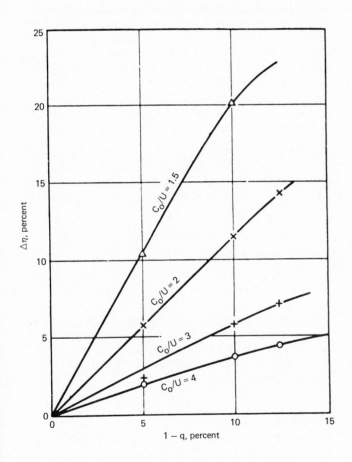

Figure 3.23 Effect of steam wetness on the loss in turbine stage efficiency. The parameter C_0/U is proportional to the energy flow per unit area. Abscissa is wetness fraction (1 − steam quality). *(Courtesy Escher-Wyss, Ltd.)*

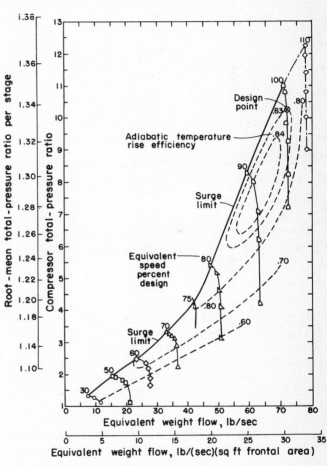

Figure 3.24 Overall performance of an eight-stage, axial-flow research compressor designed and tested at the NASA Lewis Laboratory. *(Courtesy Prof. G. K. Serovy.)*

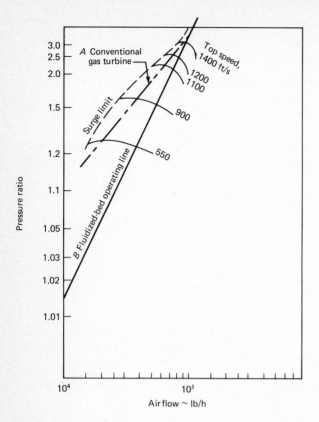

Figure 3.25 Compressor map with typical system operating lines for both a conventional gas turbine and a turbine-compressor–generator unit for a supercharged fluidized-bed furnace and boiler designed to give a steaming rate of 10^5 lb/h.

Surge Limit

An important compressor characteristic defined by a diagram such as Fig. 3.24 is the *surge limit*. Note that each curve for pressure ratio versus flow at constant speed is only half of the similar parabolic curve for a turbine (such as those in Fig. 3.5). The reason for this can be understood readily by considering Fig. 3.25, which shows a compressor performance map with curves for the compressor pressure ratio at a series of constant speeds as a function of flow rate. On this there have been superimposed two curves for system pressure drop as a function of the flow rate: one is for the usual gas turbine system, in which the pressure drop varies as the square of the flow, and one is for a gas turbine utilizing a fluidized-bed combustor in which the pressure drop does not vary as the square of the flow but rather has different characteristics to be discussed in Chaps. 5, 11, and 13. The right-hand portions of the two different system pressure-drop lines intersect the curve for the compressor ratio to the right of the point of maximum efficiency to give stable operating conditions. This is because any perturbation giving a small increase in flow leads to a drop in the compressor outlet pressure, and so the increase in flow cannot be sustained. Similarly, a perturbation causing a decrease in the flow through the system leads to an increase in the compressor outlet pressure, and so its output is greater than the pressure drop through the system and the flow increases back to its original condition. On the other hand, if the pressure drop through the system at a given flow is substantially greater so that it is similar to the top of curve *B* of Fig. 3.25, a flow perturbation could make the flow drop to zero. This gives an unstable condition, and the flow tends to jump to a high value. The high flow rate cannot be sustained by the compressor; hence the flow again drops rapidly and the cycle is repeated. The resulting flow and pressure fluctuations are usually violent and are likely to damage elements of the system such as ducts and expansion joints. Anyone who has experienced compressor surge in an operating system will accept without question the dictum that the condition must be avoided. For maximum efficiency one would like the system to operate just to the right of the surge point, but problems of matching system components over a wide range of loads, provisions for increases in pressure drop through heat exchangers as they become partially clogged with dirt, etc., make it necessary to design for operation at a point well to the right of that for maximum compressor efficiency.

Machine Size and Capacity

The size of a compressor or turbine is closely related to its capacity by the fact that the axial velocity through a stator or rotor is limited by efficiency considerations to a Mach number of about 0.3. For good proportions, the hub diameter must be at least 60 percent of the tip diameter; hence the maximum effective flow-passage area of a stage is roughly 64 percent of the rotor-disk area in axial-flow machines. (In centrifugal machines it is less than 20 percent of the disk area.)

Cavitation

In hydraulic turbines and pumps operating with a relatively low inlet pressure but high velocities, the drop in static pressure associated with liquid-stream acceleration at the inlet may be sufficiently great that the local static pressure will be below the vapor pressure, in which case vapor bubbles will form. These bubbles reduce the density and weight flow rate of the liquid through the rotor, causing the output of the machine to drop. The controlling parameters are the fluid flow rate, the rpm, and the difference between the static pressure of the liquid entering the machine and its vapor pressure. Points plotted in Fig. 3.26 for a series of tests that the author ran on a pump using the full range of possible water-glycol mixtures show that data for a wide range of fluid boiling points can be correlated by means of a set of parameters that follow directly from the equations presented in this chapter.[3] Note that flow under cavitating conditions becomes increasingly unstable as cavitation conditions become more severe, causing the experimental points to scatter over an increasingly wide range.

Cavitation in a pump or turbine is a serious matter not only because it causes a loss in output but, more importantly, because implosion of the vapor bubbles as they move from a low-pressure to a higher-pressure region leads to severe local impact loads on rotor blade surfaces, and this may cause severe erosion. To avoid erosion, local velocities in water pumps and turbines should be kept below 20 m/s, and preferably below 17 m/s if cavitating conditions are likely to be encountered in service.[3]

TURBOMACHINE DESIGN TECHNIQUES AND PROBLEMS

Space has not permitted a treatment of turbomachines beyond the basic principles of their operation, a discussion of the principal losses, and their performance characteristics. For those interested in information on design more detailed and specific than that in Refs. 1 to 3, Ref. 10 presents an excellent technique for arriving at the aerodynamic proportions of an extremely wide range of gas and vapor turbines, while Ref. 13 is widely used for making detailed accurate estimates of the efficiency of large steam turbines for a wide range of steam bleed-off conditions. Reference 11 is similar to Ref. 10 but is concerned with compressors. Reference 14 gives a nice treatment of methods for arriving at a good set of proportions for Kaplan turbines, while Ref. 15 does a similar service for those interested in Francis turbines.

Rotor Dynamics

A major consideration in the design of high-speed turbomachines is shaft dynamics. The basic problem is that no rotor is perfectly balanced, especially after slight warpage under high-temperature operation. If the rotor is operated at a high enough speed so that the exciting force from the rotating unbalanced load is at the natural frequency of the shaft in bending, large

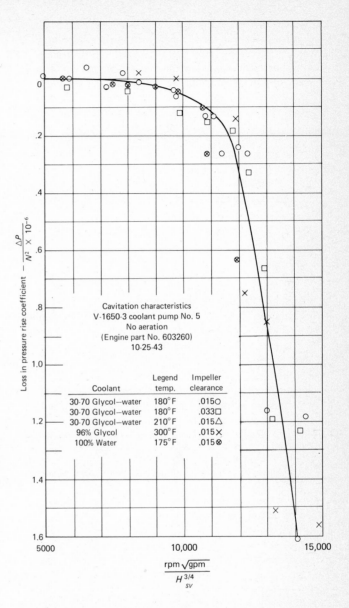

Figure 3.26 Cavitation characteristics of an engine coolant pump for a wide range of water-glycol solutions and temperatures. The pump pressure rise parameter is proportional to the pump pressure rise ΔP divided by the square of the rpm N. The cavitation parameter is proportional to the product of the rpm and the square root of the flow Q divided by the cavitation suppression head $H_{sv}^{3/4}$, where H_{sv} is the difference between the inlet static pressure and the vapor pressure. (The flow data were in gpm and the pressures in psi. The vapor pressure of 96% glycol at 300°F is 560 mmHg.)

amplitudes of vibration will occur as a consequence of shaft whip, or whirl. If the speed is increased further, the vibration amplitude will usually drop back to a low level; the trick is to keep the unbalance small and incorporate enough damping in the system so that it will limit the amplitude to a tolerable level for the short time required to accelerate through the critical speed region. A sufficient further increase in shaft speed will bring the machine into the second critical speed region, which is that for the second mode bending vibration. The problems are complex, in part because the bearings and their characteristics are involved. In any case, increasing the length of the rotor increases its flexibility; the critical speed is inversely proportional to the square of the rotor length for a constant rotor cross section.

Thrust Bearings

Very large thrust forces can be induced by the pressure differential between the inlet and outlet of an axial-flow machine because the rotor acts as if it were a piston. This situation can give not only a very difficult thrust-bearing design problem but also substantial frictional losses. In gas turbines these forces can be balanced in large measure by placing the compressor and turbine on the same shaft with their respective high-pressure regions at the center. In large steam turbines the problem is handled by employing double-flow units with the high-pressure steam entering the turbine casing at the center and splitting into two equal streams that flow through the rotor in opposite directions.

PROBLEMS

1. Steam enters a convergent-divergent nozzle at zero velocity, 6 bars, and 100°C superheat and leaves at 1 bar. The nozzle efficiency is 85 percent. Using a Molier diagram, determine the velocity, temperature, and quality (or superheat) at the nozzle exit if the nozzle has been designed for that set of operating conditions.

2. Gas enters an impulse turbine wheel at 600 m/s at an angle of 20 degrees to the plane of the wheel. The wheel tip speed has been selected to give the maximum efficiency with symmetrical blades. Neglecting losses, determine the following:

   (a) Wheel tip speed
   (b) Force acting on wheel in newtons per kilogram per second of gas flow
   (c) Power input to the wheel in kilowatts per kilogram per second of gas flow
   (d) Efficiency

3. Sketch vector diagrams for both the above case and its practical counterpart, indicating very briefly the nature of the difference between the two and its approximate magnitude.

4. Sketch velocity diagrams for one stage of a reaction turbine through which air flows at the rate of 10 kg/s given the following:

$$\text{Temperature into stage} = 1000\,°C$$
$$\text{Pressure into stage} = 10 \text{ bars}$$
$$\text{Pressure out of stage} = 6 \text{ bars}$$
$$\text{Velocity into stationary blades} = 75 \text{ m/s}$$
$$\text{(Axial flow into stationary blades)}$$

Assume that half the drop in pressure occurs in the stationary blades and half in the rotating blades, and neglect changes in density. Determine values for the principal angles and velocities.

REFERENCES

1. Stodola, A., and L. C. Loewenstein: *Steam and Gas Turbines*, McGraw-Hill Book Company, New York, 1927.

2. Lee, J. F.: *Theory and Design of Steam and Gas Turbines*, McGraw-Hill Book Company, New York, 1954.

3. Csanady, G. T.: *Theory of Turbomachines*, McGraw-Hill Book Company, New York, 1964.

4. Keator, F. W.: "Benoit Fourneyron (1802–1867)," *Mechanical Engineering*, vol. 61, no. 4, April 1939, pp. 295–301.

5. Gabriel, D. S., and L. R. Carman: "The Performance of a Single-Stage Impulse Turbine Having an 11.0-in Pitch Line Diameter Wheel with Cast Airfoil Shaped and Bent Sheet Metal Nozzle Blades," NACA-ARR-E5H31, September 1945.

6. Emery, J. C., et al.: "Systematic Two-Dimensional Cas-

cade Tests of NACA 65-Series Compressor Blades at Low Speeds," NACA Technical Report No. 1368, 1958.

7. Brown, L. E., and F. G. Groh: "Use of Experimental Inter-stage Performance of Multistage Axial Compressors," *J. Eng. Power, Trans. ASME*, vol. 84, April 1962, p. 187.

8. Eckert, B.: *The Influence of Physical Dimensions (Such as Hub:Tip Ratio, Clearance, Blade Shape) and Flow Conditions (Such as Reynolds Number and Mach Number) on Compressor Characteristics*, vol. 3, BUSHIPS-338, (Wright Aeronautical Corporation Engineering Translation No. 22), May 1946, Bureau of Ships, U.S. Navy.

9. Pfenninger, H.: "Operating Results with Gas Turbines of Large Output," *J. Eng. Power, Trans. ASME*, vol. 86, 1964, pp. 29–49.

10. Balje, O. E.: "A Study on Design Criteria and Matching of Turbomachines: Part A — Similarity Relations and Design Criteria of Turbines," *J. Eng. Power, Trans. ASME,* ser. A, vol. 84, January 1964, p. 83.

11. Balje, O. E.: "Axial Cascade Technology and Application to Flow Path Designs," *J. Eng. Power, Trans. ASME*, ser. A, vol. 90, October 1968, p. 309.

12. Serovy, G. K.: "Recent Progress in Aerodynamic Design of Axial-Flow Compressors in the U.S.," *J. Eng. Power, Trans. ASME*, vol. 88, July 1966, p. 251.

13. Baily, F. G., et al.: "Predicting the Performance of 1800-rpm Large Steam Turbine-Generators Operating with Light Water-Reactors," General Electric Company, GET-6020.

14. de Siervo, F., and F. de Leva: "Modern Trends in Selecting and Designing Kaplan Turbines," *Water Power and Dam Construction*, December 1977, pp. 51–56.

15. de Siervo, F., and F. de Leva: "Modern Trends in Selecting and Designing Francis Turbines," *Water Power and Dam Construction*, August 1976, pp. 28–31.

4

CHARACTERISTICS OF HEAT–TRANSFER AND HEAT–TRANSPORT SYSTEMS

The ideal Carnot, Rankine, and Stirling thermodynamic cycles are assumed to be closed circuits with the working fluid recirculated through a sequence of machines and heat exchangers so that it is recycled repeatedly, with no fluid being added to or removed from the system; only energy in the form of heat or work enters or leaves the components in the system. Thus, the heat-transfer and heat-transport characteristics of the fluid and the heat-transfer characteristics of the heat exchangers are vital elements in the performance of the complete system. Even in an open-cycle system such as an internal combustion engine, the heat-transfer and heat-transport properties are important in such vital auxiliaries as the engine cooling system. Similarly, high-temperature gas turbines make use of a blade cooling system to permit the highest permissible turbine inlet temperature, and the heat-transfer and heat-transport properties of both the working fluid and the coolant employed become vital factors in limiting the performance obtainable from the power plant. The heat-transfer equipment required in a power plant is also a particularly important factor in determining its capital cost. In a conventional steam plant, for example, the heat exchangers represent more than half of the total cost of the equipment, commonly running about 10 times the cost of the turbine-generator unit. Comparable cost ratios are found for many other types of power plant. The performance characteristics of the many types of heat-transfer matrix used in power plants differ widely, being very much dependent on the detail design and geometry employed. While the design of heat-transfer matrices is a subject in itself[1-5] that is beyond the scope of this book, the problems of selecting, sizing, and estimating the cost of heat exchangers are such vital factors in power plant system design and/or evaluation that a chapter on the heat-transfer and heat-transport characteristics of the principal fluids employed, together with the heat-transfer characteristics of heat exchangers, seemed much in order. This is particularly true in attempting to evolve designs

for advanced power plants in which there are many possibilities for the use of different types of heat-transfer and heat-transport systems. Not only are conventional forced-convection fluid systems—possibly with boiling or condensing elements in the system—likely possibilities, but also worthy of consideration in many instances is the use of fluidized-bed heat-transfer or heat-transport elements, heat pipes, and unconventional surfaces designed to enhance heat transfer through various special turbulence promotion schemes.[3-6]

PLANT REQUIREMENTS

From the standpoint of the power plant, the heat-transfer fluid and the exchangers should give as high a plant efficiency as is cost-effective (e.g., the heat exchanger should have lots of surface area to minimize the temperature losses), the system should operate with low flow velocities to minimize the pumping-power requirements (which also implies a large heat exchanger and connecting piping), it should be highly reliable (which implies large, heavy units of expensive materials), and it should be low in cost and easy to maintain (which implies small, light units of inexpensive materials). Clearly, these conflicting requirements require compromises that pose difficult questions.

Not only are the requirements of a particular application highly varied and difficult to quantify, but also there is a wide diversity in the choice of fluid—ranging from conventional fluids such as air or water to organic liquids and fluids such as molten salts or liquid metals—and there is a wide diversity in the heat-exchanger geometries that might be employed. Together, these present a terribly complex and confusing array of possibilities. For example, some twelve independent variables are involved in defining the heat-transfer characteristics of a heat-exchanger matrix: the flow-passage equivalent diameter and length for each of the two fluids, the surface and flow-passage areas for each of the two fluids, the degree to which the surface geometry promotes turbulence for each of the two surfaces, the temperature distribution factor for the matrix (which is dependent on whether the heat exchanger is crossflow, counterflow, or parallel-flow), and finally the thermal resistance of the wall separating the two fluids (which includes the fin efficiency if extended surfaces are employed). When this array of geometric parameters is coupled with the corresponding array of physical-property parameters for the various working fluids that might be employed, with their widely differing densities, specific heats, thermal conductivities, viscosities, heats of vaporization, and vapor pressures, and on these are superimposed the power plant requirements, including the temperature and pressure limitations associated with such phenomena as corrosion and creep strength, the number of possible combinations and permutatons of the variables becomes astronomical. Fortunately, most of these countless combinations and permutations are clearly not of interest. Hence the problem becomes one of narrowing the field to the regions of interest while yet being careful not to exclude promising possibilities. This is the crux of the art of choosing heat-transfer fluids and heat-exchanger matrix geometries to best meet overall plant requirements.

BASIC RELATIONS FOR HEAT-TRANSFER AND HEAT-TRANSPORT PROCESSES

Before attempting to handle the problems of selecting a heat-transfer fluid and heat-exchanger matrix geometry, it seems well to review briefly the heat-transfer and heat-transport processes involved in power plant equipment. It is assumed that the reader is acquainted with these relations from previous formal courses; hence the emphasis here is on the special characteristics which these relations imply for heat-transfer systems.

Types of Heat-Transfer Process

The characteristics of a heat-transfer system differ widely with the heat-transfer processes involved. Forced convection, boiling, condensing, thermal radiation, etc., each has its own special characteristics.

Forced Convection

Forced convection is the most common heat-transfer process employed in power plants. The principal variables involved are incorporated in the usual relationship for the heat-transfer coefficient h; that is,[1,2,3]

$$h = \frac{k^{0.6} c_p^{0.4}}{\mu^{0.4}} \frac{G^{0.8}}{D^{0.2}} \text{ (constant*)} \qquad (4.1)$$

The heat-transfer characteristics of the fluid are defined by its thermal conductivity k, its specific heat c_p, and its viscosity μ. The fluid velocity is defined by the mass flow rate of the fluid, G, and the effects of the size of passages in the heat-transfer matrix are defined by the symbol D. This relationship includes the effects of Reynolds number on the characteristics of the flow, particularly the thickness of the laminar boundary layer through which heat must pass by conduction. Note particularly that for the usual situation in which the controlling resistance to heat flow is a forced-convection heat-transfer coefficient, the heat flow rate is almost directly proportional to the fluid velocity through the heat-transfer matrix, and so doubling the velocity almost doubles the heat-transfer rate and thus the output of the heat exchanger is almost doubled and its capital cost per unit of output is almost halved.[3] Thus in forced-convection heat-transfer systems there is a strong incentive to use as high a velocity as possible within the limitations imposed by the system.

Thermal Conduction

Thermal conduction plays a role in virtually every heat exchanger in that heat must be transferred from one fluid to another through the wall separating the fluids. The heat flux Q/A for thermal conduction is given by

$$\frac{Q}{A} = \frac{\kappa \, \Delta T}{l} \qquad (4.2)$$

where Q is the amount of heat transferred, A is the area of the cross section through which the heat flows, k is the thermal conductivity of the material, l is the length of the conduction path, and ΔT is the temperature difference between the inlet and outlet faces of the heat-conduction path. The thickness of the wall separating the two fluids is usually small and its thermal conductivity relatively large; hence the temperature drop involved in conduction through the wall is ordinarily small in relation to the temperature drops through the two fluid films of the heat-transfer surfaces. However, if fins are employed to increase the amount of heat-transfer surface area, the heat-flow path becomes relatively long and thermal conduction through the fin

---

*The constant for Eq. (4.1) is 0.023 for k in Btu/(h·ft·°F), c_p in Btu/(lb·°F), μ in lb/(h·ft), G in lb/(h·ft²), and D in ft, while for SI units its value is 0.131 for k in W/(m·°C), c_p in kJ/(kg·°C), μ in Pa·s, G in kg/s·m², and D in m.

may entail a substantial temperature drop. To reduce this effect, finned surfaces ordinarily employ a high thermal conductivity material such as aluminum or copper in the fins.

Heat Transfer in Boiling and Condensing

The large heat of vaporization makes the heat-transfer coefficients for boiling and condensing much higher than for forced convection, but the magnitude and mechanisms vary somewhat with the operating conditions. The most easily visualized set of conditions are those prevailing in a nominally static pool of liquid with a submerged heated surface. Streams of tiny vapor bubbles are emitted from a host of nucleation sites scattered over the heated surface. These bubbles coalesce as they rise to the surface, forming large bubbles whose buoyancy effects induce strong churning and thermal convection in the pool. This is called *nucleate boiling* because the bubbles originate at nucleation sites that, on microscopic inspection, are found to be tiny pits or crevices in the metal surface in which a small amount of air or inert gas is trapped. The wetting forces and surface tension of the liquid are sufficient to inhibit vapor bubble formation except at the free surfaces provided by the nucleation sites. If the nucleation sites are not present, the surface tension may prevent the formation of bubbles until the surface temperature exceeds the boiling point of the liquid by as much as 20°C or more. In the vicinity of the nucleation site, however, the temperature need rise only a few degrees above the boiling point of the liquid before the surface tension is overcome by the vapor pressure and bubbles form rapidly and then are broken away by buoyancy forces. Heat-transfer coefficients are high—of the order of 30 kW/(cm²·K) [5284 Btu/(h·ft²·°F)]—and so the temperature difference between the hot metal surface and the boiling point of the liquid is quite small, commonly only 5 to 10°C (9 to 18°F).

If the boiling fluid is circulated, the potent effects of nucleate boiling dominate so that the heat-transfer coefficient is relatively insensitive to the mass flow rate, though it does increase somewhat as the fluid velocity is increased.

In once-through boiler tubes the character of the boiling flow changes as the water rises through the tubes. Nucleate boiling near the bottom of the tube creates so many bubbles that the flow regime quickly shifts from a bubbly bulk flow to an annular flow condition in which the liquid moves in an annular film upward along the tube walls while the vapor with some suspended droplets of moisture moves at a much higher velocity up the center of the tube. Depending on the fluid used, the heat flux, and the operating pressure, the annular flow regime persists until from 40 to 98 percent of the liquid has been evaporated, at which point dry spots appear on the wall, the annular liquid film

disintegrates, and subsequent heat transfer is largely by the conventional mechanism of forced convection in the vapor with some enhancement because liquid droplets in the free stream strike the wall where they flash into vapor. In the annular flow region the heat-transfer mechanism is likely to be simple conduction from the wall through the thin liquid film, with evaporation occurring directly from the surface of the liquid film without any bubble formation at nucleation sites. This type of boiling characterizes operation in all but the entrance region of liquid-metal boilers because of the high thermal conductivity of liquid metals.

Although not generally appreciated, nucleate boiling plays an important role in liquid cooling systems in which the fluid is not far below its boiling point. If there are local regions in which the heat flux is high as a consequence of a factor, such as flame impingement, that one might otherwise expect to cause a hot spot, local boiling will occur when the metal temperature rises a little above the boiling point of the liquid, and the hot spot is relieved by nucleate boiling. The vapor bubbles formed are quickly reabsorbed in the subcooled liquid. This is the controlling factor in providing uniform temperatures in the cylinder heads of internal combustion engines, where high local heat fluxes from exhaust ports, coupled with the low velocities in the complex water-cooling passages, would otherwise yield serious trouble with hot spots. Local boiling is also an important factor in preventing hot spots in the water-cooled walls of such components as rocket nozzles, combustors for coal gasifiers, and electrodes for magnetohydrodynamic (MHD) generators.

Thermal Radiation

Thermal radiation is a major heat-transfer mechanism in coal-fired steam boilers. The net heat flux from a hot to a cooler surface is given by the classical equation

$$\frac{Q}{A} = \epsilon_1 \epsilon_2 \sigma \left[\left(\frac{T_1}{100} \right)^4 - \left(\frac{T_2}{100} \right)^4 \right] \tag{4.3}$$

where T_1 is the hot surface temperature, T_2 is the cooler surface temperature, and ϵ_1 and ϵ_2 are the respective emissivities of the two surfaces (they commonly run around 0.80 for oxidized surfaces). The constant σ (in this case it is the Stefan-Boltzmann constant multiplied by 10^{-8}) has the value 5.67×10^{-4} W/(cm²·K⁴) for temperatures in degrees Kelvin, or 0.1713 Btu/(h·ft²·°R⁴) for temperatures in degrees Rankine.

Fluidized Beds

Heat transfer in fluidized beds differs from forced-convection heat transfer in that once the bed is well fluidized, the heat-transfer coefficient is essentially independent of the gas velocity through the fluidized bed. The principal interest in fluidized beds for power plant applications is their usefulness for coal combustion systems. In view of this, and in view of the fact that the fluidized-bed heat-transfer problems are intimately involved in combustion problems, fluidized-bed heat transfer is treated in the discussion of combustion systems in the next chapter. Suffice it to say at this point that the heat-transfer coefficient of a fluidized bed does not vary much with the gas velocity through the bed over the range for good fluidization. The coefficient depends primarily on thermal conduction between the particles and the metal surfaces, although there is a small secondary contribution from thermal radiation in the 800°C temperature range of interest for combustors.

Heat Transport

The problems of heat transport in fluid systems somehow seem to receive less attention than those of heat transfer, possibly because so many different design considerations are involved. For a given energy flow rate through the system, the principal parameters are the pressure drop, the fluid temperature rise or temperature drop, the total fluid flow rate, the pumping-power requirement, and the flow-passage cross-sectional area required.

Pressure Drop

The pressure drop through a long straight passage is given by the expression

$$\Delta P = fq\frac{l}{d} = f\frac{\varrho v^2}{2g}\ \frac{l}{d} \tag{4.4}$$

where the pressure drop ΔP is a function of the friction factor f (which depends on the Reynolds number), the dynamic head q, and the length-diameter ratio of the passage, l/d. The dynamic head q is of course equal to the quantity $\varrho v^2/2g$, where ϱ is the density of the fluid, v its velocity, and g is the acceleration of gravity. The friction factor f varies somewhat with the Reynolds number but is commonly equal to about 0.02. Note that f is also sometimes defined in terms of the hydraulic radius r_h, which is the flow-passage area of noncircular passages divided by their wetted perimeter.[3] For circular passages $r_h = \pi r^2/2\pi r = r/2$ so that the equivalent diameter $d_{eq} = 4r_h$.

The greater part of the pressure drop in a fluid system usually stems from turbulence losses associated with discontinuities such as bends, entrance or exit regions, or crossflow over tube banks. Each of these disturbances to the flow commonly causes a pressure loss roughly equal to 1 dynamic head on the basis of the maximum velocity through the restriction or discontinuity.

Fluid Flow Rate and Temperature Rise or Drop

The volumetric flow rate of the fluid is directly proportional to the quantity of heat to be transported divided by the product of the fluid's specific heat, the fluid density, and the fluid temperature rise or temperature drop in the course of its flow through the heat source or heat sink. The flow-passage cross-sectional area required is simply the volumetric flow rate of the fluid divided by the fluid velocity through the passage.

Pumping Power

Perhaps the most important parameter associated with a heat-transport system is the power required to circulate the fluid. This pumping-power requirement for either a particular component or the system as a whole is simply the product of the pressure drop for the component or the system and the volumetric flow rate V (the product of the fluid velocity v and flow-passage area A). That is,

$$\text{Pumping power } = \Delta PV = \Delta PvA \sim v^3 \tag{4.5}$$

The pumping power required has proved to be a good basis for comparing the heat-transfer performance characteristics of different heat-transfer matrix geometries with various devices for promoting turbulence and thus enhancing heat transfer at the expense of increased pressure drop.[7]

Pumping-Power–Heat-Removal Ratio

The most important figure of merit for a heat-transfer and heat-transport system is the pumping-power–heat-removal ratio.[3,4] To avoid a serious loss in thermal efficiency, this parameter is ordinarily kept to around 1 percent. This means that if the overall thermal efficiency of the power plant is 33 percent, the power required for the heat-transfer and heat-transport system is 3 percent of the net electric power output, a substantial penalty. Thus, while on the one hand there is a strong incentive to increase the fluid velocity through the equipment in the system in order to reduce its size and hence its capital cost, on the other hand the value of the power consumed by pumping increases as the cube of the velocity of the fluid, and hence this loss must be balanced against the savings in capital costs.

Thermal Convection

Although up to this point the discussion has been directed mainly at forced-convection fluid systems, in many instances simple thermal convection is an economically attractive alternative.

The combustion gas system for a household furnace is a good example; a stack is required under any circumstances in order to get the fumes above the level of living quarters, and the difference in density between the hot gases in the relatively tall stack and the ambient air provides a pressure differential that is sufficient to overcome the pressure drop induced by the gas flow through the furnace and the stack. This gives a simple system in which the somewhat higher costs associated with generous flow passages through the furnace and stack are more than offset by the savings effected by eliminating the blower and drive motor that would otherwise be required. The greater reliability of the simple system is probably an even more valuable advantage. The same conditions may hold for fairly large furnaces, depending in part on the stack height required to lift the combustion gases above the buildings in the area. Other familiar examples of thermal convection systems include cooling coils in refrigerators, hot-water heating systems for residences, and the old Model T Ford engine-radiator system. (Designers of later automobiles have found that the extra cost and complexity of a water-circulating pump is more than justified by the reduced cost and weight of the radiator.)

Steam power plants in regions where supplies of condenser cooling water have not been adequate have used cooling towers. For small plant capacities air has commonly been circulated through the cooling towers by natural thermal convection, but for medium-size plants it has usually been found that the extra cost of fans and their power requirements could be more than justified by the reduced size and cost of the cooling towers. For large plants in which the cooling towers reject ~ 500 MWt or more, the tower size required is sufficiently great that natural convection towers in the form of reinforced-concrete shells that are hyperboloids of revolution have proved to be the most economical system.[3]

Water circulation in small and medium-size steam boilers is commonly accomplished by thermal convection, the large difference in density between the bubble-free liquid in the downcomers and the bubbly liquid in the boiler tubes giving a quite adequate circulation rate for boiler designs in which the tubes are vertical or steeply inclined.[3] For horizontal tubes or for high heat fluxes, forced circulation is commonly required, especially in high-pressure boilers in which the difference in density between the liquid and vapor is much reduced, thus reducing the driving force for thermal convection.

Boiling-Condensing Refluxing Systems and Heat Pipes

Cooling systems for internal combustion engines and some processes have sometimes employed refluxing systems in which pool boiling occurs in the region being cooled, the vapor rises in-

to an air- or water-cooled condenser, and then the condensate drains back into the boiling pool.

A *heat pipe* is a compact version of this type of system, one end of the heat pipe being in the region from which heat is to be extracted and the other in the region from which heat is to be removed. The interior wall of the heat pipe is lined with a capillary structure such as a wick made of fine wire cloth or a set of fine, closely spaced grooves. The heat pipe is evacuated and charged with just enough of a suitable liquid to fill the interstices in the capillary structure.[8,9,10] In operation, the liquid evaporates at the hot end, the vapor flows to the cold end and condenses, and the condensate is returned to the hot end by capillary action in the wick structure lining the interior wall. The capillary action makes it possible for the heat pipe to function in the horizontal position or in the zero-*g* conditions in spacecraft. The action of gravity is not required for refluxing.

The heat-transport capacity of a heat pipe is high, the prime limitation usually being sonic velocity in the vapor flow from the hot to the cold end. By choosing a liquid with a suitable vapor pressure, a heat pipe can be designed to give a high heat-transport capacity with little temperature drop (< 20°C) at the desired temperature level. Freons are used for low-temperature applications such as regenerative heat exchange between the fresh-air inlet and building-air exhaust for air-conditioning systems, water is used for somewhat higher-temperature applications, and liquid metals such as mercury, cesium, and potassium are used for relatively high-temperature applications. Figure 4.1 gives a set of curves showing the sonic velocity limitation on the heat-transport capacity of heat pipes utilizing the alkali metals.

At the higher heat-transport capacities, the pressure drop in the liquid flow through the capillary passages or entrainment of the returning liquid stream by the high-velocity vapor may also be limiting, the value of these limits depending on the form of the capillary structure employed. The usual upper limit on the heat-transport capacity for a readily fabricated capillary structure is indicated in Fig. 4.1.

Calculation of Flow Rates for Thermal Convection and Heat Pipe Systems

Space does not permit inclusion of sample calculations for the flow rates induced in thermal convection systems, but the procedure is relatively simple and straightforward. A convenient method is to calculate first the pressure differential available from the density differences in the hot and cold legs for a set of system temperature distributions in the range expected, and then the fluid flow rate for each of these pressure drops can be calculated from the available pressure differential. The resulting flow rates can then be plotted as a function of a measure of the temperature differential, e.g., the stack gas temperature for a furnace. After the flow rate has been estimated, the Reynolds number should be calculated for each passage to obtain a better estimate of the friction factor; this may show that a second iteration is in order. The same procedure can be used for estimating the natural thermal-convection circulation rate in a reactor cooling system—an important consideration in the analyses of certain types of hypothetical reactor accidents, such as the failure of the electric power supply for the circulating pumps.

The calculation of the performance of heat pipes involves the latent heat of vaporization of the working fluid, its viscosity, surface tension, vapor and liquid densities, and sonic velocity. Excellent wetting of the wall and wick are, of course, essential. The application of the basic relations involved is straightforward, though a bit tedious and heavily dependent on the geometry of the capillary structure, i.e., the fabrication processes to be used.[9,10]

CHARACTERISTICS OF HEAT-TRANSFER FLUIDS

Power plants deriving their energy from combustion heat sources ordinarily employ conventional working fluids, usually air or steam. The plants using air as the working fluid commonly operate on an open cycle so that the waste heat is rejected in the form of the products of combustion that are discharged up a stack. Both the special requirements of nuclear power plants and the special boundary conditions that are applicable have led to an interest in a much wider range of working fluids. The demonstrated advantages of these fluids have in turn aroused interest in their use for such special applications as cooling gas-turbine blades, providing heat-transfer fluids for gasifiers and certain types of heat-recovery systems (Chap. 17), and serving as working fluids or heat-transfer media for low-temperature thermodynamic cycles coupled to geothermal or solar-energy heat sources (Chaps. 17 and 21). While the heat-transfer and heat-transport characteristics of these various fluids are the prime figures of merit, factors such as the fire hazard, toxicity, neutron activation cost, and compatibility of the fluid with the materials of the system must be considered.

Comparison of Heat-Transfer and Heat Transport Properties

There are so many parameters involved that it is exceedingly difficult to make a good comparison of the full range of working fluids that might be employed in a heat-transfer and -transport system.[11-14] One common approach has been to take a fixed geometry for a nuclear reactor core and determine, with some arbitrarily imposed temperature limits, the maximum reactor power that can be obtained with each of a number of different

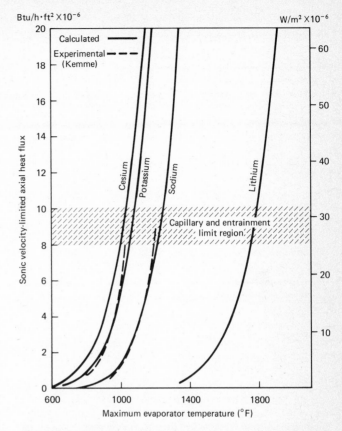

Figure 4.1 Comparison of calculated and experimental sonic limits on axial heat flow for cesium, potassium, and sodium heat pipes. Calculated values only for lithium.[10]

coolants. Because of the enormous differences in density between liquids and gases, such comparisons are usually confined either to liquids or to gases. Another common basis for comparison is to look only at the temperature difference between the heat-transfer matrix and the coolant; again such comparisons are usually made for either a group of liquids or a group of gases. A typical comparison of coolants for gas-cooled reactors is summarized in Table 4.1, which includes columns indicating the relative suitability of four gaseous coolants from the standpoints of nuclear and materials compatibility considerations, as well as heat transfer and heat transport.[12]

Rationale for a Comprehensive Comparison

After reviewing the literature in an unsuccessful effort to find a comprehensive basis for comparing all the heat-transfer fluids

TABLE 4.1   Some Characteristics of Selected Potential Gaseous Coolants[12]

	ϱ, density, g/cm³	c_p, specific heat, cal/(g · °C)	μ, viscosity, g/(cm · s)	k, thermal conductivity, cal/(s · cm · °C)	ϱc_p, volumetric specific heat, cal/(cm³ · °C)	Pumping power index*	Cooling efficiency index†	σ_T, thermal neutron absorption cross section, barns/nucleus
Hydrogen (1 atm)	6.6×10^{-5}	3.43	10^{-4}	5.33×10^{-4}	2.26×10^{-4}	4.4×10^{6}	17.5×10^{-8}	0.33
Helium (1 atm)	1.4×10^{-4}	1.25	2.2×10^{-4}	4.0×10^{-4}	1.75×10^{-4}	2.09×10^{7}	3.83×10^{-8}	~0
Air (1 atm)	9.5×10^{-4}	0.240	2.1×10^{-4}	7.57×10^{-5}	2.28×10^{-4}	6.28×10^{7}	1.25×10^{-8}	1.5
Carbon dioxide (1 atm)	1.5×10^{-3}	0.218	1.8×10^{-4}	5.0×10^{-5}	3.27×10^{-4}	3.7×10^{7}	2.34×10^{-8}	0.003

* $\dfrac{\mu}{\varrho^2 c_p^2 k \, (c_p \mu/k)^{0.4}}$ (lowest is best).

† $\varrho^2 c_p^3$ (highest is best).

of interest, the author reached the conclusion that the most important parameters are really the pumping power, the size of the fluid-flow passages required, and the heat-transfer coefficient. (Figure A4.1 in the Appendix gives heat-transfer coefficients for typical fluids.) In view of the fact that overall power-plant design considerations generally lead one to employ the maximum permissible pumping-power–heat-removal ratio, it seemed in order to fix that parameter and determine the other two. In this way one would have a direct indication of the relative merits of the various fluids. The diameters of the piping and pressure vessel are determined by the flow-passage-area requirement, while the length of a heat-transfer matrix is inversely proportional to the heat-transfer coefficient. These two factors, together with the system pressure level, largely determine the cost of the equipment required. This rationale provided the basis for the quantitative comparison of a typical set of heat-transfer and heat-transport fluids listed in Table 4.2.

In the detailed calculations of Table 4.2 it was necessary to fix a number of key parameters. Values for these were chosen to be representative so that, while absolute values of the parameters evolved might vary, the ranking of the various heat-transfer fluids would not be changed by a change in the value of the param-

eters that were fixed. Thus the heat-transfer matrix was considered to be built of 2.5-cm-ID (1-in-ID) tubes, and the temperature rise in the system was taken to be 111°C (200°F) for fluids in which no boiling took place, whereas for fluids that would boil in the heat-transfer matrix the temperature rise was taken as only 17°C (30°F), just enough so that the fluid would be definitely subcooled before entering the heat-transfer matrix. In each case the pressure and temperature conditions chosen were representative of those of interest for that fluid in one or another of the many power plant applications that might be considered. In keeping with the thesis that, for this comparison, everything would be pushed to the limit, it was decided that the dynamic head could be as much as one-tenth of the total pressure drop in the system. That is, the pressure drop in each of the two heat-transfer matrices (the heat source and the heat sink) might be kept to as little as two dynamic heads, the inlet and exit losses from each matrix might be kept to as little as two dynamic heads, and the pressure drop in each of the two connecting pipes might be kept to one dynamic head. On this basis, the overall system pressure loss would amount to ten dynamic heads.

The first step in carrying out the calculations of Table 4.2 was to fill in the columns for the physical properties using data from

Induced activity	Approximate loss in reactivity	Chemical activity (corrosion)	Radiation stability
None	0.0007	?	Good
H^3; 0.018-MeV β	0	Inert	Good
N^{16}; 7.3 s, 6-MeV γ A^{41}; 1.8 h, 1.4-MeV γ	0.003	Oxidizing	Good
N^{16}; 7.3 s, 6-MeV γ	0	Oxidizing	Some dissociation

Refs. 3, 4, and 14–20, and Refs. 17–19 of Chap. 2. The heat removed per unit weight of working fluid was then calculated from the specific heat and temperature rise (or the latent heat of vaporization for cases in which boiling occurred). The volumetric flow rate of the fluid was then calculated on the premise that the energy throughput would be 100 MWt. The pumping power was taken as 1 MWt and it, together with the volumetric flow rate, determined the value for the system pressure drop. The dynamic head was taken as one-tenth of the system pressure drop, and this determined the fluid velocity. The fluid velocity and volumetric flow rate then determined the flow-passage area required. The heat-transfer coefficient for flow through 2.5-cm-ID tubes was then determined from charts in Ref. 3 for those cases involving forced-convection heat transfer. Heat-transfer coefficients for boiling were obtained from Refs. 3, 21, and 22.

Comparison Chart

The key figures of merit yielded by the above calculations of Table 4.2 are shown in the two right-hand columns of numbers. In each column the values differ by as much as a factor of 10^4. To help visualize the implications of this confusing array of numbers, it proved helpful to plot them on logarithmic coordinates to yield Fig. 4.2. Perhaps the most surprising result is that virtually all the cases plot within a factor of ± 3 of a mean curve drawn through the scatter band. The best heat-transfer and heat-transport fluids lie at the upper left of the chart and yield both very high heat-transfer coefficients and very small flow-passage cross-sectional area requirements. The poor heat-transfer fluids, on the other hand, all plot in the lower right corner of the chart and are characterized by both low heat-transfer coefficients and large flow-passage-area requirements. Not surprisingly, all of these latter are gases. Further, the chart shows the marked improvements both in the heat-transfer coefficient and the flow-passage-area requirements obtained by operating heat-transport gases at high pressures. Boiling water and boiling liquid metals yield high heat-transfer coefficients but, except for the high-pressure boiling water, the flow-passage-area requirements are relatively large because of the low density of the vapor evolved in boiling. A relatively low heat-transfer coefficient for boiling is characteristic of any of the organic or fluorocarbon fluids because of the low values for their thermal conductivities and latent heats of vaporization.

A number of key points emerge from an inspection of the up-

TABLE 4.2 Comparisons of the Characteristics of Typical Heat-Transfer Fluids

	Density, lb/ft³		Specific heat, Btu/ (lb·°F)	Thermal conductivity, Btu/(h·ft·°F)	Viscosity, lb/(h·ft)	Melting point, °F	Boiling point @ 1 atm, °F	Heat of vaporization, Btu/lb	Temperature rise, °F
	Liquid	Gas							
Water: no boiling, 200°F	60		1.0	0.393	0.738	32			200
boiling, 1 atm, 212°F	60	0.0373	1.0				212	970	30
20 atm, 416°F	53	0.636	1.0					811	30
200 atm, 692°F	30	11.1	2.8					222	30
Air: 1 atm, 200°F		0.0602	0.241	0.0184	0.0519				200
1 atm, 1000°F		0.0272	0.263	0.0332	0.0884				200
20 atm, 1000°F		0.544	0.263	0.0332	0.0884				200
200 atm, 1000°F		5.54	0.263	0.0332	0.0884				200
He: 1 atm, 200°F		0.0083	1.25	0.0985	0.0545				200
1 atm, 1000°F		0.00376	1.25	0.176	0.099				200
20 atm, 1000°F		0.0752	1.25	0.176	0.099				200
CO_2: 1 atm, 200°F		0.0915	0.218	0.0127	0.0433				200
1 atm, 1000°F		0.0415	0.2793	0.0352	0.0827				200
20 atm, 1000°F		0.830	0.2793	0.0352	0.0827				200
Na: no boiling, 1540°F	47.3		0.305	30	0.355	208			200
NaK: no boiling, 1540°F	45		0.254	16	0.30	63			200
Li: no boiling, 1540°F	30		0.9	12	0.65	355			200
K: no boiling, 1540°F	39		0.19	17	0.38	144	1425		200
boiling, 1540°F/29 psia	39	0.060	0.19	17	0.38	144		760	30
Cs: boiling, 1540°F/56 psia	92	0.41	0.060	10	0.42	83	1274	205	30
Flinak (LiF-NaF-KF), 1100°F	132		0.437	2.66	12.6				200
LiF–BeF₂–ZrF₄–UF₄: 1200°F	141		0.47	0.83	19	813			200
HTS (NaNO₃, KNO₃, KNO₂), 600°F	115.8		1.85	0.35	7.02				200
Dowtherm A: no boiling, 600°F	49.3		0.70	0.1037	0.727		495		200
boiling, 600°F/49.3 psia	49.3	0.7237	0.579					114.1	30

*The high cost of the separated isotopes ⁷Li and ³⁹K can sometimes be justified.

Heat removed, Btu/lb	Volumetric flow rate, ft³/s	Dynamic head, psi	Velocity, ft/s	Flow-passage area for 100 MWt, ft²	Heat-transfer coefficient, Btu/(h·ft²·°F)	Relative magnitude of radioactivity from neutron activation
200	7.9	16.2	98	0.08	17,500	
1000	2,540	0.0504	220	11.6	2,000	Low
841	177	0.723	202	0.87	6,500	
306	27.9	4.59	122	0.23	6,500	
48.2	25,700	0.00498	54	475	11.2	
52.6	66,200	0.00193	50	1324	6.6	Low
52.6	3,310	0.0387	50	662	77	
52.6	331	0.387	50	66	490	
250	45,600	0.0028	110	415	21.3	
250	101,000	0.00127	110	917	12.8	Very low
250	5,050	0.0253	110	46	140	
43.6	23,800	0.00537	46	518	11.5	
55.9	40,900	0.00313	52	785	10.2	Low
55.9	2,450	0.0522	52	39	114	
61	32.8	3.9	54	0.61	14,000	High
50	42.1	3.04	49	0.86	8,700	High
180	17.5	7.31	93	0.19	26,000	Low (very low for ^{7}Li)*
38	64	2.0	43	1.5	6,200	Medium (low for ^{39}K)*
766	2,060	0.062	192	10.7	6,000	
207	1,116	0.1147	100	11.2	4,000	Medium
87.4	82.1	1.56	21	3.9	8,000	High
94	71.4	1.79	21	3.4	5,400	High
370	24.2	5.29	40	0.61	6,400	High
140	137	0.934	26	5.3	2,000	Low
132	992	0.129	80	12.4	300	

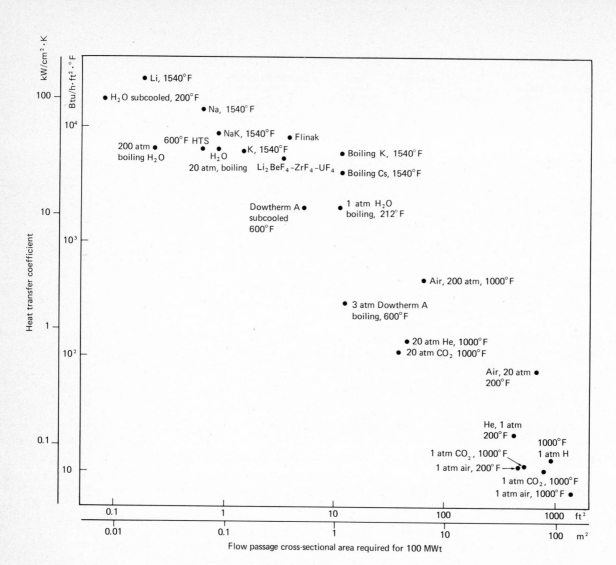

Figure 4.2 Comparison of the heat-transfer coefficients obtainable and the pipe sizes required for typical heat-transfer fluids used to remove 100 MWt from a heat-transfer matrix in a representative system in which the pumping-power–heat-removal ratio is limited to 1 percent.

per left corner of Fig. 4.2. One of the most important is that sub-cooled water in the room-temperature range gives the lowest flow-passage-area requirement of any of the coolants; thus it is no coincidence that the fission reactor that develops the highest power density of any reactor in the world, the HFIR at the Oak Ridge National Laboratory, is cooled with water at around 100°C and the water is pressurized to suppress boiling. To reach

the high temperatures required for a good thermodynamic cycle efficiency, sodium has been chosen as the coolant for fast-breeder reactors. Lithium is a better coolant and far less radio-active, but above 500°C it is not compatible with stainless steel, the best material available for the construction of fast reactors. NaK has been considered as a coolant for fast reactors because its melting point is below room temperature, whereas the melting point of sodium is 98°C (208°F). Figure 4.2 shows, however, that NaK gives a lower heat-transfer coefficient for a given pumping-power–heat-removal ratio and a greater flow-passage-area requirement. In addition, NaK is less desirable from the neutron economy standpoint.

CHARACTERISTICS OF HEAT-EXCHANGER GEOMETRIES

Most of the heat exchangers in commercial use consist of arrays of simple round tubes or round tubes fitted with fins. The bulk of the balance are made either of stacked alternately flat and corrugated plates or stacks of flat plates through which pass an array of round or flattened tubes—as in an automobile radiator. Finned surfaces of this type are ordinarily used where the heat-transfer coefficient is much lower on one side than the other (as in the water-to-air radiator for an automobile). These basic heat-transfer matrix geometries are widely used because they lend themselves to fabrication processes that can be carried out at a reasonable cost, and the tubular heat exchangers are well suited to withstand the pressure stresses imposed by high-pressure fluids such as steam.

Operating conditions in the field ordinarily fall far short of the ideal, and so corrosion, fouling, and erosion by particulate matter often impose limitations that are not obvious. As a consequence, the usual practice is to choose a heat-transfer matrix for a power plant on the basis of extensive experience with a suitable matrix, a practice that usually narrows the field to one or a few heat-transfer matrix geometries and establishes fluid velocity and operating temperature limits. Thus the principal problem of the power plant designer is one of selecting a set of heat-exchanger matrix proportions that yield a good compromise between capital costs on the one hand and operating costs on the other, with due allowance for the maintenance problems to be expected. Methods for handling analyses of this sort are presented in the next section.

HEAT-EXCHANGER PERFORMANCE CHARTS

As pointed out near the beginning of this chapter, in any heat-exchanger application there are so many variables that an explicit solution to meet a given set of requirements is usually not possible. Such quantities as tube diameter, spacing, length, and number of passes, along with fluid velocities, are the independent variables, while the dependent variables are the fluid temperature rises or drops, the temperature differences between the fluids and the metal heat-transfer surfaces, the fluid-pressure drops, and the pumping-power requirements. It is possible to set up the basic heat-transfer relations and obtain solutions by

trial and error, but the process is tedious if done by oneself and tricky if done by computer. The author has found that the best and quickest approach is to prepare charts that not only make it possible to get a solution quickly, but also show graphically the effects of each of the principal variables. In this way one can easily see the region of interest, together with the prospective trade-offs in that region, and thus make a well-informed choice of the key parameters.

In preparing a performance chart, one can usually get performance test data on a commercially available heat-transfer matrix that is reasonably representative and can use these data as the basis for constructing a chart. If the matrix chosen does not yield an attractive system with equipment that is at hand or readily available, one at least has a better idea of what to look for in another heat-transfer matrix better suited to the application.

Performance Parameters

The performance characteristics of heat exchangers are often expressed more conveniently in terms of figures of merit other than the basic heat-transfer coefficient and heat flux used in the preceding discussion.

Temperature Distributions in Typical Heat Exchangers

The heat-transfer performance of a heat exchanger is heavily dependent on the temperature distribution through the heat-transfer matrix. The principal heat exchangers employed in power plants are of four basic types: boilers, condensers, regenerative or recuperative gas heaters, and heat-recovery units. Typical axial-temperature distributions in these four types are shown in Fig. 4.3. There are three basic types of temperature distribution: a virtually constant metal-wall temperature, as in boilers and condensers; an essentially constant temperature difference, as in air preheater and recuperator units; and both varying metal temperatures and varying temperature differences, as in coolers for gas-turbine engines. Note that feedwater heaters are really high-pressure condensers and that coolers in closed-cycle gas-turbine systems have the same basic temperature distribution as in units for recovering waste heat from stack gases. Note, too, that the temperature distribution shown for the coolers and waste-heat-recovery units differs from that in regenerative gas heaters in that the temperature rise in the cold fluid is commonly much less than the temperature drop in the hot gas instead of being the same. It is for this reason that the temperature difference to drive the heat flow varies along the length of the heat exchanger, and this complicates the calculations.

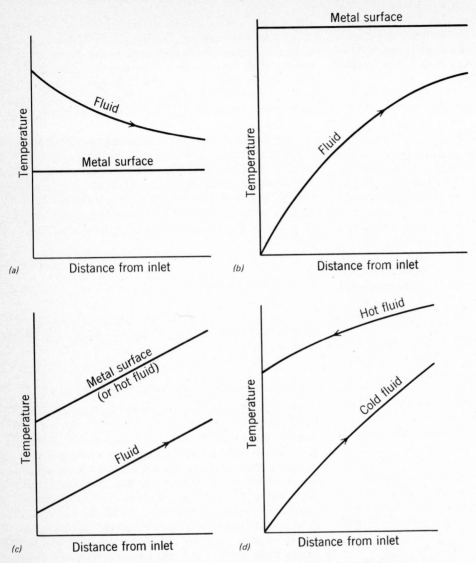

Figure 4.3 Temperature distributions in typical heat-transfer matrices. (*a*) Uniform surface temperature (as in a gas-fired boiler). (*b*) Uniform surface temperature (as in an air-cooled condenser). (*c*) Uniform temperature difference or uniform heat flux (as in a recuperator). (*d*) Counterflow heat exchanger with fluid streams having different temperature rises (as in a waste heat recovery unit).

Types of Heat-Transfer Surface

The type of heat-transfer surface employed differs with the application. In coal-fired steam boilers the high heat-transfer coefficient on the water side is used with a relatively small temperature difference between the tube wall and the water, while the relatively low heat-transfer coefficient on the combustion gas side is accompanied by a very large temperature difference; thus a high heat flux and good heat-exchanger proportions can be obtained with simple round tubes. In water-reactor steam plants the heat-transfer coefficient on the high-temperature water side of the steam-generator tubes is not as high as the coefficient for nucleate boiling, but the mismatch is not bad, and again simple round tubes can be employed. The situation is similar in condensers, feed-water heaters, and tubular air preheaters. However, the situation is different in the special case of air preheaters; lower costs can be obtained by using a rotating heat-transfer matrix of alternate layers of flat and corrugated plates with the axes of the corrugations aligned with the direction of the gas flow. The heat capacity of the metal plates is much higher than that of the gas streams so that the heat-transfer matrix can be rotated to expose a given surface element first to the hot exhaust gas on one side of the unit and then to the cold incoming air on the other side with little change in metal surface temperature. The rotating units present a serious leakage problem if used as recuperators in gas turbines. Because the two gas streams are at widely different pressures, air leakage through clearances in the rotor region bypasses the turbine, thus causing an important loss in power. As a consequence, regenerators for gas turbines usually employ either a tubular heat-transfer matrix or a plate-fin unit; in either case the gas streams are sealed from each other so that essentially no leakage can occur.

The heat-recovery units employed in gas-turbine systems are sometimes waste-heat boilers, in which case the temperature distribution is that characteristic of a boiler, but instead they may serve to heat a liquid for a chemical process or pressurized water for building heating. In either case the heat-transfer coefficient on the gas side is usually so much lower than that on the process fluid side that the tubes are usually finned on the gas side to reduce the cost and size of the heat-transfer matrix.

Heating Effectiveness

A parameter commonly referred to as *heating effectiveness* is widely used to define the performance of a given heat-transfer matrix.[3] The heating effectiveness may be defined as the ratio of the fluid temperature rise through a given matrix to the difference in temperature between the two inlet fluid streams; that is,

$$\text{Heating effectiveness} = \frac{\text{cold-fluid temperature rise}}{\text{hot-fluid inlet} - \text{cold-fluid inlet}} \quad (4.6)$$
$$\phantom{\text{Heating effectiveness} = }\quad\;\; \text{temperature} \quad\;\; \text{temperature}$$

This has proved to be an extremely convenient parameter in heat-exchanger selection work. The similar term *cooling effectiveness* is used in heat-exchanger applications in which the heat-transfer matrix is looked on as a cooler rather than as a heater.

Performance Chart Construction

An easy way to construct charts for the heating effectiveness can be deduced by examining Fig. 4.3b for a steam condenser. This gives a simple case in which not only is the steam-side temperature constant but also the condensing-heat-transfer coefficient is very high compared to the forced-convection-heat-transfer coefficient for the coolant, and so the overall heat-transfer coefficient depends mainly on the coolant velocity. If the ratio of the local temperature difference to the temperature difference at the inlet for the case covered by Fig. 4.3b is plotted on semilogarithmic coordinates against the distance the heated fluid travels through the heat-transfer matrix, a straight line is obtained, as shown in Fig. 4.4a. If data are available for one heat-exchanger passage length, the line is defined, because the value at a fluid passage length of zero is also known. The difference in temperature between the cold fluid at any given point in the heat-transfer matrix and the saturation temperature on the condensing side is related to the heating effectiveness for the matrix up to that point in a very simple fashion, namely:

$$\begin{aligned}\text{Heating effectiveness} &= \frac{\text{fluid temperature rise}}{\text{inlet temperature difference}} \\[2mm] &= \frac{\begin{array}{c}\text{inlet temperature}\\\text{difference}\end{array} - \begin{array}{c}\text{local temperature}\\\text{difference}\end{array}}{\text{inlet temperature difference}} \\[2mm] &= 1 - \frac{\text{local temperature difference}}{\text{inlet temperature difference}} \quad (4.7)\end{aligned}$$

Thus a line can be drawn as in Fig. 4.4b to define the heating effectiveness for the heat-transfer matrix by taking the same coordinates and line of Fig. 4.4a, inverting the ordinate scale, and labeling the scale to read in terms of heating effectiveness by using Eq. (4.7).

It should be noted that the straight lines of Fig. 4.4a and b are for one particular fluid mass flow rate through the heat-transfer matrix. If the cold-fluid velocity is increased, the heat-transfer coefficient h for the cold fluid will increase as the 0.8 power of the weight flow rate w, while the much higher condensing coefficient will remain essentially constant. This will result in a reduction in the heating effectiveness with an increase in fluid flow. This can be seen readily from the following simple derivation:

$$\text{Total heat transferred} = wc_p \text{ (fluid temperature rise)}$$
$$= hA \text{ (log mean temperature difference)}$$

where A is a constant for the heat-transfer surface area. Further,

$$h = Bw^{0.8}$$

where B is a constant for the heat-transfer matrix geometry and the fluids used. Equating and rearranging these expressions gives

$$\frac{\text{Fluid temperature rise}}{\text{Log mean temperature difference}} = \frac{Bw^{0.8}A}{wc_p}$$
$$= \frac{B'}{w^{0.2}}$$

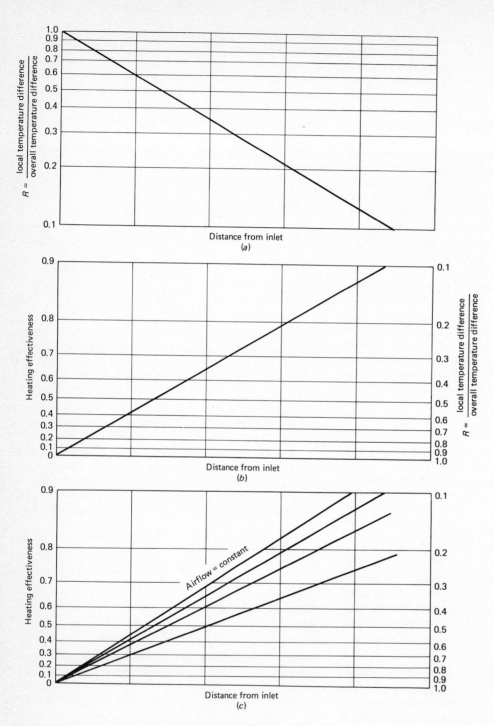

Figure 4.4 Basis for the construction of heating effectiveness charts.[3] (*Courtesy John Wiley & Sons, Inc.*)

where B' represents a compilation of the constants. *For similar temperature distributions in the heat-transfer matrix, the log mean temperature difference in each case will be proportional to the overall temperature difference, and hence the above equation applies also to the heating effectiveness.* Thus,

$$\text{Heating effectiveness} = \frac{B''}{w^{0.2}} \qquad (4.9)$$

where B'' is proportional to B'. This relation can be applied as in Fig. 4.4c to present heating effectiveness as a function of heat-transfer matrix length and fluid flow rate. Thus some good experimental points for a particular condenser make it possible to plot a chart which defines the heating effectiveness of that basic heat-transfer matrix geometry for a wide range of flow-passage lengths and a series of cold-fluid flow rates.

In reexamining Fig. 4.3b, it is evident that the heating effectiveness will remain constant with a given temperature distribution, but, as the fluid flow is increased, the temperature distribution of Fig. 4.3b will change, and hence the logarithmic mean temperature difference (LMTD) will vary. If the surface area is increased by increasing the length of the matrix, and if the fluid flow is increased simultaneously in such a way that the heating effectiveness is held constant, the log mean temperature difference will also be held constant, and thus the effects of variations in fluid flow rate can be determined directly. In order to attain the same heating effectiveness, the surface area required will increase as $w^{0.2}$ if the fluid flow is increased, and hence the matrix length required will vary accordingly.

From data for a single given set of conditions—i.e., a given basic heat-transfer matrix geometry, fluid passage length, and flow rate per unit of area—it is possible to estimate the performance of the same basic heat-transfer matrix for various fluid passage lengths and flow rates by the above technique. Of course, the whole approach is predicated on the premise that the principal barrier to heat transfer lies in one (or two) fluid films and that the heat-transfer coefficient for this film (or films) varies as the 0.8 power of the fluid flow rate. Deviations from these ideal conditions sometimes lead to a reduction in the exponent for w from 0.8 to as little as 0.5. The value is determinable with confidence only through experiments because of the difficulty in predicting degrees of turbulence and the effects of Reynolds number for irregular flow passage geometries.

General Applications

While the above discussion concerns condensers, it also applies directly to boilers, and the same basic technique can be used to give a fair approximation of the effectiveness of any type of heat-transfer matrix. It even gives a good approximation for a crossflow heat exchanger for a limited range of conditions.

Where both fluid streams change in temperature as in Fig. 4.3d, the error involved in the approximation depends on the relative temperature changes and heat-transfer coefficients in the two fluid streams. In heat exchangers in which one of the fluids is a liquid metal or a boiling or condensing liquid, the fluid film of the other liquid will account for the major portion of the local temperature difference between the two fluid streams. If a gas film temperature difference is 90 percent or more of the local total temperature difference between the two fluid streams, the metal surface temperature can be taken as approximately equal to the temperature of the fluid having the high heat-transfer coefficient. If the fraction is less than 90 percent, the effects of the film drop on the side with the high heat-transfer coefficient usually can be approximated nicely by using an exponent of less than 0.8 for the heat-transfer-coefficient–flow relationship for the low-coefficient film. Similarly, the same technique can be employed if the temperature drop through the tube wall is substantial, or if a finned tube (or other extended surface) is used and the fin efficiency runs less than 90 percent. In each instance a suitable exponent for the heat-transfer-coefficient–flow relationship can be estimated by calculation, or, better, it can be defined from experimental data. In cases where this is necessary, the heat-transfer coefficient usually will vary as the 0.5 to 0.8 power of the mass flow in well-proportioned units.

The above discussion gives an insight into the use of charts for heating or cooling effectiveness and outlines a procedure for constructing performance charts that are rough but useful approximations. For a much more detailed and rigorous treatment of these problems, the reader is referred to Ref. 3.

Charts for the heating or cooling effectiveness of two widely used heat-transfer matrices are included as Figs. A4.2 and A4.4 in the Appendix. One chart is for a tubular air heater in which the gas stream outside the tubes flows parallel to the axes of plain round tubes, while the other chart is for a waste-heat-recovery unit in which the gas flows across banks of round finned tubes.

Steam Generators

Steam generators present a particularly complex set of problems, because the steam side operates in three quite different heat-transfer regimes. The feed water must first be heated to the boiling point and then boiled, and finally the steam must be superheated and probably reheated. Thus both the temperature distribution and the performance characteristics of a steam generator are quite complex just from the standpoint of the heat transfer on the steam side. The problems are rendered still more complex in fossil-fuel-fired units, because in the combustion zone much of the heat transfer from the combustion gas to the tube wall is by thermal radiation. In a nuclear plant the problems are more straightforward and can be envisioned quite well by examining the diagram in Fig. 4.5, which shows the temperature distribution through the steam generator for a gas-cooled reactor.[23] Note that the abscissa is not the distance through the unit but rather the fraction of the heat transferred from the hot gas stream to the steam system. This abscissa is used because it defines the temperature distribution independently of the heat-transfer surface geometry.

Pressure Drop

For the regions of practical interest the fluid flow through the heat exchangers is turbulent; hence the pressure drop varies approximately as the square of the flow rate through a given heat-transfer matrix, and it increases linearly with the length of the matrix at a given flow rate. The effects of changes in gas density are readily handled because the pressure drop at a given mass flow rate is inversely proportional to the gas density.

A chart for the pressure-drop characteristics of a heat-transfer matrix can be constructed quickly using the above relations and just one good experimental point by drawing a family of straight lines with a 2:1 slope on log-log paper as in Figs. A4.3 and A4.5 of the Appendix. Note that gas density effects are handled by using the parameter $\sigma \, \Delta p$ for the ordinate rather than just Δp (σ is the ratio of the gas density to that of air at standard atmospheric temperature and pressure).

Selection of Heat-Exchanger Proportions

All too frequently people engaged in conceptual design work assume both a high effectiveness and a low pressure drop for a heat exchanger, such as a regenerator for a gas turbine, without making any effort to estimate the size and cost implied by these assumed values. Fortunately, by using charts such as those described above, it is not too difficult to estimate the size and cost of a representative unit and thus get an idea of what one might reasonably expect the result to be. Vendors can then be contacted for firmer information on specific commercial or near-commercial designs, and in these negotiations the vendor can be given a much clearer idea of the requirements once the power plant systems analyst has worked out a good set of proportions for a typical heat-transfer matrix and made the design compromises required.

Costs

Not only are there many types of heat-transfer matrix which inherently differ in cost, but also costs vary with the type of material required for strength or corrosion resistance and with the fluid pressure that must be accommodated. Several charts to aid in cost estimates are given in the Appendix. Note that the material cost factors given in the table accompanying Fig. A10.3 indicate that employing stainless steel rather than plain carbon steel, for instance, will increase the cost by a factor of 2 to 4. Special quality-control requirements such as those for nuclear plant equipment can easily double the cost. Allowances must also be made for cost escalation in periods of inflation. For purposes of crude preliminary estimates, one can get a rough idea of the cost of low-pressure carbon steel heat exchangers by taking a value of $10/ft² of heat-transfer surface as the cost of large units in 1978. The effects of cost escalation with time for all types of industrial equipment are indicated in Fig. A10.1, taken from *Chemical Engineering Progress*, which updates its equipment cost index at frequent intervals.

Figure 4.5 Typical axial-temperature distribution and heat-transfer coefficient values versus percentage of heat transferred for an axial-flow bare-tube steam generator for a helium-cooled reactor. Steam inside 0.4-in-ID, 0.5-in-OD tubes; pumping-power–heat-removal ratio = 0.002; gas at 69 bars (1000 psia); steam at 172 bars (2500 psia); gas in at 732°C (1350°F); gas out at 343°C (650°F); water in at 271°C (520°F); steam out at 566°C (1050°F); mass flow rate of steam = 104 kg/s (222.9 lb/ft²). *(Courtesy Oak Ridge National Laboratory.)*

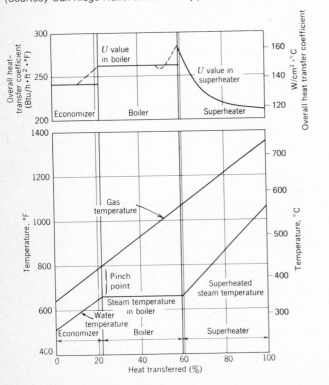

EXAMPLES OF CALCULATIONAL PROCEDURES

Several examples of representative calculations associated with heat-exchanger selection in the development of power plant designs will serve to illustrate some of the concepts outlined in this chapter.

Regenerator for a Gas Turbine

A study of proportions for a regenerator for a closed-cycle gas turbine coupled to a fluidized-bed coal combustor provides a typical case.[6]

For a closed-cycle gas turbine, the weight flow of air on the shell side is equal to that on the tube side of the heat exchanger. Thus, the air temperature rise on the tube side, δt, is equal to the air temperature drop on the shell side, and the temperature difference ΔT between the two streams for pure counterflow conditions is constant throughout the length L of the heat exchanger. The heat added to either fluid stream may be equated to the heat transferred through the available surface area $A'L$ to give

$$G'c_p 3600\, \delta t = uA'L\, \Delta T$$

$$\frac{\delta t}{\Delta T} = \frac{uA'L}{G'c_p 3600} = \frac{uA'L}{G'900}$$

where the specific heat c_p of the air is taken as 0.25 Btu/(lb·°F), G' is the mass flow rate in pounds per second per square foot, and u is the heat-transfer coefficient in Btu per hour per square foot. Using the mean surface area value for 0.50-in-OD tubes given in Table 4.3 and letting $c_p = 0.25$ gives

$$\frac{\delta t}{\Delta T} = \frac{u115L}{G'900} = \frac{0.1275uL}{G'}$$

The effectiveness is given by the following equation:

$$\text{Effectiveness} = \frac{\delta t}{\Delta T + \delta t} = \frac{1}{\Delta T/\delta t + 1}$$

The overall heat-transfer coefficient u can be determined for the heat-exchanger proportions of Table 4.3, which give a flow-passage area on the shell side 2.5 times that on the tube side. Thus, for 0.50-in-OD tubes, a tube-side mass flow rate of 10 lb/(s·ft²), and the heat-transfer coefficients obtained from Eq. (4.1),

$$u = \frac{1}{1/38 + 1/15.6} = \frac{1}{0.0904} = 11.06$$

$$\frac{\delta t}{\Delta T} = \frac{0.1278 \times 11.06}{10} = 0.141L$$

The effectiveness for a tube length of 20 ft becomes

TABLE 4.3 Geometric Data on Reference Design Tube Matrix for a Closed-Cycle Gas-Turbine Regenerator

Tube OD, in	0.50
Tube ID, in	0.444
Internal flow area, in²/tube	0.1548
Inside surface area, ft²/ft of tube	0.123
Cross-sectional area of tube wall, in²/tube	0.0415
Mean surface area of matrix, A', ft²/ft³ of internal flow passage	115
Number of tubes for 1 ft² of internal flow passage	930
Ratio of shell-side to tube-side flow-passage areas	2.5
Ratio of shell-side volume to total matrix volume	0.6635
Tube centerline spacing (triangular), in	0.815
Equivalent diameter of shell-side flow passage, in	1.00
Volume of heat-transfer matrix, ft³/ft³ of tube-side flow passage	3.768

$$\text{Effectiveness} = \frac{1}{1/2.82 + 1} = 0.738$$

The calculations for a typical case and the resulting chart are given in Table 4.4 and Fig. A4.4 of the Appendix.

Calculational Procedure for the Construction of Charts for Pressure Drop

The procedure for carrying out calculations on the pressure drop for the various flow passages is considerably simpler than that for charts for heating or cooling effectiveness. For flow inside round tubes or for flow outside tubes but parallel to them, the friction factor $f = 0.2\text{Re}^{-0.2}$ and the pressure drop given by Eq. (4.4) were employed. The Reynolds number for the flow regime of interest was checked in all cases and was found to be comfortably within the turbulent regime. If it were below ~ 3000, the matrix would probably require a few turbulators and consequently a small increase in pressure drop, but this increment may be neglected at this stage. The pressure drop for a typical condition was computed for a series of airflows for each case, and then a chart was constructed on the premise that the pressure drop would be directly proportional to the airflow passage length. Thus, a set of straight lines was drawn for the series of mass flow rates.

A single chart, that in Fig. A4.5, can be used for calculating the air pressure drop inside the tubes for the regenerator tube matrices. To allow for differences in air temperature and consequently in air density, the charts were all prepared using the pressure-drop parameter $\sigma \Delta P$, where σ is the ratio of the air density to that at standard conditions. Thus the effects of a difference in air density are easily handled. In effect, the charts were all drawn for a value of $\sigma = 1.0$. In using the charts, the value of σ for the prevailing temperature and pressure conditions in the system can be calculated to establish the pressure-drop parameter $\sigma \Delta P$.

Balancing Capital Costs Against Operating Costs

The relative importance of heat-exchanger effectiveness, pumping-power requirements, and capital cost vary widely from one situation to another. However, a perhaps oversimplified case will serve to illustrate an effective approach to a typical problem. Consider an industrial application in which a waste-heat boiler is to be installed in the exhaust of an existing simple open-cycle gas turbine that discharges hot gas at 800°F at a rate of 200 lb/s. Steam is to be generated at 300°F and 67 psia; this low-temperature steam will have a value of $0.50/10^6$ Btu. The gas-side pressure drop must not exceed 0.2 psi because of surge limit considerations for the gas turbine. Uncertainties in market conditions have led management to impose the requirement that the value of the steam generated must be sufficient so that the cost of the installation can be written off in just one year.

Neglecting the small effect of feed-water heating, the inlet temperature difference is 500°F and the heat energy ideally recoverable is the product of 500°F, 200 lb/s, and the specific heat

TABLE 4.4 Summary of Calculations for the Regenerator Heating Effectiveness Chart for 0.50-in-OD Tubes for a Shell-Side Flow-Passage Area 2.5 Times the Tube-Side Flow-Passage Area

G_1', lb/ft² · s	4.0	6.0	8.0	10.0	12.0	14.0	16.0	20.0	25.0	30.0
G_2', lb/ft² · s	1.6	2.4	3.2	4.0	4.8	5.6	6.4	8.0	10.0	12.0
h_1, Btu/(h · ft² · °F)	17.51	24.86	31.61	37.9	43.6	49.4	55.75	67.0	80.5	93.5
h_2, Btu/(h · ft² · °F)	7.45	10.25	13.0	15.6	18.0	20.34	22.7	27.0	32.4	37.4
$q_1 @ W = 0.0765$ lb/ft³	0.0224	0.0503	0.0896	0.140	0.201	0.275				
q_2	0.00358	0.00807	0.0144	0.0224	0.0323	0.044				
$U = 1/(1/h_1 + 1/h_2)$	5.226	7.258	9.212	11.051	12.740	14.408	16.3	19.30	23.06	26.77
f_1	0.030	0.029	0.0285	0.028	0.0275	0.0272	0.0270	0.0265	0.0261	0.0258

Tube length, ft					$\delta t/\Delta t = 0.1275\ UL/G_1'$					
20	3.331	3.084	2.936	2.818	2.707	2.624	2.571	2.461	2.352	2.275
40	6.664	6.169	5.872	5.636	5.415	5.249	5.141	4.922	4.704	4.551
60	9.995	9.254	8.809	8.454	8.122	7.873	7.712	7.382	7.056	6.826
80	13.327	12.338	11.745	11.272	10.829	10.497	10.283	9.843	9.408	9.102
100	16.659	15.423	14.681	14.090	13.537	13.122	12.854	12.304	11.761	11.377
120	19.991	18.507	17.617	16.908	16.244	15.746	15.424	14.765	14.113	13.653
140			20.554	19.727	18.951	18.370	17.995	17.225	16.465	15.928
160							20.566	19.686	18.817	18.204

Tube length, ft					$\eta = (\delta t/\Delta t)/(1 + \delta t/\Delta t)$					
20	0.769	0.755	0.746	0.738	0.730	0.724	0.720	0.711	0.702	0.695
40	0.870	0.860	0.854	0.849	0.844	0.840	0.837	0.831	0.825	0.820
60	0.909	0.902	0.898	0.894	0.890	0.887	0.885	0.881	0.876	0.872
80	0.930	0.925	0.922	0.919	0.915	0.913	0.911	0.908	0.904	0.901
100	0.943	0.939	0.936	0.934	0.931	0.929	0.928	0.925	0.922	0.919
120	0.952	0.949	0.946	0.944	0.942	0.940	0.939	0.937	0.934	0.932
140			0.953	0.952	0.950	0.948	0.947	0.945	0.943	0.941
160							0.954	0.952	0.950	0.948

$M_1/M_2 = 1.0$; $P_2/P_1 = 0.25$

TABLE 4.5   Summary of Calculations for Selecting a Waste-Heat Boiler for a Gas Turbine

Turbine airflow = 200 lb/s

Turbine exit temperature = 800 °F

Turbine exit pressure = 15 psia

Air pressure drop available = 0.20 psi

Heat-exchanger installed cost = $24/ft²

Value of heat recovered = $0.50/10⁶ Btu @ 5000 h/year

Boiler temperature = 300 °F

Boiler pressure = 67 psia

Finned tube matrix = 58.1 ft²/ft³

Cost for new ducts and support structure, steam system instrumentation, piping, feed pump, etc. = $120,000

Gas mass flow rate, lb/(s·ft²)	6	7	10	12
Matrix length, ft	2.27	1.77	0.91	0.66
Effectiveness, %	90.3	78.6	50	44
Value of heat recovered, $/yr	206,175	176,850	112,500	99,000
Heat-exchanger cost, $	105,510	61,700	25,380	15,340
Steam system cost, $	225,510	181,700	184,380	135,340

of ~0.25 Btu/(lb · °F), that is, 25,000 Btu/s, or 90 × 10⁶ Btu/h. The value of the heat recovered as steam is, then, the heat exchanger effectiveness × (90 × 10⁶) × $0.50/10⁶ Btu, which equals the effectiveness × $45/h. Assume that the finned-tube heat-transfer matrix of Figs. A4.2 and A4.3 will be employed and that the hot gas will have a mean temperature of 550 °F and a mean pressure of 15 psia (allowing for a small pressure drop in the stack). This gives a mean gas density of 0.040 lb/ft³, a value of σ = 0.52, and a value of $\sigma \Delta p$ = 0.104 psi. The intercept of the line for $\sigma \Delta p$ = 0.104 psi in Fig. A4.3 with each mass flow rate line defines the permissible heat-transfer matrix length from the pressure-drop standpoint for each mass flow rate. The values for the effectiveness defined by each of these combinations of mass flow rate and matrix length can then be obtained and multiplied by 45 to obtain the value of the heat recovered in dollars per hour. These values can then be plotted to give a curve for the value of the heat recovered as a function of the mass flow rate through the heat-transfer matrix. Then, assuming a heat-

exchanger matrix cost of $12/ft² and installation charges that double that value, one can calculate the heat-exchanger volume, surface area, and capital cost for a series of mass flow rates using the value of 58.1 ft²/ft³ for the heat-transfer matrix and the appropriate matrix length. (The heat-exchanger inlet face area is given by the total turbine airflow rate divided by the mass flow rate through the heat-transfer matrix.) The capital cost of the complete steam-system installation can then be computed by taking the cost of the new gas ductwork, support structure, steam piping, feed pump, instrumentation and control equipment, etc., as $120,000.

The results of these calculations are summarized in Table 4.5. Note that the size of the waste-heat boiler can be increased to somewhere between the two middle points in Table 4.5 and the cost of the complete installation written off in only a year. If the value of the heat recovered were greater or a longer capital cost write-off period could be justified, a still larger heat exchanger would be in order.

PROBLEMS

1. A design for a small toroidal thermonuclear reactor has been proposed for materials test purposes. The power density is to be as high as possible. Layout studies favor construction of the critical first wall of a layer of tubes placed side by side, using tubing having an ID of 6.35 mm (0.25 in) and an OD of 8.0 mm (0.315 in), with each tube having a total length of 3 m. Half of the length will lie in the first wall, where one side will be exposed to intense x-ray heating. Determine the maximum allowable energy flux deposited in the side of the tube facing the plasma (Fig. 5.32) for a pumping-power–heat-

removal ratio of 0.02, using water at 93°C as the coolant. Sufficient overpressure would be maintained to suppress vapor formation in the bulk-free stream with the water temperature rise limited to 50°C. (Suggest a graphical solution with pumping power on the left ordinate and heat removal for a 50°C water temperature rise on the right ordinate, with each of these two parameters plotted against flow rate on the abscissa, using log-log coordinates. Choose ordinate scales such that the pumping-power scale is shifted by a factor of 50 in relation to the heat-removal scale.)

2. Repeat the above but for 20-mm-ID tubes cooled with helium at 33 atm for a helium inlet temperature of 520°C and a 100°C temperature rise.

3. Calculate the forced-convection-heat-transfer coefficient and film temperature drop for Probs. 1 and 2 above and compare. Also calculate the first-wall metal temperature near the coolant outlet end of the heated length for each case.

4. Repeat Probs. 1 and 3 for another fluid of your choice and compare with water and helium, using the coordinates of Fig. 4.2.

5. A fluidized-bed coal combustor operated at 900°C is to be used as the heat source for a closed-cycle gas turbine with air at 10 atm as the working fluid. The heater tube matrix is to consist of 13-mm-OD (0.50-in-OD) stainless-steel tubes, the mean temperature of the air being heated is 700°C, and the bed-side heat-transfer coefficient is estimated to be 0.0284 W/(cm$^2 \cdot$ °C)[50 Btu/(h·ft$^2 \cdot$ °F)]. Calculate the tube-side heat-transfer coefficient for a tube length of 3 m and an air-mass flow rate of 5 kg/(s·m^2). Using this value, plot a set of curves similar to those of Fig. 4.4. Add curves for air-mass flow rates of 4, 8, and 10 kg/(s·m^2) to the chart for heating effectiveness versus tube length.

6. Using the heating effectiveness and pressure-drop charts of Figs. A4.4 to A4.6, construct a design chart for a regenerator to be used with the gas turbine of Prob. 5. Plot curves for the mass flow rate as a function of tube length for a series of heating effectivenesses, that is, 65, 75, 85, and 95 percent (where Figs. A4.4 to 4.6 permit that range). Superimpose on these a second set of curves for a series of constant pressure drops on the tube side (high pressure) of 0.5, 1.0, and 1.5 percent $\Delta P/P$. Compare the tube lengths required to obtain heating effectivenesses of 75 and 90 percent respectively for a 1.0 percent $\Delta P/P$.

REFERENCES

1. McAdams, W. H.: *Heat Transmission*, 3d ed., McGraw-Hill Book Company, New York, 1954.

2. Knudsen, J. G., and D. L. Katz: *Fluid Dynamics and Heat Transfer*, McGraw-Hill Book Company, New York, 1958.

3. Fraas, A. P., and M. N. Ozisik: *Heat Exchanger Design*, John Wiley & Sons, Inc., New York, 1965.

4. Kays, W. M., and A. L. London: *Compact Heat Exchangers*, McGraw-Hill Book Company, New York, 1958.

5. *Standards of Tubular Exchanger Manufacturers Association*, Tubular Exchanger Manufacturers Association, Inc., New York, 1959.

6. Fraas, A. P., et al.: "Use of Coal and Coal-Derived Fuels in Total Energy Systems for MIUS Applications," vol. II, Oak Ridge National Laboratory Report No. ORNL/MIUS-28, April 1976.

7. Kays, W. M., and A. L. London: "Heat Transfer and Flow Friction Characteristics of Some Compact Heat Exchanger Surfaces," *Trans. ASME*, vol. 72, 1950, p. 1075.

8. Grover, G. M., et al.: "Structures of Very High Thermal Conductance," *Journal of Applied Physics*, vol. 35, 1964, pp. 1990–1991.

9. Kemme, J. E.: "Ultimate Heat Pipe Performance," *IEEE Trans. on Electron Devices*, Ed-16, 1969, pp. 717–723.

10. LaVerne, M. E.: "Performance Characteristics of Cylindrical Heat Pipes for Nuclear Electric Space and Undersea Power Plants," Oak Ridge National Laboratory Report No. ORNL-TM-2803, January 1971.

11. Fraas, A. P.: "ORNL Aircraft Nuclear Power Plant Designs," Oak Ridge National Laboratory Report No. ORNL-1721, Jan. 28, 1955.

12. "The ORNL Gas-Cooled Reactor," Oak Ridge National Laboratory Report No. ORNL-2500, Mar. 1, 1958, pp. 3.17–3.23.

13. Fraas, A. P.: "Comparative Study of the More Promising Combinations of Blanket Materials, Power Conversion Systems, and Tritium Recovery and Containment Systems

for Fusion Reactors," Oak Ridge National Laboratory Report No. ORNL-TM-4999, November 1975.

14. Seifert, W. F., L. L. Jackson, and C. E. Sech: "Organic Fluids for High-Temperature Heat-Transfer Systems," *Chemical Engineering*, Oct. 30, 1972, pp. 96–104.

15. Lyon, R. N., et al.: *Liquid Metals Handbook*, Atomic Energy Commission and Department of the Navy Report No. NAVEXOS P-733 (rev.), June 1952.

16. Weatherford, W. D., Jr., et al.: "Properties of Inorganic Energy Conversion and Heat Transfer Fluids for Space Application," U.S. Air Force Report WADD TR 61-69, Aeronautical Systems Division, Wright-Patterson Air Force Base, 1961.

17. Ewing, C. T., et al.: "High Temperature Properties of Potassium," U.S. Naval Research Laboratory Report NRL-6233, September 1965.

18. Ewing, C. T., et al.: "High Temperature Properties of Cesium," U.S. Naval Research Laboratory Report NRL-6246, September 1965.

19. Cantor, S., et al.: "Physical Properties of Molten Salt Reactor Fuel," Oak Ridge National Laboratory Report No. TM-2316, August 1968.

20. *The Dowtherm Heat Transfer Fluids*, Dow Chemical Co., 1967.

21. Hoffman, H. W., and A. I. Krakoviak: "Convective Boiling with Liquid Potassium," *Proceedings of the 1964 Fluid Mechanics Institute*, Stanford University Press, 1964.

22. Martz, J. L.: "Organic Fluid Boiler Investigation," A.E.C. Rankine Cycle Technology Program, Sundstrand Corp. Report No. SAN-651-94, 1969.

23. Fraas, A. P., and M. N. Ozisik: "Steam Generators for High-Temperature Gas-Cooled Reactors," Oak Ridge National Laboratory Report No. ORNL-3208, Apr. 8, 1963.

5

CHARACTERISTICS
OF HEAT SOURCES

The two principal ways of increasing the thermal efficiency of thermodynamic cycles were indicated by Carnot over a century ago: increase the peak temperature in the cycle, or find ways to make the actual cycle more closely approach the ideal Carnot cycle. This chapter is concerned primarily with the limitations on the peak cycle temperature imposed by the characteristics of the heat source. Allowances must of course be made for the practical limitations imposed by materials, capital costs, and reliability considerations. Particular attention must be paid to the type of fuel to be used. Coal, for example, contains corrosive impurities which impose more severe peak-temperature limitations on surfaces exposed to combustion gases than a very clean fuel such as natural gas.

Before examining the major elements of the problems involved, it is worthwhile to gain some perspective by examining the chart of Fig. 5.1, which shows practicable temperature ranges for both nuclear and combustion heat sources for typical energy conversion systems. In addition, columns are provided to show the practicable creep strength limit for various types of structural alloy materials together with their melting points and sintering temperatures.

FOSSIL FUEL SYSTEMS

Combustion reactions occur in a wide variety of ways, ranging in rate from the slow, low-temperature reactions involved in spontaneous combustion to the extremely rapid reactions in detonation waves that travel at several times the speed of sound. There are many different types of combustion in power plants, including those in gas and diesel engines, combustors in gas turbines, oil and pulverized-coal burners in steam boilers, traveling grate furnaces burning coal and/or solid wastes, and fluidized-bed combustors. These differ in many ways, but all involve related problems in ignition, combustion efficiency, and emissions of air pollutants; hence a comprehensive survey is helpful in relating the many phenomena and putting them in perspective.

Fuels

Coal displaced wood as the principal source of controlled energy in England in the eighteenth century, when England began to run out of forests to fuel its economy. Wood was the dominant source of fuel in the United States up until after the middle of the last century, when coal displaced it. Petroleum subsequently became an important factor, and that, along with natural gas, displaced coal as the principal fuel in the United States by 1920. Because of their low cost, these three fossil fuels have yielded the bulk of the energy employed in this country, and their cost was essentially the cost of production up until the middle 1970s when fuel shortages made it possible for the Organization of Petroleum Exporting Countries (OPEC) to exercise monopolistic practices, and then the price of fuels became a matter of a complex interplay of monopolistic forces and government regulation. The increasing cost of fossil fuels, coupled with the realization that the readily available reserves of oil and gas are likely to be exhausted by the turn of the century, renewed interest in other fuels. In particular, wood and solid wastes, lignite, peat, and alcohol have begun to receive attention as fuels. Each of these fuels has its own characteristics and properties that influence the combustion system employed and the thermodynamic cycle chosen to convert thermal energy to electric power. Table 5.1 summarizes the principal properties of interest to the power plant engineer for typical fuels, while Fig. 5.2 gives a more detailed picture of the especially important matter of the volatile content of the different types of coal.[1,2]

In addition to the obvious importance of the heating value and sulfur content of the fuel, the fractions of the fuel in the form of volatiles and ash both represent important considerations in the design and operation of combustion systems. Furnace design and performance, for example, are affected by the fraction of coal in the form of volatile material, and the substantial amount of alkali metals—mostly potassium—present in wood ash strongly influence the character of hot corrosion.

The amount of moisture in solid fuels is also important, because it influences their tendency to agglomerate and clog

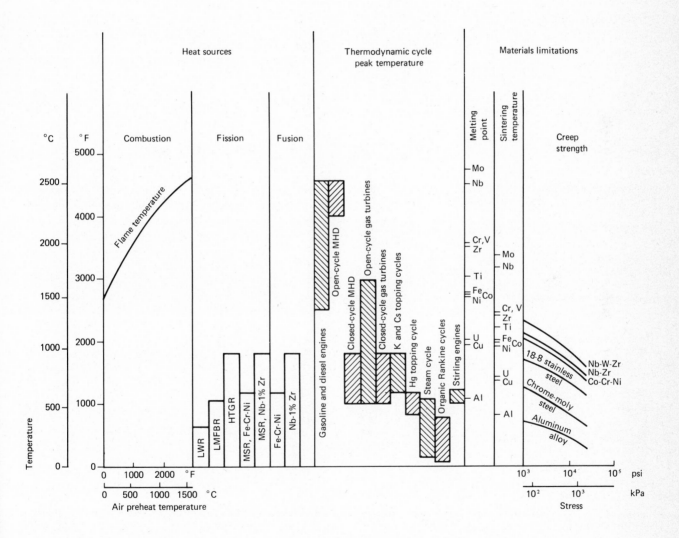

Figure 5.1   Temperature limitations imposed on thermodynamic cycles by both heat sources and the physical properties of materials.

TABLE 5.1   Composition and Characteristics of Representative Fuels*

	Penn. anthracite	Pittsburgh seam bituminous	Illinois No. 6 bituminous	Montana sub-bituminous	North Dakota lignite	Coke breeze	Char	Air-dried peat	Oak	Pine bark
Proximate analysis, wt. %										
Moisture	4.5	2.0	13.2	10.5	33.7	7.3	0.7	9.6	24.0	
Volatile matter	1.7	30.5	36.0	34.7	26.6	2.3	14.7	82	65.5	72.9
Fixed carbon	84.1	58.2	41.8	43.6	32.5	79.4	70.4		9.5	24.2
Ash	9.7	9.3	9.0	11.2	7.2	11.0	14.2	7.9	1.0	2.9
Ultimate analysis, wt. %										
H	2.1	4.7	4.9	4.9	4.2	0.3			7.2	5.6
C	93.9	75.8	70.8	66.8	62.9	80.0			37.9	53.4
S	0.7	2.2	3.4	1.2	0.9	0.6	4.1	0.5	0	0.1
N_2	0.3	1.5	1.3	1.5	1.0	0.3			0.1	0.1
O_2	2.3	6.3	9.3	13.1	20.1	0.5			53.8	37.9
H_2O	4.5					7.3				
Ash	9.7	9.5	10.3	12.5	10.9	11.0			1.0	2.9
Higher heating value, Btu/lb	12,750	13,620	11,080	10,550	7070	11,670	12,100	7861	6300	9030
Bulk density (specific grav.)			0.80							
Viscosity, cp @ 100°F										
Ignition temp., °F	1000	765		870						
Stoichiometric air-fuel ratio		10.17								
Ash composition, wt. %										
SiO_2		44.8	39.8	30.7						39.0
Fe_2O_3		28.3	32.9	18.9						3.0
TiO_2				1.1						0.2
Al_2O_3		21.4	15.1	19.6						14.0
Mn_3O_4										Trace
CaO		3.3	8.4	11.3						25.5
MgO		0.7	0.8	3.7						6.5
Na_2O		1.5	3.0	1.9						1.3
K_2O				0.5						6.0
SO_3				12.2						0.3
Cl										Trace
Ash fusion point, °F		2300	2200							2300

*Data selected from Refs. 1, 2, and 12

Oak bark	Bagasse	Municipal solid waste	No. 6 fuel oil	No. 2 fuel oil	Gasoline	Shale oil	H-coal syncrude	Low-Btu gas	Natural gas
	52.0	35.1			0				
76.0	40.2	48.9		>98	>98				
18.7	6.1								
5.3	1.7	16.0			<0.1				
5.4	2.8		9.5–12	12–14	15.5	10.7	11.8	1.9	24.8
49.7	23.4		86–90	86–88	84.5	83.8	87.1	15.9	74.4
0.1			0.7–3.5	0.05–1.0	<0.1	0.68	0.17		0
0.2	0.1			0–0.1		2.1	0.17	58.5	0.6
39.3	20.0							23.7	0.2
	52.0								
5.3	1.7		0.01–0.5		<0.1				
8370	4000	8580	18,000	19,500	20,700				23,500
0.5		0.1–0.15	0.92–1.02	0.82–0.89	0.72	0.94	0.87		
			250–750	1.7–2.7	1.1				
					800				1346
			12.5	14.5	15			1.7	17.2

Oak bark
11.1
3.3
0.1
0.1
Trace
64.5
1.2
8.9
0.2
2.0
Trace
2750

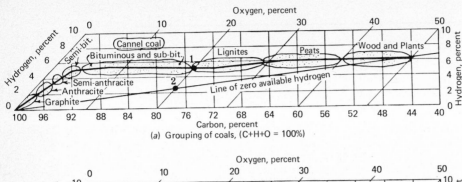

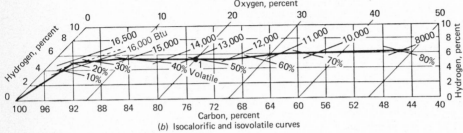

Figure 5.2 Trilinear charts for the hydrogen, carbon, and oxygen contents of moisture and ash-free solid fuels together with their volatile contents and higher heating values. *(Barnard, Ellenwood, and Hirshfeld, 1935.)*

equipment used to move, meter, and feed the coal. The amount of bound nitrogen in the fuel, particularly in coal, is an important consideration, because this nitrogen is likely to form oxides of nitrogen which can go up the stack and constitute an important pollutant. Note that coal commonly contains a significant fraction of nitrogenous compounds.

Typical values are given in Table 5.1 for synthetic fuels derived from coal. These range from low-Btu gas, which would consist mostly of carbon monoxide and hydrogen heavily diluted by nitrogen, all the way to high-quality methane, which would have essentially the same characteristics as natural gas. However, the cost of the more highly refined fuels derived from coal is clearly a step function of the degree to which they must be refined. Estimates vary, but it appears that the cost per unit of heating value is likely to be doubled in choosing methane rather than low-Btu gas or in selecting a high-grade fuel oil, comparable in characteristics to a No. 2 fuel oil derived from petroleum, rather than a very low grade tarry liquid fuel.

The composition of solid waste obviously varies widely with the source and with the methods of preparing the solid waste for combustion. A most difficult set of problems is posed by the presence in municipal solid wastes of polyvinylchloride (PVC) plastics, which tend to produce HCl, a highly corrosive material, in the flue gas. There seems to be no good way to

remove this type of material economically, and hence the best practicable approach seems to be to dilute the solid waste with a fuel such as coal so that the HCl concentration in the flue gas is kept to an acceptable level.

Combustion

Mechanism of Combustion

Combustion is an exceedingly complex phenomenon that is still something of a mystery. The medieval notion that there is something so unique in a flame that a special quantity must be present was closer to the truth than most people realize. Experiments have shown that simply raising the temperature of a combustible gas mixture will not necessarily make it burn; a certain type of chemical activity peculiar to a flame must be initiated. Although the high speed and complexity of the reactions make analysis difficult, it has been demonstrated that even so apparently simple an oxidation process as that in an oxygen-hydrogen flame does not occur to any appreciable extent by the direct combination of oxygen and hydrogen molecules but rather must take place as the result of ionization and a series of intermediate reactions involving a set of ion species. Thus, while the conventional chemical equation for the reaction is

$$2H_2 + O_2 \rightarrow 2H_2O$$

this represents an oversimplification, as it serves only to relate the initial composition to the products of combustion. While the intermediate reactions might appear to be of only academic interest, some conception at least of their occurrence is essential to an understanding of many practical combustion problems. A typical series of these intermediate reactions is as follows:

$$H_2 + OH^- \rightarrow H_2O + H^+$$

$$H^+ + O_2 \;\; \rightarrow OH^- + O^{--}$$

$$O^{--} + H_2 \rightarrow OH^- + H^+$$

In examining these equations, two important features are evident. First, a free active ion of OH, H, or O is necessary to initiate the series. In the second place, once such an ion enters into a reaction, it produces *both the oxidation product and a number of new active ions.* A reaction of this sort is called a *branched-chain* reaction, because the original active ion produces a number of new active ions. Each of these will in turn tend to start a new chain of reactions, and thus the reaction rate in the gas as a whole will tend to increase rapidly until it reaches an equilibrium rate.

The predominant reactions occurring in a flame depend on the concentration of the active ions required to produce them. Dissociation of the fuel molecules, as well as H_2O and CO_2, into simpler forms becomes significant at temperatures around $550°C$ ($1020°F$) and occurs at an increasing rate as the temperature is raised; this gives an increase in the number of active ions. The matter is best visualized in terms of the kinetic theory of gases. In a combustible gas mixture, in which many different types of intermolecular collision occur, only certain types of collision are likely to result in a reaction, the probability of any given reaction depending on such factors as the temperature and the composition of the burning gases. Under some conditions one type of reaction will be favored; under others, another. An extremely complex set of equilibrium relations governs these reactions and determines the rate of each and hence which predominate.[3]

The oxidation of hydrocarbons involves an even more complex group of reactions than the oxidation of hydrogen. Cracking of the larger molecules to yield hydrogen and simpler forms of hydrocarbon and the partial oxidation of some hydrocarbons to form active intermediate compounds, such as aldehydes, alcohols, peroxides, ketones, and organic acids, are some of the reactions that contribute to the complexity of the problem. The aldehydes, in particular, appear to play a very important part in hydrocarbon flames; spectroscopic analysis always indicates their presence. The oxidation mechanism for paraffinic molecules, for example, appears to start with the partial oxidation of the methyl group at the end of the longest straight chain, thus converting the molecule into an aldehyde. The oxidation process then progresses along the chain.

Ignition Temperature

The ignition temperature might be defined as that temperature above which the rate of formation of active ions is greater than the rate of their destruction by impingement on the walls, collision with heavy molecules, or other means. Thus to bring about ignition, the temperature may be raised locally with a hot bulb, the gas may be ionized with an electric spark, or some other means of forming a sufficient number of active ions to initiate a flame must be employed. Once started, the flame will usually sustain itself by the continuous production of a plentiful supply of active ions.

Ignition of fuel-air mixtures is a basic problem in heat engines, and is one of a wide range of combustion problem areas studied extensively by the NACA. (This work included excellent studies of combustion in gasoline and diesel engines, gas turbines, ramjets, and rocket engines.[4-7]) If the temperature of a mixture of fuel and air is increased gradually, some oxidation reactions start to take place, their rate increasing with temperature until the rate of heat generation may exceed the rate of heat loss to the surroundings. In this event the temperature will rise until ignition occurs, at which point the temperature rises rapidly and a flame develops. The ignition temperature varies with the structure of the fuel molecule, the paraffins having relatively low ignition temperatures while the aromatics have relatively high ignition temperatures[6] (Fig. 5.3 and Table A5.1).

The Stoichiometric (or Chemically Correct) Mixture

The amount of fuel ideally required to react with all the oxygen in a given quantity of air to yield H_2O and CO_2 is always an important reference condition. The fuel-air ratio of such a mixture is called the *chemically correct*, or *stoichiometric*, fuel-air ratio.

The chemically correct mixture varies somewhat from one hydrocarbon to another, because it depends on the ratio of the weight of hydrogen to the weight of carbon in the compound, i.e., the *hydrogen-carbon ratio*. Hydrocarbons having the general formula C_nH_{2n} yield chemically correct fuel-air ratios 0.0677. The highest and lowest chemically correct fuel-air ratios for any hydrocarbons commonly found in petroleum fuels are those for benzene and methane, respectively; namely, 0.0754 and 0.0580. Commercial fuels are normally mixtures of hydrocarbons; hence the chemically correct mixture depends on

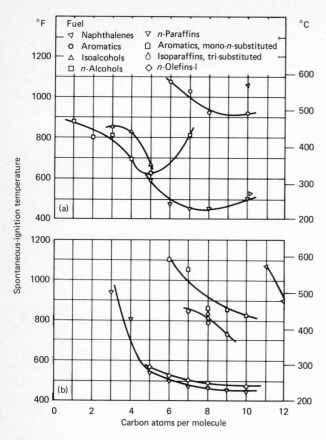

Figure 5.3 Effect of number of carbon atoms per molecule on the spontaneous-ignition temperatures of various liquid fuels.[6]

the hydrogen-carbon ratio of the fuel. A general expression for the chemically correct fuel-air ratio is

Chemically correct $\dfrac{F}{A} = \dfrac{0.0874(1 + \text{hydrogen-carbon ratio})}{1 + 3(\text{hydrogen-carbon ratio})}$

Most light fuel oils in the United States have a hydrogen-carbon ratio of very nearly 0.167. Thus the chemically correct fuel-air ratio for these fuels may be taken as 0.0677 with negligible error for practical purposes. Coal contains a substantial amount of hydrocarbons, ranging from methane to heavy tars. The volatile content of coal varies widely, but for a bituminous coal it commonly runs around 35 percent. This yields a stoichiometric fuel-air ratio of about 0.10 if the coal weight represents moisture- and ash-free coal. This is a higher stoichiometric fuel-air ratio than the value for pure carbon (0.0867) because there is a substantial amount of combined oxygen in the organic matter of coal.

Excess Air

In general, heat engines operate with fuel-air ratios that differ from the stoichiometric value. Boiler furnaces, for example, normally operate with 10 to 20 percent excess air, both to assure complete combustion and to reduce corrosion (a problem treated in the next chapter). Gas turbines and diesel engines ordinarily operate with a large amount of excess air, but maintain near-stoichiometric mixtures in the combustion zone. Gasoline engines often operate with mixtures a bit richer than stoichiometric to get smoother combustion.

Combustion Products

The products of combustion consist predominantly of N_2, O_2, H_2O, CO_2, and CO with small amounts of SO_2 and NO_x, depending on the fuel composition and combustion conditions.

While the composition of the exhaust gas from the combustion of hydrocarbons at any given fuel-air ratio varies somewhat with the hydrogen-carbon ratio of the fuel, the curves in Fig. 5.4 are representative. They give the average composition of the exhaust gases from three spark-ignition engines as a function of fuel-air ratio.[8]

Three vertical lines are drawn on Fig. 5.4 to indicate critical or important fuel-air ratios. One is for the chemically correct fuel-air ratio. The other two are for the rich and lean combustibility limits for a typical gasoline. While it is possible to ignite homogeneous fuel-air mixtures in a bomb at fuel-air ratios outside of the combustibility range indicated in Fig. 5,4, the time required for combustion is so great that combustibility limits determined in this fashion have little significance from the standpoint of heat-engine operation. For example, a short-duration spark occurring in a mixture leaner than the lean limit or richer than the rich limit will not ignite the fuel vapor. Combustion of mixtures leaner than the lean limit is possible only if the fuel vapor is not uniformly distributed throughout the air. Such a condition prevails in a diesel engine where only in the immediate vicinity of the fuel spray is the mixture in the combustion chamber rich enough to burn. A similar situation prevails in a gas turbine. Thus gas turbines and diesels can, and normally do, operate with an average overall fuel-air ratio that is leaner than the lean combustibility limit. The average mixture in a spark-ignition engine, on the other hand, must be in the combustibility range if the engine is to run under its own power, unless a stratified charge condition is obtained by direct fuel injection.

Explosive Mixtures

The combustibility limits shown in Fig. 5.4 raise the important question of the fire and explosion hazard potential associated with the use of gaseous liquid fuels that are sufficiently volatile to form a combustible mixture in air at room-temperature conditions. Interesting enough, the gasoline vapor–air mixture over the liquid fuel in a gasoline tank is too rich to burn, thus reducing the fire and explosion hazard somewhat. Some of the early fuel supplied for aircraft jet engines was just enough less volatile to allow an explosive mixture to form inside a partly empty tank. The fact was recognized and the specification on volatility set so that this would no longer be the case.

While the combustibility limits of Fig. 5.4 are for gasoline vapor–air mixtures, they are representative of all the hydrocarbons. Hydrogen, however, is much more flammable, giving a much wider range between the lean and rich mixture flammability limits, that is, from 9% H_2 to 74% H_2 in air (by

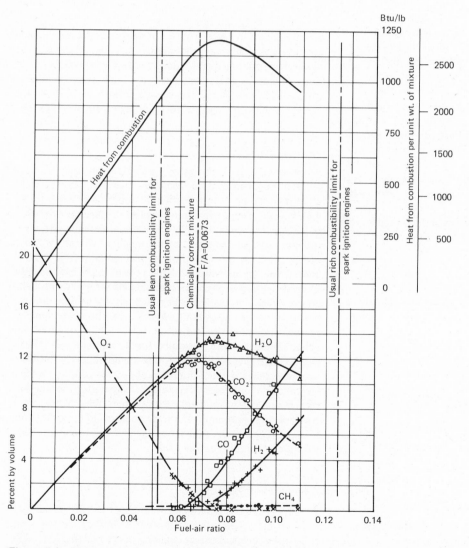

Figure 5.4 Effect of fuel-air ratio on the composition of the products of combustion of gasoline and air, and on the heat released in Btu per pound of mixture. Hydrogen-carbon ratio = 0.176. *(Calculated from data of D'Alleva and Lovell.[8])*

volume) at standard temperature and pressure.[7] It is for this reason that hydrogen represents a much more serious explosion hazard than any of the hydrocarbons, including those that are sufficiently volatile to give combustible mixtures in air at room temperature (i.e., the hydrocarbons ranging from methane to the components of gasoline).

The matter of hydrogen explosions brings up an additional problem: that of detonation. If a hydrocarbon-air mixture is sufficiently hot and at a sufficiently high pressure, a detonation wave may be initiated in the flame front, giving the sharp "knock" often heard when an automobile engine is accelerating. The reason for the knock is that the detonation wave traverses the mixture at a velocity of ~2000 m/s instead of 10 to 20 m/s,[4] and the pressure in the wave is of the order of four times the normal combustion pressure. When this happens, a severe impact load is imposed on the wall of the combustion chamber.[3] Although at atmospheric pressure detonation never occurs in methane- or gasoline-air mixtures, it may occur in hydrogen-air mixtures, thus greatly increasing the damage potential of a hydrogen explosion. Further, the range of hydrogen-air mixture ratios over which detonation may occur is quite wide, i.e., from 18 to 58 percent hydrogen in air (by volume). This, coupled with the wide flammability range for hydrogen-air mixtures (9 to 74 percent), implies that if hydrogen were to become a major fuel as some have advocated, the incidence of deaths from fires and explosions in the United States would be greatly increased over the 7500 per year associated with present fuels.

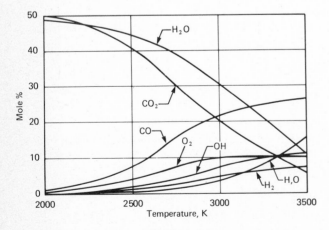

Figure 5.5 Composition of stoichiometric JP4-O_2 combustion gases versus temperature at atmospheric pressure.[9]

Heat Released by Combustion

A number of qualifying factors affect the amount of heat released by combustion. For one thing, the heat of vaporization of the water formed by combustion has never been made available in any type of heat engine. A second factor is dissociation. As can be seen in Fig. 5.5, this becomes important at temperatures above ~1650°C (3000°F), a region of interest in respect to gasoline and diesel engines and to open-cycle MHD systems. A third factor is the hydrogen-carbon ratio of the fuel. Inasmuch as the heating value of hydrogen is 141,990 kJ/kg (61,045 Btu/lb) while that of carbon is 337,735 kJ/kg (14,520 Btu/lb), the greater the proportion of hydrogen in a fuel, the higher its heating value. However, from the practical standpoint, the fuel-air ratio is in many ways more important than all three of the above items put together. Under favorable lean-mixture conditions, practically all the heat ideally available from combustion can be obtained. But if the fuel-air ratio is richer than the chemically correct value, the fuel cannot be burned completely.

Flame Temperature

A little of the energy released by the combustion process is emitted as ultraviolet radiation from the active ions, but practically all of it goes initially into heating the products of combustion, including the nitrogen in the air supplied. The temperature after combustion depends primarily on the initial air temperature and the amount of heat released per unit weight of mixture, which in turn is primarily a function of the fuel-air ratio (Fig. 5.4). In considering the magnitude of the effects, one finds that the flame temperature falls off only about half as rapidly with an increase in fuel-air ratio above the stoichiometric value as does the heat obtained per pound of fuel because of the greater amount of fuel present per pound of air. The author computed the values for Fig. 5.6 with allowances for the variable specific heats and dissociation to show the idealized temperature of the products of combustion as a function of fuel-air ratio for several typical sets of conditions. The higher temperatures for the gas-turbine and combustion-engine cases stem from the higher air temperatures resulting from the air compression prior to combustion. Experimental data have shown excellent agreement with such calculated curves.

Combustion Efficiency

An important consideration in evaluating the performance of a combustion system is the combustion efficiency, i.e., the fraction of the chemical energy available in the fuel that is actually released in the combustion process. The combustion efficiency may be reduced, for example, by incomplete combustion caused by flame quenching and evidenced by the presence of substantial amounts of CO in the stack gas. There may be other losses in the form of partially oxidized hydrocarbons or particles of unburned carbon leaving as soot in the stack gas or entrained in the ash from a coal-fired furnace. To reduce these losses to a minimum, the usual practice is to operate furnaces with some excess air and provide sufficient time in the combustion zone to give ample opportunity for all the carbon present to react with the oxygen. (The hydrogen in hydrocarbon fuels is oxidized quickly, the carbon more slowly.)

Effects of Excess Air on Stack Losses

It is essential to reject the products of combustion to the stack at a temperature in the range of 100 to 150°C (212 to 300°F) both to lift the plume of stack gas well above ground level and to avoid corrosion in the flue gas system by sulfuric acid, which will form if there is any sulfur in the fuel and the gas temperature drops below the dew point (Chaps. 6 and 9). The heat required to raise the stack gases from ambient air temperature to the stack gas temperature required commonly represents 10 to 20 percent of the heat available from the fuel. As the amount of excess air is increased beyond the ideal stoichiometric amount required for reaction of the fuel, the amount of heat lost in the stack gases is increased. That is, if excess air is increased from 0 to 100 percent, the stack losses will double. Thus, there is a strong incentive to make use of a combustion system that will give complete combustion with a minimum amount of excess air, preferably no more than 10 percent.

NO$_x$ Emissions

The type of combustion system chosen affects the amount of NO$_x$ to be expected in the stack gases. As of 1980, there has been some difficulty in enabling power plants to meet air quality standards.[10] Some insight into the problem is given by Fig. 5.7, which shows the equilibrium NO$_x$ concentration as a function of flame temperature for a range of excess-air flow rates. In point of fact, the NO$_x$ content of the combustion gases often will be less than indicated by Fig. 5.7, because it was calculated for equilibrium conditions. Depending on the flame temperature, the time required for the formation of NO$_x$ is ordinarily of roughly the same order as the transit time for the gases through the hot portion of a short flame; hence equilibrium conditions are not reached. As indicated in Fig. 5.7, the NO$_x$ concentration in the stack gas can be reduced by reducing the flame

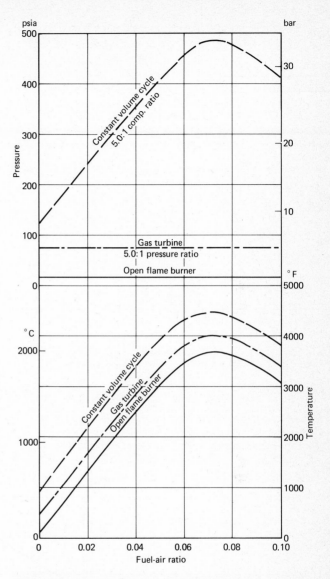

Figure 5.6   Effect of fuel-air ratio on the ideal temperature and pressure after combustion for an octene-air mixture in an open flame, a constant-pressure (gas-turbine) cycle, and a constant-volume cycle. Full allowances for variable specific heats, dissociation, etc., were made in the calculations.

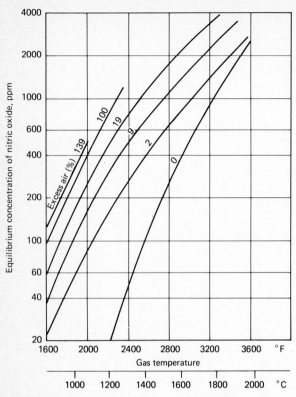

Figure 5.7 Effects of gas temperature and the amount of excess air on the calculated equilibrium nitric oxide concentration in the combustion products. *(Courtesy Oak Ridge National Laboratory.)*

temperature or reducing the amount of excess air. The latter approach has been found effective for certain special types of burner fueled with natural gas, but, for the fuels of prime interest in this text, the most effective approach is to reduce both the amount of excess air and the flame temperature. For open-flame burners, this can be accomplished by delaying the admixture of the secondary air until a substantial amount of heat has been lost by radiation from the primary combustion region. It is also possible to reduce the flame temperature by recirculating some of the relatively cool flue gas, and this approach is being used in some steam power plants.[11] Under any circumstances it is important to minimize the excess air in the combustion zone.

Air Preheat

The amount of heat-transfer surface area required in a boiler furnace can be reduced by increasing the difference in temperature between the hot gas and the boiler tubes. This can be accomplished by preheating the combustion air to raise the flame temperature. In large steam plants the air is commonly preheated to 260 to 315°C (500 to 600°F) to give flame temperatures around 1540°C (2800°F). Note that air preheating has the further advantage that it improves the combustion and thermal efficiencies (Fig. 5.8) and reduces the tendency of the flame to blow out when quenched by a high-velocity stream of cold air.

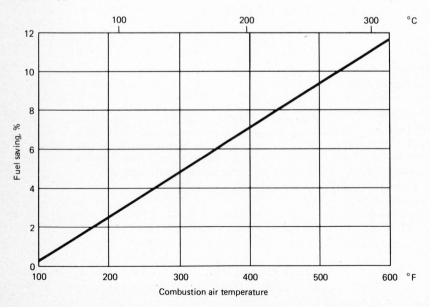

Figure 5.8 Reduction in fuel consumption associated with preheating of the combustion air for steam-boiler furnaces.[12] *(Courtesy Babcock & Wilcox Co.)*

Types of Combustion

Many types of combustion have been investigated. These types of combustion might be grouped into the classifications described in the following subsections.

Open Flames Supplied by a Gas Stream in Which the Flow Is Laminar   The bunsen burner provides a familiar example of this type of flame. The fuel-air mixture entering the flame (Fig. 5.9) is richer than the stoichiometric value; i.e., it is deficient in oxygen and so the base of the flame consists of a thin conical surface, the primary zone, in which the oxygen initially in the mixture combines with the combustible to form H_2O and CO. Additional oxygen from the air surrounding the flame diffuses into the burning gas stream from the side at a uniform rate to react with the CO. As this diffusion takes place, further oxidation occurs along any given streamline in the flame from the primary combustion zone to the point where combustion is complete or the gases are chilled to a temperature too low for combustion. The primary combustion zone is kept at the burner tip by mixing an inadequate supply of primary air with the combustible so that the velocity of the primary flame is lower than the velocity of the incoming gas stream. The boundary layer together with tiny eddies at the tip of the burner where the gas stream emerges into the atmosphere provide a low-velocity zone in which the flame is maintained and from which it can propagate toward the higher-velocity core. The flame boundaries are smooth and sharply defined. The ionization from the chemical activity in the flame is responsible for its luminosity.

Flames in Enclosed Quiescent Mixtures or Enclosed Gas Streams in Which the Flow Is Laminar   The most easily visualized flames of this type are those in coal mine explosions (usually involving methane-air mixtures) or in laboratory bombs. Since oxygen has already been mixed with the combustible gas, combustion is not delayed by the time required for the admixture of oxygen. Because of this, combustion in any given element of gas mixture is so rapid once it has been touched by the flame that the reactions proceed to completion before the front surface of the flame has progressed more than a microscopic distance beyond the element. Thus the luminous zone of intense chemical activity is only a few thousandths of an inch in thickness. In a chamber containing a quiescent mixture, the flame takes the form of a thin, smooth shell much like a rapidly expanding soap bubble. Once started, the flame progresses radially outward from its origin until it reaches the walls of the chamber and no combustible material remains. The velocity of the flame relative to the gas is moderate—of the order of 30 cm/s—and depends on the composition of the mixture and its temperature and pressure.

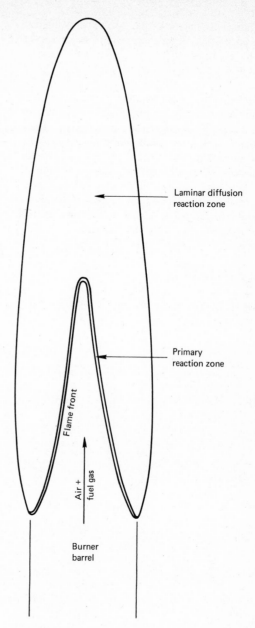

Laminar diffusion reaction zone

Primary reaction zone

Flame front

Air + fuel gas

Burner barrel

Figure 5.9   Structure of the laminar flame from a bunsen burner.

Flames in Enclosed Turbulent Mixtures or Gas Streams
Flames in engine cylinders are of this type and are similar to that described above except that the smooth, thin, shell-like flame is broken up by turbulence into a highly irregular surface in a relatively thick region in which the rapid expansion of pockets

of the burning gas act like microexplosions that greatly increase the intensity of the original turbulence, giving flame velocities of as much as 20 m/s in the high-temperature, high-pressure, highly turbulent mixture in a gasoline engine cylinder.[4]

Open Flames Supplied by a Gas Stream in Which the Flow Is Turbulent   The flames in a gas-turbine combustion chamber and a plumber's gasoline blowtorch are of this type. These flames are similar to the bunsen burner flames described above except that admixture of air takes place by turbulence, as well as by diffusion, so that the combustion rate per unit volume of flame is much higher and the eddies in the turbulent stream of burning gases break up the flame boundaries into highly irregular and unstable surfaces. The velocity of the flame depends on the degree of turbulence, the initial temperature, the fuel-air ratio, and the type of fuel. Figure 5.10 shows data for a typical case.

The behavior of turbulent flames depends mainly on the velocity of the combustible mixture, the intensity of turbulence, the fuel-air ratio of the primary air-fuel mixture, and the initial air temperature. If the velocity of the combustible gas mixture emerging from the barrel of the burner is too low, the flame may flash back down the burner barrel; if the velocity is too great, the flame will blow out. Depending on conditions, there may be a regime at velocities a little below that for blowout in which the flame will become detached from the tip of the burner barrel and burn in a metastable condition, lifted off the burner by several burner barrel diameters. The effects of the burner jet velocity and equivalence ratio (i.e., the ratio of the fuel flow rate to the stoichiometric value) on both the flashback and blowoff limits are shown in Fig. 5.11 for several initial mixture temperatures for a typical burner.

Charcoal Beds   Combustion in a bed of charcoal entails gas-solid reactions. These proceed much more slowly than the reactions in a gas flame, because the reaction rate is limited by the rate of O_2 dissociation and reaction at the solid surface. The bed is ordinarily operated with a deficiency of oxygen and so CO is evolved. This burns in a flame over the bed as secondary air diffuses into the hot combustible gas rising from the glowing bed of charcoal.

Types of Combustor

Many types of combustion systems are employed, depending in part on the fuel used. These systems differ widely in their configurations, heat release rates, excess-air requirements, provisions for ignition, etc. The characteristics of the principal types are outlined in this section.

Gas Burners

The simplest and least expensive type of combustor is a gas burner, because the fuel is in gaseous form so that it can react rapidly with air with the general combustion characteristics outlined in the previous section. The burner designs are basically similar to that of a bunsen burner, with some primary air mixed with the gas to give a mixture richer than the stoichiometric value to give a stable flame (Fig. 5.11) and with secondary air diffused into the flame from the side to complete combustion. Turbulence-promoting devices are incorporated in the burners to increase the combustion rate, reduce the flame size, and improve the combustion efficiency. To increase the inlet gas velocity as much as possible without getting into difficulty with flame blowoff or blowout, a *flameholder* is commonly inserted into the gas stream so that its wake provides a low-velocity, highly turbulent region that keeps the flame attached to the burner. A metal ring or a ceramic cup is often placed in the center of the entering stream of gas–primary air mixture for

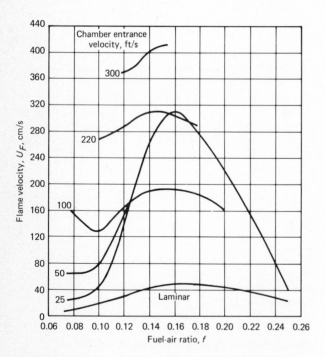

Figure 5.10   Effect of fuel-air ratio on the turbulent flame velocity of Cambridge city gas-air mixtures measured from confined V-flames at various chamber entrance velocities.[6]

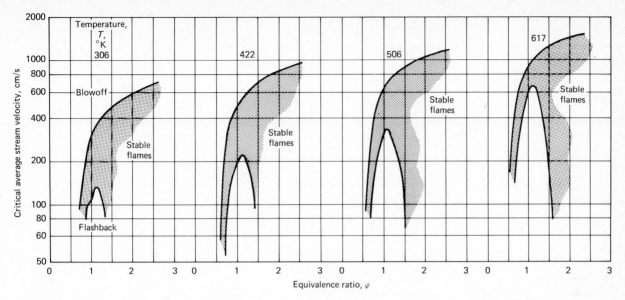

Figure 5.11   Effect of initial mixture temperature on the stable flame region for propane-air flames on a 15.6-mm burner at atmospheric pressure.[6]

this purpose. Such flameholders usually run at a sufficiently high temperature so that they serve as an igniter if flame blowoff occurs momentarily as a consequence of an airflow fluctuation.

Oil Burners

Oil burners are commonly similar in construction and operation to gas burners except that the fuel is injected through an atomizing nozzle, emerging in a fine spray that will evaporate rapidly into a vapor form in which it will burn quickly. Interestingly, the fuel emerges from the spray nozzle in the form of thin liquid jets that are broken up into droplets by the shear forces between the high-velocity liquid stream and the turbulent airstream into which it is injected; the higher the injection pressure and the lower the viscosity of the liquid, the smaller the droplets.[13] The rate of evaporation of the droplets depends on their size, the ambient air temperature, and the volatility of the liquid, typically running from a few milliseconds for isooctane to several seconds for a heavy fuel oil.[6] Figure 5.12 shows some typical evaporation rates as obtained for single droplets injected into airstreams. In furnaces, the fuel spray is carried into a large turbulent flame in which thermal radiation to the droplets from the flame greatly increases the evaporation rate. Thus heavy fuel oil can be used to give a good combustion efficiency in a large furnace for a steam boiler, whereas a light, fairly volatile fuel oil must be used in household furnaces and in the compact combustion chambers of aircraft-type gas turbines.

The larger the burner, the more difficult the admixture of secondary air from the region surrounding the flame and the greater the tendency for the flame to blow out. As a consequence, the usual procedure is to employ many burners—the large number of small holes in the gas burners for cooking stoves exemplifies this approach. (The stove flames, of course, are laminar.) A furnace for a large power plant may employ over 100 burners, each operating like a giant blowtorch burning 1 to 8 tons of fuel per hour to release as much as 50 MW of heat. Not only is the flow into these burners highly turbulent, but the fuel and airstreams are commonly injected in the form of intersecting jets both to promote rapid mixing and to stabilize the flame.

Each burner is equipped with an electric spark igniter or a gas pilot light to ignite the burner. Safety considerations require that the fuel flow be shut off immediately if the flame should blow out. Hence a sensing element is employed—usually an ultraviolet light–sensitive photocell commonly called a *purple peeper*. This is coupled to a fuel shutoff valve to ensure that an explosive mixture will not be formed in the furnace in the event that the flame blows out.

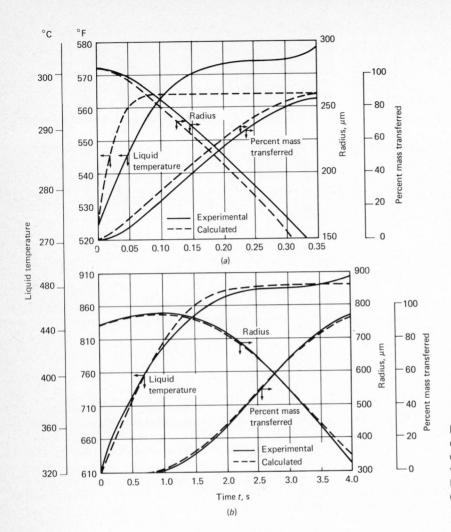

Figure 5.12 Comparison of experimental and calculated histories of droplets: (a) hexane droplet, air temperature 438°C (1280°R), air velocity 300 cm/s, initial diameter 560 μm; (b) hexadecane droplet, air temperature 666°C (1230°R), air velocity 200 cm/s, initial diameter 1520 μm.[6]

Stationary Grates for Coal Furnaces

The earliest steam power plants made use of a stationary grate over which coal was spread. Air was admitted from below through openings in the grate, and the coal was burned in large, open, flat beds. This system functions best if coal is added to the bed in the form of lumps of an appropriate size — usually 2 to 7 cm—because fines of most bituminous coals are inclined to fuse and cake into tarry masses that prevent a uniform airflow distribution through the bed and cause uneven combustion. The system is commonly hand-fired to maintain a uniform bed thickness, and ashes are shaken down through the grate.

Automatic firing systems have been devised; they usually involve pushing or scattering fresh coal out onto the stationary grate. The stationary-grate systems do not lend themselves well to close control of the fuel-air ratio (an important factor to be discussed later) and work well only with selected grades of lump coal. The excess air is commonly 60 percent or more, and the combustion efficiency may be only about 80 percent, partly because many of the fines in the coal fall out with the ash, and partly because poor mixing of the secondary air chills some of the combustion gases so that some of the volatiles escape up the stack unburned. Local temperatures in the burning bed of fuel commonly run at ~1100°C (~2000°F).

Traveling Grates for Coal Furnaces

Efforts to automate coal firing of boilers and permit the use of a wider range of grades of coal led to the development of traveling-grate stokers. In these a bed of coal 15 to 30 cm (6 to 12 in) thick is fed with an automatic stoker from one end of the traveling grate and ashes are dumped off the other end. Movement of the grate agitates the bed a bit, tending to break up regions in which caking occurs. The system is currently used not only for burning coal but also for burning mixtures of coal and solid waste. A disadvantage is that the high operating temperature of the traveling grate, coupled with abrasion by the ash, leads to a substantial amount of maintenance and a limited life. Control of the fuel-air ratio is better than for a stationary-grate system, but it still leaves much to be desired because of variations in the airflow resistance of the bed from one local region to another. The excess air usually amounts to 40 percent or more.

Pulverized-Coal Burners

One of the major improvements in steam power plants in the 1920s was the development of burners in which finely powdered coal was blown into an open-flame burner to give a flame essentially similar to that in an oil burner. This made it possible to design and build a large steam boiler so that it could operate with whatever fuel was most economical or available at any particular time simply by changing the burners, and the fuel-air ratio could be controlled closely to give ~ 99 percent combustion efficiency with only 10 to 20 percent excess air. Thus, a large fraction of modern steam plants, particularly on the east coast, have been designed and built so that they can burn either residual fuel oil or powdered coal almost equally well, and can be converted to use natural gas with only a small loss in performance. This loss in performance stems from the fact that in these furnaces much of the heat is transmitted from the hot gases to the furnace walls by thermal radiation from hot, glowing carbon particles in a bright yellow flame if residual fuel oil or powdered coal is employed. However, if natural gas is used, it burns with a pale blue flame from which relatively little energy is emitted by thermal radiation. As a consequence, in a conventional furnace designed for operation on coal or residual fuel oil with much of the heat going into the water-cooled walls of the furnace (which serve as the boiler portion of the steam generator), the capacity of the unit may be reduced when gas is employed as the fuel.

A pair of burners designed to operate on gas, oil, or pulverized coal is shown in Fig. 5.13. The pulverized coal, when used, is conveyed into the burner with the primary airstream. The secondary air is directed tangentially into the burner region through an array of adjustable doors, or vanes, similar to the gates of Francis hydraulic turbines. Its velocity as it leaves the burner may be as much as 30 m/s. The oil atomizer is located in a cool zone at the air inlet end of the burner so that it will not get hot and give trouble with coke formation irrespective of the fuel used. When operating on coal it is necessary to start up with oil and bring the furnace up to temperature before starting to feed coal to the burner while cutting back on the oil feed rate.

For good combustion the coal must be crushed and then ground to a fine powder, the degree of fineness usually specified in terms of a major fraction (e.g., 85 percent) passing through a standard mesh sieve having a certain number of wires per inch (e.g., 200). Figure 5.14 shows the size distribution for a typical sample. Note that a 200-mesh sieve has apertures of 73 μm, or ~0.003 in. Coal particles of this size are small enough to burn completely in a few seconds so that there is little or no carbon left in the ash carried out of the furnace.

Effects of Fuel on Furnace Design   The reduced availability of high-quality bituminous coal has led to changes in operating

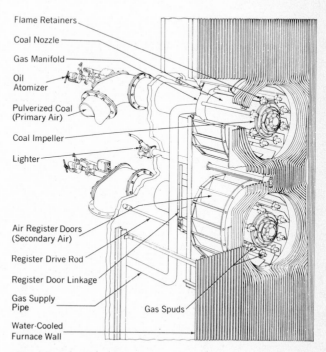

Flame Retainers
Coal Nozzle
Gas Manifold
Oil Atomizer
Pulverized Coal (Primary Air)
Coal Impeller
Lighter
Air Register Doors (Secondary Air)
Register Drive Rod
Register Door Linkage
Gas Supply Pipe
Water-Cooled Furnace Wall
Gas Spuds

Figure 5.13   Cell burner for pulverized-coal, oil, and natural-gas firing.[12] *(Courtesy Babcock & Wilcox Co.)*

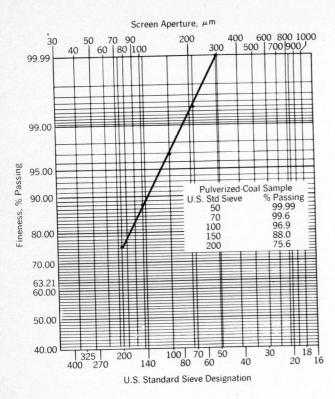

Pulverized-Coal Sample	
U.S. Std Sieve	% Passing
50	99.99
70	99.6
100	96.9
150	88.0
200	75.6

Figure 5.14 Rosin and Rammler chart for plotting pulverized-coal-sample sieve analyses.[12] *(Courtesy Babcock & Wilcox Co.)*

procedures for existing pulverized-coal-fired steam generators and changes in furnace design for new plants. In some instances the ash content of the coal available has been as high as 25 percent, and this and other factors have led to slagging and fouling problems that have forced reductions in the peak load output of as much as 20 percent in plants that had operated well on better coal. To accommodate these effects, it has been found necessary to increase the furnace size in designing for the lower-grade coals. Figure 5.15 shows the furnace sizes for various coals relative to that for a representative bituminous coal.[14] The increasing moisture and ash contents reduce the flame temperature and hence both the radiant heat flux from the combustion zone and the mean temperature differences available throughout the steam generator. Note, too, that the reduced combustibility of the lower-grade coals makes it necessary to provide a greater volume for the combustion region.

Cyclone Furnaces

For coals containing ash that gives a relatively low slag viscosity (e.g., < 250 poise at $1425\,°C$), a still more compact flame can be obtained in the intensely turbulent vortex of a cyclone furnace of the type shown in Fig. 5.16. The walls are water-cooled so that a thin layer of frozen slag forms to cover the metal surface and reduce the heat flux to ~ 16 W/cm² [50,000 Btu/(h · ft²)]. Around 70 percent of the ash is centrifuged to the wall and drains off in liquid form, and so the amount of fly ash in the combustion gas leaving the furnace is much less than for the

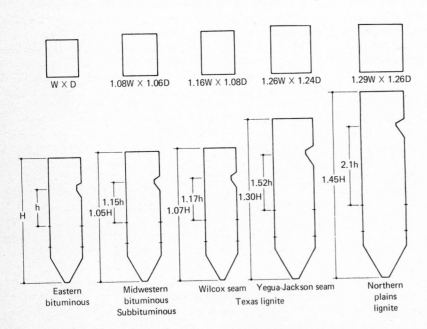

Figure 5.15 Effects of type of coal on furnace size for large steam generators.[14] *(Courtesy Power Engineering.)*

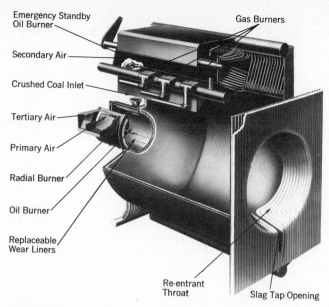

Figure 5.16 A cyclone furnace type of burner for pulverized-coal firing. Droplets of molten ash are centrifuged to the water-jacketed wall, from which molten slag is drained off, while the low-ash-content gas from the center of the vortex flows through the re-entrant throat at the right into the main part of the furnace.[12] *(Courtesy Babcock & Wilcox Co.)*

Labels on figure: Emergency Standby Oil Burner, Gas Burners, Secondary Air, Crushed Coal Inlet, Tertiary Air, Primary Air, Radial Burner, Oil Burner, Replaceable Wear Liners, Re-entrant Throat, Slag Tap Opening

Figure 5.17 Ash deposits on secondary superheater tubes. *(Courtesy Babcock & Wilcox Co.)*

pulverized-coal burners discussed above. This eases slagging problems in the superheater, which can severely reduce the heat-transfer rate and hence the capacity (Fig. 5.17). It also reduces the load on the stack-gas particulate-removal equipment, thus permitting the use of higher-ash (and lower-cost) coals. The intense turbulence makes possible a good combustion efficiency with much coarser coal-particle size ($-\frac{1}{4}$-in-mesh screen) so that the substantial cost and power requirements of pulverizers can be avoided. A key factor making this possible is that the larger coal particles are centrifuged into the liquid ash swirling around the bore of the combustion chamber, where the carbon is oxidized by the oxygen dissolved in the liquid ash. Heat release rates are very high—as much as 0.8 W/cm³ [800,000 Btu/(h · ft³)].

Cyclone burners are widely used where the coal available has an ash composition suitable for the special conditions in the cyclone burner.

Pneumatic Conveyance of Coal Pneumatic conveyance of crushed coal to a cyclone burner presents somewhat more difficult problems than is the case for pulverized coal, because the coarser particles are more inclined to settle out. There are many special considerations that, while logical when pointed out, are not obvious.[12,15] Suffice it to say here that the gas velocity is usually in the 6- to 12-m/s range, the pipe size must be at least eight times the largest size of particle conveyed in order to avoid bridging and blocking of the channel, and the superficial moisture content of the coal must be less than 1 percent.

Fluidized-Bed Combustion of Coal

The most promising new coal combustion system in the 1970s has been the fluidized-bed combustor, or FBC. The development of fluidized beds for chemical processes such as sulfide ore roasting and petroleum catalytic cracking and reforming processes began in the 1930s and has led to their extensive use for processes requiring the reaction of solids and gases (and sometimes liquids).[15-18] Over 1000 such units are in use in the United States at the time of writing. Fluidized-bed operation depends on the fact that when the velocity of fluid flowing upward through a bed of small particles is increased sufficiently, the particles begin to float. At the threshold velocity for

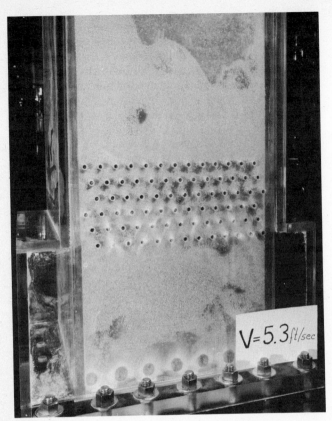

Figure 5.18 View of a cold-flow model of a fluidized bed in operation. Note the large gas bubble in the lower right corner, and near the center of the top of the bed, note the cavity left by a bubble that has just erupted. *(Courtesy Oak Ridge National Laboratory.)*

bed (Fig. 5.18). The large surface area of the solid particles in the bed and the continuous agitation give a high reaction rate per unit of volume and a uniform distribution of both the reaction rate and the temperature throughout the bed.

The work on fluidized beds for exothermic chemical processes that required heat removal from the bed to hold its temperature constant led to the discovery that the heat-transfer coefficient in fluidized beds is high—of the order of 280 to 570 W/(m² · K) [50 to 100 Btu/(h · ft² · °F)]. This in turn suggested the possibility of designing a fluidized-bed coal combustion system for a steam boiler, the idea being that the boiler size and cost could be much reduced by taking advantage of the high heat-transfer coefficient characteristic of fluidized beds. This concept was investigated experimentally beginning in the 1950s, and work continued at a low level because of problems with control, until the pressure to reduce sulfur emissions became a major consideration. The possibility was recognized of employing a coal combustion system based on a fluidized bed of limestone that would react with the SO_2 immediately on combustion to form calcium sulfate, and experiments in England in the latter 1960s showed that ~90 percent of the sulfur could be retained in a limestone bed operated in the 800 to 900°C (1470 to 1650°F) range.[20] The process usually involves calcination of the $CaCO_3$ to CaO when it is heated to bed temperature, after which the CaO absorbs SO_2 and oxygen to give $CaSO_4$.

Concurrently with this, it was found that the fluidized-bed combustion system could be used to burn gaseous fuels, char, high-sulfur residual fuel oil, solid wastes, or even sludge containing as much as 35 percent water to give a combustion efficiency of as much as 99 percent with the excess air running as little as 10 percent.[20]

Superficial Velocity Limitations The gas velocity through a fluidized bed is usually expressed in terms of the nominal average velocity upward through the freeboard above the bed and is referred to as the *superficial velocity*. For a fluidized bed to operate properly, the airflow through the bed must be high enough to float the particles and produce the agitation and circulation alluded to above. However, the airflow rate ought not be so high as to blow a large fraction of the smaller-diameter particles out of the bed so that they are carried off with the stack gas. Some insight into these limitations is given by Fig. 5.19, which shows the relation between particle diameter and the superficial velocity of the gas leaving the bed for these two limits.[17] Curves are given for three different densities of the settled bed, i.e., the bed density prior to fluidization, which depends mainly on the density of the particles. For the usual coal ash–sulfated lime mixture of interest here, the settled density is ~ 1.3 g/cm³. Inasmuch as the particle size in the bed will vary

fluidization, the bed expands upward a little and behaves as if it were a viscous liquid. A further small increase in velocity causes the bed to expand about 30 percent as bubbles form, and the bed behaves much like a turbulent boiling liquid. It does not have a sharply defined free upper surface, but, as the solid fraction decreases in the upper regions of the bed, the gas velocity also decreases and the fluid dynamic forces tending to lift the particles are reduced and the particle concentration drops.[19] The large gas bubbles that form move upward at perhaps 10 times the average gas velocity, and in so doing displace bed material so that some of it circulates downward to the side of the bubble path. When the bubbles rise to the surface of the bed, they erupt as if they were small volcanoes and blow some particles out of the bed at sufficient velocity to make the particles arc upward through trajectories well above the nominal upper surface of the

substantially about a mean value, the range of superficial velocities for satisfactory operation is substantially less than is implied by Fig. 5.17. For example, if one were to assure a uniform particle diameter of 2500 μm (0.10 in) with a settled bed density of 1.6 g/cm³, the superficial velocity of the combustion gas leaving the bed could be varied by a factor of about 20 without giving difficulty with particle carry-over or settling of the bed. However, if smaller particles about one-sixth the 2500-μm size are present and are not to be blown out of the bed, the superficial velocity could be varied by only a factor of 3 between the minimum for fluidizing the bed and the maximum usable without excessive carry-over, or elutriation.

Air Distributor Many air inlet port, or *tuyere*, designs have been tested for fluidized beds in an effort to find an inexpensive, simple system that will give a good airflow distribution across the bed. Arrangements considered have included those shown in Fig. 5.20: a simple pattern of drilled holes in a plate, porous plates, grids made by stacking alternate layers of flat and sinusoidally bent bars, vertical spigots projecting up into the bed with their ends covered with fine-mesh screen or a porous disk, and stub tubes extending up into the bed but capped and provided with holes extending horizontally or inclined at a small angle downward. Note that the tuyeres of Fig. 5.20*d*, *e*, and *f* have been designed particularly to avoid settling of particulate matter downward through the tuyere into the plenum chamber when the airflow through the tuyere is stopped. However, tests with the other types of tuyere have shown that relatively little sifting of this sort takes place even with particle sizes substantially smaller than the open holes in the tuyeres.

Inspection of the region in the vicinity of air tuyeres after hot operation has disclosed that designs *a*, *b*, and *c* give difficulty

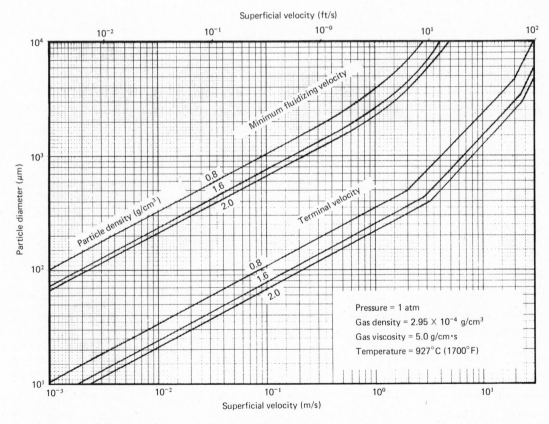

Figure 5.19 Effects of particle size on both the minimum gas velocity leaving the bed for full fluidization and the maximum velocity without elutriation.[17] *(Courtesy Oak Ridge National Laboratory.)*

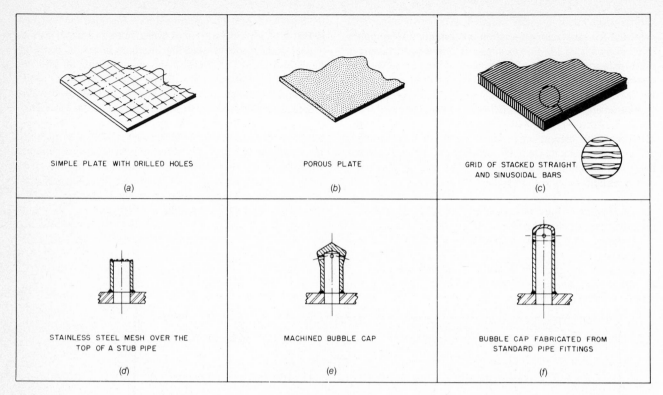

Figure 5.20   Typical air tuyere geometries that have been used in fluidized-bed coal combustion systems.[17] *(Courtesy Oak Ridge National Laboratory.)*

with warping, distortion, and sometimes cracking of the bed plate when a hot bed settles on it after an abrupt forced outage. This problem results from the abrupt heating of the plate, which is normally cooled by the air supplied to the tuyeres. The problem has also been noted by the author in plates employing the design of Fig. 5.20e when the distance between the air orifices and the plate was only 3 to 5 cm. In some designs this difficulty has been avoided by using tuyeres that project above the metal base plate by 20 to 25 cm and by protecting the plate from settling of the hot bed with a layer of firebrick or castable ceramic 10 to 15 cm thick. Analyses indicate that coal ash has a sufficiently low thermal conductivity so that it can be depended on to provide adequate thermal insulation if an air tuyere of the type shown in Fig. 5.20e is used and is made ~20 cm high, permitting a static layer of ash ~15 cm thick to settle on and protect the plate.

Past experience with fluidized beds has shown that one should include sufficient flow resistance across the air tuyeres to assure a uniform distribution of airflow over the entire cross section of the bed and to inhibit the amplitude of bed pulsations, which can be quite severe and lead to severe vibration in the tube matrix.[19] It has been found empirically that this pressure drop should represent at least 20 to 30 percent of the total pressure drop across the bed region.

Coal Feed Ports   It is possible to feed coal to a fluidized bed simply by dumping it on top. However, this has a number of serious disadvantages, probably the most important of which is that the fine particles in crushed coal tend to be blown out without burning completely or giving the SO_2 formed an opportunity to react with the lime. Hence it is much better to introduce the coal near the bottom of the bed. If this is done, the coal is most conveniently transported into the bed by using between 5 and 10 percent of the full-power airflow to convey the coal through tubes to the coal delivery ports.

As discussed in the next chapter, the most important factor influencing the rate of corrosion of metal parts in a fluidized-bed coal combustion system is the local fuel-air ratio. If slightly reducing conditions (or even stoichiometric conditions) prevail, rapid sulfidation occurs, and if conditions vacillate locally from oxidizing to reducing, very rapid oxidation and corrosion will result. The critical zone in a bed is in the vicinity of the coal feed ports, where the supply of fuel greatly exceeds the supply of air. In this region the coal particle is heated—in the course of which the bulk of volatiles will be evolved and will burn rapidly (these represent ~35 percent of the particle by weight)—and then the glowing carbon particle will oxidize at a relatively slow rate that will depend on its size and porosity. The problem is similar to that of oil droplet evaporation in an oil burner (Fig. 5.12), but it is more complex, because parameters such as the thermal conductivity and the porosity of the particle vary widely in the course of the emission of the volatiles. The problems involved are not only too complex to be handled analytically except in a qualitative fashion, but they are also difficult to handle experimentally. The limited data available indicate that by injecting the coal horizontally to give good lateral dispersion and by spacing the feed ports on roughly 1-m-square centers, good dispersion of the coal in the bed can be obtained and reducing zones can be confined to small regions close to the coal feed ports where the rapid vaporization of volatiles gives a locally fuel-rich mixture. Of course, if char rather than coal were used as a fuel, a larger coal-feed-port spacing could be employed.

Combustion Efficiency   A scrutiny of fluidized-bed coal combustion experience indicates that, in beds having a depth of ~1 m, as much as 25 percent of the carbon is blown out of the bed with the fines before it is burned if the superficial velocity is ~4 m/s, although only ~1 percent will pass through unburned if the superficial velocity is ~1 m/s. Examination of the elutriated material indicates that most of the unburned carbon is in particles larger than 40 μm; the surface-volume ratio of smaller particles is apparently high enough so that the heating, volatilization, and combustion processes take place rapidly and all the carbon is burned in the short transit time available (a few seconds). An indication of the time required for combustion is given by analysis of samples of bed material that are quenched with inert gas when taken from the bed. These analyses show that less than 1 percent by weight of the bed material is unburned carbon while the balance is roughly 20 percent ash and 80 percent sulfated lime, depending on the operating conditions, the coal and limestone used, etc. In general, the coal ash is friable and decrepitates fairly rapidly so that the fines are elutriated, whereas the lime is relatively hard and is more resistant to decrepitation.

Effects of Bed Depth and Furnace Pressure   To keep the pumping power for the combustion air to a reasonable value—i.e., around 2 percent of the net plant power output—the depth of the bed ought not exceed about 1 m (40 in) if it is operated at atmospheric pressure. This bed depth in turn favors the use of horizontal tubes for the heat-transfer matrix. If the furnace is pressurized and the combustion-air pumping-power requirement is kept to a fixed percentage of the plant output, the depth of the bed can be increased in direct proportion to the furnace operating pressure. Increasing the bed depth has the advantage from the combustion standpoint of providing more time for fine particles of coal to burn and for the SO_2 to react with the lime so that both the amount of unburned carbon in the fines blown out of the bed and the SO_2 emissions are reduced (Figs. 5.21 and 5.22). Increasing the bed depth also makes possible the use of vertical tubes, which is particularly desirable for steam boilers where it is advantageous to employ natural thermal convection for recirculation of the liquid in the boiler. Vertical tubes have the additional advantages of giving a somewhat higher heat-transfer coefficient on the bed side[16] and appearing to be less subject to erosion by the particles in the bed. Unfortunately, the pumping power required to pressurize the furnace is excessive unless the energy in the hot, high-pressure combus-

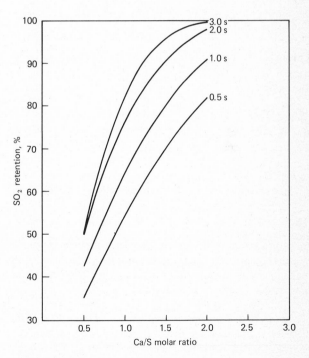

Figure 5.21   Effects of gas transit time on SO_2 retention versus Ca-S ratio in tests of a pressurized fluidized bed using dolomite as the sorbent.[21]

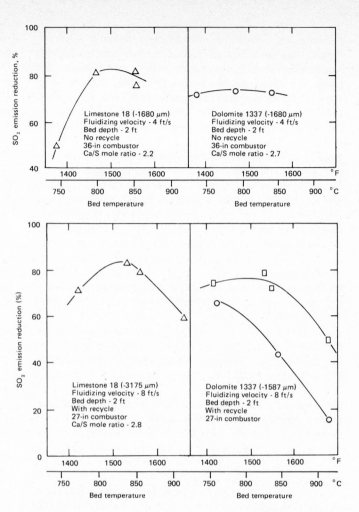

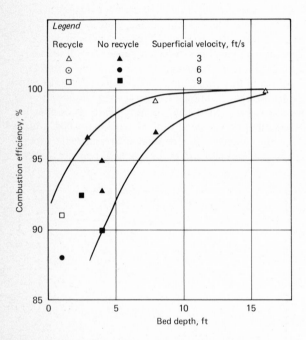

Figure 5.22 Effects of bed depth on combustion efficiency. Data plotted from tests at Exxon, BCURA, Curtiss Wright, and B & W.

Figure 5.23 Effects of the temperature of a fluidized-bed combustor on the sulfur retention for Pittsburgh coal under four different conditions.[20]

tion gas leaving the furnace is recovered in a gas turbine used to drive the compressor, and this presents difficult problems with erosion, deposits, and corrosion in the turbine. These problems are treated in Chaps. 12 and 13.

Effects of Bed Temperature and the Calcium-Sulfur Ratio
The three most important factors affecting the degree to which the sulfur in the fuel is retained in the bed as $CaSO_4$ are the calcium-sulfur ratio, the residence time, and the fluid-bed operating temperature.[20] The effects of two of these parameters on SO_2 emissions are shown in Figs. 5.23 and 5.24. The mat-

ter is complicated by the fact that both coal and limestone from different strata vary substantially in their characteristics, and these factors influence the effectiveness of the fluidized bed in removing sulfur.

Dolomite is often used instead of limestone and seems to give somewhat better sulfur retention per mol of calcium. However, the dilution of the $CaCO_3$ by $MgCO_3$ acts to give about the same weight flow of stone to the bed for a given degree of sulfur capture.

Regeneration of the Lime   It would be advantageous to

regenerate the lime in the bed and thus reduce both the consumption of limestone and the quantity of ash that must be hauled away. Processes have been investigated that would yield elemental sulfur, a salable product. Some depend on roasting calcium sulfate under mildly reducing conditions at ~1065°C (1950°F) to evolve a gas rich in sulfur dioxide. Others depend on reduction of the calcium sulfate by a gas containing hydrogen or carbon monoxide to yield calcium sulfide, and on subsequent reaction of the calcium sulfide with steam and CO_2 to produce $CaCO_3$ and H_2S, from which sulfur may be produced in elemental form more readily than from SO_2. Irrespective of the process, the regenerated lime generally has less reactivity than fresh lime. This, plus the effects of decrepitation in the turbulent bed, makes it necessary to supply fresh stone at a rate equivalent to 20 percent of the sulfur to be captured, on a stoichiometric basis, and a comparable amount of lime must be withdrawn for sale or disposal.

Exxon Research and the Argonne National Laboratory (ANL) have carried out government-funded programs in this area. They have been able to regenerate lime and recirculate it for 3 to 5 cycles by using a fairly high recirculation rate so that the lime is not sulfated more than 50 percent before recycling.[21,22] At the time of writing, the Exxon process for regenerating the lime and extracting elemental sulfur appears to require about 8 percent of the energy available in the coal fed to the fluidized bed, whereas the ANL process designed to accomplish the same purpose gives promise of requiring only 2 percent of the energy in the coal.

NO_x *Formation* The relatively low temperature that is characteristic of fluidized-bed combustion acts to keep the formation of NO_x to a low level, but the gas transit time through the high-temperature region is sufficiently long that the equilibrium concentration of NO_x can be reached (Fig. 5.7). This makes the NO_x concentration in the stack gas quite sensitive to the amount of excess air,[10] and this in turn places a premium on holding the amount of excess air to a low level. Test experience indicates that the NO_x concentration in the stack gas is higher than indicated by Fig. 5.7, apparently because there is a substantial amount of nitrogenous material in coal that tends to oxidize to NO_x. In spite of this, the NO_x concentration in the stack gas is normally half the current Environmental Protection Agency (EPA) limit of 0.7 lb $NO_x/10^6$ Btu heat release.

Heat-Transfer Coefficient The dominant heat-transfer mechanism in a fluidized bed is direct conduction from the hot particles in the bed to the heat-transfer surface, although thermal radiation also makes a minor contribution.[23] As a consequence, as shown in Fig. 5.25 the heat-transfer coefficient is almost independent of the gas superficial velocity over the bed operating range of interest. At very low velocities where the bed is barely fluidized, the heat-transfer coefficient is low, but this region is unsuitable for good combustion because the coal mixing rate is too low. At high superficial velocities, the heat-transfer coefficient also falls off because of excessive void formation in the bed, but this region is also of little interest because of excessive elutriation. The scatter in the points in Fig. 5.25 appears to have been caused by changes from one metastable mode of bed bubbling to another.

The heat-transfer coefficient is essentially independent of the furnace pressure, but it does increase ~25 percent as the mean particle size is reduced from 2 mm to 0.5 mm. It also increases ~20 percent as the temperature rises from 100°C to 900°C.

Summary

A major consideration in the choice of a burner or combustion system for any application is its suitability for use with a wide range of fuels to give a high combustion efficiency with relatively little excess air. Further, in boiler furnaces, it is important to employ a combustion system that will burn high-sulfur coal and yet reduce the sulfur emissions in the stack gases to a level consistent with both present and probable future EPA standards. As will be discussed in Chap. 11, there is no process available

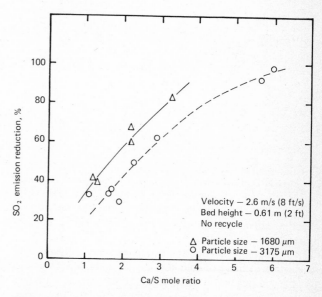

Figure 5.24 Effect of additive particle size on reduction of the stack gas SO_2 content for Limestone 18 at 1560°F.[20]

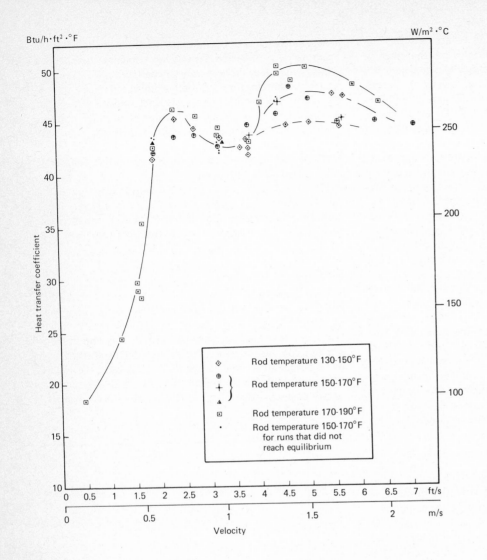

Figure 5.25 Fluidized-bed heat-transfer coefficient as a function of superficial gas velocity for an electrically heated rod in a 10-in-square Lucite model of a fluidized bed. *(Courtesy Oak Ridge National Laboratory.)*

commercially at the time of writing that is effective in removing sulfur from stack gases following combustion of high-sulfur coal and yet is economically attractive, particularly when operational, maintenance, and waste-disposal requirements are considered.

In attempting to give perspective to the many problems associated with a choice of combustion system, Table 5.2 was prepared to indicate the suitability of the different types of combustor for typical fuels. Note that only the fluidized-bed combustion system appears to show promise for handling the full range of fuels of interest, i.e., high- and low-sulfur bituminous coal, lignite, anthracite, gaseous and liquid fuel derived from coal, mixtures of residential or industrial waste with coal, or mixtures of coal and sewage sludge from filter beds in sewage plants (for which coal has proved to be an excellent filter medium). Further, the FBC may be able to accomplish this with nearly as little excess air as pulverized-coal or cyclone burners. If pressurized to 5 atm or more, the furnace horizontal cross-sectional area would be about the same or less than for the other types of combustor, and the height would be less. These matters are treated further in Chaps. 11, 13, and 14 in sections dealing with the coupling of FBCs to particular types of power plant.

TABLE 5.2 Characteristics of Typical Combustion Systems and Their Suitability for Use with Different Types of Fuel for Installations Designed to Meet EPA Standards without Equipment for Removing Sulfur from the Stack Gases

Type of fuel	Stationary grate	Moving grate	Open burner	Fluidized bed
High-sulfur bituminous coal				X
Low-sulfur bituminous coal	X	X	X	X
Lignite	X	X	X	X
Anthracite	X	X	X	X
Char (from coal conversion plants)			?	X
Natural and high-Btu gas			X	X
Medium-Btu gas			X	X
Low-Btu gas			X	X
No. 2 fuel oil			X	X
Residual fuel oil—low sulfur			X	?
Residual fuel oil—high sulfur				?
Liquid fuel from coal			X	?
Domestic solid waste + coal		X		X
Sewage sludge + coal				X
Wood waste + coal		X		X

Combustion parameters				
Typical heat release rate,* MWt/m²	1.25	2.2	5.6	1†
Btu/(h · ft²)	400,000	700,000	1,800,000	300,000
Excess air required, %	30–60	15–50	5–20	10–30
Combustion efficiency, %	90–99	90–99	98–99	92–99

*Heat release from bituminous coal per unit of furnace horizontal cross-sectional area.
†Atmospheric bed. Higher furnace pressures give proportionally higher outputs.

Coal Gasification and Liquefaction

The problems posed by burning high-sulfur, high-ash coals discussed in the previous section introduce the closely related question of converting the coal to a clean gaseous or liquid fuel before supplying the fuel to a power plant. There are two basic approaches. The first integrates a coal gasification plant with the utility plant so that heat losses from the coal gasification process can be used to advantage in the steam cycle; the second locates the coal gasification or liquefaction plant and the utility plant at different sites. In either case most of these plants are de-signed to convert as much as possible of the free carbon in the coal into gaseous fuel or hydrocarbons via the water-gas reaction, i.e., reacting steam with carbon to give CO plus H_2. The CO and H_2 can be used directly as fuel or they can serve as feed stock for methanation processes to produce hydrocarbons. The fuel conversion efficiencies claimed for the various coal conversion systems range from 55 to 90 percent.[24-27] There are differences in definitions of efficiency, but the author has been unable to find any test data obtained from an actual pilot or demonstration plant that gives an efficiency of conversion of the chemical energy in the coal into chemical energy in the product in excess of 67 percent after allowances for energy inputs represented by steam, compressed air, oxygen (if used), and circulating pumps, together with losses in the gas clean-up train for removing sulfur and particulates. The problems involved have much in common with those of the combustion systems of the previous section. Hence it seems appropriate to treat them briefly here.

Inherent Losses in Coal Gasification and Liquefaction Processes

In appraising the potential of the various methods for coal gasification and liquefaction, it is helpful to consider first the inherent losses in the basic processes, such as pyrolysis, the water-gas reaction, hydrogenation, methanation, and methanol synthesis, following an approach suggested by D. M. Eissenberg.

Pyrolysis To remove the volatile constituents from coal by pyrolysis, its temperature must be raised to 500 to 1000°C (932 to 1832°F). The sensible-heat input to the coal, whose specific heat is ~1.25 J/(g · °C)[~0.3 Btu/(lb · °F)], thus is at least 600 J/g (~260 Btu/lb). The yield of volatiles (other than water) is commonly ~20 percent with a heating value of 35,800 J/g (16,000 Btu/lb), or 7150 J/g of coal (3200 Btu/lb). Thus the heat input to the pyrolysis process runs 8 percent or more of the chemical energy in the volatiles, and much of this heat cannot be recovered unless the pyrolysis is carried out as an integral step in the course of feeding coal to a furnace–steam generator where the coal would have to be heated anyway (as in the system discussed at the end of Chap. 13). Pyrolysis of oil shale entails the same sort of loss, but its value is about twice as great because the hydrocarbon yield per ton is about half as much.

Hydrogenation The hydrocarbons produced by both pyrolysis and liquefaction processes commonly include a substantial fraction of unsaturated compounds such as olefins. These tend to be unstable and polymerize even at room temperature to form gums and tars, in some cases rendering the raw liquid product too viscous to pump in as little as a few days. This problem is handled by *hydrogenation,* commonly in a reactor at 50 atm or more and a temperature around 370°C (700°F). The energy required for the process includes that to produce the hydrogen (often including cryogenic separation from other gases), pumping, and heating. These requirements entail energy losses that commonly total 5 to 10 percent of the chemical energy in the liquid product.

The Water-Gas Reaction The basis for most coal gasification and liquefaction processes is the water-gas reaction; that is,

$$2C + 2H_2O \rightarrow 2CO + 2H_2$$

The reaction is endothermic, and so the chemical energy available from the CO and H_2 is greater than that in the carbon that reacts with the water. To supply this increment in energy, however, nearly one atom of carbon must be burned to CO_2 for each two atoms of carbon involved in the water-gas reaction. Thus there is a substantial heat loss to the CO_2 and N_2 diluents. As a result, the overall *energy efficiency* for the ideal process is no more than 90 percent if the total energy input in the form of carbon and superheated steam is considered, together with stack losses. Further, if air is used and nitrogen must be removed from the product gases, there is an additional energy loss, whereas if oxygen is fed to the process to avoid a nitrogen removal step, the energy required for the cryogenic separation of the oxygen from air must be considered, and this alone amounts to 1680 J/g of O_2 (420 kWh/ton of O_2). The latter item is roughly 14 percent of the chemical energy in the CO and H_2 produced if one allows for a 33 percent thermal efficiency for producing the electricity to make the oxygen, and this reduces the ideal energy efficiency of the process to 76 percent. If waste heat from other processes in the plant were employed to generate this electricity, the energy charge to the other process would be reduced, but the loss would be greater because of the lower thermal efficiency in utilizing the lower-temperature waste heat. If sulfur is present in the coal, necessitating the removal of H_2S from the product gas, the additional energy loss will be roughly equal to the sulfur content of the coal percentagewise.

Methanation The heat content of the low-Btu gas from the usual water-gas reaction is too low for economical storage or transport of the gas over substantial distances; hence additional processes may be employed to increase the heat content. One of these is methanation; that is,

$$2CO + 2H_2 \rightarrow CH_4 + CO_2$$

This is an exothermic process in which the chemical energy in the two gram mol each of CO and H_2 amounts to 569 kJ and 484 kJ, while the chemical energy available from combustion of the CH_4 product is 800 kJ, giving an energy efficiency for the methanation process of $\frac{800}{1053}$, which equals 76 percent. The heat given up in the process can be employed to make steam, but equipment design considerations make it necessary to keep the steam system temperature and pressure to modest levels, and so the thermal efficiency for the generation of power is limited to less than 25 percent. Thus the overall energy efficiency of converting carbon to methane can be expressed in the following equation, if we assume an oxygen-blown water-gas process and credit the methanation process for two-thirds of the heat re-

TABLE 5.3   General Comparison of Coal Gasification Reactor Types[27]

Function	Moving bed		Fluidized bed	Entrained flow
	Dry ash	Slagging bottom		
Capacity	Low	High	Intermediate	High
Ability to handle caking coals without pretreatment	Moderate	Moderate	Poor	Excellent
Temperature of operation, °F	2000–800	2800–800	1600–1900	3000–1700
Temperature control	Poor	Poor	Good	Moderate
Refractory problems	Moderate	Poor	Moderate	Poor
By-product tar formation	Yes	Yes	Possibly	Probably not
Ability to extract ash low in carbon	Moderate	Good	Moderate	Good
Ability to consume fine carbon particles	Poor	Good	Probably poor	Good

leased in it to allow for the reduced value of that heat relative to a utility boiler:

(Water-gas efficiency)(methanation efficiency)
$$= \text{(overall conversion efficiency)}$$
$$= 0.76(0.76 + 0.16) = 0.70$$

This represents the upper limit that one might hope to obtain with no allowance for pumping power, sulfur removal, miscellaneous heat losses, etc.

Methanol Synthesis   A favored route for making liquid fuel from coal is the synthesis of methanol using the CO and H_2 from the water-gas reaction, i.e.,

$$CO + 2H_2 \rightarrow CH_3OH$$

This is also an exothermic reaction, the energy available from the mol of CO and the 2 mol of H_2 being 284.5 kJ and 484 kJ, while that from the gram mol of methanol produced is 381.8 kJ, giving an energy conversion efficiency of only 50 percent for the process. As is the case for methanation, the chemical energy converted to heat in the process can be salvaged as intermediate pressure steam, but the heat content of the fuel produced would be only half that of the water gas supplied, and only 37 percent of that of the carbon that was the source of the water gas. Further, these are idealized values and do not include allowances for sulfur removal, pumping power, and miscellaneous heat losses.

Characteristics of Typical Coal Gasification Processes

The general subject of coal gasifiers suitable for coupling to a combined cycle is beyond the scope of this report, but an outline of the characteristics and problems of typical gasifiers is needed to provide some insight into the limitations that the various types of gasifier impose when used to supply fuel to a power plant. As pointed out in an excellent survey article on this set of problems,[27] the various gasifier types under development can be classified as moving-bed, fluidized-bed, or entrained-flow units (Table 5.3).

Moving-Bed Gasifiers   In the moving-bed gasifiers, raw coal is fed into the top, while steam and air (or oxygen) enter near the bottom and pass up through the bed in counterflow with the coal. Low-Btu fuel gas leaves at the top at ~ 500°C (932°F), and ash is discharged from the bottom. The bed is stirred mechanically by a slowly rotating paddle, and this requires that a noncaking coal be used with a minimum of fines, because otherwise the force required to stir the bed becomes excessive. Temperatures drop from ~ 1200°C (2200°F) in the combustion zone in the lower part to ~ 500°C (932°F) at the top. Many units of this type of gasifier (e.g., the Lurgi) have been used commercially and give a cold-gas conversion efficiency of 55 to 60 percent. The process inherently yields a gas with a high tar content that may be removed with a water scrubber. However, thorough cleaning is difficult, and the tars not removed in the scrubber have given trouble with fouling of heat exchangers and other equipment downstream. A variation of this type is the slagging-bottom gasifier developed by the British Gas Company. Oxygen rather than air is used to give a temperature of ~ 1500°C (2730°F) at the bottom of the gasifier so that the ash is above the fusion point and is removed as molten slag. The high-temperature molten slag is strongly oxidizing, thus reducing carbon losses in the ash and making it possible to handle fines in the coal. The gas produced has a heat content more than double that of the

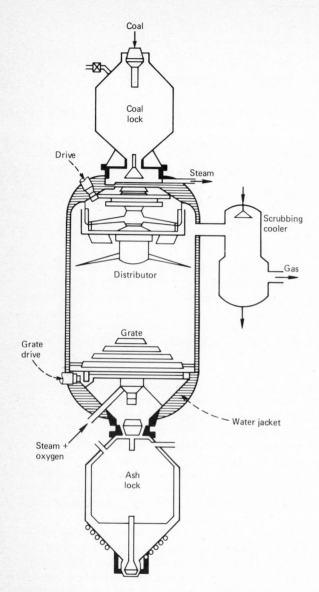

Figure 5.26 Diagram showing the construction of a Lurgi moving-bed gasifier.

air-blown type because it is not diluted by nitrogen. This is an important consideration, because the low Btu content of the fuel from gasifiers fed with air has such a low density, because of nitrogen dilution, that it cannot be stored or piped economically but must be used at once in a furnace close to the coal gasification plant.

Figure 5.26 shows the basic elements in a typical moving-bed gasifier. The power plant integration problems are treated in Chap. 13, and Fig. 13.16 gives a schematic diagram for a combined-cycle power plant utilizing this type of gasifier closely integrated into the power plant cycle.

Fluidized-Bed Gasifiers   The fluidized-bed gasifiers operate with a fluidized bed of ash or sand held at a constant temperature of 870 to 1040°C (1600 to 1900°F). Coal or char is fed into the bed, which is fluidized with steam and air or oxygen. Operation of the bed is similar to that of the fluidized-bed combustion system of the previous section except that the heat release rate is relatively small, the dwell time of the coal particles is much longer, and the atmosphere is strongly reducing instead of oxidizing; thus sulfidation of metal surfaces is a major problem. The high temperature in the gas leaving the bed serves to eliminate tars from the fuel gas produced. At the time of writing there is little commercial operating experience with this type of gasifier. The overall power plant integration problems are also discussed in Chap. 13, and Fig. 13.17 shows how this gasifier might be utilized in a power plant.

Entrained-Flow Gasifiers   The entrained-flow gasifier systems supply pulverized coal, steam, and air or oxygen to the gasifier. These swirl through the gasifier at sufficiently high velocities that the coal and ash particles are entrained in the gas throughout its transit through the reactor vessel. The throughput is high, which helps to reduce capital costs (Table 5.3). About 30 percent of the coal, together with the steam and preheated air or oxygen, is injected tangentially into a combustion chamber region at the bottom, where the coal is burned essentially stoichiometrically to give a temperature of ~ 1870°C (3400°F). The hot gases then pass upward into the gasifying zone, where the balance of the coal feed is injected and much of the sensible heat in the hot gas is absorbed in endothermic reactions to yield the fuel gas. The high velocities reduce the transit time so that the carbon is not all consumed in the first pass; hence the entrained ash in the exit gas is removed and recirculated. Ash collected as molten slag on the walls of the combustion chamber is drained off, quenched, and removed as a water slurry. Figure 5.27 shows the construction of a typical unit.

Losses Detracting from Gasifier Efficiency   It is instructive to look closely at the energy input and output of a gasifier to get a quantitative notion of the losses that detract from its conversion efficiency. One of the highest fuel conversion efficiencies of record is that for the Texaco entrained-bed gasifier of Ref. 28. For a typical set of conditions when a tarry residue from a coal liquefaction plant was used as feed, the heat content of the gaseous product was 83.5 percent of the heat content of the feed material—seemingly an attractive value. However, allowance should be made for the power required to make the oxygen supplied to the gasifier [basic oxygen steel furnaces require 420 kWh/ton (717 Btu/lb) of O_2]. If the thermal efficiency of the thermodynamic cycle for generating this power is taken into account, the corrected conversion efficiency is reduced from 83.5 percent to ~72 percent. A further allowance for the energy in the steam fed to the gasifier reduces the fuel conversion efficiency to 70 percent. A sulfur-removal system would degrade it further by about two points, and pumping losses would amount to another point or two. Thus, the fuel conversion efficiency after correction for all these factors would be ~66 percent—not 83.5 percent. Note that the energy losses representing the difference between 83.5 and 66 percent would yield heat at too low a temperature to be of much value in a steam cycle.

FISSION POWER REACTOR SYSTEMS

Fission reactors impose a very different set of constraints on the design of the heat source for a power conversion system than hold for fossil fuels.[29] In the first place, there is no nuclear limitation on the power density or temperature at which the nuclear reactor can operate; the limits on these quantities are imposed by mechanical engineering considerations. The characteristics of the reactor coolant chosen, the heat-transfer matrix geometry employed, the fuel element design and the characteristics of the fuel material, and corrosion limits are all factors in determining the temperature and power density of the fission reactor. Of these quantities, probably the most important is the type of coolant employed, and, as a consequence, fission reactors are commonly characterized as being *water-cooled* and moderated, *gas-cooled*, or *liquid-metal-cooled*. A second important nuclear consideration is the average energy at which fissions take place. If the bulk of the neutrons slow down to thermal energies before they induce fissions, the reactor is called a *thermal reactor*, and it is relatively easy to control. If there is little or no material present of low atomic number such as hydrogen or carbon, most of the neutrons will not be slowed down very much before they will induce fissions, and the reactor is called a *fast reactor*. The neutron generation time is much shorter—of the order of 10^{-6} s, as opposed to 10^{-4} s for a thermal reactor—and so the reactor responds more rapidly to any nuclear perturbation. Hence the control system must be more sensitive and more elegant to give good control characteristics for the power plant. For a good general treatment of these problems, the reader is referred to a basic text such as Ref. 29.

Water-Cooled and Moderated Reactors

Hydrogen, having the lightest atom, is the most effective element in slowing down neutrons to thermal energies where there is a high probability that they will induce fissions. As water has a high atomic density of hydrogen, it is one of the most effective moderators for fission neutrons. It happens that normal hydrogen has an appreciable neutron capture cross section, whereas heavy hydrogen, or deuterium, does not. Hence it is possible to achieve criticality with a natural uranium fuel element matrix flooded with heavy water, whereas some enrichment of the ^{235}U content (from the 0.7 percent of natural uranium to ~3 percent) is required if the neutron moderation is to be accomplished with

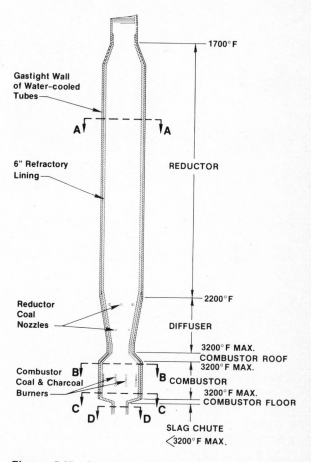

Figure 5.27 Simplified side elevation section of an entrained-bed gasifier. (*Courtesy Combustion Engineering Corp.*)

SEQUOYAH

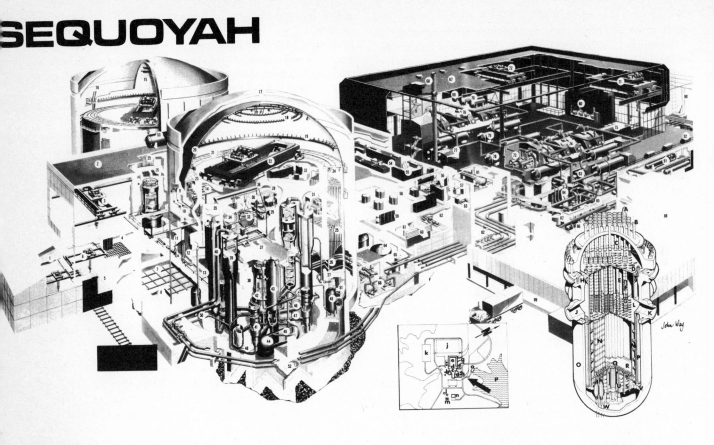

LEGEND

1. Waste packaging area
2. Auxiliary building
3. Auxiliary building crane
4. New fuel storage area
5. Railroad delivery area
6. Waste handling crane
7. Spent fuel pit
8. Fuel cask loading
9. Hoist for fuel transfer system
10. Fuel transfer canal - spent fuel pit gate
11. Fuel transfer canal valve
12. Fuel transfer conveyor up-ending frame reactor 2
13. Fuel canal to reactor 1
14. Spent fuel pit bridge and hoist
15. Reactor building 2
16. Access ladder to dome
17. Reactor building 1
18. Containment spray pipes
19. Crane collector rail
20. Steel containment vessel
21. Ice condenser top deck
22. Reactor building polar crane
23. Ice condenser system bridge crane
24. Ice condenser system air handling units

25. Ice baskets
26. Ice condenser system lower inlet doors
27. Ice condenser system floor drains
28. Ice machines
29. Ice storage bin
30. Borax solution mixing tanks
31. Package chillers
32. Control rod drive equipment room
33. Equipment hatch - reactor building
34. Personnel hatch - reactor building
35. Steam generator containment
36. Manipulator crane
37. Control rod drive missile shield
38. Gate to refueling cavity
39. Steam generators (4)
40. Main steam pipes
41. Reactor coolant pumps (4)
42. Pressure Vessel - unit 1
43. Pressurizer
44. Pressurizer relief tank
45. Accumulators (4)
46. Reactor - steam generator main coolant piping

47. Pump - reactor main coolant piping
48. Steam generator - pump main coolant piping
49. Pressurizer surge pipe
50. Feedwater pipes to steam generators
51. Ventilation fan
52. Access to sump beneath reactor
53. Raw water tanks
54. Main control room
55. Unit 1 control boards
56. Shift engineer's office
57. Kitchen and lunch room
58. 480 V shut-down board transformers
59. 480 V shut-down boards
60. Air intake housing
61. Filter units
62. Auxiliary building lighting board
63. Mechanical equipment room
64. Hold-up tanks (2)
65. Gas decay tanks
66. Component cooling pumps
67. Turbine building
68. Fresh air intakes
69. Gland seal water tank

70. Potable water tanks
71. Turbine building crane - turbine 1
72. Turbine building crane - turbine 2
73. H.P. turbine - unit 1
74. L.P. turbines - unit 1
75. Generator - unit 1
76. Turbine - unit 2
77. Auxiliary boilers
78. Reheaters - turbine 1
79. Reheaters - turbine 2
80. Heating and ventilating equipment
81. Turbine oil tank
82. Feedwater control station reactor 1
83. Heaters - low pressure
84. Turbine by-pass pipes
85. Feedwater pump turbines
86. Feedwater pump turbine condenser
87. Condenser
88. Service building
89. Service building loading dock
90. Switchyard
91. Heaters - high pressure
92. Exhaust fan housing

REACTOR

A. Control rod drive mechanism head adaptors
B. Instrumentation ports
C. Thermal sleeves
D. Upper support plate
E. Support column
F. Control rod drive shaft
G. Control rod guide tube
H. Internals support ledge
J. Inlet nozzle
K. Outlet nozzle
L. Upper core plate
M. Baffle and former
N. Fuel assemblies
O. Reactor vessel
P. Thermal shield
Q. Access port
R. Lower core plate
S. Core support
T. Diffuser plate
U. Lower support column
V. Radial supports
W. Instrumentation thimble guides

MAP

a. Reactor building 1
b. Reactor building 2
c. Auxiliary building
d. Control building
e. Turbine building
f. Service building
g. Office building
h. Primary water and refueling tanks
j. 500 kV switchyard
k. 161 kV switchyard
l. Cooling towers (auxiliary)
m. Fuel oil storage
n. Diesel generator building
o. Intake structure
p. Chickamauga reservoir

Figure 5.28 Cutaway view of the TVA Sequoyah PWR plant, including a cutaway detail of the reactor.[31] (*Courtesy Nuclear Engineering International.*)

light water. It is advantageous for a nation that does not have uranium isotope separation facilities to use natural uranium as fuel; hence the Canadians have developed the heavy-water-moderated type of reactor. They have twelve large commercial heavy-water reactor power plants in operation at the time of writing and nine more under construction. (A typical unit is described in Ref. 30.) In addition, they have sold a few reactors to other countries, but the high cost of the heavy water ($250/kg) leads to high capital costs which more than offset the extra cost of the enriched uranium fuel for the light-water reactors. As a consequence, the light-water reactors are far more widely used, and hence are of principal interest here.

Construction of a Typical Light-Water Reactor

The construction of a typical light-water reactor is shown in Fig. 5.28. A massive pressure vessel about 22 cm (8.6 in) thick, 4.4 m (14.4 ft) in diameter, and weighing ~400 tons is the primary structural element. The reactor core consists of a set of fuel element assemblies such as that of Fig. 5.29. The core is concentric with the pressure vessel and is surrounded by a region about 50 cm thick that serves to attenuate the fast neutron flux sufficiently to keep the loss of ductility of the pressure vessel to an acceptable level throughout its operating lifetime. (This problem is discussed in the next chapter.) A thick steel thermal shield just inside the pressure vessel absorbs the bulk of the intense gamma radiation emitted from the core so that it will not be absorbed in the pressure vessel and cause locally severe temperature gradients and thermal stresses. Most of the array of tubes shown penetrating the top head of the pressure vessel are used for control-rod drives, but some provide access to the core for instrumentation. The control rods contain a neutron absorber so that, as they are inserted into the reactor, they reduce the neutron population and thus reduce the rate of power generation. Note that the system is fail-safe: the rods are held up by magnets, and if the electric power were to fail, they would fall under the action of gravity, thus shutting down the reactor.

A description of the power plant coupled to this reactor is given in Chap. 11. Key design data are given in Table 11.4.

Neutronics

In water-cooled reactors the volume fraction of water in the cooling channels is sufficient so that the amount of neutron moderation that inherently takes place inevitably yields a *thermal* or *epithermal* reactor. This means that the neutron economy is sensitive to the amounts of structural material present that will absorb thermal neutrons, and this in turn means that, even with the most careful design and with large sacrifices in the capital cost, it is still not really possible to build a breeder reactor because of neutron losses as a consequence of parasitic absorptions in both the cooling water and the structural material.[32] As a result of these losses, water reactors as a general class burn more fissionable material than they produce (they commonly produce about 70 percent as much plutonium as they burn ^{235}U or plutonium), the value of the *conversion ratio* from nonfissionable ^{238}U to fissionable material increasing with the amount of capital investment that the builder is willing to make. Within this general nuclear framework, the reactor designer has a number of options in attempting to obtain a heat source that would

Figure 5.29   Typical fuel element assembly for a light-water reactor. *(Courtesy Oak Ridge National Laboratory.)*

yield as high a temperature and pressure in the thermodynamic cycle as practicable while at the same time keeping the capital costs to a minimum level. These factors are considered in the following sections.

Nuclear Reactivity, Stability, and Control

In water reactors the cooling water serves also as the moderator, and if it boils, the resulting voids in the form of steam bubbles affect the reactivity. If the fuel channels are spaced widely apart so that the bulk of the volume of the reactor is made up of liquid water, the reactivity effect of a steam bubble is small. If, however, a maximum number of fuel elements are installed per unit of volume to reduce the capital cost per unit of power output, the reactor is undermoderated and the formation of a steam bubble will act to reduce the reactivity. As a consequence, a major class of water reactors operates with sufficient overpressure so that almost no steam bubbles are ever produced; these are referred to as *pressurized-water reactors*, or PWRs. If the pressure is reduced and the water velocity through the reactor is sufficiently high so that the steam bubbles represent a small percentage of the cooling-water volume, it is possible to make the reactor serve as a boiler and still obtain good control characteristics with a more sophisticated control system. These reactors are called *boiling-water reactors*, or BWRs. The PWRs commonly operate at a pressure of 155 bars (2250 psi) with a water outlet temperature from the reactor of about 330°C (625°F). This high-temperature water is circulated through a steam generator that produces saturated steam at about 282°C (540°F). The BWRs commonly operate at a lower pressure of 72 bars (1050 psi), thus reducing the weight and cost of the pressure vessel, yet deliver steam directly to the turbine at about the same temperature: 282°C (540°F).

Efforts to obtain higher-temperature steam from water-cooled reactors have followed several different lines. First, much thought has been given to the use of a superheating section in a reactor, but this has the disadvantage that the relative amounts of energy going into boiling and into superheating change with the steam-system operating pressure, and under some conditions this leads to the possibility of inadequate cooling of the fuel elements used for superheating and possible damage to those fuel elements. Consideration has been given to the use of separate reactors, one for boiling and one for superheating, but this has appeared to involve more expense than it is worth. Another approach is to make use of an oil-fired superheater, but again this has not appeared attractive in terms of the overall operating costs. Yet another approach would be to make use of a steam-cooled reactor in what is known as a Benson boiler system, in which the steam would enter the reactor at saturat-

ed conditions and be superheated. A portion of the superheated steam would flow to the turbine, while the balance would be directed to a steam drum where the enthalpy of superheat would be used to boil water in the steam drum to supply saturated steam to the reactor. Again, this type of system has not proved to be attractive economically.

Capital Cost and Heat Transfer

To minimize the capital cost of a fission reactor, there is a strong incentive to increase the power density at which it operates (Chap. 10). This in turn means that the heat flux from the fuel elements is made as high as is consistent with safety considerations. Deviations from ideality in the coolant flow distribution make it necessary to limit the peak heat flux to ~65 percent of that which might otherwise appear acceptable. Extensive tests have been carried out to determine the maximum permissible level of the heat flux in both pressurized- and boiling-water reactors if local heat-transfer instabilities and burnout conditions are to be avoided. In both cases the maximum local heat flux at full power is normally considered to be around 158 W/cm^2 [500,000 Btu/(h · ft^2)], somewhat higher than in a coal-fired steam generator.

Fuel Elements

The most crucial factor in the design of a reactor is the fuel element, and its life is a major factor in the operating cost of a fission reactor plant. Many types of uranium fuel-element materials and geometries have been tested in reactors, including metallic uranium, uranium alloys, and many different uranium compounds. Of these by far the most satisfactory has been found to be uranium oxide fabricated in the form of dense ceramic pellets roughly 1 cm in diameter. These are encased in metal capsules. The metal most commonly employed is stainless steel, but metallic zirconium is also used because it has a lower neutron absorption cross section and hence yields a somewhat higher fuel conversion ratio. In either case the thickness of the capsule wall is kept to a minimum value consistent with strength considerations, and this is usually between 0.25 mm and 0.75 mm (0.010 in and 0.030 in). The capsules are sealed with welded end caps so that the fission products are tightly contained within the capsule. Excellent quality-control practices have reduced the incidence of leaks from capsules to approximately 1 in 10^9 capsule operating hours. Note that one of the advantages to the use of UO$_2$ pellets is that the bulk of the fission products are contained within the UO$_2$ crystal lattice, and only a small fraction of the inert gases and volatile fission products, such as iodine and cesium, are released into the interstices between the

UO_2 and the metal wall. However, even this small amount of fission product gas is sufficient to build up a substantial pressure within the metal capsule, and the buildup of this pressure is a principal factor in limiting the life of the fuel element. The rate of fission product diffusion out of the UO_2 crystal lattice increases rapidly as the UO_2 approaches its melting point. Hence to keep the release rate to an acceptable level, the peak centerline temperature of the UO_2 is normally kept to ~ 2300°C (4172°F). Inasmuch as the temperature drop between the center and the surface of the fuel element increases linearly with either the diameter or the surface heat flux, the fuel-element diameter is determined if the peak heat flux is chosen on the basis of surface heat-transfer considerations, the coolant temperature fixed by system design considerations, and the fuel-element centerline temperature fixed by fission-product retention requirements.

Another major factor limiting the life of the fuel element is embrittlement of the metal by fast neutron irradiation (another problem discussed in the next chapter). In practice, it has been found possible to obtain roughly the same limitations on fuel-element life imposed by fission-product gas buildup and loss of ductility in the metal capsule wall as are imposed by the loss in reactivity stemming from burnup of the fissionable material.

The fuel capsules are assembled with spacers to form *fuel rod clusters*, or *assemblies*, such as that of Fig. 5.29, that can be handled readily during reactor fuel loading and unloading operations. The top head of the pressure vessel is removed and the reactor kept flooded with water during the fueling operations to provide a gamma shield that is both fluid and transparent, a procedure that permits the operators to see what they are doing in the reactor core while they manipulate tools, fuel elements, and control rods.

Pressure-Vessel Limitations

As discussed in the next chapter, the reactor size and operating pressure are limited by pressure-vessel fabrication and handling considerations. Equipment for forming steel plate permits the use of plate thicknesses up to ~ 25 cm (10 in), and cranes can handle loads up to ~ 500 tons. For a given design pressure these parameters limit the diameter of the reactor pressure vessel to ~ 4 m and the power output from a reactor to ~ 1000 MWe.

Gas-Cooled Reactors

One of the principal approaches followed in an effort to increase the temperature of fission reactors, and hence the peak temperature and efficiency of the thermodynamic cycle, has been to employ a gas as the reactor coolant and in that way avoid the problems associated with vapor formation and the high cooling-system pressures in water reactors. Gaseous coolants that have been employed include air, nitrogen, CO_2, helium, hydrogen, and mixtures of helium and high-atomic-weight inert gases. A major design constraint is the pumping power required for circulating the gas through the reactor, and this in turn makes helium and CO_2 the outstanding candidates, because they give a lower pumping-power requirement than any of the other gases except for hydrogen (Chap. 4). Hydrogen has not been used except in nuclear rockets because of the fire hazard. This hazard is rendered particularly troublesome by the fact that hydrogen tends to permeate through structural metals at high temperature and thus presents a difficult containment problem. CO_2 has given fine service in the British Magnox gas-cooled reactors, in which the fuel elements are made of metallic uranium encased in magnesium alloy and the reactors are moderated by graphite. However, the peak steam temperature obtainable in these systems is not much higher than in the water reactors for two reasons: First, the melting point of the magnesium alloy limits the reactor gas outlet temperature to ~ 390°C (734°F). Second, local hot spots in the graphite lead to reactions between the graphite and the CO_2 to form CO, and this in turn leads to a variety of materials difficulties. Helium can be employed as the coolant to permit operation to much higher temperatures; with stainless-steel–UO_2 fuel elements the advanced British gas-cooled reactor produces steam at about 538°C (1000°F), i.e., about the same as obtained in fossil fuel plants. Still more advanced systems employ uranium carbide fuel encased in graphite to permit higher fuel temperatures, which in turn permit the use of greater temperature differences in the heat exchangers. Since such systems require less heat-transfer surface, their capital costs can be reduced.

The high gas temperatures obtainable with graphite–uranium carbide fuel elements have led to many studies and some experiments with gas-cooled reactors coupled directly to gas turbines. This approach has the disadvantage that the fission products are not contained perfectly in the graphite fuel elements, and as a consequence, the various components in the gas circuit become contaminated with radioactive material, leading to difficult maintenance problems, particularly in the turbine. This drawback, coupled with various safety problems, such as emergency core cooling, have prevented this approach from gaining general acceptance. Briefly, the improvement in thermal efficiency obtainable with the high-temperature gas-turbine system is not sufficient to offset the increased capital and operating costs associated with the problems just cited. It should be noted at this point that, while it was originally hoped that very high temperatures could be used in a helium-cooled fission reactor system and the turbine built of a refractory metal such as molybdenum or niobium, it has been found that trace amounts of moisture

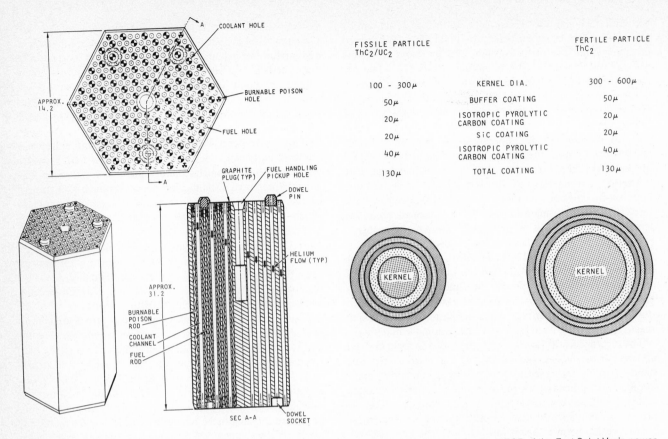

FISSILE PARTICLE ThC$_2$/UC$_2$		FERTILE PARTICLE ThC$_2$
100 - 300μ	KERNEL DIA.	300 - 600μ
50μ	BUFFER COATING	50μ
20μ	ISOTROPIC PYROLYTIC CARBON COATING	20μ
20μ	SiC COATING	20μ
40μ	ISOTROPIC PYROLYTIC CARBON COATING	40μ
130μ	TOTAL COATING	130μ

Figure 5.30 Diagrams showing the construction of the uranium carbide–graphite fuel elements for the HTGR of the Fort Saint Vrain power plant near Denver, Colorado. The distance across the flats of the fuel element is 36 cm (14.2 in), the holes for the coolant channels have a bore of 15.8 mm ($\frac{5}{8}$ in), and the holes packed with fuel microspheres have a bore of 12.4 mm (0.491 in).[33] *(Courtesy General Atomic Corp.)*

and other contaminants in the helium lead to severe corrosion of the refractory metals if they are operated at temperatures above about 650°C (1200°F). Thus, aside from reactor safety and maintenance considerations, as discussed in the next chapter, materials problems limit the peak practicable temperature in gas-cooled reactors to ~650°C (~1200°F) if helium is used to cool stainless-steel–UO$_2$ fuel elements. However, helium outlet temperatures up to ~1000°C (~1832°F) can be obtained with graphite–uranium carbide fuel elements with only mild contamination of the system by fission products leaking from the graphite encapsulation, and this permits operation of the steam system at the same temperatures as in fossil fuel plants.

HTGR Construction

High-temperature gas-cooled reactors (HTGRs) utilizing uranium carbide contained in graphite as the fuel are perhaps the most interesting type of gas-cooled reactor. The fuel elements may be in the form of spheres and the reactor core made a simple pebble bed to simplify the fuel-handling operations, or they may be in the form of hexagonal prisms to give aerodynamically cleaner flow passages through the core and a greater power output from a given size of core. The spherical fuel element approach has been pursued in Germany, where an experimental *pebble-bed* reactor utilizing ~5-cm-diameter graphite spheres has been built and operated. The prismatic fuel-element approach has been pursued in both Great Britain and the United States. A 300-MWe commercial power plant of this type built and operated near Denver, Colorado, is described in Chap. 11. The fuel-element construction employed in this reactor is shown in Fig. 5.30. Microspheres of uranium carbide and thorium carbide are coated with layers of pyrolytic graphite and silicon carbide to keep practically all the fission products within the microspheres. These microspheres are loaded, with powdered carbon packing, into holes drilled in hexagonal graphite blocks that serve as fuel elements. Another set of holes drilled in the graph-

ite blocks provides the cooling passages through which helium is circulated.

The graphite–uranium carbide fuel elements are assembled to form a reactor core as shown in Fig. 5.31. Cores constructed of graphite in this way give a good neutron economy, because relatively few neutrons are absorbed in the graphite, which serves as both structure and moderator. As a consequence, they can be designed to give a conversion ratio approaching unity by including thorium as the fertile material and converting it to ^{233}U as it absorbs surplus neutrons from fission. The design compromises are complex and subtle, but it normally proves best from the cost

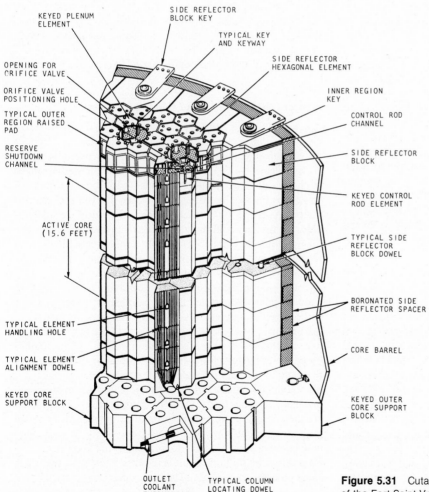

Figure 5.31 Cutaway view showing the construction of the core of the Fort Saint Vrain reactor.[33] *(Courtesy General Atomic Corp.)*

standpoint to minimize capital costs and obtain a long fuel-element life that reduces the fuel-reprocessing costs rather than try to improve the conversion ratio.[32]

The construction of an HTGR pressure vessel differs from that of the water reactor of Fig. 5.28 in that the system pressure is usually lower (commonly ~ 50 bars, or 725 psia) so that the pressure vessel can be larger and its wall thickness reduced. In fact, in some cases a steel-lined prestressed concrete vessel is used with the concrete serving as both the vessel and the reactor shield (Fig. 11.14). Thermal insulation keeps the vessel wall temperature far below the gas temperature.

Fast-Breeder Reactors

Only 0.7 percent of natural uranium is the fissionable isotope ^{235}U. Although uranium is one of the more common elements in the earth's crust, relatively little of it occurs in deposits of sufficiently high concentration to be mined profitably for use in light-water reactors that operate with a conversion ratio of around 0.7. Even with some conversion, less than 1 percent of the uranium can be burned in light-water reactors to give a useful power output. By using fast-breeder reactors, however, it becomes possible to burn virtually all the uranium by converting most of the ^{238}U into plutonium, which is fissionable. This greatly increases the price that the utilities can afford to pay for natural uranium, and makes it possible to utilize the vastly larger reserves of low-concentration uranium ore, such as the granites. Although there are substantial differences of opinion as to the amount of high-grade uranium ore yet to be discovered, there seems to be fairly general agreement that it will be necessary to start to employ fast-breeder reactors by about the year 2000 if we are to avoid a serious shortage of uranium fuel for nuclear plants. (See the latter part of Chap. 1.)

Choice of Coolant

It happens that an average of 2.07 neutrons are released per thermal neutron absorbed in ^{235}U, which is not sufficient to convert more than one fertile atom such as ^{238}U into a fissionable atom such as ^{239}Pu, but if a fast neutron is absorbed in a ^{235}U atom, an average of ~ 2.5 neutrons are released and the amount of new fissionable material produced can be greater than the amount consumed. (^{233}U and ^{239}Pu both give somewhat higher yields of neutrons.) To obtain a fast-neutron-breeder reactor, the number of low-atomic-weight atoms, such as hydrogen and carbon, in the core must be kept very small so that the fast neutrons produced by fission will not slow down too much before they are absorbed to generate more fissions. Of the various fluids having an acceptable atomic weight that might be used as

coolants in fast reactors, sodium appears to be the best—in part because it has a low neutron-capture cross section, in part because it is an exceptionally good heat-transfer and heat-transport fluid (points discussed in Chap. 4), and in part because it is compatible with the Fe-Cr-Ni alloys that are best suited for use as the reactor structure.

Fuel Investment

Although the number of neutrons produced per fission is greater for fast than for thermal neutrons, the fission cross sections (i.e., the probability of a neutron causing a fission) for ^{233}U, ^{235}U, and ^{239}Pu are much lower for fast neutrons. As a consequence, the density of fissionable atoms must be roughly 10 times greater to provide the required critical mass. This greatly increases the capital investment in the fuel and further increases the incentive to achieve the highest practicable power density in the reactor core. In practice, the liquid-metal-cooled fast-breeder reactors (LMFBRs) are commonly designed to give a power density about three times that of PWRs.

Reactor Design Features and Design Limitations

A review of the LFMBR designs evolved by many different groups around the world discloses a surprising degree of similarity with the water reactors. Virtually all of them use the same basic fuel element—UO_2 encapsulated in stainless steel arranged in clusters of long, slender fuel rods—with sodium instead of water flowing axially through the closely packed fuel-element matrix. As in the water reactors, the fuel-element capsule provides the primary containment of fission products so that these highly radioactive materials will not contaminate the primary reactor coolant circuit. Inasmuch as sodium has a much higher boiling point than water—about 888°C (1630°F) at atmospheric pressure—it is possible to operate the primary reactor circuit at a much higher temperature than is possible with water reactors without getting into serious difficulty with pressure stresses in the pressure vessel. As will be discussed in the next chapter, it happens that nickel is slightly soluble in sodium and this solubility increases fairly rapidly with temperature so that in practice the solution corrosion of stainless steel by high-temperature liquid sodium becomes a serious matter if temperatures are raised much above about 600°C (1112°F). Further, the stainless steels are subject to chloride corrosion on the water side of the steam generator so that it would be advantageous to use a chrome-molybdenum alloy. If this is done, a somewhat lower peak temperature in the sodium circuit is required because of the lower creep limit for these alloys. These considerations together with temperature drops in the fluid circuits and in the intermediate heat exchangers normally lead to a peak temperature in the steam system of around 482°C (900°F) in LMFBR plants.

Some consideration has been given to the use of refractory-metal fuel elements in LMFBR reactors. However, the neutron absorption in niobium becomes a serious problem if niobium is used in place of stainless steel, and molybdenum appears to be much more sensitive to embrittlement by fast-neutron radiation than stainless steel. As a consequence, virtually all the effort on LMFBRs is being directed at the use of stainless-steel–encapsulated-UO_2 fuel elements with temperature limitations similar to those outlined above for water-reactor fuel elements.

Design Compromises

The design compromises that must be made in an LMFBR pose much more difficult problems than in the water reactors. This stems first from the fact that the investment in fissionable material is inherently much greater in an LMFBR than in a water reactor, partly because the fissionable material is not diluted with the low-cost ^{238}U. Thus, if capital costs for the fuel are to be kept to attractive levels, it is necessary to obtain a higher *specific power*, i.e., a higher thermal energy output per kilogram of total uranium present in the reactor core. This in turn leads to the use of smaller-diameter fuel elements—7 to 10 mm diameter instead of the 10 to 13 mm commonly used in water reactors—and somewhat higher surface heat fluxes. These requirements present a problem in that the smaller the fuel-element diameter the greater the cost of fabrication per kilogram of uranium, and, for a fixed capsule-wall thickness, the volume fraction of stainless steel is increased and so is the volume fraction of sodium. Unfortunately, this result is counterproductive from the breeding standpoint because, by definition, the reactor must breed and this in turn means that the volume fractions of stainless steel and sodium should be kept to a minimum to increase the breeding gain. The situation is complicated further by yet a third major set of requirements, namely, those of reactor safety, which favor a low-power density and a relatively large set of volume fractions for the sodium and stainless steel. Inasmuch as the reactor safety considerations cannot be compromised, the tendency has been to make design compromises that have led to reductions in the breeding gain and increases in the capital costs. It happens that the character of these compromises also tends to reduce the permissible peak operating temperature of the reactor and hence of the thermodynamic cycle.

"Pipe versus Pot"

A major question of design philosophy is presented by the problem of choosing a configuration for the reactor primary cooling system.[34] Differential thermal expansion in the piping between the major components of a high-temperature fluid system poses serious thermal stress problems (Chap. 6), particularly in the junctures between large-diameter pipes and pressure vessels. These problems are aggravated by the thermal stresses developed in rapid thermal transients such as occur in the course of a rapid emergency shutdown when the temperature rise through the reactor may drop abruptly from perhaps 150°C (270°F) to almost zero. The thermal stress problems can be greatly eased by placing all the primary reactor cooling circuit within a single large vessel (or pot) that is isolated by baffles from the rapidly flowing sodium streams, thus protecting this vital containment envelope from serious thermal stresses under any operating conditions. Slip joints can be used between the components mounted within this large vessel, because small amounts of leakage through these joints would have no ill effects other than a slight increase in the pumping power. The principal disadvantage of this approach is that it increases the inventory of liquid sodium. There is no clear-cut basis for choosing between these two approaches; the Europeans generally favor the tightly integrated "pot," while the U.S. Atomic Energy Commission (AEC) and subsequently the Department of Energy (DOE) have favored the "pipe" configuration.

Construction of a Typical Reactor

The only full-scale commercial breeder reactor under construction in the world at the time of writing is the French Creys-Malville reactor, shown in Fig. 5.32. This is of the "pot" type with the reactor core, primary sodium circuit pumps, and intermediate heat exchangers (IHX) contained in a common outer vessel.[35] For safety, this is a double-walled unit so that if the inner vessel were to leak, the sodium would then be contained by the outer "safety" vessel. Access to the reactor core is provided through a large rotating plug that contains a smaller rotating plug mounted eccentrically. By rotating these two plugs to appropriate positions, the remote handling mechanism in the smaller, inner rotating plug can be positioned over any point in the reactor for fuel handling or servicing operations.

The principal design parameters for the reactor are included in Chap. 11 along with information on the rest of the power plant. (See Table 11.6.)

Molten-Salt Reactors

A substantial program on fluid-fuel reactors was carried out at the Oak Ridge National Laboratory (ORNL) during the 1950s and 1960s. By dissolving the uranium fuel in a fluid that can also serve as a moderator, it is possible not only to eliminate much of

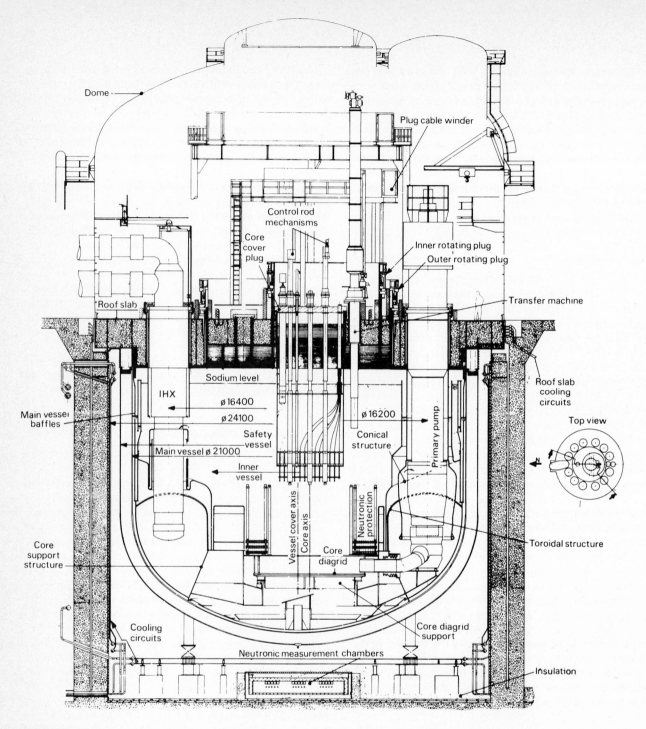

Figure 5.32 Vertical section through the 1200-MWe Creys-Malville LMFBR on the Rhône River in France. The entire primary sodium circuit is contained within one double-walled vessel. Dimensional and performance data for the plant are given in Table 11.6.[35] (*Courtesy Nuclear Engineering International.*)

134

the expense associated with fuel fabrication and reprocessing but also to improve the conversion ratio and control characteristics through the continuous removal of some of the fission-product poisons that would otherwise accumulate in the reactor. The work initially concentrated on the aqueous homogeneous reactor, an aqueous solution of uranyl sulfate whose corrosivity limited the peak temperature to ~300°C. This program was terminated in 1961, and the effort was concentrated on the molten-salt reactor (MSR), which employed UF_4 dissolved in various eutectic or near-eutectic melts of fluoride salts, including NaF, KF, LiF, BeF_2, and ZrF_4. Probably the most attractive system included ThF_4 with the UF_4 in an LiF-BeF melt (melting point 500°C) which made possible a thermal breeder utilizing thorium rather than ^{238}U as the fertile material.[32,36] The reactor fluid circuit was generally designed to be constructed of Fe-Cr-Ni alloys with a reactor outlet temperature of ~650°C for coupling to a conventional steam cycle with a peak temperature of ~565°C (1050°F). If a Nb–1% Zr alloy were used as the structural material, the system would also appear attractive for use with a potassium vapor topping cycle with a peak cycle temperature of ~838°C (1540°F).[37] An experimental reactor was operated successfully for the equivalent of 13,000 h at full power with a reactor outlet temperature of 650°C, but the program was terminated by the AEC in the early 1970s in order to concentrate all available funds on the LMFBR program.

Summary

The complex interrelationships of the numerous factors involved are confusing. A helpful perspective is provided by the overall comparison of the various types of fission reactors in Fig. 22.1 and the accompanying discussion of the history of their development in the second section, ''The Nature of R&D Work,'' of Chap. 22.

FUSION REACTORS

Major efforts to develop fusion reactors have been under way in the United States, the United Kingdom, and the U.S.S.R. since 1950, and more recently have been initiated in several other nations, yet, as of 1980, no one has demonstrated experimentally that controlled thermonuclear reactions can be obtained on anything like the scale required for a full-scale fusion reactor. However, many reduced-scale models have been designed, built, and tested to investigate the plasma physics problems. Progress in these experiments in recent years has been encourag-

ing,[38] and it now appears not only that the plasma physics problems probably can be solved, but also that it may be possible to build a full-scale electric power plant that would produce electricity at a fuel cost of only 0.01 mil/kWh, or about 1 percent of the cost of the basic fuel for fission reactors. Further, the potential supply of fuel for fusion reactors is enormous. There is sufficient heavy hydrogen in the world's oceans so that if only half of it were employed, it would serve to provide all the world's energy needs at 10 times the level anticipated for the year 2000 for a period much longer than our sun is expected to last. Inasmuch as water—which contains ~0.01 percent deuterium—is abundantly available, a practicable fusion power system would greatly reduce competition for the world's limited reserves of fossil fuel and thus should help greatly to reduce international tensions. Further, since the product of the fusion reaction is helium, there would be no long-lived fission products. There would, of course, be activated structure as a consequence of the copious production of neutrons, and it would be necessary, at least for the first few generations of fusion power plants, to employ tritium with deuterium as the fuel. This will introduce some difficult radiological safety problems in handling the volatile, radioactive tritium. However, conceptual design studies indicate that it may be possible to keep the inventory of volatile radioactive material in a fusion reactor sufficiently low that it will represent a biological hazard potential about one-millionth that in a fission reactor, and, by proper design, activation of the structure might be kept sufficiently low so that the inventory of nonvolatile radioactive material would be of the order of 1 percent of that in a fission reactor.[39] Thus, there is a whole set of extremely important incentives to obtain viable engineering solutions to the design of fusion reactors provided that the plasma physics problems can be resolved satisfactorily.

Fusion Reactions

The fusion reactions that might be employed are summarized in Table 5.4. The bulk of the effort is being directed toward the use of the deuterium-tritium (D-T) fusion reaction, because this can be ignited at a temperature roughly one-tenth that for the deuterium-deuterium (D-D) reaction, and it is proving difficult enough to ignite the D-T reaction. The D-T reaction has the disadvantages that it requires fueling with tritium and yields a very penetrating 14-MeV neutron, but both the D-D and D-^{3}He reactions are also accompanied by side reactions that generate both tritium and neutrons. Thus, in any case the fuel cycle will entail handling material containing a substantial fraction of tritium, and the rate of emission of 14-MeV neutrons from D-D or D-^{3}He plasmas will be at least one-tenth that for a D-T plasma. Fur-

TABLE 5.4 Energies Associated with Fusion Reactions

Reaction equation			Plasma temperature	Energy gain per fusion
D + T $\longrightarrow$	^{4}He (3.5 MeV)	+ n (14.1 MeV)	10 keV	1800
D + D	^{3}He (0.82 MeV)	+ n (2.45 MeV)	50 keV	70
	T (1.01 MeV)	+ p (3.02 MeV)		
D + ^{3}He $\longrightarrow$	^{4}He (3.6 MeV)	+ p (14.7 MeV)	100 keV	180

ther, D-D and D-^{3}He reactions will lead to radiation damage effects from fast neutrons and an amount of structural activation within a factor of roughly 2 of the corresponding values for the D-T fuel cycle.

Basic Reactor Concepts

Half a dozen different concepts for confining the high-temperature plasma of a fusion reactor are under active development. Confinement can be accomplished either by a powerful magnetic field or, in a pulsed system, by inertia. The engineering problems differ in that magnet design considerations and neutral beam ignition systems are major factors in the design of magnetically confined systems, whereas the laser ignition system and provisions for absorbing the blast wave from an inertially confined system dominate its engineering design.

Magnetic Confinement

In magnetically confined fusion reactors, the magnetic field must be sufficiently strong and so shaped that the plasma of deuterium and tritium ions can be maintained at an ion density of the order of 10^{14} ions/cm^3 and a temperature of over 10^8 K. These quantities are interdependent: If the ion temperature can be increased, the ion density required for the desired fusion reaction rate is reduced.

The simplest confinement scheme makes use of a solenoid magnet with extra coils at the ends to serve as magnetic mirrors

and inhibit ion leakage out the ends. The end-leakage problem can be avoided by bending the solenoid to form a torus. In either case the magnet system may be designed for steady-state operations of the plasma, or, to avoid certain instability problems, the magnetic field can be pulsed at intervals ranging from milliseconds to hundreds of seconds.

Inertial Confinement

The development of high-power laser beams and electron guns has introduced the possibility of putting a sufficiently intense burst of energy into a frozen pellet of deuterium and tritium that it can be vaporized, ionized, and heated to the temperature required to ignite the fusion reaction in such a short time (roughly 10^{-10} s) that the material in the pellet will not have an opportunity to expand very much; i.e., the plasma will be confined by its own inertia.[40] The resulting energy release will probably have to be of the order of that from the explosion of around 100 kg of high explosive if the net energy output is to be sufficient to give an economically attractive system. Thus, the difficult problems of designing large, high-field magnets are replaced by the problems of designing a pressure vessel and energy recovery system that will contain the blast.

To absorb the blast energy the author has favored the injection and ignition of the pellet in a cavity formed by a swirling vortex of lithium.[40] Gas bubbles can be injected into the lithium to cushion the blast wave and keep the stresses in the pressure vessel to a modest level. Another approach makes use of a porous wetted wall to form a cavity perhaps 100 cm in radius inside a lithium-filled pressure vessel about 300 cm in radius.[41] This arrangement avoids the hydrodynamic problems of generating and maintaining a vortex, but it introduces a set of severe problems associated with the porous wall, notably high dynamic stresses, severe radiation damage by fast neutrons, and severe activation and afterheat removal problems.

Representative Engineering Design

Although some of the problems mentioned above are quite specialized and peculiar to only one or a few concepts, most of the engineering problems to one degree or another are common to all types that currently appear promising. (Exceptions are the dynamic stresses in the inertial-confinement systems and the problems of superconducting magnets for the magnetic-confinement systems.) It is extremely difficult at this stage to assess the relative probability of success in solving the several scientific feasibility questions associated with any of the fusion reactor concepts. It must be emphasized that it will not suffice for the solutions to serve only to demonstrate scientific feasibility.

They must also permit the development of an engineering design of a power plant that will be economically attractive, and this means a long life with a high degree of reliability. It currently seems likely that economically attractive solutions to the scientific problems of magnetic confinement are closer than economically attractive solutions to the construction of high-power lasers of the required characteristics, together with the other key components of a laser-fusion system. As a consequence, for the purpose of this text it seemed in order to simplify this presentation by choosing just one of these concepts to illustrate the engineering problems involved in the development of a full-scale fusion power plant. Because it is receiving the most attention all over the world, the tokamak type of fusion reactor was chosen for this purpose. This choice has the additional advantage that tokamaks have a great deal in common with stellarators and a number of other toroidal reactors.

Tokamaks use a set of magnet coils arranged, as in Fig. 5.33, to form a toroidal cavity in which an ionized plasma of deuterium and tritium can be confined by a magnetic field if a strong electric current circulates around the torus through the plasma. This current can be generated and the plasma can be heated part way to the ignition temperature by pulsing a secondary poloidal field so that a current is induced in the plasma, because it acts as the single-turn secondary winding of a transformer[42] (Fig. 5.33). However, the electrical resistance of the plasma falls off so rapidly at ion temperatures above about 1 keV that an additional means must be employed for heating the plasma further. Currently one of the most promising approaches appears to be the use of neutral beams injected into the plasma with energies of the order of 40 keV. When the plasma temperature reaches around 10 to 20 keV, if the plasma density is sufficiently high, the fusion reaction rate should become great enough so that the reaction will be self-sustaining, and the energy output will exceed the energy input.

A cross section through the torus of a representative conceptual design for a full-scale fusion-reactor power plant of the tokamak type is shown in Fig. 5.34. Note that the plasma temperature is estimated to be 140 million degrees Celsius. The plasma is surrounded by a 1000°C lithium blanket designed to absorb over 99 percent of the fusion energy released by the plasma. The energy from the D-T fusion reaction is divided between a 14.3-MeV neutron and a 3.7-MeV alpha particle. The bulk of the energy from the 14.3-MeV neutrons is converted into heat as the neutrons penetrate deeply into the blanket, but the energy from the 3.7-MeV alpha particle induces x-ray and synchrotron radiation, which is converted into heat in the metal wall separating the lithium from the plasma. This wall is also subject to severe radiation damage from the 14.3-MeV neutrons, as well as the possibility of sputtering damage from ions that escape from

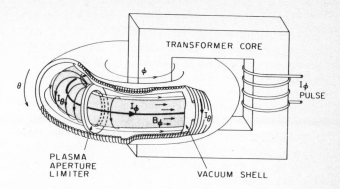

Figure 5.33 Configuration of the initial U.S.S.R. Tokamak Fusion Experiment.[42] *(Courtesy Oak Ridge National Laboratory.)*

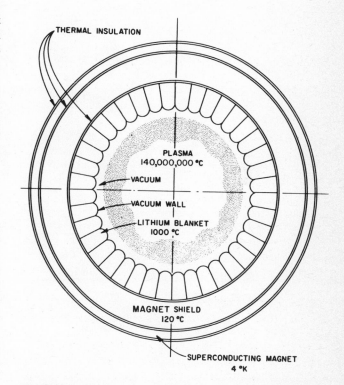

Figure 5.34 Section through the minor diameter of a toroidal fusion reactor showing the temperatures in the plasma, blanket, magnet shield, and magnet regions.[43] *(Courtesy Oak Ridge National Laboratory.)*

the plasma in spite of the confining magnetic field. Thus, it is the most critical structural element in the reactor.

A conceptual design for a full-scale fusion reactor employing the blanket region of Fig. 5.34 is shown in Fig. 5.35. Niobium was chosen for the blanket structure in this design because it has good strength and is compatible with lithium at 1000°C, thus providing a high-temperature heat source for a thermodynamic cycle.

The blanket would be cooled by an array of potassium boiler tubes immersed in the lithium blanket.[43] The potassium vapor would drive a potassium turbine in a binary vapor cycle of the type treated in Chap. 15.

Major Developmental Problems

The most crucial problems of magnetically confined fusion reactors are those concerned with the plasma physics. First, a high-density plasma must be confined for a sufficiently long time to yield a useful energy output, and this must be done with a magnetic field that is not too expensive to achieve. Second, a means for heating the plasma to the ignition temperature must be developed that will require only a modest fraction of the useful power output of a full-scale plant.[44] Third, the confinement of the plasma both before and after ignition must be so effective that under no circumstances (including peculiar transients) will the ions or the electrons in the plasma impinge on the first wall in sufficient concentrations to cause serious damage.[45] (Even with the relatively low plasma temperatures and densities obtained in the plasma physics experiments, there have been repeated instances in which local melting of the wall has occurred.) None of these three requirements has been met in any experiments up to the time of writing. However, progress toward meeting the first and second of these requirements has been impressive—a fact shown by Fig. 5.36.[38,46]

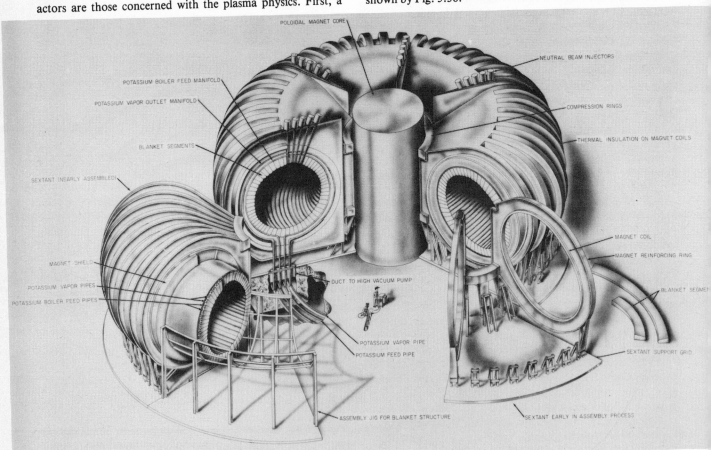

Figure 5.35   Conceptual design for a full-scale fusion reactor of the tokamak type designed to produce 1000 MWt.[43] (*Courtesy Oak Ridge National Laboratory.*)

As mentioned early in this section, demonstration of the scientific feasibility of obtaining a suitable plasma is a necessary but not sufficient condition for obtaining an economically attractive commercial fusion power reactor, because to a far greater degree than for any other type of power plant, a bewildering array of boundary conditions must be met in the engineering of an operable power reactor. Some notion of the difficulties involved is given by the following list of boundary conditions. This list, while prepared for tokamak reactors, is typical of the boundary conditions for others as well.[44,45]

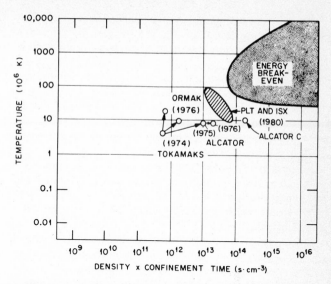

Figure 5.36 Diagram showing the progress in plasma physics experiments in terms of the plasma temperature obtained and the plasma confinement parameter, i.e., the product of the ion density and the average ion confinement time.[46] *(Courtesy Oak Ridge National Laboratory.)*

Basic Requirements to Be Satisfied in the Design of a Tokamak Reactor

1. The magnetic field configuration must give a stable plasma.
   (a) Superconducting magnets are required to keep the power cost reasonable.
   (b) The structure should be nonmagnetic.
2. The D–T cycle will be used; hence the unit must breed tritium via n-2n reactions of 14-MeV neutrons with Li or Be, and this must be recovered efficiently.
   (a) The blanket must contain metallic Be and/or Li, LiF-BeF$_2$, or other compounds of Li and Be.
   (b) The tritium concentration in the blanket must be small because of capital investment, availability, reactor safety, and structural embrittlement considerations.
3. The thickness of the high-temperature blanket region should be sufficient to remove ~99 percent of the nuclear energy released to give a high thermal efficiency.
   (a) If heat is removed from the blanket by cooling coils, these should be far enough from the first wall for low activation of the coolant.
4. The thickness of the shield for the magnet should be sufficient to keep nuclear radiation heating of the cryogenic coils to $< 10^{-6}$ of the total power output. (It takes ~500 kW of electric power to pump 1 kW of heat from 4 K to 300 K.)
5. The reactor concept should be suitable for operation at a blanket temperature of at least 400°C to give a reasonably good thermal efficiency for the thermodynamic cycle.
6. The structural material in the first wall must satisfy the following requirements:
   (a) Acceptable effects on the neutron economy.
   (b) Compatibility with the coolant.
   (c) Satisfactory life, allowing for radiation damage from fast neutrons.
   (d) Acceptable life under the sputtering action of ions from the plasma.

(e) Ductility and weldability with good ductility in the weld zone. Retention of at least 3 percent ductility after irradiation to the projected life.
(f) No serious embrittlement by hydrogen from the plasma or the blanket fertile material.
(g) Adequate strength at elevated temperatures.
(h) The lowest-practicable activation by fast neutrons to ease the problems of afterheat removal and recycling of structural material and/or radioactive waste disposal.
7. The blanket in general and the first wall in particular should be adequately cooled with a modest pumping-power requirement.
   (a) If the coolant is electrically conducting, the flow should be approximately parallel to the lines of force.
   (b) Allowances should be made for the low eddy diffusivity in liquid-metal streams flowing in a magnetic field when the heat-transfer coefficient and possible local hot-spot temperatures are being estimated.
   (c) The blanket-cooling passages should be designed with extra cooling capacity to accommodate local transient hot spots resulting from plasma instabilities that may dump large amounts of ion and/or electron energy into local regions of the first wall.

8. The potential hazards to the public should be as nearly zero as possible.

    (a) The structure must retain its integrity in a severe earthquake.

    (b) The design should be such that no single failure could lead to a release of radioactive material or to a series of events that could cause a release of radioactive material to the environment.

    (c) Provision must be made for removal of heat from the decay of radioactive material.

9. Adequate structure must be provided for the support for all elements of the system against both gravitational and magnetic forces, and for the containment of all fluids with stresses well below the elastic limit and with acceptable distortion.

    (a) Provision should be made for the forces on the magnet coils if one or more coils quenches and goes from superconducting to normal conduction.

    (b) The first wall must be stable against creep buckling.

10. The design should permit replacement of any part of the reactor in the event of unforeseen failures, and there should be provision for remote handling if this should be required.

11. Differential thermal expansion between various regions must be accommodated elastically.

    (a) Structural support of the various regions and the clearances between them must be adequate.

    (b) Temperature rises in liquid systems should be low enough to avoid serious thermal stresses.

    (c) Temperature gradients in structural supports should be low enough so that they do not induce serious local stresses.

12. Penetrations through the major regions must be provided.

13. Thermal insulation is required at the interface between the blanket and the magnet shield, at the interface between the magnet shield and the magnet coil, and around the outside of the magnet coils.

14. The capital cost of the complete power plant should not exceed 150 percent of that for the most economical fission power plant.

    (a) The basic fuel cost in a fission reactor is ~ 25 percent of the power cost, but it would be less than ~ 1 percent in a fusion reactor.

    (b) The cost of the fusion reactor itself can be several times that for a fission reactor if the cost of the rest of the plant is the same, and it can be even greater if—as seems likely—the costs associated with reactor safety can be lower.

15. The power plant should have a life of at least 20 years with one scheduled shutdown of about 2 weeks required no more than once per year.

16. Operating experience has shown that high-temperature fluid systems are extremely unforgiving of errors in design, fabrication, system cleanliness prior to filling, purity of the fluid installed, and the effectiveness of instrumentation for detecting impurities. Hence these systems should be designed accordingly.

17. Leak-tightness problems in cryogenic and vacuum systems more often than not have severely hampered operations of this type of equipment. Full consideration of these problems should be a prime factor in the design.

18. The systems and geometry chosen should inherently have good reliability. Estimates of the probability of a leak in the blanket system, for example, should be made on the basis of generic data and should indicate a mean time between leaks of at least 10^5 h with full allowances for thermal stresses in transients associated with misoperation, abrupt shutdowns, etc.

Fusion-Fission Hybrid Reactors

The large uncertainties with respect to the amount of power required to ignite and maintain the plasma raise the possibility that, for many years after it becomes possible to build a fusion reactor, the net electric output would be small even for very large thermal power outputs. To make such a system economically attractive, one school of thought has been to use the large excess supply of neutrons available from the fusion reactor to drive a subcritical fission reactor and breed plutonium.[47] Several studies have indicated that the value of the extra power plus the plutonium would make the overall system attractive. Unfortunately, these studies have been concerned with the physics and have not included a comprehensive look at all of the engineering boundary conditions that must be satisfied. Not only must one meet essentially all the boundary conditions given in the previous section, but one must also meet nearly all the truly difficult boundary conditions for a fission breeder reactor. The author has been involved in a number of critical reviews of the conceptual designs for fusion-fission hybrid reactor plants and is highly doubtful that an economically attractive system could be evolved, however attractive the objectives of both cheap power and cheap plutonium.

PROBLEMS

1. The heat content of coal includes the chemical energy available from the oxidation of the sulfur present. If a fluidized-bed combustion system is employed to remove the sulfur as $CaSO_4$ when a coal containing 5 percent sulfur, 15 percent ash, and 6 percent moisture is burned, calculate the reduction in heating value stemming from the conversion of the sulfur to $CaSO_4$ rather than SO_2. Assume 90 percent sulfur retention in the bed, a hydrogen-carbon ratio of 0.06, and half of the sulfur in the form of FeS and half in organic compounds.

2. Estimate the heat loss to the stack for a stack gas temperature of 149°C (300°F) caused by the moisture in the coal of Prob. 1, and estimate the heat loss associated with rejection of the ash and spent lime at the bed temperature of 874°C (1600°F). Use a Ca/S ratio of 2.

3. If a fluidized-bed furnace is to be used to generate steam at 122 bars and 565°C (1800 psia and 1050°F) with one reheat at 27 bars and 565°C (400 psia and 1050°F), what percentage of the superheating must be carried out in the bed and what percentage in the hot gases leaving the bed? Assume that the temperature of the bed is 874°C (1600°F), that all the reheating is done with combustion gases, and that the temperature of the gases leaving the superheater and reheater region to go to the economizer is 538°C (1000°F). Assume a feed-water temperature of 253°C (488°F) entering the economizer.

4. Estimate the percentage reduction in the economizer, boiler, and superheater surface area obtained in a pulverized-coal-fired furnace by air preheating to give an air temperature rise of 100°C (180°F), assuming the steam conditions of Prob. 3 but with no reheat. For simplicity, assume a feed-water temperature of 121°C (250°F) and neglect the variable specific heat of the combustion gas and plot a set of curves similar to Fig. 4.5, assuming a flame temperature of 2800°F and a stack temperature of 300°F with no air preheat. Repeat for an air temperature rise in the preheater of 222°C (400°F). Assume a mean temperature difference in the air preheater of 111°C (200°F). Calculate the LMTDs for the economizer, boiler, and superheater for each case. Neglect thermal radiation effects. Note that the combustion gases give up heat first to the boiler then to the superheater, and finally to the economizer.

5. If the cost per unit of surface area in the air preheater of Prob. 4 is 10 percent of the average value for the economizer, boiler, and superheater of the steam generator, compare the decrease in the cost of the economizer, boiler, and superheater with the cost of the air preheater, using the mean temperature differences given by the curves drawn in Prob. 4. As a first approximation neglect the temperature drops through the walls and the water-side film.

6. In selecting the design pressure for a BWR, it is necessary to balance the increases in capital costs stemming from an increase in reactor pressure against the reductions in capital costs stemming from an increase in thermal efficiency which gives a greater electric output from a given size of reactor core, set of cooling towers, etc. As a rough approximation, assume a fixed size of reactor core and hence a fixed thermal power input. Assume that, in the vicinity of conditions for minimum capital costs, 10 percent of the plant costs are directly proportional to the electric output (e.g., transformers and switchgear), 80 percent is directly proportional to the thermal power input, and 10 percent is directly proportional to the pressure (reactor pressure vessel, high-pressure piping, etc.). Take the steam condenser temperature as 38°C (100°F) and the electric output as directly proportional to the temperature drop in the steam turbine. Determine the reactor outlet steam pressure and temperature giving the minimum capital cost per kilowatt of electric output, assuming saturated-steam conditions at the turbine inlet and neglecting the pressure drop from the reactor to the turbine. (Suggestion: Tabulate values for the steam pressure for a set of six temperatures from 450°F to 700°F, and from these calculate values at each temperature for the relative electric output and the relative unit cost per kilowatt of each of the three major sets of capital costs.) From these data choose a design pressure for the reactor.

7. Using the curves of Fig. 2.5, estimate the increase in thermal efficiency obtainable in going from the minimum-capital-cost BWR of Prob. 6 to an LMFBR that would give 103 bars (1500 psia) steam at 510°C (950°F).

REFERENCES

1. National Coal Association: *Bituminous Coal Data—1970 Edition*, March 1971.

2. U.S. Bureau of Mines: "Coal—Bituminous and Lignite in 1972," *Mineral Industry Surveys*, Nov. 15, 1973.

3. Lewis, B., and G. von Elbe: *Combustion, Flames, and Explosion of Gases*, Academic Press, Inc., New York, 1951.

4. Miller, C. D.: "The Roles of Detonation Waves and Autoignition in Spark Ignition Engine Knock as Shown by Photographs Taken at 40,000 and 200,000 Frames per Second," *SAE Quarterly Abstract*, vol. 1, no. 1, 1947.

5. Miller, C. D.: "Slow Motion Study of Injection and Combustion of Fuel in a Diesel Engine," *SAE Trans.*, vol. 53, 1945.

6. Graves, C. C., et al.: "Basic Considerations in the Combustion of Hydrocarbon Fuels in Air," NACA Technical Report No. 1300, 1957.

7. Drell, I. L., and F. E. Belles: "Survey of Hydrogen Combustion Properties," NACA Technical Report No. 1383, 1958.

8. D'Alleva, B. A., and W. G. Lovell: "Relation of Exhaust Gas Composition to Air-Fuel Ratio," *SAE Trans.*, vol. 38, 1936.

9. Brogan, T. R.: "Electrical Properties of Seeded Combustion Gases," *Progress in Aeronautics and Astronautics*, vol. 12, Academic Press, Inc., New York, 1963, pp. 319–345.

10. Kurylko, Lubomyr: "Control of Nitric Oxide Emissions from Furnaces by External Recirculation of Combustion Products," ASME Paper No. 71-WA/PID-6, Nov. 28, 1971.

11. Sonderling, H. H., et al.: "Operation of Scattergood Steam Plant Unit 3 under Los Angeles County Air Pollution Control District Rule 67 for NO_x Emissions," *Proceedings of the Sixth Intersociety Energy Conversion Engineering Conference*, Aug. 3–5, 1971.

12. *Steam—Its Generation and Use*, The Babcock and Wilcox Co., New York, 1972.

13. Lee, D. W., and R. C. Spencer: "Photomicrographic Studies of Fuel Sprays," NACA Technical Report No. 454, 1933.

14. Olds, F. C.: "Trends in Power Boilers," *Power Engineering*, 1978, pp. 42–52.

15. Zenz, F. A., and D. F. Othmer: *Fluidization and Fluid Particle Systems*, Reinhold Publishing Corporation, New York, 1960.

16. Davidson, J. F., and D. Harrison: *Fluidization*, Academic Press, Inc., New York, 1971.

17. Fraas, A. P., et al.: "Use of Coal and Coal-Derived Fuels in Total Energy Systems for MIUS Applications," Oak Ridge National Laboratory Report No. ORNL/HUD/MIUS-28, April 1976.

18. Reh, Lothar: "Fluid Bed Combustion in Processing, Environmental Protection, and Energy Supply," paper presented at International Fluidized Bed Combustion Symposium of the American Flame Research Committee, Boston, Apr. 30–May 1, 1979.

19. Thompson, A. S.: "Instabilities in a Coal-Burning Fluidized Bed," Oak Ridge National Laborary Report No. ORNL/HUD/MIUS-38, November 1978.

20. *Pressurized Fluidized Bed Combustion Research and Development Report No. 85, Interim No. 1*, prepared for Office of Coal Research, Department of the Interior, by National Research Development Corporation, London, England, 1974.

21. Hoke, R. C., et al.: "Miniplant Studies of Pressurized Fluidized-Bed Coal Combustion: 3rd Annual Report," Exxon Research and Engineering Co., EPA-600/7-78-069, April 1978.

22. Jonke, A., G. Vogel, and I. Johnson: "Sulfated Limestone Regeneration and General FBC Support Studies," paper presented at Fluidized Bed Combustion Technology Exchange Workshop, McLean, Va., Apr. 13–15, 1977.

23. Holcomb, R. S.: "Heat Transfer Performance of a Tube in a Fluidized Bed Furnace," Oak Ridge National Laboratory Report No. ORNL/HUD/MIUS-49, February 1979.

24. Braunstein, H. M., et al.: "Environmental, Health, and Control Aspects of Coal Conversion: An Information Overview," vol. 1, Oak Ridge National Laboratory Report No. ORNL/EIS-94, April 1977.

25. Bresler, S. A., and J. D. Ireland: "Substitute Natural Gas: Processes, Equipment, Costs," *Chemical Engineering*, Oct. 16, 1972, pp. 94–108.

26. Patterson, R. C.: "Low-Btu Gasification of Coal: A C-E Status Report," paper presented at Fourth International Conference on Coal Gasification, Liquefaction, and Conversion to Electricity, University of Pittsburgh, Aug. 2–4, 1977.

27. Papamarcos, J.: "Combined Cycles and Refined Coal," *Power Engineering*, December 1976, pp. 34–42.

28. Robin, A. M.: "Hydrogen Production from Coal Liquefaction Residues," EPRI Report No. AF-233, December 1976.

29. Weisman, J.: *Elements of Nuclear Reactor Design*, Elsevier Scientific Publishing Co., Amsterdam, 1977.

30. Renshaw, R. H., and E. C. Smith: "The Standard CANDU 600 MWe Nuclear Plant," *Nuclear Engineering International*, vol. 22, no. 258, June 1977, pp. 45–53.

31. Eicheldinger, C.: "Sequoyah Nuclear Steam Supply System," *Nuclear Engineering International*, vol. 16, October 1971, p. 864.

32. Perry, A. M., and A. M. Weinberg: "Thermal Breeder Reactors," *Annual Review of Nuclear Science*, vol. 22, 1972, pp. 317–354.

33. Walker, R. E., and T. A. Johnston: "Fort St. Vrain Nuclear Power Station," *Nuclear Engineering International*, vol. 14, December 1969, pp. 1069–1073.

34. Murray, Peter: "The LMFBR Concept for Utility Systems," *Nuclear Energy Digest*, Westinghouse Nuclear Energy Systems, no. 3, 1971.

35. "Construction of the World's First Full-Scale Fast Breeder Reactor," *Nuclear Engineering International*, vol. 23, June 1978, pp. 43–60.

36. "A Review of Molten Salt Reactor Technology," *Nuclear App. & Tech.*, vol. 8, no. 2, February 1970, pp. 1–217.

37. Fraas, A. P.: "A Potassium-Steam Binary Vapor Cycle for a Molten-Salt Reactor Power Plant," *Trans. ASME*, vol. 88, 1966 (1), pp. 355–364.

38. Furth, H. P.: "Progress toward a Tokamak Fusion Reactor," *Scientific American*, vol. 241, no. 2, August 1979, pp. 50–61.

39. Fraas, A. P.: "Environmental Aspects of Nuclear Power Plants," ANS, *Proceedings of International Conference on Nuclear Solutions to World Energy Problems*, Nov. 13–17, 1972, pp. 261–273.

40. Lubin, M. J., and A. P. Fraas: "Fusion by Laser," *Scientific American*, vol. 224, no. 6, June 1971.

41. Williams, J., et al.: "A Conceptual Laser-Controlled Thermonuclear Reactor Power Plant," *Proceedings of First Topical Meeting on the Technology of Controlled Nuclear Fusion*, CONF-740402, vol. 1, 1974, p. 70.

42. Fraas, A. P., and D. J. Rose: "Fusion Reactors as Means of Meeting Total Energy Requirements," ASME Paper No. 69-WA/Ener-1, November 1969.

43. Fraas, A. P.: "Conceptual Design of the Blanket and Shield Region and Related Systems for a Full-Scale Toroidal Fusion Reactor," Oak Ridge National Laboratory Report No. ORNL-TM-3096, May 1973.

44. Fraas, A. P.: "Comparative Study of the More Promising Combinations of Blanket Materials, Power Conversion Systems, and Tritium Recovery and Containment Systems for Fusion Reactors," Oak Ridge National Laboratory Report No. ORNL-TM-4999, November 1975.

45. Fraas, A. P., and A. S. Thompson: "ORNL Fusion Power Demonstration Study: Fluid Flow, Heat Transfer, and Stress Analysis Considerations in the Design of Blankets for Full-Scale Fusion Reactors," Oak Ridge National Laboratory Report No. ORNL/TM-5960, February 1978.

46. Steiner, Don: "Fusion Power Development: Status and Prospects," *Mechanical Engineering*, vol. 102, no. 6, June 1980, pp. 48–53.

47. Bethe, H. A.: "The Fusion Hybrid," *Nuclear News*, vol. 21, no. 7, May 1978, pp. 41–44.

6

LIMITATIONS IMPOSED BY MATERIALS

The lives of power plant components are often limited by corrosion, and the thermal efficiency of the power plant system depends on the peak temperature of the cycle, which in turn is limited by either corrosion or the strength of the structural material at elevated temperatures. Thus the tremendous improvements in steam plants, internal combustion engines, and gas turbines in the course of the past century have stemmed from and been dependent on developments in materials. However, by 1980 the performance of these power conversion systems appeared to have risen to about as high a level as the basic properties of practicable materials would permit, and proposed future developments or systems should now be looked at critically from the materials standpoint in appraising their feasibility and potential. This chapter summarizes the principal problems, together with the characteristics of the materials of interest. Corrosion effects and allowable stresses are dominant factors and the main concern of this chapter. Fabrication, lubrication, and the effects of nuclear radiation are also important, hence the more significant considerations in these areas are treated briefly in the latter part of this chapter.

Steel alloys are by far the most widely used materials in power plants because of their high strength, relatively low cost, and reasonably good resistance to corrosion. In fact, 10 times as much iron is produced in the United States each year as all the other metallic elements combined. Although a tremendous variety of alloys is available, relatively few are used to any great degree in power plants, and these alloys are commonly identified by terms that may not be familiar to the reader. As a point of departure, and to save space in referring to these alloys in subsequent sections, Table 6.1 is included here to relate the class of alloy and its common designation to its composition and principal uses. In general, the more severe the corrosion conditions and/or the higher the operating temperature,

the greater the content of expensive elements such as Cr, Ni, Mo, Co, Ti, etc., in the alloy.

CORROSION

Corrosion—particularly of steel, the most commonly used metal in the United States—is a familiar problem. We know that there is essentially no rusting in a clean, dry atmosphere but that moisture induces fairly rapid corrosion, particularly if the water contains salts, acids, or alkalis, because these provide a plentiful supply of ions, which accelerate the chemical attack. Corrosion is especially rapid if there are dissimilar metals present creating electrical potential differences, which accelerate the corrosion rate of the more active metal. The availability of oxygen is also an important factor, as evidenced by the especially high rate of attack on steel pilings where the waterline fluctuates with the tide in harbors along the seacoast. Yet in very pure water essentially free of both oxygen and salt, the corrosion rate of steel is virtually nil. These familiar examples indicate something of the nature of corrosion problems and their great complexity.

A good indication of the tendency of a metal to be subject to corrosion is given by its position on the electrochemical scale. Table 6.2 lists the principal elements of interest here, together with their electrode potentials and the heats of formation of their oxides, to give an idea of the relative driving forces acting to produce corrosive chemical reactions. Note that these are only two factors in the complex surface chemistry involved. Very active metals such as aluminum and titanium, for example, can form dense, adherent oxide films that provide highly effective protective coatings. A complex oxide coating is also responsible for the excellent corrosion resistance of stainless steel, a point discovered quite by accident when a metallurgist

TABLE 6.1   Typical Alloys Used in Power Plants

Type	Common commercial designation	Specification	Alloying elements, %	Comments on use
Plain carbon steels		AISI–1020	0.2 C	General purpose. High strength and wear resistance.
		AISI–1045	0.45 C	
		A212 Grade B	0.31 C, 1.0 Mn	Pressure vessel and firebox
Ferritic alloy steels		AISI4140	0.4 C, 1 Cr, 0.9 Mn, 0.2 Mo	High strength parts
		A202/56		Pressure vessels
		SA–210A1	0.27 C, 0.8 Mn	Pressure vessels
		A533 Grade B	0.25 C, 1.3 Mn, 0.6 Ni, 0.5 Mo	Pressure vessels
	Croloy $2\frac{1}{4}$	SA–213T22	$2\frac{1}{4}$ Cr, 1 Mo, 0.5 Mn	Boiler tubes
	Croloy 5	SA–213T5	5 Cr, $\frac{1}{2}$ Mo	Boiler tubes
	Croloy 9	SA–213T9	9 Cr, 1 Mo	Boiler tubes
Ferritic stainless steels (Type 400 SS)	Type 410	A 268–47, TP410	12 Cr, 0.5 Ni, 1 Mn, 0.75 Si	Oxidation resistance at high temperatures, poor weldability
	Type 431		16 Cr, 2 Ni, 1 Mn, 1 Si	
	Type 446	A 268–47, TP446	26 Cr, 1 Mn, 0.5 Si, 0.3 Ni, 0.2 N_2	
Austenitic stainless steels (Type 300 SS)	Type 304 SS	A240 Type 304	18 Cr, 8 Ni	High strength and oxidation resistance at high temperature. Used as plate and sheet, same basic alloys used for tubing, pipe, and forgings. Good weldability.
	Type 310 SS	SA240 Type 310	25 Cr, 20 Ni, 1 Mn	
	Type 316 SS	A240 Type 316	18 Cr, 11 Ni, 2.5 Mo	
	Type 321 SS	SA240 Type 321	18 Cr, 8 Ni, 0.6 Ti	
	Type 347 SS	SA240 Type 347	18 Cr, 11 Ni, 1 Nb	
		A286	15 Cr, 26 Ni, 1.4 Mn, 2.1 Ti, 1.25 Mo, 1 Nb, 0.4 Si, 0.2 Al	
Nickel-base alloys	Incoloy 600		15 Cr, 8 Fe	Nuclear-steam-generator tubes
	Incoloy 625		21 Cr, 9 Mo, 4 Fe, 4 Nb	High strength and corrosion resistance
	Incoloy 800		46 Fe, 20.5 Cr, 0.75 Mn, 0.3 Al	Boiler tubes
	Inconel		13 Cr, 6.5 Fe	Wrought alloys, excellent strength and oxidation resistance at high temperatures. Good weldability.
	Inconel X		15 Cr, 6.8 Fe, 2.5 Ti	
	Hastelloy X		22 Cr, 18.5 Fe, 9 Mo, 1.5 Co, 0.6 W	
	Nichrome		80 Ni, 20 Cr	
	Nimonic 90		17 Co, 20 Cr, 2.6 Ti, 1.6 Al	
	Monel		30 Cu, 1.4 Fe, 1 Mn	Resistant to corrosion of aqueous solutions
	T.D. Nickel		2.4 ThO_2	MHD electrodes
	Udimet 700		18.5 Co, 15 Cr, 5 Mo, 4.3 Al, 3.5 Ti	Cast gas-turbine blades
Cobalt-base alloys	Haynes 25		20 Cr, 15 W, 10 Ni, 1.5 Mn	High temperature tubing, etc.
	Haynes 188		22 Cr, 22 Ni, 15 W, 2 Fe	High strength and corrosion resistance
Copper alloys	Commercial Cu			Electric conductors
	Admiralty metal		29 Zn, 1 Sn	Condenser tubes
	Cupronickel	ASTM B111	10 Ni	Condenser tubes
	Berylium Cu	ASTM B196	1.9 Be, 0.25 Ni	Strong, hard, wear-resistant
	AMX-MZC		0.6 Cr, 0.1 Zr, 0.04 Mg	Good strength up to 550°C
	Glidcop AL-20		0.4 Al_2O_3	Good strength up to 550°C

TABLE 6.1 Typical Alloys Used in Power Plants (*Continued*)

Type	Common commercial designation	Specification	Alloying elements, %	Comments on use
Aluminum alloys	3003	ASTM B221	1.2 Mn, 0.12 Cu	Good corrosion resistance and weldability
	2024	ASTM B211	4.5 Cu, 1.5 Mg, 0.6 Mn	High-strength plate, extrusions
	5052	ASTM B211	2.5 Mg, 0.25 Cr	Good strength, corrosion resistance, weldability
	7075	ASTM B211	5.6 Zn, 2.5 Mg, 1.6 Cu, 0.3 Cr	High strength and corrosion resistance
Titanium alloys	Commercial Ti	ASTM B265–58T		Chemical plant piping
	Ti–6Al–4V	ASTM B265–58T–5	6 Al, 4 V	High-strength compressor blades
	Ti–4A1–4Mn		4 Al, 4 Mn	Forgings, compressor parts
Zirconium alloys	Commercial Zr		1.9 Hf	Chemical process equipment
	Zircaloy 2		1.46 Sn, 0.15 Fe, 0.2 Hf	Water reactor core elements
Refractory metal alloys	TZM		99 Mo, 0.5 Ti, 0.1 Zr	High-strength high-temperature forgings, poor weldability
	Nb–1% Zr		1 Zr	High-strength high-temperature
	T–111		90 Ta, 8 W, 2 Hf	tubing, plate, and forgings; good weldability

TABLE 6.2 Electrode Potentials and Heats of Formation of Typical Oxides of Elements of Interest in Power Plants (Values in volts referred to the hydrogen-hydrogen ion half-cell as zero at 25°C, heats of formation are for 18°C.)

	Volts	kg · cal/mol		Volts	kg · cal/mol		Volts	kg · cal/mol
Li	− 3.045	− 142	Mn	− 1.18	− 230	Mo	− 0.2	− 133
K	− 2.925	− 87	V	− 1.18	− 342	Sn	− 0.136	− 69
Sr	− 2.89	− 142	Nb	− 1.1	− 382	Pb	− 0.126	− 52
Ca	− 2.87	− 152	S	− 0.92	− 71	H₂	0.000	− 57
Na	− 2.714	− 100	Se	− 0.78	− 53	Cu	+ 0.337	− 38
Mg	− 2.37	− 145	Zn	− 0.763	− 85	I₂	+ 0.536	− 43
Th	− 1.90	− 294	Cr	− 0.74	− 275	Hg	+ 0.789	− 22
Be	− 1.85	− 144	Fe	− 0.44	− 200	Ag	+ 0.799	− 7
U	− 1.80	− 260	Cd	− 0.403	− 63	Pt	+ 1.2	− 17
Al	− 1.66	− 404	Tl	− 0.336	− 118	Cl₂	+ 1.36	+ 33
Ti	− 1.63	− 586	Co	− 0.277	− 207	Au	+ 1.50	− 2
Zr	− 1.53	− 263	Ni	− 0.250	− 57	F₂	+ 2.85	+ 5

noticed the bright, clean appearance of experimental specimens of armor plate in a scrap yard after World War II.

Corrosion in Air

Perhaps the most straightforward corrosion problems deal with the corrosion of the various steels in clean, dry air. As shown in Fig. 6.1, the oxidation rate increases fairly rapidly with temperature over the range of interest in power plants. The resistance to oxidation is especially improved by adding chromium to the alloy. Additions of aluminum, nickel, and silicon are also beneficial, because they also help form protective oxide films that are highly adherent. The aluminum and silicon tend to embrittle the steel; hence the amounts added must be kept small. Nickel, however, improves the ductility,

toughness, and weldability of steel and so is widely used in spite of its rather high cost. Where particularly good corrosion resistance is desired and the higher cost can be justified, a nickel-base alloy may be used rather than an austenitic stainless steel (Table 6.1).

Catastrophic Oxidation

The importance and complexity of the protective oxide film on austenitic stainless steels is shown vividly by the phenomenon of catastrophic oxidation that has occasionally been a problem with certain steels, notably Type 316 SS. This steel contains molybdenum, which can act as a catalyst to promote oxidation. If access of air to the surface is limited in some way, e.g., by thermal insulation, the low-melting and/or oxygen-deficient

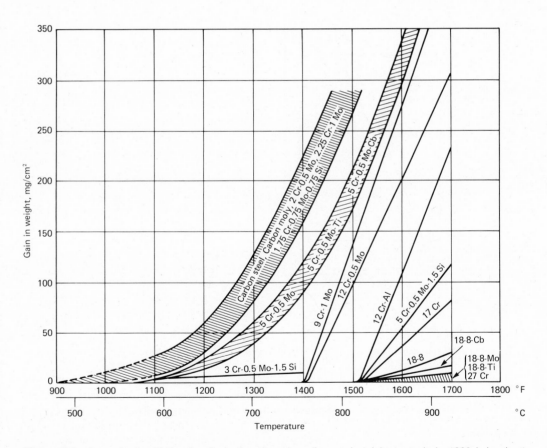

Figure 6.1 Amount of oxidation (scaling) of carbon, low-alloy, and stainless steels in 1000 h in air at temperatures from 1100 to 1700°F.[1]

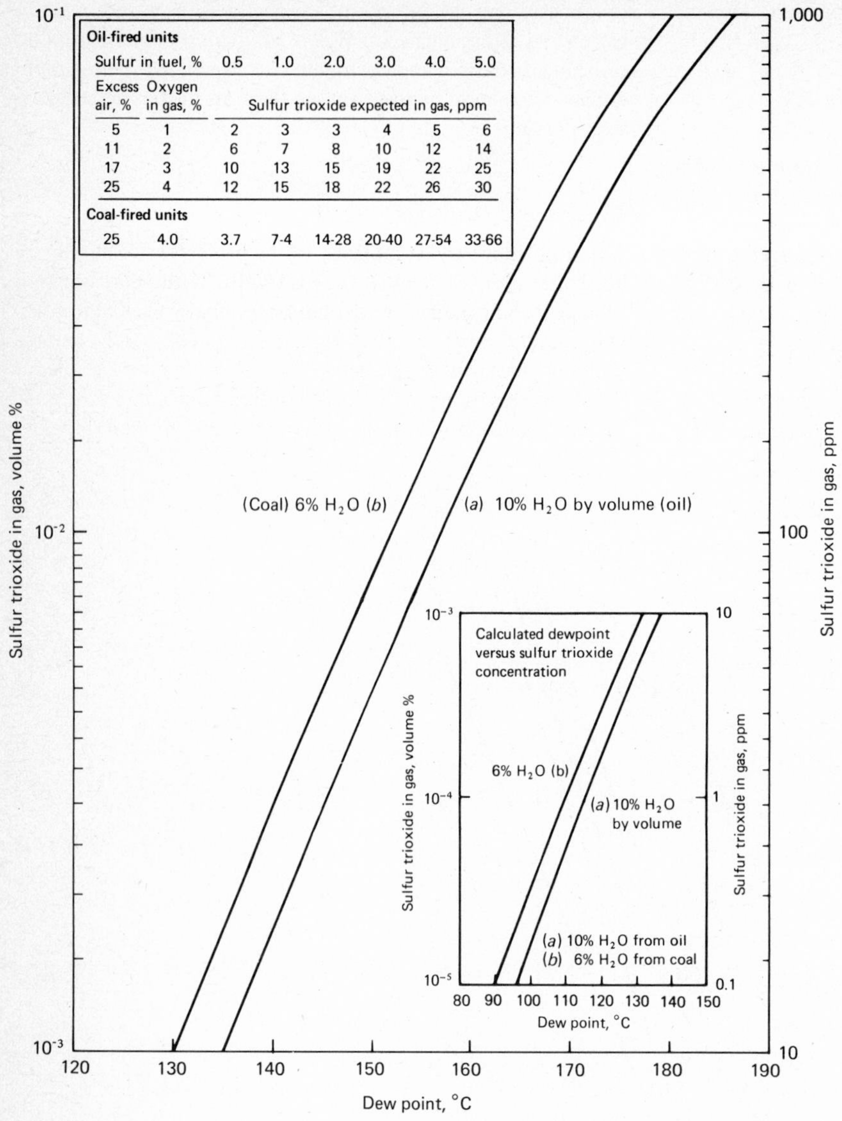

Oil-fired units
Sulfur in fuel, % 0.5 1.0 2.0 3.0 4.0 5.0
Excess air, % Oxygen in gas, % Sulfur trioxide expected in gas, ppm
5 1 2 3 3 4 5 6
11 2 6 7 8 10 12 14
17 3 10 13 15 19 22 25
25 4 12 15 18 22 26 30
Coal-fired units
25 4.0 3.7 7-4 14-28 20-40 27-54 33-66
(Coal) 6% H_2O (b)
(a) 10% H_2O by volume (oil)
Calculated dewpoint versus sulfur trioxide concentration
6% H_2O (b)
(a) 10% H_2O by volume
(a) 10% H_2O from oil
(b) 6% H_2O from coal
Sulfur trioxide in gas, volume %
Sulfur trioxide in gas, ppm
Dew point, °C
10^{-1}
10^{-2}
10^{-3}
10^{-4}
10^{-5}
1,000
100
10
1
0.1
80 90 100 110 120 130 140 150
120 130 140 150 160 170 180 190

Figure 6.2 Relation between SO_3 content of stack gas and the dew point of sulfuric acid for typical coal- and fuel oil–firing conditions.[8]

oxides of molybdenum locally destroy the otherwise protective film, leading to oxidation reactions that proceed rapidly and may penetrate a pipe wall in less than 100 h.

Corrosion by Combustion Products

If a clean fuel such as natural gas is employed, and if the metal temperature is kept above the dew point of the water vapor in the combustion products, there is little difference between the corrosion rates in air and in the products of combustion as long as the fuel-air ratio remains either above or below the stoichiometric value. If, however, the atmosphere fluctuates between oxidizing and reducing conditions, the scale formed during oxidizing conditions tends to flake off during a shift to reducing conditions and rapid attack can occur [as much as 1 cm of Type 310 SS in 21 days at temperatures approaching 700°C (1300°F)].[2] For good fuel utilization, of course, the usual procedure is to provide only a small excess of air so that oxidizing conditions prevail at all times yet stack losses chargeable to the excess air are minimized.

Effects of Sulfur

Sulfur is commonly present in coal in the form of both metal sulfides and organic compounds, in petroleum as organic compounds and H_2S, and in "sour" natural gas as H_2S. Sulfur alone does not greatly increase the corrosion rate at intermediate temperatures, but in combination with other elements in the ash, for example, it is a major source of corrosion problems.[3-7] Further, it gives serious troubles at temperatures below 150°C (300°F) because of the high dew point of the sulfuric acid formed by the combination of SO_3 and H_2O in the stack gas (Fig. 6.2). Although the bulk of the sulfur in the fuel forms SO_2 in the combustion process, depending on the amount of excess air the equilibrium relations are such that a small fraction of the sulfur forms SO_3. The table in the upper left corner of Fig. 6.2 gives the amount of SO_3 to be expected for both oil-

and coal-fired furnaces for a range of amounts of both sulfur in the fuel and excess air. Note that the resulting dew point in the stack gas depends on the percentage of moisture, and this is substantially higher for oil than for coal firing. At temperatures above the boiling point of water, the liquid condensate contains relatively little water, e.g., at 140°C (298°F) the sulfuric acid concentration is 82.5 percent by weight. Hot concentrated sulfuric acid is highly corrosive not only to metal but also to concrete, brick, mortar, plastics, etc. Hence if sulfur is present in the fuel, both the temperature of the stack gas and that of any heat-transfer surfaces used to recover heat from the stack gas should be kept above the dew point.

If the metal temperature is raised above ~815°C (~1500°F), an insidious phenomenon known as *sulfidation* becomes a problem. Metal sulfides form in the grain boundaries, and because solid-state diffusion proceeds progressively more rapidly above ~800°C, the sulfur migrates deeply into the metal. This both weakens and embrittles the metal. Sulfur has a particular affinity for nickel, and so the higher nickel alloys are especially subject to sulfidation.

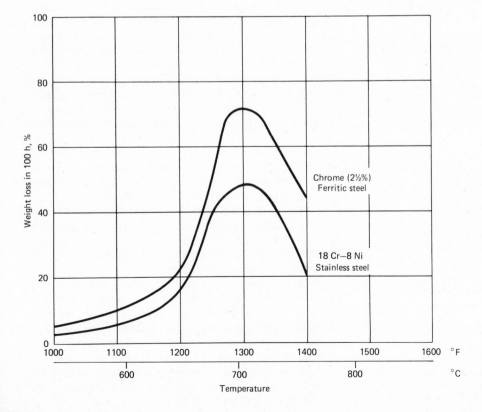

Figure 6.3 Effects of metal temperature on relative corrosion rates of superheater tubes in coal-fired furnaces. *(Courtesy Babcock & Wilcox Co.)*

Effects of Na, K, Cl, V, *and* Pb

A variety of minerals are commonly present in both coal and petroleum, and these in combination with sulfur aggravate corrosion. Sodium and potassium sulfates and vanadates have proved particularly troublesome because they have sufficiently high vapor pressures to emerge from flames in vapor form and then condense on metal surfaces that operate at temperatures below ~815°C (~1500°F), e.g., on superheater tubes in steam generators or blades in gas turbines. These sulfates and vanadates are liquids at temperatures down to ~650°C (~1200°F), and in liquid form react with the oxide films that would normally protect the alloy surface. Fairly rapid attack occurs under these films. This sort of corrosion is accelerated if chlorides and/or lead are also present.

Figure 6.3 shows a typical set of curves for operation with high sulfur coal.[3] Note that the rate of removal of material from the surface by corrosion is at a maximum around 1300°F and then falls off at higher temperatures. Sulfur migration into

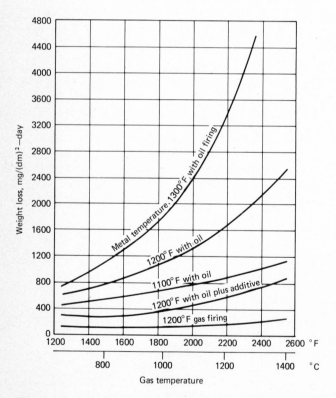

Figure 6.4 Effect of gas and metal temperatures on corrosion of Type 304, 316, and 321 stainless-steel alloys in a unit fired with oil containing 150 ppm vanadium, 70 ppm sodium, and 2.5 percent sulfur. Test duration 100 h. *(Courtesy Babcock & Wilcox Co.)*

the grain boundaries builds up less rapidly with temperature but becomes dominant at temperatures above about 815°C (1500°F), although its effects do not show up as either a gain or a loss in weight of the test specimens. Figure 6.4 shows a set of curves somewhat similar to those of Fig. 6.3 but for oil and gas fuels. Vanadium is usually not present in coal in sufficient quantities to be a problem, but it is a major problem in fuel oils and gives serious trouble with corrosion in both steam plants and gas turbines. Its effects can be reduced by adding magnesium compounds to the fuel oil, a point evident in the curves of Fig. 6.4.

Effects of Type of Fuel

Experience with conventional coal-fired steam boilers provides some valuable insights into the extreme variability in corrosion rates from one fuel to another. Table 6.3 shows the corrosion rates found in a series of tests carried out by inserting test specimens into boiler furnaces in the superheater region where the local gas temperature was approximately 1350°F, i.e., in the temperature region in which corrosivity would be expected to be nearly maximum.[6] Data were obtained for a variety of alloys operated in furnaces using widely different fuels. The composition of most of the alloys used is given in Table 6.1.

If a fluidized-bed coal combustion system were employed, it would be expected that the alkali metal sulfates would not be completely vaporized, as they are in pulverized-coal burner flames, but instead would be largely retained in the ash in the relatively low-temperature (~900°C) fluidized bed. This thesis is supported by British Coal Utilization Research Association (BCURA) tests.[7] However, sulfates have an appreciable vapor pressure at 900°C and will be present in the ash particles deposited in the gas turbine if one is used, so that some difficulty with sulfidation of gas-turbine blades appears likely where the turbine blade temperatures are above ~650°C. The limited data available from tests at BCURA on cascades of blades indicate that sulfidation is definitely a problem at metal temperatures of ~815°C (~1500°F). To determine the extent to which corrosion will prove to be a problem in either the superheater or a turbine will require extensive, carefully run tests with a wide variety of coals and limestones.

Wood has a low sulfur content. It contains, however, ~0.1 percent of potassium by weight, and thus, when used as a fuel, it forms K_2O, which is very corrosive above ~650°C (~1200°F). The same problem arises in burning municipal solid wastes. Interestingly, it has been found that by burning a properly chosen mixture of moderately high sulfur coal and solid wastes, the potassium and sulfur combine to produce potassium sulfate, thus avoiding both the sulfur emission problem and trouble with K_2O corrosion.

TABLE 6.3 Corrosion Data for Material Exposed to Various Combustion Atmospheres[6]

Alloy	High-sulfur, high-vanadium fuel oil			High-alkali content coke and blast-furnace gas		
	Weight loss, g/in²/mo	Decrease in diameter ipm*	Max depth subsurface attack, in	Weight loss g/in²/mo	Decrease in diameter ipm*	Max depth subsurface attack, in
	Sewaren Station			Penwood Station		
Aluminized steel	1.451	§	None	2.535	§	0.008
Illium G†	0.835	0.0062	0.002	1.290	0.0093	0.036
Inconel	0.455	0.0032	0.009	1.762	0.0134	0.035
17-14 Cu-Mo	‡	‡	†	1.808	0.0139	0.010
AISI 316	2.029	0.0156	0.006	0.785	0.0061	0.008
AISI 308	0.543	0.0042	0.001	0.618	0.0048	0.001
AISI 321	3.452	0.0266	0.002	0.945	0.0075	0.005
AISI 314	0.867	0.0068	None	0.078¶	0.0006	None
AISI 309	0.681	0.0052	None	0.493	0.0038	0.002
AISI 304	1.439	0.0110	None	0.783	0.0061	0.005
AISI 347	1.492	0.0114	0.001	0.643	0.0049	0.003
AISI 310	0.310	0.0023	None	0.048¶	0.0004	0.005
AISI 309S	0.183	0.0013	None	0.210¶	0.0016	0.001
16-25-6	‡	‡	‡	0.810	0.0062	0.008
	Natural gas and Illinois coal Venice No. 2 Station			High-sulfur oil and gas Bayway Refinery		
Aluminized steel	0.099	§	None	0.724	§	None
Illium G†	...	...	...	...	...	...
Inconel	0.210	0.0015	None	1.860	0.0134	None
17-14 Cu-Mo	0.098	0.0008	None	‡	‡	‡
AISI 316	0.007	0.0001	None	1.216	0.0094	None
AISI 308	0.098	0.0008	None	0.063	0.0005	0.004
AISI 321	0.068	0.0005	None	0.896	0.0070	0.013
AISI 314	0.048	0.0004	None	0.607	0.0048	0.003
AISI 309	0.203	0.0015	None	0.127	0.0010	0.005
AISI 304	0.195	0.0015	None	0.544	0.0042	0.013
AISI 347	0.128	0.0010	0.001	0.141	0.0011	0.003
AISI 310	0.000	0.0000	None	0.171	0.0013	0.001
AISI 309S	0.100	0.0007	None	0.067	0.0005	0.003
16-25-6	0.000	0.0000	None	‡	‡	‡
	Low-sulfur anthracite coal Sunbury Station			Sulfur-free natural gas Baton Rouge Station		
Aluminized steel	0.62	*	None	0.565	§	None
Illium G†	...	...	...	...	...	...
Inconel	0.0000	0.0000	None	0.0008	0.0001	None
17-14 Cu-Mo	0.0000	0.0000	None	0.0279	0.0002	None
AISI 316	0.0000	0.0000	None	0.1508	0.0012	None
AISI 308	0.0000	0.0000	None	0.0307	0.0002	None
AISI 321	0.0000	0.0000	None	0.2520	0.0020	0.001
AISI 314	0.0084	0.0001	None	0.0195	0.0002	None
AISI 309	0.0000	0.0000	None	0.0923	0.0007	None
AISI 304	0.0218	0.0002	None	0.0279	0.0002	None
AISI 347	0.0082	0.0001	None	0.0307	0.0002	None
AISI 310	0.0000	0.0000	None	0.0081	0.0001	None
AISI 309S	0.0000	0.0000	None	0.0106	0.0001	None
16-26-5	0.0455	0.0003	None	0.0162	0.0001	None

*ipm = inches per month

†No specimens available for 4 of 6 stations.

‡Specimens completely corroded away.

§Due to uncertainty of density figure required in calculation, conversion to inches of radius lost per month could not be accomplished.

¶Slight pitting occurred.

151

Corrosion in Reducing Atmospheres

Corrosion in coal gasification equipment, in which reducing atmospheres are inherent, presents problems with both hydrogen embrittlement and attack by H_2S. The chrome-moly alloys are particularly subject to hydrogen embrittlement, and ferritic steels are attacked by H_2S roughly 10 times more rapidly than the austenitic steels.[2] The corrosion rate varies with both the metal temperature and the partial pressure of the H_2S. The corrosion rate is highest at $\sim 540\,°C$ ($\sim 1000\,°F$), running ~ 2.5 mm (0.1 in) per year for an H_2S partial pressure of 12 mm (0.25 psia) and ~ 0.25 mm per year at an H_2S partial pressure of 3 mm (0.05 psia) for a 0.5% Cr ferritic steel.[2] At metal temperatures above $\sim 815\,°C$ (1500 °F), sulfidation by H_2S is also a serious problem.

Corrosion in Inert Gases

Helium has many advantages for use as a coolant in nuclear reactors, either for making steam or for use directly as the working fluid in a closed-cycle gas turbine.[9] At first thought one would expect that the use of helium as the coolant should eliminate oxidation as a problem and should even permit the use of high-strength refractory alloys of molybdenum, niobium, and tungsten. However, J. H. DeVan has pointed out that, in practice, operation of iron-chrome-nickel alloys in a helium environment has shown that impurities in helium tend to be more aggressive than air.[10,11,12] This characteristic stems from trace levels of active impurity gases (CO, CO_2, O_2, H_2O, and CH_4) in the helium; the partial pressure of these impurities is too low, typically 50 to 200 $\mu m/atm$, to form a protective oxide film on the alloy surface. As a consequence, the corrosion reactions tend to concentrate below rather than at the gas-metal interface. Obviously, metal consumption at the surface will be minimal, but the internal reactions which occur between the more reactive alloying elements and the dissolved gases reduce the strength of the alloy.[13,14] Superalloys which derive strength from gamma and gamma-prime precipitates are especially prone to internal oxidation in helium turbine environments above 800 °C (1472 °F). Further, trace amounts of oxygen in helium are sufficient to give serious corrosion of niobium alloys if the peak temperature in the system exceeds about 600 °C (1112 °F). Coatings have been suggested, but none have been found to be effective in spite of extensive test programs. As a consequence, if helium were employed with niobium in a nuclear gas-turbine cycle, the peak temperature probably should be limited to about 600 °C.

The difficulties with the attack of refractory alloys by trace amounts of oxygen and water vapor in inert gases proved so troublesome in development tests of refractory alloy systems for nuclear electric power plants for spacecraft that it was found best to conduct such tests in vacuum chambers in which the pressure was reduced to 10^{-8} torr or less.[15] This high vacuum could be easily monitored with reliable instrumentation, whereas the oxygen and moisture contents of inert gases could not be monitored readily at the very low concentrations required. These tests are significant with respect to fusion reactors, because a niobium alloy appears to be especially attractive as the structural material for the blanket surrounding the plasma, and plasma physics considerations require that this region operate with an extremely pure hydrogen atmosphere or a high vacuum, either of which should be compatible with a niobium alloy.[16]

Protective Coatings

A wide variety of protective coatings have been applied to steam-generator tubes and gas-turbine blades in an effort to protect them from oxidation and other types of corrosion. Although many have proved beneficial over the short term, no coating has proven sufficiently effective over the long term to gain widespread acceptance for long-lived equipment. Plated metal coatings tend to develop pinholes, and ceramic coatings tend to crack and spall off under thermal stress or react with ash particles, in which case they may actually accelerate corrosion. The most effective approach appears to be the choice of a proper combination of basic alloy and design conditions.

Corrosion in Steam Systems

Several types of corrosion have proved to be problems inside steam systems. These include oxidation by high-temperature steam, local attack by trace contaminants such as chlorides, solution corrosion in high-temperature feed-water heaters, and galvanic-type corrosion in condensers. The basic problems involved in each of these are discussed below, and measures for handling them are treated in Chap. 11, "Steam Power Plants."

Corrosion in High-Temperature Steam

The major mechanism causing corrosion in high-temperature steam is the increasing degree to which H_2O dissociates as the temperature is raised above $\sim 315\,°C$ (600 °F). This, coupled with the increasing diffusion rate of hydrogen through metal walls in the same temperature range, frees oxygen to attack the metal wall. A classic experiment run to investigate this phenomenon made use of an evacuated jacket around sealed tubu-

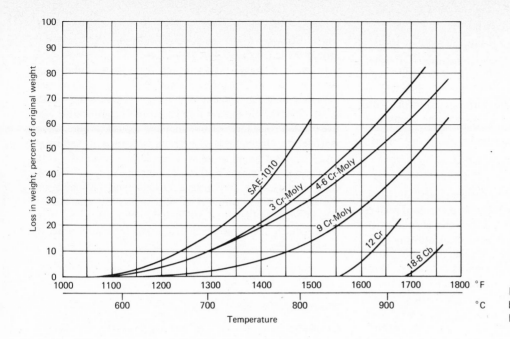

Figure 6.5 Corrosion of steel bars in contact with steam for 500 h at various temperatures.[18]

lar capsules partially filled with oxygen-free distilled water. These capsules were operated at a temperature of 315°C (600°F). The results showed excellent correlation between the rate of hydrogen effusion through the wall as measured during the test with the oxidation attack of the interior tube wall by the steam as measured after the test.[17] This is the reason that hydrazine (N_2H_2) is added to the feed water in steam plants; it serves to maintain a high pH and provides an excess of hydrogen in the steam. This increases the back reaction rate that keeps the free oxygen to a very low level.

As is evident from the above, the corrosion rate of Fe-Cr-Ni alloys in steam is very much dependent on the water chemistry. However, for a given state of the water chemistry, the corrosion rate is also dependent on both the temperature and the particular alloy. Figure 6.5 shows the results of a typical series of tests, the alloys having the higher contents of Cr and Ni giving the lower corrosion rates.[18]

Localized Corrosion Caused by Chlorides

Chlorides are associated with three types of localized corrosion: *pitting, crevice corrosion,* and *stress-corrosion cracking*. Pitting and stress corrosion result from the local breakdown of otherwise protective passive films. The mechanism involves diffusion of chloride ions into the localized region, the higher resulting chloride ion concentration preventing repassivation.

The local corrosion rates may be 1000 times the rate found for a passive protective film. In the presence of stress above a threshold level, rates of stress-corrosion crack propagation may be very high.

Trace amounts of chlorides in the feed water tend to be concentrated in small pits and crevices in regions where boiling occurs. When this happens, the relatively high concentration of chlorides can cause severe attack along grain boundaries and crack formation in some alloys. The austenitic stainless steels are particularly susceptible, whereas the nickel-base and ferritic alloys are much less so. In fossil-fuel steam plants the austenitic stainless steels are ordinarily used only in the upper stage superheater, where there is no mechanism to concentrate the chlorides such as by evaporation. In water-reactor plants, however, there is a strong incentive to keep overall corrosion to a minimum in the reactor coolant circuit, and hence the austenitic stainless steels are preferred over the less expensive ferritic chrome-moly steels. To avoid the possibility of chloride corrosion on the steam side of the steam generator for water-reactor systems, a nickel-base alloy, Incoloy 600, is commonly used to avoid dissimilar metal corrosion that might occur in the transition from the austenitic to the ferritic alloy. As will be discussed in a subsequent section, the nickel-base alloys are more subject to corrosion in sodium. Therefore, in the LMFBR it appears desirable to employ a chrome-moly alloy in the steam generator and throughout the steam system.

Solution Corrosion in Feed-Water Heaters

In some of the earlier supercritical pressure steam plants, Monel Metal was employed in the high-temperature feed-water heaters. This gave difficulty with copper deposits in the steam turbines because small amounts of copper were dissolved in the high-temperature, high-pressure water passing through the feed heaters. The problem was solved by changing to chrome-moly steels.

Corrosion in Steam Condensers

The corrosion problems on the cooling-water side of steam condensers in coastal areas are particularly difficult to predict because they are dependent on the local water chemistry. In one instance in the author's experience two plants only 8 km apart on the same estuary of the Atlantic had the same materials in their condensers; one was afflicted with severe corrosion, and the other was not. It appeared that another plant in the area was discharging a chemical in its waste water that caused the trouble, although it is possible that there was a difference in the local solids content which induced a combined erosion-corrosion effect.

To reduce corrosion in the condenser to an acceptable level, the usual practice is to employ Admiralty Metal, a copper-base alloy developed by the British Admiralty. In some instances in freshwater areas, stainless steel has given excellent service. Where particularly difficult corrosion conditions have been encountered along the seacoast, a titanium alloy has given outstanding service.

Although titanium is a very active metal chemically, it forms a thin tenaciously adherent oxide film that is highly effective in preventing corrosion in a wide variety of aqueous solutions including salt water. Extensive endurance tests in seawater by the U.S. Navy have given particular attention to those areas of the equipment where corrosion or erosion had been found before.[20,21] In all cases, the parts made of titanium were found to be remarkably free of both erosion and corrosion attack. The areas under washers and bolt heads are typical examples of regions commonly subject to attack in stainless-steel, Monel, copper, and bronze parts, but no trace of attack in these regions was found with titanium. The inside of condenser tube walls just downstream from the tube inlet are often subject to erosion in Admiralty Metal, cupronickel, and Monel, but no attack was found in the titanium condenser tubes. Similarly, attack is commonly found just downstream of valve ports, but again no attack was found in the titanium parts. An all-titanium pump tested in seawater for 4 years at Harbor Island was unattacked. In short, the experience at the U.S. Navy Marine Engineering Laboratory at Annapolis indicates that titanium is much superior to any other material tested there for use in salt water in the temperature range of $-2°C$ ($28°F$) to around $38°C$ ($100°F$).[20,21]

In addition to the long-term exposure tests cited above, the heat-transfer performance of titanium condenser tubes has been determined and found to be competitive with that of Admiralty Metal.[21] Tests have also been run to compare the relative erosion resistances of titanium and typical turbine bucket materials to high-velocity wet steam, and titanium and its alloys have been found superior to the 13% chrome alloy commonly used for turbine buckets. The metal loss was $\frac{1}{9}$ to $\frac{1}{4}$ of that found with the 13% chrome steel in a typical 45-day test in wet $250°C$ ($480°F$) steam.[22] Somewhat similar tests run in a turbine at Imperial Metal Industries, Birmingham, England, have shown much the same results, as have tests performed by Westinghouse.[23]

Corrosion in Water Reactors

Corrosion in water reactors poses particularly exacting requirements. This is partly because the intense ionizing radiation in the nuclear reactor excites chemical activity. Another reason is that any activated atoms of structural material in corrosion products entering the water stream, either in dissolved form or as particulates, are likely to deposit elsewhere in the system outside the reactor and greatly complicate maintenance by giving a high level of radioactivity in a component that would not otherwise be radioactive. As a consequence, to minimize corrosion, water-reactor systems are normally built of austenitic stainless steels, high-nickel alloys, and zirconium alloys. All of these have excellent resistance to corrosion in water and have proved compatible when used together in water-reactor systems. Some corrosion still occurs, however, but its effects are minimized with an elaborate bypass clean-up system that includes filters and ion exchange resins. In pressurized-water reactors the pH is adjusted between 9 and 11 with ammonia to minimize both corrosion and the entry of any corrosion products into the coolant stream, as well as to suppress the radiolytic decomposition. Hydrogen is also added to maintain its concentration at 25 to 50 cm^3/kg of water.[24] In boiling-water reactors, boiling in the reactor gives a different situation in which the aforementioned corrective measures have been found inadvisable. Much of the hydrogen and oxygen formed in the core is not recombined in back reactions but leaves the core with the steam. These gases are subsequently stripped as noncondensibles in the condenser and are vented to the atmosphere. The condensed steam is passed through a full-flow ion exchange resin bed to minimize corrosion and scale deposition in the reactor core, where deposits on the fuel elements would inhibit heat transfer.[24]

In small power plants there is a strong incentive to reduce the system complexity and operating crew size. The author suggested in 1966 that this might be done if the system were hermetically sealed and made of an alloy of titanium. This appeared to be a good possibility in view of experience gained first in the homogeneous reactor program and subsequently in a seawater desalination program.[24] In those programs extensive endurance tests were carried out with three small titanium systems in which pumps circulated water containing first uranyl sulfate with pH values of 1.0 to 2.5 (about as corrosive as the sulfuric acid in storage batteries) and subsequently in simulated seawater with a nearly neutral pH. The temperature in these corrosion-test loops ranged from 150 to 325°C with uranyl sulfate, and up to 200°C (392°F) with the seawater. After tens of thousands of hours in each of these corrosive conditions, the only sign of corrosion was a microscopically thin film of titanium oxide. Experts in water chemistry outlined a program to investigate this approach,[24] but the program for a hermetically sealed water-reactor system was not funded.

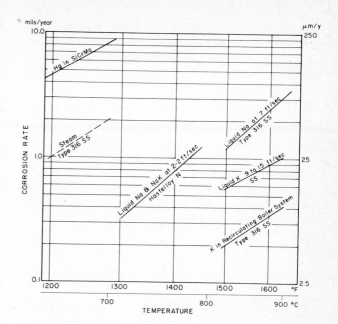

Figure 6.6   Effects of temperature on the corrosion rate for typical combinations of materials for Rankine cycle systems. *(Prepared by J. H. DeVan, ORNL.)*

Corrosion and Mass Transfer in Liquid-Metal and Molten-Salt Systems

High-temperature liquid systems have been found to be subject to corrosion and mass transfer if one (or more) of the metals in the alloy of which the system is built is slightly soluble in the fluid circulated. The solubility generally increases with temperature; hence metal tends to be dissolved from the hot zone and deposited in the cold zone, producing a phenomenon known as *mass transfer*.[25] Difficulties of this sort have been particularly severe with the heavy metals lead, bismuth, and mercury at temperatures above about 482°C (900°F). Flocculent deposits of dendritic crystals tend to plug the small-diameter passages in the colder portion of the system. While filters are helpful, their effectiveness is limited by the fact that material tends to precipitate out of solution wherever there is a drop in temperature, and so plugs may form either ahead of or downstream of the filter. An indication of the degree to which corrosion and mass transfer occur is indicated in Fig. 6.6, which shows corrosion rates as functions of temperature for several typical materials combinations.

Extensive tests have shown that sodium and potassium give very low mass transfer and corrosion rates if the oxygen content is kept down to around 20 ppm or less.[15] A large further reduction can be effected in alkali-metal-vapor systems by using a recirculating boiler so that the fluid moves from the hot

zone of the system as a pure vapor and thus will not transport dissolved solids. While this vapor will tend to dissolve a small amount of material from the walls on which it condenses, the low solution rate at the lower temperatures in the condenser makes it likely that the amount of material involved is extremely small. This has been demonstrated at ORNL by over 160,000 h of tests with potassium-vapor systems, some built of stainless steel and some of refractory metal.[26] The corrosion rates for refractory alloys are essentially zero up to ~1100°C (2000°F). A once-through boiler presents greater problems; material that dissolves in the liquid in the first portion of the boiler is either deposited toward the boiler outlet or is carried along in particulate form with the vapor and may deposit in the turbine.

In molten-salt systems the situation is more complex because chemical attack can occur as well as solution corrosion. Fused nitrate-nitrite melts such as HTS (Table 4.2 in Chap. 4) can be used in Fe-Cr-Ni alloys up to ~482°C (900°F), but they tend to decompose at higher temperatures.[27] The molten alkali fluorides are the most stable heat-transfer liquids at higher temperatures, having good compatibility with high-nickel alloys up to ~815°C (1500°F), but their melting points are high.[28] A great deal of work was carried out with the molten fluorides during the 1950–1970 period with a view to their use as a vehicle for uranium in high-temperature fluid-fuel reactors,[29] but this effort was discontinued because of a Washington budgetary decision to concentrate the AEC breeder reactor effort on the LMFBR.

LIMITATIONS IMPOSED BY STRENGTH

Engineering courses in the strength of materials and stress analysis are concerned mainly with structures that operate near room temperature at which the elastic limit of the material is well defined, thermal and fatigue stresses are not a problem, and buckling is elastic. In power plants, however, the need for the highest practicable operating temperature to improve the thermal efficiency and for the highest practicable heat fluxes to reduce capital costs have made creep, stress-to-rupture, thermal stresses, thermal-strain cycling, and sometimes creep buckling the dominant factors in limiting plant performance and life. Further, safety considerations and potential failure modes make ductility and toughness in structural materials far more important for power plants than for conventional structures. Thus this section was prepared on the premise that the reader is familiar with the room-temperature properties of alloys and their use in ordinary stress analysis, but will find it helpful to survey the specialized materials and stress problems that are so vital under the high-temperature conditions that prevail in advanced energy systems.

In considering the behavior of alloys at high temperature and the basic limitations of any of these materials, it is helpful to think in terms of the basic characteristic of metals which gives them ductility. If the shear stress acting on a plane through the crystal lattice of a grain of metal exceeds a critical value, slip occurs, resulting in shear glide between sections of the crystal adjacent to the glide plane. This occurs by generation and motion of lattice imperfections called *dislocations* within the crystal lattice. These processes and the interaction with the dislocations of the grain boundaries between crystals tend to increase the resistance to further slippage, giving a *work-hardening*, or strengthening, effect. If alloying elements are added to a metal, they may be present in the form of a solid solution, in which case they interact with the dislocations, increasing the critical shear stress for slip. Alloying elements may form discrete crystals of intermetallic compounds that also act to resist slip; in either case they strengthen the alloy. These effects vary with heat treatment and with the amount of hot and cold working of the alloy, processes that alter the form and dispersion of the alloying constituents. As the temperature increases, thermally activated diffusion processes alter the structure with time and change the temperature-dependent mechanical properties.

The strength of an alloy at elevated temperatures is related to its melting point. As the temperature is raised, atoms can shift their positions in the crystal lattice more readily so that the resistance to stress is reduced. Further, when the temperature is increased to 75 percent of the melting point, the mobility of atoms becomes so great that adjacent particles of metal tend to weld together, or *sinter*, when forced against each other. This tendency is the basis for powder metallurgy techniques and a variety of clever fabrication processes. The increasing mobility of atoms in the crystal lattice places an upper limit to the temperature at which alloys of a given base metal can have a useful strength no matter what additions of alloying elements may be made. Thus, even though over a billion dollars has been devoted to metallurgical research on the development of new alloys for high-temperature service in jet engines and gas turbines since World War II, the permissible metal temperature in gas-turbine blades has been increased only 100°C: from ~ 700 to ~ 800°C (~ 1292 to ~ 1472°F). This temperature limitation stems from the fact that the melting points of Fe-Cr-Ni alloys run around 1370°C (2550°F), that is, 1640 K, giving a sintering temperature of ~ 1230 K (~ 1750°F).

Creep Stress and Stress-to-Rupture

As the temperature of an alloy is raised above some threshold value, its elastic limit and yield tensile strength begin to fall off somewhat in short-time tests, but, if fairly high steady-state stresses are maintained, much more pronounced effects are noted in the form of creep, i.e., gradual plastic deformation of the metal under load. These effects are shown in Fig. 6.7 for one of the most commonly used materials in high-temperature steam plants, a $2\frac{1}{4}$ % Cr–$\frac{1}{2}$ % Mo steel. Creep strength data are commonly given (as in Fig. 6.7) for two creep rates: 0.1 percent creep in 1000 h and 0.1 percent creep in 10^4 h, that is, 0.0001 percent per hour and 0.00001 percent per hour, respectively. In addition, the stress required to cause rupture in 10^4 h is also a key consideration. Somehwat higher stresses and creep rates may prove more acceptable in some types of structure than in others, e.g., a 1 percent dilation of a pipe might be acceptable in the course of a 20-year life, but a dilation of much less than 0.1 percent in a turbine rotor would probably produce interference with the casing and cause severe damage.

Creep Buckling

Creep can have disastrous effects for structures that can buckle under load, e.g., a column loaded in compression. A circular shell under external pressure may become unstable and buckle

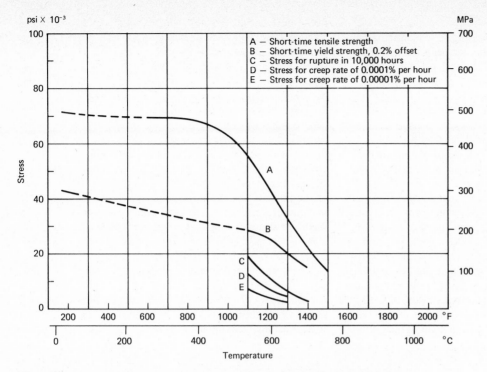

psi X 10⁻³ ... MPa

A — Short-time tensile strength
B — Short-time yield strength, 0.2% offset
C — Stress for rupture in 10,000 hours
D — Stress for creep rate of 0.0001% per hour
E — Stress for creep rate of 0.00001% per hour

Figure 6.7   Effects of temperature on the strength of $2\frac{1}{4}$ % Cr–$\frac{1}{2}$ % Mo (Ref. 1).

inward (as, for example, in submarine hulls). The problem is aggravated if the structure is operated at a temperature high enough that creep may occur, in which case the shell may be subject to creep buckling. One example is given by concentric ducts for closed-cycle gas turbines in which the outer annulus carries the higher-pressure, lower-temperature gas while the inner duct carries lower-pressure, substantially higher-temperature gas. Blanket structures for fusion reactors offer another example with a substantial pressure outside the thin shell surrounding the plasma and a high vacuum inside the shell. The creep buckling phenomenon stems from the fact that even the most carefully fabricated shells are not perfectly circular, and even a slight deviation from circularity induces substantial bending stresses. The gradual creep of the locally overstressed material acts to increase the out-of-roundness and hence the bending stresses until a point is reached at which the shell abruptly buckles inward, or implodes. Fortunately, extensive tests at ORNL of both cylindrical and spherical shells subjected to external pressure at elevated temperatures provide a sound basis for design.[29,30] Briefly, the amount of structural material required to provide the requisite greater stiffness is greatly increased over that implied by first-order hoop stress calculations. To avoid this sort of trouble, it is best, where possible, to design a structure so that there are no elements in what appears to be simple compression.

Failure Modes in High-Temperature Structures

If a failure occurs in a high-temperature structure, it is likely to progress much more slowly than in a conventional structure that operates near room temperature. The reason for this is that creep and creep-rupture are more likely failure modes than the more familiar yielding or burst-type rupture failure modes typified by a short-time tensile test; i.e., the ratio of the ultimate tensile stress to the design stress is much higher than for conventional structures. This effect is shown in Fig. 6.8 for 2% Cr–$\frac{1}{2}$ % Mo steel. The values are given for both the stress for creep-rupture and the code stress of the American Society of Mechanical Engineers (ASME), i.e., the usual design stress. Because the ultimate tensile stress is many times the design stress at high temperatures, burst-type failures initiated by fatigue cracks or (in rare cases) as a consequence of short periods of overpressurization caused by a malfunction or an operator error are far less likely in a high-temperature structure than in one designed for operation near room temperature. Rather, it is more likely that small cracks will develop from thermal-strain cycling and progress slowly until fluid leakage occurs and leads to a controlled shutdown.

Thermal Stresses and Thermal-Strain Cycling

Probably the most difficult and subtle—and one might even say the most insidious—set of problems in the design of high-temperature systems are those posed by thermal stresses and thermal-strain cycling. Everyone knows that a brittle material such as glass must be heated gradually and uniformly to avoid cracking. On the other hand, a metal pan can take severe thermal mistreatment without cracking, at least for a few times. However, anyone who has worked with combustion chambers for jet engines knows that the fluctuating temperatures in these chambers can lead to cracks. These are akin to those that one can induce by bending a piece of wire back and forth until it breaks. A ductile metal can withstand a number of severe strain cycles, but only a limited number. In point of fact, it has

been found that the degree to which a metal such as a 300 series stainless steel can withstand plastic strain cycling is actually reduced if its temperature is raised to a red heat. This was contrary to the hopes of those of us who first attempted to design liquid-metal systems for operation at 815°C (1500°F). Tests, however, disclosed that our worst fears were realized. Systems designed to test components under severe thermal-strain cycling conditions to see if they would break, broke. Extensive controlled tests showed that the total integrated amount of thermal-strain cycling that a given material will withstand increases as one reduces the amount of strain per cycle.[31] A typical curve showing the amount of strain per cycle as a function of the number of cycles to failure is given in Fig. 6.9.[32] Note that the proceedings of the symposium of Ref. 32 include a wealth of strain cycling data on different alloys.

Designers of high-temperature systems must force themselves to try to envision unfavorable off-design conditions that will produce thermal stresses and strains. Some typical conditions likely to give difficulty can be envisioned by inspecting the geometries of Fig. 6.10. The first and simplest of these, a simple round tube with a radial heat flux, is subject to thermal stresses arising from the difference in temperature between the

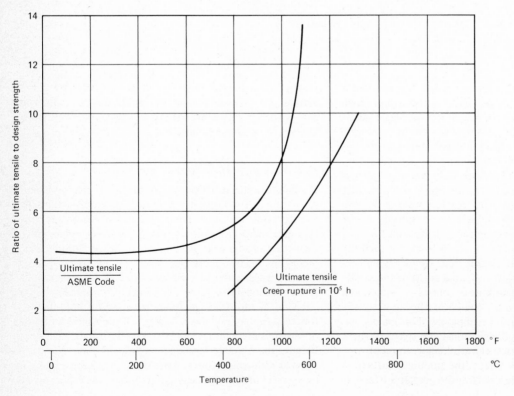

Figure 6.8 Effects of metal temperature on the ratio of the ultimate tensile stress to typical design stresses for 2% Cr–$\frac{1}{2}$% Mo ferritic steel showing the much greater margin for severe transient pressure stresses for high-temperature design conditions. *(Courtesy Oak Ridge National Laboratory.)*

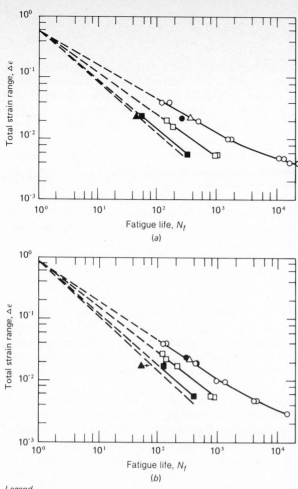

Figure 6.9 Total strain range versus cycles to failure for Incoloy 800 at (a) 665°C (1200°F) and (b) 760°C (1400°F).[32]

Legend

○ No hold time
□ 10-min hold time in tension only
△ 10-min hold time in both tension and
   compression
● 10-min hold time in compression only
■ 60-min hold time in tension only
▲ 300-min hold time in tension only

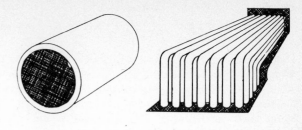

(A) SIMPLE ROUND TUBE   (B) HOCKEY STICK HEAT EXCHANGER

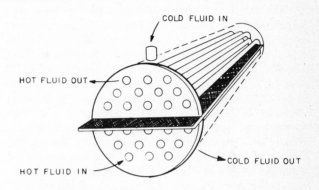

(C) U-TUBE SHELL-AND-TUBE HEAT EXCHANGER

Figure 6.10 Typical geometries in which thermal stresses may
be a problem.[16]

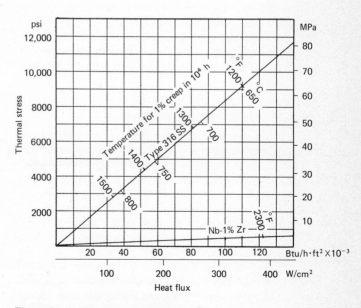

Figure 6.11 Effects of heat flux on the thermal stress in the wall of a 1-in-
OD, 0.8-in-ID tube for two different alloys.[16]

inner and outer layers, and these stresses are directly proportional to the heat flux up to the point where they are relieved
by creep. They are also dependent on physical properties other
than the strength, a point brought out by Fig. 6.11, which
shows these effects for two 1-in-OD, 0.8-in-ID tubes, one of
stainless steel and the other of the refractory alloy Nb–1% Zr.
The thermal stress is proportional to $E\alpha \, \Delta T$ where E is the
modulus of elasticity, α is the coefficient of thermal expan-

sion, and ΔT for the case of Fig. 6.11 is the temperature difference between the tube surface and the midplane of the tube wall. Figure 6.11 shows the marked reduction in thermal stress for Nb–1% Zr as compared to stainless steel because it has a much lower coefficient of thermal expansion and a lower modulus of elasticity. The temperature difference is also lower because Nb has a high thermal conductivity. It should be noted that thickening the tube wall would increase the temperature drop through the wall and thus aggravate the problem. If a double-wall tube is fabricated by drawing one tube down over another in an effort to reduce the possibility of a leak from one fluid to another, as might be done in a sodium-heated steam generator, the temperature difference between the inner and outer tubes of the double-wall design is inherently indeterminant, and the differential diametral and axial expansions are potentially large and thus could be responsible for severe thermal-strain cycling and tube failures. Such a construction certainly would be expensive. There are large temperature variations in the furnace waterwall of Fig. 11.3 as a consequence of the asymmetric heat input. These induce shear stresses in the webs between tubes and bending stresses in the tubes.

The hockey-stick tube configuration at the upper right of Fig. 6.10 is subject to severe thermal stresses because differential thermal expansion between the tubes and the casing must be accommodated by bending deflections in the short straight sections of the tubes near the header sheet. The U-tube shell-and-tube heat exchanger at the bottom of Fig. 6.10 is subject to shear stresses in the shell near the midplane baffle as a consequence of the temperature rise (or drop) in the shell-side fluid. Similarly, there are shear stresses near the midplane of the header sheet, and their effects are magnified by bending stresses in the ligaments between the holes in the header sheet.

Liquid-Metal-Heated Steam Generators

The high heat-transfer coefficients characteristic of both liquid metals and boiling water lead to exceptionally high heat fluxes in steam generators when liquid metal is on one side of a tube wall and water is on the other. While this is advantageous from the heat-transfer standpoint, it is likely to give large temperature differences in tube walls and even greater temperature differences in thick tube sheets and shells. The problems are difficult indeed when one considers the changes in temperature distribution in going from zero to full power and back. But the problems are even worse if consideration is given to off-design temperature transients associated with abrupt shutdowns, malfunctions of automatic controls, and inadvertent errors of operators. For example, anything that would cause an abrupt drop in steam pressure would lead to a step change in the

water-side temperature in the boiler equal to the change in the saturation temperature between the two pressure levels. A drop in boiler pressure from 55 bars (800 psi) to 20 bars (300 psi), for example, would mean a drop in boiler temperature of 55°C (100°F). This would induce severe thermal stresses as a consequence of the abrupt increase in the difference between the tube wall and shell temperatures, as well as the temperature difference across the tube header sheet. It would also lead to unstable film boiling conditions with the alternately high and low heat-transfer coefficients associated with the local formation and destruction of steam blankets. Anyone who has heard this type of boiling (steam hammer) will realize that severe thermal stresses must accompany it.

Effects of Surface Temperature Fluctuations

Fairly rapid fluctuations in surface temperatures may occur under unstable boiling conditions of the sort mentioned above or as a consequence of turbulence in a fluid stream having large differences in the local temperature, as in the region downstream of a combustion chamber, e.g., in the channel of an MHD generator. Extensive experience with low-level temperature excursions indicates that no damage will occur if the temperature variation can be kept below some limit characteristic of the particular material in question, even for millions of thermal cycles. Above that limit, as the temperature excursion per pulse is increased, the number of thermal cycles required to produce damage is reduced. For Inconel or stainless steel at 600°C, for example, the amplitude of the temperature excursion per pulse required to produce cracks from thermal-strain cycling in a relatively short time is of the order of 60°C, with some variation from one alloy to another. A typical example of a surface cracked in this fashion is shown in Fig. 6.12. In this instance molten-salt streams at two different temperatures were fed into a Y, and the flow in the two streams was pulsed so that the wall of the passage downstream of the Y was exposed alternately to fluid at the two different temperatures.[33] Once the temperature difference between the two streams was increased beyond about 50°C, a further increase in the temperature difference caused the number of cycles to produce cracking to fall off rapidly. This effect is shown in the curve in Fig. 6.13. Note that the material used in this case, Inconel, is an exceptionally tough and ductile material.

The effect of pulse frequency on the temperature range required to induce thermal-strain cracking is not great until the pulse frequency exceeds about 10 Hz. At higher frequencies the time becomes too short for full relaxation of the thermally induced stresses so that less plastic flow of the metal occurs; hence somewhat greater temperature amplitudes per cycle can

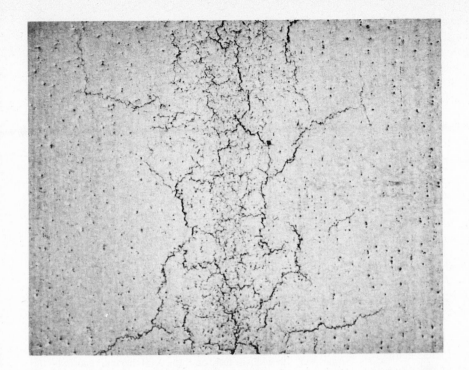

Figure 6.12 Photo of the interior surface of an Inconel tube subjected to thermal-strain cycling at 600°C by a temperature variation of 60°C in a stream of molten salt.[33] The number of cycles to failure was about 10^5. (*Courtesy Oak Ridge National Laboratory.*)

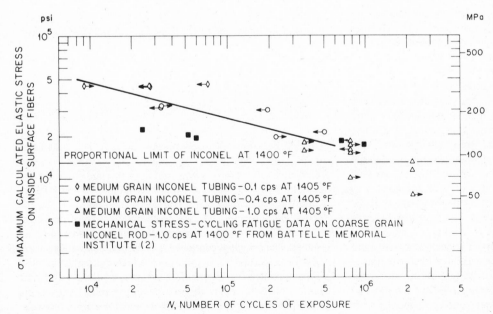

Figure 6.13 Correlation of calculated stress with the number of cycles to failure in thermal-stress cycling tests of Inconel at 763°C (1405°F) in a fused-salt environment: NaF-ZrF_4-UF_4 (56-30-5 mole percent).[33]

be tolerated for a given number of cycles to damage. At pulse frequencies of 1000 Hz, for example, one might be able to double the temperature amplitude for a given amount of damage from a given number of cycles. This, of course, depends on the particular material and the mean temperature of the test. The data of Figs. 6.9 and 6.13 show these effects for two typical alloys.

Estimation of the Life of Parts Subject to Thermal Stresses

As discussed further in later chapters on specific types of power plant, severe thermal stresses are induced by high heat fluxes in such parts as the water-cooled walls of steam-boiler furnaces and solar-energy collectors, in water-cooled blades of gas turbines, in the electrodes in MHD channels, in components of liquid-metal-cooled reactors, and in the first wall of a fusion reactor. These are likely to limit the life of the parts involved; hence estimates of the effects of both operating conditions and the choice of material are often required.

Fatigue tests, in which the number of rapidly repeated cycles of stress to failure are determined, provide important experimental information on which to base the design of power apparatus subjected to repeated conditions of stress caused by startup and shutdown of the apparatus and by vibrations. For many steels there is a fatigue limit, a stress below which test data indicate that the material will not fail no matter how many times the stressed condition is repeated. The number of cycles commonly associated with this fatigue limit is $\sim 10^6$ cycles. The larger the stress above the fatigue limit, the less the number of repeated cycles at which the material will fail. Note that many materials, including copper and aluminum, appear not to have a definite fatigue limit.

Many design problems which are related to thermal stress involve the possibility of failure at a relatively low number of repeated cycles (tens to thousands of repetitions rather than millions). It is this situation with which we are concerned here. It may be observed that the extreme case of failure at a low number of cycles is a single applied stress to failure in rupture (one-quarter cycle). Under thermal-strain cycling the material commonly yields and deforms plastically so that the thermal stresses are largely relieved. Thus, it is more appropriate to speak of "thermal strain" rather than "thermal stress." However, many stress analysts find it convenient to speak in terms of the equivalent thermal stress in an ideal elastic material.

Many factors influence the ability of a material to withstand thermal-strain cycling. These include its ductility, the temperature at which it is operated, and the range of plastic strain per cycle. Although good test data are not available for most of the materials and conditions of interest here, a paper by Coffin[34] reviews an extensive set of experimental data on the low

cycle fatigue of materials at high temperature and shows that good correlation can be obtained through the use of the empirical equations:

$$\Delta\epsilon_p \nu^{B(k-1)} = C N_f^{-B} \tag{6.1}$$

$$\Delta\epsilon_e \nu^{-k'} = \frac{A'}{E} N_f^{-B'} \tag{6.2}$$

where $\Delta\epsilon_p$ is the range of plastic strain (the sum of the negative and positive amounts by which the strain goes into the plastic region during a cycle) and $\Delta\epsilon_e$ is the range of the elastic strain (positive and negative). N_f is the number of cycles to failure under the conditions imposed, and E is the modulus of elasticity. A', C, k, k', B, and B' are experimentally determined quantities for each material, and ν is the frequency with which the stress application is repeated.

The total range of the strain, $\Delta\epsilon$, is the sum of the plastic and elastic ranges, or

$$\Delta\epsilon = \Delta\epsilon_p + \Delta\epsilon_e = C \frac{N_f^{-B}}{\nu^{B(k-1)}} + \frac{A'}{E} \frac{N_f^{-B'}}{\nu^{-k}} \tag{6.3}$$

For the range of interest here, the effects of frequency are small so that $k = 1$ and $k' = 0$, B is approximately 0.5, B' is nearly zero, A' is close to the yield stress σ_y, C is half the elongation $\epsilon_{1/4}$ in a tensile test, $\Delta\epsilon$ is the total thermal strain which is given by the coefficient of thermal expansion α, the temperature range ΔT, and a function of Poisson's ratio μ. That is,

$$\Delta\epsilon = \left(\frac{1+\mu}{1-\mu}\right) \alpha \, \Delta T$$

Substituting these values in Eq. (6.3) gives the useful approximation

$$N_f = \frac{1}{\left[\frac{2}{\epsilon_{1/4}}\left(\frac{1+\mu}{1-\mu}\right)\alpha \, \Delta T - \frac{2\sigma_y}{\epsilon_{1/4}E}\right]^2} \tag{6.4}$$

This has been applied to obtain the curves of Fig. 6.14 using the physical property data of Table 6.4.[35] Note that copper at 150°C gives about the same resistance to thermal-strain cycling as stainless steel at 600°C and that Haynes 188, even at 1500°C, is more resistant than either copper or stainless steel. Experimental data for IN 718, an exceptionally strong high-temperature alloy, indicate an unusually high resistance to strain cycling. The principal characteristics desired in an alloy for resistance to thermal-strain cycling are a high thermal conductivity to minimize the temperature differential for a given heat flux, a low coefficient of thermal expansion and a low modulus of elasticity to minimize the thermal stress for a given temperature differential, and a high elastic limit and a high ductility at

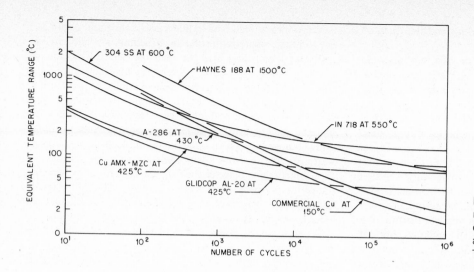

Figure 6.14 Equivalent temperature range as a function of the number of cycles to failure as calculated for typical alloys using the physical properties of Table 6.4.[35]

TABLE 6.4 **Material Properties Useful for Estimating Thermal Stresses, Strains, and Low-Cycle-Fatigue Life** (Ref. 35)

	Specific heat c_p, cal/(g·°C)	Density ϱ, g/cm³	Thermal conductivity k, cal/(s·cm·°C)	Coefficient of thermal expansion $\alpha \times 10^6/°C$	Modulus of elasticity $E \times 10^{-6}$, psi	Yield point σ_y, psi	Elongation $\Delta\epsilon_{1/4}$	Poisson's ratio μ
Type 304 SS, 600°C	0.105	8.0*	0.041	17	21	10,000	0.4	0.30
Haynes 188, 500°C	0.12	8.95*	0.0475	14.8	27.6	43,800	0.7	0.30
In 718, 550°C	0.104	8.2*	0.049	14.2	16.7	16	0.20	0.28
A-286, 430°C			0.045	16.7	25	60,000	0.20	
Commercial oxygen-free Cu, 150°C	0.092	8.96*	0.923	16	17*	5,000	0.4	0.36
Glidcop AL-20, 425°C			0.81	19.6	16*	26,000	0.07	0.36
AMAX-MZC, 425°C	0.094	8.8*	0.74	20	20*	55,000	0.09	0.36
Nb-1% Zr, 1000°C	0.065	8.5	0.165	7	10	20,000	0.4	0.30
TZM, 1000°C	0.066	10.0	0.250	5	50	18,000	0.4	0.32
W, 150°C	0.032	19.3	0.476	4.5	50	40,000	0.01	0.28
V, 800°C	0.152	6.11	0.095	10.9	18	38,000	0.2	0.30
Ti-6Al-4V, 500°C	0.124	4.1	0.023	9.6	11	45,000	0.27	0.30

*Value at room temperature; elevated-temperature data not available.

the desired operating temperature to withstand the thermal-strain cycling.

In using Fig. 6.14 it should be remembered that in strain-cycling laboratory test work alloy specimens are held at a constant temperature and are subjected to mechanical rather than thermal-strain cycling. The strain ordinate in the figure is replaced by the temperature range that would induce a thermal strain equivalent to the mechanical strain required to produce failure in the mechanical strain-cycling tests. Thus Fig. 6.14 is a bit artificial, especially for large strains and few cycles, because the properties of the alloys vary with temperature. At first thought, one might employ a second scale for the mechanical strain per cycle, but this cannot be done because the conversion from thermal to mechanical strain involves the modulus of elasticity and the coefficient of thermal expansion, both of which vary from one alloy to another. Thus in trying to make a direct comparison of different alloys from the thermal-strain cycling standpoint, the scales of Fig. 6.14 are by far the most convenient. One simply must keep in mind that they are an approximation made with the simplifying assumption that they are for a set of ideal materials having the temperature-invariant properties of the materials specified at the temperatures specified. For the region of interest here, the peak temperature in the thermal cycle is that for which the curves were drawn.

Effects of Radiation Damage

Materials in the core of fission reactors are subject to damage because fast-neutron collisions frequently displace atoms in the crystal lattice and sometimes cause transmutation of elements —a process that for some elements such as nickel may produce

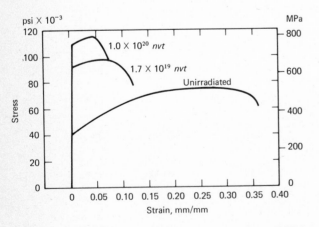

Figure 6.15 Stress-strain behavior of an irradiated ASTM A-212 Grade B carbon-silicon steel.[37]

helium. These effects lead to a loss of ductility and swelling of the irradiated material (especially if helium is produced), together with reductions in the thermal and electrical conductivities.[36] Typical curves showing the effects of irradiation on the stress-strain curves for a steel pressure vessel are shown in Fig. 6.15. The parameter *nvt* in Fig. 6.15 is the total number of fast neutrons per square centimeter that have passed through the material. The embrittling and swelling are particularly important, because they act to limit the life of parts in the cores of water reactors to a few years. Depending on the thickness of the blanket region between the core and the pressure vessel, the life of the pressure vessel may be limited to 20 to 40 years. If and when the plasma physics problems of fusion reactors are solved, one of the most difficult engineering problems of fusion power will be posed by the 14-MeV neutrons released by the D-T fusion reaction, which cause several times more severe damaging effects than the 1- to 2-MeV neutrons from fission.[35,37]

Ceramic Structures

It is often suggested that ceramic structural parts could be employed to permit higher operating temperatures than are practicable with metal. Such structures present many problems, not the least of which involve thermal stress. The basic difficulty can be illustrated by considering the possibility of constructing glass bridges. Glass is a light, strong, cheap, abundant material. However, it is not used for bridges because it has zero ductility, and any local stress concentration will initiate a crack that will propagate and induce a general failure. A similar problem is presented by a heat-transfer matrix. A complex geometry and temperature distribution arising from large temperature differences leads to large local thermal stresses. The problem is rendered more acute if the structure must withstand large tensile stresses from internal pressure. A simplistic view of the problem may indicate that for a particular design these two conditions can be satisfied for the nominal pressure and thermal stresses that might be expected for the full-power design condition. However, one must provide for thermal transients under off-design conditions, and these are severe. Further, considerations of fracture mechanics indicate that, first, one must expect macroscopic stress concentrations in regions in which the geometry changes (such as in the vicinity of header sheets) and, second, that these will be aggravated by microscopic stress concentrations in the form of small defects and cracks that are invariably associated with the fabrication process. The combined effects of these two factors will serve to increase local stresses by a factor of 5 or 10 even in a carefully designed joint. But this is not all; the Occupational Safety and Health Administration (OSHA) standards require that all engineering structures be capable of withstanding an earthquake, and this complicates the support problem. In a heat exchanger, for example, provisions for accommodating the shaking forces to be expected in an earthquake make it essential to employ an array of tube supports that form a redundant structure with a multiplicity of restraints on the tubes, and these make it impossible to satisfy the thermal stress and strain requirements for a brittle material if thought is given to the variety of off-design conditions that must be expected in transients in the course of emergencies or misoperation of the system.

If a crack once starts in a brittle material, extensive experience indicates that it will progress rapidly if the pressure stresses are high, and shattering of the structure will result. This will occur even if the stresses are nominally compressive, as shown by an investigation of glass hemispherical shells for use as crew compartments in deep submersibles. This type of structure appeared to be an extremely attractive possibility in the middle 1960s, and was widely accepted at that time as the most promising direction for development because of the light weight and high strength of structural glass. However, Martin Krenzky of the Naval Ship Engineering Command Test Laboratory at Carderock pointed out that even small defects in the glass would be likely to cause failures under external hydrostatic pressures far less than the design values, even though the spherical shells would be nominally in pure compression. Over a dozen glass hemispherical shells were fabricated with a thickness of about 12.5 cm (5 in) and shipped to Carderock for testing. A number of tests were run; in every instance failure occurred at much less than the design pressure in spite of the fact that the transparency of the glass made examination for small defects exceptionally easy. Clearly, even a small defect in a brittle component in compression is likely to cause a disastrous failure if the part is fairly highly stressed. Not surprisingly, enthusiasm for glass crew compartments for deep submersibles completely evaporated following these tests.

Brittle-Ductile Transition Temperature

The brittleness problems of ceramics point up the importance of ductility in metal structures and the emphasis on this property in the ASME pressure vessel code. Of particular importance is the abrupt loss in ductility that occurs in carbon and low-alloy steels as the temperature is dropped—an effect shown in Fig. 6.16. The temperature at which this transition from ductile to brittle failure occurs is known as the *transition temperature*. As shown in Fig. 6.16, irradiation increases the transition temperature, as do small differences in the melting and fabrication practices. The material in weld zones sometimes has a transition temperature even higher than 0°C. The

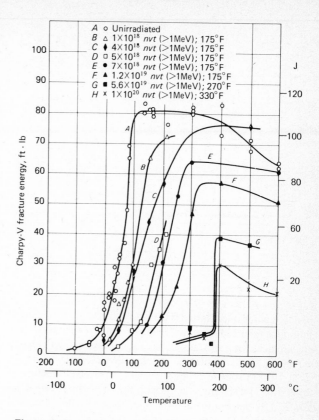

Figure 6.16 Charpy V-notch impact test behavior of an irradiated ASTM A-285 grade A carbon steel (normalized).[37]

implications of this phenomenon were not appreciated until after World War II, when there was an intensive effort to find the reason why a number of Liberty ships suddenly broke in two for no apparent cause.[37] This phenomenon also explained a whole series of previously mysterious failures in welded structures, including the failures of some bridges and storage tanks in Belgium during winter in the early 1920s. Note that increasing the thickness of steel plate tends to increase the brittle-ductile transition temperature, a problem vital to heavy-pressure vessels and treated further in Chap. 8.

Fracture Mode

The manner in which a crack in a material propagates and possibly fractures is an important consideration from the safety standpoint. In tough materials the crack will progress until it enters a region of lower stress, where it is arrested. In less tough

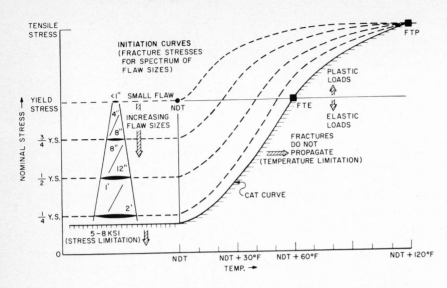

Figure 6.17 Generalized fracture analysis diagram as referenced by the NDT temperature.[37]

materials a crack once started is likely to continue to propagate to complete failure. Figure 6.17 helps to visualize the behavior of these cracks for a typical steel at various stress and temperature regions. The curves for the stress at the Crack Arrest Temperature (CAT) start at a low stress and at the Nul Ductility Temperature (NDT) and rise through the Fracture Transition Elastic (FTE) temperature and level out at the Fracture Transition Plastic (FTP) temperature.

LIMITATIONS IMPOSED BY FABRICATION

Alloys differ widely in their suitability for fabrication in a practical engineering sense. While perhaps obvious, these considerations are so important that a brief summary appears in order.

Fabricability

One of the key considerations in the design of a component is the method to be used in its fabrication: casting, forging, forming or pressing from plate, tube drawing, bending, or machining. The method used affects the strength, reliability, and cost. A material having good ductility that does not work harden rapidly is likely to be suitable for tube-drawing operations; some alloys do not have these properties. Others do not lend themselves to forging but may be suitable for casting. Under any circumstances the fabricability characteristics affect the cost, life, and reliability of the part.

Weldability

One of the most important engineering requirements for a power plant fluid circuit is that the structural material must contain the working fluid with a high degree of leak-tightness to avoid both loss of the fluid and its contamination. This in turn means that the structural material must be weldable and ductile with good ductility in the weld zone because there will inevitably be substantial thermal stresses as a consequence of temperature differences in the welding operation. Extensive experience indicates that at least a few percent elongation in the weld zone is required if cracks are to be avoided. Welding stresses can be reduced by preheating, and this is often necessary for thick sections in high-strength carbon steel pressure vessels.

Cost and Availability

Cost, an obviously important consideration, will be discussed in Chap. 10. While less obvious, availability is also an important factor. There is also a strong incentive to employ a commercial material that is routinely available in the form of sheet, plate, and tubing, and for which quality-control procedures have been well developed. This allows a high level of confidence in the consistency of the material's physical properties.

LIMITATIONS IMPOSED BY LUBRICATION

Most moving parts in familiar equipment operate near room temperature and can be lubricated in a conventional fashion, but some such as pistons in high-temperature steam engines (Chap. 2) and control rod drives in nuclear reactors present formidable lubrication problems. Rubbing surfaces in high-temperature water, helium, sodium, or a vacuum are often essential elements in reactor servicing equipment. Not only must the moving parts in these machines continue to function with a high degree of reliability, but also they must continue to operate accurately so that the various positioning operations will not deteriorate with wear; otherwise, jamming in one form or another might occur. This in turn means that the moving parts should function with little wear and a low coefficient of friction. Lubrication of the moving parts in reactors poses some exceptionally difficult problems because in some instances parts of the machine must operate at a temperature of 350 °C (662 °F) or more under intense gamma radiation. Aside from temperature effects, the radiation dose rate at the face of an operating reactor is such that, under operating conditions, the fast-neutron dose alone would cause severe damage to petroleum-base grease in about 6 min. Even after shutdown, when the neutron flux drops to zero, the gamma-ray dose is high. Although gamma rays do not damage metals and ceramics, they are as damaging to organic materials as neutrons, and, as the temperature increases, the deterioration occurs even more rapidly. Some of the synthetic lubricants, such as the silicones, are better suited to operation at high temperature than petroleum-base lubricants, but they are substantially more sensitive to radiation, and hence would not be satisfactory. Petroleum-base lubricants ordinarily may not be used at temperatures above approximately 450 °F, and even at that temperature their life is limited. If a circulating-oil system with the oil reservoir in a low-radiation-level region is used, substantially greater life can be obtained, but this poses serious limitations on the design of the mechanism. Under any circumstances the lubricant must have a very low vapor pressure to avoid evaporation and the formation of tarry deposits that might lead to such difficulties as fuel elements' sticking in channels.

Most of the bearing loading conditions involved entail operation in what one would normally consider the boundary lubrication regime, i.e., frequent starts and stops in which a fully developed hydrodynamic fluid film would not be present. Lubrication in conventional mechanisms would be provided by adsorbed films of moisture or organic materials such as oils or greases that adhere tenaciously to the rubbing surfaces and minimize the contact of high spots that can lead to wear, scuffing, galling, or even seizing of the parts involved. Only a few lubricants appear to be suited to the more hostile applications. The most promising are dry lubricants, such as molybdenum disulfide, graphite, and phthalocyanine. The lubricity of graphite depends on films of adsorbed moisture; hence it is not suitable for use in a vacuum, with a dry inert gas such as helium, or at temperatures above ~ 200 °C (400 °F), above which it loses its adsorbed moisture. There has been a substantial amount of work on all three materials not only for high-temperature gas-cooled reactors but also for aircraft power plant mechanisms designed to operate at temperatures up to 1200 °F and for spacecraft mechanisms designed to operate in a vacuum. This work indicates that reasonably low coefficients of friction and low wear rates can be obtained if the mechanisms are carefully designed for this sort of service, and that molybdenum disulfide appears to be more promising for this application than either graphite or phthalocyanine. Relatively large clearances must be employed, and a means for renewing the dry lubricant films must be provided. Gas or liquid jets carrying a suspension of sulfide powder have been used with some success. Thin coats of varnish containing molybdenum disulfide powder have given good service, especially if baked on in several thin layers. A sintered compact containing approximately 10 percent molybdenum disulfide powder, 5 percent copper, and the balance silver appears promising.[38] Compacts of this character may be employed as blocks in the separator cages of ball or roller bearings, as bushings for simple journal bearings, or as idlers in gear trains.

For shafts that must run at a substantial speed, it is possible to use the ambient fluid as a lubricant in hydrodynamic bearings. Good success has been achieved with both liquid-metal- and gas-lubricated bearings. In all cases the unit loads have been kept low and the machines have been designed to avoid edge-loading of journal bearings. Very hard materials such as tungsten or titanium carbide have proved to be best for bearing surfaces. Rubbing velocities should be kept moderate to minimize heat generation, which can lead to self-welding.

PROBLEMS

1. A 2-in-OD, $2\frac{1}{4}$ % Cr-$\frac{1}{2}$ % Mo tube in the reheater of a coal-fired steam plant is designed to operate at 565 °C (1050 °F) with an internal pressure of 3 MPa (440 psi). Using Fig. 6.7, estimate the tube-wall thickness required if the allowable stress is that for a creep rate of 0.00001 percent per year.

2. If the fire-side corrosion rate for the tube of Prob. 1 is estimated to be 0.012 mm (0.005 in) per year and the design life is to be 20 years, by how much should the tube-wall thickness be increased to allow for corrosion?

3. The heat flux through the lithium-cooled first wall of a fusion reactor (Fig. 5.34) caused by x-ray and plasma heating is 100 W/cm^2 [3.17 × 10^5 Btu/(h · ft^2)], and the thickness of the IN 718 wall is 2.5 mm. Determine the temperature drop through the wall. If the reactor experiences 10 power pulses per hour with a mean wall temperature of 550°C, estimate the life of the wall as limited by the thermal-strain cycling induced by the radial temperature difference through the wall. (Consider that the compound curvature of the wall prevents a change in shape that would relieve the stresses from the radial temperature difference.)

4. The Type 304 SS heater for a gas turbine is built of $\frac{3}{4}$-in-OD tubes and operates at 10 atm. It might be operated at a higher temperature to give a higher cycle efficiency if it were made of Hastelloy X instead of stainless steel. Investigate the cost effectiveness of this change, taking the peak metal temperature as 25°C greater than the turbine air-inlet temperature for turbine air-inlet temperatures of 815°C (1500°F) and 900°C (1652°F), respectively, for the Type 304 stainless steel and the Hastelloy X. Using the curves of Figs. 10.14 and A6.1, with a wall thickness allowance of 0.050 in for corrosion, estimate the wall thickness required in each case and the increased capital cost in dollars per kilowatt of electric output for the Hastelloy X, assuming an air-heater surface requirement of 0.2 m^2/kWe. Estimate the increased thermal efficiency obtainable from Fig. 2.26 and the consequent fuel saving per year, assuming a fuel cost of $1/10^9 J for a total power output per year equivalent to continuous operation at 60 percent of full load. Compare with the increased capital charge at 16 percent per year based on the increased capital investment. (Neglect both the reductions in capital charges associated with the higher output per unit weight of air handled and the increases in duct and turbine costs stemming from the higher gas turbine inlet temperature.)

REFERENCES

1. *Steels for Elevated Temperature Service*, United States Steel Corporation, ADV-18566, 1949.

2. Probert, P. B., and L. Katz: "Corrosion in Reducing Atmospheres—A Designer's Approach," *International Symposium on Corrosion and Deposits, New England College, Henniken, New Hampshire*, June 26–July 1, 1977.

3. *Steam—Its Generation and Use*, 38th ed., The Babcock and Wilcox Co., New York, 1972.

4. Goldberg, S. A., et al.: "A Laboratory Study of High-Temperature Corrosion on Fireside Surfaces of Coal-Fired Steam Generators," *Trans. ASME*, vol. 90, April 1968, p. 193.

5. Sedor, P., et al.: "External Corrosion of Superheaters in Boilers Firing High-Alkali Coals," *Trans. ASME*, vol. 82, July 1960, p. 181.

6. Blank, H. A., et al.: "Behavior of Superheater Tubing Material in Contact with Combustion Atmospheres at 1350°F," *Trans. ASME*, vol. 74, 1952, p. 813.

7. *Pressurized Fluidized Bed Combustion Research and Development Report No. 85, Interim No. 1*, prepared for Office of Coal Research, Department of the Interior, by National Research Development Corporation, London, England, 1974.

8. Pierce, R. R.: "Estimating Acid Dewpoints in Stack Gases," *Chemical Engineering*, Apr. 11, 1977, pp. 125–128.

9. Fraas, A. P., and M. N. Ozisik: "A Comparison of Gas-Turbine and Steam-Turbine Power Plants for Use with All-Ceramic Gas-Cooled Reactors," USAEC Report ORNL-3209, Oak Ridge National Laboratory, July 1965.

10. Roberts, D. I., et al.: "Behavior of Structural Materials in the High-Temperature Gas-Cooled Reactor Systems," *Symposium on Materials Performance in Operating Nuclear Systems, Nucl. Met. 19*, CONF-730801, August 1973, pp. 325–358.

11. Hedgecock, P. D., and P. Patriarca: "Materials Selection for Gas-Cooled Reactor Components," *ANS Topical Meeting Gas-Cooled Reactors: HTGR and GCFBR, May 7–10, 1974*, CONF-740501, 1974, pp. 525–559.

12. Wunderlick, J. W., and N. E. Baker: "Exposure of HTGR Candidate Core Plate and Thermal Insulation Materials to Impure Helium at 1650°F to 1850°F for 3000 Hours," General Atomic Company GAMD-7377, Dec. 29, 1966.

13. Huddle, R. A. U.: "The Influence of HTGR Helium on the Behavior of Metals in High Temperature Reactors,"

Effects of Environment on Material Properties in Nuclear Systems (Proceedings of the International Conference on Corrosion, London, July 1–2, 1971), Institution of Civil Engineers, London, 1971, pp. 203–212.

14. Wood, D. S., M. Farrow, and W. T. Burke: "A Preliminary Study of the Effect of Helium Environment on the Creep and Rupture Behavior of Type 316 Stainless Steel and Incoloy 800," *Effects of Environment on Material Properties in Nuclear Systems (Proceedings of the International Conference on Corrosion, London, July 1–2, 1971)*, Institution of Civil Engineers, London, 1971, pp. 213–228.

15. Harms, W. O., and A. P. Litman: "Compatibility of Materials with Alkali Metals for Space Nuclear Power Systems," *Nuclear App. & Tech.*, vol. 5, September 1968.

16. Fraas, A. P.: "Comparative Study of the More Promising Combinations of Blanket Materials, Power Conversion Systems, and Tritium Recovery and Containment Systems for Fusion Reactors," Oak Ridge National Laboratory Report No. ORNL-TM-4999, November 1975.

17. Bloom, M. C., et al.: "Corrosion Studies in High Temperature Water by a Hydrogen Effusion Method," *Corrosion*, National Association of Corrosion Engineers, vol. 13, May 1957, pp. 297–302.

18. Hawkins, G. A., et al.: "The Corrosion of Alloy Steels by High Temperature Steam," *Trans. ASME*, vol. 66, 1944, pp. 291–295.

19. Latanision, R. M., and R. W. Staehle: "Stress-Corrosion Cracking of Fe-Ni-Cr Alloys," *Proceedings of Conference on the Fundamental Aspects of Stress Corrosion Cracking, Ohio State University, Department of Metallurgical Engineering*, 1969.

20. Basil, J. L.: "Corrosion of Materials in High Velocity Sea Water," Naval Engineering Experiment Station Report 910160A, Annapolis, December 1960.

21. Basil, J. L.: "Performance of Plain and Duplex Condenser Tubes at High Water Velocities," Naval Engineering Experiment Station Report 040020C, Annapolis, December 1957.

22. "Elevated Temperature Properties of Titanium and Titanium Alloys," Naval Engineering Experiment Station Report 4A066876, Annapolis, March 1951.

23. DeCorso, S. M.: "Erosion Tests of Steam Turbine Blade Materials," *ASTM Proceedings*, vol. 64, 1964, pp. 782–796.

24. Jenks, G. H., and J. C. Griess: "Water Chemistry in Pressurized and Boiling Water Reactors," USAEC Report ORNL 4173, Oak Ridge National Laboratory, November 1967.

25. R. N. Lyon et al., *Liquid Metals Handbook*, Office of Naval Research, NAVEXOS P-733 (Rev.), June 1952.

26. DeVan, J. H.: "Compatibility of Structural Metals with Boiling Potassium," *International Conference on Liquid Metal Technology in Energy Production*, CONF-760503-Pl, pp. 418–426, Champion, Pa., May 1976.

27. Silverman, M. D., and J. R. Engel: "Survey of Technology for Storage of Thermal Energy in Heat Transfer Salt," Oak Ridge National Laboratory Report No. ORNL/TM-5682, January 1977.

28. Grimes, W. R.: "Molten Salt Reactor Chemistry," *Nuclear App. & Tech.*, vol. 8, no. 2, February 1970, pp. 137–155.

29. Tong, K. N.: "Buckling and Creep Buckling of Spherical Shells Under Uniform External Pressure," Syracuse University Research Institution Report No. ME 922-764S, July 1964.

30. Corum, J. M.: "An Investigation of the Instantaneous and Creep Buckling of Initially Out-of-Round Tubes Subjected to External Pressure," USAEC Report ORNL 3299, Oak Ridge National Laboratory, Jan. 16, 1963.

31. Coffin, L. F., Jr.: "A Study of the Effects of Cyclic Thermal Stresses on a Ductile Metal," *Trans. ASME*, vol. 76, 1954.

32. Ostergren, W. J.: "Correlation of Hold-Time Effects in Elevated Temperature, Low Cycle Fatigue Using a Frequency-Modified Damage Function," *1976 ASME-MPC Symposium on Creep-Fatigue Interaction*, ASME MPC-3, 1976, pp. 179–202.

33. Keyes, J. J., and A. I. Krakoviac: "High Frequency Surface Thermal Fatigue Cycling of Inconel at 1406°F," *Nuclear Science and Engineering*, vol. 9, no. 4, April 1961.

34. Coffin, L. F. Jr.: "Fatigue at High Temperature," *The Institution of Mechanical Engineers Proceedings, London*, vol. 188, September 1974, pp. 109–127.

35. Fraas, A. P., and A. S. Thompson: "ORNL Fusion Power Demonstration Study: Fluid Flow, Heat Transfer, and Stress Analysis Considerations in the Design of Blankets for Full-Scale Fusion Reactors," Oak Ridge National Laboratory Report ORNL/TM-5960, February 1978.

36. Fraas, A. P.: "Materials Problems in the Design of Magnetically Confined Fusion Reactors," *Nuclear Technology,* vol. 22, April 1974.

37. Whitman, G. D., et al.: "Technology of Steel Pressure Vessels for Water-Cooled Nuclear Reactors," Oak Ridge National Laboratory Nuclear Information Center, Report No. ORNL-NSIC-21, December 1967.

38. Johnson, R. L., et al.: "Friction, Wear, and Surface Damage of Metals as Affected by Solid Surface Films," NACA Technical Report 1254, 1956.

7

INSTRUMENTATION AND CONTROL

Organizations engaged in the design, development, construction, and operation of power plants virtually always have a department called *Instrumentation and Control* (I&C), yet in organizing the material for this chapter the author realized that the term *Control and Instrumentation* would be more appropriate. This is because the prime objective of the discipline is control; the instrumentation is simply a means to that end, no matter how impressive the huge panels of instruments in the control room. Perhaps the reason for the inversion of terms is that the kind of expertise required for the control system is that found in the instrumentation field. In the author's view this is a symptom of a deficiency on the part of mechanical engineers who are so preoccupied with the heat transfer, fluid flow, and mechanical design of the major components that they fail to give control problems the major consideration they deserve. These problems ought not be regarded as peripheral and the instrumentation and control equipment as a sort of major appendage to be tacked on after the rest of the plant has been designed. Thus this chapter is concerned mainly with control. It is intended to delineate the control problems in power plants, put them in perspective, and give some insights into steps that can be taken in component selection and design, as well as in system layout, to improve the control characteristics of the complete integrated plant. Minimizing the amount of instrumentation and control equipment required should improve the plant reliability; as indicated in the next chapter, roughly half of the outages of a power plant are caused by malfunctions in the instrumentation and control systems. Reducing and simplifying the control functions required would serve to reduce the incidence of these outages, and at the same time reduce the capital investment required for instrumentation and control equipment.

BASIC FUNCTIONS OF INSTRUMENTATION AND CONTROL SYSTEMS

Instrumentation and control systems have three major functions:

1. Control

2. Diagnosis of incipient or serious trouble

3. Protection of personnel, equipment, and the public

All three functions usually depend on sensors such as thermocouples or strain gages to indicate temperatures, pressures, liquid levels, etc. The three functions are related, and some of the sensors may be used for all three functions, but the processing of the signals is quite different. Signals from sensors used for control functions are fed directly into the control system, which responds with some action to regulate one or more components of the power plant, usually with some proportional control action. Most sensors serve diagnostic purposes, and their signals may simply actuate an instrument on a display panel, or they may be recorded on a strip chart or on magnetic tape in a computer, or both. Signals used for operation of the protective system are ordinarily monitored continuously and automatically with no response unless they fall outside of prescribed limits, in which case they actuate warning lights or buzzers for less serious indications, or they initiate control action automatically when prompt action is in order.

However big and impressive a power plant may be, it must function as a slave to the load demand—hence the common control terms *slaved-to-the-load* and *load-following.* In a sense it functions with blind obedience and with no advance warning that the load will go up or down; the plant must respond immediately and automatically to supply the demand whatever it may be. For example, when an increase in load on a steam turbine causes it to slow down slightly, a governor (or other equipment having a similar function) must act at once to open a throttle valve and admit more steam to the turbine. The resulting drop in steam pressure and flow calls for an increase in the fuel and airflow to the furnace and in the feed-water flow to the boiler. In modern plants these actions are automatic. The function of the operating personnel is to carry out some fine tuning to improve the plant efficiency, minimize emissions, and watch for trends or other signs of an incipient or developing malfunc-

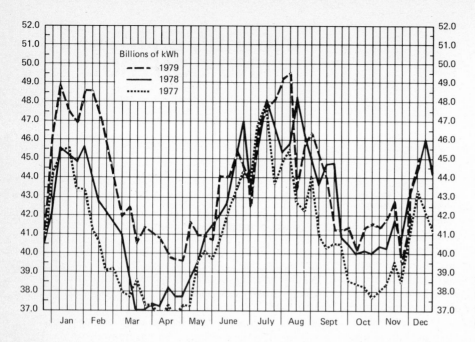

Figure 7.1   Total weekly output of U.S. electric utilities as a function of the time of the year for 1977, 1978, and 1979.[1] (Courtesy Electrical World.)

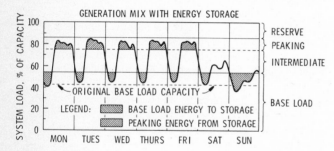

Figure 7.2   Average hourly output of a large electric utility system with energy storage capacity for a typical week. (Courtesy Argonne National Laboratory.)

more complex functions of furnace firing rate were not automated until a century later.

CHARACTERISTICS OF LOAD ON ELECTRIC UTILITIES

The electric load on a typical utility varies diurnally, weekly, and seasonally with the lightest loads coming at night on spring and fall weekends and the greatest loads occurring during hot summer weekdays when air-conditioning loads peak, or during winter cold spells when heating loads peak.

Figure 7.1 gives the weekly electric power output of U.S. utilities in units of 10^9 kWh for 1977, 1978, and 1979, showing the seasonal variation in demand. Figure 7.2 gives the hourly output during a typical week for a large utility system with a well-balanced combination of *base, intermediate,* and *peaking-power* capacity.[2] In this instance the system has some conventional hydroelectric capacity but not enough to handle the full amount of peak load; hence the utility has also installed some pumped hydro storage capacity. Depending on other uses of a river, conventional hydroelectric units are ordinarily constrained to maintain some minimum river flow for the fish downstream and for water supplies, thus they cannot be shut down completely in periods of low demand. However, their

tion that might be corrected, or at least checked and delayed until a shutdown would be convenient. Note that the concept of automatic control is not new: the flyball governor invented by Christiaan Huyghens about 1657 for use in clocks was patented in England for the automatic control of windmills by Thomas Mead in 1787 and was adopted a little later by James Watt early in his development of the steam engine. Significantly, the far

great flexibility makes them prized elements in a utility system; at least 10 percent of the system capacity in hydro units is generally considered highly desirable. Part of this hydro capacity may be pumped storage, which helps further to reduce the range of load variation by providing an extra load for the base-load units during off-peak hours. This is especially advantageous if the system includes some nuclear power plants, for which the incremental cost of the fuel for the extra off-peak power is quite small.

Load Allocation

The load dispatcher in a utility system tries to keep his *base-load* units fully loaded all the time in order to take advantage of their low fuel costs obtained at the expense of high capital costs. The *intermediate-load* units are called on next as the load builds up. These are usually coal- or residual-fuel-oil-fired units in which the plant design has been simplified and compromised to give reduced capital costs at the expense of a higher fuel consumption, usually accomplished by using lower steam pressures and temperatures, together with a less elaborate regenerative feed-water heating system and perhaps no reheat. Often the intermediate-load units are old base-load units designed in former years when lower thermal efficiencies were usual. Although their fuel costs are higher than for new base-load units, their capital charges are relatively low, because the capital costs have been largely written off. Finally, when the load increases beyond the capacity of the base- and intermediate-load units, the *peaking-power* units are coupled into the system. These units are often gas turbines that have low capital costs but both a relatively high specific fuel consumption and high unit costs for a premium fuel (natural gas or a distillate fuel oil). (The problems of balancing capital and operating costs are treated in Chap. 10.)

Magnitude of Load Variations

The cumulative effects of load variations differ from one utility to another, mainly as a consequence of the fraction of the total load represented by the fluctuating residential as opposed to steady industrial loads. In fact, industrial power users such as aluminum smelters often have clauses in their contracts calling for them to drop a substantial part of their load during periods of high residential demand. In spite of such efforts to flatten load profiles such as those in Figs. 7.1 and 7.2, diurnal variations in load for some utilities may be ± 30 percent, and the maximum summer or winter load and the minimum load during a spring or fall weekend night may differ by a factor of 4. This in turn means that for some utilities even base-load plants may have an *annual capacity factor* of only 65 percent,[3] i.e., their average output may be as little as 65 percent of their rated capacity. Usually it is higher, a point implicit in the curve of Fig. 7.3 which gives the average load factor for all U.S. utilities as a function of year.[4] Note the optimistic FEA extrapolation to 70 percent.

Incidence of Starts

During periods when the electric load on a utility system is low, it is desirable to shut down units that are not needed in order to save fuel and cut costs. However, the large temperature changes associated with startups and shutdowns induce substantial thermal stresses and some thermal-strain cycling, which limit the life of plant components. Thus an important consideration in design work is the number of starts and stops expected in the life of the unit. Figure 7.4 gives data for three types of fossil-fuel steam plant. The first column is for a conventional coal-fired base-load plant that will spend the latter part of its life in intermediate-load service. The second is for a plant designed and built for peaking service; i.e., it will be shut down every night and during weekends. (To facilitate startups it would ordinarily employ gas or oil as the fuel.) The third column is for an intermediate-load type of plant which is coal-fired to minimize

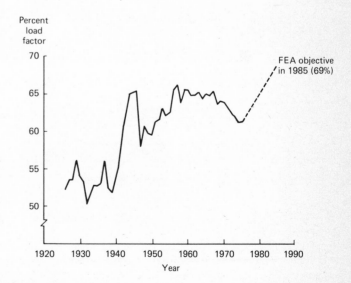

Figure 7.3 Improvement in the average annual load factor of U.S. electric utilities in the 1925–1975 period.[4]

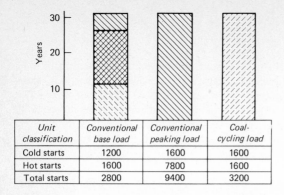

Unit classification	Conventional base load	Conventional peaking load	Coal-cycling load
Cold starts	1200	1600	1600
Hot starts	1600	7800	1600
Total starts	2800	9400	3200

Base-load mode   Peaking mode

Weekend mode   Coal-cycling mode

Figure 7.4 Loading cycles for base-load steam plants, peaking gas turbines, and coal-fired steam plant units designed for peaking service.[5]

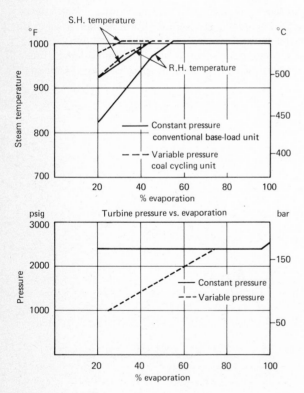

Figure 7.5 Effects of load on superheater and reheater outlet temperatures for both constant pressure and variable pressure control modes.[5]

fuel costs and is designed to give a reasonably good efficiency at low loads while operating at reduced steam pressures and temperatures (e.g., following the lower curves shown in Fig. 7.5) to increase the life of the equipment. The most interesting point is that the intermediate-load plant which was designed to operate over wide ranges of load is subject to only a relatively small incremental number of starts over those for a coal-fired plant designed for base-load service, in part because the latter portion of the life of the base-load plant will be spent in intermediate-load service. At least equally important is the fact that the base-load plant must undergo not only the rather large number of 1200 cold starts but also a total of ~2800 cycles between an "idle" condition and high-power operation. These are major considerations in establishing permissible amounts of thermal-strain cycling in parts in which the key temperature differences depend on the heat flux, e.g., boiler furnace walls, water-cooled gas-turbine blades, and water-cooled electrodes in MHD generators.

Power Plants in Process Industries

The above discussion has been concerned mainly with plants in electric utility service. About 10 percent of the electric power in the United States is produced in plants operated by industry as integral elements in their processes, usually supplying both heat and electricity. These plants normally operate at full load without any shutdowns except for about one per year for scheduled maintenance and occasional forced outages caused by malfunctions. Such plants present much less difficult control problems than those in conventional utility service, and the drastically reduced incidence of both power cycling and shutdowns leads to major improvements in component life.

CONTROL FUNCTIONS

When confronted with a set of problems as complex as those associated with the control of any type of power plant, it is worthwhile to begin with a review of the fundamental characteristics of the power plant components, the control functions inherently involved in coordinating these components, and the overall power plant requirements from the operators standpoint.

Basic Components and Their Characteristics for a Steam Rankine Cycle

In view of its widespread use, it is probably best to begin with a conventional steam power plant. The basic components of a

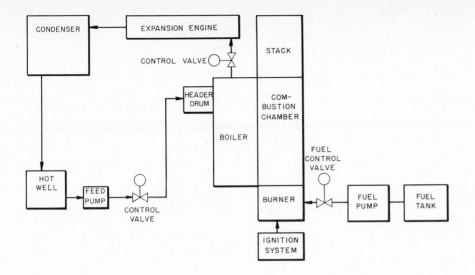

Figure 7.6 Schematic diagram of a steam power plant with a recirculating boiler and the simplest system possible.

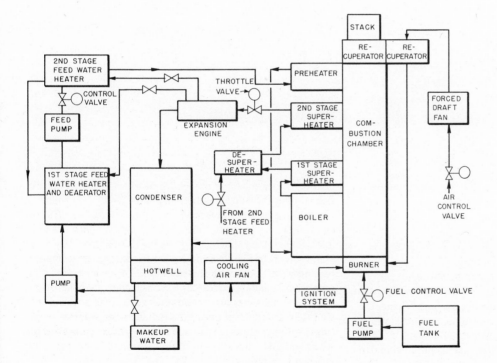

Figure 7.7 Schematic diagram of a steam power plant with a once-through boiler and the principal components usually required. In this case an air-cooled condenser is employed.

simple Rankine cycle system are indicated in Fig. 7.6. Heat added in the boiler generates vapor that flows through a throttle valve to a turbine and then to a condenser. Condensate is scavenged from the hot well of the condenser by a feed pump and, in the simpler systems, returned to the boiler where the

boiler drum serves as a reservoir in which the bulk of the liquid inventory in the system is retained. The basic, simple system of Fig. 7.6 is commonly elaborated as in Fig. 7.7 to include a recuperator, a feed-water preheater or economizer, a super-heater, a set of regenerative feed-water heaters to increase the

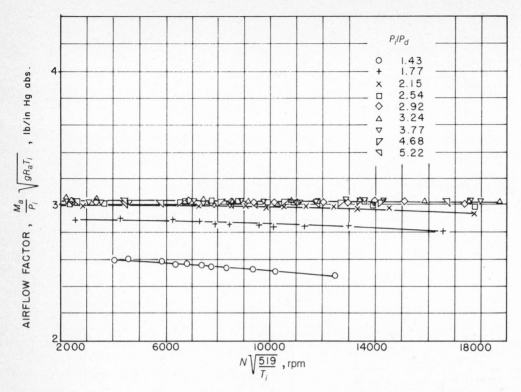

Figure 7.8 Effects of turbine rpm on the gas flow rate through a single-stage impulse turbine at a constant pressure ratio.[6]

thermal efficiency, a deaerator to reduce corrosion, and a de-superheater to control the amount of superheat.

Turbine Performance Characteristics

The principal operating variables affecting the output and efficiency of a turbine are the inlet pressure, the pressure ratio, the vapor flow rate, and the rpm. Most turbines of interest here would be designed for stage pressure ratios of about 2; hence the vapor flow will be directly proportional to the turbine inlet pressure and independent of the turbine rpm[6] (Fig. 7.8). The torque input to the turbine wheel will be directly proportional to the vapor weight flow rate, and, to a first approximation, the aerodynamic efficiency at design speed will be independent of load, because the volumetric flow will be essentially constant. The effects of turbine rpm can be deduced readily from velocity diagrams (Chap. 3), and these effects for a typical machine are shown by Figs. 3.4 and 7.8. Note that as the turbine speed in-creases, the torque input to the turbine wheel drops off linearly from double the value at the design point to zero at the runaway condition at nearly double the design rpm. The turbine efficiency peaks at the design turbine wheel tip speed, which is a little less than half the vapor velocity leaving the inlet nozzle.

The curves of Figs. 7.8 and 3.4 are for a single-stage impulse turbine wheel. Although the situation is more complex if the number of stages is increased, the basic characteristics of the multistage machines are essentially similar. The principal advantage of a multistage machine is that increasing the number of stages with a given pressure ratio reduces the turbine wheel tip speed and this reduces compressibility losses and improves the efficiency somewhat.

Steam-Generator Characteristics

One of the major problems in the control of boilers stems from changes in the boiler temperature distribution with changes in load. Some insight into these problems is given by Fig. 7.9, which shows the local temperature of the water and steam as a function of the fraction of heat added in the course of its transit through a monotube, or once-through, steam generator for a nuclear plant.[7] If the steam conditions are to be held constant at reduced load, the temperature of the reactor coolant must be reduced as the power output is reduced.

In fossil fuel plants the combustion gas temperature is essentially independent of the power output, but for control modes such as the variable pressure mode of Fig. 7.5 the fraction of heat added at the boiling point and the temperature at the boiling point will vary with the boiler operating pressure.

As a point of interest, in small plants it is possible to operate an oil or gas burner intermittently; that is, if the load demand is less than the boiler output, both the burner and the feed-water flow can be stopped and the turbine can be operated by the steam in the reservoir. This approach has the disadvantage that a relatively large reservoir is required, and this increases the size, cost, weight, and the steam explosion hazard potential. Such an approach is obviously unsuited to a system designed to operate on superheated steam, but it has been used in both boats and steam automobiles employing saturated-steam systems, in which case extra steam for an abrupt increase in load can be obtained by flashing from the superheated water in the header drum.

If an attempt is made to vary the power output of a Rankine cycle system while holding a constant steam temperature and pressure at the exit of a monotube boiler, the heat-transfer coefficients both on the gas side and on the steam side in the superheater will vary as the 0.6 to 0.8 power of the fluid flow rate.[7] The heat-transfer coefficient on the water side in the boiler will be substantially independent of the feed-water flow rate. As a consequence of these effects, the superheater outlet temperature will tend to increase as the load is reduced. It is for this reason that it is common to split the superheater into two parts with a desuperheater between them. The desuperheater usually consists of a water spray whose flow rate can be adjusted to hold a constant superheater outlet temperature irrespective of the heat flux distribution in the steam generator (Fig. 7.7).

Condenser Performance Characteristics

The principal barrier to heat transfer in a condenser is the cooling-water-side fluid film. The overall heat-transfer coefficient varies approximately as the 0.6 power of the cooling-water flow through the condenser because much of the temperature drop is through the tube wall and condensate liquid film. With a given water flow rate, the temperature difference between the condenser cooling water and the condensing steam is directly proportional to the heat load.

Burner and Combustion Air Systems

One of the most subtle and difficult sets of problems in the design of a Rankine cycle power plant is concerned with the burner and combustion chamber. The operating characteristics of these components are far more complex than they at first appear, and these complexities are reflected in the requirements imposed on the control system. The chamber must provide initial ignition of the flame with equipment that avoids an explosion hazard by preventing fuel injection unless an electric spark

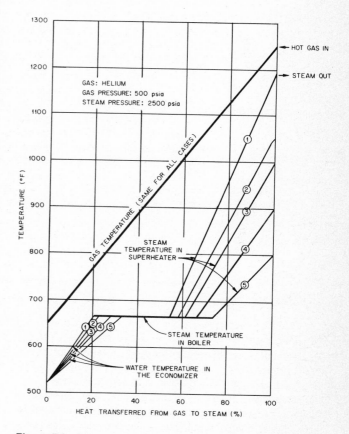

Figure 7.9 Temperature distribution in a steam generator for a gas-cooled reactor with fixed gas inlet and outlet temperatures for five power output conditions with no desuperheater.[7] *(Courtesy Oak Ridge National Laboratory.)*

TABLE 7.1 Principal Parameters Related to Control of Typical Rankine Cycle Systems

	System of Fig. 7.6	System of Fig. 7.7
Feed water	Feed flow rate	Feed flow rate
	Pump discharge pressure	Pump discharge pressure
	Liquid level in makeup tank	Liquid level in makeup tank
		Steam flow to air ejector
		Deaerator temperature
		Temperature leaving feed heater
		Steam flow to first-stage feed heater
		Steam temperature to first-stage feed heater
		Steam flow to second-stage feed heater
		Steam temperature to second-stage feed heater
Steam generator	Discharge pressure	Superheater outlet temperature
	Header drum liquid level	Superheater outlet pressure
		Desuperheater water flow rate
Condenser	Steam pressure	Steam pressure
	Steam temperature	Steam temperature
	Hot well temperature	Hot well temperature
	Coolant temperature	Coolant temperature
	Coolant flow rate	Coolant flow rate
		Cooling fan rpm
		Cooling fan power
Condenser scavenging pump		Discharge pressure
		Discharge flow rate
Fuel system	Fuel tank liquid level	Fuel tank liquid level
	Fuel pump discharge pressure	Fuel pump discharge pressure
	Fuel flow rate	Fuel flow rate
Burner, combustion chamber, and stack	Airflow rate	Airflow rate
	Stack gas exit temperature	Fan speed
		Stack gas exit temperature
		Air temperature leaving recuperator
		Speed of rotary recuperator
Ignition system	Current to igniter	Current to igniter
	Temperature of flameholder	Temperature of flameholder
Heat engine	Inlet pressure	Inlet pressure
	Engine power	Engine power

or pilot flame is present, smoke-free operation and maintenance of the proper ratio of the fuel and airflow rates over the full range of load conditions, avoidance of hot spots in the boiler and consequent possibilities of tube burnout, and provision of an adequate gas flow capacity over the full load range. All of these impose control requirements, some obvious and some subtle. If the system is to have a good overall thermal efficiency, it should include an air preheater to reduce the heat lost to the gases leaving the stack, and this also may require control.

The characteristics of the burner vary widely with the detail design, but under any circumstances the low volatility of fuel oil imposes a vaporization and ignition problem. To obtain good atomization, the fuel injection pressure must be substantial, which makes the fuel- and air-metering problem for a wide range of loads much more difficult than is the case for the carburetor of an automobile engine. This, coupled with problems in maintaining good combustion over a substantial range of firing rates, has led many designers of small boilers to use an on-off burner control similar to that in household furnaces so that orifices can be adjusted to give good combustion at the full-power fuel and airflow conditions. With this approach it is sufficient to get a fuel-metering system, burner, and combustion chamber combination that will give good combustion and no trouble with hot spots at the design point only. For large boilers, however, the fuel and airflow rates must be modulated to suit the load, and the extra complexities must be handled properly both in the control system and in the details of the burner and furnace design.

Fossil Fuel Systems

The fuel system for an oil-fired plant requires as a minimum a fuel storage tank, a fuel pump, and some means for regulating the fuel flow rate. Similar functions are handled by the corresponding equipment in a coal-fired plant. For safety reasons the fuel-metering equipment should include provisions to prevent pumping fuel into the burner unless the flame is burning properly. Most of the explosions that have occurred in gas- and oil-fired furnaces have stemmed from flooding the combustion chamber with fuel before actual ignition has been obtained. Serious explosions caused by bursting of the steam boiler have been much less frequent.

Feed-Water System

Most Rankine cycle power plants suffer an appreciable loss of working fluid as a consequence of leaks. To avoid fouling the boiler and other system components, it is essential to treat the makeup water and store enough treated water to assure an adequate reserve supply of clean water for makeup and for use under emergency conditions (such as a major leak in the boiler tube array).

Overall System Characteristics

The relationship of the principal parameters for the overall system can be deduced by relating the characteristics of the individual components. The relationships will, of course, depend in substantial measure on the control mode chosen. However, under any circumstances, the local temperatures of the combustion gases will be essentially constant, and the steam temperature in the condenser will be nearly constant, varying only a little with changes in the condenser cooling-water temperature available and the heat load.

Control Parameters, Functions, and Requirements

To provide some perspective on the control problems, it seems best to list the principal parameters involved in the control of a Rankine cycle power plant, the major control functions, and the principal requirements from the standpoint of the operator.

Table 7.1 lists the principal parameters involved in the control of both the simple system of Fig. 7.6 and the more complex system of Fig. 7.7, while Table 7.2 lists the corresponding principal control functions. Examination of Tables 7.1 and 7.2 indicates that the many operating parameters and control functions imply a complex control system. This point of view is reinforced by an examination of the performance characteristics of the various components discussed in the previous sections. Note that most of these may be either independent or dependent

TABLE 7.2   Principal Control Functions for Typical Rankine Cycle Systems

System of Fig. 7.6	System of Fig. 7.7
Ignite the burner.	Ignite the burner.
Limit fuel flow to a low level until combustion chamber heats up.	Limit fuel flow to a low level until combustion chamber heats up.
Regulate the feed-water flow rate to maintain the proper liquid level in boiler header drum.	Regulate the water flow rate to maintain the desired boiler pressure. Add makeup water to maintain the proper liquid level in hot well.
Regulate the fuel flow rate to maintain the boiler pressure constant.	Regulate the fuel flow rate to maintain the boiler pressure constant.
Regulate the throttle valve to give the desired power output.	Regulate the throttle valve to give the desired power output. Regulate the water flow rate to the desuperheater to maintain the superheater outlet temperature.
Proportion the fuel and airflow rates to maintain a constant fuel-air ratio.	Proportion the fuel and airflow rates to maintain a constant fuel-air ratio.
Pressure relief valve to blow off steam to avoid an explosion if steam pressure becomes excessive.	Orifices in steam bleedoff lines maintain proper ratio between feed-water and bleed steam flow rates. Pressure relief valve to blow off steam to avoid an explosion if steam pressure becomes excessive.

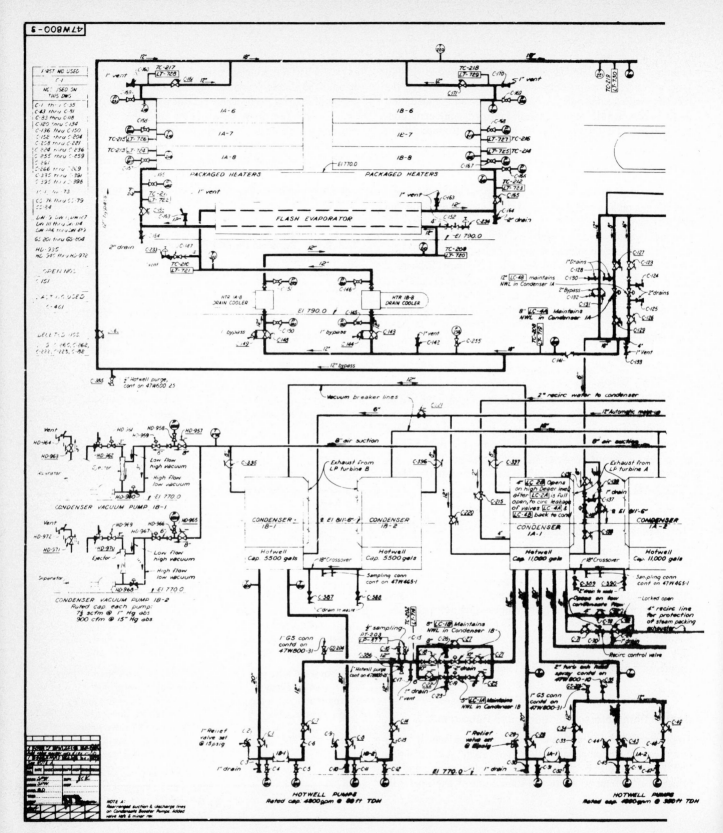

Figure 7.10 The flowsheet for the condensate system of the TVA Bull Run power plant.[8] (*Courtesy TVA.*)

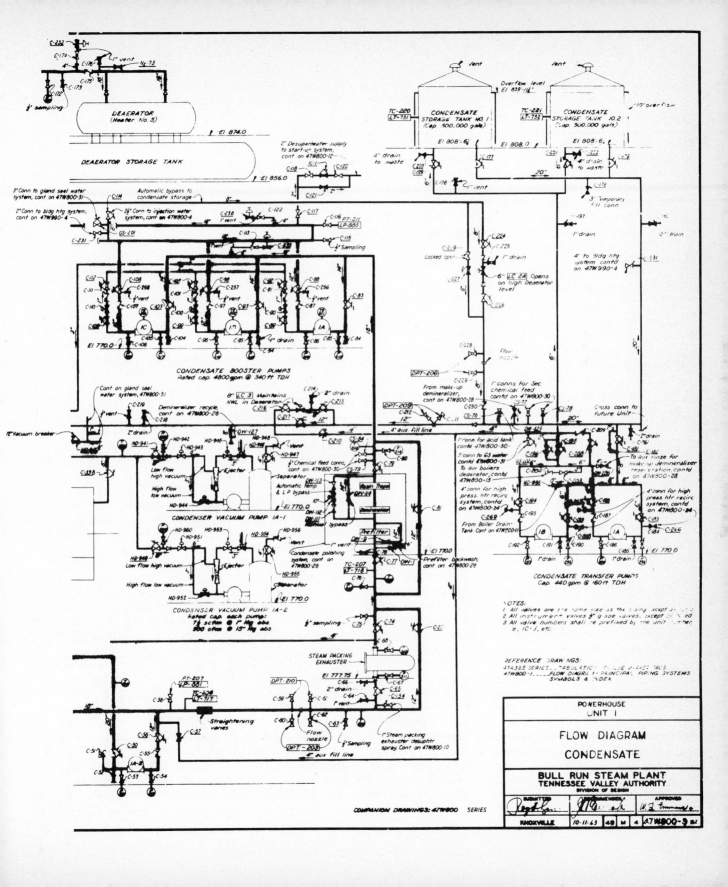

POWERHOUSE
UNIT I

FLOW DIAGRAM

CONDENSATE

BULL RUN STEAM PLANT
TENNESSEE VALLEY AUTHORITY
DIVISION OF DESIGN

COMPANION DRAWINGS: 47W800 SERIES

variables, depending on the type of control scheme employed. A few of these need not be measured for routine operation of old established systems, but all of them must be considered when the control system is being designed.

Some of the principal control functions summarized in Table 7.2 are likely to be quite complex. For example, ignition of the burner includes provision of adequate fuel pressure, opening a valve to admit the fuel, generation of a spark for ignition or introduction of a pilot flame, and limitation of the fuel flow to a small value until a sensing device such as a "purple peeper" (which detects the ultraviolet emission from flames) indicates that the burner is properly ignited.

Complete Power Plant

The above outline of major control parameters and functions may be regarded as a gross oversimplification; an actual plant is enormously more complex. Some notion of the degree of complexity is given by Fig. 7.10, which is just one of the 72 flowsheets and wiring diagrams required to define the various systems for the 950-MWe unit of the Tennessee Valley Authority (TVA) Bull Run power plant.[8] The complexity of this flowsheet for the condensate is typical of the others in the set. Suffice it to say that just the right half of this flowsheet includes ~ 150 valves, 28 pressure sensors, and 9 temperature sensors. In the plant as a whole there are so many pressure, temperature, and flow sensors, valves, and other instrumentation and control devices that a substantial book is required just to define their functions.

Characteristics of Other Types of Power Plant

The fossil-fuel-fired steam plant has been treated above at some length, in part because it has provided the bulk of the U.S. electric power and in part because it presents the same basic problems as many of the advanced types of plant treated in later chapters. However, it is in order to note that hydroelectric, diesel, and gas-turbine units present far less complex control problems, a characteristic that is reflected in the smaller operating crew sizes required. Also, diesels and gas turbines involve lower capital costs. Nuclear reactors for steam plants actually present less complex control problems than fossil-fuel-plant burners and furnaces, but public safety considerations usually entail a great deal of special nuclear instrumentation, control, and monitoring equipment. However, for small power plants the basic possibilities for system simplification exist if some sacrifice in performance is accepted. An extreme example indicating what can be done is given by a boiling-water reactor design that the author evolved for a 100-KWe unmanned nuclear undersea power plant. One conclusion of the study was

that the control functions of this simplified nuclear plant would be simpler than those for a fuel-cell system of similar output!

DYNAMIC RESPONSE OF THE SYSTEM

The discussion up to this point has been concerned with essentially steady-state operation at any of a wide range of loads. However, the dynamic response characteristics of power plants are quite as important and certainly even more complex.

Types of Load Change

The response requirements for a power plant unit depend in part on the role that the unit is called on to play in a system and in part on the rate at which the system may be called on to change load. The habits of residential users, for example, lead to a ramped increase in system load in the morning when people get up, turn on lights, start cooking, and use hot water. A second, more rapid, ramped rise in system load occurs as machines are started up in industrial plants—effects evident in Fig. 7.2. Large step changes in load can occur when city street lights are turned on or off, unless—as is normally the case—the street lighting system switchgear is designed to phase small blocks of load in succession to avoid large step changes. Step changes in system load may also occur in starting very large machines, such as supersonic wind tunnels. The location of the U.S. Air Force AEDC facility at Tullahoma, Tennessee, was chosen because it was one of the few places in the country where the electric utility could handle the huge step change in load when the wind-tunnel motor was started. In that instance the hydroelectric units of TVA had the ability to handle the load. In a case of this sort the wind-tunnel operator would, of course, work closely with the TVA load dispatcher to ensure that the system was operating with enough spinning reserve in the hydro units to pick up the load when applied. Even more severe step changes in load have occurred through electric grid interconnections when a failure of some sort has thrown a heavy load through tie lines on an adjacent system. It was this sort of event that led to the great 1965 blackout of the entire U.S. northeast when a progressively more severe overload was thrown from one utility to the next, causing a chain of outages from New York to Maine. Switchgear for handling such intergrid loads is now designed to limit the magnitude of such step changes in load and thus to prevent this sort of event.

The most difficult yet unavoidable step change in load that can occur with an individual unit is a load trip at full power. (Lightning is a common cause of a load trip.) The first concern is to prevent excessive overspeed of the turbine generator—which would cause a probably ruinous rotor failure. Rapid closing of the throttle valves actuated by the governor will normally limit the overspeed to less than 10 percent. The heat capacity of the boiler is usually so great that the steaming rate cannot be halted abruptly; hence valves must be opened so that steam bypasses the turbine and is dumped through a desuperheating water spray into the condenser or to the atmosphere—all of which call for further automatic control actions. The situation is complicated by the fact that the power plant unit must be kept ready to be coupled back onto the line when the fault in the electric system has been cleared—which may be accomplished in a few seconds, or it may take many minutes.

Response Rates

Once a steam or hydraulic turbine generator is up to synchronous speed, the rate at which it can respond to a change in load depends on the rate at which the flow of working fluid can be changed, and this can be accomplished in roughly a second. Large load swings can be handled by hydro units in a few seconds, but steam turbines are subject to damage by thermal distortion as a consequence of changes in temperature distribution if the rate of change in load is greater than ~ 2 percent per minute. This limitation stems in part from the fact that the casings and the central portion of the rotor are massive and so their temperature cannot be changed rapidly, whereas the blades are relatively thin with a high surface-volume ratio and hence their temperature closely follows that of the steam flowing through them. Some notion of the mass of material involved is given by the 850-MWe unit of the TVA Bull Run plant: the high-pressure turbine weighs 627,000 kg (1,380,000 lb) while the low-pressure turbine weighs 1,485,000 kg (3,275,000 lb).[8] Similarly, there is an enormous amount of material in the steam generator—in the Bull Run unit ~ 5,450,000 kg (~ 12,000,000 lb) in heat-transfer surface alone—so thermal-stress consideratons favor limiting the rate of change in load to about the same as for the turbine.

Gas turbines operate at only 10 to 20 atm, hence are much lighter in weight and, in simple open-cycle units, require no heat-transfer surface. Changes in load are accomplished by varying the fuel flow while the airflow remains fixed. The combustion chambers and turbine casings are relatively thin and are designed to tolerate rapid changes in temperature so that they can accept a load swing from low to full power in as little as 30 s.

Startup Time

Hydro units are commonly kept running even at no load as spinning reserve, an operational mode called *condensing,* because it permits the units also to act like condensers (capacitors) in the electric system to correct a lagging power factor and thus im-prove the efficiency of the system. Medium-size diesel-generator units can start, be brought to speed, and pick up load in less than a minute. Hence they are excellent as emergency power supplies used to take care of essential circuits for instrumentation, controls, and certain critical equipment items such as lubricating-oil pumps. (Some big steam turbines suffered ruinous failures in the great northeast blackout of 1965 because they kept spinning while their electric motor-driven lubricating-oil pumps stopped when the power line went dead. They lacked the instrumentation and controls to cope with such an event.) Gas turbines for peaking service in an emergency can be taken from a cold start to full load in as little as 3 min with some models[9] (Fig. 7.11), though a warm-up period of 10 min or more is preferable. Intermediate-load, coal-fired steam units can be started and brought on stream in 1 to 2 h after an overnight shutdown, while base-load units normally require at least 2 to 4 h. The time required after an extended shutdown is likely to be much longer, often taking a day or more to bring the water purity to the proper level.

Example

A good idea of the effects of the system heat capacity on the rates at which temperatures can change can be obtained by some estimates for the 950-MWe Bull Run unit cited above. Consider, for example, the time required to drop the system temperature 50°C if the fuel flow to the burners were stopped abruptly and the steaming rate continued at the full design rate. The amount of steel in heat-transfer surface in the steam generator is ~ 5.45 × 10⁶ kg, that in high-temperature piping and other components closely coupled thermally to the steam circuit is ~ 0.95 × 10⁶ kg, and the weight of steel in the high- and low-pressure turbines is 2.1×10^6 kg, giving a total of 8.5×10^6 kg of steel whose temperature is closely coupled to the high-temperature portion of the steam system. The heat capacity of this steel is 0.42 J/(g·°C), or a total of 3.57×10^9 J/°C for the system. At a power output of 850 MWe, the thermal efficiency is ~ 40 percent with stack and furnace losses running roughly 10 percent and so the heat input to the steam is approximately 2.25 × 850 = 1912 MWt, or ~ 1.9×10^9 J/s. For a 50°C drop in temperature the heat capacity of just the steel in the system (i.e., neglecting the heat capacity of the water) would yield ~ 180 × 10⁹ J, while the heat required to sustain the design steaming rate is ~ 1.9×10^9 J/s. Thus it would take about 95 s for the 50°C temperature change. If instead of being abruptly stopped, the fuel flow were reduced 10 percent and the steaming rate were maintained, it would take ~ 15 min for the system temperature to drop 50°C. This crude example gives a good notion of the enormous ''thermal inertia'' of a steam system and shows why

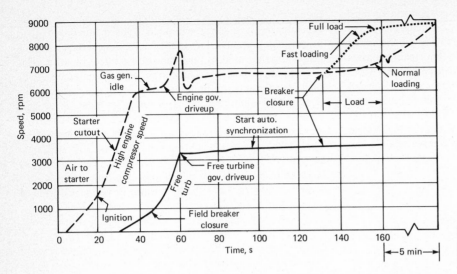

Figure 7.11   Sequence of events in a fast start of a gas turbine operating on liquid fuel.[9]

changes in the output of base-load steam plants are commonly limited to 2 percent per minute.

STABILITY

The stability problems of individual components, power plant units, and interconnected power plant systems are much too complex to treat in the space available here other than by giving some notion of their nature and implications. Fortunately, the basic concepts involved—and actually the mathematical relations as well—apply to both fluid flow and electric systems. In fact, it is this commonality that makes it possible to build relatively inexpensive electric simulators of power plant systems for use in studies of stability and control problems of plants that are in the conceptual design stage, as well as much more complex and expensive simulators for subsequent training of operating personnel.

For a component or a system to operate in a stable fashion, it should be so proportioned that it responds to a perturbation by tending to return to its original condition. For example, the steering gear of an automobile is designed so that if the front end is aligned to give the correct camber, castor, and toe-in, the front wheels will tend to position themselves to point straight ahead. In this case, as in most systems, it is desirable to proportion the system to give only a small restoring force if the operator wishes to turn the car; too great a degree of stability may be almost as undesirable as an unstable system. (In a car, the steering would be unstable if a small deflection of the steering wheel generated a force that acted to increase the turning

angle rather than reduce it.) The same basic considerations apply in power plants. Both the individual components and the manner in which they are coupled should be such that they form a stable system.

Stability of Components

The stability characteristics of the principal components are treated briefly in previous chapters. The discussion here is therefore limited to the minimum needed to convey an appreciation for overall system stability problems.

Turbines

The linear drop in turbine torque with increasing shaft speed (Fig. 3.4) makes for stable turbine operation. That is, a perturbation that acts to slow the turbine down yields an increase in torque—which acts to speed the turbine up again. At the other extreme, a complete loss of load will cause a turbine to speed up to roughly twice its design speed, at which point the torque developed will be zero. (This assumes, of course, that the overspeed does not lead to a rotor failure from excessive stresses, bearing overheating, or a shaft whipping problem.) In short, turbines are inherently stable. Further, as indicated by Fig. 7.8, variations in turbine speed have essentially no effect on the fluid flow rate through the turbine; hence a perturbation in turbine speed will not affect the flow rate through the fluid circuit.

Compressors

As discussed in Chap. 3, centrifugal and axial-flow compressors present a stability problem in that the pressure rise developed at a constant speed peaks at some flow rate and drops off if the flow is either increased or decreased from that value. This leads to violently unstable operation (or "surging") at low flows but permits stable operation at flows higher than that for the peak pressure rise. Thus gas-turbine systems must be designed so that the flow through the compressor is always greater than that for the peak pressure rise (and peak efficiency), because otherwise oscillations in the system flow will be induced. Note that in the stable region a small increase in flow will reduce the compressor pressure rise and this will act to reduce the system flow; hence the system is stable.

Boilers

The large increase in specific volume that occurs when water is vaporized leads to complex two-phase flow regimes in the boiler tubes. The much greater volume of the steam leads to a greater pressure drop, and so under some conditions flow oscillations, or "chugging," may occur. One example of this type of problem is discussed in Chap. 11; for a more general treatment the reader is referred to Ref. 10. Suffice it to say here that boiling-flow stability is a major consideration in boiler design, and great care is taken to choose proportions. Thus stability is not often a source of trouble in conventional boilers. Generally an increase in steam flow rate with a constant firing rate will lead to a drop in the boiler outlet temperature and pressure, and this will tend to cause a reduction in steam flow—a stabilizing effect.

Condensers

The cooling-water rate through a condenser is normally kept constant. An increase in steam flow into the condenser leads to an increase in the temperature drop across the heat-transfer surfaces, and this causes an increase in the local steam pressure. Note that a quite small increase in condenser pressure will give a large increase in the available temperature drop, e.g., typically an increase in condenser pressure of 10 mm Hg (0.4 in Hg) will give a 5°C, or 50 percent increase in both the temperature drop and the heat flux. Thus the condenser is also a stable element in the system.

Combustors

Burners that give high heat-release rates per unit of volume inherently involve both high gas velocities and intense turbulence.

Inasmuch as the gas velocities are of the same order as the flame speed, the flame may tend to "blow out," producing a highly unstable condition (Chap. 5). Thus one of the major problems in the design and development of furnaces for steam generators and combustors for gas turbines is the achievement of stable operation in the burners, an essential requirement for the power plant.

Nuclear Reactors

Nuclear reactors are designed to be stable. Fortunately, an increase in the reactor temperature tends to reduce the neutron scattering and fission cross sections, and this decreases the reactivity and the power. An increase in temperature also causes the reactor to expand. The reactor expansion tends to increase the neutron leakage, and this also acts to reduce the reactivity and the power. (In one reactor, the EBR-I, thermal expansion acted to change the geometry of the fuel in such a way as to increase the reactivity—a destabilizing effect that, when noted, was eliminated by design changes.) Thus reactors can be, and are, designed to have a *negative temperature coefficient* of reactivity so that their temperature is inherently stable.

A *negative power coefficient* is also desirable in reactors, and it is usually obtainable with no special effort because of the large temperature drop between the central portion of a fuel element and the surface; an increase in power leads to an increase in the average fuel temperature and thus to a reduction in reactivity and power even though the reactor coolant inlet and outlet temperatures are held constant. However, there can be "too much of a good thing." In boiling-water reactors the negative power coefficient is very large, because an increase in power leads to an increase in the volume fraction of steam in the core, and this markedly reduces reactivity. The effect is sufficiently great that it requires special attention in the design of the control system so that the control rod positions are shifted as soon as a power change occurs in order to avoid large overshoots in the action of the controls.

System Stability and Controllability

Not only must the individual components of a power plant unit be stable themselves but also the integrated system must be stable, and this includes the control system. To simplify the system it is desirable to have the control actions roughly proportional to the demand signals. This requirement ordinarily leads to a much more complex mechanism than the simple "on-off" controls used for a household refrigerator, air conditioner, or furnace, which operate in systems with such large heat capacities that the on-off action occurs perhaps only once per hour.

Operation of a Single Unit

If a single power plant unit serves a building or a plant, a small addition to the electric load will lead to a drop in both the voltage and the frequency if no control action is taken. The drop in speed will give a small increase in torque and in generator current output, while the drop in voltage will reduce the electric power input to the previously connected set of loads in parallel, and that decrement in power will go to the new increment in system load. The system will stabilize at the new lower voltage and frequency unless control actions are taken to restore them to their former values. The basic relations are simple, and the interrelated electric load and power plant systems are stable.

Operation of Multiple Power Plant Units in Parallel

If two or more power plant units are operated in parallel, they must be synchronized so that they generate the same frequency and their control systems must function so that the electric load is shared between them with the load distribution desired.[3,11-14] The problems are far too complex to be treated here other than by citing the control specifications for individual power plant units as adopted by the North American Power Systems Interconnection Committee (NAPSIC) to assure compatibility between interconnected electric utility systems.[3,11] These specifications are summarized in Table 7.3, taken from Ref. 3. They are intended for all types of power plant, including hydraulic, steam, and gas-turbine generator units where applicable.

Several points in Table 7.3 are particularly important. Item *d* under the first performance need calls for linearity in response to perturbations in frequency. This would ensure that for nearly the full operating range of a turbine, its controls will act to produce a change in power that will be at a constant rate fixed between 10 and 67 percent per 1 percent change in frequency, depending on the unit. Note that changes in the system frequency are usually less than 0.05 percent and rarely as much as 0.1 percent.

Item *b* under this same performance need calls for a maximum dead-band width of 0.06 percent (that is, ±0.03 percent) to assure that units will not be too sluggish in response to drifts in the system frequency. Although not stated in Table 7.3, it is also important that the turbine and power plant unit controls not be subject to "hunting," i.e., power oscillations about the midpoint. Some overshoot in the reaction to a perturbation is acceptable, but the resulting oscillation should be highly damped and fade out in a few cycles.

To meet the above control requirements for steam turbines requires much more than operation of the throttle valve by the governor. As outlined in Table 7.2, the fuel, air, and feed-water flow rates must be varied, and they are ordinarily "slaved" to the steam flow rate with provisions for modulation by other parameters, such as an oxygen analyzer in the flue gas.

INSTRUMENTATION

Instrumentation requirements vary widely with the type of power plant from relatively little for hydro units, to modest amounts for diesels and gas turbines, to almost incredible amounts for coal-fired and nuclear steam plants. Only a few of the sensors are used for primary control functions; most are for diagnostic purposes or control of auxiliary systems such as feed-water clean-up. In all cases it is desirable to have linearity of response, freedom from drift in the calibration of sensors, and high reliability. For some applications, remote reading and perhaps strip-chart recording which favor electrical types of pickup are important. For others, simple thermometers, Bourdon tube pressure gages, or manometers may be satisfactory—and much less expensive in sensor cost, installation time, and maintenance. Some data are important for monitoring plant performance, and these are fed into a computer to provide an immediate output of such vital information as the heat rate or stack gas emissions. It is sometimes desirable to use a computer to store a large variety of diagnostic information. This storage capability is highly advantageous for new types of equipment, because it provides a fine reservoir of data that can be reviewed if difficult-to-diagnose problems arise.

Reliability of Instrumentation

The reliability and performance of instrument sensors and controls are best if the sensors are simple and rugged, and have a nearly linear response characteristic. Thermocouples are perhaps the best sensing elements available in these respects, yet even they are sensitive to installation conditions. Although they can be obtained in very small wire sizes, fine wires and thin electrical insulation are easily damaged; hence relatively large wire sizes are preferable. For example, if a sheathed thermocouple is to be employed, it is best to use a ~7-mm-OD ($\frac{1}{4}$-in-OD) sheathed unit if practicable, even though much smaller sizes are available. Extended operation at high temperature is likely to cause a drift in calibration, and there may also be some differences in calibration from one production lot to another. While more expensive, platinum resistance temperature detectors (RTDs) are less subject to drift in calibration. Most other types of sensor are more complex and hence more subject to problems that affect their reliability. So many types are available that no effort to treat them further will be made here.

TABLE 7.3   Suggested Performance Requirements[3]

Performance need	Response required
Frequency governing in normal operation	(a) Respond $+1.3\%$, -0.7% of unit nameplate megawatt rating in 2 s in prompt, stable fashion. (b) Maximum dead band of 0.06% frequency. (c) Overall steady-state regulation of 5%. (d) Linearity – per standards.
Normal daily load following	Able to go from 100 to 50% of nameplate MW rating at rates up to 2%/min over a 2-h period, stay at 50% load for 4 to 6 h and return to 100% in 2 h at rates up to 2%/min. Periods of zero response rate are permissible during loading and unloading while plant components are being added or removed. In addition, peaking units should be able to load or unload over 70% of nameplate load in 10 to 20 min.
Normal startup and shutdown	(a) Startup of base-load units following a brief shutdown in 2 to 4 h; 6 to 10 h following a more extended shutdown. Intermediate units at full load in 1 to 2 h from startup. Peaking units at full load in 30 min. (b) Shutdown rates same as startup are acceptable.
Tie-line backup	Rates of response as for daily load following, over spinning reserve range.
System emergency—off-nominal frequency	Ability of steam supply and auxiliaries to maintain operation at full load with off-normal frequencies for maximum permissible times as specified by turbine and generator manufacturers. A typical limit of 1% change from rated frequency for sustained operation has been cited. The permissible time of operation at greater frequency deviations decreases, until immediate trip is required for deviations of approximately 5% or 3 Hz. Rapid response under governor control from 100% to some lower value and return to 100% in 20 min. The larger the total possible excursion the better, but unit controls should be coordinated in a fashion so as to keep the unit on the line. A minimum generation of 70% under these abnormal conditions would be a desirable objective.
System emergency—off-nominal voltage	Capability for continuous operation at rated load at any terminal voltage within $\pm 5\%$ of rated. Capable of maintaining auxiliaries supporting load with auxiliary bus voltages in the range of 80 to 110% of normal voltage. Generator operation below 95% of rated voltage is possible with suitable reduction of load as defined by the generator manufacturer. Generator voltage may be restricted by volts per hertz considerations.
Unit emergency—total load rejection	Response is desired in a manner which would permit reloading to 100% power in 20 to 30 min after prompt resynchronizing, particularly of peaking and cycling plants.

Automatic Warning, Setback, and Shutdown Systems

Sensing elements may be coupled into a plant control system in several ways, depending on the seriousness of the indication. If a pressure or temperature exceeds normal limits by a modest amount, a circuit may be arranged so that the signal from the sensor not only continues to give a reading on an indicator but also can actuate a warning light, a buzzer, or a whistle. If the condition gets worse and the signal exceeds some critical level, a circuit can be arranged to take some automatic setback action to relieve the situation, e.g., reduce the firing rate to the boiler, or, in extreme cases, actually shut down the system by stopping the fuel flow or *scramming* the reactor (driving the control rods in at the maximum design rate for an abrupt shutdown). The latter type of action should not be programmed lightly, not only because of problems in handling the electric load but also because of possible damage to the power plant unit itself. An abrupt shutdown is likely to give serious thermal stresses and possibly local overheating or overcooling of some components. Thus, to minimize the need for such drastic action, systems should include adequate fluid reservoirs and standby equipment. This in turn places on the plant design personnel the responsibility for envisioning the various types of malfunction and failure that may occur, evolving a scenario for the probable course of events in each case, and modifying the system design to minimize the ill effects.

Coincidence Circuits

The relatively high incidence of malfunctions in instrumentation sensors coupled with the ill effects of abrupt shutdowns have led to the use of coincidence circuits for vital measurements. If, for example, it is decided that a fission reactor should be scrammed when the reactor coolant outlet temperature exceeds a certain level, such an action is too serious to be dependent on the indication of a single thermocouple. Three thermocouples can be hooked up in a coincidence circuit so that at least two of them must indicate an excessive temperature before the control system will scram the reactor.

Setback Actions

The automatic control actions taken in the event of an automatic setback should be designed to facilitate a resumption of the load. A load trip caused by lightning, for example, will probably mean a loss in the load for only seconds or at most a minute; hence the initial action should be only enough to protect the turbines from an overspeed. There may be no immediate cutback in the furnace firing rate. Steam may be simply dumped to the condenser for many seconds, giving the operator time to judge what further action should be taken.

SYSTEM DESIGN PHILOSOPHY

Different types of power plants vary so widely in the characteristics of their components that it is difficult to indicate the possibilities for improving and simplifying their control systems. At the very least, the engineers responsible for the system design should strive to get nearly linear response characteristics in each major component, simple and nearly linear relations between components, a minimum of requirements for control actions, and dependence where possible on simple, rugged sensors having roughly linear characteristics and adequate sensitivity, yet not subject to drift. Heat capacities, fluid inventories, fluid circuit transit times, etc., should be chosen to ease control problems and give the I&C engineers as much latitude as possible in their design work. This implies that I&C expertise should be brought to bear on the system design at an early stage so that the system can be simulated electronically in simplified form and major control problems defined during the conceptual design phase when changes in the system design can be made inexpensively.

Computers have proved to be enormously useful in the operation and control of power plants, so much so that some wonder why there is any need for a human operator. This question was answered beautifully by J. Bronowski in his book *The Origins of Knowledge and Imagination.* His thesis is that computing machines follow some one basic system of logic, and, if that system of logic encounters an inconsistency, it is helpless to change. Human beings, on the other hand, are not only better able to recognize an inconsistency, but are also able to throw out the whole set of axioms, postulates, and theorems with which they have been working and construct a completely new system of logic that is consistent with the realities with which they are working. Bronowski cites as a classic example Einstein's development of the theory of general relativity to cope with the inconsistencies between Newton's system and the physical world when these inconsistencies became evident. On the more mundane scale of electric power plants, not only does Murphy's law hold—i.e., "anything that can go wrong will go wrong" —but also we are unable to think of every possible contingency and combination of contingencies. Nor can we predict just how the power plant system itself will behave when some of these troubles arise. This means that we cannot program a computer accurately and that, just as airplanes need pilots, power plants need intelligent human operators.

PROBLEMS

1. A sodium-cooled fission reactor is to be operated at a constant mean reactor core temperature of 500°C. The temperature difference between the primary and secondary sodium circuits is designed to be 30°C and the temperature rise in each circuit is to be 200°C at full power, with about half of the temperature difference between circuits represented by the temperature drop in the tube wall. Neglecting the relatively small change with flow rate in the

forced-convection heat-transfer coefficient for sodium, plot curves for the temperatures of the primary and secondary sodium circuits at their respective inlets and outlets to the intermediate heat exchanger as a function of load for the following:

(a) Constant flow rates in both circuits
(b) Flow rates directly proportional to the reactor power output in both circuits
(c) Constant flow rate in the primary circuit and a flow rate directly proportional to power in the secondary circuit

2. Estimate the heat capacity of both the UO_2 in the fuel elements and the water inventory in the pressure vessel of the PWR of Fig. 5.28 and Table 11.4. Compare these values, and estimate the time required to raise their mean temperatures 100°C if all the reactor output at 100 percent power were to go into heating (a) just the UO_2 and (b) both the fuel elements and the water in the pressure vessel.

3. Estimate the mean transit time around the circuit for the air in a small closed-cycle gas-turbine system if the compressor pressure ratio is 3:1, the flow rate is 8 kg/s, the peak pressure is 10 atm, the volume of the low-pressure portion of the system is 3 m³, of which one-third is near room temperature and the balance at an average temperature of 300°C, and the volume of the high-pressure portion of the system is 1 m³ with an average temperature of 800°C.

4. A fluidized-bed combustion furnace operates at 3.5 atm and 800°C, has a bed depth of 3 m, has an average bed density of 1 g/cm³ when fluidized, and at full power operates with a superficial combustion gas velocity of 1.8 m/s leaving the bed. If the specific heat of the bed material is 0.5 kJ/kg and the excess air is 20 percent, estimate the heat release per square meter of bed cross-sectional area and the rate at which the bed temperature will rise at full power if the flow of coolant to the tube matrix in the bed is stopped abruptly so that no heat is removed from the bed.

5. Estimate the bed temperature overshoot above normal for Prob. 4 if the transit time through the coal feed and metering system is 10 s from the point where the coal flow can be cut off, assuming further that the average time for combustion of a coal particle in the bed is 10 s, and that the control of the coal feed flow depends on sensing the bed temperature with a dead-band width of ± 10°C relative to the design value for the bed operating temperature.

6. Count the number of sensors called for in the flowsheet of Fig. 7.10 and estimate the installed cost of this instrumentation, assuming that for temperature-measuring equipment it would be $200 per sensor and that for pressure indication it would be $200 per sensor for Bourdon tube gages mounted on the pipeline. Sensors are indicated by a circle or rectangle at the dead end of a side leg. The letters P and T inside these symbols indicate pressure or temperature, and the following number identifies that particular sensor.

REFERENCES

1. "World News Beat," *Electrical World,* Jan. 1, 1980, p. 14.

2. Yao, N. P., and J. R. Birk: "Battery Energy Storage for Utility Load Leveling and Electric Vehicles: A Review of Advanced Secondary Batteries," *Proceedings of the Tenth IECEC,* August 1975, pp. 1107–1119.

3. Ewart, D. N.: "Power Response Requirements for Electric Utility Generating Units," *Proceedings of American Power Conference,* 1978, vol. 40, pp. 1139–1150.

4. *Public Utilities Fortnightly,* June 27, 1976, p. 28.

5. Brown, R. C., and D. A. Harris: "Large Coal-Fired Cycling Units," *ASME-IEEE-ASCE Joint Power Generation Conference,* Sept. 28–Oct. 2, 1975.

6. Gabriel, D. S., and L. R. Carman: "The Performance of a Single-Stage Impulse Turbine Having an 11.0-in Pitch-Line Diameter Wheel with Cast Airfoil Shaped and Bent Sheet Metal Nozzle Blades," NACA-ARR-E5H31, September 1945.

7. Fraas, A. P., and M. N. Ozisik: "Steam Generators for High-Temperature Gas-Cooled Reactors," USAEC Report ORNL-3208, Oak Ridge National Laboratory, Apr. 8, 1963.

8. "The Bull Run Steam Plant," Tennessee Valley Authority Technical Report No. 38, 1967.

9. *P & WA Turbojet Power Pac, 122 MW,* Pratt and Whitney Aircraft Bulletin.

10. Fraas, A. P., and M. N. Ozisik: *Heat Exchanger Design,* John Wiley & Sons, Inc., New York, 1965.

11. North American Power Systems Interconnection Committee (NAPSIC): "Minimum Criteria for Operating Reliability," Feb. 25, 1970; also NAPSIC "Minimum Performance Criteria—A Supplement to the Operating Manual," June 1, 1972.

12. "Deep Freeze Forces Rotating Blackouts in Six States—First Ever in Winter," *Electrical Week,* Jan. 24, 1977, pp. 1–4.

13. Davidson, D. R., D. N. Ewart, and L. K. Kirchmayer: "Long Term Dynamic Response of Power Systems—An Analysis of Major Disturbances," IEEE Paper No. T74 519-5, *IEEE Trans.,* vol. PAS-94, no. 3, May/June 1975, pp. 819–826.

14. Joint ASME-AIEE Committee: "Recommended Specification for Speed-Governing of Steam Turbines Intended to Drive Electric Generators Rated 500 kW and Larger," IEEE Standard 122 (AIEE No. 600), December 1959.

8

RELIABILITY
AND SERVICE LIFE

The greater part of an engineer's formal education is concerned with design for performance. As a result, many engineers can, on a purely analytical basis, design turbines, pumps, blowers, and heat exchangers, and predict their performance within 10 to 30 percent. To design for a long life with a high degree of reliability is vastly more difficult. A traditional approach has been to "design, build, and bust," and repeat the process. Years of engineering test experience, coupled with that intangible "engineering judgment," has seemed to be the best—if not the only—basis for this aspect of design. Some indication of how difficult the problems are is given by the fact that, to the writer's knowledge, in only two instances during the past 50 years have aircraft engines of a new design been put on the test stand and run through a 150-h endurance test successfully in the first attempt. Large steam turbines and boilers present still more difficult problems, because they are so big that they cannot be tested by the manufacturer before shipment. Thus their design must be much more conservative and, wherever possible, represent a minimum extrapolation of extensive field-operating experience. Although the penalties for a forced outage in a stationary power plant are less serious than those that would result from such an outage in an airplane, they still entail formidable costs. Not only are the capital charges on the idle equipment high, but also, if the utility must purchase power from another source in order to accommodate its load, the cost of the purchased power normally entails a substantial premium. As a consequence, for a 1000-MWe unit, the cost of an outage is commonly from \$250,000 to \$500,000 per day, depending on local fuel and other charges.[1] For the usual incidence of forced outages, an improvement in the availability of the turbine-generator unit by 1 percent was estimated in 1979 to be worth from 11 to 21 million dollars in the initial capital investment!

Shortcomings in reliability can have even greater repercussions. The severe system overload that led to the major blackout of New York City in the summer of 1977 resulted not only in widespread disruption of services but also in extensive looting, running the cost of the blackout to over a billion dollars. Similarly, the combination of equipment failures and human errors that produced the Three Mile Island nuclear plant accident has entailed not only an enormous direct cost to the utility operating the plant, but also a probably far greater cost to the nation as a consequence of the public reaction. Clearly, it is particularly important to have the highest possible reliability in each and every component (and in the personnel) of a nuclear plant because of both public safety considerations and the high costs of repair where nuclear radiation may be a problem.

TERMINOLOGY

A number of special terms are commonly used in treating reliability problems in central station plants.[2] Of first importance is the amount of time that the unit is on the line producing power per year, commonly referred to as *service hours*. The time during which a unit is shut down because the load on the system is much less than the system capacity is referred to as *reserve shutdown hours*. The sum of these two quantities is the number of hours per year that the unit is "available." The *availability* of the unit, usually expressed as the *availability factor*, or the percentage of time that the unit is available, is obtained by dividing the total available time in the course of a year by 8760, the number of hours in a year.

Plants are commonly taken out of service once a year for regularly scheduled maintenance; these periods are considered *planned outage hours*. In addition, during the course of a year it may become evident that parts need to be replaced, but this is almost always detected early enough that the parts in question are able to continue functioning until a period of low power demand when the plant may be shut down more conveniently. The time spent in this maintenance work is referred to as *maintenance outage hours*. The most troublesome problems result from *forced outages*, which occur when a unit develops prob-

TABLE 8.1   Average Operating Experience for Typical Fossil-Fuel-Steam Units [2]

	Unit-year average	
	Hours	Days
Service hours *(SH)*	6387	266
Reserve shutdown hours *(RSH)*	1563	65
Available hours *(AH)*	7950	
Maintenance outage hours *(MOH)*	198	
Planned-outage hours *(POH)*	376	
Scheduled-outage hours *(SOH)*	574	24
Forced-outage hours *(FOH)*	192	8
Period hours *(PH)*	8716*	363
No. of forced outages	4.42	

*Less than 8760 because some units less than a year
Source: Edison Electric Institute report 69–33

TABLE 8.2   Average Operating Experience of Typical Gas-Turbine Units [2]

	Unit-year average, h
Service hours	641
Reserve shutdown hours	5763
Available hours	6404
Maintenance outage hours	169
Planned-outage hours	117
Scheduled-outage hours	286
Forced-outage hours	205
Period hours	6895
No. of forced outages	3.87

Source: Edison Electric Institute report 69–33

lems so serious that it must be shut down in spite of the fact that its output is needed under the prevailing load conditions. The time spent per year in forced outages is alluded to as *forced outage hours,* and this parameter divided by the sum of the service and forced outage hours is called the *forced outage rate.* The incidence of forced outages is also an important consideration, and it is referred to as *mean time between forced outages,* or MTBFO. It is determined by dividing the number of hours per year that the unit is in service by the number of forced outages. A similar term is widely used for failure rates of components in general industrial service; this is the *mean time between failures,* or MTBF.

Another term related to reliability problems is the *capacity factor,* the ratio of the total power produced by a unit per year divided by the amount of power that it would have produced if operated throughout the year at full capacity. The capacity factor is of consequence because generally the stresses on a piece of equipment are lower at reduced capacity, a matter of particular importance for electric motors, generators, and gas turbines.

Another factor affecting reliability is the number of starts and stops. The thermal cycles associated with starting and stopping a unit are often of greater consequence than the number of hours of operation in some types of equipment, such as gas turbines used for peaking-power operation.

Typical values for the above parameters differ widely with the type of service. A representative set of data for a fossil-fuel-fired steam plant employed for base-load operation is given in Table 8.1 while similar data for peaking-power gas-turbine units

are given in Table 8.2. In spite of determined efforts to improve plant reliability, increasing environmental restrictions have made necessary progressively more encumbering front and back end equipment that reduces the overall plant reliability[3, 4] (Fig. 8.1 and Table 8.3).

TYPES OF FAILURE

There are many types of equipment in a power plant, any one of which may fail and induce a forced outage. For example, difficulties in the Three Mile Island nuclear plant accident began with the failure of two water-circulating pumps, and in the great northeast blackout of 1965 there were a number of bearing failures in big steam turbines because the electric motors driving the lubricating-oil pumps lost their power supply. This section outlines the principal types of failure in order to give some insight into the failure modes that may be expected in power plant systems.

Ruptures

Perhaps the first possibility brought to mind by the term *failure* is a burst type of rupture in a pressure vessel. Failures of this type were common during the early days of steam power. In the years from 1862 to 1879 there were over 10,000 boiler explosions in England, and in the 40 years from 1880 to 1919 there were 14,281 boiler explosions in the United States, resulting in over 10,000 deaths and 17,000 injuries.[5] These terrible tragedies led to the development of the ASME pressure-vessel code, which requires conservative design along with the use of a construction material having a good ductility in order that serious local stresses can be relieved by yielding. This code, coupled with good quality control in the course of manufacture and the use of reliable safety valves, has made burst-type failures of boilers and other pressure vessels very rare, although some do occasionally occur. For example, in Pennsylvania, a 12-in main steam line operating at about 800°F burst in August 1977, killing six men.

An investigation disclosed that the failure was caused by graphitization and complete loss of ductility in the weld zone adjacent to a valve after 33 years of service, including 220,000 h of operation and 1830 startups.[6] Graphitization is an insidious phenomenon that is initiated in weld zones in carbon and carbon-moly steel. It develops slowly with aging at high temperatures (above 468°C, or 875°F, for carbon-moly steel), occasionally leading to severe embrittlement.[7] Nondestructive testing techniques are available for checking susceptible regions in old equipment, but the importance of such tests had not been appreciated by the utility involved.[6]

The consequences of a failure in the pressure vessels of a nuclear reactor are so serious that particularly stringent requirements are imposed on their design and quality control, or "quality assurance" (QA). In addition, periodic inspections are required using ultrasonic techniques to detect the development of any small cracks. These techniques are not only sensitive but are additionally attractive because they can be employed with remotely operated equipment after the pressure vessels have become radioactive.

A truly impressive series of tests was carried out to investigate the efficacy of the above precautions.[8,9] A series of small and then progressively larger model pressure vessels were tested to investigate the failure mode of typical vessels and the regions in which they would be most sensitive to small defects and cracks. These tests showed that, as expected, an axial crack in the outside of the vessel near the midline was the most serious type of defect. It happens that the ductility of steel plate falls off with increasing thickness because of the effects on the grain structure of the slower cooling rates in the thicker sections. Further tests were therefore carried out with quarter-scale pressure vessels having 15-cm-thick (6-in-thick) walls. As discussed in Chap. 6, concerning materials, the ductility of steels commonly falls with a reduction in temperature. These effects combine to give very

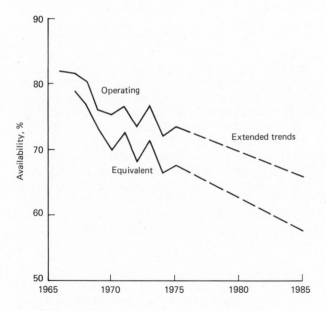

Figure 8.1 Actual operating and equivalent availability of units rated 400 MW and higher, with extended trends of these factors. Equivalent availability is operating availability with allowances for partial outages.[3]

TABLE 8.3 Forced-Outage Rate, Availability Factor, and Capacity Factor for All U.S. Nuclear Units and Fossil Steam Units of 600 MWe and Larger[4]

	1972	1973	1974	1975	1976
Forced-outage rate, %					
All nuclear units	12	18	17	15	13
PWR units	10	21	18	13	13
BWR units	12	10	13	18	13
Fossil units	21	18	18	14	—
Availability factor, %					
All nuclear units	73	70	69	72	69
PWR units	74	66	66	76	71
BWR units	73	75	72	68	68
Fossil units	68	75	73	73	—
Capacity factor, %					
All nuclear units	61	58	57	60	59
PWR units	60	55	57	67	62
BWR units	62	62	58	50	64
Fossil units	57	62	54	54	—

Source of data on nuclear units: For 1974–1976, NRC Grey Books (1/75, 1/76, 1/77); for 1972–1973, AEC Reports 1/74 #OOE-ES-001, 5/74 #OOE-ES-004, and 12/74 –OOE-ES-004 and ERDA Report –76/125

Source of data on fossil units: FEA Data Base and EEI Equipment Availability Report

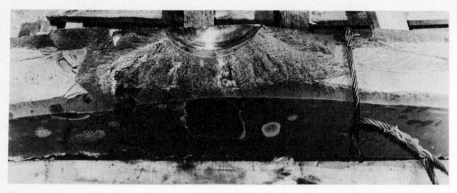

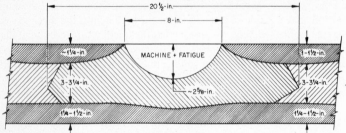

Figure 8.2 Close-up of the fracture surface in the vicinity of the machined defect and fatigue-induced crack in a quarter-scale steel pressure vessel that was pressurized to failure. The fatigue-induced crack represents about 25 percent of the depth of the defect.[8,9]

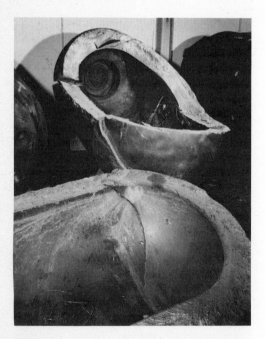

Figure 8.3 View of the two sections of a quarter-scale pressure vessel after pressurizing to failure at 0°C. The extensive cracking is characteristic of a brittle failure.[8,9]

little ductility in thick sections of pressure-vessel steel at temperatures around 0°C; hence a series of tests was run at temperatures ranging from 0°C (32°F) to 88°C (190°F). The quarter-scale pressure vessels were fabricated to the full nuclear pressure-vessel standards, and then a severe crack was simulated by cutting a Woodruff key type of axial slot with a slitting wheel in the outer surface of the pressure vessel at the midplane. A fitting resembling a vacuum-cleaner nozzle was clamped over this slot so that the slot could be pressurized with hydraulic oil to 1350 atm (20,000 psi) and the pressure relieved every few seconds for thousands of cycles in order to produce a fatigue crack that progressed into the vessel from the base of the slot. The progress of the fatigue crack during this cycling operation was followed with ultrasonic equipment until the crack had progressed approximately halfway through the vessel. The vessel was then installed in a pit and pressure-tested hydraulically until it failed. A close-up photo of the simulated crack after the test in a typical case is shown in Fig. 8.2, in which the textures of the surfaces of the initial cut with the slitting wheel, the fatigue crack induced by pressure cycling with hydraulic oil, and the fracture surface in the final rupture are easily distinguished. Figure 8.3 shows the way in which the vessel tested at 0°C fractured into two major pieces as a consequence of the brittle type of failure

Figure 8.4 Quarter-scale pressure vessel after testing to failure at 88°C (190°F). The crack propagation was limited by the ductility and toughness of the steel.[8,9]

that occurred. Figure 8.4 shows the fracture in a vessel tested at 88°C (190°F), demonstrating the important effects of ductility and toughness on the failure mode. In this instance, the crack progressed only a short distance before it was arrested, and opened only enough to relieve the high internal pressure. The most significant result of the tests is that, irrespective of the test temperature, none of the vessels failed at a pressure less than double the design operating pressure in spite of the extreme severity of the cracks, which were far larger than cracks that can be readily detected with the ultrasonic techniques available.

Another catastrophic burst type of failure is that of large dams used for hydroelectric and water-storage projects. As with boilers, the incidence of failures dropped rapidly as a consequence of improved engineering from about 1815 to about 1930, but there has been little further improvement[10] (Fig. 8.5). About a third of these failures since 1930 have stemmed from severe floods which exceeded the spillway capacity, but about a quarter resulted from foundation problems caused by insufficiencies in design, and the balance stemmed from other design problems or improper construction, including use of inferior materials.[11,12] As one might expect, the latter two categories are likely to lead to failure in the first year or two after completion of construction, and this has indeed been found to be the case[10] (Fig. 8.6).

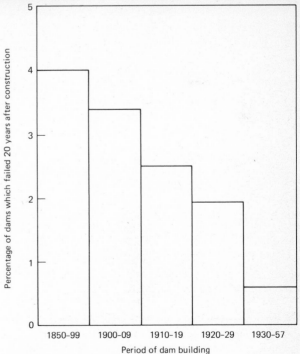

Figure 8.5 Failures of large dams within 20 years of construction.[10]

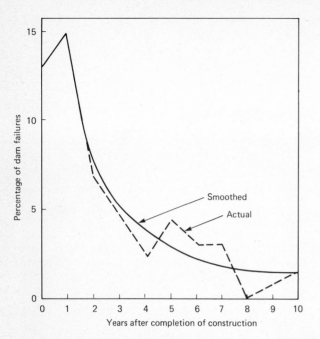

Figure 8.6   Approximate timing of dam failures.[10]

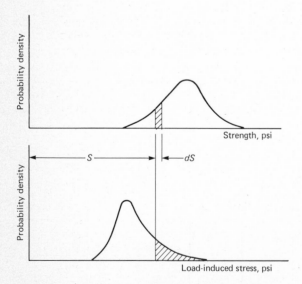

Figure 8.7   The shaded rectangle in the upper diagram is the geometric representation of the probability of encountering a material strength between S and $S + dS$. The shaded area in the lower diagram is the geometric representation of the probability of encountering a load-induced stress greater than S in magnitude.[13]

Cracks and Fractures

Cracks and fractures sometimes occur in metal components as a consequence of fatigue stresses or thermal-strain cycling. They commonly begin at stress concentrations caused by defects in the material or poor design. Equipment designs are usually conservative enough that such failures occur only occasionally as a consequence of unfavorable combinations of the effects of variability in both the strength of parts and the loads imposed. If a large number of seemingly identical parts are tested, they will be found to fail at substantially different loads, as indicated by the upper curve of Fig. 8.7 or, if fatigue-tested, they will last for substantially different times. Further, the loads that metal parts withstand in actual service vary as indicated by the lower curve of Fig. 8.7. Although an electric motor driving a fan is subject to a predictable load, in a coal-handling system both the motor and the various elements in the mechanism may be subject to widely varying loads as a consequence of changes in the size and hardness, etc., of the lumps of coal and rock that pass through the system. This is one of the reasons why the coal- and ash-handling systems require a large fraction of the total maintenance effort of a coal-fired power plant. As indicated in Fig. 8.7, the combined effects of variations in the loading from one component to another and variations in material strength lead to occasional failures in parts that are less strong than the average and are subjected to higher than normal loads.[13] For some applications, such as motor-driven fans, it is possible to obtain a high degree of reliability by a modest amount of overdesign. For others, however, such as coal-handling equipment that may ingest a large hard rock, the extra cost of overdesign becomes excessive, and so the lowest overall costs are obtained by replacing overloaded parts when necessary.[14] This of course applies only when the failure of a part could not have catastrophic or even severe effects. A machine element in the coal-handling equipment of a coal yard is a very different matter from a nuclear reactor pressure vessel. A somewhat similar situation exists in steam generators, in which thermal stresses vary widely with both the fuel being fired and the rate at which the load may be changed. There are so many possible combinations of these conditions that there is no way to assure that thermal-strain cycling will not cause cracks under unforeseen combinations of unfavorable conditions. However, cautious design and proper selection of materials can assure that if a crack develops, it will progress slowly and form a leak rather than result in a burst type of rupture. In a steam plant the loss of water from the leak will serve as an advance warning of more serious trouble.

One of the most common problems is development of cracks, usually from thermal-strain cycling, which lead to leaks. These cracks occur most often in welds, and it is for this reason that great attention is given to weld quality control in order to assure a high degree of integrity, particularly in piping joints.

TABLE 8.4   Observed Working Fluid Leakage Rate from Typical Systems

Plant	Leakage rate, %/day
Coal-fired steam plant (3500-psi, 950-MW TVA Bull Run)	0.5
Pressurized-water reactor (160-MW Yankee)	0.2*
Boiling-water reactor (52-MW Humboldt Bay)	0.2*
Helium-cooled reactor (50-MW Peach Bottom)	0.2
Reactor containment shell (normal test requirement)	0.1
Boiling heavy-water reactor (200-MW Marviken)	0.00055
Molten-salt reactor (8-MW MSRE)	0.000024
Boiling-potassium power plant (0.4-MW ORNL-LPS)	0.000024

*Most of this leakage is caught and recycled.

Fluid System Leakage

For every instance in which a part cracks or breaks there must be at least 100 instances in which a gasketed flanged joint, a shaft seal, a valve bonnet, a tube fitting, or some such joint develops a troublesome leak. Though not spectacular, such leaks not only can be exasperating and terribly time-consuming, but they also can be dangerous if the leaking fluid is radioactive, constitutes a fire or explosion hazard, is highly corrosive, or is expensive (e.g., helium or heavy water). Thus having a system that is reliably leak-tight is often not simply important but absolutely vital to the feasibility of a power plant concept.

Typical Degrees of Leak-Tightness

A good degree of leak-tightness is important in the condenser of a steam plant, particularly if cooled by seawater. A still higher degree of leak-tightness is required in a nuclear plant where fluids may be radioactive. In appraising these problems, it is instructive to examine Table 8.4, which summarizes data obtained in the course of operation for a wide variety of typical systems. Note that in a conventional steam plant, where the incentive to reduce leakage is not very high because the cost of treating makeup water is low, the leakage runs around 0.5 percent per day. Somewhat lower leakage rates have been obtained in boiling-water and pressurized-water reactor systems and in the Peach Bottom helium-cooled reactor. The specifications normally imposed on reactor containment shells are met routinely; these require a leakage rate of less than 0.1 percent per day.

In heavy-water reactors the high cost of heavy water ($250/kg in 1979) has led to much more stringent efforts to obtain a leak-tight system; hence the very low leakage rate for the third-last item of Table 8.4. Because of rigorous attention to details in the design and construction, a 160-MWe conventional power plant in Sweden demonstrated that it is possible to build and operate a boiling heavy-water reactor system with a leakage rate of only 0.2 percent per year, or 0.00055 percent per day.[15] A report on it indicated that 98 percent of this small leakage would be through the large shaft seals in the turbines and feed pumps and through the threaded joints for instrument sensors.[16] Most of the balance of the fluid losses would stem from flushing operations during the annual shutdown.

Experience with liquid-metal and molten-salt systems indicates that it is possible to reduce the leakage of the working fluid to essentially zero (i.e., less than the limit of detection, which in a run of 3000 h or more is about 10^{-6} percent per hour). The next to the last item of Table 8.4, the molten-salt reactor experiment, was operated for over 25,000 h with no trace of a salt leak from the operating system. The last item of Table 8.4 is for a small liquid-metal system—in this case a potassium system designed to simulate a 400-kWt space power plant operating on a potassium vapor Rankine cycle similar to that of the topping cycle in Fig. 15.1.

System Costs as a Function of Leak-Tightness

Some indications of the incremental cost of attaining a high degree of leak-tightness in a system are given by the author's experience with high-temperature reactor systems. In general, it appears that the cost of fabricating an all-welded, hermetically sealed, stainless-steel system for use with liquid metals or molten salts at high temperatures runs about three times as much as building a geometrically similar system using conventional plain carbon-steel piping and flanged joints. Another indication is given by experience with containment vessels. Experience indicates that conventional commercial practice yields vessels that with their flanged closures will leak about 1 percent per day. At an incremental cost of about 10 percent, it is possible to obtain a factor of 10 increase in leak-tightness through better detail design and quality-control procedures.

Design for Leak-Tightness

A high degree of leak-tightness cannot be maintained in flanged joints in high-temperature piping systems because thermal cycles will result in relaxation of the metal in local, highly stressed regions, the joints will loosen up, and trace leaks will develop. To avoid difficulties of this sort, all joints should be welded with high-quality heliarc or tungsten–inert gas (TIG) welds. The number of flanged joints should be kept to a minimum, and they should be used only in relatively cool zones, i.e., where temperatures are below 150°C (300°F) and high-class ASME oval ring metal gaskets can be employed. In general, tube fittings in instrument and gas pressurizing lines should be avoided, although if the incentive is worth the risk, it is possible to employ Swagelok (or equivalent) fittings in low-temperature zones with a fair probability that they will yield an adequate degree of leak-tightness.

Where possible, welds should be designed so that they can be made in the shop; the number of field welds should be kept to an absolute minimum because of difficulties in maintaining good quality control. Although elaborate formalized welding and inspection procedures have been worked out, there is absolutely no good substitute for having conscientious, highly skilled welders and inspectors with extensive experience.

Provision for leak testing during the shakedown operations should be included in the design. One requirement in the leak-test work should be that the system be evacuated and that the pressure rise in the system be no more than a few microns when the evacuated system is left sealed overnight. A variety of leak-testing equipment can be used if the vacuum test reveals the presence of a leak. Of these, helium leak detectors appear to be the most effective means of finding trace leaks.[17]

The system design should also include provision for detection of any leaks that may develop during system operation. Oxygen leakage into sodium or NaK can be detected with a plug indicator.[18] This device is sufficiently sensitive so that a buildup of oxygen in sodium or NaK can be detected long before any serious increase in corrosion rate will occur. This, of course, assumes that the leak will stem from a low-cycle fatigue crack or material defect in a moderately stressed part and hence will develop slowly. Similarly, a leak of liquid metal into the atmosphere surrounding the liquid-metal system can be detected with instruments sensitive to the presence of alkali metal oxides, although in practice this has been difficult to accomplish. A leak of a liquid metal into a steam system, or of water or steam into a liquid-metal system, can be detected by providing a gas trap with a nickel window kept at a sufficiently high temperature so that hydrogen will diffuse through the window into an evacuated zone where it can be sensed with a vacuum gage.[19]

It is also important to include some means for purifying the fluid in the system to clean up any contamination that may enter either during the charging operation or as a consequence of a leak. This may be accomplished with cold traps, hot traps, or a distillation system.

Wear

Wear of rubbing parts always limits their useful life. If fully developed hydrodynamic lubrication conditions prevail and no dirt is present in the lubricant, there should be no wear. Boundary lubrication conditions are usually unavoidable, however, during starting and stopping of the equipment, and so most of the wear occurs at these times. Once a surface is scuffed, its load capacity is reduced and further deterioration is likely to occur. If it is not possible to design for fully developed hydrodynamic lubrication conditions, or if dirt is unavoidable, as in the case of coal- or ash-handling equipment, wear rates can be reduced by the proper choice of lubricant, rubbing surface materials, operating temperature, use of seals to keep out dirt, and proper maintenance with adequate lubrication. However, for the more difficult applications, some wear is unavoidable, and hence the best solution is to replace parts at proper intervals.

Wear is a mild form of abrasion. Coal and ash sliding down chutes or carried pneumatically through pipes can abrade away a pipe wall in a few hours. The problems are too specialized to treat here, but one set of these problems—turbine blade erosion—is treated in Chap. 12. That section includes some basic data on erosion.

Corrosion

As discussed at some length in Chap. 6, concerning materials limitations, corrosion is particularly likely to give problems under humid or wet conditions or in combustion atmospheres combining dirty fuels with metal temperatures above about 300°C. Corrosion rates can be reduced through the use of more expensive materials, but a balance between outage and maintenance costs on the one hand and materials costs on the other usually favors moderately expensive, rather than very expensive, materials. The usual procedure is to operate until an inspection during a routine shutdown indicates that part replacement is in order or until a failure occurs.

Deposits

Deposits in the form of ash, sludge, or scale often limit the operating time possible without maintenance. Deterioration of lubricants because of high-temperature conditions may lead to the formation of deposits that are likely to limit the operating life of machines such as internal combustion engines in which deposits in piston-ring grooves, on valve stems, and the like, gradually build up and lead to a deterioration in performance, a malfunction, or an actual failure of parts.

Ash deposits on tubes in steam generators are particularly troublesome and often unpredictable. The ash may deposit in a fairly loose form, in which case it can be removed by soot blowers, or it may be fairly adherent (Fig. 5.17). In some instances a glassy slag forms that is so adherent that it can be removed only by heavy blows with a sledge hammer. Difficulties of this sort can be avoided through the use of a better and more expensive fuel, if such a fuel is available.

Electrical Failures

Perhaps the most common source of trouble is the failure of some element in the electric circuitry, and this may come about as a consequence of a short, arcing, or an open circuit. The results of the failure vary widely with the particular type of equipment, but there are so many components in the instrumentation and control circuitry that failures of one sort or another are one of the most common causes of power plant forced outages. Fortunately, electric-equipment components are usually readily accessible and easily replaced, and so the duration of the outage is ordinarily short.

Failures in electric equipment are induced by electric stresses (overvoltage), reduced resistivity of insulators caused by above-normal temperature and/or humidity, and corrosion. Solid-state electronic components are particularly sensitive and for this reason are commonly kept in clean, air-conditioned rooms with controlled humidity.

GENERIC DATA ON FAILURE RATES

Before one can attempt to assess the reliability of a system, one must have generic data on the reliability of the principal components. Unfortunately, the data available are limited because neither component manufacturers nor power plant operators care to publicize their problems. The best and most extensive data are from the aerospace and nuclear power plant fields where the combined effects of the public interest and the serious consequences of a failure have led to the accumulation of vast amounts of data. Table 8.5 summarizes a set of these data for

TABLE 8.5 Failure Rates for Typical High-Quality Components Operated under Favorable Conditions with Maintenance in the Aerospace and Nuclear Power Fields

Component	Failure rates (failures/10^6 h)
Electric motors	1.0
Controllable valves	1.7
Blowers	2.4
Pumps	2.7
Generators	0.9
Circuit breakers	0.3
Fuses	0.5
Relays	0.5*
Switches	0.5*
Electric instruments (voltmeters, etc.)	1.3
Pressure sensor	3.5
Temperature sensor	1.5
Wire-wound resistors	0.5
Condensers	0.2
SCRs	1.0
Transistors	0.1

*Assuming 1 cycle per hour.

the principal components used in auxiliary systems in power plants. Note that the failure rates are given in terms of the number of failures per million hours of component operation. These data are for well developed, high-quality components operated under favorable conditions with appropriate maintenance. Note that the failure rates generally run roughly one per million hours of operation. More detailed data on components are given in Table A8.1 of the Appendix.

DETAILED ANALYSIS OF STEAM-TURBINE PLANT RELIABILITY

As pointed out in the beginning of the chapter, there are strong incentives to improve the reliability and availability of steam power plants, and the major component manufacturers have put forth great efforts to achieve the relatively high levels of reliability that have been obtained. A detailed insight into these problems is important not only as a means toward improving the reliability of existing equipment, but also as a means for assessing the reliability to be expected of advanced energy systems for which few data are available. Good data are hard to find, in part because operators are more concerned with keeping their equipment running than with compiling the sort of detailed records required to make a sound statistical analysis of failures. Further, not many people have a clear idea of how best to handle the data available. As a consequence, it seems desirable at this point to include a detailed analysis that gives a good insight into both the reliability problems of steam turbines and methods for handling failure data from plant operations.

An excellent example is provided by a set of data assembled by Millard Myers of ORNL on the operating experience with 14 turbine-generator units having capacities from 10 to 25 MW. These units were installed in 1944 and were in operation up to 1962 in a power plant at Oak Ridge. In attempting to analyze these data, the author found it difficult to see how best to organize the information. It was decided to ignore all except forced shutdowns and to separate these into two groups: those chargeable to the boiler and those chargeable to the turbine generator.[20] Further, inasmuch as 7 of the 14 turbine-generator units were definitely less subject to trouble than the others, only

these 7 best units were considered. Because there was a higher incidence of forced outages in the early years of the power plant, it seemed advisable to plot the incidence of outages as a function of the year in which they occurred. The resulting graph showed that the shakedown period for the plant lasted for 4 or 5 years, by which time the incidence of outages reached a minimum that lasted for 7 years. The effects of wear and tear then led to a gradual increase in the incidence of outages. The balance of the study was then concentrated on the 7 best years of the lives of the 7 best units of the plant—seemingly as auspicious a basis as could be chosen.

The first step was to tabulate the number of forced outages per 500-h period as a function of the number of operating hours after the previous shutdown. The forced outages associated with the turbine-generator unit and its related equipment were separated from those chargeable to the boiler. In keeping with the usual procedure followed in probability studies of this sort, the data for forced outages of the turbines were plotted on semi-logarithmic coordinates to give Fig. 8.8. Where the number of outages per 500-h operating period averaged less than one, the time interval was increased, and, in plotting the points, the number of outages was divided by the number of 500-h operating periods and the point plotted at the middle of the time interval involved. As can be seen in Fig. 8.8, the scatter band was not very wide, and it defined rather well a straight line fitting the conventional probability relationship:

$$N = Ae^{-Bt}$$

where N = number of outages per unit of operating time divided by the total number of outages considered
A = a constant which determines the value of N at $t = 0$
B = a constant defining the slope of the curve
t = operating time

Since the total number of outages must be equal to the sum of the outages in the various 500-h time intervals, the integral under the curve must be equal to unity. Thus we may write

$$\sum_{t=0}^{t=\alpha} N = 1 = \int_{t=0}^{t=\alpha} Ae^{-Bt}\, dt = \frac{A}{B}$$

It is evident that A must equal B, and hence the slope of the line in Fig. 8.8 must be related to the left-hand intercept.

Since the data plotted were in terms of the number of outages per 500-h interval of operation, for further analysis it was easier to express t in terms of the number, n, of 500-h operating periods. Several values for the constant A were tried and that appearing to give the best fit was chosen. The equation for the curve of Fig. 8.8 then became

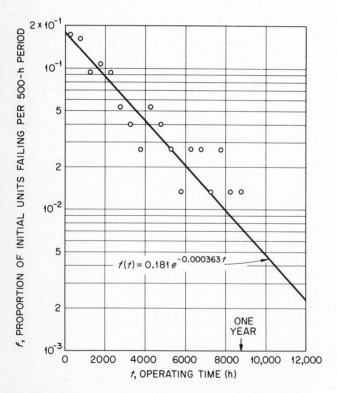

Figure 8.8 Incidence of forced shutdowns of seven turbine-generator units in the power plant of the Oak Ridge Gaseous Diffusion Plant during the 1950–1955 period. There were 75 forced outages during a total of 175,000 unit operating hours.[20]

$$N = 0.181e^{-0.181n}$$

where N = number of outages per 500-h period divided by the total number of outages

n = number of 500-h periods up to and including the time interval in question

If the preliminary form of the equation is used with the time expressed in hours, the equation that best fits the data of Fig. 8.8 becomes

$$N = 0.000363e^{-0.000363t}$$

where t is in hours.

The average probability of a unit operating for any given length of time is given by the number of units started up minus the number of units forced out of operation divided by the number of units started up. Since the number of units forced out of operation is the integral under a curve such as that of Fig. 8.8, the probability P that a unit will operate successfully up to any given time t is given by

$$P = 1 - \int_{t=0}^{t=t} Ae^{-At}dt = e^{-At}$$

For the case cited in Fig. 8.8, where t is hours, the probability P that a unit will operate satisfactorily up to a time t becomes

$$P = e^{-0.000363t}$$

This equation has been plotted in Fig. 8.9 to show graphically the probability of successful operation as a function of operating time. It is striking to note from Fig. 8.9 that only about 5 units out of 100 could operate for a year without a forced outage.

The surprisingly high incidence of forced outages in these turbines led to a similar analysis of forced-outage data for TVA equipment. A study of experience with 60 large turbine-generator units showed similar results. Further study showed that the reliability of over 100 hydro units was only about twice as good as that of the steam-turbine–generator units.

The log books were then reviewed in an effort to categorize the different types of failure. It was found that the electric equipment was the biggest single source of difficulty, brush troubles in the exciter being particularly frequent. Sticking in the governor and/or the throttle mechanism with consequent tendency of the unit to overspeed when the load was reduced was the next most common cause of forced outages. These two items, in fact, accounted for about half of the forced outages chargeable to the steam-turbine–generator units. The balance of the troubles were mainly in the auxiliaries.

Unfortunately, statistical studies of this sort are very tedious and time-consuming, and much qualifying information is re-

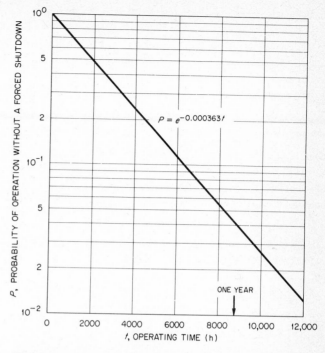

Figure 8.9 Probability of operation without a forced shutdown as calculated from data of Fig. 8.8.[20]

quired to make them meaningful. Yet the low reliability indicated by Fig. 8.9 for conventional steam-turbine–generator units indicates that far more work of this sort is badly needed.

Effects of Steam-Turbine Inlet Temperature on Availability

Extensive data compiled by the General Electric Company[21,22] on the availability and forced-outage rates for high-pressure, high-temperature steam turbines have been summarized in Fig. 8.10. These data show one of the key reasons for limiting the turbine inlet temperature in steam plants to the 538 to 565°C (1000 to 1050°F) temperature range.

Tube Leaks in Coal-Fired Steam Generators

Tube leaks are the most common cause of forced outages chargeable to steam generators, particularly in the larger units. To investigate this problem, in the early 1970s the author collected data on the incidence of tube leaks experienced in about 60 coal-fired steam generators operated in U.S. central stations. Seventeen different models by several vendors were repre-

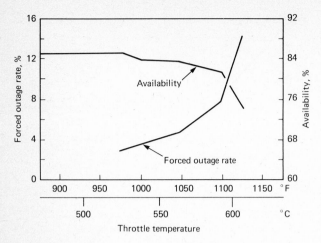

Figure 8.10   Forced outage rate and availability for steam turbines with elevated throttle temperatures.[21]

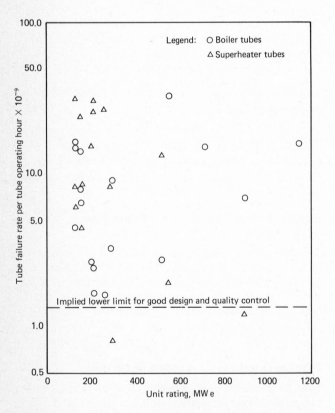

Figure 8.11   Summary of steam-generator failure experience in
typical utility units for 4,300,000 unit-hours of operation giving 95
× 10⁹ tube-hours of operation. Data are plotted as a function of
steam-generator unit design output for 17 different models of
steam generator. *(Courtesy Oak Ridge National Laboratory.)*

sented. The number of boiler and superheater tube leaks
reported for each model was divided by the product of the
number of tubes per unit and the number of unit hours operated
to give the incidence of tube leaks per tube operating hour. The
results are plotted as a function of the unit power rating to give
the broad scatter band shown in Fig. 8.11.

The wide range of tube failure rates at first seemed confusing.
However, further reflection led to the conclusion that the wide
spread (about two orders of magnitude) simply indicates the
range of the effects of variations in both the refinement of the
designs and in the quality control exercised in the course of materials procurement, fabrication, and erection. Thus, a dashed
line was drawn across the lower portion of the scatter band to indicate the tube failure rate to be expected in reasonably well-
designed units fabricated to the better quality-control standards
in current use for coal-fired steam plants. It seems reasonable to
believe that lower failure rates might be obtained with more refined designs achieved through extensive component testing,
tighter quality control, and cleaner fuels.

It was thought initially that the tube failure rate might be a
function of unit size; hence the data were plotted on the coordinates of Fig. 8.11. However, there does not seem to be any definite size effect. The poorer reputation of large units evidently
just stems from the fact that the greater the number of tubes, the
greater the likelihood of a tube leak.

The incidence of tube failures always appears to be higher in
new plants during the first year or two of operation. This is particularly true if equipment of a new design is employed. The
incidence of failures of this sort can be reduced by extensive
component testing if this is practicable—as is the case for liquid-
metal pumps and heat exchangers for LMFBR plants. Incisive
analyses and conservative design, particularly with respect to
thermal stresses, should also help to avoid trouble.

Tube Leaks in Steam Generators Made for Sodium-Cooled Fast Reactors

An attempt to carry out a reliability study in considerable depth
has been made by the British for the sodium-cooled fast-breeder
reactor system.[23] The prime concern in this study was the incidence of tube failures in the liquid-metal-heated steam generator. One of the most important elements of the study was a review of failure experience in some 80 steam generators with
fossil fuel firing having capacities of 100 to 500 MWe. This work
showed that the incidence of tube failures was approximately 1
in 10,000 tube operating years, or 10^{-8} failures per tube operating hour. Although one would have expected that the incidence
of failures in the British nuclear steam generators would be
lower than in the fossil fuel plants, experience with these
generators has varied widely. In some plants the incidence of

leaks has been high as a consequence of design errors that have led to failures from tube vibration, while in others tube defects were troublesome. However, steps taken to eliminate the design shortcomings and institute good quality-control procedures should bring failure rates well below those for fossil fuel plants, and this seems to be the situation in the later plants. Unfortunately, the data available to the author for the newer plants are not sufficient at the time of writing to be definitive.

The very tight controls exercised over leaks in reactor fuel elements provide another set of data that appears to be relevant to the tube leakage problem in steam generators. Reactor fuel elements are made of stainless steel or zirconium tubing 7 to 13 mm in diameter. The wall thickness is only 0.25 to 0.5 mm (0.010 to 0.020 in). Metallographic experience shows that these walls are only a few grains thick, and so small defects in the grain boundaries of just two adjacent grains may lead to a leak. Further, the fuel elements are commonly only 1 to 3 m (3 to 10 ft) long with an end cap welded to each end, so the number of welded joints is large.

Extensive experience with U.S. water reactors shows a failure rate per operating hour of about 10^{-9} leaks per weld and 10^{-10} leaks per foot of tubing. Essentially similar data have been obtained by the British in their AGR (Advanced Gas-Cooled Reactor) with fuel-element operating temperatures of 300 to 600°C (600 to 1100°F).

RELIABILITY OF IDEALIZED AND ACTUAL SYSTEMS

Some insight into the application of generic data of the type shown in Table 8.5 to system reliability appraisals can be obtained by setting up a series of idealized systems. The simplest set of cases to consider is a series of systems having various numbers of components of essentially similar reliabilities as in Table 8.6. For a relatively simple system with only 10 major components, for example, the mean time to failure would be only 1000 h if it were built of ordinary commercial components having a failure rate of 10^{-4} failures per hour. If, on the other hand, one assumes the highest reliability that there is reason to hope might be obtainable in mechanical components—i.e., a failure rate of 10^{-6} per component operating hour[20]—the mean time to failure for a 10-component system would be 100,000 h. Many people prefer to speak in terms of a 90 percent probability of satisfactory operation for some specified period. This reduces the operating period by a factor of 10, so that a 10-component system with a mean time to failure of 100,000 h would have a 90 percent probability of operating for 10,000 h without a repair or a forced outage.

If we turn our attention from mechanical systems to electronic systems, it is necessary to think in terms of much larger numbers of components—at least 100 and commonly 1000 or more. Thus the mean time to failure of an electronic system such as an amplifier made up of 100 components having a reliability of 10^{-7} failures per component operating hour would have about the same mean time to failure as a mechanical system made up of 10 components each having a failure rate of 10^{-6} per component operating hour.

Reliability of Familiar Systems

It is interesting to compare the idealized systems of Table 8.6 with some actual systems for which there is extensive operating experience. Table 8.7 presents a simplified analysis of the relia-

TABLE 8.6   Mean Time to a Forced Outage for a Series of Idealized Systems Differing in the Number of Components and Average Component Reliability (The time given in the table indicates number of hours. The number should be reduced by a factor of 10 to obtain the time for a 90% probability of operation without a forced outage.)

Number of components in system	Average failure rate for individual components (failures/10^6 h)				
	100	10	1	0.1	0.01
1	10,000	100,000	1,000,000	10^7	10^8
5	2,000	20,000	200,000	2,000,000	2×10^7
10	1,000	10,000	100,000	1,000,000	10^7
25	400	4,000	40,000	400,000	4,000,000
100	100	1,000	10,000	100,000	1,000,000
500	20	200	2,000	20,000	200,000
1000	1	100	1,000	10,000	100,000
5000	0.2	20	200	2,000	20,000

TABLE 8.7   Mean Time to Failure of Components in Typical Systems in Widespread Use

System	Number of components		Mean time to a forced outage from field experience	Average component failure rate per 10⁶ h
Household refrigerator	5	(motor, compressor, thermostat, solenoid switch, vapor system)	130,000 h (15 years)	1.5
Automatic washing machine	15	(motor, agitator, spin dry mechanism, 2 water inlet valves, drain valve, thermostat, fill limit switch, drain limit switch, timer, cycle selector switch, cycle shift mechanism, start-stop switch, detergent dispenser, water pump)	500 h (2 years at 5 h/week)	130
Automobile engine	25	(starter, starter solenoid, ignition switch, ignition wiring, spark plugs, distributor, coil, generator, voltage regulator, water pump, oil pump, oil pressure regulating valve, oil filter, radiator, thermostat, valve mechanism, pistons, crankshaft and rods, timing gears, fuel pump, fuel filter, carburetor, fuel tank, oil pressure gage, water temperature gage)	1000 h (30,000 mi)	40
Turbojet engines	25	(starter, ignitor, ignition switch, ignition wiring, compressor, turbine, accessory drive gear box, oil lines, oil filter, oil pump, oil pressure regulating valve, oil cooler, oil temperature regulator, 8 burners, fuel control, fuel pump, fuel lines)	4000 h (Extensive checks and minor maintenance at no more than 2000 h)	10
Molten-salt pump test loop (500 gpm)	35		3000 h	10
Transistor radio (small)	100		5000 h (4.5 h/day for 3 years)	2
Household TV set (black and white)	1000		5000 h (4.5 h/day for 3 years)	0.2

bility experience with a series of familiar systems. It is assumed that in each case the reliability of all the components is the same for any given system. Actually, such systems would not be built like Oliver Wendell Holmes "wonderful one-hoss shay," in which all parts lasted without fault for "a hundred years to a day" and then all crumbled at once, but rather would have substantial differences in reliability from one component to another. Although detailed data on the reliabilities of individual components are lacking, it is still informative to assume as a first approximation that the reliabilities of the various components are approximately equal in each particular system.

The first example in Table 8.7 is a household refrigerator, a machine that has become a watchword in matters of reliability. In this instance, in addition to each of the major items of mechanical or electric equipment, the fluid containment system is listed as a component on the basis that a leak at any point in that system would constitute a failure. According to one of the major manufacturers of refrigerators, the mean time to failure is 15 years. Using this number coupled with a value of 5 for the number of components makes it possible to compute backwards to obtain an average failure rate per component, and this turns out to be approximately 1.5×10^{-6}. This happens to be essentially the same as the lowest failure rate that the large amount of data cited above indicate can be achieved with highly developed mechanical components operated under benign conditions.

A similar procedure was carried out for another familiar household appliance, namely, an automatic washing machine. It appears that not only are there many more components in the system but also the reliability of the individual components is much lower. This difference probably stems from the fact that the components in the refrigerator system are hermetically sealed, hence are not subject to deterioration or damage from corrosion or dirt as are the components of an automatic washing machine. It is worth noting that about half of the manufacturers that have built clothes washers and driers in a single unit have discontinued these items, largely because of the high incidence of service troubles. Not only are such combined machines more complex, but also the range of temperature and humidity conditions is greater, and thus the conditions are more stressful.

In turning to the next item in Table 8.7, an automobile engine, it appears that even after enormous amounts of experience it is still not possible to get a high reliability in ordinary service. While not generally realized, there are data to show that a high price does not necessarily solve the problem: Because of the enormous amount of service experience behind it, a standard Ford is likely to require less servicing than a Rolls-Royce for operation under rough road conditions.

Experience with high-temperature liquid systems at ORNL gives some insight into the effects on reliability of shifting the operating temperature base from near room temperature to a red heat. Close to 200,000 h of operating experience have been accumulated with rather simple systems designed for performance and endurance testing of liquid-metal and molten-salt pumps. Unfortunately, the operating schedule on these systems has ordinarily called for a shutdown every 3000 h (approximately) to permit recalibration of the instrumentation, because the sensors are subject to a drift in calibration with extended operation at high temperatures. However, the data indicate that the mean time between forced outages has been of the order of 3000 h, and this figure was used in Table 8.7 as a basis for estimating the mean effective time to failure for an average component.

It should be noted that a detailed analysis of the actual failures experienced in the liquid-metal and molten-salt test loops indicates that roughly 90 percent of the forced outages were associated with the electric equipment. At first glance, this appears surprising in that many would expect the conventional electric components to have a much higher reliability than the high-temperature mechanical components. In fact, after initial shakedown runs, there were no failures in the high-temperature pumps in 165,000 h, but 52 failures in the electric equipment. What is not obvious is that there are usually numerous junctions and a fair number of contact surfaces in most items of electric equipment, and there are many places in which the electrical insulation can break down. As a consequence, there are likely to be many different failure modes in a component which, on the surface, may appear to be quite simple.

Two good examples of well-developed electronic systems are given by the last two items in Table 8.7, i.e., a small transistor radio and a black-and-white TV set. Good statistics on the reliability of typical units were not available; hence the values used in Table 8.7 for the mean time to failure were obtained by questioning friends on their personal experience. The resulting values are believed to be representative even if not based on a broad statistical base. The value derived in Table 8.7 for the mean failure rate for the average component in commercial electronic equipment of this sort is $\sim 10^{-6}$ failures per component operating hour. Very tight quality control for spacecraft has given MTBFs as low as 10^{-8}/h.

The author has not found good data on the reliability of electronic control systems for power plants, but some idea of how formidable the problem really is can be seen from consideration of a relatively simple item, a voltage regulator for a diesel generator set. To the author's knowledge by far the least complex design recommended for this application by a major electric equipment manufacturer includes 85 electronic components even with no provisions for redundancy, and it was estimated by the manufacturer to have a mean time to failure of $\sim 10,000$ h.

Forced-Outage Rates in Conventional Power Plant Units

The fraction of forced outages in fossil-fuel-fired steam plants caused by each of the major components in typical plants is indicated by Table 8.8.[22] As one would expect, the boiler and turbine are the principal sources of outages, but note that the category "other" for miscellaneous causes of forced outages represents about 10 percent of the total. It is also interesting that the incidence of outages increases with the size of the unit, the 600-MW and larger units having more than double the outage rate of the 200- to 390-MW units. This probably stems in part from the smaller background of operating experience that has been obtained with the larger units.

A set of forced-outage information for gas turbines is shown in Fig. 8.12. In this instance the compressor-combustor-turbine generator unit is a smaller factor in the overall incidence of forced outages than the auxiliary equipment. Table 8.8 indicates that the boiler-turbine-generator unit is responsible for 90 percent of the forced outages in steam units. This is not surprising because the gas-turbine and compressor units are simpler than the corresponding components in the large steam-generator–furnace-turbine-generator combination. Another factor is that the smaller size of the components in the gas turbines makes it easier to repair these components or replace them, so the time associated with each forced outage is reduced. Since the time in a forced outage is charged to the component requiring the longest period to repair, and since maintenance work is carried out on a variety of components during any given shutdown, the relatively small size and ease of replacement of gas-turbine components may make their repair time shorter

Mean time between failures (MTBF), h

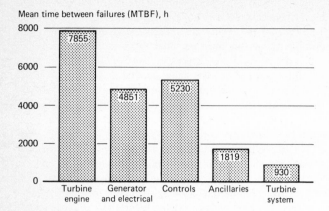

Figure 8.12   Gas-turbine system reliability results.[24]

than that of complex electric components controlling the fuel feed and metering system. The situation is quite complex, and much more detailed information is needed before a proper assessment of the reliability problems can be made.

Causes of Outages in Coal-Fired Steam Generators

Data showing the relative incidence of forced outages, outages for maintenance, and planned outages for annual overhaul work in 435 steam plants are summarized in Table 8.9. A detailed insight into the causes of a major set of these outages is given by an excellent set of data on the coal-fired steam-generator outage experience of 39 utilities operating 82 units of similar design during the 1977–1979 period. The 2254 forced outages reported in that study were charged to one of five categories: the economizer, the waterwall, the superheater/reheater, the coal pulverizers, and miscellaneous (or "other") items. The fraction of the total number of outages chargeable to each of these is shown in Fig. 8.13, and detailed breakdowns of the causes of trouble for the waterwall and superheater/reheater, the two components giving 70 percent of the difficulties, are summarized in Tables 8.10 and 8.11. A review of these tables indicates that, in the cases where the cause of the failures could be determined, about one-third were chargeable to the ash in the fuel, roughly another one-third might be amenable to refinements in design, about 10 percent stemmed from defective welds, and the balance were caused by a variety of factors. As one might expect, the incidence of defects in welds made by maintenance personnel was roughly five times that of defective field welds, while shop welds were responsible for only about one-fourth the number of failures resulting from field welds. These ratios are still more significant when one remembers that the number of

TABLE 8.8   EEI Data for Average Forced-Outage Rates and Availabilities for Components in Stations with GE Large Steam-Turbine-Generator Sets for the 10-Year Period 1965–1974 (Ref. 21)

	200–389 MW		390–599 MW		600 MW and Above	
	F.O.R.%	Avail.%	F.O.R.%	Avail.%	F.O.R.%	Avail.%
Boiler	3.7	88.4	5.7	85.1	8.8	83.9
Turbine	0.9	93.2	1.2	90.9	2.1	90.2
Condenser	0.1	96.5	0.1	95.4	0.2	96.9
Generator	0.2	95.8	0.2	94.9	1.3	93.7
Other	0.4	96.8	1.0	95.1	1.4	96.3
Unit	5.3	85.8	8.0	81.1	13.0	76.7

TABLE 8.9   Incidence of Outages and Outage Time per Year for 435 Steam Power Plant Units of 100 to 199 MW in the 1967–1976 Period[24] (Service factor = 81%; capacity factor = 65%)

	Availability	Forced		Maintenance		Planned	
	Percent	Per unit events	Year-hours	Per unit events	Year-hours	Per unit events	Year-hours
Boiler	90.2	5.2	219.4	1.95	158.6	0.91	475.6
Turbine	93.7	0.69	68.5	0.84	84.2	0.63	390.7
Condenser	97.2	0.22	6.5	0.46	43.1	0.31	190.7
Generator	96.2	0.24	20.0	0.39	57.9	0.36	254.1
Other	97.8	0.56	20.2	0.55	42.1	0.23	123.0
Unit	86.8	6.9	330.4	3.1	209.1	1.3	611.0

shop welds is many times the number of field welds, which in turn is many times the number of maintenance welds. The relative incidence of these failures probably reflects more on the degree of adversity of the conditions under which the welds are made than the relative skills of the welders.

The high proportion of outages caused in one way or another by ash in the fuel suggests that possibly the extra expense of reducing the ash content of the fuel by coal washing might be justified by reductions in outage costs. Table 8.9 indicates that boilers are responsible for ~1400 h per year of outage, and Tables 8.10 and 8.11, together with Fig. 8.13, indicate that ash is responsible for about one-third of these outages, or ~18 days per year. To give a rough idea of the cost-benefit relation for reducing the ash content of the coal, consider a 1000-MW plant with a load factor of 60 percent and an availability of 80 percent. The amount of coal burned per year would be $\sim 2 \times 10^6$ tons. If the outage time chargeable to ash could be halved by coal washing and if the cost of outages runs $250,000 per day, the savings in reduced outage time would justify ~$2.25 per ton for coal

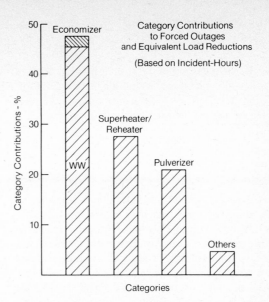

Figure 8.13 Contributions to forced outages and equivalent load reductions in coal-fired steam generators broken down by categories.[25] (*Courtesy Combustion Engineering Power Systems.*)

TABLE 8.10 Steam Generator Waterwall Tubing Failure Causes and Their Incident Frequency Ranking by Percentages (Oct. 1, 1977 through Dec. 31, 1979)

Category	Percent
Indeterminate	22.9
Sootblower erosion	12.8
Slag fall erosion	11.5
Tube/attachment failures	10.7
Defective maintenance weld	6.6
Furnace ash hopper tube quenching	5.7
Overheating (indeterminate duration)	3.8
Internal corrosion	3.8
Fly ash erosion	3.6
Center wall flexibility	3.3
Short-term overheating	3.0
Header/nipple flexibility	2.2
Sootblower system failures	1.6
Defective original field weld	1.6
Thermal expansion (seal-to-seal tube)	1.6
Abrasion (tube-to-tube)	1.4
Internal obstruction	1.1
Overheating (internal deposits)	0.8
Outage oxygen pitting (related to improper lay-up)	0.5
External corrosion	0.5
Defect in roof casing seal weld	0.2
Casing leak erosion	0.2
Defective shop weld	0.2
Embrittlement	0.2
Maintenance procedures	0.2
	100.0

TABLE 8.11 Steam Generator SH/RH Tubing Failure Causes and Their Incident Frequency Ranking by Percentage (Oct. 1, 1977 Through Dec. 31, 1979)

Category	Percent
Indeterminate	17.9
Header/nipple flexibility	14.3
Fly ash erosion	12.1
Sootblower erosion	9.4
Overheating (indeterminate and long-term duration)	8.5
Defective maintenance weld	7.2
Abrasion	5.4
Dissimilar weld failures	4.5
Tubing/attachment related problem	4.5
Outage oxygen pitting (related to improper lay-up procedures)	4.0
Overheating (internal deposits)	3.6
Vibration fatigue	1.7
Defective original field weld	1.3
Incorrect material erroneously installed	1.3
Sootblower system failure	0.9
Slag fall/erosion	0.9
External tube wastage by chemical additive system	0.5
Internal obstruction	0.5
External corrosion	0.5
Defective shop weld	0.5
Maintenance procedures	0.5
	100.0

washing, about half of the cost of the washing process. Some or all of the rest of the coal washing cost might be written off against reduced emissions, reduced maintenance of the stack gas clean-up system, and other such cost-saving results.

REDUNDANCY

The reliability of a system usually can be increased by providing spare components that can be brought on stream whenever required to take the place of units that have failed. Until the 1920s, steam power plants were ordinarily built with a multiplicity of boilers feeding a common steam header from which steam was drawn to supply a multiplicity of turbines. As pressures and temperatures were increased, the reliability of the valves required in the steam lines fell off to such a degree that directly coupling a single boiler to a single turbine was found to yield a more reliable system, and this approach is used universally today. Redundancy is obtained by having at least 10 steam-turbine units coupled to a utility system so that no one unit represents more than 10 percent of the system capacity. Thus, even if two units are out at the same time—say, one for routine maintenance and the other because of a forced outage—the system can still handle a load equal to 80 percent of its nominal full capacity. Obviously, it would be desirable to have a still greater margin between the peak load to be expected and the total nominal system capacity. Statistically, one must expect to have a number of units shut down occasionally as a consequence of forced outages, and when this occurs a specie of redundancy is obtained by drawing power from adjacent utilities that have spare power available.

Wherever possible, redundancy is provided in auxiliary equipment, particularly for such vital components as instrumentation and controls, feed-water pumps, and lube-oil pumps. These latter components are installed in such a way that they can be valved off and defective units removed while the plant continues to operate. For components such as coal pulverizers or burners, this may mean a reduction in the power plant output obtainable, but a reduction is much less serious than a complete shutdown.

DETECTION OF INCIPIENT TROUBLE

Much can be done to detect incipient trouble before it becomes serious. The most important measures of this sort commonly taken in steam plants are concerned with the control of water quality. This entails maintaining a high degree of purity in the feed water as indicated in the chapter on materials. Even seemingly tight controls may not suffice; a number of steam genera-

tors in nuclear plants have experienced extensive tube failures as a consequence of corrosion between the baffles and tubes. The formation of iron oxide on the surface of the holes in the carbon-steel baffles in these units led to pinching and local necking-down of the tubes where they passed through the baffles. These regions acted as stress concentrations subject to thermal-strain cycling as a consequence of the differential thermal expansion between the tubes and the shell, and the thermal-strain cycling led to extensive cracking of tubes. More elegant monitoring of the steam-system feed water to detect the presence of chloride in-leakage from the brackish condenser cooling water might have prevented these problems.

Vibration test equipment is commonly used on both steam and gas turbines to detect the buildup of deposits, possible deterioration of bearings, and the like so that more serious failures, such as a rub between the rotor and the casing, can be avoided.[26] Visual examination of the interior of the furnace can be helpful in the detection of difficulties with slagging and ash deposits or problems with the burners. Buildup of ash or slag deposits in the superheater and reheater can be detected by monitoring the temperature difference between the steam and the flue gas in these components. Most important of all, an alert and experienced crew making full use of the diagnostic instrumentation available is often able to catch trouble before it becomes serious.

HUMAN ERRORS

Roughly 25 percent of forced outages apparently stem from human errors in either operation or maintenance. These errors may result from carelessness, negligence, poor instructions, or a poor comprehension of the task at hand.[27] It is significant that 90 percent of aircraft accidents stem from pilot error and that all the reactor accidents experienced to date, including the worst reactor accident experienced in England—that at Windscale in 1957—and the Three Mile Island accident in the United States, were caused by operator errors. The reactor accidents have generally stemmed from an inadequate understanding by some operating personnel of the complex system with which they were working. Although quantitative data are hard to obtain, it seems likely that in all types of plants a large fraction of the minor forced outages are caused by human errors, resulting mostly from an inadequate understanding and working knowledge of the system but partly from carelessness. There are no substitutes for competence, training, experience, diligence, sense of responsibility, intelligence, and esprit de corps in the operating and maintenance personnel. No amount of bureaucratic formalities and red tape can take the place of these basic characteristics of a good staff.

THE SPECIAL RELIABILITY PROBLEMS OF NUCLEAR PLANTS

The consequences of a malfunction in a nuclear power plant are potentially so serious that a great deal of thought and effort has been given not only to the reliability of individual components but even more to the possibility that one failure might lead to a series of events that would be disastrous. The space exploration program confronted a basically similar problem, and, particularly in the Apollo missions, demonstrated dramatically that analysis of even extremely complex systems can reduce the probability of a serious failure to an extremely small value through judicious use of redundancy. In both fields the approach followed is the same, and is referred to as *fault tree analysis*.[28] In principle, each component, subsystem, and system is studied, the various possible types of failure and failure mode are delineated, and the resulting chains of events are worked out by several different, essentially independent experts to minimize the probability that any eventuality is overlooked. These analyses of chains of events are summarized diagrammatically in the form of *fault trees*. To be most effective, this work should be initiated in the early stages of the system design so that the system configuration is amenable to provision of redundant components to limit the extent of the ill effects of any failure. Just as commercial aircraft have a multiplicity of engines and are designed so that they can be flown safely after the failure of any one of the engines, nuclear plants have a multiplicity of reactor cooling pumps so that they can continue to operate safely in the event of a pump or drive motor failure. This basic philosophy is followed throughout the plant with all vital components, such as valves and instrument sensors, provided with parallel or back-up units. The failure probabilities of each of these elements in the system can then be used to estimate the probabilities and consequences of each failure mode envisioned, and the system design can be modified until the probability of a serious failure becomes exceedingly small.[28]

PROBLEMS

1. Consider a coal conveyor for which a statistical analysis shows that 1 piece in 100 of a key part will have a defect which will serve to initiate a fatigue crack and the crack will progress fairly rapidly if the part is subjected to a stress of 100 MPa (45,000 psi). Service experience indicates that such a stress will be experienced about once in 10 years for each part in service. Estimate the mean time between failures (MTBF) for the machine as a whole caused by this sort of failure if there are 5000 of these parts in the conveyor chain of each machine (Fig. 8.7).

2. Stress analysis shows that thermal stresses will be induced in the vicinity of a penetration in a Type 304 stainless-steel pressure vessel for an LMFBR operating at 600°C when going from 10 percent to full power, and these stresses will be equivalent to a temperature difference of 100°C. From Fig. 6.14, estimate the number of such cycles that could be sustained before a crack would be initiated in a unit of this design.

3. Assume that each component of the system of Fig. 7.7 has a mean time to failure of 10^{-5}/h, and from this estimate the mean time between forced outages of the system.

4. Repeat Prob. 3, assuming that each of the five automatic control valves, together with its instrumentation and control circuitry, has a mean time to failure of 10^{-4}/h.

REFERENCES

1. Retalick, F.: "Preliminary Assessment of the Requirements and Potential of Open Cycle MHD as an Electric Utility Power Plant," paper presented at the Eighteenth Symposium on Engineering Aspects of MHD, Butte, Mont., June 18–20, 1979.

2. Albrecht, P. P., and W. D. Marsh: "Gas Turbines Require Different Outage Criteria," *Electrical World*, Apr. 27, 1970, pp. 38–39.

3. "Fossil Plant Performance and Reliability," *EPRI Journal*, vol. 3, no. 8, October 1978, pp. 53–56.

4. Niebo, R. J.: "Power Plant Productivity Trends and Improvement Possibilities," *Combustion*, vol. 50, no. 1, January 1979, pp. 12–21.

5. Flynn, G., et al.: "GMR Sterling Thermal Engine—Part of the Sterling Engine Story—1960 Chapter," *SAE Trans.*, vol. 68, 1960, pp. 665–684.

6. "Summary Report: Williamsburg Station Accident Investigation," *Penelec GPU*, Dec. 14, 1977.

7. Nuchols, J. B., and J. R. McGuffey: "Graphitization Failures in Piping," *Mechanical Engineering*, May 1959.

8. Derby, R. W., et al.: "Test of 6 in Thick Pressure Vessels, Series 1: Intermediate Test Vessels V-1 and V-2," Oak Ridge National Laboratory Report No. 4895, February 1974.

9. Merkle, J. G., et al.: "An Evaluation of the HSST Program Intermediate Pressure Vessel Tests in Terms of Light Water Reactor Pressure Vessel Safety," Oak Ridge National Laboratory Report No. TM 5090, November 1975.

10. Germond, J. P.: "Insuring Dam Risks," *Water Power and Dam Construction*, June 1977, pp. 36–39.

11. Johnson, F. A., and P. Illes: "A Classification of Dam Failures," *Water Power and Dam Construction*, December 1976, pp. 43–45.

12. *Lessons from Dam Incidents*, International Commission on Large Dams, Paris, 1973.

13. Mischke, C.: "A Method of Relating Factor of Safety and Reliability," *J. Eng. Industry, Trans. ASME*, 1970, pp. 537–542.

14. Nelson, N. W., and K. Hayashi: "Reliability and Economic Analyses Applied to Mechanical Equipment," *J. Eng. Industry, Trans. ASME*, 1974, pp. 311–316.

15. "Direct Cycle Heavy Water Turbine for Marviken," *Nuclear Engineering International*, vol. 12, no. 126, January 1967, p. 33.

16. Christensen, J., and C. B. Olesen: "Predicting Performance of D_2O Steam Turbines in Boiling Heavy Water Reactors," *Nucleonics*, vol. 24, no. 5, May 1966, p. 67.

17. Steckelmacher, W.: "Leak Detection," *Nuclear Engineering*, vol. 4, no. 43, December 1959, pp. 450–453.

18. Collins, G. D.: "Plugging Indicator Operation," AEC Research and Development Report GEAP-10048, September 1969.

19. Hayes, D. J., and G. Horn: "Leak Detection in Sodium-Heated Boilers," *Journal of the British Nuclear Energy Society*, vol. 19, no. 1, January 1971, pp. 41–48.

20. Fraas, A. P.: "Reliability as a Criterion in Nuclear Space Power Plant Design," paper No. 661A presented at the SAE-ASME Air Transport and Space Meeting, New York, April 1964.

21. Elston, C. W.: "Needs and Directions for Improving Power Plant Reliability: A Turbine Manufacturer's View," paper presented at the ANS Executive Conference on Improving Power Plant Reliability, Sept. 27–29, 1976.

22. Downs, J. E.: "Forced Outages of Large Steam Turbines—An Analysis of Causes," *Proceedings of The American Power Conference*, vol. 32, 1970.

23. Bolt, P. R., and H. M. Carruthers: "Some Comments on Sodium-Water Reaction Problems from the Viewpoint of a Power Station Purchaser and Operator," paper presented at a symposium on Progress in Sodium-Cooled Fast Reactor Engineering held by the IAEA in Monaco, Mar. 23–27, 1970.

24. Spencer, D. F., et al.: "Development of High Reliability Combustion Turbine Combined-Cycle Power Plants," paper presented at the ANS 1979 Annual Meeting, Atlanta, Ga., June 3–7, 1979.

25. Thomas, G. C., et al.: "C-E Availability Program," *American Power Conference*, April 1980.

26. Bannister, R. L., et al.: "Modern Diagnostic Techniques Improve Steam Turbine Reliability," *Power*, vol. 124, no. 1, January 1980.

27. Finnegan, J. P., et al.: "The Role of Personnel Errors in Power Plant Equipment Reliability," EPRI Report No. AF-1041, prepared for the Electric Power Research Institute by Failure Analysis Associates, April 1979.

28. Fussell, J. B., and J. S. Arendt: "System Reliability Engineering Methodology: A Discussion of the State of the Art," *Nuclear Safety*, vol. 20, no. 5, September–October 1979.

9

ENVIRONMENTAL EFFECTS

From the early 1960s to the time of writing, there has been a tremendous increase in political activity concerning the environment, the political activists directing their attacks particularly at electric power plants. This has led to a vast increase in governmental regulation of the power industry. At the time of writing, impassioned attacks on the industry continue with demands for ever more stringent controls on emissions and a host of lawsuits to prevent the construction of new power plants. The activists claim that the industry shows a callous disregard for the public welfare by building plants that will "poison the atmosphere and cause people to die of cancer by the hundreds of thousands."[1] They further claim that this is particularly reprehensible because the sun pours enormous amounts of energy on the earth, and we are only prevented from getting this perfectly clean energy practically "for free" by the machinations of the big oil companies.[2] Opposing the activists are many utility executives, engineers, and scientists who state that economical electricity from "natural" sources lies far in the future and that, for at least the next few decades, most of the electric energy must be obtained from coal and fission power plants. They acknowledge that troublesome environmental problems are associated with both types, but they claim that governmental regulations are unreasonably stringent, greatly increasing the cost of electricity without yielding commensurate benefits, thus creating "economic pollution" with severe inflationary effects. Most serious of all, it is charged that restrictions on new power plant construction will lead to major power shortages by the latter 1980s which will dislocate the economy and cause massive unemployment. The activists respond that industrial growth must be halted to prevent further injury to the environment, and they accuse the industrialists of greed for obscene profits. However, the imputation of motives cuts both ways, and many of the "professional activists" are accused of being publicity seekers and deriving handsome incomes from lectures, book royalties, lobbying, and particularly legal fees.[3]

The situation is confusing indeed, with numerous disturbing facts intermixed with plausible hypotheses, highly emotional arguments, and myriad conflicting "conclusions." How can one decide which "conclusions" are valid and which are sophistries? Unfortunately, there is no simple answer. A person who wants to understand the situation has no choice but to spend hours of stiff mental effort to assimilate the principal facts of the situation. This chapter has been prepared in an effort to aid those genuinely interested in getting a reasonably good perspective on the basic problems and choices involved.

TECHNOLOGY, INDUSTRY, AND THE QUALITY OF LIFE

The terribly confusing situation in the environmental arena cannot possibly be understood until it is recognized that perhaps the most vital difficulty in the acrimonious debates over "power versus the environment" is that the two sides commonly begin with such completely different axioms and postulates that no compromise is possible. The question could be more correctly stated as one of comparative life-styles: Should the United States proceed with further technological and economic development or should it go back to the technology, economy, and life-style of some earlier period, e.g., that at the end of the nineteenth century? Thus the first point to be considered seems to be the flat statement that our lives are being horribly poisoned by industry and technology. In attempting to appraise its validity, one significant parameter is the average life span. Data in Table 9.1 show the great increase in life span brought about by technology and industrialization,[4] an increase that is still continuing in spite of the increased number of deaths stemming from abuses of affluence such as smoking, rich diets, insufficient exercise, and drugs.[5,6]

A return to the life-style of the latter nineteenth century would not eliminate occupational deaths; the number of U.S. farmers killed just by kicks from horses or mules was apparently about 2000 per year around 1900.[7] This may be compared with a

TABLE 9.1 Average Life Span from Neolithic Times to the Twentieth Century

Neolithic period	20
Classical Rome	27
Medieval England	32
United States	
1760	35
1860	45
1900	49
1920	56
1940	63
1960	70
1978	73

Sources: *World Almanac* and *Encyclopedia Britannica*

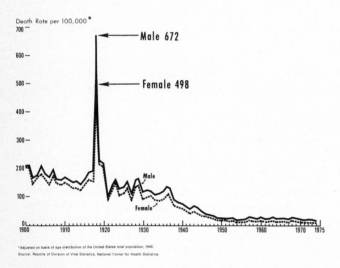

*Adjusted on basis of age distribution of the United States total population, 1940.
Source: Reports of Division of Vital Statistics, National Center for Health Statistics.

Figure 9.1 Mortality from influenza and pneumonia in the United States from 1900 to 1974. *(Courtesy Metropolitan Life Insurance Co.)*

total occupational death rate of ~ 13,000 from all causes in 1977 for a population about six times that of U.S. farms in 1900.[7] If one looks back to still less affluent nineteenth-century farming societies, one might contemplate the facial expressions of the people in the original of Van Gogh's painting the "Potato Eaters" to get a vivid impression of the brutalizing effects of the utter drudgery and misery of the bare subsistence existence implicit in such a life-style. In point of fact, the great increase in population now makes such a return even less attractive, because the productivity of the nineteenth-century-type farm is

low and even our present relatively small farm population has been forced to expand into progressively less productive land, as indicated by a recent study of Mennonite and Amish farmers.[8] The large families and increasing numbers of these thrifty, hard-working people have forced many to move out of their ancestral areas in rich farming country to less productive land, and for those less fortunate families this has markedly reduced even their simple standard of living. Other major blessings of technology and industry include plumbing systems (only 25 percent of U.S. homes had flush toilets in 1920), electric washing machines, kitchen appliances, prepared foods, and inexpensive ready-made clothing that have freed women from household drudgery and made the women's liberation movement possible. The luxuries of fresh fruits, vegetables, and meat year round can only be appreciated if one lives for some years in societies without them. Culturally, the greatest benefits of the affluence brought by technology and industry have gone to what is now our largest single "industry": education. The arts have also benefited immensely; e.g., more tickets to ballet performances were sold in the United States in 1978 than tickets to professional football games.

One could go on and on, but the items cited here should suffice to make the point that technology and industry have increased our life span and have enormously enhanced the quality of life. While there have been numerous abuses, many have been corrected, and the sensible course is to distinguish which of those remaining are the more serious and determine how best to cope with them rather than to try to go back to an idyllic past that did not really exist.

HEALTH EFFECTS

Concern for the effects of environmental pollution generated by the human race is not new. Seneca, Pliny, and other Roman writers were distressed by the foul smells and noxious fumes of Rome, and as early as the fourteenth century the fumes from coal fires in London led Edward I to decree the death penalty for anyone who burned coal in England.[9] Every major city in the United States had placed restrictions on smoke emissions by the beginning of the twentieth century; e.g., anthracite was the only form of coal that could be burned for home heating in most cities in the northeast. It is significant that the unwholesome effects of fumes from coal fires in England in the centuries following the decree of Edward I were more than offset by their benefits, for, as the English forests were consumed, England shifted to coal for fuel in the course of the fifteenth and sixteenth centuries, and as a result, led Europe in the industrial revolution.[10] Whatever its faults, industrialization must have been beneficial to the health of the society, for the population of

England began to expand rapidly at the beginning of the seventeenth century, evidently because people were at last getting enough nutritious food to improve their health and thus their resistance to disease.[11] This population explosion did not stem from advances in sanitation, medicine, or public health measures because more than a century passed before these came into being as an outgrowth of the advances in technology, education, and research made possible by the unprecedented prosperity of the industrial revolution. These advances have now largely eliminated infectious diseases as a cause of death, and the incidence of pneumonia (the most serious remaining communicable disease) is continuing to decline (Fig. 9.1).

Some valuable insights into the current health situation in the United States can be gained from Table 9.2, which gives the proximate causes of deaths from disease and accidents for the U.S. population as a whole,[4,12] while Fig. 9.2 shows cancer mortality data as a function of age group.[5] One of the most striking points evident in these data is that cardiovascular and cancer problems are responsible for about 70 percent of the total deaths, both mainly afflicting the past-45 age group. In considering ways in which the death rate might be reduced, studies of the detailed data indicate that a large fraction of deaths from cardiovascular problems and two-thirds of the cancer deaths are caused either by smoking (e.g., lung cancer) or by too rich a diet (e.g., cancer of the large intestine).[6] It can be seen from Fig. 9.2 that the only form of cancer for which the mortality rate has increased is lung cancer, and this is almost entirely among tobacco smokers. It is also significant that deaths from alcoholism (e.g., cirrhosis of the liver), dope, and homicide each greatly exceed deaths from occupational accidents, and deaths from congenital defects exceed deaths from communicable diseases. It is also noteworthy that from 70 to 85 percent of the traffic fatalities are caused by the use of alcohol or drugs.

TABLE 9.2   Deaths and Death Rates in the U.S. for Selected Causes[4] (Rates per 100,000 population)

1977* Cause of death	Number	Rate	1977* Cause of death	Number	Rate
All causes	1,898,000	877.4	Acute bronchitis and bronchiolitis	680	0.3
Enteritis and other diarrheal diseases	1,510	0.7	Influenza and pneumonia	49,960	23.1
Tuberculosis, all forms	2,960	1.4	Influenza	1,060	0.5
Syphilis and its sequelae	130	0.1	Pneumonia	48,900	22.6
Other infective and parasitic diseases	3,840	1.8	Bronchitis, emphysema, and asthma	22,220	10.3
Malignant neoplasms, including			Chronic and unqualified bronchitis	4,310	2.0
neoplasms of lymphatic and			Emphysema	16,320	7.5
hematopoietic tissues	387,430	179.1	Asthma	1,590	0.7
Diabetes mellitus	33,570	15.5	Peptic ulcer	6,320	2.9
Meningitis	1,620	0.7	Hernia and intestinal obstruction	5,660	2.6
Major cardiovascular diseases	960,090	443.8	Cirrhosis of liver	31,260	14.5
Diseases of the heart	717,320	331.6	Cholelithiasis, cholecystitis, and cholangitis .	2,770	1.3
Active rheumatic fever and chronic			Nephritis and nephrosis	8,130	3.8
rheumatic heart disease	12,560	5.8	Infections of kidney	3,650	1.7
Hypertensive heart disease and			Hyperplasia of prostate	910	0.4
renal disease	3,700	1.7	Congenital anomalies	12,540	5.8
Ischemic heart disease	637,670	294.8	Certain causes of mortality in early infancy . .	23,310	10.8
Chronic disease of endocardium and			Symptoms and ill-defined conditions	32,150	14.9
other myocardial insufficiency	3,930	1.8	All other diseases	128,620	59.5
All other forms of heart disease	52,800	24.4	Accidents	104,000	48.5
Hypertension	5,310	2.5	Motor vehicle accidents	50,380	23.3
Cerebrovascular diseases	182,840	84.5	All other accidents	54,640	25.3
Arteriosclerosis	29,040	13.4	Suicide	28,390	13.1
Other diseases of arteries,			Homicide	21,090	9.7
arterioles, and capillaries	25,580	11.8	All other external causes	4,230	2.0

*Provisional. Owing to rounding estimates of death, figures may not add to total. Data based on a 10% sampling of all death certificates for a 12-month (Jan.-Dec.) period.

Source: National Center for Health Statistics, U.S. Department of Health, Education, and Welfare

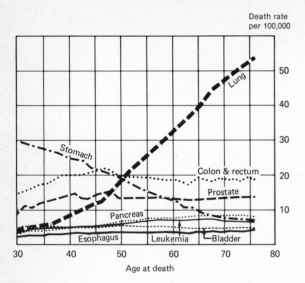

Death rate per 100,000

Figure 9.2 Male cancer death rates per 100,000 of the U.S. male population after adjustment for age on the basis of age distribution of the 1940 U.S. male population.[5]

The various types of accidents far more often cause disability than death. Table 9.3 compares data for deaths, permanent impairment, and temporary total disability for major types of accident, together with some major costs associated with these injuries. Table 9.3 indicates that in 1977 these costs totaled 62×10^9 of a gross national product of 1690×10^9, or roughly 3.5 percent of the GNP.

The beneficial effects of technology and industry on the health and life expectancy of the society as a whole are implicit in Table 9.1 and Fig. 9.1. There are some adverse effects, however, and the nature of these must be understood if effective steps are to be taken to reduce them.

Statistical Studies

The effects of various environmental factors on health are difficult to determine because individuals may vary widely in their susceptibility (e.g., allergic reactions or their age at exposure). There may be unsuspected synergistic effects, or the adverse effects may take a long time to develop. Controlled experiments

with humans are usually out of the question, and the various animal species differ widely in their responses. The problems are particularly difficult when looking for small effects. It is not hard to determine how great a dose of a toxic material will result in the death of half of a given group of rats, because good statistics can be obtained with test groups of a few dozen, but to obtain good statistics for the dose that would cause one death in a thousand would require test groups of at least ten thousand. Further, the same test would have to be repeated a number of times to make sure that some other factor, such as a mild flu epidemic, was not the cause of the deaths in the first test. The problems are also complicated by the fact that some materials such as NaF and NaI are lethal in large doses but actually beneficial in small doses. In human beings, NaF and NaI are deliberately added to water and salt, respectively, to improve the formation of teeth and the function of thyroid glands. Chlorine gas is poisonous, but when added to municipal water supplies in small amounts it appears to have no adverse effects on humans and is highly effective for controlling bacterial concentrations that would otherwise cause widespread difficulties and hundreds of thousands of deaths annually from intestinal ailments. Seawater is also toxic if drunk regularly, because the kidneys cannot eliminate the salt at the rate it is ingested; yet some salt in the diet is essential to health. The subtle balancing of such effects is particularly evident with respect to drugs. Aspirin is an obvious example of a widely used drug that is beneficial in small doses but lethal in large doses. Although it is not really common knowledge, trace amounts of arsenic appear to be essential to human life, and arsenic can also be enormously beneficial medicinally, as the author found after contracting amoebic dysentery in South America. (Doses of arsenic that are disagreeable but sublethal for humans are totally lethal for the amoebas, which would otherwise establish themselves in the small intestine and gradually cause the death of their host.) Plants also are often adversely affected by large doses of a material but favorably affected by small doses. This is true not only for conventional fertilizers but also for about 20 other trace elements (Table 9.4) whose presence in small amounts has been found essential to plant and/or animal health because they act as catalysts in enzyme systems. As will be discussed later, many of these are on the EPA 1979 list of elements that must be kept to extremely low levels in aqueous emissions from power plants.[13]

As in the case of NaF, NaI, and chlorine in drinking water, a key question that must be considered with regard to possibly lethal substances is whether there is a threshold below which no ill effects will occur. Because of the difficulty in establishing whether such a threshold exists, the usual procedure—probably pessimistic—is to assume a linear relation between the total accumulated dose and the adverse effects (such as a reduction in

TABLE 9.3   Accidental Injuries in the U.S. by Severity of Injury [4]

1977 Severity of injury	Total*	Motor vehicle	Work	Home	Public nonmotor vehicle
Deaths*	104,000	49,500	13,000	24,000	22,000
Disabling injuries*	10,400,000	1,900,000	2,300,000	3,600,000	2,800,000
Permanent impairments	380.000	150,000	80,000	100,000	70,000
Temporary total disabilities	10,000,000	1,750,000	2,200,000	3,500,000	2,700,000
	Certain costs of accidental injuries, 1977 ($ billions)				
Total*	$62.0	$30.5	$20.7	$8.6	$5.4
Wage loss	18.3	8.8	4.1	3.2	3.2
Medical expense	7.5	2.5	2.2	1.7	1.3
Insurance administration	12.3	9.1	3.0	0.1	0.1

*Duplication between motor vehicle, work, and home are eliminated in the total column.
Source: National Safety Council

life span) found at relatively high dosage levels, and then impose permissible dose limits well below those given by the same or similar materials naturally present in the environment.

Toxic Materials

The character of the effects of toxic materials differs widely from one type to another.[14] Acids and alkalis actually destroy living tissue, causing effects much like those of a burn. If small amounts of soluble compounds containing heavy metals, such as lead or mercury, are ingested at a higher rate than they can be eliminated by the bile gland and kidneys, they accumulate in the body and affect the function of various glands, causing among other possible effects anemia, loss of hair, and even loss of potency. (The author has found that pointing out the latter can have a salutary effect on the care with which technicians and students treat the use of mercury in laboratory instruments.) Many organic compounds can have very severe specific effects. For example, in a statistical study in Europe an alert observer noted a high incidence of bladder cancer in rubber workers. All the cases were traced to a chemical used in a particular process for a specialty rubber, and it was found that everyone who worked on that process for 2 years or more died of bladder cancer within 7 years. The corrective action was obvious and was taken immediately: production of the rubber was terminated.

Chlorinated hydrocarbons such as those used in pesticides and herbicides are common causes of dysfunction of glands or the nervous system. Again, it is difficult to determine permissi-

TABLE 9.4   Elements Essential or Probably Essential to Plant and Animal Life

Elements basic to life		Trace elements essential to life		Trace elements probably essential to life	
H	P	Cr	Se	F	Cd
C	S	Mn	Mo	V	Ba
N	Cl	Fe	Sn	Ni	Sr
O	K	Co	I	As	
Na	Ca	Cu	Zn	Br	

Source: E.J. Underwood, *Trace Elements in Human and Animal Nutrition*, Academic Press, New York, 1971.

ble levels of exposure to these substances from inhalation, ingestion, and skin contact because of wide ranges in susceptibility between individuals, and between animals and humans, coupled with the large number of individuals required in test groups to investigate trace effects. Note that serious health effects and deaths caused by a toxic material before its degree of toxicity is realized, together with the effects of heavy doses associated with accidental spills, often provide vital data on the effects on humans to supplement data obtained from animals. Similarly, data from other countries with less stringent standards than the United States are often valuable, although their significance may be reduced because levels of exposure, numbers of people involved, etc., are not well documented.

From the standpoint of the general public, the most serious

toxic materials produced by power plants are sulfur compounds derived from fossil fuel. Sulfur is emitted as SO_2, which, while toxic in large concentrations, does not appear to have appreciable short-term adverse effects in concentrations below ~ 200 ppb.[15] Probably the most comprehensive data have been obtained in Great Britain, where major programs both to study the problem and to reduce sulfur emissions were instituted following the terrible "black fog" of London in 1952 that was charged with around 4000 deaths from pulmonary trouble. (The smog was so thick that cinemas closed, because it obscured the view of their screens!) Figure A9.1 includes data from British experience to indicate some of the major effects of sulfur emissions in relation to flue gas emissions and EPA regulations. A particularly significant point is that in studies of bronchitic patients during the 25-year period in which SO_2 levels in Britain were dramatically reduced (1952 to 1977), no statistically significant ill effects were found for 24-h exposures to SO_2 concentrations of less than 500 $\mu g/m^3$ (0.18 ppm) accompanied by 250 $\mu g/m^3$ of particulates.[15] However, SO_2 reacts in the atmosphere to form H_2SO_4, some of which is adsorbed on ash particles, and these, when ingested into the lungs, appear much more likely to have serious long-term effects, as discussed in the next section. Data for some other pollutants of interest are also included in Fig. A9.1.

Particulates

Fine airborne particles of dust are naturally present in the air, sometimes in high concentrations, as a consequence of dust storms or volcanic eruptions. Smoke emissions from furnaces have been sufficiently obnoxious to have been the subject of governmental regulation for centuries. Hard-rock mining has led to silicosis, and coal mining has brought on the "black lung disease." Asbestos fibers have been troublesome also, with the synergistic effects of heavy smoking being a major contributing factor. Whatever the type of material involved, the trouble is caused by the deposition on delicate lung tissue of substances insoluble in body fluids and the subsequent formation of scar tissue. This may gradually become so widespread that it severely reduces the effective lung surface area available for the exchange of O_2 and CO_2. The result is drastically limited capacity for physical activity, the effects being generally the same as the effects of emphysema, caused by smoking tobacco. Particular attention is being given at the time of writing to the ill effects of submicron-size particles emitted from the stacks of coal-fired power plants and to questions as to whether the composition or form of these particles differs from natural dust in ways that make their effects more serious (e.g., adsorbed H_2SO_4).

Ionizing Radiation

The long-term ill effects of continuing small doses of ionizing radiation first received widespread public attention in the early 1920s. Radium in paint for luminous watch dials was being applied in a small plant in Boston by women who pointed their brushes with their tongues and lips and thus gradually ingested appreciable quantities of radium that became concentrated in their bone marrow and severely interfered with the production of red corpuscles. General physical deterioration and death occurred, resulting in a dramatic court case. This and subsequent events, particularly the atom bombs of Hiroshima and Nagasaki, have led to an emotional reaction on the part of many people that has been without parallel for any other hazard of the twentieth century. The problems are real, but actually present a far less serious threat than most of the other hazards of our environment—a fact implicit in Table 9.2. This section on the effects of radiation on living organisms was prepared in an effort to provide the basic background required for a quantitative appreciation of the problems caused by radioactivity and a rational assessment of their magnitude.

Types of Radiation

The four types of radiation of prime interest from the standpoint of public health are x-rays, gamma rays, beta rays, and alpha particles. X-rays are photons having shorter wavelengths and higher energies than ultraviolet radiation, but they are similar in origin: both are photons emitted when electrons circling an atomic nucleus drop from one orbit to another in the course of decay from an excited state. Usually x-ray emission is stimulated in a material by electrons generated in a high-voltage electromagnetic device similar to a TV picture tube but operated at a higher voltage. The stream of high-voltage electrons gives up energy to electrons in the target atoms, putting them in an unstable excited state, from which they quickly decay. The x-rays produced are much more penetrating than ultraviolet rays (which will penetrate only about 1 mm of tissue), the most energetic being able to pass through many centimeters of steel. Gamma rays are basically similar to x-rays, but have still higher energies and hence are even more penetrating. They normally originate in the atomic nucleus as a consequence of radioactive decay from one isotopic state to another. For situations of interest here, beta rays (high-energy electrons) and alpha particles (the nuclei of helium atoms) are both emitted in the course of the decay of the nuclei of radioactive isotopes. These isotopes include naturally occurring elements such as radium, uranium, and thorium. Potassium is another of the naturally radioactive elements, though because its half-life is long (10^9 years) and the natural concentration of the radioactive isotope is only 0.01 percent, the activity per unit weight is small. However, it is an essential element in all living organisms, representing 0.1 percent by weight of wood and forming a vital constituent of animal bones.

Nuclear power plants generate radioactive material in the form of fission products (the residual fragments from fission of uranium and plutonium), and small amounts of these materials may leak into the environment.[16,17] In addition, uranium absorbs neutrons to form plutonium, which, like uranium, emits alpha particles together with gamma rays. Structural materials in the reactor core such as the Fe, Cr, and Ni in stainless steel also absorb neutrons to form radioactive isotopes with various half-lives, and reactor cooling water absorbs neutrons to give tritium and nitrogen 16 (^{3}H and ^{16}N), both of which are radioactive.[12]

Neutron and proton radiation will not be discussed here. They are not normally a problem from the public health standpoint because they are present only inside the shield when the reactor is in operation.

Radiation Measurement

The first unit of measurement of radiation intensity was developed for x-rays and is called the *roentgen* (so named after the inventor of the original x-ray tube). This unit is defined in terms of the ionization produced in air, 1 R being the amount of radiation required to ionize air sufficiently to produce one electrostatic unit of charge per cubic centimeter of air at standard conditions. The same unit of measurement is directly applicable to gamma rays. For all types of radiation and for materials in general (i.e., not just x-rays in air) the radiation dose is measured in *rad* which has been defined as the amount of radiation that will deposit 100 ergs/g at the point of interest. (A roentgen gives 87.7 ergs of energy deposited per gram of air.) The damage to animal and particularly human tissue varies both with the type of radiation and the type of tissue, e.g., the cornea of the eye is especially sensitive to neutrons. From the standpoint of damage to human tissue, a radiation dose unit called the *rem* (roentgen *e*quivalent, *m*an) is employed. Inasmuch as 1 rem represents a relatively large amount of radiation, the unit usually cited is the millirem; 1 rem = 1000 mrem. Both the *dose rate*—for example, 10 mrem/h—and the total accumulated dose are common measures of irradiation.

The amount of radioactivity associated with a radioactive material is expressed in *curies*. Although a curie was originally defined as the amount of activity in one grain of radium, the more fundamental definition of 1 Ci = 3×10^{10} disintegrations

TABLE 9.5 Typical Radioactive Elements and Their Half-Lives

Naturally radioactive		Fission products		Structural materials activated in reactors	
^{14}C	5700 yr	^{85}Kr	10 yr	^{55}Fe	2.6 yr
^{40}K	10^8 yr	^{90}Sr	28 yr	^{59}Fe	45 d
^{226}Ra	1620 yr	^{131}I	8 d	^{54}Mn	303 d
^{228}Th	1.9 yr	^{133}I	21 h	^{56}Mn	2.6 h
^{230}Th	10^5 yr	^{133}Xe	5 d	^{58}Co	71 d
^{235}U	7×10^8 yr	^{137}Cs	29 yr	^{60}Co	5 yr
^{238}U	5×10^9 yr	^{140}Ba	13 d	^{95}Nb	35 d

per second is now the international standard. Note that as time passes, a given quantity of a radioactive material decays and its activity is reduced by a factor of 2 in a time interval known as the *half-life*. Each radioactive isotope has its own characteristic emissions and half-life; Table 9.5 lists the principal radioisotopes of interest from the public health standpoint.

Effects of Radiation

The effects of various amounts of radiation on mammals are indicated in Table 9.6. For total doses below about 25,000 mrem the short-term effects of dose rate appear to be minor, and the overall long-term effects are indistinguishable from those of aging. That is, radiation causes cell mutations that are degenerative: graying of hair, reduced elasticity of the skin, etc. The two most important deleterious changes that may occur are the transformation of normal cells into cancerous cells and effects on the rapidly evolving cells of a fetus or a young child. The latter is the reason that a particular effort is made to avoid irradiation of pregnant women—and this includes the use of x-rays.

There have been many dire predictions of the horrible disfigurations that will stem from the genetic effects of radiation, two-headed babies being a favorite example of this sort of speculation. Certainly radiation during pregnancy can induce birth defects, but the magnitude of the genetic hazard (which is the long-term concern) is indicated by the fact that 34 years of searching for signs of genetic effects in the progeny of the survivors of Hiroshima and Nagasaki have failed to turn up statistically significant evidence of genetic damage.[18] The lack of evidence apparently stems from the fact that serious mutations are usually lethal, i.e., they result in miscarriages or still births. (Roughly 15 percent of conceptions in humans result in miscar-

TABLE 9.6   Biological Effects of Radiation[12]

Dose level, mrem*	Biological effect
Less than 1000	Data at low dose rates and total doses, even in test animals, are inadequate to assess effects at low levels. Therefore, effects are conservatively assumed to be directly proportional to total dose received regardless of dose rate (linear hypothesis).
1000	100 leukemia cases per million persons exposed.
9000 (lifetime dose from natural background radiation) ($\sim$ 125 mrem/year)	Normal human life span and diseases.
Less than 25,000	Detectable only by laboratory examination. No clinically observable effects below about 50,000 mrem.
Less than 100,000	Little or no life shortening.
Less than 250,000†	Few or no deaths, but acute radiation sickness and significant life shortening.
450,000†	Death of 50% of those exposed in less than 30 days; the other 50% recover but with some permanent impairment.
1,000,000†	Death in less than 30 days.

*Once in a lifetime dose received in a short time (i.e., a few hours or less.)
†Radiation sickness includes vomiting, diarrhea, loss of hair, nausea, hemorrhaging, fever, loss of appetite, and general malaise. Recovery (if no complications) occurs in about 3 months.

TABLE 9.7   Average Annual Doses from Radiation Sources in the United States[12]

Type of radiation	Some sources	Average annual dose, mrem/year*
Alpha and gamma	Natural radioactivity (e.g., uranium) in solids, rocks, minerals	30
Beta	Natural radioactivity (e.g., potassium 40) in soils, rocks, minerals, bones	20
	Television (an average of 1 h per day)	$\frac{1}{2}$
	Natural radioactivity in the air (e.g., tritium)	2
Gamma	Medical and dental x-rays	20
	Cosmic radiation at sea level	40

*Dose to reproductive organs.

riages, mostly at very early stages and so are not recognized as such by the mothers.) Controlled experiments with mice have yielded similar results. At Los Alamos, New Mexico, J. R. Spalding exposed each of 83 generations of mice to 200,000 mrad of ionizing radiation. Individual mice were adversely affected, but no long-term genetic effects were observed in this 22-year experiment—which would have taken 2500 years if 83 generations of humans had been similarly exposed.

Sources of Radiation

The sources of most of the 9000-mrem total radiation dose received during an average lifetime are listed in Table 9.7. The bulk of this radiation is from natural sources—e.g., uranium in rocks, cosmic rays, etc. A substantial amount (about 15 mrem per year) is from potassium in our bones, where this element must be present for good bony structure. Medical and dental x-rays contribute the bulk of the balance. Radiation to the public from nuclear power plants either directly or through the food chain is almost negligible.[16,17] Even in the highly publicized Three Mile Island nuclear plant accident in 1979, the maximum dose that anyone could have received would have been ~ 40 mrem even if that person had stood at the worst point along the perimeter fence and had remained in that spot throughout the crucial week in which activity was emitted. This was the equivalent of about three chest x-rays, a radiation dose vastly less serious than the impression given by the sensational treatment of the incident by the news media. One cannot help but feel that journalists and political figures have a basic responsibility to be sufficiently well informed to put matters in good perspective in emergencies of this sort rather than to generate a highly emotional drama that induces mass hysteria and public panic.

The most toxic of the radioactive materials associated with fission reactors is plutonium. Its toxicity is compared with that of more familiar poisons in Table 9.8. It is an α-emitter, which means that its radiation is not penetrating; hence it must be ingested to have large effects. Statements that it is the most toxic substance known to human beings are simply not true. Table 9.8 shows that its toxicity is less than that of several other familiar poisons, and of the same order as many others. Significantly, not a single human death has been caused by plutonium poisoning up to the time of writing.

Further perspective on the relative effects of various common environmental factors on life span is given by Table 9.9. Table 9.10 gives data on the estimated incidence of cancer from the principal sources of radiation in our environment assuming linear effects, i.e., that there is no threshold. Note that present and projected nuclear power plants give a virtually negligible contribution—a fact very different from the fearful picture painted by antinuclear activists.

TABLE 9.8 Comparison of Reactor Plutonium* with Highly Toxic Materials [12]

Toxin or poison	Lethal dose (mg)	Time to death
Ingested (swallowed)		
Anthrax spores	under 0.0001	—
Botulism	under 0.001	—
Lead arsenate	100	hours to days
Potassium cyanide	700	hours to days
Reactor plutonium	1,150	over 15 years
Caffeine	14,000	days
Injected		
North American coral snake venom	0.005	hours to days
Indian King cobra	0.02	hours to days
Reactor plutonium	0.078	over 15 years
Diamondback rattler	0.14	hours to days
Inhaled		
Reactor plutonium‡	0.26	over 15 years
Reactor plutonium‡	0.7	3 years
Nerve gas (Sarin)	1.0	few hours
Reactor plutonium‡	1.9	1 year
Reactor plutonium‡	12	60 days
Benzpyrene (1 pack/day of cigarettes for 30 years)	16.0	over 30 years

Toxin or poison	Lethal concentration† (mg/ft³)	Time to death
Inhalation atmosphere		
Reactor plutonium‡	0.026	over 15 years
Reactor plutonium‡	1.3	60 days
Cadmium fumes	10	few hours
Mercury vapor	30	few hours
Phosgene	65	few hours

*Mixture of plutonium isotopes as plutonium oxide (more than five times more hazardous than plutonium 239)

†Four-hour exposure.

‡Exposure at different levels.

TABLE 9.9 Life-Shortening Effects of Various Factors in Human Experience [12]

Factors tending to decrease average lifetime	Decrease of average lifetime
Overweight by 25%	3.6 years
Male rather than female	3.0 years
Smoking	
1 pack per day	7.0 years
2 packs per day	10.0 years
City rather than country living	5.0 years
Actual radiation from nuclear power plants in 1970	less than 1 min
Estimate for the year 2000 assuming hundredfold increase in nuclear power production	less than 30 min

TABLE 9.10 Estimated Annual Cancer Mortality in the United States from Radiation and Other Causes—1970–2000 [12]

Cause of cancer	Cumulative cancer deaths to 2000	Average annual cancer deaths
Radiation		
Natural background	200,000	6,700
Medical x-rays	100,000	3,300
Jet airplane travel	7,000	230
Weapons fallout	7,000	230
Nuclear power industry	90	3
Total from all radiation sources	314,090	10,500
All other causes	11,686,000	389,500
Total from all causes	12,000,000	400,000

Maximum Permissible Concentrations

The maximum permissible concentrations of representative toxic materials in air and drinking water are summarized in Tables 9.11 and 9.12. The values for radioactive materials have been established by an international commission and are based on extensive experimental data. The values for the other more conventional materials in some instances have an equally firm base—e.g., heavy metals such as lead and mercury—but in general the values used in the regulation of these materials are less firmly based than those used for radioactive materials because they have not been the subject of as much research.

It should be remembered that the toxicity of a material depends on how it is ingested and its residence time in the body. This varies widely with the material, and is easily determined for radioactive materials. Tritium, for example, is eliminated rapid-

TABLE 9.11 Typical Upper Limits on Concentrations of Air Pollutants [15, 16, 17]

Combustion products, ppm		Industrial chemicals, ppm		Radioisotopes, Ci/cm^3	
CO	60 (1 h)	Ammonia	50	^{3}H	2×10^{-7}
	15 (8 h)	Arsenic	0.3	^{85}Kr	3×10^{-7}
NO$_x$	0.1 (1 h)	Benzene	25	^{90}Sr	3×10^{-11}
SO$_2$	1	Bromine	0.1	^{131}I	1×10^{-10}
H$_2$S	0.1 (1 h)	Butyl alcohol	100	^{135}Xe	1×10^{-7}
		Carbon tetrachloride	65	^{137}Cs	2×10^{-9}

Particulates, particles/m$^3 \times 10^{-6}$	
Asbestos	175
Mica	700
Silica	5250
Soapstone	700
Portland cement	175

Industrial chemicals (cont.)		Radioisotopes (cont.)	
Chlordane	0.5	^{228}Th	3×10^{-12}
Chloroform	240	^{235}U	2×10^{-11}
DDT	1	^{238}U	3×10^{-12}
Ethyl alcohol	1000	^{239}Pu	6×10^{-14}
Gasoline	500		
HCl	5		
Ozone	0.1		
Phosgene	1		
Turpentine	100		

TABLE 9.12 Mineral Content in ppm of Water from Typical Sources Together with Standard Tolerances. (Values for emissions from power plants have been underlined where they exceed drinking water tolerances by a factor of ten or more.)

	Sea-water	New Zealand geothermal well [23]
Aluminum	1	
Antimony	0.0004	
Arsenic	0.003	3
Barium	0.03	
Boron	4.6	19
Bromine	6.5	4
Cadmium	0.0001	
Calcium	400	12
Carbon	30	
Chlorine	19,000	1460
Cobalt	0.0004	
Copper	0.003	
Fluorine	1.3	0.5
Germanium	0.00006	
Lead	0.00002	
Lithium	0.17	9
Magnesium	1275	0.003
Manganese	0.002	
Nickel	0.002	
Nitrogen	0.5	
Phosphorus	0.07	
Potassium	380	135
Selenium	0.004	
Silicon	3	440
Silver	0.0002	
Sodium	10,500	850
Strontium	8	
Sulfur	885	17
Thorium	0.00005	
Uranium	0.003	
Vanadium	0.002	0.2
Zinc	0.01	
Fe	0.01	
Chromium	0.0004	
Mercury	0.00002	0.0001
Be		

Handbook of Tables for Applied Engineering Science, 2d ed., CRC Press, 1973.

Leachates from coal ash[21]				Actual public water sources[24]		Permissible concentrations		Trace elements required by plants and animals
Ash pond	FGD sludge pond	FBC ash	FBC spent stone	Port Arthur, Tex., water supply	Lower Mississippi River	Irrigation water*	EPA primary drinking water[13]	
						1		
0.03				0.3				
0.08	0.3	2.5	5			1	0.05	✔
40	2						1	
17	40	0.6	0.8			0.75	1	
								✔
0.01	0.05	0.01	0.2			0.005	0.01	✔
				19	135			✔
				25	11			✔
						0.2		✔
0.09	0.56		0.1			0.2	1	✔
17							2	✔
0.1								
0.24		2.5	2.4			5	0.05	
						5		
0.002	0.04			9	53			
						2	0.05	✔
0.05	0.05					0.5		✔
				5	13		10	✔
				1	1			✔
				1	1			✔
0.47	0.54					0.05	0.01	✔
				6	7			✔
							0.05	
				17	29			✔
								✔
				12	84		80	✔
0.2						10		✔
0.2	4.2					5	5	✔
							0.3	✔
	0.25	0.1	0.1			5	0.05	✔
0.015	0.07	6	13				0.002	
0.003	1		0.01			0.5		

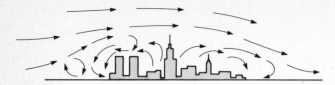

Figure 9.3 Air convection currents producing a heat bubble over a city.

ly; only half the amount entering the body remains at the end of a week. Strontium, on the other hand, is chemically similar to calcium and if it enters the bloodstream, it tends to be absorbed in the bones so that its average residence time in the body is a matter of years rather than days. Particulates may become embedded in the lungs and, if insoluble, remain there for years.

ATMOSPHERIC POLLUTION

The biggest sources of pollution of the earth's atmosphere as a whole are natural and agricultural. Volcanoes, desert windstorms, and decaying vegetation contribute roughly double the total amount of sulfurous gases, dust, and organic vapors emitted to the atmosphere by industry. A glance out my window at the majestic Great Smoky Mountains reminds me that the blue haze enveloping them—the haze which gave both them and the Blue Ridge Mountains of Virginia their names two centuries ago—is formed largely by atmospheric reactions with the organic vapors rising from decaying vegetation on their forest floors and from terpenes emitted by evergreens. A major problem with homogenic emissions is that they are concentrated in urban areas, a situation that is aggravated by both temperature inversions that develop during the night and heat bubbles that prevail over large cities (Fig. 9.3). These conditions lead to recirculation of the air over cities, a buildup in the concentration of pollutants, and the formation of all-too-familiar smogs. Although most of the smog comes from motor vehicle exhaust, about 10 percent is commonly from coal- or oil-fired steam power plants and hence a matter of prime interest here.

Dispersion of Pollutants

The modes of dispersion of pollutants are readily visualized by considering the plume emerging from a smokestack. The smoke is carried horizontally by the wind while being dispersed laterally by atmospheric turbulence to give a plume that grows in diameter until it becomes so diffuse that it is no longer visible.[19] The higher temperature of combustion gases makes them lighter

than the surrounding air, and so the plume rises until the adiabatic temperature drop in the gas brings its density into equilibrium with that of the surrounding air. This may occur abruptly at a relatively low elevation if there is an atmospheric temperature inversion, or at a much higher elevation if the normal atmospheric temperature lapse with increasing altitude prevails. The wind velocity, the terrain, the volume flow rate of the stack gas, and the difference in temperature between the stack gas and the surrounding air affect the rate of lateral dispersion.

In attempting to assess the potential for adverse effects of a smoke plume on any given location in the surrounding area, it is necessary to consider the probability that the wind will carry the plume in that direction. This probability is conveniently expressed diagrammatically by a *rosette* such as that of Fig. 9.4. The probability that the wind will be in a particular direction, coupled with the degree of plume dispersion into the air at ground level, provide a basis for estimating the effects of the stack emissions on human health. Note that areas close to a tall stack may be subject to little or no pollution because the plume envelope does not expand to ground level. For the tall stacks of coal-fired steam plants, the peak concentration at ground level is commonly found at distances of 2 to 15 km (depending on the terrain and atmospheric conditions), the dilution factor at that point being of the order of 10^4 in relation to the initial stack-gas emission (depending on stack height and wind conditions). Consider, for example, a case in which the sulfur content of the coal used is 2 percent, giving a sulfur content in the stack gas of 4 g/m^3, or ~ 1440 ppm. If the ground level tolerance is taken to be 200 $\mu g/m^3$ (0.072 ppm), the dilution factor required is ~ 20,000. Generally, it is possible to get sufficient dilution to meet ground level standards by using a sufficiently tall stack, the approach followed in Great Britain. However, in the United States the EPA has chosen to place limits on the sulfur content of the flue gas leaving the stack irrespective of stack height. Political forces from high-sulfur-coal-mining areas have asserted that this imposes an unfair penalty on their coal and thus have obtained regulations mandating the use of SO_2 scrubbers irrespective of the sulfur content of the coal used. Consequently, a utility finds little advantage either in burning low-sulfur coal or in using tall stacks. Thus politics rather than public health or cost-benefit effects have come to dominate the regulation of sulfur emissions. It is significant that the factor-of-10 reduction in total sulfur emissions achieved in the New York metropolitan area between 1965 and 1979 has certainly reduced the smog problem, but contrary to expectations, it has not been accompanied by statistically significant reductions either in pulmonary problems or in the rate of deterioration of paint, marble, limestone, etc.

Emissions other than smoke plumes may also pose problems.

The exhaust from stationary diesel engines and gas turbines is usually discharged near ground level but is ejected at a high enough temperature and velocity to lift it well above the stack outlet. The wet vapor from cooling towers exits at low velocity and at a temperature only a little above the ambient air; hence there is little rise in the discharge plume. Leaks of toxic chemicals from piping or a damaged tank car often constitute ground-level point sources, resulting in peak concentrations immediately downwind. The ventilation systems of nuclear plant buildings are designed so that any accidental leaks of radioactive gas into the building are diluted by the building air, which is filtered and discharged up a stack perhaps 70 m high to minimize the potential ground concentration. Here again the most unfavorable atmospheric conditions are given by a low wind velocity and a temperature inversion. Under these conditions the ground-level concentration in the plume drops as the square of the distance from the source—as is the case along the centerline of a smoke plume.

Reduction of Gaseous Emissions

The amount of objectional material emitted may be reduced by any of several methods, each generally most effective if applied before there is any more dilution than can be helped. The concentration of particulates can be reduced by means of cyclone separators, electrostatic precipitators, filters, and/or water scrubbers (spray towers). Some gases and vapors can be removed chemically in scrubbers; SO_2 gas, for example, can be reacted with lime in the scrub water to form $CaSO_3$, which is removed as suspended solid particles. The technology and costs involved are treated in Chap. 11, which discusses steam plants.

In dealing with the much lower volume flow rates of potentially contaminated air that might be released from nuclear plants, more elegant equipment can be employed. Silver and copper mesh are used to remove radioactive iodine that might contaminate air vented from nuclear plants, and activated carbon or fluorocarbons can be used to adsorb or absorb the radioactive fission product gases xenon and krypton. These measures commonly can be used to reduce the amount of activity released by a factor of 100 to 1000.

WATER POLLUTION

Questions to be answered about water pollution concern the acceptable concentration of each of the potential contaminants in various bodies of water, methods for restricting the release of contaminants, together with the amounts involved and the associated costs, as well as the movement of the contaminants with water circulating in the hydrosphere.[20-31] Once this information has been acquired, one has a basis for determining the best methods of handling water-pollution problems.

In former years water quality was judged by turbidity, bacterial content, hardness, and acidity or alkalinity. The suspended solids and much of the bacteria can be filtered out, the remaining bacteria can be kept to acceptable levels by chlorination, the hardness can be reduced by chemical treatment with a material such as zeolite, and the acidity or alkalinity can be neutralized. The principal objectionable bulk contaminant of municipal waste water has been organic material, and this is removed by digestion with bacteria, a process requiring aeration — hence the use of the parameter *biological oxygen demand* (BOD) as a measure of the organic material content. However, the rapid growth of the chemical industry and the proliferation of new commercial chemical compounds present a new and complex set of potential contaminants. Many of these present problems far more serious than those posed by wastes from power plants, but they are beyond the scope of this text. The prime concerns here are with trace elements in coal ash and radioactivity from nuclear plants that might enter drinking

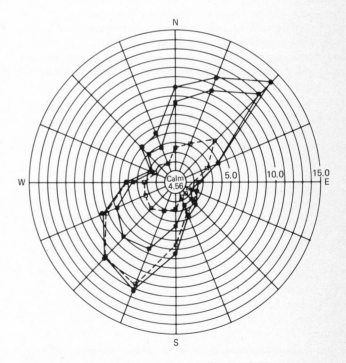

Figure 9.4   On-site wind rose for October 20 to December 31, 1972. Facility is on a site at 620 ft MSL. The wind instrument is 33 ft above ground. *(Courtesy TVA.)*

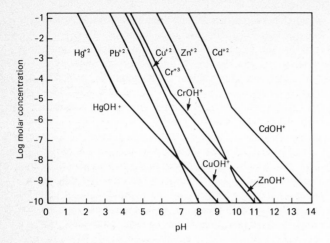

Figure 9.5 Effects of pH on the solubility of metals in coal ash leachate. *(From "Conceptual Design of an Atmospheric Fluidized-Bed Combustion Electric Power Generating Plant," prepared by Burns and Roe, Inc., for U.S. DOE.)*

water supplies. The trace elements and radioactive materials may be released to the environment in the course of or following mining operations, they may escape with the stack gas, or they may be leached from the ash or residue resulting from flue-gas cleaning processes such as wet-limestone flue-gas desulfurization (FGD).

Water Discharges from Coal and Geothermal Plants

Table 9.12 includes columns for the permissible concentration of minerals considered objectionable in drinking and irrigation water, together with a summary of the mineral content of typical seawater and freshwater supplies. Data are also given for the water discharged from a geothermal power plant. Note that, where the concentration of an element in the leachate or geothermal plant effluent exceeds the drinking water tolerance by a factor of 10, the value has been underlined to indicate that even after dilution with ground or river water it might be a problem. There are 15 numbers underlined in Table 9.12, the elements involved being arsenic, barium, boron, cadmium, lead, mercury, and selenium. For all seven elements the permissible concentration in drinking water is 1 ppm or less (for mercury it is only 0.002 ppm); for such low concentrations quite elegant, exacting, and expensive techniques are required to check for compliance. It is interesting to observe that, as one would expect, the elements whose concentrations are considered unhealthful and thus are being limited have low concentrations in seawater, the substance in which life first evolved.

The quantity of a trace element in the leachate from an ash or sludge settling pond is approximately proportional to its concentration in the coal;[25] hence if a problem develops it may be handled by a change in the source of the coal used. Another approach is to treat the ash or sludge to change the solubility of the element in question. Changing the pH, for example, has been found effective[21] (Fig. 9.5).

The leachates from both ash and scrubber sludge are usually quite alkaline, with the value of the pH commonly ranging from 8 to 12. The water draining from western coal and coal mines is also commonly alkaline. However, in the east the drainage from piles of coal, mine tailings, and from mines themselves is usually quite acid because the eastern coal contains substantial amounts of inorganic sulfur—largely iron pyrite—that oxidizes readily on exposure to air and moisture to form sulfuric acid. This is a major source of stream pollution. To protect fish and other aquatic life, the pH of water draining from mining areas is not supposed to be less than 6.5. Drainage from alkaline sources such as ash ponds should have a pH of no more than 8.5. (Pure water has a pH of 7, a phenothaline-neutral pH is 8.2, and the pH of rain is normally 5.6 because of dissolved CO_2.)

Limiting Releases of Contaminated Water

In terms of sheer quantity, the most serious water-pollution problem associated with power plants is acid drainage from mined areas, coal-washing plants, and coal piles. Using average values for the Appalachian area, in which about 35 percent of the coal is surface-mined, a recent estimate shows that a 1000-MWe plant supplied with an average mix of coal would lead to the annual release of 66,000 tonnes (t) of sulfuric acid from the surface-mined areas and 55,000 t of acid from the deep mines.[26] About 90 percent of this could be neutralized with lime if proper measures were taken.

Although formerly the ill effects of high concentrations of inorganic material in waste water were handled by dilution in riverine systems, the enormous increase in industrial activity since the end of World War II has led to excessive buildups of dissolved solids in many rivers. This has been particularly bad for cities downstream of Pittsburgh and St. Louis, for example, because the dissolved solids content of the river water increases as the river passes each urban area. The water passing through a municipal water and sewage system undergoes an increase in dissolved solids content of ~250 ppm, which compares with the usual upper limit of 500 ppm for potable water. Although the water diverted from a river through a city water system is usually a modest fraction of the total river flow, the cumulative effects of a series of cities become large, e.g., the Mississippi at New Orleans contains 350 ppm of dissolved solids. In view of the seriousness of this problem, the author has suggested that if fusion reactor plants of the next century could be located within 20 km of the city, the reactors could be run at full power all the time irrespective of the electric load, and the waste heat might be used for a distillation purification of all the city's sewage. Such a step apparently would cost an urban area less than current EPA requirements for FGD systems because of the low cost of fuel for fusion reactors. Systems of this type are discussed in Chap. 17.

Radioactivity in Waste Water from Nuclear Plants

All emissions from nuclear plants are tightly controlled. The amount of radioactivity permitted in the waste water is carefully kept to low levels. As a check, the water in the surrounding area is continually monitored to ensure the efficacy of the control measures. In addition, detailed exhaustive surveys have been conducted in the vicinity of plants at typical sites to trace the paths that radioactive isotopes may take through the food chain. Extensive work has been carried out around all the U.S. government plants since their construction in the 1940s[16] and around some of the commercial power plants.[17] Extremely sensitive instrumentation is used in these studies, hence very slight amounts of radioactivity are detected. In fact, the dominant activity is commonly of natural origin. In general, it is found that the concentrations of radioactive isotopes are highest in the lowest forms of life, such as algae, rather than increasing progressively up through the hierarchy of the food chain.[16,17]

It is interesting to review a typical study made in the vicinity of the 640-MWe Oyster Creek boiling-water reactor plant, which began operation in 1969.[17] The plant is located on Barnegat Bay, a popular site for fishing, swimming, and sailing about 75 km south of New York City. The bay is ~50 km long and 5 km wide, with an average depth of only 1.5 m and a volume of ~2.4 × 10^8 m^3. Commercial fishing yields ~30,000 kg per year of fish, 200,000 kg per year of clam meat, and 30,000 kg per year of blue crabs. The flushing action of tides causes an exchange of roughly 14 percent per day of the water in the bay with the ocean. The condenser cooling-water flow rate is typically 1700 m^3/min, or 10^9 m^3 per year, with a maximum temperature rise of 12.8°C. The waste water containing radioactive material is released periodically from storage tanks to the condenser cooling-water flow after monitoring to ensure that the activity is below tolerance levels, the release rate running ~17,000 m^3 per year. For details of this fine study the reader is urged to read Ref. 17. The most important results are reproduced here in Table 9.13, which summarizes the results from the standpoint of the radiation dose to a person living in the vicinity with a taste for seafood, assuming that such an adult would eat 21 kg of fish and 5 kg of shellfish and crab per year. Note that the principal dose measured would be from ^{210}Po, a naturally present isotope not contained in the nuclear plant wastes, and that the activity actually measured in seafood was lower than that calculated from a computer model for the conditions given. Note also that the total dose to bathers spending 67 h per year on the beach was found to be only 7×10^{-4} mrem.

Hydrology

One of the most important considerations for appraising the environmental effects of power plants is the circulation of water in the hydrosphere.[28,29,30] A remarkably detailed quantitative insight into this circulation has been provided by monitoring the movement of the tritium released in H-bomb explosions and following its course first through the atmosphere and then through the hydrosphere.[30] (Tritium is a beta-emitter with a half-life of 12.36 years. The nature of its radiation makes possible exceedingly sensitive instruments that can detect even the one tritium atom in 10^{18} atoms naturally present in hydrogen as a consequence of cosmic-ray activity in the upper atmosphere.) The bulk of the tritium from an H-bomb explosion initially appears in the troposphere, where its mean residence time is about a month, with about one-third of it eventually flushed out by rain and two-thirds by vapor exchange with the oceans. About half of this occurs in the temperate region between 30 and 50° latitude. Of the rainfall on land, on the average about two-thirds returns to the atmosphere by evaporation and plant transpiration. This ratio varies widely from one region to another and is heavily dependent on the precipitation rate. During a heavy rain, the runoff is a large fraction of the rainfall, because there is not time for evaporation to occur. However, if a pool of water is maintained in a leak-tight basin, the annual evaporation rate will exceed the rainfall over most of the western United States. The net balance between rainfall and evaporation for the United States as a whole is shown in Fig. 9.6. A study of the Mississippi River valley area indicates that about two-thirds of the rainfall is water evaporated from the oceans (largely the Gulf of Mexico) and about one-third is water reevaporated from the land.[30] Thus of the 0.77 m per year rainfall, about 0.28 m returns to the ocean by runoff and 0.49 m per year is evaporated, of which 0.24 per year returns to oceans through the atmosphere.

Depending on the structure of the soil and the rock strata in the region, the bulk of the runoff may occur on the surface or a large fraction may percolate through the soil and/or flow through aquifers (porous regions above or between impermeable rock strata). The velocity of groundwater flow varies widely from ~2 m per year to as much as ~100 m per day.[30] In some areas in the U.S. west and the Sahara Desert, the rock

TABLE 9.13 Internal Radiation Dose from Fish and Shellfish Consumption in the Vicinity of the Oyster Creek BWR[17]

Radionuclide	Food	Ingestion dose factor, mrem/pCi	Critical organ	Concentration, pCi/kg		Dose equivalent rate, mrem/year		Exposure source
				Calculated	Measured	Calculated	Measured	
^{32}P	Fish	1.93×10^{-4}	Bone	1600	<200	6.2	<0.8	S
	Invertebrate			1700	<400	1.5	<0.4	S
^{55}Fe	Fish	2.75×10^{-6}	Bone	1500	< 80	0.09	<0.005	S
	Invertebrate			9800	<100	0.14	<0.002	S
^{60}Co	Fish	4.02×10^{-5}	GI-LLI	76	6	0.06	0.005	S
	Invertebrate			760	180	0.15	0.04	S
^{65}Zn	Fish	1.54×10^{-5}	Liver	30	< 30	0.01	<0.01	S
	Invertebrate			4200	< 25	0.3	<0.002	S
^{90}Sr	Fish	7.58×10^{-3}	Bone	0.06	2.3	0.01	0.37	S
				0.6		0.096		F
	Invertebrate			0.6	< 12	0.02	<0.6	S
				6		0.2		F
^{95}Nb	Fish	2.10×10^{-5}	GI-LLI	690	< 10	0.3	<0.005	S
	Invertebrate			2.3	< 80	0.0002	<0.009	S
^{131}I	Fish	1.95×10^{-3}	Thyroid	2.6	< 20	0.10	<0.0.7	S
	Invertebrate			13	< 20	0.12	<0.2	S
^{137}Cs	Fish	1.09×10^{-4}	Liver	33	30	0.08	0.07	S
				16		0.04		F
	Invertebrate			20	30	0.011	0.02	S
				10		0.005		F
^{210}Po	Invertebrate	2.74×10^{-3}	Spleen		380		5.2	N

Notes:
1. Annual ingestion rate for maximum exposure individual (adult) is 21 kg fish and 5 kg invertebrates ("other seafood") according to Ref. 20, p. 1.109–40; ingestion dose factors are for the critical organ in adults, from the same reference, p. 1.109–56, except that the value for ^{210}Po is from Ref. 4; the interval between catching and eating seafood is assumed to be 1 day.
2. GI-LLI = gastrointestinal (tract)—lower large intestine; S = station; F = fallout; N = nature.
3. Invertebrate = clam or crab. ^{60}Co and ^{210}Po were found in clam meat; ^{137}Cs was found in crab meat.

strata form basins filled with sand and gravel containing "fossil" water that has remained from the period of heavy rainfall of the last ice age (as determined from carbon 14 analysis of the dissolved CO_2).

The oceans are the final repository for minerals dissolved in water, and their volume of 1.35×10^{18} m³ represents 97.3 percent of the earth's water (not including bound water in the earth's interior).[29] Glaciers and polar ice represent another 29×10^{15} m³, and water in underground aquifers about 8.4×10^{15} m³. Lakes and rivers total only 0.2×10^{15} m³, the atmosphere still less—0.013×10^{15} m³—and the biosphere a mere 0.0006×10^{15} m³.

The rate of mixing in large bodies of water varies widely.[30] In the oceans, wave action produces relatively rapid mixing to a depth of about 40 m, with less rapid mixing down to the thermocline at a depth of ~ 100 m. Estimates of the mixing time down to the thermocline range vary from ~ 3 to ~ 20 years. The upper 100 m of the oceans represents an enormous reservoir—$\sim 40 \times 10^{15}$ m³—whose mineral content is vastly greater than the mass of any wastes that might be discharged into it. For example, the total annual world's production of mercury, the most toxic of the materials in Table 9.13, is equivalent to ~ 1 percent of the mercury contained in the top 100 m of the oceans.

Mixing rates in other large bodies of water are much faster. For example, Lake Michigan has complete vertical mixing each year,[30] and tidal action leads to flushing of the Delaware River estuary at a rate that gives about 50 percent dilution per month. Mixing rates in rivers are quite rapid. Essentially full lateral and

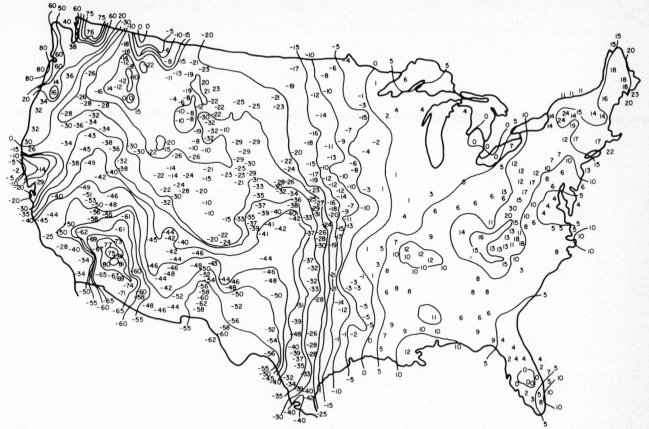

Figure 9.6 Average net precipitation in the United States. Figures represent differences in inches between precipitation and evaporation.[27] *(Source: U.S. Weather Bureau.)*

vertical mixing occurs in 6 to 10 km for a river such as the Clinch[30] (in Tennessee), which has a mean flow rate of 130 m³/s.

The oceans are a sink for CO_2, as well as for dissolved solids; roughly half of the CO_2 released by combustion of solid fuel diffuses from the atmosphere into the oceans. Much of the SO_2 entering the atmosphere follows a similar course.

Reducing Emissions

One of the most important factors influencing the choice of measures to minimize water pollution by power plants is the prevailing rainfall. As shown in Fig. 9.6, for much of the United States the annual rainfall is less than the evaporation losses from the surface of a pond; hence in these regions sludge and ash slur-

ries can be pumped into a pond to dry out, and there will be no runoff. Seepage through the soil to groundwater can be largely prevented by lining the pond with an impermeable barrier of clay, plastic film, asphalt, etc.

For regions where the rainfall exceeds the evaporation rate, the drainage from the ash or sludge in a settling pond may be treated to bring the contamination within acceptable limits. This may be accomplished by running the acid drainage from the coal pile into the ash settling pond, at once neutralizing the acid and reducing the pH of the effluent from the ash pond. The amount of suspended solids in the effluent can be kept below the EPA requirement of 100 ppm maximum, 30 ppm average, by staging the water flow through sections of the settling pond by means of inverted weirs, etc. The contamination

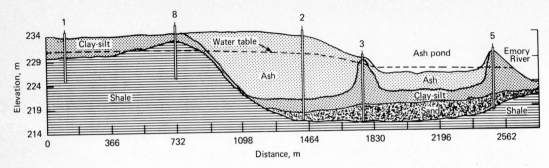

Figure 9.7 Topographic profile through an ash pond in relation to the water table, wells, and a river. *(Courtesy TVA.)*

of groundwater can be minimized by using a barrier of asphalt or plastic as described above (Fig. 9.7). These problems are treated in the latter part of Chap. 11.

In many instances the overall economics or the lack of water supply favor complete recycling of the water. This requires treatment to keep the level of dissolved solids to an acceptable level—a subject treated further in Chap. 11. Organic materials in the form of grease, fuel, lube oil, and cleaning materials represent a biological oxygen demand, as well as forming unsightly and damaging scums on the water surface. EPA requires that the total of these materials in waste water be kept to 20 ppm maximum, 15 ppm average.

SOLID WASTES FROM FOSSIL FUEL PLANTS

The amount of solid waste associated with operation of a coal-fired plant is large and presents a major disposal problem. Extensive efforts have been made to find uses for these materials, and about 20 percent of the 68×10^6 tons of ash produced by electric utilities in 1977 was utilized in some way, e.g., in concrete, in the manufacture of portland cement, as fertilizer, or as landfill. Some of the sludge from FGD processes is used for making wallboard and similar building materials, but the market is much less than the supply in most localities. If coal mines are not far from the plant, both ash and FGD sludge may be dumped back into the mines if the rock structure is favorable. This can be helpful in reducing or eliminating both acid drainage and subsidence in the mined area. However, in most cases the material is simply impounded in settling ponds. For a 1000-MWe plant using 12 percent ash, 3.5 percent sulfur coal, the pond capacity for a 35-year plant life would typically be 6.1×10^6 m³ (~5000 acre-ft) for the ash and 26×10^6 m³ (~21,000 acre-ft) for the FGD sludge.[26,31]

Mine tailings and the residue from coal-washing operations

represent huge volumes of waste material. Assuming that these are stacked to a depth of 12 m (40 ft), the area required per year for a 1000-MWe plant using 65 percent strip-mined coal would be ~4 ha (10 acres). The surface area disturbed in mining this Appalachian coal would include ~60 ha (150 acres) affected by subsidence of underground-mined coal, 960 ha (2400 acres) strip-mined, and 2200 ha (5600 acres) disturbed by strip-mining operations.[26] Under laws passed in the 1970s the bulk of this area could and would be restored within a few years time. The thicker seams of northwestern coal and much smaller slopes of the ground surface entail much less area (~200 ha) mined each year, but the much drier climate seriously slows reclamation by replanting.

RADIOACTIVE WASTES

Probably the most serious charges leveled against nuclear power are that the problems of radioactive-waste disposal have not been solved, that the wastes must be guarded for millions of years to prevent hundreds of millions of deaths from cancer, and that this is a horrible burden to place on the thousands of generations to come. The extent to which these charges are misrepresentations may be judged by several facts. First, the fission-product activity decays so that in 500 years it is less than that of the original uranium. Second, consider the often-repeated statement that the number of deaths from cancer that *could* be caused by fission-product wastes is 10^{10} deaths per year, which happens to be roughly double the total population of the earth. This figure is derived by assuming that all the radioactive material would be encapsulated and administered in just the right dosage to all the people on earth plus as many imaginary people as would be required to completely consume it. Following this "logic," it might be reasoned that the arsenic trioxide imported annually into the United States similarly administered would cause roughly as many deaths,[32] or that one man could supply enough sperm to impregnate every fertile woman on the whole earth. Clearly, such an approach to a potential hazard is irrational, and a more realistic basis for an appraisal is in order.

Reducing Emissions

A variety of ways to dispose of radioactive wastes has been proposed, depending on their individual character. By far the most difficult materials to dispose of are the fission-product wastes from fission-reactor fuel elements. A high-level group commissioned by the President concluded in a 1979 report[33] that "Confidence has now increased to the point that the majority of informed technical opinion holds that the capability now exists to characterize and evaluate media in a number of geological environments for possible use as repositories built with conventional mining technology and that successful isolation of radioactive wastes from the biosphere appears feasible for thousands of years." The report summarizes extensive public hearings and makes the specific recommendation that disposal should be started in mined repositories in several geological environments using a variety of emplacement media. Conditions in the disposal area should be monitored, and the casks containing the radioactive wastes should be retrievable in the event any difficulties should develop. The committee further recommended that secondary efforts should be carried out to investigate disposal in deep ocean sediments and very deep holes drilled by conventional techniques. Thus, in effect, the committee endorsed the disposal procedure that has been the principal candidate of nuclear experts since the early 1960s, i.e., disposal in deep mines with the radioisotopes as constituents of ceramics or glasses having inherently low solubilities in water and encased in shielded casks that would constitute an additional barrier to leaching by groundwater. Political, not technical, considerations have prevented the implementation of this approach not only in the United States but also elsewhere in the world, in part because the incremental cost of continuing aboveground temporary storage is small in relation to the large front-end costs of starting up a completely new system. However, the United Kingdom, France, Germany, Japan, and the U.S.S.R. all have plans for storage in mined repositories.

A favored geological formation for a repository is a salt deposit. In a typical case a deposit $\sim 10^8$ years old would be employed, its freedom from groundwater percolation problems testified by its age. Salt has the additional advantage of a relatively high thermal conductivity, and so the heat release by radioactive decay would be conducted away from disposal casks with a minimal rise in the local temperature. The principal objection to this method of disposal is that somehow groundwater might get to the fission products and become saturated with them, and that if this were to occur, the results would be catastrophic. As pointed out by B. L. Cohen,[32] the low solubility of the stored fission-product compounds would give a leaching time of $\sim 10^4$ years plus a further time delay of at least hundreds of years for the contaminated water to percolate from a deep mine through an aquifer to the surface. Further, ion exchange with rock and soil materials in the long course of percolation would attenuate the concentrations by large factors, typically 10^4 for those of most concern (e.g., plutonium). The combined effects of these factors is to introduce a delay time of the order of millions of years before the wastes would reach surface waters, by which time they would have decayed to minor levels. A striking demonstration of the validity of this rationale and the extremely low rate of leaching to be expected is given by the natural reactor that went critical in Gabon $\sim 2 \times 10^9$ years ago. Detailed analyses of material in the area show that most of the fission products are still in the rock matrix in which they were formed.[34]

Another approach to the problem has also been investigated by B. L. Cohen.[32] He suggests that the probability of an atom of this waste reaching and being ingested by people should certainly be no greater than for an atom of radium in the rock and soil above it. Inasmuch as the average concentration of uranium in rocks (2.7 ppm) is known, the number of cancer deaths that could be caused by radiation doses from the ingestion of the radium yielded by the decay of this uranium can be calculated. Cohen assumed that the fission-reactor wastes would be buried at a depth of 600 m, and estimated that if the amount of radium generated per year in the rock layers down to 600 m in the United States were ingested in just lethal doses, it could induce 3×10^{13} cancer deaths in humans, or roughly 1000 times the number from the fission products that would be produced each year if all the electric power in the United States were generated by fission reactors. Analyses indicate that the amount of radium actually ingested from natural sources gives the average person in the United States a dose of ~ 10 mrem per year, to which can be ascribed 12 cancer deaths per year. Inasmuch as the probability of an atom getting into food or water ingested by humans must decrease rapidly with the depth of the rock stratum from which it originates, the probability of an atom from the buried wastes getting into the groundwater must be lower than for the average radium atom in the first 600 m of rock. Further, the rock strata chosen for a repository should have a much lower probability than average of yielding leached material to groundwater, and the method of encapsulation should further reduce this probability. Thus the second approach also yields a negligibly small probability of radioactive material escaping from fission-product wastes in glass or ceramic form in a deep mine repository and, by entering the drinking water or food chain, causing deaths from cancer. In fact, after a decay period of 500 years, the radioactivity of the wastes is actually less than that of the original uranium because of their more rapid decay rate. Therefore, if one wishes to reduce the amount of radioactivity in the environment, one might advocate that as much uranium as possible be fissioned in reactors to reduce the radiation dose to future generations! Of course there is a good probability that cures for cancer will be found if our economy continues to have a surplus for such research—rendering moot the question of the number of deaths from cancer caused by trace amounts of radioactivity in the environment.

Mine Tailings

The radioactivity associated with the trace amounts of uranium in mine tailings is substantial and may be able to enter the biosphere more readily than from uranium in the original rock formation unless the dumps of tailings are sealed off with impermeable layers of clay to isolate them from groundwater. This precaution was not taken in earlier mining operations but is now required.

Low-Level Activity Solutions and Sludges

Low concentrations of radioactive wastes (i.e., very dilute solution of small amounts of radioactive material) have been disposed of by mixing with cement and pumping the cement slurries into deeply drilled holes that are cased down to the region where they penetrate shale formations. By using high pressures, the rock layers can be forced apart (a process called *hydrofracturing*) and the slurry pumped into the interstices, where it hardens to form rocklike material.

Activated Structure

Neutron activation of reactor structural materials—principally stainless steel—presents yet another type of disposal problem. The usual procedure is to bury this material in a sort of cemetery dedicated to this purpose. The solubility of the material is so low that the amount of activity picked up by groundwater is negligible and the half-life is sufficiently short (mostly a matter of a few years) that the activity will be largely gone in 100 years.

EFFECTS OF WASTE-HEAT REJECTION

Up to the 1960s the bulk of the waste heat from U.S. central power stations was rejected to rivers, lakes, or estuaries. However, rapid growth of the power industry finally reached the point that the heat absorption capacity of some rivers was not sufficient to accommodate additional plants. At the same time the environmental movement raised serious questions with respect to the long-term ecological effects of the heat rejected from the power plants, particularly on riverine and estuarine biota. Concurrently, studies of urban climates showed that their average temperature is commonly warmer by ~ 0.5 to $1\,°C$ than for surrounding rural areas, and under low wind velocity conditions may run $\sim 5\,°C$ higher. This raised the specter of a global change in climate caused by power plants, but one finds that the total energy consumption in the United States represents only ~ 0.002 percent of the solar flux. In fact, even in urban areas it appears that changes in albedo associated with closely spaced buildings give a difference in the balance between the solar energy absorbed and that reradiated that is greater than the energy released by the consumption of fossil fuel.[35] Thus the prime problem is the determination of the permissible temperature rise in the cooling water flowing through the plant condensers, together with the permissible temperature rise in the river or estuary.

Ecological Field Studies

The field studies up to the time of writing have not been sufficient to give a clearly defined picture, in part because the effects of chemical pollution appear to be far greater and mask the thermal effects of power plants. There have been many large fish kills, for example, but the great majority have been definitely traced to chemical rather than thermal effects. Simple observations show that in fall, winter, and spring months fish like the warm water in the vicinity of power plant cooling-water outlets, and fishermen congregate in these areas. It has been argued that this leads to "overfishing" and hence is bad for the fish population in the area. TVA studies show, however, that the fish populations per acre of lake are greater for TVA lakes having large steam power plants than for nearby lakes without steam plants, and the ratio of game fish to "rough" fish is consistently higher.[36] Other studies have yielded similar results.[37]

Laboratory Studies

Controlled experiments in laboratories have yielded much information, and this has served as the principal basis for strict state and federal regulations on thermal discharges issued in the 1960s and 1970s which have led to the widespread use of cooling towers.

A primary factor in considering thermal effects is the fact that the solubility of oxygen in water drops as the temperature is

TABLE 9.14 Minimum and Maximum Temperatures for 50% Survival of Freshwater Fish*

Species	Acclimated to °F	Minimum temperature		Maximum temperature	
		°F	Time, h	°F	Time, h
Largemouth bass	68.0	41.0	24	89.6	72
	86.0	51.8	24	93.2	72
Bluegill	59.0	37.4	24	87.8	60
	86.0	51.8	24	93.2	60
Channel catfish	59.0	32.0	24	86.0	24
	77.0	42.8	24	93.2	24
Yellow perch	41.0	—	—	69.8	96
(winter)	77.0	39.2	24	86.0	96
(summer)	77.0	48.2	24	89.6	96
Gizzard shad	77.0	51.8	24	93.2	48
	95.0	68.0	24	98.6	48
Common shiner	41.0	—	—	80.6	133
	77.0	39.2	24	87.8	133
	86.0	46.4	24	87.8	133
Brook trout	37.4	—	—	73.4	133
	68.0	—	—	77.0	133

*From *Industrial Waste Guide on Thermal Pollution,* Pacific Northwest Water Laboratory, Federal Water Pollution Control Administration, September 1968.

raised. However, a drop in the oxygen content as a consequence of a small increase in temperature will not occur unless the water is saturated with oxygen before heating. Offsetting the reduced solubility caused by the rise in temperature, however, is an increase in the rate at which the water will absorb oxygen at the air-water interface. Further, bacterial digestion of organic material in water proceeds more rapidly as the temperature is raised, thus improving the water quality. This effect can be particularly advantageous if the condenser cooling water is drawn from the cold, nearly stagnant region near the bottom of a lake or estuary, where the water tends to be low in oxygen. When heated and discharged on the surface at a temperature equal to or a little greater than the surface water temperature, it will remain near the air-water interface and pick up oxygen.

Studies of fish show that the different species vary widely in their tolerance to both short- and long-duration exposure to extremes in temperature. Table 9.14 shows data obtained for typical adult freshwater fish. The fish larvae are much more sensitive to temperature changes, and it is because of this that regulations on cooling-water temperature changes are commonly set to much lower values than could be justified by data on larger fish.

Studies of the myriad forms of smaller biota are rendered much more complex because their interrelationships in an ecosystem involve so many subtleties. If one examines the bottom of the food chain, one finds that diatoms develop most rapidly in the temperature range of 15 °C (59 °F) to 25 °C (77 °F), green algae at 25 °C (77 °F) to 35 °C (95 °F), blue-green algae at 35 °C (96 °F) to 40 °C (104 °F), while the rate of bacterial growth increases up to a temperature of ~ 32 °C (90 °F).[36] These growth rates are, of course, also heavily dependent on the presence of nutrients.

The ecological problems are complex, but at least one comprehensive examination of all the overall effects has indicated that the loss of water by evaporation from cooling towers, coupled with the higher fuel consumption and greater capacity required, largely offset the small advantages of eliminating thermal discharges to bodies of water—so much so that the overall environmental effects are often counterproductive.[37]

Costs

From the standpoint of capital and operating costs the optimum temperature rise in the condenser cooling water for a steam plant is 5.5°C (10°F) to 11°C (20°F); state and federal regulations commonly limit the temperature rise to 1°C (2°F) to 2.6°C (5°F). Compliance with these regulations increases plant capital costs by about 3 percent unless cooling towers are required, in which case costs are increased about 5 percent and the thermal efficiency is reduced about 2 percent, depending on the site and the load factor.[37]

CATASTROPHIES

Natural phenomena such as floods, earthquakes, volcanic eruptions, tidal waves, and avalanches may cause great loss of life and enormous amounts of property damage. Over 10^6 people have died in earthquakes in the past 50 years. Failures in energy systems can also have disastrous effects. For example, a leak from a large storage tank of liquefied natural gas led to a tremendous fire in Cleveland, Ohio, in 1944, causing 137 deaths and millions of dollars in property damage. The principal public hazards presented by electric power plants are failures in hydroelectric dams or nuclear reactors, the possibility of climatic changes induced by combustion of fossil fuels, and the possibility of a major war induced by conflicts over the rapidly diminishing supplies of petroleum. At least some of these risks can be quantified and some interesting comparisons can be made. This section summarizes a number of efforts in this direction.[38-45]

Dam Failures

Dams have been used for thousands of years to impound water supplies, drive machines, and, since 1889, to generate electric power. Serious failures have occurred: 33 dams failed in the United States between 1918 and 1958, five of which caused the loss of 1680 lives.[38] Worldwide, there have been over 300 dam failures, two of the worst disasters being the failure of the Vajont dam in Italy in 1963 which caused ~ 3000 deaths and the failure in 1979 of a dam near Morvi, India, that caused ~ 10,000 deaths.[39-42] Two failures in the United States in 1977 alone—the Teton and Toccoa dams—caused 50 deaths. Thus adequate statistics are available, and these indicate that, although the incidence of dam failures dropped from 1850 to 1930, the failure rate has been fairly constant in the past 50 years (see page 195), with about 0.5 percent of major dams failing during the first 20 years of their lives, giving a failure probability of $\sim 2 \times 10^{-4}$ per dam-year. In relating the failures of hydroelectric power dams to the resulting loss in life, Inhaber's analysis of the statis-

tics yields a value of 0.0011 to 0.0016 life lost per megawatt-year of electric output.[42] (This value was calculated before the dam failure near Morvi, India.) It is interesting that worldwide there have been eight major failures of hydroelectric dams in the 20-year period between 1960 and 1979 with a loss of ~ 15,000 lives. During the same period, commercial nuclear power plants produced about one-eighth as many kilowatthours without a single accident that caused a death among the general public, and there were no deaths to plant employes caused by radiation. This yields a fatality rate per unit of electric output for nuclear plants that is less than one-thousandth that for hydroelectric plants.

The hazard potential presented by a dam failure is much greater than indicated by the number of casualties experienced in any accident to date. One estimate for an existing dam in the event of its failure is as high as 250,000 deaths;[38] this estimate stemmed from a study in California of the possibility that an earthquake might cause the failure of a large dam when filled to capacity. An even greater hazard potential of 750,000 deaths has been estimated for the proposed Auburn dam to be located north of Sacramento.[38] The estimates of potential property damage are vague, but the amounts of money involved are on the scale of national disasters.

Earthquakes

People commonly associate earthquakes with locations in California or Alaska, but the most severe earthquake in the history of the United States occurred in 1811 in the vicinity of New Madrid, Missouri. One of its effects was that a large area of land dropped to form Reelfoot Lake, which is 15 km long, 4 km wide, and 6 m deep. Other examples include an earthquake in Charleston, South Carolina, that killed 120 people in 1886, and one in Boston in 1775. There is no region that is completely free of the possibility of this phenomenon, the question being mainly one of the probability that an earthquake will exceed a given intensity (Fig. 9.8).

The shaking forces associated with these events can rupture structures such as dams or shake equipment such as transformers or heat exchangers loose from their foundations. As a consequence, regulations require that all structures and vital equipment in nuclear power plants be designed to withstand earthquakes of the magnitudes indicated in Fig. 9.8. Lateral accelerations of 0.25 to 0.5 g must be accommodated not only by the foundations and mounting structure but also by minor components such as heat-exchanger tubes and tube bundles, cable trays, and reactor core internal elements, including control rods.

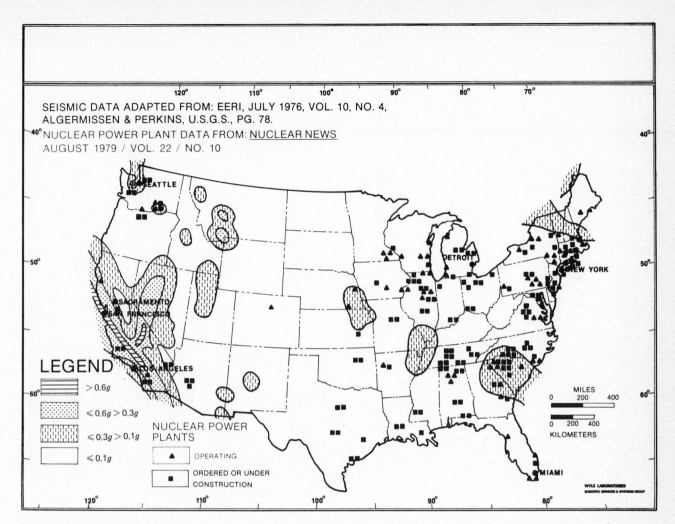

SEISMIC DATA ADAPTED FROM: EERI, JULY 1976, VOL. 10, NO. 4,
ALGERMISSEN & PERKINS, U.S.G.S., PG. 78.
NUCLEAR POWER PLANT DATA FROM: <u>NUCLEAR NEWS</u>
AUGUST 1979 / VOL. 22 / NO. 10

LEGEND

≡	> 0.6g
⣿	≤ 0.6g > 0.3g
⣿	≤ 0.3g > 0.1g
□	≤ 0.1g

NUCLEAR POWER PLANTS

▲ OPERATING

■ ORDERED OR UNDER CONSTRUCTION

MILES
0 200 400

0 200 400
KILOMETERS

WYLE LABORATORIES
SCIENTIFIC SERVICES & SYSTEMS GROUP

Figure 9.8 Seismic map of the United States, showing the zones of various degrees of damage probable in an earthquake, together with nuclear plant locations.[12]

Disasters in Mining and Construction

Underground coal mining is more dangerous than other types of underground mining because both coal dust suspended in the air and methane diffusing out of the coal seam can form explosive mixtures in the shafts and tunnels. To minimize this hazard, large amounts of fresh air are continuously circulated through the mines by powerful fans, and great precautions are taken to avoid sparks or flames that could ignite a combustible mixture if one were to accumulate locally because of poor flow distribution in the ventilating air. But even with these precautions, in

addition to the roughly 300 workers killed per year in many small coal-mining accidents in the United States, about every other year there is a major coal mine disaster in the form of an explosion or fire, in which sometimes more than a hundred workers lose their lives. It is also worthy of note that the MHSA standards instituted in the early 1970s in an effort to improve mine safety have led to a small reduction in the number of miner deaths per 10^6 man-hours, but the standards made it necessary to employ more workers underground so that the number of deaths per 10^6 tons of coal mined actually increased. Hence the regulations have proved counterproductive while substantially

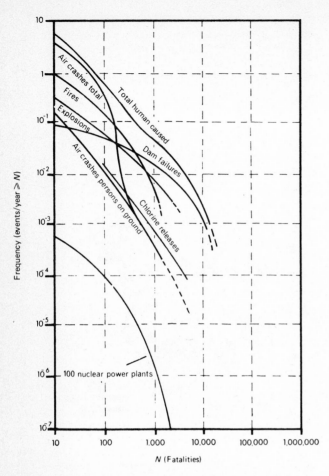

Figure 9.9 Frequency of fatalities owing to human-caused events.[43] Notes:

1. Fatalities due to auto accidents are not shown because data are not available. Auto accidents cause about 50,000 fatalities per year.

2. Approximate uncertainties for nuclear events are estimated to be represented by factors of $\frac{1}{4}$ and 4 on consequence magnitudes and by factors of $\frac{1}{5}$ and 5 on probabilities.

3. For natural and human-caused occurrences the uncertainty in probability of largest recorded consequence magnitude is estimated to be represented by factors of $\frac{1}{20}$ and 5. Smaller magnitudes have less uncertainty.

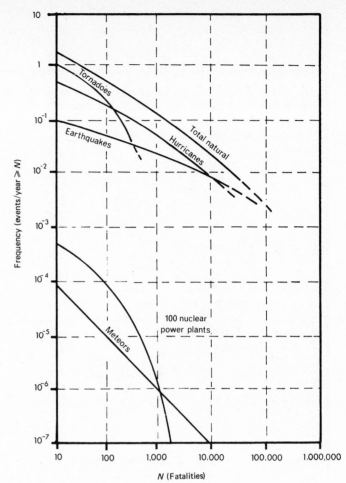

Figure 9.10 Frequency of fatalities owing to natural events.[43] Notes:

1. For natural and human-caused occurrences the uncertainty in probability of largest recorded consequence magnitude is estimated to be represented by factors of $\frac{1}{20}$ and 5. Smaller magnitudes have less uncertainty.

2. Approximate uncertainties for nuclear events are estimated to be represented by factors of $\frac{1}{4}$ and 4 on consequence magnitudes and by factors of $\frac{1}{5}$ and 5 on probabilities.

increasing costs. Another interesting point is that twice as many coal miners are killed in accidents in personal automobiles as in mine accidents.

Accidents that occur in power plant construction work usually involve only one or a few people, although 51 workers died in the collapse of the scaffolding on a cooling tower being built in West Virginia in 1978.

Possibility of a Nuclear Power Plant Disaster

A substantial number of people have become preoccupied with the terrible prospect of a nuclear power plant somehow belching forth an enormous quantity of deadly radioactive material that would spread across the countryside, killing thousands of people and dooming tens of thousands to horrible deaths by cancer. The position held by the antinuclear activists is that, if the potential is there—and it cannot be proved beyond a doubt that such an event could not happen—no nuclear power plant operation should be permitted. Thus it becomes necessary to assess not only the probability of a terrible accident but also the consequences if such an event were to occur, no matter how low the probability.

Probability of a Major Accident

The probability of a major nuclear accident was assessed by a high-level committee in 1975 using generic data for the reliability of components and systems in the basic fashion outlined in the previous chapter. The results presented in what was generally called the "Rasmussen Report," or WASH 1400, indicated that the probability is exceedingly low: of the order of one failure causing 100 or more fatalities in 10^7 reactor operating years.[43] A 1978 review of that study by another prestigious committee led to an endorsement of the WASH 1400 methodology but suggested that the combined effects of uncertainties might give a higher probability of a major accident by a factor of from 2 to as much as 10.[44] On the other hand, a similar Swedish study yielded a probability lower than that of WASH 1400 by a factor of ~ 100.[45] Taking WASH 1400 as the basis, if all the U.S. electric power in 1978 had been obtained from 1000-MWe nuclear reactors, ~ 500 would have been required, and this implies that a failure causing 100 or more fatalities might be expected once every 200 to 2000 years. This is a small probability when compared with other natural disasters or events of homogenic origin, such as a dam failure or an airliner crash into a stadium filled with people. The probabilities of such events as functions of the number of fatalities as given by the Rasmussen Report are shown in Figs. 9.9 and 9.10, while Fig. 9.11 shows similar

estimates of property damage. The soundness of the Rasmussen Report is indicated by the fact that it also considered the probability of an accident similar to that of the Three Mile Island plant and predicted that such an event was likely to occur in the course of the number of plant operating years that have been accumulated in the United States up to 1980.

Character of a Major Nuclear Accident

Many people associate nuclear reactors with atom bombs and somehow imagine that an explosive rupture could occur. In

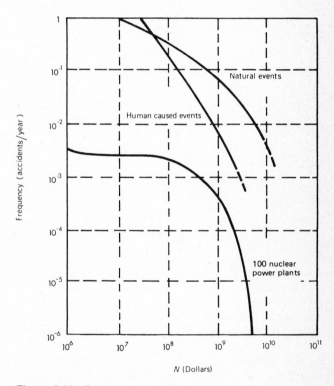

Figure 9.11 Frequency of property damage owing to natural and human-caused events.[43] Notes:

1. Property damage due to auto accidents is not included because data are not available. Auto accidents cause about 15×10^9 damage each year.
2. Approximate uncertainties for nuclear events are estimated to be represented by factors of $\frac{1}{5}$ and 2 on consequence magnitudes and by factors of $\frac{1}{5}$ and 5 on probabilities.
3. For natural and human-caused occurrences the uncertainty in probability of largest recorded consequence magnitude is estimated to be represented by factors of $\frac{1}{20}$ and 5. Smaller magnitudes have less uncertainty.

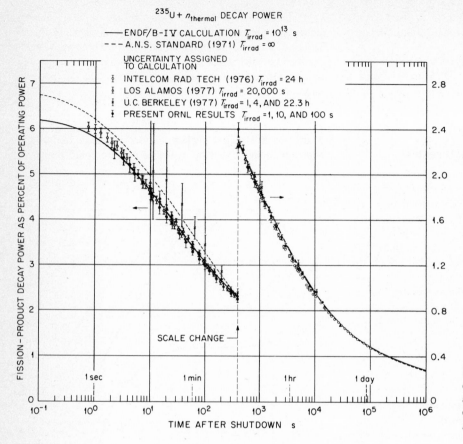

^{235}U + $n_{thermal}$ DECAY POWER

—— ENDF/B-IV CALCULATION T_{irrad} = 10^{13} s
--- A.N.S. STANDARD (1971) T_{irrad} = ∞

UNCERTAINTY ASSIGNED TO CALCULATION
INTELCOM RAD TECH (1976) T_{irrad} = 24 h
LOS ALAMOS (1977) T_{irrad} = 20,000 s
U.C. BERKELEY (1977) T_{irrad} = 1, 4, AND 22.3 h
PRESENT ORNL RESULTS T_{irrad} = 1, 10, AND 100 s

SCALE CHANGE

1 sec   1 min   1 hr   1 day

FISSION-PRODUCT DECAY POWER AS PERCENT OF OPERATING POWER

TIME AFTER SHUTDOWN s

Figure 9.12 The energy release by decay of the fission products of ^{235}U as a function of the time after reactor shutdown. *(Courtesy Oak Ridge National Laboratory.)*

point of fact it is exceedingly difficult to cause a nuclear explosion, and the science of nuclear reactor physics has reached the point that reactors can be and are designed so that a nuclear explosion positively could not occur. Thus the only mechanism that might lead to a major dispersal of radioactive material is a reactor core meltdown as a consequence of a failure to remove the afterheat from fission-product decay—a much more severe case of the sort of accident that occurred at the Three Mile Island plant in April 1979. The amount of heat available to cause such an accident is indicated in Fig. 9.12, which shows the heat generation rate from fission-product decay as a function of time after an abrupt shutdown from full power. The heat generation rate is highest if virtually all the usable fuel in the core has been consumed (as for Fig. 9.12), but under any circumstances the heat release rate falls off rapidly with time at first and then progressively more slowly. The energy release rate P for a core near the end of its life is approximated well by the expression

$$\frac{P}{P_0} = 0.06\, t^{-0.2}$$

where P_0 is the reactor full-power output and t is in seconds.

If the reactor coolant flow is stopped, the fuel temperature will rise until the fuel reaches a temperature in equilibrium with its surroundings. Possible courses of events can be envisioned with the help of the drawing of a nuclear plant shown in Fig. 5.28. Note that the fuel is contained in metal capsules which have demonstrated a high degree of leak-tightness, that the reactor cooling system constitutes a second containment envelope, and that the reactor system is housed in a large steel vessel (the *containment shell*, item No. 20 in Fig. 5.28) which constitutes a third barrier to the escape of the fission products, and this in turn is surrounded by the reinforced-concrete shell of the reactor building (item No. 17 of Fig. 5.28). In the event that a *loss of coolant accident* occurs in spite of the extensive set of *engineered safeguards* that normally come into play automatically, the coolant inventory in the reactor pressure vessel will begin to boil off, carrying with it the afterheat. In the Three Mile Island accident this occurred as a consequence of a nearly incredible series of equipment malfunctions and operator errors (including a deliberate 2-h shutdown of the emergency core cooling system by an operator). With no coolant flow, water boiled off the core for over 2 h until the water level dropped to perhaps 0.6 m below the top of the core. At this point the upper portion of the fuel elements was apparently cooled only by thermal radiation and by water droplets splashing up from the free liquid surface of the boiling water in the lower part of the core. The fuel elements did not melt, but some of the volatile fission products such as I, Cs, Xe, and Kr diffused through the overheated metal of the fuel capsule walls and contaminated the cooling water. In the Three Mile Island accident the operators finally restarted the core cooling circuit and brought the core temperature down to normal, but it can be imagined that on some occasion nothing would be done by the operators to prevent further loss of water boiled out of the reactor and so the fuel elements would eventually melt, releasing most of the volatile fission products. The uranium oxide fuel and the refractory fission products remaining would melt at ~ 2800°C (5000°F) and form a white-hot melt consisting mainly of uranium oxide, and this molten mass could then melt its way down through the reactor pressure vessel and thence, following the so-called China Syndrome, presumably melt its way down through the foundation into the earth beneath—in fantasy, all the way through the earth to China (hence the term *China Syndrome*). In point of fact, heat losses to the surroundings from the molten oxide would be high, fission products intermediate in volatility between uranium and cesium would distill off, and these would condense out on relatively cool surfaces within the containment vessel. This would disperse the heat generation over a large region and yield a high rate of heat loss to the surrounding air. Various auxiliary cooling systems within the reactor containment shell would act to remove this heat.

The bulk of the volatile fission products in the atmosphere of the reactor building would be confined by the containment shell. If somehow large amounts of activity did leak out as a consequence of a rupture of the containment shell, they would form a plume that would drift downwind and disperse laterally and vertically. (One of the several operator errors in the course of the Three Mile Island accident permitted the release of some activity by venting from the containment system.) The velocity with which the plume would move downwind would probably be that of a conventional weather front—around 16 km/h (10

mph)—and so there would be time for evacuation. Further, the material escaping would be mainly the volatile fission products. The refractory uranium, plutonium, and strontium oxides have low vapor pressures and hence would remain in the reactor building in solid form. Thus the gas and vapor emitted would present a much less serious fallout problem than the debris from an atom bomb.

Comparison of a Major Nuclear Accident with Rupture of a Tank Car of Liquid Chlorine

It is instructive to compare the consequences of a nuclear accident far worse than that at the Three Mile Island plant with the consequences of rupturing a tank car of liquid chlorine. Suppose that in such a reactor accident a failure occurred in the reactor primary coolant system causing fission products to escape into the containment shell, and then yet another failure occurred in the containment shell causing fission products to escape into the atmosphere. Analysis of the gases in the reactor cooling system after the Three Mile Island accident disclosed that, as one would expect, a substantial percentage of the volatile fission products I, Cs, Kr, and Xe escaped from the UO_2 crystal lattice, diffused through the red-hot stainless fuel capsules, and entered the reactor coolant system. However, it was found that over 99 percent of the I and Cs that escaped were dissolved and retained in the water so that only the noble gases Kr and Xe were in gaseous form. If it is assumed that 10 percent of the total Kr and Xe that had accumulated in the fuel escaped into the containment shell, and that 10 percent of this escaped from the containment shell to the atmosphere a few hours after the emergency reactor shutdown, a cloud of dangerously radioactive gas would develop in the form of a plume that would drift downwind from the plant. Assume further that the wind velocity would be low so that it would disperse the radioactive gas in such a way as to give the maximum volume with a lethal concentration. The total inventory of Kr and Xe in the reactor would amount to about 0.1 Ci/W a few hours after shutdown, or ~ 3 × 10⁸ Ci for a 3000-MWt reactor. The amount leaking to the atmosphere following the above rationale thus would be ~ 3 × 10⁶ Ci.

The toxic effect of the radioactive Kr and Xe can be estimated from the standards for occupational exposure to radiation, which give the concentration in air that, if breathed for a year of 40-h weeks (i.e., for 2000 h), would give a dose of 5 rem (Table A9.5). This concentration for both ^{85}Kr and ^{133}Xe is 10^{-5} μ Ci/mL, or 10^{-5} Ci/m³ (Table A9.5). The concentration giving an LD-50 dose (500 rem) to people exposed for 2 h to a cloud containing this gas would be 100,000 times greater, i.e., 1

Ci/m^3. Thus the 3×10^6 Ci of Kr and Xe escaping to the atmosphere could be dispersed in $3 \times 10^6 m^3$ to give people breathing it for 2 h an LD-50 dose.

In considering the potentially lethal effects of the rupture of a tank car of liquid chlorine, one finds that a typical tank car carries 55 tons, or about 50,000 kg. Once the pressure on the chlorine was relieved by rupture of the tank, one would expect all of it to boil off and escape to the atmosphere. At standard temperature and pressure, chlorine gas has a density of 3.3 kg/m^3; hence the volume of chlorine released would be 15,000 m^3. The lethal concentration of chlorine in air is given by Ref. 14 as 1000 ppm (Fig. A9.1); hence dilution by $15 \times 10^6 m^3$ of air would still give a lethal concentration. Thus the volume of the lethal cloud of chlorine would be about five times that of the cloud of radioactive Kr and Xe that would give an LD-50 dose of radiation.

The actual clouds—were they to occur—would of course not be uniform in concentration with sharply defined boundaries, but some idea of their size can be obtained by assuming that the toxic material would be dispersed in the form of a half-cone with the surface of the ground lying in a plane through the axis. If the volume of the half-cone were $3 \times 10^6 m^3$ and the cone were long and slender so that its length was nine times the radius, the length would be almost 1000 m, and the area covered would be 0.07 km^2, much less than one would at first expect.

Some further comparisons between these two cases can be drawn. The toxic effects of chlorine—choking, etc.—would be felt immediately and would not only interfere with efforts to evacuate the area, but also would be lethal in much less than 2 h. More important, the probability of the rupture of a tank car of liquid chlorine is surely vastly greater than that of the sequence of events assumed for the reactor case—a point indicated by the fact that three tank cars of liquid chlorine have been ruptured in the 1960–1980 period. It is also significant that in two of these cases there were no fatalities and there were only two deaths in the third case. It might be argued that all shipment of liquid chlorine should be halted immediately, but liquid chlorine is widely used for purifying water supplies, and, if shipments were halted, people would die by the hundreds of thousands of intestinal ailments, as was the case before municipal water supplies were routinely treated with chlorine.

This brings up the question of the time required to evacuate an area in an emergency. An excellent example is given by an actual case in which a train hauling tank cars of liquid chlorine, propane, butane, styrene, and NaOH was wrecked in Mississauga, a suburb of Toronto, Canada, on Nov. 11, 1979. A quarter of a million people were evacuated from the area in a few hours in an orderly fashion with no deaths or serious injuries.[46]

Measures to Reduce Operator Errors

The 1979 accident at the Three Mile Island (TMI) nuclear plant has raised serious questions regarding the probability of operator error. On the one hand, it has been argued that the TMI accident demonstrated the effectiveness of the engineered safeguards and that it has greatly reduced the probability of another such accident by giving the electric utility industry an unforgettable lesson in the enormous financial loss that a utility stands to sustain if its operators lack competence. It seems likely that this tremendous financial incentive will prove more effective than any governmental regulation in reducing the likelihood of future accidents in which operator error plays a major role. Significantly, the president of the utility responsible for the TMI plant was forced to resign following the accident. Obvious steps for utility management to take include selection of personnel of high aptitude, giving them more thorough training including extensive experience with electronic simulators in which the operator can be confronted with all sorts of possible accident situations (similar to those in airline pilot training programs), and providing higher-pay incentives.

A more comprehensive and fundamental approach to the problem of assuring a high level of expertise at nuclear power plant sites has been urged by A. M. Weinberg.[47] This entails the clustering of nuclear power plants in large "power parks" located in remote sites. Having four or more reactors at each site instead of only one or two (the situation in 1980) would give a sufficiently large operating budget at each site to support a full complement of real experts who could be called on immediately if any emergency arose. Further, a much greater reservoir of operating experience would be accumulated at the site so that symptoms of incipient trouble would be recognized more readily in the early stages when they could be handled promptly and effectively. About 80 of the 90 sites at which nuclear power plants were in operation or under construction in the United States in 1980 have less than 25,000 people within an 8-km radius. Hence, further construction of new nuclear plants might be confined to these sites so that they could become "power parks" that would provide the broad range of expertise that is so important in reducing the probability of operator error.

Wars

The growing competition for the earth's dwindling supplies of petroleum is leading to increasingly tense geopolitical situations that are raising the probability of a major war. In fact, restricting the construction of both nuclear power plants and nuclear-fuel-reprocessing facilities on the premise that this will reduce the proliferation of nuclear weapons is probably counterproductive, for it increases the pressures for petroleum and thus both international tensions and pressures within other countries to develop and stock nuclear weapons. An excellent treatment of this hazard makes a convincing case for the thesis that easing the competition for scarce petroleum through the extensive use of nuclear power is the least hazardous course to follow.[48]

Change in Climate

As shown in Fig. 9.13, the CO_2 content of the earth's atmosphere has been increasing steadily for over a century because the high consumption of fossil fuels is generating CO_2 more rapidly than it is being absorbed by plants and the oceans.

This reduces the rate of heat loss from the earth because CO_2 absorbs infrared radiation much more strongly than visible light and thus tends to reduce thermal radiation from the earth and raises the mean annual temperature via the "greenhouse effect."[49] This effect is being offset in part by an increase in particulate matter in the atmosphere stemming partially from increases in industrialization but largely, it is thought, from the high incidence of volcanic eruptions in recent years.[50] The particulates in the atmosphere reflect sunlight into space, preventing it from reaching the earth's surface. The energy balance is difficult to predict, but some estimates indicate that continuing our high rate of fossil fuel consumption well into the twenty-first century might lead to melting of the polar icecaps. While at

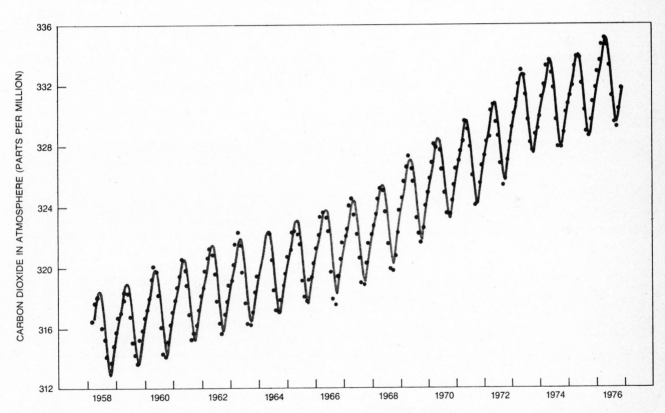

Figure 9.13 Trend in atmospheric carbon dioxide concentration as measured since 1958 at the Mauna Loa Observatory on the island of Hawaii by Charles D. Keeling of the Scripps Institution of Oceanography. The dots indicate the monthly average concentration of carbon dioxide. The seasonal oscillations are caused by the removal of carbon dioxide by photosynthesis during the growing season in the Northern Hemisphere and the subsequent release during the fall and winter months. The Mauna Loa measurements and those made elsewhere show that the average carbon dioxide content of the atmosphere has risen more than 5 percent since 1958. The rate of increase has varied from year to year from causes not yet known. The current rate is 1 ppm per year, equivalent to 2.3×10^{15} g of carbon (Ref. 49).

first thought one might expect the icecap melting to occur very slowly—requiring perhaps a millenium—geological evidence indicates that in at least some instances our ice ages have ended abruptly with increases in sea level of the order of 30 m (100 ft) occurring in as little as 40 years.[51] Such a rise in sea level would inundate a substantial fraction of the world's most fertile agricultural land in alluvial deltas and valleys such as those of the Nile, the lower Mississippi, and the great rivers of China and India. The risk is difficult to quantify, but surely the probability is not zero and the number of people that would be affected is enormous.

Energy use can definitely affect the local climate and possibly the earth's climate by affecting the heat and moisture retention of foliage. Deforestation to obtain fuel has been proceeding for millenia and is continuing; nearly half the forest products being harvested in the latter part of the twentieth century are being used for fuel. The Mediterranean area has been largely deforested, and this has led to a hotter, drier local climate with an agricultural output much reduced from Roman times. Northern India, Indonesia, and Brazil are among the areas that appear to be particularly threatened with local changes in climate from deforestation as the twentieth century nears its close. Thus measures are urgently needed to reduce the dependence of the less developed areas of the world on wood for fuel in order to avoid changes at least in local climates and possibly in the worldwide climate. Note that a major mechanism for removing CO_2 from the air is carbon fixation by the world's forests, and reducing their extent will accelerate the rate of increase in the CO_2 of the atmosphere.

Acid Rain

One of the major charges of some environmentalists in the 1970s was that the use of high-sulfur coal and residual fuel oil, together with the smelting of pyrite ores, has led to an increase in SO_2 and NO_2 emissions in the northeastern United States and Europe, and some reports indicate that the average acidity of the rainfall has been increased seriously in many areas.[52,53,54] The data cited to show marked increases in the acidity of the rainfall in the United States are for many sites, but most of the sites for the various periods are not the same. If the only data used are those taken at the same sites over the 1955–1972 period, no increase in the acidity of the rain is apparent.[54] This is consistent with estimates that the total emissions of SO_2 and NO_2 from U.S. plants did not increase over this period. The problems are complex, the data fragmentary, and interpretations of the information differ widely; hence, the information available must be examined closely.

To begin with, the CO_2 present in the earth's atmosphere

dissolves in atmospheric moisture to give a weak carbonic acid concentration, so a pH of 5.6 is normally found in rain or snow. In industrial areas in one extreme case SO_2 and NO_2 so increased the acidity that a pH of 2.4 was recorded for a rain in Scotland in 1974.[52] (This is the acidity of vinegar.) While strongly acid rains are rare, in one of the most badly affected areas in the northeastern United States the average value for the 1964 to 1974 period was a pH of 4.1.

The most pronounced adverse effect of acid rain has occurred in areas in which the soils and rock outcroppings are acidic and so the acidity of the rain is not neutralized and the runoff makes lakes and rivers acidic. Organic acids from the decaying litter on forest floors contribute to this process, as well as to the acidic compounds in the atmosphere. In areas with acid soils it appears that if the average pH in a lake or river drops below 4.5, it becomes sufficiently acidic so that the fish populations do not reproduce and soon disappear—a distressing phenomenon that seems to have developed in the Adirondacks and Norway since 1950. Other biota are also adversely affected, though they generally appear to be less sensitive than fish.[52] The problem may be more serious than that of human health effects as far as SO_2 emissions are concerned.

Sources

Analyses of acid rain indicate that SO_2 is from 3 to 10 times as large a factor in the acidity as NO_2.[52] In attempting to establish the sources of the sulfur in the acid rainfall, Table 9.15 was prepared to provide an interesting comparison of the sulfur cycle in the biosphere on both a global basis and for the northeastern United States (the section east of the Mississippi between the southern boundary of Kentucky and a line through Buffalo, New York, an area that contains 75 percent of the United States urban population and an even higher fraction of heavy industry).[53] Some of the sulfur is precipitated in rain, while some falls out as sulfates in the form of fine dust. For the earth as a whole the sum of the two equals the emissions, but for the northeastern United States apparently half of the sulfur emitted is carried off and is probably deposited mostly in the ocean. It is also true that some of the sulfur deposited, particularly in the Adirondack area, is probably carried southward from the big copper and nickel ore smelter at Sudbury, Ontario, when winds blow from the northwest. This smelter complex includes a 400-m stack, and emits about 3 percent as much sulfur as all the power plants in the northeastern area of the United States covered by Table 9.15.[52,53] Summation of data on stack gas emissions of sulfur indicate that roughly 65 percent is from power plants and most of the balance is from smelting operations.[52]

Sulfur in the Soil

The importance of sulfur to plant life has been known since the mid eighteenth century, and was dramatically demonstrated by Benjamin Franklin, who applied gypsum to a hillside near Philadelphia to stimulate grass growth in a pattern that spelled THIS LAND HAS BEEN PLASTERED to provide a vivid lesson in the value of fertilizing with gypsum. The natural recycling of sulfur in agricultural areas has been disturbed by removing crops for marketing in urban centers, so that in many areas sulfur must be added to the soil in fertilizers to maintain productivity. Thus sulfur deposition from power plant effluents can be advantageous if the associated increase in acidity is not excessive.[15,54] The extent to which sulfur is used for this purpose is indicated in Tables 9.15 and 9.16. Larger applications of sulfur in fertilizers would be required if it were not for the sulfur contributed by industrial emissions, a situation clearly demonstrated in Germany following World War II when sulfur emissions from plants in the Ruhr were severely restricted by law for some years. Yellowing of tomato plants gave a clue to the reason for the reduced agricultural productivity that was being experienced, and agitation by farmers led to the removal of governmental restrictions on sulfur emissions. Yellowing of tomato plants stopped and agricultural productivity returned to normal. There are indications that NO_2 emissions may have similar though less pronounced beneficial effects, but this has not been clearly demonstrated.

Soluble Sulfates in Groundwater

Acid mine drainage was mentioned earlier in the chapter, together with leaching of sulfates and sulfites from sludge ponds for stack gas scrubbers. Contributions from these and other industrial sources to groundwater in the northeastern United States are summarized in Table 9.17. The contribution shown for stack gas scrubbers is small because the data are for 1975, when few such scrubbers were in service.

Liming Acidic Lakes

One solution to the acid rain problem is the addition of lime. In Sweden it has been found that the addition of 50 to 75 kg per year of $CaCO_3$ per hectare of watershed is effective in coping with a pH of 4.0 to 4.3 in rain accompanied by a comparable contribution by dry dust sulfates to acidity in the area. The total cost of liming 700 lakes and rivers in the 1977–1979 period was found to be $50 to $70 per ton where surface equipment could be used and somewhat higher where it was necessary to employ

TABLE 9.15   Comparison of Global Land Sulfur Cycle with the Sulfur Cycle of Northeast United States* (Ref. 53)

Cycle flux	Global	By land-area fraction	Est by data
		Study region (northeast U.S.)†	
SO₂ emissions	150	0.88	25.0
Bacteriogenic prodn.	210	1.23	2.0
Precip. + dry depos.	360	2.11	13.0
Fertilizer	33	0.19	1.0
Weathering	42	0.25	1.0
Acid-mine draining	?	?	2.0
Industrial operation	?	?	0.0
Stream load	225	1.32	15.0
Atmos. transport out	0	0.00	10.0

*10⁶ tons of equivalent SO₄/year.
†Figures given as "by land-area fraction" are calculated by using the global flux, multiplying by the study region land area, and dividing by the global land area. Figures listed as "est by data" are the actual fluxes as estimated from the data presented in Ref. 53.

TABLE 9.16   Deposition of Sulfate in Rainwater and Fertilizer Applications in the Northeastern U.S. [53]

State	Sulfate in rainfall		Fertilizer applications* (10³ tons/year)
	tons/ (mi²·year)	10³ tons/year	
Mass.	10.3	81	5.7
R.I.	8.6	9	1.2
Conn.	10.8	53	4.6
N.J.	17.4	131	14.9
Del.	12.8	25	9.5
Md.	12.3	121	27.4
Pa.	13.4	604	68.1
Va.	8.6	341	72.3
W. Va.	10.8	261	5.5
Ky.	12.1	486	60.3
Ohio	14.0	574	138.0
Ind.	17.4	630	151.2
N.Y.†	12.1	326	33.8
Mich.†	14.3	220	25.8
Ill.†	17.4	619	107.2
	Av 13.3	Total 4481	Total 725.0

*Extrapolated from data supplied by the Sulphur Institute.
† Data given for portion of state within study region.

TABLE 9.17 Ground-Level Fluxes of Soluble Sulfate to the Soil and Groundwater in the Northeastern U.S.* (Ref. 53)

	Total sludge	Overflow	Seepage	Deposited sludge
Sulfate sludge				
Industry	12.0	0.3	0.03??	11.94
Scrubbing†	1.0	0.003	0.003??	1.0
Total	13.0	0.303	???	12.94
			Total assumed ≪ 0.5	
Other sources				
Normal erosion				0.5
Excess erosion (from strip mining)				0.5
Acid-mine drainage				2.0‡
Total				3.0

*Expressed as 10^6 tons of SO_4/year.
† Figure for early 1975, should rise rapidly through the 1980s as more flue-gas desulfurization units come on line.
‡ U.S. Department of Interior estimate.

helicopters.[55] The marine plant and animal populations generally were well recovered in about 2 years following liming.

EPA studies indicate that a program to lime the portions of the Adirondacks where ~ 100 lakes have been seriously affected by acid rain would cost $150,000 per year, with an additional $70,000 per year required for monitoring.[56] This approach appears to be highly cost-effective when compared with an estimated cost of 10 billion dollars per year for universal stack gas scrubbing (in old as well as in new power plants) to reduce emissions by 90 percent, a course urged by key EPA officials at the time of writing. These alternatives suggest that a tax on emissions to cover the cost of liming watersheds affected by acid rain would be an inexpensive and sensible way to handle the problem.

Mass Starvation

Much of the underdeveloped world is engaged in a struggle to rise from a low-technology-subsistence existence. These countries have a high birthrate and a rapid population growth. There is therefore a major question as to whether they can increase their standard of living fast enough to bring their population growth rate under control before they have reached the limit of their agricultural production. Petroleum is a vital factor in their race to avert mass starvation, and it has been pointed out that the most important contribution that could be made by the developed countries would be to shift from petroleum to nuclear energy and thus make available the supplies of petroleum the poor countries need so badly. Such a shift can be made by the developed countries, because they have the capital, expertise, and industrial base required, whereas the underdeveloped countries do not.

COMPARISONS OF THE ENVIRONMENTAL EFFECTS OF VARIOUS ENERGY SYSTEMS

A number of attempts have been made to compare the environmental effects of various energy systems.[26,38,42-45,57-64] Some of these have entailed comparisons of just the effects on human health of emissions from fossil fuel and nuclear power plants, while others have included a greater variety of energy systems and other effects, including land use and cost-benefit relations. The major uncertainties in the basic data are compounded by variations in the choice of methodology for making comparisons; hence it is not surprising that there are substantial differences in the resulting estimates of the relative impact of the various energy conversion systems. Possibly chief among the uncertainties are those associated with the feasibility and monetary costs of advanced energy systems such as solar cells; these questions are addressed in Chaps. 21 and 22. Suffice it to say here that the consensus of the technically well-informed (e.g., the National Academy of Sciences[64]) is that the bulk of the electric power produced in the United States through the year 2000 will have to come from coal and fission reactors, and therefore the prime question is the relative environmental impact of these two types of system. On this basis, a study by Resources for the Future[61] concentrated on the relative impacts of coal and nuclear power and organized the data available on injuries, fatalities, and land use associated with these two systems. One of the principal results of this study is summarized in Fig. 9.14, which shows graphically the relative impacts in terms of occupational deaths per 2×10^{12} kWh (the total electric power output of the United States in 1975). Figure 9.15 gives a similar comparison for the total fatalities in the population as a whole based on the National Energy Plan (NEP), which projects an electricity production rate of $\sim 6 \times 10^{12}$ kWh in the year 2000. Of this, $\sim 3.2 \times 10^{12}$ kWh would be from coal and $\sim 1.9 \times 10^{12}$ kWh would be nuclear. The increased fatalities resulting from a nuclear power plant moratorium are shown at

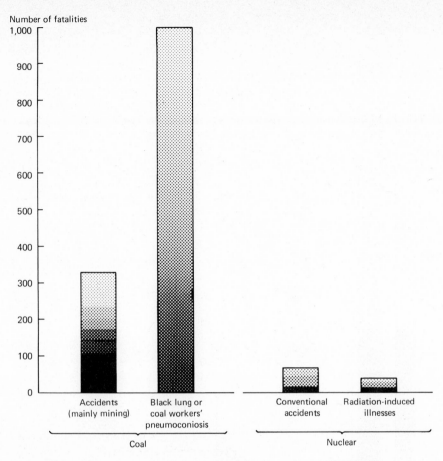

Number of fatalities

Figure 9.14   Estimates of workers' occupational deaths per 2×10^{12} kWh of power production.[61] *(Courtesy Resources for the Future.)*

the right of Fig. 9.15. They are based on the assumption that the output from coal would be increased to 4.4×10^{12} kWh while that from nuclear plants would be reduced to 0.74×10^{12} kWh. An estimate of the relative land use for these two cases is shown in Fig. 9.16.

Similar results for health effects were obtained in a study conducted by the American Medical Association (AMA). This study indicated that, for the same total electric output, both the occupational death and injury rates of coal run about 10 times those for a nuclear system, and the nonoccupational death rates about 100 times higher for coal than for a nuclear system.[60]

A more ambitious attempt to compare the probable health effects of the principal energy systems that have been proposed

with little regard to their economical feasibility was carried out by H. Inhaber[42] and is summarized in Fig. 9.17. This study inherently entailed a multitude of detailed assumptions and judgments, and these have drawn criticism from environmentalists (e.g., Ref.62), who, for their part, in taking exception to Inhaber's statistics make optimistic assumptions with respect to "natural" energy systems. Probably the most serious questions are concerned with means for supplying electricity to consumers when the sun doesn't shine or the wind doesn't blow; Inhaber's position was that by far the least expensive option would be coal-fired back-up power plants. (These questions are treated in Chaps. 20, 21, and 22.) The use of coal-fired plants for back-up power as advocated by Inhaber leads to almost as many deaths

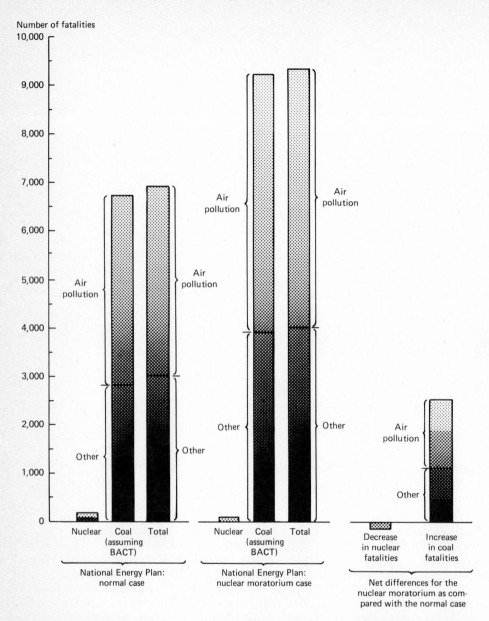

Figure 9.15 Fatalities for the year 2000 from coal-fired and nuclear plants, based on the NEP's 1985 prediction and two alternative projections. The first assumes annual electric power production of 3.2 × 10¹² kWh from coal and 1.9 × 10¹² kWh from nuclear reactors. The second projection assumes a nuclear moratorium, giving 4.4 × 10¹² kWh from coal and 0.74 × 10¹² kWh from nuclear reactors.[61] *(Courtesy Resources for the Future.)*

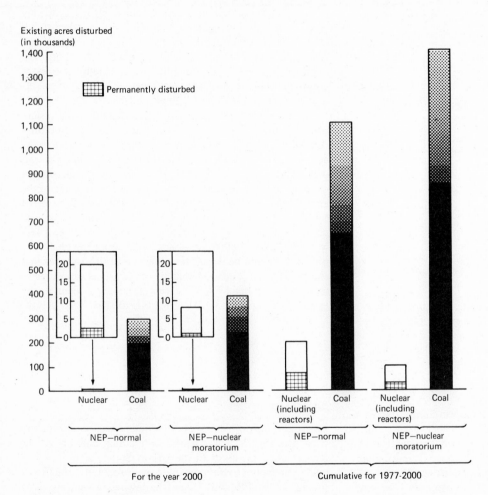

Figure 9.16 Land use impacts in the year 2000 and cumulative land use impacts (assuming certain cycles of reclamation) for 1977 to 2000, based on the same projections as Fig. 9.15.[61] *(Courtesy Resources for the Future.)*

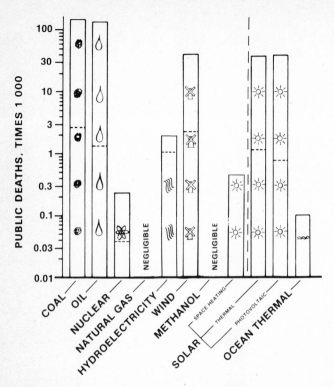

Figure 9.17   Public deaths, times 1000, per megawatt-year, as a function of energy system. Much of this risk is produced by emissions created after fuel is gathered (for the case of conventional technologies) or in the course of producing the materials for the power system (for nonconventional technologies). This explains in part the relatively high values of the latter group in comparison with coal and oil.[42] (Courtesy Atomic Energy Control Board, Canada.)

per unit of electric output for solar and wind power plants as result from the use of conventional coal-fired plants. This is partly because most of the time the power must come from the coal-fired back-up plants and partly because there are many fatalities associated with the mining and smelting of metals used for the construction of capital-intensive plants such as windmills, solar power plants, etc. Thus the occupational deaths associated with the construction of solar plants, for example, are higher than for conventional plants. In addition, weather conditions, together with such factors as the poor performance of solar plants in the morning and late afternoon, as well as at night, place the bulk of the load on the coal-fired back-up

plants. The coal-fired plants could be avoided through the use of very expensive energy storage systems, but the potential fatalities associated with both the energy storage systems and the greatly increased solar plant capacity required to provide the energy going into storage would inevitably be substantial. It can also be deduced from Figs. 9.15 and 9.17 that, after allowing for the differences in their bases, Inhaber's estimate that the total fatalities from coal run nearly 100 times those from nuclear power is much the same as obtained completely independently by Ramsay of Resources for the Future.[61] Further, other studies by high-level professional committees, such as that of the AMA, have also found a similar ratio.[43,60] In short, the technical studies by the leading professionally competent groups indicate that nuclear energy gives far smaller adverse effects than coal on both health and land use.

An interesting set of data supporting this thesis was assembled by the United Kingdom Atomic Energy Authority (UKAEA). In comparing death rates for those working in the atomic energy field with those for the population as a whole, it was found that, for each of the major causes of death except road accidents, in all age groups the death rate was lower for the atomic energy employees. This was true for the various forms of cancer, the incidence of deaths from leukemia running only 48 percent of that for the general population.[63] The statistical base was quite large: 350,000 man-years at risk during the 1962–1974 period. It should be noted, too, that the British have taken a considerably more relaxed view of occupational exposures to radiation than the USAEC, a fact strikingly evident to the author during visits to UKAEA plants in the 1957–1971 period.

SUMMARY

Environmental problems are both serious and enormously complex, with electric power plants representing only one of the many factors involved. In the United States sulfur and particulate emissions to the atmosphere were reduced by factors of roughly 2 from 1964 to 1975, and the downward trend appeared to be continuing at the time of writing.[65] The key difficulty appears to be one of getting adequate public perspective on these problems so that the United States as a nation will not become so preoccupied with one set of problems that much larger problems are ignored, in effect "straining at a gnat and swallowing a camel."

PROBLEMS

1. The life-shortening effects of accumulated doses of low-level nuclear radiation have been estimated to be as much as 8 h per 1000 mrem. Estimate the life-shortening effect of the Three Mile Island incident for a person who stood at the highest dose level point at the plant perimeter where the total dose during the 3 weeks of the accident ran ~40 mrem. Compare this with the effect of smoking just one pack of cigarettes (Table 9.9).

2. From Ref. 3 and Table 9.2 it appears that the average age at death from automobile accidents is about 20 years, while the average age at death from the various forms of malignancy is ~60 years. Taking the normal life expectancy to be 73 years, estimate the number of man-years of life lost, respectively, to automobile accidents and to all types of malignancy per 100,000 people in the United States. Use the data of Table 9.2.

3. The fatality rate for coal mining in the United States was $0.35/10^6$ tons mined in 1976. If the United States derived all its energy from coal at the energy consumption rate of 70 quads per year, estimate the resulting number of coal-mining fatalities per year.

4. A good indication of the danger to the public associated with the release of a toxic material is given by the amount of air required to dilute the quantity released to standard tolerance levels. Using the tolerance levels given in Fig. A9.1, estimate the volume of air required to dilute to tolerance concentrations the chlorine leaking from a 55-ton tank car of liquid chlorine.

REFERENCES

1. Tamplin, A. R., and J. W. Gofman: *Population Control through Nuclear Pollution,* Nelson-Hall Co., 1970.

2. Hayes, D.: *Sun Day/Solar Action,* letter widely circulated in a fund solicitation campaign, April 1978.

3. Tucker, W.: "Environmentalism and the Leisure Class," *Harper's Magazine,* December 1977, p. 49.

4. *The World Almanac and Book of Facts, 1979,* Newspaper Enterprise Association, Inc., New York, 1978.

5. Schwartz, H.: "A Look at the Cancer Figures," *The Wall Street Journal,* Nov. 15, 1979.

6. American Cancer Society: *Annual Report,* March 1979.

7. Fraas, A. P., and G. Samuels: "Power Conversion Systems of the 21st Century," *Journal of the Power Division, ASCE,* vol. 104, no. P01, Proc. Paper 13545, February 1978, pp. 83–97.

8. Johnson, W. A., et al.: "Energy Conservation in Amish Agriculture," *Science,* vol. 198, no. 4315, Oct. 28, 1977, pp. 373–378.

9. *World Book Encyclopedia,* 1959, vol. 3, p. 1523.

10. Harris, J. R.: "The Rise of Coal Technology," *Scientific American,* August 1974, pp. 92–97.

11. McKeown, T.: *The Modern Rise of Population,* Academic Press, Inc., New York, 1979.

12. *Nuclear Power and the Environment: Questions and Answers,* rev. ed., ANS, La Grange Park, Ill., April 1976.

13. *EPA Human Health Water Quality Criteria,* September 1978.

14. Sax, H. I.: *Dangerous Properties of Industrial Materials,* Reinhold Publishing Corporation, New York, 1957.

15. Ross, F. Fraser: "A 1977 Approach to Sulfur Oxide Emissions," ASME Paper No. 77-JPGC-Pwr-1, September 1977.

16. Nelson, D. J.: "Radioactive Waste in the Fresh Water Environment, Energy and the Environment-Cost-Benefit Analysis," *Proceedings of Conference Sponsored by Georgia Institute of Technology,* June 23–27, 1975, pp. 262–272.

17. Blanchard, R. L., and B. Kahn: "Abundance and Distribution of Radionuclides Discharged from a BWR Nuclear Power Plant," *Nuclear Safety,* vol. 20, no. 2, March-April 1979, pp. 190–205.

18. Rados, B.: "Primer on Radiation," *FDA Consumer,* vol. 13, no. 6, July-August 1979, pp. 4–9.

19. Briggs, G. A.: *Plume Rise,* AEC Critical Review Series, U.S.A.E.C., 1969.

20. *Code of Federal Regulations, Energy,* sec. 10, parts 0–199, App. B, pp. 202–211, rev. Jan. 1, 1978.

21. *Assessment of the Cost Impact of Resource Conservation and Recovery Act of 1976,* report prepared for U.S. Department of Energy, Environmental Issues Committee, prepared by Energy and Environmental Analysis, Inc., Nov. 30, 1978.

22. "Control of Waste and Water Pollution from Coal-Fired Power Plants," Second R&D Report, U.S. EPA, Industrial Environmental Research Laboratory, EPA-600/7-78-224, November 1978.

23. Axtmann, R. C.: "Environmental Impact of Geothermal Power Plant," *Science,* vol. 187, no. 4179, March 7, 1975.

24. Krisher, A. S.: "Raw Water Treatment in the CPI," *Chemical Engineering,* vol. 85, no. 19, Aug. 28, 1978, pp. 78–98.

25. "Controlling SO_2 Emissions from Coal-Fired Steam-Electric Generators: Solid Waste Impact," vol. 1, *Executive Summary,* U.S. EPA, Industrial Environmental Research Laboratory, EPA-600/7-78-044, March 1978.

26. Pigford, T. H., et al.: *Comprehensive Standards: The Power Generation Case,* U.S. EPA, Office of Health and Ecological Effects, EPA-600/9-78-013, June 1978.

27. Duvel, W. A., Jr., et al.: *FGD Sludge Disposal Manual,* EPRI Report No. FP-977, prepared by Michael Baker, Jr., Inc., January 1979.

28. Wenk, E., Jr.: "The Physical Resources of the Ocean," *Scientific American,* September 1969, pp. 167–176.

29. Peixoto, J. P., and M. A. Kettani: "The Control of the Water Cycle," *Scientific American,* vol. 228, no. 4, April 1973, pp. 46–61.

30. Jacobs, D. G.: *Source of Tritium and Its Behavior on Release to the Environment,* Health Physics Division, Oak Ridge National Laboratory, USAEC Division of Technical Information, 1968.

31. Steiner, G. R., and D. G. Jahnig: "Role of Ponds in Treatment of Wastewater Expands at TVA Plants," *Power,* July 1977, pp. 54–58.

32. Cohen, B. L.: "Environmental Hazards in Radioactive Waste Disposal," *Physics Today,* January 1976, pp. 9–13.

33. *Report to the President by the Interagency Review Group on Nuclear Waste Management,* NTIS, March 1975.

34. Cowan, G. A.: "A Natural Fission Reactor," *Scientific American,* vol. 235, no. 1, July 1976, p. 36.

35. Rotty, R. M., et al.: "Atmospheric Considerations Regarding the Impact of Heat Dissipation from a Nuclear Energy Center," Oak Ridge National Laboratory Report No. ORNL/TM-5122, May 1976.

36. *Problems in the Disposal of Waste Heat from Steam-Electric Plants,* FPC Bureau of Power, 1969.

37. Reynolds, J. Z.: "Power Plant Cooling Systems: Policy Alternatives," *Science,* vol. 207, no. 4429, Jan. 25, 1980, pp. 367–372.

38. Okrent, D.: "Risk-Benefit Analysis for Large Technological Systems," *Nuclear Safety,* vol. 20, no. 2, March-April 1979, pp. 148–164.

39. Johnson, F. A., and P. Illes: "A Classification of Dam Failures," *Water Power and Dam Construction,* December 1976, pp. 43–45.

40. Germond, J. P.: "Insuring Dam Risks," *Water Power and Dam Construction,* June 1977, pp. 36–39.

41. Ayyaswamy, P., et al.: "Estimates of the Risks Associated with Dam Failure," USAEC Report UCLA-ENG-7423, University of California, Los Angeles, March 1974.

42. Inhaber, H.: "Risk of Energy Production," Atomic Energy Control Board Report AECB-1119/Rev-1, Ottawa, Canada, May 1978.

43. *Reactor Safety Study: An Assessment of Accident Risks in U.S. Commercial Nuclear Power Plants,* U.S. Nuclear Regulatory Commission, WASH-1400, October 1975.

44. Lewis, H. W., et al.: *Risk Assessment Review Group Report to the U.S. Nuclear Regulatory Commission,* NUREG/CR-0400, September 1978.

45. Greenhalgh, G.: "The Safety and Siting of Nuclear Power Plants," *Nuclear Engineering International,* vol. 19, no. 223, December 1974, pp. 1015–1027.

46. Knox, J. B., "Mississauga—A Milestone in Emergency Response Planning," Technical Note, *Nuclear Safety,* vol. 21, no. 5, September-October 1980, pp. 569–571.

47. Weinberg, A. M.: "Salvaging the Atomic Age," *The Wilson Quarterly,* Summer 1979, pp. 88–112.

48. Rose, D. J., and R. K. Lester: "Nuclear Power, Nuclear Weapons, and International Stability," *Scientific American,* vol. 238, no. 4, April 1978, pp. 45–57.

49. Woodwell, G. M.: "The Carbon Dioxide Question," *Scientific American,* vol. 238, no. 1, January 1978, pp. 34–43.

50. Frazier, K.: "Earth's Cooling Climate," *Science News,* vol. 96, Nov. 15, 1969, pp. 458–459.

51. Hollin, J. T.: "Wilson's Theory of Ice Ages," *Nature,* vol. 208, 1965, p. 12.

52. Likens, G. E., et al.: "Acid Rain," *Scientific American,* vol. 241, no. 4, October 1979, pp. 43–51.

53. Shinn, J. H., and S. Lynn: "Do Man-Made Sources Affect the Sulfur Cycle of Northeastern States?" *Environmental Science and Engineering,* vol. 13, no. 9, September 1979, pp. 1062–1067.

54. "Acid Rain and the Politics of Fear," *Electrical World,* Sept. 1, 1980, pp. 27–28.

55. Bengtsson, B.: "Liming Acid Lakes in Sweden," *Ambio,* vol. 9, no. 1, February 1980, pp. 34–36.

56. "EPA Acid Rain Analysis: Adverse Effects Are Proven and Proliferating," *Inside EPA,* Apr. 25, 1980, pp. 15–16.

57. Terrill, J. G., Jr., et al.: "Environmental Aspects of Nuclear and Conventional Power Plants," *Industrial Medicine and Surgery,* vol. 36, no. 6, July 1967, pp. 412–419.

58. Fraas, A. P.: "Environmental Aspects of Fusion Power Plants," *Proceedings of the International Conference on Nuclear Solutions to World Energy Problems*, Nov. 13–17, 1972, pp. 261–273.

59. Brouns, R. J.: "Environmental Impacts of Nonfusion Power System," Battelle Pacific Northwest Laboratories Report No. BNWL-2027, UC-20, September 1976.

60. AMA Council on Scientific Affairs: "Health Evaluation of Energy Generating Sources," *Journal of the American Medical Association,* vol. 240, no. 20, Nov. 10, 1978.

61. Ramsay, W.: *Unpaid Costs of Electrical Energy,* published for Resources for the Future by The Johns Hopkins Press, Baltimore, 1979.

62. Holdren, J. P., et al.: "Energy: Calculating the Risks," *Science,* vol. 204, no. 4393, May 11, 1979, pp. 564–568.

63. Hill, J.: "Nuclear Power Comes of Age," *Journal of the British Nuclear Energy Society,* vol. 15, no. 2, April 1976, pp. 105–109.

64. *Study of Nuclear and Alternative Energy Systems,* National Academy of Sciences, Washington, D.C., 1978.

65. *Energy/Environment Fact Book,* DOE/EPA Decision Series, EPA-600/9-77-041, March 1978.

10

COSTS

The prime criterion for the selection of a power plant type or unit size is minimum cost, if for no other reason than that there is usually a statutory requirement that the utility supply electric power to the consumer at the lowest practicable cost. The cost problems involve not only engineering complexities and a host of arbitrary (but generally reasonable) cost accounting formalities, but also a confusing tangle of laws and political considerations.[1] The latter stem in part from the fact that there are major reductions in cost associated with increasing scale. As a consequence, throughout the United States and the world the electric utilities have been granted a monopoly position: each utility organization is sufficiently large to operate at low costs and thus is given a monopoly in its area of operation. In some instances a local, state, or federal nonprofit organization operates the utility, but most electric power in the United States (~ 75 percent) is produced by investor-owned (i.e., privately financed) utility companies. The distinction between public and private ownership is important from the cost standpoint, first because the private utilities must pay both property and income taxes, and second because in this capital-intensive business there is a major difference in capital charges. Municipal, state, and Rural Electric Association utilities obtain their capital from bonds that are not only tax-free but also are backed by the tax base of the community; thus they involve almost no risk, and they bear a much lower interest rate than required for corporate bonds or for the dividend return rate for corporate stocks if these are to be competitive in the financial markets. (The ratio of bonded debt to equity capital for investor-owned utilities is about 65:35.) It is this difference in taxes and capital charges that is responsible for the lower rates charged by publicly owned utilities as compared to those charged by investor-owned utilities; e.g., average charges were $0.0241/kWh and $0.0416/kWh, respectively, for the United States as a whole in 1978.[2] As is the case in so many areas, superficial comparisons, however tempting politically, can be highly misleading; a sound analysis requires rigorous and tedious cost accounting.

MAJOR ELEMENTS OF POWER COSTS

The simplest and most widely used cost breakdown is shown graphically in the bar chart of Fig. 10.1, in which the costs of power generation at the plant output bus bars are divided into three categories, i.e., capital, fuel, and operation and maintenance. Note the large differences in these costs for the three different types of plant, the nuclear plant entailing the highest capital and lowest fuel costs whereas the reverse holds for the oil-fired plant.

A quite different cost breakdown is shown in Fig. 10.2, which gives the total expenditures of all United States electric utilities for the 1970–1978 period with the percentages in each of six major categories.[3] The category labeled "Production" corresponds to the costs represented in Fig. 10.1. Of particular interest is the item for "Transmission and Distribution," which represents only 5.6 percent of the total cost to the consumer, or about double the cost of handling customers' accounts. Those who claim that huge savings in distribution costs are possible through the use of many small, individual, widely distributed power plants apparently have not examined publicly audited actual costs for both public and private utilities. In this connection it should be mentioned that the power losses in long-distance electric-power transmission lines for most utilities run between 1 and 2 percent[1] (Chap. 20). Thus transmission losses and costs are small compared to the economies of scale for large central stations.

CAPITAL COSTS

Each power plant, whether fossil-fuel steam or nuclear steam, usually consists of one to four units, though occasionally of as many as 10. In thermal power plants each unit is a tightly integrated heat-source-turbine–generator unit with its own controls and instrumentation. Units are usually not built simultaneously at a given plant site, but in sequence as the expanding load

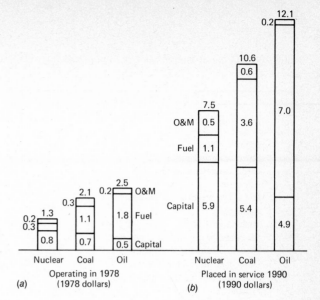

Figure 10.1 Principal cost components for electric power generation in cents per kilowatthour for nuclear, coal, and oil fuels for (a) plants operating in 1978 and (b) plants assumed to be placed in service in 1990.[1]

Figure 10.2 Total expenditures of U.S. utilities with a breakdown into six major expense categories for the 1970–1978 period.[3] *(Courtesy EBASCO Services Inc.)*

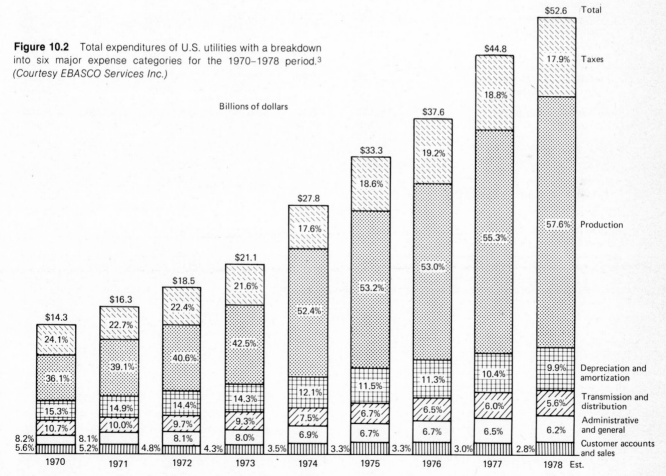

251

TABLE 10.1   Order-of-Magnitude Cost Estimates for 1000-MW Initial Units at New Sites[5]   (All Amounts in $1000—Mid-1975 Price Level)

Ebasco Account No.	Description	Nuclear			Coal-Fired		
		Total	Material	Installation	Total	Material	Installation
1.	Improvements to site	$ 1,690	$ 780	$ 910	$ 2,030	$ 1,000	$ 1,030
2.	Earthwork and piling	10,150	2,350	7,800	4,050	1,250	2,800
3.	Circulating water system	15,850	7,350	8,500	11,600	5,300	6,300
4.	Concrete	44,900	18,300	26,600	9,550	3,500	6,050
5.	Structural steel, lifting equipment, stacks	43,600	26,200	17,400	28,600	16,600	12,000
6.	Buildings	15,100	6,500	8,600	10,600	4,600	6,000
7.	Turbine generator	41,570	39,000	2,570	29,550	27,700	1,850
8.	Steam generator and accessories	—	—	—	128,900*	93,200	35,700
9.	Nuclear steam supply system	72,300	65,100	7,200	—	—	—
10.	Other mechanical equipment	20,900	18,300	2,600	14,700	12,600	2,100
11.	Coal- and ash-handling equipment	—	—	—	18,800	13,600	5,200
12.	Piping	45,900	24,700	21,200	17,300	9,600	7,700
13.	Insulation and lagging	3,700	1,040	2,660	10,900	3,300	7,600
14.	Instrumentation	2,870	2,000	870	2,680	2,300	380
15.	Electrical equipment	43,200	24,300	18,900	25,400	15,800	9,600
16.	Painting and finishing	2,100	600	1,500	1,790	500	1,290
17.	Off-site facilities	—	—	—	19,030	630	18,400
18.	Substation	2,290	1,820	470	2,290	1,820	470
	Total direct construction cost	$366,120	$238,340	$127,780	$337,770†	$213,300	$124,470
	Indirect construction cost	38,300	—	38,300	28,600	—	28,600
	Subtotal for contingencies	$404,420	$238,340	$166,080	$366,370	$213,300	$153,070
	Contingencies	77,300	35,750	41,550	51,900	21,300	30,600
	Total specific construction cost	$481,720	$274,090	$207,630	$418,270	$234,600	$183,670
	Home office services and fees	62,580			41,830		
	Total construction cost	$544,300			$460,100		
	Interest during construction and other client charges	Not included			Not included		
	Total project cost	$544,300			$460,100		

*Includes electrostatic precipitator and SO_2 removal equipment. Direct cost of SO_2 removal equipment included in Account No. 8 is $52,700,000.

†Total cost of SO_2 removal system including cost of required structures, foundations, off-site facilities, piping, electrical equipment, wiring, etc., for installation and operation of this system that are covered under the other appropriate accounts is $81,900,000.

| | Oil-Fired | |
Total	Material	Installation
$ 1,620	$ 750	$ 870
1,820	550	1,270
11,600	5,300	6,300
4,060	1,380	2,680
11,800	6,700	5,100
6,930	2,930	4,000
29,550	27,700	1,850
38,800	26,300	12,500
—	—	—
16,380	13,800	2,580
—	—	—
14,800	8,200	6,600
2,190	600	1,530
2,150	1,850	300
16,400	10,100	6,300
1,320	370	950
—	—	—
2,290	1,820	470
$161,710	$108,410	$53,300
12,300	—	12,300
$174,010	$108,410	$65,600
23,900	10,800	13,100
$197,910	$119,210	$78,700
19,790		
$217,700		
Not included		
$217,700		

of a utility requires additional capacity. Both the unit output and the capital costs commonly increase from the first to the latest unit installed. To facilitate rate setting, the Federal Power Commission (FPC)—subsequently incorporated into the DOE as the Federal Energy Regulatory Commission (FERC)—standardized the major cost categories involved in the construction and operation of each power plant unit and organized a standard format for presentation of these costs.[4] Cost summaries of these data are presented in Table 10.1 for typical nuclear, coal-fired and oil-fired power plant units.[5] The complete detailed cost breakdowns from which these summaries were prepared commonly total around 50 pages for each unit with each detailed item bearing a standard FPC account number; e.g., account number 234.11 is for closed feed-water heaters and account number 232.151 is for the chlorine injection system for the condenser cooling-water system. Table 10.2 has been included both to show the standard format for summarizing the capital costs and to indicate the relative size of the principal cost items. Note particularly that although the heat source, steam generator, and turbine generator are the major cost items, both Table 10.1 and Table 10.2 indicate that their combined cost is still only about 25 percent of the total cost of the plant.

Major Factors Affecting Capital Costs

Capital costs vary widely from one unit to another. The extent of this variation is indicated by Fig. 10.3, which shows a scatter band of actual costs for coal-fired power plant units as a function of the year in which they began commercial operation.[6] A similar scatter band for light-water reactors is displaced upward 10 to 30 percent, that for oil-fired steam units is displaced downward about 40 percent, and that for gas-fired steam units is displaced downward at least 50 percent. In all cases the width of the scatter band is large, the lowest costs running about half of the highest. The reasons for these variations include the size of the unit, whether the unit is the first to be installed at the site or is an addition (i.e., whether site preparation costs have already been charged off), the soundness of foundation conditions, the amount and type of excavation required, and local labor rates.

Size of Unit

One of the biggest factors affecting the capital cost per kilowatt is the size of the unit, the unit cost of a 1000-MWe unit being only ~ 60 percent as much as that for a 200-MWe unit[1] (Fig. 10.4). This stems in large measure from the fact that many costs, such as those for engineering and for instrumentation and control equipment, increase relatively little with the size of the unit, and the cost of any particular item of equipment does not go up in

TABLE 10.2 Plant Capital Investment Summary

BASIC DATA

	Seabrook Station		
Name of plant	Unit #1 + $\frac{1}{2}$ Common	Cost basis: at start of construction	
Net capacity	1150 MWe		
Reactor type	Westinghouse PWR	**Type of cooling**	
Location	Seabrook, N. H.		
		Run of river	

Design and construction period

Natural draft cooling towers

		Mechanical draft cooling towers	
Month, year NSSS order placed	January 1973	Other (describe)	Atlantic Ocean
Month, year of commercial operation	April 1983		
Length of workweek	40 hours, 5 days hours		
Interest rate, interest during construction	$9\frac{1}{2}$ or compound?		

COST SUMMARY

Account number	Account title	Total cost (thousand dollars)
DIRECT COSTS		
20	Land and land rights	$ 1,250
	PHYSICAL PLANT	
21	Structures and site facilities	198,092
22	Reactor plant equipment	170,232
23	Turbine plant equipment	68,264
24	Electric plant equipment	48,930
325, 352, 353	Misc. plant equipment	20,950
	Subtotal.................................	$ 507,718
	Spare parts allowance	3,236
	Contingency allowance	38,835
	Subtotal.................................	$ 549,789
INDIRECT COSTS		
91	Construction facilities, equip't. and services	$ 40,502
92	Engineering and const. mg't. services	141,316
93	Other costs	46,341
94	Interest during construction	371,000
	Subtotal.................................	$ 599,159
	Start of construction cost	$1,148,948
	*Escalation during construction (8%/yr)	88,673
	Total plant capital investment ($1077/kW)	$1,237,621

*Indicate separate escalation rates for site labor, site materials, and for purchased equipment, if applicable. Escalation rate is 8%/yr, simple.

Note: Cost data above are for Unit #1 plus $\frac{1}{2}$ of the Common facilities. Date of latest construction cost estimate is January 1979.

(Table submitted to the U.S. Nuclear Regulatory Commission June 22, 1979 in connection with a request for a license, Docket Nos. 50-443 and 50-444.)

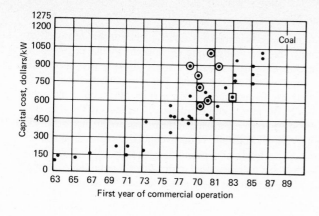

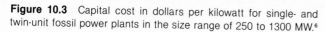

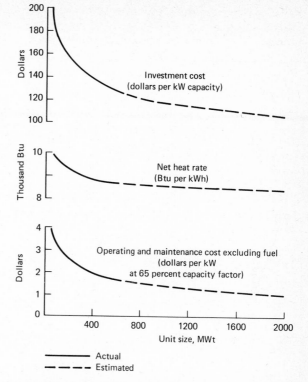

Figure 10.3 Capital cost in dollars per kilowatt for single- and twin-unit fossil power plants in the size range of 250 to 1300 MW.[6]

Figure 10.4 Effects of size on the 1964 cost and the performance of coal-fired steam-electric plants consisting of two units.[7]

proportion to its size. For example, the cost of the coal-handling equipment per ton of coal handled drops off rapidly as the capacity of the system is increased up to 1000 tons per hour (which is equivalent to ~2000 MWe), an effect shown in Fig. 10.5.[8]

Effects of Fuel Used

The amount and cost of the auxiliary equipment required varies widely with the type of fuel, coal requiring a large coal yard, conveyors, crushers, pulverizers, precipitators, and flue-gas desulfurization, soot-cleaning, and ash-handling equipment, as well as an ash-disposal area. Most of these costs can be avoided if oil is used, and all of them can be avoided if gas is the fuel. Further, the burners are simpler and less expensive for oil and even more so for gas. Costs also vary with the type of coal burned and its ash content. The furnace size must be increased for lignite, for example (Fig. 11.11). Nuclear plants require a

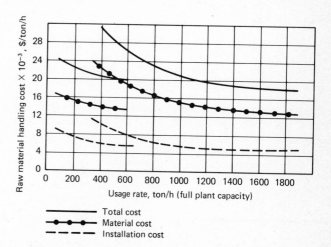

Figure 10.5 Effects of size on the unit price of systems for handling coal, limestone, and dolomite from the point of delivery by rail to silo storage at the furnace.[8]

large capital investment in the reactor plus much nuclear instrumentation and complex facilities for spent fuel and radioactive waste handling as dictated by nuclear safety regulations.

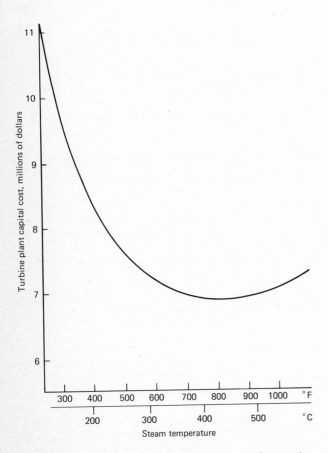

Figure 10.6 Effect of steam-turbine inlet temperature on the capital cost of the steam turbine, condenser, and feed-water system (but not the boiler, electric generator, or other elements of the plant).[9] Steam conditions chosen:

Steam cycle		Feed-water system	
Pressure, psig	Temperature, °F	Heaters	Temperature, °F
175	377	0	92
140	500	1	200
265	525	1	240
400	746	3	280
500	750	3	280
565	800	3	280
850	900	5	400
1250	950	5	435
1450	1000	5	450

Plant Complexity

The thermal efficiency of a steam plant can be improved by refining the thermodynamic cycle through the inclusion of reheating and progressively more elaborate provisions for regenerative feed-water heating. The more elegant these details in the thermodynamic cycle, the greater the amount and complexity of the piping and equipment; increases in thermal efficiency are therefore obtained at the expense of an increase in the capital cost per kilowatt. Thus, in coal-fired plants designed for intermediate loads and peaking service the system design is simplified and compromised in the direction of reduced capital costs at the expense of a somewhat reduced efficiency.

Effects of Steam Temperature and Pressure on Steam-Turbine Costs

Increasing the steam pressure and temperature increases the cycle efficiency, and this reduces the capital charges per unit of electric output for auxiliaries such as the coal-handling, feed-water, and condenser-cooling systems. It also reduces the unit costs for the steam generator and turbine up to the point where an increase in the temperature requires the use of much more expensive materials. This effect is shown by Fig. 10.6, which gives a curve for the cost of the turbine, condenser, and feed-water system for a 100-MW unit as a function of the steam temperature into the turbine.[9] Although the curve of Fig. 10.6 was prepared in 1953, the same basic trends still hold at the time of writing, because the effects of design temperature on materials costs have not changed significantly since 1950. Recent data for a higher temperature range are given in the chapter on steam plants (Fig. 11.12).

The increases in efficiency and reductions in cost associated with the increasing temperature and pressure, along with increases in scale, were major factors in the big reduction in electric power rates in spite of inflation in other costs through the period from 1900 to 1970. For example, as can be seen in Fig. 10.7, the cost of electricity to consumers in then-current dollars dropped by a factor of 2 between 1930 to 1970, and by a factor of 5 after adjustment for inflation.[10] This is a good measure of the effectiveness of the continuing effort by the electric-utility industry to reduce costs to the consumer.

Effects of Head on the Costs of Hydraulic Turbines

The head on a hydroelectric turbine generator is analogous to the temperature "head" on a steam turbine, and increasing the head reduces the cost per kilowatt of output. This stems from the fact that the physical size of the turbine is determined by the volumetric flow rate while the power output is directly propor-

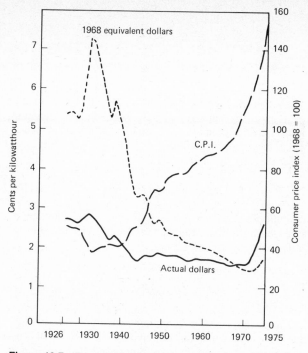

Figure 10.7 Price of electricity to ultimate consumers.[10]

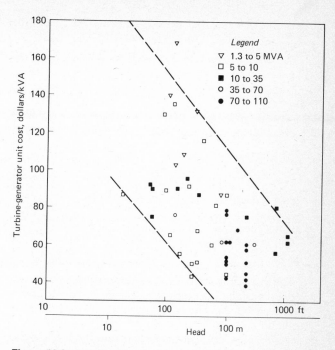

Figure 10.8 Effects of head and power capacity on the unit costs of 117 hydroelectric turbine-generator units. The data were taken from Ref. 11 for the 1950–1958 period, and the costs were converted to 1975 dollars by using the cost escalation factors of Fig. A10.1. A single point was plotted for multiple units purchased on the same contract.

tional to the head. Figure 10.8 shows a scatter band of cost data for 117 turbine-generator units purchased by the U.S. Bureau of Reclamation during the 1950–1958 period.[11] In preparing Fig. 10.8 the author corrected the original data for inflation to give equivalent 1975 dollars by using Fig. A10.1 in the Appendix. Note that there are points in Fig. 10.8 for power capacities ranging from 1.3 to 110 MVA (megavoltamperes). Inspection of the scatter band indicates that the unit cost is inversely proportional to the head. For outputs above 6 MVA there is no significant effect of capacity on the unit cost, but all the points for outputs below 6 MVA lie in the upper part of the scatter band. Note, too, that only 6 of the 117 cases are for heads below 30 m (100 ft), and none are for a head below 12 m (40 ft) because in that region unit costs were too high to be attractive. Although the fact is not readily shown, a review of the basic data disclosed that the lower points in the scatter band were for orders placed during periods of economic recession—a condition to be expected, because fabricators will bid low during such periods in order to avoid laying off their employes and so will accept a loss on the job. While the scatter band of Fig. 10.8 may appear wide, even the bids for a particular contract commonly differ by as much as a factor of 2, a reflection of differences between manufacturers in the availability of designs, model test data, and tooling applicable to the particular specifications provided.

A sophisticated effort to relate the costs of hydraulic turbines to the design parameters was prepared for publication just as this book was being completed. The analysis was based on data for 29 Kaplan and 17 Francis turbine contracts (each involving from 1 to 14 turbines) let by the U.S. Army Corps of Engineers during the 1950–1979 period.[12] The cost of the generator was not included; the generators were procured under separate contracts. A multiple-regression computer program was employed to obtain empirical values for the constants in the equations used. Eight major cost items were considered, of which the turbine cost typically represented ~83 percent of the total, while the balance was for model tests, tools, spare parts, and installation. The resulting equations for the major cost item—the design, construction, and delivery of the turbine itself—for Kaplan and Francis units are as follows:

Kaplan: $W = 7.8215 \times 10^{-5} D^{3.3407} n^{0.064} H_s^{1.363} \text{hp}^{-0.7338}$

(10.1)

$$\text{Cost (\$)} = (\text{HCIT})^{1.0101} (nW)^{0.9104} \qquad (10.2)$$

Francis: $W = 355 \times 10^{-5} D^{1.9566} H_s^{0.331}$ (10.3)

$$\text{Cost (\$)} = 38.1725 (\text{HCIT})^{1.0394} D^{2.0008} n^{0.8203} \quad (10.4)$$

where W = weight, tons
     D = throat diameter, in
     n = number of units in the contract
     H_s = static head, ft
     hp = horsepower
     HCIT = hydro cost index for turbines

This cost index is compiled by the Water and Power Resources Service (formerly the U.S. Bureau of Reclamation) and published quarterly in *Engineering News Record*. It is approximately proportional to the *Engineering News Record* Construction Cost Index plotted in Fig. A10.1, running ~ 83 percent of the latter in 1979.

If one examines these equations to appraise the effects of the prime design variables (the output and head) on the unit cost in dollars per kilowatt, cost escalation and the number of units can be dropped and the cost for the turbines can be divided by the power output which is proportional to the head and the water flow rate. For Kaplan turbines

$$\text{hp} \sim D^2 V H_s \qquad (10.5)$$

where V is the axial flow velocity through the throat, and

$$\frac{\text{Cost}}{\text{hp}} \sim \frac{D^{3.34} H_s^{1.36} \text{hp}^{-0.73}}{\text{hp}} = \frac{D^{3.34} H_s^{1.36} (D^2 V H_s)^{-0.73}}{D^2 V H_s} \quad (10.6)$$

$$\frac{\text{Cost}}{\text{hp}} \sim \frac{1}{D^{0.12} V^{1.73} H_s^{0.37}} \qquad (10.7)$$

For Kaplan turbines which normally operate at relatively low heads (< 30 m, or < 100 ft), the rotational speed is usually not limited by cavitation considerations. Hence the wheel speed and the throat velocity vary as $H_s^{1/2}$ (Chap. 3), and the unit per cost per kilowatt becomes:

$$\text{Unit cost (\$/kW)} \sim \frac{1}{D^{0.12} H_s^{1.23}} \qquad (10.8)$$

Thus the cost data show that the unit cost of Kaplan turbines is not very sensitive to the diameter but increases rapidly with a reduction in the head, both vital factors when considering the possibilities of exploiting small, low-head hydro sites.

A similar rationale for Francis turbines yields

$$\text{Cost} \sim D^2$$

$$\text{Unit cost (\$/kW)} \sim \frac{D^2}{D^2 V H_s} \sim \frac{1}{V H_s} \qquad (10.9)$$

For the higher head region (above ~ 30 m) in which Francis turbines are normally employed (Fig. 3.16), cavitation considerations limit the tip speed to around 36 m/s (120 ft/s) and hence the axial velocity through the throat. Thus V is essentially constant for the range of interest, and

$$\text{Unit cost (\$/kW)} \sim \frac{1}{H_s} \qquad (10.10)$$

which indicates that the unit costs of Francis turbines are only a little less sensitive to the head than for Kaplan turbines, and are essentially independent of the size.

Cooling Towers

As discussed in the previous chapter on environmental effects, cooling towers are commonly required for new power plants on lake and river sites, and these are expensive.[13] The commonly used wet cooling towers entail incremental capital costs over those required for the simpler direct use of lake or river water of ~ $30/kW in 1978 dollars. To save water in arid regions, dry cooling towers using water-to-air heat exchangers have been employed in a few cases, but their incremental cost ran ~ $100/kWe over wet towers in 1975.[13] (See also the last section of Chap. 11.)

Time for Construction

The time required for the construction of a power plant unit is substantial, running as much as 10 years for a nuclear plant in the latter 1970s. The sequence of events and the time required for each phase of a project can be envisioned by examining Fig. 10.9, which outlines the major steps between the time a contract is awarded and the time commercial operation is commenced.[14] Of course, a substantial period of work precedes the contract award, including site selection, core drilling to establish foundation conditions if a new site is used, and conceptual design studies to establish firm design specifications. Note that a cost estimate is made at the time the design contract is awarded, and that Fig. 10.9 shows the cost estimate being revised at three subsequent points as the design and subsequent construction progress. Note, too, that construction is started before the

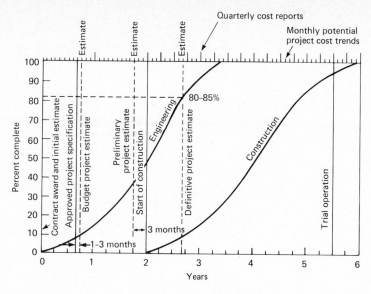

Figure 10.9 Typical schedule for estimates and cost reports—fossil project.[14]

detail design work is completed in order to expedite the project. However, this can lead to trouble if the design work falls behind schedule (usually because of changes in government regulations). In some instances construction workers are forced to stop work for lack of drawings, and in one instance in the author's experience, piping was installed without drawings and the drawings were subsequently corrected to conform to the actual installation! In any event such delays increase costs substantially.

A major step toward minimizing the time for construction has been the development of techniques for estimating the time required for preparing drawings, procuring materials and equipment, and carrying out the various construction and installation jobs at the site. The detailed estimates are organized in a chart that shows the flow of items into the site, and a computer program is commonly used to assist in detailed surveys and analyses. The chart and computer data are revised regularly as work progresses, mainly to highlight items that are falling behind schedule and may cause serious delays, but also to keep a close check on costs.

Costs of Interest During Construction and Escalation

In recent years the combined effects of construction delays stemming from government regulations, high interest rates, and inflation have increased the cost of interest on the investment during construction to as much as 35 percent of the total construction cost. This cost is called *Interest During Construction* (IDC), or *Allowance for Funds Used During Construction* (AFUDC). Figures 10.10 and 10.11 show that this cost plus escalation can total nearly double the cost of the plant estimated on the basis of the prevailing costs at the time the commitment for the plant construction was made.[1,15,16]

Additional costs to the consumer are likely to result from delays in construction if the utility finds itself short of power and must purchase power at a premium from another utility or operate high-cost oil-fired peaking units for intermediate load service.

The 1970s have seen a rate of increase in construction costs without precedent in history. Much of this has stemmed from inflation, but costs have also increased because of reduced labor productivity and increased regulation. Thus construction costs have escalated at a higher rate than inflation in spite of advances in technology, such as the widespread use of automatic welding equipment and larger earthmoving equipment.

Licensing

The direct costs associated with licensing a power plant were running roughly one-third of the cost of new plants in the latter 1970s. An incredible amount of paper work is required: from 10 to 20 tons for a new nuclear power plant! Many local, state, and

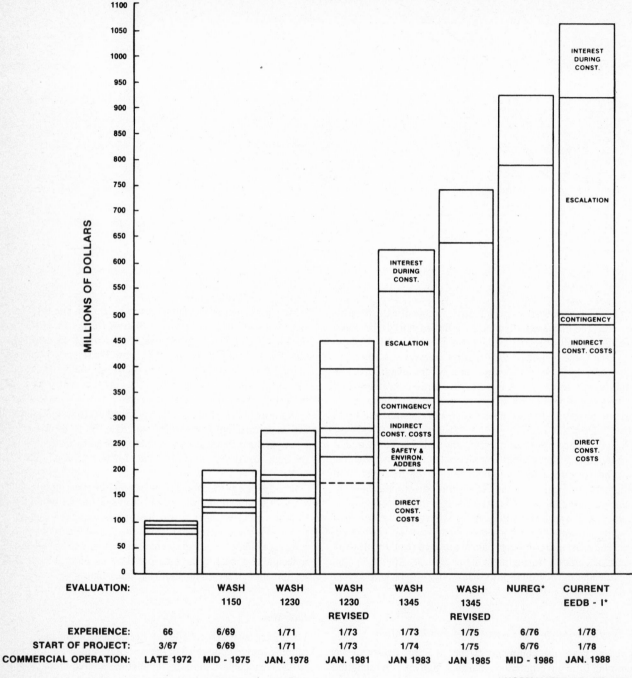

Figure 10.10 Comparison of coal-fired plant cost estimates (total investment cost for 1000-MWe units.).[15]

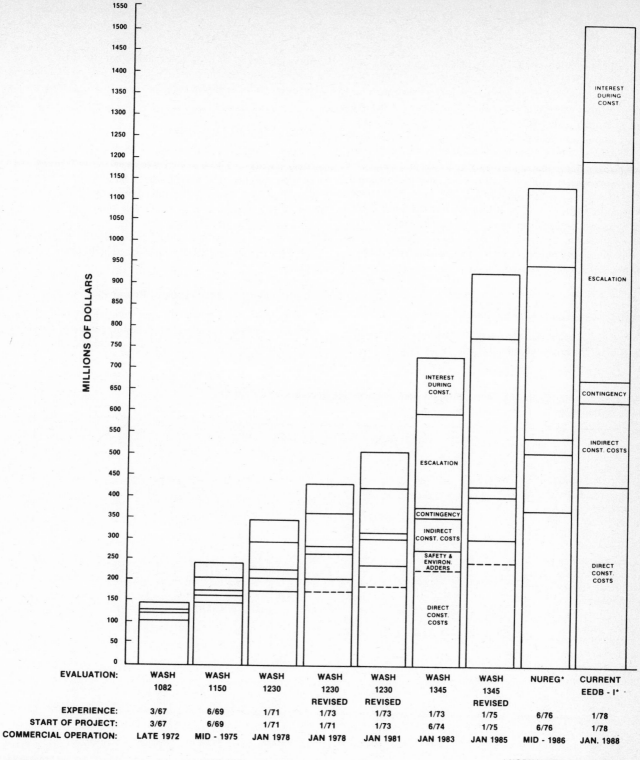

Figure 10.11   Comparison of nuclear plant cost estimates (total investment cost for 1000-MWe units).[15] (*Courtesy United Engineers and Constructors, Inc.*)

* NORMALIZED TO 1000 MWe

261

federal agencies (including, on the federal level, the FERC, EPA, OSHA, NRC) each have voluminous requirements that must be met with reports prepared according to their own particular formats and content specifications, often inconsistent with each other and sometimes actually contradictory. The resulting amount of engineering work and the attendant costs are staggering, commonly running over 30 percent of the direct cost of a nuclear plant. Some notion of the character and magnitude of this work is given by Figs. 10.10 and 10.11 and by Table 10.3, which also show the enormous increase in these requirements between the 1960s and 1970s. It is not clear that there has been any increase in public health or safety as a consequence of these requirements, but the increase in the cost of electric power to the consumer unquestionably has been high, especially when both direct and indirect costs are included.

Relative Amounts of Major Capital Cost Items

The relative amounts of the direct and indirect costs of new power plants have changed greatly in the 1970s, an effect shown graphically in Fig. 10.12.[15] Note that this bar chart also gives an excellent insight into the relative size of the various components of the direct and indirect costs.

Effects of Extended Workweek

Efforts to "make up time" and stay on schedule by going to an extended workweek lead to increased costs, not only because of the wage bonuses for overtime and Sunday work but also because of reduced worker productivity as a consequence of fatigue. This effect is indicated in Fig. 10.13 and applies to both craft labor and professional workers. The effect was first determined quantitatively in England during World War II, and essentially similar data have been obtained in the United States.

Capital Costs of New Types of Power Plants

Detailed cost breakdowns for conventional fossil fuel and nuclear power plants are available, and so good cost estimates can be made except for uncertainties stemming from possible

TABLE 10.3 Changes in the Documentation Requirements for Nuclear Units from the 1960s to the 1970s[17]

Item	1960s	1970s
Engineering man-hours	500,000	3,500,000
Job duration, years	8	13
Engineering personnel	125	350
Correspondence		
Transmitted	12,400	55,000
Received (not including vendor dwgs. trans.)	8600	17,000
PSAR volumes	2	13
Environmental reports		
Construction permit stage	0	10 (volumes)
Operating license stage	1 (volume)	12 (volumes)
Public hearings	2 (days)	16 (days)
		(ACRS & ASLB)
Quality manuals	1	6
Calculation pages	1400	70,000
Hanger engineering calculation	0	12,000
Pipe stress computer calculations	500	500
Civil/structural seismic computer calculations	16	200
Drawings (total)	2200	45,000
Purchase orders	240	400
Subcontracts	25	60
Specifications	230	490
Vendor drawings	37,000	90,000

Source: AIF January 1978

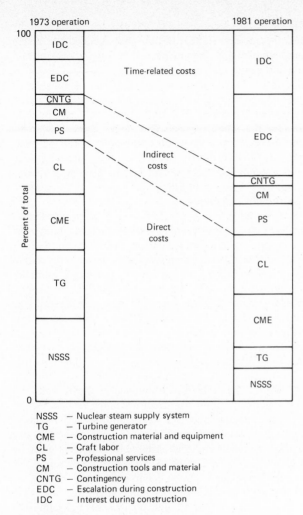

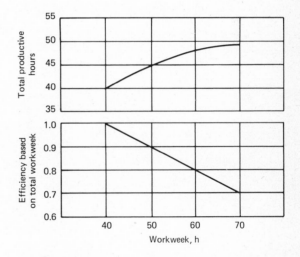

NSSS — Nuclear steam supply system
TG — Turbine generator
CME — Construction material and equipment
CL — Craft labor
PS — Professional services
CM — Construction tools and material
CNTG — Contingency
EDC — Escalation during construction
IDC — Interest during construction

Figure 10.12 Shifts in distribution of nuclear power plant capital costs.[15] (*Courtesy United Engineers and Constructors, Inc.*)

Figure 10.13 Effects of sustained overtime on productivity of site labor.[15] (*Courtesy United Engineers and Constructors, Inc.*)

delays in licensing and the amount of inflation. The situation is much more difficult with respect to new types of power plants, and the uncertainties are inherently greater. Cost estimates must be made, however, and this is ordinarily done by estimating the cost of each individual piece of equipment required by using generic data for similar components for both power plants and the chemical process industries. Compilations of these cost data can be found in books, such as Ref. 17, where costs are given in terms of such parameters as the power for motors, the surface area for heat exchangers, the diameter and length for piping, the capacity and head for compressors, fans, and pumps. (Some

useful data of this sort are given in Figs. A10.3 to A10.7.) Cost estimates for unusual mechanical equipment, such as mountings for solar mirrors designed to track the sun, can be made on the basis of weight, type of material, and the amount and precision of the machining required by using costs for roughly similar machinery. In fact, component weight is probably the best single index of cost if the appropriate cost factors for the type of equipment and the material involved are employed. Under any circumstances, for unusual equipment it is best to make estimates on two or three different bases to provide some degree of assurance that the cost estimates are meaningful.

Heat-Exchanger Costs

Heat exchangers often constitute the largest item in the cost of a power plant, particularly in advanced types. Hence a generalized survey of factors determining their costs appears in order, particularly because they give a good illustration of the various design factors affecting costs. To keep the cost of equipment at as low a level as possible, it is desirable to employ as inexpensive a material as can be used after allowances are made for corrosion considerations and pressure stresses. This, in turn, implies that the peak temperature and pressure chosen for the thermodynamic cycle must represent a compromise between cycle efficiency and the pressure differentials across the walls of piping, tubing, heat-exchanger shells, etc., so that thick tube walls of an expensive alloy will not be required.

Example   The effect of the choice of material and tube-wall thickness on the cost of tubing is given in Fig. A10.3 for 0.75-in-

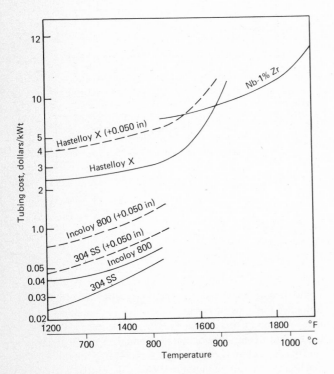

Figure 10.14   Effects of design temperature on the cost of tubing for a 500 psi pressure differential and a heat flux of 50,000 Btu/(h·ft²) assuming the estimated costs of Fig. A10.3 and ASME code allowable stresses (or equivalent). Two curves are given for the Fe-Cr-Ni alloys, one with and one without an extra thickness allowance of 0.050 in for corrosion in a coal-fired furnace.[18]

OD tubing made to a conventional ASTM specification for nonnuclear applications.[18] The basic data were obtained via conversations with tubing vendors in May 1975 assuming only modest production quantities of about 3048 m (10,000 ft). The effects of tube-wall thickness on costs were available only for Incoloy 800. As a consequence, in constructing Fig. A10.3, the writer assumed that the effects of wall thickness on cost would be similar for the other materials and simply drew in a set of parallel curves through the points obtained for a wall thickness of 1.65 mm (0.065 in). If the tubing must meet nuclear quality assurance standards, the extra cost for quality control and inspection in 1975 ran about $10/m ($3/ft), essentially irrespective of the alloy employed.

The tube-wall thickness required for any given installation depends in part on the pressure differential that must be accommodated across the tube wall and in part on the extent to which the wall must be thickened to allow for corrosion in the course of the life of the heat exchanger. In attempting to appraise the relative advantages of different materials after allowing for differences in strength at elevated temperatures, the curves of Fig. 10.14 were prepared using the data of Fig. A10.3 and assuming a heat exchanger that would have a pressure differential across the tube walls of 34 atm (500 psi) and a temperature difference sufficient to yield a heat flux of 15.8 kW/cm² [50,000 Btu/(h·ft²)]. Allowable stresses were taken from the ASME pressure vessel code for the three iron-chrome-nickel alloys, while the allowable stress for the niobium alloy was taken as 60 percent of the stress for 1 percent creep in 10,000 h, a value which is consistent with the relation between the ASME code stresses for the other three materials and their stresses for 1 percent creep in 10,000 h. In examining Fig. 10.14 and applying a factor of 2 to 2.5 to convert the tubing cost in dollars per kilowatt thermal to dollars per kilowatt of electric output, it is interesting to see that if the heat flux can be as high as 15.8 kW/cm² [50,000 Btu/(h · ft²)], the cost of the tubing is not a serious charge even for the expensive alloys. Note, too, that above 870°C (1600°F) the higher strength of Nb-1% Zr gives a lower cost than Hastelloy X, and that for an advanced nuclear plant employing a 982°C (1800°F) boiler for a potassium vapor cycle with a thermal efficiency of about 50 percent, if allowances are made for the lower pressure the cost of the Nb-1% Zr tubing would run about $18/kWe, a not unreasonable figure.

One of the most important steps that can be taken to reduce the cost of equipment is to design it so that it can be fabricated and assembled in the shop rather than in the field. This, in turn, implies that it is desirable to design components such as a large heat exchanger so that they consist of a multiplicity of basic modules, each of which can be fabricated and inspected independently before final assembly. If possible, these modules should be assembled into complete units before shipment to the

site, although if they are too large for shipment of assembled units to be feasible, they can at least be shipped as subassemblies.

A major factor in the cost of heat exchangers is the fabrication of the headers including making the tube-to-header joints. If a high degree of leak-tightness must be obtained, as in a helium or an alkali metal system, the joints must be welded and the detail design of the weld at the header sheet deserves much attention both to facilitate the welding operation and to ease inspection. For a nuclear quality heliarc weld in Fe-Cr-Ni alloys, the cost in 1975 was around $30 per tube-to-header joint including inspection, while for Nb-1% the cost was about double that because the welds must be made in a dry box.

It is usually necessary to make use of bent tubes. Where possible, all the bends should be kept in the same plane, because this not only simplifies the fabrication operation but also eases inspection. Springback problems make it difficult to maintain close tolerances on the shape of the tubes. Hence it is desirable to design the heat exchangers so that the tubes will be flexible enough for their shape to be determined by spacers and close tolerances on the bends will not be required.

FUEL COSTS

Fuel costs are a major fraction of the cost of power generation (Fig. 10.1), running about half the total cost for coal-fired plants. But the absolute and relative costs have changed markedly since 1970, with the price of oil in particular escalating rapidly. The world's fuel resource base is such that coal and nuclear fuel are clearly the lowest-cost fuels of the future (Fig. 1.16); hence the fuel cost trends of Fig. 1.8 can be expected to continue well into the twenty-first century.[19] There is little reduction in fuel costs gained by an increase in scale beyond ~ 600 MWe because the thermal efficiency increases relatively little with a further increase in the size of the plant (Fig. 10.4). The biggest uncertainties lie in the questions of changes in the relative costs associated with the environmental and safety regulations attending the use of these fuels.

Flue Gas Desulfurization

The environmental effects of sulfur emissions are treated in the previous chapter, while the equipment required and its performance characteristics are treated in the next chapter on steam power plants. The discussion here is thus limited to a brief look at the overall cost effects. A typical estimate of the incremental capital cost of flue gas desulfurization when incorporated in new coal-fired plants in the design stage has been made from actual installation costs by Battelle Columbus Laboratories.[20]

This study yielded an incremental capital cost in 1980 dollars of $101/kW for a 175-MW unit and $70/kW for a 300-MW unit on the basis of 1978 EPA requirements (subsequently made more stringent in July 1979). The study indicated that costs would be reduced ~ 10 percent if the design value for the coal sulfur content for which the system was designed were less than 1 percent, while capital costs would be increased by ~ 10 percent if the sulfur content were above 3.5 percent. In addition, there would be a capital cost for sludge disposal running from as little as $2/kW to as much as $66/kW, depending on the plant site. Retrofitting an FGD system to an existing power plant gives costs from 30 to 50 percent greater than for systems incorporated in a plant in the design stage.

The direct operating costs of FGD systems are substantial. A typical 1978 estimate was 3 to 8 mills per kilowatthour, or 40 to 60 percent of the total costs for operation and maintenance.[20] In addition, there have been major costs stemming from the increased plant downtime caused by corrosion and clogging of FGD system components.

Nuclear Plant Fuel Costs

The overall costs for the fuel cycle of nuclear plants are complicated by the large capital charge for the investment in the fuel and the expenses associated with its preparation, together with political uncertainties with respect to reprocessing and radioactive waste disposal. An excellent yet brief treatment of these cost factors for LWRs is given in Ref. 1, from which Table 10.4 has

TABLE 10.4 Projected Nuclear Fuel Cycle Cost for 1250-MW Plant Placed in Service in 1990 (Ref. 1)

Fuel cycle cost, cents/kWh	No recycle	Recycle
Levelized for first 5 years	1.3	1.1
Levelized over 30 years	2.0	1.5
Components (based on 30 years), %		
Uranium	68.0	58.0
Conversion	1.6	1.4
Enrichment	22.0	21.0
Fabrication	5.9	9.4
Transportation	0.6	1.2
Reprocessing	—	5.0
Waste	1.9	4.0

Note: Financing of fuel cycle is included in above. Costs are expressed in 1990 dollars. Recycle is believed more appropriate than nonrecycle for fuel cost predictions in 1990, and is used for prediction of generating cost.

TABLE 10.5   Staff Requirement for Coal-Fired Plants with FGD Systems [19]

| | 400–700 MWe unit | | | | 701–1300 MWe unit | | | |
| | Units per site | | | | Units per site | | | |
	1	2	3	4	1	2	3	4
Plant manager's office								
Manager	1	1	1	1	1	1	1	1
Assistant	1	2	3	4	1	2	3	4
Environmental control	1	1	1	1	1	1	1	1
Public relations	1	1	1	1	1	1	1	1
Training	1	1	1	1	1	1	1	1
Safety	1	1	1	1	1	1	1	1
Administrative services	13	14	15	16	13	14	15	16
Health services	1	1	1	2	1	1	1	2
Security	7	7	9	14	7	7	9	14
Subtotal	27	29	33	41	27	29	33	41
Operations								
Supervision (excluding shift)	3	3	5	5	3	3	5	5
Shifts	45	50	60	65	45	50	60	65
Fuel and limestone handling	12	12	12	18	12	12	12	18
Waste systems	15	30	45	60	15	30	45	60
Subtotal	75	95	122	148	75	95	122	148
Maintenance								
Supervision	8	8	10	12	8	8	10	12
Crafts	90	115	135	155	95	120	140	160
Peak maintenance annualized	33	66	99	132	35	70	105	140
Subtotal	131	189	244	299	138	198	255	312
Technical and engineering								
Waste	1	2	3	4	1	2	3	4
Radiochemical	2	2	3	4	2	2	3	4
Instrumentation and controls	2	2	3	4	2	2	3	4
Performance, reports, and technicians	14	17	21	24	14	17	21	24
Subtotal	19	23	30	36	19	23	30	36
Total	252	336	429	524	259	345	440	537

**TABLE 10.6   Maintenance Materials Cost Factors as a
Percentage of Maintenance Labor Cost*** (Ref. 19)

	Fixed	Variable	Total
LWR	100	0	100
Coal	50	17	67
Coal with FGD	53	29	82

*Estimated at 80% plant capacity factor.

been taken. Two columns of numbers are given, one for "no recycle" (the practice up to the time of writing) and the other for "recycle" (the approach adopted by nearly every country having nuclear power plants). Kennedy in Ref. 1 addressed the charges by antinuclear activists that the costs of Table 10.4 do not include all the charges that should be made, and he quite effectively refutes these claims. The biggest questions are environmental, and these are treated in the previous chapter.

For breeder reactors, the fuel cycle costs would be much lower than for the LWRs of Table 10.4, because the cost of fresh uranium—the biggest item in Table 10.4—would be almost nil. The costs and advantages of the breeder reactor fuel cycle are treated in Ref. 21. This paper is representative of the views of those who have made a thorough engineering and economic study of the problems involved and who have generally concluded that basic economics and the relative availability of fuels of various types will eventually drive the world energy economy to the widespread use of breeder reactors early in the twenty-first century.

OPERATION AND MAINTENANCE

The operation and maintenance (O&M) costs of power plants vary substantially with both the size of the individual units and the number of units in a plant.[22] The biggest cost is for personnel. As shown in Table 10.5, the number of people required increases much less rapidly than the number of units or their size; hence here again there are major economies of scale that have contributed to the trend toward ever larger power plants.

Nuclear Versus Coal-Fired Plants

As one might expect, the manpower requirements for fossil fuel and nuclear plants are similar. However, nuclear plants require an additional 50 to 100 employees in the security force but only about half as many in the operating crew and roughly two-thirds as many in the maintenance force; they do not entail the extra work and the troubles with dirt and corrosion in the coal- and ash-handling equipment that coal-fired plants involve. Oil- and gas-fired plants require substantially smaller crews than coal-fired plants because the fuel- and ash-handling problems are greatly eased with fuel oil and eliminated with natural gas.

Utility systems normally have special crews that move from plant to plant to expedite maintenance work during major shutdowns, particularly during the annual shutdowns for general maintenance and overhaul. Allowances for these employees were included in Table 10.5 and the above discussion.

Hydroelectric Plants

Hydroelectric units and plants are generally much smaller than the 400-MW size that is the smallest in Table 10.5. An interesting comparison can be made, however, with the 24 plants totaling 3000 MW of hydro operated by TVA. These 24 plants have about the same number of operation and maintenance personnel as a single nuclear steam plant with four units each in the 400 to 700 MW range.

Materials and Supplies

The expenses for materials and supplies vary somewhat with the type of plant and its age, as well as the scale of the plant. It has been found empirically that a reasonable estimate of these costs can be obtained by multiplying the personnel costs of the plant by the factors given in Table 10.6.[22]

COST OF ELECTRICITY

In view of the substantial differences in the costs of the distribution of electricity as a consequence of the fixed costs for the lines and differences in the amount consumed by the customer, it seems best to confine the discussion here to the cost of electricity at the plant output bus.

Capital Charges and Taxes

Capital costs for new investor-owned electric utility plants that formerly ran around 6 percent increased to about 12 percent for either stock or bonds in the 1970s. The factors involved and their implications are complex[1] and consequently not generally understood or appreciated by the general public. Ad valorem taxes levied by state and municipal governments are also roughly proportional to the capital investment and commonly run 1 to 2 percent. Depreciation is ordinarily computed on a straight-line basis for an assumed 30-year life of the plant, and it is based on the initial rather than the replacement cost, giving a first-year capital charge of $3\frac{1}{3}$ percent. Federal corporate income taxes are normally about 50 percent of the net income before calculation of the income tax; hence if all the capital investment were in the form of corporate stock, to obtain a 12 percent net yield on this stock would require earning an additional 12 percent to cover the income tax. If the ratio of bonded debt to stockholders' equity is 2:1, the capital charge to cover the income tax would be 4 percent instead of 12 percent. (This shows the incentive to obtain as much capital through bonds as possible.) The total of these charges for a 2:1 ratio of bonded debt-to-stockholders' equity thus becomes about 21 percent of the initial capital investment (not including charges for Interest During Construction). These costs are essentially independent of the amount of operation and hence are inversely proportional to the capacity factor, usually ~ 60 percent.

Fuel Costs

The fuel cost component of the electricity produced depends on both the thermal efficiency and the cost of the fuel itself, together with the costs related to the fuel consumed—such as the transportation, reprocessing, and waste charges for nuclear fuel and the limestone required for flue gas scrubbing of sulfur from coal. These costs are roughly directly proportional to the amount of power produced rather than a function of the capacity factor.

Operation and Maintenance

The operating staff must be on duty whether the unit is running or not; hence the bulk of the operating costs are inversely proportional to the capacity factor, as is the case for the capital charges. Some of the maintenance costs may be reduced if the unit is not operating or is at low loads; the extent to which this affects the O&M charge depends on the type of unit, the fuel used, etc. Note that deterioration from corrosion in many com-

ponents is likely to proceed more rapidly during shutdown when acidic moisture can collect than during hot operation when the parts are dry.

GENERAL PROBLEMS

The general problems associated with setting electric utility rates are both complex and vitally important to the nation. An enormous amount of capital—over 500 billion dollars—will be required in the 1980s for constructing new or replacement electric utility plants, and about two-thirds of this will have to be new capital (the balance will come from depreciation accounts for old plants).[1] Since this investment will represent ~ 25 percent of the new capital raised by private industry in financial markets, reasonable returns in the form of bond interest and stock dividends must be available. Probably the prime objection to these returns is that raised by no-growth advocates who also overlook the need for replacing worn-out plants. It is on the basis of this point of view that they approach such complex questions as whether to include in electric power rates a portion or all of the interest incurred during construction of new and replacement facilities. The inclusion of these charges in rates as a "Construction Work in Progress" (CWIP) item has been a major public issue that has involved a confusing mixture of statements and much misrepresentation. Unfortunately, the long-term costs to the consumer stemming from inadequate rates to finance the construction needed are likely to be much greater than those associated with adequate rates to finance an orderly construction program. It makes little difference whether this is done by providing an adequate profit for the utility so that its stock and bond offerings bring in all the required capital or whether a portion of it comes from inclusion in the rates of a charge for interest during construction of new facilities. Experience has shown that if these investments are not made, the consumer will eventually pay substantially more because of the high costs of power from gas turbines or premium power purchased from other utilities.

Cost Estimating

Estimating the costs for a projected plant has been rendered particularly difficult by the recent high rates of inflation with the escalation rate varying from one cost component to another. As a result, it is not surprising that estimates from different cost-estimating groups may differ substantially. A typical set made on the basis of similar ground rules is given in Table 10.7. It shows differences of ± 20 percent in some items, and ± 10 percent in the cost of the complete plant.[23]

TABLE 10.7 Three Independent Estimates of 500-MW Pulverized-Fuel Station Costs in Dollars per Kilowatt Using Similar Ground Rules

| Plant item | Estimator | | |
	General Electric	Bechtel	United Engrs. and Constrs.
Coal handling	10.6	6.7	13.2
Ash handling	7.5	2.8	8.0
Boiler	108.4	120.3	78.1
Turbogenerator	43.7	54.9	47.2
Electrostatic pptr.	13.3	31.1	13.4
Scrubber	78.9	98.5	61.3
Cooling tower	10.3	*	13.7
Pumps, pipes, condenser, etc.	77.7	80.1	82.0
Electrical, instrumentation, etc.	51.9	59.3	74.0
Civil, yardwork, etc.	56.2	83.3	61.7
Total	458.5	543.4	452.5

*Not given separately.

Probably the best and most convenient source of up-to-date cost information is the *Engineering News Record*, which gives quite detailed data for the costs of craft labor and construction materials for 20 cities in the United States, as well as some overall indices for the country as a whole. Data for one of these, *plant construction costs,* are plotted in Fig. A10.1 for the 1913–1979 period. Space has been provided so that the reader can fill in later data to provide an up-to-date basis for estimating the effects of inflation.

An important aid to cost estimators is a comprehensive set of eight reports of 200 to 500 pages each covering the capital, fuel, and operating costs of coal and nuclear plants prepared for the Nuclear Regulatory Commission (NRC).[24] Reference 25 is a similar but much less detailed study prepared for the Electric Power Research Institute (EPRI).

Relation between Energy Costs and Capital Costs

The rapid increase in the market prices of fuels in the 1970s led many to expect a shift in the relative importance of capital and fuel costs in the production of electricity. This has not occurred because, as pointed out by H. W. Parker,[26] energy is the driving force in our economy, and its cost directly affects the cost of everything including equipment and construction. A striking illustration of this is provided by the plot of Fig. 10.15 showing the industrial fuel cost index plotted against the process plant construction cost index for the 50-year period of 1926 to 1978. These data show a one-to-one relation between fuel and construction costs except in the 1958–1974 period, when a world oil glut temporarily depressed the price of petroleum and greatly stimulated the economy of the entire world.

The one-to-one relation between bulk energy costs and capital costs shown in Fig. 10.15 has a number of implications. First, it is apparent that increases in fuel cost apparently have not justified increasing the capital cost to get an increase in thermal efficiency and hence a lower fuel consumption. Secondly, while the effect is not obvious, it means that the cost of producing fuel oil and gasoline from coal increases about as rapidly as the cost

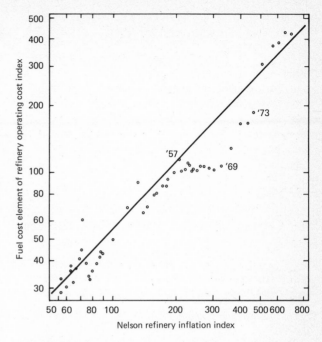

Figure 10.15 Relation between industrial fuel costs and process plant construction costs for the 1926 to 1978 period.[26] (*Courtesy ESCOE.*)

of bulk industrial fuel, because both the cost of the input material and the capital cost of the processing plant are roughly proportional to the industrial fuel cost. It is for this reason that the estimated cost of synfuel from coal seems to increase at almost the same rate as crude petroleum, thus somehow remaining substantially more expensive. Thirdly, the costs of "natural" sources of energy also continue to escalate at about the same rate as fossil fuels, because they require large capital investments; hence the expected "crossover" in costs does not materialize and the economical feasibility of "natural" energy sources continues to shift into the future.

PROBLEMS

1. Compare the capital cost data for the four types of plant for which data are given in Tables 10.1 and 13.5, consolidating some of the cost categories to facilitate comparisons. Comment briefly on what appear to be consistencies and inconsistencies and also indicate possible explanations for disparities.

2. Using the capital cost data of Table 10.1 and the fuel cost data of Fig. 1.8, estimate the cost of electricity for both coal-fired and nuclear plants for private financing. Assume a 70 percent load factor, capital charges of 18 percent per year, and express in 1975 dollars. (Reference to Fig. A10.2 is suggested.)

3. Repeat Prob. 2 for a publicly owned utility for which the capital charges are 9 percent per year.

4. Recalculate the costs of the coal-fired plant of Table 10.1, using the amounts of materials given in Table 10.1 but the unit costs of labor and materials given in a recent issue of *Engineering News Record.*

5. Estimate the current capital cost of a coal-fired plant built

for $185/kW in 1962, using Fig. A10.1 and data in a recent issue of *Engineering News Record.*

6. Estimate the cost of the tube matrix for a heat exchanger for intermediate pressure service for an installation requiring tight quality control. Use $\frac{3}{4}$ 110-in-OD, 0.10-in-wall tubes, a tube length of 4 m, Type 304 SS, and 10,000 tubes. Express in terms of current dollars, using the cost data of Fig. A10.3 and the escalation factors of Fig. A10.1.

REFERENCES

1. Kennedy, W. J. L.: "The Economics of Nuclear Power," paper presented to Atomic Industrial Forum, Inc., Conference on Nuclear Power and the Public, Feb. 26, 1979.

2. Weslowski, J. D.: "50 Locally-Owned Utilities in 1978 Outperformed Investor-Owned Counterparts," *Electric Light and Power,* vol. 57, no. 7, July 1979, pp. 10–12.

3. *1978 Business and Economic Charts,* Ebasco Services, Inc., 1979.

4. *Uniform Systems of Accounts Prescribed for Public Utilities and Licensees,* U.S. Federal Power Commission, April 1, 1973.

5. *Fossil and Nuclear 1000 MW Central Station Power Plants Investment Estimates,* prepared for EPRI by EBASCO Services, Inc., September 1975.

6. Budwani, R. N.: "Fossil-Fired Power Plants: What It Takes to Get Them Built," *Power Engineering,* May 1978, pp. 36–42.

7. *National Power Survey,* U.S. Federal Power Commission, 1964, p. 70.

8. Beecher, D. T., et al.: *Energy Conversion Alternatives Study—ECAS—Westinghouse Phase I Final Report,* vol. 1, *Introduction and Summary and General Conclusions,* NASA CR-134941, vol. 1, 1976.

9. *Nuclear Power Project Report,* Foster Wheeler Corp. and Pioneer Service and Engineering Co., October 1958.

10. *Factors Affecting the Electric Power Supply, 1980-85, Executive Summary and Recommendations,* U.S. Federal Power Commission, Dec. 1, 1976.

11. *Estimating Reclamation Instructions,* App. A, Estimating Data, Series 150, U.S. Bureau of Reclamation.

12. Sheldon, L. H.: "Cost Analysis of Hydraulic Turbines," *Water Power and Dam Construction,* June 1981.

13. Guyer, E. C., et al.: "Optimization-Simulation Methodology for Wet-Dry Cooling," EPRI Report No. FP-1096, May 1979.

14. Hallberg, L. K.: "Of Time and the Estimate," *Combustion,* September 1977, pp. 16–17.

15. "Power Plant Capital Investment Cost Estimates: Current Trends and Sensitivity to Economic Parameters," USDOE Report DOE/NE-0009, June 1980.

16. "1000 MWe Central Station Power Plant Investment Cost Study," vols. I, II, and III, USAEC Report WASH-1230, United Engineers and Constructors, Inc., June 1972.

17. Guthrie, K. M.: *Process Plant Estimating, Evaluation, and Control,* Craftsman Book Company of America, New York, 1974.

18. Fraas, A. P.: "Heat Exchangers for High Temperature Thermodynamic Cycles," ASME Paper No. 75-WA/HT-102, December 1975.

19. Brandfon, W. W.: "Comparative Costs for Central Station Electricity Generation," *Atomic Industrial Forum Conference on Energy for Central Station Electricity Generation,* Apr. 18, 1978.

20. Bloom, S. G., et al.: "Analysis of Variations in Costs of FGD Systems," prepared by Battelle Columbus Laboratories for EPRI, EPRI Report No. FP-909, October 1978.

21. Stauffer, T. R., et al.: "To Breed or Not to Breed," *Mechanical Engineering,* February 1977, pp. 32–41.

22. Myers, M. L.: "A Procedure for Estimating Nonfuel Operation and Maintenance Costs for Large Steam-Electric

Power Plants," Oak Ridge National Laboratory Report No. ORNL/TM-6467, January 1979.

23. Johnston, R.: *The Economics of Coal-Based Electricity Generation by Conventional Methods and by Fluidized Bed Combustion,* International Energy Agency, Economic Assessment Service, Working Paper No. 33, September 1978.

24. *Commercial Electric Power Cost Studies,* prepared for the Nuclear Regulatory Commission and ERDA by United Engineers and Constructors, Inc., June 1977. Set of eight reports (200 to 500 pp. each):

　1. *Capital Cost: Pressurized Water Reactor Plant,* NUREG-0241, COO-2477-5

　2. *Capital Cost: Boiling Water Reactor Plant,* NUREG-0242, COO-2477-6

　3. *Capital Cost: High and Low Sulfur Coal Plants—1200 MWe,* NUREG-0243, COO-2477-7

　4. *Capital Cost: Low and High Sulfur Coal Plants—800 MWe,* NUREG-0244, COO-2477-8

　5. *Capital Cost Addendum: Multi-Unit Coal and Nuclear Stations,* NUREG-0245, COO-2477-9

　6. *Fuel Supply Investment Cost: Coal and Nuclear,* NUREG-0246, COO-2477-10

　7. *Cooling Systems Addendum: Capital Total Generating Cost Studies,* NUREG-0247, COO-2477-11

　8. *Total Generating Costs: Coal and Nuclear Plants,* NUREG-0248, COO-2477-12

25. *Technical Assessment Guide,* Electric Power Research Institute Report No. EPRI PP-877-SR, June 1978.

26. Parker, H. W.: "The Energy Component of Future Energy Costs," *ESCOE Echo,* The Engineering Societies Commission on Energy, vol. 3, no. 24, Nov. 19, 1979.

11

STEAM
POWER PLANTS

In view of the fact that there has been no increase in the temperature and pressure employed in steam power plants since about 1950 and no improvement in the thermal efficiency, one might ask why a chapter on steam plants is included in a text on advanced power plants. This has been done because steam plants are truly highly developed systems, they provide the bulk of our power, and they provide a standard against which any advanced concept must be compared to determine whether it is sufficiently attractive to justify its selection.

HISTORICAL DEVELOPMENT

It is instructive to trace the development of steam power from its inception with the Newcomen engine nearly 300 years ago. Interestingly enough, that engine was so inefficient (less than 1 percent) that it was competitive with horses in only one field: pumping water from mines. Even there it gained only limited acceptance until James Watt in 1765 made the first major improvement: separation of the steam condenser from the cylinder to cut heat losses and increase the thermal efficiency by a factor of 4—to a few percent! People were so skeptical that this big improvement could be effected that for the first 20 years the Watt engines were not sold; they were erected at the site, and Watt's company was paid royalties on the basis of the coal savings that were effected through the use of the more efficient engine that Watt had patented.[1] This gave an opportunity for cheating on the part of the engine operator, who of course got more power from a given size of engine and hence more useful work with the former fuel consumption—a situation that led Watt to develop the Prony brake for measuring horsepower and defining that quantity.

Reciprocating-Steam-Engine Development

Refinements in the engine mechanisms and improvements in machine tools to give better fitting parts followed rapidly in the next 100 years. Thus the reciprocating steam engine was developed to essentially its natural limits by the time Edison invented the electric light bulb. As discussed in Chap. 2, the reciprocating steam engine is very different from gasoline or diesel engines in that the condensing-heat-transfer coefficient for steam is of the order of 5.7 W/(m^2 · °C) [10,000 Btu/(h · ft^2 · °F)], about 100 times as high as the corresponding value in a gasoline or diesel engine cylinder. As a consequence, the upper cylinder wall in a steam engine is heated by steam condensation to very nearly the saturation temperature of the steam supplied to the cylinder. The best lubricants available will keep wear rates low only if the rubbing surfaces are below about 204°C (400°F). With care in piston design, it has proved possible to keep the temperature at the piston rings about 40°C below that of the inlet steam and thus to operate with a steam saturation temperature and pressure of 242°C and 34 bars (467°F and 500 psia). Significantly, no improvements in this temperature limit have been achieved in the past 100 years; repeated efforts to exceed it in automotive steam engines have led to high wear rates and short engine life.

The development of steam plants for utility service in the United States began with the construction of the Brush Electric Light Company plant in Philadelphia in 1881.[2] Construction of other plants followed rapidly throughout the country to satisfy the need for electric power for lighting and for electric street railways. Initially, single-expansion reciprocating steam engines were employed. By 1890, double-expansion engines and shortly thereafter triple-expansion engines became common. Note that because of their low cost and simplicity of construction, triple-expansion steam engines were used in the huge fleet of Liberty ships built during World War II.

Steam-Turbine Development

By 1900, the advantages of a steam turbine in permitting higher temperatures and pressures with greater expansion ratios became apparent, and in 1904 the first steam turbine in U.S. utility service went into operation, giving a higher thermal efficiency than any reciprocating engine in service.[3-8] Although pressures up to 41 bars (600 psia) were used in steam engines for mobile power plants in order to reduce the engine size and weight while sacrificing engine life, overall expansion ratios of

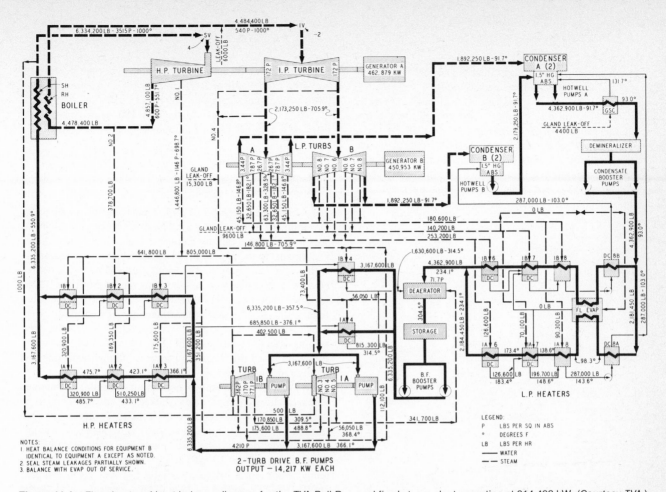

Figure 11.1 Flowsheet and heat balance diagram for the TVA Bull Run coal-fired steam plant operating at 914,402 kW. (*Courtesy TVA.*)

more than 100 were not practicable even in triple-expansion reciprocating steam engines, and hence there was no incentive from the standpoint of thermal efficiency to go to steam pressures in excess of about 10 bars (145 psia). A steam turbine could employ much higher expansion ratios to good advantage, thus leading to fairly rapid increases in both temperature and superheat.[5,6,7]

Increasing the steam pressure and the resulting expansion ratio in steam turbines increased the number of stages and rotor length required to the point that bearing and shaft dynamics problems became limiting. (As discussed at the end of Chap. 3, the critical speed of a shaft is inversely proportional to the square of its length.) This limitation led to the use of first two and finally as many as four turbines in tandem. In a supercritical-pressure steam plant, for example, there may be a superhigh-pressure, a high-pressure, an intermediate-pressure, and a low-pressure turbine. Other factors favoring the use of a multiplicity of turbines in series are the large differences in cas-

ing thickness required to withstand the internal pressure and the large differences in rotor diameter stemming from the enormous change in steam specific volume associated with the steam expansion (a factor of over 2000). In fact, the flow-passage area required for the volume of steam leaving the lower-pressure turbines becomes so great in units of large capacity that double-flow turbines are employed with the steam entering in the center and splitting into two streams that flow in opposite directions toward the outboard ends of the rotor. As mentioned at the end of Chap. 3, this has the additional advantage that the axial forces on the two sets of rotor blades balance each other, thus avoiding a difficult thrust-bearing problem. For units of more than 300 MWe, the size of the low-pressure turbine becomes so great that two double-flow turbines are used in parallel, and so there are actually a total of four stages of rotor blades operating in parallel. Note that the last-stage turbine rotor size required is greatly reduced by bleeding off steam for regenerative feed-water heating. (Examination of the flowsheet of Fig. 11.1 shows that in this case the steam flow to the condenser is only 60 percent the flow into the high-pressure turbine.)

The high-, intermediate-, and low-pressure turbines may be coupled in series on a single shaft having a total length (including the generator) of as much as 77 m (250 ft) for a 1300-MW unit. An alternative arrangement is to reduce the size of the higher-pressure units by designing them to operate at 3600 rpm and mounting them on one shaft and mounting the inherently large low-pressure units on a separate shaft that operates at 1800 rpm. The latter arrangement is referred to as a *cross-compound* turbine-generator unit. The steam pressure available from PWR or BWR nuclear plants is enough lower than in fossil fuel plants—i.e., 65 bars (~950 psia) as opposed to 240 bars (~3500 psia)—that an 1800-rpm single-shaft machine is employed with a number of turbines in tandem.

Steam Generator–Furnace Units

The rapidly increasing size of early steam power plants led to the development of traveling grate and other types of mechanical stoker, and these rapidly displaced hand-fired boilers.[3] Improvements in metallurgy made possible higher-strength steels with greater resistance to corrosion, thus leading to further increases in pressure and temperature throughout the 1920s together with concurrent increases in unit size.[4] At the same time, refinements in the thermodynamic cycle were made to improve the thermal efficiency by reheating,[9] regenerative feed heating, and air preheating. Improved methods of feed-water treatment eased the corrosion problems on the steam side and reduced maintenance.[2]

By 1930, developments in both electric welding techniques and x-ray inspection of welds enabled the advancement from riveted header drums with rolled tube-to-header joints and flanged joints in the piping to virtually all-welded construction of the header drums, tube-to-header joints, and plant piping.[2] This, in turn, permitted marked further increases in the system temperature and pressure to around 170 bars (2450 psi) and 565°C (1050°F), so that by 1950 the new central station steam plants had virtually reached the operating conditions current at the time of writing.[10,11] Concurrently, the reliability of boilers and turbines increased sufficiently that the initial practice of manifolding a multiplicity of boilers to a multiplicity of turbines was dropped in favor of independent units with each boiler coupled to a particular turbine. This simplified the piping (particularly provisions for thermal expansion) and eliminated many large shutoff valves which gave a great deal of trouble when used at high temperatures and pressures. In the 1950s, the once-through boiler came into use and was almost immediately applied to produce supercritical-pressure steam, commonly 240 bars (3500 psi) and 565°C (1050°F).[12-16] A simplified flowsheet showing the major components in a typical modern plant is presented in Fig. 7.7, and a more detailed flowsheet for a supercritical-pressure steam system is shown in Fig. 11.1.

A section through a pulverized-coal-fired-furnace–steam generator unit is shown in Fig. 11.2. Combustion takes place in the large open chamber at the lower right, which is surrounded by vertical boiler tubes joined together with short thick fins as shown in Fig. 11.3 to form a continuous "waterwall" that ab-

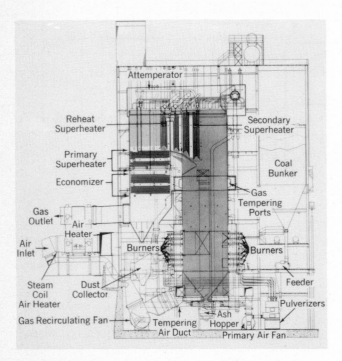

Figure 11.2 Radiant boiler for pulverized-coal firing. Design pressure 2875 psi; primary and reheat steam temperatures 1000°F; maximum continuous steam output 1,750,000 lb/h.[2] (*Courtesy Babcock & Wilcox Co.*)

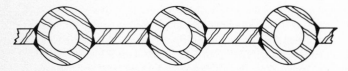

Figure 11.3 Section through the boiler tubes in a waterwall for a boiler furnace such as that of Fig. 11.2.

TABLE 11.1 Summary of Principal Developments in Steam Power Plants for Utility Service

Period	Typical parameters for new plants				Principal developments
	Peak temp., °C (°F)	Peak press., bars (psia)	Unit output, MW	Full-load thermal efficiency, %	
1880–1890	173 (344)	8.6 (125)	0.3	5	Generators belt-driven from reciprocating steam engines with hand-fired boilers.
1890–1900	173 (344)	9.0 (130)	0.7	7	Double- and triple-expansion engines with Corliss poppet valve gear introduced.
1900–1910	181 (358)	10 (145)	1	10	Steam turbine with steam superheat came into common use with mechanical coal stokers.
1910–1920	293 (560)	19 (275)	10	15	Steam pressures and temperatures increased somewhat and sizes and efficiency of turbines much increased.
1920–1930	385 (725)	38 (550)	25	25	Large pulverized-coal-fired boilers introduced with new furnace designs employing water-cooled walls with steam reheat, regenerative feed-water heating, and combustion air preheat. Turbine sizes increased to as much as 100 MW. Feed water treated.
1930–1940	482 (900)	62 (900)	50	32	Arc-welded construction and new alloys permitted marked increase in pressures and temperatures: large steam generators and turbines combined into single units (to eliminate troublesome valves) replaced former multiplicity of small boilers (as many as 60 in one powerhouse) manifolded to a multiplicity of turbines. Control of feed-water chemistry improved.
1940–1950	538 (1000)	83 (1200)	150	36	Higher pressures and temperatures came into use with further improvements in alloys and control of feed-water chemistry. Further increases in unit size.
1950–1960	566 (1050)	165 (3500)	300	38	Once-through and supercritical steam pressure systems came into use. Turbine designs improved to tolerate and remove much more moisture, thus permitting use of saturated steam from nuclear plants. Unit sizes increased further, and still better control of feed-water chemistry.
1960–1970	566 (1050)	240 (3500)	500	39	Unit size further increased. Nuclear reactors come into use. Cost of electricity continues to drop so that by 1970 the inflation-adjusted cost was $\frac{1}{5}$ that in 1930.
1970–1980	538 (1000)	240 (3500)	1000	36	EPA requirements lead to widespread introduction of cooling towers, stack gas scrubbers, and improved particulate removal equipment, all causing losses in plant efficiency and increased costs. Further increases in unit size served to reduce costs, but increasing shortages of gas and petroleum lead to large increases in the overall cost of electricity.

sorbs the intense thermal radiation heat flux from the burning fuel. The heat flux is high: ~ 63 W/cm² [200,000 Btu/(h · ft²)]. Thus a large fraction of the heat of combustion is removed in a few seconds as the hot gases pass upward through the furnace into the second-stage superheater, then across through the reheater and downward through the first-stage superheater and economizer, and finally laterally through the air preheater to the stack.

Chronological Development

The highlights[18] of this long developmental history are summarized in Table 11.1 and Fig. 11.4. Note the steady increases in pressure and temperature up to 1957, with a concomitant increase in thermal efficiency, but with no further significant increases in any of these three parameters in the subsequent 20 years. Efforts in the 1960s and early 1970s were directed

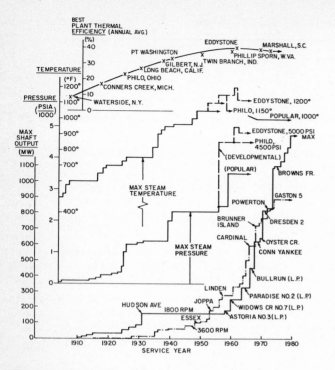

Figure 11.4 Historical development of advanced turbine generators.[18] (*Courtesy General Electric Company.*)

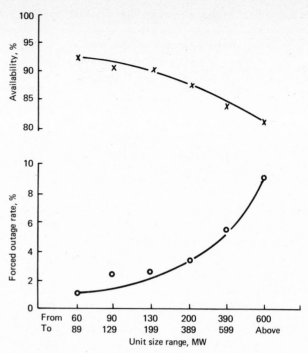

Figure 11.5 Averages for 1965–1974 EEI availability data for fossil-fuel steam-turbine units as a function of unit size.[19]

primarily toward increases in both reliability and in unit size; the latter effect is also shown in Fig. 11.4.[18] However, the reliability and availability[19] of units seem to have declined with increasing size, as indicated in Fig. 11.5. This stems in part from the increase in the number of boiler tubes; for a given probability of a tube failure (see Chap. 8 and Fig. 8.10), the probability of a forced outage from a boiler tube failure increases with unit size, so that boiler tube failures become a large fraction of the total outages. Partly for this reason the trend toward even larger units seems to have tapered off in recent years (Fig. 11.6).[19]

Emissions

Since passage of the Environmental Protection Act of 1968, the prime emphasis in fossil-fuel steam power plant development has been on the reduction of stack gas emissions and minimizing the environmental effects of waste-heat rejection; both efforts have led to reductions in thermal efficiency. The environmental effects involved and something of the cost/benefit aspects of

the EPA regulations were treated in Chap. 9, whereas this chapter is concerned with the equipment involved, the problems it poses, and the prospects for future development.

Work on the reduction of stack gas emissions and their effects has included the use of stacks as much as 366 m (1200 ft) tall to reduce concentrations at ground level, marked improvements in electrostatic precipitators to remove particulates, bag houses for even better removal of particulates with fabric filters, stack gas scrubbers to remove SO_2, and improvements in furnace and burner design to reduce the emission of oxides of nitrogen, e.g., via staged combustion (see Chap. 5).[20,21,22] Most of this work has been carried out by superimposing new equipment on existing types of power plant and selection of low-sulfur fuels. In addition, work has been under way to remove sulfur from the fuel before it is fed to the burners, either by solvent extraction of sulfur from coal or by a coal gasification or liquefaction process that includes removal of the sulfur.[23] Work has also been under way to reduce the sulfur content in fairly low sulfur coals by coal-washing operations that appear to be effective in removing perhaps half of the ash and ~ 15 percent of the sulfur.[24]

As discussed in Chap. 5, several efforts to develop fluidized-bed coal combustion systems have been under way since the middle 1950s.[25,26,27] This process is the most promising development in sight for the economical and efficient use of high-sulfur fuels in coal-fired steam plants at the time of writing. A series of both generalized and specific design studies of fluidized-bed coal combustion systems for utility, as well as industrial and institutional applications, have been carried out to show the potential of fluidized-bed furnaces coupled to steam boilers.[16,28-36]

The characteristics of the principal means for reducing the amount and effects of stack gas emissions are summarized in Table 11.2. Of these, the implications of the use of low-sulfur coal need no further discussion, the coal solvent refining, gasification, and liquefaction are treated in Chap. 5, and particulate removal equipment is treated in the next chapter. The special problems of stack gas scrubbers and fluidized-bed fur-

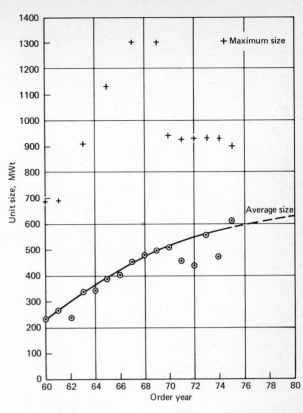

Figure 11.6 Average and maximum size of fossil-fuel steam-turbine units as a function of the year in which they were ordered.[19]

TABLE 11.2 Summary of Characteristics of the Principal Approaches to Reducing the Amount and Effects of Sulfur Emissions from Steam Power Plants

System	Relative capital cost	Efficiency of coal utilization	Relative plant avail-ability	Provisions for retrofit	Estimated date for commercial operations	R&D uncertainties
Low-sulfur coal	1.0	0.99	0.80	No problem	Current	None
Fluidized-bed comb.						
AFB	1.2–1.5	0.9–0.98	0.8	Furnace vol. larger than for pulv. coal	1984–1986	Corrosion in bed, coal metering and feed, reliability
Supercharged bed	0.8–1.1	0.99	0.8	Furnace small	1986–1988	Ditto above, turbine erosion and deposits
PFB	?	0.99	?	Furnace small	?	Ditto above, hot gas cleanup system
Wet limestone stack gas scrubbers	1.2–1.3	0.93	0.7?	Site-dependent	Current	Reliability and life of components, high capital and operating costs
Solvent refined coal	1.6–2.5	0.6–0.8	0.8	No problem	1985?	Reliability and life of plant; conversion efficiency
Coal liquefaction	1.6–2.5	0.6–0.7	0.8	No problem	?	Ditto above
High-Btu gas coal gasification	1.6–2.5	0.5–0.6	0.8	No problem	?	Ditto above
Low-Btu gas coal gasification	1.4–2.0	0.6–0.8	0.6	New plant required for good integration	1980–1982	Reliability and life of plant, conversion efficiency
Tall stack	1.01–1.02	0.99	0.80	No problem	Current	None

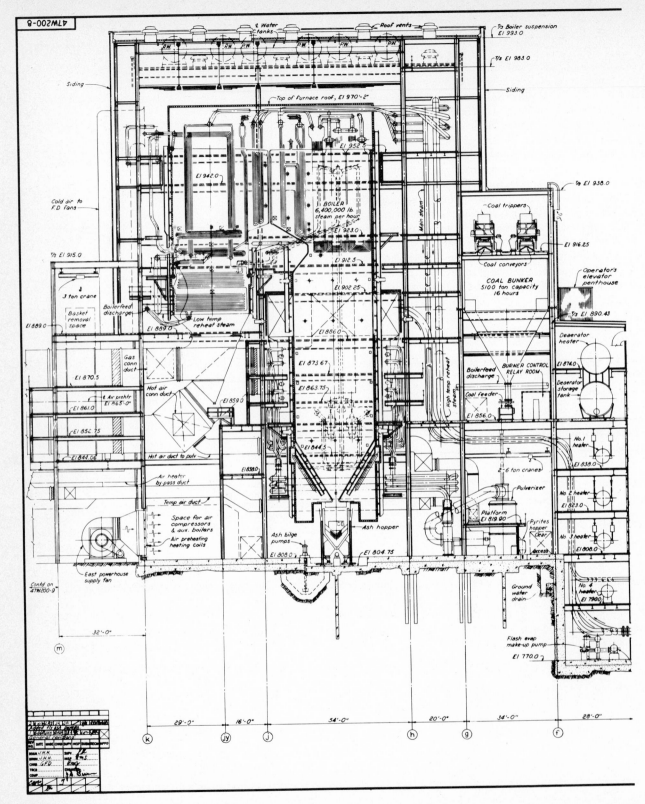

Roof vents

To Boiler suspension
El 993.0

To El 983.0

Siding

Siding

Water tanks

Top of furnace roof, El 970'-2"

El 952.5

El 942.0

To El 938.0

Cold air to F.D fans

BOILER
6,400,000 lb
steam per hour

Coal trippers

El 923.0

Main steam

El 916.25

Coal conveyors

El 912.5

To El 915.0

Operator's elevator penthouse

El 902.25

COAL BUNKER
5100 ton capacity
16 hours

3 ton crane

To El 890.43

Basket removal space

Boilerfeed discharge

El 889.0

El 889.0

El886.0

Deaerator heater

Low temp reheat steam

El 874.0

El 873.67

Burner control relay room

El 870.5

Gas conn duct

Boilerfeed discharge

Deaerator storage tank

El 863.75

Hot air conn duct

High temp reheat steam

El 861.0

& Air prehtr El 863-0"

El 859.0

No.1 heater

Coal feeder

El 856.0

El 852.75

El 838.0

No 2 heater

El 844.06

Hot air duct to pulv

El844.5

El 823.0

Air heater by pass duct

El 833.0

No 3 heater

2-6 ton cranes

Temp air duct

El 808.0

Pulverizer

Space for air compressors & aux. boilers

Platform El 819.90

Air preheating heating coils

Ash bilge pumps

Pyrites hopper clear

Ash hopper

El 808.0

El 804.75

Access

East powerhouse supply fan

Ground water drain

No 4 heater El 7900

Contd on 47W200-9

Flash evap make-up pump

El 770.0

32'-0"

m

29'-0"   16'-0"   54'-0"   20'-0"   34'-0"   28'-0"

k   JY   J   h   g   f

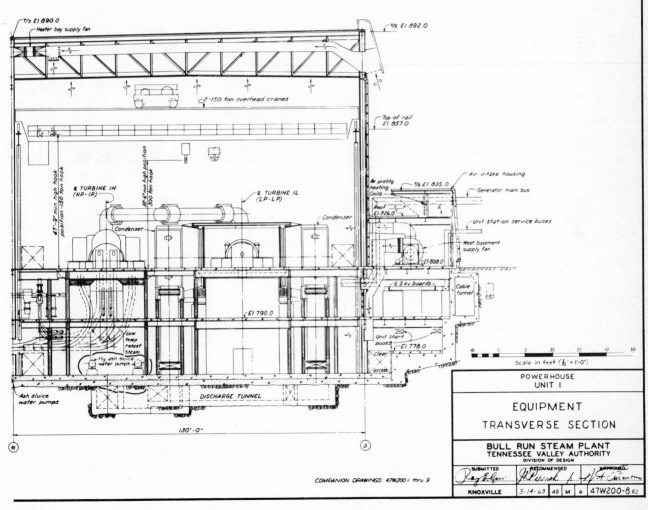

Figure 11.7 Elevation showing the arrangement of the principal components in the 950-MWe Bull Run power plant unit of TVA near Knoxville, Tenn. (*Courtesy TVA.*)

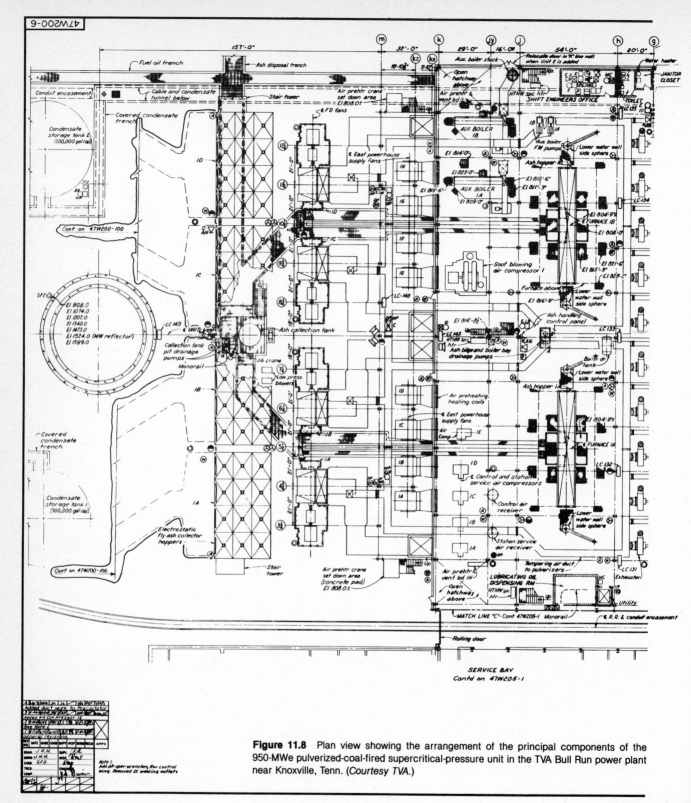

Figure 11.8 Plan view showing the arrangement of the principal components of the 950-MWe pulverized-coal-fired supercritical-pressure unit in the TVA Bull Run power plant near Knoxville, Tenn. (*Courtesy TVA.*)

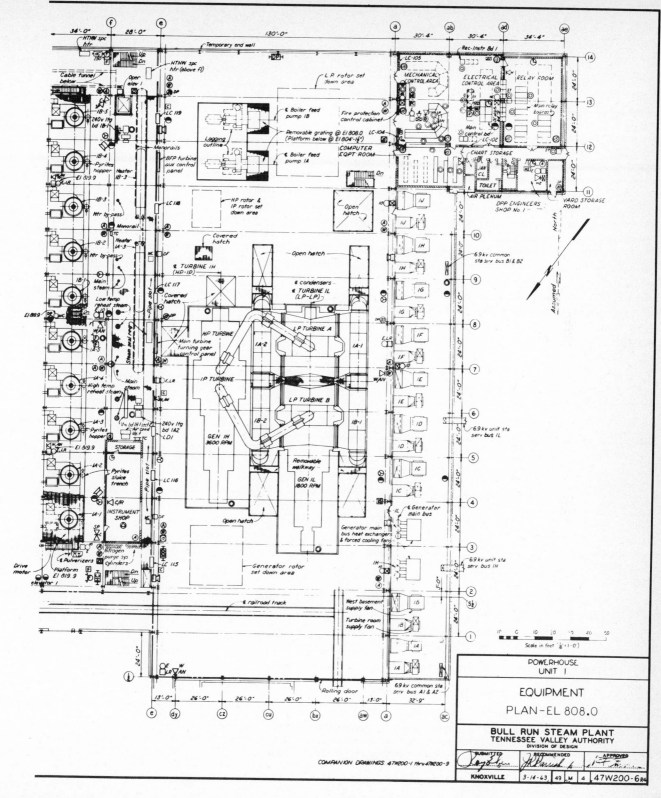

POWERHOUSE
UNIT 1

EQUIPMENT

PLAN-EL 808.0

BULL RUN STEAM PLANT
TENNESSEE VALLEY AUTHORITY
DIVISION OF DESIGN

SUBMITTED	RECOMMENDED	APPROVED			
KNOXVILLE	3-14-63	49	M	4	47W200-6 R4

COMPANION DRAWINGS 47W200-1 thru 47W200-9

nace–steam generators are treated in later sections of this chapter.

TYPICAL PLANTS

The size, complexity, and cost of modern steam power plants can be appreciated only by a fairly detailed look at the principal components in the system coupled with visits to plants during both construction and operation. In 1980 the cost of a 1000-MWe power plant unit was approaching a billion dollars, and the size and variety of the equipment required are even more impressive than the cost. Further, in developing and appraising conceptual designs for advanced power plants, it is often helpful to make comparisons between the equipment required and that of existing power plants; hence fairly detailed descriptions of some typical examples are presented in this section. The first, the TVA Bull Run plant, is one of the 10 most efficient units in the world and is widely used as an example in engineering studies and comparisons. The second example is the TVA

Sequoyah Unit No. 1 PWR nuclear plant. The third is a high-temperature gas-cooled reactor (GCR), and the fourth is a liquid-metal-cooled fast-breeder reactor (LMFBR).

Pulverized-Coal-Fired Supercritical-Pressure Plant

At the time it began commercial operation in 1967, the 950-MWe Bull Run steam plant was the largest steam-boiler-turbine-generator unit in the world with many new features in the detail design of its components. Thus it was not surprising that the solution of numerous mechanical problems required several years of shakedown operation. These problems were solved, and at the time of writing the plant is considered to be one of the most reliable and efficient coal-fired plants in the country.

A flowsheet for the plant is presented in Fig. 11.1, and both a vertical section through the plant and a plan view are shown in Figs. 11.7 and 11.8. Diagrams outlining the coal yard and fuel

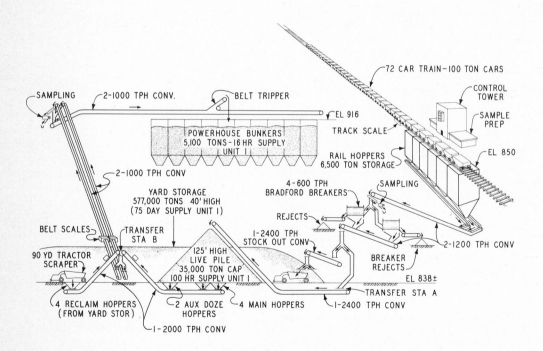

Figure 11.9 Schematic diagram for the entire coal-handling operation of the TVA Bull Run plant. The coal arrives from the coal field in 72-car unit trains at the right, is handled and processed, and is delivered to the bunkers beside the furnaces shown near the top of the diagram.[17] (*Courtesy TVA.*)

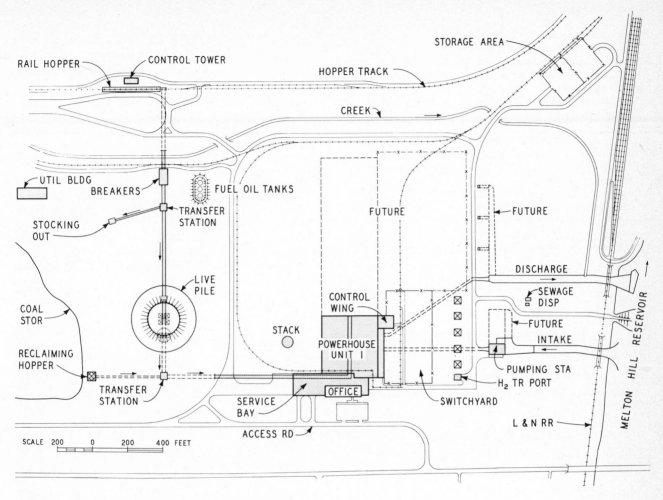

Figure 11.10 Site layout for the TVA Bull Run Power Plant. (*Courtesy TVA.*)

supply system are given in Figs. 11.9 and 11.10. Table 11.3 summarizes the principal design parameters. (This table represents 7 of 35 pages of data in Ref. 17.) The author has found that this table and these figures have provided invaluable data for making equipment comparisons in design studies and analysis work on new types of power plants.

A detailed account relating the functions of the various items in Fig. 11.1, Figs. 11.7 to 11.9, and Table 11.3 is probably not needed by most readers, would require many pages, and does not appear necessary. Readers not familiar with steam plants will probably find it more fruitful to take the time to trace through the drawings and table themselves with the help of in-

formation in previous chapters, particularly Figs. 2.3 to 2.10, Fig. 7.7, and Table 10.1, and thus become familiar with the principal components of the plant and their size, costs, and interrelationships.

In making comparisons with other fossil-fuel-fired steam plants, it should be remembered that oil- and gas-fired units do not require coal- and ash-handling equipment that consume a substantial amount of power and increase both capital and operating costs. On the other hand, the stack loss chargeable to the heat of vaporization of the H_2O in the combustion products increases with the H_2 content of the fuel so that, while coal-fired furnaces have efficiencies of ~ 90 percent, the values for oil- and

TABLE 11.3 Design and Construction Data for the 950-MWe Pulverized-Coal-Fired Bull Run Power Plant Designed, Built, and Operated by TVA near Knoxville, Tenn.[17]

TURBOGENERATOR

Number of units installed (HP and LP)—1
Manufacturer—General Electric Company
Maximum generator nameplate rating—950,000 kW
Turbine, type—horizontal, impulse reaction, cross compound, 4-flow exhaust extraction
  3600-rpm shaft—HP and IP turbines
  1800-rpm shaft—LP turbines with 4-flow exhaust and 43-in last-stage buckets, 22 stages total
Generator, type and maximum nameplate rating—2 direct connected, hydrogen-cooled rotor, water-cooled stator, 527,778 kVA, 0.9 pf, 45 psig H_2, 3 ph, 60 cps, 24,000 V, 12,696 A, 0.640 minimum test scr, Y-connected
  Temperature rise—stator 45°C; rotor 74°C
Exciter, type and capacity—2 shaft driven, HP 1880 kW, 500 V, 603 rpm, and LP 970 kW, 500 V, 892 rpm
Spacing of shafts (HP and LP)—50 ft 0 in
Overall dimensions above turbine room floor
  Length—HP 106 ft $7\frac{1}{4}$ in; LP 132 ft $3\frac{1}{8}$ in
  Width—HP 28 ft 0 in; LP 27 ft 10 in
  Height (HP and LP)—22 ft (approximate)

Low-Pressure Turbine Weights, lb

Diaphragms (upper and lower combined)

Low pressure A—210,160
Low pressure B—210,160

Oil tank

Without oil—52,240
Weight of oil—48,375

Rotors (with buckets and wheels)

Low pressure A—281,950
Low pressure B—292,910

Rotors (with 7th and 8th stage buckets removed)

Low pressure A—	258,935
Low pressure B—	269,895
Subtotal	1,624,625

Inner shells, A or B

	Unassembled	Diaphragms Stages	Diaphragms Total	Assembled
Turbine end, inner casing, upper	8,415	—	—	8,415
Turbine end, inner casing, lower	16,110	—	—	16,110
Generator end, inner casing, upper	8,415	—	—	8,415
Generator end, inner casing, lower	16,110	—	—	16,110
Turbine inner shell, upper	5,525	2-4	20,700	26,225
Turbine inner shell, lower	5,580	—	—	5,580
Generator inner shell, upper	5,525	2-4	20,700	26,225
Generator inner shell, lower	5,580	—	—	5,580
Turbine packing casing cone extension, upper	4,465	—	—	4,465
Turbine packing casing cone extension, lower	4,465	—	—	4,465
Gen packing casing cone extension, upper	4,465	—	—	4,465
Gen packing casing cone extension, lower	4,465	—	—	4,465
Subtotal				130,520

Total weight

Estimated total weight of all LP turbine parts—3,275,000

Exhaust hoods A and B

	Unassembled	Inner casing	Diaphragms	Steam guides	Assembled
Upper turbine end	86,250	—	—	—	86,250
Generator end	86,250	—	—	—	86,250
Middle	27,330	—	—	—	27,330
(Middle) inner	88,565	16,830	60,670	14,210	180,275
Lower turbine end (A)	136,365	—	—	—	136,365
Lower turbine end (B)	137,890	—	—	—	137,890
Generator end (A)	141,500	—	—	—	141,500
Generator end (B)	135,115	—	—	—	135,115
Middle	29,140	—	—	—	29,140
(Middle) inner	101,035	32,220	—	14,210	147,465
Subtotal					1,107,580

TURBOGENERATOR (Continued)

High-Pressure Turbine Weights, lb

Outer shell

	Unassembled	Inner shells	Misc.	Assembled
HP, upper	99,895	—	19,275	119,170
HP, lower	100,675	46,085	—	146,760
Reheat, upper	97,565	—	17,775	115,340
Reheat, lower	100,095	23,880	—	123,975
Subtotal				505,245

Inner shell

	Unassembled	Misc.	Assembled
HP, upper	30,685	870*	31,555
HP, lower	34,765	6180*	40,945†
No. 2, upper	5,040	—	5,040
No. 2, lower	5,140	—	5,140
No. 1 reheat turbine, upper	5,605	—	5,605
No. 1 reheat turbine, lower	5,660	—	5,660†
No. 2 reheat turbine, upper	6,240	—	6,240
No. 2 reheat turbine, lower	6,280	—	6,280†
No. 1 reheat generator, upper	5,605	—	5,605
No. 1 reheat generator, lower	5,660	—	5,660†
No.2 reheat generator, upper	6,240	—	6,240
No. 2 reheat generator, lower	6,280	—	6,280†
Subtotal			130,190

*Control, intercept, etc., when mounted on shell
†No diaphragms included

Diaphragms

High pressure—15,270
Reheat pressure—27,800

Oil tank

Without oil—48,160
Weight of oil—40,875
360-degree nozzle box, estimated—3740

Steam chest

A—89,790
B—90,100
Combined reheat stop valves—108,400

Rotors (with buckets and wheels)

HP—38,105
IP, double flow—51,445
Hydraulic enclosure (servomotors)— 24,000
Subtotal 538,785

Total weight

Estimated total weight of all HP turbine parts—1,380,000

TURBINE

Exhaust size (inside dimensions)—Eight at 20 ft 9$\frac{1}{2}$ in by 11 ft 3$\frac{1}{2}$ in
Number of steam chests—2
Number of steam control valves—8 (4 in each chest)
Number of stop valves—4
Number of combined reheat valves (stop and intercept)—2
Type—horizontal
Throttle pressure—3500 psig
Throttle temperature—1000°F
Reheat temperature—1000°F
Design back pressure—1$\frac{1}{2}$ in Hg abs
Extraction stages, five points—No. 1, 5th HP; No. 2, 8th HP exhaust; No. 4, 14th IP (exhaust); No. 6, 4th LP; No. 7, 6th LP; No. 8, 7th LP

Speed—No-load steam, 3500 psig, 800°F, at throttle and 4 in Hg abs exhaust conditions, full excitation, pph (approximate)
  speed-matching valve open—175,000
  speed-matching valve closed—95,000
Heat rate—Guaranteed performance based on extraction for feedwater heating, including all losses in the unit, also exciter and rheostat losses, rated throttle steam conditions, and 1$\frac{1}{2}$ in of Hg abs exhaust pressure, with zero makeup
  kW—914,402
  Btu per kWh—7142

TABLE 11.3   Design and Construction Data for the 950-MWe Pulverized-Coal-Fired Bull Run Power Plant Designed, Built, and Operated by TVA near Knoxville, Tenn.[17] (Continued)

TURBINE (Continued)

Turning gears

Number—2 (1 each HP and LP element)
Location—adjacent to each turbine-generator coupling
Type—electric motor-driven, transmitted through a link belt and a reducing gear train to the turbine shaft
Speed—2–3 rpm
Drive—15 hp for HP shaft, 20 hp for LP shaft, 3 ph, 60 cps, 120 rpm, 440 V ac

Main turbine oil tanks

Number—2 (1 each HP and LP elements)
Lubricating oil requirements, gal:

	HP	LP
Oil tank capacity	5450	6100
Piping runback	2100	1800
Seal oil system	300	300
Trap drain	350	350

Total lubricating oil—16,750 gal

Vapor extractors

Number—2 (1 each HP and LP tanks)
Location—one mounted externally on roof of each turbine oil tank
Type—electric motor driven, horizontal, gear-type blower
Drive—3 hp, 3 ph, 60 cps, 900 rpm, 440 V ac

Lubricating oil coolers

Number—4 (2 each for HP and LP tanks)
Manufacturer—General Electric Company
Location—supported at roof of main turbine oil tank with shell inside tank and main head outside
Type—vertical tank, straight tubes (through-type), water-cooled, 2-pass
Design data—surface 2080 ft^2, 1272 tubes, length 9 ft 11$\frac{1}{4}$ in; $\frac{5}{8}$-in OD, No. 18 Bwg inhibited admiralty, aluminum bronze tube sheets; cast-iron channel; steel cover; steel shell; 40-psig operating pressure, oil side; 125-psig operating pressure, water side
Weight of tube bundle (estimated), lb—5800
Cooling water required, raw—HP 1700 gpm; LP 1980 gpm at 90°F inlet water temperature

Shaft seals (steam)

Number of packings—8 (4 each HP and LP element)
Sealing steam required at startup, normal (including boiler feed-water pump turbine requirements)—   14,000 pph
Sealing steam leakoff, normal (including boiler feed-water pump turbine)—   13,600 pph
Air leakage to steam packing exhauster (including boiler feed-water pump turbine)—   1900 pph

Pumps, turbine accessory

Service	No.	Location	Type	Drive hp	ph	cps	rpm	V ac
Main oil HP element	1	(1)	(6)	. . . . (13)				
Main oil LP element	1	(2)	(7)	. . . . (14)				
Auxiliary oil	2[11]	(3)	(8)	250	3	60	3600	440
Booster oil	2[11]	(4)	(9)	. . . . (15)				
Turning gear oil	2[11]	(3)	(9)	30	3	60	1800	400
Emergency bearing oil	2[11]	(3)	(9)	30	—	—	1700	240 dc
Lift oil	6[12]	(5)	(10)	5	3	60	1200	440

Location:
1. Mounted on HP turbine shaft.
2. Mounted on LP turbine shaft.
3. One pump in each turbine oil tank with electric motor drive mounted externally on tank roof.
4. One in each turbine oil tank.
5. Two packages of three pumps and motors each; assembly A for bearing Nos. 1, 2, and 3 mounted on platform El. 780.5 adjacent to LP oil tank; assembly B for bearing Nos. 4, 5, and 6 mounted on floor El. 790 adjacent to No. 5 bearing of LP element.

Type:
6. Horizontal, double suction, single stage, centrifugal.
7. Horizontal, single suction, 2 stage, centrifugal.
8. Vertical centrifugal, single stage.
9. Vertical, centrifugal.
10. Horizontal, axial piston, constant volume.

Data:
11. One each HP and LP tanks.
12. LP element only. 14. LP turbine shaft.
13. HP turbine shaft. 15. Oil turbine.

TURBINE (*Continued*)

Generators

	HP generator	LP generator		HP generator	LP generator
Speed	3600 rpm	1800 rpm	Main exciters		
Total losses, kW @ 37 psig H_2 (at			Drive	gear driven	gear driven
500,000 kVA, 0.9 pf)	5732	4714	Rating—kW	1880	970
Reactances (based on 500,000 kVA)			—V	500	500
Direct axis synchronous	160%	164%	—rpm	603	892
Direct axis subtransient	24%	30.5%	Pilot exciters		
Negative sequence	16%	16.5%	Type	amplidyne	amplidyne
Zero sequence	13%	12.0%	Rating—kW	10	3
Resistance at 75°C			—V	250	250
Stator per phase	0.00295 Ω	0.00311 Ω	Motor—hp	25	10
Field	0.11600 Ω	0.22100 Ω	Neutral grounding transformer		
Negative sequence	2.10%	2.50%	kVA	75	75
Field currents			V	24,000/220	24,000/220
Full load—1.0 pf	2630 A	1400 A	Neutral resistor		
—0.9 pf	3400 A	1760 A	Amp (1 min)	345	440
Short-circuit ratio test values	0.640	0.648	Ohm	0.4	0.29
Hydrogen treatment	vacuum detraining	vacuum detraining	Voltage regulators—automatic high-speed, continuous-acting, dynamic type		
Hydrogen cooling			Surge protection—lightning arresters only		
Number of coolers	4	8			
Cooling water (gpm at 90°F)	1500	1400			
Gas space in generator casing	3600 cu ft	2500 cu ft			
Length of generator	30 ft $0\frac{1}{2}$ in	32 ft 11 in			
Weight of principal parts					
Stator	269.79 tons	258.66 tons			
Rotor	58.4 tons	145.9 tons			
Resistance temperature detectors (number installed)					
Stator windings	48	60			
Stator gas ducts	8	16			
Thermocouples (number installed)					
Stator cooling water	48	60			
Stator iron	28	28			
Stator gas ducts	8	16			

STEAM GENERATOR

General data

Manufacturer—Combustion Engineering, Incorporated

Type—supercritical, combined circulation, twin divided furnaces, reheat

Rated capacity, pph—6,400,000

Steam pressure (at superheater outlet), psig—3650

Steam temperature (at superheater outlet), °F—1003

Efficiency (guaranteed at rated load), %—90.08

Furnaces

Type—pressurized, twin, waterwall, dry bottom

Principal dimensions—60 ft $9\frac{1}{4}$ in wide by 29 ft $9\frac{3}{16}$ in deep by 92 ft high (each furnace)

Total heating surface, ft²—74,590

Total volume, ft³—382,000

TABLE 11.3 Design and Construction Data for the 950-MWe Pulverized-Coal-Fired Bull Run Power Plant Designed, Built, and Operated by TVA near Knoxville, Tenn.[17] *(Continued)*

STEAM GENERATOR *(Continued)*

Superheaters

Type—4-stage, horizontal, partition panel, platen, and pendant
Tube size—$2 \times 1\frac{1}{4}$ in OD
Heating surface, ft²—255,200
Design pressure, psi—3840
Design temperature, °F—1003

Boiler recirculating pumps (4 total, 2 spares)

Type—canned motor
Capacity, gpm—6500
Manufacturer—Westinghouse Electric Corporation
Motors—375 kW

Economizers

Type—continuous-loop, finned-tube
Tube size—2-in OD
Design pressure, psig—4340
Total heating surface, ft²—270,000

Reheaters

Type—three-stage, horizontal, inlet and outlet pendant
Tube size—$2\frac{1}{2}$- and $2\frac{1}{8}$-in OD
Rated capacity, pph—4,500,000
Design pressure, psig—725
Design temperature, °F—1003
Operating inlet pressure, psig—575
Operating outlet pressure, psig—545
Operating inlet temperature, °F—552
Operating outlet temperature, °F—1003
Heating surface, ft²—153,000

Air preheaters

Number—4
Manufacturer—Air Preheater Corporation
Type and size—Ljungstrom, regenerative, counterflow, $29\frac{1}{2}$ HX
Heating surface, each, ft²—302,800
Element length—$75\frac{1}{2}$ in (hot end 36 in; intermediate $27\frac{1}{2}$ in; and cold end 12 in)
Design temperature, gases
  Entering, °F—656
  Leaving, °F—290
Design temperature, air
  Entering, °F—110
  Leaving, °F—591

Firing equipment

Burners—80, tilting tangential
Pulverizers—10, Raymond No. 843RPS
Feeders—10, volumetric
Pilot oil torches—48
Burner oil guns—48, retractable

Boiler safety valves

Number and size
  Intermediate superheater—three 3 in; three $2\frac{1}{2}$ in
  Superheater outlet—four $2\frac{1}{2}$ in
  Low temperature reheat—twelve 4 in
  High temperature reheat—three 4 in
Manufacturer—Consolidated
Type—maxiflow

Power control valves

Number and size
  On superheater outlet—two $2\frac{1}{2}$ in
Manufacturer—Consolidated
Type—electromatic

Startup system safety valves

Number and size
  On separator spillover line—two 4 in
Manufacturer—Consolidated
Type—maxiflow

Soot blowers

Number installed
  Long retractable—20
  Waterwall deslaggers—72
Blowing medium—air, with steam standby
Drive—electric motor
Manufacturer—Copes-Vulcan
Control—automatic sequential or manual
Blowing pressure—125–200 psig

COAL FEEDERS

Number—10
Type—volumetric

Drive—variable speed motor
Manufacturer—Stock Equipment Company

COAL VALVES

Number—10
Size—24 in

Type—slide gate, rack and pinion drive, motor operated
Manufacturer—Fairfield Engineering Company

FORCED DRAFT FANS

Number—4
Manufacturer—Buffalo Forge Company
Type—double width, double inlet
Capacity, each (at test block)—545,000 cfm

Static pressure (at test block)—32 in water
Temperature (at test block)—120°F
Control—variable speed (fluid drive) and inlet dampers
Motor—3500 hp, General Electric

FLY ASH COLLECTORS

Number—4
Manufacturer—American-Standard Industrial Division
Type—electrostatic

Capacity, each—594,000 cfm
Efficiency—99 %
Pressure drop—0.5 in water

CONDENSING EQUIPMENT

Condenser

Number—1 (2 twin shells)
Manufacturer—Foster Wheeler Corporation
Type—horizontal, twin shell, modified double flow, single pass, side inlet
Location—axial, along each side of two, double flow, LP turbines
Surface, sq ft—320,000
Tube data—41,412 (3552 90-10 cupronickel; 37,860 inhibited admiralty) tubes, 33-ft $8\frac{3}{4}$-in effective length, approximate weight of tubes 683,000 lb, $\frac{7}{8}$-in OD, No. 18 Bwg
Shell-and-tube support plates—36 (9 per section), $\frac{3}{4}$-in copper bearing plates
Tube sheets—8 (2 per section), $1\frac{1}{2}$-in rolled steel
Waterboxes—nondivided, design pressure of 40 psi, steel, 2 inlet (6 by 11 ft inside) and 4 outlet (78-in diam) bottom connections
Hotwell data—full rectangular pattern, deaerating-type, storage capacity hotwell B sections 5500 gal each, A sections 11,000 gal each

Total weight (approximate), lb

Empty—2,640,000		
Operating—3,520,000		
Condenser performance		
Tube cleanliness, %	85	100
Steam load, pph	3,787,000	3,787,000
Btu rejected per lb steam	940.3	940.3
Absolute pressure, in Hg	1.30	1.18
Circulating water temperature, °F	55	55
Circulating water flow, gpm	397,500	397,500
Water velocity, fps	6.5	6.5
Friction loss through water passages, ft	10.95	10.95
Maximum oxygen content of effluent, cc per liter	0.01	0.01

TABLE 11.3   Design and Construction Data for the 950-MWe Pulverized-Coal-Fired Bull Run Power Plant Designed, Built, and Operated by TVA near Knoxville, Tenn. [17] *(Continued)*

CONDENSING EQUIPMENT *(Continued)*

Vacuum pumps

Number—4
Manufacturer—Nash Engineering Company
Type—Model H-8 Nash rotary vacuum pump with atmospheric air ejector

Operating performance
  Suction pressure, in Hg abs—1.0
  Suction temperature, °F—74.0
  Rated capacity, each—7.5 scfm
  Motor hp—50
  Motor manufacturer—General Electric Company

CONDENSATE AND FEED-WATER SYSTEMS

Deaerating heater and storage tank

Number—1
Manufacturer—Graver Water Conditioning Company
Type—nonmetering, cylindrical, spray tray, nonstorage-type, horizontal heater, and horizontal storage tank
Location—storage tank supported at El. 856.0 floor in heater bay; heater supported on storage tank but accessible from El. 874.0 floor
Performance—under conditions of service outlined below, the heater will deliver 6,335,150 pph, heated to a temperature equal to that of saturated steam in the heater and containing not more than 0.005 cc per liter of dissolved oxygen; when operated at overload conditions and delivering up to a maximum of 6,650,000 pph for short periods of peaking operation, the oxygen content of the fluid will not exceed 0.03 cc per liter of dissolved oxygen
Conditions of service—performing under the conditions of service given above, the proportions of water and steam quantities are as follows
  Condensate, pph (rated)—4,368,490
  HP heater drains, pph (rated)—1,624,460
  Extraction steam to heater, pph (rated)—342,200
Construction details

	Heater	Storage tank
Shell diameter	11-ft 0-in OD	12-ft 0-in OD
Length (overall)	37 ft 10 in	94 ft 5 in
Material	Copper bearing steel plate	
Thickness of heads	$\frac{11}{16}$ in	$\frac{11}{16}$ in spherical
Thickness of shell	$\frac{9}{16}$ in	$\frac{1}{2}$ in
Design pressure shells, psi	85	85
Design temperature, °F	650	650

Capacity of deaerator storage tank (minimum)—48,000 lb at 304.4°F
Weight, lb
  Entire unit, dry—192,500
  Unit at maximum operating level—785,000
  Complete unit, flooded—990,000

Flash evaporator

Number—1
Manufacturer—Westinghouse Electric Corporation
Type—horizontal, single stage, flash type
Location—turbine bay, on El. 790 floor
Approximate weight, lb
  Empty—60,000
  Operating—78,000
  Flooded—129,500
Design data
  Heater section (integral with evaporator condenser and flash chamber)
    Total surface, ft²—4710

	Shell	Tubes
Design pressure, psig	15.0*	50
Design temperature, °F	300	300
Test pressure, psig	22.5	75
Quantity fluid, pph	60,000 (steam)	2,625,000 (brine)
Tubes, OD, in		$\frac{3}{4}$-18 Bwg
Number		630
Average effective length		$\frac{3}{4}$-18 Bwg
Material		Admiralty
Pitch, in		1 Tri
Passes		1

*At 30 in Hg vac

CONDENSATE AND FEED-WATER SYSTEMS (*Continued*)

Condenser section (integral with evaporator heater and flash chamber)
Total surface, ft^2—3200

	Shell	Tubes
Design pressure, psig	15.0	300
Design temperature, °F	300	300
Test pressure, psig	22.5	450
Quantity fluid, pph	60,000	4,368,000
Tubes, OD, in		$1\frac{1}{2}$-22 Bwg
Number		214
Average effective length		38 ft 0 in
Material		304 SS
Pitch, in		$1-\frac{3}{4}$ Tri
Passes		1

Flash evaporator chamber (knitted wire mesh mat moisture separators between flash chamber and evaporator condenser)

Injection water coolers

Number—2
Manufacturer—The Whitlock Manufacturing Company
Type—closed, horizontal, straight tube, 2 pass
Location—turbine bay floor El. 770.0
Design performance data

	Shell	Tubes
Total effective surface, sq ft		299
Medium flowing	injection water	cooling water
Quantity, gpm	90	255
Design temperature, °F	350	350
Design pressure, psig	450	100
Test pressure, psig	675	150
Shell ID, in	10.136	
Tubes OD, in		$\frac{5}{8}$-22 Bwg
Material	steel	304 SS
Pitch, in		0.833 Tri
Number		80
Length, in		276

Shell and bundle weight, dry, lb—2800
Shell and bundle weight, flooded, lb—3650

Auxiliary deaerator

Number—1
Manufacturer—Allis-Chalmers Manufacturing Company
Type—nonmetering contact, spray tray type, vertical heater with integral storage capacity
Location—floor El. 856.0 in heater bay
Performance—The heater will deliver 180,000 pph, heated to a temperature equal to that of saturated steam at operating pressures from 1 to 10 psig with inlet distilled water at temperatures varying from 40° to 212°F, and inlet steam enthalpy of 1140 Btu per lb and containing not more than 0.005 cc per liter of dissolved oxygen
Construction details
   Shell diameter—8 ft 0 in OD
   Length (overall)—12 ft 8$\frac{7}{8}$ in
   Material—steel plate
   Thickness of heads, in—$\frac{5}{8}$
   Thickness of shell, in—$\frac{7}{16}$
   Design pressure, psi—85
   Storage capacity, gal—1800
   Weight, lb
      Entire unit, dry—11,900
      Unit at operating level—28,000
      Unit flooded—47,000

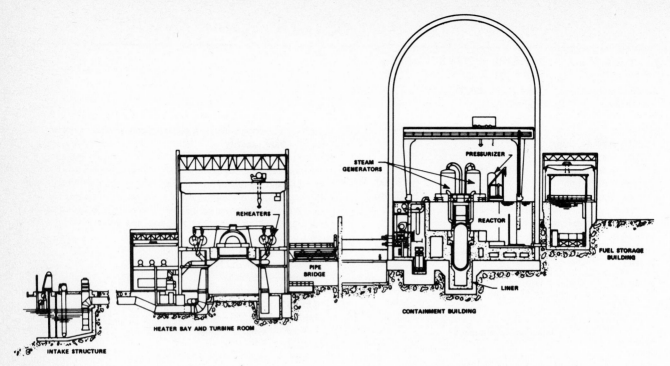

Figure 11.11   Longitudinal section through a typical PWR power plant. (*From "Pressurized Water Reactor Fundamentals Manual," Inspection and Enforcement Training Center, NRC.*)

gas-fired units are ~87 and ~85 percent, respectively, when calculated on the basis of the lower heating value of the fuel.

Light-Water Reactor Plants

TVA, the largest electric utility system in the United States, began the design and construction of nuclear plants later than many large utilities, but by 1990 about half of its power output will be nuclear. The first plant, that at Brown's Ferry, consists of three boiling-water reactor (BWR) units of 1097 MWe each. That plant is in operation at the time of writing and, in spite of high capital costs stemming in part from the facts that these units represented a new type of plant for TVA and were larger than any previous BWR units, this plant produces power at a substantially lower cost than any of the TVA coal-fired plants, and it is not vulnerable to coal shortages caused by strikes or other problems. A second, and, from the potential-accident-containment standpoint, more advanced TVA nuclear installation is the Sequoyah pressurized-water reactor (PWR) plant that consists of two 1125-MWe units, the first of which began

operation in 1980. The design of a nuclear plant is dominated by reactor safety considerations and provisions for the containment of radioactivity that might otherwise be released in the event of an accident. Because of this, and because of their interesting reactor safety characteristics, one of these units was chosen as an example for inclusion in this section.

The Sequoyah PWR units are the first water reactors to make use of an ice condenser system in a free-standing steel containment vessel as a means for absorbing the energy that might conceivably be released in the form of steam by a rupture in the primary reactor cooling system.[37] There is so much energy in the sensible heat in the inventory of high-temperature water in the reactor cooling system that, if the pressure were relieved by a rupture in the system, the steam formed would produce a high pressure in the containment building. It is for this reason that reactor containment structures are commonly made in the form of either spherical or vertical cylindrical shells with hemispherical or ellipsoidal heads. (The first such structure was a 200-ft-diameter sphere built by the General Electric Company near Schenectady in the latter 1940s for testing the prototype sodium-

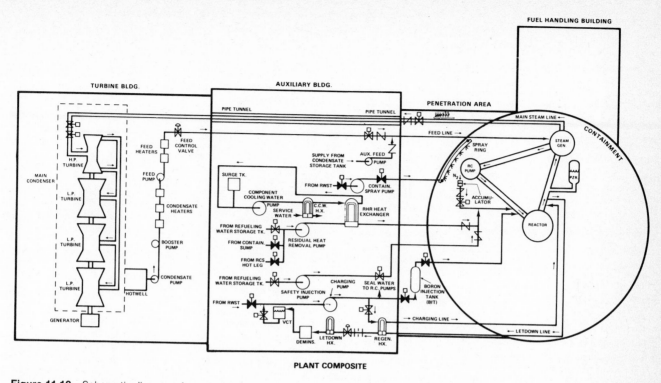

Figure 11.12 Schematic diagram of a pressurized-water reactor plant showing the way in which the principal components are coupled. (*From "Pressurized Water Reactor Fundamentals Manual," Inspection and Enforcement Training Center, NRC.*)

cooled reactor subsequently used in the Sea Wolf submarine.) The pressure that might conceivably build up in the containment vessel in the event of the maximum credible accident can be reduced drastically by using a chilled water spray or by a heat-transfer matrix made of ice. In the Sequoyah plant the ice is in the form of an array of vertical columns similar to a bank of tubes in a shell-and-tube heat exchanger. These units are arranged in the containment shell along with the other reactor components, as shown in the plant sectional view of Fig. 5.28. The size and overall cost of the containment system can be reduced substantially through the use of this ice condenser system.

A section through the reactor for the Sequoyah plant was included in Chap. 5 along with a photo of a typical fuel element for a water reactor (see Figs. 5.28 and 5.29). The longitudinal section of Fig. 11.11 and the schematic diagram of Fig. 11.12 show the way in which the principal components of a PWR plant are coupled. As is the case for coal-fired plants, numerous detailed sheets, such as Fig. 7.10, are required to show the host of lesser components. The steam system is similar to that for a

coal-fired plant except for the steam generator; a section through this component is shown in Fig. 5.28. Table 11.4 gives the principal design data for the plant.

A familiarity with and an appreciation for the relations between the principal components, their characteristics, proportions, and general construction are probably best obtained by a detailed examination of Figs. 5.28 and 5.29, together with Figs. 11.11 and 11.12 and Table 11.4. This, coupled with a visit to a nuclear plant (preferably in the latter stages of construction when the key components can be seen before they become radioactive), the reader will probably find more effective than 10,000 words on the subject—which the space available in this text does not permit.

Gas-Cooled Reactor Plants

The characteristics and relative advantages and disadvantages of gas-cooled reactors are discussed in Chap. 5. The evidence of the marketplace is that light-water reactors give lower costs but a poorer neutron economy, whereas sodium-cooled reactors

TABLE 11.4 Summary of Design Data for the 1125-MWe PWR Unit No. 1 of the TVA Sequoyah Power Plant near Chattanooga, Tenn. [37]

Power

Net electric output	1125 MWe
Gross electric output	1171 MWe
Gross thermal output	3411 MWt

Reactor core

Core diameter (equivalent)	3.40 m (133.7 in)
Core height (active)	3.66 m (144 in)
Number of fuel assemblies	193
Fuel pin lattice pitch	14.3 mm (0.563 in)
Average thermal output	589,200 kcal/(m$^2\cdot$h)
	217,200 Btu/(ft$^2\cdot$h)
Maximum thermal output	1,573,200 kcal/(m$^2\cdot$h)
	579,600 Btu/(ft$^2\cdot$h)
Weight of fuel as UO$_2$	97.6 te (215,400 lb)

Fuel assemblies

Fuel material	UO$_2$
Pellet diameter	9.29 mm (0.366 in)
Clad material	Zr-4
Clad thickness	0.61 mm (0.024 in)
Pin diameter	10.7 mm (0.422 in)
Number of pins per assembly	204
Maximum fuel central temperature	2282°C (4140°F)
Maximum clad surface temperature	247°C (657°F)
Feed enrichment (equilibrium)	3.2%
Fuel discharge burnup (equilibrium)	31,000 MWd/t

Control rods

Neutron absorber	Ag In Cd
Cladding material	S.S. type 304
Number, full length	53
part length	8
Shape	Rod cluster
Length of poison section	3.62 m (142.7 in)

Primary coolant system

Type	Forced circulation
Operating pressure	158 kg/cm^2 (2250 psia)
Reactor inlet temperature	285°C (545°F)
Reactor outlet temperature	321°C (610°F)
Coolant pumps	4
Total reactor flow	61 × 10^6 kg/h (134 × 10^6 lb/h)

Reactor pressure vessel

Inside diameter	4.39 m (173 in)
Inside height	12.6 m (495 in)
Wall thickness (core region)	219 mm (8.625 in)
Material	ASTM A-508 Class II
Design pressure	176 kg/cm^2 (2500 psia)
Design temperature	343°C (650°F)

Containment building

Type	Double (steel vessel, concrete shield)
Pressure suppression	Ice condenser
Design pressure	0.76 kg/cm^2 (10.8 psi)
Inside diameter (steel vessel)	34.4 m (115 ft)
Inside height (steel vessel)	47.5 m (156 ft)

Turbogenerator

Rating	1220 MWe
Speed	1800 rpm
TSV pressure	55 kg/cm^2 (782 psi)
TSV temperature	268°C (514°F)

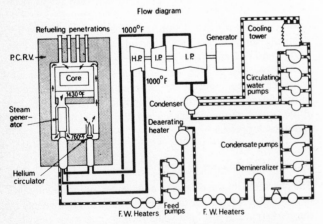

Figure 11.13 Flowsheet for the Fort St. Vrain GCR power plant. (The reactor is shown in Figs. 5.30 and 5.31.) [38] (*Courtesy General Atomic Corp.*)

show greater promise for breeding but give higher capital costs. As a consequence, although many gas-cooled reactors have been built in Great Britain, only a few have been built elsewhere. Only one large commercial gas-cooled reactor is in operation in the United States at the time of writing, and no others are planned. However, because of their excellent fuel economy potential, they may prove more attractive in the future. Hence it seems best to include one example, and thus the 330-MWe Fort Saint Vrain reactor is included here. The system is similar to that of a PWR. The principal differences are that the reactor core must be larger for a given power output, while the reactor system pressure is lower, and the reactor coolant outlet temperature is higher, permitting the steam system temperature to be as high as in a fossil fuel plant. The reactor and fuel element are described in Chap. 5 (Figs. 5.30 and 5.31). A flowsheet is shown in Fig. 11.13, and the principal design data are given in Table 11.5. For a detailed description of this plant the reader is referred to Ref. 38.

TABLE 11.5 Summary of Design Data for the 330-MWe GCR of the Fort Saint Vrain Power Plant of the Public Service Co. of Colorado near Plattville, Colo. [38]

Capacity

Net electric output	330 MWe
Gross generation	342 MWe
Overall station net efficiency	39.2%
Design life of plant	30 years

Reactor core

Reactor output	841.7 MWt
Core diameter	5.9 m (19.5 ft)
Active core height	4.7 m (15.6 ft)
Number of fuel elements	1482
Element lattice pitch	360 mm (14.2 in)

Fuel

Fuel material	Th/U^{235} (93% enriched)
Total quantity of thorium	19,458 kg
Total quantity of uranium	882 kg
Fuel form	Coated particles in cylindrical beds
Number of fuel elements per refuel region	42
Element (hexagonal across flats)	355 mm (14 in)
Element (length)	787 mm (31 in)
Fuel bed diameter	12.4 mm (0.491 in)
Coolant channel diameter	15.8 mm (0.625 in)
Burnup	100,000 MWd/t

Control

Control rods	37 pairs
Active length	4.7 m (186 in)
Absorber material	B$_4$C/Graphite
Canning material	Incoloy
Shape	Hollow cylindrical
Drive, normal	Electric motor
Scram	Gravity

Thermal data

Primary steam flow	2,305,300 lb/h
Primary steam pressure	168 kg/cm^2 (2400 psig)
Feed-water temperature	206°C (403°F)
Primary coolant flow	3.39 × 10^6 lb/h
Primary coolant pressure	49 kg/cm^2 (700 psia)
Coolant temp. reactor inlet	406°C (762°F)
Coolant temp. reactor outlet	785°C (1444°F)
Avg. heat flux	142 kW/m^2 (45,000 Btu/h · ft^2)
Max. heat flux	441 kW/m^2 (140,000 Btu/h · ft^2)
Max. fuel temperature	1260°C (2300°F)
No. of steam generator modules	12
No. of steam generators	2

Reactor vessel

Type	Prestressed concrete
Internal clearance dimension	9.4 × 23 m (31 ft ID × 75 ft IH)
Max external dimension	(21 m × 32 m) (68 ft across corner × 106 ft high)
Normal working pressure	48 kg/cm^2 (688 psig)

Circulators

Type	Axial-flow compressor with integral driver
Drive	Single-stage steam turbine
Flow control	Variable speed
Number of circulators	4 (2 per loop)
Rated steam flow	70 kg/s (155 lb/s/circulator)
Speed	9550 rpm
Compressor press. rise (helium)	1.0 kg/cm^2 (14 psi)
Compressor inlet temperature	395°C (742°F)
Power	5200 hp

Steam generators (per module)

Total heat transfer	6 × 10^{11} Cal/h (2.4 × 10^8 Btu/h)
Bulk gas inlet temperature	776°C (1427°F)
Gas mass flow	128,000 kg/h (284,170 lb/h)
Superheater steam flow	87,000 kg/h (192,110 lb/h)
Superheater outlet pressure	2512 psia
Superheater outlet temp.	540°C (1005°F)
Reheater steam flow	84,000 kg/h (187,150 lb/h)
Reheater inlet pressure	178 kg/cm^2 (650 psi)
Reheater inlet temperature	356°C (673°F)
Reheater outlet pressure	42 kg/cm^2 (600 psi)
Reheater outlet temperature	539°C (1002°F)
Number of steam generator modules	12

Turbine generator

Type	Cross-compound
Gross output	342 MWe
TSV pressure	168 kg/cm^2 (2400 psig)
TSV temp.	538°C (1000°F)
IP cylinder TSV pressure	399 kg/cm^2 (567.5 psia)
IP cylinder TSV temperature	538°C (1000°F)
Vacuum	63 mm Hg (2.5 in Hg)

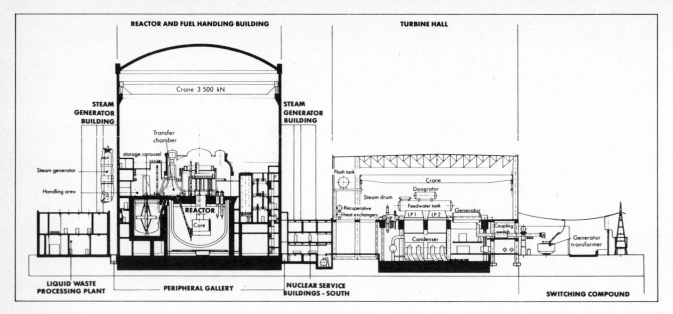

Figure 11.14 Longitudinal cross section of the Creys-Malville power station.[39] (*Courtesy Nuclear Engineering International.*)

Liquid-Metal-Cooled Fast-Breeder Reactor (LMFBR) Plant

The world's first full-scale commercial LMFBR plant is the 1200-MWe unit at Creys-Malville in southwestern France, scheduled for commissioning in 1983. The design of this plant was based on that of its prototype, the 250-MWe Phoenix (so named because, as a breeder reactor, it is designed to generate more fuel than it burns), which was started in 1973. A cross section through the reactor for this plant was included in Chap. 5 (Fig. 5.32). Figures 11.14 to 11.17 show, respectively, vertical and horizontal sections through the plant, a flowsheet for the main power circuits, and a section through the steam generator. Table 11.6 gives the principal design data.[39] The materials limitations on the reactor operating temperature turn out to be such that the plant produces steam at almost as high a temperature as a fossil fuel plant.

The plant design is typical of most of the designs for LMFBRs in that both the reactor core and the intermediate heat exchangers (IHXs) are mounted within a single pressure vessel along with the sodium pumps for the primary cooling circuit. There are eight intermediate sodium circuits that transport the heat out to four steam generators, thus providing a high degree of redundancy to assure cooling of the reactor under all conditions. The reactor is surrounded by ~4 m of concrete shielding,

and all the circuits containing radioactivity are housed within the reactor building, which is 64 m in diameter, 84 m high, and has reinforced-concrete walls 1 m thick. The concrete walls of the vaults for equipment containing sodium are lined with sheet metal so that liquid sodium could not contact concrete and react with moisture in the concrete. A nitrogen buffer gas region lies between the main vessel containing the primary sodium system and a safety tank that surrounds it, in effect providing a double-walled tank.

As was the case for the two previous plant designs presented in this section, it seems best to leave it to the reader to trace through the flowsheet, table, and the drawings to become familiar with the relationship of the various components and their characteristics. For further details refer to Ref. 39.

PRINCIPAL DESIGN PARAMETERS

Numerous compromises must be made in power plant design to balance improvements in thermal efficiency against increases in capital costs. These compromises differ, depending on the type of service expected, i.e., base load, intermediate load, or peaking power (Chap. 7). Fossil-fuel base-load plants typically employ high steam temperatures and pressures, i.e., 538 to

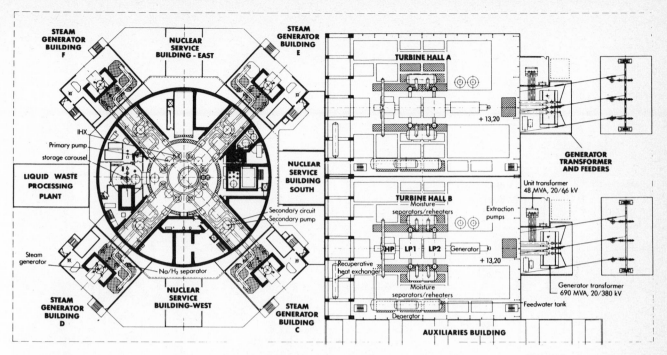

Figure 11.15 Plan view of the 1200-MWe LMFBR power plant at Creys-Malville on the Rhône River in France.[39] (*Courtesy Nuclear Engineering International.*)

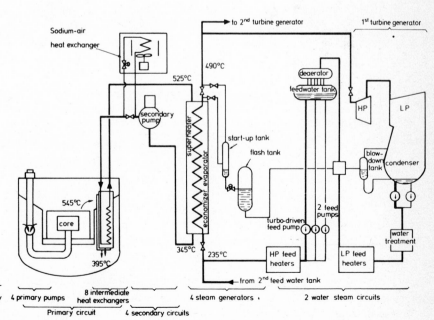

Figure 11.16 Flowsheet for the Creys-Malville LMFBR power plant.[39] (*Courtesy Nuclear Engineering International.*)

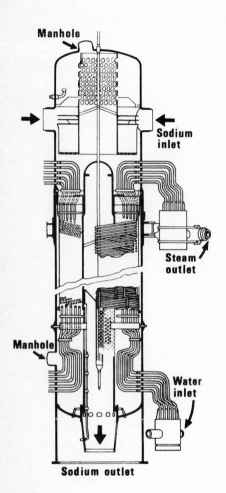

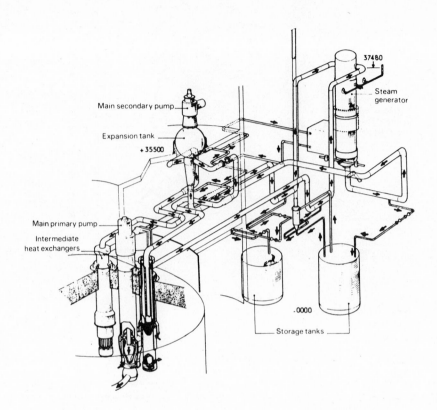

Figure 11.17 Isometric view of the secondary sodium circuit of the Creys-Malville power plant, together with a detail showing a section through the steam generator.[39] (*Courtesy Nuclear Engineering International.*)

TABLE 11.6 Summary of Design Data for the 1200-MWe LMFBR of the Centrale Nucléaire Européenne à Neutrons Rapides SA Creys-Malville Plant on the Rhône River near Creys-Malville, France[39]

Reactor type (Type de reacteur)

Sodium cooled, pool, fast breeder (à neutrons rapides, du type intégrè, á refroidissement par sodium)

Fuel (Combustible)

Composition	sintered UO_2-PuO_2 mixed oxide (oxide mixte fritté UO_2-PuO_2)
Average Pu (Enrichissement moyen)	16%
Mass of Pu (Masse du Pu)	5600 kg
Breeding gain in core (Gain de régéneration dans le coeur)	0.24
Maximum burnup (guaranteed) [Taux de combustion maximal (garanti)]	70,000 MWd/te
Target (objectif visé)	100,000 MWd/te

Fuel assemblies (Assemblages combustibles)

Assemblies in core (Assemblages dans le coeur)	364
Pins per assembly (Aiguilles par assemblage)	271
Pin length (Longueur de l'aiguille)	2700 mm
Assembly length (Longueur de l'assemblage)	5400 mm
Cladding (Matériau de gainage)	stainless steel (acier inoxydable)
Nominal maximum cladding temperature (Température maximale nominale de gaine)	620°C

Blanket assemblies (Assemblages fertiles)

Assemblies in core (Assemblages dans le coeur)	233
Pins per assembly (Aiguilles par assemblage)	91
Pin length (Longueur totale de l'aiguille)	1950 mm
Assembly length (Longueur totale de l'assemblage)	5400 mm
Cladding material (Matériau de gainage)	stainless steel (acier inoxydable)

Control rod assemblies (Assemblages de commande)

Main shutdown system (Système d'arrêt principal)	
Assemblies in core (Assemblages dans le coeur)	21
Absorber pins per assembly (Aiguilles absorbantes par assemblage)	31
Pin length (Longueur de l'aiguille)	1300 mm
Cladding material (Matériau de gainage)	stainless steel (acier inoxydable)
Backup shutdown system (Système d'arrêt complémentaire)	

Control rod assemblies (Assemblages de commande) (Cont.)

Assemblies in core (Assemblages dans le coeur)	3
Elements for assembly (Nombre d'éléments par assemblage)	3
Cladding material (Matériau de gainage)	stainless steel (acier inoxydable)

Main reactor tank (Cuve de réacteur)

Inside diameter (Diamètre intérieur)	21,000 mm
Height (Hauteur)	19,500 mm
Material (Matériau)	stainless steel (acier inoxydable)

Primary circuits (Circuits primaires)

Total mass of sodium in primary circuits (Masse totale de sodium dans les circuits primaires)	3.500 te
Nominal flow rate (Débit nominal)	4×4.1 te/s
IHX outlet temperature (Température de sortie des échangeurs intermédiares)	392°C
Core inlet temperature (Température d'entrée dans le coeur)	395°C
Core outlet temperature (Température de sortie du coeur)	545°C
IHX inlet temperature (Température d'entrée aux échangeurs intermédiaires)	542°C

Secondary circuits (Circuits secondaires)

Total mass of sodium in secondary circuits (Masse totale de sodium dans les circuits secondaires)	1.500 te
Nominal flow rate (Débit nominal)	4×3.3 te/s
Steam generator outlet temperature (Température de sortie des générateurs de vapeur) IHX inlet temperature (Température d'entrée aux échangeurs intermédiaires)	345°C
IHX outlet temperature (Température de sortie des échangeurs intermédiares) Steam generator inlet temperature (Température d'entrée aux générateurs de vapeur)	525°C

Water-steam circuits (Circuits eau-vapeur)

Water temperature/pressure at steam generator inlet (Température/pression de l'eau à l'entrée des générateurs de vapeur)	235°C/210 bars
Steam temperature/pressure at turbine stop valves (Température/pression de la vapeur à l'admission des turbines)	487°C/177 bars
Nominal flow rate (Débit nominal)	4×340 kg/s

565°C (1000 to 1050°F) and 100 to 270 bars (1500 to 4000 psi), whereas intermediate- or peaking-load steam plants commonly employ peak steam temperatures that are lower by 50 to 100°C and pressures of 70 to 120 bars. The steam temperatures and pressures used in nuclear plants are determined by the characteristics of the reactor (see Chap. 5 and Tables 11.4 to 11.6).

Effects of Unit Size on Efficiency and Cost

Increasing the size of a unit leads to reductions in capital costs and small increases in the efficiency stemming largely from the higher Reynolds numbers inherent in the larger sizes (Fig. 3.19). An important savings in operating costs stems from the reduction in the amount of manpower required per unit of output (Table 10.5); an increase in unit size by a factor of 10 commonly leads to an increase in manpower by only a factor of ~6. Other savings result from the fact that some components, such as instrumentation and control equipment, increase relatively little

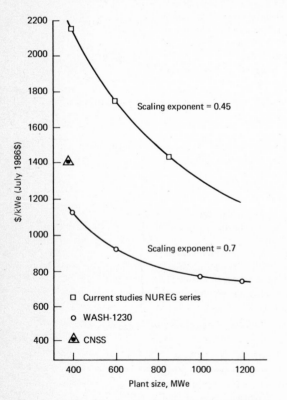

Figure 11.18 Effects of unit size on the estimated capital cost of nuclear power plants as obtained in two different studies.[40]

in cost and complexity as the unit size is increased. As shown in Fig. 11.18, these effects are especially pronounced for nuclear plants.[40] Other data on these effects are given in Chap. 10, e.g., Figs. 10.4 and 10.5.

Plant Thermal Efficiency

The overall thermal efficiency of a steam power plant is normally taken as the ratio of the useful electric output delivered to the grid divided by the chemical energy available in the fuel. The principal component efficiencies commonly run around 90 percent. This is the usual value for a coal-fired steam generator in which most of the loss is in the form of heat in the stack gases. Aerodynamic and moisture-churning losses in the turbine and pressure losses in the piping between turbines reduce the turbine efficiency to 80 to 85 percent. The heat losses inherent in the thermodynamic cycle run about 45 percent of the heat in the steam delivered to the turbine. In nuclear reactors the pumping power for the reactor cooling circuit commonly runs ~2 percent of the plant gross output.[41] Other smaller losses include electrical losses in the generator (about 1 percent) and energy to the steam-turbine-driven feed pumps (~1.5 percent of the useful output). The electric power required to drive induced- and forced-draft fans, coal feed and preparation equipment, condensate pumps, condenser cooling-water circulating pumps, miscellaneous minor equipment, and general station service and lighting totals about 3 percent of the plant net power output.[41] Table 11.7 gives data for a typical coal-fired power plant.

Effects of Peak Steam Temperature on Thermal Efficiency

Any effort to increase plant thermal efficiency by increasing the steam pressure and/or temperature entails many compromises that must be made in an effort to avoid excessive increases in

TABLE 11.7 Auxiliary Power Requirements for the TVA 950-MW Bull Run Coal-fired Plant[17]

	Fraction of net output, %
Forced-draft fans	1
Pulverizers	0.6
Condenser cooling water	0.25
Miscellaneous pumps	0.5
Station lighting and service	0.65
Total auxiliary power	3.0

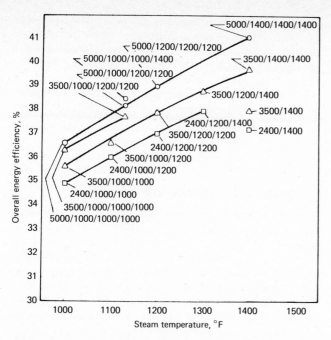

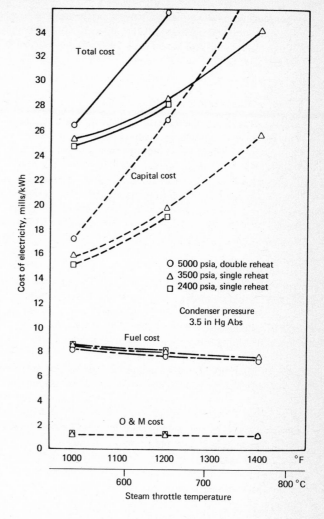

Figure 11.19 *(Above)* Effect of steam throttle conditions on overall efficiency for a 500-MWe steam plant with an atmospheric furnace. The steam conditions given on the curves are: pressure, psi/superheat, °F/first reheat, °F/second reheat, °F.[16]

Figure 11.20 *(Right)* Effect of steam-turbine throttle conditions on the cost of electricity from a 500-MWe steam plant with an atmospheric pressure furnace on the basis of 1975 dollars.[16]

capital cost. A nice set of curves showing these effects for a large fossil fuel plant is given in Figs. 11.19 and 11.20, taken from a Westinghouse study.[16] This shows the relatively small improvements in the thermal efficiency obtainable through the use of increased steam temperatures and pressures beyond 550°C/240 bars together with the relatively large increases in capital costs, so that the minimum cost of electricity is obtained with a peak steam temperature of about 538°C (1000°F). This result is consistent with numerous other studies that have been carried out in the past 25 years.

The curves of Fig. 11.20 suggest that the minimum capital cost of a plant would be found at steam pressures and

temperatures lower than those considered in preparing Fig. 11.19. A review of the whole development of steam power plants shows that increasing the steam pressure and temperature from the low levels of the 1920s has led to reduction in the size and unit cost per kilowatt of electric output for the combustion air and stack gas equipment. On the other hand, as is evident from Fig. A6.1, increasing the steam temperature beyond 482°C (900°F) leads to a fairly rapid loss in strength of the relatively inexpensive low-alloy steels. The best resolution of the various compromises that must be made in selecting a steam temperature and pressure for minimum costs is indicated by the actual sales of steam plants for intermediate- and peaking-power purposes;[38] these plants are often designed for 60 to 120 bars (900 to 1800 psi) at 482 to 510 °C (900 to 950 °F), with many having steam conditions of 160 bars (2400 psi) and 540/540 °C (1000/1000 °F).

Effects of Superheating and Reheating

Superheating leads to an increase in the peak temperature in the cycle for a given peak steam pressure, and this increases the thermal efficiency. The increase is not as great as one might expect from simple peak temperature considerations because, as discussed in Chap. 2, when the amount of superheat is increased, the temperature-entropy diagram deviates farther and farther from the ideal rectangular Carnot cycle; this effect is shown in Fig. 2.4. Reheating has the effect of giving a somewhat closer approach to the ideal Carnot cycle, as can be seen in Fig. 2.10.

One of the most important reasons for employing superheat and reheat relatively early in the development of steam power plants was that steam expansion in the turbine from a saturated-steam condition leads to moisture formation, which in turn may lead to turbine blade erosion, coupled with losses in turbine efficiency stemming from moisture-churning energy losses (Fig. 3.33). Refined design of the turbine blading to prevent the accumulation of liquid slugs in stagnation regions near the tips of stator blades, coupled with bleed-off slots in the turbine casing and the use of bleed-off steam for regenerative feed heating, has greatly reduced the moisture-churning losses and tendencies toward blade erosion. In addition, the use of erosion-resistant materials, such as stellite and titanium, in the lower-stage turbine blades has made it possible to increase the turbine tip speed to ~ 500 m/s (1640 ft/s) in the last stage without getting into difficulty with serious moisture-churning losses or turbine bucket erosion even when operating with saturated steam at the turbine inlet.[42] (The discussion on turbine blade erosion in the latter part of the next chapter includes some information on erosion by wet steam.)

Air Preheat

The principal loss from the furnace–steam generator units in fossil fuel plants is heat in the stack gas. By using a recuperative heat exchanger, a pronounced energy savings can be effected by transferring heat from the stack gas to the incoming combustion air. This is commonly accomplished with a large cylindrical matrix of alternate corrugated and flat plates placed so that one portion is in the hot gas stream and the other in the incoming air. The cylinder is rotated at low speed so that heat is picked up and stored in the metal plates on the hot side, moved over and transmitted to the incoming air on the other side, giving a heat

exchanger that has a low cost per unit of surface area. Preheating the air also serves to improve combustion conditions and increases the temperature difference in the furnace and thus the heat flux. This is particularly important because the SO_3 in the stack gas combines with moisture to give corrosive sulfuric acid if the stack gas temperature is reduced below the dew point. Depending on the sulfur content of the fuel (Fig. 6.2), this ordinarily places a lower limit on the stack gas temperature of about 150 °C (300 °F).

FLUE GAS DESULFURIZATION

Most of the developmental activity on steam power plants at the time of writing is concerned with problems stemming from environmental legislation and efforts to use lower-grade coals, particularly those high in sulfur and ash.[43-47] Chief among the steps being taken is the development of stack gas scrubbers. The basic concept is simple—i.e., spray an alkaline solution or slurry into the flue gas to react with the SO_3 and SO_2 to form a sulfate or sulfite—but in practice such systems present serious corrosion, scaling, plugging, and waste-disposal problems.

Background of Experience

The reduction of sulfur emissions by stack gas scrubbing began in England in 1925 at the Battersea station with lime used to neutralize the acidity of the scrub water, which was discharged to the Thames River in a once-through process.[43] In the late 1930s a closed-loop lime-limestone scrubber system was installed at the nearby Fulham station that burned 1 to 1.5% S coal. Scaling and corrosion proved to be severe problems, and the FGD system was removed from service in 1942. Further work in England on both stack gas scrubbing and environmental effects led to the choice of tall stacks rather than scrubbers, and that course is being followed in Great Britain at the time of writing.[43] (See also Ref. 15 of Chap. 9.)

Serious work on flue gas desulfurization began in the United States in the early 1940s, and small-scale efforts were continued on both lime-limestone scrubbing and processes for recovery of the products. In the 1960s, experiments were conducted with pulverized limestone injected into the furnace to calcine it and obtain very reactive lime particles in the flue gas, but the lime formed low-melting-point carbonate glasses that caused severe slagging of the boiler tubes, making this approach clearly unsatisfactory. Subsequent work has been directed mainly to various stack-gas scrubbing processes together with processes for disposing of the sludge produced. Extensive work has also been under way in Japan, largely for use with high-sulfur fuel oils.

Parameters and Criteria

As discussed in Chap. 9, the problem of sulfur emissions has been highly politicized and the public bombarded with a host of contradictory statements. Moreover, the courts have been flooded with suits brought by government agencies against utilities for failing to comply with regulations, by utilities against government agencies claiming that regulations are unrealistic, by environmentalist groups against government agencies charging laxity in enforcement, and by consumers against utilities to protest high rates stemming from expensive measures to reduce emissions. Much of both the acrimony and the litigation stems from preoccupation with one or a few of the parameters and criteria involved without recognition of their dependence on other parameters. The situation is enormously complex with no clear-cut solution in sight, but it is instructive to review the key technical facts, parameters, and criteria, their relationships, and the experience with operating systems. The scope of these problems is indicated in Table. 11.2.

Sulfur Content of Fuel

The sulfur content of the fuel is a major parameter. The higher the sulfur content, the greater the droplet density required in the spray tower, the more rapid the sulfur pickup, the greater the difficulties with scale formation and mist flow up the stack, the more difficult the removal of a high percentage of sulfur, and the greater the amount of waste material.

Ash Content of Fuel

Both the percentage of ash and its chemistry affect the desulfurization process. Eastern coals are generally acidic, whereas western coals are usually basic so that their alkali content contributes to the conversion of sulfurous and sulfuric acid to sulfates. In addition, the chloride content of the coal is an important factor affecting corrosion, particularly in closed-loop systems where the chloride concentration builds up.

In some systems electrostatic precipitators are installed ahead of the scrubbers to reduce the adverse effects of the ash on scaling and corrosion in the air preheaters and scrubbers.

Type of Sorbent

Lime is both more soluble in water and more reactive than limestone, but it is more expensive. Further, in either case the presence of other minerals may have adverse or beneficial effects; e.g., magnesium that may be present in locally available limestone acts to improve its effectiveness. In fact, MgO is used as the sorbent in some processes. Lime obtained as a by-product

from acetylene plants has been used at some sites and has been found to be much less prone to cause scaling because of the presence of a small amount of sodium thiosulfate.

Sodium carbonate is available at an acceptable cost at some sites in the western United States and has been found to give much less trouble with scaling and plugging. It is normally regenerated by reacting with lime or limestone and recirculated. A disadvantage is that the resulting calcium sulfite and sulfate waste has a substantial content of soluble sodium compounds that present a leaching problem, as well as requiring a makeup stream of sodium carbonate. In the arid west, where evaporation from a disposal pond exceeds the rainfall, the leaching problem can be handled; but in the east, contamination of streams by effluent from settling ponds is a serious problem.

Fraction of Sulfur Removed

The costs and problems of FGD processes increase with the fraction of the sulfur removed. Not only is it necessary to employ progressively greater amounts of sorbent to drive the reaction, but the fraction of the unreacted sorbent in the waste increases, thus increasing both the cost for the sorbent and the disposal problem more rapidly than linearly.

Water Recycle

There is a step change in difficulties with scaling, plugging, and corrosion if one shifts from an open cycle with waste-water discharge to a closed cycle in which impurities such as chloride build up. Significantly, no full-scale system handling high-sulfur coal has been operated on a closed cycle up to the time of writing.[43]

Energy Requirements

Scrubbing the flue gas reduces its temperature below the dew point, thus introducing major corrosion problems in damper mechanisms, induced-draft fans, stacks, piping, and in heat exchangers employed to raise the stack gas emission temperature to avoid moisture precipitation and high ground concentrations of stack gases immediately downwind. Reheating the stack gas after scrubbing is sometimes done with gas or oil burners, sometimes with steam, and sometimes with regenerative heat exchangers. All such measures entail energy losses equivalent to several percent of the energy consumption of the plant. In addition, extra power is required for the induced- and forced-draft fans, pumps, etc., so losses total 3 to 7 percent of the plant energy input. Additional energy is required if the waste is converted to a salable product such as H_2S or elemental sulfur. In the latter case the energy consumption of the FGD system has run as much as 25 percent of the energy input to the plant.[43]

Product

In principle, the sulfur removed from the stack gas can be made into a salable product, and efforts to do this have been under way in the United States and other countries since the 1930s. The only really successful commercial operations have been in Japan, where there is a good market for gypsum. Other potential products for which the waste might be used include portland cement, building and road base materials, H_2S, and elemental sulfur. However, the total market for these products is far too small to utilize the output from the U.S. utility industry if all its coal-fired plants were fitted with FGD systems, and even for the limited number of FGD installations at the time of writing (a total of ~ 15,000 MWe of capacity), it has been found that the lowest-cost method is to use either a settling pond or a landfill after dewatering the sludge and treating it chemically to give a stable material. The calcium sulfite is gelatinous and tends to flow under load. Oxidation to calcium sulfate gives a stable product. Depending on the chemistry of the ash, mixing with fly ash may also give a structurally stable material.

Capital Cost

The capital cost of an FGD system depends on all the above parameters, generally running from $60 to $130/kWe (giving annualized capital costs of 3 to 6 mills per kilowatthour) for new plants in 1975 dollars according to an EPA study.[46] Cost estimates by industrial organizations are generally higher.[47] Operating costs vary even more widely, depending on maintenance problems, running as much as 26 mills per kilowatthour in some plants. Another indication of costs is given by a 1979 compromise agreement on new FGD installations between TVA and EPA, which by one government estimate will increase TVA's cost of electricity generation by 30 percent in coal-fired plants.

Availability

The continual maintenance required to cope with scaling, plugging, corrosion, etc., is expensive not only in man-hours, replacement parts, and supplies but also in reduced availability of the plant. In some cases the normal load on the plant is sufficiently low at night so that dampers can be operated to close off perhaps half of the units to permit cleaning and servicing, and in some plants this is done every night.[43] For a base-load plant it is necessary to install extra FGD capacity to reduce losses in the availability of the full capacity.

Life of Equipment

Operating any type of equipment in a wet flue gas atmosphere is bound to give trouble because the gas is inherently corrosive (Chap. 6), and the erosive effects of slurries aggravate the corrosion problems. Many different methods of coping with these problems have been tried, including lining pipes with rubber, plastic, or special coatings, but up to 1980 none has proved satisfactory when employed over a period of several years.[43] At the time of writing, expensive high-nickel alloy parts are being tried.

Operating Experience

The Clean Air Act was amended for the third time in August 1977 to require that all new U.S. power plants employ the "best available technology" to reduce emissions, "taking into consideration the cost of achieving such emission reduction, any nonair quality health and environmental impacts and energy requirements." There have been widely different interpretations of the Congressional language. Those demanding tight controls on emissions can point to satisfactory service of flue gas scrubbers in a plant that has operated a short time on relatively low-sulfur coal with a low-cost supply of a sorbent having good characteristics and a plant load pattern that permits operation at low nighttime loads with maintenance of the FGD equipment every night. Others can point to the high costs that everyone has experienced, the short life of the equipment, the severe loss in availability, and the inability of any plant to operate on high-sulfur coal with a closed system. A detailed and comprehensive review of all the U.S. and Japanese experience up to 1979 is given in Ref. 43 and clearly indicates that the technology is not well in hand. In fact, none of the systems installed by 1976 was still being offered for sale by May 1978—another indication that the technology had not yet been developed.

Efforts to apply cost-benefit criteria have foundered on the refusal of some activists to accept such studies as long as any medical authorities express concern over possible health effects. At the time of writing it is not clear what systems will prove viable, how great costs will prove to be, or what standards will emerge, though the trend to date has been a continual tightening of EPA requirements irrespective of the costs. At the time of writing (1981) it appears that this trend is changing, in part because of rapidly escalating costs and in part because of recent studies by such organizations as the National Academy of Sciences indicating that earlier inferences of severe health effects were exaggerated or in error.[48] Significantly, at a symposium conducted by the New York Academy of Medicine on the health effects of sulfur oxides and related particulates in

December 1978, 72 percent of the participants felt that then-current standards were too stringent.[48]

FLUIDIZED-BED COAL COMBUSTION FOR STEAM GENERATION

The most promising approach to future improvements in steam plants at the time of writing appears to be through the development of fluidized-bed combustors (FBCs) that will give lower costs and higher efficiencies for coal-fired operation than conventional pulverized-coal-fired furnaces (PCFs) fitted with FGD equipment.[49] As discussed in Chap. 5, experiments with fluidized-bed combustion systems have been under way since the middle 1950s, yet none is in regular commercial service at the time of writing. Some insight into the problems involved is given by Table 11.8, which summarizes the principal design data for three TVA plants having conventional PCFs and eight FBC plants.[49-52] The first fairly large-scale FBC pilot plant to be built, the 30-MWe experimental unit at Rivesville, West Virginia, is the first of the four atmospheric fluidized-bed combustors (AFBCs) in the table, the other three being design studies prepared for TVA. Data from design studies for one supercharged fluidized-bed combustor (SFBC) and three pressurized fluidized-bed combustors (PFBCs) are also included, the latter three being combined-cycle plants in which part of the electric output would be delivered by steam turbines and part by gas turbines. Note that the use of an FBC affects only the furnace and stack gas clean-up equipment; the rest of the steam system is unaffected. In fact, the steam conditions chosen for the FBC plants of Table 11.8 range from 6 MPa/440°C (800 psi/825°F) to 26 MPa/540°C (3800 psi/1005°F), a substantially wider range than for the three PCF plants shown.

A good notion of the furnace geometry envisioned for AFBCs is given by Fig. 11.21, while Fig. 11.22 shows the principal components required in the system. Table 11.9 gives design data for this furnace.

Capital Cost Parameters

For FBCs to be attractive, their capital costs ought to be less than for conventional plants. Costs for new systems are always difficult to estimate, but several parameters in Table 11.8 provide a basis for appraising some of the major cost factors.

Heat-Transfer Surface

One of the most important factors affecting the capital cost is the heat-transfer surface area required per unit of output. Note that the heat-transfer surface area parameter is about the same for the AFBC as for the PCFs, but that increasing the furnace pressure reduces the surface area requirements for the PFBCs. The reason is that, while the gas-side heat-transfer coefficient in fluidized beds is high and acts to reduce the surface area requirements, the average temperature difference available is less than for a conventional furnace, so that the heat-transfer surface area requirements are about the same for PCF and AFBC furnaces. For PFBCs, increasing the furnace pressure improves the heat-transfer coefficient in the heat-transfer matrices above the bed and thus reduces their size, weight, and cost—a major advantage of furnace pressurization.

Floor Space

The amount of floor space required for the furnace is another important parameter from the cost standpoint, particularly if an AFBC is being retrofitted to an existing PCF plant. Table 11.8 indicates that this parameter is twice as great for full-scale AFBCs as for the PCF plants, but that supercharging the FBC to 3 atm gives about the same floor area requirement, and pressurizing the furnace to 10 atm cuts the floor space required to roughly half that for a conventional furnace. (Note that the value for the Rivesville furnace is not consistent with the others because it does not include ancillary equipment.) The crux of the problem is that the heat release per unit of fluidized-bed cross-sectional area is directly proportional to the airflow rate, and, as indicated by Fig. 5.19, the superficial gas velocity leaving the bed is usually limited to ~1.8 m/s (6 ft/s) to avoid excessive elutriation of fine particles. The shallow depth (~1 m) and large cross-sectional area of an AFBC gives an excessive floor space requirement unless a series of beds is stacked one above the other in tiers, as in the AFBC furnace of Fig. 11.21, but even this appears to give about double the floor space requirement of a conventional furnace. In addition, the floor space required for auxiliaries, particularly the coal feed equipment, seems higher than for conventional plants.

Number of Fuel Feed Points

The cost, complexity, and maintenance problems of a furnace increase with the number of fuel feed points, a parameter treated in the bottom line of Table 11.8. Pressurizing a fluidized-bed furnace reduces the bed area, and—other considerations being equal—should make the number of coal feed points inversely proportional to the pressure. The wide range of design values given in Table 11.8 indicates that the information available to the designers was insufficient to form a firm basis for the designs. There is no question but that, to minimize both the cost and complexity of the coal feed system for a fluidized-bed fur-

TABLE 11.8   Major Design Parameters for Both Typical Pulverized Coal-Fired Power Plants and Conceptual Designs for Fluidized-Bed Combustion Systems [49]

Plant	Conventional pulverized coal			Atmospheric fluidized bed			
	Bull Run	Colbert #5	Kingston #9	Rivesville	Foster-Wheeler	Combustion engineering	Babcock & Wilcox
Design steam pressure and temperature, MPa/°C/°C (psig/°F/°F)	24/538/538 (3500/1000/ 1000)	17/566/538 (2400/1050/ 1000)	12/566/566 (1800/1050/ 1050)	9/496/538 (1350/925/ 1000)	26/540/540 (3800/1005/ 1005)	18/540/540 (2600/1005/ 1005)	18/539/539 (2581/1003/ 1003)
Design output, MWe	850	550	200	30	150	200	200
Overall plant thermal efficiency, %	40	38.6	26.7				
Furnace arrangement	Conventional	Conventional	Conventional	Single bed 4 compart-ments	Stacked 4 main 1 CBC†	Ranch 7 main 1 CBC	Stacked 4 main 2 CBC
Furnace pressure, MPa (atm)	0.10 (1)	0.10 (1)	0.10 (1)	0.10 (1)	0.10 (1)	0.10 (1)	0.10 (1)
Surface area, m² (ft²)	70,000 (753,000)		20,500 (221,000)			22,000 (237,000)	
Surface area, m²/MWe (ft²/MWe)	82 (886)		103 (1105)			110 (1185)	
Furnace plan area, m²/MWe (ft²/MWe)	0.051 (5.5)		0.052 (5.6)	0.14* (1.52)	0.97 (10.4)	1.24 (13.4)	1.80 (19.4)
Heat-transfer surface weight, kg (lb)		2,270,000 (5,000,000)	1,000,000 (2,200,000)				
Total furnace weight, kg (lb)	10,900,000 (24,000,000)		3,800,000 (8,400,000)				
Heat-transfer surface weight, kg/MWe (lb/MWe)		4140 (9100)	5000 (11,000)				
Total furnace weight, kg/MWe (lb/MWe)	12,800 (28,230)		19,000 (42,000)				
Number of burners or fuel feed points	176		56	44	16	118	640
Number of burners or fuel feed points/100 MWe	21		28	147	11	59	320

*Does not include space for coal feed and other equipment.
†Carbon burnup cell

Supercharged fluidized bed	Pressurized fluidized bed		
ORNL	GE—ECAS	Westinghouse ECAS	Curtiss-Wright
12/566/566 (1800/1050/ 1050)	24/538 (3515/1000)	24/538 (3515/1000)	6/440 (800/825)
200	904	679	500
	39.2	39.0	38.8
Single bed vertical tubes	Stacked 7 bed	Stacked 4 bed	
0.30 (3)	1.0 (10)	1.0 (10)	0.7 (7)
8000 (86,700)	8300 (89,400)	5900 (62,980)	
40 (434)	9 (99)	9 (93)	
0.632 (6.8)	0.055 (0.59)	0.17 (1.85)	
404,500 (890,000)			
	1,825,000 (4,016,000)		
2000 (4450)			
	2000 (4400)	1872 (4120)	
144	335	64	
72	37	9.4	

TABLE 11.9 Fluidized-Bed Design Parameters for Fig. 11.21

Superficial velocity	
Main beds	12 ft/s
Carbon burnup cells	9 ft/s
Excess air	
Main beds	20%
Carbon burnup cells	25%
Temperature	
Main beds	1550 F
Carbon burnup cells	2000 F
Bed depth	
Static	24 in
Full load	48 in
Grid plate pressure drop	16 in H_2O
Bed pressure drop	30 in H_2O
Coal/limestone feed spacing	18 ft^2
Freeboard	6 ft
Limestone feed rate	
Design coal—0.9% S	2.3 Ca/S mol ratio
3.25% S coal	4.0 Ca/S mol ratio
Heat-transfer coefficient	
Vertical tubes	50 Btu/(h · ft^2 · °F)
Horizontal tubes	45 Btu/(h · ft^2 · °F)
Combustion efficiency	
Main beds	90%
Carbon burnup cells	90%
Elutriation	
Carbon	10% of heat input weight equivalent
Coal ash	100% of input coal ash weight
Limestone	40% of solid weight after calcination

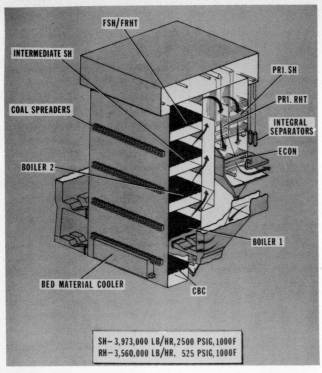

Figure 11.21 Schematic isometric view of an atmospheric fluidized-bed steam generator with five beds arranged in vertical tiers for a 570-MWe power plant.[50] (*Courtesy Foster Wheeler Energy Corp.*)

SH – 3,973,000 LB/HR, 2500 PSIG, 1000 F
RH – 3,560,000 LB/HR, 525 PSIG, 1000 F

nace, one should minimize the number of coal feed ports. However, a number of other considerations, such as tube corrosion, improved sulfur retention, and carbon and lime utilization, favor close spacing of the coal feed ports. Of these, corrosion of metal tubes in the bed is probably controlling; rapid corrosion of metal in the fluidized bed will occur if there are local regions in which there are vacillations between oxidizing and reducing conditions—e.g., in fuel-rich zones above coal feed ports. Tendencies toward such conditions can be reduced by increasing the nominal amount of excess air, reducing the coal feed port spacing, increasing the depth of the mixing region between the coal feed ports and the heat-transfer matrix in the bed, reducing the amount of volatile matter in the coal, increasing the amount of air preheat, and modifying the geometries of both the air tuyeres and the coal feed ports to increase the rate of lateral mixing of the coal injected into the bed.

Major Problem Areas

Hundreds of fluidized-bed combustion systems are in use commercially to burn low-grade solid waste, such as sawdust and

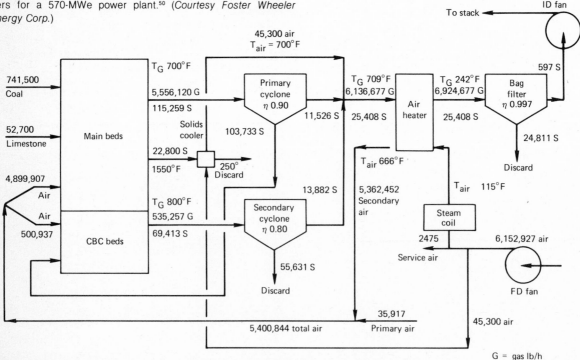

Figure 11.22 Flowsheet for the fluidized-bed furnace system of Fig. 11.21.[50] (*Courtesy Foster Wheeler Energy Corp.*)

sludge from sewage plants, while other hundreds of units are in commercial operation roasting pyrite ores. In view of this, it is surprising that there are no fluidized-bed coal combustion systems in commercial operation with steam boilers, although quite a number of experimental units have been operated. The principal reasons for this situation are treated in this section.

Boiler Stability and Control

There have been serious problems in controlling the combination of a fluidized-bed coal combustion system and a steam boiler with its tubes immersed in the bed. For good sulfur removal the bed must be operated in the 816 to 927°C (1500 to 1700°F) range, and to maintain this temperature about two-thirds of the heat released by combustion of the coal must be removed from the fluidized bed by a tube matrix immersed in the bed. Herein lies the crux of the problem. The heat-transfer coefficient between the bed and the tube wall is high—around 300 J/(s · m² · °C) [50 to 60 Btu/(h · ft² · °F)]. At the same time the heat-transfer coefficient from the tube to the boiling water under nucleate boiling conditions is very high—of the order of 30,000 J/(s · m² · °C) [5000 Btu/(h · ft² · °F)]. Thus, the tube wall tends to run close to the water temperature. Both the fluid-bed-side and water-side coefficients are essentially independent of the combustion gas and water flow rates, respectively. Inasmuch as $Q/A = u\,\Delta T$ and the overall heat-transfer coefficient u is constant, if one attempts to reduce the rate of heat release in the bed, the bed temperature will drop until it reaches about 540°C (1000°F), at which point the bed will quench; i.e., "the fire will go out." If the coal feed is continued, coke will form and clog the bed. If the water flow to the boiler is reduced in an effort to keep the bed temperature constant at reduced outputs at 870°C (1600°F), the water-side heat-transfer mechanism shifts from a nucleate boiling condition to a vapor-film blanket condition. This effect is illustrated by Fig. 11.23, which shows a typical curve for the heat flux to boiling water as a function of the temperature difference between the boiling point of the water and the metal wall from which the heat is being conducted.[33] If the fluid-bed temperature is held at 870°C (1600°F) and the metal tube wall is held at 316°C (600°F), the heat flux through the tube wall under normal nucleate boiling conditions would run about 300,000 J/(s · m²) [50,000 Btu/(h · ft²)]. This point has been plotted in Fig. 11.23 to indicate the normal operating point to be expected. Note that operation in this region is stable, because any small increase in the tube-wall temperature will result in a greatly increased heat flux from the tube wall to the boiling water. If one attempts to reduce the heat flux from the bed to the tube wall by allowing the tube-wall temperature to increase to, say, 540°C (1000°F), the

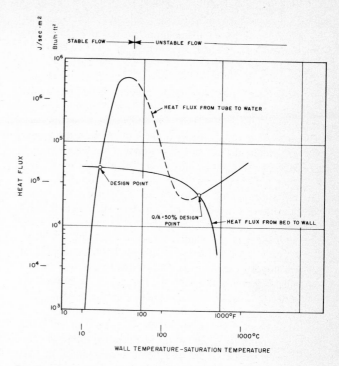

Figure 11.23 Curve for the calculated heat flux from the bed to the tube in a fluidized-bed steam boiler with a bed temperature 555°C (1000°F) above the saturation temperature of the steam is shown superimposed on a typical curve for the heat flux from the tube wall to boiling water. Both heat fluxes are plotted against the difference between the temperature of the tube wall and the saturation temperature of the boiling water.[33]

heat flux will be cut roughly in half, but the heat-transfer mode will be via vapor-blanket boiling, which will be very unstable and will lead to wide fluctuations in the local metal temperature. Even worse, the water flow rate through the tube will become highly unstable, and this in turn will lead to even more severe temperature fluctuations and thermal stresses. Thermal-strain cycling failure of the tubes will result if the tubes do not overheat so much that their loss in strength leads to failure from pressure stresses. Note that this unstable operating region is indicated in Fig. 11.23. Note, too, that intermediate conditions between these two regions would be unstable because any small increase in the tube-wall temperature would lead to a reduction in the heat flux from the tube wall to the boiling water.

For good reliability, operation must be constrained to the nucleate boiling regime with a fixed bed temperature of around 870°C (1600°F). Basic heat-transfer considerations indicate that the only good way in which the heat input to the boiler can be reduced is to reduce the amount of surface area of the tube matrix immersed in the bed. This, in turn, implies that either the bed should be compartmentalized so that the output can be changed in quantum jumps or the level of the fluid bed should be varied. The former approach presents difficulties in the relative phasing of the flow of red-hot "sand" and feed water to a compartment being activated or deactivated. Unless this phasing is carried out in just the right fashion, one runs the danger either of chilling the bed in that region and building up heavy coke deposits or of getting into a vapor-blanket boiling region, generating hot spots in the tubes, and of failure of the tubes after a series of perhaps 10 to 100 thermal cycles.

Varying the bed depth presents a problem in that it is difficult to tell just where the surface of the bed is during operation, because in the regime of interest the fluid bed will be bubbling violently (Fig. 5.18). Experience at the BCURA Laboratory in England indicates that the bed depth can be determined and controlled within 7 to 10 cm (3 or 4 in) by measuring the pressure drop across the bed. For an atmospheric fluidized bed for which the maximum bed depth with an acceptable pumping-power loss would be around 1 m (40 in), the uncertainty in bed depth would be around 10 percent of the bed depth at full power and progressively greater if the bed depth were reduced at part power. If, on the other hand, the bed depth at full power is made around 3 m (10 ft), the uncertainty in bed depth becomes only a few percent and provides a good basis for control of the power output. That is, the power output will not be too sensitive to modest additions or removal of hot "sand" to or from the bed. Thus, just from the control standpoint, there is a strong incentive to supercharge the bed so that the bed depth can be made great enough to permit good control and yet not be so great as to entail excessive pumping-power requirements for the combustion air.

Coal Feed and Metering System

Pulverized-coal and fluidized-bed coal feed and metering systems differ in two important respects; namely, the requirements on particle size and requirements on the uniformity in fuel distribution in the furnace. Pulverized-coal firing requires not only crushing and screening but also pulverizing to a size less than 2 μm. The pulverizers are expensive, require considerable maintenance, and consume a substantial amount of electric power—typically 0.5 to 1.0 percent of the gross plant output. On the other hand, the open furnace can tolerate substantial variations and fluctuations in the fuel flow to individual burners, whereas in the fluidized bed such variations

and fluctuations affect the local fuel-air ratio and may lead to local fluctuations between oxidizing and reducing conditions in the vicinity of the coal feed ports. (This would be much less of a problem if char were the fuel rather than raw coal with its volatile content of ~35 percent.) Getting a high degree of uniformity in coal feed rate and distribution to a large number of feed points in a fluidized bed is an extremely formidable problem.[53]

Forcing the coal into the furnace against the furnace pressure becomes progressively more difficult as the furnace pressure is increased from atmospheric to the 10 atm projected in some designs. The pressure barrier can be handled with lock hoppers of the sort employed for solids handling in many chemical processes (and in blast furnaces), but to get good sealing in the valves of lock hopper systems has generally proved troublesome. A variety of helical screw feeders have been developed, and these have proved satisfactory for feeding against small pressure differentials. Gas backflow through the material in the screw has been a problem in some types during operation against pressures of more than 1.5 atm, though at least one reportedly has functioned well in bench tests against 4 atm. Rotary feeders with vanes or pockets have been used to pump the crushed coal into a pressurized furnace in several experimental FBCs, but these machines seem to be subject to rapid wear unless the solids feed rate is kept well below their capacity. This experience indicates that they are suitable for metering and might be used as a pressure lock, but not for both functions.

Dividing the coal stream so that it is uniformly distributed to dozens or hundreds of coal feed points is difficult for any given flow rate. It is even more difficult to get the same set of devices to perform well over a wide range of flow rates and do so consistently on both a short-time and long-time basis, but such performance is required in order to prevent the local fluctuations that might cause serious fire-side corrosion. Experience in bench tests shows that the only way that it can be achieved is to dry the coal to less than 1 percent superficial moisture so that the granular solid flows freely.[53]

After feeding the coal through a pressure lock, it can be metered and then divided before it is entrained in an airstream for pneumatic transport to the furnace, or the solid particles can be entrained in an airstream and the stream divided. In either case, the airstream velocity required depends on the particle size, generally running about 6 m/s for −3 mm ($-\frac{1}{8}$ in) particles. At this velocity erosion in straight pipes has not been a problem, but severe erosion has been experienced in bends and elbows. Interestingly, use of a tee with a blocked leg facing the inflowing stream has proved to be an entirely satisfactory replacement for elbows. Solid particles deposited in the dead leg of the tee act as a cushion to absorb the impact of particles in the

TABLE 11.10 Comparison of Relative Advantages and Disadvantages of Atmospheric and Pressurized Fluidized-Bed Coal Combustion Systems

Furnace pressure	Atmospheric 0.10 MPa (1 atm)	Pressurized (turbosupercharged) 0.3–0.5 MPa (3–5 atm)	Pressurized (gas turbine–generator) 1.0 MPa (10 atm)
Advantages	Fewest feasibility problems Greatest amount of test experience Coal feed against only 0.014 MPa (~2 psi) Solid waste disposal much less difficult than for wet scrubbers Plant efficiency higher than for wet scrubbers by ~ 1 point Capital cost modestly improved over conventional plant with scrubbers	Capital cost of furnace and steam generator appears lower than for conventional pulverized-coal unit Coal feed points reduced by a factor of 3–5 over AFB Large bed depth improves combustion efficiency and SO_2 removal, and permits vertical boiler tubes with simpler furnace structure Plant thermal efficiency about 1 point higher than for AFB Control characteristics appear to be excellent	Potential cycle thermal efficiency is about 3.5 points higher than AFB Capital cost of furnace and steam generator appears lower than for conventional pulverized-coal unit Coal feed points reduced by about a factor of 10 relative to AFB Large bed depth gives better coal utilization and SO_2 removal as well as permitting vertical tube array and simpler furnace structure
Disadvantages	Large number of coal feed points gives complex coal feed system Shallow beds require horizontal tubes with support structure at bed temperature Startup and control problems appear to be complex	Feasibility of acceptable turbine erosion and deposits with 538°C (1000°F) turbine inlet temperature seems likely but has not yet been proven Coal feed against 0.3–0.5 MPa (3–5 atm) poses some reliability problems Test experience much less than for AFB	Feasibility of acceptable turbine erosion and deposit rates is highly doubtful Coal feed against 1.0 MPa (10 atm) poses difficult reliability problems Test experience much less than for AFB Large heat capacity of granular bed filters for hot gas cleanup gives very slow response times Particulate removal equipment may entail excessive capital cost

airstream so that they bounce around the bend without eroding the metal wall.

Advantages and Disadvantages of Pressurized Furnaces

As indicated by Table 11.8, pressurizing the furnace for a fluidized bed reduces the furnace size and capital costs, and it permits an increase in bed depth which, as indicated in Figs. 5.21 and 5.22, improves both sulfur retention and combustion efficiency. However, pressurization requires such a large power input to the compressor that it can be accomplished economically only if a gas turbine is employed to recover power from the hot pressurized gas leaving the furnace, and this, in turn, presents problems with gas-turbine blade erosion and deposits caused by ash particles entrained from the bed. These problems are treated in the next chapter. Combined gas turbine–steam power plants, including those employing FBCs, are treated in Chap. 13. However, a brief summary of the principal advantages and disadvantages of the three basic types of plants is presented in Table 11.10. A further comparison is made in Chap. 13.

COOLING TOWERS

Heat rejection from a steam power plant to rivers, lakes, or estuaries may have adverse effects on the marine biota (Chap. 9). To avoid possible ecological damage, natural-draft cooling towers with an intermediate coolant circuit have been required by the EPA on new plants as the least expensive alternative. The incremental capital cost of wet cooling towers has proved to be ~ $40/kWe in 1978 dollars. The higher condenser temperature reduces the thermal efficiency ~3 percent, and these factors coupled with added maintenance and troubles such as freezing and solids buildup from evaporation increase the cost of electricity by ~7 percent. Efforts to investigate the ecological justification for these costs have shown that in many cases there is no significant ecological advantage to the use of cooling towers, and, in fact, mandating their use may be counter-productive.[54]

Cooling towers function by evaporating water and hence have a substantial water consumption, which is a problem in arid areas. They also introduce a fog problem under some conditions at certain sites. Both of these problems can be avoided through the use of "dry" cooling towers, which employ heat-transfer matrices in the air inlet region. In principle, these cooling towers could act as direct condensers for the steam, but the steam volume flow is so great that, to reduce the size and cost of piping, an intermediate fluid is ordinarily employed—usually either water or a water-glycol solution to avoid difficulties with freezing. A few such installations have been built, one of them a 300-MWe plant in the Dakotas that uses forced-draft air cool-ing. The dry cooling towers have proved very expensive, increasing the plant capital cost by ~30 percent. Further, corrosion has caused trouble. In an experimental dry cooling tower installation near Manchester, England, the aluminum heat exchangers corroded and leaked so badly that the system was removed from service after only 2 years.

An effort to design a special finned tube tailored to suit the special requirements of dry cooling towers indicates that costs might be reduced somewhat. Depending on the amount of pumping power and the temperature differences between the air and cooling water considered acceptable, a design study has yielded an estimate of $100 to $136/kWe for the proposed type of extruded-machined aluminum finned-tube system in 1978 dollars.[55]

PROBLEMS

1. Using the heat balance diagram of Fig. 11.1, find the following:
   (a) The fraction of the steam condensed in the main condenser
   (b) The fraction of the steam condensed in the feed heaters
   (c) The overall efficiency of the turbine–feed pump unit, i.e., actual pump work/isentropic drop in the steam.

2. Make a rough estimate of the change in overall steam plant thermal efficiency in going from pulverized-coal firing to an atmospheric fluidized bed by simply assuming as a first approximation that the principal differences will be elimination of the coal pulverizers and the additional power for fans to overcome the static pressure drop through a 40-in-deep bed. (Fluidized density = 65 lb/ft³.)

3. Estimate the increase in the thermal efficiency for the Bull Run system obtainable by going to steam-turbine inlet temperatures of 621 and 677 °C (1150 and 1250 °F), respectively, assuming the curves of Fig. 11.19. Estimate the resulting percentage increases in the capital cost of the superheater and reheater associated with (a) increasing the wall thickness (assume a metal temperature 55 °C higher than the steam temperature) and (b) using Incoloy 800 in place of Type 304 stainless steel. (See tubing cost data in Fig. A10.4.)

4. Estimate the effect on the thermal efficiency of the Bull Run plant of using natural-draft cooling towers giving 32 °C (90 °F) condenser cooling water instead of river water at 15 °C (60 °F). (See last portion of Chap. 2.)

5. Estimate the capital cost in dollars per kilowatt of plant output for heat exchangers for cooling stack gas from 150 °C (300 °F) to 65 °C (150 °F) for scrubbing and then reheating to 121 °C (250 °F). (See Fig. A10.3.) Estimate the increase in the cost of electricity chargeable just to this heat exchanger. (See Chaps. 4 and 10.)

6. Estimate the amount of desuperheating water to be injected between the first and second stages of the superheater at 25 percent load for a 200,000 kg/h steam generator heated by a gas-cooled reactor. Assume that the inlet and exit gas temperatures for the reactor are to be held constant, irrespective of load, by varying the gas circulation rate so that the ratio of the heat added in the superheater to the heat added to the boiler remains constant. Assume also that the feed water is 30 °C below the boiling point, that no desuperheating is required at full load, and that the steam delivery pressure and temperature at full power are 180 bars and 550 °C (2610 psig and 1022 °F), whereas at 25 percent power they are to be 45 bars and 400 °C (650 psig and 752 °F).

REFERENCES

1. Orrok, G.A.: "James Watt, 1736–1819," *Mechanical Engineering,* vol. 58, 1936, pp. 75–80.

2. *Steam: Its Generation and Use,* 38th ed., The Babcock & Wilcox Company, New York, 1972.

3. Reynolds, H. B., et al.: "New Boiler Equipment at the Interborough Rapid Transit Co.'s Fifty-Ninth Street Power Station," *Trans. ASME,* vol. 48, 1926, p. 1369.

4. Greene, A. M., Jr.: "The Tale of Two City Stations: A Half Century of Progress in Steam-Power Generation," *Mechanical Engineering,* vol. 63, 1941, p. 109.

5. Keller, E. E., and F. Hodgkinson: "The Steam Turbine in the United States, I: Developments by the Westinghouse Machine Company," *Mechanical Engineering,* vol. 58, 1936, p. 683.

6. Christie, A. G.: "The Steam Turbine in the United States, II: Early Allis-Chalmers Steam Turbines," *Mechanical Engineering,* vol. 59, 1937, p. 71.

7. Robinson, E. L.: "The Steam Turbine in the United States, III: Developments by the General Electric Company," *Mechanical Engineering,* vol. 59, 1937, p. 239.

8. Kimball, D. S.: "The Century's Great Inventions," *Mechanical Engineering,* vol. 59, 1937, p. 507.

9. Blowney, W. E., and G. B. Warren: "The Increase in Thermal Efficiency Due to Resuperheating in Steam Turbines," *Trans. ASME,* vol. 46, 1924, p. 563.

10. Christie, A. G.: "Development and Performance of American Power Plants," *Mechanical Engineering,* vol. 58, 1936, p. 539.

11. Kerr, H. J.: "Once-Through Series Boiler for 1500 to 5000 lb. Pressure," *Trans. ASME,* RP-54-1a, vol. 54, 1932.

12. Gastpar, Jacques: "European Practice with Sulzer Monotube Steam Generators," *Trans. ASME,* vol. 75, 1953, p. 1345.

13. Rowland, W. H., and A. M. Frendberg: "First Commercial Supercritical-Pressure Steam Generator for Philo Plant," *Trans. ASME,* vol. 79, 1957, p. 409.

14. Dauber, C. A.: "Avon No. 8-A Supercritical-Pressure Plant," *Trans. ASME,* vol. 79, 1957, p. 727.

15. Campbell, C. B., C. C. Franck, Sr., and J. C. Spahr: "The Eddystone Super-Pressure Unit," *Trans. ASME,* vol. 79, 1957, p. 1431.

16. Wolfe, R. W.: *Energy Conversion Alternatives Study (ECAS), Westinghouse Phase I Final Report,* vol. XI: *Advanced Steam Systems,* NASA CR-134941, Feb. 12, 1976.

17. "The Bull Run Steam Plant," Tennessee Valley Authority Technical Report No. 38, 1967.

18. Elston, C. W.: "Design and Development Philosophy to Achieve High Reliability and Long Life in Large Turbine Generators," Paper No. AAS73-055, presented to the American Astronautical Society, September 1976.

19. Davis, C. M., et al: "Large Utility Boilers—Experience and Design Trends," *Proceedings of the American Power Conference,* vol. 38, 1976, p. 280.

20. Henke, W. G.: "The New 'Hot' Electrostatic Precipitation," *Combustion,* October 1970.

21. Bazelmans, C. L., et al.: "Study of Options for Control of Emissions from an Existing Coal-Fired Electric Power Station," Oak Ridge National Laboratory Report No. ORNL-TM-4298, September 1973.

22. Burchard, J. K., et al.: "Some General Economic Considerations of Flue Gas Scrubbing for Utilities," *Proceedings of Conference on Sulfur in Utility Fuels: The Growing Dilemma, Drake Hotel,* Chicago, Oct. 25–26, 1972.

23. Elliot, M. A.: *Chemistry of Coal Utilization,* 2d suppl. vol., Wiley, New York, 1981.

24. "Multi-Stream Coal Cleaning System Promises Help with Sulfur Problem," *Coal Age,* vol. 81, January 1976, pp. 86–88.

25. Anson, D.: "Fluidized Bed Combustion of Coal for Power Generation," *Progress in Energy and Combustion Science,* vol. 2, 1976, pp. 61–82.

26. Squires, A. M.: "Applications of Fluidized Beds in Coal Technology," *Alternative Energy Sources,* Academic Press, Inc., New York, 1976, chap. 4, p. 49.

27. Stringfellow, T. E., and J. G. Branam: "Start-Up and Initial Operation of the Rivesville 30 MW Fluid Bed Boiler," *Proceedings of Fifth International Conference on Fluidized Bed Combustion,* Washington, D.C., Dec. 12–14, 1977.

28. Spencer, D. F., O. D. Guildersleeve, and R. A. Loth: "Initial Comparative Analysis of the Market Penetration Potential of Coal and Coal Derived Fuels in the United States Utility Industry (1985–2005)," presented at the IEEE PES Summer Meeting, Mexico City, Mex., July 17–22, 1977.

29. Brown, D. H., et al.: *Energy Conversion Alternatives Study (ECAS), General Electric Phase II Final Report,* vol. II: *Advanced Energy Conversion Systems—Conceptual Designs,* part 2, "Closed Cycles," NASA-CR 134949, SRD-76-064-2, December 1976.

30. Becker, T. W.: "Application of Atmospheric Fluidized Bed Combustion for Electric Power Generation," *Proceedings of Fifth International Conference on Fluidized Bed Combustion,* Washington, D.C., Dec. 12–14, 1977.

31. Corell, R. B.: "Conceptual Design of a 570 MW Combustion Engineering, Inc. Atmospheric Fluidized Bed Steam Generator," *Proceedings of Fifth International Conference on Fluidized Bed Combustion,* Washington, D.C., Dec. 12–14, 1977.

32. Reed, K. A., and R. L. Gamble: "Conceptual Design of a 570 MW Foster-Wheeler Energy Corp. Atmospheric Fluidized Bed Steam Generator," *Proceedings of Fifth International Conference on Fluidized Bed Combustion,* Washington, D.C., Dec. 12–14, 1977.

33. Fraas, A. P., G. Samuels, and M. E. Lackey: "A New Approach to Fluidized Bed Steam Boiler," ASME Paper No. 76-WA/Pwr-8, December 1976.

34. Farmer, M.: "Application of Fluidized Bed Combustion Technology to Industrial Boilers," paper presented at the Fluidized Bed Combustion Technology Exchange Workshop, McLean, Va., Apr. 13–15, 1977.

35. Webb, R.: "Natural Versus Forced Circulation in Fluidized Bed Combustion," paper presented at the Fluidized Bed Combustion Technology Exchange Workshop, McLean, Va., Apr. 13–15, 1977.

36. Miller, W.: "Fluidized Bed Boiler at Georgetown University," paper presented at the Fluidized Bed Combustion Technology Exchange Workshop, McLean, Va., Apr. 13–15, 1977.

37. Iredale, A. J. F., and N. P. Grimm: "Ice Condenser Reactor System Containment," *Nuclear Engineering International,* vol. 16, October 1971, pp. 864–867.

38. Walker, R. E., and T. A. Johnston: "Fort Saint Vrain Nuclear Power Station," *Nuclear Engineering International,* vol. 14, December 1969, pp. 1069–1073.

39. "Construction of the World's First Full-Scale Fast Breeder Reactor," *Nuclear Engineering International,* vol. 23, June 1978, pp. 43–60.

40. Crowley, J. H.: "Power Plant Cost Estimates Put to the Test," *Nuclear Engineering International,* July 1978, pp. 39–43.

41. Leung, P., and K. A. Gulbrand: "Power System Economics: An Evaluation of Plant Auxiliary System Incremental Kilowatt Consumption," *J. Eng. Power,* July 1977, pp. 419–423.

42. Wood, R. A.: "Status of Titanium Blading for Low Pressure Steam Turbines," EPRI Report No. AF-445, February 1977.

43. Rush, R. E., and A. V. Slack: *Status Report on Flue Gas Desulfurization at Coal-Fired Power Plants,* Utility Air Regulatory Group, Jan. 15, 1979.

44. "Shifting SO_2 from the Stack," *EPRI Journal,* July/August 1979, pp. 15–19.

45. *Interagency Flue Gas Desulfurization Evaluation,* vol. 1, first draft report, Nov. 30, 1977.

46. Klett, M. G.: *Typical Costs for Electric Energy Generation and Environment Controls,* Interagency Energy/Environmental Protection Agency Report No. EPA-600/7-79-026, January 1979.

47. Bloom, S. G., et al.: *Analysis of Variations in Costs of FGD Systems,* Electric Power Research Institute Report No. FP-909, October 1978.

48. "Symposium on Health Effects of Sulfur Oxides and Related Particulates," *Bulletin of the New York Academy of Medicine,* vol. 54, no. 11, December 1978.

49. Fraas, A. P., et al.: "Assessment of the State of the Art of Pressurized Fluidized Bed Combustion Systems," Oak Ridge National Laboratory Report No. ORNL/TM-6633, June 1979.

50. Reed, K. A., and G. G. Cervenka: "Conceptual Design of a Foster Wheeler Energy Corporation Atmospheric Fluidized Bed Steam Generator," *Proceedings of Fifth International Conference on Fluidized Bed Combustion,* Dec. 12–14, 1977, vol. II, MITRE Corp., pp. 285–310.

51. Becker, R. W.: "The Application of Atmospheric Fluidized Bed Combustion for Electric Power Generation," *Proceedings of Fifth International Conference on Fluidized Bed Combustion*, Dec. 12–14, 1977, vol. II, MITRE Corp., pp. 267–282.

52. Covell, R.B.: "Conceptual Design of a 570-MW Combustion Engineering, Inc. Atmospheric Fluidized Bed Steam Generator," *Proceedings of Fifth International Conference on Fluidized Bed Combustion,* Dec. 12–14, 1977, vol. II, MITRE Corp., pp. 326–340.

53. Lackey, M. E.: "Design and Performance Testing of a Coal Feed and Metering System for the MIUS Fluidized Bed Combustor," Oak Ridge National Laboratory Report No. ORNL/HUD/MIUS-47, December 1977.

54. Reynolds, J. Z.: "Power Plant Cooling Systems: Policy Alternatives," *Science,* vol. 207, no. 4429, Jan. 25, 1980, pp. 367–372.

55. Haberski, R. J., and J. C. Bentz: *Conceptual Design and Cost Evaluation of a High Performance Dry Cooling System,* U.S. Department of Energy Report No. COO-4218-1, Curtiss-Wright Corp. Report CWC-WR-78-001, Mar. 1, 1978.

12

OPEN-CYCLE
GAS TURBINES

By the latter 1800s it was recognized by quite a number of engineers both in Europe and the United States that a gas turbine offered the possibility of a simpler, lighter, smaller, and less expensive power plant than the Rankine cycle because the turbine could be operated directly on the products of combustion, thus eliminating the boiler. Further, the enormous gas-handling capacity of the gas turbine gives it a major advantage over reciprocating internal combustion engines. Thus efforts to design and build gas turbines began 100 years ago concurrently with similar efforts on steam turbines.[1-5] By 1904 a gas turbine developed by Armengaud and Lemale in France became the first gas turbine to produce a useful net output.[4] Significantly, Brown Boveri and Co. built a 25-stage centrifugal compressor in 1906 for a later Armengaud-Lemale machine that produced 75 hp with a thermal efficiency of ~2 percent, and subsequently many units of this design of compressor were built, making it the world's first commercial centrifugal compressor.[5] Although its efficiency was surprisingly good, ~70 percent, it was still not sufficient to yield an attractive gas turbine. Because of this, the Armengaud-Lemale and other early experiments with gas turbines—unlike those with steam turbines—gave such disappointingly low power plant thermal efficiencies that no commercial gas turbines were built until over 30 years later.

As pointed out in Chap. 2, the Brayton cycle is basically a more difficult cycle with which to work than the Rankine cycle because of a fundamental difference between them: The work required to drive the feed pump in the Rankine cycle is only ~1 percent of the turbine output, whereas for a gas turbine even in an ideal cycle the work required to drive the compressor is around half of the work output of the turbine. This effect can be envisioned readily by examining the pressure-volume (P-V) diagrams in Fig. 12.1, remembering that the ideal work input for the compression process and the ideal work output from the expansion process are directly proportional to the area enclosed by their respective P-V diagrams. Aerodynamic losses in the

compressor and turbine represent a large fraction of the ideal net work, as do the pressure losses between the compressor and the turbine. As a consequence, high aerodynamic efficiencies in the compressor and turbine and a high turbine inlet temperature are essential, for otherwise little or no useful net work will be obtained. A high turbine inlet temperature (over 650°C) is required because the greater the amount of heat added between the compressor and turbine, the greater the work output obtainable from the turbine and the less sensitive the cycle becomes to aerodynamic losses in the compressor and turbine and in the ductwork and burner between them.

Developments in aerodynamics in the late 1920s on airfoil theory for wings and propellers, coupled with the application of vortex theory to the design of wind-tunnel fans (by T. H. Troller—first at Aachen, Germany, and then at the Daniel Guggenheim Airship Institute in Akron, Ohio), made it possible to obtain axial-flow compressor efficiencies in excess of 80 percent by 1932.[6] Further, metallurgical developments in iron-chrome-nickel alloys in the 1920s produced stainless steels and nickel alloys capable of operating at 650°C (1200°F) or more, and this made it possible to build simple gas turbines for a number of important applications.

COURSE OF DEVELOPMENT
Exhaust Turbosuperchargers

The first practical application of the gas turbine was for supercharging diesel engines for locomotive service and gasoline engines for aircraft.[2,3] In 1905, after working on one of the early gas turbines, A. J. Buchi recognized the difficulties posed by the high gas temperatures required to get a good system efficiency. As a means of circumventing these problems, he proposed that a gas turbine could perform a useful function using the energy in the exhaust of a diesel engine to provide the power to supercharge the diesel and thus increase both its output and its efficiency. He finally interested the Sulzer Company in Switzerland and in 1911 began the development of an engine-exhaust-gas-driven gas turbine coupled to a centrifugal compressor for supercharging diesel engines.[2] In the United States about the same time S. A. Moss saw the potential for the concept in the aircraft field, and in 1918 he operated a Liberty aircraft engine with a turbosupercharger.[3] The excellent potential of this approach for improving the high-altitude performance of aircraft engines led to the design, construction, and installation of an improved version of exhaust turbosupercharger on a Curtiss D-12 engine in a Curtiss Hawk fighter in 1927.[7] Further aerodynamic and metallurgical developments, coupled with the great incentive for major improvements in high-altitude engine performance, led to the initiation in 1938 of substantial programs at the General Electric Company in Schenectady and at the Wright Aeronautical Corporation in New Jersey (where the author worked on the installation of the first unit). These programs were aimed at testing a wide variety of iron-chrome-nickel-cobalt alloys suitable for operating at turbine inlet temperatures of 815 to 1100°C (1500 to 2000°F). Much of this work was directed toward providing some air cooling of the blades to reduce the metal temperature well below the turbine inlet gas temperature. The development programs were successful, and the resulting exhaust turbosuperchargers were widely used in high-altitude U.S. aircraft in the latter part of World War II.

This work led to improved versions that would yield a net useful power output so that they could be coupled into the engine through a gearbox and thus not only increase the engine output but also reduce its fuel consumption.[8] This type of unit—in effect a bottoming cycle—was fairly widely used in transport aircraft such as the DC-7 and the Super Constellation.

Velox Boilers

The development of high-efficiency axial-flow fans for wind tunnels and iron-chrome-nickel alloys suitable for high-temperature turbines, coupled with the successful application of turbosuperchargers to internal combustion engines, led to the development of the first combined gas-turbine–steam cycle, the Velox boiler. In 1934 the combustion products from the first Velox boiler were expanded through a gas turbine to drive an axial-flow compressor that supercharged the oil-fired furnace for a steam boiler.[9,10] This served to reduce the amount of heat-transfer surface area required in the furnace, eliminated the parasitic electric load for driving induced- and forced-draft fans, and in some subsequent cases actually yielded a small

amount of net power output from the gas turbine which was used to drive an electric generator, thus increasing the overall power plant thermal efficiency. These advantages were partly offset by the additional costs of a pressure-tight shell for the furnace and limitations on the ash content of the fuel oil to ~550 ppm to avoid difficulties with the gas turbine. Because of its higher ash content, coal was not employed. One of the most successful applications of this concept has been to a set of boilers used in 17 U.S. destroyers built in the 1960s, giving excellent service at the time of writing with furnaces pressurized to about 5 atm. These boilers are about half the size and half the weight of conventional marine boilers for that application, cost less, and have given better service.[11] This concept is discussed further in the next chapter with particular reference to its application to a fluidized-bed coal combustion system.[12]

Gas Turbines for Catalytic Cracking Units

The success of the gas turbines built for the Velox boiler led to their application to Houdry catalytic cracking units for the pro-

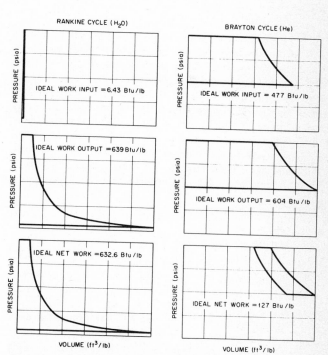

Figure 12.1 Pressure-volume diagrams for typical ideal thermodynamic cycles for steam and gas turbines.

duction of high-octane gasoline in petroleum refineries. The first unit went into operation at the Marcus Hook, Pennsylvania, plant of the the Sun Oil Company in 1936.[13] In this application, the air from the compressor is fed to a bed of catalyst particles when the effectiveness of the catalyst has been reduced by coking. The carbon is burned out of the catalyst, thus heating the air and yielding hot combustion gases that are fed to the turbine. A small net useful electric power output is obtained. This process was so eminently successful for providing high-octane gasoline that it soon became widely used and made possible the huge U.S. output of 100-octane gasoline in World War II.

First Industrial Gas Turbine

The fine performance of the gas-turbine units built for the Velox boilers led Brown-Boveri to construct a 4000-kWe gas-turbine–generator unit that was tested extensively in Neuchatel, Switzerland, in 1939.[14] This unit showed a compressor efficiency of 84.6 percent and a turbine efficiency of 88.4 percent, giving an overall cycle thermal efficiency of 18 percent, with a turbine inlet temperature of only 552°C (1025°F).

Locomotive Gas Turbines

The outstanding success of the Brown-Boveri gas turbine in 1940 led to the construction of a similar turbine for use in the Swiss railway system in 1942. In 1944, this in turn led to the launching of a substantial program to develop a coal-fired gas turbine for locomotive service in the United States.[15] The program was finally dropped in the early 1960s after nearly 20 years of unsuccessful efforts to resolve the difficulties with turbine bucket erosion and deposits caused by fine ash particles from the combustion process that could not be removed satisfactorily by multistage cyclone separators.[16] These special problems of coal-burning, open-cycle, gas turbines are treated later in this chapter.

Closed-Cycle Gas Turbines

Ackeret and Keller[17] invented the closed-cycle gas turbine and built a demonstration unit that produced electric power beginning in 1939. This system has the advantage that dirty fuels such as coal can be used because the products of combustion do not go through the turbine; their sensible heat is transferred through a heat exchanger to the gaseous working fluid (usually air or helium) of the closed-cycle gas turbine. The closed cycle has the further advantage of giving a high efficiency irrespective of the load. The efficiency of an open-cycle unit, on the other hand, falls off rapidly as the load is reduced, partly because of the reduction in the turbine inlet temperature and partly because of a progressively greater aerodynamic mismatch between the compressor and turbine. (The latter effect can be largely corrected by employing variable-pitch stator vanes, but this increases the complexity, cost, and maintenance of the engine.) The special problems of closed-cycle gas turbines are treated in Chap. 4.

Turbojet Engines

In the middle 1930s, the possibilities of applying the gas turbine to the jet propulsion of aircraft were recognized in both England and Germany, and programs were launched to develop jet engines. Interestingly enough, in both programs there was little difficulty in developing the compressor and turbine but great difficulty in developing compact combustion chambers having a good operating life and not subject to flameout (blowing out of the flame in the burner). The success of these efforts led to major engine development programs in the United States during and following World War II, the cost of which totaled 10×10^9 in the 1940–1980 period.

Commercial Gas Turbines

The success of the jet engine developments in the 1940s and 1950s led to the widespread application of the resulting jet engine technology to gas turbines for industrial power plants and for use as peaking-power units in central stations. By the latter 1970s, some 50,000 MW of gas-turbine capacity was installed in electric utility plants, mostly for peaking-power service, and ~ 16,000 MW of equivalent power output in industrial plants and pipeline pumping stations. The thermal efficiency of these units commonly runs from 24 to 28 percent at full load. Table 12.1 gives performance data for some typical gas turbines employed in commercial service in the 1934–1965 period, during which most of the development for commercial applications took place.

By 1980 the severe shortages of natural gas and fuel oil that developed following the Yom Kippur War had cut the amount of clean fuels available for industrial and utility gas turbines to less than half of that in the early 1970s. Both fuel costs and governmental pressure have acted to force industry and the public utilities to phase out their use of oil and gas and shift to coal or nuclear fuel. This caused a big drop in orders for new gas turbines; as indicated in Fig. 12.2, the power capacity of new units installed in 1978 was only about 10 percent of the peak

TABLE 12.1 Characteristics of Typical Industrial and Electric Utility Gas Turbine Units Installed and Operated for Extended Periods during the 1934–1965 Period†

Location or application	Cycle type*	Start of operation	Turbine inlet temperature, °F	Turbine adiabatic efficiency, %	Compressor adiabatic efficiency, %	Regenerator effectiveness, %	Cycle thermal efficiency, %
Velox steam generator		1934	900	83	73		
Marcus Hook, Pa. (Houdry reactor)		1936	950	85	79		
Zurich, Switzerland	5	1939	1268	89	83	90	12
Neuchatel, Switzerland	1	1940	1025	88.4	84.6		31.6
Swiss Federal Railways	2	1941	1110	86	86	50	18
Beznan, Switzerland	4	1948					
Lester, Pennsylvania		1948			87.5		30.5
G.E. pilot model locomotive	1	1948	1290		82.8		
Chimbote, Peru	1	1949					16
Alexandria, Egypt	2	1949					19.5
Pertizalete, Venezuela	2	1949	1112				22.9
Lima, Peru	4	1949					25
Beznan, Switzerland	4	1949	1200				28
Westinghouse electric test locomotive	1	1949	1350	84	82.5		26.5
British Railways	2	1949	1112	86	86	42	16.9
Baracalldo, Spain	2	1951					15.8
Dudelange, Luxemburg	2	1951					25
Filaret, Rumania	3	1951					21.5
Oklahoma City, Oklahoma	6	1951	1340				23.5
British Railways	1	1952	1300				18
Allis-Chalmers Loco. Dev. Division	2	1952	1300	87.7	82.8	52.7	21
Union Pacific Railways	1	1952	1290		82.8		16
St. Dizier, France	1	1956	1380				19.2
Ravensburg, Germany	5	1956	1220				25
Duisburg, Germany	1	1957	1380				
Otterbacken, Sweden	3	1957	1160				
Toyomi, Japan	5	1957	1220				26
Oberhausen, Germany	5	1960	1328				26
Coburg, Germany	5	1961	1256				29.5
H.M.S. Ark Royal	1	1961	1390				28
Nippon Kokan, Japan	5	1961	1256				16.4
Kashira, Russia	5	1962	1256				29
Haus Aden, Germany	5	1963	1256				28
Orenda test engine	2	1965		85	84.5	85.5	29.5

*Cycle types: (1) Open —without intercooling or regeneration
(2) Open —without intercooling and with regeneration
(3) Open —with intercooling and without regeneration
(4) Open —with intercooling and regeneration
(5) Closed —with intercooling and regeneration
(6) Gas turbine–steam turbine combined cycle

†From M. E. Lackey, "Second Iteration Analysis of a Fossil Fuel-Fired Gas Turbine-Potassium-Steam Combined Cycle," Oak Ridge National Laboratory Report No. ORNL-NSF-EP-39, July 1973.

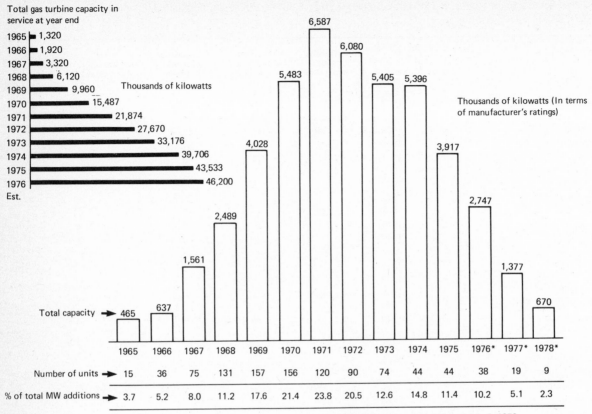

Total gas turbine capacity in service at year end

Year	Value
1965	1,320
1966	1,920
1967	3,320
1968	6,120
1969	9,960
1970	15,487
1971	21,874
1972	27,670
1973	33,176
1974	39,706
1975	43,533
1976 Est.	46,200

Thousands of kilowatts

Thousands of kilowatts (In terms of manufacturer's ratings)

	1965	1966	1967	1968	1969	1970	1971	1972	1973	1974	1975	1976*	1977*	1978*
Total capacity	465	637	1,561	2,489	4,028	5,483	6,587	6,080	5,405	5,396	3,917	2,747	1,377	670
Number of units	15	36	75	131	157	156	120	90	74	44	44	38	19	9
% of total MW additions	3.7	5.2	8.0	11.2	17.6	21.4	23.8	20.5	12.6	14.8	11.4	10.2	5.1	2.3

*Installed and/or scheduled as of July 1, 1976

Source: Based on F.P.C.; E.E.I. Data; Estimate by Ebasco.

Figure 12.2 Gas-turbine capacity placed in service in the total electric utility industry. (*1976 Business and Economic Charts, Ebasco Services, Inc.*)

value reached in 1971. Thus the role of the gas turbine in U.S. electric utility service in the 1980–2000 period will depend on developments designed to increase the thermal efficiency, permit operation on low-cost fuels, or take advantage of special situations such as attend the use of cogeneration (Chap. 17).

PRINCIPAL PROBLEMS

The principal objectives in the development of gas turbines for industrial and utility service during the past 40 years have been to obtain a thermal efficiency, service reliability, and system life competitive with steam systems. The high degree of refinement in the aerodynamic, detail design, and fabrication techniques for turbojet engines have made it possible to produce relatively

lightweight, compact, low-capital-cost gas turbines for industrial and central station service. However, their thermal efficiency is low and the fuel costs are high as compared to those for steam plants. The key problems involved in making gas turbines more efficient and longer lived are reviewed in this section from the standpoints of thermodynamics, aerodynamics, materials, and engine design.

The progress in the technology affecting gas-turbine performance, particularly the thermal efficiency, is indicated in Fig. 12.3. Data for five key performance parameters as obtained from operating experience with commercial gas turbines are plotted in Fig. 12.3 as functions of the year of initial operation. The basic problems involved in each of these are discussed in the following sections.

Thermodynamic Considerations

The thermal efficiency of a simple gas-turbine cycle designed for maximum work is shown in Fig. 12.4 as a function of the cycle temperature ratio for a cycle having component efficiencies of 90 percent—about as high as can be achieved in view of inherent losses (Chap. 3). This cycle efficiency is compared with that of the ideal (Brayton) gas-turbine cycle designed for maximum work ($\eta = 1.00$) and with that of the theoretical Carnot cycle. Here the large difference in efficiency of the cycle caused by changing either the peak temperature or the component efficiency is readily apparent. Note that the peak practicable tur-bine inlet temperature of ~1100°C (2000°F), coupled with a typical atmospheric air temperature, gives an overall cycle temperature ratio of between 4 and 5. Note also that Fig. 12.4 does not include the effects of pressure losses between the com-pressor and the turbine; these commonly run about 5 percent, largely in the combustion chamber.

It is evident from Fig. 12.4 that the two principal parameters affecting the cycle efficiency are the peak turbine inlet temperature permissible and the component efficiencies. In ad-dition, it is possible to increase the cycle efficiency by employing a regenerative heat exchanger that removes heat from the tur-bine exhaust gas and transfers it to the air leaving the com-

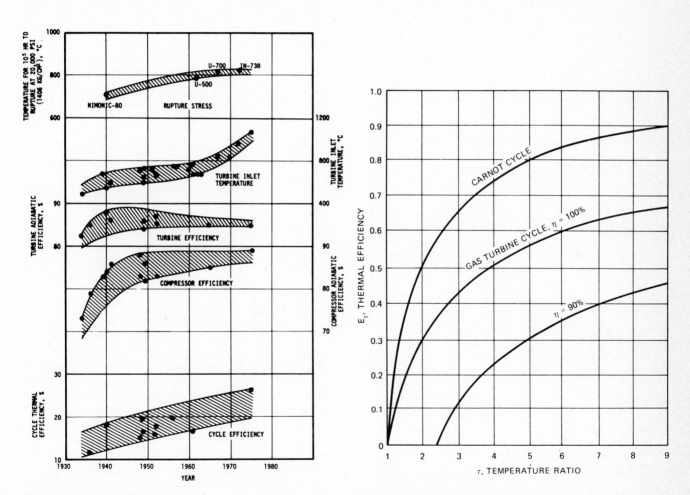

Figure 12.3 Operating charateristics of gas-turbine power systems installed for continuous operation as a function of the year of initial operation. (*Courtesy Oak Ridge National Laboratory.*)

Figure 12.4 Ideal thermal efficiency of gas turbines for com-pressor and turbine efficiencies, η, of 90 and 100 percent. (*Courtesy Oak Ridge National Laboratory.*)

pressor en route to the combustion chamber. It is also possible to reduce the compression work through the use of intercoolers between stages in the compressor. The advantages of these measures were recognized in the 1930s, and the second utility-type gas turbine, which was tested in Switzerland in 1939 and 1940, incorporated both intercoolers and regenerators.[17] However, the cost of the heat exchangers required is high—much more than the cost of the basic gas turbine—and hence this equipment has been used in relatively few gas-turbine installations. Marked increases in fuel costs in recent years are likely to favor the increased utilization of regenerators and intercoolers. Figure 12.5 shows the effect on the thermal efficiency for some typical utility installations of both simple open-cycle gas turbines and open cycles with regeneration to indicate the advantages associated with the use of regenerators. Figure 12.5 indicates that in practice there is little or no advantage to

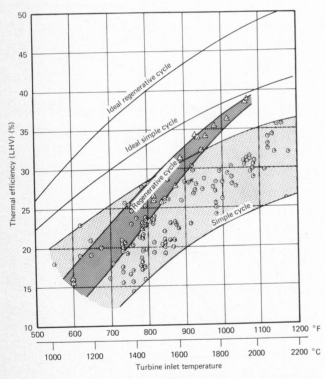

Figure 12.5 Reported turbine inlet temperature and thermal efficiency for units installed during the 1950–1976 period compared to the ideal efficiency of a simple open and a regenerative cycle with a compressor efficiency of 88 percent and a turbine efficiency of 90 percent. (*Courtesy Oak Ridge National Laboratory.*)

the use of a regenerator for open-cycle gas turbines having inlet temperatures below about 900°C. This stems from the fact that the extra pressure losses in the ductwork and regenerator heat-transfer matrix largely offset the thermal gain unless large and expensive ducting and heat exchangers are employed or unless the temperature difference available is large.

Aerodynamics

The initial British effort on turbojet engines employed centrifugal compressors, whereas the initial German effort employed axial-flow compressors. The centrifugal units are easier to design and fabricate, and in small sizes their efficiency approaches that of axial-flow units. However, in larger sizes the axial-flow compressors can be made more efficient than centrifugal compressors, their size can be much reduced, and hence their cost is less. As a consequence, all the gas turbines designed for shaft power ouputs of more than about 2000 kWe employ axial-flow compressors. The pronounced effect of size was discussed in Chap. 3 and is indicated in Fig. 3.7, which shows the efficiency for a series of axial-flow compressors as a function of Reynolds number.

It can be seen in Fig. 12.3 that improvements in turbine and compressor efficiency have been small in the past 40 years as have increases in the creep strength of the oxidation-resistant alloys suitable for gas-turbine buckets. Little further improvement in these parameters can be expected. As discussed in Chap. 3, the aerodynamic losses in the compressor and turbine stem mainly from friction in the boundary layer of the high-velocity gas stream and from the inevitable losses associated with the wakes downstream of cascades of blades. These wakes lead to deviations from ideality in the gas flow through subsequent stages of rotor and stator blades.

As also discussed in Chap. 3, the pressure rise per stage in a compressor and the work output per stage in a turbine are directly proportional to the square of the tip speed. Thus, there is a strong incentive to run machines at as high a tip speed as possible within the limitations imposed by either the stresses from centrifugal force or the aerodynamic losses stemming from shock waves, which rapidly become more serious as the tip speed approaches the velocity of sound. In a good design of a practical gas turbine operating on air (but not helium), the tip speed limitation imposed by centrifugal stress considerations in the turbine is about the same as that imposed by sonic velocity considerations. In the compressor, where the temperature is lower and allowable stresses are higher, compressibility effects are limiting. The sonic velocity in helium is 2.6 times that in air, hence rotor stress is always limiting in helium machines.

Throughout the history of the development of the gas tur-

TABLE 12.2 Chemical Composition of Alloys Used in Gas Turbines

Alloy	Cr	Ni	Co	Fe	W	Mo	Ti	Al	Cb	V	C	B	Ta	Mn	Si	Na
Composition, %																
Bucket alloys tested in turbosuperchargers 1938–1943 (Data courtesy of R. Cole, Curtiss-Wright Corp.)																
S-495 ht. 22792	13.87	19.89			3.54	4.40			3.51		0.41			0.53	0.47	
N-155 low carbon ht. 11534	21.50	18.99	17.60		2.08	2.90			1.06		0.15			1.68	0.31	0.06
N-155 high carbon ht. 11535	21.12	18.97	17.70		2.17	2.90			1.12		0.31			1.58	0.33	0.07
19-9 DL BIO658	19.34	8.51			2.10	1.28	0.22		0.11		0.29			0.89	0.54	
K42B KB476	16.78	42.68	21.23	15.04			2.70				0.05			1.16	0.36	
S-590 ht. 42883	17.80	19.64	18.37		3.94	3.82			2.75		0.43			0.69	0.09	
16-25-6 ht. 11542	14.98	25.40				6.00					0.08			1.79	0.58	0.15
Refractory ht. M406	20.64	20.71	31.41	12.95	4.46	8.40					0.15			1.21	0.29	
Cast 61 alloy	24.66		67.73		5.18						0.43					
Cast vitallium	28.28		63.84			5.63					0.25					
Cast 6059 alloy	24.61	33.15	33.05			5.34					0.39					
Cast 422-19 alloy	27.56	15.51	47.61			6.59					0.51					
Nimonic 80	20.00	73.00		4.00			2.30				0.05					
Alloys used in gas turbines for utilities in 1975 [18]																
Buckets																
S816	20.00	20.00	BAL	4.00	4.00	4.00			4.00		0.40					
NIM 80A	20.00	BAL		4.00			2.30	1.00			0.05					
M-252	19.00	BAL	10.00	2.00		10.00	2.50	0.75			0.10					
U-500	18.50	BAL	18.50			4.00	3.00	3.00			0.07	0.006				
RENE 77	15.00	BAL	17.00			5.30	3.35	4.25			0.07	0.02				
IN 738	16.00	BAL	8.30	0.20	2.60	1.75	3.40	3.40	0.90		0.11	0.01	1.75			
CTD 111	14.00	BAL	9.50		4.00	1.50	3.00	5.00			0.11	0.01	3.00			
Partitions-1st																
X-40	25.00	10.00	BAL	1.00	8.00						0.50	0.01				
X-45	25.00	10.00	BAL	1.00	8.00						0.25	0.01				
FSX-414	29.00	10.00	BAL	1.00	7.00						0.25	0.01				
N-155	21.00	20.00	20.00	BAL	2.50	3.00					0.20					
Turbine wheels																
Cr-Mo-V	1.00	0.50		BAL		1.25				0.25	0.30					
A-286	15.00	25.00		BAL		1.20	2.00	0.30		0.25	0.08	0.006				
M-152	12.00	2.50		BAL		1.70				0.30	0.12					
IN 706	16.00	41.00		BAL			1.70	0.40	3.00		0.06	0.006				

bine, a considerable effort has gone into the improvement of component efficiencies, particularly the compressor. As discussed in the section "Surge Limit" near the end of Chap. 3, there is an instability inherent in the dynamic compression process in axial-flow or centrifugal compressors. It is caused by the fact that the static pressure increases in the direction of flow except at the walls, where it decreases because of skin friction, and this tends to cause flow separation, eddies, and aerodynamic "stall," or backflow. Thus the design of an efficient compressor is more difficult than the design of an efficient turbine where the static pressure continues to drop throughout the length of the channel in the direction of the flow through the turbine.

Materials Limitations

From the beginning, the development of high-temperature materials has played a major role in the gas-turbine program. A wide range of iron-chrome-nickel-cobalt alloys was investigated in the development of exhaust turbosuperchargers for aircraft applications in the latter 1930s and early 1940s. The upper portion of Table 12.2 lists some of these alloys and gives their composition. The lower portion of Table 12.2 summarizes the materials commonly used in production gas turbines at the time of writing.[18] Note the similarity in the compositions even after 40 years and a billion dollars in metallurgical development. Note also at the top of Fig. 12.3 that permissible metal temperatures increased only about 100°C (180°F) from 1940 to 1980. Extensive research on the properties of materials indicates that all metals begin to soften so much that they sinter at a temperature about 75 percent of the absolute temperature at which they melt. The melting points of nickel and cobalt, the metals that give the best bases for high-strength, oxidation-resistant alloys, lie in the 1450 to 1500°C (2600 to 2700°F) range. The melting points of their alloys are lower (much lower for eutectic compositions) and fall around 1370°C (2500°F). The sintering point of these alloys is about 1000°C (1832°F). Thus, there seems to be little hope of getting further improvements in the permissible operating temperature of oxidation-resistant alloys suitable for combustion gas turbines. Ceramics have been considered for the past 40 years but their inherent brittleness presents formidable thermal-stress problems that will be discussed in a later section.

Progressively more elaborate provisions for air cooling the turbine blades have been a far more important factor in making possible increases in turbine inlet temperature than the development of alloys having greater strength at high temperatures. Actually, major metallurgical developments have been involved in making possible the fabrication of nickel- and cobalt-base alloys into blades having complex internal cooling passages. These blades utilize the cooling airstreams as effectively as possible and thus minimize the pumping losses associated with diverting a portion of the airflow from the compressor to cool them. The major incentive to develop them has been the increase in power output thus obtainable from a given size and weight of jet engine—an even more important consideration than fuel economy for military aircraft service. The scatter band of data in Fig. 12.3 indicates that the point of diminishing returns is being approached in efforts to increase the thermal efficiency of commercial gas turbines through the use of further improvements in air cooling, although further increases in turbine inlet temperature and, consequently, in the output from a given size of jet engine are obtainable.

The alloy compositions of Table 12.2 are evidence of the major effort that has been made in the 1960s to minimize the cobalt content because of supply limitations and costs. In addition, alloys have been favored that are suitable for fabrication in complex blade forms for blades with internal cooling passages.

The design of a gas-turbine blade involves consideration of a combined stress situation that includes creep caused by tensile and bending stresses and fatigue caused by the rapidly fluctuating aerodynamic loading of rotor blades as they pass through the wakes of stator blades, as well as the effects of stress concentrations and corrosion. A correlation of the expected operating lifetime of such a blade with the metal temperature and the imposed level of applied stress is shown in Fig. 12.6 for a typical set of design conditions. (See Chap. 6 for further background.) It can be seen that the operating temperature and, therefore, the thermal efficiency of the engine depend on the desired operating lifetime of the turbine blade.

Combustion Chamber

Gas-turbine combustion chambers are designed so that the liquid or gaseous fuel is burned in a central region in which the fuel-air mixture is somewhat richer than stoichiometric to give rapid combustion and minimize the tendency of the flame to blow out (see Fig. 5.11), but high flame temperatures are produced. This high-temperature central gas stream is then tempered by the admixture of 200 to 300 percent of excess air to give the desired turbine inlet temperature. The combustion chamber is carefully designed so that the excess air cools the combustion chamber walls and blankets them with a film of relatively cool air, keeping the walls far below the flame temperature[19] (Fig. 12.7). However, to keep the combustion chamber compact, high turbulence levels are required, and this turbulence leads to combustion reaction and heat release rates vastly higher than those in conventional boiler furnaces (Chap. 5). It also causes some elements of the combustion chamber wall to fluctuate in

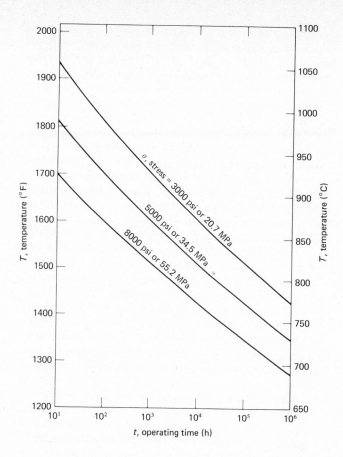

Figure 12.6 *(Left)* Effect of the design operating life on the allowable temperature for representative allowable stresses for a typical gas-turbine bucket alloy. *(Courtesy Oak Ridge National Laboratory.)*

Figure 12.7 *(Below)* Section through the combustion chamber liner for the P&W-Stal Laval FT-51 gas turbine showing the louvers used for cooling the liner and the axial temperature distribution along the liner for a turbine inlet temperature of 1204 °C (2200 °F).[34]

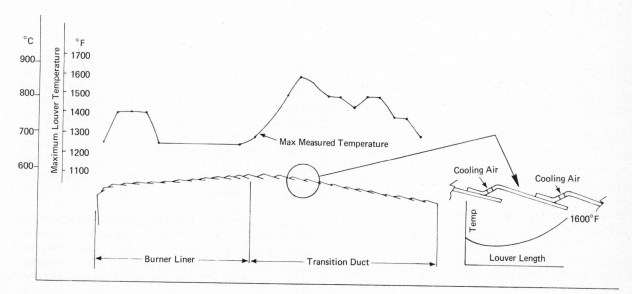

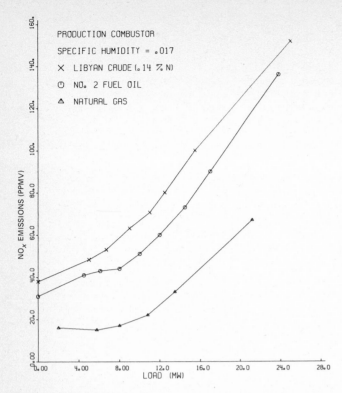

Figure 12.8 The effect of fuel (Libyan crude, No. 2 fuel oil, and natural gas) on NO$_x$ emissions from a 25-MW gas turbine.[20]

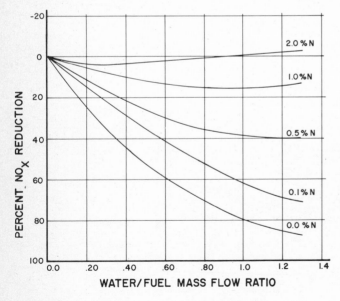

Figure 12.9 Predicted decrease in the reduction of NO$_x$ emissions through water injection for a wide range of bound nitrogen contents in fuel oil.[20]

temperature, and this in turn leads to severe thermal stresses and eventually to failures from thermal-strain cycling. This has proved to be one of the more difficult problems in developing gas turbines. Any design change that increases either the air inlet temperature or the combustion chamber exit temperature increases the difficulty in getting good life from the combustion chamber.

From the standpoint of utility service, one of the most important characteristics of a combustor is the amount of NO$_x$ emitted in the engine exhaust. As discussed in general terms in Chap. 5, the amount depends on the temperature of combustion and the residence time in the hot zone, together with the fuel-air ratio (Fig. 5.7), and these factors depend on both the combustion chamber design and the load. The fuel used is also a factor, particularly its nitrogen content. A typical set of curves showing the effects of both load and the type of fuel are shown in Fig. 12.8, the NO$_x$ emission rate increasing with load (i.e., combustion gas temperature) and with the fuel nitrogen compound content, as one would expect.[20] The tendency toward NO$_x$ formation can be reduced by water injection into the inlet airstream (which reduces the flame temperature), but, as can be seen in Fig. 12.9, it does little good in reducing the NO$_x$ produced from nitrogen compounds in the fuel. A substantial effort is underway at the time of writing to reduce NO$_x$ emissions through improvements in combustion chamber design. These, together with other problems of combustion chamber design and development, are both difficult and important. They are not, however, the prime factors that limit the thermal efficiency of gas-turbine plants, and they involve so many factors peculiar to specific engines that they will not be treated further here.

Life and Reliability

The total operating time that has been obtained with gas turbines in utility and industrial service is impressive indeed: the engines of just one manufacturer had accumulated over 900 million h of operation as of the end of October 1976. Under favorable conditions the reliability of industrial and central station gas turbines has been good (Chap. 8): in one study of 35 units over a period of 10 years the mean operating time between forced outages was 750 h.

The time between major overhauls is heavily dependent on the fuel employed. For a given turbine inlet temperature the best results are obtained with natural gas. With No. 2 fuel oil the time between major overhauls is reduced by ~ 20 percent, and with residual fuel oil the time between overhauls is reduced by around 60 percent as compared to operation with natural gas. These reductions in life have stemmed from the corrosive effects of sulfur, vanadium, alkali metals, and chlorides in the fuel.

The effects are complex and heavily dependent on the particular combination of these contaminants. In practice, the usual procedure is to reduce the peak turbine inlet temperature for operation with the dirtier fuels. The operating life also depends on the incidence of starts and stops and is much greater for gas turbines that operate continuously in industrial service than for peaking power units employed by utilities.

GAS-TURBINE BLADE COOLING

A variety of methods of cooling gas-turbine blades have been employed. The first and obvious step, taken in the 1930s, was to take advantage of thermal conduction in the turbine rotor buckets and remove heat from them by directing a cooling airstream over the rotor disk to which they were attached. This had the advantage that it reduced the blade temperature in the vicinity of the blade root, where the stresses from centrifugal force are the highest. This approach was used in the development of exhaust turbosuperchargers in the early 1940s. It was also obvious at that time that more effective cooling could be accomplished by employing hollow turbine buckets supplied with cooling air from the rotor hub, an approach used in the early German turbojet engines and in some of the U.S. exhaust turbosuperchargers for aircraft engines.

Other cooling arrangements that have been considered include the fabrication of hollow turbine buckets partially filled with an alkali metal such as potassium and sealed in much the same way as the stems of sodium- or potassium-cooled exhaust valves for internal combustion engines. With this arrangement, as in a heat pipe (Chap. 4), heat would be removed from the blade tip by boiling of the alkali metal and transferred in vapor form to the blade root, where the vapor would condense. This tends to give a uniform temperature over the length of the blade, which in turn means a higher root temperature for a given heat load on the airstream employed to cool the rotor. Hence this approach is less desirable than other simpler systems.

The bulk of the effort on cooling of turbine blades has been concerned with bleeding air through internal passages in the blade and exhausting it through ports into the hot combustion gas stream passing over the blade. In addition, there has been a substantial amount of work on water cooling. These two sets of work are summarized in the following sections.

Air Cooling of Gas-Turbine Blades

A tremendous background on turbine blade cooling has been accumulated in the course of developing air-cooled blades for turbojet engines, and this also provides a splendid background for studies of water cooling. In fact, a familiarity with the work

on air cooling is vital to an understanding of the problems of water cooling the blades of gas turbines for utility service.

Principal Arrangements for Air Cooling

The principal methods that have been used for cooling gas-turbine rotor and stator blades are indicated in Fig. 12.10, which was taken from a summary of NACA and NASA work on blade cooling.[21] The simplest, most obvious approach is indicated in Fig. 12.10a, where the cooling air flows radially outward through passages inside the blade. Somewhat higher heat-transfer coefficients can be obtained with a given cooling airflow but a higher pressure drop by making use of the impingement cooling arrangement of Fig. 12.10b, where air is supplied to the blade through a central perforated duct and small air jets from the duct impinge on the outer shell of the blade. This arrangement has the advantage that the coolant flow distribution can be varied readily to suit the differing requirements of local blade regions; the nose of the blade, for example, is subject to a higher external heat-transfer coefficient

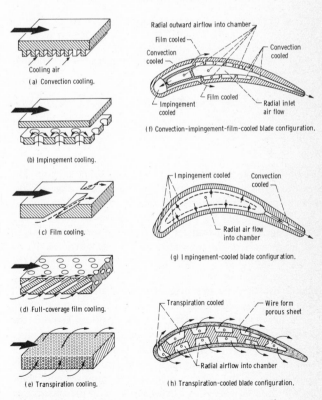

Figure 12.10 Methods of air cooling of turbine blades.[19]

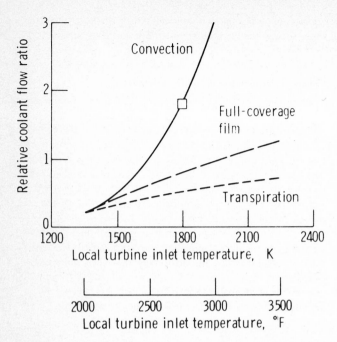

Figure 12.11 Potential effectiveness of cooling methods. Outside metal temperature, 1255 K (1800°F); coolant temperature, 811 K (1000°F); wall thickness, 0.127 cm (0.05 in); gas pressure, 20 atm; convection efficiency, 0.7; film cooling efficiency, 0.6; transpiration cooling efficiency, 0.8.[21]

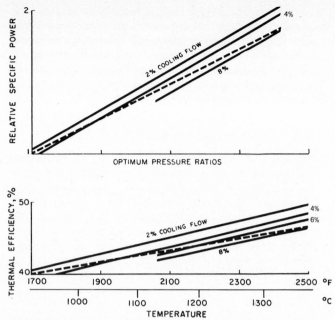

Figure 12.12 Analytical estimates for an idealized cycle of the effects of the amount of cooling air bled from the compressor discharge on the thermal efficiency as a function of turbine inlet temperature and on the specific power output as a function of the optimum pressure ratio. The dashed lines indicate the minimum amount of cooling air ideally required to maintain an acceptable blade temperature. The cooling airflow in an actual engine is usually about double that shown.[30]

and hence requires more cooling than any other portion. The third approach to blade cooling is indicated in Fig. 12.10c, which employs slits in the blade that inject cooling air into the boundary layer passing over the outer surface of the blade and thus reduce the heat-transfer rate from the combustion gas to the blade surface. This approach can be made more effective by using a large number of small holes rather than a few slits in the blade, as indicated in Fig. 12.10d. If carried to the limit, it makes use of a porous blade shell—an arrangement called *transpiration cooling* (Fig. 12.10e). Because of the variations in local heat-transfer coefficient with position around the perimeter of the blade, the blade cooling air can be used more efficiently if some portions of the blade are cooled by convection of airstreams flowing radially outward, while other portions, such as the nose, are cooled by impingement, some regions are film-cooled by exhausting some of the air through small holes or slits in the blade surface, and the thin trailing edge

region of the blade is cooled by a slit as indicated in Fig. 12.10f. Note that ejection of the cooling air through the trailing edge of a rotor blade might be expected to give a Hero turbine type of jet reaction that would contribute to the efficiency of the turbine. In practice, it has been found that at best this arrangement avoids aerodynamic losses sustained as a consequence of disturbance of the main combustion gas stream by the ejected cooling air. Figure 12.10g shows a blade in which the main portion is cooled by impingement and only the trailing edge is cooled by chordwise convection. Figure 12.10h shows a transpiration-cooled blade that might be formed by winding layers of fine wire steeply pitched in alternating directions on a central strut, which constitutes the main structural element of the blade, and sintering the assembly to give a strong, rigid body.[22] Other variations of these basic cooling methods include elegant passages to conduct the air radially outward, return it radially inward, and then pass it radially outward again.

Extensive tests of many different geometries have been carried out by NASA and by the gas-turbine manufacturers. This work has had to be closely coordinated with the development of fabrication techniques for making these geometries of the difficult-to-work iron-chrome-nickel-cobalt alloys required to obtain good strength and oxidation resistance.

Gains and Losses Associated with Air Cooling

As the amount of air bled from the compressor discharge and used for blade cooling is increased, the turbine inlet temperature can be increased, thus increasing both the output from a given size engine and the thermal efficiency. However, both of these factors are offset by the pumping-power losses involved, because the air bled from the compressor does not yield work from the turbine stage into which it is injected or from turbine stages upstream of the stage into which it is injected. Consequently, increasing the amount of cooling airflow eventually brings one to the point of diminishing returns. Figure 12.11

shows this effect for three typical methods of blade cooling (i.e., convection, impingement, and transpiration cooling).[21] The overall effect on plant performance is indicated in Fig. 12.12, which shows the effects on both thermal efficiency and specific power output as functions of the turbine inlet temperature obtainable from elegantly designed cooling provisions for several different percentages of cooling air bled from the compressor discharge.[23]

Local Heat-Transfer Coefficients

The local heat-transfer coefficient varies with position around the perimeter of the airfoil. Figure 12.13 shows the local heat-transfer coefficient as a function of distance from the leading edge as calculated by an elegant analysis.[24] Also shown are local surface temperatures for an uncooled blade with a turbine inlet total temperature of 1850°F. An indication of the validity of the analysis technique is given in Fig. 12.14, which shows a curve for the calculated surface temperature along the pitch line of a

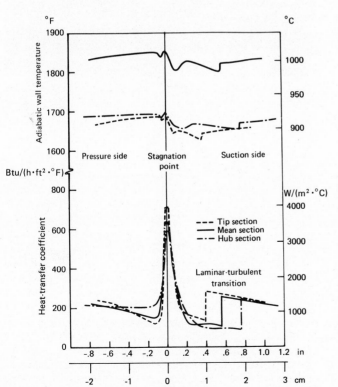

Figure 12.13 *(Left)* Typical gas-side heat-transfer coefficients and adiabatic wall temperatures in a gas-turbine blade.[24]

Figure 12.14 *(Below)* Comparison of an analytical estimate of the blade temperature distribution with engine test data.[24]

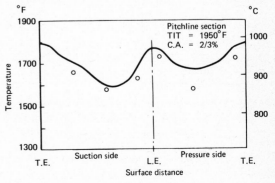

cooled blade together with some experimentally measured points.

The local heat-transfer coefficient around the perimeter of a turbine blade varies somewhat with the airfoil shape. Figure 12.15 shows another curve for the local heat-transfer coefficient as a function of position around the perimeter of the blade for a different airfoil section.[25] Note the similarity between Figs. 12.13 and 12.15 in that the highest heat-transfer coefficient is at the leading edge, but note also that in Fig. 12.15 the heat-transfer coefficient is also high at the 45 percent chord point on the suction surface of the blade and is substantial on the pressure side near the trailing edge. As is implicit in Fig. 12.13, secondary flows lead to differences in the chordwise position of the transition from laminar to turbulent flow on the suction side of the blade; in this instance, the transition is delayed substantially near the hub, while it occurs quite early near the tip. If slots are provided in the blade for film cooling as in Fig. 12.10f, aerodynamic conditions and the temperature in the boundary layer are obviously affected and can substantially reduce the nominal local heat-transfer coefficient.

Local Temperature Distribution

It is interesting to examine some local temperature distributions both as calculated and as measured in experiments. Figure 12.16 shows an interesting case for which both calculated and experimentally determined temperatures are available for a transpiration-cooled turbine rotor blade.[22] Note that in spite of

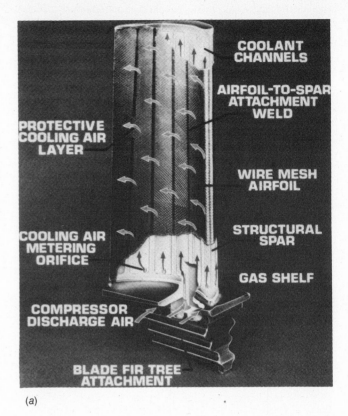

(a)

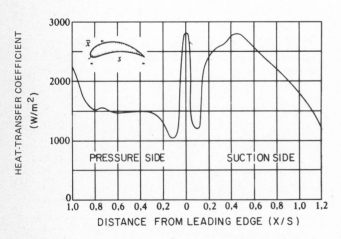

Figure 12.15 Typical curve for the local heat-transfer coefficient at the blade surface.[24]

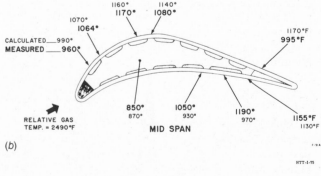

(b)

Figure 12.16 (a) Schematic view of the blade; (b) calculated and measured airfoil temperatures in a transpiration-cooled turbine rotor blade with a 1371°C (2500°F) turbine inlet temperature.[22] (*Courtesy Curtiss-Wright Corp.*)

the difficulty in estimating the complex effects of cooling airflow through a porous blade surface into the boundary layer of the high-velocity airstream over the airfoil, it apparently was possible in this case to calculate the local temperature within $\sim 110\,°C$ ($\sim 200\,°F$) of the values found in experiments.

Stress Distribution

Variations in the local temperature through the blade section lead to thermal stresses as a consequence of differential thermal expansion. These thermal stresses are superimposed on the stresses induced by centrifugal force coupled with the bending stresses on the blades stemming from the aerodynamic load. Figure 12.17 shows both the temperature and the stress distributions at the midsection for a typical air-cooled blade. Note that in this instance the hottest element of the blade occurs on the suction side at about 30 percent of the chord, while the coolest section of the blade is the trailing edge. These temperatures and stresses were calculated for the local heat-transfer coefficient distribution shown in Fig. 12.15. Hence it is not surprising that the peak temperature region in the blade is as indicated.

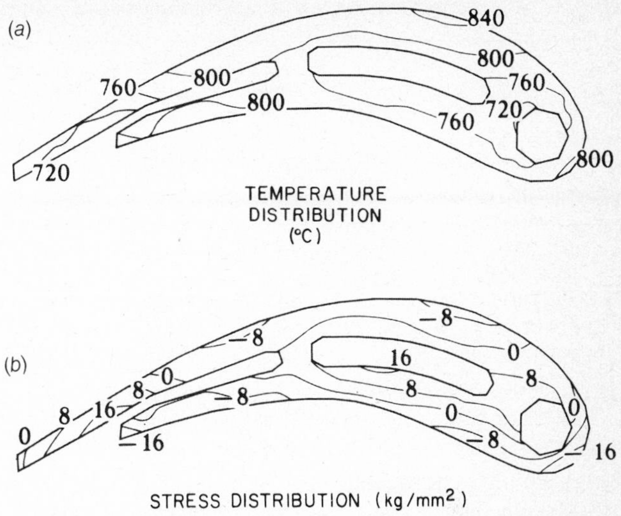

Figure 12.17 Calculated distribution of temperatures (a) and stresses (b) at the midsection of a cooled rotor blade under steady-state operating conditions with a gas temperature of 986 °C and a cooling-air inlet temperature of 380 °C.[25]

Water Cooling of Gas-Turbine Blades

Water-cooled gas-turbine blades offer such intriguing possibilities for achieving turbine inlet temperatures as high as 2000 °C ($\sim 3600\,°F$) that the concept appeared to deserve substantial space in this text, particularly because it involves an exceptionally fine case study demanding that practically every discipline in power plant engineering work be strained to its limit. Water cooling has not been applicable to aircraft engines, because the weight of water required would be prohibitive; however, water cooling has appeared attractive for stationary gas-turbine applications.

Experience with Water-Cooled Gas Turbines

A number of efforts involving water cooling of high-temperature gas-turbine blades have been carried out in the past. In fact, the very earliest gas turbine to produce a net power output, the Lemale and Armengaud turbine mentioned at the beginning of this chapter, employed water cooling in 1904.[4] Professor K. Bammert at the Braunschweig Technischen Hochschule included both air- and water-cooled turbines in the gas-turbine work that he initiated there in 1938, and this work stimulated the initiation of British work on water-cooled gas turbines beginning in 1945 and continuing off and on up to the time of writing.[26] NACA built and tested many water-cooled gas turbines in the latter 1940s,[27,28,29] work that was covered by some 15 reports. Data for both the 1904 and the NACA tests show the heat losses to the cooling water seriously degraded the turbine output. Fortunately, these heat losses become a smaller fraction of the system output as the size of the turbine is increased, because the surface-volume ratio becomes more favorable. In the latter 1950s, the Solar Aircraft Company, under the Bureau of Ships of the U.S. Navy, designed and built a small water-cooled gas turbine with a closed cooling system. The turbine was operated a total of 219 h but was troubled with thermal-stress cracking of the blades.[30] Subsequently, a Siemens project involved operation of a 1200-kW unit in Germany in the early 1960s.[31] A closed water-cooling circuit was used with the water circulated out to the blades and then back to the center of the rotor. This arrangement made it necessary to use thick-walled passages to contain the high pressure generated by centrifugal force and gave trouble with deposits of solids centrifuged out of the water near the blade tips. Work was discontinued after ~ 1000 h of testing because of difficulties with plugging of coolant passages, instabilities in the cooling water flow, leaks, and, more particularly, severe thermal stresses in the buckets.

TABLE 12.3 Water-Cooled Gas-Turbine Combined-Cycle Operating Parameters Given in Ref. 23

	Flow, lb/s	Pressure, psia	Temperature, °F		Flow, lb/s	Pressure, psia	Temperature, °F
AIR				**BUCKET COOLANT**			
Compressor inlet	700	14.62	59	Turbine inlet			
Bleed air				1st stage	19.00	250	100
8th stage	3.05	46.81	346.9	2d stage	11.27	250	100
13th stage	13.91	96.84	523.5	3d stage	4.86	250	100
Compressor discharge				Coolant lost to gas path			
(*a*) Turbine cooling	3.28	200.36	696.8	1st stage	5.70	103.84	337.3
(*b*) Bearing seal	2.71	200.36	696.8	2d stage	2.82	49.14	291.8
Combustor inlet	677.05	200.36	696.8	3d stage	0.97	21.70	253.5
FUEL				Recovered coolant			
Fuel heater inlet	29.48		125	1st stage	13.30	103.84	337.3
Combustor inlet	29.48		250	2d stage	8.45	49.14	291.8
				3d stage	3.89	21.70	253.5
NO$_x$ SUPPRESSION STEAM				Feed-water heater discharge	25.64	14.70	130.5
Combustor inlet	8.12	300.36	417.5	Makeup water	19.39	14.70	59.0
				Polisher inlet	35.13	14.70	99.7
COMBUSTION PRODUCTS				Steam-turbine condenser inlet	9.90	14.70	99.7
Combustor discharge	714.65	192.34	2877.1				
Turbine exhaust	744.38	15.42	1384.6	**STEAM CYCLE**			
HRSG bypass	7.44	15.42	1384.6	Feed-water pump inlet	177.97	21.7	222.3
HRSG inlet	736.94	15.42	1384.6	Economizer inlet	177.97	1485	225.0
HRSG outlet	736.94	14.88	300.0	Boiler blowdown	1.78	1415	588.5
Stack	744.38	14.70	311.7	Evaporator inlet	176.19	1415	588.5
				Superheater inlet	176.19	1415	588.5
NOZZLE COOLANT				Steam-turbine inlet	176.19	1250	950.0
Turbine inlet	98.83	1500	225	Condenser inlet	176.19	1.23	108.7
Turbine discharge	98.83	1250	512.4	Feed-water heater inlet	186.09	24.11	108.8
Flash tank inlet	60.63	1250	512.4	Deaerator inlet	186.09	21.7	134.4
Fuel heat exchanger inlet	38.20	1250	512.4				
Fuel heat exchanger outlet	38.20	1250	463.9	Gas-turbine output, MW	176.24		
Flash tank return	52.51	300	417.3	Steam-turbine output, MW	82.16		
Deaerator inlet (nozzle return)	38.20	21.7	395.0	Net plant output, MW	258.40		
Deaerator inlet (flash tank return)	52.51	21.7	395.0	Gas-turbine efficiency, %	0.3244 (HHV)		
				Combined-cycle thermal efficiency, %	0.4757 (HHV)		

A different approach employing a once-through water circuit discharging a steam-water mixture from the blade tips was initiated by General Electric in the early 1960s. The first experiments were carried out with a 24.7-cm-diameter (9.7-in-diameter) simple turbine wheel with turbine inlet temperatures ranging from 1540 to 1925°C (2800 to 3500°F) and, as in the Lemale-Armengaud and NACA turbine tests, the weight flow rate of water was only a little less than that of the fuel.[4,27,32] Results of the initial tests in 1964 were encouraging in that there were no difficulties with water flow instabilities or passage plugging as a consequence of deposits in the short period of the tests. This

work was followed a few years later by the design, construction, and operation of a second turbine in which the design took advantage of the lessons learned from the first unit. This unit was operated for some 20 h with 50 starts and stops in a series of tests in which the maximum gas temperature into the turbine was as high as 1925°C (3500°F). A local heat flux in the turbine rotor of 440 W/cm^2 [1.4 × 10^6 Btu/(h · ft^2)] was measured. A third turbine was then built and tested. About 100 h of operation had been obtained on these three turbines as of February 1979. There has been no effort to run an endurance test; the first objective was to determine whether the new approach eliminated the problems encountered in the Siemens work. Subsequently, the effort was directed mainly toward assuring adequate cooling with acceptable thermal stresses, but included studies of corrosion in the cooling-water passages and erosion of the casing and stator by liquid droplets thrown off the rotor.[32-38] The effort has also included the preparation of conceptual designs for full-scale systems. Table 12.3 gives the principal parameters for a reference design system.

Methods of Water-Cooling Blades

At first glance, one might consider quite a variety of water-cooled blades in a turbine rotor. Perhaps the first logical possibility would be to circulate water radially outward through one set of passages of the blade and then return it to the rotor through another set. This approach was employed in NACA tests of two aluminum turbine wheels and one stainless-steel wheel in the 1948–1952 period. The wheel diameters were about 30 cm (1 ft). The bulk of this work was carried out with rotor tip speeds of 214 to 305 m/s (700 to 1000 fps) with turbine inlet temperatures of 540 to 1095 °C (1000 to 2000 °F). Both natural thermal convection and forced-convection flow of the water through the rotor were investigated, and a wealth of heat-transfer data were obtained.[27-29] Siemens, in Germany, subsequently built a full-scale 1200-kW gas turbine employing this method of blade cooling.[31] The testing included 700 h at temperatures above 1000 °C (1830 °F). However, the blade tips operated in a centrifugal field of 10,000 to 20,000 g, leading to a radial pressure differential between the rotor hub and the blade tip of hundreds of atmospheres; this gave severe pressure stresses in the blades which were superimposed on the stresses from centrifugal and aerodynamic loads, as well as severe thermal stresses. To avoid the pressure stresses, the water can be admitted to the blade root and allowed to flow radially outward to a discharge port near the blade tip, with boiling occurring along the length of the passage.[32] The water flow rate must be limited to avoid an excessive water consumption and quenching of the hot gas stream. Ideally, one would like to have a thin film of water flowing on the wall, while steam would flow through the central portion of the cooling passage. This approach is complicated by the fact that the water film not only is subject to centrifugal force but also to Coriolis forces associated with the tangential acceleration of the water as it flows radially outward along the blade.[35] Consider the cross section through such a passage as shown in Fig. 12.18. As a consequence of the Coriolis force, the liquid film is swept to the trailing side of the circular water passage to give the crescent-shaped section shown in Fig. 12.18. The lower radial velocity in the boundary layer reduces the Coriolis forces in that region and leads to the secondary circulation shown with the small arrows in Fig. 12.18. Evaporation

takes place from the surface of the thin liquid film rather than by the formation of tiny bubbles at the heated surface as in conventional nucleate boiling.

To provide a good water flow distribution around the perimeter of the airfoil with the limited water flow rate available, it appears best to make use of an array of small-diameter passages arranged around the perimeter of the airfoil as shown in Fig. 12.19. The flow to each passage must be individually metered to restrict the water flow to the proper level.

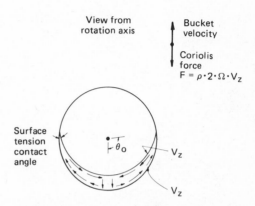

Figure 12.18 Schematic view of the lateral circulation in the cooling passage for a water-cooled turbine rotor blade.[35]

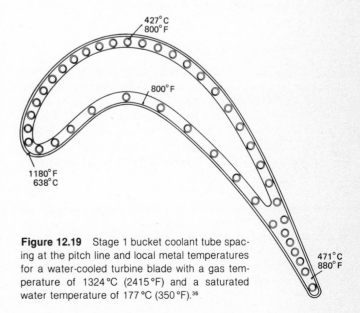

Figure 12.19 Stage 1 bucket coolant tube spacing at the pitch line and local metal temperatures for a water-cooled turbine blade with a gas temperature of 1324 °C (2415 °F) and a saturated water temperature of 177 °C (350 °F).[36]

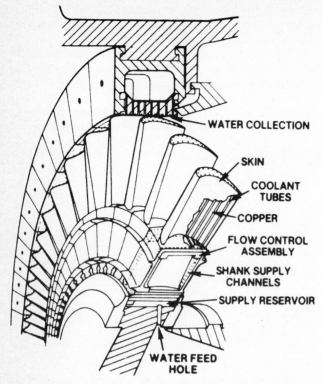

Figure 12.20 Conceptual design for the water-cooling system for a water-cooled gas turbine.[37]

Water Flow Supply Arrangements

Getting water into the blade root presents a problem. A promising configuration entails the use of a groove near the rim of the rotor into which a water jet can be directed, as in Fig. 12.20, so that it enters the rotor at a pressure substantially equal to that of the local hot gas.[36,37] Small complex passages in the blade root distribute the water to the radial passages from a system of weirs that meter the flow to provide the proper flow rate to each passage.

Limiting Heat Flux

Extensive experience with heat transfer under boiling conditions has shown that the two principal parameters are the local heat flux and the temperature difference between the hot metal surface and the saturation temperature of the boiling liquid for the pressure at which it is operating.[39] As the heat flux is increased, the temperature difference between the metal surface and the saturation temperature of the boiling liquid increases relatively little until a critical heat flux is reached; after this, further increases in the metal surface temperature lead to a drastic reduction in the heat flux because a vapor blanket forms at the surface (Fig. 11.25). The peak heat flux is commonly alluded to as the *critical* or *burnout* heat flux, or *departure from nucleate boiling* (DNB). For conventional stationary boiler tubes, the burnout condition is commonly reached when the vapor quality is 30 to 70 percent with heat fluxes of the order of 30 to 100 W/cm^2 [100,000 to 300,000 Btu/(h · ft^2)].

The strong centrifugal field makes the situation in a water-cooled turbine blade substantially different from that in a stationary boiler tube; hence experiments have been run in the General Electric program to investigate the situation.[35,36] Figure 12.21 shows a typical set of data for the heat flux to a 1.1-mm-diameter 12.5-cm-long passage in an electrically heated copper block mounted on a rotor that gave a centrifugal field of 8000 g. Analyses showed that the critical, or burnout, heat flux (i.e., the heat flux at which the metal temperature began to increase rapidly) nominally was about what one would expect in static boiler tubes. However, a detailed analysis carried out to determine the fraction of the passage perimeter wet by the water film indicated that only about 20 percent of the perimeter had been wet; hence the local heat flux under the wet film was about five times what one would expect in static boiler tubes. This probably stems from the high radial velocity and the large Coriolis forces acting to keep the film tightly against the heated surface. Subsequent tests indicate that with approximately 2-mm-diameter passages, special treatment of the internal surface of the passage, and canting the passage to reduce the Coriolis forces, it appears possible to increase the fraction of the wet perimeter to ~45 percent and the total heat removal capacity correspondingly,[36-38] but canting poses serious fabrication and stress problems.

Temperature Distribution in a Water-Cooled Blade

At first, one might expect to be able to employ the array of small water-cooling passages of Fig. 12.19 in otherwise solid stainless-steel or nickel alloy blades. A brief look at the heat transfer, temperature distribution, and thermal-stress situation, however, quickly shows that in the first stage of the turbine the heat flux is so high that this approach would give completely unacceptable thermal stresses, although it can be used in the third stage and possibly in the second stage where the heat fluxes are lower. This stems from the fact that with hot gas temperatures of 1370 to 1650°C (2500 to 3000°F) and cooling-water saturation temperatures of 170 to 235°C (340 to 455°F), corresponding to gas pressures of 8 to 30 atm, and with blade surface temperatures of 540 to 650°C (1000 to 1200°F), the gas film temperature drop will be around 850°C (1500°F). From Figs. 12.13 and 12.15 one can see that for conventional 1000°C gas turbines the local heat-transfer coefficient will run as much as 2800 W/(m² · °C) [500 Btu/(h · ft² · °F)], which gives a local heat flux of about 240 W/cm² [(750,000 Btu/(h · ft²)]. At the higher pressure ratios required to take advantage of turbine inlet temperatures in the 1370 to 1650°C (2500 to 3000°F) range, the heat-transfer coefficient would be still higher. As an example, in the small (24.7-cm-diameter) water-cooled turbine run by General Electric at 16 atm, a heat flux of 440 W/cm² [1.4 × 10⁶ Btu/(h · ft²)] was measured experimentally.[32] Such high heat fluxes lead to much steeper temperature gradients in the metal blade between the heated surface and the cooling passage than in air-cooled blades.

Thermal Stresses

In the extensive studies of water-cooled gas turbines carried out by NACA around 1950, the severe thermal stresses associated with the high heat fluxes and steep thermal gradients caused serious concern even with solid aluminum alloy blades. The thermal conductivity for aluminum alloys is nearly 10 times that of the iron-chrome-nickel alloys ordinarily used in gas turbines. For example, in a homogeneous iron-chrome-nickel alloy blade for a heat flux of 330 W/cm² [10⁶ Btu/(h · ft²)], the temperature drop across a 2.5-mm thickness would be ~400°C (720°F). Inspection of Fig. 12.19 indicates that the bulk of the metal would be at a temperature perhaps 50°C above that of the saturation temperature of the water [i.e., around 230°C (450°F)]. Hence, a great deal of elastic and plastic deformation would take place in the hot surface skin, particularly in the hot streak along the leading edge and probably an additional hot streak on the suction surface at 30 or 40 percent of the chord. Changes in power level for the plant would lead to changes in the gas inlet temperature and hence in the blade temperature, so that these power changes, coupled with startups and shutdowns, would subject the hot-streak regions to severe thermal-strain cycling (Chap. 6). Note that for the 400°C temperature differential cited above, Fig. 6.14 indicates that, in a homogeneous stainless-steel blade, failure would occur in about 300 cycles if the stresses were uniaxial and no other stresses were present. In a turbine rotor blade, the thermal stresses will be biaxial and will be superimposed on large stresses from the centrifugal load on the blade. Thus failure should be expected in a much smaller number of cycles—probably between 10 and 100. Note that this rough analysis is for an ideal simple slab geometry, whereas in

an actual blade (Fig. 12.19) the heat flow path from the surface to the portion of the passage cooled by the liquid film would be longer than 2.5 mm. In the critical region close to the leading edge, the limited metal cross-sectional area effective in heat conduction from the hot surface to the reduced area of the cooling liquid meniscus would roughly double the heat flux between adjacent cooling passages. This would lead to temperature differences greater than used in this example. Further, the cooling passages constitute stress concentrations, the thermal stresses in a turbine blade in the course of a temperature transient are biaxial, and the surface stresses are tensile for an air-blast quench under flameout conditions.

The temperature differential between the outer surface and the coolant passages can be reduced by employing copper in the blades. Inasmuch as copper could not withstand the centrifugal stresses, the oxidizing conditions on the hot gas side, or the corrosive attack of high-pressure, high-temperature water, it is necessary to clad the outer surface of the copper with a high-strength, oxidation-resistant material and make the central structural spar and the cooling passage liners of a high-strength iron-chrome-nickel alloy that has a higher coefficient of thermal expansion.[35,36] The General Electric choices of Refs. 35 and 36 for these two materials were, respectively, Inconel 671 (also designated A-286)—a 25% Cr-15% Ni alloy having an exceptionally high strength and high coefficient of thermal expansion and containing small amounts of Mo, Mn, V, Ti, and Si—for the central spar, and IN 718—another exceptionally high-strength

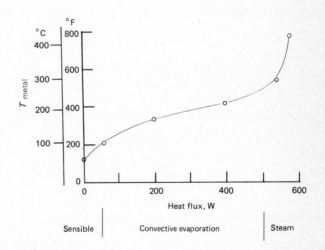

Figure 12.21 The metal temperature measured in the wall of a 0.043-in-ID cooling passage for a series of heat fluxes in the rotor of an electrically heated simulated blade in a test rig.[35]

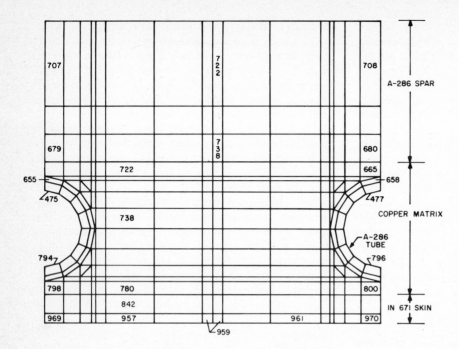

Figure 12.22 The temperature distribution in a section of a water-cooled turbine blade as given by a computational model for stage-1 bucket thermal-stress analysis (temperatures in °F).[36]

alloy of ~50 percent nickel having a relatively low coefficient of thermal expansion—for cladding. These alloys must be thermally bonded to the copper, and the bonds must withstand severe thermal-strain cycling.

Elegant analyses of the temperature and thermal-stress distribution in the rotor and stator blades have yielded estimates of the temperature distribution in a blade element such as that shown in Fig. 12.22. The central spar is massive in relation to the rest of the blade, and so differential expansion and hence local thermal stresses depend on the difference between the local temperature and the mean temperature of the central spar. For example, note that the 475°F in an element of the A-286 coolant passage liner gives a temperature differential of ~240°F between this local region and the mean temperature of the spar. The resulting thermal stress would be ~58,000 psi, which is below the yield point and hence acceptable. The peak temperature in the cladding is ~255°F above the mean temperature of the spar, but this is partially compensated by the lower coefficient of thermal expansion. The net effect is to give a thermal stress in the cladding of ~30,000 psi, well within the elastic limit. However, these rough estimates of the thermal stresses were made for simple uniaxial conditions with no allowances for three-dimensional effects, the stresses from centrifugal force and aerodynamic loads, and stress concentration

effects associated with changes in section in the vicinity of the blade root. For example, stretching the central spar under the centrifugal load will act to alleviate the axial thermal stress in the cladding, but Poisson's ratio effects will act to increase the tangential thermal stress in the cladding. The effects are exceedingly complex, but a satisfactory solution to the stress problems may be possible in view of the very clever and sophisticated design analyses that have been made.[33-38]

Heat Losses

In the 1904 Lemale-Armengaud and 1950 NACA tests of water-cooled turbines, the heat lost from the hot gas to the cooled walls substantially reduced the power output. This type of loss would be less pronounced in a large turbine because of the more favorable surface-volume ratio, but it is not trivial. In the NACA tests, the heat loss per stage amounted to ~5 percent of the energy input in the fuel with a temperature difference of 950°C (1700°F) between the hot gas and the metal surfaces.

The total heat loss to the cooling water for a large three-stage turbine having blades similar to those shown in Fig. 12.19 was estimated as 155 kJ/kg (66.8 Btu/lb) of gas, which is equivalent to 9.3 percent of the heat input from combustion. Almost one-third of this was estimated to occur in the first-stage stator, and over 20 percent would be in the first-stage rotor. The author estimated that the total heat loss to the cooling water would reduce the gas-turbine output by ~6 percent and the steam turbine output by ~7 percent. This effect would be partly offset by the increase in mass flow through the gas turbine represented by the steam escaping from the rotor blade tips, which would increase the gas-turbine output by ~0.5 percent. The heat flowing to the cooling water for the stator and casing, coupled with the heat in the liquid water thrown off the rotor and collected by slots in the casing, could be used for feed-water heating; this would represent an increase in plant output by ~2 percent. Thus the net overall effect would be a loss of a little over 5 percent. Note that this assumes no water cooling of the combustion chamber and includes no allowance for water churning losses associated with the liquid water thrown off the rotor or the pumping-power requirement for the cooling-water system. Note too, that if the liquid water thrown off the turbine rotor is not collected, its vaporization will act to quench the hot gas stream and further reduce its temperature.

It is interesting to compare these losses with those estimated for air-cooled turbines in Fig. 12.12. The ideal thermal efficiency indicated by Fig. 12.12 for 1370°C (2500°F) is ~50 percent for zero cooling airflow, whereas the dashed curve that indicates the minimum cooling airflow ideally required gives an ideal thermal efficiency of ~45 percent or a 10 percent drop in power plant output. In comparing this with the losses estimated above for a water-cooled turbine, it appears that water cooling might give half the losses of an air-cooled turbine.

It has been suggested that the heat losses might be reduced by applying a ceramic coating to the stator and rotor blades.[36] The temperature drop through this coating would reduce the temperature difference between the hot gas and the blade for a given peak metal temperature and thus reduce the heat loss. This concept has been investigated repeatedly for air-cooled blades, the key problem being to find a ceramic coating that will not crack and spall off under the severe thermal cycling conditions that it must withstand. In addition, recent tests of such coatings at United Technologies Corporation have shown that, to be effective, these coatings must be polished. In one test with a coating having a surface roughness of 600 μin, the increased heat-transfer coefficient essentially offset the thermal insulation effect, and even polishing to a surface roughness of 125 μin did not completely eliminate this effect.

Erosion

The mixture of water and steam leaving the blade will have a quality no higher than 70 percent if satisfactory cooling is to be assured.[33-36] The droplets of water emerging from the tips of the blades will be thrown off at high velocity and will tend to impinge on either the turbine casing or the downstream stator blades; in either case erosion will be a major problem. Methods for coping with this at the low-temperature end of steam turbines have been developed, but they have entailed bleeding off substantial quantities of steam from the main stream flowing through the turbine stage and using it for feed-water heating; this would represent a major loss if it were done near the inlet of a gas turbine. The erosion problems in a water-cooled gas turbine will also be rendered more severe by the high-temperature environment.

Deposits in the Water-Cooling Passages

Trace amounts of solids dissolved in the feed water may lead to deposits in the small-diameter cooling passages in the blade and blockage of the channels. By keeping the quality of the steam leaving the blade below 70 percent, it seems likely that this kind of difficulty can be avoided. However, the small cooling passages in the turbine blades will give a situation far more sensitive to trace amounts of deposits than in any steam boiler built to date. This will be particularly important in regard to the cooling passages in the stator and casing because they entail closed-circuit flow with subcooled boiling. Any nonuniformity in solids deposition from one passage to another can imbalance the flow distribution. This could lead to boiling in some channels, flow instabilities, and a rapidly progressing failure condition.

Reliability

In attempting to appraise the reliability of the water-cooled turbine system design represented by Figs. 12.19 and 12.20, five points appear to be particularly important:

1. The cooling passages must be small in diameter (between 1 and 2.6 mm) and hence are subject to plugging.

2. To keep heat losses, etc., to acceptable levels, the water flow rate through the rotor blades must be kept to a fairly low level so that the vapor exit quality will be ~70 percent.

3. In the first stage, in particular, the heat flux is close to that for burnout.

4. The water flow rate to each of many thousands of passages must be metered and kept within close limits to avoid excessive heat losses on the one hand and burnout on the other. Any deposits of solids in the metering weirs or orifices would spoil the water flow distribution

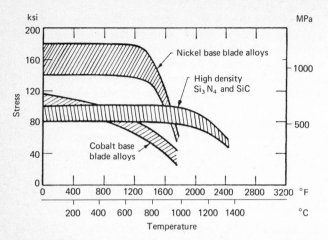

Figure 12.23 Four-point flexural strength of high-density Si_3N_4 and SiC and ultimate tensile strength of metals as a function of temperature.[41]

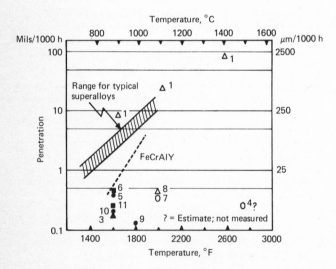

Figure 12.24 Effects of temperature on the corrosion behavior of ceramics exposed to hot combustion gases.[42]

Symbol	Ceramic	Conditions	Author/Source
△1	H.P. Si_3N_4	Oil burner rig; Na, V. 80 h	Brooks/CEGB
▲3	H.P. Si_3N_4	Diesel + 125 ppm Na; GT rig; 7200 h	GE/GTPD
○4?	Sintered SiC	Ng plus NaCl; burner flame 60 h	GT/CRD
●5	Sintered SiC	Diesel + 125 ppm Na; GT rig; 6000 h	GE/GTPD
■6	Silcomp	Diesel + 125 ppm Na; GT rig; 6000 h	GE/GTPD
○7	H.P. SiC	Diesel + 5 ppm Na + 0.5 wt% S; V, Mg 250 h	Singhal/W
△8	H.P. Si_3N_4	Diesel + 5 ppm Na + 0.5 wt% S; V, Mg 250 h	Singhal/W
●9	Sintered SiC	Diesel + 125 ppm Na; GT Rig (ni-10) 10,000 h	GE/GTPD
●10	Sintered SiC	Diesel + 125 ppm Na; GT Rig (ni-8) 10,000 h	GE/GTPD
■11	Silcomp	Diesel + 125 ppm Na; GT Rig (jm-1) 6,000 h	GE/GTPD

Notes: Open symbols (○) under 1000 h; (●) above
? estimated temperature; no penetration data taken; no apparent attack.

5. If the cooling-water flow to a region of a blade is interrupted, the local surface temperature will rise at a rate of ~ 280°C/s (~ 500°F/s) in high flux regions and local melting will occur in a few seconds. Such a failure is likely to spread rapidly and could easily lead to a general failure of the turbine.

Thus the crux of the reliability problem is the probability that the water flow rate to any one of many thousands of small cooling passages will drop more than ~ 30 percent below the design value for even a few seconds. That probability must be kept exceedingly low if the water-cooled turbine is to be commercially viable.

Summary

In view of the degree to which every phase of the technology involved must be stretched to the limit to take advantage of water cooling, it seems unlikely that this approach to high turbine inlet temperatures will prove practicable.[40]

CERAMIC TURBINES

The high melting point and excellent oxidation resistance of ceramics have made them tantalizingly attractive candidates for gas-turbine applications for the past 40 years. Their high strength at elevated temperature and low density make them particularly attractive for turbine wheels, where the most important stresses stem from the loads imposed by centrifugal forces. Figures 12.23 and 12.24 show these characteristics in comparison with values for typical metal alloys developed for high-temperature applications.[41,42] On the one hand, the apparent potential is great indeed, but on the other hand, the problems are not only extremely formidable, but also subtle. Hence they merit examination in sufficient detail to convey an appreciation for their character and magnitude.

Development Work from 1939 to 1957

Efforts to develop a ceramic gas turbine were initiated in Germany in 1939 by Prof. K. Bammert at the Braunschweig Technischenhochschule.[43] He was assisted by Erich Soehngen, who was responsible for the fabrication and testing of the ceramic blades. The first test of a gas turbine with ceramic blades in a metal wheel was carried out in 1940 with steatite blades. The turbine ran only a short time before blade failures occurred. A fairly comprehensive program to fabricate and test blades made of silicon carbide and aluminum oxide was initiated, and numerous tests were carried out in static test rigs in an effort to obtain good resistance to thermal shock in a geometry that would give relatively low stresses in the juncture between the blade root and the wheel. Concurrent with the materials development, a number of designs were prepared in an attempt to keep the blades in compression by supporting them from the outer periphery. The design work led to construction of a rotor drum made up of a series of rings, one for each stage. In parallel with this effort, a second approach followed was to develop cermet (or ceramel) blades made by powder metallurgy techniques using a mixture of metal and ceramic powders, a technology originally developed for making tool bits for lathes. The composition of the blade material was varied from largely aluminum oxide at the outer tip of the blade to entirely metal at the root. This arrangement ensured that there would be ductile material in the region of the juncture between the blade and the rotor disk to accommodate small differences between the blade and rotor profiles in the joint between them and, consequently, stress concentrations. The end of the war terminated these programs before an engine was run employing any of these three approaches. Some of the work was moved to Farnborough, England, however, where a turbine with cermet blades was assembled and operated with turbine inlet gas temperatures as high as 1200 °C (2192 °F). Erich Soehngen went from Germany to Wright Field and participated in the U.S. Air Force ceramic gas-turbine effort there.

After World War II, the German work on cermets intrigued engineers both at the NACA and WADC, the Wright Air Development Center in Dayton, Ohio. Comprehensive development programs were initiated both within those organizations and under subcontracts with General Electric, Westinghouse, Pratt & Whitney, the Wright Aeronautical Corporation, and a number of materials research laboratories. The main emphasis was on basic materials work as a search was made for the best possible combination of physical properties. A wide range of material compositions was investigated in the course of these programs. Concurrently with this work, efforts to design turbine wheels were carried out with particular attention given to minimizing the stress concentrations at the root of a ceramic blade where it was joined to the turbine wheel. Special efforts were made to minimize the stress concentrations, both by the choice of the basic geometry and by inserting compliant pads of metal between the turbine bucket root and the recess in the wheel into which it fitted. This work was closely coupled with the development of fabrication techniques suitable for producing blades of a consistently high quality. This proved to be very difficult because small defects tend to be common in ceramics, and they are hard to find by any conventional inspection technique. In fact, it is not possible to produce a ceramic part completely free of defects; the problem proved to be one of keeping the maximum size of the defects present below a tolerable level.

Quite a number of both cold and hot spin tests were carried out in the course of the 1945-1957 development work. Probably the most extensive of these were hot spin tests carried out at Pratt & Whitney with a rotor designed to fit in the J-57 engine.[44] Other tests were carried out in burner rigs using cascades, and General Electric tested a stator with ceramic blades in a J-47 engine. Failures were common in all these tests, and in the spin tests any blade failure resulted in the failure of all the blades in the rotor. (Such a failure would, of course, be catastrophic in a full-scale engine.)

The only tests of rotors in full-scale engines were carried out at the NACA Lewis Laboratory using a J-33 engine.[45-49] Profiting from the hot spin tests, only four or six ceramic blades were installed in the turbine rotor for each test while the rest were metal. In all, 27 sets of cermet blades were tested. The sets differed both in composition and in the detailed design of the root. The metal used to bind the ceramic particles together was nickel or cobalt, and the ceramic material in most cases was titanium carbide. The weight fraction of the metal in the blade generally ran from 20 to 30 percent, which gave a ductility at 900 °C (1552 °F) of 1 or 2 percent. Failures occurred in all the tests except one in operating times of 1 to 144 h. In the one test in which a failure did not occur, the rotor ran for 150 h, the original test objective. (Note that a statistical treatment of the test results would indicate that if a full complement of blades had been installed in the "successful" 150-h test instead of only six, it probably would have not been possible for the rotor to run for the full 150 h.) The total turbine operating time accumulated in all 27 of these tests was 1700 h. The principal advance made in the program was the reduction in the stress concentrations at the blade root by improving the design in that region.

The NACA work was phased out in the latter 1950s after correlation of the data on the range of metal contents in the cermet showed that the ductility must exceed 2 percent if a turbine rotor is to withstand the thermal stresses to be expected in routine operation. It was found that when the metal content of the cermet was increased sufficiently to give this amount of ductility, its creep strength was no better than that of the better metallic alloys available, and they gave much greater ductility and resistance to mechanical and thermal shock.

Programs Initiated by ARPA in 1971

About 15 years later, when it became evident that the point of diminishing returns was being reached in the improvement in metal alloys and in blade cooling for m. ·v turbojet engine applications, ARPA (Advanced Research Projects Agency, Department of Defense) in 1971 instituted a new program aimed at the development of ceramic turbine blades. This work included both a comprehensive program of basic materials development and specific engine design and development programs.[50] The extensive earlier work carried out by NACA and WADC served as the point of departure. The materials examined have included aluminum, magnesium, beryllium, yttrium, and zirconium oxides; silicon and titanium carbines; and silicon and boron nitride. The greatest emphasis in the effort has been on silicon nitride, with most of the balance on silicon carbide.

Fabrication

The properties of ceramics are heavily dependent on the fabrication methods employed. In the ARPA program two processes have been investigated for fabricating Si_3N_4: hot pressing and reaction sintering.[50] The hot-pressing process uses Si_3N_4 powder as the starting material. Additives such as calcia are sometimes used to allow sintering at a lower temperature, but this degrades the high-temperature strength. Hot pressing is carried out utilizing graphite dies in a nitrogen atmosphere furnace. (Si_3N_4 disintegrates at high temperatures in a vacuum.) Hot pressing is not very adaptable to large or complex shapes, but for small simple shapes it produces near-theoretical density and higher strength than reaction sintering. Diamond grinding to the final dimensions is usually necessary and can be accomplished, but it is obviously more costly than similar work on metals.

Reaction sintering usually entails forming a compact of silicon powder by cold pressing and heating it in a nitrogen atmosphere furnace. Some reaction is allowed to take place while the silicon is well below its melting point of 1430 °C (2600 °F), giving grains that have a shell of Si_3N_4 containing silicon. Then the temperature is raised above the melting point of silicon and the reaction carried to completion. (The melting point of Si_3N_4 is 1900 °C, or 3450 °F.) The volumetric change from the silicon powder compact to the finished ceramic body is of the order of $\frac{1}{2}$

percent, and so reasonably good dimensional control is possible without final machining. A variation in the fabrication process is to react the body partially so that it is weakly bonded. It can then be machined with conventional metal working tools, and subsequently the reaction sintering process is completed. The density achievable is substantially less than the theoretical value, because nitrogen must penetrate the body to reach the unreacted silicon. Therefore, the strength is generally less than for the hot-pressed material, and the pores constitute a set of microscopic defects that are stress concentrations.

Three processes are commonly used to fabricate SiC bodies: sintering of SiC powder either with or without graphite hot-pressing equipment, and reaction-sintering a powder compact of SiC and carbon. For the latter, the firing takes place in the presence of silicon vapor and/or liquid to yield a body that contains about 10 percent free silicon. This excess silicon tends to reduce the strength at high temperature and makes such a body less desirable for turbine applications. Some limited work has been done on the fabrication of bodies of SiC using chemical vapor deposition, but the resulting parts have been plagued by defects.

Design, Construction, and Testing

The fabrication process development in the ARPA program has been closely coupled with work on the detail design of parts suitable for gas turbines and testing of the parts produced.[50] It is relatively easy to design a die for hot-pressing a simple cylindrical shape, and, if the length-diameter ratio of the cylinder is around unity, a uniform density can be obtained in the part. However, it is very difficult to design tooling to produce a uniform density in a hot-pressed airfoil in which the section thickness and the airfoil chord vary, the airfoil is twisted, and there is a massive bulb at one end. Similarly, it is difficult to fabricate large, relatively thin-walled, cylindrical shells for use as combustion chamber liners and achieve physical properties approaching those obtainable in small, simple, cylindrical test pieces. Thus the design effort must be tightly integrated with the development of fabrication methods.

The research laboratory of Ford Motor Company, working under an ARPA contract, designed an all-ceramic gas turbine intended to produce about 200 hp for automotive service.[51-54] The turbine rotor was about 127 mm (5 in) in diameter. The design employed an injection-molded, reaction-sintered, silicon nitride blade ring which was diffusion bonded to a rotor hub made of hot-pressed silicon nitride. Elegant computer programs for three-dimensional stress analysis yielded a blade form and rotor shape that gave promise of withstanding the centrifugal and aerodynamic loads, together with the thermal stresses associated with acceleration from idle to full power in a few seconds. The latter requirement was imposed to permit the rapid increase in power required for starting up and accelerating a vehicle after a stop at a traffic light.

Other components were developed, and some of these were subjected to simulated engine operation in burner test rigs. These parts include the combustor, stator, and rotor tip shrouds. The longest operation of these parts in bench tests without failure was a 245-h test of a set of rotor tip shrouds at 1055°C (1930°F) and 26 h at 1370°C (2500°F). The running time on ceramic parts in test rigs totaled approximately 2000 h. In the most impressive test, a single-stage turbine with a ceramic stator and rotor was operated for 35 h with a turbine inlet temperature of approximately 1232°C (~2250°F). The temperature was then increased to ~1370°C (~2500°F) for 1.5 h, at which point it was noted that a metallic component in the rotor shaft assembly had overheated to 1038°C (1900°F). A gradual shutdown was then initiated to minimize thermal stresses, but the rotor failed at ~1038°C (~1900°F) in the course of this shutdown operation. Examination of the failed parts indicated that the failure stemmed from seizure between a metal coupling and the ceramic rotor where slippage was supposed to occur to accommodate differential thermal expansion. Note that the Fe-Cr-Ni alloys have coefficients of thermal expansion about three times the values for ceramics, thus posing difficult joint design problems.

Westinghouse has worked on the development of a ceramic stator under both ARPA and EPRI contracts.[55] The design chosen made use of ceramic blades, the ends of which were fitted into the ceramic housings that formed the walls of the annular gas passage entering the turbine. An eight-blade sector of a full-scale first-stage stator was subjected to about 100 h of testing in a burner test rig with a peak hot-gas inlet temperature of 1232 to 1465°C (2250 to 2670°F). When the peak temperature was 1370°C, the average hot-gas temperature was 1065°C (1950°F). Although this deviation from ideality in the burner test rig may seem large, it is similar to that in an actual engine. These cascade tests entailed over 200 cycles simulating startup, ramps up to and down from full power, and shutdown. Three of eight vanes survived operation to peak temperatures of 1465°C (2670°F). The blades were inspected after 25 and 60 cycles, and defective blades were replaced. Temperature changes were ramped at 25°C/s when simulating controlled emergency shutdown conditions, and the fuel flow was cut abruptly to simulate a load trip condition for a few temperature cycles.

General Electric has worked on the development of silicon carbide liners for combustion chambers.[56] Small-scale parts have been tested in a high-temperature, high-pressure burner test rig. Effects of transient test conditions such as those pro-

duced under flameout and restart conditions have been investigated. This work was directed toward engine designs in which the gas pressure of the combustion chamber would be contained by an outer metal housing; thus the silicon carbide combustion chamber liner would be subject only to the pressure differential within the combustion chamber and the fluctuating pressures associated with the intense turbulence inherent in any compact combustion chamber design. The investigation has included study of the corrosive effects of such elements as sulfur to be expected in less than perfectly clean fuels that will be derived from coal.

Pratt & Whitney, under NASA and ARPA funding, investigated two methods for mounting ceramic blades in metal turbine wheels.[57] The first of these makes use of a compliant platinum foil to give a good local stress distribution in fairly conventional blade root designs. The second approach makes use of a split rotor hub, the halves of which are brought together to enclose the bulb-shaped blade roots, and the rotor hub halves are diffusion-bonded together. The rotor hub has an electroplated platinum layer at the blade roots. Some of these components have been subjected to hot spin tests at temperatures up to 1200°C (2200°F).

In related programs, other development work has been carried out at the Solar Division of International Harvester[58] and at AiResearch, the latter entailing the design and construction of a ceramic version of the T-76 engine for use in a Navy patrol boat.[59] None has yielded any substantial amount of turbine operation up to the time of writing.

Principal Stress Problems

A number of stress problems arise as a consequence of the brittle character of ceramics. First and foremost is the sensitivity of ceramics to thermal shock. Second is the difficulty of avoiding stress concentrations in the complex geometries inherently required for gas turbines. Third is the impossibility of making parts in production that are perfect; hence the problem becomes one of producing parts with stress-raising defects that are as small as possible and ensuring that the defects are tolerable by inspection techniques. A fourth major problem area is concerned with the long-term stability of ceramic parts in a strongly oxidizing atmosphere with substantial amounts of contaminants that can prove corrosive. (See also Chap. 6, page 164.)

Thermal and Mechanical Stresses

It is easy to see that the thermal stresses associated with temperature transients are superimposed on large steady-state stresses in a turbine rotor. It is less obvious that great difficulty

TABLE 12.4 Basic Properties of Some Selected High-Temperature Ceramics and Superalloy Metals[57]

Material	Melting point, °F	Coefficient of thermal expansion × 10⁶/°F	Density, lb/in³	Thermal conductivity, Btu/(in·ft²·h·°F)	Specific heat, Btu/(lb·°F)	Young's modulus, psi × 10⁶ at 70°F
Ceramics						
Al₂O₃ Alumina (commercial grade)	3660	3.7 (70–500°F)	0.130	74	0.19	40.2
SiO₂ Fused silica (slip cast)	3100	0.29 (70–400°F)	0.067	2	0.15	3.8
C (ATJ graphite)	6360 (sublimes)	2.2	0.062	725	0.17	1.1
SiC Silicon carbide (recrystallized)	4080 (dissociates)	2.8 (70–400°F)	0.090	115	0.20	23.0
SiC Silicon carbide (hot pressed)	4080 (dissociates)	2.8 (70–400°F)	0.112	700	0.20	56.0
Si₃N₄ Silicon nitride (reaction sintered)	3450 (dissociates)	1.4 (70–1800°F)	0.090	28	0.26	25.0
Si₃N₄ Silicon nitride (hot pressed)	3450 (dissociates)	1.4 (70–1800°F)	0.112	105	0.26	31.0
Lithium-alumina-silicate (glass ceramics)	2400	−6 (750–1650°F)	0.08	7	0.20	12.0
3 Al₂O₃·2 SiO₂ Mullite (pressed and sintered)	3320	2.8 (70–212°F)	0.097	28.5	0.23	22.5
Metals						
Mar M 246	2450	7.24	0.305	117	≃ 0.10	29.8
Hastelloy-X	2470	9.0 (70–1600°F)	0.298	76	0.116	28.6
Inco 713C	2350	8.3 (80–1600°F)	0.286	146	0.10	29.9

has been experienced with blade vibration, leading to the failure of metal blades in both rotors and stators, and that vibratory stresses from fluctuating aerodynamic loads will be a major problem in nominally static components such as stators or combustion chamber liners; thermal stresses will be superimposed on these vibratory stresses. Thus, in highly stressed zones thermal cycling appears likely to cause small cracks that propagate slowly under vibratory stresses until they become large enough to cause an abrupt major failure under centrifugal or aerodynamic loads.

Physical Properties The thermal stress in a component depends on its physical properties. For a given temperature differential and part geometry, the thermal stresses are directly proportional to the elastic modulus and the coefficient of thermal expansion. Under transient temperature conditions, they are also inversely proportional to the thermal conductivity and directly proportional to the volumetric specific heat. Table 12.4 gives typical physical properties for the principal ceramics of interest here.

Allowable Stress The allowable stress in ceramics is much lower than the average flexural or tensile stress would imply because of the wide scatter of test data for brittle materials. A good idea of the problem is given by Fig. 12.25, from Ref. 60, which shows the failure probability of hot-pressed Si_3N_4 as a function of the short-time flexural strength at failure for 1200°C (2200°F) operation after either $\frac{1}{2}$ or 100 h of exposure to 1200°C. Although exposures up to 100 h reduced the failure probability for a given stress, probably because of a reordering of microcracks or formation of a surface oxide film in compression, longer exposures caused the failure probability to increase. Note that to reduce the failure probability to 1 percent implies an allowable stress about half that for the mean stress to failure. If a ceramic turbine includes 100 blades and the probability of a blade failure is to be kept to 1 percent, the design stress must be low enough so that a curve such as Fig. 12.25 gives a failure probability of only 0.0001, and this stress would be only about half of that for the lowest point plotted in Fig. 12.25. Analysis of extensive test work such as that of Fig. 12.25 led the author of Ref. 52 to limit the allowable design stress at 1200°C for a typical reaction-sintered Si_3N_4 to 75 MPa (10,900 psi).

Analytical Estimates of Stresses Produced in Fast Thermal Transients The heat-transfer coefficient in the high-velocity gas stream flowing through turbine stators and rotors is of the order of 5680 W/(m² · °C) [500 to 1000 Btu/(h · ft² · °F)] at the high pressure ratios (12 to 40) envisioned for turbine inlet temperatures > 1200°C. The abrupt drop in the temperature of the gas stream flowing through the stator and rotor in the event of a flameout will be of the order of 830°C (1500°F), giving a heat flux of ~ 4 × 10⁶ W/m² [~ 10⁶ Btu/(h · ft²)]. The distance from the combustion chamber to the turbine is ordinarily only ~ 1 m and the velocity is ~ 100 m/s. Thus the temperature change in the gas stream at the rotor inlet occurs in a few milliseconds. The surface of the blade begins to chill immediately, and the resulting temperature distribution as a function of distance into the blade from the surface is indicated in Fig. 12.26 for each of a series of time intervals following the flameout. Tensile stresses

are induced in the chilled surface region that reach a maximum a short time after the initiation of the temperature transient. Then, as the temperature change penetrates to the center of the blade from both sides, the temperature differential between the core and the surface of the blade is reduced and the thermal stresses are eased. However, this process takes place at different rates, depending on the chordwise position in the blade, with the temperature distribution reaching equilibrium in the trailing edge long before it reaches its new equilibrium value in the thick central portion. Thus, cracks can be expected to develop in the surface of the blade, particularly at the trailing edge, and cracks of this character are commonly found in tests. The effect of blade thickness on the maximum stress to be expected from a typical temperature transient of this sort is indicated in Fig.

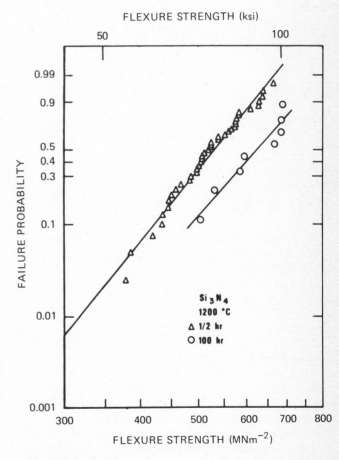

Figure 12.25 Failure probability as a function of breaking stress for hot-pressed silicon nitride.[60]

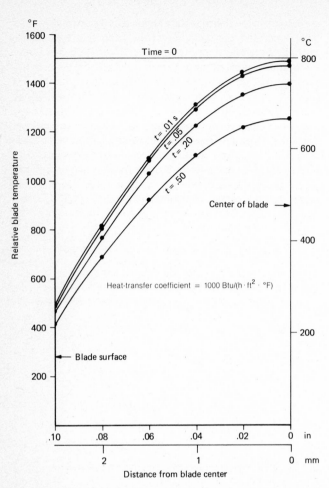

Figure 12.26 Transient temperature profiles in an infinite plate 5 mm (0.20 in) thick of reaction-sintered silicon nitride. (*Courtesy Oak Ridge National Laboratory.*)

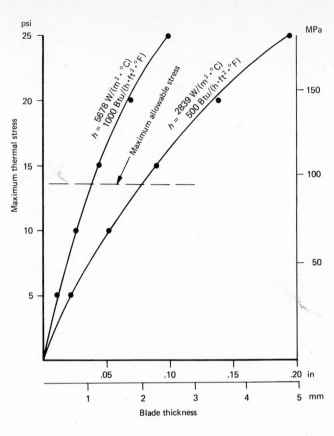

Figure 12.27 Calculated thermal stress following a flameout as a function of blade thickness for reaction-sintered silicon nitride for an initial temperature difference between the blade and the compressor air outlet of 834°C (1500°F) for two assumed heat-transfer coefficients. (*Courtesy Oak Ridge National Laboratory.*)

12.27, which was prepared for reaction-sintered silicon nitride. Note that the thermal stress would exceed the stress at which a tensile failure would normally be expected in the reaction-sintered silicon nitride favored for turbine blade applications if the blade thickness exceeds about 1.5 mm ($\frac{1}{16}$ in). Note also that in any operating turbine the thermal stress in rotor blades would be superimposed on aerodynamic, centrifugal, and vibratory stresses, so that failures probably would occur with substantially thinner blade sections than implied by a simplistic interpretation of Fig. 12.27.

The extreme severity of the air-blast quench following a

flameout can be judged by considering the classic curve for the heat flux to boiling water as a function of the temperature of the water (Fig. 11.25). The peak heat flux at the point of DNB runs ~110 W/cm² [350,000 Btu/(h · ft²)] with a temperature difference of ~50°C (~90°F). For higher metal surface temperatures a vapor blanket forms, and for a temperature difference of ~1000°C the heat flux drops to ~16 W/cm² [~50,000 Btu/(h · ft²)]. This has been confirmed by metallurgists in measuring the cooling rates of steel parts quenched in water from a red heat. Thus the air-blast quench to which ceramic turbine blades would be subjected in the event of a flameout is roughly 20 times more severe than a water quench.

Incidence of Severe Thermal Transients or Mechanical Impacts Good statistical data on the incidence of severe thermal transients for gas turbines in utility service are not available. However, data on the forced-outage experience of a major utility having 35 gas turbines give some insight on the matter. Over a 10-year period the mean operating time between forced outages was ~ 750 h, and ~ 25 percent of the outages stemmed from faults or failures in the control system. It seems likely that a fair percentage of these involved an abrupt interruption of the fuel supply. (Data from some other utilities indicate about three such incidents per 10,000 h.) Experience with gas turbines operating on blast furnace gas indicate an appreciable incidence of flameouts because of the poor flammability of this fuel. The low-Btu gas from a coal gasification plant is likely to present similar problems.

In another study it was found that 70 percent of the forced outages in gas turbines in utility service stemmed from human errors, many of them in maintenance.[61] An appreciable fraction involved small parts or tools inadvertently left in the air duct system. Any such object entering a ceramic turbine would have catastrophic effects. In this regard many feel that such an event would occur frequently enough so that just this consideration alone makes ceramic gas turbines impractical.

Thermal Stresses in Combustion Chamber Components Although the above analysis indicates that there is little probability of obtaining commercially useful ceramic turbine rotors and stators, the thermal-stress conditions in combustion chambers can be substantially less severe. The gas velocities are much lower, and hence the heat-transfer coefficient for an air quench following a flameout is much lower. On the other hand, during a start or an increase in power, the radiant heat flux can be quite high—as much as 160 W/cm^2 [500,000 Btu/(h · ft^2)]. At least some of the components would be rather thin: ~ 3 mm. For some air-cooled parts the operating temperature would be well below the exit gas temperature, and this would reduce the heat flux associated with the air-blast quench following a flameout. Perhaps most important of all is the fact that combustion chambers can be designed so that ceramic liners or flameholders will not be subject to substantial tensile stresses because the pressure loads can be carried by an outer, air-cooled, metal can. Briefly, it appears that the prospects for employing ceramic components in the combustion chamber are better than for stator or rotor blades.

Creep Most ceramics of interest undergo creep and creep-to-rupture at the potential application temperature of 1200 to 1370°C (2200 to 2500°F). There has been some confusion of these inelastic creep strains with the ductility of metals. Such inelastic strains develop over long periods of time and therefore are not available for the mitigation of impact and thermal shock stresses imposed in very short time periods. In other words, if a ceramic were employed at a temperature sufficiently high that creep could take place rapidly enough to relieve thermal or mechanical shock effects, the creep rate would be too high for the ceramic part to serve as an important structural element. Silicon nitride, for example, displays good resistance to thermal shock at 1370°C (2500°F), but its creep strength is so low at that temperature as to make it marginally useful as a structural element. Further inspection of Fig. 12.26 shows that in a flameout the highly stressed skin of a ceramic airfoil will be cooled to a temperature of ~ 650°C (1200°F). At this temperature, creep would not relieve the thermal stress, while the bulk of the blade—although at high temperature—will be subject to such a low thermal stress that creep could have little effect.

Accurate determination of thermal and other strain-controlled stresses is often confounded by the creep behavior. Such analyses tend to be carried out on the assumption of completely elastic behavior. The elastic strains are very small in view of the high modulus of elasticity of ceramic materials, and it follows that creep strains of the order of elastic strains can change the stress picture in a gross fashion.

Consequences of a High Modulus of Elasticity The principal difference in the impact resistance of metals and ceramics arises from the fact that the energy absorption in ductile deformation is much greater than the strain energy associated with elastic deformation. Further, even in the elastic range, ceramic materials have the disadvantage that they have a modulus of elasticity of around 344,000 MPa (50 × 10^6 psi). The elastic strain energy per unit volume can be written as $\sigma^2/2E$, where σ is the stress level. The deleterious effect of a high value for E is obvious.

Size Effects The size effect in brittle fractures can be significant, and neglecting it can be misleading. The statistical treatment of brittle fracture as originated by Weibull[62] in 1939 gives a good approximation of the size effect. The simplest form of the scaling relation may be written as

$$\frac{\sigma_L}{\sigma_S} = \left(\frac{V_S}{V_L}\right)^{1/m}$$

where σ = fracture stress
V = stressed volume
subscripts $_S$ and $_L$ = small and large
m = empirical constant, typically about 10

Thus, if one estimates the strength of a full-scale structural element from a specimen that has one-tenth the size (i.e., one-thousandth of the volume) of the intended stressed part, the small specimen turns out to be twice as strong as the intended part.

Stress Concentrations Where two parts are joined, as in the highly stressed region at the juncture between the blade root and the rotor, it is difficult to get perfect matching of the parts so as to obtain a perfectly uniform load distribution and avoid a stress concentration caused by local high spots.[63] When such stress concentrations occur with ductile materials, local yielding takes place and the loads are redistributed so that the stress concentration is relieved. This is the secret to the load-sharing characteristics and hence to the effectiveness of bolted and riveted joints in steel or aluminum structures, but yielding cannot occur in a brittle ceramic part. This is why we do not build truss bridges or other structures of glass in spite of the fact that glass is light, strong, cheap, and abundant. Similarly, this inability of a brittle material to accommodate deviations from ideality in loading is the reason for ASME pressure vessel code requirements for a high degree of ductility in vessel materials. See Chap. 6 for a more detailed discussion.

To cope with the load distribution problem, it is necessary to hold exceedingly close tolerances on the mating surfaces in ceramic parts,* a problem that is rendered more difficult by the necessity for using curved surfaces to avoid stress concentrations in corners. The problem can be eased somewhat by the use of a compliant layer such as the platinum shims used by Pratt & Whitney between the blade root and the rotor in its development test work.[57] However, the problem is aggravated by the fact that the coefficient of thermal expansion of the ceramics is low (Table 12.4) compared to that of the iron- or nickel-base alloys with which the ceramic parts must mate.

Small randomly distributed defects are present in any material. In wrought metal alloys these defects are greatly reduced by hot and cold working. This is, of course, not possible in ceramics. Also, the fabrication processes for ceramics are inherently more likely to yield parts with defects than would be found in a cast metal ingot.

Fiber Reinforcement To reduce the sensitivity of ceramic parts to the stress concentrations at small defects, it has been suggested that fine metal wires be incorporated in the ceramic part. The excellent properties of plastic-bonded carbon filament composites are well known, and they are used in some

special applications such as golf club shafts, fishing rods, and aircraft wing spars. A less well known example is the attempt by Rolls-Royce Aero Engines, Inc. to use epoxy-bonded carbon filament fan blades in the development of RB211 engine (the engine model later used in the Lockheed 1011 aircraft). Development was completed satisfactorily until failures were experienced in the bird ingestion test. Concurrently, the development program included limited commercial service of similar fan blades in the Rolls-Royce Conway engine. That service test went well until a plane in Nigeria went through an exceptionally heavy tropical rain that caused the epoxy-bonded carbon fiber blades to delaminate. These troubles made it necessary to shift to heavier titanium blades. These catastrophic failures in the engine development work were also catastrophic financially for the Rolls-Royce company.

In view of the fact that neither rain nor birds could reach the turbine, it seems possible that wire reinforced ceramic blades might be used in turbine blades. A key question is whether a filament-reinforced composite material can be fabricated to give adequate high-temperature strength and oxidation resistance, as well as the ability to withstand the thermal shock of flameouts. Carbon-bonded carbon filament composites have adequate mechanical properties at high temperature, but they are not usable at high temperatures in the presence of oxygen. The development of more appropriate composites using some of the many high-temperature filamentary materials has looked interesting.[64-68] Superalloy-bonded tungsten filament materials have been investigated,[64] but they have the shortcomings of high density and poor oxidation resistance for the refractory metal component. The Naval Air Systems Command supported a 4-year program at United Technologies Corporation for improving the impact resistance of hot-pressed silicon nitride by reinforcing it with refractory metal fibers.[65] Tungsten wire was tried but was dropped in the first year because of a reaction between the tungsten and the silicon nitride. Tantalum wire greatly increased the impact resistance of the bodies tested at both room and elevated temperatures, but the tantalum was subject to oxidation. Efforts to cope with the oxidation problem included the application of silicide and aluminide coatings on the tantalum, but these attempts were unsuccessful because the coatings either reacted with the silicon nitride or embrittled the tantalum.

A fairly comprehensive treatment of the basic problems is given in a British publication[68] which points out that reinforcing a brittle ceramic with ductile metal fibers inherently gives discontinuities that represent stress concentrations, and these lead to trouble with microcracking of the ceramic matrix. The microcracks tend to propagate under strain-cycling conditions because they constitute stress concentrations, and thus failure of the part results.

---

*AiResearch has held the blade root contours in its ceramic turbine work to within 0.005 mm (0.0002 in) in the latter 1970s.

Nondestructive and Proof Testing   The most effective means of detecting defects in ceramic parts for gas turbines appears to be the use of ultrasonic devices. A major difficulty with this approach is that density variations in the ceramic confuse the inspection procedure. Dye checks can be used but are less effective.

One of the most effective methods used to date for weeding out slightly defective parts for experimental units has been to subject individual parts to stress situations similar to those expected in a full-scale engine.[60] Turbine buckets, for example, can be individually spin-tested, but it is best not to test a large number in a single wheel, at least in preliminary sorting tests, because experience has shown that a single blade failure will wipe out all the blades in the wheel.

Corrosion

Although ceramics are much less subject to attack by oxidation than metals, all the ceramics that have been tested are subject to some degradation after prolonged operation in high-temperature combustion gases, particularly if contaminants such as sulfur are present. The ceramics that have stood up best in long-term endurance tests are silicon carbide and silicon nitride, and, in both cases, the reason for the good resistance to attack appears to be that oxidation acts to form a protective coating of SiO_2.[50] Small cracks that may develop in such a coating are inherently self-healing, much like the protective aluminum oxide coating on aluminum and the chrome oxide coating on stainless steel.

There are indications that other long-term effects and types of corrosion may be operative. Anyone who has employed silicon carbide electric heating elements in high-temperature furnaces knows that they seem to lose strength and become fragile after a few thousand hours of operation. It is possible that trouble of this sort might develop with silicon nitride because the partial pressure of nitrogen required over this material to prevent decomposition increases rapidly above 1100°C (2012°F). Another cause for concern is that any ash present in particulate form in the high-temperature airstream will probably be molten at temperatures above 1100°C and will tend to form films of slag on the surface of the blades. This slag is likely to be quite reactive with the SiO_2 protective coating, so that it then may react with the base material. If local temperatures rise to 1400°C (2552°F), the SiO_2 tends to break down into SiO in local zones in which the oxygen partial pressure is low, and, once the protective layer of SiO_2 is reduced to SiO, fairly rapid oxidation of the base material may take place. Any of these problems may be aggravated by the fact that solid-state diffusion processes tend to take place fairly rapidly at temperatures over 1000°C, and diffusion barriers tend to be ineffective. The cumulative effects of these considerations have led some experts in the ceramic field to state that the upper limit imposed by corrosion in gas turbines may be only 100 to 200°C higher for ceramics than for the nickel-chrome-cobalt alloys.

Summary

A review of 40 years of experience gained in efforts to build ceramic gas turbines, together with analyses of the implications of their basic brittle character, leads one to conclude that ceramics are inherently unsuitable for use in gas-turbine stators and rotors because of thermal stresses in a flameout and mechanical shocks from stray foreign objects. They may, however, serve for some combustion chamber parts.

COAL-FIRED GAS TURBINES

The advantages of using low-cost coal instead of high-cost petroleum distillates to fuel open-cycle gas turbines have led to a series of programs since 1944 that have been designed to exploit this possibility. Each of the earlier programs foundered on difficulties with turbine blade erosion and deposits and was eventually terminated. As discussed in the next chapter, developments in fluidized-bed combustion have generated great interest in the use of this type of combustor in a combined gas turbine–steam cycle, and this has led to new efforts initiated in the 1970s. The problems are so complex and subtle that they deserve a comprehensive survey, which properly includes not only the work with coal-burning gas turbines but also experience such as that with wet vapor erosion in steam turbines, dust erosion in helicopter engines, etc. (More detailed reviews of one phase or another of these problems may be found in Refs. 69–71.)

Operating Experience

As with water-cooled and ceramic gas turbines, the coal-fired gas turbine is an attractive concept that appears plausible, but the seriousness of the problems can be appreciated only by a review of operating experience. Hence all the experience available deserves a critical review with particular attention to problems that proved to be serious barriers.

Locomotive Development Program

The success of the Swiss oil-fired gas-turbine locomotive in the early 1940s, mentioned earlier (see the fifth item in Table 12.1), led to a series of efforts to develop coal-fired gas turbines for railroad locomotive service in the United Kingdom and the

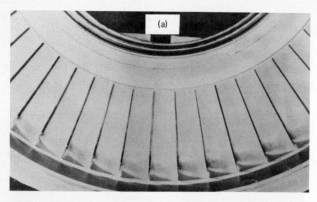

Figure 12.28   Photos of the third-stage stator after two tests in the turbine of the Locomotive Development Committee. Stator in (a) showed no measureable erosion after 300 h with a turbine inlet temperature of 570° (1060°F) and properly functioning ash separators (<1 percent plus 20 μm, <10 percent plus 10 μm). Stator in (b) was severely eroded at the trailing edges toward the blade roots after 757 h at a turbine inlet temperature of 663°C (1255°F) with periods in which the ash separator did not function properly. (*Courtesy of R. T. Sawyer.*)

United States immediately after World War II. There were strong incentives to employ the railroads' traditional fuel—coal—in an engine with a higher thermal efficiency than the 6 to 8 percent obtained in steam locomotives, and an engine that did not require more water than fuel was particularly attractive. Diesel locomotives had come into use, and oil-fired steam-turbine locomotives were under development, both giving much better thermal efficiencies than the conventional steam locomotive. The coal-burning gas turbine showed promise of giving a good thermal efficiency, a more compact and less expensive power plant, and would operate on a lower-cost fuel; hence an effort was initiated by Bituminous Coal Research, Inc., and supported by some of the major railroads and coal companies. They set up and funded the Locomotive Development Committee (LDC) to carry out a program to develop a coal-burning gas turbine for locomotive service. Interestingly, in a prescient December 1945 ASME paper,[72] K.A. Browne of the LDC stated that one of the major reasons for the program was the fact that U.S. petroleum deposits would be seriously depleted in 30 years, a situation that would require the United States to import much of its oil supply and thus would make the nation highly vulnerable in time of war.

The LDC effort began with a series of experiments on what were correctly envisioned to be the key problems, namely, the combustion chamber, particle separation from ash-laden hot combustion gas, turbine blade erosion, and ash deposition. The results were encouraging: A high volumetric heat release rate of over 5 J/(s · cm³) [5 × 10⁵Btu/(h · ft³)] was obtained with a vortex combustor, 90 percent of the fly ash in the size range above 10 μm was removed with cyclone separators, and erosion and/or deposition rates on turbine blade specimens in cascade tests with hot-gas temperatures up to 700°C (1300°F) appeared to be tolerable. Tests were then initiated using a turbine built for a Houdry catalytic cracking unit in a petroleum refinery.[13] This unit was run with a turbine inlet temperature of 482°C (900°F) with no serious problems.

A prototype coal-burning gas turbine designed for locomotive use was then obtained and, between 1951 and 1958, it was operated for more than 4000 h with turbine inlet temperatures of 675°C (~1250°F). Severe blade erosion necessitated three partial replacements of the blades in spite of a series of improvements that increased the effectiveness of the particle separation equipment. A review of the operating experience discloses that most of the erosion occurred in the upper stages where the gas temperature was above 565°C (1050°F) (Fig. 12.28). The importance of gas temperature is also indicated by the fact that the LDC operated the turbine with an inlet temperature of 565°C (1050°F) for approximately 500 h with negligible erosion and deposits. Blade erosion and deposits with a turbine inlet temperature of 675°C (1250°F), the minimum value for a useful power plant, proved to be such difficult obstacles to the coal-burning turbine development that the LDC program was terminated in 1959. The cumulative cost of the program in 1950 dollars was about $5,500,000.[76,77]

Concurrently, a similar though smaller effort in England was carried out with no greater success. In Canada the turbine erosion and deposit problems of the coal-burning gas turbine were circumvented by employing a heat exchanger to heat the air flowing from the compressor to the turbine. The fairly high temperature turbine exhaust air was then fed to a coal-fired cyclone burner that served as the heat source for the system.[78] There were no problems with the turbine, but, even with the much lower combustion gas velocities in the heat exchanger, ash deposits led to a substantial loss in performance in a few hundred hours, and severe corrosion of the ferritic alloy tubes occurred. The turbine air inlet temperature generally ran ~650°C (~1200°F).

The Union Pacific Railroad, wishing to utilize its large coal reserves, commissioned Alco and General Electric to modify an oil-fired gas-turbine engine so that it could be fired with coal and operated in a locomotive to carry the LDC program to its logical conclusion.[79] Experience with this unit in rail service was disappointing; in the initial operation, the power fell off badly in 200 h. Improvements were made in the cyclone separation system used to remove the dust, and a second test was run in which operation up to 400 h was obtained before the loss in power output became so severe that the project was terminated.

Concurrently, a program was initiated at the Bureau of Mines Research Center in Morgantown, where the gas turbine run initially by the LDC at the American Locomotive Works in Dunkirk, New York, was operated after it had been modified in an effort to reduce the severity of turbine bucket erosion.[80,81] A new set of blades of an improved design was installed. Those in the first-stage rotor and second-stage stator were omitted, however, to provide an annular space for centrifuging ash from the airstream out to the casing and thus prevent recirculation of the ash particles, an important factor contributing to erosion damage. Titanium carbide wear strips were installed at key points (particularly where severe erosion had occurred at the base of the rotor blades) to improve resistance to abrasion. The first test carried out in 1963 was terminated after 878 h as a consequence of ash deposits, which caused a severe loss in power in spite of the fact that the turbine was cleaned at intervals of not more than 81 h throughout this initial test. A number of modifications were made in an effort to reduce the difficulties with deposits, and a second test of 1085 h was carried out. In this second series of tests, the dust content in the gas entering the turbine was around 130 ppm by weight, with over 90 percent of the particles having a size below 10 μm. Conditions were improved, but it was still necessary to clean the turbine to remove ash deposits at intervals of no more than 89 h. In the second test the ash deposits could be removed completely by cleaning and so further operation would have been quite practicable. Estimates of turbine bucket erosion rates led to the conclusion that a 5000-h life could be obtained for the stator blades and a 20,000-h life for the rotor blades.

In discussing the whole complex of tests as outlined above with J. P. McGee (who was responsible for the program), the writer learned that the program was terminated because it was felt that the particulate content would have to be kept below 20 ppm and because the personnel involved decided that there was

no practical way to do this with a high-temperature gas stream. For example, concurrent work carried out by General Electric indicated that electrostatic precipitators were ineffective in the temperature range required for a viable gas turbine because conductive dust accumulated on the electric insulators, preventing maintenance of an adequately high voltage on the collector plates.

Coal-Burning Gas-Turbine Development at ARL in Australia

A development program was carried out in the 1960s at the Aeronautical Research Laboratory (ARL) in Melbourne, Australia, at a total cost of about 1 million dollars.[69,82] This program culminated in the operation of a Ruston and Hornsby engine with a nominal rating of about 1000 hp at 650°C (1200°F). This was preceded by tests with both an aircraft exhaust turbosupercharger and a 60-hp Solar gas turbine. Extensive tests of components such as combustion chambers, cyclone separators, and blade cascades were carried out to give many important insights into the phenomena encountered. The fuel used was brown coal from Victoria in the initial tests and subsequently bituminous coal from New South Wales. Early experience with turbine bucket erosion and deposits was similar to that in the program of the LDC as outlined above. Subsequent work was carried out with the turbine redesigned for much reduced turbine wheel tip speeds, i.e., 244 instead of 335 m/s (800 instead of 1100 ft/s). In the last test the erosion problem was essentially eliminated, but deposits still presented difficulties and required cleaning with water injected along with crushed nutshells at frequent intervals (less than 100 h). On the basis of the erosion experience in the final test of about 100 h, the blade life was estimated to be about 5000 h.[82]

The total running time with the Ruston and Hornsby engine was about 500 h. The work as a whole indicated that erosion and deposits depend heavily on both the type of coal used and the engine design. There was not sufficient operation to give significant data on corrosion.

Tests with Fluidized-Bed Combustors

Experiments at BCURA with a pressurized fluidized-bed coal combustion system have included a number of runs in which the hot gases leaving the bed at about 6 atm have been directed through a nozzle into a cascade of blades representing a turbine rotor.[83,84] Tests run with inlet gas temperatures of around 843°C (1550°F) showed relatively little in the way of deposits or erosion for periods of 500 h of operation (but it should be remembered that LCD experience showed that cascade tests are not definitive). One test was carried out with an inlet gas

Figure 12.29 Photograph of a two-vane section of the first-stage stator from an LM2500 marine gas turbine after 200 h of cycled operation at the Philadelphia Navy Yard. The turbine inlet temperature was up to 1150°C (2100°F) about half the time. Note that deposits on the blades have partially blocked the cooling-air discharge ports near the leading edge of the vane at the left, and blockage of the ports near the leading edge of the vane at the right led to severe overheating and burning of a slot through what was the cooling-air discharge port region. (*Courtesy of the Philadelphia Navy Yard.*)

temperature of 870 to 925°C (1600 to 1700°F). This yielded appreciable deposits with little erosion, but photomicrographs indicated appreciable corrosion. In general, the corrosion was comparable to that found with oil-fired gas turbines operating at the same temperature, but the deposits were substantially greater and would require periodic cleaning.

The Combustion Power Company began work on coupling a gas turbine to a fluidized bed burning solid wastes in 1969 and subsequently shifted in 1974 to operation with coal.[85-88] By 1976, the unit had accumulated approximately 600 h of turbine operation at an inlet temperature of approximately 760°C (1400°F). The hot-gas clean-up system for these tests was operated at temperatures of 815 to 970°C (1500 to 1600°F) with cooling downstream of the last stage of particle separation so as to not exceed the design turbine inlet temperature of 788°C (1450°F). The hot-gas clean-up devices used in this series of tests consisted of tangential entry cyclones and 15.2- and 7.6-cm (6- and 3-in) multicyclones. In each test the multicyclones became ineffective after a few hours of operation owing to ash

plugging, which resulted in heavy fouling and erosion in the turbine. Subsequent work has been directed toward the use of granular-bed filters.[88]

Catalytic-Cracking Units

The use of gas turbines to provide combustion air to burn out the coke that accumulates on the Al_2O_3 pellets used as the catalyst in petroleum refinery catalytic-cracking units[13] was confined to operation with turbine inlet temperatures largely below 425°C (800°F) until the 1950s, and hence little or no net useful power was obtained.

In 1950, the Shell Development Co. made the first serious effort to recover power from a petroleum catalytic-cracking unit by introducing a higher inlet temperature turboexpander into the flue gas stream directly downstream of the regenerator.[89] Because there was nothing ahead of the expander to reduce the particulate level, the performance deteriorated rapidly, and by the end of 750 h the turbine was virtually useless.

A review of the Shell Development Co. experience and the LDC work indicated that not only should the particulate content of the gas be reduced, but also every effort should be made to keep the particles uniformly dispersed in the gas stream to avoid severe local erosion, particularly at the roots and tips of the blades. This led to the decision to use a single-stage turbine with a long, straight inlet passage. This decision, together with extensive development work on particle separators, led to a test with three stages of cyclone separators conducted in 1957 by the Shell Development Company.[89,90] After 4000 h of continuous operation, inspection of the blading revealed no serious erosion. Commercial installation of these units began in 1963 and was extended to eight refineries by 1973. The total power recovery capacity of the eight installations was approximately 62 MW with an estimated turbine life of 25,000 to 40,000 h.[89,90,91] These units operate with a turbine inlet temperature of about 594°C (1100°F). Turbine inlet temperatures as high as 675°C (1250°F) have been used with Inconel X blades coated with tungsten carbide to give a turbine life of 2 to 3 years.[90] Extensive experience with these units indicates that erosion can be kept to a tolerable level with two stages of cyclone separators followed by a third-stage multicyclone (which gives a particulate loading of ~100 ppm*) for operation at turbine inlet temperatures up to 620°C (1150°F) if a single-stage turbine is used with aerodynamically clean inlet conditions permitting a uniform distribution of the particles across the inlet face of the turbine blading to prevent severe local erosion, and if the

---

*1 ppm = 1.22 mg/m³ = 0.0345 mg/ft³ = 0.00054 gr/ft³; 1 mg/m³ = .000442 gr/ft³

spouting velocity from the nozzle box into the wheel is kept below 488 m/s (1600 ft/s). (See Refs. 89–96.)

Gas Turbines Operated on Low-Btu Gas from Blast Furnaces and Gasifiers

A substantial amount of experience has been obtained in the United States and in Europe with gas turbines operating on low-Btu gas from blast furnaces with turbine inlet temperatures around 730°C (1350°F). After extensive testing of hot-gas clean-up equipment, it has been found best to cool this gas and clean it with a two-stage wet scrubber and often an electrostatic filter before burning it in a combustion chamber ahead of the turbine.[97-102] The particulate content in the gas fed to the turbine is commonly kept below 1 ppm by weight. The same approach and requirement have been imposed on the low-Btu gas supplied from Lurgi gasifiers to the gas turbine of the combined-cycle plant at Lünen, Germany.[103,104]

A typical example of turbine performance with blast furnace gas is a Sulzer gas turbine employed for supercharging a blast furnace at the Hainaut-Sambre works in Belgium. The turbine began operation in 1955, producing 7500 kWe with a gas inlet temperature of 710°C (1310°F). Erosion of the first-stage turbine buckets progressed until some loss in efficiency occurred after 81,000 h of operation, at which point these blades were replaced. However, the second- and third-stage blades were still in satisfactory condition after 136,000 h of operating time.

Deposits in Conventional Gas Turbines from Ash in the Fuel and Dust in the Inlet Airstream

There has been a substantial amount of trouble with losses in efficiency in commercial gas turbines caused by erosion and deposits from ash in the fuel oil and dust in the inlet air (Figs. 3.22 and 12.29). For this reason, gas-turbine manufacturers specify that the ash content of the fuel oil should be less than 0.003 percent and the particulate content of the inlet airstream should be less than 1 ppm.[105,106] (Air dilution makes these two values essentially the same limitation on the particulate content in the hot gas.) Air filters are recommended and are commonly used because the dust content in many environments is often above this level.

Basic Studies of Erosion

There have been many studies of erosion phenomena that give some insight into the mechanisms involved and the principal parameters affecting erosion rates.[107-114] These studies indicate that the erosion rate depends on the kind of particle, its size, the angle of incidence relative to the surface on which it impinges, the hardness of the surface, the velocity of impingement, and the particulate content of the gas stream. The effects of each of these are discussed in the following sections.

Effects of Particle Character

Extensive tests carried out with two extremely different types of particle, i.e., angular grains of silica sand and water droplets, give some valuable insights. Both analyses and experiments indicate that the damage mechanism is fundamentally different for these two types of particle. The maximum damage caused by a particle of sand occurs at an incident angle of about 35 degrees relatively to the surface, and the resulting damage appears to be primarily a scoring of the surface[109] (Figs. 12.30 and 12.31). With liquid droplets, on the other hand, the maximum damage occurs when the angle of incidence of the particle is 90 degrees and the form of the damage appears to be either of two types. There may be plastic indentation of the blade material and subsequent failure by low cycle fatigue if the material is soft and ductile. In hard materials, the damage may be similar to that in the races of ball bearings; i.e., high shear stresses may be induced below the surface by the impact, and fatigue may cause cracks to develop and spread in a plane parallel to but below the surface until material spalls off and leaves pits. Further, liquid droplet erosion in wet vapor turbines also differs in a fundamental way from solid particle erosion in gas turbines in that the micron-size droplets of moisture that form in the wet vapor do

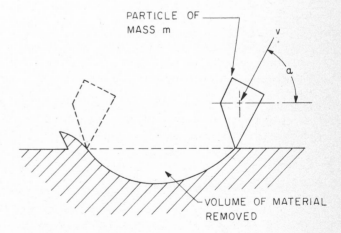

Figure 12.30 Mechanism of material removal by an impacting particle.[109]

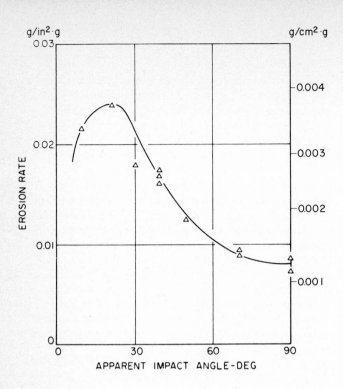

Figure 12.31 *(Left)* Erosion loss in weight per unit area per unit weight of incident particle plotted as a function of the apparent impact angle (tempered C-1050 steel, 25 ppm or 0.013 gr/ft³ dust concentration; 0-74 μm silica flour; 255 m/s or 835 fps airstream velocity).[109]

Figure 12.32 *(Below)* Erosion versus number of impacts on metal specimens rotated through a water jet. Impact velocity = 125 m/s, jet diameter = 1.3 mm.[112]

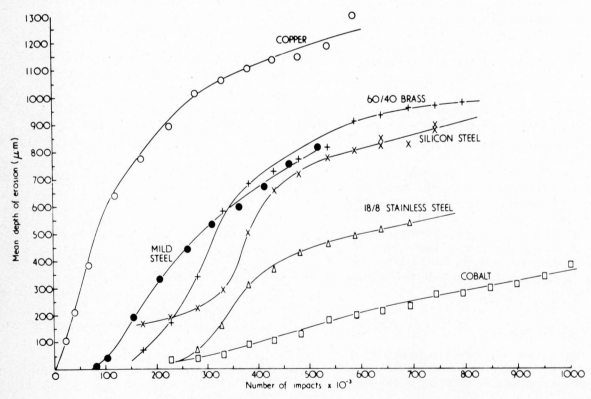

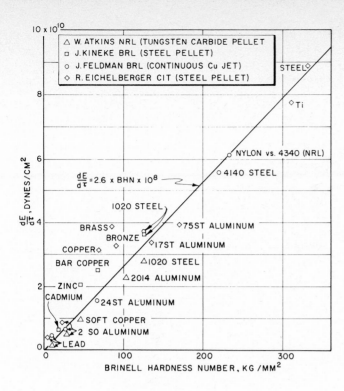

Figure 12.33 Effects of target material hardness on the ratio of the kinetic energy in the incident particle to the volume of material removed in hypervelocity impacts. (*Eichelberger and Gehring, Ballistics Research Labs, Report No. 1155, Dec., 1961.*)

not cause erosion directly. Rather, they impinge on stator vanes, the liquid film that forms flows to the trailing edge, and large droplets are then shed from the trailing edges of stator blades.[110,111] It is these large droplets that are responsible for erosion in the rotor. This is partly because a thin liquid film, which tends to form on the surface of the rotor blade, acts as a cushion to protect the blade from very small droplets whose diameter is not many times greater than the thickness of the liquid film. As a consequence, the measures taken by both the designer and the operator to cope with turbine blade erosion in wet vapor turbines are quite different from those in gas turbines that ingest solid particles.

It has been suggested by H. R. Hoy[83] that the erosion of blades in a gas turbine coupled to a fluidized-bed coal combustion chamber might be much less serious than if a pulverized-coal burner were employed. The reason given was that the ash formed in a fluidized bed tends to be soft and friable, having been formed well below the fusion point, whereas the ash particles in a pulverized-coal burner are vitreous cinders that are formed well above the fusion point of the ash and subsequently chilled by the secondary air to give vitreous particles. The writer has not yet found any clear-cut experimental demonstration that this hypothesis is valid, and, although it appears reasonable, as will be discussed later, the substantial percentage of angular particles of refractory oxides present in the ash from FBCs can cause serious erosion.

Effects of Blade Hardness

The choice of blade material influences the erosion rate. Figures 12.32 and 12.33 show the effect of metal target hardness on erosion rate and support one's intuitive feeling that increasing the hardness of the blades should reduce the erosion rate. It also supports Hoy's thesis that the softer particles from a fluidized bed should be less erosive than ash from conventional burners.

Experience with steam turbines has indicated that a very hard cobalt-base alloy, such as Stellite, is exceptionally resistant to erosion by wet vapor,[113] and experience with gas turbines is consistent with the steam-turbine experience[80,114] (Fig. 12.34). Inasmuch as cobalt-base alloys are also exceptionally good from both the high-temperature-strength and scuff-resistance standpoints, and are commonly used in gas-turbine blades, at least some of the usual materials of construction for turbines are of about as erosion-resistant an alloy as one might find.

It should be noted that, in the work carried out by the Bureau of Mines with the turbine from the LDC program, the use of titanium carbide inserts near the blade roots largely eliminated erosion with no changes in the particulate content of the gas or in the turbine wheel tip speed.[80] This, together with the Shell experience with tungsten carbide coatings for blades in units for cat cracker service,[90] also supports the intuitive feeling that increasing the turbine bucket hardness will reduce the erosion rate.

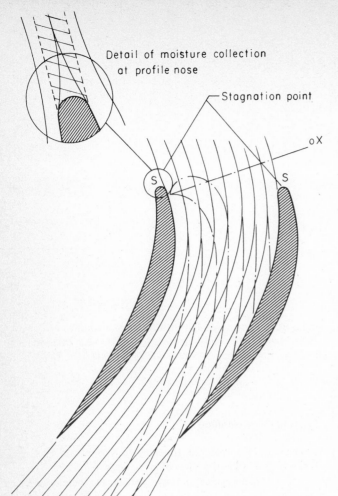

Detail of moisture collection at profile nose

Stagnation point

oX

S   S

Figure 12.34 *(Left)* Steam and water droplet paths between stator blades. *(Courtesy Westinghouse Electric Corp.)*

Figure 12.35 *(Below)* Erosion loss for five dust particle size ranges [Pearlitic C-1050 steel; 25 ppm (0.013 gr/ft³) dust concentration; 255 m/s (850 fps) airstream velocity; 40 degrees apparent impact angle].[109]

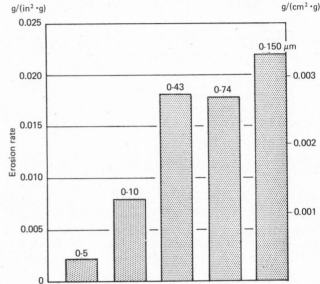

Effects of Particle Size

Both analyses and experiments indicate that the smaller the particle, the more nearly it tends to follow the airstream and the less inclined it is to be centrifuged out so that it will impinge on a blade. This effect is shown in Fig. 12.34. The data available indicate that particles smaller than about 10 μm tend to follow the airstream sufficiently well so that, as the particle size is reduced further, there is progressively less difficulty with turbine blade erosion. Note that the smaller particles are more inclined to erode the blade near the trailing edge than near the leading edge.

The effects of particle size on the erosion rate are indicated by Fig. 12.35. These data were obtained with silica particles in the same series of tests as the data of Figs. 12.31 and 12.32. Somewhat similar data are plotted in Fig. 12.36 for units employed with fluidized-bed catalytic-cracking systems in a petroleum refinery.[115] The data of Fig. 12.36 indicate that the

erosion rate is inversely proportional to the particle size for a given particulate content. Note that these curves were extrapolated linearly beyond the data points for the smallest particle size—i.e., about 5 μm—and that Fig. 12.35 indicates that the damage would be less serious for the smaller particle sizes than implied by the curve extrapolations of Fig. 12.36.

Effects of Tip Speed

Turbine bucket erosion is strongly affected by the turbine wheel tip speed, i.e., by the velocity of impact of the particle on the turbine blade.[116] Experiments indicate that the erosion rate varies as about the fifth power of the tip speed (Fig. 12.37),[109] and this suggests that the erosion problem can be drastically eased by reducing the design tip speed of the turbine.[108] This thesis has been validated by the Australian work in which reduc-

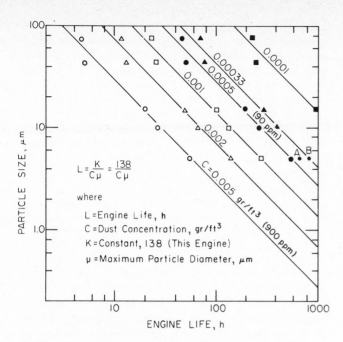

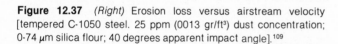

Figure 12.36 *(Above)* Particle size versus usable safe gas turbine life in cat cracker service (maximum weight loss of 21 gr).[114]

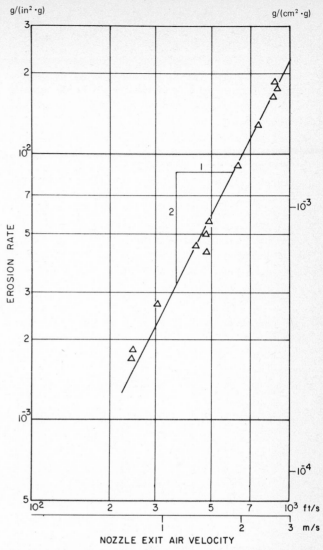

Figure 12.37 *(Right)* Erosion loss versus airstream velocity [tempered C-1050 steel. 25 ppm (0013 gr/ft³) dust concentration; 0-74 μm silica flour; 40 degrees apparent impact angle].[109]

ing the tip speed from 335 to 244 m/s (1100 to 800 ft/s) essentially eliminated erosion as a problem.[69] However, this approach has the disadvantage that the work output per stage varies as the square of the tip speed. Hence cutting the tip speed from 335 to 244 m/s would in effect require increasing the number of stages by a factor of about 2, thus increasing the cost of the turbine by nearly as great a factor.

Deposits

As the size and concentration of particulates entering coal-fired gas turbines have been reduced, thus easing erosion problems, difficulties with deposits have increased. These deposits have formed under a wide variety of conditions with widely differing types of particle, and their character has varied accordingly. It is significant that in military gas turbines, where the dust consists primarily of silica and feldspars (having melting points in the 1100 to 1700°C range, i.e., 2012 to 3092°F), the dust is not inclined to stick; thus deposits build up only in low-velocity regions where local aerodynamic conditions favor deposition.[107] Coal ash, on the other hand, deposits in high-velocity regions, and its character is different. In the LDC and Australian tests, the combustors were similar to conventional gas-turbine combustors (Fig. 12.7). The pulverized coal was burned under near-stoichiometric conditions, giving flame temperatures well above the fusion point of the ash, so that the sulfates were vaporized and consequently passed in vapor form into the turbine, where some could condense on the blades. The ash, when quenched by the secondary air, emerged from the combustor as vitreous particles with melting points much lower than silica or alumina. In the BCURA and Combustion Power tests with FBCs, the temperature in the combustion zone was 800 to 900°C (1472 to 1652°F), well below the fusion point of the ash. Some of the sulfates were not vaporized, and the bulk of the particles were soft and friable calcined minerals (though microscopic analyses have shown that roughly 25 percent of the fine particles elutriated from a fluidized bed are dense angular particles of silica).[87]

Effects of Low-Melting Constituents

With either pulverized-coal or fluidized-bed combustion, both the low-melting-point sulfates and some of the glassy components of the ash are still sticky well below their melting points, and so they tend to adhere to blade surfaces and form a sticky, flypaper-like film, particularly if they impact at high velocity. [The kinetic energy associated with an impact at 610 m/s (2000 ft/s) is sufficient to raise the temperature of a dust particle

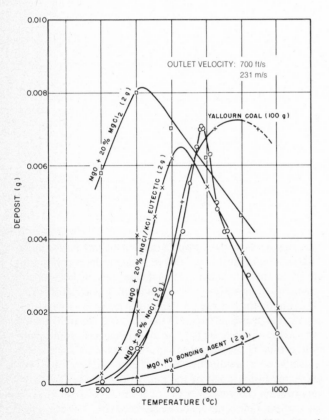

Figure 12.38 Effects of temperature on the deposition rate of coal ash and various mixtures of MgO and chloride salts fed into the hot-gas stream of a turbine simulation test rig in Australia. The particle size used was nominally 5 μm. (NaCl melts at 801°C.)[69]

220°C (396°F).] Higher-melting-point materials, such as silica and alumina, tend to become trapped in the sticky film, which acts as a binder. Note, too, that the sintering temperature of a solid is 75 percent of its melting temperature in absolute units, and the softening points of glasses have a similar relation to their melting points, so that an ash particle having a melting point of 1400 K (2025°F) might sinter under pressure (e.g., from an impact load) at 1950 K (1395°F). Thus, considering the effects of impact energy, sintering, and the low melting points of some constituents of coal ash such as sulfates and chlorides, the formation of hard adherent deposits on turbine blades is not surprising.

Evidence that low-melting components of coal ash act as a cement between higher-melting-point particles is given by an Australian investigation conducted in a test rig designed to simulate turbine conditions.[69] Tests were carried out both with ash from a typical Australian coal and with MgO containing low-melting salts that would serve as bonding agents (i.e., give a "flypaper" effect of the sort noted in the Bureau of Mines tests[80]). Figure 12.38 shows the results of this set of controlled experiments. Note that the amount of material deposited from the Yallourn coal ash dropped rapidly with a reduction in temperature, becoming practically zero at 500°C (932°F). Note also that pure MgO, which has a melting point of 2800°C (5072°F), when run with no bonding agent gave almost no deposits at 600°C (1112°F). However, the addition of low-melting salt in the form of sodium, potassium, or magnesium chloride, or eutectic mixtures of these materials led to both a much heavier deposit and a shifting of the temperature for the onset of deposition to a lower value. In an FBC, a mixture of calcium and magnesium sulfates will form a glass that is soft and plastic at about 700°C (1300°F) and may be expected to give the same mechanism for the buildup of deposits as the alkali metal sulfates. Deposits formed in the first-stage cyclone separators of the Combustion Power and BCURA units were apparently caused by this calcium-magnesium sulfate glass.[85]

Thermophoresis

It appears that an esoteric phenomenon known as thermophoresis may be an important factor in the formation of gas-turbine blade deposits.[117] One of the first strong indications of this was encountered in a 1974 gas-turbine acceptance test at the Philadelphia Navy Yard. Substantial deposits were noted after only a few hundred hours of operation under cycling conditions in which the average turbine inlet temperature was in the 1150°C (2100°F) range about half the time.[118] The cooling-air discharge ports in the first-stage stator blades were affected, and some blades were damaged by overheating (Fig. 12.29). Investigation disclosed that the bulk of the deposit consisted of submicron-

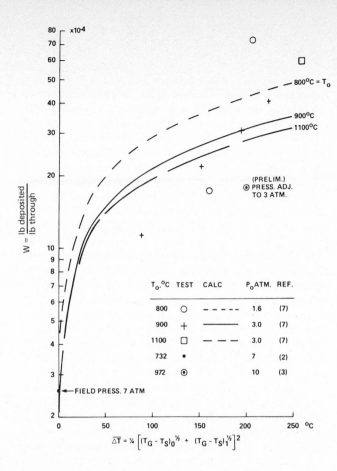

Figure 12.39 Deposition rates in a cooled cascade (T_G = gas temperature, T_S = surface temperature, subscripts 0 and 1 refer to cascade inlet and outlet conditions respectively).[117]

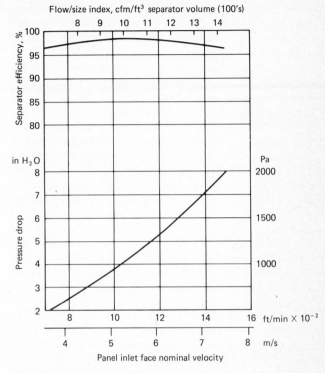

Figure 12.40 STRATA Panel multiple cyclone separator performance characteristics for 1.5 × 6 in tubes in parallel series. Performance is based on a scavenging airflow requirement of 15 percent of the airflow rate.[116]

size Fe_2O_3 particles that entered the engine with the inlet air that had a particulate content of only 0.06 ppm. Further investigation disclosed that Philadelphia air was not dirtier in this respect than the air in most localities, and that the deposits did not form when the nominal turbine inlet temperature was limited to 982°C (~1800°F). Note that the melting point of the Fe_2O_3 is 1570°C (2860°F), which implies a sintering temperature of 1110°C (~2030°F). This appears to be the reason that this type of deposit did not prove a problem until operations at a turbine inlet temperature of 1150°C (2100°F) were initiated. Note, too, that coal ash commonly contains ~5 percent Fe_2O_3.

Work by NASA,[119] Westinghouse,[120] and United Technologies Corporation (UTC) seems to be consistent with the Navy experience. One of the interesting analyses of the problem was evolved by G. Vermes at Westinghouse to explain the markedly heavier deposits found on both the pressure and suction surfaces in engines with strongly cooled blades.[117] He has shown that the temperature differential between hot particles in the gas stream and the cooler metal blade surfaces induces a force that drives the particles toward the blade surface and causes them to

adhere—the phenomenon known as *thermophoresis*. As shown in Fig. 12.39, his analytically derived relations correlate a substantial amount of experimental data suprisingly well and support his thesis that the greater the temperature difference and the smaller the particle, the greater the deposition rate.[117]

Particle Separators

In most of the coal-burning gas-turbine programs, large conventional cyclone separators have been employed to remove ash particles from the hot gases flowing out of the combustion chamber. These are effective in removing most of the dust down to about the 20-μm-size particles. Two stages are commonly employed with the large particles being removed in the first stage and the smaller particles in the second. The most effective method found for removing still smaller particles has been the use of a large number of small-diameter (~4-cm) cyclone separators operating in parallel. These units were used in the LDC program and were further developed for aircraft and cat cracker service.[90,121] Figure 12.40 shows the particle removal efficiency of a unit of this type and the pressure drop as a function of the airflow rate per unit of inlet face area of the separator bank.[121] These data were for dust in which 25 percent by weight of the particles had equivalent diameters of less than 10 μm.

Removal of particles of less than about 5 μm is best accomplished with some form of filter if the gas temperature is below around 260°C (200°F), while electrostatic precipitators can be used at temperatures up to about 538°C (1000°F).[122] As indicated previously, at higher temperatures the coal ash becomes sufficiently conducting that deposits formed on the electrical insulators produce high-tension shorts that prevent the buildup of an adequate voltage to give effective electrostatic precipitation. Further, low velocities are required, and hence the size and cost of the large pressure shell required to house the large volume of precipitator plates present serious problems for gas-turbine applications.

Several types of granular-bed filters have been used to remove fine particulates. These may confine the granular material with fine wire screens, or they may use a set of louvers arranged as shown in Fig. 12.41. In operation they characteristically develop a filter cake of particles covering the inlet face, which greatly increases the pressure drop and must be removed. For example, in tests of an FBC at Exxon Research, a set of granular-bed filters was operated downstream of cyclone separators.[123] The first set of these filters was run with hot-gas downflow through a 50-mesh inlet screen and a bed of 250- to 600-mesh particles of crushed quartz. The inlet screens clogged quickly, so that it was necessary to clear them at intervals of 5 to 10 min by blowing back first one and then another of several units operating in

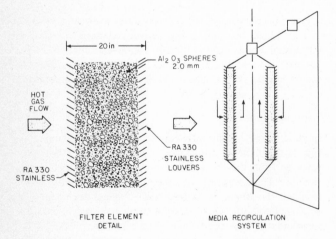

Figure 12.41 Schematic diagram of the Combustion Power Company design for a granular-bed filter. (*Courtesy Oak Ridge National Laboratory.*)

parallel. However, the backflow operation was only partially effective in clearing the units and reducing the pressure drop to the design range; the pressure drop became excessive within 24 h of running. The filters were then modified by removing the inlet screens; this helped but led to excessive losses of the fine silica particles from the filter bed during the blowdown operation. The problem was corrected by changing to coarser particles of crushed alumina (840 to 1400 μm). Tests with these at the time of writing indicate that when a clean bed is used at the start, the particulate content in the exit gas stream can be held to the EPA specification for stack emissions of 0.05 gr/ft³ (gas volume at standard conditions). However, the fine dust tends to intermix with the granular bed when it is fluidized during the blowback operation, and after a few hours of running, the particulate content of the gas leaving the bed rises to the point where it exceeded the target limit of 0.05 gr/ft³ (90 ppm) for this program. A review of the extensive series of tests shows that the most satisfactory particle-removal equipment was a series of cyclones similar to those used in cat cracker service; these gave particulate contents of ~30 ppm over periods of hundreds of hours.

The Exxon tests included cascades of blades downstream of the particle separators to investigate erosion and deposit problems. Figure 12.42 shows the ash deposits in one of these cascades after 274 h of operation with one of the best particle separators employed. The fact that these deposits were much heavier than in some of the more favorable BCURA tests probably stems from differences in the test apparatus. The relatively large length-diameter ratio and the cooler walls of the hot gas piping ahead of the cascades in the BCURA tests may have removed the bulk of the material prone to form deposits before it reached the cascade. Thus conditions in the BCURA tests may not have been representative of those in full-scale engines.

A Combustion Power Company granular-bed filtration system tested was in the form of a cylindrical annulus about 11.8 in (30 cm) thick contained between two sets of louvered plates with the louvers sloped inward toward the bed so that the granular material was contained (Fig. 12.41).[124] The concept entails gradual movement of the granular material downward through the bed so that the filter cake on the inlet face is broken up and fine particles that get into the bed can be removed in the course of recycling the granular material from the bottom to the top of the bed through an external circuit. A test with a full-scale hot-flow unit was terminated in the initial shakedown by a creep-buckling type of failure. Data from a cold-flow model tested to determine the effects of bed thickness, filter media size, and flow rates on filtration performance indicated that particulate loadings of as low as 20 ppm were obtained in the exit gas stream.[86] It is not clear, however, whether a level this low can be achieved under hot operation and sustained for thousands of hours. As in the Exxon tests, one problem was that fines from the inlet filter cake tended to mix in with the granular bed when it was moving or during blowback operations.

SUMMARY

The gas turbine provides a splendid example of a case history for an advanced energy conversion system that has been carried through a long period of development to yield a mature system playing a major role in the U.S. economy. The technology is now mature, with 48,000 MW in gas-turbine power capacity of the 600,000 MW total capacity of the U.S. electric utility industry in 1980. Most utility gas turbines are used for peaking-power purposes, for which their low capital cost offsets their relatively high fuel costs (premium fuels are required and the thermal efficiency is lower than for steam plants). The process industries employ many gas turbines for cogeneration of electric power and process heat, and some are used in combined gas turbine–steam cycle systems. The amount of gas-turbine power capacity installed per year peaked at 6500 MW in 1971 and dropped abruptly with the steep increase in fuel prices following the Yom Kippur War and was only about 670 MW for 1978.

Efforts to reduce the high fuel costs by increasing the turbine inlet temperature and hence the thermal efficiency have involved the development of both improved high-temperature materials and improved turbine blade cooling. The limits of oxidation-resistant metal alloy development appear to have been reached. The peak permissible metal temperature from the creep strength standpoint is about 60 percent of the melting point in absolute temperature units (which compares to ~75 percent for the sintering temperature), and the melting points of Fe, Cr, Ni, and Co, the best of the elements from the strength and oxidation-resistance standpoints, impose immutable limits. Improvements in air cooling of turbine blades permit operation with blade temperatures as much as 300 °C below the inlet gas temperature. The point of diminishing returns appears to have been reached, however, because the pumping-power losses involved in increased air cooling largely offset the incremental gain in thermal efficiency obtainable at the higher temperature. Eighty years of effort to water cool turbine blades indicate that high heat losses, severe thermal stresses, and complex construction make that route to higher temperatures impractical— however challenging and intriguing.

The light weight, excellent strength and oxidation resistance at high temperature, availability of the constituent elements, and the low cost of ceramics have long made them intriguing possibilities for high-temperature gas turbines. However, a review of the many efforts to use them during the past 40 years in

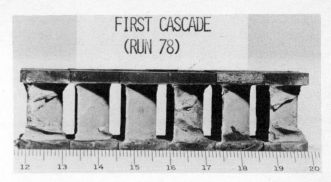

Figure 12.42 Ash deposits in a blade cascade operated for 274 h downstream of a series of cyclone separators that reduced the hot-gas particulate content to 30 ppm in tests with a 3.3-m-deep FBC. The bed temperature was ~ 900 °C (1652 °F), the superficial velocity ~ 1.8 m/s (6 ft/s), and the furnace pressure 9.25 bars. (*Courtesy Exxon Research Engineering Co.*)

Germany, England, and the United States discloses that no one has yet succeeded in running a gas turbine with a ceramic stator and rotor (the vital high-temperature parts) for more than 40 h without a catastrophic failure. The basic problem is that ceramics are inherently brittle. This makes them subject to failure from mechanical or thermal shock, particularly because of stress concentration effects of small defects, which are always present. The problems are rendered more acute because any turbine in commercial operation will be subject to catastrophic failure in severe temperature transients, which appear to be unavoidable. Analysis discloses that the best physical properties obtained in laboratory samples of simple geometry are hardly sufficient, even ideally, to withstand the air-blast quench following a flameout or an abrupt interruption of the fuel supply. Surprisingly, such an air-blast quench is of the order of 20 times as severe as would obtain if a white hot part were quenched in a bucket of water. Thus the development of a commercially useful turbine with a ceramic rotor and stator seems highly unlikely.

Efforts to reduce gas-turbine operating costs by changing to direct coal firing have been under way since 1945. None has been successful in avoiding serious trouble with blade erosion and/or deposits caused by fine particles of ash. Evidence both from these efforts and from operating experience with oil and gaseous fuels containing small amounts of ash shows that the particulate content of gas entering the turbine in the 800 to 900 °C (1472 to 1652 °F) range must be kept below 1 ppm, yet the best that any high-temperature particle separation equipment can give with an acceptable pressure drop is 100 times as high.

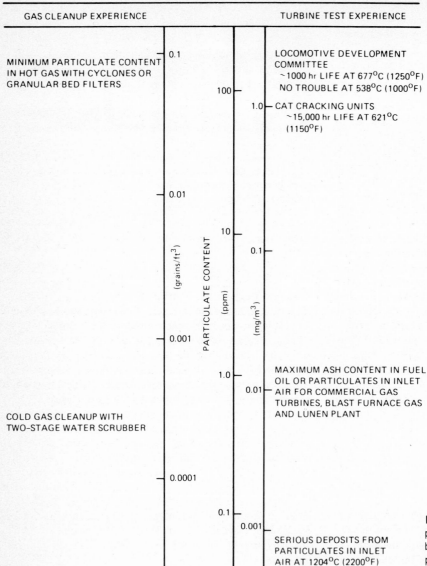

Figure 12.43 Summary of experience with particulate-removal equipment and gas turbines operating with particulates in the gas. All points are for the particulate content in the hot gas.

Figure 12.43 shows this tremendous gap and gives good perspective on the problem. This same body of experience shows that the turbine blade erosion and deposit problems are strongly dependent on the temperature, and that dropping the turbine inlet temperature below 600°C (1110°F) gives an acceptable life with gas particulate contents of 100 ppm, i.e., the level demonstrably obtainable with cyclone separators. Thus it seems likely that the use of gas turbines with coal as the fuel will be limited to Velox boiler installations, in which the turbine inlet temperature can be kept below 600°C (as discussed in the next chapter), or to closed-cycle installations (as discussed in Chap. 14).

PROBLEMS

1. Estimate the effects of load on the fuel flow to a gas turbine, its exhaust temperature, and the thermal efficiency. Assume a 10:1 pressure ratio, operation at synchronous speed, a compressor efficiency of 85 percent, a turbine efficiency of 90 percent, a 900°C turbine inlet temperature at full load, and a 5 percent pressure drop from the compressor to the turbine. Calculate conditions for 100, 75, 50, and 25 percent of full load, using basic thermodynamic relations and the properties of air at room temperature to simplify the calculations. Neglect the effects of shifts in the operating point on the compressor and turbine characteristic curves. Plot the resulting values as functions of load.

2. A new model gas turbine is to be coupled in a simple open cycle to a PFBC. The engine is designed to produce 100 MWe with a turbine inlet temperature of 850°C, a compressor inlet temperature of 20°C, and a pressure ratio of 10:1. About two-thirds of the air will be heated by tubes in the bed, one-third will be used as combustion air in the pressurized furnace, and the hot streams will be combined before entering the turbine. Estimate the following:

    (a) Coal-flow and airflow rates (moisture- and ash-free basis)
    (b) Diameter of compressor and turbine inlet and outlet ducts (assuming a Mach number of 0.20)

3. The particulate content of combustion gases going from the furnace to the turbine of the above system is to be reduced to 3 ppm by two stages of cyclone separator followed by a bag filter system in which the cylindrical bags are hung from header sheets in equilateral triangular arrays. These arrays would be arranged in tiers with each tier having a configuration something like that of the granular-bed filters of Fig. 12.41. Hot dirty gas would enter the top of 5-m-diameter vessels, flow down through an outer annulus, radially inward, down through the bags, and then the clean gas would flow radially inward to and down through the central duct, from which it would pass out the center of the bottom of the shell. Estimate the following:

    (a) The volume of bag filter matrix required, assuming 10-cm-diameter cylindrical bags with an inlet face velocity of 1 m/min and a maximum axial velocity over the bag surfaces of 1 m/s
    (b) The size of vessel required, assuming five tiers of bags and gas velocities of 70 m/s in the outer annulus and central duct
    (c) The weight and cost of just the vessel (with no internals), using cost data given in the Appendix and assuming a layer of thermal insulation lining the carbon steel vessel so that it would operate at ~300°C
    (d) The ratio of the weight of gas in the filter vessel to the gas flow rate per second to give an indication of the extra lag time to be expected in controlling the system.

REFERENCES

1. Stodola, A.: *Steam and Gas Turbines,* McGraw-Hill Book Company, New York, 1945.

2. Buchi, A. J.: "Supercharging of Internal-Combustion Engines with Blowers Driven by Exhaust-Gas Turbines," *Trans. ASME,* vol. 59, 1937, p. 85.

3. Moss, S. A.: "Gas Turbines and Turbosuperchargers," *Trans. ASME,* vol. 66, 1944, pp. 351–371.

4. Sawyer, R. T.: *The Modern Gas Turbine,* 2d ed., Prentice-Hall, Inc., Englewood Cliffs, N.J., 1947.

5. Pfenninger, H.: "The Evolution of the Brown-Boveri Gas Turbine," paper presented to visitors at ASME International Gas Turbine Conference and Products Show, Zurich, Mar. 13–17, 1966.

6. Troller, T.: Zur Berechnung von Schauben-ventilatoren, Abhandlungen aus dem Aerdynamischen Institut, Aachen, 1931.

7. Taylor, C. F.: *Aircraft Propulsion, Smithsonian Annals of Flight,* The Smithsonian Institution, vol. 1, no. 4, 1971.

8. Welsh, H. W., and E. F. Pierce: "Engine Compounding for Power and Efficiency," *S.A.E. Quart. Abstract,* vol. 2, 1948, p. 316.

9. Seippel, Claude: "Gas Turbines in Our Century," ASME Gas Turbine Progress Report, 1952, *Trans. ASME,* vol. 75, 1953, pp. 1–2.

10. Meyer, Adolph: "The Velox Steam Generator—Its Possibilities as Applied to Land and Sea," *Mechanical Engineering,* vol. 57, 1935, p. 469.

11. Fritz, W. A., Jr., and T. P. Tursi, Jr.: "Development and Experience with a Supercharged Steam-Generating System," *Marine Technology,* January 1967.

12. Fraas, A. P., et al.: "A New Approach to a Fluidized Bed Steam Boiler," ASME Paper No. 76-WA/Pwr-8, December 1976.

13. Pew, A. E., Jr.: "Operating Experience with the Gas Turbine—Report on Use as Propulsion Equipment for Axial Air Compressors at Sun Oil Refineries," *Mechanical Engineering,* vol. 67, 1945, pp. 594–598.

14. Stodola, A.: "Load Tests of a 4000-kW Combustion-Turbine Set," *Power,* February 1940, pp. 62–63.

15. Brown, K. A., et al.: "Gas Turbine Progress—Railroad," *Trans. ASME,* vol. 75, 1953, pp. 161–168.

16. Smith, J., et al.: "Bureau of Mines Progress in Developing Open and Closed-Cycle Coal-Burning Gas Turbine Power Plants," *Trans. ASME,* ser. A, vol. 88, 1966, p. 313.

17. Keller, C.: "The Escher-Wyss-AK Closed Cycle Turbine, Its Actual Development and Future Prospects," *Trans. ASME,* vol. 68, 1946, p. 791.

18. Lordi, F. D.: *Gas Turbine Materials and Coatings,* General Electric Company, GER-2182J, 1976.

19. de Biasi, V.: "FT51 Design Shortcut to 1980 Technology," *Gas Turbine World,* November 1975.

20. Hung, W. S. Y.: "A Diffusion-Limited Model that Accurately Predicts the NO_x Emissions from Gas Turbine Combustors Including the Use of Nitrogen-Containing Fuels," *J. Eng. Power, Trans. ASME,* vol. 98, 1976, pp. 320–326.

21. Livingood, J. N. B., H. H. Ellerbrock, and A. Kaufman: "1971 NASA Turbine Cooling Research Status Report," NASA TM X-2384, September 1971.

22. Moskowitz, S. L.: "2750°F Engine Test of a Transpiration Air-Cooled Turbine, 70-WA/GT-1," *Trans. ASME,* vol. 93, no. 1, 1971, p. 238.

23. Barnes, J. F.: "The Role of High Temperature Gas Turbines in Power Generation," *Combustion,* September 1968.

24. Matchett, J. D., J. N. Colborn, and A. F. Ahles: "A Comparison of Calculated and Measured Temperature Distributions in Forced-Convection Air-Cooled Gas Turbine Airfoils," ASME 67-WA/GT-4, July 31, 1967.

25. Murkerjee, D. K.: "The Cooling of Gas Turbine Blades," *Brown Boveri Review,* vol. 67, January 1977.

26. Bayley, F. J., and B. W. Martin: "A Review of Liquid Cooling of High Temperature Gas Turbine Rotor Blades," *Proceedings of the Institute of Mechanical Engineers,* vol. 185, 18/71, 1970–1971.

27. Freche, J. C., and A. J. Diaguila: "Heat Transfer and Operating Characteristics of Aluminum Forced-Convection and Stainless-Steel Natural Convection Water-Cooled Single-Stage Turbines," NACA RM E50D032, June 30, 1950.

28. Freche, J. C.: "A Summary of Design Information for Water-Cooled Turbines," NACA RM E51A03, Mar. 9, 1951.

29. Livingood, J. N. B., and W. B. Brown: "Analysis of Temperature Distribution in Liquid-Cooled Turbine Blades," NACA Technical Report 1066, 1952.

30. Alpert, S., et al.: "Development of a Three-Stage Liquid-Cooled Gas Turbine," ASME Paper No. 59-GTP-1, 1959.

31. Friedrich, R.: "A Gas Turbine with Cooled Blades for Temperatures over 1000°C," *Brennstoff-Waerme-Kraft,* vol. 14, 1962, pp. 368–373.

32. Kydd, P. H., and W. H. Day: "An Ultra High Temperature Turbine for Maximum Performance and Fuels Flexibility," ASME 75-GT-81, Dec. 2, 1974.

33. "Assessment of a Water-Cooled Gas Turbine Concept," EPRI Report No. 234-1, Final Report, August 1975.

34. "Water Cooled Gas Turbine Development," EPRI Report No. AF-231, Project 234-2, Phase 1A, Annual Report, January 1976.

35. *UHT Turbine Program,* General Electric Company, Contract RP234-3, Phase 1A Semiannual Report, January 1–June 30, 1976.

36. "Reference Turbine Subsystem Designs," ERDA-HTTT Topical Report, General Electric Company, 1977.

37. Day, W. H.: "An Ultra High Temperature Gas Turbine for Coal-Derived Fuels with Maximum Fuels Flexibility," *Proceedings of the American Power Conference,* vol. 39, 1977.

38. Day, W. H., and M. W. Horner: General Electric Co., personal communication to A. P. Fraas, Nov.14, 1977, and Dec. 7, 1977.

39. Fraas, A. P., and M. N. Ozisik: *Heat Exchanger Design,* John Wiley & Sons, Inc., New York, 1965.

40. Van Fossen, G. J., Jr.: "Review and Status of Liquid-Cooling Technology for Gas Turbines," NASA Lewis Research Center Report No. NASA-RP-1038, April 1979.

41. Brown, D. H., and J. C. Corman: *Energy Conversion Alternatives Study, General Electric Phase I Final Report,* vol. II, *Advanced Energy Conversion Systems,* Pt. I, "Open Cycle Gas Turbines," NASA-CR 134948, General Electric Company.

42. Sims, C. T., and J. E. Palko: "Surface Stability of Ceramics Applied to Energy Conversion Systems," *Proceedings of Workshop on Ceramics for Advanced Heat Engines,* Jan. 24–26, 1977, CONF-770110, p. 287.

43. Soehngen, Erich: personal communication to A. P. Fraas, January 1977.

44. Wright Air Development Center: *Proceedings of the WADC Ceramic Conference on Cermets,* Oct. 6–8, 1952, WADC Technical Report 52-327.

45. Deutsch, G. C., A. J. Meyer, and G. M. Ault: *A Review of the Development of Cermets,* AGARD Report No. 185, North Atlantic Treaty Organization, March–April 1958.

46. Deutsch, G. C., A. J. Meyer, Jr., and W. C. Morgan: *Preliminary Investigation in J33 Turbojet Engine of Several Root Designs for Ceramel Turbine Blades,* NACA RM E52K13, Lewis Flight Propulsion Laboratory, Cleveland, Ohio, Jan. 26, 1953.

47. Meyer, A. J., Jr., G. C. Deutsch, and W. C. Morgan: *Preliminary Investigation of Several Root Designs for Cermet Turbine Blades in Turbojet Engine,* II, *Root Design Alterations,* NACA RM E53G02, Lewis Flight Propulsion Laboratory, Cleveland, Ohio, Oct. 13, 1953.

48. Pinkel, B., G. C. Deutsch, and W. C. Morgan: *Preliminary Investigation of Several Root Designs for Cermet Turbine Blades in Turbojet Engine,* III, *Curved-Root Design,* NACA RM E55J04, Lewis Flight Propulsion Laboratory, Cleveland, Ohio, Dec. 28, 1955.

49. Morgan, W. C., and G. C. Deutsch: *Experimental Investigation of Cermet Turbine Blades in an Axial-Flow Turbojet Engine,* NACA Technical Note 4030, Lewis Flight Propulsion Laboratory, Cleveland, Ohio, October 1957.

50. *Proceedings of Workshop on Ceramics for Advanced Heat Engines,* ERDA Division of Conservation Research and Technology, Orlando, Fla., Jan. 24–26, 1977, CONF-77010.

51. McLean, A. F.: "The Application of Ceramics to the Small Gas Turbine," ASME Paper No. 70-GT-105, presented at the ASME Gas Turbine Conference and Products Show, Brussels, Belgium, May 24–28, 1970.

52. McLean, A.F.: "Automotive Ceramic Turbines and Directions for Future Ceramic Technology," *Proceedings of Workshop on Ceramics for Advanced Heat Engines,* ERDA Division of Conservation Research and Technology, Orlando, Fla., Jan. 24–26, 1977, CONF-77010.

53. deBiasi, V.: "Ford Runs Uncooled Ceramic Turbine Engine at 2500°F," *Gas Turbine World,* July 1977.

54. "Ceramics for High Performance Applications— II," *Proceedings of the Fifth Army Materials Technology Conference,* Mar. 21–25, 1977, Brook Hill Publishing, Chestnut Hill, Mass.

55. Bratton, R.: "Ceramic Developments for Industrial Gas Turbines," *Proceedings of Workshop on Ceramics for Advanced Heat Engines,* ERDA Division of Conservation Research and Technology, Orlando, Fla., Jan. 24–26, 1977, CONF-77010.

56. Day, W. H.: "Recent Developments in the Application of Ceramics to Gas Turbine Combustors," *Proceedings of Workshop on Ceramics for Advanced Heat Engines,* ERDA Division of Conservation Research and Technology, Orlando, Fla., Jan. 24–26, 1977, CONF-77010.

57. Calvert, G., and B. H. Walker: "Ceramic Blade Attachment Programs at P&W Aircraft," *Proceedings of Workshop on Ceramics for Advanced Heat Engines,* ERDA Division of Conservation Research and Technology, Orlando, Fla., Jan. 24–26, 1977, CONF-77010.

58. Napier, J. C., A. G. Metcalfe, and T. E. Duffy: "Application of Ceramics to the MERADCOM 10 kW Engine," *Proceedings of Workshop on Ceramics for Advanced Heat Engines,* ERDA Division of Conservation Research and Technology, Orlando, Fla., Jan. 24–26, 1977, CONF-77010.

59. Wallace, F. B., and N. R. Nelson: "ARPA/Navy Ceramic Engine Progress," *Proceedings of Workshop on Ceramics for Advanced Heat Engines,* ERDA Division of Conservation Research and Technology, Orlando, Fla., Jan. 24–26, 1977, CONF-77010.

60. Wiederhorn, S. M., and N. J. Tighe: "Application of Proof Testing to Silicon Nitride," *Proceedings of Workshop on Ceramics for Advanced Heat Engines,* ERDA Division of Conservation Research and Technology, Orlando, Fla., Jan. 24–26, 1977, CONF-77010.

61. Donahoe, P. G.: "Preventive Maintenance for Gas Turbines," *Chemical Engineering Progress,* vol. 68, no. 7, July 1972.

62. Weibull, W.: "Statistical Theory of the Strength of Materials," *Ingenious Vetenskaps Akademieu Handlinger* no. 151, 1939 (in English).

63. *Brittle Materials Design, High Temperature Gas Turbine, Seventh Semi-Annual Progress Report Covering Period Ending Dec. 31, 1974,* AMMRC-CTR-75-8, April 1975.

64. Weeton, J. W.: "Design Concepts for Fiber-Metal Matrix Composites for Advanced Gas Turbine Blades," ASME 70-GT-133, May 1970.

65. Peterson, G. P.: "Advanced Composites—A State-of-the-Art Assessment," ASME 70-GT-120, May 1970.

66. Metcalfe, A. G.: "Current Status of Titanium-Boron Composites for Gas Turbines," ASME Paper No. 69-GT-1, March 1969.

67. Brennan, J. J.: "Evaluation of Ceramic Fiber Reinforced Si_3N_4, Final Report," Report No. NADC-75207-30, United Technologies Corp., NADC Contract No. 62269-75-CO137, April 1, 1977.

68. "Review-Ceramic Matrix Composites," *Journal of Materials Science,* vol. II, May 1976.

69. "The Coal-Burning Gas Turbine Project," Report of the Interdepartmental Steering Committee, Department of Minerals and Energy, Department of Supply, Commonwealth of Australia, 1973.

70. Fraas, A. P.: "Survey of Turbine Bucket Erosion, Deposits, and Corrosion," ASME Paper No. 75-GT-123, March 1975.

71. Lackey, M. E.: "Summary of the Research and Development Effort on Open-Cycle Coal-Fired Gas Turbines," Oak Ridge National Laboratory Report No. ORNL/TM-6253, October 1979.

72. Browne, K. A.: "Future Use of Coal in Railway Motive Power," *Mechanical Engineering,* vol. 68, 1946, p. 547.

73. *Progress Report No. 1 to The Locomotive Development Committee, May 1 to August 1, 1945,* Bituminous Coal Research, Inc., Dunkirk, N.Y., Aug. 25, 1945.

74. Hazard, H. R., and F. D. Buckley: "Experimental Combustion of Pulverized Coal at Atmospheric and Elevated Pressures," *Trans. ASME,* vol. 70, 1948, p. 729.

75. Fisher, M. A., and E. F. Davis: "Studies on Fly-Ash Erosion," *Mechanical Engineering,* vol. 71, June 1949, p. 481.

76. Broadley, P. R., and W. M. Meyer: *1957 Annual Report of the Locomotive Development Committee,* Locomotive Development Committee, Bituminous Coal Research, Inc. (1957).

77. "Coal-Fired Gas Turbine," *Mechanical Engineering,* vol. 81, May 1959, p. 79.

78. Mordell, D. L., and R. W. Foster-Pegg: "Test of an Experimental Coal-Burning Gas Turbine," *Trans. ASME,* vol. 78, 1956, pp. 1807–1833.

79. "Coal-Fired Gas Turbine Is Road Tested in Freight Service," *Railway Locomotives and Cars,* January 1963.

80. Smith, J., D. C. Strimbeck, N. H. Coates, and J. P. McGee: "Bureau of Mines Progress in Developing Open and Closed-Cycle Coal-Burning Gas Turbine Power Plants," *J. Eng. Power Trans. ASME,* ser. A, vol. 88, no. 4, October 1966, p. 313.

81. Smith, J., et al.: *Bureau of Mines Coal-Fired Gas Turbine Research Project: Tests of New Turbine Blade Design,* BM-RI-6920, U.S. Bureau of Mines, 1967.

82. *Annual Report 1970–1971,* Aeronautical Research Laboratories, Department of Supply, Melbourne, Australia.

83. Hoy, H. R., and J. E. Stantan: "Fluidized Combustion Under Pressure," paper presented at the Joint Meeting of the Chemical Institute of Canada and the Division of Fuel Chemistry of the AIChE in Toronto, May 24, 1970.

84. *Pressurized Fluidized Bed Combustion Research and Development Report No. 85, Interim No. 1,* prepared for Office of Coal Research, Department of the Interior, by National Research Development Corporation, London, England, 1974.

85. Furlong, D. A., and G. L. Wade: "Use of Low Grade Solid Fuels in Gas Turbines," ASME Paper No.

74-WA/ENER-5, presented at the ASME Winter Annual Meeting, Nov. 18, 1974.

86. *Energy Conversion from Coal Utilizing CPU-400 Technology—Monthly Report for August 1976,* FE-1536-M37, Combustion Power Company, Inc., September 1976.

87. Stevens, W. G., and A. R. Stetson: *Corrosion and Erosion Evaluation of Turbine Materials in an Environment Simulating the CPU-400 Combustor Operating on Coal,* DOE Report FE/1536-3, April 1977.

88. Van Grouw, S. J.: "Corrosion Studies and High Temperature Filtration," *Proceedings of the Fluidized Bed Combustion Technology Exchange Workshop*, Apr. 13–15, 1977, NTIS CONF-770447-P-2, vol. II.

89. Stettenbenz, L. M.: "Power-Recovery Gas Expander Gains in FCC Cycle," *The Oil and Gas Journal,* Dec. 18, 1972, p. 60.

90. Wilson, J. G., and J. C. Dygert: "Separator and Turbo-Expander for Erosive Environments," *Proceeding of the Seventh World Petroleum Congress, Mexico City*, April 1967, p. 95.

91. Balfoort, J. P.: "Improved Hot-Gas Expanders for Cat Cracker Flue Gas," *Hydrocarbon Processing,* March 1976, p. 142.

92. Dygert, J. C.: "Power-Recovery Gas Turbines for Fluid-Bed Process," *The Oil and Gas Journal,* Apr. 20, 1959, p. 94.

93. Wilson, J. G.: "Energy Recovery Pays Off at Three Shell Refineries," *The Oil and Gas Journal,* Apr. 18, 1966, p. 77.

94. Franzel, H. L., and D. W. Miller: "Cleaner Stack Gases Are a Bonus," *The Gas and Oil Journal,* Mar. 24, 1969, p. 95.

95. Stettenbenz, L. W.: "Minimizing Erosion and Afterburn in the Power-Recovery Gas Turbines," *The Oil and Gas Journal,* Oct. 19, 1970, p. 65.

96. Brown, S. S.: "Power Recovery Pays Off at Shell Oil," *The Oil and Gas Journal,* May 21, 1973, p. 128.

97. Jaumotte, A. L., and J. Hustin: "Experience Gained from a Ten-Year Operation of a Gas Turbine Working with Blast Furnace Gas," ASME Paper No. 66-GT-97, 1966.

98. Strub, A.: "Field Experience with Industrial Gas Turbine Installation," *Sulzer Technical Review,* BST Brown Boveri-Sulzer Turbomachinery Ltd., vol. 48, no. 3, 1966, p. 129.

99. Pfenninger, H.: "Operating Results with Gas Turbines of Large Output," *J. Eng. Power, Trans. ASME,* January 1964, p. 29.

100. Aguet, E., and J. Von Salis: "Three Years' Operating Experience with 7500 kW Gas Turbine Plants in Belgian Steelworks," ASME Paper No. 60-GTP-3, March 1960.

101. Felix, P. C.: personal communication to A. P. Fraas, Apr. 3, 1975, summarizing Brown Boveri blast furnace gas turbine operating experience in Europe.

102. Frieder, A.: personal communication to A. P. Fraas, July 12, 1976, summarizing Sulzer blast furnace gas turbine operating experience in Europe.

103. Bund, K., K. A. Henney, and K. H. Krieb: "Combined Gas/Steam Turbine Generating Plant in the Kellermann Power Station at Lünen," *STEAG, Anlagentechnik,* 1970.

104. Kahrweg, H. Meyer: discussion with A. P. Fraas at Lünen plant, Sept. 27, 1973.

105. Foster, A. D., H. Doerning, and J. W. Hickey: *Fuel Flexibility in Heavy-Duty Gas Turbines,* GER-2222J, General Electric Company, 1974.

106. "Gas Turbine Inlet Air Treatment," General Electric Gas Turbine Reference Library, GER-2490.

107. Horton, J. W.: "Environmental Factors in Engine Design for Military Applications," ASME Paper 65-GTP-1, February 1965.

108. Smeltzer, C. E., and W. A. Compton: "Mechanisms of Sand and Dust Erosion in Gas Turbine Engines," Solar Division, IHC, U.S. AvLabs Contract DAAJO2-68-C-0056, AD 680 299, September 1968.

109. Wood, C. D., and P. W. Espenschade: "Mechanisms of Dust Erosion," *SAE Trans.,* vol. 73, 1965, p. 515.

110. Wood, B.: "Wetness in Steam Cycles," *Proceedings of the Institution of Mechanical Engineers,* vol. 174, no. 14, 1960, pp. 491–511.

111. Evans, D. H., and W. D. Pouchot: "Flow Studies in a Wet Steam Turbine," NASA CR-134683, prepared for

NASA by Westinghouse Electric Corporation, Aug. 13, 1974.

112. Fraas, A. P., H. C. Young, and A. G. Grindell: "Survey of Information on Turbine Bucket Erosion," USAEC Report ORNL-TM-2088, Oak Ridge National Laboratory, July 1968.

113. Brunton, J. H.: "Liquid Impact and Material Removal Phenomena," *Summary of Turbine Erosion Meeting,* Jet Propulsion Laboratory Technical Memorandum no. 33-354, June 1967, pp. 173-191.

114. Bulkley, E. L.: "New Studies Shed Light on Erosion Problems," *Petro/Chem Engineer,* March 1962, p. 218.

115. "Dust Erosion Parameters for Gas Turbine," *Petro/Chem Engineering,* December 1962, p. 198.

116. Grant, G., and W. Tabakoff: "Erosion Prediction in Turbomachinery Resulting from Environmental Solid Particles," *J. Aircraft,* vol. 12, no. 5, May 1975, p. 471.

117. Vermes, G.: "Thermophoresis—Enhanced Deposition Rates in Combustion Turbine Blade Passages," *J. Eng. Power, Trans. ASME,* vol. 101, October 1979, pp. 542–548.

118. Weinert, E. P.: Naval Ship Engineering Center, personal communication to A. P. Fraas in March 1975 and February 1978.

119. Deadmore, D. L., and C. E. Lowell: "Plugging of Cooling Holes in Film-Cooled Turbine Vanes," NASA TMX-73661, April 1977.

120. "High Temperature Turbine Technology Program, Phase I—Program and System Definition, Topical Report—Fuels Cleanup and Turbine Tolerance," prepared by Westinghouse Electric Corp. for ERDA, FE-2290-27, February 1977.

121. Mund, M. G., and Hanspeter Guhne: "Gas Turbine—Dust—Air Cleaners: Experience and Trends," ASME Paper No. 70-GT-104, January 1970.

122. Henke, W. G.: "The New Hot Electrostatic Precipitation," *Combustion,* October 1970.

123. Hoke, R. C. et al.: *Miniplant Studies of Pressurized Fluidized Bed Coal Combustion: 3rd Annual Report,* Exxon Research and Engineering Co., EPA-600/7-78-069, April 1978.

124. Van Grouw, S. J.: "Corrosion Studies and High Temperature Filtration," *Proceedings of Fluidized Bed Combustion Technology Exchange Workshop*, McLean, Va., April 13–15, 1977, vol. II, CONF-770447-P-2, p. 235.

13

COMBINED-CYCLE, GAS-TURBINE–STEAM POWER PLANTS

The term *combined cycle* refers to any power plant system in which a higher-temperature thermodynamic cycle rejects its heat to a lower-temperature thermodynamic cycle, ordinarily employing a different working fluid. The Buchi compound turbocharged diesel mentioned in the previous chapter and the binary vapor cycles of Chap. 15 are examples of combined cycles. However, the term *combined-cycle plant* has now become the accepted shortened form of the term *combined-cycle, gas-turbine–steam plant*, and it is this type of combined-cycle plant that is the subject of this chapter.

DEVELOPMENTAL BACKGROUND

As mentioned in the previous chapter, the gas turbine was first applied commercially to supercharge steam boilers and thus reduce their size and cost, a development carried out by Brown Boveri in Switzerland.[1] Over 100 of these oil-fired units, called *Velox boilers*, were built by Brown Boveri, and many are still in operation at the time of writing.[2] Foster-Wheeler built 35 oil-fired boilers of this type for U.S. Navy destroyers in the 1960s;[3] one of these units is shown in Fig. 13.1. Note the cylindrical furnace shell to sustain the internal pressure. The combustion air is supplied to the furnace at 6 atm by a compressor that is driven by a turbine operating on the exhaust gas from the furnace–steam-generator unit. The gas enters the turbine at

Figure 13.1 Section through a pressurized furnace–steam generator of the Velox boiler type for marine service. The flames from three oil burners at the top are directed downward into the combustion chamber. The hot gas flows radially outward at the bottom, then upward, back down, and upward again through the boiler and superheater tube matrices around the perimeter, and into the gas turbine at the top left at 500 °C. The compressor to the right of the turbine delivers air to the furnace. The unit generates 56,800 kg/h of steam at 538 °C (1000 °F) and 82.7 bars (1200 psi).[3] *(Courtesy Foster Wheeler Energy Corp.)*

a temperature of ~450°C (840°F), i.e., just high enough for it to drive the turbine-compressor unit with no other useful output.

The first gas turbine that produced electric power and rejected its heat to a steam-turbine system was installed in 1949 by the Oklahoma Gas and Electric Company.[4] The company superimposed a 3500-kW gas turbine on an existing medium-pressure steam system to increase the plant output and raise the overall plant cycle efficiency from 22 to 26 percent. However, it was not until the 1960s that commercial gas turbines reached the development stage at which they could be built large enough to provide an economically attractive option as a large power unit. The low cost of gas and distillate fuels in the 1960s, coupled with a rapid increase in the need for peaking-power units, led to the installation of many peaking-power gas turbines, together with a substantial number of combined-cycle gas-turbine–steam power plants, through the sixties and early seventies. This trend was halted abruptly by the sharp in-crease in the cost of natural gas and distillate fuels following the Yom Kippur War in 1973.[5] The installation of new combined-cycle plants had dropped to a low level by 1979, by which time 179 units had been installed in the United States (Fig. 13.2). The need to shift to coal as a fuel for utilities, coupled with stiff limitations on air pollution, have turned the development effort for combined cycles to the use of either a fluidized-bed coal combustion system as the heat source or a coal gasification plant tightly integrated with the combined cycle. Of course, there still is a substantial market for combined-cycle plants employing distillate fuels for use in the oil-producing countries, such as those in the Middle East, because of the low capital investment involved. There still is a market also for the Velox boiler, because it can employ a residual fuel and operates with turbine inlet temperatures below 500°C, a temperature region in which the ash and sulfur contents of a residual fuel may be ~20 times the maximum level permitted in the No. 2 fuel oil for high-temperature turbines without causing serious difficulties with the turbine. (See the latter part of the previous chapter.)

THERMODYNAMIC CYCLE CONSIDERATIONS

There is a wide range of possible divisions of the energy in the cycle between the gas turbine and the steam system. On the one hand, it is possible to employ a power plant unit designed to operate primarily as a gas turbine with an auxiliary low-pressure steam system that utilizes heat from the gas-turbine exhaust. At the other extreme, it is possible to design a steam plant of the Velox boiler type in which the gas turbine is an auxiliary that serves just to pressurize the steam boiler to reduce its size and cost. An intermediate type is a unit designed for a high-pressure and -temperature steam system to give a high Rankine cycle efficiency, with the gas turbine serving as an auxiliary designed to extract some extra power from the hot combustion gas.

A temperature-entropy diagram for the first type of combined cycle is shown in Fig. 13.3. The only heat going into the steam cycle is that derived from the gas-turbine exhaust. Note that the peak steam pressure obtainable with a given turbine exhaust temperature is substantially lower than one might otherwise expect because most of the heat added to the steam in the boiler must be added at the boiling point. This means that the saturation temperature in the boiler must be close to the temperature of the hot gas leaving the boiler rather than that of the hot gas leaving the turbine (Fig. 4.5).

The efficiency of the steam cycle can be improved by taking advantage of the fact that less than half of the oxygen in the air supplied to the gas turbine is consumed in combustion. The re-

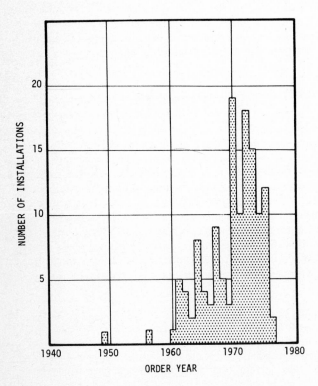

Figure 13.2 Gas-turbine–steam combined-cycle power plant installations made each year for electric utility power generation (1949–1976). *(Courtesy Oak Ridge National Laboratory.)*

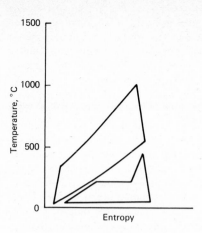

Figure 13.3 Temperature-entropy diagram for a combined cycle in which the steam cycle is an auxiliary that operates on the waste heat from the gas turbine.

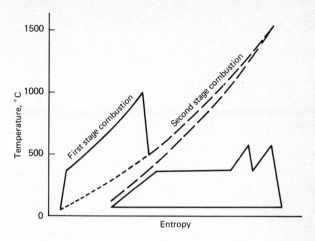

Figure 13.4 Temperature-entropy diagram for a combined cycle designed for maximum efficiency with a high-pressure, high-temperature steam system.

maining oxygen can be utilized by a burner installed downstream from the gas-turbine exhaust to increase the temperature of the gas before it enters the boiler and thereby increase the temperature and pressure for which the steam cycle can be designed. The resulting temperature-entropy diagram is shown in Fig. 13.4. Note that the addition of the burner also gives a marked increase in the power output of the plant for a given size of gas turbine. With such an arrangement the principal power output is obtained from the steam turbine; the gas turbine simply provides some extra power at relatively little incremental capital cost, and with a steam cycle efficiency of ~39 percent it improves the overall system thermal efficiency to as much as 42 percent. Further, since the flue gas leaving the boiler furnace does not pass through the turbine, it is possible to use lower-cost residual fuel oil or coal for firing the boiler as in the system shown in Fig. 13.5.[6]

Range of Gas-Turbine–Steam Cycle Combinations

An outline of the range of possible combinations of gas-turbine and steam cycles that might be employed is given by Fig. 13.6.[6] This includes curves for the two firing modes discussed above: one with the steam cycle designed to recover heat from the gas-turbine exhaust and the other with the gas turbine designed to produce some useful power while serving mainly to supply preheated air to the boiler-furnace for a high-temperature and -pressure steam system. Depending on the design conditions chosen for the gas-turbine and steam systems, Fig. 13.6 indicates that the waste-heat boiler type of

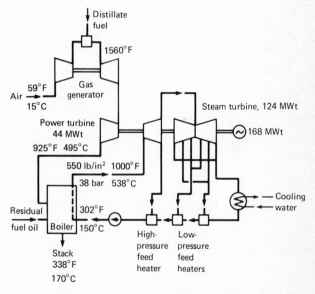

Figure 13.5 Combined cycle designed for maximum plant efficiency with most of the power produced by the steam plant.[6]

system gives a steam turbine output about half that of the gas turbine when system conditions are chosen for the best thermal efficiency. On the other hand, if the steam system is made the primary prime mover, the best efficiency is obtained with the gas turbine producing only about 14 percent of the total out-

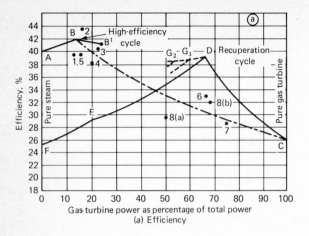

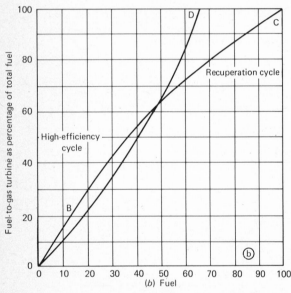

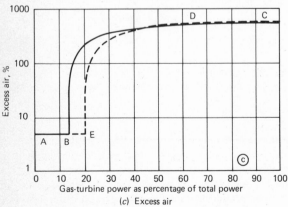

put. These effects are reflected in the two lower curves of Fig. 13.6 for both the fuel consumption distribution and the amount of excess air in the stack gas. A low value for the latter makes possible a reduced heat loss to the stack, an important factor contributing to the higher thermal efficiency for the system in which steam-cycle considerations are dominant.

Note that Fig. 13.6 shows the relation between the steam cycle with an auxiliary gas turbine and the Velox boiler cycle. The latter is represented by the condition at the extreme left of the set of curves. Note also that there are other combinations of gas- and steam-turbine cycles not shown in Fig. 13.6. For example, all the fuel could be burned in the combustor for a supercharged boiler and the flue gas discharged from the boiler furnace to the gas turbine at a temperature of perhaps 800°C so that it would give a substantial net power output. This system differs from the Velox boiler discussed earlier in that it requires clean fuel because the turbine inlet temperature is high —generally 800 to 1000°C versus the 400 to 500°C employed in the Velox boiler. Systems of this type are discussed in later sections.

Distribution of Losses

A nice comparison of the energy flows through both a conventional steam cycle and a combined cycle designed with the steam system as the prime element is given in Fig. 13.7.[6] There are some subtle differences, such as the smaller amount of steam bled from the turbine for feed-water heating in the combined-cycle system because of the large amount of heat available at low temperatures from the turbine exhaust gas.

Effects of Gas-Turbine Inlet Temperature on the Efficiency of Combined Cycles

The gas-turbine inlet temperature not only affects the efficiency of the gas-turbine cycle, but it also affects the peak temperature and efficiency of the steam cycle employed to recover the waste heat from the gas-turbine exhaust. As can be seen from the lower curve in Fig. 13.8, increasing the gas-turbine in-

Figure 13.6 Effects of the fraction of the total plant output produced by the gas turbine at the design point on the calculated performance of combined-cycle plants. The numbers beside the points plotted for the thermal efficiency of actual plants identify these plants as follows: (1) Horseshoe Lake, (2) Hohe Wand, (3) San Angelo, (4) Altbach, (5) Vitry, (6) Korneuburg, (7) Neuchatel, (8a) Socolie with supplementary firing, (8b) Socolie without supplementary firing.[6]

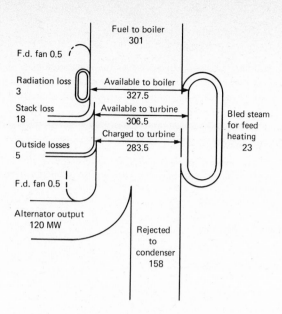

Conventional steam cycle 120 MW for 301 MWt input
(= 39.9 percent efficiency)

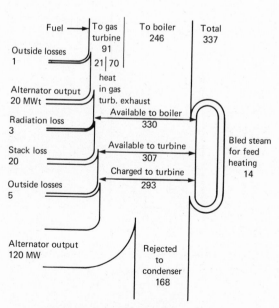

Combined cycle 140 MW for 337 MWt input
(= 41.6 percent efficiency)

Figure 13.7   *(Left)* Sankey diagrams comparing a steam cycle and a high-efficiency combined cycle. The numbers in the diagram are for megawatts of energy.[6]

Figure 13.8   *(Below)* Effects of gas-turbine inlet temperature on the turbine outlet temperature and the overall thermal efficiency of a combined cycle with oil firing and component efficiencies of 90 percent. *(Courtesy Oak Ridge National Laboratory.)*

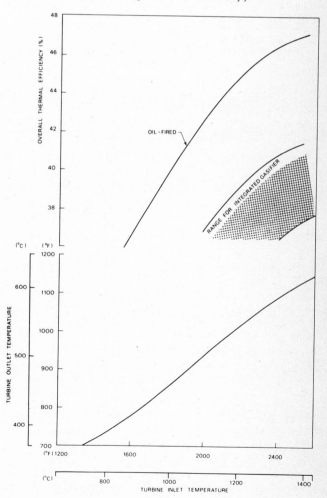

let temperature also increases its exhaust temperature, thus increasing the peak temperature and the efficiency of the steam cycle as well. Note that the upper curve is for the efficiency of the combined cycle with an oil-fired system. Below it a broad band indicates the range in which the overall efficiency of a combined-cycle plant closely integrated with a coal gasifier may be expected to lie. The gasifier systems will be discussed in a later section, but it may be mentioned here that the losses inherent in the coal gasification process are the reason for the much lower thermal efficiency indicated in Fig. 13.8 for such systems.

CHARACTERISTICS OF TYPICAL COMBINED-CYCLE PLANTS

There are so many possible variations of the combined-cycle concept that it seems best to present typical examples in some detail, outlining their major characteristics, rather than to treat the subject in generalities.

Brown Boveri Velox Boiler

In the mid-1970s Brown Boveri offered a modernized version of their Velox boiler.[2] This unit makes use of an efficient

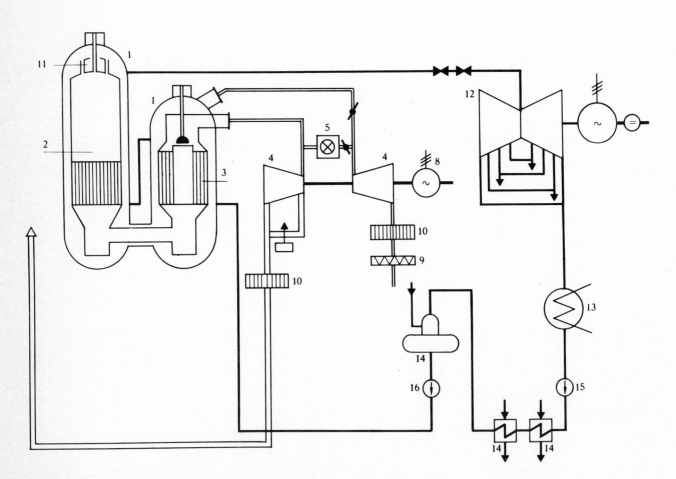

Figure 13.9 Flowsheet for a Velox boiler system.[2] *(Courtesy Brown Boveri and Co., Ltd.)*

1. Furnace
2. Superheater
3. Boiler
4. Gas turbine
5. Furnace bypass valve
8. Starter motor
9. Air filter
10. Silencers
11. Superheater bypass valve
12. Steam turbine
13. Steam condenser
14. Feed-water pump and heaters
15. Deaerator
16. Feed-water control valve

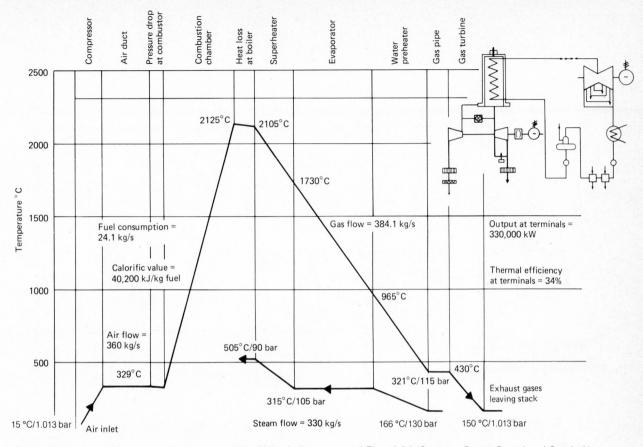

Figure 13.10   Temperature distribution through the Velox boiler system of Fig. 13.9.[2] *(Courtesy Brown Boveri and Co., Ltd.)*

modern gas turbine with a pressure ratio of 10:1 and a turbine inlet temperature of 400°C (752°F). The steam system is designed to deliver 175-bar (2540-psia) steam at 540°C (1000°F). A flowsheet for the plant is shown in Fig. 13.9, a typical temperature distribution is presented in Fig. 13.10, and the compact plant layout is shown in Fig. 13.11. The thermal efficiency as a function of load is given in Fig. 13.12 for two different control modes, the higher efficiency mode requiring a more complex control system.

Several variations of the plant using the same gas turbine have been offered, the differences lying in the complexity and cost of the steam system. Table 13.1 shows four cases, the one at the right including a reheater. Note the increases in output obtainable, and particularly the high overall thermal efficiency of 42.8 percent obtainable with reheat. At first glance this appears to be much higher than estimates for the combined cycles

integrated with coal gasifiers designed for large net power outputs from the gas turbine as presented in the next section, but it is based on the lower rather than the higher heating value of the fuel and does not include stack gas scrubbers for sulfur removal.

Gas Turbine with an Auxiliary Steam Cycle

One type of combined-cycle plant is that in which the waste heat in the gas-turbine exhaust is the sole source of energy for a steam cycle. A typical example is the 26-MWe plant at Clarksdale, Mississippi.[7] To utilize the heat in the gas-turbine exhaust as efficiently as possible, a *dual-pressure steam cycle* is employed similar to that used in the British Magnox gas-cooled reactors, which operate with a relatively low reactor gas outlet

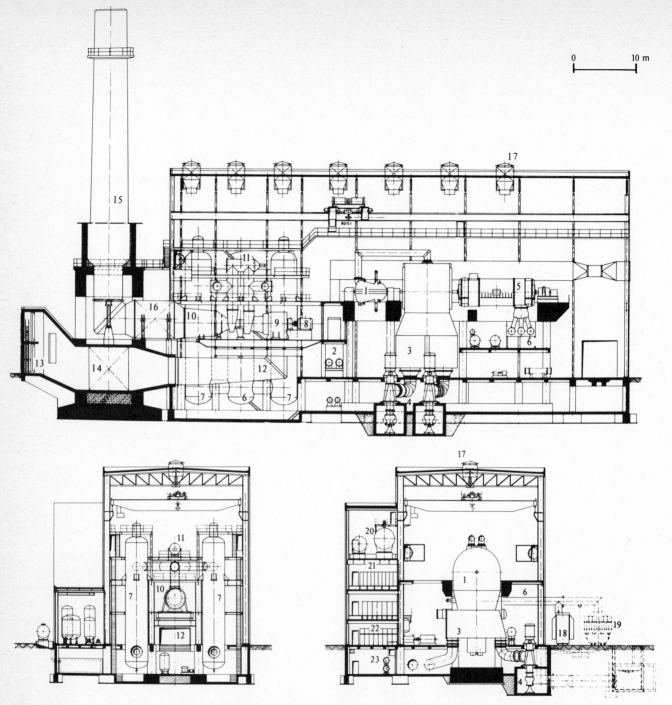

Figure 13.11 Longitudinal and lateral cross sections through a 330-MWe plant for a 10-atm Velox boiler furnace.[2] *(Courtesy Brown Boveri and Co., Ltd.)*

1. Steam turbine, 330 MW
2. Oil cooler
3. Condenser
4. Main cooling water pumps, 2×50%
5. Generator
6. Generator lead
7. Supercharged Durr-Benson steam generator (1120 t/h, 90 bars, 550°C), four heated containers, diameter 3.6 m, height 20 m
8. Starter motor for gas turbines
9. Axial compressor
10. Gas turbine
11. Startup combustion chamber
12. Air intake
13. Air filter
14. Silencer in air intake duct
15. Stack
16. Exhaust-gas silencer
17. Vent

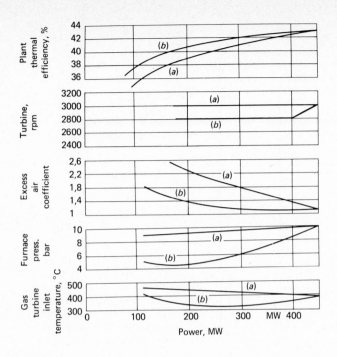

Figure 13.12 Part-load performance characteristics of a super-charged boiler steam power plant for two gas turbine operating modes: (a) with constant combustion airflow through the gas-turbine set and (b) with the turbine speed reduced to decrease the airflow and improve the thermal efficiency at part load.[2] *(Courtesy Brown Boveri and Co., Ltd.)*

TABLE 13.1   Characteristics of a Series of Power Plants Utilizing the Same Model of Gas Turbine to Supercharge Velox Boilers Designed for Four Different Steam Conditions [2]

Design no.		1	2	3	4 (with reheater)
Combustion air flow	kg/s	360	360	360	360
Fuel-air coefficient	kg/s	2.41	24.1	24.1	24.1
Pressure ratio of the compressor	λ	10	10	10	10
Temperature at gas-turbine inlet	°C	400	400	400	400
Excess power of gas-turbine set	λ	0	0	0	0
Type of fuel		Heavy fuel oil	Heavy fuel oil	Heavy fuel oil	Heavy fuel oil
Live steam conditions	bar/°C	80/500	109/525	109/525	175/540, 36.5/540
Vacuum	bar	0.104	0.08	0.04	0.03
Number of bleed steam reheaters including deaerator		3	4	5	7
Output at terminals	MW	333	360	375	428
Thermal efficiency at terminals (without feed pump)	%	34.3	37.1	38.6	44.1
Auxiliary services		3	3	3	3
Net power	MW	323	349	364	415
Net thermal efficiency*	%	33.3	36	37.5	42.8

*Based on LHV of the fuel.

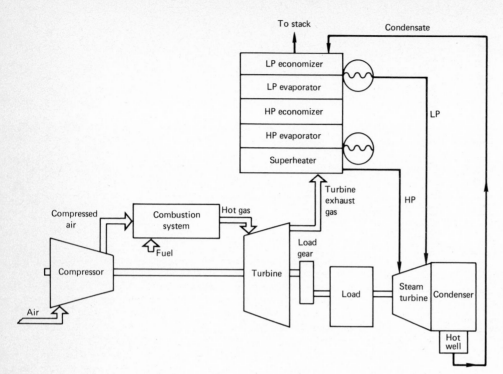

Figure 13.13 Schematic diagram for a combined-cycle gas and steam turbine system in which all the heat for the steam system is obtained from a waste-heat recovery boiler in the turbine exhaust.[4] *(Courtesy Encotech, Inc.)*

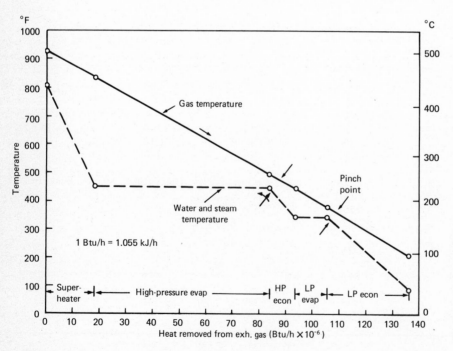

Figure 13.14 Temperature distribution in the steam generator for the dual-pressure steam system of Fig. 13.13. The steam pressures are 29.3 bars (425 psig) and 6.27 bars (91 psig), and the steam flows 38,828 and 6486 kg/s, respectively.[4] *(Courtesy Encotech, Inc.)*

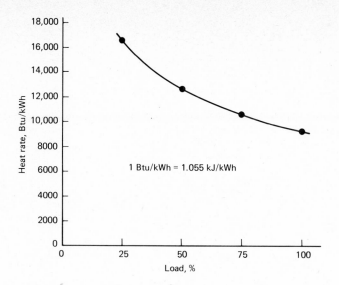

Figure 13.15 Part-load heat rate for the plant of Figs. 13.12 and 13.13 operating on No. 2 fuel oil with an ambient air temperature of 27°C (80°F).[4] *(Courtesy Encotech, Inc.)*

temperature of around 390°C (734°F) (Chap. 5). A flowsheet for the steam system of the Clarksdale combined-cycle plant is presented in Fig. 13.13, and the temperature distribution through the steam generator is shown in Fig. 13.14. The outputs of the gas and steam turbines are, respectively, 15 MW and 11 MW, giving an overall net thermal efficiency of 36.6 percent at full load for operation on No. 2 fuel oil. The part-load heat rate is given in Fig. 13.15.

The Lünen Combined-Cycle Plant Coupled to Lurgi Coal Gasifiers

One of the shortcomings of coal gasification processes (as discussed in Chap. 5) is that over 15 percent of the chemical energy in the coal fed to the process is lost to cooling water, generally at temperatures below ~200°C (392°F). One way to employ this heat to good advantage is to couple the coal gasification plant directly to a combined-cycle power plant. The clean low-Btu gas from the gasifier is burned in the combined-cycle plant to give a high-temperature heat source for the gas turbine, and the heat lost to cooling water in the gasifier is employed for feed-water heating in the steam cycle. A further advantage of close integration of the gasifier with the combined-cycle plant is that the energy conversion efficiency of coal gasification processes drops with increasing refinement of the gaseous fuel produced; hence the most energy-efficient and lowest-cost processes for converting coal into gaseous or liquid fuel are those yielding low-Btu gas. Since the volume of

the low-Btu gas is too great for the gas to be stored or transported economically (Chap. 20), gasifiers yielding low-Btu gas must be located close to the plant in which the gas is consumed.

The only combined-cycle power plant closely integrated with a coal gasification plant that had been built and operated anywhere in the world by 1980 is a unit at Lünen, near Essen in West Germany, which employs five Lurgi gasifiers in parallel.[8,9]

The attached flowsheet in Fig. 13.16 shows the principal components in the plant. Air from the compressor of the gas turbine at 10 atm is split into two streams, with 10 percent going through a second compressor for delivery to the Lurgi units at 20 atm. This air, along with prime steam bled from the boiler, goes into the bottom of each Lurgi unit, while coal is fed through a lock hopper at the top. Low-Btu gas emerging from the Lurgi units is passed through two scrubber stages to remove tar and ash (but not sulfur), and then it is expanded through a small turbine which drives the small, high-pressure compressor that boosts air from the main compressor to 20 atm to feed the Lurgi units. The low-Btu gas from the scrubbers, along with the main airstream from the compressor, is fed to burners in two large furnace–steam-generator units that operate in parallel. Over half of the heat of combustion is transferred to steam in the steam generators, and the balance as sensible heat in the combustion gas is fed to the main gas turbine at 820°C. The gas-turbine exhaust at 400°C passes through an economizer for the steam plant and then is delivered to the stack at 170°C. The design output of the steam

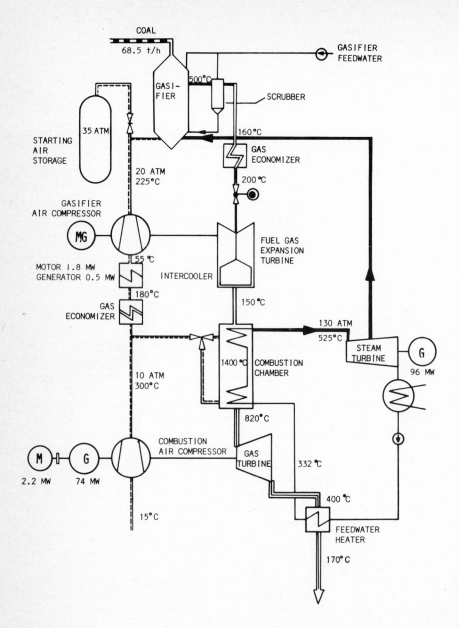

Figure 13.16 Flowsheet for the integrated coal-gasification–combined-cycle Lünen plant with Lurgi moving-bed gasifiers.[8]

turbine is 100 MWe, and the output of the gas turbine is 70 MWe. The only steam bled from the steam turbine for regenerative feed heating is that used to preheat the feed water fed to the deaerator. Thus the feed water entering the economizer in the exhaust stream from the gas turbine is up to the boiling point, and as a consequence, there is no tendency for condensation of sulfurous acid in the economizer if SO_2 is present in the turbine exhaust gas. The overall thermal efficiency at full load was estimated to be about 37 percent from the existing plant. This estimate is based on the lower heating value of the fuel and does not include losses in a sulfur-removal system.

The furnace–steam-generator units are designed to deliver steam at 130 atm and 525°C (1910 psia and 977°F). The two units operating in parallel are each 3.6 m in diameter and 20 m high. They are shop-fabricated (as is the case with the Lurgi units) and can be shipped by rail. Fuel and combustion air are fed through a burner at the top, and combustion takes place in a cylindrical cavity with a waterwall formed by welding spiral tubes 22 mm in diameter to give a continuous water-cooled shell around the combustion chamber. Steam from the header drum at the top of the waterwall is fed to superheater tubes arranged in concentric, annular tube banks in the lower portion of the vessel.

To avoid excessive temperatures in the combustion-product gas stream entering the main gas turbine, an air bypass valve is provided in the main air duct from the compressor to the combustion chamber to permit tempering the turbine air inlet temperature.

One of the features of the combined cycle as embodied in this plant is that the steam from the gas scrubbers and the water evaporated from the coal increase the mass flow through the gas turbine sufficiently to increase its net output from about 50 MWe to 70 MWe. Thus, the large amount of heat absorbed by this water is not wasted but serves a useful function. The plant is started up on fuel oil and can be up to full power in 30 min.

The plant began operation in January 1973. Early operations encountered difficulties with fouling of components by tar from the Lurgi gasifiers and with inadequate solid particle removal from the fuel gas by the two-stage water scrubber (i.e., initially the particulate concentrations were in the 50-ppm range instead of below 1 ppm as specified by the turbine manufacturer). These and other problems, such as difficulties in developing the sulfur-removal process planned for the plant, continued to handicap operation, so that at the end of 1979 the plant was still not in regular service. This must be considered significant, because the Lurgi gasifier is widely regarded as the most reliable of the various coal gasification systems, and it is the most efficient to be used commercially as of 1981.

Conceptual Designs for Advanced Coal Gasifier–Combined-Cycle Plants

Efforts to obtain a high thermal efficiency from coal gasifiers closely integrated with combined-cycle plants in order to effect a major reduction in sulfur emissions while yet obtaining a good thermal efficiency have been premised on two major advances. These are (1) a coal gasification process giving a higher fuel conversion efficiency than the Lurgi, and (2) a gas turbine with inlet temperatures higher than any used to date in utility base-load service.[5,10-13] A number of large-budget programs to obtain the advances required in both the gasifiers (Chap. 5) and the high-temperature gas turbines (Chap. 12) were under

way during the 1970s, and these programs seem likely to continue in the 1980s.

Combined-Cycle System Integrated with a Coal Gasification Plant

A conceptual design of a combined-cycle system closely integrated with a coal gasification system is presented in Ref. 10 by R.W. Foster-Pegg, one of the earliest and leading advocates of systems of this type. The flowsheet for the conceptual design given in Fig. 13.17 is similar to that in Fig. 13.16 for the Lünen plant, but there are many differences. The most important is the use of fluidized-bed gasifiers operating at 18.27 bars (250 psig) and 1038°C (1900°F) with a gas residence time of 10 to 15 s so that the tars and oils will be thermally cracked, thus eliminating fouling of heat-exchanger surfaces and other components in the fuel gas stream (major problems in the Lünen plant). In addition, the coal feed stream is treated for ~30 min at ~427°C (800°F) by partial oxidation with air to minimize the tendency to agglomerate into tarry masses when it enters the fluidized-bed gasifier during heating (another major problem in Lurgi units). The heat released by this oxidation is utilized as a portion of the heat input to the boiler of the steam cycle. This pretreatment of the coal stream, together with the high temperature of the gasifier, should make it possible to avoid the most serious problems experienced with the Lünen plant. Another important difference is the much higher gas-turbine inlet temperature. Having an inlet temperature of ~1204°C (2200°F) rather than the 820°C (1508°F) in the Lünen plant, the gas turbine has a 547°C (1016°F), instead of a 400°C (752°F), exhaust temperature. This, together with some high-temperature heat recovery from the gasification process, make it possible to carry out the entire combustion process ahead of the gas turbine while still obtaining relatively high steam turbine inlet conditions, i.e., 101 bars/526°C (1450 psig/978°F). The high turbine inlet temperature requires the use of an elegant air-cooling system for the turbine blades. The success of the latter depends on the effectiveness of the water scrubber and sulfinol tower in giving an extremely low particulate content in the hot gas entering the turbine in order to avoid clogging of the cooling air exit ports in the blades (Fig. 12.29) and/or a buildup of deposits on the blades which might cause a loss in gas-turbine efficiency.

Figure 13.17 also shows the use of lower-steam-pressure stages in the counterflow steam generator to produce intermediate- and low-pressure steam for the coal gasification process, thus matching the steam temperature requirements to the availability of heat at lower temperatures in the flue gas. For additional details, the reader is referred to the original paper.

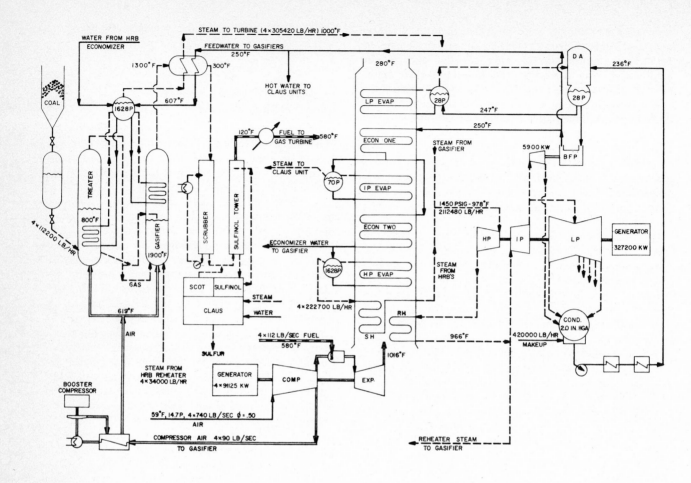

$$112200 \times 12470 \times 4 \div [91125 \times 4 + 327200 - 31700] = 8480 \text{ Btu/kWh H.H.V.}$$

Figure 13.17 Energy balance schematic diagram of a combined-cycle power plant coupled to a coal gasifier.[10] *(Courtesy Westinghouse Electric Corp.)*

It contains more information on making the most effective use possible of all the heat released in the coal gasification process than any other proposal the author has examined. Note that in this case the overall thermal efficiency was estimated to be 40.2 percent with a heat rate of 8946 kJ/kWh (8480 Btu/kWh) based on the higher heating value of dry coal, or 39.7 percent with a heat rate of 9037 kJ/kWh (8600 Btu/kWh) based on coal containing 13 percent moisture.

EPRI Study of Seven Advanced Gasifier Plants

The DOE development program that evolved in the latter 1970s for combined-cycle systems coupled to coal gasifiers (see

first portion of Chap. 22) was based on the two major studies of Refs. 12 and 13. These in turn were based on an assumed gasifier conversion efficiency that was prescribed by the Energy Research and Development Administration (ERDA) as a basis for the study. The value used—80 percent— is higher than that of any system tested (Chap. 5). A more comprehensive study sponsored by EPRI and carried out by the Fluor Corporation[11] took as its point of departure the best information available on seven different coal gasification processes that had been utilized in pilot plants and simply assumed that a gas-turbine inlet temperature of 1315 °C (2400 °F) could be obtained. Attention was focused on the many details of the equipment required for the coal gasification and fuel gas clean-up systems. Enough detailed data are included in the report in tables and flowsheets to provide good quantitative estimates of the principal losses for each of the seven systems studied. Table 13.2 summarizes key data for these seven systems for which design conditions were chosen so that the results should be comparable. For example, the gas-turbine inlet temperature and pressure ratio, the steam-cycle conditions, and the stack gas temperatures were the same in all cases. The design power output varied somewhat, but only by ± 10 percent. It is interesting to note that, when the principal losses are taken into account, the highest overall thermal efficiency for any of the systems was 40.6 percent (based on the fuel higher heating value, HHV). This efficiency is essentially the same as the 40.2 percent of the study by Foster-Pegg cited above, which was based on a 1204°C (2200°F) gas-turbine inlet temperature instead of the 1315°C (2400°F) of the EPRI study. On

TABLE 13.2 Summary of Performance Estimates for Combined-Cycle Plants Closely Integrated with Representative Coal Gasification Plants [11]

	Lurgi MACW	BGC slagger MXSC	Foster Wheeler EAHC	Foster Wheeler EXHC	Combustion engineering EALC	Texaco EXTC (slurry feed)	Texaco EXTC (dry feed)
GASIFICATION AND GAS CLEANING SYSTEM							
Coal feed rate, lb/h m.f.	1,014,814	798,333	798,333	798,333	798,333	798,333	798,333
Oxygen or air/coal ratio,[a] lb/lb m.f.	1.562	0.481	2.857	0.609	4.37	0.858	0.806
Oxidant temperature, °F	340	214	800	335	437	300	300
Steam/coal ratio, lb/lb m.f.	0.758	0.31	0.150	0.624	0	0	0.610
Slurry water/coal ratio, lb/lb m.f.	NA	NA	NA	NA	NA	0.503	NA
Gasifier exit pressure, psig	340	320	360	360	[b]	600[c]	600[c]
Crude gas temperature, °F	861	820	1700	1700	1700	2300–2600	2300–2600
Crude gas HHV (dry basis),[d] Btu/SCF	189.1	379.0	174.1	315.4	113.0	281.1	280.7
Temperature of fuel gas to gas turbine, °F	425	580	800	800	1200	781	781
POWER SYSTEM							
Gas turbine inlet temperature, °F	2400	2400	2400	2400	2400	2400	2400
Pressure ratio	17:1	17:1	17:1	17:1	17:1	17:1	17:1
Gas turbine exhaust temperature, °F	1137	1128	1127	1133	1147	1140	1140
Steam conditions, psig/°F/°F	1450/ 900/ 1000	1450/ 900/ 1000	1450/ 900/ 1000	1450/ 900/ 1000	1450/ 900/ 1000	1450/ 900/ 1000	1450/ 900/ 1000
Condensing pressure, in Hg abs.	2.5	2.5	2.5	2.5	2.5	2.5	2.5
Stack temperature, °F	275	275	275	275	275	272	272
Gas turbine power,[e] MW	590	857	751	803	886	745	763
Steam turbine power,[e] MW	430	385	504	384	307	448	425
Power consumed, MW	32	30	42	38	55	36	46
Net system power, MW	988	1212	1213	1149	1138	1157	1142
OVERALL SYSTEM							
Process and deaerater makeup water, gpm/1000 MW	2207	834	497	1031	157	362	1072
Cooling tower makeup water, gpm/1000 MW	5698	5882	6125	6003	7439	7588	7255
Cooling water circulation rate, gpm/MW	366	307	341	321	343	347	352
Cooling tower heat rejection, % of coal HHV	33.9	33.8	36.8	33.2	37.6	38.7	35.6
Air cooler heat rejection, % of coal HHV	6.5	4.7	3.2	7.2	4.9	5.2	4.6
Net heat rate, Btu/kWh	9762	8410	8428	8876	8959	8813	8928
Overall system efficiency (coal→power), % of coal HHV	34.96	40.6	40.5	38.5	38.1	38.7	38.2

[a]Dry basis, 100% O_2 for oxygen blown case.
[b]Gasifier exit pressure is − 0.5 in H_2O.
[c]Average gasifier pressure.
[d]Excluding the HHV of H_2S, S, and NH_3.
[e]At generator terminals.

the other hand, the design of the Foster-Pegg system entailed a greater effort to utilize all the waste heat to the best possible advantage.

As is usual in comparisons of this sort the coal flow rates are given on a moisture-free (m.f.) basis because the moisture content of coals varies widely. Often the effects of varying ash content are also included, and the coal flow rates are given on a moisture-and-ash-free (m.a.f.) basis.

Effects of Coal-Conversion Process Efficiency and Heat-Recovery Temperature

The above EPRI study, together with the doubtful prospects of achieving the 1315°C (2400°F) turbine inlet temperature (a point implicit in the previous chapter), raises a serious question as to whether the combined-cycle plant integrated with a coal gasifier will prove to be competitive with a conventional coal-fired supercritical-pressure steam plant with a wet limestone stack gas scrubber, particularly in view of the large capital investment required for the gasifier system. This question is so vital that it deserves more general scrutiny. Close integration of a coal gasification plant with a combined cycle that utilizes the waste heat from the gasifier gives a system whose overall thermal efficiency is difficult to appraise. To get some perspective, one can draw a coal gasification plant enclosed in an envelope and in effect prepare a diagram similar to that of Fig. 13.7 to delineate the eventual destination of all the energy coming into the system. To a first approximation, the chemical energy in the coal flowing into the plant will be equal to the chemical energy in the fuel gas leaving the plant, plus the waste heat recovered in the form of steam at a useful temperature, plus the unrecoverable heat losses to the surroundings or to low-temperature cooling streams. One would prefer to have the fuel gas from the gasifier flow to the burner of the combined cycle at a high temperature so that the sensible heat in this fuel gas would also be part of the energy output. But in most gasification processes for combined cycles, the gas must be cooled in order to remove sulfur and particulates. Hence, the sensible heat available in the fuel gas is usually small, and the parameter *cold-gas conversion efficiency*, the ratio of the chemical energy in the product gas to that in the original coal (both at room temperature), is commonly used to define the efficiency of the gasification process.

Some losses are hard to estimate without detailed drawings for the plant considered. However, losses to the surroundings, plus allowances for pumping power (including the thermal efficiency of the system for generating that power), commonly run at roughly 7 percent. The fact that much of the sensible heat recovered from the coal gasification process can be utilized either for feed-water heating or for a low-pressure steam cycle is confusing. A series of ''Btu-chasing'' exercises leads to a simplifying conclusion that is supported by basic thermodynamic considerations: the overall plant thermal efficiency will be the same whether the heat recovered is used for feed heating or for a separate steam cycle. Thus the intermediate-temperature heat may be treated as if it were utilized in a simple steam cycle which (with allowances for conventional component efficiencies) would give cycle efficiencies of 12, 25.5, 32, and 33.5 percent for heat available in steam at temperatures of 93, 204, 316, and 427°C (200, 400, 600, and 800°F), respectively. Thus, there are just three major independent variables affecting the overall plant thermal efficiency: the coal-to-fuel gas conversion efficiency, the temperature of waste-heat recovery from the coal gasification plant, and the gas-turbine cycle inlet temperature. This gives too many variables for a simple chart.

One way of treating the problem is to calculate first what can be called an *effective cold-gas thermal efficiency*, which is the cold-gas conversion efficiency plus a quantity representing the practical value of the waste heat from the gasifier. The latter is the summation of the products of the fractions of the chemical energy in the coal recovered as steam at various temperatures and the ratio of the thermal efficiency of the steam cycle for each of these temperatures to the thermal efficiency of the combined cycle at the specified gas-turbine inlet temperature (Fig. 13.8). The resulting values are shown in the lower part of Fig. 13.18 for a series of cold-gas thermal efficiencies.[5] There is a marked increase in the effective cold-gas thermal efficiency with an increase in the temperature at which heat is recovered from the gasifier. The dependent variable of prime interest is the overall thermal efficiency of the plant complex, which can be found from the chart at the top of Fig. 13.18 by using the effective cold-gas thermal efficiency and the gas-turbine inlet temperature. For the highest gas-turbine inlet temperature likely to be permitted by the blade deposit problems of Fig. 12.39—i.e., 1100°C (2010°F)—an optimistically high cold-gas conversion efficiency of 80 percent (much better than 67 percent, the best value demonstrated up to 1980), and recovery of waste heat from the gasifier at 204°C (400°F), the overall thermal efficiency of the plant complex is only ~36 percent, about the same as for a conventional coal-fired steam plant with stack gas scrubbers. In view of both thermal efficiency and capital cost considerations, it appears that a higher gasifier efficiency, a higher waste-heat-recovery temperature, and/or a gas-turbine inlet temperature of over 1100°C (2010°F) will be needed to make the system attractive. The situation would also change if even tighter restrictions were placed on sulfur emissions, because this would impose an even greater penalty on stack gas scrubbing.

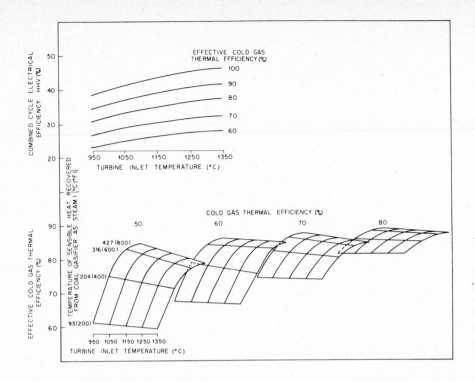

Figure 13.18 Effects of cold-gas efficiency, sensible heat rejection temperature, and turbine firing temperature on integrated coal-gasification–combined-cycle power plants. *(Courtesy of Oak Ridge National Laboratory.)*

Operational Problems

One of the most important advantages of open-cycle gas turbines is that they can be started up in minutes and can respond in seconds to abrupt changes in the electric load. When a steam bottoming cycle is added, good response to load changes can still be obtained, but the startup time increases to ~ 45 min. If a coal gasification plant is tightly integrated with a combined gas-turbine–steam system, the situation becomes much more complex. The volumetric flow rate of the low-Btu gas is so great that it is impractical to store the gas, and it must be used about as rapidly as it is generated. The coal gasifiers have a high heat capacity and so take a long time to heat up. Also, the efficiency of the coal gasification process is sensitive to pressure, temperature, and flow rate conditions. Hence the gasifiers operate well over only a limited range. These characteristics vary substantially with the type of gasifier (i.e., fixed-bed, fluidized-bed, or entrained-bed), but it appears that plants of this type may require 12 to 20 h to get on line from a cold start, are relatively inflexible, and thus are suitable only for baseload service.

Integration of a gas-turbine–steam combined-cycle plant with a coal gasification plant presents the utilities with the unfamiliar problems of the operation of a relatively complex chemical plant in addition to the electric generation plant. This major increase in the ability and experience required of the operators will probably result in a lower reliability and availability for the integrated plants, at least during their formative years, and a longer period for shakedown than has been experienced with the oil-fired combined-cycle plants.

The fuel quality from an integrated coal gasification plant will probably be such that the gas-turbine life, reliability, and/or the allowable turbine inlet temperature will be reduced, because the ash, alkali metal, and sulfur contents will tend to run higher than for No. 2 fuel oil, especially during startup conditions. The overall reliability and availability of the integrated plant must include the coal gasification plant, whose complexity will inherently have adverse effects, as will the severe corrosion conditions in the gasification system.

Combined-Cycle Coupled to a Fluidized-Bed Combustor

As discussed in Chap. 5, the FBC provides an attractive approach to the combustion of high-sulfur, high-ash coals of all types—including lignites—with a minimum of coal preparation and no need for stack gas scrubbers. However, as discussed in Chap. 11, if the furnace is operated at atmospheric pressure, the bed depth is limited to ~ 1 m by pumping-power considerations. This makes the bed area awkwardly large,

TABLE 13.3 Comparison of Design Studies for Three Combined-Cycle Plants Fueled, Respectively, with Gas from a Low-Btu Gasifier, Semiclean Liquid Fuel from Coal, and Coal Burned in a Pressurized Fluidized Bed [16]

Item	Plant 1 Combined gas-steam turbine plant with an integrated low-Btu gasifier	Plant 2 Combined gas-steam turbine plant burning semiclean fuel from coal	Plant 3 Advanced steam plant with pressurized fluidized-bed boiler
Plant size (net power), MW	785.978	872.840	679.207
Gas-turbine–generator gross power, MW	543.320	584.360	142.002
Steam-turbine–generator gross power, MW	285.054	313.547	562.636
Auxiliary power requirements and transformer losses, MW	42.396	25.067	25.431
Gas turbine-generator			
Number	4	4	2
Pressure ratio	16:1	16:1	10:1
Turbine inlet temperature, °F	2500	2500	1760
Inlet airflow per turbine, lb/s	749.82	750.09	760
Steam generator type	HRSG	HRSG	PFBB
Number	4	4	4
Total steam generated $\times 10^{-6}$, lb/h	1.5167	1.7601	3.6000
Throttle pressure, psig	2400	2400	3500
Throttle temperature, °F	1000	1000	1000
Reheat temperature, °F	1000	1000	1000
Condenser pressure, in Hg abs.	2	2	2.5
Performance			
Power plant efficiency, %	46.8	52.1	38.99
Overall efficiency (coal to switchyard), %	46.8	38.55	38.99
Plant heat rate, Btu/kWh	7293	6551	8753

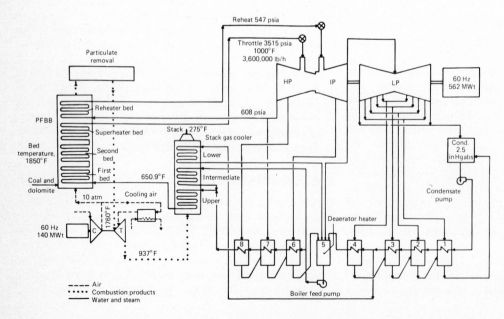

Figure 13.19 Energy balance diagram for a combined-cycle plant coupled to a fluidized-bed coal combustor.[16] *(Courtesy Westinghouse Electric Corp.)*

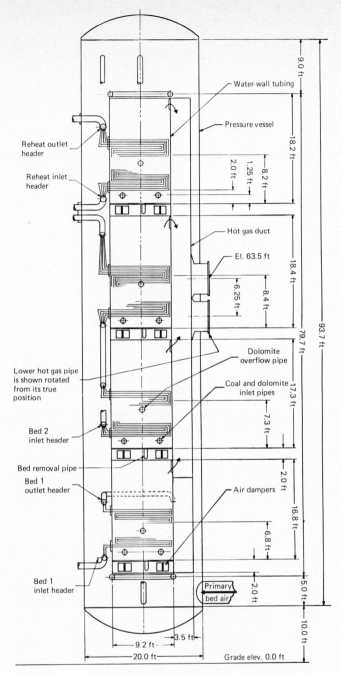

Labels on figure (top to bottom, left and right):

Reheat outlet header

Reheat inlet header

Lower hot gas pipe is shown rotated from its true position

Bed 2 inlet header

Bed removal pipe
Bed 1 outlet header

Bed 1 inlet header

Water-wall tubing

Pressure vessel

Hot gas duct

El. 63.5 ft

Dolomite overflow pipe

Coal and dolomite inlet pipes

Air dampers

Primary bed air

Dimensions: 9.0 ft, 18.2 ft, 1.25 ft, 2.0 ft, 8.2 ft, 18.4 ft, 6.25 ft, 8.4 ft, 93.7 ft, 79.7 ft, 17.3 ft, 7.3 ft, 2.0 ft, 16.8 ft, 6.8 ft, 2.0 ft, 5.0 ft, 10.0 ft, 3.5 ft, 9.2 ft, 20.0 ft, Grade elev. 0.0 ft

Figure 13.20 Pressurized fluidized-bed steam generator for the plant of Fig. 13.19.[16] (*Courtesy Westinghouse Electric Corp.*)

tional to the furnace pressure,[14] and this has led to numerous design studies and some laboratory-scale experiments with beds pressurized to 6 to 10 atm.[14-20] The latter have shown much improved combustion efficiencies and sulfur retentions, because the greater the bed depth, the longer the residence time for both coal particles and the gases evolved from them (Figs. 5.21 and 5.22). The prime problem has been that of particulate removal from the hot gases leaving the bed. As discussed near the end of the previous chapter, the best of the processes tested up to 1980 has given particulate concentrations about 100 times the level that appears acceptable for gas turbines that operate in the range of 815 °C (1500 °F) to 982 °C (1800 °F). However, extensive programs have been under way since the mid-1970s to develop improved particulate-removal equipment for operation in this temperature range, and a number of conceptual designs for pressurized fluidized-bed power plants have been prepared on the assumption that one or another of these particle separators will prove economically attractive. A review of a typical study gives some perspective on the problems involved in these large budget efforts, e.g., approximately 30 million dollars of DOE funds in fiscal year 1979.

Westinghouse Conceptual Design for ECAS-II

One of the most detailed and complete conceptual design studies of a pressurized fluidized-bed combustor (PFCB) is presented in the 380-page report of Ref. 16. The design study assumed a combustor temperature of 955 °C (1750 °F) and a design net power output of ~680 MW. The principal design data are summarized in Table 13.3, along with directly comparable values from two other parallel studies of combined-cycle plants which used, respectively, a coal gasifier[12] and a liquid fuel derived from coal.[21] [The conceptual design for the coal gasifier case is similar to that given in the previous section but followed more optimistic DOE assumptions, including ceramic blades in the upper stages of the gas turbine with no blade cooling-air power requirement and a higher turbine inlet temperature, i.e., 1371 °C (2500 °F) versus 1204 °C (2200 °F).]

A heat balance diagram showing the principal elements in the system is given in Fig. 13.19. The layout is similar to that of Fig. 13.16 for the plant with the low-Btu Lurgi gasifiers in that there is just one combustion chamber which is placed ahead of the gas turbine, with most of the heat released flowing into the steam generator. The hot gas leaving the gas turbine at 473 °C (884 °F) provides heat for the economizer (or upper-stage feedwater heater).

The furnace–steam generator, shown in Fig. 13.20, utilizes the configuration commonly employed in conceptual designs for both atmospheric and pressurized fluidized beds, i.e., a

giving a high capital cost that is marginally competitive with conventional PCFs fitted with stack gas scrubbers. It was recognized in the 1960s that pressurizing the furnace with a gas turbine would reduce the bed area by a factor directly propor-

TABLE 13.4 Auxiliary Power Requirements for the 680-MWe (net) Combined-Cycle Pressurized Fluidized-Bed Plant Shown in Fig. 13.19[16]

	Units	Power/unit, kWe	Total power, MWe	Percent auxiliary power
Gas-turbine auxiliaries	2	448	0.896	4.06
Pressurized-fluidized-bed boiler auxiliaries			9.349	42.46
Transport compressor	2	794		
Multiclone	4	282		
Auxiliary compressor	2	1746		
Impact mills	4	238		
Miscellaneous	4	210		
Waste disposal	LS	1349		
Steam-turbine auxiliaries			3.183	14.46
Condensate pumps		2063		
Miscellaneous		1120		
Circulating water auxiliaries and cooling towers			6.350	28.84
Circulating water pumps		3175		
Cooling-tower fans		2321		
Miscellaneous		854		
Coal and dolomite handling			1.120	5.09
Station auxiliaries (misc.)			1.120	5.09
Total auxiliary loads			22.018	100.0
Main transformer losses			3.413	
Total station auxiliary power and power losses			25.431	

TABLE 13.5 Capital Cost Estimates for the 680-MWe Combined-Cycle Plant of Fig. 13.19[16]

Acct. no.	Description	Unit of measure
1.00	Land and land rights	
1.10	Land cost	acres
1.20	Site development	LS
1.30	Access and site railroad	miles
2.00	Structures and improvements	
2.10	Plant island foundations	yd³
2.20	Station structural steel	tons
2.30	Chimney	pair
2.40	Structural features	LS
2.50	Station buildings	ft³
2.60	Other buildings	ft³
2.70	Process waste disposal	LS
3.00	Heat rejection system	
3.10	Condenser	ft²
3.20	Circ. H_2O system structure, piping and auxiliary	LS
3.30	Cooling towers	cell
4.00	Material handling, storage and preparation	
4.10	Coal-handling system	ton/h
4.20	Dolomite-handling system	
4.30	Fuel oil or coal distillate handling system	LS
4.40	Coal and dolomite preparation, feed and silo storage	LS
4.50	Process-waste-handling system	LS
5.00	Energy conversion	
5.10	Fluidized-bed steam generator	
5.11	PFBB	pair
5.12	Gas cleaning & waste handling	pair
5.20	Gas-turbine/generator	each
5.30	Steam-turbine/generator	each
5.40	Heat-recovery apparatus	
5.41	Stack gas cooler	each
6.00	Auxiliary mechanical equipment	
6.10	Piping systems	
6.11	Hot-gas piping systems	LS
6.12	Steam piping systems	tons
6.13	Feed-water piping systems	tons
6.14	Misc. service piping systems	tons
6.20	Feed-water heaters	LS
6.30	Misc. mechanical equipment	
6.31	Boiler feed pump and drive	LS
6.32	Other pumps	LS
6.33	Auxiliary boiler	each

series of rectangular beds arranged one above the other in a tall cylindrical pressure shell. Horizontal serpentine tube banks in the beds are supported on a rugged structure to resist the shaking forces of bed pulsations (Chap. 11).

The power requirements of the various auxiliaries in a power plant often represent a substantial fraction of the gross electric output, thus degrading the thermal efficiency. It is interesting to note from Table 13.4 that the power saved by eliminating the coal pulverizers and using a gas turbine in place of induced- and forced-draft fans is roughly offset by the extra power required for pneumatic transport of the coal feed stream and ash from the cyclone separators. Thus the requirement for auxiliary power is ∼3.1 percent, about the same as the amount required for a conventional pulverized-coal-fired steam plant (Table 11.7).

A summary of the estimated capital cost of the plant in 1975 dollars is given in Table 13.5. In reviewing these estimated costs, it is worth noting that the biggest cost item is the fluid-

Quantity	Unit price — Material — Mjr. comp.	B.O.P.	Inst. cost	Total price × 10⁻³ — Material — Mjr. comp.	B.O.P.	Inst. cost	Total ext. cost × 10⁻³, $	Percent of total	Subtotal
					2,743	3,200	5,943		3.315
380		2500			950		950	0.530	
1		883,500	2,705,800		884	2,706	3,590	2.002	
7.9		115,088	62,494		909	494	1,403	0.783	
					6,291	6,174	12,465		6.952
24,500		131	175		3,210	4,288	7,498	4.182	
1950		700	300		1,365	585	1,950	1.088	
2		140,000	80,000		280	160	440	0.245	
1		490,000	163,000		490	163	653	0.364	
2,700,000		0.18	0.18		486	486	972	0.542	
28,000		15.00	13.00		420	364	784	0.437	
1		40,000	127,500		40	128	168	0.094	
					5,824	3,341	9,165		5.112
249,200		5.85	1.10		1,458	274	1,732	0.966	
1		2,014,300	1,876,680		2,014	1,877	3,891	2.170	
14		168,000	85,000		2,352	1,190	3,542	1.976	
				5,471	10,926	9,422	25,819		14.400
276		19,200	9,360		5,299	2,583	7,882	4.396	
30% of 4.10					1,590	775	2,365	1.319	
1		225,000	190,000		225	190	415	0.231	
2	2,735,463	80,080	1,553,002	5,471	160	3,106	8,737	4.873	
1		3,651,800	2,768,000		3,652	2,768	6,420	3.581	
				63,963	4,544	16,500	85,007		47.410
				21,955	4,544	12,195	38,694	21.580	
2	4,685,124	1,261,318	2,644,598	9,370	2,522	5,280	17,172		
2	6,292,330	1,011,282	3,457,386	12,585	2,022	6,915	21,522		
28,358,900	8,258,900		676,700	16,718		1,353	18,071	10.079	
1	18,950,000		1,321,875	18,950		1,322	20,272	11.306	
2	3,169,830		815,000	6,340		1,630	7,970	4.445	
				15,179		6,438	21,617		12.056
				9,697		5,018	14,715	8,207	
1	3,389,278		338,928	3,389		338	3,727		
800		4600	2400		3,680	1,920	5,600		
510		3600	3600		1,836	1,836	3,672		
220		3600	4200		792	924	1,716		
1		970,000	48,500		970	49	1,019	0.568	
					3,340	1,072	4,412	2,461	
1		1,225,000	110,000		1,225	110	1,335		
1		696,000	111,000		696	111	807		
1		517,000	211,500		517	212	729		

TABLE 13.5 Capital Cost Estimates for the 680-MWe Combined-Cycle Plant of Fig. 13.19[16] (*Continued*)

Acct. no.	Description	Unit of measure	Quantity	Unit price Material Mjr. comp.	B.O.P.	Inst. cost
6.34	Misc. service system equip.	LS	1		901,700	639,000
6.40	Water treatment equipment					
6.41	Makeup demineralizer and pretreatment equipment	LS	1		362,000	104,000
6.42	Condensate polishing	LS	1		810,000	195,000
7.00	Auxiliary electrical equipment					
7.10	Auxiliary transformers and motors	LS	1		1,127,000	167,000
7.20	Switchgear and control boards	LS	1		1,496,000	434,000
7.30	Cable, conduit, trays & ducts	linear ft	2,970,000		1.35	1.48
7.40	Buswork	LS	1		781,000	454,300
7.50	Lighting and communications	LS	1		558,000	837,000
7.60	Instrumentation and controls	LS	1		1,675,000	729,000
8.00	Transmission plant station equipment					
8.10	Main transformers	LS	1	2,562,120		60,000
	Total direct accounts					
10.00	Indirect construction costs					
	Subtotal					
11.00	Professional services					
	Subtotal					
12.00	Contingency (10%)					
	Subtotal					
13.00	Escalation during construction (5 years)					
	Subtotal					
14.00	Interest during construction					
	Total capitalization					

ized-bed steam generator, which suggests that a different approach to its design might reduce its size, weight, and cost. The second largest cost item is the gas-cleaning and waste-handling equipment, for which the uncertainty is definitely large, because no suitable prototype has ever been operated. Note also that the estimated cost of these two items is about seven times the total cost of the "structures and improvements," including all the building foundations.

The close integration of the coal gasifier with the combined-cycle plant of Fig. 13.18 presents a complex set of control problems. A good indication of the manner in which these might be handled is given by Fig. 13.21.

Supercharged Fluidized Beds

The relatively high costs of atmospheric fluidized-bed furnace–steam generators stemming from their awkward proportions, coupled with the low probability of developing a hot-gas clean-up system for a pressurized fluidized-bed furnace that

Total price × 10⁻³			Total ext. cost × 10⁻³, $	Percent of total	Subtotal
Material					
Mjr. comp.	B.O.P.	Inst. cost			
	902	639	1,541		
	1,172	299	1,471	.820	
	362	104	466		
	810	195	1,005		
	9,646	7,017	16,663		9.293
	1,127	167	1,294	0.722	
	1,496	434	1,930	1.076	
	4,009	4,396	8,405	4.687	
	781	454	1,235	0.689	
	558	837	1,395	0.778	
	1,675	729	2,404	1.341	
	2,562	60	2,622		1.462
	2,562	60	2,622	1.462	
69,434	57,715	52,152	179,301	100.00	
		26,598			
69,434	57,715	78,750	205,899		
			20,590		
			226,489		
			22,649		
			249,138		
			50,326		
			299,464		
			71,004		
			370,468		

would prove both economically attractive and give a good gas-turbine life with a turbine inlet temperature of 800°C or more, led the author to propose a different approach to pressurized fluidized-bed system design.[22] Earlier work on coal-fired gas turbines has indicated that if the gas-turbine inlet temperature were dropped to ~540°C (1003°F), cyclone separators can be used to reduce the particulate content sufficiently that there should not be serious difficulties with erosion and deposits in the turbine (Fig. 12.43). There would be little or no useful electric output from the gas-turbine–compressor unit, but it would serve to supercharge the furnace to 3 to 10 atm, thus making it possible to employ a bed depth of around 3 to 6 m (10 to 20 ft). This not only permits the use of a much smaller fluidized bed, but it also makes it possible to use a less expensive boiler tube configuration, i.e., vertical rather than horizontal tubes. This in turn permits the use of natural thermal convection recirculation of the water in the boiler tubes, and makes it possible to eliminate the tube-bank support structure required for the horizontal serpentine tubes of Fig. 13.20. (That tube-bank support structure is inherently un-

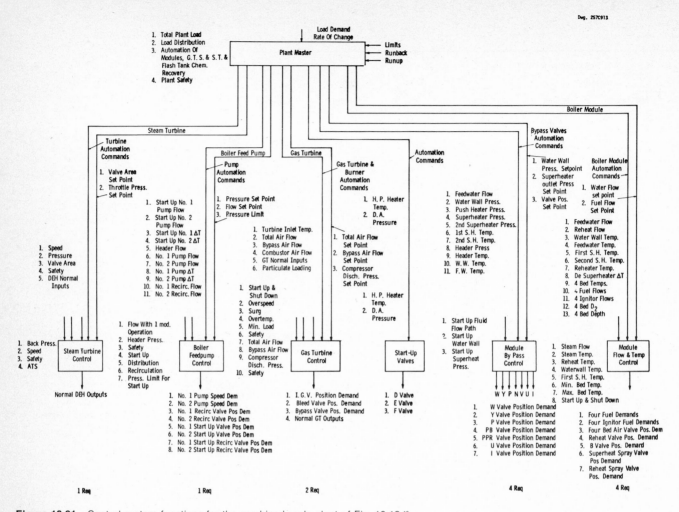

Figure 13.21 Control system functions for the combined-cycle plant of Fig. 13.19.[16]

cooled, corrosion-prone, and expensive.) The vertical tubes also are advantageous because they give a favorable configuration for controlling the steaming rate by varying the bed depth and thus the fraction of the surface area of the boiler tube matrix that is immersed in the bed. (See Fig. 11.23 and the related discussion.)

A conceptual design evolved from these considerations is shown in Fig. 13.22. The vertical boiler tube configuration makes it possible to use a round rather than a rectangular bed so that it fills virtually the entire cross-sectional area of the cylindrical pressure vessel, thus avoiding the complications of the multitier construction of Fig. 13.20 and markedly reducing the vessel fabrication and piping costs. For a high-pressure, high-temperature steam system, such a large fraction of the heat

goes to the superheater and reheater that heat balance considerations make it necessary to place the first stage of the superheater in the bed. The second-stage superheater and the reheater tube banks are located above the freeboard. Hot gases leaving the bed are cooled by counterflow through these tube banks to ~538°C (1000°F) before flowing to the gas turbine. The gases leaving the turbine are then cooled further by giving up their heat to the boiler feed water before passing to the stack.

Scale Effects

In scaling up this type of furnace to large outputs, the relations between furnace steaming rate, shell diameter, operating pressure, and unit cost are vital considerations. Design studies indicate that the cost of the furnace shell per unit of output drops rapidly if the furnace pressure is increased from 1 atm to 3 atm because the minimum shell thickness for structural stability is about that required to confine a pressure of 3 atm. Further increases in pressure require corresponding increases in shell thickness; hence the cost per unit of output does not drop much further. A second major consideration is the bed support grid. The controlling structural problem is supporting the weight of the bed after an abrupt hot shutdown and at the same time accommodating differential expansion between the support grid and the shell while preventing uncontrolled air leakage past the air distributor plate. These structural requirements lead to a rapid increase in the weight and cost of the support grid for bed diameters greater than ~ 10 m. At the same time cost considerations strongly favor shop fabrication of major furnace shell subassemblies, and these should be no more than 10.67 m (35 ft) in diameter, the upper limit for shipment by barge (a practicable course for most plant sites). Thus there is a strong incentive to limit the diameter of a furnace unit to ~ 10.67 m (35 ft). This in turn means that the design output per unit will depend on the furnace pressure chosen, i.e., ~ 125 MWe for 3 atm and ~ 400 MWe for 10 atm. If commercial turbochargers for large diesel engines are used, the upper limit on the pressure is ~ 4 atm, and the same value holds for the commercial turboexpanders that have been employed in petroleum refinery catalytic cracker service. However, for a design output of ~ 400 MWe, the gas turbine employed in the Brown Boveri Velox boiler described early in this chapter (Figs. 13.9 to 13.12) appears suitable. It is a refined design with a high efficiency in both the compressor and turbine so that for an oil-fired Velox boiler the turbine output is sufficient to drive the compressor and give a pressure ratio of 10:1 with a turbine inlet temperature of only 400°C (752°F). The pressure drop across the fluidized-bed furnace–steam generator would be greater than for the oil-fired Velox boiler, but this could be accommodated with a small increase in the gas-turbine inlet temperature.

Control

One of the most important advantages of the proposed system is that it eases the difficult control problems posed by operation of a fluidized-bed combustor (FBC). The lower limit on the gas flow velocity through a fluidized bed is imposed by getting sufficient agitation for good coal mixing and combustion; the upper limit is imposed by excessive elutriation of the fines. The velocity range available turns out to be roughly a factor of 3 (Fig. 5.19). On the other hand, for good control of a power plant from startup to full power it would be highly desirable to vary the combustion air weight flow rate by a factor of 10. As

discussed in Chap. 5, limitations on the superficial gas velocities through the bed for good coal mixing and acceptable elutriation rates are essentially independent of furnace pressure. If the bed is started up at atmospheric pressure with the minimum air velocity for good fluidization, the pressure and superficial velocity each can be increased by a factor of 3 or

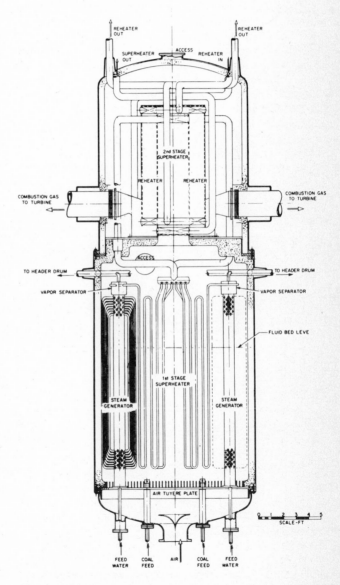

Figure 13.22 Vertical section through a pressurized fluidized-bed steam-generator unit.[22]

more in going to full power, thus allowing the air weight flow to be varied by a factor of 10 or more while still staying within the limits imposed at the lower end by good combustion and at the upper end by elutriation.

Inasmuch as no electric power would be generated by the gas turbine, there would be no need to maintain a constant speed (to give a constant frequency). The turbine-compressor speed could be allowed to vary, greatly easing control problems. This would require some auxiliary power input for startup and part-load operation, but the deficit between the compressor power requirement and the turbine output is small (Fig. 13.23) and might be accommodated with a variable-speed electric motor drive or an auxiliary motor-driven blower in the air duct ahead of the compressor.

There would be two main control functions involved in the operation of the furnace. The furnace temperature would be kept essentially constant by varying the airflow and coal feed rate while maintaining a small amount of excess air. The boiler pressure would be kept constant by varying the height of the fluidized bed, thus controlling the effective heat-transfer sur-

face area in the bed. The sensing elements required are standard reliable units, and the relations are practically linear.

Hydrocarbon Fuels Obtained by Pyrolysis of the Coal Fed to an FBC Plant

Occasionally one encounters a favorable combination of system characteristics with important possibilities. An examination of the losses inherent in all the coal gasification and lique-faction processes that have been under development indicates that substantial heat losses at moderate temperatures are inherent, and even close integration of the gasifier with a combined-cycle plant does not give a really attractive conversion efficiency (Fig. 13.16 and Table 13.2). It has often been suggested that the losses associated with the water-gas reaction could be avoided if one attempted only to skim the bulk of the volatiles off the coal enroute to a furnace in which the resulting char containing the heavy tars would be burned. To get a high yield of the lighter hydrocarbons, it has been found best to employ a flash pyrolysis process in which the coal is heated

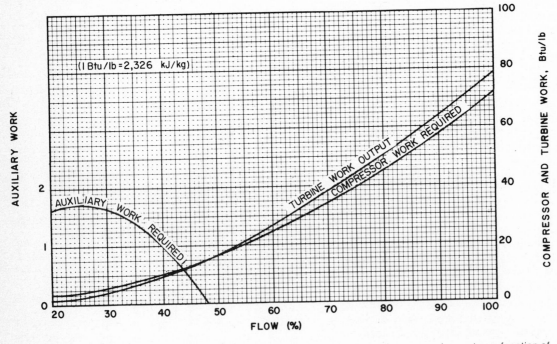

Figure 13.23   Turbine power output, compressor power input, and electric motor power requirement as a function of combustion airflow for a supercharged fluidized-bed furnace, assuming compressor and turbine efficiencies of 80 percent and a system pressure drop equal to 20 percent of the compressor pressure ratio.[22]

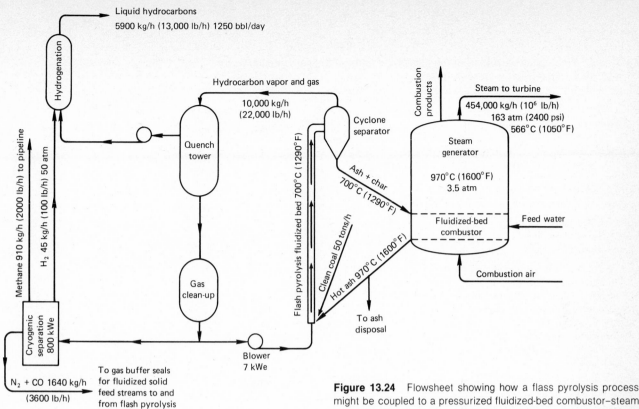

Liquid hydrocarbons
5900 kg/h (13,000 lb/h) 1250 bbl/day

Hydrogenation

Methane 910 kg/h (2000 lb/h) to pipeline

H_2 45 kg/h (100 lb/h) 50 atm

Cryogenic separation 800 kWe

N_2 + CO 1640 kg/h (3600 lb/h)

To gas buffer seals for fluidized solid feed streams to and from flash pyrolysis bed

Quench tower

Gas clean-up

Blower 7 kWe

Hydrocarbon vapor and gas

10,000 kg/h (22,000 lb/h)

Flash pyrolysis fluidized bed 700°C (1290°F)

Cyclone separator

Ash + char 700°C (1290°F)

Clean coal 50 tons/h

Hot ash 970°C (1600°F)

To ash disposal

Combustion products

Steam to turbine
454,000 kg/h (10^6 lb/h)
163 atm (2400 psi)
566°C (1050°F)

Steam generator

970°C (1600°F)
3.5 atm

Fluidized-bed combustor

Feed water

Combustion air

Figure 13.24 Flowsheet showing how a flash pyrolysis process might be coupled to a pressurized fluidized-bed combustor–steam generator.[26]

very rapidly and held at ~538°C (1000°F) for only a few seconds, then separated from the gas and vapor evolved, and the hydrocarbon vapors quenched with an oil spray to prevent further cracking.[23-25] The quenched hydrocarbons include a substantial fraction of unsaturated molecules that must be treated quickly by hydrogenation to prevent breakdown or polymerization to form gums, resins, and tars.

A particularly energy-efficient flash pyrolysis process proposed by the author entails diverting a stream of hot ash from a fluidized bed operating at 3 to 10 atm, mixing it with a stream of fresh coal in a small-diameter entrained fluidized bed that empties into a cyclone separator, and then returning the ash to the main FBC along with the char.[26] A flowsheet for the proposed process is given in Fig. 13.24. Note that the only heat and combustible material leaving the system (other than the usual amounts for the steam power plant) is that in the gas and vapor from the flash pyrolysis process. Hence, to a first approximation, the energy loss is only the latent and sensible heat given up by the gas and vapor in the quench following the flash pyrolysis. This represents only a few percent of the heat content of the hydrocarbons produced. Additional energy is re-

quired for pumping in the hydrogenation and cryogenic separation processes, etc., but the energy losses involved are roughly the same as for refining crude petroleum. The yield of gaseous and liquid hydrocarbons would run ~20 percent of the ash- and moisture-free weight of the coal, and the hydrogen produced in the flash pyrolysis process should be sufficient for the hydrogenation process.

Potential Yield of Flash Pyrolysis to the U.S. Energy Economy

An interesting indication of the potential contribution of the proposed coal pyrolysis process to the U.S. energy economy can be obtained by considering what might have been accomplished if all the fuel fed to fossil-fuel central stations in 1976 had been in the form of coal which was first subjected to a flash pyrolysis process such as that of Fig. 13.24. Table 13.6 summarizes the key numbers associated with this possibility. Note that the total U.S. energy consumption in 1976 was 71 quads, of which 16.1 quads were used by electric utilities in the form of fossil fuels. Oil and gas constituted 5.7 quads of the total 16.1 quads of fossil fuel employed in steam-electric utility

TABLE 13.6   Potential Contribution of Coal Pyrolysis to the U.S. Energy Economy as Indicated by a Perturbation to Data for 1976 [26]

	Actual consumption 1976, quads	Projected consumption with proposed flash pyrolysis system, quads
Total U.S. energy consumption	71.1	71.1
Petroleum consumption	29.2	23.1
Coal consumption	13.5	23.7
Natural gas consumption	20.2	16.1
Total fossil fuel consumed by electric utilities	16.1	16.1
Gas and oil consumption	5.7	0
Coal consumption	10.4	16.1*
Reduction in petroleum consumption if steam-electric utilities used only coal, of which 20% by weight was removed by pyrolysis to supply liquid and gaseous hydrocarbons to other sectors of the economy		
Savings in oil and gas used by utilities		5.7
Liquid and gaseous fuels produced		4.5
Reduction in U.S. consumption of natural gas and petroleum		10.2†

*Equivalent to ~ 800 × 10⁶ tons/year of coal.

†Equivalent to ~ 5 × 10⁶ bbl/day of petroleum.

plants. Some 18.5 quads energy were employed in fuels for motor vehicles. If all the fossil fuel plants had been fueled with char from a flash pyrolysis process, some 5.7 quads of energy from petroleum would have been saved and 7 quads of gaseous and liquid fuels (equal to ~ 3.3 = 10⁶ bbl/day of crude oil) would have been produced from the coal. This would have in effect reduced the total U.S. petroleum consumption almost in half. In addition, although a small factor from the energy or mass balance standpoints, coal pyrolysis yields small amounts of high-value products that can be selectively extracted in the refining process at some additional cost.

Comparison of Coal Flash Pyrolysis with Production of Shale Oil

It is interesting to compare the proposed coal flash pyrolysis process with shale oil production. The hydrocarbon yield from coal flash pyrolysis appears to be about twice as great— ~ 60 gal/ton as compared to ~ 30 gal/ton for oil shale—while the capital equipment cost appears to be much smaller. The solid-waste-disposal problem is much reduced, because the volume of ash plus spent limestone would be far less than the volume of coal mined whereas the volume of the oil shale waste after retorting is ~ 130 percent of the volume of shale mined. Further, coal from seams in the eastern United States can be employed so that the costs of transportation to load centers would be greatly reduced, and water supplies in the east are far less of a problem than in the oil shale areas of the west. In addition, about 75 percent of the capital and labor costs for mining and handling the coal and ash should be charged to the production of electric power, further reducing the cost of producing the hydrocarbons from coal as compared to oil shale. Perhaps equally important, the octane rating is much higher for the hydrocarbons from coal than for those from oil shale.

PROBLEMS

1. If a coal gasification system having a cold-gas conversion efficiency of 60 percent is coupled to a combined cycle that would have a thermal efficiency of 41 percent if operated on natural gas, estimate the overall thermal efficiency for the integrated plant, assuming that half the energy lost in the gasification process can be recovered at 200°C for use in the steam cycle. (Suggest use of Fig. A2.9.)

2. A combined-cycle gas turbine with a pressure ratio of 10:1 is to be coupled to an FBC-steam generator operated with a combustor temperature of 850°C. The hot-gas clean-up system employs three stages of cyclone separator followed by a ceramic fabric filter. The pressure losses for each stage of the cyclone separators and the bag filter run at 1 percent of the inlet pressure, and the blowdown requirements for removing trapped particles also run at 1 percent of the flow for each of the four units. Estimate the cycle efficiency with the hot-gas clean-up equipment if the efficiency without that equipment were 41 percent. Use data for plant 3 of Table 13.3 to estimate the fraction of the total plant output obtained from the gas turbine.

3. Estimate the reduction in auxiliary power requirements and consequent increase in thermal efficiency in going from a conventional PCF-steam generator to a Velox boiler type of FBC supercharged with a gas-turbine-driven compressor that would not generate any electric power (Tables 11.7 and 13.4).

REFERENCES

1. *The Evolution of the Brown Boveri Gas Turbine*, Brown Boveri & Co., Ltd., Baden, March 1966.

2. Pfenninger, H., and G. Yannakopoulos: "Steam Power Stations with Pressure-Fired Boilers," *Brown Boveri Review*, vol. 66, July/August 1975, p. 309.

3. Daman, E. L.: "Supercharged Boiler Development," *Heat Engineering,* November–December 1955.

4. Blake, J. W.: "Combination Gas Turbine-Steam Turbine Unit—Performance of Installation Burning Natural Gas at Belle Isle Station," *Mechanical Engineering*, vol. 73, January 1951, pp. 14–16.

5. Lackey, M. E.: "Summary of the Development of Open-Cycle Coal-Fired Gas Turbine-Steam Cycles," Oak Ridge National Laboratory Report No. ORNL/TM-6252, September 1980.

6. Wood, B.: "Combined Cycles: A General Review of Achievements," *Combustion*, vol. 43, no. 10, April 1972, pp. 12–22.

7. Kindl, F. H., and J. Haley: "Design and Operation of a 26-MW Stag Plant," ASME Paper No. 75-GT-18, March 1975.

8. Bund, K., K. A. Henney, and K. H. Krieb: "Combined Gas/Steam-Turbine Generating Plant with Bituminous-Coal High-Pressure Gasification Plant in the Kellermann Power Station at Lünen," STEAG Anlagentechnik, 1970.

9. Kahrweg, H. Meyer: personal discussion with A. P. Fraas during visit to Lünen plant, Sept. 27, 1973.

10. Foster-Pegg, R. W., and H. L. Jaeger: "Low Btu Gas Powering of Combined Cycle Plants," American Power Conference, Chicago, Ill., Apr. 20, 1976.

11. *Economic Studies of Coal Gasification Combined Cycle Systems for Electric Power Generation*, EPRI Rep AF642, January 1978.

12. Amos, D. J., R. M. Lee, and R. W. Foster-Pegg: *Energy Conversion Alternatives Study (ECAS) Westinghouse Phase I—Final Report, vol. V, Combined Gas-Steam Turbine Cycles,* NASA CR-134941, vol. V, Westinghouse Electric Corporation Research Laboratories, February 1976.

13. Corman, J. C., and G. R. Fox: *Energy Conversion Alternatives Study (ECAS),* General Electric Phase II Final Report, NASA-CR-134949, vol. 1, December 1976.

14. Fraas, A. P., et al.: "Assessment of the State of the Art of Pressurized Fluidized-Bed Combustion Systems," Oak Ridge National Laboratory Report No. ORNL/ TM-6633, June 1979.

15. Hoy, H. R., and J. E. Stantan: "Fluidized Combustion under Pressure," paper presented at the American Chemical Society Meeting, Toronto, Canada, May 24, 1970.

16. Beecher, D. T., et al.: "Summary and Advanced Steam

Plant with Pressurized Fluidized-Bed Boilers," *Energy Conversion Alternatives Study—ECAS—Westinghouse Phase II Final Report*, NASA CR-134942, vol. III, Nov. 1, 1976.

17. Harris, L. P., and R. P. Shah: "Open Cycle Gas Turbines and Open Cycle MHD," *Energy Conversion Alternatives Study—ECAS—, General Electric Phase II Final Report*, vol. II, pt. 3, NASA CR-134949, December 1976.

18. Markowsky, J. J., and B. Wickstrom: "170 MW Pressurized Fluidized Bed Combustion Electric Plant," paper presented at the Sixth Energy Technology Conference and Exposition, Feb. 26–28, 1979.

19. Fraas, A. P., "A Comprehensive Overview of the Effects of Furnace Pressure and Turbine Inlet Temperature on the Principal Design Parameters for Fluidized-Bed Combustion Systems," (to be published).

20. Hoy, H. R., et al.: "Further Experiments on the Pilot-Scale Pressurized Combustor at Leatherhead, England," *Proceedings of the Fifth International Conference on Fluidized-Bed Combustion,* Dec. 12–14, 1977, MITRE Corp., McLean, Va.

21. Beecher, D. T., et al.: "Summary and Combined Gas-Steam Turbine Plant Using Coal-Derived Liquid Fuel," *Energy Conversion Alternatives Study—ECAS—Westinghouse Phase II Final Report*, NASA CR-134942, vol. II, Nov. 1, 1976.

22. Fraas, A. P., et al.: "A New Approach to a Fluidized Bed Steam Boiler," ASME Paper No. 76-WA/Pwr-8, December 1976.

23. Howard, J. B., "Fundamentals of Coal Pyrolysis and Hydropyrolysis," in M. A. Elliot (ed.), *Chemistry of Coal Utilization*, suppl. vol. 2, Wiley, New York, 1981, pp. 665–784.

24. Anthony, D. B.: "Rapid Devolatilization and Hydrogasification of Coal," doctoral thesis submitted to the Department of Chemical Engineering, Massachusetts Institute of Technology, January 1974.

25. Smith, I. W., "New Approaches to Coal Pyrolysis—CSIRO," *EPRI Conf. Coal Pyrolysis*, Palo Alto, 1981.

26. Fraas, A. P.: "Production of Motor Fuel and Methane from Coal via a Flash Pyrolysis Process Closely Coupled to a Fluidized Bed Utility Steam Plant," ASME Paper No. 80-WA/Fu-7, November 1980.

14

CLOSED-CYCLE GAS TURBINES

The closed-cycle gas turbine was patented by Ackeret and Keller in Switzerland in 1936, and one of the first gas turbines to produce significant amounts of electric power was a closed-cycle system built and operated in Switzerland in 1939.[1,2] Since that time, 14 fairly large commercial units have been built and operated abroad, and a number of small special-purpose units have been built and operated in the United States.[3-5]

ADVANTAGES

The closed-cycle gas turbine system was originally conceived as a means of effecting good control of the system over a wide range of loads with essentially no loss in overall system thermal efficiency at reduced loads down to around 25 percent of the design output.[2] This is accomplished by varying the pressure in the closed-cycle system; the compressor power input and the output of the turbine are both directly proportional to the mass flow of gas if the system pressure level is varied while the turbine inlet temperature is held constant. This is in sharp contrast to the performance characteristics of conventional open-cycle gas turbines, in which the system pressure level is kept constant and control is accomplished by varying the turbine inlet temperature. That not only reduces the ideal cycle efficiency as the turbine inlet temperature is dropped to reduce the load, but it also leads to a mismatch between the turbine and compressor. The volumetric flow rate through the turbine is reduced, the relative velocity of the gas entering and leaving the turbine buckets changes, the operating points on the turbine and compressor characteristic curves shift, and the aerodynamic efficiencies of the turbine and compressor vary as a consequence, thus degrading the plant thermal efficiency. As shown in Fig. 14.1, this does not happen in a closed-cycle gas turbine, where the temperatures throughout the system remain substantially constant irrespective of load, as does the volumetric flow rate of gas.

As discussed in Chap. 12, open-cycle gas turbines are not well suited to operation on dirty fuels; even trace amounts of solid particles lead to turbine bucket erosion and deposits in the rotor and stator that damage the blades, obstruct the gas flow, and detract from the aerodynamic efficiency. Thus, a major second advantage of the closed-cycle gas turbine is that it can be operated with dirty fuel. This has been a particularly important advantage in the past in Europe, where there has been a strong incentive to make use of readily available low-grade coal or dirty gas because of limited amounts of foreign exchange.

A third major advantage of the closed-cycle turbine is that,

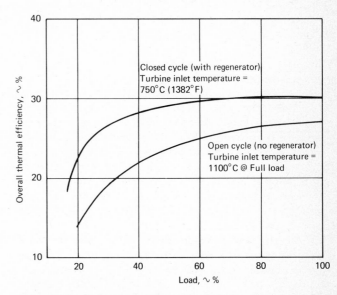

Figure 14.1 Comparison of the part-load efficiencies of open- and closed-cycle gas turbines.

TABLE 14.1   Summary of Closed-Cycle Industrial and Utility Power Plants That Have Been Built and Operated[3]

Manufacturer	Application	Continuous output, MWe	Heat supply, MW	Plant efficiency, %	Commissioning date	Running time, h	Fuel
Escher-Wyss	Power	2		32.6	1940	6000	Oil
E-W + GHH	Power and heat	2.3	2.3–4.1	25	1956	120,000 to 6-76	Coal or oil
GHH	Power and heat	6.6	8–16	28	1961	100,000 to 6-75	Coal
GHH	Power and heat	13.75	18.5–28	29.5	1960	100,000 to 6-76	Coal/coke oven gas
GHH + EVO	Power and heat	50	53	31.3	1975	3000 to 5-76	Coke oven gas
GHH	Power and heat comp. drive	6.4	7.8	29.5	1963	100,000 to 6-75	Mine gas + coal
GHH	Power and heat	17.25	20–29	30	1967	75,000 to 6-76	Bl. furn. gas and oil
	Power	12.5			1951	5000 since 1956	Oil
Fuji Elect. & E-W	Power	2.0		26	1957	90,000	Natural gas
Fuji Elect.	Power	12		29	1961	85,000 to 12-70	Bl. furn. gas
	Power	2.0			1960	~1000	Coal slurry
	Power	2.2			1959	~1000	Peat
E-W	Power and heat	12	9–12	28	1961		Brown coal
BBC/E-W	Power and heat	30–22	29–58	31–24	1972	5000	Oil or gas
E-W & LaFleur	Cryogenic gas production	2			1966		Natural gas
Corps of Engineers (Stratos et al.)	Army power requirements	0.5		18.6	1959		Oil
Corps of Engineers (Stratos et al.)	Army power requirements	0.500		16.7	1964		Oil

if a regenerator is introduced to increase the cycle efficiency, or if a waste-heat-recovery unit is to be installed in the turbine exhaust, the size and cost of these heat exchangers can be reduced in relation to those of an open-cycle gas-turbine installation because in the closed-cycle system the hot gas leaving the turbine is at a pressure of 3 to 30 atm. The high gas density improves the heat-transfer coefficient for a given pumping power and thus reduces the heat-transfer surface area requirements by a factor of 2 to 20. This is extremely important because even for closed-cycle units the cost of the heat exchangers is commonly several times that of the gas-turbine–compressor unit. In practice, therefore, it is economically attractive to use a regenerator with a closed-cycle gas turbine, whereas this is not often the case for open-cycle units. Figure 12.5 shows the overall thermal efficiency at full load for both a closed regenerated cycle and a simple open cycle as a function of turbine inlet temperature to show the quantitative advantage of regeneration.

A fourth important advantage of the closed-cycle gas turbine is that even in sizes as small as 100 kWe, the Reynolds number is sufficiently high to give a good turbine efficiency, whereas the efficiency of steam turbines falls off rapidly below about 10 MWe. As a consequence, gas turbines yield a definite efficiency advantage over steam turbines for power outputs under about 5 MWe, and this advantage becomes very pronounced for power outputs below 1 MWe, particularly if a high-molecular-weight gas is used instead of air for outputs below 100 kWe.

DISADVANTAGES

The principal disadvantage of the closed-cycle gas turbine for fossil-fuel-fired applications is the cost of the heat exchanger required to transfer heat from the combustion products to the turbine working fluid. The heat-transfer coefficients and temperature differences in these units are relatively low, and so the

Turbine inlet temperature, °C	Compressor inlet pressure, bars	Remarks
700		First test plant retired
660	7.2	
680	7.3	
710	8	
750	10.5	First use of helium as working fluid in electric power plant
680	9.3	
711	10.2	
660		First big plant/double pressurized heater
660	7.2	
680	6.7	
660		Stopped due to mine closure
660		Stopped due to mine closure
680	7	Achieved guarantees
720		Achieved guarantees, dismantled
680		Helium fluid, dismantled
650	8.1	Experimental (fluid, N_2)
650	8.1	Experimental (fluid, N_2

surface areas required are large and these heat exchangers are consequently large, heavy, and expensive. Secondly, the metal temperature in these heat exchangers must be somewhat higher than the gas temperature entering the turbine; hence the strength and corrosion resistance of the metal used limit the peak operating temperature and thus the thermal efficiency. The open-cycle gas turbine, on the other hand, can operate with cooled blades and casings so that the turbine inlet gas temperature can run hundreds of degrees Celsius above the temperature of the hottest metal. These effects are implicit in Fig. 12.5, which shows actual operating data for the thermal efficiency as a function of temperature for both simple open-cycle and closed-cycle plants with regeneration. Note also that the pressure drop and pumping power losses in the various components of the closed-cycle system, together with the somewhat higher compressor inlet temperature, act to degrade the thermal efficiency of the closed-cycle system.

A third disadvantage of the closed-cycle gas turbine for some applications is that a substantial amount of cooling water is required to cool the recirculated gas before it returns to the compressor. In some instances, this heat exchanger can be integrated into the process heat system so that this requirement does not represent a penalty, but in other instances it may constitute an important disadvantage of the closed-cycle gas turbine.

OPERATING EXPERIENCE

Table 14.1 lists the closed-cycle gas turbine units that have been operated commercially abroad.[3] All have been relatively small and have employed dirty fuels. The demonstrated reliability has been high, but the capital costs have also been relatively high. For this reason, most installations have been made where there has been both a low-cost source of dirty fuel and a need for process heat at a temperature above 100°C.

The principal operating experience in the United States with closed-cycle gas-turbine systems has been with two relatively small, special-purpose units. The first of these was an Army package power unit employing a high-temperature nitrogen-cooled fission reactor as the heat source. The turbine was designed for an output of 415 kWe.[5] Turbines were first operated with combustion-fired heaters in the course of the component development program. The total operating time obtained with closed-cycle gas-turbine systems with both fossil and nuclear heat sources in the course of the program was 8000 h.

One of the NASA nuclear electric space power units developed in the 1960s made use of a closed-cycle gas turbine designed to operate with an isotope heat source.[6] By 1978 units of this model had operated a total of 39,395 h, 26,040 h of which had been on the principal system tested.[7] The longest uninterrupted run was 7200 h. The working fluid was a mixture of helium and either argon or krypton, the mixture being chosen to give a good compromise between a high heat-transfer coefficient and a high molecular weight that would give a relatively high Reynolds number and thus a good efficiency in the compressor and turbine. The overall system thermal efficiency was ~25 percent in spite of the small size—only 7.5 kWe.

DESIGN STUDIES FOR ADVANCED POWER PLANTS

Closed-cycle gas turbines have been employed in many studies of both nuclear and fossil-fuel power plants of advanced types. In fact, most of the papers in the literature on closed-cycle gas turbines fall in this area. A brief survey of this work is helpful in appraising the potential role of closed-cycle gas turbines in the future.

Gas-Cooled-Reactor Central Stations

There has been a great deal of interest in the use of closed-cycle gas turbines with high-temperature gas-cooled reactors for central stations, and many design studies have been carried out in the past 20 years.[8] Central stations are an obvious application: It is necessary to circulate gas through the reactor to remove the heat generated, the gas commonly chosen is helium because of its good heat-transfer characteristics, and hence the logical thermodynamic cycle to employ is a gas turbine.[9] Unfortunately, all-ceramic high-temperature fuel elements are a trifle porous so that they emit a few parts per million of the more volatile radioactive fission products, vastly more than is the case for the metal-encapsulated fuel in the lower-temperature, water-cooled reactors. As a consequence, the plant layout is largely determined by reactor safety and maintenance considerations which favor the use of a steam turbine that is isolated from the reactor coolant circuit rather than a gas turbine.[10] Thus the only significant operating experience with a nuclear closed-cycle gas-turbine system is the small Army package power unit cited above, and this employed metal sandwich plate fuel elements having better fission product retention characteristics than the all-ceramic fuel.

Fusion Reactor Concepts

The combination of the severe heat loading and cooling problem posed by the first wall surrounding the plasma of a fusion reactor, coupled with the magnetohydrodynamic problems posed by schemes to pump a liquid metal through the high magnetic field, has led to many design studies of helium-cooled fusion reactor blankets. Some of these have envisioned the use of gas turbines as a means of circulating the helium coolant and at the same time extracting useful power.[11-14] The feasibility of these conceptual design studies cannot be appraised properly until the plasma physics problems of fusion reactors are more firmly in hand.[14]

Fluidized-Bed Coal Combustion Systems for Electric Power and Process Heat

At the time of writing, a major application of interest for closed-cycle gas turbines in the United States is for use with fluidized-bed coal combustion systems. Industrial plants and institutions currently using natural gas or distillate fuels for on-site plants that generate both electricity and process heat are being faced with a need to shift to coal as fuel. (In 1978 more natural gas was used in these plants than for residential heating.) The power capacity of these plants is generally in the range of 1 to 50 MWe.[15,16] Stack gas scrubbers for these small

plants are far too expensive; hence the fluidized-bed coal combustion system appears to be an attractive possibility.

For new chemical processes, commercial competition makes it necessary to procure, install, and bring on line new power sources in periods of 2 to 3 years.[15] The amounts of heat required and the temperatures of utilization usually vary with the process involved, and so the power plant must be of a type that can be tailored to meet the particular needs. For institutional and commercial building heating and air-conditioning applications, the requirements for both heat and electricity vary diurnally, weekly, and seasonally, and so a flexible system is required.

Closed-Cycle Plants for Cogeneration

In 1972 the author proposed a closed-cycle gas turbine coupled to a fluidized-bed coal combustion system and showed through a conceptual design study that such a power plant appears to be especially well suited to these applications,[17,18] which are responsible for a large fraction of U.S. energy consumption. After long delays, a program to investigate this type of system with an experimental unit was funded in 1979. Figure 14.2 presents a flowsheet for the basic system proposed. The heat source, in the form of a fluidized-bed coal combustion system, is located a little to the right of the center of the flowsheet, and the closed-cycle gas-turbine system lies below it. Air heated by the fluidized bed goes to the gas-turbine inlet at 815°C (1500°F) and about 10 atm at full power. The hot air exhausted from the turbine at about 580°C (1075°F) enters the regenerator, where it gives up much of its heat to air en route from the compressor to the fluidized-bed coal combustion system. The turbine airstream leaving the regenerator for the furnace passes first through an economizer section, which removes a substantial amount of heat from the combustion gas leaving the fluidized bed. The low-temperature, low-pressure turbine exhaust stream leaving the regenerator passes first to a waste-heat-recovery heat exchanger to provide process heat, in this instance for the heating and air-conditioning system for a building complex. The low-pressure turbine airstream, before going to the compressor, is cooled further to 27°C (80°F) by giving up its heat to potable water for the domestic hot-water system. The combustion air system on the right side of the flowsheet takes air in at atmospheric conditions and preheats it to about 535°C (997°F), partly to reduce stack losses and partly to improve combustion conditions in the fluidized bed. The combustion gas leaving the system is cooled in the recuperator to 150°C (300°F) and is discharged through an induced-draft fan and a bag house to remove ash particles. (The bag house and a forced-draft fan at the air inlet are not shown in the flowsheet.) Note that an air compressor and air reservoir are

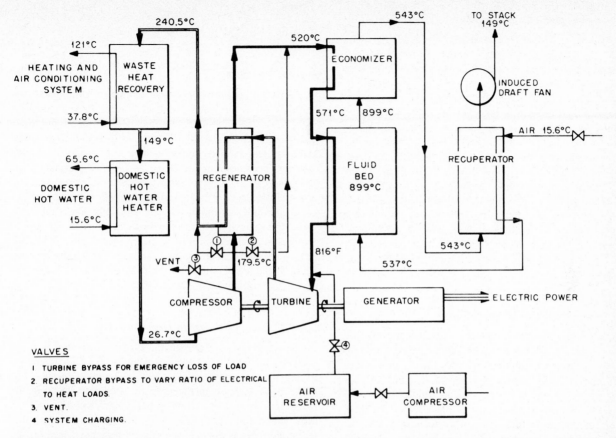

Figure 14.2 Flowsheet for a closed-cycle gas turbine coupled to a fluidized-bed coal combustion system.[18] *(Courtesy Oak Ridge National Laboratory.)*

shown at the bottom of the flowsheet; with these the pressure in the turbine air system can be increased rapidly to accommodate an increase in load. Similarly, a vent valve (No. 3) just above the compressor provides for a rapid reduction in system pressure to accommodate a drop in load. Provision has also been made for bypassing a portion of the air from the compressor around the regenerator to reduce the amount of heat recovered from the hot turbine exhaust. In this way the ratio of the energy recovered as heat to the energy produced as electricity can be changed with changing load conditions. Finally, valve No. 1, near the lower left corner of the regenerator, makes it possible to bypass high-pressure air around the furnace and turbine if a mismatch between the compressor and turbine should occur under any abnormal load conditions.

The design shown is intended to give a relatively high thermal efficiency in the conversion of chemical energy in the fuel into electricity (about 30 percent), while about 50 percent is recovered as heat at about 120°C (250°F). The regenerator could, however, be replaced with an intermediate-temperature steam generator to yield higher-temperature steam at about 260°C (500°F) for a chemical process if desired. The thermal efficiency for production of electricity would then run about 25 percent.

DESIGN PARAMETERS AFFECTING THERMAL EFFICIENCY

Closed-cycle gas turbines present design problems somewhat different from those of open-cycle gas turbines. The more important problems and parameters involved are discussed in this section.

Closed Cycles for Central Stations

Phase I of the Energy Conversion Alternatives Study (ECAS), a major government-sponsored comparison of advanced energy systems (described along with similar studies in the first part of Chap. 22), included studies of closed-cycle gas-turbine plants of 500 to 1000 MWe with helium as the working fluid.[19,20] A closed-cycle peak temperature of 815°C (1500°F) was chosen for this work by both General Electric and Westinghouse in view of the anticipated limit that would be imposed by fire-side corrosion. The capital costs of the resulting systems were too high to appear attractive. In Phase II, at ERDA's request a more optimistic view was taken by General Electric and a peak helium temperature of 1010°C (1850°F) was assumed, together with steam and organic Rankine bottoming cycles (which became interesting possibilities at the higher temperature level). The much higher cost of the materials required for the higher-temperature system essentially offset the fuel savings gained from the higher thermal efficiency.[21]

Turbine Inlet Temperature Limitations

Increasing the turbine inlet temperature increases the thermal efficiency of the thermodynamic cycle, but it also increases the rate of oxidation and/or sulfidation attack of the heater tubes in the furnace of a combustion heat source. The limited data available in 1980 from corrosion tests in fluidized-bed coal combustion systems, the most promising combustion heat source at the time of writing, are not completely consistent, probably as a consequence of differences between operating conditions in the various tests. The indications seem to be that 300-series stainless-steel alloys show the most promise.[22-25] The ORNL-FluiDyne tests and the BCURA tests indicate that metal temperatures in the 815 to 870°C (1500 to 1600°F) range will give acceptably long lives [corrosion rates of 0.025 mm (0.001 in) in 4500 h],[22,23] but the EPRI-sponsored tests at Stoke Orchard in England indicate that the metal operating temperature with these materials may be limited to a lower value.[24,25] Close inspection of all this work indicates that reducing conditions must be avoided by obtaining a good fuel distribution and operating with sufficient excess air to assure that fuel-rich zones do not occur in the vicinity of the coal feed ports. It shows that high, even catastrophic, corrosion rates can stem from local fluctuations between oxidizing and reducing conditions. The basic corrosion problems are discussed in Chap. 6. Suffice it to say here that many furnace design variables are involved, and so extensive test work will be required to resolve this vital question for FBCs. The importance of such investigations cannot be overemphasized.

In nuclear fission or fusion reactors the upper temperature limit imposed by the materials available also appears to be in the 850°C range even if helium is used as the working fluid. The reasons for this are treated in Chap. 6.

System Pressure

As the system pressure is increased, the size of components for closed-cycle gas turbines is reduced. This results in a reduction in cost up to the point where the thickness of casings and shells becomes too great and the thickness of heat exchanger tubes becomes great enough that the temperature drop through the tube wall becomes a major factor. In practice, a peak system pressure of 20 to 30 atm is usually favored except for large fission or fusion reactor heat sources, for which pressures of 30 to 50 atm are commonly chosen.[8-14] Note that Table 14.1 shows that all the systems that have been built make use of compressor inlet pressures between 7 and 10.2 atm. While not shown, the compressor pressure ratios commonly run between 3 and 4, the usual range for best efficiency in either open or closed gas-turbine cycles with recuperators (Chap. 12).

Ratio of Pumping Power to Heat Removal

The size and cost of components such as ducts and heat exchangers can be reduced by increasing the gas velocity and reducing the flow passage cross-sectional area. However, this leads to an increase in the pressure drop and hence in the pumping losses. The problem is particularly important in the most vital component in the plant: the gas heater. Many different approaches to analyzing this question have been carried out with generally similar results: the pumping-power–heat-removal ratio for the heater is ordinarily kept below ~2 percent.[10,23] This means that, if the overall thermal efficiency of the plant is 33 percent, the equivalent of about 6 percent of the gross electric output must go into the pumping power for the heater. In addition, of course, pressure losses in other components such as the ducts, regenerator, and cooler commonly increase the overall pumping energy requirements to 10 percent of the net electric output.

It happens that in design studies it is often more convenient to state the permissible pressure drop across a component in terms of the ratio of that pressure drop to the average absolute pressure in the component, usually expressed as $\Delta P/P$. The summation of these values for $\Delta P/P$ for the system as a whole is ordinarily kept to the 5 to 10 percent range.

Regenerator Effectiveness

To a first approximation, increasing the regenerator size for a given pumping-power–heat-removal ratio in the regenerator will increase the cycle efficiency by recovering a greater fraction of the heat in the turbine exhaust. However, as the compressor air temperature leaving the regenerator approaches the turbine outlet temperature, i.e., the heating effectiveness (Chap. 4) approaches unity, one quickly reaches a point of diminishing returns because the size and capital cost of the regenerator increase rapidly with little increase in heat recovery. This effect is shown in Fig. 14.3, which was prepared for a gas-cooled reactor system.[10] Studies at Escher-Wyss[9] on plants such as those in Table 14.1 and studies by the author on closed-cycle gas turbines coupled to either nuclear reactors[10,11] or FBCs[18] have indicated that a heating effectiveness of around 0.80 represents a good balance between capital costs and fuel costs as influenced by the thermal efficiency.

Recuperator Effectiveness

The heat in the hot combustion gases leaving the heat-transfer matrix of the turbine air heater can be utilized to good advantage by using it to preheat the combustion air with a heat exchanger called a *recuperator*. Preheating the air also serves to improve the combustion efficiency. The pressure difference between the two fluid streams is small; hence a rotary recuperator similar to those employed in conventional steam plants can be used. As is the case for the regenerator, a good balance between heating effectiveness and capital costs is obtained with a heating effectiveness of ~80 percent. Corrosion problems usually make it necessary to keep the stack gas temperature above 120°C (250°F), and so this temperature, rather than capital costs, is usually the design limitation.

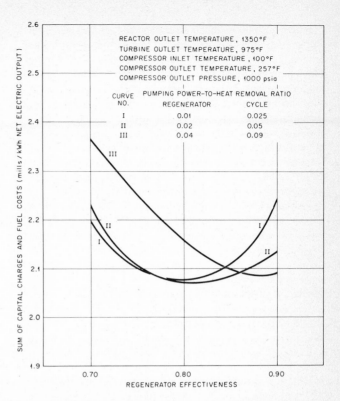

Figure 14.3 Sum of capital charges and fuel, regenerator, and cooler costs as a function of regenerator effectiveness.[10] *(Courtesy Oak Ridge National Laboratory.)*

Choice of Working Fluid

Helium is commonly chosen in design studies as the working fluid for closed-cycle gas-turbine systems. The primary reason is that the high thermal conductivity of helium, coupled with its high specific heat, make it a better heat-transfer fluid than any other gas except hydrogen. (Hydrogen has sometimes been considered but has never been used as a working fluid in a closed-cycle gas turbine because of the explosion hazard.) A major disadvantage of helium is its cost, which in turn poses a requirement for an exceptionally high degree of leak-tightness throughout the system. This requirement presents a particularly troublesome problem with respect to the turbine-generator shaft seal, which will be considered in the next section. Carbon dioxide and nitrogen are also often candidates, primarily because they, like helium,—on the surface at least—appear to reduce the oxidation and corrosion problems. However, it has been found that iron-chrome-nickel alloys owe their high-temperature oxidation resistance to the formation of a dense,

protective oxide film much like that of the oxide film on anodized aluminum. For these iron-chrome-nickel alloys, strongly oxidizing conditions are required not only to form this film initially but also to maintain it in the face of tendencies to form minute cracks with thermal-strain cycling. As discussed in Chap. 6, it has been found that, at temperatures above and about 600°C, ordinary air is a more favorable environment for the iron-chrome-nickel alloys than is helium. Thus there is really no incentive from the structural materials standpoint to use helium rather than air.

The output of a closed-cycle gas-turbine system is normally controlled by varying the system pressure. The most difficult control condition that must be met is the abrupt loss of load as a consequence of a circuit-breaker trip. For this condition, the control problem is greatly eased if the working fluid can simply be vented to the atmosphere,[26] a step that presents no problem with air but would represent an important expense if helium were employed.

Helium has the advantage that it makes possible smaller turbine and compressor units, but the number of stages of these units is increased.[9] If an appreciable number of similar machines were produced, there would undoubtedly be an important cost savings. However, no such machines are in production, and hence there is a strong incentive to employ air as a working fluid so that existing gas-turbine machinery can be employed without carrying out a substantial design and development effort. As a consequence of these factors, only one of the closed-cycle gas-turbine systems in operation in 1979 employed helium as the working fluid; it was located in the plant at Oberhausen in West Germany.[27] It is interesting to note that the helium inventory in this system was ~ 1000 kg, which entailed a cost of only ~ $10,000. The capital cost of the special tanks and compressors required for storing the helium was surely much greater.

The thermal properties of SO_2 have aroused an interest in the use of this material as a working fluid in gas-cooled reactors because it gives a lower pumping power for cooling a given geometry of reactor core than either CO_2 or He.[28] The results of a design study are shown in Table 14.2. It was found that the component costs should be low because the size of the piping, turbine, and heat exchangers would be less than for either CO_2 or He. However, the corrosion problems have appeared so formidable that no development effort has been initiated up to 1979, and none seems likely in the future.

Shaft Seal

The shaft seal between the turbine and the generator presents some difficult problems, particularly if helium is employed. It is sobering to note that of the several different seal arrangements that have been employed in helium, nitrogen, and CO_2 gas-cooled reactor systems, a substantial percentage have encountered serious difficulty with shaft seal leakage. These instances have included oil leakage into the system from the cir-

culating blowers of the Calder Hall reactors, oil leakage into the nitrogen circuit of the Army package power reactor, oil leakage into the helium circuits of both the EBOR maritime reactor and the Oberhausen helium turbine system, and water leakage through the water-buffered blower shaft seal in the Fort Saint Vrain helium-cooled reactor. Rather surprisingly, in only one of these systems was there instrumentation designed to detect such leakage before it became great enough to give serious trouble. Such instrumentation can and should be provided, as indicated by the serious consequences of the leaks in every one of the cases mentioned. The oil polymerizes at high temperature to form tars that spoil the heat-exchanger performance and cause reactor fuel elements to stick in their channels, making removal a monumental task. In one instance (the test unit for the Army package power reactor) the oil that leaked into the nitrogen system ignited when air got into the system, and a ruinous fire resulted.

THE SUPERCRITICAL CO_2 CYCLE

The supercritical CO_2 thermodynamic cycle is a special type of closed-cycle gas-turbine system that takes advantage of the low specific volume in the vicinity of the critical pressure and temperature to reduce the compressor work and hence make the cycle less sensitive to the aerodynamic efficiency of the compressor. (See Chap. 2 and Fig. 2.18.) The first study of record of this cycle was made by G. Samuels in 1958.[24] He concluded that the high peak pressure (about 260 bars) required in the system led to difficult problems in the design of the pressure vessels and piping, and this in turn led to high capital charges that more than offset the improvement in thermal efficiency obtainable.

In the 1960s E. G. Feher proposed the supercritical CO_2 cycle for use in small power plants both for terrestrial use and for space nuclear electric power applications.[29,30,31] A number of design studies were carried out, and a small system mock-up was built and operated.

In the middle 1970s the cycle was considered for use in fossil-fuel-fired central stations on the premise that it would give a higher thermal efficiency than the closed-cycle gas turbine.[20]

System Description

Figure 14.4 shows a flowsheet for a supercritical CO_2 cycle designed to operate with a fluidized-bed coal combustion chamber as the heat source.[20] The three fluidized-bed coal combustion systems at the left operate in parallel, heating the CO_2 coming from the high-temperature regenerators. The hot, high-pressure CO_2 leaving the heat-transfer matrices in the fluidized beds flows to two turbines in series, the first of which

TABLE 14.2 Performance Parameters for SO_2, CO_2, and He Cycles Coupled to a Gas-Cooled Fast Reactor System[28]

Reactor coolant	SO_2	CO_2	He
Peak pressure, bars	92	88	68
Reactor gas outlet temperature, °C	490	620	620
°F	914	1148	1148
Turbine working fluid	SO_2	H_2O	H_2O
Pumping power/net power	0.05	0.13	0.13
Overall thermal efficiency, %	31.8	38.7	38.7

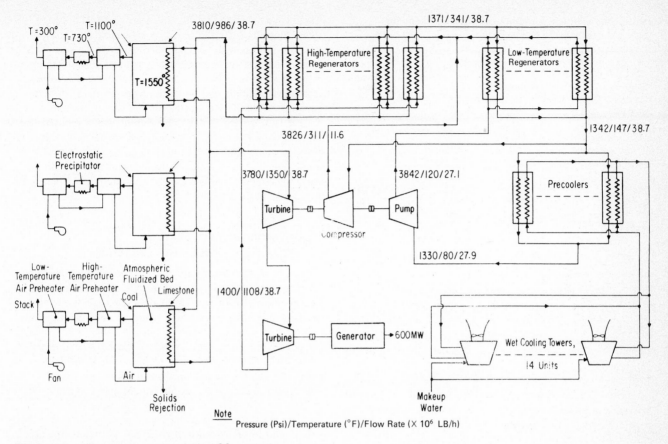

Figure 14.4 Flowsheet for a supercritical CO_2 cycle.[19]

drives the pump and compressor and the second the generator. Note that both a pump and a compressor are used in conjunction with the low-temperature regenerators and precoolers to give a little better thermodynamic cycle efficiency than if a pump only were employed.

A temperature-enthalpy diagram for the cycle is shown in Fig. 2.18. The critical pressure and temperature for CO_2 are 73 bars and 31 °C, respectively (1072 psia and 88 °F). To stay close to the critical point without causing condensation and the extra problems that it would entail, the minimum pressure must be kept somewhat above 73 bars and, for a pressure ratio giving a good cycle efficiency, it is necessary to have a peak pressure of the order of 260 bars (3780 psia). The strength of iron-chrome-nickel alloys suitable for the heater falls off so rapidly at temperatures above 600 °C that the peak cycle temperature must be limited to about 732 °C (1350 °F) because of creep-strength considerations.[20] Thus, there is little flexibility in the choice of cycle operating conditions.

Note in Fig. 2.18 that a great deal of heat must be added in the regenerator system between the compressor outlet (point No. 2 in Fig. 2.18) and the heater inlet (point No. 3). In fact, this amount of heat is of the order of three times that added in the heater, and the heat addition must be at a small temperature difference if the large amount of heat available from cooling the turbine exhaust (the path from point No. 5 to point No. 6) is to be employed to good advantage in obtaining a high overall efficiency. These factors lead to a surface area in the regenerator four or five times that in the heater, making the regenerator an expensive piece of equipment.

Advantages and Disadvantages

The prime advantage of the supercritical CO_2 system is that it gives a cycle much less sensitive to losses in the compressor. It also reduces the size of the components, but at the expense inherent in the construction of pressure vessels and piping for

TABLE 14.3 Critical Constants of Working Fluids for Supercritical-Pressure Cycles[30]

Name	Formula	Critical temperature, °F	Critical pressure, psia
Ammonia	NH_3	271.2	1636
Carbon dioxide	CO_2	87.8	1072
Hexafluorobenzene	C_6F_6	460	402
Perfluoropropane	C_3F_8	161.4	388
Sulfur dioxide	SO_2	315.5	1143
Sulfur hexafluoride	SF_6	114	546
Water	H_2O	705	3206
Xenon	Xe	61.9	853

high system pressures. Header sheets, pressure vessel penetrations, and flanged joints become particularly awkward to handle, and disproportionately high costs result. (Note that the pressure involved is a little higher than that commonly used in supercritical-pressure steam plants, and the temperatures are substantially higher, necessitating greater wall thicknesses.) Because of these limitations, the peak temperature in the cycle cannot be as high as in a closed-cycle gas turbine that would operate at a pressure 25 percent or less of that required in the supercritical CO_2 system. Thus, the increase in the practicable peak temperature for a conventional closed-cycle gas turbine in large measure offsets the reduced losses in the compressor of the supercritical CO_2 system from the thermal efficiency standpoint.

The advantages of the supercritical CO_2 cycle stem from careful selection of the key pressures and temperatures in the cycle to give the maximum thermal efficiency at full power. As a consequence, operation at reduced loads entails a substantial loss in thermal efficiency, whereas in a closed-cycle gas turbine the thermal efficiency remains substantially constant from full load down to around 25 percent load—an important advantage in a utility system, because loads vary throughout the day.

Both open- and closed-cycle gas turbines reject their heat at a fairly high temperature, and so they are well suited to applications requiring the cogeneration of heat and electricity. However, the supercritical CO_2 cycle rejects its heat at a relatively low temperature—64°C (149°F) and below—and hence is not suited for applications requiring both electricity and heat at moderate temperatures.

The supercritical CO_2 system has the further disadvantage that the turbine and compressor would require special design.

There is no question that suitable units could be designed and their compact size could reduce costs once they were in production. But, as with the helium closed-cycle gas turbine, the special character of the machinery required constitutes something of a hurdle. This, of course, is not the case for open or closed cycles employing air as the working fluid. Partly as a consequence of the special costs of innovation, the only supercritical CO_2 system that has been built was a reduced-scale mock-up of a 250-kWe system designed for the Army Corps of Engineers and built by McDonnell Douglas. This 15-kWt heat-input system did not include a turbine generator; a throttling valve was employed to simulate that component. The system was operated for 25 h.

OTHER WORKING FLUIDS FOR SUPERCRITICAL-PRESSURE SYSTEMS

If one wishes to take advantage of the reduced work of compression for a closed-cycle gas-turbine system that operates with the compression portion of the cycle close to the critical pressure and temperature but wishes a working fluid that might be better than CO_2, the fluids of Table 14.3 have been considered promising candidates.[30] Water has been included in Table 14.3, although in practice it has been found best, with water, to operate the system on a Rankine cycle so that the pump work is carried out in the liquid phase and thus is even less than if it were in the gas phase. The fluorocarbons, SO_2, and SF_6 do not appear to be good candidates because they would tend to break down or dissociate at high temperatures and give serious problems with corrosion. These problems could be avoided with xenon, but this material is very expensive and has properties little different from CO_2. Ammonia could be employed, but both its critical temperature and its critical pressure are higher than for CO_2, making CO_2 the more promising candidate. Thus, CO_2 seems to be the best candidate of any of the fluids that have been considered for a supercritical-pressure cycle in which there would be no condensation and no problems with two-phase flow such as instabilities and difficulties in controlling free liquid surfaces. (This was one of the advantages of the supercritical CO_2 cycle for space power plant applications where control of free liquid surfaces under 0-g conditions poses major problems.)

Capital Costs

Capital costs are difficult to estimate even for modest variations in conventional systems. Inasmuch as most of the closed-cycle systems involve developmental uncertainties, it has seemed best to consider only the effects of the design conditions inherent in roughly similar systems on the relative cost of the most expensive components, i.e., the high-temperature heat exchangers. Extensive experience indicates that the cost of these units is approximately proportional to both the surface area required and the pressure if the operating pressure is above about 10 bars. Further, the cost increases rapidly with the design temperature for temperatures above about 500°C, the point at which the strength of low-alloy steel begins to fall off rapidly. More expensive alloys must be considered as the design temperature is increased above 500°C because of both creep-strength and fire-side corrosion considerations (Fig. 10.13).

With the above considerations in mind, it is interesting to consider the design temperature and pressure conditions for the high-temperature heat exchangers required for typical cycles, together with the temperature difference available, the heat flux, and the heat energy that must be transferred per unit of the electric output. The latter is a particularly important factor for the closed-cycle gas turbine and the supercritical CO_2 cycle because both depend on regenerators in which the amount of energy transferred is two to four times the useful

power output of the cycle. Values for some typical cases designed on the basis of similar criteria were taken from the author's files[32] to yield the numbers in the first five columns of Table 14.4, and this made it possible to compute the sixth column. Data on tubing cost as a function of temperature and wall thickness then gave a basis for calculating the relative costs given in the extreme right-hand column. (See Fig. A10.4.)

The relative cost values appear to be consistent with a number of previous studies and provide some insights into the reasons for differences between systems. For example, the heat-transfer matrix costs for the closed-cycle gas turbine of Table 14.4 are roughly half again as high as for the steam plant. The still higher cost for the supercritical CO_2 cycle is reasonably consistent with the results of Ref. 20, which, for the same cycle pressure and temperature conditions as used in Table 14.4, yielded heat-exchanger costs six times as high for the supercritical CO_2 system as for the helium closed-cycle gas turbine. Almost half of this difference stems from the much greater amount of heat that must be transferred in the regenerator per unit of electric output. The balance apparently stems from the thicker tubes and the awkward header design problems associ-

TABLE 14.4   Comparison of Factors Affecting the Size and Cost of the High-Temperature Heat-Transfer Matrices for some Typical Power Plants Employing Fluidized-Bed Furnaces

	Pressure, bars	Peak temperature, °C (°F)	Mean temperature difference, °C (°F)	Ratio of heat transferred to electric output	Heat flux, W/cm² [Btu/(h·ft²)]	Relative surface area	Temperature-cost factor	Relative cost
Fluidized-bed steam plant	246	565 (1050)	~ 400 (720)	2.5	3.15† (10,000)	1.0	1.0	1.0
Closed-cycle gas turbine								
Heater	68	815 (1500)	167 (300)	2.5	3.15 (10,000)	1.0	2.0	0.6*
Regenerator	68	538 (1000)	56 (100)	2.1	0.63 (2,000)	4.2	0.8	$\frac{0.9}{1.5}$
Supercritical CO_2								
Heater	260	733 (1350)	250 (450)	2.5	4.4 (14,000)	0.7	2.0	1.5
Regenerator	260	594 (1100)	56 (100)	4.6	1.9 (6,000)	3.1	0.8	$\frac{2.6}{4.1}$
Alkali-metal-vapor cycle								
Alkali-metal boiler	2	815 (1540)	61 (110)	2.1	2.5 (8,000)	1.1	2.0	0.9*
Alkali-metal condenser	246	594 (1100)	167 (300)	1.7	15.7 (50,000)	0.13	0.9	$\frac{0.12}{1.02}$

*Minimum wall thickness requirements when allowing for fire-side corrosion cause the tube wall thickness and cost to vary nonlinearly and to be independent of pressure for pressures less than ~ 10 bars.

†Average heat flux for both the boiler tubes in the fluidized bed and superheater and reheater tubes in the hot-gas stream leaving the bed.

ated with the very high pressures inherently required in the supercritical CO_2 cycle.

DISSOCIATING GAS CYCLES

A number of dissociating gas cycles have been proposed which are somewhat similar to the supercritical CO_2 cycle in that they fall somewhere between a Brayton cycle and a Rankine cycle. These cycles employ a working fluid that dissociates during the heating phase of the cycle, greatly increasing the volume of gas to be expanded through the turbine, and then reassociates in the cooler, reducing the volume of gas that must be compressed and thus increasing the net work from the cycle.

Aluminum Chloride

One of the most attractive of the dissociating gas cycles is based on the use of aluminum chloride.[33] The pressures and temperatures are chosen so that the gas will be mostly in the form of Al_2Cl_6 during compression, while during the expansion process it will be mostly $AlCl_3$. This, in effect, will cut the compression work roughly in half and thus produce a marked improvement in cycle efficiency over a conventional Brayton cycle operating between the same temperature limits.

An investigation of this cycle by the author in 1959[33] disclosed that it is necessary to go to a peak cycle temperature of about 815°C (1500°F) and a temperature of around 225°C (440°F) at the cold end of the cycle to take full advantage of the unusual properties arising from the dissociation of alumi-

TABLE 14.5   Principal Parameters for a Dissociating Gas Topping Cycle using Aluminum Chloride Coupled to a Steam Cycle [33]

Ideal mass ratio = 0.18487 lb of water per pound of aluminum chloride
Actual mass ratio = 0.19585 lb of water per pound of aluminum chloride
Ideal cycle efficiency = 53.2%
Actual cycle efficiency = 41.4%

Condition	Temperature, °F	Enthalpy, Btu/lb	Entropy, Btu/°F	Pressure, psia	Specific volume, ft³/lb	Weight fraction dissociated or steam quality
			Aluminum chloride			
Compressor inlet	440	142	0.02530	5	7.2476	0.00176
Compressor outlet (isentropic)	570	164	0.02530	100	0.41506	0.00259
Compressor outlet (80% efficiency)	615	169.5	0.0311	100	0.4344	0.00447
Turbine inlet	1540	478	0.22584	100	1.4614	0.818
Turbine outlet (isentropic)	1150	406	0.22584	5	23.07	0.781
Turbine outlet (80% efficiency)	1175	420.4	0.2343	5	23.92	0.818
			Steam*			
Pump inlet	91.72	59.71	0.1147	0.7368	0.01611	Saturated liquid
Pump outlet (isentropic)	91.72	66.14	0.1147	2400	0.01600	Compressed liquid
Pump outlet (50% efficiency)	91.72	72.57	0.1243	2400	0.01600	Compressed liquid
Turbine inlet	1050	1494	1.5554	2400	0.3373	Superheated vapor
Turbine outlet (isentropic)	91.72	855	1.5554	0.7368	339.5	0.763
Turbine outlet (80% efficiency)	91.72	983	1.790	0.7368	394.2	0.886

*The bases for calculating the enthalpy and entropy of aluminum chloride and steam were not the same. Comparison of the absolute values of these properties between the two fluids is therefore meaningless.

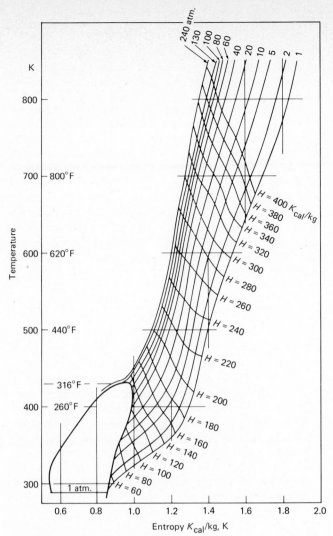

Figure 14.5 Effective temperature-entropy diagram for the dissociating mixture $N_2O_4 \rightleftarrows 2NO_2 \cdot Q_1 \rightleftarrows 2NO + O_2 \cdot Q_2$. (*From Nesterenko & Bubnov data.*)

TABLE 14.6 Comparison of the Thermophysical Characteristics of Gas-Cooled Fast Neutron 1000-MWe Reactors Operating on H_2O and N_2O_4 (Ref. 35)

Quantity	N_2O_4	H_2O (steam)
Thermal capacity of reactor, MW	2400	2519
Electric power output of nuclear power station (MW)	1100	1000
Reactor coolant	N_2O_4	H_2O
Working substance of turbine	N_2O_4	H_2O
Gas temperature at reactor inlet, °C	370	375
Gas temperature at reactor outlet, °C	570	540
Coolant pressure, atm (abs)	135	175
Core diameter, mm	2200	2628
Core height	1400	1510
Thickness of end face and lateral blanket reflectors, mm	400	350
Max. temperature of fuel element cladding, °C	737	754
Ditto, disregarding superheating factors, °C	620	635
Hydraulic resistance of reactor core, atm (abs)	8.2	9.27
Gas velocity in reactor, m/s	45	40
Thermodynamic layout of nuclear power station	Single-loop condenser	Single-loop condensation
Coolant, loop I	N_2O_4	H_2O
Coolant, loop II	—	—
Gas parameters at turbine inlet:		
Pressure, atm (abs)	240	160
Temperature, °C	565	536
Intermediate superheating		
Pressure, atm (abs)	40	30
Temperature, °C	540	510
Efficiency of nuclear power station cycle, %	47.4	43.3
Net efficiency of nuclear power station, %	44	39.7

num chloride. The relatively low pressure in the high-temperature region of the aluminum chloride system—about 6 bars—serves to reduce pressure stresses sufficiently to make operation at 815°C practicable. For good cycle efficiency, a binary fluid cycle employing aluminum chloride in the high-temperature portion and steam in the lower-temperature region must be employed. The set of calculations given in Table 14.5 indicate that such a binary cycle might give a thermal efficiency of around 41 percent.

The N_2O_4-$2NO_2$ Cycle

A considerable amount of work has been carried out in Minsk, Soviet Union, on a dissociating gas cycle employing N_2O_4 as the working fluid.[34,35] The resulting cycle is quite complex, because it involves two stages of dissociation, the first from N_2O_4 to $2NO_2$ in the temperature range around 100°C (212°F) and a second dissociation from $2NO_2$ to $2NO + O_2$ in the temperature range around 600°C (1100°F). In effect, this increases the specific heat in these regions, and it results in an unusual temperature-entropy diagram for the cycle (Fig. 14.5).

The Russian advocates of the dissociating N_2O_4 cycle believe that it is particularly well suited to use with high-temperature, gas-cooled, fast reactors, and they have prepared a fairly complete conceptual design for such a plant.[35] Estimates of the net power plant efficiency run about 40 percent for a peak cycle temperature of 540 to 565°C at a pressure of 130 to 170 bars (e.g., see Table 14.6).

Major Problems

The dissociating gas cycles show some intriguing possibilities such as a much reduced turbine size,[35,36] but all of them share a vital practical problem in that the working fluid in each case is inherently inclined to give free oxygen or free chlorine as a dissociation product. If there is the slightest trace of moisture present, this is likely to lead to reactions with the protective surface coatings, which will cause severe attack in local zones. The difficulties are likely to be rendered substantially more serious if the system is used to cool a nuclear reactor because the intense ionizing radiation in the reactor acts to aggravate the corrosion problems.

The dissociating gas cycles have another serious disadvantage in that in each case the operating temperatures and pressures are fixed to narrow ranges by the chemical thermodynamics of the dissociation process. This drastically limits the development potential of any given fluid.

The effective specific heat and effective thermal conductivity of the gas are increased by factors of as much as 6 in the dissociating and reassociating regions, and this gives improved heat transfer in those regions. Most of the heat-transfer surface, however, does not have the benefit of this effect, and so conventional forced-convection heat-transfer relations apply. As a consequence, the surface areas and cost of the heat exchangers required are high for the regenerator and cooler at the low temperature end of the cycle, where the temperature differences are small and the gas density is low. Therefore, the mass velocity must be kept relatively low to avoid excessive pumping-power losses.

PROBLEMS

1. Using data in Table 14.2, the charts of Figs. A4.2 to 4.6, and the results of Prob. 6 of Chap. 4, estimate the tube matrix lengths, cross-sectional area, and heat-transfer surface area for the regenerator and the waste-heat boiler for the system shown in the flowsheet of Fig. 14.2

2. Calculate the weight of the tube matrix for the two heat-exchanger tube matrices of Prob. 1, and estimate the cost of the tube matrix for each unit, assuming $100/m^2.

3. Outline a series of calculations designed to determine whether the choice of parameters and conditions prescribed in the flowsheet of Fig. 14.2 yields near-minimum costs. Assume the cost of coal to be $50 per ton. Also assume that the capital cost of the casings and ductwork for the heat exchangers will be roughly equal to the cost of the heat-transfer matrices, and that the capital cost of the balance of the plant will be $300/kW irrespective of these perturbations. Use capital charges of 20 percent with a plant load factor of 60 percent.

REFERENCES

1. Stodola, A.: "Load Tests of a 4000-kW Combustion-Turbine Set," *Power*, February 1940, pp. 62–63.

2. Keller, C.: "The Escher-Wyss Closed Cycle Turbine," *Trans. ASME*, vol. 68, 1946, p. 791.

3. Harmon, R. A.: *Operational Status and Current Trends in Gas Turbines for Utility Applications in Europe*, Fossil Energy Organization of U.S. Energy Research and Development Administration, TIC-FE/WAPO-0016-1, Aug. 16, 1976.

4. Keller, C.: "Operating Experience Design Features of Closed Cycle Gas Turbine Power Plants," *Trans. ASME*, vol. 79, 1957, p. 627.

5. Crim, W. M., et al.: "The Compact AK Process Nuclear System," *Trans. ASME*, vol. 88, 1966, p. 127.

6. Brown, W. J.: "Brayton-B Power System—A Progress Report," paper presented at Fourth Intersociety Energy Conversion Engineering Conference, Washington, D.C., Sept. 22–26, 1969.

7. Personal communication from Lloyd Schurr and James Dunn, NASA Lewis Laboratory, Apr. 27 and 29, 1977.

8. *State-of-the-Art of HTGR Gas Turbine Technology*, Gulf General Atomic Corp., Report No. Gulf-GA-A12098, June 1, 1973.

9. Keller, C., and D. Schmidt: "Industrial Closed-Cycle Gas

Turbines for Conventional and Nuclear Fuel," ASME Paper No. 67-GT-10, December 1966.

10. Fraas, A. P., and M. N. Ozisik: "A Comparison of Gas-Turbine and Steam-Turbine Power Plants for Use with All-Ceramic Gas-Cooled Reactors," Oak Ridge National Laboratory Report No. ORNL-3209, July 1965.

11. Fraas, A. P.: "Problems in Coupling a Gas Turbine to a Thermonuclear Reactor," paper presented at the ASME Gas Turbine and Fluids Engineering Conference, Mar. 26–30, 1972, ASME Paper No. 72-GT-98.

12. Forster, S.: "Closed Helium-Turbine Cycle with a Fusion Reactor," *Mechanical Engineering*, August 1973, p. 13.

13. Mitchell, J. T. D., and J. A. Booth: "Wall Loading Limitations on a Helium-Cooled Fusion Reactor Blanket," CLM-R126, UKAEA Research Group, Culham Laboratory (paper 13 of Workshop on Fusion Reactor Design Problems, Jan. 29–Feb. 15, 1974).

14. Fraas, A. P.: "Comparative Study of the More Promising Combinations of Blanket Materials, Power Conversion Systems and Tritium Recovery and Containment Systems for Fusion Reactors," Oak Ridge National Laboratory Report No. ORNL-TM-4999, November 1975.

15. Anderson, T. D., et al.: *An Assessment of Industrial Energy Options Based on Coal and Nuclear Systems*, Oak Ridge National Laboratory Report No. ORNL-4995, July 1975.

16. *Use of Coal and Coal-Derived Fuels in Total Energy Systems for MIUS Applications*, vol. 1, *Summary Report*, Oak Ridge National Laboratory Report No. ORNL/HUD/MIUS-27, April 1976.

17. Fraas, A. P.: "Small Coal Burning Gas Turbine for Modular Integrated Utility Systems," paper presented at the Ninth Intersociety Energy Conversion Engineering Conference, San Francisco, Calif., Aug. 26–30, 1974.

18. Fraas, A. P., et al.: "Design Study for a Coal-Fueled Closed Cycle Gas Turbine System for MIUS Applications," paper presented at the Tenth Intersociety Energy Conversion Engineering Conference, Newark, Del., Aug. 8–12, 1975.

19. *Energy Conversion Alternatives Study (ECAS), General Electric Phase I Final Report*, NASA-CR-134948, 1976.

20. Deegan, P. B.: *Energy Conversion Alternatives Study (ECAS), Westinghouse Phase I Final Report*, vol. II, *Metal Vapor Rankine Topping-Steam and Bottoming Cycles*, NASA CR-134941, Feb. 12, 1976.

21. *Energy Conversion Alternatives Study (ECAS), General Electric Phase II Final Report*, NASA-CR-134949, 1976.

22. Cooper, R. H. et al.: *Corrosion of High-Temperature Materials in AFBC Environments,* Oak Ridge National Laboratory Report No. ORNL/TM-7734 P1 and P2, 1981.

23. British Coal Utilization Research Association, *Research and Development Report No. 85*, prepared for and released by the Office of Coal Research, Washington, D.C., September 1973.

24. Stringer, J.: "High-Temperature Corrosion in Fluidized Bed Combustion," paper presented at the 1977 Engineering Foundation Conference on Ash Deposits and Corrosion Due to Impurities in Combustion Gases, New England College, Henniker, N.H., Hemisphere Publishing Co.

25. Stringer, J.: "High-Temperature Corrosion of Metals and Alloys in Fluidized Bed Combustion Systems," *Proceedings of the International Conference on Fluidized Bed Combustion Systems,* Dec. 12–14, 1977, Washington, D.C., MITRE Corp.

26. Ball, S. J.: "Initial Simulation and Control System Studies of the MIUS Coal-Fired Turbine Experiment," Oak Ridge National Laboratory Report No. ORNL/HUD/MIUS-42, February 1978.

27. Bammert, K., et al.: "Highlights and Future Development of Closed-Cycle Gas Turbines," *J. Eng. Power, Trans. ASME,* vol. 96, October 1974, pp. 342–348.

28. Bender, M., R. S. Carlsmith, W. R. Gall, and G. Samuels: "Comparative Features of Gas-Cooled Fast Reactors," Oak Ridge National Laboratory, Reactor Study, 1958.

29. Feher, E. G., et al.: "Investigation of Supercritical (Feher) Cycle," McDonnell Douglas Astronautics Co., AFAPL-TR-68-100, October 1968.

30. Feher, E. G.: "The Supercritical Thermodynamic Power Cycle," paper presented at the 2d Intersociety Energy Conversion Engineering Conference, Miami Beach, Fla., Aug. 13–17, 1967, pp. 37–44.

31. Hoffman, J. R., and E. G. Feher: "150 KWE Supercritical Closed Cycle System," *Trans. ASME*, vol. 93, 1971.

32. Fraas, A. P.: "Heat Exchangers for High Temperature Thermodynamic Cycles," ASME Paper No. 75-WA/HT-102, Aug. 6, 1975.

33. Blander, M., et al.: "Aluminum Chloride as a Thermodynamic Working Fluid and Heat Transfer Medium," USAEC Report No. ORNL-2677, Oak Ridge National Laboratory, Sept. 21, 1959.

34. Krasnin, A. K., and V. B. Nesterenko: "Dissociating Gases: A New Class of Coolants and Working Substances for Large Power Plants," *Atomic Energy Review*, vol. 9, no. 1, 1971. IAEA, Vienna.

35. Krasnin, A. K., et al.: "The Experimental Power Installation BRG-30 with Gas-Cooled Fast Neutron Reactor and Dissociating Coolant," *Atomnaya Energiya*, vol. 30, no. 2, February 1971, p. 180.

36. Luchter, S.: "A Survey and Comparison of the Supercritical and Dissociating Gas Power Cycles," *Proceedings of the Fifth Intersociety Energy Conversion Conference*, 1970, pp. 2-1 to 2-5.

METAL-VAPOR TOPPING CYCLES

One of the major advantages of the Rankine cycle is that in practice it permits a closer approach to the efficiency of the Carnot cycle than any other cycle in general use. For example, supercritical-pressure steam Rankine cycle plants give operating thermal efficiencies a little over 39 percent with a peak temperature of 565°C (1050°F), whereas open-cycle gas turbines with peak temperatures of 1000°C (1832°F) yield overall thermal efficiencies of only ~28 percent, and a closed-cycle gas turbine with a peak temperature of 800°C (1472°F) and regeneration gives a thermal efficiency of ~33 percent. As pointed out in Chap. 11, the thermal efficiency of the steam Rankine cycle increased by a factor of ~10 from 1881 to the early 1950s (Fig. 11.4), at which point the limits of development appear to have been reached because of basic materials limitations, particularly the severe stress and creep problems arising from the high steam pressures inherently required. One approach to raising this temperature limit is to employ a lower-vapor-pressure working fluid in a Rankine topping cycle superimposed on a conventional steam Rankine cycle to which it rejects its waste heat. Although in principle a wide variety of working fluids might be used, the compounds with suitable vapor pressure are either organics (or fluorocarbons) that decompose at high temperature, giving tarry deposits or corrosive gases, or else are inorganic compounds that decompose and cause corrosion. As a consequence, in practice the only serious candidates have been the metal vapors of Hg, K, and Cs. This chapter summarizes the developmental history, assessments of system potentials, the operational experience, and the principal problems and limitations of these working fluids.

MERCURY-VAPOR TOPPING CYCLE

In a series of studies in the 1900–1903 period, Steinmetz pointed out the advantages of mercury and high-molecular-weight organic compounds as turbine working fluids, particularly as a means of reducing wheel tip speeds and the number of stages. The advantages of a binary vapor cycle for power generation were recognized by Emmet in 1913, and a total of seven mercury-vapor–steam-power plants were built between 1917 and 1948.[2-5] All were built in the United States by General Electric. The Schiller plant at Portsmouth, New Hampshire, the last of these to operate, was shut down in 1968 after 20 years of service.

The Schiller Plant

The Schiller plant, as the latest to be built, took advantage of the operating experience gained with the previous plants and hence is of the greatest interest here. When first built in 1949, it was in a unique position in that it was in an area where there was a need for a relatively small-capacity plant and the fuel costs were high. The mercury-steam unit was the first unit to be installed in that plant, and it was put into commercial operation with a design rating of 8.5 atm (125 psia) and 515°C (960°F) at the inlet to the mercury turbines.[4,5] The rated capacity was 15,000 kW for the two mercury units, which exhausted into a mercury condenser and steam boiler that fed a steam turbine having a capacity of 25,000 kW. The total plant cost at that time was a bit high: $12,700,000 or $317/kW. According to a Federal Power Commission report,[6] the heat rate averaged 9700 Btu, giving an overall plant efficiency of 36.3 percent. Significantly, the Schiller plant was enlarged in the 1950s to give a capacity of 190,000 kW by the addition of standard steam-turbine units of about 40,000 kW each, with a heat rate of about 11,000 Btu/kW, or a 31 percent overall efficiency, at a cost of $140/kW.

Principal Problems

Although the mercury-steam binary vapor cycle showed a favorable economy in the 1920s when the maximum steam pressures were of the order of 30 bars (450 psi) and the capacity of turbine-generator units was much smaller than at present, conditions changed by 1950. On the one hand, metallurgical advances have made it possible to contain steam at 540°C (1000°F) and 240 bars (3500 psi), while on the other hand, it has become clear that difficulties with mercury corrosion and mass transfer prevent its use with iron-chrome-nickel alloys at temperatures much above 480°C (900°F). (Although mercury does not react chemically with iron-chrome-nickel alloys, it tends to take them into solution at high temperatures and precipitate them out at low temperatures, plugging pipes and heat-exchanger passages. See Chap. 6.) As a consequence, the thermal efficiency of steam plants in normal operation became higher than is practicable for mercury-steam binary vapor cycle systems.

There are also some practical disadvantages. The supply of mercury is so limited that the amount required for a large modern plant would represent a major capital investment, even if it did not drive up the market price. The mercury inventory in the 40-MW Schiller plant was 140,000 kg, while that of a 400-MW plant would be about 1,500,000 kg, or about 15 percent of the total world production in 1977. The capital cost, therefore, at the January 1979 price of $5.28/kg would run about 8 million dollars, or ~ $20/kWe, and the world supply is too limited to permit the extensive use of mercury-vapor plants. Further, the poor wetting characteristics of mercury make it necessary to employ heat-transfer rates about 15 or 20 percent of those commonly employed in steam boilers, thus increasing the size and capital cost of both the mercury boiler and the mercury condenser. Although it was found in the early 1920s that the wetting difficulties could be much reduced by the addition of small amounts of magnesium and titanium, this practice is not wholly satisfactory.

The health hazard represented by the toxicity of mercury vapor in air is a serious one. An indication of the seriousness of the problem is given by considering a hypothetical accident in which the total inventory of the poisonous volatiles might escape from a plant. It would take about as much air to dilute the mercury vapor to standard tolerance levels as would be required for similar dilution of the volatile fission products accumulated in a nuclear reactor in the course of a year of operation. As mentioned in Chap. 9, the maximum allowable concentration of Hg in air is 0.01 ppm, or 0.1 mg/m³. Thus even slight leaks in a mercury-vapor power plant are serious matters.

Nuclear Space Power Program

The prime problem in high-temperature systems is materials compatibility, hence the experience in the nuclear space power program is highly significant in spite of the small size of the units used—generally in the 50 to 500 kWe range. While these seem very small by current central station standards, the early steam turbines in central stations were similar in size, i.e., in the 150 to 1000 kWe range.

In the middle 1950s the U.S. Air Force became interested in the possibilities of reconnaissance satellites that would require substantial amounts of electric power for radar, infrared, and visible-light scanning equipment. Several different types of energy conversion system were considered, including solar cells, solar-Rankine plants, radioisotope-thermoelectric units, and a nuclear-reactor–mercury-vapor cycle system. Work on a small 2-kWe version of a mercury-vapor system was initiated in 1955 with Atomics International (AI) and Thompson-Ramo-Wooldridge (TRW) as contractors.[7] In 1959 an additional effort aimed at a 50-kWe plant was initiated by NASA with AI and Aerojet as contractors and a modest in-house support effort at the NASA Lewis Lab. These power plants were designed for a peak temperature of 650°C (1200°F) with Type 300 series stainless steel as the structural material. This temperature, well above the peak levels that had been found tolerable for mercury-vapor central station plants, was chosen in an effort to bring the weight of the condenser to an acceptable value. This was because the waste heat from the thermodynamic cycle could be rejected only by thermal radiation to space, which made the weight of the radiator inversely proportional to the fourth power of the absolute temperature of the radiating surface. Thus the peak cycle temperature was chosen to be ~ 650°C (~ 1200°F) and the condenser temperature ~ 360°C (~ 680°F), because these were the minimum values that appeared to give an acceptable specific weight for the system.[8] Unfortunately, the mass-transfer problem in a mercury–stainless-steel system remained just as stubborn as in the central station plants, and no test runs of more than 1000 h were obtained with this materials combination in the course of the 10-year program in spite of an expenditure of over 10^8—an object lesson in the intractability of basic materials problems.

It happens that the refractory metals have been known since the early 1950s to give good compatibility with mercury, lead, and bismuth, but they are subject to serious attack by oxidation and, if they are to be run in air, must be clad with an alloy such as stainless steel. This requirement poses difficult and expensive problems in fabricating pipe, tubing, and plate, as well as in assembling parts with welded joints, because it results in an interior weld made with a refractory metal electrode and an exterior weld made with stainless steel. These extra costs and fabrication problems were ultimately accepted, and a SNAP-8 system was constructed with tantalum as the metal in contact with the mercury in the boiler and stainless steel as the cladding material. An 8700-h test with a peak mercury-vapor temperature of 676°C (1250°F) was successfully completed in 1967.[8] However, the system was still too heavy to be attractive and the program was terminated in 1970.

POTASSIUM-VAPOR CYCLE

The limitations imposed by solution corrosion and mass transfer in the mercury-vapor space power plant led two USAF officers involved in the program in 1958 to request the author to try to find a more promising approach. In the ensuing study many different systems that appeared to be suited to the space power plant application were examined critically. These included inert-gas Brayton cycles, dissociating gas cycles, and a number of metal-vapor Rankine cycles. It happened that the author had been involved in some studies of sodium- and rubidium-vapor cycles for use with fan-jet engines in the course of the Aircraft Nuclear Propulsion Program in the early 1950s and hence had at hand Mollier diagrams for sodium and rubidium.[9,10] Additional diagrams for potassium and cesium were prepared, and the proportions of the turbines and radiators for each of the working fluids were estimated for space power plants. The results indicated that the vapor cycles employing potassium, rubidium, and cesium were the most promising candidates, yielding much lower specific weights than mercury, as well as a much higher temperature limit imposed by materials.[11] Rubidium has properties intermediate between potassium and cesium and is less available, hence it has received no further attention. Physical properties for potassium, cesium, and water are given in Table 15.1. Cesium

was very expensive and not readily available; hence potassium was chosen as the prime candidate. Similar conclusions were reached in subsequent studies at NASA and General Electric.[12,13] Work was initiated at ORNL in 1959 on what were considered to be the two most crucial feasibility questions of a potassium-vapor cycle, namely, the compatibility of boiling potassium with stainless steel in the temperature range of 816 to 870°C (1500 to 1600°F) and the heat-transfer coefficients characteristic of boiling potassium. By 1961 both questions had been resolved favorably and a program to develop a potassium-vapor cycle space power plant was initiated. Subsequently, some 200,000 h of boiling-potassium system operation at ORNL and ~70,000 h at the G.E. Evendale plant under the space power plant program[14] resolved some 40 other important questions related to the design of potassium-vapor power plants.

Central Station Power Plants

Concurrently with the space power plant work, the author investigated the possibility of coupling a potassium-steam binary vapor cycle to a molten-salt reactor for central station use[15] and found that this might yield an overall thermal efficiency of the order of 54 percent. As work proceeded on the development of thermonuclear reactors,[16] high-temperature gas-cooled reactors,[17] and LMFBRs,[18] the results of the space power plant work were applied in a series of advanced-system studies carried out at ORNL and by various organizations in Europe. These studies indicated that in high-temperature nuclear plants thermal efficiencies of ~50 percent could be obtained with reasonable capital costs.

TABLE 15.1   Comparison of Physical Properties of Potassium, Cesium, and Water for Condensing Conditions

	Potassium		Cesium		Water	
	Liquid	Vapor	Liquid	Vapor	Liquid	Vapor
Temperature, °F	1040.0		800		115.6	
Pressure, psia	1.50		0.66		1.50	
Specific volume, ft³/lb	0.02269	267.15	0.0091	610	0.01619	228.65
Enthalpy, Btu/lb	283.0	1170.4	69	267	83.56	1111.8
Heat of vaporization, Btu/lb	887.4		232		1028.14	
Specific heat, Btu/(lb · °F)	0.1823	0.1266	0.056	0.06	0.998	0.43
Viscosity, lb/(ft · h)	0.37	0.0189	0.50	0.054	1.42	0.029
Thermal conductivity, Btu/(h · ft · °F)	21.0	0.00363	11.2	0.0055	0.371	0.012
Prandtl No., $c_p \mu / k$	0.00321	0.659	0.0025	0.589	3.82	1.04
Surface tension, lb/ft	0.0041		0.0038		0.00469	

Studies of fossil fuel plants with alkali-metal-vapor topping cycles were not made before 1970 because of the serious combustion gas-side corrosion problems at tube-wall temperatures of ~800°C posed by the relatively high sulfur content of the low-cost fuel (coal or residual fuel oil) employed in most central stations. However, it became evident in 1970 that the impending fuel shortages, coupled with restrictions on sulfur emissions, provided both new incentives to employ a potassium-vapor topping cycle for central stations and a new opportunity. Since the requirement for low sulfur emissions favored either coal gasification to remove the sulfur or a combustion process in which the sulfur would be absorbed by a more active agent than iron-chrome-nickel alloys, sulfur corrosion would be a much less serious problem in such plants than in conventional oil- or coal-fired plants. Design studies by several organizations showed that a potassium-vapor cycle plant indeed appeared attractive and that the principal uncertainties were associated with the potassium boiler.[19-22]

Conceptual Design for a Fossil-Fuel Potassium-Vapor Cycle Plant

Going from a nuclear to a fossil-fuel heat source raises two problems for a potassium-vapor cycle plant. The first is that the combustion gas-side heat-transfer coefficient in a boiler of this type is completely dominant, because the boiling-potassium heat-transfer coefficient is extremely high. As a consequence, the size and cost of the boiler are determined by the heat-transfer coefficient for the combustion gas. To increase this coefficient it is highly advantageous to increase the operating pressure of the furnace. The second major problem is that the combustion gases inherently must leave the potassium boiler at a temperature of ~930°C (~1700°F) in order to boil potassium at, say, 838°C (1540°F). This would result in exorbitant heat losses if the hot gases were discharged to the stack. The amount of heat in these gases is greater than that required for boiler feed heating; hence some other approach ought to be employed to utilize this heat. Both of these problems can be handled nicely by making the potassium boiler and furnace the combustion chamber for a gas turbine. The burners and furnace can be operated to burn an amount of fuel equivalent to something close to the stoichiometric value, i.e., about four times the fuel-air ratio ordinarily used in the combustion chamber of a gas turbine. Thus about 75 percent of the heat added to the combustion gases in the furnace would flow into the potassium-vapor cycle, and only about 25 percent would flow into the gas-turbine cycle. Further, much of the heat in the exhaust gas leaving the gas turbine can be used to take care of about half of the feed-water heating for the steam cycle, thus increasing the useful work that can be extracted from the steam in the turbine per pound of steam delivered from the steam generator.

Flowsheet

A schematic diagram of the system proposed in Ref. 19 is shown in Fig. 15.1, and Table 15.2 summarizes the principal parameters for the conceptual design. Air from the compressor of the gas turbine is supplied to the furnace that heats the potassium boiler. The exhaust gas from this furnace then flows to the turbine of the gas-turbine unit. The vapor discharged from the potassium turbine would flow directly into a heat exchanger that would serve as the potassium condenser and the steam generator. Inasmuch as energy is removed efficiently in the potassium-vapor cycle, there is no incentive to increase the peak temperature of the steam cycle other than to increase the density of the potassium vapor in the condenser and thus reduce the size of the potassium turbine and condenser. Rough estimates indicated that near-minimum costs would be obtained by condensing the potassium at 593°C (1100°F) and transferring the heat to steam at a peak temperature of 565°C (1050°F).

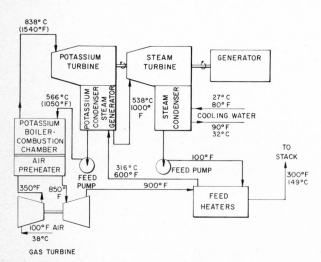

Figure 15.1 Flowsheet for the proposed potassium-vapor topping cycle system showing the relationships of the major components and the principal elements of the instrumentation and control system. *(Courtesy Oak Ridge National Laboratory.)*

TABLE 15.2   Summary of Performance Estimates for an Oil-Fired Gas-Turbine–Potassium-Vapor–Steam Ternary Cycle

Power plant		Power plant	
Power output, MWe	600	Number of passes for shell cooling gas	2
Gas turbine, MWe	105.2	Heat input to potassium in each furnace, Btu/h	4.322×10^8
Potassium turbine, MWe	134.3	Air pressure entering burner, psia	132.2
Steam turbine, MWe	360.5	Air density entering burner, lb/ft³	0.341
		Fuel-air ratio (90% stoichiometric)	0.060
Net thermal efficiency, %	51.87	Combustion gas flow leaving to turbine, lb/h	604,209
		Recirculation of combustion gases, %	50
Gas turbine		Combustion chamber	
Number of gas-turbine units	2	Number of combustion chambers per furnace	31
Airflow to each compressor, lb/h	1,710,225	Combustion chamber diameter, in	22
Fuel flow to combustion chambers, lb/h	204,805	Combustion chamber volume flow rate, ft³/s	2375
Heat input to gas flowing into each turbine, Btu/h	6.775×10^8	Combustion chamber flow area per furnace, ft²	81.8
Total heat to four combustion chambers, Btu/h	39.48×10^8	Mean gas velocity, ft/s	29
Heat to potassium boilers, Btu/h	25.93×10^8	Mean gas density, lb/ft³	0.104
Compressor pressure ratio	9:1	Mean gas dynamic head, psi	0.009
Compressor inlet temperature, °F	60	Pressure drop in combustion chamber and tube matrix, psi	1.6
Compressor outlet temperature, °F	588	Pressure drop in burner-aspirator, psi	8.0
Turbine expansion ratio	8:1		
Turbine inlet temperature, °F	1700	Gas side of tube matrix (per furnace)	
Turbine exhaust temperature, °F	976	Gas-flow-passage area in tube matrix, ft²	17.9
Useful heat in gas-turbine exhaust, Btu/h	7.398×10^8	Gas-flow geometry	1-pass, axial
		Gas-side heat-transfer coefficient, Btu/(h·ft²·°F)	44
Potassium-vapor turbine		Gas-side equivalent passage diameter, in	0.555
Potassium-vapor flow to two turbines, lb/h	30.78×10^5	Gas-side mass flow rate, lb/(s·ft²)	13.8
Vapor temperature into turbine, °F	1540	Gas-side Reynolds number	18.300
Vapor temperature leaving turbine, °F	1100	Mean gas temperature, °F	2200
Enthalpy drop (actual), Btu/lb	164.9		
Vapor quality leaving turbine, %	85.3	Potassium side of tube matrix	
Heat rejected to steam, Btu/h	21.14×10^8	Potassium-vapor temperature, °F	1540
		Potassium boiler tube OD, in	1.0
Steam turbine		Potassium boiler tube ID, in	0.8
Steam flow to high-pressure turbine, lb/h	2.27×10^6	Potassium boiler tube centerline spacing, in	1.3
Steam pressure to turbine, psia	3515	Potassium boiler tube vapor exit velocity, ft/s	114.7
Steam temperature to turbine, °F	1000	Potassium-vapor density, lb/ft³	0.058
Steam condenser temperature, °F	91.7	Potassium-vapor dynamic head, psi	0.0823
Fraction of feed heat from gas-turbine exhaust, %	56	Potassium-vapor mass flow rate, lb/(s·ft²)	6.65
Enthalpy rise in feed heater, Btu/lb	636.9	Heat added to potassium in boiler, Btu/lb vapor	842.5
Enthalpy rise in boiler and superheater, Btu/lb	726.1	Heat added to potassium in preheater, Btu/lb	74.08
		Potassium-vapor flow rate per tube, lb/s	0.0232
Furnace and potassium boilers		Potassium heat load per tube, Btu/s	19.56
Total number of units (3 per gas turbine)	6	Potassium heat load per tube, Btu/h	70,409
Airflow into each furnace, lb/h	570,075	Mean temperature difference between gas and tube, °F	385
Airflow into each furnace, lb/s	158.35		
Airflow into each furnace, ft³/s	464.7		
Shell cooling-gas-flow-passage area per pass, ft²	3		

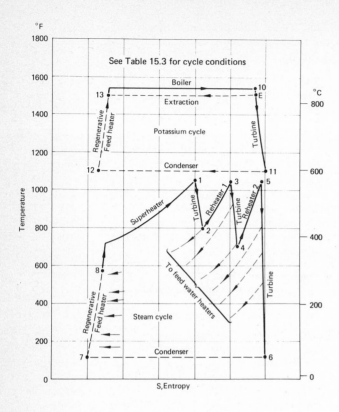

Figure 15.2 Temperature-entropy diagram for the steam and potassium cycles.[14] *(Courtesy Oak Ridge National Laboratory.)*

TABLE 15.3 Temperature and Pressure Conditions for the Steam and Potassium Cycles Shown in Fig. 15.2

		Pressure, psia	Temperature, °F	Enthalpy, Btu/lb
Steam	1	4000	1050	1446.1
cycle	2	1133	786	1374.0
	3	1043	1050	1535.0
	4	263	705	1372.7
	5	251	1050	1552.9
	6	0.50	115	1111.6
	7	0.50	115	82.9
	8	4300	558.4	560.1
Potassium	10	29.0	1540	1210
cycle	11	2.4	1100	1068
	12	2.4	1100	290
	13	30.0	1500	368

Thermodynamic Analysis

The temperature-entropy diagram for the resulting binary vapor cycle is presented in Fig. 15.2, and the principal data for the potassium and steam cycle calculations are presented in Table 15.3. The numbers for the lines in the table refer to numbered positions in the temperature-entropy diagram of Fig. 15.2. The potassium-vapor cycle calculations were carried out by using physical property data for potassium given in Ref. 23.

In preparing the flowsheet of Fig. 15.1, it was possible to employ the steam flowsheet for the Eddystone supercritical-pressure steam plant of Refs. 24 and 25 with essentially no changes. Further, the overall thermal efficiency of the steam portion of the system could be taken as that of the Eddystone plant with allowances for differences between the two systems stemming from such factors as the stack losses and the power requirements for auxiliaries.

The cycle conditions were chosen to give a recirculating boiler and saturated vapor at the turbine inlet, in part because this gives the highest thermal efficiency and in part because this makes it possible to keep all the metal surfaces in the furnace covered by liquid potassium, thus preventing them from exceeding the boiling point of potassium by more than about 30°C. Thus it is possible to obtain a boiler design that should be free of hot-spot difficulties and yet maintain a rather high average heat flux throughout the entire heat-transfer matrix. While it may appear that eliminating the potassium-vapor superheater to avoid problems with local hot spots would be likely to give difficulty with turbine bucket erosion, it should be remembered that all the light-water reactor power plants are operated with saturated steam delivered to the turbines, and these turbines operate with acceptable erosion rates. Further, studies of and experience with potassium turbines for space power plants have shown that potassium is much less likely to give difficulty with turbine bucket erosion than water. This comes about in part because of its lower density, in part because of the greater compressibility of the liquid, and in part because of its lower viscosity (which reduces the diameter of the droplets that could impinge on the turbine buckets—a point discussed in the latter part of Chap. 12).

Gas- or Oil-Fired Potassium Boiler

To facilitate fabrication and maintenance, the potassium boiler design of Ref. 19 made use of a set of boiler tube bundle and burner modules. Figure 15.3 shows the construction of the potassium boiler tube bundle module which consists of two annular rows of 1-in-diameter stainless-steel tubes arranged around a 22-in-diameter, long, vertical, cylindrical combustion chamber.[26] The flame from the burner directs hot gas upward through this combustion chamber. The tubes are bent at the top to provide flow passages between them so that the hot gas from the top of the combustion chamber can flow radially outward and then downward through an outer annulus over the outer row of tubes. Potassium would be circulated from a header tank at the top of the tube bundle through downcomers to ring-shaped manifolds at the bottom, and then vertically upward through the tubes back to a header drum and vapor separator at the top of the header tank. Around 200 tube bundles of this type would be required in a 600-MWe power plant.

The boiler tube bundles would be mounted in a furnace shell in the form shown in Fig. 15.4. In this layout the burner assemblies are independently mounted in the bottom head of the furnace shell with individual flanges so that any burner can be

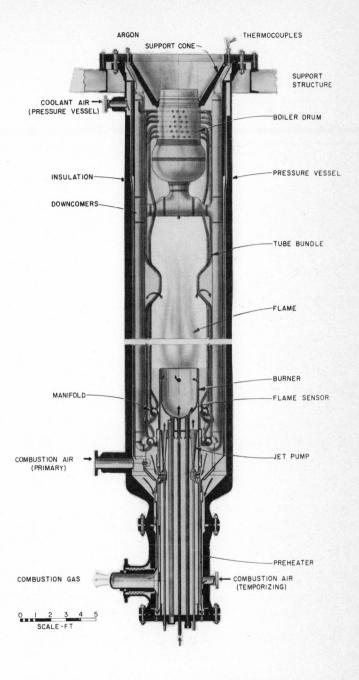

Figure 15.3 Section through an experimental gas-fired potassium boiler tube bundle and burner module installed in a furnace shell for testing.[26] *(Courtesy Oak Ridge National Laboratory.)*

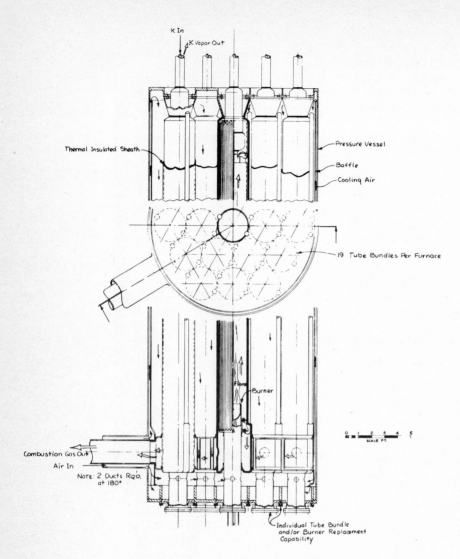

Figure 15.4 Section through a full-scale furnace-potassium boiler unit showing provisions for the burner installation at the bottom and the boiler-tube-bundle mounting arrangement at the top.[26] *(Courtesy Oak Ridge National Laboratory.)*

removed without disturbing the others. The boiler tube bundles are also independently mounted to a grid at the top which in effect serves as the top head. The design is such that any tube bundle can be replaced independently of the others by disconnecting it and withdrawing it through either the top or the bottom of the furnace.

Potassium-Vapor Turbine

The potassium turbine of Fig. 15.5 would be essentially similar to a low-pressure steam turbine except that more elaborate

shaft seals would be required. These shaft seals would employ a three-stage labyrinth seal between two face seals as in Fig. 15.6. In the labyrinth seal, argon buffer gas would be admitted to the center stage and bled from the two bleed-off regions between the three stages, one bleed consisting of air-plus-argon and one consisting of argon-plus-potassium vapor. Note that the pressure differentials are not large. This type of seal was used in the potassium-vapor turbine operated for ~2000 h by General Electric under the NASA space power plant program.

Mach number limitations make necessary a lower tip speed with potassium than for steam, and hence with an 1800-rpm

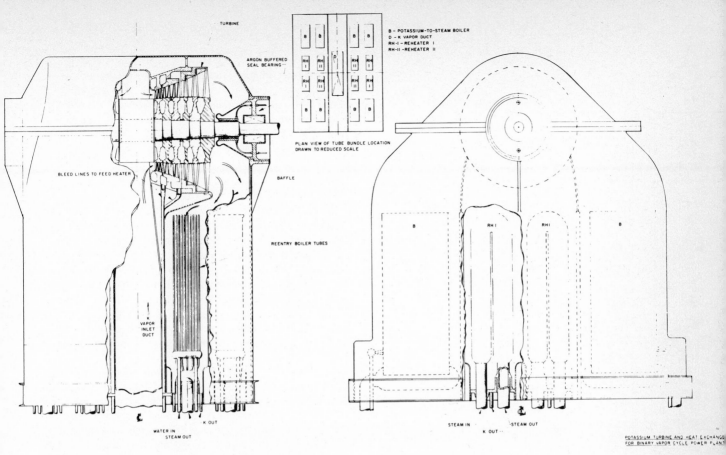

Figure 15.5 Conceptual design for a 115-MW potassium-vapor turbine built integrally with the potassium-condenser-steam generator.[14] *(Courtesy Oak Ridge National Laboratory.)*

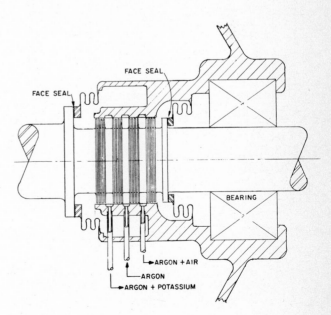

Figure 15.6 Proposed potassium-turbine shaft seal and bearing arrangement.[19] *(Courtesy Oak Ridge National Laboratory.)*

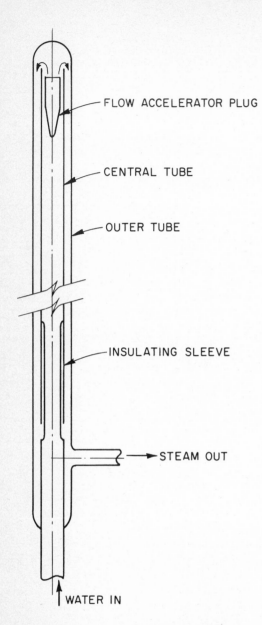

— FLOW ACCELERATOR PLUG

— CENTRAL TUBE

— OUTER TUBE

— INSULATING SLEEVE

→ STEAM OUT

↑ WATER IN

Figure 15.7 Section through a reentry-tube boiler designed for heating by condensing potassium vapor.[19] *(Courtesy Oak Ridge National Laboratory.)*

shaft and two turbines (to keep the size of each unit down) the rotor diameter at the outlet end is smaller for a given volumetric flow—in this instance the diameter was ~3 m (10 ft). Five stages are required, with the turbine rotor inlet around 2 m (6 ft) in diameter for each of the two double-flow turbines for a 600-MWe plant.

Potassium Condenser–Steam Generator

The very large volumetric flow rate of the potassium vapor out of the turbine made it desirable to install the potassium condenser directly beneath the turbine in much the same fashion as is standard practice with conventional steam-turbine condensers. To minimize the temperature gradient in the vicinity of the turbine shaft seals and bearings, and to keep the temperature of the turbine casing down to 1100°F, where its strength would not be much lower than at room temperature, the condenser was divided into two sections, with one section at either end of the double-flow turbine. The general layout is shown in Fig. 15.5.

The condensing coefficient for the potassium vapor is very high—about three times that for water—and hence it is possible to operate with a small temperature difference between the condensing potassium and the steam-generator exit temperature. As a consequence, the amount of surface area required in the potassium condenser–steam generator is much less than that ordinarily employed in a steam generator for a comparable fossil-fuel-fired power plant. A uniformly high heat flux can be obtained throughout most of the steam generator with no danger of burnout, because the peak metal temperature cannot exceed the condensing temperature chosen for the potassium, i.e., 593°C (1100°F).

In designing the potassium condenser–steam generator, it was decided that to avoid serious thermal stresses it would be best to employ a new type of reentry-tube steam generator that has been evolved in conceptual form for molten salt and other high-temperature liquid-cooled reactors.[27,28,29] This concept employs two vertical, concentric tubes in the form shown in Fig. 15.7. The water enters at the bottom through a central tube having a diameter of about 7 mm ($\frac{1}{4}$ in). Preheating and boiling occur as the water rises in this tube until evaporation is complete, after which there is some superheating. The steam emerges from the top end of the small-diameter tube, reverses direction, and flows back downward through the annulus between the inner small tube and the outer tube having an ID of around 13 mm ($\frac{1}{2}$ in). The steam annulus between the inner and outer tubes acts as a buffer to absorb the large temperature difference between the relatively low-temperature boiling-water region and the high-temperature outer tube, which operates at the condensing temperature of the potassium. In order to get the temperature of the superheated steam to approach the condensing temperature of the potassium, it is necessary to provide a little thermal insulation around the lower portion of the inner tube. A thin layer of plasma-sprayed zirconia or a static steam annulus inside a concentric stainless-steel tube have been found to be effective for this purpose (Fig. 15.7).

The potassium vapor enters the condenser shell at the top, flows downward between the tubes, condenses, and the condensate drains to the bottom. With this arrangement the header sheet separating the potassium from the atmosphere is not subject to a large pressure differential. There would be no high-pressure header sheets in this system; the tubes for the high-pressure feed water and the exit steam would be manifolded, as in the superheaters of high-pressure coal-fired steam boilers, rather than run into header sheets.

Although more complex and expensive than the usual designs for LMFBR steam generators, the reentry-tube once-through steam generator is remarkably free from thermal stresses and has the ability to operate well over a very wide range of conditions, including hot startups. For example, it was found possible in tests to bring a small unit to a red heat before beginning to admit feed water, yet there were no signs of serious thermal stresses or boiling-flow instabilities when the feed-water flow was initiated.

OPERATING EXPERIENCE WITH ALKALI-METAL-VAPOR SYSTEMS

Extensive operating experience with alkali-metal-vapor systems was gained in the 1959–1970 period under the space power plant program.[14] Further experience was obtained in the latter 1970s under a small program directed toward central station applications funded initially by the National Science Foundation (NSF) and subsequently by ERDA and DOE as programmatic responsibilities changed in Washington. The principal developments of that work are summarized here.

Experimental Program at ORNL under the Space Power Program

Experimental work on the potassium-vapor cycle began at ORNL in 1959 with a series of experiments concerned with the compatibility of boiling potassium with stainless steel at a system peak temperature of 870°C (1600°F) and also concerned with the heat-transfer coefficients and burnout heat flux obtainable with boiling potassium under forced-convection flow through tubes. The results of this work indicated that the materials compatibility was excellent and that potassium boiling in forced convection yielded heat-transfer coefficients and burnout heat fluxes at least as good as those obtainable with water operating at the same pressure [i.e., potassium boiling at 870°C (1600°F) gives heat-transfer coefficients and burnout fluxes at least as high as those obtainable with water boiling at 125°C].[30] Concurrent studies at ORNL indicated that the most promising space power plant system from the performance and reliability standpoints could be obtained with a boiling-potassium reactor coupled directly to a potassium-vapor turbine.[31] Thus a program was launched to develop the principal components for such a system with particular attention given to the following: (1) the heat-transfer matrix of the boiling-potassium reactor; (2) the vapor separator; (3) a potassium-vapor condenser designed to operate under 0-g conditions with a low probability of puncture by meteorites; (4) a free turbine-driven potassium feed pump whose performance characteristics were matched to the rest of the system so that no control action was required throughout the operating regime from zero to full power; and (5) the neutronic characteristics of a boiling-potassium reactor having temperature, power, and void coefficients such that a minimum of control actions would be required.[31]

Forty-two key problems were envisioned at the inception of the program, and most were solved by the time funding was terminated in 1966 when the nuclear space power plant program was being phased out.[31] At this point the total amount of running time accumulated at ORNL on boiling-potassium systems operating with stainless steel as the structural material was 160,000 h, of which about 8000 h was on systems with potassium turbine-driven feed pumps. The pumps operated in the cavitating mode with dry sump condensers so that the only control action required was to vary the electric power input to the electric heater rods, which simulated the fuel elements to be used in a full-scale reactor.[32]

The most significant result of the experimental work was that boiling potassium operating in a stainless-steel system with a peak temperature of about 850°C (1600°F) caused essentially no difficulty with corrosion or mass transfer. This stems from the fact that, although there is substantial differential solubility of iron, chromium, and nickel in liquid potassium between the boiler operating temperature of ~850°C and the condenser operating temperature of ~550°C, in a recirculating potassium boiler the liquid in the boiler rapidly becomes

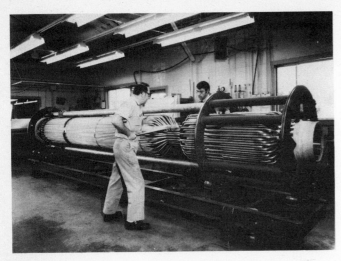

Figure 15.8 Photo of the ORNL-NSF/RANN potassium boiler under construction. The Type 304 stainless-steel tube bundle module is 7 m long and 0.7 m in diameter. It contains 134 tubes with a diameter of 2.54 cm. *(Courtesy Oak Ridge National Laboratory.)*

saturated with iron, chromium, and nickel (a few ppm). However, since only vapor moves out of the boiler through the system to the condenser, the dissolved elements are retained in the recirculating boiler and cannot be deposited out in the cooler, lower-solubility zone of the condenser. Further, the solution rate of iron, chromium, and nickel in the cold zone is so low that there is no appreciable transfer of material to the boiler in the condensate flowing from the cold zone to the hot zone and hence no appreciable deposition of material in the boiler. (See Ref. 25 of Chap. 6.)

The only serious problem encountered in the course of the development program was difficulty with nucleation in the boiler.[33] Alkali metals wet the iron-chromium-nickel alloys so tenaciously that it is difficult to initiate the bubbles required for nucleate boiling. As a consequence, the liquid may superheat by as much as several hundred degrees Celsius before a bubble begins to form. If this occurs, the sensible heat in the superheated liquid is released to the vapor extremely rapidly and leads to a phenomenon which is appropriately called *explosive boiling*. It was found that this condition could be avoided by deliberately introducing nucleation sites in the form of crevices to provide local hot spots.[33] Further, it was necessary to ensure that a small amount of inert gas in the system could collect in the crevices to provide an initial free liquid surface for initiating boiling under startup conditions. Once the problem was understood and devices of this sort were incorporated into the systems, no further difficulty with explosive boiling was experienced.

Experimental Programs at NASA Lewis Laboratory and G.E.-Evendale under the Space Power Program

System studies at NASA Lewis Laboratory[12] and the General Electric-Evendale Plant[13] favored a potassium-vapor cycle coupled to a lithium-cooled reactor, so an experimental program was launched to develop the key components, particularly the turbine. This program led to a series of experiments in which a total of ~ 55,000 h of boiling-potassium system operation was obtained, including ~ 12,000 h of turbine operation. The turbine tested was a three-stage unit of what would have been a six-stage machine for a full-scale 500-kWe power plant.[34,35] The results of the test work[36,37] confirmed the thermodynamic data of Refs. 38 and 39 and generally paralleled the experience at ORNL. There was no difficulty with mass transfer or corrosion when operating boiling potassium in stainless-steel recirculating boiler systems, and boiling and condensing heat-transfer coefficients similar to those found at ORNL were measured. There was some difficulty with unstable operation of the 3-MWt gas-fired potassium boiler which stemmed from the fact that it did not include nucleation sites, but about 12,000 h of boiler operation was obtained.[37] Six tube failures occurred; they were clearly the result of thermal stresses stemming from local superheating of liquid as a consequence of irregular failures to nucleate. There was also some difficulty with turbine bucket erosion caused by wet vapor, but evidently this stemmed from unstable boiling which led to excessive liquid carry-over from the boiler.

An important development of the program was the satisfactory operation of a labyrinth-type shaft seal on the turbine similar to that of Fig. 15.6. This work demonstrated that a seal somewhat similar to that employed in the mercury-vapor turbines built by General Electric could be employed with potassium by using an inert gas as a buffer in a region between the potassium vapor and atmospheric air portions of the labyrinth seal.

Other Experimental Space Power Programs

The AEC and NASA also funded work on the potassium-vapor cycle at Pratt & Whitney, AiResearch, Jet Propulsion Laboratory (JPL), and Rocketdyne.[40-43] The running time ac-

cumulated on boiling-potassium systems in these efforts totaled 25,000 h, 4000 h of which involved turbine operation.[15]
The experience gained in these efforts generally paralleled that
of ORNL and General Electric-Evendale.

Test of a Full-Scale Gas-Fired Potassium Boiler Tube Bundle and Burner Module

Following the favorable indications of the design study[19] of the
plant of Figs. 15.1 to 15.7, a small program was undertaken at
ORNL to design, build, and test a full-scale gas-fired
potassium boiler tube bundle module and burner assembly
similar to that of Fig. 15.3.[26] This was originally supported by
NSF and subsequently by ERDA and then DOE-FE funds. (FE
signifies the Division of Fossil Energy.) A photograph of this
unit as it was being completed in the shop is shown in Fig. 15.8.

In the initial tests of the boiler and burner module assembly,
water was used instead of potassium, because both analyses
and experience in the space power plant program indicated
that the performance characteristics with water at 100°C
should be essentially similar to those with potassium at 760°C.
Numerous difficulties with the pilot light were experienced in
the shakedown testing, but these were corrected. Good con-
trollability of both the boiler and the burner were found over
the operating range from light-off to about 20 percent power,
the maximum that could be run with the system as set up for
the shakedown tests. The boiling flow stability through this
startup and low-power-output regime was excellent. This is
vitally important because it is in this boiling regime that instability normally occurs; stable conditions should prevail at
higher outputs.

The components required for operating the system with potassium were installed, and shakedown tests with potassium in
the system began in the latter part of 1977. These tests were run
with an air-cooled potassium condenser. Performance tests
were completed and the data indicate that the unit met the design estimates. There was a problem with a small potassium
leak that developed just as the performance tests were being
completed. This was caused by a crack in a tee in which the
condensate flows from the two sections of the potassium condenser joined to return to the boiler. An imbalance in the cooling of the two sections of the condenser led to temperature differences of as much as 300°C between the two streams, and the
resulting thermal stresses in the zone where the two streams
merged caused a small crack to develop from thermal-strain
cycling. (See Fig. 6.12 for a similar case.) The design changes
to correct this condition were made, and the test resumed. The
test was terminated by a creep-buckling failure in the burner
just before funding for the program was discontinued. Note
that these two failures point up the special difficulties in the
design of equipment for operation in high-temperature liquid-
metal systems, and they show the importance of designing
these systems so that key modules can be thoroughly tested to
turn up oversights in the design work.

PRINCIPAL DEVELOPMENTAL PROBLEMS

The experience summarized in the previous section indicates
that most of the basic technology of the boiling-alkali-metal
system is well in hand. The principal problems associated with
the alkali-metal-vapor cycles are the thermal efficiency obtainable in a practical system, the capital costs of such a system,
the operating life, and the safety, reliability, and maintenance.
These depend heavily on detail design problems in the alkali-
metal boiler, turbine, and condenser-steam generator, which in
turn have been found to depend on whether cesium or potassium is used as the working fluid. If a fluidized-bed coal combustion system is employed, the problems of that system are
also involved.

Cesium versus Potassium

The first major problem to be faced in designing an alkali-
metal-vapor system is the choice between cesium and potassium for the working fluid. In conjunction with the space
power plant program for the potassium-vapor system development a small effort was directed toward the use of cesium. Although there had been some use of cesium compounds in man-
tles for gas lights and a few other special applications, there
had been no demand for cesium metal, and hence if any were
desired for experimental use it was necessary to build a special
pilot plant to produce it. Not only was the cost high—about
the same as for gold—but also the purity was uncertain. Purity
is a vital consideration in the early stages of a materials test
program in which it could have a pronounced effect on corrosion and thus on concept feasibility, especially if—as was initially the case for cesium—analytical techniques for trace impurities are not well developed. These questions were attacked
under the space program, in part because of interest in cesium
as a propellant for ion jet propulsion of spacecraft, and in part
because of the need for cesium in thermionic cells.

Compatibility with Structural Materials

Although the operating experience obtained with boiling
cesium systems was much less than for potassium, thorough

metallurgical examinations after ~20,000 h of operation showed no significant differences between cesium and potassium in their compatibility with the iron-chromium-nickel alloys.[44]

Resources and Costs

The resources for potassium are abundant, and the pure metal was available commercially for ~$2.20/kg ($1/lb) in 1975. Although the market for cesium is small, and there has been no effort to search for it, a number of large deposits have been noted.[45,46] For example, a deposit containing about 350,000 tons of pollucite, a mineral containing about 20 percent cesium, has been found in Manitoba. Other deposits include one of about 150,000 tons in Rhodesia and one of 50,000 tons in southwest Africa. Extensive deposits of a cesium biotite are found in South Dakota with a cesium content of around 3 percent. Substantial amounts of cesium are also found associated with molybdenum-vanadium ores, and there are piles of tailings in the U.S. northwest that could be processed relatively inexpensively to yield thousands of tons of cesium. The 1975 cost of these ores ran less than $2.20/kg ($1/lb) of contained cesium. Thus, ample supplies of cesium are available, and the cesium could be extracted at a reasonable price if a market were to develop. (1975 estimates ran from $4 to $22/kg or $2 to $10/lb.) Further, note that the known reserves represent about 100,000 times the current U.S. annual consumption; hence there has been no incentive to find new bodies of ore.

Effects on Boiler Design

As detail design work progressed on a full-scale potassium-boiler tube bundle module at ORNL in 1973, it became evident that the alkali-metal-boiler tube matrix presents a special set of stress problems not encountered in conventional steam boilers. These stem from the fact that the alkali-metal-vapor pressure in the boiler is only a few atmospheres, and so if a 10-atm furnace is employed, the external pressure on the boiler tubes, header drums, and vapor manifolds may induce collapse as a consequence of creep-buckling effects.[26,47-49] That is, as discussed in Chap. 6, small initial eccentricities will lead to substantial local bending stresses, gradual creep of the stressed metal will accentuate the eccentricities, and buckling and implosion will result. To avoid these difficulties, some designs for furnaces intended to operate at 10 atm have employed wall thicknesses two to three times what would otherwise be required for simple hoop stresses, and this has led to large capital costs, particularly where expensive alloys such as Hastelloy X were employed.[47,48] Thus, the capital cost of the boiler is very

dependent on the extent to which the furnace is pressurized and whether potassium or cesium is employed. The vapor pressure of cesium at 840°C is about 4 atm, whereas the vapor pressure of potassium at that temperature is about 2 atm. Thus, if the furnace pressure were 3 atm, the walls of a cesium boiler would be stressed in tension, not compression, while the walls of a potassium boiler would be subjected to an external pressure differential of only 1 atm, rather than 8 atm as would be the case if a 10-atm furnace were employed. This set of boiler tube stress considerations favors a furnace pressure of 3 to 5 atm.

Effects on Turbine Design

The initial interest in cesium in the space power plant program stemmed from the fact that its atomic weight and vapor pressure are higher than those of potassium. It raised the possibility of building a turbine that would be both smaller in diameter because of the higher vapor density and have fewer stages because of the greater stage temperature drop for a given set of vapor and blade velocities. Both effects act to reduce both the weight and the mechanical complexity of the turbine.[11,12,13] Basically, there is no difference between cesium and potassium in the thermal efficiency of the Rankine cycle for operation between the same temperature limits, but there are some differences in the design compromises that can be made.

Extensive design studies indicate that the alkali-metal-vapor turbines pose problems little different from those of mercury-vapor and steam turbines from the standpoints of aerodynamic design, moisture control, and turbine bucket erosion on the one hand, and shaft seals on the other.[11-13,47-51] The construction materials for both the rotors and the turbine buckets would be essentially the same as those for the gas turbines currently in operation. The principal problems appear to be those of scale effects. For large turbine sizes there are serious questions as to whether forgings of sufficiently large size can be made of suitable alloys in current use for gas turbines.[47,48] Thus a major feasibility question is the choice of turbine size, and this in turn is partly dependent on the choice of working fluid, because cesium will give a turbine about 60 percent the size or 60 percent the top speed that potassium provides.[52,53] Thus one set of feasibility questions can be avoided by choosing cesium rather than potassium as the working fluid.

As mentioned in a previous section on cycle efficiency, the choice of turbine inlet temperature is dependent in part on turbine rotor stress considerations. Air bled from the compressor discharge is employed in conventional gas turbines to cool the rotor to a temperature substantially below that of the hot gas stream passing through the turbine blades, at least in the higher-temperature stages. This approach cannot be used in an alkali-metal-vapor turbine because the metal vapor can readily flow into the space between rotor disks, and its very high condensing coefficient would defeat any effort to cool that disk below the static temperature of the metal vapor in that stage. Thus the static temperature drop in the first-stage nozzle box becomes particularly important in the design of alkali-metal-vapor turbines. In this respect cesium has an important advantage over potassium: The much smaller number of stages required for a cesium-vapor turbine also means that the temperature drop in the first-stage stator could be 60 to 100°C greater for cesium than for potassium and would permit a corresponding increase in the turbine inlet temperature for a given allowable rotor temperature and creep strength. If the turbine inlet temperature is increased, the diameter of the turbine rotor can be reduced. Alternatively, the lower tip speed of the cesium-vapor turbine can also be used to obtain lower stresses from centrifugal force, and this, in turn, will permit a further increase in turbine inlet temperature for a given choice of turbine rotor material. In fact, it appears that this factor alone will reduce the stresses in the turbine rotor sufficiently to permit operation of large disks of available alloys with a turbine inlet temperature of 840°C (1540°F), using the same material and rotor diameter as contemplated in a G.E. study in which rotor stress considerations limited a potassium-vapor turbine to operation at 760°C (1400°F).[53]

If turbine size limitations are considered, a cesium-vapor topping cycle offers an important subtle advantage over potassium from the thermal-efficiency standpoint. This advantage is that the combined cesium-vapor–steam cycle can be designed for a lower cesium-condenser temperature than is the case for potassium (for a given upper limit on turbine size). This reduces the amount of steam superheating and gives a more nearly rectangular temperature-entropy diagram for the steam cycle and thus the combined cycle will approach more closely the ideal Carnot cycle—an effect that can be deduced from Fig. 2.3 which shows temperature-entropy diagrams for a typical set of Rankine cycles.

Alkali-Metal-Vapor Cycle for an Advanced LMFBR

Although the problems of an LMFBR designed for a reactor outlet temperature of ~650°C (~1200°F) are formidable, it seems likely that the time will come when efforts to improve the plant thermal efficiency will lead to designs for reactor outlet temperatures of 700 to 800°C (1292 to 1472°F). This would make it possible to use a metal-vapor cycle in place of the intermediate sodium circuit (Fig. 11.16), thus reducing the temperature drop between the reactor outlet and the thermodynamic

cycle.[52] The cesium-vapor cycle appears more suitable than potassium for this application, because for practicable size limitations on the turbine, it would permit operation with a lower turbine-exit temperature, and this would help to give a favorable cycle efficiency as outlined above. It would also ease awkward problems with both maintenance and the effects of a leak in the steam generator of an LMFBR—points that will be discussed in a later section.

Alkali-Metal Boiler in a Fluidized-Bed Coal Combustion Furnace

In shifting from gas or oil as a fuel to coal in a fluidized-bed combustion chamber, it at first seemed logical to employ a 10-atm furnace supercharged with a gas turbine as in the initial work with oil or gas as the fuel.[19] Analyses indicated that this would give essentially the same thermal efficiency, although the higher pressure drop across the fluidized bed led to a tenth of a point loss in overall thermal efficiency.[20]

Fluidized-Bed Design Considerations

When it became apparent that the 10-atm furnace gave a creep-buckling problem in the tubes and headers of the 2-atm potassium boiler, it was recognized that the heat-transfer coefficient in a fluidized bed is essentially independent of the pressure in the furnace. Hence from the heat-transfer and boiler surface area standpoints, there is no incentive to employ a high pressure in the furnace. However, pumping losses in the fluidized bed are important, and the depth of the bed for acceptable pumping-power losses can be increased in direct proportion to the furnace pressure. As discussed in the section on fluidized-bed furnaces in Chap. 11, it is highly advantageous to employ vertical tubes in the furnace in order to obtain good natural thermal convection in a recirculating boiler, and since it is desirable to have fewer tube-to-header drum joints in the boiler, there is a strong incentive to employ a furnace pressure of at least 3 atm. This pressure permits nearly as great a tube length as can be used while still maintaining a vapor quality well below that for the burnout heat flux with the circulation rate provided by natural thermal convection of the boiling potassium or cesium.

Turbine Inlet Temperature Limitations

As indicated in the section in Chap. 12 on coal-fired gas turbines, past experience indicates that the problems with turbine bucket erosion, deposits, and corrosion caused by coal ash make it doubtful that gas-turbine inlet temperatures above

538 °C (1000 °F) can be employed. As is the case for steam-generator–furnace systems for combined cycles (Chap. 13), at turbine inlet temperatures of 538 °C (1000 °F) there is extensive experience to indicate that a satisfactory gas-turbine life can be obtained, and this temperature will serve to supercharge the furnace to 3 to 4 atm.

An important advantage of the fluidized-bed furnace design is that it is insensitive to tube spacing, whereas the design of Fig. 15.3 for a gas-fired potassium boiler has the disadvantage that tube warpage permits hot gas to bypass the upper portion of the tube bundle, causing a loss in performance.

Cost Factors

In the preliminary studies of potassium-vapor cycles carried out by both Westinghouse and General Electric in Schenectady under the ECAS program, the only furnace design condition considered was that of a fluidized-bed coal combustion system supercharged to 10 atm.[47,48] To ease the creep-buckling problem in the header drums and vapor piping, the boiler operating temperature was reduced to 1400 °F, which led to a loss in thermal efficiency of several points and an increase in costs. Further, the particulate-removal equipment required for the high gas-turbine inlet temperature and the high-temperature gas ducts added greatly to the capital costs. These problems are much the same as for fluidized-bed furnaces for combined cycles (Chap. 13). As is the case for steam generators, a furnace pressurized with a free gas turbine to 3 to 4 atm gives a substantially lower-cost boiler-furnace than an atmospheric unit while avoiding the severe turbine blade erosion, corrosion, and deposit problems inherent in the use of a ~900 °C (~1650 °F) gas turbine to pressurize the furnace to 10 atm. Note that the details of the cost estimates in Refs. 47, 48, and 54 indicate that major cost savings could be effected by designing for a 3- to 4-atm furnace rather than a ~10-atm furnace.

Temperature Limit Imposed
by Combustion Gas-Side Corrosion

Although the fluidized-bed coal combustion system has the advantage that it would permit operation with coal, a much more readily available and lower-cost fuel than clean gas or fuel oil, it presents a major set of problems with gas-side corrosion of the high-temperature boiler tubes. As discussed in Chap. 14, the prospects for a 10- to 20-year life with a Type 300 stainless steel at metal temperatures up to ~850 °C (1560 °F) appeared to be good in 1979, but the data available at the time of writing were not sufficient to establish the permissible temperature limit.

Alkali-Metal-Condenser–Steam Generator

Most of the studies carried out to date for alkali-metal-vapor topping cycle systems have assumed the use of a steam generator similar to that in an LMFBR system. The resulting capital costs have been quite modest—of the order of 20 percent of the cost of the alkali-metal-boiler and furnace unit. However, consideration of the difficult thermal-stress problems, coupled with stability and control problems and the possible consequences of leaks, have led to the author's proposal to employ the reentry-tube steam generator concept of Fig. 15.7. The capital costs of such a unit will be higher, but the thermal-stress problems and the consequences of a leak are alleviated. Tests of single-tube experimental units have yielded performance characteristics essentially the same as those computed before the test, and extensive operation over a wide range of conditions demonstrated excellent stability and control characteristics with no signs of serious flow instabilities.[27,28,29] However, no multitube unit has been built. Tests of such a unit are essential to establish the feasibility of this concept. These should include endurance tests to determine whether there are problems with the small layer of thermal insulation required along the lower part of the inner tube to permit good superheating.

System Control

Both conventional steam power plants and conventional open-cycle gas turbines pose quite difficult control problems—so difficult that the control system in either case represents an important capital cost item (Chaps. 7, 11, and 12). At first, the control problems posed by a binary vapor cycle would appear to be even worse because of the greater complexity of the system. The problem is compounded if a fluidized-bed coal combustion system is employed because, as discussed at the end of Chap. 11, the heat-transfer coefficient on the bed side is essentially independent of the combustion airflow through the bed and the heat-transfer coefficient on the boiling liquid side is also essentially independent of liquid flow rate. Consequently, the only two power control variables at the disposal of the designer are the temperature difference between the fluidized bed and the boiling liquid metal and the surface area of the portion of the boiler tubes immersed in the bed. The simplest control approach proposed makes use of a supercharged fluidized-bed combustion system in which the level of the fluidized bed would be varied to control the rate of heat transfer from the bed to the boiling liquid metal, while the combustion airflow and fuel flow to the bed would be controlled to hold a constant bed temperature and fuel-air ratio. (See the discussion of the similar system in Chap. 13.) The resulting control system has the advantages that all the variables have a linear relation to the power output over the full range from zero to full power and that the temperature and pressure sensing equipment required for control is simple, inexpensive, and reliable. A variable-speed electric motor-driven blower similar to that for the steam generator of Fig. 13.23 could be used at low powers to supplement the output of the turbocharger for supercharging the furnace, and it could be controlled to give the desired speed. (The turbine shaft speed would be essentially proportional to the system power output.) With a reentry-tube steam generator, it would be necessary to control only the feed-water flow rate to the steam generator so that a constant condensing temperature would be maintained for the alkali-metal-vapor system. If no throttle valve were used in the steam system, the pressure in the steam generator would automatically vary proportionally to the power output, because the first-stage steam-turbine nozzle would act as a critical pressure drop orifice with the steam flow rate being directly proportional to the pressure at the turbine inlet.[14,32]

Off-design and emergency conditions, such as an abrupt loss in electric load, pose additional problems that are not readily handled. The methods for dealing with these will depend on the detailed design of the major components in the system, as well as the methods used for coupling them and effecting control. The cleverness and thoroughness of the designer in coping with these problems will determine whether the actual system can survive emergency conditions without ill effects. Extensive analyses and tests with reduced-scale systems will be required to determine whether any given approach will be satisfactory.

Operational, Maintenance, and Safety Problems

Probably the most serious doubts people have with respect to the practicality of alkali-metal-vapor topping cycles have stemmed from the potential fire and/or explosion hazard associated with a liquid-metal leak, particularly in the steam generator. The high costs and slow pace of the U.S. program for developing an LMFBR have introduced a major mental block against any liquid-metal system into the minds of most utility personnel. In view of this situation, a fairly comprehensive look at these problems was undertaken in a study at ORNL in 1973,[55] and it was found that with proper design the safety problems of the alkali-metal-vapor cycle appear far less formidable than those of the LMFBR as currently envisioned.

The most serious questions with respect to the feasibility of an alkali-metal-vapor cycle are associated with the possibility of leaks into or out of the alkali-metal system, particularly

those that would lead to a liquid-metal–water reaction. The following section discusses these problems and their counterparts in the LMFBR together with the nature of the failures to be expected, a review of experience with liquid-alkali-metal leaks and attendant reactions, and an appraisal of the probable course of events in case of failures. Reference 56 gives a much more detailed treatment.

Anticipated Failure Modes in the Alkali-Metal-Vapor Topping Cycle

Operational and safety problems were major factors in the design philosophy on which the system in Ref. 19 was based. Every effort was made to minimize the possibility of a failure, and, if one should occur, to minimize its consequences (Chap. 8). The pressure in the alkali-metal system would be below that of either the combustion gases in the furnace–potassium boiler units or the steam in the potassium-condenser–steam generator units. As a consequence, if a leak develops, leakage will be into the alkali-metal system rather than from the alkali-metal system into the combustion chamber or steam generator.

Combustion Gas Leak into the Alkali-Metal Boiler A leak of combustion gases into the boiler will lead to the formation of alkali-metal oxide in the recirculating liquid. Gaseous nitrogen will be carried to the condenser, where it will accumulate in a region at one end. Both will provide simple, easily monitored bases for detection of a small leak (Chap. 8). Little or none of the oxide will be carried out of the boiler into the balance of the system. The low pressure in the boiler implies low pressure stresses; hence, if a leak does develop, it will probably stem from thermal-strain cycling and can be expected to progress slowly. This, in turn, means that the damage can be kept minor if the leak is detected in the early stages.

Steam Leak into the Alkali-Metal Condenser If a leak develops in the steam generator, the steam jetting into the alkali-metal condenser will react with the metal vapor to form alkali-metal oxide and hydrogen. A small leak can be detected readily by accumulation of the noncondensable hydrogen at one end of the condenser. Since the condenser will have a large vapor volume space available, there will be plenty of space for the hydrogen, even with a large leak, and no explosion or even large increase in pressure will occur. (This situation is completely different from that in a liquid-metal-heated boiler in which there is no free volume in the liquid-metal side into which the hydrogen from the reaction can expand.) As the hydrogen builds up in the condenser, it will block the flow of metal vapor into the condenser and produce a back pressure

TABLE 15.4 Summary of System Parameters Both for Major Experimental Units and Full-Scale Plant Reference Designs for Alkali-Metal-Vapor Cycle Systems

	Reference no.	Combustion system	Cycle working fluids	Plant output, MWe	Thermal efficiency, %	Furnace pressure, atm	Alkali-metal system		
							Turbine inlet temperature, °C	Boiler type	Structure material
G.E. space power plant turbine test rig	38, 43	Gas	K	3 MWt[a]		1	840	Nat. circ. vert. tube	SS
ORNL full-scale gas-fired K boiler tube bundle module	89, 90	Gas	K	1 MWt[a]		1.5	840	Nat. circ. vert. tube	SS
ORNL K vapor topping cycle plant design (1973)	35, 46	Gas or oil	Air-K-H_2O	600	51	9	840	Nat. circ. vert. tube	SS
G.E.-Evendale K vapor topping plant design	38	Coal fluid bed	Air-K-H_2O	1200	48.9	9	925	Horiz. tube forced conv.	SS
Austrian K vapor topping cycle plant design	85	Gas or oil	Air-K-or-ganic-H_2O	615	56	3.1	890	Vert. tube	SS
Westinghouse K vapor plant design, ECAS-I	39	Coal fluid bed	Air-K-H_2O	1200	42.4	10	760	Vert. tube forced conv.	Inco-800
G.E.-Schenectady, K vapor plant design									
ECAS-I	40	Coal fluid bed	Air-K-H_2O		39.6	10	760	Horiz. tube forced conv.	HA-188
ECAS-II	86	Coal fluid bed	Air-K-H_2O	995	44	10	760	Horiz. tube forced conv.	HA-188
ORNL supercharged furnace plant design (1978)	27	Coal fluid bed	Cs-H_2O	200	47.8	4	840	Nat. conv. vert. tube	Inco-800

[a] Output in MWt because no electric power was generated.
[b] Operating time in test work.
[c] Author's estimate using data of Fig. A10.4 for consistency.

on the metal-vapor turbine. Only a few pounds of steam leaking into the condenser would produce a marked change in the turbine back pressure which would be obvious to an operator or could also trip a warning signal.

Although it is extremely unlikely, if a large steam leak were to develop as a consequence of a tube rupture, the inherent nature of the double-walled tube configuration and inlet orificing of the inner tube of the reentry-tube boiler is such that vapor rather than water would be injected into the metal-vapor region (Fig. 15.7), and the rate of injection would be relatively low—about 1.8 kg/s (4 lb/s) per ruptured tube. The vapor injection would lead to an increase in the pressure in the condenser at a rate of about 0.07 atm/s (1 psi/s), or about 4

atm/min. Thus, if the condenser shell were designed for an internal pressure of 4 atm (60 psi), and if the flow of either steam or metal vapor into the condenser could be stopped within 1 min of the first evidence of the rupture, the damage would be limited to the broken tube. For the extreme case of an abrupt, complete rupture of a steam-generator tube, the condenser pressure would rise to about 0.3 atm (4 psia), or about three times its normal value, in about 3 s. This would be easily and reliably detectable and could be the basis for closing valves in the feed-water supply line. If this were done in an additional 10 s, the inventory of superheated water downstream of the shut-off valve in the boiler design proposed would be exhausted in another 15 s in the course of the "coastdown," and the peak

Boiler tubes and headers		Boiler and/ or turbine life, h	Steam generator		Boiler tubes and headers		Gas-turbine inlet tem- perature, °C	Cost of electricity, mills/kWh
Weight, kg/kWt	Cost, $/kWt		Peak temperature, °C	Peak pressure, atm	Weight, kg/kWt	Cost,[c] $/kWt		
1.06	21	10^b						
0.8	16							
0.58	12	10^5	540	240	0.17	3.4	925	
0.45	9		565	240	0.21	42	930	9
0.45	12	10^5	540	240	0.077	1.54	980	29
		10^5	540	240				40
1.4	140	10^5	540	240	0.564	11.28	930	40
0.74	20	10^5	480	136	0.29	2.1	540	

pressure in the condenser would be held to about 2 atm (15 psig). To protect against the contingency that no action might be taken, a rupture disk should be provided to blow off at per- haps 2.7 atm (40 psia).

It should be emphasized that a complete rupture of a tube appears highly unlikely; all experience to date indicates that a leak would develop gradually as a result of thermal-strain cy- cling and, if the leakage were sufficient to be detectable, the system could be shut down in an orderly fashion before any great amount of leakage had occurred. Thus, it is apparent that there is a strong premium for detecting trace amounts of steam or gas leakage into the potassium system, and instru- mentation for this is available.

SUMMARY

The future development of the alkali-metal-vapor cycle de- pends in part on the general view of the practicality of any liquid-metal system and in part on the escalation in fuel prices. When the cost and availability of fuel become sufficiently acute problems to arouse a greater interest in a higher thermal efficiency, there will be an increased interest in alkali-metal- vapor topping cycles. A major question then becomes one of capital costs, and this suggests that the principal system de- signs available be examined from the capital cost standpoint.

Table 15.4 summarizes system parameters for seven typical studies of alkali-metal-vapor topping cycle plants, together with two cases of fairly large experimental test systems that have been built to investigate problems of alkali-metal-vapor cycles. The latter two cases are included to provide a basis for comparing the operating temperatures and specific weights of the boilers in these systems with the corresponding values for the design studies.

Combustion and Power Conversion Systems

Both of the experimental systems listed have been fueled with gas, as have two of the full-scale plant studies. However, the bulk of the plant studies have employed fluidized-bed coal combustion systems because it appears that coal is the fuel that probably would be used. All the full-scale studies employ a furnace pressure of 3 to 10 atm in order to reduce the capital costs of the alkali-metal-boiler and furnace unit. All but two of the full-scale systems accomplish the furnace pressurization with a gas-turbine circuit that also produces a substantial amount of electric power. In the other two systems, the gas turbine serves simply to drive the compressor that supercharges the furnace. In one of these two cases an additional organic-fluid Rankine-cycle circuit was interjected between the potassium-vapor topping cycle and the steam cycle to avoid any possibility of a violent reaction between the potassium and steam in the potassium-condenser–steam generator. The seriousness of such an event was reduced in some of the other designs primarily by reducing the liquid-alkali-metal inventory and by providing a large space in the alkali-metal-vapor condensing region to absorb the pressure buildup associated with a possible alkali-metal–water reaction. This led to some increase in the cost of the steam generator, but it was only a small fraction of the plant total cost.

Power Plant Output

In all but one of the design studies, the projected plant output was 600 to 1200 MWe. The exception, the last entry in the table, is the most recent design and was intended primarily as a demonstration unit especially suitable for industrial power plant installations, rather than as a fully mature central station.

Thermal Efficiency

The estimated thermal efficiency for the seven cases ranges from 39.6 to 51 percent. It should be noted that the 39.6 percent efficiency case was later revised to yield a thermal efficiency of about 44 percent and the 51 percent efficiency was for a gas-fired unit with no sulfur problem; hence the range is really from 44 to 51 percent. A large portion of the revision

from 39.6 to 44 percent stemmed from redesign of the boiler to eliminate an excessive pumping power requirement for potassium recirculation. Much of the difference between the 44 percent efficiency of that revised design and the 51 percent thermal efficiency of the most optimistic study lay in the choice of peak temperature for the alkali-metal-vapor cycle, which was 760 °C for the lower value and 890 °C for the higher thermal efficiency. Other factors included the degradation associated with the losses inherent in going from a clean, high-quality fuel to coal with a substantial ash and sulfur content, the use of cooling towers, and differences in allowances for parasitic power requirements (whose cumulative effect amounted to ~4 points in efficiency). In view of extensive operating experience with potassium-vapor systems under the space power plant program, it seems likely that, if the fire-side corrosion problem in fluidized-bed coal combustion systems will permit, an alkali-metal-vapor turbine inlet temperature of 840 °C should prove practicable. This should permit an overall power plant efficiency of 48 percent for an installation employing cooling towers in a reasonably refined design in which parasitic power losses are kept to a practicable minimum.

The most expensive component in the entire plant is the alkali-metal boiler and furnace assembly. This cost, in turn, depends in part on the design—whether a natural-convection, vertical-tube, recirculating boiler or a horizontal-tube, forced-circulation boiler is chosen, with the former giving a less complex assembly and hence somewhat lower costs. The second major consideration is that of stresses in the boiler tubes, header drums, and vapor piping inside the furnace; these depend on the choice of alkali metal and the furnace pressure. The two studies of Refs. 47 and 48 assumed potassium as the alkali metal with a 10-atm furnace pressure; this led to severe creep-buckling stresses because of the large pressure differential across the tube and header walls. The stresses, in turn, led in the Ref. 47 study to the choice of Hastelloy 188, a very expensive alloy. In both the potassium-vapor power plant study of Ref. 48 and the cesium-vapor cycle studies of Ref. 54, Inco 800 or Type 304 stainless steel were considered likely to be satisfactory and were employed in the design.

The weight of the boiler in pounds per kilowatt of heat transferred to the alkali metal is probably the best indication of the cost of the unit. Table 15.4 shows that the specific weights range from 0.45 to 1.4 kg/kWt. This range stemmed from variations in the temperature difference between the fluidized-bed or combustion gases and the boiler outlet temperature, differences in the pressure differential across the tubes and hence in the creep-buckling stress limitation, and differences between horizontal- and vertical-tube boiler units. The gas-fired units, for example, operate with a much larger temperature difference between the combustion gases and the alkali-metal-boiler temperature, thus reducing the alkali-metal-boiler surface area requirement.

In an attempt to estimate the boiler cost (not including the furnace), the cost data of Fig. A10.4 for alloy boiler tubing were used. The cost of 304 stainless steel was found to be $2.70/lb, Incoloy 800 H was $4.72/lb, and Hastelloy X was $26.21/lb. The cost of the fabricated unit was estimated by taking the cost of the tubing in dollars per pound; multiplying it by the weight of the finished boiler assembly in pounds; multiplying by a factor of 1.5 to allow for material lost in the shop to machining, cutoff, and similar operations; and adding $5.00/lb for the fabrication operations, irrespective of the type of material. Results were converted into dollars per kilogram and are given in Table 15.4. Note that these yielded fabricated costs of $20.00/kg, $26.00/kg, and $100/kg for the stainless steel, Incoloy 800 H, and Hastelloy 188, respectively. Note also that the cost of the Hastelloy 188 was taken to be the same as that of Hastelloy X because their chemical compositions are similar. Resulting boiler costs ranged from $9.00 to $140/kWt. The differences would be even greater if the thermal efficiency were factored in and values quoted in dollars per kilowatt of net electric output from the plant. This spread in costs by more than a factor of 10 stems mostly from differences in the cost of the material (notably Hastelloy), the extra wall thickness required to resist creep buckling in the potassium units, the magnitude of the temperature difference available for heat transfer, and to some degree the choice of boiler configuration. Note that all the boilers were designed for the same nominal boiler operating life (around $\sim 10^5$ h).

The last column in Table 15.4 lists the cost of electricity for those studies in which such a cost was estimated. The lowest estimate was for the General Electric-Evendale study made in 1973 before the severe escalation in costs became a serious factor to be reckoned with. The large differences between the Westinghouse- and General Electric-ECAS studies of Refs. 47 and 48 in the estimated costs of major components give some indication of the large uncertainties in design precepts at this early stage in the development of alkali-metal-vapor systems. The cumulative effects of all the uncertainties are probably larger than those implied by the numbers given in Table 15.4.

It follows from the above discussion that there is a good possibility of obtaining an alkali-metal-vapor topping cycle system at an attractive cost, but that there are major uncertainties that can be resolved only by a substantial program of development tests.

PROBLEMS

1. Superimpose a potassium-vapor cycle on the heat-balance diagram for a steam plant similar to that for the Bull Run plant of Fig. 11.1. Assume an 838°C (1540°F) potassium-turbine inlet temperature, a 621°C (1150°F) potassium-turbine outlet temperature, an 85 percent potassium-turbine efficiency, and a furnace supercharged to 9 atm with a gas turbine operating with a 927°C (1700°F) inlet temperature. Assume a fuel energy input of 1180 MWt, 10 percent excess air, and 11 percent pressure loss between the compressor and gas turbine, a 10 percent stack loss, 98 percent generator efficiency, and 90 percent gas turbine and compressor efficiencies. Estimate the power output and thermal efficiency for each cycle and the overall thermal efficiency of the system.

2. Repeat the above for a 760°C (1400°F) potassium-turbine inlet temperature.

3. Using the results of Prob. 1, estimate the thermal efficiency if feed-water heating by the stack gas is greatly reduced so that the stack gas temperature is 343°C (650°F) instead of 121°C (250°F).

4. An advanced high-temperature LMFBR is to be coupled to a cesium-vapor cycle with a cesium-turbine inlet temperature of 677°C (1250°F) and an outlet temperature of 427°C (800°F) which rejects its heat to a 371°C (700°F), 138-bar (2000-psia) steam cycle that operates with regenerative feed heating and a steam temperature in the condenser of 40°C (102°F). Estimate the thermal efficiency of the Cs and steam cycles and the overall thermal efficiency, neglecting pumping and heat losses. (There are no stack losses.)

REFERENCES

1. Pummer, W. J.: *The Kinetics and Mechanism of the Pyrolytic Decomposition of Aromatic Heat Transfer Fluids, Final Report NBS Project No. 3110541*, National Bureau of Standards, April 1970.

2. Emmet, W. L. R.: "The Emmet Mercury-Vapor Process," *Trans. ASME*, vol. 46, 1924, p. 253.

3. Emmet, W. L. R.: "Mercury Vapor for Central Station Power," *Mechanical Engineering*, vol. 63, 1941, p. 351.

4. Smith, A. R., and E. S. Thompson: "The Mercury Vapor Process," *Trans. ASME*, vol. 64, 1942, p. 625.

5. Hackett, H. N., and D. Douglass: "Modern Mercury-Unit Power Plant Design," *Trans. ASME*, vol. 72, 1950, p. 89.

6. Federal Power Commission: *Steam-Electric Plant Construction Cost and Annual Production Expenses*, 1950, p. 50.

7. Wallerstedt, R. L., et al.: *Final Summary Report–SNAP 2/ Mercury Rankine Program Review*, vol. 1, NAA-SR-12181, Atomics International Division, Rockwell International, June 15, 1967.

8. Derow, H.: "Tantalum as a Mercury Containment Material in Mercury Rankine Cycle System Boilers, and SNAP-8 Power Conversion System, Breadboard Assembly—Materials Evaluation after 8700 hr Operation," Energy 70, *Proceedings of the 1970 Intersociety Energy Conversion Engineering Conference*, vol. 1, pp. 11-18 to 11-27.

9. Inatomi, T. H., and W. C. Parrish: "Thermodynamic Diagrams for Sodium," NAA-SR-62, July 1950.

10. "Boiling Rubidium as a Reactor Coolant: Preparation of Rubidium Metal, Physical and Thermodynamic Properties, and Compatibility with Inconel," Oak Ridge National Laboratory, unpublished internal document, August 1954.

11. Fraas, A. P.: "Fission Reactors as a Source of Electrical Power in Space," *Proceedings of Second Symposium on Advanced Propulsion Concepts*, vol. III, Avco-Everett Research Laboratory, Oct. 7–8, 1959.

12. Stewart, W. L.: "Analytical Investigation of Multistage-Turbine Efficiency Characteristics in Terms of Work and Speed Requirements," NACA-RM E57K22b, Lewis Flight Propulsion Laboratory, February 1958.

13. Schnetzer, E.: "Comparison Study of Cesium and Potassium for Rankine Cycle Space Power Systems," TMS Report No. 67-1, General Electric Space Power and Propulsion Section, July 1966.

14. Young, H. C., and A. G. Grindell: "Summary of Design and Test Experience with Cesium and Potassium Components and Systems for Space Power Plants," USAEC Report ORNL-TM-1833, Oak Ridge National Laboratory, June 1967.

15. Fraas, A. P.: "A Potassium-Steam Binary Vapor Cycle for a Molten-Salt Reactor Power Unit," *Trans. ASME*, vol. 88, 1966, pp. 355–366.

16. Fraas, A. P., and D. J. Rose: "Fusion Reactors as Means of Meeting Total Energy Requirements," ASME Paper No. 69/WA-Ener-1, November 1969.

17. Collier, J. G., et al.: "A Potassium-Steam Binary Turbine Cycle for a High Temperature Gas-Cooled Reactor," AERE-R-7038, December 1972.

18. Wood, B.: "Alternative Fluids for Power Generation," *Proceedings of the Institution of Mechanical Engineers*, vol. 184, part 1, 1969–1970.

19. Fraas, A. P.: "Preliminary Assessment of a Potassium-Steam-Gas Vapor Cycle for Better Fuel Economy and Reduced Thermal Pollution," Oak Ridge National Laboratory Report No. ORNL-NSF-EP-6, August 1971.

20. Fraas, A. P.: "A Fluidized Bed Coal Combustion System Coupled to a Potassium Vapor Cycle," *AIChE Symposium Series*, vol. 70, no. 137, November 1972, pp. 238–244.

21. Rossbach, R. J.: *Final Report: Study of Potassium Topping Cycles for Stationary Power*, GESP-741, NASA Contract No. NAS3-17354, General Electric Company, Energy Systems Programs, Nov. 13, 1973.

22. Rajakovics, G. E.: "Energy Conversion Process with about 60% Efficiency for Central Power Stations," *Proceedings of the Ninth Intersociety Energy Conversion Conference*, 1974, p. 1100.

23. Weatherford, W. D., et al.: "Properties of Inorganic Energy-Conversion and Heat-Transfer Fluids for Space Applications," Wright-Patterson Air Force Base, WADD Technical Report 61-96, November 1961.

24. Harlow, J. H.: "Engineering the Eddystone Plant for

5000 lb 1200-deg Steam," *Trans. ASME*, vol. 79, 1957, p. 1410.

25. Trumpler, E. W., Jr., et al.: "Development Associated with the Superpressure Turbine for Eddystone Station Unit No. 1," *J. Eng. Power, Trans. ASME*, ser. A, vol. 82, no. 4, October 1960, pp. 286–292.

26. Fraas, A. P., et al.: "Design of a Potassium Boiler Tube Bundle Module," Oak Ridge National Laboratory Report No. ORNL/NSF/EP-46 (unpublished).

27. Fraas, A. P.: "A New Approach to the Design of Steam Generators for Molten Salt Reactor Power Plants," Oak Ridge National Laboratory Report No. ORNL/TM-2953, June 1971.

28. Bruens, N. W. S., et al.: "Modeling of Nuclear Steam Generator Dynamics," paper presented at the International Conference on Materials for Nuclear Steam Generators, Gatlinburg, Tenn., Sept. 9–12, 1975.

29. Viresema, B.: "Aspects of Molten Fluorides as Heat Transfer Agents for Power Generation," doctoral thesis, Technische Hogeschool Delft, February 1979.

30. Hoffman, H. W., and A. I. Krakoviak: "Forced Convection Saturation Boiling of Potassium at Near Atmospheric Pressure," *Proceedings of the 1962 High Temperature Liquid Metal Heat Transfer Technology Meeting*, BNL-756, pp. 182–203.

31. Fraas, A. P.: "Summary of the MPRE Design and Development Program," Oak Ridge National Laboratory Report No. ORNL-4043, June 22, 1967.

32. Yarcsh, M. M., and P. A. Gnadt: "The Intermediate Potassium System—A Rankine Cycle, Potassium Test Facility," Oak Ridge National Laboratory Report No. ORNL-4025, October 1968.

33. MacPherson, R. E.: "Techniques for Stabilizing Liquid Metal Pool Boiling," II-B/11, *Proceedings of the Conference Internationale sur La Surete des Reacteurs a Neutrons Rapids,* Sept. 22, 1967, Aix-en-Provence, Commissariat a l'Energie Atomique, France.

34. Zimmerman, W. F.: "Two-Stage Potassium Turbine: IV-Materials Support of Performance and Endurance Tests," NASA CR-925, February 1968.

35. Schnetzer, E., and G. M. Kaplan: "Erosion Testing of a Three-Stage Potassium Turbine," ASME Preprint 70-AV/SPT-37, June 1970.

36. Peterson, J. R.: "High Performance Once-Through Boiling of Potassium in Single Tubes at Saturation Temperatures of 1300 to 1750°F," NASA CR-842, August 1967.

37. "Facilities of Space Power and Propulsion Section," SPPS 6-113, General Electric Company, Feb. 25, 1966.

38. Ewing, C. T., et al.: "High-Temperature Properties of Potassium," NRL Report 6233, September 1965.

39. Ewing, C. T., et al.: "High-Temperature Properties of Cesium," NRL Report 6246, Sept. 24, 1965.

40. "Corrosion Studies of Refractory Metal Alloys in Boiling Potassium and Liquid NaK," *Proceedings of AEC-NASA Liquid Metals Information Meeting*, CONF-650411, April 1965.

41. *SNAP 50/SPUR Program, Nuclear Mechanical Power Unit, Experimental Research Development Program*, Final Report, APS-5249, AiResearch Mfg. Co., December 1966.

42. Davis, J. P., et al.: "Lithium-Boiling Potassium Refractory Metal Loop Facility," Jet Propulsion Laboratory Technical Report No. 32-508, Aug. 31, 1963.

43. Spies, R., and A. H. Cooke: "Investigation of Variables in Turbine Erosion," paper presented at ASTM Sixty-Ninth Meeting, June 1966.

44. Jansen, D. H., and R. L. Klueh: "Effects of Liquid and Vapor Cesium on Structural Materials," USAEC Report ORNL-TM 1813, Oak Ridge National Laboratory, June 1967 (AEC Interagency Agreement 40-98-66, NASA Order W-12, 353).

45. Dean, K. C., et al.: "Cesium Extractive Metallurgy; Ore to Metal," *Journal of Metals*, November 1966.

46. Heindl, R. A.: *Cesium, Mineral Facts and Problems*, U.S. Bureau of Mines Bulletin 650, 1970, p. 650.

47. Bass, R. R., et al.: *Energy Conversion Alternatives Study (ECAS), General Electric Phase II*, Final Report: *Research and Development Plans and Implementation Assessment*, NASA CR-134949, vol. III, SRD-76-064-3, December 1976.

48. Deegan, P. B.: *Energy Conversion Alternatives Study (ECAS), Westinghouse Phase I Final Report*, vol. VII: *Metal Vapor Rankine Topping-Steam and Bottoming Cycles*, NASA CR-134941, Feb. 12, 1976.

49. Fraas, A. P., D. W. Burton, and L. V. Wilson: "Design

Comparison of Cesium and Potassium Vapor Turbine-Generator Units for Space Power Units," Oak Ridge National Laboratory Report No. ORNL/TM-2024, February 1969.

50. Young, H. C., et al.: "Survey of Information on Turbine Bucket Erosion," Oak Ridge National Laboratory Report No. ORNL/TM-2088, July 1968.

51. Varljen, T. C., and C. M. Glassmire: "Estimation of Moisture Formation and Deposition and of the Threshold for Turbine Bucket Erosion in Potassium and Cesium Vapor Turbines," WANL-PR(CCC)-003, Westinghouse Astronuclear Laboratory, December 1967.

52. Fraas, A. P.: "A Cesium Vapor Cycle for an Advanced LMFBR," ASME Paper No. 75-WA/Ener-5, December 1975.

53. Fraas, A. P.: Comparison of Helium, Potassium, and Cesium Cycles, *Proceedings of the Tenth Intersociety Energy Conversion Engineering Conference*, 1975, pp. 486–495.

54. Samuels, G., et al.: "Design Study of a 200 MWe Alkali Metal-Steam Binary Power Plant Using a Coal-Fired Fluidized Bed Furnace," Oak Ridge National Laboratory Report No. TM-6041.

55. Fraas, A. P., G. Samuels, and M. E. Lackey: "A New Approach to a Fluidized Bed Steam Boiler," ASME Paper No. 76-WA/Pwr-8, Dec. 5, 1976.

56. Fraas, A. P.: "Operational, Maintenance and Environmental Problems Associated with a Fossil Fuel-Fired Potassium-Steam Binary Vapor Cycle," Oak Ridge National Laboratory Report No. ORNL/NSF/EP-30, August 1974.

MAGNETO-HYDRODYNAMIC SYSTEMS

The very word "magnetohydrodynamics" (MHD) is impressive. The news media have been fascinated by it and since the early 1960s have often reported that this system will soon generate electric power at high efficiency and low cost in large central stations. Its advocates are dedicated with an almost religious zeal. On the other hand, many experts in fluid mechanics and materials have serious doubts that a commercially viable system can be developed. The concept is sufficiently abstruse that it is difficult to judge just how formidable the problems are. The literature is extensive with an overwhelming volume of abstract mathematical analyses.[1] The author has made an extensive review of the past 40 years of experimental and theoretical efforts in this field and has reluctantly reached the conclusion that there is a large gap between the actual state of development and that required to design and build a useful power plant. This chapter summarizes the principal theoretical and experimental work accomplished to date, outlines a representative full-scale power plant concept prepared by leading advocates, and then examines the principal problems that must be resolved before a realistic power plant design can be evolved.

HISTORICAL BACKGROUND

One hundred and fifty years ago, Faraday, in working on his concept for generating electricity by moving an electric conductor through a magnetic field, considered the possibility of using a conducting fluid stream which might be forced to flow through a magnetic field rather than move a solid conductor.[2] He experimented with a stream of mercury flowing through a glass tube, and set up an inconclusive experiment using the River Thames as his flowing, conducting fluid. He also considered the possibility of using the tides in estuaries where the electrically conducting seawater ebbs and flows across the earth's magnetic field. However, once he had developed a sufficiently powerful magnet, he found the solid conductor concept

much more practical and pursued it, dropping work on the conducting fluid stream, though his diary contains speculation as to whether the aurora borealis is what we would now describe as an MHD phenomenon.

Essentially nothing further was done with this concept until the late 1930s, when developments in binary vapor cycles and gas-turbine technology led to the thought that one might replace the moving parts of a high-temperature turbine and its generator with an MHD generator that would carry out the same function in the thermodynamic cycle. By burning fuel with oxygen or highly preheated air, it is possible to obtain gas temperatures of around 2600°C (4700°F). At this temperature the products of combustion are sufficiently ionized to have an appreciable electric conductivity, though not enough to give acceptable I^2R losses. The degree of ionization can be increased about one-hundredfold by adding 1 percent of potassium or cesium, the most easily ionized of the elements, to give an electric conductivity approximately one-millionth that of copper, which is good enough to give a useful generator.[3,4,5] By carrying out the combustion at a pressure of 5 to 10 atm, the hot gases can be expanded at high velocity through a nozzle transverse to a magnetic field so that an electric current will be generated in the gases and will flow at right angles to the direction of both the high-velocity gas flow and the magnetic field (Fig. 16.1). Thus neither a turbine nor a generator would be required, and the temperature limitations of the gas turbine might be avoided. The basic concept is wonderfully simple and has appealed to many investigators all over the world, leading to numerous efforts to develop MHD systems. Table 16.1 lists some of the more significant experiments carried out both in the United States and abroad. A large-scale program has been pursued in the U.S.S.R., and roughly comparable programs were pursued in the Federal Republic of Germany, Great Britain, and France during the 1960s. In 1967 both Great Britain and France decided to close out their MHD programs because they felt that the system did not appear promising, and the Federal Republic of

TABLE 16.1   MHD Generator Performance Parameters for Eight Experimental Units and a Reference Design

Experiment (*Reference*)	Avco Mark V *(Salzburg MHD Symp. 1966)*	Shibaura Electric Co., Tokyo *(Salzburg MHD Symp. 1966)*	Avco Mark II *(5th Int. Conf. on MHD, Munich 1971)*	USSR U-02 *(9th World Energy Conf. Trans. vol 5, 1974)*
Fuel	Ethyl alcohol	Light oil	Toluene	Natural gas
Oxidant	Liquid oxygen	Oxygen	Oxygen	Air
MHD generator efficiency, %				
Enthalpy extraction, %	7	1.4	7.5	3.75
Current density, A/cm²	6.8	2.2	1 to 10	
Terminal voltage, v	2000	100	960 to 96	
Magnetic field, T	2 to 3	2	2.6 to 3.5	1.8
Gas inlet temperature, K (°F)	3100 (5000)			
Seed	Potassium hydroxide		Cesium carbonate	K_2CO_3
Operating life, h	0.0167	0.02	0.0022	300
Power output, MWe	23.6 8.4 to magnet	0.1	0.4	0.075
Goodness index‡	2.75 net 3.73 gross	0.003	0.007	84

*Magnet consumes 0.12 MW power.
†Operating time limited by size of coal hopper.
‡Goodness index, % enthalpy extraction × power output × operating time.

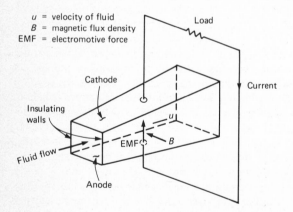

u = velocity of fluid
B = magnetic flux density
EMF = electromotive force

Figure 16.1   A basic MHD channel includes cathode, anode, magnet, and insulating walls. The flow of the conducting fluid across the magnetic field results in an induced electromotive force and a current in an external load.[5]

Germany followed suit in 1971. However, a substantial program was under way in Japan in the 1970s, and at the time of writing, valuable research efforts are under way in the Netherlands, Poland, Italy, China, and Israel.

THEORY AND EXPERIMENTS

As in most fields, one must examine both the basic theory involved and the results of experiments to develop a good appreciation for the nature and problems of MHD generators and associated power systems.

Elementary Theory

The open-cycle MHD system is commonly spoken of as a replacement for the gas turbine in that it would operate on the same basic thermodynamic cycle but at a much higher temperature so that a higher cycle efficiency could be obtained. The hot combustion gases would be expanded from a pressure of 5 to 10 atm down to about 1 atm through a divergent channel. Ideally, the expansion would be isentropic with the electromagnetic retarding force on the gas through which the electric current flows imposing a pressure drop equivalent to that imposed by banks of turbine blades. Note that unless this retarding force removes energy from the gas in the course of its expansion

USSR U-25 (9th World Energy Conf. Trans. vol. 5, 1974)	Avco Mark VI (Personal Commun., Feb. 1977)	ETL Mark V (Japan) (16th Symp. on E.A. of MHD)	UTSI (Personal commun., Feb. 1977)	HPDE Facility (7th Int. Conf. on MHD, 1980) (Ref. 17)	Full-Scale Reference Design (Ref. 27)
Natural gas	Fuel oil	Light oil	Coal	Toluene	Coal
Air	Air and oxygen	Oxygen	Oxygen	Air — O_2	Air
	Not measured			15.5	80
8	1.33 to 2	1.93	0.56	9	47.2
	0.3 to 1	0.7	0.5	0.6	0.8
600 to 700	1000 to 2000	20,000	850	1800	40,000
2 (iron magnet)	2.5	4.2	2* 3000	3.4	5
	Potassium carbonate and fly ash	KOH	K_2OH_4	KOH	1% K_2SO_4
3	98	1	1†	0.01	100,000
4.5	0.2	0.482	0.09	22	1405
108	26–39	0.93	0.05	3.9	4×10^8

through the nozzle, the expansion will not be isentropic but will be a simple throttling process, and no useful energy will be recovered. Thus to extract work from the expanding gas flow requires a substantial current density. To achieve this current density with the ionization that can be obtained at the highest practicable temperatures, it is necessary to add potassium or cesium to the hot gas. For a simple case in which the channel design gives a constant axial velocity, the basic relations involved follow directly from elementary electromagnetic theory. That is,

$$\frac{dP_x}{dx} = \frac{IB}{A} = J_y B_z \qquad (16.1)$$

where I/A = current density J, A/m^2

dP/dx = pressure drop per meter of length, N/m^2

B = magnetic field, tesla (T)

x = distance along flow path, m

y = distance along electric current path, m

z = distance along magnetic flux path, m

For good electrical efficiency, the voltage generated should be at least 10 times the voltage drop caused by the resistance of the conducting gas between the electrodes that span the channel. The voltage obtainable is given by

$$E = Bud \qquad (16.2)$$

where E = emf, V

B = magnetic field, T

u = gas velocity, m/s

d = channel thickness in direction of electric current flow, m

The voltage loss stemming from the electrical resistance of the hot gas is given by

$$E_i = \frac{I\varrho d}{A} = J\varrho d \qquad (16.3)$$

where E_i = internal voltage loss

I/A = current density, A/m^2

ϱ = gas resistivity, Ω/m

d = distance between electrodes, m

Hall Effect

In addition to the Faraday electric field imposed at right angles to the gas stream, the electric current flowing across the gas stream induces its own field in the direction of the gas stream—a phenomenon called the Hall effect.[6] The Hall field is present in all conductors, but in the copper conductors of conventional generators it is small and has no practical effect. In MHD

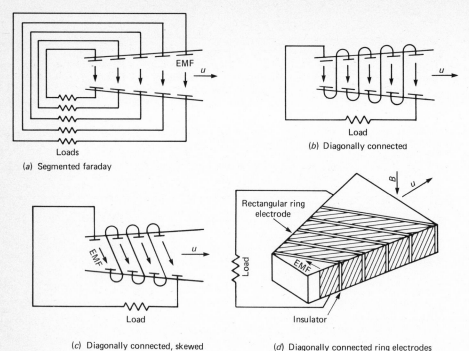

(a) Segmented faraday

(b) Diagonally connected

(c) Diagonally connected, skewed

(d) Diagonally connected ring electrodes

Figure 16.2 Representative MHD channel electrode connections.

generators with the conducting system comprising an ionized gas, it is typically as large or even larger than the Faraday field. The Hall field causes continuous electrodes to short-circuit the generator and creates a tendency toward fluid flow separation along the anode wall. These difficulties can be reduced by segmenting the electrodes as in Fig. 16.2a.[4,5] Another approach is to connect them in a staggered pattern, as in Fig. 16.2b, so that the equipotential planes lie diagonally across the channel. This has the further advantage in small units of increasing the voltage output of the generator, because the electrodes are in series. The increased voltage reduces the relative importance of the resistance losses in the external circuit. Depending on the proportions of the system and the operating conditions, the equipotential lines may be inclined to such a degree that the diagonal array of Fig. 16.2c is in order. These latter two electrode arrays can be formed of stacks of rectangular ring-shaped electrodes in the form shown in Fig. 16.2d. The top and bottom channel walls in Fig. l6.2d serve as the conductors that provide the series connections between the anodes at the right and the cathodes at the left in the rings of Fig. 16.2d.

Open-Cycle Systems

Most of the work on MHD generators has been directed toward open-cycle systems in which air is taken from the atmosphere,

compressed, heated by combustion, expanded through the MHD generator, and then exhausted to the atmosphere through a heat-recovery system. Inasmuch as this type has shown the greatest promise, this chapter is concerned mainly with the use of combustion gases as the working fluid.

Conductivity of Hot Combustion Gases

One of the serious constraints in the open-cycle MHD systems is the poor conductivity of hot combustion gases. The conductivity can be greatly increased if the hot gas is seeded with an alkali metal. Potassium is much more readily ionized than sodium and much less expensive than cesium; hence it is normally used in MHD work. The effects of both temperature and additions of potassium for the range of interest are shown in Fig. 16.3; the addition of 1 percent by weight of potassium carbonate increases the conductivity by a factor of over 100. As shown in Fig. 16.4, increasing the pressure reduces the conductivity.[7]

Conditions of Interest for a Useful System

Extensive design studies have shown that the various practical constraints that must be considered if one is to have a useful system lead to a rather narrow range of interest for the design parameters. Basic thermodynamic considerations show that, as in gas turbines, the efficiency of the MHD generator must exceed ~70 percent if it is to deliver sufficient power to supply the air required for the combustion chamber and at the same time yield sufficient additional power to give a worthwhile net power output from the system. For this case, even under ideal conditions, it appears to be necessary to obtain a hot gas temperature in the MHD generator of the order of 2540°C (4600°F) and employ as high a magnetic field as can be obtained with a practicable superconducting magnet (i.e., of the order of 5 T), the generator must be large (of the order of 1000 MW), and the gas velocity must be high (~500 m/s).

Closed-Cycle Systems

The difficulties posed by the poor conductivity of the combustion gases, even when seeded with potassium or cesium, has led to the investigation of other working fluids that might be employed at reduced temperatures in closed cycles.[8] These have included the use of cesium vapor ionized in inert gases (such as argon or neon) operating in essentially a Brayton cycle, the alkali metals potassium and cesium operating in Rankine cycles,[9] and two-phase mixtures with inert gas or metal vapor bubbles in a liquid alkali metal stream operating in essentially a

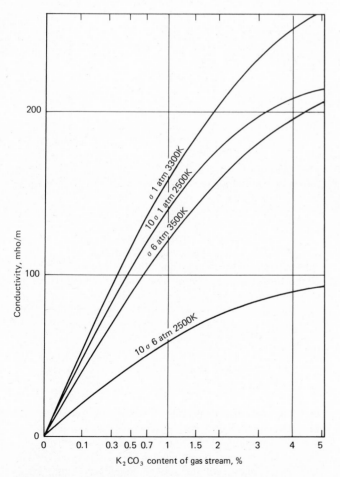

Figure 16.3 Effects of alkali concentration on the gas conductivity σ for several representative temperatures and pressures.[7]

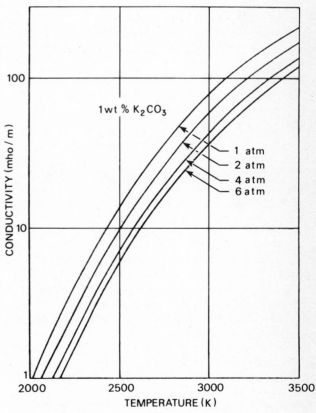

Figure 16.4 Effects of temperature and pressure on the conductivity of combustion products seeded with 1 percent by weight of K_2CO_3 (Ref. 7).

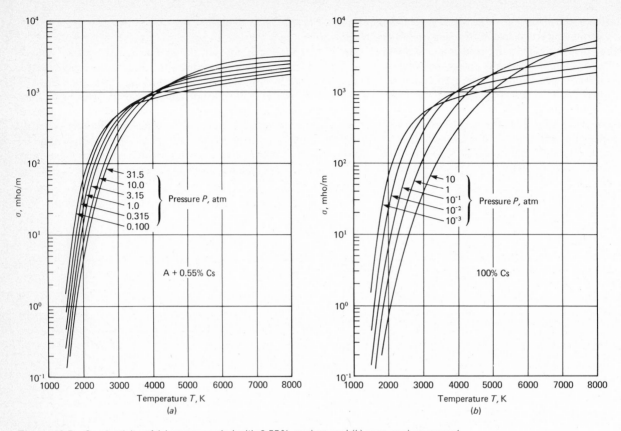

Figure 16.5 Conductivity of (a) argon seeded with 0.55% cesium and (b) pure cesium vapor.[4]

constant-temperature expansion cycle.[10] In the two-phase systems the driving force moving the fluid through the MHD generator is derived from the expansion of the gas bubbles.

Systems that make use of a two-phase mixture of an alkali metal with a metal vapor or an inert gas depend on the conductivity of the liquid metal. The bubbles of gas or vapor must be in a very finely divided form and very uniformly distributed through the cross section. The electric conductivity is provided by the ligaments of liquid between the bubbles. If a nearly perfect bubble distribution is not maintained throughout the expansion process in the MHD generator, the flow will be unstable, and the efficiency will be badly degraded.

The conductivities of typical gaseous working fluids for closed-cycle MHD systems are shown as a function of temperature in Fig. 16.5.[11] In comparing Fig. 16.5 with Fig. 16.3, it is evident that the inert gases permit operation at a temperature that is lower by around 800°C (1440°F) than is practicable with combustion gases.

All these closed-cycle systems must receive their heat through some sort of heat exchanger, and the temperature limitations on the materials in the tube wall and other elements of the heat exchanger place an upper temperature limit of around 1000°C for systems heated with fossil fuels and possibly 1300°C for systems employing a nuclear heat source. Thus the advantages of the greater electric conductivity of the working fluid for these systems at a given temperature are offset by the much lower operating temperature limit imposed by materials considerations in the heat exchanger at the high-temperature end of the MHD cycle.

Although all these closed-cycle systems have been examined experimentally on a small scale, none has appeared sufficiently attractive to lead to a major effort. In the two-phase flow system, for example, there is serious doubt that a uniform bubble distribution could be obtained in the large sizes required for full-scale systems. All the systems employing liquid metals would have to operate at relatively high temperatures—i.e., around 700°C—and in this region there is no electrically insulating material known that could be exposed to an otherwise suitable liquid metal and used between the electrodes. A thin metal can might be used to line the channel,[10,11] although in full-scale units it would pose severe thermal-stress problems in transients. In addition, the expansion process through the nozzle would yield a two-phase stream moving at high velocity, and this appears likely to cause severe erosion in elements downstream of the MHD working section, the erosion mechanism being similar to that caused by cavitation in pumps and hydraulic turbines (Chap. 3). Even if these materials problems were solved and an MHD generator efficiency (i.e., "turbine" efficiency) of ~ 80 percent could be obtained, basic thermodynamic considerations limit the cycle efficiency obtainable for a given peak temperature to the same level as for closed-cycle turbines. Thus there is no incentive to develop such a system for fossil fuels, although there might be certain advantages for some types of nuclear reactor plant. For these reasons, the closed-cycle systems have received little attention except as a field for basic research, and they will not be treated further in this text.

Brief Review of MHD Experiments

Many different experiments have been run on MHD generators,[1,12-26] but space limitations make it necessary to restrict this section to a review of only the most significant of the experiments, particularly those listed in Table 16.1. The most important considerations are the scale of the experiment, the generator efficiency obtained, the duration of the runs, the problems encountered, and the success of innovations. The really significant tests should obtain not only a good efficiency but also a long life and a high output. W.D. Jackson has proposed a parameter to measure this combination: The *goodness index,* as he terms it, is the product of the percentage of enthalpy extraction (a good indication of the generator efficiency), the electric output in megawatts, and the operating time in hours.[13] This parameter is given in the last line of Table 16.1.

Westinghouse Research

One of the first serious efforts to investigate MHD power generation was a series of experiments carried out at Westinghouse by B. Karlovitz between 1938 and 1946.[3] Ionization in the combustion gas was so poor and hence the electric conductivity so low that it was difficult to detect any sign of electric power generation. The program was dropped for a time, and then the tests were renewed in the latter 1950s with alkali-metal seeding of the combustion gas to give better results. The system still did not seem sufficiently promising to warrant further effort at the time.

Avco

The first MHD effort in the world to yield promising results was carried out at Avco with USAF funding. A nominal objective was to obtain a very lightweight, air-transportable, electric generator unit for use under special conditions where a high fuel consumption and a short life would be acceptable. Initially, Avco found it difficult, as had Westinghouse, to obtain any electric output from the generator section. However, a significant electric output was obtained in 1959, and by 1963 a large unit having a throat roughly 40 × 60 cm for the working section was built and, after extensive shakedown testing, delivered a peak of about 30 MW of electricity for a few seconds during a 1-min-duration run.[15] It was fueled by 52 kg/s (114 lb/s) of ethyl alcohol reacted with liquid oxygen from storage tanks. Note that the power requirements for making low-quality gaseous oxygen at the rate required for the combustor would have been of the order of 200 MW.* If preheated air rather than liquid oxygen had been employed, the power required for the air compressors would have been about 50 MW, still almost double the electric output of the generator. This experiment is enormously significant, in part because it represented a tremendous step forward and in part because no experiment since has yielded a substantially more efficient MHD generator.

The Avco program in the intervening years has been directed mainly at operation of units of modest size that could be built and operated at lower cost in investigations of ways to increase the operating life and the generator efficiency. By 1979 three test runs of 250 h had been obtained,[5,25] but there had been no substantial improvements in the generator efficiency.

Work at Avco in the 1970s has included an effort to use coal as the fuel. One of the experiments employed in this work involved a disk generator powered by a shock tube. An alkali-metal-seeded mixture of nitrogen, carbon dioxide, and hydrogen was used as the working fluid, because when heated by a shock, it produces chemical species essentially the same as those present when coal is burned in air with a mixture deficient in oxygen. The system used is shown in Fig. 16.6. The generator consisted of two insulating walls and a single pair of electrodes to give a more compact geometry than the linear generator discussed previously.[16] The test system is such that the generator can be operated only for very short pulses of about 2 milliseconds (ms). An output of about 8 MW has been obtained

---

* The electric energy required to make low-purity oxygen for basic oxygen steel furnaces is 400 to 430 kWh per ton.

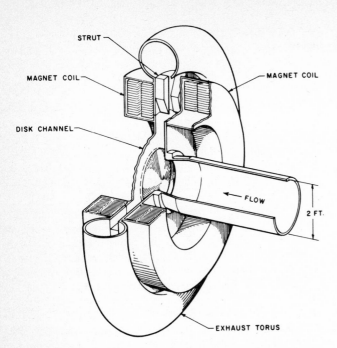

STRUT

MAGNET COIL

MAGNET COIL

DISK CHANNEL

FLOW

2 FT.

EXHAUST TORUS

Figure 16.6 Section through an MHD disk generator showing the positions of the principal elements in the unit. (*Courtesy AVCO Everett Research Laboratory.*)

periments have not mentioned the fact that the generator efficiency has never exceeded a few percent, but instead have focused attention on operation with high-sulfur pulverized coal, for which they report 90 percent recovery of the potassium seed added to the hot combustion gases. The potassium recovery was accomplished by collecting the ash from the flue gas and treating it to recover potassium sulfate.[19]

Research at AEDC (ARO)

The large air-handling facilities and exceptionally large electric power capacity at the Arnold Engineering Development Center (AEDC) near Tullahoma, Tennessee, have made it a natural site for the large-scale MHD experiments conducted there since the mid-1960s. In 1979 this laboratory (operated by the ARO Corporation for the U.S. Air Force) started up a new large MHD generator designed to produce 40 MW utilizing the most advanced MHD design techniques available. The second column from the right in Table 16.1 summarizes the performance reported for this unit in June 1980.[17] Even though the generator inlet throat is large (0.60 × 0.82 m) so that scale effects should not be large in going to a full-scale generator (e.g., 1.45 m × 1.45 m in the ECAS design of Ref. 27), the generator efficiency was only about 16 percent. Thus the electric output obtained fell short of what would be required for the air compression process in an MHD power plant. The electric losses in the boundary layer were minimized in these tests by operating with electrode temperatures as high as practicable—i.e., over 1000°C (1832°F)—and keeping the runs to short durations, such as 10 s.

Research at Universities

In relatively small but often well-directed programs, a substantial amount of basic work has been carried on at universities in the United States. The High Temperature Gas Dynamics Laboratory at Stanford University has produced much significant information on detailed problem areas such as boundary layers on the electrodes,[20] the Hall field limitations, magnetoacoustic instabilities, the effects of electric conductivity inhomogeneities,[21,22] and diagnostic techniques. This work is extremely important as a means toward understanding the large differences between the ideal performance of an MHD generator and the best obtainable in any experimental unit built to date.

A wide variety of work has also been carried out at the Massachusetts Institute of Technology (M.I.T.). This has included work explicitly directed toward the development of MHD generators (particularly the disk generator), together with much of a quite general nature, of interest to but not necessarily directed primarily at MHD development.

with a generator efficiency better than that achieved with the best linear unit mentioned above. However, while there are differences in opinion, some analytical estimates supported by some experiments indicate that, if such a unit were operated continuously, the degree of ionization would be less, and hence the generator efficiency would be lower for steady-state operation.

UTSI Research

The University of Tennessee Space Institute (UTSI), at Tullahoma, Tennessee, has been working on MHD power generation since the 1960s. The initial work at UTSI was carried out with a solid-fuel rocket propellant supplying hot gases to a generator that delivered 8 MW of electric power with a generator efficiency about half that of the Avco unit of 1963. The facility was used for a fairly extensive series of investigations of the diagonal placing of wall electrodes, as indicated in Fig. 16.2c. Most of this work was carried out with liquid propane and liquid oxygen supplied to the combustion chamber. Attention was then turned to the possibility of burning coal or char.[18] The publicity releases in connection with these ex-

Research Abroad

As mentioned earlier, the only major MHD program under way abroad at the time of writing is that in the U.S.S.R. It is interesting to note that the Soviet program might also have been closed out in the early 1960s had not a team of Russian scientists visiting the United States happened to hear from Philip Sporn, then president of the American Electric Power Co., of the successful experiment at Avco in 1963. The Russians were dubious and made a strong effort to see the experiment and obtain some detailed data on it. After numerous and extended high-level telephone calls, arrangements were made for the Russian scientists to visit the U.S. Air Force–sponsored experiment at Avco. They were greatly impressed and this led to a fresh lease on life for the Russian program. The Russians then began the ambitious program that led to the design and construction of the U-25, an MHD generator designed to produce 25 MW and deliver the power to the Moscow grid.[21] The unit was built adjacent to a specially constructed oxygen plant so that the combustion air could be enriched with oxygen to about 40% O_2 by weight to produce the desired combustion gas temperatures. This was necessary because the highest temperature that was expected to be obtainable with the air preheaters—1200°C—was not enough to give the temperature required for the MHD generator.

The U-25 was first operated in 1971 and has been modified repeatedly since then in an effort to increase the generator efficiency and operating life. At the time of writing the output has been increased to as much as 20 MW in some tests (the original design output was 25 MW), a run duration of as much as 250 h has been obtained, and the generator efficiency has been im-

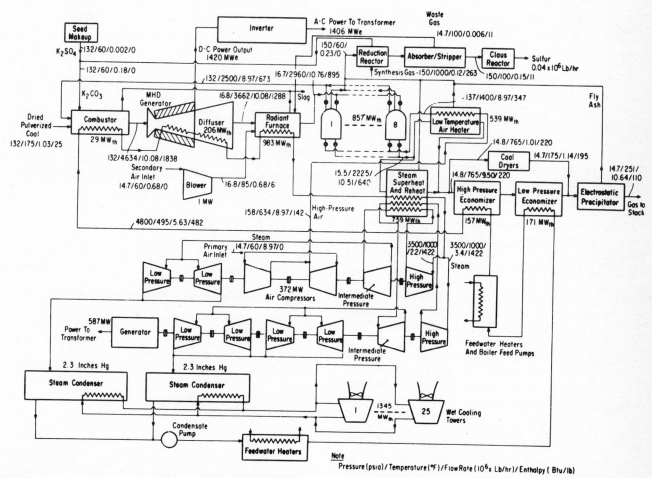

Figure 16.7   Flowsheet for the reference design MHD system used in Task II of the ECAS study.[27]

proved. However, the 20-MW output is insufficient to drive the air compressor and oxygen-manufacturing equipment required to supply the combustion chamber. The U-25 program has supplied the American advocates of MHD programs with potent arguments for increased funding, in this way enabling those in the Soviet MHD program to return a favor to their American counterparts.

A substantial MHD program in Japan and quite modest MHD research efforts in other countries are under way at the time of writing. At the Eighteenth Symposium on the Engineering Aspects of MHD in 1979, there were reports of work in Japan, the Netherlands, Poland, Australia, Israel, and Italy.

CONCEPTUAL DESIGN FOR A FULL-SCALE POWER PLANT

In order to judge the feasibility of obtaining a commercial MHD system, one must first have a reasonably complete conceptual design of a full-scale power plant so that one can compare the problems involved and levels of performance required with the current state of the art. There are many sketchy conceptual designs depicting the MHD generator in a topping cycle from which the hot exhaust serves as the heat source for a conventional steam cycle. The most complete design for a full-scale MHD power plant that the author has been able to find is one prepared by Avco.[27] A flowsheet for that system is presented in Fig. 16.7, and the principal design parameters are listed in Table 16.2. Note that the generator efficiency assumed in the U.S. government policy-making study of Ref. 27 was 80 percent, whereas the best efficiency ever obtained in any MHD experiment has been around 30 percent. Had the 30 percent value been used, the overall thermal efficiency for the complete plant, including the steam cycle, would have been less than 30 percent rather than the 50 percent calculated.

Most of the items in Fig. 16.7 are self-explanatory, but a number of points are not necessarily obvious. The low-temperature air heater employed for regeneratively preheating the combustion air with the products of combustion is a heat exchanger of fairly conventional geometry designed to operate at exceptionally high temperatures: The entering flue gas temperature is 1400°C (2550°F) and heats air to 1051°C (1900°F) for admission into the high-temperature air heaters numbered 1 and 8 in the diagram. These heaters would operate like blast furnace stoves, one of them being heated by flue gas that would enter at about 2000°C (3650°F) while the other would be heating combustion air to about 1370°C (2500°F). Another item not necessarily obvious is that the steam system is closely integrated with the MHD topping cycle because the

TABLE 16.2 Design Data for the ECAS Reference Design Open-Cycle MHD System [27]

Electrical output summary

Total prime cycle (open-cycle) MHD) output (MWe at 60 Hz ac)	1406
Total bottoming cycle (steam) output (MWe at 60 Hz ac)	587
Total gross output (MWe)	1993
Total auxiliary losses (MWe) including transformer losses	61
Net power plant output (MWe at 60 Hz ac and 500 kV)	1932

Thermal energy balance

Energy outputs, MW			
MHD power output		1420	
Combustor/channel/diffuser cooling		235	
Radiant furnace heat transfer		983	
HTAH heat transfer		857	
Secondary furnace heat transfer		1298	
Economizers		328	
Coal dryers		8	
Leaving losses		370	
Coal ash (sensible + latent)	22		
K_2SO_4 (sensible + latent)	10		
Combustion gas (sensible)	183		
Combustion gas (latent)	155		
Energy inputs other than combustion			
Air heating (857 + 539)		1396	
Air compressor power		372	
Coal heating in mills, dryers		8	
Net energy output			3723 MWt
Combustion energy input			3750 MWt
Fuel HHV at 10.788 Btu/lb		3688	
Correction for $SO_x \rightarrow K_2SO_4$ (gas)		17	
Condensation and solidification of K_2SO_4		45	
Excess energy input			27 MWt

Thermodynamic cycle parameters for MHD system

Combustion pressure, atm	9
Combustion temperature, °F	4634
Air preheat temperature, °F	2500
F/A ratio relative to stoichiometric F/A	1.07
Slag rejection	85%
Compressor pressure ratio	10.75
Diffuser outlet pressure, atm	1.14
Electric load parameter	0.80
Potassium seeding	1%

Steam cycle conditions

Steam pressure	3500 psia
Steam temperature, initial	1000 °F
Reheat	1000 °F

MHD channel

$E_{x\,max}$, kV/m	2.7
$J_{y\,max}$, A/m^2	0.74×10^4
$\sigma(UB)^2$, W/m^3	200×10^6
Electric power/area, W/m^2	5.6×10^6
Maximum Hall parameter	4.1
Electric energy/mass flow, J/kg	1.1×10^6
Maximum Hall potential, kV	43

Superconducting magnet

Magnetic field		
Channel inlet	2.496 T	
Peak (near inlet)	5.992 T	
Channel outlet	3.12 T	
Active length	82 ft	(25 m)
Current density (average)	2×10^7 A/m^2	
Ampere turns	50.8×10^6	
Ampere meters	34.2×10^8	
Stored energy	15.2×10^6 kJ	
Warm bore inlet diameter	9.42 ft	(2.87 m)
Warm bore exit diameter	21.3 ft	(6.50 m)
Dewar inlet outer diameter	30.5 ft	(9.3 m)
Dewar outlet outer diameter	44.6 ft	(13.6 m)
Dewar length	101.7 ft	(31 m)

Combustor

Type	Single stage	
Coal pulverization	70% through 200 mesh	
Moisture content dried coal	2%	
Cooling	B.F. water	
Pressure	9 atm	
Temperature	4634°F	
Residence time	50 milliseconds	
Specific heat release rate	8 MW/m^3 atm or 0.75 MBtu/ft^3 atm	
Combustor inner diameter	9.02 ft	(2.75 m)
Combustor active length	27.07 ft	(8.25 m)
Combustor weight	5.07×10^4 lb	(2.3×10^4 kg)

High-temperature heater

Matrix geometry	
Hole diameter	2.0 in
Geometric porosity	40%
Bed length	40 ft
Bed diameter	23.5 ft
Number of heaters	8 { 2 blowdown / 5 reheat / 1 spare
Operating conditions	
Blowdown/reheat cycle time	4/10 min
Combustion gas inlet temperature	2960 °F
Combustion gas outlet temperature	2225 °F
Air inlet temperature	1400 °F
Air outlet temperature	2500 °F
Combustion gas pressure drop	0.07 atm
Air pressure drop	0.026 atm

Pressure drops in the combustion gas system

High-temperature air heater	
Type	Refractory ceramic storage
Gas $\Delta p/p$	0.07
Air $\Delta p/p$	0.026
Radiant furnace	
Gas $\Delta p/p$	0.01
Water Δp, psi	570
Secondary furnace	
Gas $\Delta p/p$	0.03
Steam Δp, psi	854
Air Δp, psi	21
Economizers	
Gas $\Delta p/p$	0.02
Water Δp, psi	21

Figure 16.8 Conceptual design for a full-scale 1406-MWe MHD channel /diffuser.[27]

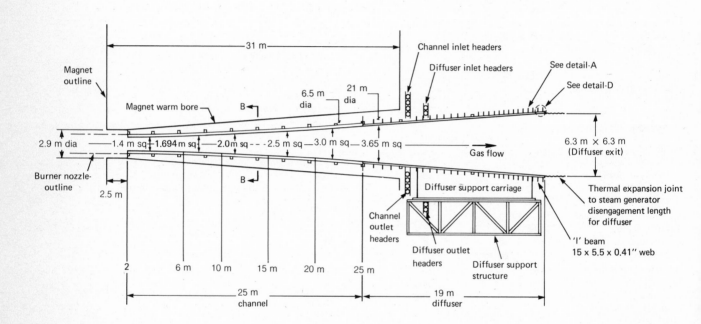

heavy heat loads involved in cooling the walls of the combustion chamber, the MHD generator, and other elements of the hot-gas system represent such a large fraction of the energy throughput that it must be employed to advantage in the steam cycle. This integration, in turn, imposes a very difficult set of problems in balancing the various elements of the steam generator so that there will be, on the one hand, adequate cooling of MHD system parts and, on the other hand, a proper distribution of the heat input to the economizer, boiler, superheater, and reheater of the steam system.

Figure 16.8 shows a section through the MHD generator of the plant of Fig. 16.7 and Table 16.2. The throat at the inlet of the generator is 1.45 m², and the overall length of the generator is about 30 m. The voltage output of the 10 sets of electrodes arranged in series is 40,000 V. As can be seen in Fig. 16.7, this

would be converted to alternating current for delivery to the grid.

MAJOR PROBLEM AREAS

The feasibility of developing an MHD power plant is best appraised by examining each of the major problem areas implicit in the plant flowsheet of Fig. 16.7. In each case the requirements implicit in the conceptual design of Fig. 16.7 and Table 16.2 may be compared with the status of the development work that has been carried out during the past 20 years. Where pertinent, development work from other fields is also cited. Those who prepared the conceptual design of Fig. 16.7 assumed component performance far beyond the levels that had been obtained in experiments. They recognized that such exceptional performance is required in order to obtain an attractive overall efficiency from the power plant, and they felt that somehow such performance goals might be achieved. In fact, most claims for the high performance of MHD power plants have called for still greater extrapolations of test experience than employed in the conceptual design used for reference purposes here.

Generator Efficiency and Enthalpy Extraction

As in any thermodynamic cycle, conversion of the energy obtained from the expansion of hot gases in passing through the MHD generator should approach 100 percent of that available from an isentropic expansion. It is generally recognized that this efficiency cannot be realized for many reasons. There are inevitably losses in the form of heat conduction to the walls, electrical resistance in the plasma, and electrical resistance losses in films on the surfaces of the electrodes and in the electrodes themselves. There is electrical leakage through the insulators between electrodes and short-circuiting of some of the output via fringing currents either through the plasma or through conducting films on the walls both between electrodes and to ground through the structure supporting the generator. There are fringing currents through the plasma both back into the combustion chamber and downstream into the diffuser, boiler tubes, and other components. And finally, there are deviations from ideality in the temperature and velocity distributions in the gas stream flowing through the generator, as well as in the magnetic field through which the hot gas is flowing. Further, turbulence in the hot-gas stream can be expected to lead to some difficulties, and there are possibilities of instabilities and further deviations from ideality under part-load conditions, not to mention compromises in the design that may be required to carry out startup and shutdown operations.

Generator Channel Design

Helpful insights into some of the above reasons for the poor efficiency of actual, as opposed to ideal, MHD generators can be obtained by considering the design of a channel for an MHD generator. Suppose one follows a common approach and decides to keep the magnetic field, gas velocity, channel height, and current density constant throughout the length with the channel diverging in width (i.e., in the direction of the electric current flow) with segmented electrodes connected as in Fig. 16.2b. The channel cross-sectional area ratio (i.e., local area divided by the inlet area) at any given station can then be related to the local pressure by assuming an isentropic expansion, because, for a constant velocity, the cross-sectional area should be directly proportional to the specific volume. Hence, from basic thermodynamics, it follows that

$$\frac{A}{A_0} = \frac{v}{v_0} = \frac{P_0^{1/k}}{P} \qquad (16.4)$$

where A = channel flow passage area, m²
v = gas specific volume, m³/kg
P = gas pressure, Pa
0 = subscript denoting channel inlet

The rate of divergence can be determined from Eq. (16.1). Inasmuch as both the current density and the magnetic field have been chosen to be constant, dP/dx should be constant. This in turn means that, because the absolute pressure drop per unit of channel length is constant, dP/dx must become a progressively larger fraction of the local absolute pressure, and thus the divergence angle must increase along the length of the channel. Note that this is the case for the channel of Fig. 16.8.

There are a number of advantages to choosing a somewhat different set of design parameters than chosen for the simple case examined above; e.g., the channel height might diverge as well as the width. This condition reduces the current density toward the outlet and hence the resistance losses—an important consideration, because the electric conductivity falls off rapidly with temperature. To avoid an increase in the emf generated as the channel width increases toward the outlet, it is possible to reduce the magnetic field. Actually this tends to be what happens, because, as the channel width increases, the magnetic field tends to fall off for a given set of coil windings. The compromises of this sort made for the elegantly designed channel of Ref. 27 and Fig. 16.8 are indicated in Figs. 16.9 and 16.10. Some

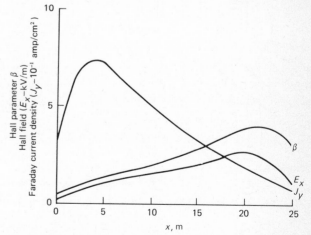

Figure 16.9 Calculated variation of the Hall parameter β, the axial electric field E_x, and the transverse electric current density, J_y (Ref. 27).

additional insights into the intricacies of channel design are given by Fig. 16.11, which shows the electric field and current functions as calculated for the U-25 in Moscow.[21]

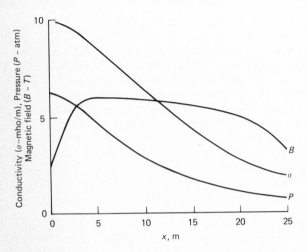

Figure 16.10 Calculated variations in the electric conductivity σ, magnetic flux density B, and pressure P, along the generator length at nominal design conditions.[27]

Enthalpy Extraction

It has been found convenient to express the effectiveness of an MHD generator in terms of the percent enthalpy extraction, i.e., the percentage of the chemical energy supplied to the combustion chamber that is obtained in the form of electricity generated in the MHD channel. In none of the cases of Table 16.1 did the enthalpy extraction in a steady-state experiment exceed 10 percent, yet to make a full-scale plant attractive a value of 47.2 percent was assumed for the reference design of Ref. 27. An appreciation of the seriousness of the problem is given by Fig. 16.12, which gives the enthalpy ideally available from an isentropic expansion from 2550°C (4620°F) for a range of pressure ratios. Also shown is a curve for the enthalpy input required to compress air with a compressor efficiency of 85 percent. If one allows for the pressure drops inherently required for the combustion chamber, ducts, etc., the compressor pressure ratio must be at least 20 percent higher than the expansion ratio across the MHD generator. Thus the enthalpy extraction required just to provide the air for the combustion chamber with no allowance for other losses would have to be at least 20 percent, or about double that obtained in an experiment to date. The importance of this point is commonly obscured by the assumption that steam turbines will be used to drive the air compressors (as in the flowsheet of Fig. 16.7), and so the steam cycle appears to make a less important contribution to the plant output than is really the case.

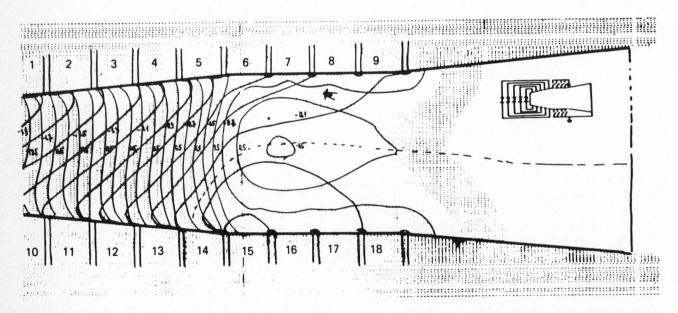

Figure 16.11 Two-dimensional distribution of potential and current functions in the discharge section of the U-25 channel for the optimal potential difference between opposite electrodes.[21]

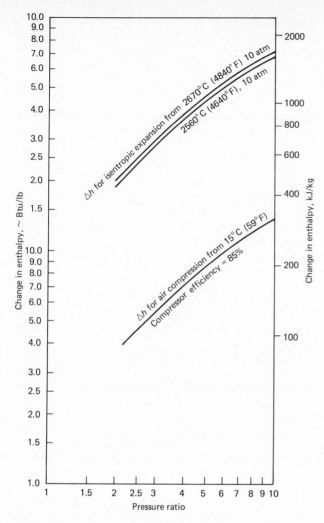

Figure 16.12 Change in enthalpy for ideal isentropic processes as a function of pressure ratio.

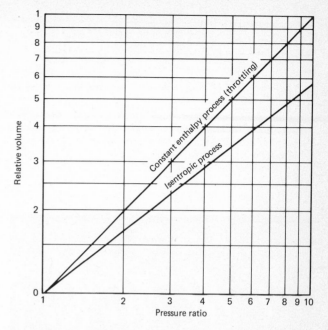

Figure 16.13 Relative specific volume as a function of pressure ratio for expansion of combustion products through an MHD channel from an initial temperature of 2830 K (4634°F).

Deviations from Ideality

The huge gap between the enthalpy extraction ideally obtainable and that actually obtained in MHD channel experiments has been given little attention in the literature, yet a high generator efficiency is vital to the practical utilization of the concept. As in other types of energy conversion systems, it is instructive to clarify the character of the losses that detract from the efficiency by considering the effects of deviations from ideality. Consider first what will happen if, instead of having a uniform velocity distribution across the channel, the velocity is higher than average in the center of the channel. The retarding force on an element of fluid is independent of the velocity and a function only of the current density and the magnetic field; hence there is no stabilizing effect that would change the local value of dP/dx and tend to flatten the velocity distribution. In the high-velocity region the expanding gas stream is not doing the work it should be doing for an isentropic expansion, and the average gas expansion process will be somewhere between an isentropic and a constant enthalpy, or throttling, process. It is roughly analogous to operation of a reciprocating engine with a small hole in the piston head allowing some of the gas to leak past the piston and expand without doing work. The magnitude of the difference in velocity between an isentropic expansion through a divergent channel of a given geometry and a constant enthalpy (or throttling) expansion can be deduced from the curves for these two expansion processes plotted in Fig. 16.13, assuming that the axial static pressure distribution along the channel is the same for both cases. For example, at a point where the pressure has dropped by a factor of 3 from the inlet, the velocity for the throttling process would be higher than the design value for an isentropic expansion by only 28 percent.

Another closely related problem is posed by variations in the temperature distribution across the channel. Note that in gas turbines temperature variations across the flow through the tur-

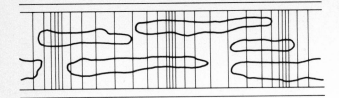

Figure 16.14 Representation of temperature inhomogeneities in the Stanford M-2 MHD facility channel. The disturbances consist of a combination of one-dimensional axial acoustic standing waves and three-dimensional convected conductivity nonuniformities.[23]

Figure 16.15 Series of high-speed photographs of an MHD channel showing luminous streamers or filaments of high current density extending across the channel between electrodes. The successive frames show that the streamers move with the flow from right to left except that the ends tend to concentrate on the upstream edge of the anode and the downstream edge of the cathode. Exposure time = 2.5 μs, electrode pitch = 25 mm, frame-to-frame interval = 2.5 μs, B = 2 T, T = 2000 K (two 4-frame sets).[24] (*Courtesy University of Technology, Eindhoven.*)

bine blades may cover a range of ~ 200°C (~ 360°F). As can be seen from Fig. 16.3, in an MHD generator this would mean a variation in the electric conductivity of nearly a factor of 3. Since the emf across the local region is essentially constant, the local current density through the high-temperature element of the gas stream will be roughly three times that through the low-temperature region, and this will produce three times the retarding force—which will act to reduce the velocity of the high-temperature element of fluid. In view of the extensive experience with the highly turbulent character of the gas flow leaving

high-performance combustors such as those for gas turbines and rocket engines, it is evident that large deviations from ideality in the temperature and velocity distribution in the flow entering an MHD generator must be expected. It is also evident that these deviations will lead to losses in efficiency of the magnitude that has characterized every MHD experiment to date.

Unfortunately, the above problems have rarely been addressed in these terms in the literature. However, several papers presented at the Eighteenth Symposium on the Engineering Aspects of Magnetohydrodynamics in Butte, Montana, in June 1979 show that experimental experience is consistent with the above characterization of this major loss mechanism. First, a paper by R. R. Rankin showed that in combustion MHD generators the operating regime is in the region in which turbulence is only mildly suppressed by MHD effects, and so the flow is still highly turbulent.[28] Secondly, V. D. Semenov, from the U.S.S.R., stated that the temperatures in the U-25 channel at a typical location varied over a range of around 250°C.[26] In discussing this matter with Semenov, the author pointed out that this meant a variation in current flow locally of the order of a factor of 4, which in turn meant that drag forces would vary accordingly and that the velocity distribution would be badly disturbed. Semenov acknowledged this situation but was unable to provide any other data regarding its influence on overall generator efficiency. He agreed, however, that the overall generator efficiency was poor. Two papers from Stanford[22,23] and one from the Netherlands[24] also showed large variations in the local current density by various techniques, including laser diagnostics and some high-speed motion pictures taken at around 2½ μs per frame. One of the Stanford papers[22] included Fig. 16.14, which shows high-temperature regions, or "islands," in the high-velocity gas stream. A paper from Eindhoven in the Netherlands[24] included the remarkable high-speed photographs of Fig. 16.15 that show the generator flow field textured by luminous streamers or filaments of high current density that correspond to the high-temperature "islands" of Fig. 16.14. These islands must inevitably cause substantial variations in the velocity distribution in the MHD working section, and their effect probably constitutes a major factor in the large differences between the ideal and actual efficiencies of MHD generators. It is significant that the Eindhoven test facility was for a closed-cycle MHD system and should have given a more uniform temperature distribution across the gas stream than can be obtained from a combustion chamber.

Temperature and Current Density Variations in the Boundary Layer

The electrodes forming the walls of the MHD generator channel must be cooled, and this in turn means that the gas in the boundary layer must be at a temperature much lower than that in the core of the stream. In fact, the gas temperature in the boundary layer close to the wall is of necessity too low for it to be ionized. Hence the electric current can be conducted through the boundary layer only by local ionization induced by arcing. This situation tends to cause current concentration because, where a local element in the boundary layer is heated by an arc, the conductivity becomes substantial compared to the surrounding gas, and so the current flow tends to concentrate in that local region rather than to distribute uniformly over the face of the electrode. The high-speed photographs of Fig. 16.15 show the discrete arc filaments of the sort one would expect in an MHD channel from the above rationale. This current concentration has the adverse effect of locally overheating the electrode, causing it to erode away. W. D. Jackson has pointed out that the problem is somewhat similar to that in the commutators of dc motors and generators, which operate well only if the current can be kept uniformly distributed over the interface between the carbon brush and the copper commutator segment. A century of experience has shown that the only satisfactory material for brushes is a special class of carbon formulations that inherently give a uniform distribution of microarcs across the brush-commutator interface. No satisfactory metallic alloy has ever been found for this service, and graphite could not withstand the high-temperature oxidizing atmosphere in the MHD channel. A similar situation prevails in breaker points and electric switchgear, in which platinum has been found to give the best service, and thus it is not surprising that platinum has proved to be the best material for the electrodes in MHD generators.

System Dynamic Instabilities

High-velocity turbulent flow is characterized by local fluctuations in flow velocity and direction which provide a mechanism for triggering large-scale and sometimes violent pulsations such as those common in liquid-fuel rocket combustion chambers, where the high heat-release rates act to amplify small perturbations. The situation is rendered much more complex with many more degrees of freedom in an MHD system, because the mass flow of the fluid is inductively coupled to the electric system,[29] and the degree of ionization varies with conditions.[30,31] As a consequence, not only can there be oscillations between energy stored as pressure and kinetic energy in the flow stream, but also there can be oscillations involving these quantities and the electric current flow and the energy stored in the inductance and capacitance of the electric system, as well as the magnetic field. Both analyses and experiments indicate that the amplitude of these fluctuations increases with system size (in part because of the greater scale of turbulence in the hot gas stream) and the

generator efficiency (because of reduced damping by fluidynamic losses).[32,33] In fact, these considerations have led some MHD experts to conclude that a large-scale system would be inherently unstable if a good efficiency could be obtained. It is the author's understanding that this was one of the reasons the British closed out their MHD development program in the 1960s, though the main reason was the government decision to shift from coal to nuclear energy.

Control

In any commercial power plant it is essential to be able to vary the power output while maintaining a reasonably good efficiency. For an MHD generator there are two basic controllable quantities: the fuel flow rate and the airflow rate. If the fuel flow rate is reduced while the airflow rate is held constant, the gas temperature will drop and the efficiency will drop rapidly because of the loss in plasma electric conductivity. If the airflow rate is reduced, the pressure ratio across the MHD channel will be reduced, and this will cause a mismatch between the channel area ratio and the pressure ratio, making an isentropic expansion impossible. Any effort to control by changing both the fuel and airflow rates would give similar effects. Thus there is no way in which an MHD generator can be controlled that will not result in a rapid loss in efficiency as the load is reduced.

Wall Cooling and Thermal Stress

The extremely high gas temperatures make it essential to cool the walls of an MHD generator, and the high gas velocities—particularly in the high-pressure region at the inlet to the channel—yield high heat-transfer coefficients. As a consequence, the heat flux to the cooled wall is inherently very high: of the order of 315 W/cm^2 [$10^6 \text{ Btu}/(\text{h} \cdot \text{ft}^2)$]. This imposes not only difficult heat-transfer problems but also thermal stresses even more severe than those in water-cooled or ceramic blades in high-temperature gas turbines (Chap. 12).

The required heat transfer can be accomplished by using high-velocity, subcooled water (i.e., water pressurized sufficiently so that its boiling point is above the metal-wall-surface temperature) or by using special turbulators as well as high velocities if boiling is to occur. In either case the passage length-diameter ratio must be kept fairly low—of the order of 100—and the pumping power is inherently substantial. The design problems are formidable but appear to be manageable for a short-lived system. The same cannot be said for a long-lived system, which would require a more complex system and a lower plant efficiency than implied by the reference design of Fig. 16.7 and Table 16.2. An idea of the complexity of just the cooling-

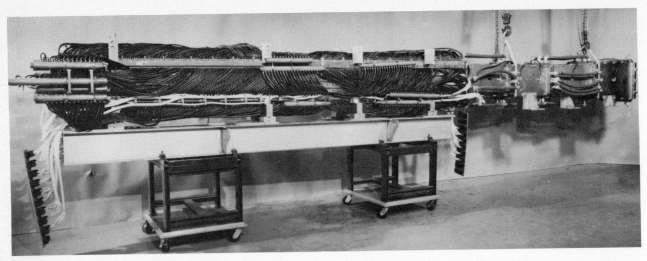

Figure 16.16   Photo of the first MHD test channel built for installation in the Component and Development Integration Facility (CDIF) of the Montana Energy and MHD Research and Development Institute, Inc. (MERDI). The generator is shrouded by the manifolds and rubber hoses of the water-cooling system for the electrodes. (*Courtesy MERDI.*)

water supply pipes required is given by Fig. 16.16. which shows a typical test channel.

Most of the MHD channel experiments have employed low-temperature water-cooled copper walls, which have usually operated without cracking from thermal-strain cycling because of the high ductility of copper and the relatively small number of thermal cycles to which the apparatus was subjected. (The basic elements of thermal-strain cycling effects are treated in the section on thermal-strain cycling in Chap. 6.) Examination of the background for the reference design of Fig. 16.7 shows that the heat flux to the walls must be recovered at a fairly high temperature for use in a high-pressure steam cycle, and so the walls must be made of a high-strength iron-chrome-nickel alloy and be sufficiently thick to withstand a high pressure, e.g., about 270 atm (4000 psi) in the reference design of Table 16.2. Assuming what is probably the best commercial structural material available for this application, Haynes 188, the wall thickness required in a 0.64-in-ID (16-mm-ID) round tube (the smallest diameter employed for a conventional steam plant) would be 2.5 mm (0.10 in). Using this thickness and a heat flux of 315 W/cm^2 [10^6 Btu/(ft^2 · h)], together with Fig. 6.14, one finds that the wall temperature drop would be 460°C (830°F) and thermal-strain cycling would cause the tube to fail in ∼1000 cycles if one neglects both the pressure stresses and the reduction in wall strength resulting from the large temperature drop. These factors must, of course, be included; hence the life to

failure would probably be ∼10 cycles. Increasing the wall thickness would not help because it would increase the temperature drop through the wall and increase the thermal stress. Actual electrode geometries would probably be more complex than a simple round tube and would therefore yield stress concentrations, further aggravating the problem.

Coating the wall with a ceramic has been suggested, but efforts to do this for gas-turbine blades (Chap. 12) have invariably led to spalling of the coating under the thermal stresses inherently involved. An advantage claimed for using coal as the fuel is that a frozen slag layer on the wall would provide some thermal insulation, but it would be effective only up to the melting point of the slag, ∼1000°C (∼1830°F). Inasmuch as the frozen layer would not reduce the heat-transfer coefficient, it could only be expected to reduce the heat flux by reducing the temperature difference between the hot gas and the slag-metal electrode interface, i.e., by perhaps 20 percent if one designs for a metal-slag interface temperature of 650°C (1200°F).

One might consider employing a liquid metal such as sodium as the wall coolant because it would give a very high heat-transfer coefficient and could be operated at a relatively low pressure, minimizing the cooling-passage wall thickness required. However, pumping it across the strong magnetic field would entail an excessive pressure drop and attendant pumping-power requirement. Further, its high electric conductivity would make it necessary to have a separate liquid-metal system for each electrode—a serious complication, especially when one considers the problems of electrically isolating each liquid-metal–steam generator from the rest of the steam system.

A close and realistic scrutiny of the wall cooling and thermal-stress problems indicates that they are so formidable that it is doubtful that it will be possible to obtain both a long life and recovery of the heat losses to the wall at a useful temperature. Although these problems have not been treated in the MHD literature, they are similar to, but more severe than, those of water-cooled gas-turbine blades (which are treated in Chap. 12).

Electrode Corrosion and Erosion

Every effort to operate MHD generators for extended periods has been frustrated by severe damage to the electrodes. The most successful effort up to the time of writing was reported by Avco at the Eighteenth Symposium on Engineering Aspects of MHD.[34] In this program, which was explicitly directed toward obtaining the longest possible life, an MHD channel completed a 500-h endurance test, twice as long as the longest previously obtained. To achieve this, it was necessary to use a very complex electrode construction with the basic water-cooled copper electrode covered by inlays of platinum, tungsten-copper alloy, and a special nickel alloy (Fig. 16.17). Even with this elaborate electrode structure the electrodes were in such poor condition after the 500-h test that they probably could not have run another 100

h (Fig. 16.18). Avco estimated that, if a 6000-h life could be obtained with these electrodes, the cost of the platinum would be only 2 mills per kilowatthour. However, unless the corrosion and erosion rates of the platinum can be reduced by a factor of 10, this estimate implied 20 mills per kilowatthour for a 600-h life for just the platinum with no allowances for downtime, fabrication, etc., Note, too, that in the full-scale MHD generator the erosive effects would be greater, because the voltage would be roughly 10 times that of this Avco 500-h endurance test.

Electrical Insulation

As in the electrode endurance test cited above, electrical insulation has been a source of many difficulties in the relatively small-scale experiments that have been run to date in which voltages have generally been 100 to 1000 V and running times have been a matter of hours. Degradation with running time and thermal cycling clearly occurs, and the higher voltages required for commercial plants (40,000 V for the reference design of Fig. 16.8) would make these problems more acute. A major factor in the problem is the fairly rapid drop in resistivity with

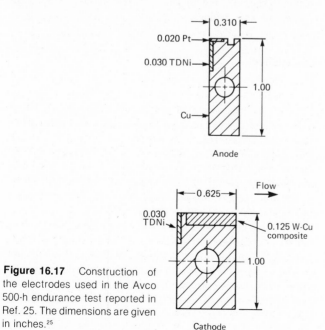

Figure 16.17 Construction of the electrodes used in the Avco 500-h endurance test reported in Ref. 25. The dimensions are given in inches.[25]

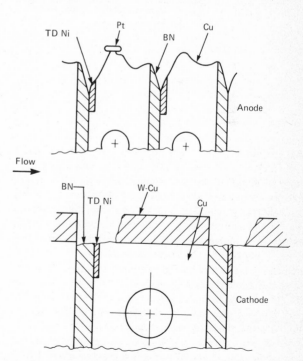

Figure 16.18 Effects of erosion after 500 h of operation of the electrodes of Fig. 16.16.[25]

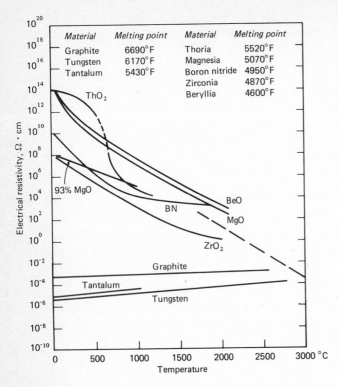

Figure 16.19 Electrical resistivity of several conductors and insulators.[3] *(From Kingery, Norton, Campbell.)*

where it will tend to form a conducting film across the insulators between electrodes and lead to shorting.

Magnet Design

The ~ 5-T magnetic field required for good performance of an MHD generator is roughly five times the field used in conventional generators. It could be obtained with water-cooled copper coils, but the electric power requirements would detract so much from the useful plant output that for a commercial power plant there is widespread agreement that a superconducting magnet will be required. The most promising superconducting material is a niobium-titanium alloy that can be used at 4 K in a field up to ~ 7.5 T. Inasmuch as the local field in the windings will be roughly half again that in the channel, the Nb-Ti material can be used, but it will be operating close to its upper limit. One of the major accomplishments of the MHD program has been the demonstration of the feasibility of building a 5-T magnet using Nb-Ti as the superconducting material. This magnet was designed and built in the United States and shipped to the U.S.S.R. for testing in the U-25. Results of tests reported in 1979 indicated that the magnet has functioned satisfactorily.[26]

Although the superconducting magnet requires virtually no electric power input to the coils, it must be cooled to 4 K with a cryogenic refrigeration system that requires ~500 kWe for each 1 kWt of heat removed. Thus much attention must be given to insulating the coils from the hot generator channel, and space is required for this insulation (Fig. 16.8).

increasing temperature for any of the ceramic insulators that might be used. As shown in Fig. 16.19, even for MgO, the best of the commercial insulating materials, the resistivity drops to ~10^8 ohm · cm (Ω · cm) at 800°C (1470°F), and so an applied voltage of 40,000 V across a 1-mm-thick layer of insulation would lead to a nominal leakage of 0.004 A/cm². The consequent heat release would be ~ 160 W/cm², and this would lead to rapid local overheating and complete breakdown of the insulation. Thus, above some threshold temperature between 650 and 870°C (1200 and 1600°F), the insulation gives an unstable condition—a major design limitation and a frequent source of trouble in MHD test work. If coal is used as the fuel, the ash attacks ceramic insulators, greatly increasing the problems. The rate of attack is increased as much as tenfold by the electrical potential across the insulating material.[34,35]

A closely related problem is that even with methane as the fuel to avoid corrosive material in the hot gas stream, the K_2CO_3 vapor that serves as the seed will condense on the cold walls,

Scale Effects

Development programs for power conversion equipment normally depend heavily on scaling parameters that serve to relate the performance of small-scale experimental units to full-scale plants. It is frequently claimed that large-scale MHD generators will be much more efficient than the smaller test units. Certainly one must expect some reductions in the percentage of the heat lost to the walls, electrical resistance losses in the lower temperature boundary layers in the flow over the electrodes, etc., but the increase in scale from the Avco Mark V or the Russian U-25 does not seem likely to produce much of a reduction in losses. Some nice theoretical work on scaling has been carried out,[36] but the author has searched the literature repeatedly and talked with leading authorities in the field without finding any published paper or report on a method for scaling that has been validated by use in correlating the extensive experimental data available. Such work is essential if one is to have a sound basis for predicting the performance of a full-scale generator. It appears that the reason for the absence of a validated scaling method is that, as indicated in the section ''Deviations from Ideality,'' there have been major energy losses in the experiments that must be accounted for properly, and probably must be drastically reduced or eliminated before scaling relations can be validated by their successful use to correlate experimental data.

Combustion Gas System

Although the MHD generator itself presents the most difficult and challenging problems, the other components in the hot-gas system also present truly formidable problems that are discussed in this section.

Combustor and Ash Problems

There seems to be fairly general agreement that fuel cost considerations will probably make it necessary to employ coal as the fuel in any MHD central station plant; hence it is necessary to cope with the difficulties posed by the ash in the fuel. At first thought one might expect that most of the ash could be removed by a cyclone separator in the combustor, and this approach is being followed.[37,38,39] However, most of the ash would be vaporized at combustor outlet temperatures.[35] Because of the high combustion air preheat temperature required, even burning the coal in two stages leads to vaporization of much of the ash in the first stage. Specifically, if all the carbon is to be burned so that there will not be a large loss of carbon with the ash, all the carbon and hydrogen must be reacted to give CO and H_2O, and this yields a hot-gas temperature of ~2000°C (3700°F), leaving the first-stage combustor.[37,38] Not only will this temperature vaporize all the alkali oxides, chlorides, and sulfates (Fig. A6.4), but also most of the MgO, 50 percent of the SiO_2 and Fe_2O_3, and even ~12 percent of the CaO will be vaporized (Fig. 16.20). These vaporized components of the ash will deposit on cooled surfaces both in the combustion chamber and downstream—most of which will be cooled by some element of the steam-generation system. There will, of course, be a sort of fractionating tower effect in that the lower-vapor-pressure materials such as CaO, SiO_2, and Fe_2O_3 will tend to deposit to a greater degree in the first portion of the hot gas passage and to a lesser degree later, but the cooled walls will be at a sufficiently low temperature that one must expect substantial amounts of all the vaporized compounds of the ash to be present in the slag even at the MHD channel inlet. This increases the corrosivity of the slag.

The high heat-transfer coefficients in the high-velocity regions in the combustion chamber and generator channel will give such high heat fluxes there (as mentioned earlier in the section on thermal stresses) that the temperature gradient through the solid slag layer will be so steep that it cannot have a thickness of more than ~0.040 in (1 mm), and the high shear forces from the gas stream will drive the liquid film of slag along the walls. However, in the lower velocity zones, such as in steam-generator tube bundles (Fig. 16.7), one must expect a building up of massive deposits of slag if even a small percentage of the ash enters the generator channel. These slag deposits on the boiler tubes will not only spoil the heat transfer but will also block the gas flow passages. Clearly, if a workable system is to be obtained, the vaporized ash must be condensed and removed before low-velocity hot gas encounters heat-transfer surfaces that are important elements of the steam generator.

The design of the combustor itself is greatly complicated by the ash problem so that, although the combustion chambers developed for liquid-fuel rocket engines can be used as the point of departure for initiating development work, major changes must be expected. Further, the life of rocket engines is normally a matter of minutes with relatively few power cycles, whereas a combustor for a central station must have a life of at least tens of thousands of hours and should withstand ~3000 power cycles (Fig. 7.4). The combustion chamber walls certainly must be water-cooled; the heat fluxes and thermal stresses in the metal walls will depend in part on the thickness of the solid and liquid layers of slag that will form. These thicknesses will depend on the convective and radiant heat fluxes from the flames, the

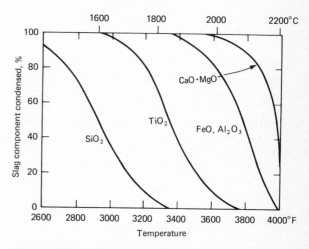

Figure 16.20   Volatility of slag constituents at 1 atm from the combustion of Pittsburgh seam HVAB coal with 10 percent excess air and 2 g-moles K_2CO_3/kg coal.[38]

operating conditions, and the coal composition—which can be expected to vary substantially from time to time even when the fuel is coal from the same seam.

The degree to which the vaporized ash will tend to condense on water-cooled walls is indicated by a test of a half-scale first-stage experimental combustion chamber in which an intense vortex was induced to centrifuge out liquid and solid particles.[39] Surprisingly, it proved possible to remove as much as 95 percent of the coal ash as liquid slag with the vortex separator in the combustion chamber in spite of the fact that the temperatures leaving this first-stage combustor were on the order of 1980°C (3600°F). However, the walls of the cyclone combustion chamber were cooled intensively with water passages, and, as a consequence, they maintained quite a low temperature. Thus vaporized slag condensed in the relatively low-temperature liquid stream collected on the walls and flowed out through the ports provided. The slag temperature probably was about 1340°C (2450°F) at the free liquid surface, and, with intense swirl in the cyclone combustion chamber, there was so much turbulence that the bulk of the ash and vapor formed came in contact with the relatively cool liquid-slag surface, where it condensed out. This approach, while effective in removing most of the ash, also led to a large heat loss: 18 to 25 percent of the heat throughput.[39] In scaling up from the half-scale test unit to a full-scale combustion chamber, the author of Ref. 39 estimated that the heat losses would be reduced to about 13 percent. This implies that the slag removal efficiency would also drop—probably permitting something of the order of 7 to 10 percent of the ash to pass on through in vapor form.

Diffuser

The kinetic energy in the high-velocity gas stream leaving the MHD generator channel must be recovered to the extent possible as a means of reducing the air compressor power requirement. Even with a good velocity distribution at the diffuser inlet and with smooth walls, extensive experience indicates that it will probably not be possible to recover more than ~45 percent of the kinetic energy as static pressure.[40] Note that the reference design of Table 16.2 assumed a 90 percent diffuser efficiency.

Air Preheater

The reference design of Fig. 16.7 and Table 16.2 assumes that the hot gases leaving the diffuser at ~1925°C (3500°F) will be employed to preheat the combustion air to ~1500°C (2730°F) by means of equipment similar to the stoves used for preheating the air for blast furnaces. The severity of the materials problems involved is indicated by the fact that blast furnace stoves ordinarily preheat the air to only 980°C (1500°F), about 1000°F less than required for an MHD system. The highest air preheat temperature obtained with blast furnace stoves to the author's knowledge has been 1300°C (2370°F) in tests in Germany. This was achieved just in experimental runs and was obtainable only by heating with clean hot gas from a natural-gas burner, because otherwise the refractory material for packing the stoves deteriorated too rapidly. Efforts to develop refractory materials compatible with the products of combustion of coal have shown that both MgO and Al_2O_3 deteriorate rapidly at temperatures above ~1300°C (2370°F). Spinels of these materials show somewhat better resistance to attack.[41] The problems are complicated by the fact that coal ash may be either acid or basic; refractories not very susceptible to one are usually rapidly attacked by the other.

Tests on slagging effects at Fluidyne indicate that beds of spheres tend to clog rapidly because of slag accumulation by impingement where the gas flow changes direction, but that perforated blocks or plates stacked to give long, straight, vertical passages seem to clear themselves by slag drainage to the bottom under both the shear forces of the high-velocity hot-gas flow and gravity.[41]

Steam Generator

The slagging and corrosion problems in the steam generator can be expected to be similar to but worse than those in conventional steam generators (Chap. 11).

Seed Recovery

The cost of the seed (usually K_2CO_3) for a commercial MHD system will be excessive unless at least 90 percent can be recovered. Most of the potassium can be expected to come out as K_2SO_4 in the slag and ash removed by the cyclone separators or deposited as solids on surfaces at lower temperature throughout the hot-gas system. The K_2SO_4 will have to be leached out of the ash and slag and converted back to K_2CO_3 for recirculation through the system.

SUMMARY

In principle, the MHD topping cycle offers the highest peak cycle temperature and the highest thermodynamic cycle efficiency of any system that has been proposed, yet none of the many MHD generators tested up to 1980 has yielded a high enough efficiency so that it would have supplied even half the power required to provide the air (or oxygen) fed to the combustor. The most vital problem seems to be that the flow through the MHD channel is inherently very sensitive to deviations from ideality in the temperature distribution, and this leads to large variations in the local electric current density which must affect the fluid velocity distribution. A poor velocity distribution will cause the expansion to be more nearly a constant enthalpy (or throttling) process than an ideal isentropic process; hence the efficiency is low. The problem is subtle and the losses involved are inherently much more serious than any losses in the expansion process for a reasonably well-designed conventional gas turbine. (Even the world's first gas turbine to produce a net output—in 1904—had a far better fluid dynamic and thermodynamic efficiency, and the character of the losses was better delineated.)

In addition to the above problem with losses in the basic MHD process, developmental efforts have been handicapped by such serious materials problems associated with the high temperatures required that the total running time obtained in all the MHD generator tests performed in the United States up to 1980 (at a cost of over 200 million dollars) has been less than 2000 h. Severe erosion, corrosion, and thermal stresses in the electrodes and insulators have limited the longest test to 500 h. For utility service, coal will have to be used as the fuel, and coal ash will aggravate the corrosion problems, particularly in the heat-transfer matrix for the regenerative heater, and will present serious slagging problems by fouling surfaces in the steam generator. Recovery of the bulk of the potassium (or cesium) seed from the ash at an acceptable cost presents another set of difficult materials problems.

In view of the inherent difficulty of the above basic problems as testified by the slow progress toward their solution during the 1960–1980 period, the prospects for obtaining a commercially viable MHD system before the year 2000 must be judged to be poor.

PROBLEMS

1. Consider an MHD generator channel with an inlet 1×1 m and a pressure ratio of 8 with an inlet temperature of 2550°C (4620°F). Assume that the combustion chamber is fed a stoichiometric mixture of kerosene and air, and that the pressure losses in the combustion air system other than the pressure drop through the generator amount to 20 percent of the compressor pressure ratio. Using Fig. 16.12, estimate the ideal enthalpy drop for an isentropic expansion through the generator and the percentage of enthalpy extraction ideally obtainable. Also estimate the compressor energy input, assuming a compressor efficiency of 85 percent.

2. Assume a gas velocity through the above generator region of 600 m/s, a current density of 1 A/cm², and a magnetic field of 5 T. Estimate the ideal voltage generated, the pressure drop in the first 1-m-long section, and the flow passage area at the 1-m station associated with the isentropic expansion ideally required to do the amount of work involved.

3. Estimate the voltage and power losses stemming from the resistance of the plasma seeded with 1% K_2CO_3 for the con-

ditions of Probs. 1 and 2. Repeat assuming that the gas temperature is reduced by 200°C (360°F).

4. Assume that as a consequence of inhomogeneities in temperature and other conditions that one-third of the total weight flow of the gas stream of Prob. 1 flows through the working section at 80 percent of the design velocity and one-third flows at the design velocity. As a first approximation, estimate the velocity of the remaining one-third, neglecting the fact that the temperatures of the three fractions would differ because their expansions would not be isentropic. Compare the volumetric flow rate at the generator outlet for a pure isentropic expansion with that for a pure throttling process.

5. Estimate the kinetic energy in the gas stream leaving the generator of Prob. 1. Assuming that the design value for the velocity is achieved (i.e., 600 m/s) and that there is a 45 percent pressure recovery in the diffuser, what would be the consequent energy loss per unit of mass flow? Compare these results with the values obtained in Prob. 1 for the ideal enthalpy extraction and the compressor energy input.

REFERENCES

1. Energy Research and Development Administration, *Magnetohydrodynamics, Power Generation and Theory—A Bibliography,* TID-3356, November 1975.

2. Pai, Shih-I: *Magnetogasdynamics and Plasma Dynamics,* Springer-Verlag, Vienna, 1962.

3. Way S., et al.: "Experiments with MHD Power Generation," *J. Eng. Power, Trans. ASME,* vol. 83, 1961, p. 397.

4. Rosa, R. J., *Magnetohydrodynamics Energy Conversion,* McGraw-Hill Book Company, New York, 1968.

5. Levi, E.: "MHD's Target: Payoff by 2000," *IEEE Spectrum,* May 1978, p. 46.

6. Harris, L.P., and J.D. Cobine: "The Significance of the Hall Effect for Three MHD Generator Configurations," *J. Eng. Power, Trans. ASME,* vol. 83, 1961, p. 392.

7. Gilbey, D. M.: "The MPD generation of D.C. Power," *I.E.E. Conference Report,* Series 4, Institute of Electrical Engineers, London, 1962.

8. *Proceedings of the Fifth International Conference on MHD Electrical Power Generation, Munich,* Apr. 19–23, 1971.

9. Elliott, D. G.: "Magnetohydrodynamic Power Systems," *Journal of Spacecraft and Rockets,* AIAA, vol. 4, no. 7, 1966, p. 842.

10. Petrick, M., and K. Y. Lee: "Performance Characteristics of a Liquid Metal MHD Generator," *Proceedings of Symposium on MHD Power Generation,* vol. 2, Paris, 1964, p. 953.

11. Pierson, E. S., et al.: "Open-Cycle Coal-Fired Liquid-Metal MHD," *Proceedings of the Eighteenth Symposium on Engineering Aspects of Magnetohydrodynamics,* June 18–20, 1979, pp. D-2.3.1 to D-2.3.8.

12. Jackson, W. D., et al.: "A Critique of MHD Power Generation," *J. Eng. Power, Trans. ASME,* vol. 92, 1970, p. 217.

13. Jackson, W. D., and P. S. Zygielbaum: "Open-Cycle MHD Power Generation: Status and Engineering Development Approach," *Proceedings of the American Power Conference, Chicago, Ill.,* Apr. 21–23, 1975, pp. 1058–1071.

14. Young, F. J.: *Summary of the Research and Development Effort on Open-Cycle MHD Systems,* Oak Ridge National Laboratory Report No. TM-6257 (to be published).

15. Sporn, P., and A. Kantrowitz: "Large Scale Generation of Electric Power by Application of the MHD Concept," *Power,* vol. 103, 1959.

16. Rosa, R. et al.: *Recent MHD Generator Testing at Avco Everett Research Laboratory, Inc.,* paper presented at ASME Winter Annual Meeting, New York, Nov. 17–22, 1974, ASME Preprint 74-WA/Ener-7.

17. Starr, R. F. et al.: "Description, Performance, and Preliminary Faraday Power Production Results of the HPDE Facility," *Seventh International Conference on MHD Power Generation,* June 1980.

18. Dicks, J. B., et al.: "Direct Coal-Fired MHD Power Generation," *J. Eng. Power, Trans. ASME,* vol. 96, 1974, p. 153.

19. Templemeyer, K., et al.: "Investigations of Factors Influencing Potassium Seed Recovery in a Direct Coal-Fired MHD Generator System," *Sixteenth Symposium on Engineering Aspects of MHD, Pittsburgh, Pa.,* May 16–18, 1977, p. IX.6.34.

20. Rankin, R. R., S. A. Self, and R. H. Eustis: "A Study of the Insulating Wall Boundary Layer," *Proceedings of the Sixteenth Symposium on Engineering Aspects of MHD, Pittsburgh, Pa.,* May 16–18, 1977, p. VI.3.13.

21. Sheindlin, A. E.: "The U-25 MHD Pilot Plant," Institute of High Temperatures, U.S.S.R. Academy of Science, Moscow, September 1974.

22. Barton, J. P., et al.: "Fluctuations in Combustion MHD Generator Systems," *Proceedings of the Eighteenth Symposium on Engineering Aspects of Magnetohydrodynamics,* June 18–20, 1979, pp. J.1.1–J.1.8.

23. Kowalik, R. M., and C. H. Kruger: "Experiments Concerning Inhomogeneities in Combustion MHD Generators," *Proceedings of the Eighteenth Symposium on Engineering Aspects of Magnetohydrodynamics,* June 18–20, 1979, pp. J.3.1.–J.3.6.

24. Hellebrekers, W. M., et al.: "Experiments in the Nonuniform Discharge Structure in Noble Gas MHD Generators," *Proceedings of the Eighteenth Symposium on Engineering Aspects of Magnetohydrodynamics,* June 18–20, 1979, pp. D-2.4.1.–D-2.4.7.

25. Demirjian, A., et al.: "Long Duration Channel Development and Testing," *Proceedings of the Eighteenth Symposium on the Engineering Aspects of Magnetohydrodynamics,* June 18–20, 1979, pp. A.3.1.–A.3.11.

26. Iserov, A. D., et al.: "Study of the U-25B MHD Generator System in Strong Electric and Magnetic Fields," *Proceedings of the Eighteenth Symposium on Engineering Aspects of Magnetohydrodynamics,* June 18–20, 1979, pp. A.5.1–A.5.14.

27. Harris, L. P.., and R. P. Shah: *Energy Conversion Alternatives Study (ECAS), General Electric Phase II Final Report,* vol. II: *Advanced Energy Conversion Systems—Conceptual Designs,* part 3: *Open-Cycle Gas Turbines and Open-Cycle MHD,* NASA-Cr 13949, General Electric Company, December 1976.

28. Rankin, R. R.: "Turbulence Damping in MHD Generators," *Proceedings of the Eighteenth Symposium on Engineering Aspects of Magnetohydrodynamics,* June 18–20, 1979, pp. E.2.1–E.2.4.

29. Hughes, W. F., and F. J. Young: *The Electromagnetodynamics of Fluids,* John Wiley & Sons, Inc., New York, 1965.

30. Velikov, B., and A. Dykhue: "Plasma Turbulence Due to Ionization Instability in a Strong Magnetic Field," *Proceedings of the Sixth International Symposium on Ionization Phenomena in Gases, Paris,* 1963.

31. Barton, J. P., J. K. Koester, and M. Mitchner: "An Experimental Investigation of Fluctuation Properties within a Combustion MHD Generator," *Proceedings of the Sixteenth Symposium on Engineering Aspects of MHD, Pittsburgh, Pa.,* May 16–18, 1977, p. VII.5.27.

32. Wu, Y. C. L., and G. Rajogopol: "Three Dimensional Current Distribution in Diagonal Conducting Wall Channels," *Proceedings of the Sixteenth Symposium on Engineering Aspects of MHD, Pittsburgh, Pa.,* May 16–18, 1977, p. VII.2.8.

33. Biturin, V. A., et al.: "A Consideration of Some Three-Dimensional Effects in MHD Channel," *Proceedings of the Sixteenth Symposium on Engineering Aspects of MHD,* May 16–18, 1977, pp. 34–37.

34. Korenaga, S., et al.: "Deterioration of Insulating Wall Materials Near Electrodes," *Proceedings of the Sixteenth Symposium on the Energy Aspects of MHD, Pittsburgh, Pa.,* May 16–18, 1977, p. IV.3.18.

35. Bowen, H. K., et al.: "Chemical Stability and Degradation of MHD Electrodes," *Proceedings of the Fourteenth Symposium on the Engineering Aspects of MHD, UTSI, Tullahoma, Tenn.,* April 8–10, 1974, p. IV.1.1.

36. Demetriades, S. T., G. S. Argyropoulos, and C. D. Maxwell: "Progress in Analytical Modeling of MHD Power Generators," *Proceedings of the Twelfth Symposium on Engineering Aspects of Magnetohydrodynamics, Argonne National Laboratory, Argonne, Ill.,* Mar. 27, 1972, p. I.5.1.

37. Thompson-Ramo-Wooldridge Defense and Space Systems Group, *TRW 25 MW(t) Staged MHD Coal Combustor Conceptual Design Study,* ERDA Contract No. W-31-109-Eng-38, TID-27145, June 4, 1976.

38. Rockwell International, Rocketdyne Division: *MHD Combustor Design Study,* ERDA Contract No. W-31-109-Eng-38, TID-27144, May 28, 1976.

39. Demski, R. J., et al.: "Development of a Slagging Cyclone Gasifier for MHD Applications," *Proceedings of the Eighteenth Symposium on Engineering Aspects of Magnetohydrodynamics,* June 18–20, 1979, pp. K.6.1–K.6.7.

40. Doss, E. D.: "Subsonic MHD-Diffuser Performance with High Blockage," *Proceedings of the Sixteenth Symposium on Engineering Aspects of MHD, Pittsburgh, Pa.,* May 16–17, 1977, p. IX.3.18.

41. Smyth, R. R., et al.: "Progress in Testing of Refractories for MHD Air Heater Applications," *Proceedings of the Sixteenth Symposium on the Engineering Aspects of MHD, Pittsburgh, Pa.,* May 16–18, 1977, p. IV.7.40.

42. Capps, W.: "Some Properties of Coal Slags of Importance to MHD," *Proceedings of the Sixteenth Symposium on Engineering Aspects of MHD, Pittsburgh, Pa.,* May 16–18, 1977, p. VIII.2.21.

17

COGENERATION
AND BOTTOMING
CYCLES

Inasmuch as all the systems used commercially for converting heat into electricity reject from 1.5 to 5 times as much energy to the environment in the form of heat as they yield in the form of electric energy, there are clearly incentives for utilizing this waste heat. As discussed in Chap. 13, steam cycles are fairly widely used to extract useful energy from the exhaust of gas turbines. As will be discussed in the latter part of this chapter, steam or other Rankine cycles can be used to recover the heat rejected from diesel engines or from chemical processes where the heat is usually available at substantially lower temperatures than the heat from gas turbines. However, a greater potential for conserving energy is through cogeneration, i.e., the design and operation of a power plant in such a way that it meets the needs for both heat and electricity in an industrial plant, a building complex, or an urban area. Cogeneration is widely used by industry in paper mills, chemical processing plants, etc., and for institutional heating systems; about 10,000 MW of electric power (~ 15 percent of U.S. industrial electricity consumption) was produced by U.S. industry in 1976 from cogeneration power plants. The concept of comprehensive urban area cogeneration offers the possibility of huge energy savings, but it is not widely used because it involves not only a formidable capital investment and a time span of decades for implementation, but also even more formidable institutional, legal, and political barriers. Thus this chapter begins with a brief look at cogeneration plants tailored to meet the needs of specific industrial or institutional applications and then turns to a comprehensive survey of the diverse problems of integrated urban energy systems in order to provide perspective on the possibilities, the technical requirements, and the institutional problems. The last section of the chapter surveys the possibilities of using working fluids other than water in Rankine bottoming cycles for special applications that are often alternatives to cogeneration.

COGENERATION PLANTS FOR INDUSTRIAL PROCESSES AND INSTITUTIONS

As soon as electric power began to come into use, it was evident that an industry that required both electricity and low-pressure steam for process heat could operate its boilers at a pressure substantially higher than the pressure at which steam was used for the process, and the steam could be expanded through an engine placed between the boiler and the process equipment. The incremental capital and fuel costs were small; hence the resulting cost of the electricity produced was low. As improvements in steam power plants made possible progressively higher boiler temperatures, the range of possibilities for cogeneration was broadened and the economic potential improved. A flowsheet for a typical modern plant is shown in Fig. 17.1 for an industry that requires steam at two pressure levels. In this case, 56 percent of the chemical energy in the fuel is converted into either electricity or process heat, and, in some plants, the fraction may be as high as 90 percent.

Cogeneration plants for institutions such as hospitals, universities, prisons, etc., are similar to those for industry except that most of the steam is used for building heating and hot water and so there is no large bleed-off of steam at moderate pressures, and the turbine discharges at a relatively low pressure—usually only a little above atmospheric pressure. This arrangement increases the fraction of the energy in the fuel converted into electricity and usually makes possible a somewhat higher overall thermal efficiency at the design point. However, the wide variations in the heat load with the weather lead to extensive periods of part-load operation in which there is a poor match between the requirements for heat and those for electricity. As a consequence, much or all of the electric power in mild weather is commonly purchased from a utility.

COGENERATION FOR URBAN AREAS

The extent to which cogeneration of heat and electricity can be employed depends on the degree of urbanization in a nation; hence a survey of this factor is in order. Urbanization, one of the most important trends of our century, has proceeded concurrently with industrialization.[1] Studies indicate that a city ordinarily has the full complement of urban advantages and problems when its population exceeds 100,000. One of the best indices of the degree of urbanization of a nation is the fraction of the population dwelling in cities of over 100,000; a nation may be classed as urban when this fraction exceeds 50 percent. Amazingly, by that definition no nation in the world could be classed as urban until England reached that point in 1850. Other nations passed that point much later, for example, the United States in 1915, Japan in 1955, and the U.S.S.R. in 1965. As a point of interest, only 2 percent of the population of Europe lived in cities over 100,000 in the year 1800, although the fraction in England was up to 10 percent. The remarkable thing is that, as time has gone on, the period of time required for the population distribution to shift from 10 percent urban to over 50 percent urban has gone from 79 years for England, 66 years for the United States, 48 years for Germany, 36 years for Japan, to 26 years for Australia. In the United States and Europe the movement to cities appears to have leveled off, with ~75 percent of the population in urban areas having populations of 100,000 or more. The accelerated rate of industrialization in underdeveloped countries will probably lead to even more rapid rates in their urbanization in the future. It is estimated that by the year 2000 half the population of the world will be urban, thus providing a huge market for cogeneration in central station plants.

As discussed in Chap. 9, the concentration of people in our large urban complexes has led to distressing air and water contamination problems, while thermal pollution of our streams, lakes, and estuaries has become a serious matter. Thus the cogeneration of heat and electricity for urban areas offers the possibility of reducing not only energy resource consumption but also environmental pollution. Inasmuch as over 75 percent of the population in developed countries lives in urban areas suited to comprehensive cogeneration systems, the potential application is enormous.

Total Energy Requirements

If one takes a comprehensive look at all these problems and considers long-term measures to alleviate them, a first logical step is to examine the total energy requirements of an urban complex. Cities currently require about twice as much energy for heating buildings as for generating electricity. Since World War II, air conditioning has come to represent an additional large energy requirement. Water shortages have begun to develop in some areas leading to a consideration of seawater distillation, which would require large amounts of heat energy. Although heretofore our attempts to meet these various requirements for energy have led to ad hoc solutions for each specific requirement, it has become evident that there is a strong incentive to obtain a single integrated public utility complex that would be a major element in the master plan for the city as a whole.

A point of departure is provided by the estimate of the United States energy requirement for the year 2000 presented in Table 17.1. Note that about 80 percent of the energy required by residences and over half of the energy employed by commercial operations is for heating buildings. The bulk of the energy used for transportation in 1979 was supplied by petroleum products, and it was assumed in preparing Table 17.1 that this would still be the case in 2000. Pressure for reductions in atmospheric pollution continues to build up; hence it may be that by the year 2000 a substantial fraction of automotive energy requirements might be met by using new types of electric storage battery (Chap. 20). At the time of writing, over 10 percent of the energy used in industrial processes was in the form of electricity; the fraction had more than doubled in the 1960–1980 period. Note, too, that about one-third of the industrial process heat is employed at relatively low temperatures so that it could be supplied by extracting steam from a turbine at a temperature of 204°C (400°F) or less. In some industries (e.g., chemical) about 90 percent of the heat required is in this temperature range. For cities as a whole the percentage is large although it differs substantially from one city to another, depending on the local industrial operations. The balance of the heat required by industry is largely for high-temperature metallurgical or ceramics work. In any event, if one projects the current use of electricity to the year 2000, it appears that electric energy consumption may total about 25.6×10^{15} Btu/year, or about 15 percent of the total energy requirements of the nation at that time.

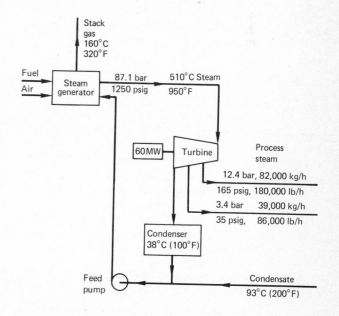

Figure 17.1 Typical cogeneration system for an industrial plant.

TABLE 17.1 A Typical Upper Limit Estimate of U.S. Energy Requirements in the Year 2000

Application	Electricity, 10^{15} Btu/year*	Heat, 10^{15} Btu/year	Total, 10^{15} Btu/year
Residential	5.0	19.5	24.5
Commercial	12.0	15.5	27.5
Industrial			
Food, paper, chemicals ($T \leq 400\,°F$)	2.7	19	21.7
Steel	1.3	12	13.3
Ceramics	0.6	5	5.6
Miscellaneous	4.0	21	25
Subtotal	8.6	57	65.6
Transportation		48	48
Total	25.6	140	165.6

*This column gives the electric energy; the thermal energy input to the plants generating electricity will be several times as great.

TABLE 17.2 Typical District Heat Consumption in 1952 (Ref. 2)

Location	Area served, mi²	Steam distributed, 10^6 lb/(year·mi²)
Baltimore	0.45	2,741,998
Boston	0.98	2,534,854
Chicago	0.20	3,361,990
Cleveland	0.88	3,239,226
Dayton	0.51	3,821,153
Detroit	1.58	2,553,496
Indianapolis	1.88	2,182,870
Milwaukee	0.87	2,190,074
New York	6.31	3,058,501
Philadelphia	1.13	1,883,422
Pittsburg	0.33	5,295,891
Rochester	0.86	3,070,672
St. Louis	0.82	2,253,461
AVERAGE	1.29	2,935,970

Heat for Buildings and Low-Temperature Industrial Processes

It is evident from Table 17.1 that heat for buildings and low-temperature industrial processes represents a large fraction of the total energy requirements, and that there is a strong incentive to employ the waste heat from the thermodynamic cycle of the electric power plants for these purposes. Fortunately, some valuable practical insights can be obtained from the more than 100 large district heating systems employed in cities in the United States, most of them providing the heat in the form of low-pressure steam.[2,3] In addition, there are also numerous smaller cogeneration systems that serve large institutions, shopping centers, and housing developments. Table 17.2 shows that nine of the larger district heating systems serve areas of 2 to 15 km² (0.8 mi² to over 6 mi²) and that the steam distributed runs approximately 0.5 kg/(year·km²) [3×10^9 lb/(year·mi²)]. Because of the distorting effects of government rate setting for utilities, the steam for most of these district heating systems is usually generated in plants that are separate from those producing electric power. Where steam is extracted from steam-turbine–generator units, the method of computing costs imposed by government regulatory agencies usually requires that each Btu of extracted steam costs the same as a Btu of prime steam. No reduction in cost is made for the value derived from the steam expansion to produce electricity. This policy reduces the rate charged residential consumers for electricity and increases the rate charged commercial establishments for heat, a politically advantageous course. To avoid the legal hassles, district heating systems are usually operated by separate companies set up for that purpose, and most of the steam for district heating systems is generated in separate steam plants that do not generate electricity. Thus, irrespective of the source, the average price of steam sold by district heating systems in 1973 was $1.38/$10^9$ J ($1.45/$10^6$ Btu).[4] Of this amount, about half the cost could be attributed to the generation of the steam and about half to the distribution system. It is especially interesting to note that the large mains were accountable for only about one-quarter of the cost of the distribution system; most of the cost derived from the small branching mains and pipes.[2]

The real cost of low-temperature heat can be reduced drastically if, instead of generating steam for heating as a separate process, steam and electricity are produced simultaneously by bleeding steam from the lower turbine stages of an electric power plant. This solution is particularly cost-effective if the power plant is a nuclear plant having low fuel costs. A typical estimate in 1969 indicated that steam for a district heating system could be produced at $0.07 to $0.30/$10^9$ J ($0.07 to $0.30/$10^6$ Btu) with a nuclear plant if it were used simply for generating steam, whereas if it were a dual-purpose plant generating both electricity and steam, the cost would run from $0.024 to $0.092/$10^9$ J ($0.025 to $0.096/$10^6$ Btu), depending on the type of reactor and the type of financing.[3] (Inflation roughly tripled these costs by 1980.) As one might expect, the lower figures in these two cases are for a high-temperature reactor such as an LMFBR, whereas the higher figures are for low-temperature water reactors. For coal-fired plants the costs run about 50 percent higher because of higher fuel costs, even though a coal-fired plant could be in the city, while a nuclear plant would have to be ~30 km outside the city according to current regulatory practice.

Air Conditioning and Refrigeration

Many of the customers of existing district heating systems employ steam during the summer months to drive air-conditioning systems. In some instances this is done by expanding the steam through low-pressure turbines that drive compressors for the refrigerant, while in others an absorption type of refrigeration system is employed. The two methods are roughly comparable in overall efficiency. The consequent load on the district heating system in northern U.S. cities in 1975 was typically about half of that represented by the building heating load in the cold-weather months,[3] but it is expected to rise to equal the heating load.

The processes described above for air conditioning and refrigeration in large commercial buildings can also be employed in residential units. The American Gas Association has sponsored the marketing of gas-heated absorption refrigeration systems, and several gas organizations have sponsored the development of small Rankine cycle systems with gas-fired boilers that provide both electricity and heat. These absorption refrigeration units could be adapted to use high-temperature water from a district heating system preferably at a temperature of 120°C (250°F) or higher.

Industrial Process Heat

Detailed data on the temperature levels at which industries utilize thermal energy are not easy to find, and the sources available give values that differ substantially, in part because of differences in ways in which the raw data have been organized or interpreted.[5,6] Figure 17.2 gives a good indication of the distributiion of energy utilization by industry, while Table 17.3 gives the temperature level at which process heat is required by the six industries that have the highest energy consumptions.[6] The electric power requirements of these industries are also included in Table 17.3. In the 1970s only ~15 percent of this electric power was produced by on-site power plants owned by the industrial user and designed to supply both process heat and electricity. The percentage has not been larger for several reasons: A cogeneration plant involves a substantial extra capital investment over that required by a simple boiler for process steam; small-scale electric power plants entail higher operating and capital costs (Chap. 10); and in many cases the process steam requirements at a particular plant are much greater than the electric power requirements, and, until 1978, most utilities would not purchase the excess

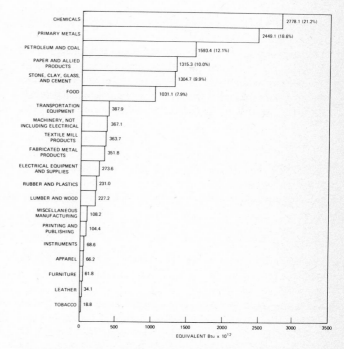

Figure 17.2 Fuels and electricity purchased by industry in 1971 (1 Btu = 1.055 kJ).[5]

TABLE 17.3   Annual Process Heat/Temperature and Electric Requirements for Six Major Industries (Ref. 6)

Industry	Process heat consumption, 10^{12} kJ (10^{12} Btu)				Electricity, 10^9 kWh	Date (reference)
	177°C (350°F)	177–593°C (350–1100°F)	593–816°C (1100–1500°F)	816°C (1500°F)		
Iron/steel	69 (65)	0 (0)	0 (0)	1806 (1712)	57.5*	1973 (6)
Petroleum	63 (60)	2958 (2804)	241 (230)	0 (0)	23.1	1973 (6)
Chemicals	322 (305)	195 (185)	0 (0)	37 (35)	80.0	1977 (5)
Paper/pulp	324 (307)	504 (478)	0 (0)	0 (0)	36.0	1973 (6)
Stone/clay/glass	77 (73)	19 (18)	12 (11)	1080 (1024)	24.0	1974 (6)
Food processing	271 (256)	64 (61)	0 (0)	0 (0)	36.0	1977 (5)

*20 percent generated in plant.

electric power that the industrial plant might produce. In fact, up until 1980, the only step taken by Congress to encourage cogeneration after the Yom Kippur War was a new law requiring electric utilities to buy such power where available. Another unfavorable factor is that the large capital investment required for electric power production can be justified only if a plant lifetime of 20 to 40 years can be expected, whereas market conditions and the technology for chemicals and many other industrial products changes so rapidly that the useful lifetime of a new processing plant may be only 3 to 5 years. Although the steam plant might be used to supply heat to another process after the first process became obsolete, the efficiency of the steam turbine is sensitive to its exhaust temperature and pressure, as well as the steam flow rate; hence a poor match is likely if the plant is used for another process. All these factors commonly favor an arrangement whereby the electric utility makes large blocks of process heat available to many industries—preferably in one or more industrial parks relatively close to the electric power plant. With a substantial number of users, the perturbations to the steam-electric plant are much reduced if a new plant process is started up or an old one discontinued.

Distillation of Seawater

Distillation of seawater to provide potable water for domestic use has been carried out in a few water-short areas for many years. At the time of writing, seawater distillation plants are in use at the U.S. naval base in Guantanamo, Cuba, and in Arabia. Extensive studies have been carried out since the early 1960s on the use of large nuclear power plants for desalting seawater for agricultural purposes with the waste heat from the generation of electric power that would be used for making aluminum, fertilizer, or other products requiring a large power input.[7,8] These agro-industrial complexes show fascinating possibilities for "making the desert bloom" in certain areas such as the upper Gulf of California, the Middle East, the Gujarat Peninsula on the northwest coast of India, and western Australia. The capital costs are huge, however, and the estimated cost-benefit ratio marginal, so that up to 1980 there had been no firm commitment to a definite project.

Distillation of Sewage

As a consequence of the growing concern over water pollution, many different efforts are being made to improve the methods of treatment used in sewage plants. Most of these efforts entail removal of organic solids. This is not sufficient for inland cities, however, because the mineral content of the water flowing through a city system is increased by about 250 ppm. Inasmuch as 500 ppm is considered to be the maximum tolerable mineral content of the water going into the city mains, it seems likely that many cities eventually must employ some means of reducing the content of the dissolved solids. If cheap low-temperature steam were available from a nuclear electric plant under off-peak conditions, one of the processes that might be developed for accomplishing this is distillation by means of multistage evaporators.[9] For the Philadelphia area with a population of 2 million people, the incremental capital cost of provisions for distilling all the sewage with heat from a centrally located nuclear electric power plant appeared in a 1969 study to be about 50 million dollars if the entire complex were properly integrated in the design stage.[9] This compared with a capital investment in the existing water and sewage systems of about 375 million dollars (in 1969 dollars). The requirements are about 237,000 J/L (850 Btu/gal) of water and about 378 L (100 gal) of water per day per person in the area served. This would amount to about 10^{19} J per year (10^{16} Btu per year) for the entire United States in the year 2000, and it would represent perhaps one-third of the waste heat available from the electric power plants.

It is believed that the costs of distilling sewage would not be greatly different from those for distilling seawater. The much lower content of dissolved inorganic solids would tend to reduce that type of fouling of heat-transfer surfaces, but, on the other hand, the presence of organic material might increase the fouling problem. In addition, it would probably be necessary to evaporate the waste solid content to dryness so that it could be burned, or it might be sterilized by heating and sold as fertilizer. Thus, with low-grade heat from a nuclear electric power plant, the entire sewage effluent from a large city could be distilled at a price equivalent to an increase in the operating cost of the water and sewage systems of about 50 percent. If waste heat from a coal-fired plant were employed, the incremental cost would be ~ 75 percent. In water-short areas the distillate could be recycled through the city water system. Preliminary estimates have indicated that, by using low-grade heat from the low-temperature end of the thermodynamic cycle of high-temperature nuclear power plants, it would be possible to distill all the sewage from the urban complexes in the United States by installing a set of systems whose incremental capital cost if designed into the set of nuclear utility systems would be only about 5 billion dollars (on the basis of 1969 costs). This would require a capital expenditure of 500 million dollars per year for 10 years in 1969 dollars, or roughly double that in 1980 dollars—a relatively modest cost. The cost of heat required for the distillation process would of course be substantial unless low-grade heat from nuclear power plants were available at a low cost under off-peak load conditions. If the value of that heat were $0.20/10^9 J ($0.20/10^6 Btu) (in 1969 dollars), the cost of the heat would be about 2 billion dollars per year—again a small fraction of the U.S. annual budget in the year 2000.

Municipal Solid Wastes as an Energy Source

Cities generate an enormous quantity of solid wastes—roughly 1 ton per year per capita, of which about one-third is from residences and the balance from commercial and industrial operations.[9-12] The heating value of municipal wastes commonly totals ~ 12 × 10^6 J/kg (5500 Btu/lb) so that in 1979 the chemical energy content approximated 10 percent that of the 1979 U.S. annual coal production of ~ 7 × 10^8 tons per year. Table 17.4 summarizes both the total amount of solid waste generated in the United States and that practicably collectible from various sources. About 5 percent of the U.S. electric power consumption might be produced from these wastes if

they could be burned efficiently in steam plants located within ~ 50 km of the source of the waste so that collection and transportation costs would be tolerable. Until the 1970s, although many municipal waste incinerators were in use and some produced useful heat and/or electric power, the extra costs of handling and burning solid wastes, together with the poor boiler efficiencies (60 to 70 percent) and potentially severe corrosion where heat recovery was attempted, led to solid-waste disposal in sanitary landfills wherever land was available for this purpose, which it usually was.[12,13] (In 1970 the on-site costs of landfill disposal ran only $1 per ton of wastes.) The rapidly rising cost of fuels in the 1970s changed the relative cost situation while the availability of sites for landfills declined. This situation led to many experimental investigations of the use of municipal solid wastes as fuels as well as the construction of some new power plants designed expressly for the use of solid wastes.[10,14,15] Even though these plants took advantage of extensive experience in Europe with the use of municipal wastes as fuel in power plants, their operation has presented a host of problems.

About half of the usable solid waste listed in Table 17.4 consists of municipal solid wastes. Although much of this consists of paper products, there are substantial amounts of metals and glass, and the value of the waste as fuel is degraded by a high moisture content (Fig. 17.3, Fig. 21.6, Tables 5.1 and 7.5). For use as fuel, the wastes are usually separated into light and heavy fractions by shredding and air flotation. These processes are usually preceded, however, by manual sorting to remove inner-spring mattresses, automobile batteries, massive metal parts, etc., that would foul or damage the shredders, and so perhaps one-third of the total must be diverted to a landfill dis-

TABLE 17.4   Dry, Ash-Free, Organic Wastes Generated and Potentially Collectible for 1971 (Ref. 10)

Source	Quantity, million tons/year	
	Total generated	Total available
Animal wastes	200	26.0
Urban refuse	129	71.0
Logging and other wood refuse	55	5.0
Agricultural and food wastes	390	22.6
Industrial wastes	44	5.2
Sewage solids	12	1.5
Miscellaneous	50	5.0
Total	880	136.3

TABLE 17.5 Expected Ranges in Mixed Municipal Refuse Composition

Component	Percent composition as received (dry weight basis)	
	Anticipated range	Nominal value
Paper	37–60	55
Newsprint	7–15	12
Cardboard	4–18	11
Other	26–37	32
Metallics	7–10	9
Ferrous	6–8	7.5
Nonferrous	1–2	1.5
Food	12–13	14
Yard	4–10	5
Wood	1–4	4
Glass	6–12	9
Plastic	1–3	1
Miscellaneous	< 5	3

Source: C. L. Wilson, "A Plan for Energy Independence," *Foreign Affairs,* July 1973, pp. 657–675.

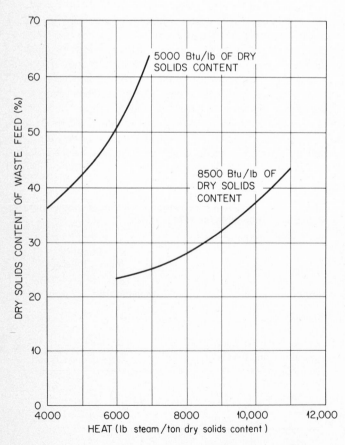

posal site. Worse, attendants must watch for cans of gasoline, solvents, gunpowder, and even dynamite. Some 97 explosions in shredders in solid-waste disposal demonstration plants in the latter 1970s represented the equivalent of one serious explosion every few months in a plant handling 1000 tons per day, and at least two fatalities have resulted, as well as much expensive damage to facilities.[10] Yet another problem is that the plastic materials mixed with paper in the combustible light fraction from the air separation process include substantial amounts of chlorinated hydrocarbons that yield HCl in the stack gas and hence serious boiler and stack corrosion problems. Yet another difficulty is that the solid-waste composition varies widely from one truckload to another, so that furnace control is difficult if the wastes are burned alone. As a consequence, combustion of municipal solid wastes is usually carried out along with some other fuel, such as coal or oil.

Extensive operating experience with demonstration plants has generally been less than satisfactory and the capital costs of the facilities have been high: typically ~ $50,000 per ton per day of capacity (in 1979 dollars).[10] For example, two plants at St. Louis and Nashville[15] burning solid wastes in boiler furnaces designed for the purpose have proved financially burdensome to these cities even with substantial funding from the federal government, and they have experienced many technical difficulties, particularly in meeting EPA requirements on particulate emissions and with HCl corrosion. Both problems have been eased by burning the solid wastes as a fuel supplementary to coal having a substantial sulfur content.[16] Dilution of the coal with the solid wastes not only reduces the effective sulfur content but also somewhat increases the retention of sulfur compounds in the ash because of reactions involving the sulfur and the chlorine in the solid wastes. These reactions also keep down corrosion of the boiler tubes by HCl so that fire-side corrosion is about the same as for furnace operation on coal alone.[16]

In Baltimore a different approach employing a pyrolysis process presented so many problems that it was almost abandoned in 1977 after 2 years of shakedown tests. The problems were finally solved sufficiently so that by 1979 the plant was generating $39,000 per week in revenues for the city.[14] It handled 700 tons per day of refuse, shredding it and partially burning it first in a rotary kiln in an oxygen-deficient atmosphere at ~ 900 °C. The gases from the pyrolysis were then burned in a conventional boiler furnace to produce steam that

Figure 17.3 Relationship between the water content of waste feed and the low-pressure saturated steam generating capability of a Copeland fluidized-bed system at autogenous combustion.[12]

was sold to the Baltimore Gas and Electric Company for $3.90/1000 lb of steam in 1979, reducing the utility's fuel oil consumption by 5×10^6 gal per year.

Recycling metals and glass from the heavy fraction of the solid-waste separation process has been an important element in most of the projects.[10,13] After the shredding and air flotation operations extract the combustibles, the iron can be removed magnetically, nonferrous metals can be removed with an eddy current generator operated in conjunction with a magnet, and glass can be separated from the residue by a water flotation process. The products of these processes are of such low grade that, while it has proved possible to market the iron in at least some cases, it has generally not been possible to find buyers for the nonferrous metals and glass. The cost of their

separation, therefore, is apparently not justified unless special markets can be developed. Possibilities include the use of glass as aggregate in concrete, in brickmaking, etc.

Combustion of municipal solid wastes greatly reduces their volume, but there is still ash to be disposed of in a landfill. The alkali content of this ash is substantial, and thus it presents a leaching and groundwater contamination problem similar to that of coal (as discussed in Chaps. 9 and 11).[11]

Matching Thermal and Electric Energy Requirements

Most of the process heat produced in the cogeneration plants of the U.S. manufacturing industry has been obtained from

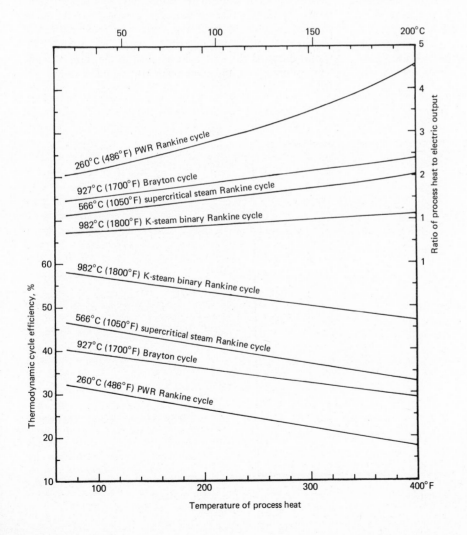

Figure 17.4 Effects of process heat-removal temperature on the overall thermal efficiency of the cycle for electric power generation and on the ratio of energy to the process heat system to the energy to electric power for some typical thermodynamic cycles.[19]

back-pressure steam turbines, i.e., turbines that discharge their steam against a substantial back pressure rather than expanding it to the low pressure obtainable with a condenser running at 38°C (100°F) or less.[17,18] In plants where steam is required at several temperatures, extraction turbines are used, and occasionally, where the requirement for electric power is large, the amount of steam extracted may be only a substantial fraction of the inlet flow. In this case, some steam is extracted and some is expanded to a conventional condenser.

The greater the temperature at which the steam is required for process purposes, the smaller the electric output per unit of energy input to the steam generator. The resulting fraction of the thermal energy input that is obtained in the form of electricity depends on both the steam extraction temperature and the peak temperature in the cycle. Figure 17.4 shows this effect for a water reactor plant, a conventional supercritical-pressure steam plant, and a concept for a plant of the twenty-first century employing a fusion reactor coupled to a potassium-vapor–steam binary vapor cycle plant. Figure 17.4 also shows a curve for a conventional gas turbine which has the inherent

advantage that it rejects its heat at a relatively high temperature and for this reason is in widespread use for cogeneration in the chemical processing industry. In addition to the curves for the efficiency of the thermodynamic cycle for electric energy production, Fig. 17.4 gives a set of curves for the ratio of the energy obtained in the form of process heat to that in the form of electricity. This is also an important parameter, and may determine the choice of a gas turbine in preference to a steam turbine, or vice versa. Figure 17.5 shows a direct comparison of these two systems in this respect.[20] It can be seen that the overall utilization of the energy input for the conditions of Fig. 17.4 would be 100 percent except for stack losses if all of both the electricity and process heat could be utilized to good advantage. There is an important proviso in this respect relative to the gas turbine; Fig. 17.4 was prepared on the assumption that the process heat would be used in the form of superheated water and so the heat could still be extracted efficiently from the turbine exhaust up to fairly high temperatures. If the heat were required in the form of high-temperature steam, the pinch-point effect implicit in Fig. 4.5

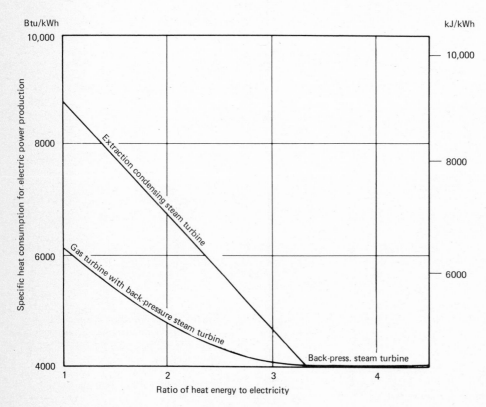

Figure 17.5 Replot of data from *Sulzer Technical Review* (April 1975) showing the heat rates for both gas and steam turbines where both process heat and electricity are required. (*Courtesy Oak Ridge National Laboratory.*)

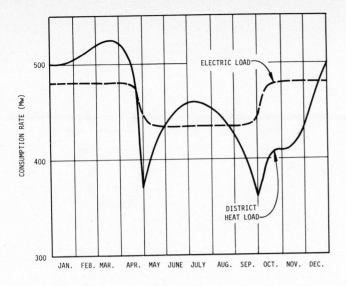

Figure 17.6 Annual electric and district heating system loads near Philadelphia.[3] (*Courtesy Oak Ridge National Laboratory.*)

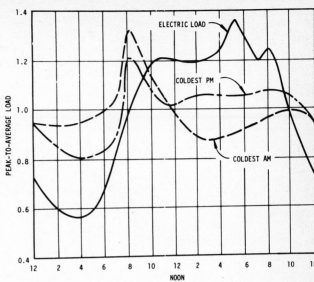

Figure 17.7 Variations in the winter diurnal electric and heating loads near Philadelphia.[3] (*Courtesy Oak Ridge National Laboratory.*)

would make it impossible to get a high overall efficiency with the gas-turbine system.

Heat Storage

One element of the matching problem cited above is possible variation in the ratio of the heat required to the electricity produced. For district heating systems, both seasonal and diurnal variations in this quantity can be quite large; e.g., the building heating load tends to be maximum during a winter night, whereas the electric load is a maximum in the late afternoon (Figs. 17.6 and 17.7). Little can be done toward utilizing all the heat available during seasons of low heat demand, but the short-term storage of heat in the form of high-temperature water can be accomplished efficiently with relatively little capital investment to accommodate diurnal variations in the ratio of heat to electric energy demand, thus giving the plant operator a valuable degree of freedom in meeting both the heat and electric loads required. The flexibility and part-load performance characteristics of the power plant then determine the overall efficiency of energy utilization, a matter that will be discussed in a later section in this chapter in regard to a particular case, the Munich municipal system.

Plant Siting

Electric energy can be transmitted substantial distances efficiently for modest capital costs by using high voltages (Chap. 20), but heat transport is another matter, in part because of pumping and heat losses and in part because of much higher capital costs. For this reason it is important that a cogeneration type of power plant be located fairly close to the heat load, the distance acceptable being dependent on the heat load and prevailing costs. Thus these factors, as well as the usual plant siting criteria, must be considered.

Plant Site Criteria

Up until the 1960s the principal considerations in steam power plant siting were the availability of cooling water, a favorable geological structure for a firm foundation, the cost of land, and the relative cost of fuel transportation and electric power transmission if coal were the fuel. Most plants were built in or on the edge of urban areas, although coal transportation costs sometimes favored riverine sites close to coal mines. The siting problem became more complicated with the introduction of nuclear plants; the AEC imposed stringent requirements that

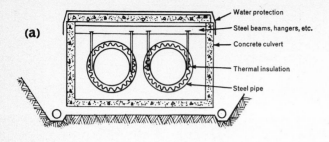

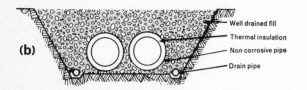

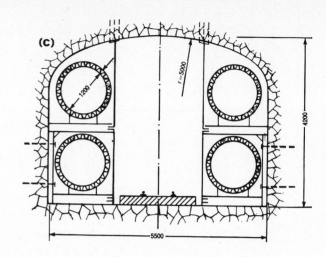

Figure 17.8 Types of construction used for the culverts and mains of district heating systems: (*a*) steel pipes in a concrete culvert, (*b*) armored plastic or prestressed concrete laid in the ground without a culvert, and (*c*) steel pipes in a tunnel.[27] (*Larsson and El Mahgary, IAEA Report No. AG-62/6.*)

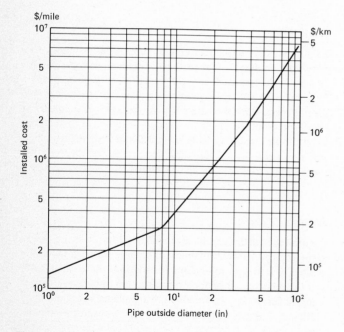

Figure 17.9 Cost of installing supply-and-return pipe system sealed with poured concrete for district heating system mains. Based on high-temperature water with an operating pressure of 28.6 bars (400 psig) and 1969 dollars.[3]

made it necessary to site plants in regions of low population density with large, fenced, exclusion areas around each plant. As discussed near the end of Chap. 9, these factors, coupled with the problems of transporting radioactive spent fuel, have led to intensive studies of large nuclear energy centers, or *power parks,* that would include fuel reprocessing facilities.[21,22,23]

The siting problems for both fossil fuel and nuclear power plants have become so strongly dominated by environmental factors (see Chap. 9 on environmental considerations) and politics that at the time of writing it was not possible to delineate any sound logical basis for plant siting that could be applied for any actual site without being subject to legal attack by environmentalist groups that are basically opposed to the construction of any new plants. As a consequence, the discussion here on the important problems of plant siting is concerned mostly with the problems of transporting heat from the electric power plant to an urban load center. It will suffice to mention that nuclear fission plants could be located in mined caverns under urban areas, as suggested by Edward Teller, and thus protect the public from any conceivable accident.[24] There is extensive experience in Sweden with this type of construction for industrial manufacturing plants, chosen in that case as a means of protecting the facilities from aircraft and missile attack in the event of war. This Swedish experience indicates that where the local geology and terrain are favorable (which is usually the case), the incremental cost of the mined cavern construction might be as little as 10 percent of the total plant cost. However, for reactors, provisions for getting large pieces of equipment, such as pressure vessels, in and out of deep caverns through vertical shafts pose severe problems that appear to require expensive solutions. Note, too, that some designs for fusion reactors, if they prove practicable, offer the possibility of a vastly less serious hazard potential than do fission reactors, and so a good case can be made for siting fusion reactors in urban areas without the need to go underground.[25]

Superheated Water versus Steam for District Heating

Most of the district heating plants in the United States were built before 1929, served relatively small areas filled with tall buildings, and made use of steam as the heat-transport fluid. Both analyses and experience indicate that if the heat must be transported over distances greater than about ~1 km, the distribution system costs can be reduced by employing high-temperature water (100 to 200°C) rather than steam as the heat-transport medium.[4] The cost reduction stems from the fact that water has ~1000 times the density of low-pressure steam, and so, even though the heat content per unit weight is lower, the diameter of the mains is greatly reduced for a given pumping-power–heat-transport ratio (Chap. 4). This is particularly important because the cost of the mains increases rapidly with diameter, in part because of the relatively high pressures and consequently large wall thicknesses required and in part because of the physical size of the trench in which the mains are installed, together with the amount of thermal insulation and concrete duct required for protecting them (Fig. 17.8).[26] Estimates in 1969 indicated that the cost of hot-water mains is largely a function of their diameter (Fig. 17.9 and Table 17.6), and the heat delivered through mains carrying high-temperature water is directly proportional to the temperature drop available.[3] For a temperature drop of ~100°C the cost of heat transport was estimated in 1969 to run about $0.02/10^9 J [$0.02/10^6 Btu/(h·mi)] for large mains handling ~1000 MWt.[3] The study of Ref. 3 indicated further

that if steam were sold at the price cited above as the 1969 average for district heating systems in the United States—i.e. at $1.45/10^6 Btu—and the steam were obtained from a nuclear plant at a cost of $0.10/10^9 J ($0.10/10^6 Btu) rather than $0.76/10^9 J ($0.76/10^6 Btu), and if half of the difference were considered as offsetting the additional cost of a distribution system, it would be possible to transport heat economically as much as 30 km (20 mi) from the electric power plant to the heat load center.[3]

The size of the pipe required to supply hot water for heating a single-family residence is actually less than that normally used for its conventional water supply. Even after allowing for the extra costs of thermal insulation, the heavier pipe required to withstand the higher pressure, and heat losses from the distribution system, preliminary calculations made by the writer indicate that heat could be distributed economically even to single-family residences in built-up areas on the outskirts of a city if the heat were supplied from a nuclear plant. European experience confirms this, and some examples will be cited later in this chapter.

Reliability Considerations

While not necessarily obvious, close integration of the various elements of the utility system for an urban complex carries with it the requirement for an exceptionally high degree of reliability. As serious as an electric power shortage may be, a failure of the heating system of an urban complex would be even more serious because there would be nothing comparable to the countrywide electric power grid to fall back on. Fortunately, there is a tremendous amount of heat capacity in the heating systems for an urban complex, so that an outage for an hour or two would probably not be serious. The heat required for the distillation of seawater or sewage need not be continuously available; in fact, with ample storage capacity for both water and sewage, it appears possible in a cold spell to operate

TABLE 17.6   Percentage of Total District Heating Pipe System Cost by Component

Component	Percentage of total cost						
	Pipe diameter (IPS)						
	4 in	8 in	12 in	16 in	24 in	30 in	36 in
Excavation	19.1	16.3	14.3	12.6	11.3	10.3	9.6
Concrete	29	26.1	24.5	22.4	21.8	20.6	20.2
Pipe	35.5	42.1	44.7	50.1	52.7	54.8	55.7
Insulation	16.4	15.5	16.5	14.9	14.2	14.3	14.5

TABLE 17.7 Estimated Size Distribution of the Electric Power Requirements of United States Metropolitan Areas in the Year 2000

Electric load, MW	Number of cities in range	Total power, MWe	Percent of total power
0–200	1	173	0.042
200–300	15	3,831	0.93
300–400	23	7,927	1.93
400–500	31	14,140	3.44
500–700	29	17,249	4.19
700–1000	38	32,055	7.79
1000–1500	24	28,416	6.90
1500–2000	13	22,341	5.43
2000–3000	18	44,163	10.73
3000–4000	10	36,387	8.84
4000–6000	7	32,279	7.84
6000–10000	7	51,386	12.49
10,000	6	121,189	29.45
Total	222	411,536	100.00

for as much as 30 days without any distillation.

In reviewing the above and considering all the fluctuations that are necessarily to be expected in both electric power demand and the demand for heat for buildings, together with the requirements for a very high degree of reliability, it seems likely that the heat load for a city should be shared by at least three, and preferably four, power plant units to assure the requisite reliability. To investigate this requirement, Table 17.7 was prepared to provide a basis for judging the size of power plant required. If most of the people in the U.S. urban complexes are to be served, it will be necessary to utilize power plant units with design outputs of no more than about 100 MWe for the smaller urban areas. Thus nuclear plants may not be well suited for use in the smaller demand areas because of their relatively high cost per kilowatt in sizes below ~500 MWe.

Typical Cogeneration Systems for Urban Areas

Although some U.S. electric utilities provide back-pressure steam for district heating systems, no U.S. city has a comprehensive integrated system supplying the greater part of the city's needs for both heat and electricity. There are a number of such examples abroad, however, such as Malmö and Västerås in Sweden, Munich in Germany, and Sapporo in northern Japan. The Sapporo case is interesting in that up until the latter 1960s the buildings were heated mainly with coal, and the release of coal smoke at low level from a multitude of small chimneys led to severe smogs under the frequent temperature inversion conditions prevailing in the winter because of the local topography. When the city was chosen as the host for the 1970 Winter Olympics, the city council decided that something had to be done, and a major program was launched to build a district heating system coupled to a large steam power plant with tall stacks.[27] The system was designed by a U.S. firm, built, and has proved to be eminently successful both economically and environmentally. The systems of the Swedish cities of Malmö and Västerås are particularly interesting in that they show that it is possible for small cities with populations of 100,000 to 250,000 to move in an orderly and economically attractive fashion from having no municipal district heating system to having one with cogeneration in a comprehensive integrated system.[26,28] In these instances the orderly buildup of the heat load (Fig. 17.10) was nicely paced with the construction of steam generation facilities especially designed for cogeneration—and consequently the central station is more complex and expensive than a conventional electric power plant. As in Sapporo, the systems have proved economically viable and environmentally attractive not only for commercial and multifamily building areas but also for areas of single-family residences. Further, as can be seen in Fig. 17.11, they have drastically reduced air pollution, in part because of the far better control of combustion conditions and better combustion efficiency in large power plants. They thus refute the Lovins hypothesis that "small is better."

Street Heating in Västerås

A special added feature in Västerås is the extensive use of street heating to free the streets of ice and snow in the winter.[30] By 1973 the street heating system had been extended to provide 1,300,000 m² of snow- and ice-free pavement heated by water entering 25-mm polyethylene tubes on 25-cm centers about 10 cm below the surface. The tubes are arranged in 200-m-long grids and supplied with water at 35°C which cools to 20°C in the circuit. Efforts to quantify the costs and benefits have yielded the bar chart of Fig. 17.12, which seems to give good justification for the Västerås street heating system.

The Munich Municipal Power System

The Munich Municipal Power System (München Stadtwerke) provides an excellent example of the application of cogeneration in a city of about 1 million people to supply both electricity

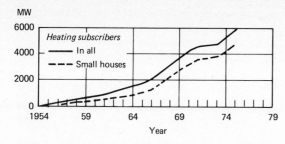

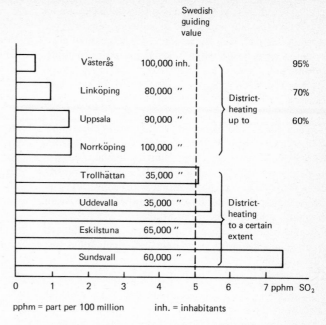

Figure 17.10 Growth of the heat load on the district heating system of Västerås, Sweden, from its inception in 1954, when the population was 80,000 to 1975, when the population reached 120,000.[29]

pphm = part per 100 million   inh. = inhabitants

Figure 17.11 Effects of district heating on the SO_2 content in parts per hundred million of the air of typical Swedish cities in February 1971.[29]

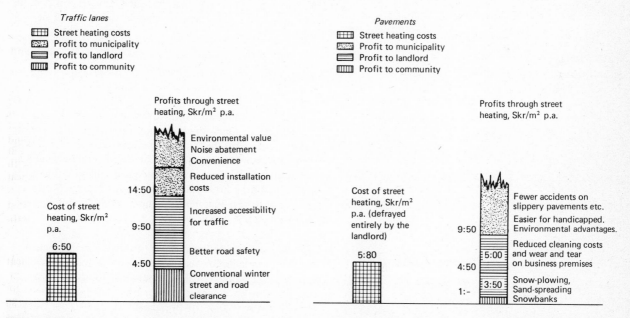

Figure 17.12 Estimated costs compared with the benefits from street heating in Västerås, Sweden. The bars at the left are for automotive traffic lanes, and those at the right are for sidewalks. At the time of the estimate in 1973 the exchange rate was 4.7 Skr per dollar.[30]

and heat.[31] The system is not allowed to supply heat or power to anyone outside the city limits, and since it is under serious governmental restrictions with respect to the purchase of electric power from private utilities, it is not connected into a utility grid. Municipally owned and operated, the electric power capacity was 600 MWe when the author visited the system in 1972. There were four steam plants and one gas-turbine plant, with a sixth plant, a 172-MWe gas-turbine plant, under construction and due for completion in 1974. There is no heavy industry in Munich, yet ~ 50 percent of their heat load is industrial and 50 percent is commercial and residential. The base heat load is 25 percent of the plant's full capacity and is mostly industrial. One of the steam plants is fired with rubbish and garbage with auxiliary fuel oil burners. Another of the steam plants—that on the upwind side of Munich—is gas-fired with gas from a Bavarian natural gas field. The system also includes a number of pumped-storage hydro units in the mountains just south of Munich.

Gas-Turbine Plant   The first gas-turbine plant was intended as a peaking power unit when initially built in 1961. However, it proved to be not only more efficient than the design estimates had indicated, but also nicely flexible in handling varying ratios of heat to electric loads. Thus it has actually become essentially

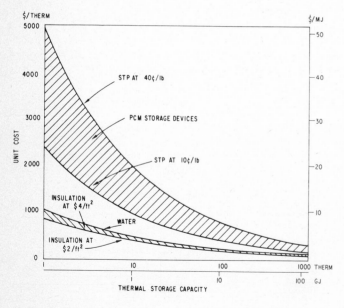

$/THERM $/MJ

UNIT COST

THERMAL STORAGE CAPACITY

Figure 17.13 Comparative costs per therm (10^5 Btu) of thermal-energy storage devices using water versus phase-change material (PCM) for storage. (STP = sodium thiosulfate.)[32]

a base-load plant with one or both turbines in operation over 6000 h per year. The gas-turbine plant has demonstrated an excellent reliability and an availability of 98 percent while using a light distillate fuel oil and operating with a turbine inlet temperature of 720°C.

The gas-turbine plant design was intended to give a good overall efficiency for any of four modes of operation. If the electric load is dominant, the plant is operated with a regenerator extracting heat from the gases leaving the gas turbine and putting it into the air leaving the compressor before it reaches the burners. When the remaining heat in the exhaust gas leaves the regenerator, it is given up to water in the heat exchanger for the superheated-water distribution system. With this arrangement the thermal efficiency of the plant for generation of electricity alone is over 30 percent, and the overall thermal efficiency counting the heat given up to the superheated-water distribution system is about 90 percent. If the heat load on the hot-water distribution system becomes sufficiently high to justify it, the hot gases leaving the turbine may bypass the regenerator and go directly to the heat exchanger for the hot-water distribution system. A third mode of operation lies between the two just mentioned: it involves modulating from one mode to the other and operating with some regeneration. For unusually high heat loads the fourth mode of operation entails lighting an afterburner at the inlet to the heat exchanger for the hot-water distribution system. Note that the heat exchanger costs for this plant ran about four times the cost of the turbine-generator units.

Heat Storage   An important element in the system both for the gas-turbine plant and for the steam power plants is the use of large tanks about 5 m in diameter and ~ 35 m tall for hot-water storage. These tanks are operated full of water all the time and function much like a domestic hot-water tank, with the hot water segregated at the top and cold water at the bottom. With these tanks it is possible to store heat during periods of low heat loads and then recover it and deliver it to the distribution system during high-peak-heat-load periods. Figure 17.13 indicates that the cost of storing heat in high-temperature water is low—definitely less than in fusible salts.[32]

Hot-Water Distribution System   In 1972 the Munich heat distribution system covered an area of 16 km² and had a capacity of 500 gigacalories (Gcal) per hour (2×10^{12} J/h, or 2×10^9 Btu/h). This gave a heat-load density of approximately 30 Gcal/(km² · h) [120 × 10⁶ Btu/(km² · h)]. In 1972 the heat was sold at 21 deutsche marks (DM) per gigacalorie (about $1.90/10⁶ Btu). The capital cost for the water-main distribution system for a new, single-family residential area where the heat distribution system was installed before the construction of the streets and houses amounted to 10^5 DM/(Gcal · h) ($25/kWt).

The high-temperature water distribution system makes use of two basic water supply temperatures. The water for industry is maintained at 160 °C as it goes into the hot-water mains. It returns to the power house at a temperature of around 50 °C. The water supplied for building heating is normally at about 90 °C, but in cold weather the temperature is increased to as much as 110 °C. The building heating system is normally operated by taking the 50 °C water being returned from industry and adding to it 160 °C water to give the desired temperature level. The hot-water distribution system in Munich is an interconnected network, and so the outage of any one of the plants will not lead to a loss in supply of hot water to either the industrial or residential consumers.

Experience in Munich has shown that if a new residential section is built and all the houses are to be heated with hot water from a central system, a system using hot water from power plants gives the lowest overall capital costs and much lower fuel costs than any other system. Two large single-family residence areas of about 1 km² each have been built with spacious yards; the capital cost for the hot-water distribution system was less than $300 per residence for construction about 1963. This low cost was possible only if the hot-water mains were installed before the streets were paved and the houses built. Further, for low costs for single-family residences, it is essential that all the houses on each street served should use the hot-water system in order to give an adequate heat-load density.

The size of the mains is adjusted to the heat load of the area served. The smallest mains are 10 cm in diameter with a 5-cm-thick layer of thermal insulation. The hot-water mains are installed inside a small concrete duct with clearance around the main to reduce heat conduction losses to the earth (Fig. 17.8). The return water main is thermally insulated but is not surrounded with a concrete duct. The mains are usually installed at a depth of about 2 m. The design accommodates a velocity of about 4 m/s in 60-cm-diameter mains and about 2 m/s in the 10-cm-diameter mains. The capital cost of installing a set of mains where there are no obstacles to their installation (i.e., in a new area) in 1972 ran about 2300 DM/m ($250/ft) for a 50-cm-ID main and about 750 DM/m ($82/ft) for a 10-cm-ID main. The thermal insulation on the larger-diameter mains was about 10 cm, but it was only about 5 cm thick on the 10-cm-ID mains.

Plant Proposed for the Minneapolis–St. Paul Area

A major problem in getting new cogeneration projects underway in cities where there is no district heating system is that special turbines and higher capital costs are required, yet it will take some years to build up the heat load for the new steam plant. During this period the plant will be handicapped by both higher capital costs and the inherently poorer efficiency of the turbines designed for high steam-bleed rates. These problems have been explored in a set of design studies of 400- to 800-MWe turbine-generator units to give good efficiency for operation over a wide range of ratios of electric to heat outputs, e.g., from 550 MWe and zero heat output to the district heating system to 471 MWe and 705 MWt for district heating.[33] The corresponding thermal efficiencies based on the heat in the steam supplied to the turbine at 165 bars, 538 °C/538 °C (2400 psi, 1000 °F/1000 °F) would be 43.2 and 82.9 percent, respectively. Further, the loss in turbine efficiency at zero heat load would be only ~0.3 percent and the incremental capital cost would be only ~5 percent when compared with a conventional turbine. Thus the cost and efficiency penalties for providing the power plant to serve as the basis for a new cogeneration system appear to be small.

An excellent insight into the operation of the proposed system is given by Figs. 17.14 to 17.16. In warm weather, when there would be little heat load except for hot water, the water circulation rate in the district heating system would be maintained at a constant relatively low rate with a hot-water supply temperature of 71 °C (160 °F). In colder weather, the water circulation rate would be increased to the full design flow rate and the water temperature would be increased sufficiently to meet the heat demand (Fig. 17.14). Figure 17.15 shows the fraction of the time that one would expect any given heat load.

The thermal efficiency of the plant based on the heat in the steam supplied to the turbine as influenced by the heat and electric loads is delineated in Fig. 17.16. This shows the energy savings made possible by cogeneration.

BOTTOMING AND OTHER SPECIAL RANKINE CYCLES

In many cases, heat is available from an industrial process at moderate temperatures (e.g., 60 to 100 °C), together with a heat-sink temperature in the 10 to 20 °C range, but the use of steam turbines is not attractive because the excessively large machines needed to handle the low-density vapor make the unit cost high. In fact, in conventional central stations the limitations on steam-turbine size generally make it uneconomical to design for steam exhaust temperatures below ~32 °C (90 °F) because the density of steam drops rapidly at lower temperatures. In cold climates the use of a working fluid having a higher vapor pressure than water offers the possibility of extracting more work from the system via a low-temperature *bottoming cycle*. It happens that lower temperature cycles such as this require essentially the same working fluid characteristics as cycles for

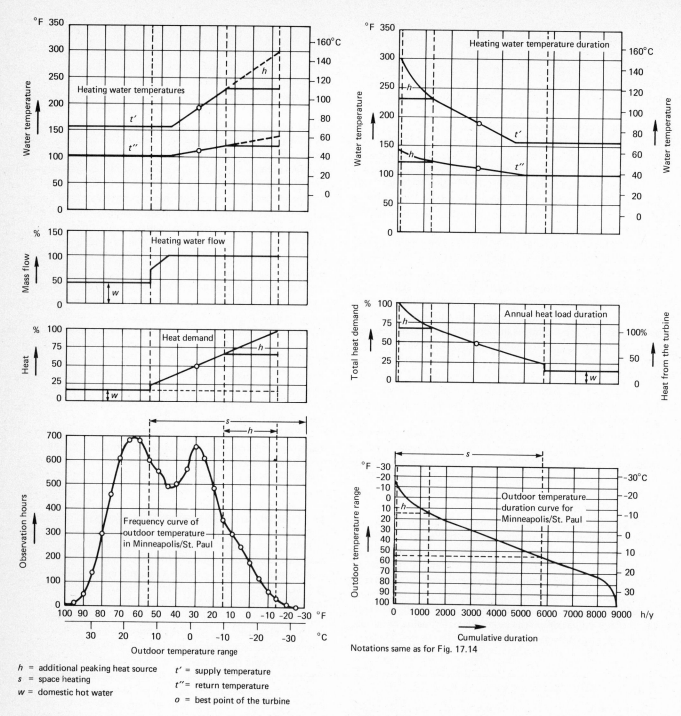

h = additional peaking heat source
s = space heating
w = domestic hot water

t' = supply temperature
t'' = return temperature
o = best point of the turbine

Figure 17.14 Typical heat load data for a U.S.-district heating installation.[33]

Figure 17.15 Duration of heat load conditions for a U.S.-district heating installation.[33]

use with solar, geothermal, and ocean-thermal-difference energy systems (treated in Chap. 21); hence the material presented here is also applicable to those systems.

The use of fluids other than water in Rankine cycles has also received considerable attention for special applications in which the system output would be relatively small so that Reynolds number effects make possible a substantially higher turbine efficiency with a high-molecular-weight fluid than obtainable with steam in a small turbine (Fig. 2.22). Examples include small nuclear electric power systems for space or undersea applications[34,35] and small systems for certain terrestrial applications.[36,37,38] Other fluids appear attractive where there are incentives to avoid high pressures at moderately high temperatures, as in certain chemical processes (Refs. 37, 38, and 39 and Ref. 14 of Chap. 4). Still other fluids are of interest to keep the turbine size down for large outputs at low temperatures, as for geothermal and Ocean Thermal Energy Conversion (OTEC) applications (Chap. 21).[40,41]

Working Fluids

The many organic compounds, together with the fluorocarbons and chlorinated hydrocarbons, appear to offer an overwhelmingly wide range of candidate compounds for special Rankine cycles, but practical considerations narrow the field to a relatively small number. The first consideration is the vapor pressure in the temperature range of interest. For favorable proportions in the turbine, heat exchangers, and vapor passages, it is desirable that the vapor pressure at the lowest temperature in the cycle be ~0.3 bar or more, although a pressure as low as 0.03 bar may be acceptable. (The pressure in the condenser of a conventional steam power plant is about as low as is practicable, i.e., commonly 0.03 to 0.1 bar.) As discussed in Chap. 6, a second major consideration is the compatibility of the fluid with the other materials in the system, together with its long-term chemical stability, particularly freedom from a tendency to polymerize to form gums, resins, or tars. If a hermetically sealed system is employed, it is essential that the fluid not decompose to give noncondensable gases, such as H_2 or CH_4. It is desirable that the fluid be nontoxic and nonflammable, that the melting point be well below temperatures that might be reached when the system is shut down, and that the cost be reasonable. The fluids that have proved of greatest interrest meet most of the above criteria; their principal properties are summarized in Table 17.8, and their vapor pressures are plotted in Fig. A2.2. Of these, NH_3 has the highest vapor pressure and is of interest mainly for the lowest temperature cycles, while Dowtherm has the lowest vapor pressure and is well suited to the higher temperature cycles. Water is included in Table 17.8 not only for

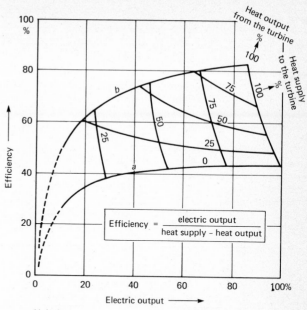

Unit size	: 400, 550, 800 MW
Steam conditions	: 2400 psig (1000°F/1000°F ⇒2.5″ Hg)
Heating water temperature:	113°F ⇒ 194°F

a = Pure condensing operation
b = Pure extraction operation (back-pressure operation)

Figure 17.16 Net thermal efficiency of the cogeneration turbine as a function of heat and electric loads.[33]

purposes of comparison but also because, after all the considerations are weighed, it is usually the best candidate even for special applications.[39] A major reason for its suitability is that its heat-transfer characteristics, including a high heat of vaporization and relatively good thermal conductivity, are exceptionally good for both boiling and condensing conditions.

Heat-Transfer Considerations

The heat exchangers in a system ordinarily cost several times as much as the turbine. Hence the boiling and condensing heat-transfer characteristics are likely to prove to be the dominant factors in the choice of a working fluid. The column at the right end of Table 17.8 gives a parameter that is proportional to the forced-convection heat-transfer coefficient for the liquid—an important factor if a regenerator is employed.

The forced-convection heat-transfer parameter in the right-hand column of Table 17.8 also influences the condensing heat-

TABLE 17.8   Properties of Typical Rankine Cycle Working Fluids at 25°C or 1 atm

| Fluid | Molecular weight | Properties of liquid | | | | | | $\dfrac{c_p^{0.4}k^{0.6}}{\mu^{0.4}}$ |
		Density, g/cm³	Specific heat, J/(g·°C)	Viscosity, cP	Thermal conductivity, W/(m·°C)	Heat of vaporization, J/g	Freezing point, °C	
Ammonia	17	0.602	4.81	0.124	0.50	1370	−78	2.85
Benzene	78.1	0.876	1.74	0.60	0.14	390	5	0.47
Dowtherm A	165.6	1.05	2.5	1.08	0.12	286	54	0.39
Ethanol	46	0.787	2.45	1.095	0.17	846	−114	0.48
Freon 12	121	1.315	0.97	0.27	0.07	165	−158	0.34
Methanol	32	0.789	2.55	0.56	0.22	1102	−98	0.54
Monochlorobenzene	113	0.876				323	−45	
Toluene	92.1	0.865	1.72	0.55	0.13	363	−95	0.46
Water	18	1.000	4.19	0.89	0.61	2262	0	1.47

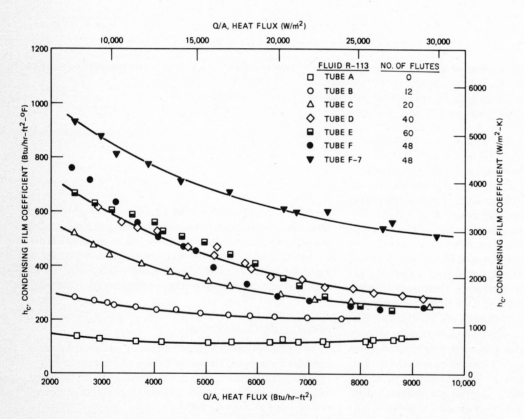

Figure 17.17   Effects of the flute geometry on the condensing heat-transfer performance of 1-in-OK, 4-ft-long, vertical condenser tubes operating with R-113 (CCl₂F-CClF₂). Tube F-7 was fitted with seven neoprene drain-off skirts (or collars) to reduce the thickness of the condensate film to a minimum.[43] (*Courtesy Oak Ridge National Laboratory.*)

transfer coefficient, because the film of liquid condensate on the surface of a condenser acts as a barrier to heat transfer. This is such an important factor that the success of power plants designed to operate on ocean thermal differences and low-temperature geothermal heat sources depends in large measure on minimizing this temperature loss. One method that has been found effective is to employ fluted vertical condenser tubes. The liquid condensate film is concentrated in the flutes by surface tension and runs down the tiny channels formed by the grooves, leaving only an extremely thin film covering the ridges between the grooves (see Ref. 43 and see Ref. 3 of Chap. 4). Figure 17.17 shows that such a geometry can increase the condensing heat-transfer coefficient by as much as a factor of 7 over that ob-

tainable with a plain tube. As shown in Fig. 17.18, the magnitude of the enhancement depends on the physical properties of the condensed liquid, particularly the thermal conductivity, surface tension, and viscosity. Note the clear superiority of ammonia from this standpoint.

The high condensing heat-transfer coefficient obtainable with fluted condenser tubes gives a situation in which the heat-transfer coefficient on the cooling water side becomes controlling so that there is little incentive for further increases in the condensing coefficient. This effect is shown in Fig. 17.19, which gives the overall heat-transfer coefficient as a function of the condensing (or boiling) coefficient for units operating with seawater flowing through the tubes.[43]

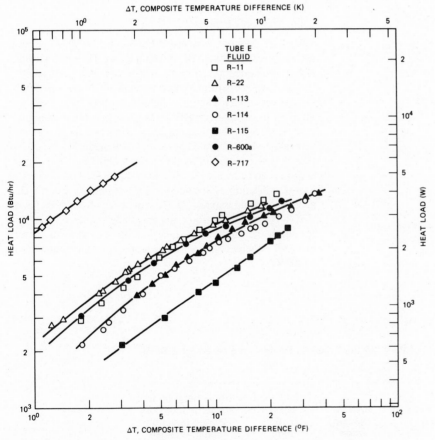

Figure 17.18 Effects of the physical properties of the working fluid on the condensing heat transfer rate for Tube E of Fig. 17.17 (surface area = 1.05 ft²). Data are plotted for five Freons, isobutane (R-600a), and NH₃ (R-717). The "composite temperature difference" includes the temperature drop through the tube wall.[43] (*Courtesy Oak Ridge National Laboratory.*)

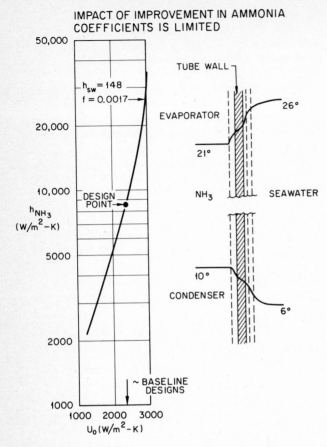

Figure 17.19 Effects of the boiling or condensing heat-transfer coefficient on the overall heat-transfer coefficient for operation with seawater as the tube-side fluid giving a heat-transfer coefficient of 812 W/(m² · K) [143 Btu/(h · ft² · °F)].[43] (*Courtesy Oak Ridge National Laboratory.*)

Turbine Design Considerations

The principal proponents of working fluids other than water for special low- and moderate-temperature Rankine cycles are turbine designers who are mainly interested in improving the proportions and/or increasing the Reynolds number in the turbine through the use of a high-molecular-weight fluid.[34,37,38] Table 17.9 shows the effects of the choice of working fluid on some of the principal design parameters for a small turbine. Tip leakage losses associated with a 0.16-mm blade height for the full admission steam turbine would be so severe that it would be better to go to partial admission, though this would still give a lower turbine efficiency than for full admission with the greater blade heights of the units designed for operation with chlorinated hydrocarbons.

Cycle Efficiency Effects

Table 17.8 does not include a column indicating differences in cycle efficiency between the working fluids. The reason for this is that, fundamentally, there is no difference in the efficiency of the ideal cycles operating between the same temperature limits; differences in efficiency for the actual cycles stem from differences in component efficiencies attributable to Reynolds number effects, design compromises, and the usable temperature range which is limited by factors such as pinch point effects (Fig. 4.5).[40] An indication of the magnitude of the latter consideration is given by Fig. 17.20, which shows the effectiveness of energy utilization as a function of the temperature at which a geothermal fluid is available.[43] It can be seen from these curves that R-115, a fluorocarbon, appears to be the best choice from the purely thermodynamic standpoint for a source temperature of 140°C, whereas NH_3 would be best for a source temperature of 300°C.

TABLE 17.9 Effects of the Choice of Working Fluid on the Principal Design Parameters of a Small Turbine Designed to Produce 285 W (Ref. 37)

	Boiler temp., °C	Condenser temp., °C	Rpm for 140-mm wheel	Total throat cross section, mm²	Blade height (mm) for full admission, 20 deg injection angle, impulse
Steam	110	45	62,500	5.8	0.16
Monochlorobenzene	110	45	26,500	13.2	0.61
Dichlorobenzene	110	45	25,000	144.0	5.00

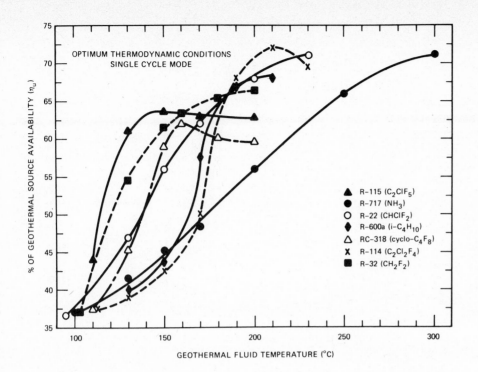

Figure 17.20 Effect of working fluid characteristics on the fraction of the thermal energy available that can be put into the thermodynamic cycle as a function of the geothermal source temperature.[43] (*Courtesy Oak Ridge National Laboratory.*)

Bottoming Cycles for Conventional Steam Plants at Dry, Cold Sites

The size and cost of steam turbines and condensers increase rapidly if one tries to increase the cycle efficiency by reducing the condenser temperature. At 120°C (250°F), for example, the specific volume of steam is 0.86 m³/kg (13.8 ft³/lb), whereas at 38°C (100°F) it is 21.8 m³/kg (350 ft³/lb) and at 15°C (60°F) it is 75 m³/kg (1200 ft³/lb). For an 1800-rpm steam turbine, it is difficult to make use of a diameter greater than 4.2 m (14 ft) because of tip speed effects on both the stresses in the turbine blades and compressibility losses in the steam. As a consequence, the velocity of the steam leaving the last stage of the turbine must be so high that it represents a large fraction of the energy that one might expect to be available from expansion from 38°C (100°F) down to 15°C (60°F). One way to avoid the losses in cycle efficiency that this entails and also effect a marked reduction in the turbine cost, would be to make use of a working fluid with a much higher vapor pressure than steam in this temperature range. Hydrocarbons such as butane, some of the Freons, and particularly ammonia have looked attractive for this purpose.[44]

The principal barrier to the use of a bottoming cycle in steam plants at favorable sites is that it inherently requires a heat ex- changer between the lower end of the steam cycle and the upper portion of the bottoming cycle. Analyses indicate that the temperature loss across the heat exchanger can be kept to as little as 7 to 10°C with acceptable costs. Even this low temperature difference inevitably leads to a loss in cycle efficiency unless the heat-sink temperature will run below around 10°C for most of the year. For such conditions the bottoming cycle has the additional major advantage that it could be used with dry cooling towers without danger of freezing of the working fluid. This is especially important in cold areas where there is not an adequate supply of cooling water.

Small Systems in the Intermediate-Temperature Range

A typical special situation favoring the use of an organic working fluid is offered by an industrial operation in which sufficient heat to produce ~100 kWe is available at ~350°C (662°F) and there is a need for process heat at ~100°C (212°F). One of the best of the working fluids in Table 17.8 for this application is Dowtherm because its chemical stability is the best among the fluids listed,[35,45] its vapor pressure is in the right range, and it yields a more efficient turbine than steam in this size range (Fig. 2.22).

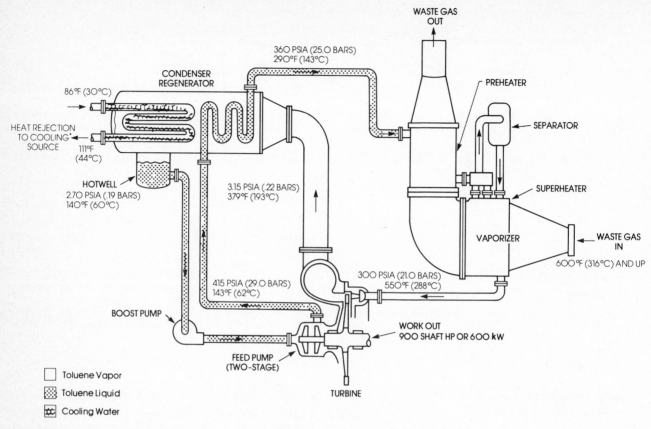

Figure 17.21 Flowsheet for an organic Rankine cycle system utilizing toluene as the working fluid for recovering heat from diesel engine exhaust gas or industrial process fluid streams to produce 600 kWe. (*Courtesy Sundstrand Corp.*)

To obtain a good efficiency from a Dowtherm cycle, it is necessary to cope with a special problem introduced by the shape of the temperature-entropy diagram (Fig. 2.5). The vapor expansion from saturated conditions at the turbine inlet entails a path into the strongly superheated region. This is advantageous in that it avoids moisture formulation in the turbine and the consequent problems with moisture churning losses and possible blade erosion, but for good cycle efficiency it is necessary that the heat removed from the superheated vapor in cooling it to the saturated condition must be employed for regenerative heating of the liquid feed to the boiler with a counterflow heat exchanger. This makes it necessary to remove the heat by forced convection with a much poorer heat-transfer coefficient than for condensing conditions; hence the amount of heat-transfer surface area required is much greater than for a condenser.[35]

Commercial Applications

Although there have been many studies of potential applications of working fluids other than water for use in low- and intermediate-temperature Rankine cycle systems, none of these has gone beyond the experimental and demonstration stages to any substantial commercial application. Thus these systems are likely to continue to be of interest only for quite special heat-recovery applications,[42] and for low-temperature heat sources such as in geothermal, solar, and OTEC applications. An excellent example of such a system for obtaining electric power where heat is available at intermediate temperatures is a 600-kWe unit built by Sundstrand for heat recovery from diesel exhaust and industrial processes.[38] As of the latter part of 1979, three of the systems were in commercial use and three more were under construction with some support funding from the conservation division of DOE. A flowsheet giving the operating conditions for this system is shown in Fig. 17.21. The system is being used where the thermal energy is in effect free, and so the cost of power depends mainly on the capital and maintenance costs. Utilizing a simple system with a fluid having good thermal stability so that corrosion and deposits are not a problem helps reduce both capital and operating costs. In this instance toluene was chosen as the working fluid because it made possible a good efficiency in a single-stage turbine, whereas steam would have required a multiplicity of stages and hence a higher cost.[38] This basic system is also suited to solar thermal power systems in the 500- to 1000-kWe range.[38]

The thermal efficiency of the system of Fig. 17.21 is modest, about 9 percent, because of the low peak temperature and the high heat-rejection temperature for a process heat application; these conditions give an ideal Carnot cycle efficiency of 25 percent. More significant is the ratio of the overall cycle efficiency to that of the Carnot cycle, a ratio that in this case is 37 percent. Part of the loss stems from the fact that small, single-stage turbines of the sort used here have a turbine efficiency of 70 to 75 percent and generators in this size range have efficiencies of 93 to 95 percent. The rest of the losses stem from differences between the actual cycle and the ideal Carnot cycle.

PROBLEMS

1. Make a rough estimate of the fraction of the chemical energy in the fuel converted to electricity, the fraction obtained as process heat at 200°C (392°F), and the fraction lost up the stack if a 172-bar/538°C (2400 psi/1000°F) steam plant were designed for cogeneration with a back-pressure turbine discharging all its steam to the process heat system. Assume that the thermal efficiency of the plant, if it is operated with a 30°C (86°F) condenser temperature, would be 35 percent and that the actual cycle efficiencies for the two cases are directly proportional to their Carnot cycle efficiencies.

2. Estimate the fractions of the energy in the fuel converted into electricity and recovered as process heat at 121°C and hot water at 66°C in the coal-fired closed-cycle gas-turbine plant of Fig. 14.2 operated at a peak pressure of 10 atm with a 3.5:1 pressure ratio and a turbine inlet temperature of 815°C (1500°F). A regenerator having an effectiveness of 85 percent is employed in the gas-turbine circuit, and the stack gases are cooled to 150°C (302°F) in a recuperator that preheats the combustion air for furnace operation with 20 percent excess air.

3. Assume that 20 percent of the steam is bled from the turbine at 204°C (400°F) for industrial process heat in a plant similar to that of Fig. 11.1. Estimate the reduction in plant output in the form of electricity and the consequent break-even charge for the bled-off steam in dollars per 10⁶ Btu, assuming that the value of the electricity is $0.04/kWh.

4. A large industrial complex located 10 km from the center of a city has 7×10^{12} J/h of heat available at 77°C (170°F) on an around-the-clock basis with a water return temperature of 38°C (100°F). A study indicates that the peak load for building heating of the 40,000 residential units in the city would approximately match the heat available from the industrial complex. Estimate the size and cost of a pair of mains to circulate the hot water between the industrial complex and the city, together with the pumping power required, using the 1969 construction costs of Fig. 17.9 and assuming $0.02/kWh as the 1969 cost of electricity. Escalate the costs to current levels.

5. The year-round average total heating load for the above case is estimated to be 14.4×10^{11} J/h (including domestic hot-water requirements). Estimate the prorated capital cost of the mains per residential unit and the capital and pumping charges, together with the sum of these two charges, for the heat. (Note that there would be additional charges for the capital and maintenance costs of the distribution system and for the heat supplied.) As above, make the cost estimates initially on the basis of the 1969 data of Fig. 17.9, and then escalate to current levels and compare with the current costs of both gas and electricity for residential heating.

6. A chemical plant has a hot gas stream leaving one process at 400°C (752°F) and needs to supply heat to another process at 100°C. It has been suggested that the energy be cascaded through an organic Rankine cycle so that it may be used to obtain some electric power. Estimate the thermal efficiency for electric power generation with a Dowtherm system operated with a boiler temperature of 343°C (650°F) and a condenser temperature of 143°C (290°F). Assume a turbine efficiency of 75 percent, a regenerator effectiveness of 90 percent, and a generator efficiency of 95 percent.

REFERENCES

1. *Cities,* special issue of *Scientific American,* vol. 213, no. 3, September 1965.

2. *District Heating Handbook,* 3d ed., National District Heating Association, 1951.

3. Miller, A. J., et al.: "Use of Steam-Electric Power Plants to Provide Thermal Energy to Urban Areas," Oak Ridge National Laboratory Report No. ORNL-HUD-14, January 1971.

4. Anderson, T. D., et al.: "An Assessment of Energy Options Based on Coal and Nuclear Systems," Oak Ridge National Laboratory Report No. ORNL-4995, July 1975.

5. Graves, R. L., et al.: "Assessment of an Atmospheric Fluidized Bed Coal Combustion Gas Turbine Cogeneration System for Industrial Applications," Oak Ridge National Laboratory Report No. TM-6626, vol. 1, October 1979.

6. Sindt, H. A., et al.: "Costs of Power from Nuclear Desalting Plants," *Chemical Engineering Progress,* vol. 63, no. 4, April 1967, pp. 41–45.

7. Probstein, R. F.: "Desalination: Some Fluid Mechanical Properties," *J. Basic Eng., Trans. ASME*, vol. 94, 1972, pp. 286–313.

8 Spiewak, I.: "Investigation of the Feasibility of Purifying Municipal Waste by Distillation," Oak Ridge National Laboratory Report No. ORNL-TM-2547, April 1969.

9. Anderson, L. L.: "Energy Potential of Organic Wastes: A Review of the Quantities and Sources," U.S. Bureau of Mines Information Circular IC-8549, 1972.

10. Sherwin, E. T., and A. R. Nollet: "Solid Waste Resource Recovery: Technology Assessment," *Mechanical Engineering,* vol. 102, no. 5, May 1980.

11. Fife, J. A.: "Incineration: Steam Generation from Solid Wastes," *District Heating,* vol. LVI, no. 2, Fall 1970, pp. 18–24.

12. Boegly, W. J., et al: "MIUS Technology Evaluation—Solid Waste Collection and Disposal," Oak Ridge National Laboratory Report No. ORNL-HUD-MIUS-9, September 1973.

13. Stabenow, G.: "The Chicago Northwest and Harrisburg Incinerators: A Proven Method of Energy Recovery and Recycling of Ferrous Metals," *Proceedings of the 1976 National Waste Processing Conference, Boston, Mass., ASME,* 1976, pp. 81–96.

14. "Baltimore's Resource Recovery Plant now Working Well City Official Says," *Environmental Reporter,* Oct. 12, 1979, pp. 1348–1349.

15. McEwen, L., and S. J. Levy: "Can Nashville's Story Be Placed in Perspective?" *Solid Waste Management,* vol. 19, no. 8, August 1976, pp. 24–60.

16. Krause, H. H., et al.: *Corrosion and Deposits from Combustion of Solid Wastes,* part VI: *Processed Refuse as a Supplementary Fuel in a Stoker-Fired Boiler,* ASME Paper No. 78-WA/Fu-4, December 1978.

17. Kovacik, J. M.: "Guidelines for Future Cogeneration," *American Power Conference,* Apr. 23–25, 1969.

18. Wilson, W. B., and W. J. Hefner: "Economic Selection of Plant Cycles and Fuels for Gas Turbines," *Combustion,* April 1974, pp. 7–16.

19. Fraas, A. P., and G. Samuels: "Power Conversion Systems of the 21st Century," *Journal of the Power Division, ASCE,* vol. 104, no. PO1, Proceedings Paper 13545, February 1978, pp. 83–97.

20. Frei, D.: "Gas Turbines for the Process Improvement of Industrial Thermal Power Plants," *Sulzer Technical Review,* vol. 4, 1975, pp. 195–200.

21. *Considerations Affecting Steam Power Plant Site Selection, A Report Sponsored by the Energy Policy Staff,* Office of Science and Technology, U.S. Government Printing Office, 1969.

22. Piper, H.G., and G. L. West, Jr.: "Siting of Nuclear Reactors," Oak Ridge National Laboratory Report No. ORNL-HUD-11, 1971.

23. "Nuclear Energy Center Site Survey—1975, Practical Issues of Implementation," U.S. Nuclear Regulatory Commission, NUREG—0001, part IV, January 1976.

24. Beck, C.: "Engineering Study on Underground Construction of Nuclear Power Reactors," USAEC Report AECU-3779, Apr. 15, 1958.

25. Fraas, A. P.: "Environmental Aspects of Fusion Power Plants," *Proceedings of the Plenary Sessions International Conference on Nuclear Solutions to World Energy Problems,* Nov. 13–17, 1972, pp. 261–274.

26. *Mälmo Kraflvärmeverk,* The City of Malmö, Technical Division, Spring 1973.

27. Raisic, N.: "The Most Promising Area of Future Nuclear Heat Utilization?" *Nuclear Engineering International,* vol. 22, no. 260, August 1977.

28. Malfitani, L.: "District Heating System Warms Heart of Japanese City," *Power,* April 1970, pp. 50–53.

29. Some articles about townplanning in Västerås, City of Västerås, 1973.

30. Västerås District Heating Power Station, Springfeldt Annonsbyrå AB, Västerås, Sweden, 1975.

31. Heizkraftwerk Sendling, Stadtwerke München, Munich, West Germany, November 1972.

32. Segasser, C. S., and J. E. Christian: "Low Temperature Thermal Energy Storage," Oak Ridge National Laboratory Report No. ANL/CES/TE 79-3, March 1978.

33. Oliker, I., and H. J. Muhlhauser: "Technical and Economic Aspects of Coal-Fired District Heating Power Plants in U.S.A." *American Power Conference,* April 1980.

34. Wigmore, D. B., and R. E. Niggemann: "The Specification of an Optimum Working Fluid for a Small Rankine Cycle Turboelectric Power System," *Proceedings of the Seventh Intersociety Energy Conversion Engineering Conference,* September 1972, pp. 303–314.

35. Samuels, G., and R. S. Holcomb: "Reference Design for an Isotope Power Unit Employing an Organic Rankine Cycle," Oak Ridge National Laboratory Report No. TM-2960, March 1971.

36. Morgan, D. J., and J. P. Davis: "High Efficiency Gas Turbine/Organic Rankine Cycle Combined Power Plant," ASME Paper 74-GT-35, March 1974.

37. Tabor, H., and L. Bronicki: "Establishing Criteria for Fluids for Small Vapor Turbines," *SEA Trans.,* vol. 73, 1965, pp. 561–575.

38. Niggemann, R. E., et al.: "Fluid Selection and Optimization of an Organic Rankine Cycle Waste Heat Power Conversion System," ASME Paper No. 78-WA/Ener-6, Dec. 10, 1978.

39. *The Dowtherm Heat Transfer Fluids,* Dow Chemical Co., 1967.

40. Milora, S. L., and J. W. Tester: *Geothermal Energy as a Source of Electric Power,* The M.I.T. Press, Cambridge, Mass., 1976.

41. McGowan, J. G., et al.: "Conceptual Design of a Rankine Cycle Powered by the Ocean Thermal Difference," *Proceedings of the Intersociety Energy Conversion Engineering Conference,* Aug. 13–16, 1973, pp. 420–427.

42. Luchter, S.: "Power Recovery from Gas Turbines—A Review of the Limitations, and an Evaluation of the Use of Organic Working Fluids," ASME Paper No. 70-GT-113, May 1970.

43. Michel, J. W.: "Heat Transfer Considerations in Utilizing Solar and Geothermal Energy," paper presented at the Miami International Conference on Alternative Energy Sources, Dec. 5–7, 1977, Miami Beach, Fla.

44. Set, R. G. and W. Steigelmann: *Binary-Cycle Power Plants Using Dry Cooling Systems,* part 1: *Technical and Economic Evaluation, Final Report F-C3023,* The Franklin Institute Research Laboratories, January 1972.

45. Pummer, J. W.; *The Kinetics and Mechanism of the Pyrolytic Decomposition of Aromatic Heat Transfer Fluids, Final Report NBS Project No. 3110541,* National Bureau of Standards, April 1970.

18

FUEL CELLS

Fuel cells offer an intriguing means of generating electric energy by utilizing directly the chemical energy released in the reaction between a fuel and oxygen.[1] The operation is similar to that of a conventional wet-cell or dry-cell battery except that the electrodes and electrolyte are not consumed; instead the reactants are fed continuously in gaseous or liquid form to porous electrodes so that—in principle—the fuel cell could have an infinite life. The basic concept is so simple in principle, its theoretical efficiency so high, and its apparent freedom from a need for noisy and troublesome machinery so attractive that it has captured the imagination of many engineers. In fact, the advantages and appealing features of fuel cells for all sorts of special applications constitute the subject of a large fraction of the papers presented on fuel cells at engineering society meetings. Unfortunately, their great attractiveness is balanced by a set of subtle yet truly formidable problems that have greatly complicated an apparently simple system and for nearly a century have frustrated efforts directed toward its commercial development.

HISTORICAL BACKGROUND

The fuel-cell concept is almost as old as that of wet-cell batteries. In 1839, Sir William Grove found that an electric current could be generated by bubbling hydrogen and oxygen, respectively, over two platinum electrodes in a sulfuric acid bath.[2] By the 1890s, it was recognized that carbon and air might be employed instead of hydrogen and oxygen and that a high thermal efficiency could be realized, at least in principle, because a fuel cell is not a heat engine. That is, ideally, most of the energy available from the chemical reaction should appear as electric voltage and current. By 1896, when electric power was coming into use for operating the gun turrets and ammunition hoists on the early armored naval vessels, W. W. Jacques, in the United States, proposed a battleship power plant based on experimental fuel cells that he had operated.[3] A typical cell about 1.2 m high and weighing about 225 kg developed about a kilowatt with a current density of around 100 mA/cm² (approximately 100

A/ft²). Each cell consisted of a cast-iron pot containing potassium hydroxide with a coal anode and provisions for bubbling air up through the molten KOH. He built many of these cells and operated as many as 100 of them in series, but their life was limited to only about 100 h.

Research objectives in the fuel-cell field have differed with the application toward which they were directed. In most of the early work the fuel cell was envisioned as a competitor with the wet-cell battery. Fuel-cell research was given particular emphasis from 1930 to 1950 because of its potential for naval submarine applications,[4] and, though very few people knew of it, small, intensive, highly secret efforts were pursued in the United States, Great Britain, France, and Germany. The possibility of applying some of these fuel cells to automotive vehicles became of interest in the latter 1940s and subsequently led to dozens of private research and development (R&D) efforts.[5-8] Support for the naval effort fell off in the latter 1950s because of the high cost and limited life of fuel-cell systems on the one hand and the success of the nuclear power plant program for submarines on the other. However, at just that time the requirements for a short-lived electric power source that would operate reliably under 0-g conditions for the manned space flight program and have a much lower weight than batteries led to a large influx of U.S. Air Force and NASA funds. The emphasis was, of course, on low weight with little regard to cost. For example, there was no objection to the use of relatively large amounts of platinum catalyst in the electrodes. This effort was greatly reduced with cutbacks in the U.S. manned space flight program in the latter 1960s; at the same time, disappointment in the high cost, short life, and relatively large specific weight of fuel cells led to the closing out of virtually all the private efforts to develop fuel cells for automotive and other commercial applications. Of the many private companies working on fuel cells under the space program, by the latter 1970s the only remaining large-scale effort on commercially marketable units was carried out by United Technologies Corporation (formerly Pratt & Whitney). This development has been pursued mainly with the objective of obtaining a system for relatively small electric utility power plants using hydrogen from natural gas or reformed fuel oil as the fuel. Starting in 1967, this effort was funded jointly by Pratt & Whitney and the gas utilities.[10] Subsequently, when it became apparent that a shortage of natural gas was developing, Pratt & Whitney turned for additional support to the Edison Electric Institute (EEI) and EPRI and, with the same type of cell, shifted the emphasis from natural gas to hydrogen from distillate petroleum as the fuel.[11] In 1975, ERDA began to provide additional support.[12] Privately funded efforts on a much smaller scale have been pursued at Energy Research, Exxon, the Institute for Gas Technology, General Electric, and Westinghouse, the latter two as outgrowths of their space power programs. At the time of writing, special projects were also being funded by NASA (space shuttle), the U.S. Navy, ARPA, the U.S. Army, NSF, and the U.S. Air Force.

Over 80 years have elapsed since Jacques built the first large fuel cells. A great deal of progress has been made and the specific weight of fuel cells has been cut by a factor of about 20, but the electrochemical current densities are still in the range of 50 to 300 mA/cm² (50 to 300 A/ft²) and limited life is still a vital problem. Extensive highly satisfactory operating experience for periods of weeks has been obtained under the space program, and in the past 30 years many different materials combinations have been investigated.[1-16] Numerous studies of the application of fuel cells to the production of electricity for utility service have shown that they have many attractive features and would be very useful if a relatively low cost and a long life (approximately 40,000 h) could be obtained.[1,12,13,16-24] The phosphoric acid cells would be particularly attractive for use in end-of-the-line substations for intermediate- and peak-load service where they have the advantages of good efficiency at part load and fast response to changes in load.

OPERATING PRINCIPLES

As in conventional batteries, fuel cells employ a pair of electrodes separated by an electrolyte that does not conduct electrons but allow ions to migrate from one electrode to the other. In a typical cell with an acid electrolyte, gaseous hydrogen is supplied to the anode, where it is ionized and releases electrons to the external circuit. The hydrogen ions migrate to the cathode, which is supplied with oxygen. The oxygen is ionized by electrons flowing into the electrode from the external circuit, and the ionized oxygen reacts with the hydrogen ions to form H_2O. If an alkaline electrolyte is employed, the operation is similar except that hydroxyl ions migrate from the cathode to the anode. A schematic diagram showing the mode of operation of a phosphoric acid fuel cell is given in Fig. 18.1. For substantially more detailed information on the electrochemistry involved, nice presentations prepared for those not familiar with the subject are given in Refs. 1, 2, and 12.

The electrochemical operation of a fuel cell, simple in principle, is complicated by a number of factors in practice. First, there are substantial losses in the cell voltage output as a consequence of activation and diffusional polarization at the electrodes, together with microscopic irregularities in surface chemistry that in effect give local short circuits. Second, there are appreciable electrical resistances in the electrodes, and especially the electrolyte, which lead to I^2R losses. These result in a

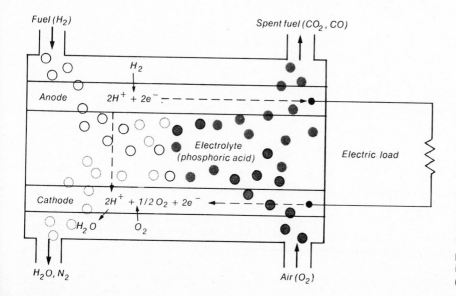

Figure 18.1 Operation of a fuel cell is based on the electrochemical reactions between a fuel (in this case a hydrogen-rich gas made from coal or oil) and an oxidant (in this case oxygen in air). At the anode of the cell, hydrogen molecules are oxidized (an electron is removed from each of the two hydrogen atoms in the molecule) to form hydrogen ions. The ions are transported through the electrolyte to the cathode, and the electrons flow through an external circuit to the cathode, producing power. At the cathode hydrogen ions, electrons, and oxygen form water.[1] (*Courtesy Scientific American.*)

substantial loss in voltage output, and the I^2R loss produces heat that must be removed through a cooling system. The voltage losses, together with the power required for pumping the coolant, detract from the overall thermal efficiency of the cell. The I^2R losses can be reduced by minimizing the thickness of the electrolyte and the active portion of the electrodes, but the degree to which this can be done is limited by the requirement that the electrodes be fairly uniformly spaced and not touch each other to avoid shorting of the electric current.

The polarization and I^2R losses — as in conventional batteries

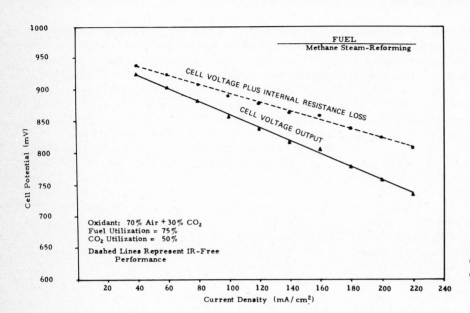

Figure 18.2 Effects of current density on cell output voltage for a molten-carbonate fuel cell.[26]

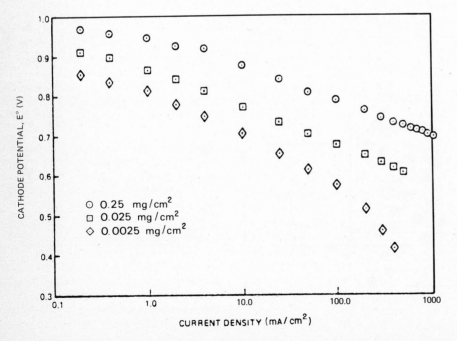

Figure 18.3 Cathode performance on oxygen for various loadings of platinum supported on Vulcan XC-72 at 160°C in 96% H_3PO_4 (Ref. 31).

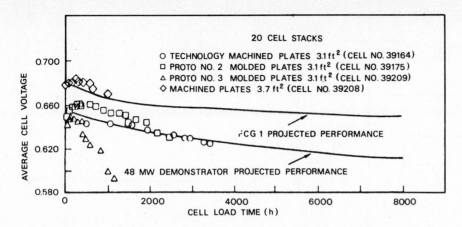

Figure 18.4 Twenty-cell stack performance histories.[43]

—increase with the reaction rate (i.e., an increasing current), and this decreases the potential difference between the anode and cathode.[25] Thus, the cell output falls off with increasing load as shown in Fig. 18.2, which gives a typical curve for cell voltage as a function of current density. The reaction rate can be greatly increased for any given potential difference by employing a catalyst to aid in the ionization of the reactants. Figure 18.3 shows the beneficial effects of introducing platinum as a catalyst into one or both of the electrodes.[26] Unfortunately, the performance of the catalyst tends to deteriorate so that the output of a fuel cell falls off with running time;[27-43] some typical curves are shown in Fig. 18.4. The prime reaction products may have an adverse effect on the cell; for example, water from the hydrogen-oxygen reaction will dilute acid or alkaline electrolytes. Thus, one of the complications of aqueous cells is that provisions must be made for maintaining the electrolyte at the proper concentration.

PRINCIPAL TYPES OF FUEL CELLS UNDER DEVELOPMENT

Five types of fuel cells have been under active development for commercial service. Three of these make use of relatively low-temperature aqueous solutions: one, KOH; the second, phosphoric acid; and the third, a sulfonated solid polymer that also contains water. Note that the operating temperature of aqueous cells is limited by loss of water vapor from the electrolyte; hence they must be pressurized sufficiently to avoid boiling. One high-temperature fuel cell employs molten carbonate as the electrolyte at a temperature of about 600°C (1112°F), while the other employs solid ZrO_2 at a temperature of about 1100°C (2012°F), which is sufficiently high that the ZrO_2 crystals permit the migration of negative oxygen ions. The principal features and

characteristics of each of these types are discussed briefly here. A much more complete picture is given in Ref. 12.

Alkaline Fuel Cells

The work of Bacon in Britain during the 1930s and 1940s developed the alkaline fuel cell to the point where it became a credible power source (and hence is often referred to as the Bacon cell).[14] This type of cell has served as the electric power source for U.S. manned spacecraft. The Apollo fuel cells made use of an 85% KOH electrolyte operating at 250°C (482°F) at a pressure of about 3 atm with a weight of about 45 kg/kW and using very pure gaseous hydrogen and oxygen as the fuel and oxidant.

A major disadvantage of the alkaline fuel cells is their sensitivity to trace amounts of carbon dioxide; CO_2 leads to carbonate formation in the cell and a progressive loss of output with operating time. This presents a serious problem, because the carbon dioxide content of atmospheric air is sufficient to cause difficulty in a matter of days, and hydrogen obtained from hydrocarbons or coal inherently contains CO_2. Reducing the CO_2 content to an acceptable level is quite expensive.

Phosphoric Acid Fuel Cells

Phosphoric and sulfuric acids have been used as electrolytes in fuel cells, but phosphoric acid is greatly preferable because it is more stable chemically. This type of cell is insensitive to CO_2 in the fuel or oxidants and works well using air as the oxidant. However, these cells are sensitive to small amounts of CO. Hence CO must be removed from H_2 obtained from reformed hydrocarbons.

An operational fuel-cell system is extremely complex, unlike the simple lead-acid battery that most people imagine it to resemble. The principal elements in a typical phosphoric acid system are shown in the schematic diagram of Fig. 18.5 for the

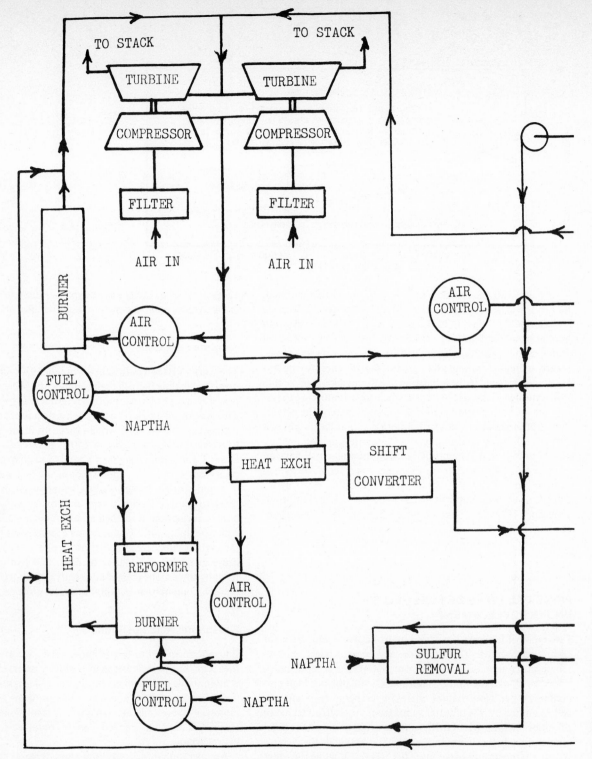

Figure 18.5 Flowsheet for a phosphoric acid fuel-cell power plant, including the reformer and gas purification equipment and cell cooling system.[43] (*Courtesy Electric Power Research Institute.*)

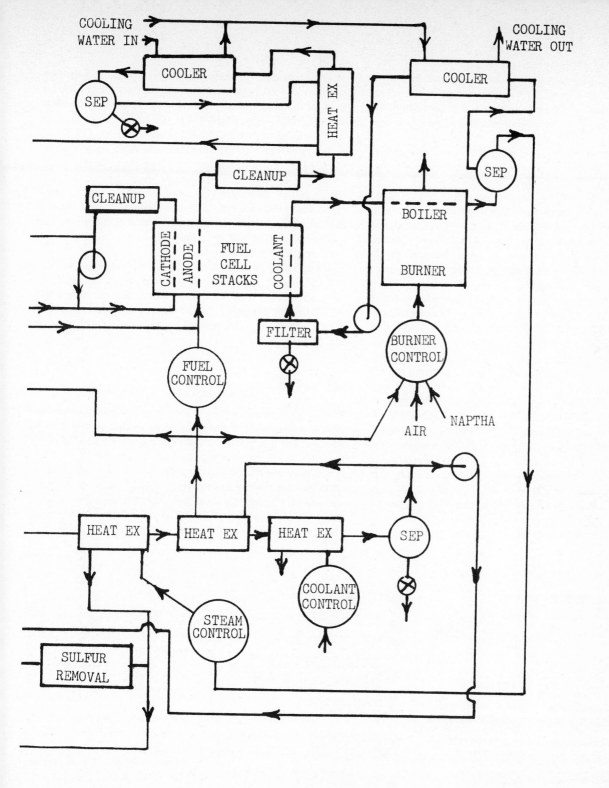

phosphoric acid fuel-cell system under development by United Technologies Corporation.[17,18] The fuel-cell stacks are represented by the small rectangle just above and to the right of the center of the flowsheet. The components of the reformer system for converting naphtha to hydrogen and purifying the product are in the lower half and upper right. The upper left quarter shows the gas turbine required to circulate the reactants and provide cooling. Note that the instrumentation and control problems are quite complex; the electrolyte must be kept at the right concentration with an evaporator to remove the water produced by the oxidation of hydrogen, the flow rates of the H_2 and air must be regulated, the temperature of the cell must be controlled, etc. The electrical equipment required to invert the dc output of the cell to alternating current at the desired frequency and voltage is not included in Fig. 18.5.

As is the case with the alkaline cells, it is possible to operate phosphoric acid fuel cells without using a precious metal catalyst (in spite of its high cost, platinum is the most cost-effective catalyst for aqueous cells), but unless some platinum is employed, the cell output per unit of area is not sufficiently high to be attractive. The capital cost of the platinum required is easily justified for spacecraft but becomes an important consideration for utility service. The NASA fuel cells, for example, have employed from 3 to 10 mg/cm^2, or 15 to 50 g/kW. At \$14.50/g (the 1980 market price of platinum), this would give a capital investment just for the platinum of \$218 to \$725/kW. Current work on phosphoric acid cells for utility service[17,18] is directed toward cells with platinum requirements of approximately 0.75 mg/cm^2, which corresponds to a capital investment for platinum of \$70 to \$140/kW, depending on the rated cell current density.

Significantly, the Pratt & Whitney fuel cells built for the Apollo program were of the alkaline type, because they operated on pure hydrogen and oxygen, whereas the subsequent work of the same organization (now the Power Systems Division of United Technologies Corporation) has been concerned mostly with phosphoric acid cells. Because the cost of hydrogen and oxygen would be excessive, the cells must operate on reformed hydrocarbons and air. This type of cell also has been under development for the U.S. Army Mobility Equipment Research and Development Command for use in mobile equipment, where there is a strong incentive to employ air as the oxidant and a liquid hydrocarbon as the source of hydrogen. The mobile units can be built to operate with a low noise level, and a long service life is not required.

For an application in which there is a use for the waste heat (such as cogeneration of electricity and heat for a housing complex), some of the waste heat can be recovered from phosphoric acid cells. For example, at full load, about 12 percent of the energy in the fuel can be recovered in the form of steam at 1 to 4 atm (15 to 60 psi).[21]

Molten-Carbonate Fuel Cells

Molten-carbonate fuel cells have been under development for 30 years.[16,18,26] Schematic diagrams showing the construction of both a cell of this type and a cell stack assembly are included in Fig. 18.6. In this construction the molten-carbonate electrolyte is contained in a porous ceramic matrix that serves to maintain a small uniform spacing between electrodes. These cells will operate with carbon, hydrogen, carbon monoxide, or hydrocarbons as the fuel and hence are well adapted to the use of liquid or gaseous fuels derived from coal. The high operating temperature

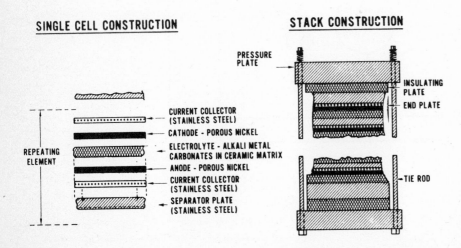

Figure 18.6 Schematic diagrams showing (a) the arrangement and functions of the components of a molten-carbonate cell and (b) the details of construction of a cell stack.[17]

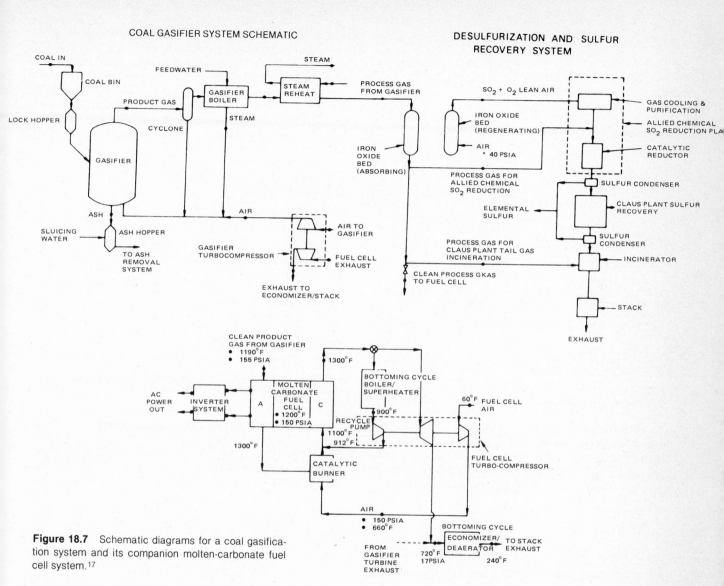

Figure 18.7 Schematic diagrams for a coal gasification system and its companion molten-carbonate fuel cell system.[17]

(about 600°C) permits the chemical reactions to proceed rapidly enough that a precious metal catalyst is not required. Further, the waste heat from the cell can be used to advantage in a thermodynamic cycle or for process heat. However, the high operating temperature has presented difficult problems with sealing, corrosion, and differential expansion in the brittle, low-strength nickel oxide plates, problems that have yet to be solved as of 1981 in the various efforts to develop long-lived units.

Figure 18.7 is a schematic diagram showing the relation of a molten-carbonate fuel cell to the coal gasifier from which it gets

its fuel and to a steam bottoming cycle employed to raise the overall plant efficiency to values estimated to be as high as 49 percent.[17] This type of cell has been the subject of major development programs at the Institute of Gas Technology, United Technologies Corporation, and the Energy Research Corporation. It is widely viewed as the most promising of all the fuel-cell systems for the production of electric power for intermediate- and base-load electric utility applications. It was the only type of fuel cell considered in the Energy Conversion Alternatives Study (ECAS) (Chap. 22) that looked as if it might be more promising than a conventional steam system for operation with coal as the basic fuel.[17,27,28]

Solid Oxide Electrolyte

This type of cell depends on the fact that a compact of ZrO_2 containing about 15% CaO permits fairly rapid migration of oxygen ions at approximately 1000°C (2012°F) while remaining a poor conductor of electrons.[13] These cells operate well with a fuel gas consisting of hydrogen and carbon monoxide using air as the oxidant. The cell operates at a high temperature, and so the waste heat can be used to advantage in a thermodynamic cycle or for process heat. The major disadvantage is that the internal electrical resistance losses are fairly high, partly because of the resistance of the solid electrolyte and partly because of contact resistance problems between the solid electrolyte and the metal electrodes. Much of the work on this type of cell has been carried out at Westinghouse, initially under contract with the Office of Coal Research. At the time of writing, efforts were being directed primarily toward materials combinations that would withstand the severe corrosion conditions at 1000°C sufficiently well to keep degradation of the interconnectors in the cell stack to an acceptable level for a cell life of at least 30,000 h.

Solid Polymer Fuel Cells

Solid polymer fuel cells make use of a sulfonated fluorocarbon similar to Teflon as the electrolyte.[25] Units of this type have been under development by General Electric for the space program for nearly 20 years (they provided the electric power for the Gemini spacecraft). A cell of this type has been operated on pure hydrogen and oxygen for 40,000 h at a current density of approximately 100 mA/cm² and 82°C with a platinum loading of approximately 8 mg/cm² (which corresponds to a platinum investment of $550/kW). In this test the degradation in cell output was only about 4 percent. The principal disadvantages of this type of cell for utility service are the high capital cost of the platinum required, intolerance to CO, and a lower energy conversion efficiency than other types.

Comparison of the Characteristics of Typical Fuel Cells

The many different parameters and criteria involved for the various types of fuel cell are confusing. To help clarify the situation, Dr. H. G. Corneil of Exxon drew up Table 18.1 and suggested its use to the author.[29] Note that the efficiencies given in Table 18.1 are based on the chemical energy in the fuel gas supplied to the cell and do not include losses in the preparation of that gas. Note also that the electric power required to make gaseous oxygen with no special purification (i.e., for the basic oxygen steel production process) is approximately 400 kW per ton. This is equivalent to approximately 40 percent of the lower heating value of the hydrogen fuel after allowing for the overall thermodynamic efficiency for conversion of heat into the electricity required to drive the compressors to make the oxygen.

PRINCIPAL PROBLEMS AND PARAMETERS

There is no question that fuel cells will produce electric power; the question is one of economics. The economics involves both the power output obtainable per unit of surface area and the

TABLE 18.1 Summary of Fuel-Cell Operating Characteristics (Ref. 42)

	Electrolyte				
	Solid polymer	KOH	Phosphoric acid	Molten carbonate	Solid oxides
Fuel	H_2	H_2	Syn. gas	Syn. gas	Syn. gas
CO tolerant	No	No	Yes	Yes	Yes
CO_2 tolerant	Yes	No	Yes	Yes	Yes
Operating temperature, °C	50–100	50–100	150–200	650	1000–1200
Cell efficiency	0.45	0.55	0.45	0.55	0.58
Heat rate, Btu/kWh	9300	7500	9000	7500	6900
Platinum required	Yes	No	Yes	No	No
Developers (in 1978)	G.E.	UTC, Alsthom-Exxon	UTC	UTC	Westinghouse

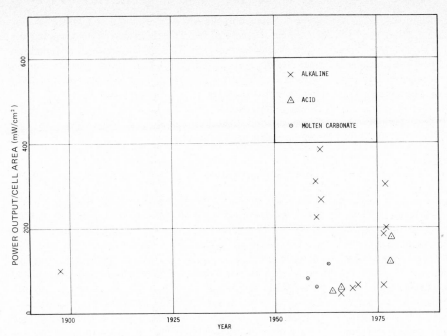

Figure 18.8 Reported outputs of fuel cells as a function of year of tests. (*Courtesy Oak Ridge National Laboratory.*)

cost of the cell, because the greater the output from a given size of cell, the lower its cost in dollars per kilowatt. The literature includes numerous curves showing the voltage obtainable from cells as a function of the output per unit of area (such as Fig. 18.3) with particular attention to the effect of the amount of platinum catalyst required per unit of area in aqueous cells. This brings out a third consideration: the length of the cell's operating life. There seems to be widespread agreement that a fuel-cell lifetime of at least 30,000 h will be necessary if costs are to be kept reasonable.[17] This stems from the fact that the deterioration in any type of fuel cell tends to be of such a character that the material in the fuel cell must be completely reprocessed, much as is the case for lead-acid batteries. Thus, the capital cost of the cell must be written off against an operating life of only 4 to 6 years.

Overall Efficiency

The high efficiency of the ideal fuel cell is commonly confused with the actual efficiency of a fuel-cell power plant. Even within a single fuel cell, the electric losses from polarization and internal resistance are substantial, so that individual fuel cells operating on H_2 and O_2 commonly give an efficiency of approximately 55 percent for alkaline cells and approximately 45 percent for phosphoric acid cells when the plant is operating at full power. Resistance losses in connections are appreciable, and

there are power losses in the inverter (which is likely to have an efficiency of approximately 96 percent). Losses in reforming the hydrocarbon fuel to give hydrogen are substantial, as are pumping losses for circulating the fuel and airstreams and cooling the cell stacks. As a consequence, the overall thermal efficiency for producing electricity from the chemical energy in liquid hydrocarbon fuel has generally been estimated to be poor—less than 25 percent according to General Electric[28] and Westinghouse.[27] With natural gas as the fuel, the naphtha reforming losses are avoided so that the fuel-cell power units in one extensive test program gave an efficiency of approximately 30 percent.[30] Values around 36 percent have been estimated by United Technologies Corporation for their advanced system under development at the time of writing.[24]

Fuel-Cell Unit Output

Some indication of the progress in improving the fuel-cell output per unit of area can be gained from Fig. 18.8, which shows experimental points on fuel-cell performance taken at random from papers that have appeared in the literature over the past 20 years. It should be remembered that in the 1890s Jacques built many cells that produced about 70 mW/cm^2 from an alkaline type of cell with no platinum catalyst.[3]

Cell performance depends in substantial measure on fuel purity and detailed fabrication techniques affecting such factors as the size, character, and dispersion of pores in the electrodes, platinum form and dispersion, plate spacing, etc.[31-44] Further, the output for short periods can be much higher than for steady-state operation. In view of this, the spread of points for the cells in Fig. 18.8, while it may at first seem large, is rather less than might otherwise be expected. It should be remembered also that the cell operating temperature, the electrolyte concentration, the flow rates of fuel and oxidant through the cell, etc., also have important influences on the cell output.

Platinum Loading

It is extremely difficult to estimate the cost of a fuel cell because it is a function not only of the quantity of expensive materials required per unit of output but also of the fabrication processes. The cost of obtaining a particular porous structure in nickel, for example, may be many times the cost of the nickel required. However, most investigators have concluded that in spite of its high cost per unit of weight, platinum is more effective on a cost basis than any other catalyst for aqueous cells.

The cost of a scarce material such as platinum depends on the demand. If phosphoric acid fuel cells were to be employed for peaking service capacity and installed at a high enough rate to justify their development (e.g., at the rate that prevailed for gas turbines during the 1970–1975 time interval), about 6000 MW of new capacity would be required per year. This would represent approximately 20 percent of the world's annual production of platinum if one uses a value of 0.75 mg/cm² as the platinum requirement, and it would probably have a large effect on the price. Inasmuch as the cost of platinum is so high that it may dominate the cost of a cell, it is instructive to examine the amount employed per unit of area as a function of time. The scattered data in the literature indicate that the platinum loading was reduced by a factor of about 3 for phosphoric acid cells in the 1960s, but there was little further reduction in the 1970s. Of course, a reduction in the amount of platinum per unit of area is likely to reduce the cell life as well as the cell output. These effects are indicated in Fig. 18.9.

Fuel-Cell Electrocatalysis

The highly subtle and involved phenomena involved in the electrocatalysis processes in fuel cells are vital[31,32] but beyond the scope of this report. In simplistic terms, the key reactions in a typical cell of the types of interest here occur where the electrolyte contacts both the crystallites of catalyst in the capillary passages of the electrode and the fuel gas or air. Because diffusion rates in liquids are relatively slow, most of the reactions occur in the liquid meniscus region in the capillary passages, a volume of microscopic thickness. The greater the total surface area in these meniscus regions, the greater the reaction rate and hence the greater the cell current output for a given voltage output. Thus, the art of fabricating aqueous fuel cells involves applying a microscopically thin layer of platinum to the inner surfaces of a fine matrix of capillary passages, and the character of the resulting surface should give just enough wetting to provide a maximum of liquid meniscus wetted surface area, but not so much wetting that the electrode pore structure becomes flooded

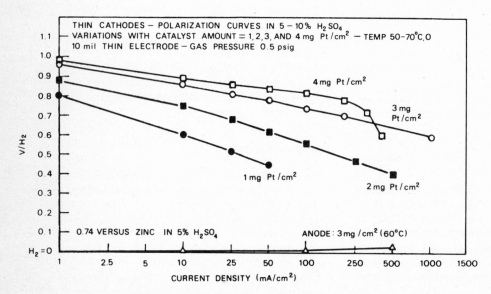

Figure 18.9 Effects of the amount of platinum catalyst on the performance of an acid H_2-O_2 fuel cell.[6]

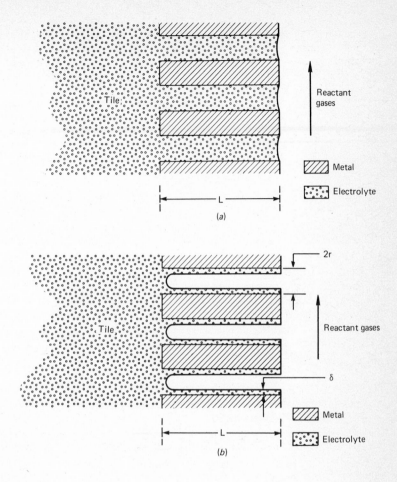

Figure 18.10 Schematic representation of an electrode (a) with flooded pores and (b) with film-covered pores (Ref. 44).

with electrolyte (Fig. 18.10). Clearly this entails a delicate balance that is sensitive to many different subtle factors that depend on the surface chemistry, particularly after long periods of running during which trace amounts of impurities in the feed gas (e.g., sulfur or CO) can contaminate the surface and lead to a deterioration in cell performance. Another factor that has proved to be troublesome in phosphoric acid cells is dissolution of the fine crystallites of platinum and their redeposition in the form of a coarser structure that presents less surface area and hence gives a lower cell output (Fig. 18.11). These problems are discussed in Refs. 31 and 32. Somewhat similar problems have been experienced with alkaline fuel cells.[20,33]

In view of the extreme complexity of the surface chemistry involved in the electrodes and its sensitivity to trace impurities in the fuel, it is not surprising that so much talent, money, and years of effort have been expended in efforts to produce long-lived cell stacks with only limited success.

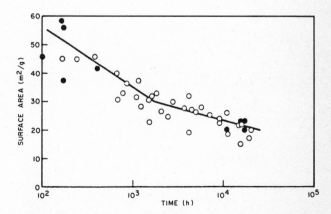

Figure 18.11 Surface area loss with time at 190°C for platinum supported on Vulcan XC-72 (o determined electrochemically; ● determined using transmission electron microscope.)[31]

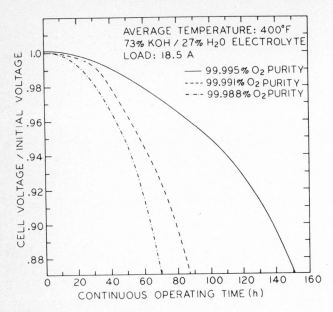

Figure 18.12 Performance degradation as a function of running time for an alkaline cell running on oxygen contaminated with small amounts of CO_2 (Ref. 34).

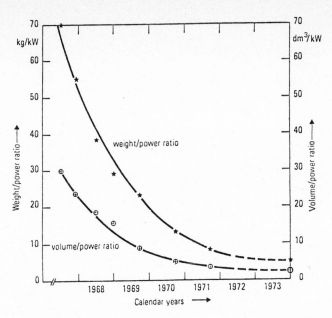

Figure 18.13 Status of volume/power ratio of alkaline fuel-cell stacks as a function of calendar year.[39]

TABLE 18.2 Life Expectancy of $\frac{1}{4}$-in Flat Carbon Electrodes (Current Density, 40–50 mA/cm²)*

Date of testing	Percentage of cells surviving			
	500 h	1000 h	2000 h	4000 h
1959	30	10		
1960	50	20	10	
1961	80	60	10	
Early 1962	90	80	50	10
Mid-1962	95	90	80	20*

*Data from Ref. 7; tests were still in progress at the time the reference was prepared.

Fuel-Cell Stack Life Expectancy

There are scattered bits of data in the literature on the useful life of individual fuel cells built for research purposes. These units are ordinarily simple disks or cylinders of rugged construction, and the life data are commonly for operation at low current densities. The engineering compromises involved in building fuel-cell stacks or modules with many large cells to build up the output power and voltage lead to a host of problems, such as warping and cracking of large, thin, brittle plates under the thermal stresses inherent in power cycling and nonuniformities in both current and coolant distribution, development of internal shorts, electrolyte leakage, increases in the contact resistance between elements of the cell stack, etc. The rare bits of data on the life expectancy of fuel-cell stacks given in the literature usually cite operating times of from 500 to 2000 h. One of the problems seems to be that small differences in the fabrication or operation of fuel cells lead to marked differences in life span from one cell to another in a set of seemingly similar cells. For example, the effects of oxygen purity on the performance of an alkaline H_2-O_2 cell are indicated in Fig. 18.12.[34] Another indication of random variations is given by Table 18.2, which summarizes statistical data on endurance tests of individual alkaline fuel cells built in plate form to give the same geometry as would be employed in a full-scale fuel-cell stack or module.[7] As one might expect, some cells failed early in the test, while a few lasted up to eight times as long. The failure of one cell in a stack, will, at the minimum, affect the output of the stack as a whole, and it may completely disable a module.

Specific Weight and Volume

There have been major advances in making fuel cells more compact and lighter to meet the stringent requirements of the space program, as indicated in Fig. 18.13. The incentives to obtain low specific weight and volume are also substantial in the automotive and the Navy DSRV (Deep Submergence Research Vehicle) programs, but are not pronounced for utility service. For vehicle applications, however, if an inverter is used, its weight will run approximately 9 kg/kW (approximately 20 lb/kW), and the specific weight of the complete power plant is much more than that of the cell stacks.

Fuel Reforming

For utility service, it was clear at the inception of the Pratt & Whitney program on fuel cells for commercial use that oxygen and gaseous hydrogen would be far too expensive. Hence their fuel-cell power plant was designed to operate on air with a reformer to convert a hydrocarbon fuel to hydrogen. Hydrocarbon reformers commonly employ a catalyst such as nickel operating at approximately 600°C to convert a mixture of steam, hydrocarbons, and air into a mixture of H_2, CO, CO_2, H_2O, and N_2 with as little CO as possible. For good efficiency it is desirable to maximize the output of H_2 and minimize dilution by steam, nitrogen, and CO. However, fouling of the catalyst by carbon is a problem if the ratio of steam to carbon is reduced in an effort to improve the H_2 content of the product. Fouling of the catalyst by sulfur is a problem if a material such as No. 2 fuel oil is employed, because some fuel oils contain as much as 2000 ppm of sulfur. For a long life of the catalyst in the reformer, it will probably be necessary to keep the sulfur content of fuel oil feed stock below 100 ppm.[35] A still lower level of sulfur must be maintained in the gas fed to the fuel cells to avoid deterioration of the catalyst in the course of long periods of operation.[34] There are indications that the sulfur content of the fuel for phosphoric acid cells must be under approximately 10 ppm, and it must be under approximately 3 ppm for molten-carbonate cells.[36] (Note the ZnO bed in the lower center of the flowsheet of Fig. 18.5; this is required for H_2S absorption.) In practice, the reformer has been found to present a formidable set of problems. The kinetics of reforming also depend on the feed (e.g., fuel oil is more difficult to handle than natural gas). The complexity of the system required is indicated in Fig. 18.5. Keeping the capital cost to acceptable levels while yet achieving a long life, high reliability, and high fuel conversion efficiency has proved to be difficult.

Inverter

Another major set of problems is presented by the inverter for converting the dc output of the cell stacks into ac current.[37] The weight, volume, and cost of these units goes down as the voltage of the fuel-cell stacks is increased, but an increase in voltage also

increases the tendency toward breakdown of the insulation for the interconnectors in the cell stacks and between cells and ground. The UTC FCG-1 Module Demonstrator cell stacks of the system of Fig. 18.5 are designed to give an output of 5000 V.

PERFORMANCE OF TYPICAL FUEL-CELL PLANTS

A great deal of work on fuel cells has been carried out in the past 30 years, and much of it is well documented (see the comprehensive bibliography in Ref. 38 and the informative survey paper of Ref. 39). Much progress has been made and many small units have been built and operated. Table 18.3 summarizes this operational experience with fuel-cell stacks operated in the United States at outputs of 1 kW or more. (A typical automobile storage battery can deliver about 1 kW for a short period.)

Numerous performance projections and system cost estimates have been prepared on the assumption of an attractively

TABLE 18.3 Summary of Operating Time on Fuel-Cell Stacks as of June 1977

	Total time, h	Longest life, h
NASA manned space flight program (1 to 2 kW stacks)(4 to 10 mg Pt/cm²)		
Johnson Space Center (alkaline)	16,500	5,000
Allis Chalmers (10 mg Pt/cm²) (alkaline)	20,000	3,300
Pratt & Whitney (alkaline)	53,400	5,000
Apollo flight operation (alkaline)	10,900	5,000
G.E. (4 mg Pt/cm²)(solid polymer, 350-W stacks)	80,000	1,938
Navy DSRV program (20 to 30 kW stacks)(alkaline)	2,000	
Union Carbide (automotive units)(alkaline)	50,000	2,000
United Technology Corporation (utility units) Phosphoric acid		
65 PC-11 (12.5-kW modules)	205,000	7,700
1 PC-18 (40-kW module)	8,000	8,000
3 20-cell stacks (3.1 ft² cells)	7,160	3,647
1 20-cell stack (3.7 ft² cells)	1,001	1,001
Molten carbonate		
10-cell stack	2,000	1,400
20-cell stack	1,500	1,000

long life with reasonable costs; often there is an assumption of a smaller amount of platinum catalyst per unit of area than employed in the cells that have been tested. Table 18.4 presents projections of this sort. It is difficult to judge the validity of these estimates because of scaling problems. Increasing the size of cells to many square feet leads to deviations from ideality in the local composition of the fuel and airstreams flowing through the electrodes, because the H_2 and O_2 contents change in the direction of flow, and crossflow conditions inherently prevail.[26] Pressure stresses stemming from the pressure drop in the passages for the reactant gases and the coolant become important, as does even the static head associated with the greater cell height. The large voltage drop to ground ($\sim$ 5000 V) associated with the large number of cells per stack required for good efficiency introduces severe electrical stresses. These and other scaling problems introduce formidable difficulties with irregu-

larities in temperature distribution, differential expansion, thermal stresses, sealing, and electrical insulation.

Phosphoric Acid Fuel Cells

The only aqueous fuel-cell stacks under development for utility service at the time of writing are the phosphoric acid cells built by United Technologies Corporation (UTC). In the early 1970s UTC built and delivered 65 complete self-contained power units of 12.5 kW output with integral natural gas reformers.[40,41] These units were operated unattended by 35 different organizations with an average requirement for servicing of only one call per 1000 h of operation. The total running time on these units was 205,000 h. Most units showed some loss in performance at the end of the 2000-h life for which they were rated, but one ran for 7700 h. A 40-kW unit was then built and operated for 8000 h, and subsequently a 1-MW unit was built and operated for 1069 h. The main effort at the time of writing is centered on a 4.8-MW unit that employs 456 cells per stack with 3.7 ft[2] per cell and 5000 V output per stack.[42,43] (A flowsheet for the power plant is given in Fig. 18.5.) This unit was installed in Manhattan by the Consolidated Edison Company for field tests in 1980. It is envisioned that six of these units would be used to form 27-MW power plants suitable for substations. The fuel used is a petroleum distillate. Ten intermediate-size experimental 20-cell stacks of the type planned for the 4.5-MW unit have been built and endurance-tested. One ran for 8200 h, three ran for 3300 to 3500 h, three ran for about 2000 h, and three ran for only $\sim$ 1000 h.

A somewhat similar program funded by the Gas Research Institute, DOE, and UTC entails the development of a 40-kW power unit designed to operate on natural gas. A complete unit has been operated for 15,000 h, with the first cell stack replaced after 8000 h. Reduced-scale cell stacks of about 1.2 kW have been performance- and endurance-tested to develop improved fabrication methods. One of these was run for 3000 h and two for 2000 h. A number of units were being built for field testing at the time of writing.

TABLE 18.4   Typical Design Parameters for Fuel-Cell Power Units (Data from Ref. 27)

	Electrolyte type	
	85 wt. % H_3PO_4	Paste of Li, Na, K carbonates and alkali aluminates
Power output, MWe	23.4	22.6
Fuel-cell rating, MW	25	25
Fuel		
High-Btu gas	X	X
Medium-Btu gas		
Methanol		
Oxidant		
Air	X	X
Oxygen		
Fuel-cell life, 10^3 h	30	30
Voltage degradation, %	5	5
Temperature, °C	190	650
Electrolyte thickness, cm	0.05	0.1
Anode		
Type	Pt/C	Ni
Catalyst loading: mg Pt/cm^2	1.0	
Cathode		
Type	Pt/C	Lithiated NiO
Catalyst loading: mg Pt/cm^2	1.0	
Current density, mA/cm^2	200	200
Average cell voltage, V	0.7	0.7
Power plant efficiency, %	35.5	48.8
Overall efficiency, %	23.9	32.9

Molten-Carbonate Cells

Table 18.5 presents an excellent chronological summary of the principal developments in molten-carbonate fuel-cell research in the past 60 years, with the bulk of that work having been carried out during the past 20 years.[16,26,38,44] There were $\sim$ 40 items in the original table (taken from an exceptionally fine and informative survey made by the Institute of Gas Technology, which presents the developmental history and key problem

TABLE 18.5 Compilation of Molten-Carbonate Fuel-Cell Developments (Ref. 44)

Year	Investigator(s) (affiliation)	Electrolyte (composition)	Support or "free"	Temperature, K	Anode (fuel gas)	Cathode (oxidant)	Performance at 0.8 V* Power flux,† mW/cm²	Performance at 0.8 V* Life, hours	Comments
1921	E. Baur et al. (Eidgenossische Technischen Hochschule, Zurich)	$(Na, K)_2CO_3$ (equimolar)	MgO	~1100	Fe/MgO (H_2)	Fe_3O_4/MgO (air)	3	4	Power flux of 4 mA/cm² at 0.75 V.
1946	O. K. Davtyan (Academy of Sciences, Moscow, Russia)	Na_2CO_3 + oxides of Ca, Ce, P, Sn, Th, W	Sand, uralmonazite, clay, quartz (35%)	973	$Fe_2O_3/Fe/clay$ $(H_2 + CO)$ 1:1	$Fe_2O_3/Fe_3O_4/clay$ (air)	24	15	OCV = 0.85. A linear current-potential curve is shown. Power flux of 32 mA/cm² at 0.75 V.
1958	G. H. J. Broers (Univ. of Amsterdam, Netherlands)	$(Li, Na, K)_2CO_3$ (eutectic)	MgO sinter (50%)	923	Ni pdr/Ag gze $(H_2 + H_2O)$ 1:1	Ag pdr/Ag gze (air + CO_2)	40	500	Loss of electrolyte by decomposition and vaporization of CO_2, Li_2O, Na_2O and K_2O. Cells experienced much gas leakage and gasket problems.
1959	E. Gorin (Consolidated Coal Co., Pa.) and H. L. Recht (Atomic International, California)	$(Li, Na)_2CO_3$ (equimolar)	MgO (72%)	1023	Ni-porous $(H_2 + H_2O)$	Ag-gauze (air + CO_2)	48	300	A lithiated NiO-sinter is also successfully used, but gives higher cathode resistance. Found much electrolyte loss and high contact resistance.
1960	H. H. Chambers and A. D. S. Tantram (Sondes Place Res. Inst., Dorking, England)	$(Li, Na)_2CO_3$ (eutectic)	MgO (50%)	923	ZnO-porous/Ag pdr (H_2)	ZnO-porous/Ag pdr (air)	55	1,000	Severe corrosion of cell components. Dual porosity electrodes in "free" carbonate version of cell. Kerosene vapor gives almost same performance as H_2.
1962	Y. L. Sandler (Westinghouse Electric Corp., Pittsburgh, Penn.)	$(Li, Na, K)_2CO_2$ (3:4:3)	MgO (55%)	853	Ni mesh/Ni pdr (H_2)	Ag mesh/Ag pdr $(O_2 + CO_2)$ 1:2	110	170	Secondary reactions at fuel electrode produce CO and CH_4. Short life due to decomposition and loss of carbonates. Cells had improved electrode/electrolyte contact.
1963	I. Trachtenberg et al. (Texas Instruments, Inc., Dallas, Tex.)	$(Li, Na)_2CO_3$ (eutectic)	MgO	873	Ag (H_2)	Ag (air)	65	500	Thermal cycling and 0.2 vol% H_2S in fuel had no degrading effects in 10-day test.
1963	L. G. Marianowski and B. S. Baker (Inst. of Gas Technology, Chicago, Ill.)	$(Li, Na)_2CO_3$ (eutectic)	MgO (70%)	873	Pd-Ag foil (H_2)	Ag paint (air)	13	>4,000	Performance not seriously affected by presence of CO, CO_2, CH_4 in fuel. CO_2 and H_2O generated at anode cannot be vented through foil.
1966	S. E. Chuck (Illinois Institute of Technology, Chicago, Ill.)	$(Li, Na, K)_2CO_3$ (eutectic)	Li AlO₃ (paste)	1030	Ni-fiber mat (CO, 100%)	Ag paint/Ag screen $(O_2 + CO_2)$ 1:1	24	—	SS-encased cell shows considerably lower level of performance compared to ceramic-encased cell, possibly due to catalytic activity.
1967	A. Clauss and G. Genin (SERAI, Brussels, Belgium)	$(Li, Na)_2CO_3$ (eutectic)	MgO	900	Ni-screen (74μ) in 4 layers $(CO_2 + H_2 + H_2O)$	NiO screen $(O_2 + CO_2)$	64	486	Concludes structure of Ni electrodes only of secondary importance at least within the limited range of experimental data offered. Flooding and tile fissures were reasons for cell failure.

*Where performance is not at 0.8V cell output, this is stated in the comments.

†1 mW/cm² = 0.93 WSF (W/ft²).

TABLE 18.5 Compilation of Molten-Carbonate Fuel-Cell Developments (*continued*)

Year	Investigator(s) (affiliation)	Electrolyte (composition)	Support or "free"	Temperature, °K	Anode (fuel gas)	Cathode (oxidant)	Power flux,† mW/cm²	Life, hours	Comments
1967	A. Salvadori et al. (Gaz de France, France)	(Li, Na)₂ CO₃ (eutectic)	Li AlO₂ (50%)	923	Ni-porous (H₂)	Ag-porous (air + CO₂) 7:1	50	2,000	Planar cell at 0.75 V output. Use of LiAlO₂ recognized as important milestone in molten-carbonate fuel-cell work, affording greater stability and less corrosion.
1967	A. D. S. Tantram (National Research Development Corp., England)	(Li, Na) (CO₂, OH) 1:1 8:1	—	873	ZnO-porous/Ag film (H₂)	ZnO-porous/Ag film (air + H₂O) 5:2	30	—	Output of 0.75 V with 40 mA/cm². Hydroxide in electrolyte may be formed *in situ* by operating with H₂O on the cathode side.
1967	D. R. Warren (Aeronautical Research Lab., Melbourne, Australia)	(Li, Na, K)₂CO₃ (eutectic)	MgO (53% paste)	973	Ag-grid/catalyst (H₂)	Ag-grid/catalyst (air + CO₂) 7:3	110	—	Output of 0.9 V with 120 mA/cm². Some results with Al₂O₃ and ZrO₂ matrix material indicate better performance with the latter.
1968	B. S. Baker et al. (Inst. of Gas Technology, Chicago, Ill.)	(Li, Na, K)₂CO₃ (eutectic)	LiAlO₂ (paste)	980	Ni-porous (Reformed JP-4 + H₂O)	NiO-porous (air + CO₂ + H₂O)	40	2,600	Fuel reforming operation and electrochem. reaction occur in an "integral" reactor operated on JP-4 and air. Small (3 cm²) cells were operated up to 13,000 h.
1969	R. P. Hamlen et al. (General Electric Co., Schenectady, N.Y.)	(Li, Na, K)₂ CO₃ (eutectic)	MgO	873	3 Pd-Ag foil (H₂)	Ag-porous (O₂ + CO₂) 1:2	10	—	A dual porosity matrix allows reaction products to pass internally from anode to cathode side. Anode bulging was observed.
1976	B. S. Baker (Energy Research Corp., Danbury Conn.)	(Li, K)₂CO₃ (1:1)	LiAlO₂	973	Ni-sinter (H₂ + CO₂) 4:1	NiO-porous (O₂ + CO₂) 1:2	144	—	Perf. from initial peak characteristics. Problems of gas sealing, tile reinforcement, and electrolyte preparation.
1976	J. M. King (United Technologies Corp., South Windsor, Conn.)	Molten carbonates (unspecified)	—	923	—	—	100	5,000	Performance shown was obtained on subscale cells. Immediate design goals for fuel cells in commercial operation are 45% power-plant efficiency (7500 Btu/kWh heat rate) at $200/kW initial cost (1975 dollars) and 5-year operating lifetime.
1976	L. G. Marianowski et al. (Inst. of Gas Technology, Chicago, Ill.)	Molten carbonates (unspecified)	—	—	—	—	160	40,000	Long-term runs with laboratory-sized cells.
1976	S. Szymanski (United Technologies Corp., South Windsor, Conn.)	Molten carbonates (unspecified)	—	—	—	—	120	—	IR-free cell voltage of 0.8 V with 80% fuel utilization. A model is presented for predicting electrode performance over a wide range of variables.

*Where performance is not at 0.8V cell output, this is stated in the comments.

†1 mW/cm² = 0.93 WSF (W/ft²).

areas, particularly on materials, of molten-carbonate fuel cells[44]). Figure 18.14, from the same report, summarizes the results of experiments with single cells. This work has progressed sufficiently that in the past few years some effort has been expended on fuel-cell stacks. One cell stack has operated for 1400 h and another for 1000 h.[35] The corrosion, materials degradation, distortion, etc., problems of these cells appear to be inherently much more formidable than those of the phosphoric acid cells. The development of large nickel oxide plate electrodes has been particularly frustrating. These plates are fabricated by pressing and sintering fine metallic nickel powder and then oxidizing it to give the desired porous nickel oxide structure. However, the resulting material is both weak and brittle. It thus tends to crack when subjected to stresses caused by differential expansion in the course of power cycling or as a result of irregularities in the power and/or coolant flow distribution.

SUMMARY

There has been great progress in fuel-cell development in the past century. The chief remaining problems are cost and service life. In particular, the importance of the complete absence of a large body of statistical data on multiyear operation with fuel-cell stacks of 1 kW or more cannot be overemphasized if one is to make a meaningful effort to project the rate of commercial use of fuel cells. All the data on the endurance tests of fuel-cell stacks of 1 kW or more that could be found are in Table 18.5. The significance of this aspect of R&D work is brought out by reviewing papers and reports that have given optimistic projections of early commercial applications despite the lack of data on fuel-cell stack endurance tests. For example, in a report prepared by a group of nine Harvard Business School students in 1959, over a page of conclusions gave a glowing picture of the many attractive applications for fuel cells once development was completed.[45] The authors were even concerned as to the major industrial dislocations that would occur when fuel cells were introduced. Other reports and papers have proved to be similarly optimistic. However, commercial acceptance must await the

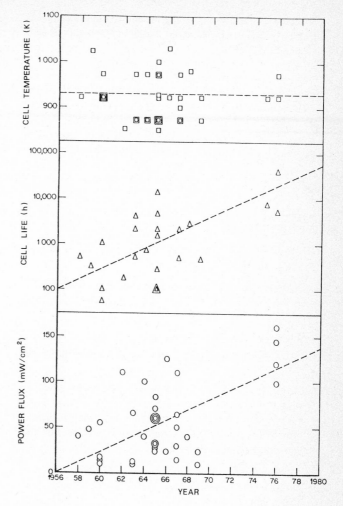

Figure 18.14 Summary of single-cell tests of molten-carbonate fuel cells.[44]

demonstration of many long endurance tests of cell stacks of substantial size, and as of the time of writing every effort to obtain a large, consistently long-lived fuel cell stack has failed.

PROBLEMS

1. For a molten-carbonate fuel cell having an output voltage of 0.75 V, a current density 125 mA/cm², and a cell thickness of 5 mm, estimate the number of cells, the cell area, and the cell stack height for a module designed to deliver 100 kW at 250 V.

2. The above fuel-cell module is to be fueled with a low-Btu synthetic gas containing 22.5% H_2, 15% CO, 11% CO_2, 8% H_2O, and 44% N_2 by volume. Estimate the gas volume flow rate to the module and the flow-passage area required for a 10 m/s feed velocity at 650°C at both 1 and 10 atm, assuming

80 percent consumption of the H_2 and CO and a cell stack thermal efficiency of 40 percent based on the higher heating value of the fuel.

3. Ten prototype fuel-cell stacks of 20 cells each are subjected to an endurance test. Deterioration or failures in cells lead to termination of the tests of these cell stacks after 10,000, 5000, 4000, 3500, 3200, 3000, 2500, 2200, 2000, and 1000 h, respectively. Full-scale cell stacks are designed with 460 cells in series to form a module. Assume that there would be no in-crease in the rate of deterioration of the cells with an increase in the number of cells in the stack and the consequent increase in voltage and thermal stresses. Estimate the mean life of the 460-cell stack, assuming: (*a*) that the above tests were terminated by cell failures, and failure of any one cell will make it necessary to take the cell stack out of service, and (*b*) that individual cell failures can be tolerated, but deterioration of cells to the point that the module loses 10 percent of its design output (i.e., 10 percent of its cells) will require that the module be removed from service.

REFERENCES

1. Fickett, A. P.: "Fuel Cell Power Plants," *Scientific American*, December 1978, p. 70.

2. Bockris, J. O., and S. Srinivasan: *Fuel Cells: Their Electrochemistry,* McGraw-Hill Book Company, New York, 1969, p. 575.

3. Wyczalek, F. A., D. L. Frank, and G. E. Smith: "A Vehicle Fuel Cell System," Automotive Engineering Congress and Exposition, Detroit, Mich., Jan. 9–13, 1967.

4. Carter, J. M.: "Batteries as Sources of Power for Undersea Warfare," *Problems in Power for Propulsion Applying to Submarines*, prepared for The Panel on Power for Propulsion, Washington, D.C., serial no. NRC: CUW: 00775D, app. D, September 1950.

5. Johnson, D. M., J. McCallum, and W. T. Reid: *Report on Fuel Cell Activities to Fuel Cell Research Group*, BMI-TM-8, Jan. 31, 1964.

6. Kordesch, K. V.: "City Car with H_2-Air Fuel Cell/Lead Battery (One Year Operating Experiences)," Society of Automotive Engineers, Inc., New York, Paper No. 719015, August 1971.

7. Kordesch, K. F., in B. S. Baker: *Hydrocarbon Fuel Cell Technology*, Academic Press, Inc., New York, 1965, p. 17.

8. Cook, N. A.: "Analysis of Fuel Cells for Vehicular Applications," Society of Automotive Engineers, Inc., Paper No. 680082, Jan. 8–12, 1968.

9. Austin, L. G.: *Fuel Cells, A Review of Government-Sponsored Research, 1950–1964*, NASA SP-120, January 1967.

10. King, J. M., Jr., and S. H. Folstad: *Electricity for Developing Areas via Fuel Cell Powerplants*, Eighth IECEC, American Institute of Aeronautics and Astronautics, Paper No. 739056, 1973, p. 111.

11. Fickett, A. P.: "An Electric Utility Fuel Cell: Dream or Reality?" *Proceedings of the American Power Conference,* Chicago, vol. 37, 1975.

12. Roberts, R.: *Needs and Recommendations for a National Fuel Cell Program*, MITRE Technical Report MTR-7933, January 1977.

13. Sverdrup, E. F., et al.: *1970 Final Report, Project Fuel Cell*, R&D Report No. 57, Westinghouse Research Laboratories no. 70-9E6-273-R8, August 1970.

14. Bacon, F. T.: "The Medium Temperature Hydrogen/Oxygen (Air) Cell," *Fuel Cells*, American Institute of Chemical Engineers, New York, 1963.

15. Evans, G. E.: "Status of the Carbon Electrode Fuel Cell Battery," *Fuel Cells*, American Institute of Chemical Engineers, New York, 1963.

16. Broers, C. H. J.: "High Temperature Cells with Carbonate Paste Electrolytes," *Fuel Cells*, American Institute of Chemical Engineers, New York, 1963.

17. King, J. M.: *Integrated Coal Gasifier/Molten Carbonate Fuel Cell Power-Plant Conceptual Design and Implementation Assessment*, United Technologies Corporation, Phase II Final Report, ECAS; NASA, ERDA, and NSF; NASA CR 134955, FCR-0237, October 1976.

18. *Phase I—Improvement of Fuel Cell Technology Base, Final Report, Jan. 19, 1976–March 31, 1977*, United Technologies Corporation, FCR-0477.

19. Wood, W., et al.: *Economic Assessment of the Utilization of Fuel Cells in Electric Utility Systems*, EPRI-336, November 1976.

20. Elzinga, E. R., et al.: *Application of the Alsthom-Exxon Alkaline Fuel Cell System to Utility Power Generation*, EPRI-384, January 1977.

21. Bolan, P., and E. Hall: "Industrial Applications of On-Site Fuel Cell Power Plants," National Fuel Cell Seminar, Boston, June 21–23, 1977.

22. Steitz, P.: "Assessment of the Fuel Cell's Role in Small Utilities," National Fuel Cell Seminar, Boston, June 21–23, 1977.

23. *National Benefits Associated with Commercial Application of Fuel Cell Power Plants*, ERDA 76-54, prepared for ERDA by United Technologies Corporation, Feb. 27, 1976.

24. Lueckel, W. J., Jr.: "Fuel Cells for Utility Service," *Proceedings of the Third Energy Technology Conference*, Washington, D.C., March 29–31, 1976.

25. McElroy, J. F.: "Status of H_2/O_2 Solid Polymer Electrolyte Fuel Cell Technology," National Fuel Cell Seminar, Boston, June 21–24, 1977.

26. Marianowski, L. G., et al.: "Fuel Cell Research on Second-Generation Molten Carbonate Fuel Cells," National Fuel Cell Seminar, Boston, June 21–23, 1977.

27. Warde, C. J., et al.: *Energy Conversion Alternatives Study, ECAS*, Westinghouse Phase I Final Report, vol. XII—*Fuel Cells*, NASA-CR-134941, Feb. 12, 1976.

28. Corman, J. C., et al.: *Energy Conversion Alternatives Study, ECAS*, General Electric Phase I Final Report, vol. II, *Advanced Energy Conversion Systems* NASA-CR-134948, SRD-76-011, February 1976.

29. Corneil, H. G.: Exxon Enterprises, Inc., personal communication to A. P. Fraas, Oct. 28, 1977.

30. Personal communication from Ray Huse, Public Service of New Jersey, to A. P. Fraas, Apr. 19, 1977.

31. Kunz, H. R.: "The State-of-the-Art of Hydrogen-Air Phosphoric Acid Electrolyte Fuel Cells," *Proceedings of the Symposium on Electrode Materials and Processes for Energy Conversion and Storage*, The Electrochemical Society, April 1977.

32. Fickett, A. P.: "Fuel Cell Electrocatalysis: Where Have We Failed?" paper presented at the Electrochemical Society Meeting, Philadelphia, April 1977.

33. Giner, J., et al.: *Fabrication and Evaluation of Electrodes for Alkaline Fuel Cells, Final Report*, EPRI RP 728, June 1977.

34. Jones, J. C., and J. E. Cox: "Experimental Determination of the Effect of Oxygen Supply Impurities on Fuel Cell Performance," *Proceedings of the Intersociety Energy Conversion Engineering Conference*, Miami Beach, Florida, 1967.

35. King, J. M., et al.: "Advanced Fuel Cell Technology Program—EPRI RP114-2," National Fuel Cell Seminar, Boston, June 21–23, 1977.

36. Cusumano, J. A., and R. B. Levy: "Fuel Processing for Fuel Cells," National Fuel Cell Seminar, Boston, June 21–23, 1977.

37. Phillips, G. A., et al.: "Inverters for Commercial Fuel Cell Power Generation," *IEEE Trans. on Power Apparatus and Systems*, vol. PAS-95, no. 3, May/June 1976.

38. *Fuel Cells, A Bibliography*, ERDA TID-3359, June 1977.

39. Kordesch, K. V.: "25 Years of Fuel Cell Development (1951-1976)," *Journal of Electrochemical Society*, March 1978.

40. Morse, W. F.: "Target On-Site Fuel Cell Program," National Fuel Cell Seminar, Boston, June 21–23, 1977.

41. Orlofsky, S.: "Development of a 12.4 kW Natural Gas Fuel Cell," *Transactions of the Eighth World Energy Conference*, vol. II, Bucharest, June 28–July 2, 1971.

42. *Improved FCG-1 Cell Technology Final Report*, United Technologies Corporation, EPRI-EM-1730, March 1981.

43. Handley, L. M., et al.: "4.8-MW Fuel Cell Module Demonstrator," *Proceedings of the 12th Intersociety Energy Conversion Engineering Conference*, Washington, D.C., 1977.

44. Maru, H. C., et al.: *Fuel Cell Research on Second-Generation Molten-Carbonate Systems*, report prepared for the Argonne National Laboratory by the Institute of Gas Technology, January 1977.

45. Lockwood, G. S., Jr.: *Fuel Cells: Power for the Future*, Fuel Cell Research Association, 1960.

19

HYDROELECTRIC POWER

Just as the highly developed coal-fired steam plant has served as a convenient standard against which other thermal power plants can be compared, so conventional, reliable hydroelectric plants provide a basis for evaluating the potential of "natural" sources of energy such as solar, wind, tides, ocean thermal differences, etc., because in each case one is confronted with high capital costs which are site-dependent and are sensitive to the head and energy flux available and to cyclic variations characteristic of the site. As indicated in Ref. 1, approximately 11 percent of the electric power produced in the United States in 1977 was provided by hydroelectric installations. However, this fraction varies widely with the region of the country as indicated by Fig. 19.1, which shows the relative amounts of hydroelectric and thermal power plant capacities for the principal geographical regions of the country, indicating a much greater availability of hydroelectric power in mountainous areas.[1]

HISTORICAL BACKGROUND

As mentioned in the first chapter, the oldest documented use of a machine to replace human or animal power was a waterwheel described in a Byzantine manuscript written a little after 600 A.D. This machine employed a paddle wheel similar to that of a Mississippi stern-wheeler, with the wheel mounted above a fast-flowing stream. The outboard end of the rotor was supported on a pontoon that was positioned by a system of booms and guys. The power was transmitted through pulleys with rope belts to drive a millstone. Returning crusaders apparently carried the concept to Europe, where waterwheels gradually came into use during late medieval times. These machines provided the basis for the beginning of the industrial revolution and were used not only for grinding grain but also for a wide variety of industrial operations, such as driving the bellows for ironworks. For example, the first ironworks in the United States was built by Governor Bradford's son just north of Boston in 1648 and required eight large waterwheels for driving the bellows for the

furnaces and for driving rolls and slitting shears. Thus, an important consideration in finding a suitable location for a plant was the availability of waterpower.

Undershot and overshot waterwheels, operating with heads of 2 to 4 m, were the source of waterpower up until the latter part of the eighteenth century, when small, vertical-shaft, turbine-type machines began to come into use not only in Spain, France, and Italy but also in the United States. These machines attracted the attention of men such as Euler, who evolved a theoretical basis for their design (Chap. 3), so that by 1830 a number of vertical-shaft turbines were in use with efficiencies of over 80 percent.[2] The use of these machines increased rapidly, but they did not fully displace the overshot and undershot waterwheels until the end of the century. Incidentally, the largest of these machines was an 1839 overshot waterwheel, 60 ft in diameter and 22 ft wide, employed in an ironworks on the Hudson River just above Troy, New York. This machine developed 1200 hp.

The use of waterpower for generating electricity began in the United States on the Fox River in Wisconsin in 1882, a year after the first steam-driven electric utility plant began operating in Philadelphia.[3] The use of hydroelectric generators grew rapidly so that by the 1930s some 30 percent of the nation's generating capacity and 40 percent of the electric energy produced was from hydro units.

HYDROELECTRIC POTENTIAL OF THE UNITED STATES

The United States is blessed with one of the largest sets of hydroelectric power sources in the world; only China, the U.S.S.R., the Congo, and Brazil have greater national hydroelectric potential. The United States has developed a larger fraction of its hydroelectric potential than almost any other major nation and hence gives an excellent indication of the extent to which the hydroelectric potential both here and abroad can be utilized.

The total hydroelectric potential of the continental United States is estimated by the Federal Power Commission[1] to be about 168×10^6 kW of capacity, of which the principal undeveloped potentials are in the north Pacific and Alaskan areas (Fig. 19.2). The capacity that has been developed or is under construction amounts to 66×10^6 kWe, roughly half of the total potential excluding the 33×10^6 kWe undeveloped potential of Alaska. At first thought one might construe this to indicate that a huge amount of hydroelectric capacity remains to be developed, but in point of fact most of the attractive sites have already been developed, and the balance would entail progressively more expensive installations. Not only are the sites naturally less well suited to development, but also the cost of land has been rising steeply and environmental considerations are leading to vociferous opposition to construction of new facilities at almost any of the remaining sites. An indication of these effects is given by Fig. 19.3, which shows the developed capacity as a function of year. It is evident from Fig. 19.3 that the construction of new facilities has already tapered off, and, in light of the above considerations, it is unlikely to increase rapidly in the future.[4]

TYPES OF SITE FOR HYDROELECTRIC POWER PLANTS

Some insight into the effects of a site on the cost of a hydroelectric power plant can be obtained by considering the half dozen different types of site commonly developed. The principal costs involve the amount of land that must be acquired, replacement of assets submerged by the reservoir, and the quantity of material required for the construction of the dam and powerhouse. The useful power output depends on the head and water flow available and the extent to which these may vary with the season of the year. Seasonal variations in water flow constitute a particularly important factor in determining the size of reservoir best suited to the site.

Falls and Short, Steep Rapids

One of the best and lowest-cost hydroelectric sites in the world is Niagara Falls. Land requirements are minimal; a relatively short canal around the falls serves to supply a huge volume of water with a high head to the powerhouse located below the falls with little seasonal variation in the flow rate. The site is close to a major load center, and so no long transmission lines are required. Interestingly, it provided one of the first cases in which a

conflict developed between environmentalists and the need for low-cost power. In this case, the conflict was resolved sensibly by compromising on the percentage of the water flow used for power, which was kept below the point at which it detracted noticeably from the aesthetic value of the falls. There is no other comparable site in the United States, but the great falls on the Parana River at Sete Quedas in Brazil provides a roughly similar situation. The water flow at Sete Quedas is much greater but varies more widely through the course of the year and the head is only half as great. More importantly, it is hundreds of miles from the nearest load centers, and so it was not until the 1970s that construction of a power plant was undertaken at this site. When fully developed, the Itaipu plant will be the largest hydroelectric plant in the world with an output roughly six times that at Niagara.

Low-Dam, Run-of-the-River Sites

The oldest and still one of the most common types of hydroelectric site entails the construction of a relatively low dam in a

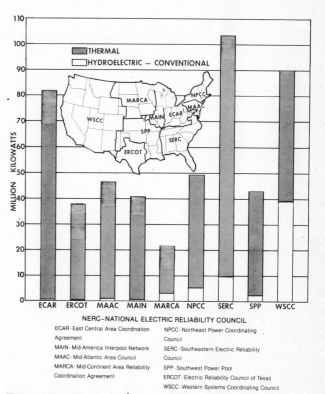

Figure 19.1 The power capacity of U.S. thermal and hydroelectric power plants in 1978.[1] (*Courtesy Federal Energy Regulatory Commission.*)

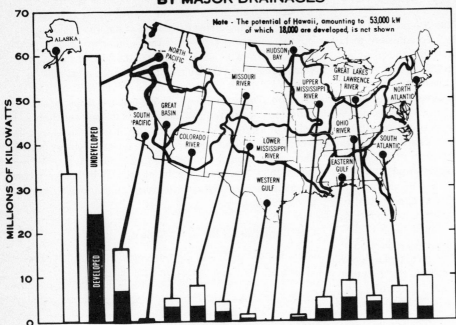

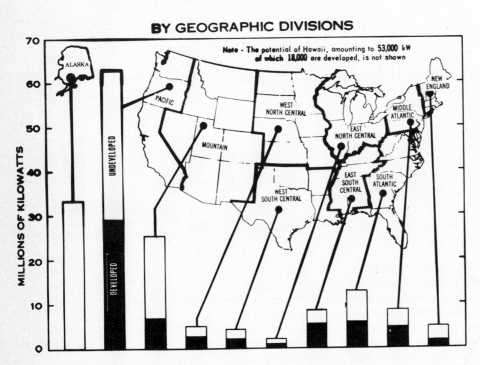

Figure 19.2 Developed and undeveloped hydroelectric power capacity of the United States in 1978 by drainage areas.[1] (*Courtesy Federal Energy Regulatory Commission.*)

stretch of river in which the stream level drops fairly rapidly. Such dams commonly yield heads of 6 to 18 m (20 to 60 ft), are often relatively small, and have little or no storage capacity. A relatively large example of this type is Wilson Dam at Muscle Shoals on the Tennessee River. Note that dams of this type are often multipurpose dams in that they include locks to improve the navigability of the river in which they are located, as is the case at Wilson Dam. When locks are included, a substantial fraction of the cost of the dam can be charged to navigation of the waterway.

High Dams

In mountainous areas where rivers drop fairly rapidly in long gorges, it is often advantageous to place a dam at a particularly narrow spot in the gorge, in part to give a high head and in part to provide a large storage capacity. Inasmuch as most rivers are characterized by large seasonal variations in flow, these dams are often multipurpose units, providing both flood control and a means of giving a more nearly uniform water flow rate downstream. The latter point is particularly important in ensuring adequate water supplies to downstream cities during seasons of low rainfall. Typical examples of such dams are TVA's Norris Dam in Tennessee and Shasta Dam in California.

Low Dams in Mountainous Areas

In mountainous areas it is sometimes possible to build a low dam in the upper reaches of a river with a fairly rapid drop around a U-bend or between one river and another so that a tunnel can be bored laterally through the ridge separating the two stretches of river and a much higher head can be delivered to the powerhouse than developed at the dam. An example of such a dam is the Santeetla Dam of Alcoa located in North Carolina. This dam has a maximum height of 66 m (216 ft) but yields a head at the powerhouse of 202 m (665 ft). A truly impressive possibility for a dam of this sort is offered by the great bend in the Brahmaputra River in India where the river flows out of the Himalaya Mountains. It is estimated that this project, if developed, would yield 20,000 MW.

Occasionally a low dam may be built on a river near the edge of an escarpment to give a high head for a plant at the foot of the escarpment. An example of this sort is the Alcoa Thorpe Dam (originally called Glenville) in North Carolina, which has a maximum dam height of 46 m (150 ft) but gives a head at the powerhouse at the foot of the escarpment of 368 m (1207 ft). Probably the most spectacular example of this type of site is

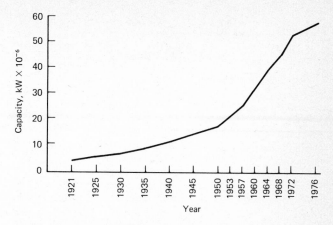

Figure 19.3 Developed hydroelectric capacity of the United States as a function of time.[4]

given by Cubatão, a plant at Santos in Brazil. In this area the Brazilian landmass is characterized by an escarpment ~ 1000 m high which rises up within a few kilometers of the Atlantic coast. Water drains inland, with the rivers draining southwestward into the Parana-Plata, which empties into the Atlantic near Buenos Aires. The prevailing winds carry the warm moist air from the South Atlantic over the escarpment, giving a rainfall of over 5 m per year (17 ft per year). In the early 1920s Billings, an American engineer, realized that a river up on the plateau could be dammed so that water would be backed up to a point near the edge of the escarpment and penstocks could be run down the very steep slope to sea level to drive Pelton wheels with a head of 713 m (2340 ft). This brilliant concept was kept secret, and options on the land required were obtained on the pretext of finding a good site for a British golf and country club. The options were all exercised, the plant was built, and the first power produced in the early 1920s. São Paulo was then a city of only 50,000 people engaged mainly in the coffee trade. The cheap power revolutionized the economy. São Paulo developed rapidly into an industrial metropolis which, at the time of writing, was a city of around 7 million people with a huge appetite for electric power. By adding small dams downstream on the rivers flowing inland and pumping water back to the edge of the escarpment, the power capacity of the Cubatão site has been greatly increased, and other, roughly similar, though less remarkably favorable, sites along the Brazilian coasts have been developed.

Similar but much smaller-capacity sites have been developed in Scandanavia, where—as in Brazil—the escarpment consists mainly of granite batholiths which make possible a special type

TABLE 19.1 Summary of Data on Hydroelectric Installations in the TVA System

Project	Owner	Year of closure	River	Max. height, ft	Overall crest length, ft	Max. spillway capacity, cfs	Volume of concrete, yd³	Volume earth and/or rock fill, yd³	No. units	Rated capacity, kW
Tennessee River Basin										
Kentucky	TVA	1944	Tennessee	206	8422	1,050,000	1,356,000	5,582,100	5	175,000
Pickwick Landing	TVA	1938	Tennessee	113	7715	650,000	679,100	3,081,000	5	218,000
Wilson	TVA	1924	Tennessee	137	4535	671,000	1,766,200	0	21	629,840
Wheeler	TVA	1936	Tennessee	72	6342	542,000	1,099,400	0	11	356,400
Guntersville	TVA	1939	Tennessee	94	3979	478,000	514,100	874,900	4	97,200
Nickajack	TVA	1967	Tennessee	81	3767	360,000	546,900	989,200	4	97,200
Chickamauga	TVA	1940	Tennessee	112	2960	470,000	506,400	2,793,500	4	108,000
Watts Bar	TVA	1942	Tennessee	112	2960	560,000	480,200	1,210,000	5	150,000
Fort Loudoun	TVA	1943	Tennessee	122	4190	390,000	586,700	3,594,000	4	131,190
Tims Ford	TVA	1970	Elk	175	1484	108,000	85,400	2,530,000	1	45,000
Appalachia	TVA	1943	Hiwassee	150	1308	136,000	237,800	0	2	75,000
Hiwassee	TVA	1940	Hiwassee	307	1376	112,000	800,600	0	2	117,100
Chatuge	TVA	1942	Hiwassee	144	2850	11,500	25,700	2,348,000	1	10,000
Ocoee No. 1	TVA	1911	Ocoee	135	840	45,000	160,000	0	5	18,000
Ocoee No. 2	TVA	1913	Ocoee	30	450	—	0	0	2	21,000
Ocoee No. 3	TVA	1942	Ocoee	110	612	95,000	82,500	82,000	1	27,000
Blue Ridge	TVA	1930	Toccoa	167	1000	55,000	—	1,500,000	1	20,000
Nottely	TVA	1942	Nottely	184	2300	11,500	21,700	1,552,300	1	15,000
Melton Hill	TVA	1963	Clinch	103	1020	122,000	246,800	0	2	72,000
Norris	TVA	1936	Clinch	265	1860	93,400	1,002,300	181,700	2	100,800
Tallico	TVA	1975	Little Tenn.	129	3238	135,000	78,000	1,883,000	—	—
Chilhowee	Alcoa	1957	Little Tenn.	91	1373	182,000	91,500	307,000	3	50,000
Calderwood	Alcoa	1930	Little Tenn.	232	916	260,000	—	0	3	121,500
Cheoah	Alcoa	1919	Little Tenn.	225	750	200,000	—	0	5	110,000
Fontana	TVA	1944	Little Tenn.	480	2365	134,300	2,815,500	760,600	3	225,000
Santeetlah	Alcoa	1928	Cheoah	212	1054	76,100	—	0	2	45,000
Nantahala	Alcoa	1942	Nantahala	250	1042	59,000	—	1,829,000	1	43,200
Thorpe·	Alcoa	1941	Tuckasegee	150	900	56,000	—	1,060,000	1	21,600
Douglas	TVA	1943	French Broad	202	1705	342,000	556,400	127,900	4	113,500
Nolichucky	TVA	1913	Nolichucky	94	480	—	—	0	4	10,640
Cherokee	TVA	1941	Holston	175	6760	694,200	894,200	3,304,100	4	120,000
Fort Patrick Henry	TVA	1952	S Fork Holston	95	737	141,000	72,500	0	2	36,000
Boone	TVA	1952	S Fork Holston	160	1532	137,000	198,400	714,000	3	75,000
South Holston	TVA	1950	S Fork Holston	285	1600	116,200	97,500	5,897,400	1	35,000
Wilbur	TVA	1912	Watauga	77	375	34,000	—	0	4	10,700
Watauga	TVA	1948	Watauga	318	900	73,200	80,400	3,497,800	2	50,000
Pumped-Storage Project										
Raccoon Mt.	TVA	1974	Tennessee	230	8500	none	135,000	9,400,000	4	1,530,000
Cumberland River Basin										
Great Falls	TVA	1916	Caney Fork	92	800	150,000	—	0	2	31,860
Barkley	C of E	1963	Cumberland	157	10,180	570,000	1,258,400	3,335,600	4	130,000
Center Hill	C of E	1948	Caney Fork	250	2,160	349,600	995,000	3,609,900	3	135,000
Cheatham	C of E	1953	Cumberland	75	981	90,000	290,400	136,000	3	36,000
Cordell Hull	C of E	1967	Cumberland	93	1,306	155,000	338,800	280,600	3	100,000
Dale Hollow	C of E	1943	Obey	200	1,717	59,450	581,710	—	3	54,000
J. Percy Priest	C of E	1967	Stones	147	2,718	182,000	226,400	1,788,100	1	28,000
Laurel	C of E	—	Laurel	282	1420	86,000	32,500	3,225,800	1	61,000
Old Hickory	C of E	1954	Cumberland	98	3,750	200,000	460,800	451,400	4	100,000
Wolf Creek	C of E	1950	Cumberland	258	5,735	434,800	1,422,000	10,319,200	6	270,000

Area at full pool El., acres	Total volume below top of gates, acre-ft	Useful controlled storage, acre-ft	Length of shore line, miles	Back-water length, miles	Full pool (El.)	Top of gates (El.)	Min. expected pool level (El.)	Avg. tail-water level (El.)	Head, ft	Cost	Project
											Tennessee River Basin
160,300	6,129,000	4,008,000	2380	164.3	359	375	354	310	47	$117,984,000	Kentucky
43,100	1,105,000	417,000	496	52.7	414	418	408	362	50	45,605,000	Pickwick Landing
15,500	641,000	59,000	154	15.5	507	507	504.5	414	92	107,585,000	Wilson
67,100	1,071,000	351,000	1083	74.1	556	556	550	507	48	87,655,000	Wheeler
67,900	1,052,000	172,300	949	75.7	595	595	593	557	37	51,054,000	Guntersville
10,730	252,400	32,300	192	46.3	634	635	632	596	34	74,942,000	Nickajack
35,400	739,000	347,000	810	58.9	682	685	675	634	45	42,065,000	Chickamauga
39,000	1,175,000	379,000	783	72.4	741	745	735	682	56	36,065,000	Watts Bar
14,600	393,000	111,000	360	55	813	815	807	740	70	42,374,000	Fort Loudoun
10,700	617,000	323,000	246	34	888	895	860	752	138	50,900,000	Tims Ford
1,100	57,800	8,800	31	9.8	1290	1280	1272	840	380	24,051,000	Apalachia
6,090	434,000	362,200	180	22	1524	1526	1415	1275	254	24,440,000	Hiwassee
7,050	240,500	222,100	132	13	1927	1928	1860	1804	126	9,122,000	Chatuge
1,890	86,500	33,800	18	7.5	837	837	816	724	113	2,963,000	Ocoee No. 1
—	—	silted	—	—	1115	1115	—	843	252	3,035,000	Ocoee No. 2
621	4,040	3,770	24	7	1435	1435	1413	1119	313	8,997,000	Ocoee No. 3
3,290	195,900	183,900	60	10	1690	1691	1590	1543	147	5,507,000	Blue Ridge
4,180	174,300	161,600	106	20	1779	1780	1690	1612	174	8,081,000	Nottely
5,690	126,000	31,900	144	44	795	796	790	742	51	36,250,000	Melton Hill
34,200	2,549,000	2,260,000	800	72	1020	1034	930	826	196	33,368,000	Norris
16,500	447,300	126,000	310	33.2	813	815	807	740	70	69,000,000	Tellico
1,890	49,250	6,564	30	8.9	874	874	870	812	80	—	Chilhowee
536	41,180	1,570	—	8	1087	1087	1084	869	209	—	Calderwood
595	35,030	1,850	—	10	1276	1276	1273	1087	187	—	Cheoah
10,640	1,443,000	1,145,000	248	29	1708	1710	1525	1276	429	78,448,000	Fontana
2,863	158,250	133,300	85	7.5	1939	1939	1863	1275	597	—	Santeetlah
1,605	138,730	126,000	—	4.6	3012	3012	2881	2007	944	—	Nantahala
1,462	70,810	67,100	—	4.5	3491	3491	3415	2284	1200	—	Thorpe
30,400	1,475,000	1,394,700	555	43.1	1000	1002	920	873	129	47,029,000	Douglas
797	9,850	—	—	—	1245	1245	—	—	68	1,747,000	Nolichucky
30,300	1,544,000	1,460,000	463	59	1073	1075	980	925	149	36,805,000	Cherokee
872	26,900	4,200	37	10.3	1263	1263	1258	1195	75	12,289,000	Fort Patrick Henry
4,400	193,400	148,400	130	17.3	1385	1385	1330	1264	123	27,766,000	Boone
7,580	764,000	642,600	168	24.3	1729	1742	1616	1490	239	31,428,000	South Holston
72	—	—	3	1.7	1650	1650	1645	1585	62	2,503,000	Wilbur
6,430	677,000	624,700	106	16.7	1959	1975	1815	1850	309	32,590,000	Watauga
											Pumped-Storage Project
528	37,930	36,340	—	—	1672	—	1530	633	1040	155,000,000	Raccoon Mt.
											Cumberland River Basin
2,110	51,300	48,300	120	22	805	805	762	655	150	9,030,000	Great Falls
57,920	2,082,000	1,472,000	1417	118.1	359	375	354	312	44	143,750,000	Barkley
18,220	2,092,000	1,254,000	422	64	648	685	618	486	162	44,491,000	Center Hill
7,450	111,000	26,800	320	67.5	385	386	382	361	22	30,185,000	Cheatham
11,990	280,500	75,700	381	71.9	504	505	499	457	47	71,700,000	Cordell Hull
30,990	1,706,000	849,000	620	51	651	663	631	514	137	25,949,000	Dale Hollow
22,720	652,000	384,000	265	31.9	490	504	480	396	94	51,625,000	J. Percy Priest
6,060	435,000	185,000	206	19.2	1018	1018	1018	765	253	34,600,000	Laurel
22,500	467,000	110,000	440	97.3	445	447	442	398	45	48,684,000	Old Hickory
50,250	6,089,000	4,236,000	1255	101.3	723	750	573	561	137	79,082,500	Wolf Creek

of construction. That is, the expense of penstocks can be avoided by boring tunnels downward through the granite and lining them with concrete to give a smoothly finished surface. In fact, the casings for the turbines themselves can be made in the same way by carving appropriately shaped cavities in the solid rock. Another area favored with this type of site is the central portion of the west coast of India. Bombay derives a large fraction of its power from installations such as the Khapoli plant at the edge of the escarpment, which rises up about 300 m (1000 ft) some 50 km east of Bombay.

Strings of Power Plants

In some instances a series of dams can be located one after another along a river to give a system such as that built by Alcoa and TVA on the Little Tennessee River in eastern Tennessee and western North Carolina. The 146-m-high (480-ft-high) Fontana Dam in the headwaters provides a large storage capacity and a steady water flow into a series of four dams and powerhouses downstream which have heads of 20 to 60 m (67 to 200 ft).

DATA FOR TYPICAL HYDROELECTRIC SITES

At the time of writing there were over 1400 hydroelectric power plants in the United States. Data for a typical set of these is presented in Table 19.1 to give some useful insights into the principal parameters. This set, which is for the TVA-Alcoa hydroelectric power system, includes examples of almost every type of site discussed above. Most of the TVA units are multipurpose dams, those on the main river being designed to provide navigation and flood control as well as hydroelectric power, while most of the tributary projects are designed to provide both flood control and hydroelectric power. Additional benefits include provision of adequate mainstream flow during the dry summer months to assure satisfactory water supplies to cities along the river system and a wealth of recreational opportunities for fishing, boating, and swimming. The Alcoa dams, on the other hand, were built primarily to provide electric power with little or no effort to provide flood control, although they provide fine recreational opportunities.

One of the most significant points implicit in the table is the fact that in no instance has it appeared appropriate to build a dam in the TVA system that yielded a head of less than 12 m (40 ft), and in only three instances is the power capacity less than 20 MW. The reasons of course were economic in spite of the fact that the dams have been built with federal funds, which have been available at low interest rates, and substantial portions of the cost of the dams have been written off against flood control

or navigation. These and other points implicit in the table will be referred to frequently in the subsequent sections.

Storage Capacity

One of the major advantages of hydroelectric units is the rapidity with which they can pick up load. If a unit is kept on the line but is delivering essentially no power while operating in the condensing mode (to improve the power factor in the system), it can accept load in a matter of seconds, far more rapidly than any thermal unit. In addition, it is very important for the system to be able to pick up peak loads for a few hours a day (Fig. 7.1). As a consequence, electric utility systems tend to have at least 10 percent of their power capacity in hydroelectric units. However, in order to take advantage of these characteristics of hydroelectric turbine generators, it is essential to have storage capacity in the dam. This should be at least enough to take care of diurnal load swings in power, and preferably sufficient to take care of weekly variations. In addition, it is highly desirable to be able to handle seasonal variations such as heavy air-conditioning loads in the summer and heavy heating and heat-pump electric loads in the winter. A convenient chart for relating the number of kilowatthours of electric power production obtainable to the water storage capacity of a reservoir is provided by Fig. 19.4. The enormous water storage capacity required to accommodate seasonal variations is indicated by taking data from Table 19.1 for the Norris power plant, which has a useful controlled storage of 2.7×10^8 m³ (2.2×10^6 acre-ft). Neglecting the loss in head associated with dropping from full pool to the ordinary minimum level [i.e., from a lake elevation of 311 m (1020 ft) above sea level to 283 m (930 ft)], this storage capacity with a head of 61 m (200 ft) will give a total output of about 4×10^8 kWh. This would amount to about 4000 h at the full-power output of around 100 MW from the Norris turbines. Thus the Norris Dam serves to collect the heavy rainfall in the winter and spring months and has sufficient capacity to provide power through the summer and autumn months, when the river flow would otherwise be relatively small. (There are 8760 h in a year.) Note also in Table 19.1 that there is only one dam in the TVA system with a higher storage capacity than that of Norris, and only five of the dams of the system have useful controlled storage capacities in excess of 8×10^7 m³ (650,000 acre-ft), i.e., the equivalent of 500 h at 100 MW with a head of 61 m (200 ft).

Neglecting the loss in head with reservoir drawdown, as was done above, involves much less of an error than one might at first think. Most reservoirs have volume characteristics similar to those of an inverted pyramid, so that the volume of water stored varies as the cube of the water height above the tail water. As a consequence, a 20 percent reduction in the level from full pool means that half of the volume at full pool will have been removed.

Geological Considerations

One of the important cost factors in constructing a dam, aside from the volume of material required, is the geological structure underlying both the dam site and the lake to be impounded. Extensive test borings must be made to assure a good foundation for the dam and freedom from cracks or permeable strata that might lead to leaks and a possible weakening of the foundation. For example, poor foundation conditions were a factor in the catastrophic failure of the Teton Dam in 1976. Several failures caused by poor foundation conditions have occurred, and this is one reason why many otherwise superficially attractive hydroelectric sites are not suitable for development. Aside from the possibility of catastrophic failure, porosity in the underlying rock strata can lead to serious leakage and a marked loss in the expected hydroelectric power capacity. One example of this sort of problem is given by the Aswan Dam: leakage through rock strata and losses to evaporation have reduced its hydroelectric power capacity by about one-third of the value originally expected.

Another major factor affecting the cost of the plant is the availability at the site of suitable material for construction. For a concrete dam this means the availability of a quarry for obtaining suitable aggregate and sand pits to provide the principal materials for the concrete. For an earthen dam or an earth-and-rock-filled dam, there must be an adequate supply of suitable earth and/or rock close to the dam site.

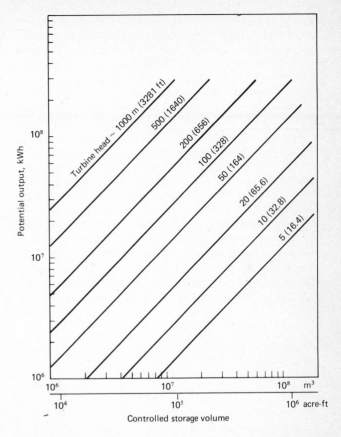

Figure 19.4 Potential electric energy output from a water storage reservoir for a 90 percent turbine-generator efficiency.

TYPES OF DAMS

The fifth and sixth columns under "Dams and Appurtenances" in Table 19.1 indicate the type of construction used in the TVA and Alcoa dams. One of the most common types is the concrete gravity dam with earth embankments. Note that this type of dam has a roughly triangular cross section with the broad base of the triangle resting on and keyed into bedrock so that it is held in place by its own weight. The construction is simple but the volume of concrete required per unit of length is large. The concrete volume required can be reduced by employing a concrete arch dam, the shoulders of which can rest against sound rock. The Calderwood and Hungry Horse dams are examples of this construction. The forces generated by the hydrostatic pressure of the impounded water act against the convex face of the dam and are carried in compression along the arc of the dam into the rock shoulders at either end. This type of structure can be used

only in deep, narrow gorges with sound rock foundations on either side. The most common type is the earth dam—which may have several types of earth in the cross section—with rock riprap facing. The facing may be heavy, as in an earth-rock dam, such as the TVA Cherokee Dam, or the dam may be made largely of rock.

TYPES OF TURBINES

Four types of turbines are employed in hydroelectric power stations. For low heads—up to as much as 55 m (180 ft)—propeller turbines are employed, primarily because they give relatively high rotor velocities for relatively low water through-flow velocities. These are usually Kaplan turbines, i.e., propeller-type turbines with adjustable blades. The blade pitch can be varied to give a good efficiency over a wide range of loads,[5,6] a second and important reason for using them. For higher heads—30 m (100 ft) to as much as 300 m (1000 ft)—Francis turbines are employed. These are radial inflow units in which the water enters the rotor through a set of variable angle inlet guide vanes and flows radially inward and axially downward, with a substantial pressure drop taking place within the turbine wheel itself; i.e., the unit is a reaction turbine (Chap. 3). The Deriaz turbine is a little-used intermediate variety with adjustable rotor vanes whose axes lie on the surface of a cone. For heads in excess of ~305 m (1000 ft), Pelton wheels are ordinarily employed. These are impulse turbines in which all the static head available in the turbine is converted into velocity in needle valve-controlled nozzles, and all this velocity energy is absorbed in the wheel so the water leaves at a negligible absolute velocity (Chap. 3). The ranges of application of the principal types of turbine are indicated graphically in Fig. 19.5.[7]

For electric power generation, the turbine speed must be kept constant to hold a constant output frequency. This poses problems for part-load operation. The changes in the angles with which the flow enters and leaves the turbine can be accommodated in the Kaplan and Francis turbines by varying either the rotor blade or the inlet guide vane angles, but this cannot be done in the Pelton wheel. As a consequence, the efficiency at part load suffers. Figure 19.6 shows typical curves for the efficiency of these three types of turbine as a function of load. Note that the Kaplan turbine with its very low pitch propeller blades yields the best performance over a wide range, the adjustable inlet guide vane Francis turbine is next best, and the Pelton wheels have a wider range for good performance than one might expect.

It is advantageous to increase the size of the turbine generator, partly because this reduces the hydrodynamic losses but mainly because it reduces the cost per kilowatt of the complete installation. However, increasing the size also increases the weight, the problem of handling the components for shipment and installation, and the bearing loads. Continuing improvements in design have made it possible to produce progressively larger units; Fig. 19.7 indicates the progress in this area.[8] Note that head is an important factor in making possible high out-

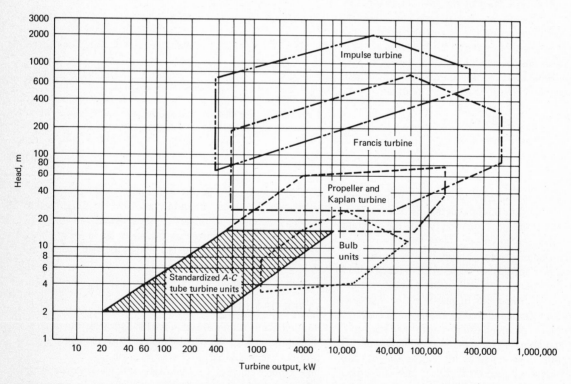

Figure 19.5 Ranges of application of the various types of hydraulic turbine units. Tube turbines and bulb units are specialized forms of propeller turbine in which the shaft is horizontal to reduce inlet and exit losses for low head sites. (*Courtesy Allis-Chalmers.*)

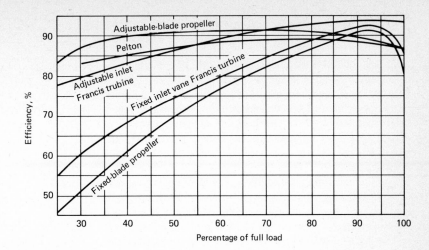

Figure 19.6 Effects of load on the efficiency of typical turbines.

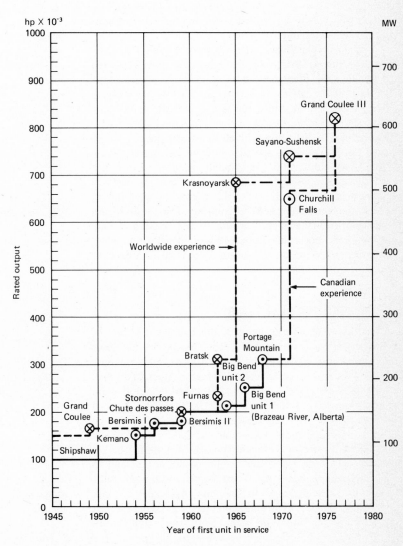

Figure 19.7 Increase in the maximum rating of hydraulic turbines in the 1945–1976 period.[8]

puts. The 600-MW Grand Coulee turbine in Fig. 19.7 was designed for a head of about 107 m (350 ft). Because of their inherently lower head, propeller-type turbines have been limited to outputs of about 100 MW, and, while they operate with very high heads, Pelton wheels have been limited to outputs of about 30 MW because of the restricted inlet nozzle area available in the wheels and the need to have a free exit for the water leaving the wheel.

To minimize fluid losses in the inlet and exit passages, propeller turbines are usually mounted within the dam itself. Francis turbines may be mounted in the dam or in a separate powerhouse. Figure 19.8 shows a typical installation for a Francis turbine, while Fig. 3.9 shows a more detailed view of a similar installation for a Kaplan turbine. For low-head installations the inlet and exit losses can be reduced by inclining the shaft from the vertical. This is particularly true for propeller turbines, which are often mounted with the shaft horizontal.

To minimize exit losses, particularly in low-head installations, a diffuser or draft tube is commonly employed between the turbine wheel discharge and the tailrace. This improves the efficiency of the turbine by recovering much of the velocity energy at the turbine outlet, but is likely to reduce the static pressure at the turbine to a level below atmospheric. The reduction may lead to difficulties with air in-leakage and possibly to cavitation. The latter has proved to be a problem in some of the higher-speed Kaplan turbine installations. Difficulties with cavitation can be avoided by installing the turbine at a level well below the level of the tailrace so that the turbine is operating under a substantial static head. This approach has the disadvantage of preventing easy draining of the turbine during maintenance. The turbine can be readily drained if it is mounted above the level of the tailrace, one of the original advantages to the use of a draft tube.

HYDROELECTRIC PLANT COSTS

As mentioned at several points in this chapter, the cost of hydroelectric plants varies widely, depending on the site. Some indication of this is given by Table 19.1, which summarizes data for the TVA and Alcoa hydroelectric plants. Much additional data can be obtained from tables published by the Federal Energy Regulatory Commission (FERC). Summaries of data for four typical plants are presented in Table 19.2. Note the wide variation in the cost of both land and the structure, mainly the dam. However, a reasonably consistent pattern emerges with respect to the cost of the turbine-generator unit and its associated equipment. A plot of cost data for turbine-generator units built for the Bureau of Reclamation was presented in Fig. 10.8, which showed that the principal factor affecting the cost of the turbine-generator is the head. The cost of the complete in-

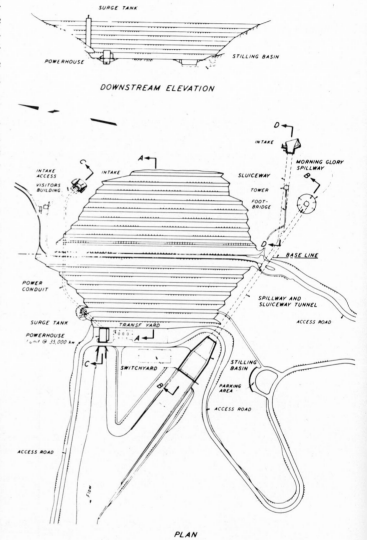

Figure 19.8 General plant elevations and sections through a typical hydroelectric power plant, the TVA South Holston project (*Courtesy TVA.*)

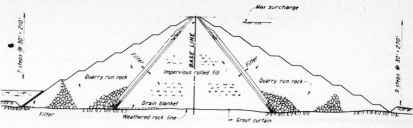

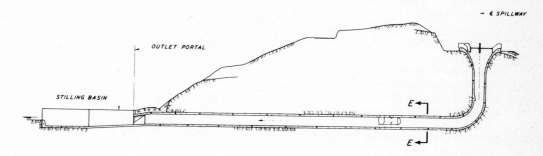

SECTION THRU UNIT

SECTION A-A

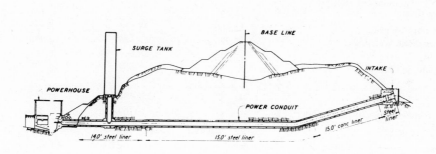

SECTION B-B

SECTION E-E

SECTION C-C

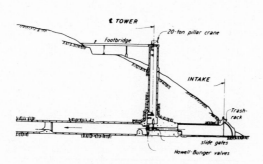

SECTION D-D

TABLE 19.2 Hydroelectric Plant Construction Cost and Annual Production Expenses, 1973[12]

	Name of Utility	U. S. DEPARTMENT OF THE ARMY, CORPS OF ENGINEERS 1/			
	Name of Plant	Albeni Falls	Dworshak 3/	Barkley	Wolf Creek
	Project				
Line	**Post Office**	Newport, Wash.	North Fork, Idaho	Grand River, Ky.	Jamestown, Ky.
No.	**County and State**	Bonner, Idaho	Idaho	Lyon, Ky.	Russell, Ky.
	River	Pend Oreille	N. F. Clearwater	Cumberland	Cumberland
	Region and Power Supply Area	VII-41	VII-41	III-20	III-20
	Licensed Project No.	—	—	—	—
1	**Installed Generating Capacity—Generator Nameplate – MW**	42.6	400.0	130.0	271.2
2	Pumping energy, Million kWh (Pumped storage)	—	—	—	—
3	**Net Generation, Million kWh**	228.1	163.1	889.9	1,271.1
4	**Plant Factor, Percent, Based on Nameplate Rating**	61	—	78	54
5	**Net Peak Demand on Plant, MW (60 Minutes)**	43.0	NR	164.0	344.0
6	**Capability Under Most Favorable Operating Conditions – MW**	49.0	460.0	166.0	300.0
7	**Capability Under Most Adverse Operating Conditions – MW**	0	38.5	0	180.0
8	**Hours Connected to Load**	7,998	2,998	8,597	8,447
9	**Planned Ultimate Generating Capacity – MW**	42.6	—	130.0	271.2
10	**COST OF PLANT (Thousands of Dollars)**				
11	**Land and Land Rights**	3,421	30,421	5,163	11,113
12	**Structures and Improvements**	11,966	18,574	18,835	9,690
13	**Reservoirs, Dams, and Waterways**	7,795	206,326	2,932	24,911
14	**Equipment Costs**	8,733	20,077	18,588	13,552
15	**Roads, Railroads, and Bridges**	—	5,954	38	682
16					
17	**Total Cost**	31,915 2/	281,352 2/	45,556 2/	59,849 2/
18	**Cost per kW, Installed Capacity (Nameplate)** $	749	703	350	222

19	**PRODUCTION EXPENSES**	$1000	Mills kWh	$1000	Mills kWh	$1000	Mills kWh	$1000	Mills kWh
20	**Operation Supervision and Engineering**	39	.17	7	.04	32	.04	36	.03
21	**Water for Power**	—	—	—	—	—	—	—	—
22	**Hydraulic Expenses**	26	.11	—	—	2	—	—	—
23	**Electric Expenses**	116	.51	28	.17	123	.14	122	.10
24	**Misc. Hydraulic Power Generation Expenses**	2	.01	—	—	14	.02	26	.02
25	**Rents**								
26	Joint Operating Expenses (Allocated)	92	.40	107	.66	19	.02	88	.07
27	**Maintenance Supervision and Engineering**	29	.13	—	—	39	.04	40	.03
28	**Maintenance of Structures**	91	.40	—	—	38	.04	6	—
29	**Maintenence of Reservoirs, Dams, and Waterways**	—	—	—	—	—	—	—	—
30	**Maintenance of Electric Plant**	40	.18	—	—	74	.08	177	.14
31	**Maintenance of Misc. Hydraulic Plant**	20	.09	—	—	6	.01	8	.01
32	Joint Maintenance Expenses (Allocated)	133	.58	81	.50	4	—	5	—
33	**Total Production Expenses**	588	2.58	223	1.37	351	.39	508	.40
34	**Production Expenses per kW (Nameplate)**	13.80		.56		2.70		1.87	
35	Cost of Pumping Energy (Pumped Storage) $	—	—	—	—	—	—	—	—

TABLE 19.2 Hydroelectric Plant Construction Cost and Annual Production Expenses, 1973[12] (*Continued*)

Name of Utility		U. S. DEPARTMENT OF THE ARMY, CORPS OF ENGINEERS 1/			
Line No.	**Name of Plant** Project Post Office County and State River Region and Power Supply Area Licensed Project No.	Albeni Falls Newport, Wash. Bonner, Idaho Pend Oreille VII-41 —	Dworshak 3/ North Fork, Idaho Idaho N. F. Clearwater VII-41 —	Barkley Grand River, Ky. Lyon, Ky. Cumberland III-20 —	Wolf Creek Jamestown, Ky. Russell, Ky. Cumberland III-20 —
36	Average Number of Employees	15	12	17	22
37	Type of Operation	Manual		Manual	Manual
38	Initial Year of Plant Operation	1955	1973	1966	1952
39	**HYDRAULIC DATA**				
40	Drainage Area, Square Miles	24,200		17,598	5,789
41	Area of Pond at Normal Full Pond Level, Acres	94,000		57,920	56,250
42	Storage or Pondage from Maximum Draw-down, Acre-Feet	1,153,000		258,900	2,142,000
43	Gross Head (Pond Elevation Minus Tailwater Elevation), ft	28.3		46	163
44	(CFS per kW)—Full Station Load	NR		0.298	0.0815
45	**GENERAL DATA—FOR PLANTS ADDED IN 1973**				
46	Kind of Development				
47	Type of Dam				
48	☐ Pumped Storage				
49	Type of Powerhouse Construction:	☐ Conventional	☐ Outdoor	☐ Semi-Outdoor	
50	Project Use:	☐ Single Purpose	☐ Multiple Purpose		

EQUIPMENT CHARACTERISTICS—FOR PLANTS OR UNITS ADDED IN 1973

WATERWHEELS						GENERATION				
No. Units	Design Head (Feet)	Max. H P* (1,000)	Type (Hor. or Vert.)	Type** Runner	Year Installation	No. Units	Nameplate Rating (MW)	P F %	Voltage k V	Year Installed
2	560	142.0	Vert.	F	1973	2	90.0		13.8	1973
1	560	346.0	Vert.	F	1973	1	220.0		13.8	1973

3/   New Plant-Commercial Operation: 1973

*At Design Head;   **F-Francis;   FP-Fixed Propeller;   AP-Automatic Adjusted Propeller;   I-Impulse.

NOTES: 2/   Includes tentative multiple purpose plant allocation.

Fiscal Year Ending: 1/   June 30th.

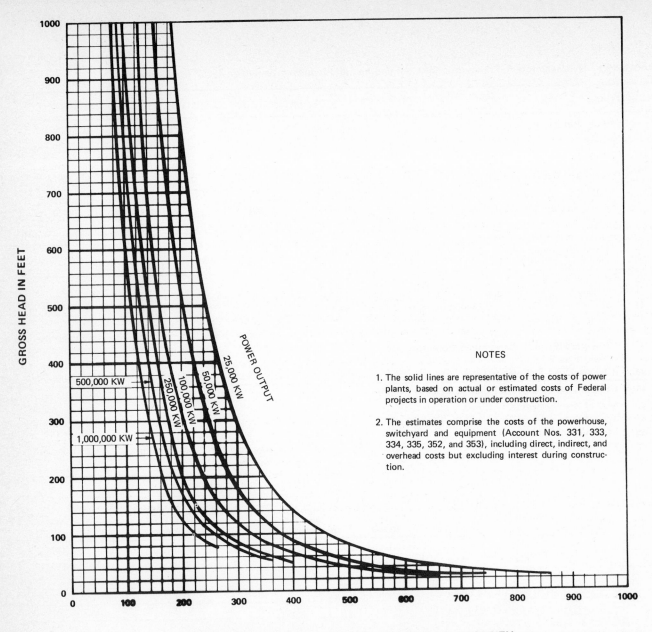

Figure 19.9 Federal Power Commission guide for estimating the cost of hydroelectric power plants (excluding the dam and reservoir) as of 1978.[1] *(Courtesy Federal Energy Regulatory Commission.)*

stallation of the turbine-generator unit with its associated control equipment and electric switchgear goes up less rapidly than linearly with power output; hence the cost per kilowatt of installed capacity falls off somewhat at a constant head with an increase in capacity. This effect is shown in Fig. 19.9.

UTILIZATION OF SITES FOR SMALL HYDROELECTRIC UNITS

The drastic increase in the cost of petroleum, natural gas, and coal following the 1973 Yom Kippur War has led to a reexamination of the possibility of utilizing small hydroelectric sites. In a study of this problem submitted to the President by the Corps of Engineers,[4] it was concluded that the most promising approach would be to install turbine-generator units in relatively small dams that had been abandoned because they were not competitive costwise with thermal power plants while the supplies of fossil fuels could be obtained inexpensively. The findings are summarized in Table 19.3, which indicates that rehabilitating existing hydroelectric dam sites could provide an additional hydroelectric power capacity of about 5000 MW, expansion of the capacity of existing dams in service could provide an additional 16,000 MW, the potential at existing nonhydroelectric dams (for irrigation, city water supplies, etc.) might be as much as 7000 MW, and the potential at existing nonhydro dams capable of less than 5000 kW of output might total as much as 26,000 MW. The total hydroelectric potential thus obtainable amounts to almost as much as that already developed and in service. Most of this would be in the form of low-head, low-power-capacity units, and hence the turbine-generator and associated equipment would inherently have a relatively high unit cost (Figs. 10.8 and 19.9). Further, the cost of maintenance and of coupling these units into a grid would be relatively high. The question is whether these high unit cost components would still yield an attractive system in view of the fact that the cost of the dam and associated structures should be low. In addition, other problems arise that could entail large legal and oher costs. One example cited was the Occoquan reservoir in northern Virginia, which supplies 220,000 m³ per day (58 × 10⁶ gal per day) of water to the Washington metropolitan area, yielding $12 × 10⁶ per year in gross revenues. If this dam were rebuilt to increase the hydraulic head available by a factor of 4 so that it would yield a capacity of 17,500 kW, it could produce 47 × 10⁶ kWh of electricity per year, which would have a sales value of $1.8 × 10⁶. Clearly the existing $12 × 10⁶ per year in revenues from supplying water to the urban market is roughly seven times the revenues that could be obtained from electric power production after a large new investment both in improving the dam and installing the hydroelectric unit. Further, there would be a serious conflict regarding the operation of the dam for water supply purposes as opposed to operation for hydroelectric power production. There is also the question as to whether the reliability of the urban water supply would be compromised by less than perfect reliability of the turbine-generator system.

The problems are much more complex than they appear on the surface. Certainly some of the small hydroelectric sites that have been abandoned because they were uneconomic can be rehabilitated to give economically attractive power, and this is being done in some cases. However, the sixfold increase in fuel costs from 1973 to 1979 has been accompanied by nearly as great an increase in capital costs (Fig. 10.14), so that the economics of hydroelectric plants relative to thermal plants has not changed as much as one might first think. Further, it seems clear that the construction of numerous new low-head dams on small streams does not offer a promising source of electric power at a competitive cost. A specific example that illustrates the problem quite well is given by an article that appeared in *The New York Times* in May 1978 giving an enthusiastic report on a small hydroelectric turbine-generator unit being marketed by a French concern. A 5-kW unit selling for $7800 was designed to operate on a head of about 3 m (10 ft). This gives a capital cost of $1500/kW, to which must be added the cost of its installation. It is difficult to estimate operating costs; certainly a substantial amount of maintenance would be required just to keep the trash racks at the turbine inlet clear of debris. Such a system would make a nice hobby for someone so inclined, but would probably not be attractive to most people.

A typical case in which changing economics have justified the redevelopment of an old hydro plant that had fallen into disuse is given by the Cornell plant in Michigan. In this instance the dam originally supplied power to a paper pulp mill, with 10 of the 12 turbines used for direct drives of pulp mills while two were used for generating electricity. The dam, which yields a head of 11 m (36 ft), has been modified to take three hydroelectric turbines of 10^4 kW each at a cost of $500/kW. Contracts for the bulk of the equipment were let in 1972, and the installation

TABLE 19.3   Conventional Hydroelectric Capacity (Constructed and Potential) at Existing Dams[4]

	Capacity, millions of kW	Generation, billions of kWh
Developed	57.0	271.0
Under construction	8.2	16.8
Total installed	65.2	287.8
Potential rehabilitation of existing hydro dams	5.1	24.4
Potential expansion of existing hydro dams	15.9	29.8
Potential at existing nonhydro dams greater than 5000 kW	7.0	20.4
Potential at existing nonhydro dams less than 5000 kW	26.6	84.7
Total potential	54.6	159.3
TOTAL (developed and undeveloped)	119.8	447.1

TABLE 19.4 Principal Cost Items for the Installation of a New 1-MW Turbine-Generator Unit in a Typical Existing Dam Giving a Head of 15 m (49 ft) (Ref. 9)

	Custom	Percent of total	Standardized	Percent of total
Equipment	$ 530,000	44	$450,000	45
Engineering	180,000	15	85,000	9
Civil Construction	500,000	41	460,000	46
Total	$1,210,000	100	$995,000	100

was completed in 1977, hence the $500/kW cost is probably in 1973 or 1974 dollars.[8]

Costs of Standardized versus Custom-Designed Installations

The major cost components in low-head, low-power hydro installations in existing dams are equipment, engineering, and construction in the field. All three items can be minimized by employing standardized turbine-generator units that include the controls.[7] Table 19.4 compares these costs for a typical case, first for a custom-designed, and secondly for a standardized, turbine-generator package unit.[7]

Estimating the Flow Potential of a Site

In attempting to estimate the hydroelectric potential of a site on a stream whose flow has not been well monitored, one can make a good estimate of the flow from the area of the drainage basin upstream of the site and the annual rainfall. A chart that makes it easier to carry out such estimates for the New England area is shown in Fig. 19.10. Note that the same chart can be used for other areas by applying a factor for the difference in rainfall between the other area and New England. If this is done, an additional factor to allow for differences in evaporation losses between the New England area and the area in question may be necessary. These evaporation losses may be quite high; they represent about two-thirds of the total rainfall in the TVA area, for example.

TIDAL POWER

Tidal energy has been used since the eleventh century to drive waterwheels, and "tidal mills" were fairly common along the French and English coasts up to the early twentieth century. The possibility of using the ebb and flow of tides to drive hydroelectric units has been an intriguing subject of discussion since 1890, and seems again to be a live issue at the time of writing. The range of sea level variation varies widely from relatively small values of about 1 m in regions near the equator to values of as much as 20 m along the northern coast of North America and Europe. There are normally two high and two low tides a day, depending on the phase of the moon. Maximum tides, called the *spring* tides, occur when the sun and the moon are both on the same side of the earth, while the *neap* tides occur when the sun and moon are at 90 degrees with respect to the earth. The neap tides are commonly about half as great as the spring tides.

The characteristics and problems of a tidal power station are well illustrated by examining the experience with the tidal power station built in the estuary of the river La Rance in Brittany, France, completed in 1966.[12] This plant has an installed capacity of 240 MW and produces power about 6 h per day, utilizing both the inflow and the outflow of water with changes in the tide. Figure 19.11 shows the variation in the sea level for a typical day, together with the variation in level of the water in the tidal basin behind the dam across the mouth of the estuary. In this instance, the turbines are operated as pumps for a short period, when the level in the basin and sea level are the same, to accelerate filling or emptying the reservoir. In that way they derive some extra power during the period when the difference in level is great enough to obtain useful power from the tidal flow. The turbines are large—their diameter is 5.35 m—and 24 units are required. These are mounted in the central part of the 750-m-long dam, in which the power station occupies about half of the length.

The principal characteristics of the power plant are summarized in Table 19.5, together with the corresponding data for the TVA Nickajack Dam, which was completed about the same time.[13] Note that the peak difference in water level in spring tides for the La Rance dam is about the same as the head available from the Nickajack dam, and the length of both dams is about the same. The turbines are designed for a lower head, and the total installed power capacity is more than $2\frac{1}{2}$ times as great in the LaRance power plant. As a consequence, it is not surprising that the cost of the La Rance power plant was about three times that of the TVA Nickajack Dam and powerhouse. In a sense, both projects are multipurpose, because the Nickajack dam also provides improved navigation and some flood control capability, while the La Rance dam serves as a bridge across the mouth of the estuary and is heavily traveled. Note that in the table two values for the La Rance power level are given: 60 and 240 MW. This is because the turbines develop power for only 6 h per day; the equivalent output on a sustained basis is only one-

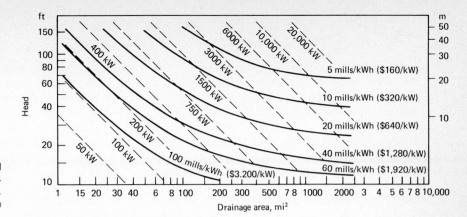

Figure 19.10 Typical capacities and minimal capital charges for small hydroelectric plants as estimated for rivers in the northeastern United States.[10]

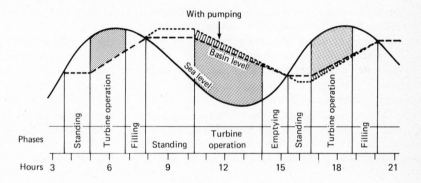

Figure 19.11 Typical operating regimes of the pump-turbines at La Rance. Note the additional head that can be gained by pumping, an option sometimes taken at off-peak times.[11]

TABLE 19.5 Comparison of Tidal Power Projects with a Conventional Low-Head Riverine Hydroelectric Project Including Actual Costs

Type of project	Riverine	Tidal	Tidal
Name of project	TVA—Nickajack[11]	La Rance[10]	Passamaquoddy[12]
Location	Tennessee River Tenn., U.S.A.	La Rance Estuary Brittany, France	Passamaquoddy Bay Maine, U.S.A.
Year of completion	1968	1966	
Design power output, MW	96	240 (60*)	166 (80*)
Dam length, m	1135	750	3660
Dam maximum height above foundation, m	24.7		45.7
Area of pool, km²	43.5	22	96
Operating head, m	13.6	5.5	3
Mean range for spring tides, m		13.5	7
Project cost, $	42,100,000	120,000,000†	
Cost per kilowatt of average output, $/kW	343‡	2000†	

*Output averaged over 24 h.
†Cost does not include the cost of the site survey and design or credit for value of the dam as a bridge.
‡Cost does not include credit for flood control and navigation.

quarter of the peak output, or about 60 MW. Although the total water flow available at Nickajack is not sufficient to permit full-power output throughout the year, the availability of the power capacity at any time is worth a great deal. Thus the 60-MW value for La Rance is reasonable for comparative purposes. Using this 60-MW value yields a project cost of about $2000/kW for the La Rance project, as opposed to about $435/kW for the Nickajack dam, if no allowance is made in either case for the benefits of the other uses of the dam. This large difference in cost is not surprising in view of the much greater turbine-generator capacity required for a tidal project. Further, many proposed projects, such as Passamaquoddy, entail exposure of dams to ocean storms, and the much more rugged construction required of such a dam would entail much higher costs.

In the United States the most promising site for a tidal power project has been the Passamaquoddy Bay in Maine. The project was originally proposed about 1920, and quite thorough design studies were carried out in the depression of the 1930s with the expectation that it would be constructed as a means of providing work for over 40,000 unemployed in the state of Maine.[13] Cost estimates proved to be so high that the project was shelved. The reason for this can be appreciated by examining the last column in Table 19.5, in which the design data for the Passamaquoddy project have been summarized. Note that the head available would have been about half that for the La Rance project, the

effective output would have been only a little greater than for the La Rance plant, the dam length would have been about five times as great, and some of that would have been exposed to wave damage in ocean storms. No data were available for the maximum height above the foundation for the La Rance dam, but note that this parameter for Passamaquoddy is 45.6 m (150 ft), while the corresponding value for the Nickajack dam is 24.7 m (81 ft). Thus the volume of material per unit of length would be four times as great for Passamaquoddy as for Nickajack, and hence one would expect the cost of the three-times-longer Passamaquoddy dam to run at least 12 times that for the Nickajack dam. The average head on the turbines when operating would be less than one-fourth as great, and they would operate only about half the time; hence their capital cost would run roughly eight times as much. Further, the availability of any tidal power unit is a function of the timing of the tides, and since this would commonly not be in phase with the load demand on the power system, the value of the power capacity would be reduced by an additional factor. Thus it appears that electric power from the most attractive tidal power project in the United States would cost roughly 10 times as much as that from a typical hydroelectric installation on a river. There would be no other benefits, such as improved navigation or flood control, and there might be serious adverse ecological effects on estuarine marine life.

PROBLEMS

1. A typical small hydraulic turbine designed for a low-head installation has a design rating of 570 kW with a head of 10 m and a flow of 25 m³/s. The draft tube outlet diameter is 3.75 m. Calculate the overall efficiency, the fraction of the total head lost as velocity head at the draft tube outlet, and the turbine-generator efficiency if the turbine is not charged with the velocity head at the draft tube outlet.

2. Choose an actual site that is familiar to you and that appears to be promising for a small, low-head, hydroelectric unit. Estimate the cost of the land required, the dam (or modifications to an existing dam), and the complete turbine-generator installation, including controls and switchgear with appropriate charges for design, overhead, etc. Obtain or estimate stream flow rates and estimate the capital charges

for the electricity that could be generated, assuming a 30-year life. Also estimate operating costs, assuming maintenance work at 60 man-days per year.

3. Using the reservoir areas for the TVA dams of Table 19.1, estimate the total annual capital charge for land use, assuming 1979 land costs of $1000 per acre for the mainstream reservoirs (which have flooded rich agricultural bottom-lands) and $500 per acre for the headwater reservoirs (which have flooded lands that usually sloped too steeply for good cultivation). Using a 16 percent capital charge and the total 1979 TVA hydroelectric output of 21 × 10⁹ kWh, calculate an appropriate capital charge per kilowatthour for land utilization on the basis of 1979 land values.

REFERENCES

1. *Hydroelectric Power Evaluation,* Federal Energy Regulatory Commission, DOE/FERC-0031, August 1979.

2. Keator, F. W.: "Benoit Fourneyron (1802–1867)," *Mechanical Engineering*, vol. 61, no. 4, 1939, pp. 295–301.

3. *Hydroelectric Power Systems,* Energy Research and Development Administration, Argonne National Laboratory, Report No. 107, 1977.

4. *Estimate of National Hydroelectric Power Potential at Existing Dams,* U.S. Army Corps of Engineers, Institute for Water Resources, July 20, 1977.

5. Hackert, H.: "Victor Kaplan: His Life and Work," *Water Power and Dam Construction,* November 1976, p. 39.

6. Terry, R. V.: "Development of Automatic Adjustable-Blade-Type Propeller Turbine," *Trans. ASME,* vol. 63, 1941, p. 395.

7. Haydock, J. L., and J. G. Warnock: "Towards 2,000,000 Horsepower for Giant Turbines," *Energy International,* vol. 7, no. 4, April 1970, p. 28.

8. Eberhardt, A.: "Cornell Hydro Plant Redevelopment," *Water Power and Dam Construction,* June 1976, p. 35.

9. Mayo, H. A., Jr.: "Modern Low Head Hydro Equipment," *1980 Joint Power Generation Conference,* Sept. 30, 1980.

10. O'Brien, E.: "Small Hydroplants for the Northeast," *Electrical World,* Aug. 15, 1977, p. 61.

11. "Ten Years of Tidal Power," *Water Power and Dam Construction,* December 1976, p. 55.

12. *Hydroelectric Plant Construction Cost and Annual Production Expenses,* Federal Power Commission, FPC S-256, 17th Annual Supplement, 1973, May 1976.

13. Casey, H. J.: "The Passamaquoddy Tidal-Power Project," *Mechanical Engineering,* vol. 57, 1935, p. 580.

20

ENERGY STORAGE
AND TRANSMISSION

Energy storage is an important element in many utility systems. It provides a means for easing load peaking problems and improving the load factor on base-load plants. Some types of storage system may also be designed to give a greatly improved system response rate to abrupt changes in load.[1] Energy storage has become particularly attractive since 1973 because the rapidly increasing availability of power from nuclear plants makes the incremental cost of off-peak power low, while at the same time the manyfold increase in fuel oil cost has made power from gas-turbine peaking units much more expensive.[2,3] Before 1961 there were only four pumped-storage hydro units in use in the United States, but by 1974 there was a total installed capacity of 33,600 MW in 51 plants, with many more under construction or planned.[1-5] The pumped-storage hydro plants are the only type of energy storage in commercial use by U.S. utilities, although electric storage batteries are employed as emergency power supplies for vital auxiliaries, and electrically heated hot-water tanks in customers' residences are in limited use for load leveling. One new type of plant that employs underground storage of compressed air for gas turbines has been installed in Germany. Other new types proposed involve the use of the heat of fusion of molten salts, improved storage batteries, and the production of hydrogen.

The engineering problems of energy transmission are generally beyond the scope of this book except insofar as they have major effects on the choice of power plant type or location and/or energy storage system. Thus the concern here is only with the costs and energy efficiency as functions of the amounts of energy handled and distances of transmission. These become crucial matters if, for example, one considers proposals to generate solar power in the U.S. southwest and transmit it to the middle west, of if one wishes to compare the overall costs for supplying Chicago with energy from a mine-mouth coal gasification plant in Montana or a mine-mouth coal-fired electric power plant. The relative costs of rail or slurry pipeline transport of coal as compared to the corresponding costs for high-Btu gas or electricity are typical of other considerations treated below.

PUMPED-STORAGE HYDRO

Pumped-storage hydroelectric installations have proved to be economically attractive where suitable sites have been available. The basic considerations outlined in the previous chapter apply. Typical examples of pumped-storage units are the Jocassee plant of the Duke Power Company in South Carolina[3] and the TVA Racoon Mountain project on the Tennessee River near Chattanooga, Tennessee.[4] In both these installations (as in most cases for pumped-storage hydro plants), high ground above a deep river valley makes it possible to employ a low-head dam to feed a high-head turbine with adequate reservoirs both on the heights and in the river valley throughout the year. Key data on these and some other typical pumped-storage projects in the United States are given in Table 20.1. Note that the head in most cases is at least 100 m and the power capacity over 100 MW. These are commonly the minimum values for economically attractive systems.

The efficiency of a pumped-storage hydro system is good, the principal losses being those in the turbine when operating as either a pump or turbine. Thus, if the efficiency of this component is 90 percent for either mode of operation, the efficiency for the complete cycle would be 81 percent. In addition, frictional losses in the pipes, gates, etc., are present and reduce the overall efficiency to 65 to 75 percent.

As is the case for any hydroelectric installation, the capital cost of pumped-storage systems is highly site-dependent. Table 20.1 gives an indication of the range to be expected for good sites.[5]

Reversible Pump Turbines

Early pumped-hydro energy storage systems employed separate pumps and turbines. Major reductions in the capital cost can be effected by employing reversible pump-turbine units, and this course has been followed for most installations made since about 1940. To do this, some design compromises must be made. As a result, the efficiency in either mode of operation is somewhat less than for units designed explicitly as pumps or turbines.[6,7] The usual practice in the design is to make compromises favoring pump operation because the most critical consideration is to avoid flow separation in regions of adverse pressure gradients.[6] Note that one of the problems is that, for constant-speed machines, the head for peak efficiency for turbine operation is inherently higher than the head for peak efficiency for pump operation.

Propeller-type turbines with adjustable rotor and stator blades are obviously best suited to low-head installations up to 50 m (164 ft), while Francis-type units are most effective for high heads. For intermediate heads, diagonal-flow adjustable-blade machines (Fig. 19.5) have some advantages and are sometimes used for heads of 20 to 100 m (Ref. 8). (See Table 20.1.)

Control

Reversible pump-turbines usually have a narrower range for good efficiency than units designed as either pumps or turbines; hence there is a strong incentive to operate them at their design points. This ordinarily is not a problem, in part because there are usually a number of units operating in parallel, permitting one or more to be shut off, and in part because the periods of high power demand are shorter than the periods when power is available for pumping. Thus it is often advantageous to have one or more units that are used only as turbines and hence can be designed for good performance over a wide range of flows. Of couse, for installations in dams in a river system where turbines in one dam discharge into the backwater of a downstream dam, there must be more turbine capacity than pump capacity be-

TABLE 20.1   Typical U.S. and Canadian Pumped-Storage Hydro Installations with Some or All Units in Operation in 1979

	Total plant output, MW	Turbine head, m	No. of units	Type of turbine	Reversible	Capital cost, $/kW	Year of initial operation
Rocky River, Conn. Lt. & Power	7	70	1	Francis	No—separate pump	215	1928
Buchanan, Lower Colo. Riv. Auth.	11	41	1	Francis	No—separate pump	309	1951
Flatiron—Bur. Rec.	9	88	1	Francis	Yes		1954
Hiwassee—TVA	60	58	1	Francis	Yes		1956
Sir Adam Beck (Canada)	272	25	8	Deriaz	Yes		1957
Lewiston, Power Auth. N. Y.	240	23	12	Francis	Yes		1961
Taum Sauk, Union Elec.	408	241	2	Francis	Yes	112	1963
Brazeau (Canada)	24	13.7	2	Kaplan	Yes		1964
Smith Mt., Appalachian Power	532	55		Francis	Yes	119	1965
Yards Creek, Jersey Central P & L	389	200	3	Francis	Yes	95	1965
Cabin Creek, Pub. Service of. Colo.	300	363	2	Francis	Yes	113	1967
Muddy Run, Phila. Elec.	800	108	8	Francis	Yes	98	1967
San Luis, Bur. Rec.	424	60		Francis	Yes	145	1967
Salina, Grand Riv. Power Auth.	520	71.6		Francis	Yes	114	1968
Racoon Mt., TVA	1530	305	2	Francis	Yes		1978
Thermalito, State of Cal.	85	25.9	3	Francis	Yes		1978
Edwd. G. Hyatt, State of Cal.	290	152	3	Francis	Yes	243	1978
Kinzua, C.E.I. & Pa. Elec. Co.	422	243		Francis	Yes	153	1969
DeGray, Army Engineers	28	52		Francis	Yes	334	1969
Mormon Flat, Salt River Project	49	39				269	1971
Horse Mesa, Salt River Project	100	80.5				155	1972
Northfield Mt., N. E. Util.	1000	227	4	Francis	Yes	463	1972
Blenheim-Gilboa, Pow. Auth. N. Y.	1000	305					1973
Ludington, Cons. Power Co.	1980	110			Yes	154	1973
Jocasee, Duke P & L	610	90	4	Francis	Yes (3 units)	295	1973
Bear Swamp, N. Eng..Power	600	223					1974
Carter's, Corps of Engineers	250	105	2	Francis	Yes		1976
Castaic, L. A. Dept. of W & P	1250	305					1972
Grand Coulee, Bur. Rec.	300	81	6	Francis	Yes		1973
Fairfield, S. Car. E & G Co.	480			Francis			1978

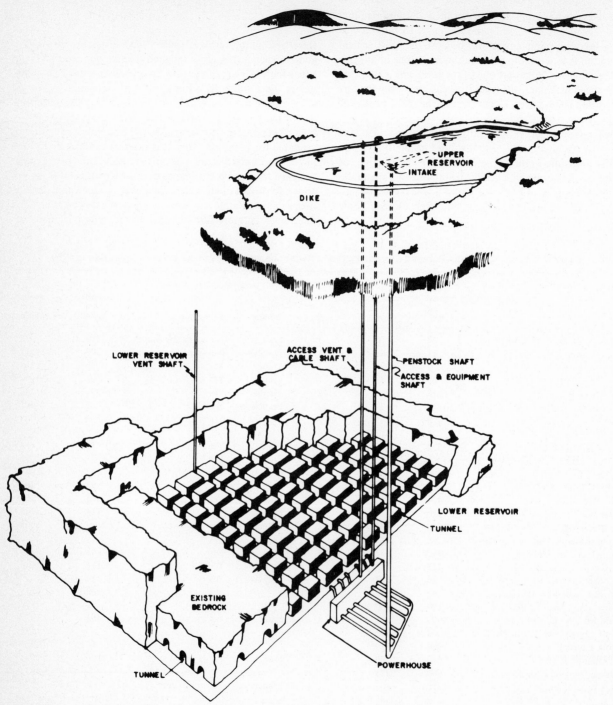

Figure 20.1 Conceptual design for an underground pumped-storage hydro power plant.[2] As in conventional mines the rectangular blocks between tunnels support the roof. *(Courtesy Chas. T. Main, Inc.)*

cause of the net flow downstream. This is the case for the Jocassee installation, for example, where one conventional turbine of 160 MW operates in parallel with three reversible pump-turbine units of 150 MW each.

Control considerations are an additional factor favoring high-head installations, because, the greater the head, the smaller the percentage change in head with drawdown of the reservoir, and hence the smaller the loss in efficiency stemming from deviations from the head for peak efficiency.

Startup and turnaround time vary from 1 to 40 min, depending on the head, the site, and the size and type of machine. The inertia in a long water column in the penstocks of a high-head machine, for example, is a major factor; shifting from pump to turbine operation can be accomplished much more quickly than vice versa, because gravitational forces aid in reversing the direction of the water flow. This is a favorable situation, because unexpected load peaks requiring a quick shift in operating mode are quite likely during pumping, whereas shifts from turbine to pump operation need not be accomplished quickly and there is a strong incentive not to throw too heavy a load on the transmission lines. Thus, in general, the system design compromises are made to permit a fairly rapid initiation of turbine operation. Turbine startup or a shift from the pumping to power-delivery mode can be accomplished in as little as 1 min, while pump startup or shifting from the power output to the pumping mode commonly requires 5 min or longer.

Environmental Effects

The environmental effects of dams for storage reservoirs are the same for pumped hydro as for conventional hydro systems. However, one effect unique to pumped-storage systems appears if a high-head system utilizes as the lower reservoir an estuary that is an important breeding area for fish. Although fish are usually not harmed when pumped up to a reservoir, in a high-head system the rapid decompression as fish pass through the turbine in the discharge mode may cause rupture of the swim bladder, killing the fish. Although the evidence is debatable, this was the deciding factor in the court case that resulted in the cancellation of the Consolidated Edison Storm King project on the Hudson River.

Underground Storage

More often than not a utility finds itself with no satisfactory site for a conventional pumped-hydro storage plant, in which case it may consider the use of an underground storage reservoir. Although no installation of this type had been built up to the time of writing, the concept was getting much serious study in the 1970s. The underground reservoir might be a natural cavern or an abandoned mine, but it would more likely be a specially excavated cavity as in Fig. 20.1. The excavation might be accomplished in rock at what appears to be reasonable cost if the regional geology is favorable. (See the map in Fig. 20.2.) If underground storage is included as a possibility, geological studies indicate that most U.S. utilities could include pumped-hydro plants in their systems.[10] Fairly extensive (and expensive) exploratory drilling is likely to be required to make certain that the rock formations at a site are in fact suitable.

Costs

As in conventional hydro systems, the costs of pumped-storage systems are highly site-dependent, but unit costs generally drop with an increase in the head available. An indication of the costs involved is given by Fig. 20.3, which presents the estimated costs of the equipment required to convert an existing conventional hydroelectric plant to a pumped-storage facility for sites in which the upper and lower reservoirs have adequate capacities.[9] These costs assume the installation of new, reversible pump-turbine units, a step that may be justified for old plants just from the standpoint that an improvement in turbine efficiency may be obtainable. Total costs for the installation including modifications to the dam are much higher and are subject to considerable variation, as indicated by the spread in the set of estimates for which points are plotted in Fig. 20.4.

Underground storage systems cost more than surface storage systems that are at favorable sites. For favorable geological structures that can be excavated at a cost of $10/yd³, the lower reservoir was estimated in 1975 to cost $33/kWh of capacity for a 1000-ft head and $17/kWh for a 2000-ft head.[10] In the same study the estimated cost for a surface reservoir was about $1.50/kWh. These cost estimates were for 1975 conditions and assumed a 4-h peak load condition with a 6-h pumping period. A rough comparison of the principal costs estimated for underground storage systems with conventional surface systems is given in Fig. 20.5.

COMPRESSED-AIR ENERGY STORAGE

The concept of underground storage for pumped-hydro systems has led to studies of similar systems employing compressed air,[1,11,12] and one such system had been completed at the time of writing.[12,13] This unit, the Nordwestdeutsche Kraftwerke A.G. plant at Huntorf, on the Hunte River in the area between Bremen and Oldenburg, West Germany, is designed to produce 290 MW of peaking power for 2 h. The air is compressed in two machines, first to 6 atm in a unit with a 25,000-kW rated power input, and then to 70 atm in a unit with a 40,000-kW rated power input.[13] Both intercooling and after-cooling are employed, so that the air is stored at 50°C in a 280,000-m³ cavity that was solution-mined in a salt dome at a depth between 600 and 800 m. The system operates in a con-stant-volume mode with the pressure in the reservoir running from 45 to 65 atm and with the charging accomplished in 8 h. The total cost of the plant was 92 × 10⁶ DM ($40 × 10⁶ U.S.), or about $138/kW. Of this, 68 × 10⁶ DM was for equipment, 18.5 × 10⁶ DM was for the solution-mined cavity, and 5.5 × 10⁶ DM was for engineering and miscellaneous (1975 to 1978).

In comparing the above plant with a conventional gas-turbine plant, it is evident that there is a substantial heat loss involved in the aftercooler and that the motors required for the air com-pressors represent a major additional capital cost. These disad-vantages are more than offset in this case by the availability of low-cost nuclear power for compressors during off-peak periods and the ready availability of natural gas from a North Sea field for fueling the gas turbine. Further, the solution-mined cavity in a conveniently located salt dome was inexpen-sive, but it could not have been used for pumped-hydro storage unless lined somehow to prevent further enlargement by solu-tion effects. Thus, although the energy efficiency of this system is only about 10 percent, much less than for the pumped hydro, the capital cost is relatively low and the air compression is ac-

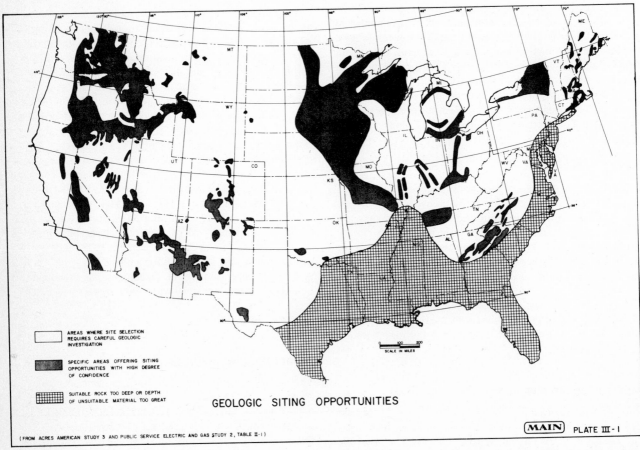

Figure 20.2 Suitability of geological conditions in the United States for siting underground compressed-air or pumped-hydro energy stor-age systems.[2] *(Courtesy Chas. T. Main, Inc.)*

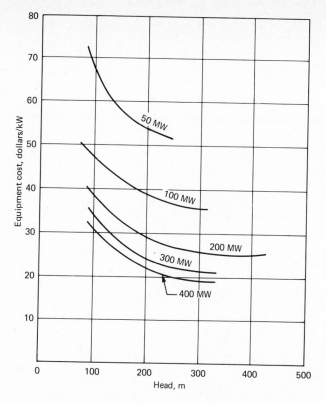

Figure 20.3 Major equipment costs as a function of head. (Equipment includes single-stage pump/turbine, valve, and generator/motor.)[10]

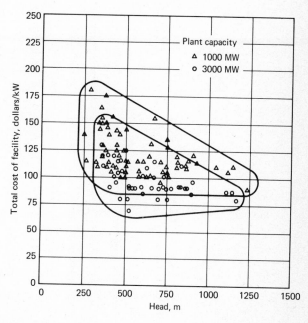

Figure 20.4 Total costs of proposed installations as a function of head.[10]

complished with cheap nuclear power. Note that the electric output of the gas turbine is about twice as great for a given fuel consumption rate as would be the case if it had to drive its own compressor.

A number of steps might be taken to improve the efficiency of the above system.[11,12] First, the heat removed after the compression process might be stored and employed for reheating the air for turbine operation, or a regenerator could be employed to transfer heat from the turbine exhaust to the air coming from storage. Second, if the storage cavity were in a rock formation, it could be pressurized by coupling it to a water reservoir on the surface so that the system would operate as a constant-pressure air reservoir instead of a constant volume. This would avoid shifts in the operating points on the compressor and turbine characteristic curves and make it possible for both units to operate at essentially their peak efficiency throughout the

charge-discharge cycle. Further, hydrostatically pressurizing the cavity would essentially triple the volume of air obtainable from its discharge. Third, if the cavity were pressurized hydraulically, a pumped-storage hydro unit could be coupled into the system, because the water flowing into the cavity during gas-turbine operation can also drive a hydraulic turbine-generator unit. This requires that the cavity be at a greater depth; i.e., the total hydrostatic head must equal the sum of the static head required to pressurize the cavity for the gas turbine and the head required for the hydraulic turbine.

It has been suggested that aquifers or depleted gas fields be employed for compressed-air or pumped-hydro underground storage. However, the permeability of the porous strata involved is low, hence large pressure losses associated with the airflow into and out of the porous rock would occur for the relatively short charge-discharge time periods of interest.

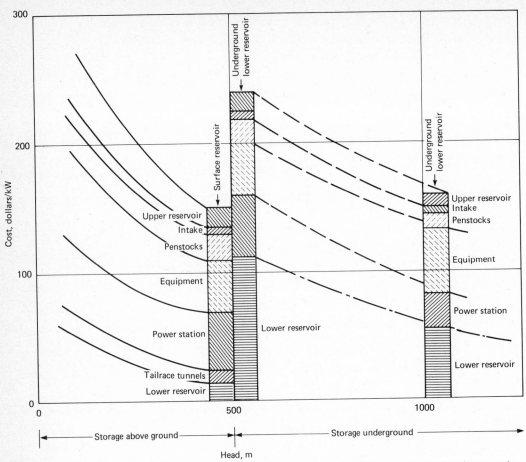

Figure 20.5 An order-of-magnitude cost comparison between surface and underground pumped storage.[9]

ELECTRIC STORAGE BATTERIES

An obvious way to provide energy storage capacity in an electric power system is to employ electric storage batteries,[14,15,16] but up to the time of writing the cost of these has been too high for their use by utilities except for very limited in-house emergency use. Batteries are employed by power consumers for emergencies in special installations where relatively small amounts of standby power must be available at all times. Typical examples of such requirements are for lights in hospital operating rooms, certain vital instrumentation and control equipment in nuclear power plants, and communications systems. If substantial amounts of emergency power are required in the event of an outage, the usual procedure has been to employ diesel-generator sets equipped with automatic controls rather than batteries. These may be kept running continuously at no load so that they can pick up the emergency power load within a few seconds, or they may be fitted with automatic startup equipment so that they can be started up, brought to speed, and pick up their load in less than 1 min.

The changing cost situation for electric utilities since 1970 has provided new incentives to develop storage batteries suitable for peaking power applications. This, coupled with the strong incentive to develop improved storage batteries for automotive vehicles, has led to substantial programs not only to improve the familiar lead-acid storage battery but also, more important, to develop new types of cells that show promise of reduced costs, longer life, greater efficiency, and lighter weight.

The minimal requirements for a suitable battery are defined by the characteristics of the principal competing systems: gas turbines and pumped-hydro installations. On this basis there seems to be general agreement that the requirements of batteries for electric utility service range about as follows:[14,15,16]

1. Cost: $30/kWh (in 1978 dollars)

2. Life: 2500 cycles (10 years)

3. Energy efficiency: 70 percent

4. Duty cycle: 2 to 10 h for discharge and 5 to 7 h for charge

The requirements for vehicular power applications are much the same with the additional limitations that the specific weight and volume should be less than 10 kg/kWh and 5 L/kWh (100 Wh/kg and 200 kWh/L).[15,17]

Types of Batteries

The characteristics of a wide variety of batteries are summarized in Fig. 20.6 and in Table 20.2.[14-24] Lead-acid batteries such as those used in automobiles can be employed, but their life is short (3 to 5 years) and they are good for only a limited number of full charge-discharge cycles (about 50). The greater the degree of discharge, the shorter the life. The nickel-iron batteries that have been used for years for industrial applications give a longer life but are heavier and more expensive. Among the more promising of the many types of electric storage battery on which research is proceeding are those that make use of an alkali metal, usually sodium or lithium, reacting with sulfur or chlorine.[18,20] Some of the work is quite encouraging; e.g., single cells using lithium and sulfur have operated for as much as a year with a full cycle every day with little loss in performance. However, the current density obtainable is relatively low, and, as in fuel cells (Chap. 18), the problems of assembling the individual cells into compact stacks are truly formidable. Electrical insulators, warpage, and corrosion present very difficult problems, especially because the cells must operate at temperatures high enough to ensure that the electrolyte be liquid if it is a molten salt or, if β alumina is the electrolyte, both of the reactants must be liquid or gaseous. Thus a lithium-sulfur cell must run at a temperature above 200°C (lithium melts at 186°C and sulfur at 119°C). Note that these batteries must be kept hot enough to avoid freezing under standby conditions. This can be accomplished with good thermal insulation. In one case for an automobile installation it was found that an inch of superinsulation would serve to keep the battery at temperature for a month if it used its own capacity to operate some small, built-in, electric heaters. Examination of Table 20.2 shows, as one would expect, that the output per unit weight is roughly 20 times higher for a lithium-sulfur cell than for a lead-acid cell, in part because of the lower atomic weight of the reagents and in part because of the much greater energy release per atom reacting.

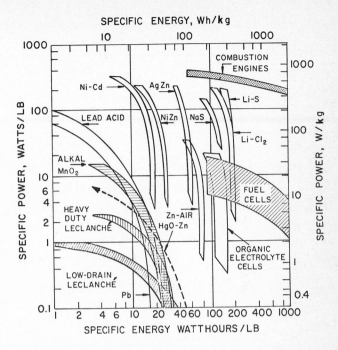

Figure 20.6 Performance of battery systems including Leclanché, or dry cells.[15] (*Courtesy Argonne National Laboratory.*)

Characteristics of Lead-Acid Batteries

An appropriate point of departure for further discussion of the characteristics and problems of electric storage batteries is a review of the characteristics of lead-acid batteries. Data for one of the most recently developed models are shown in Figs. 20.7 to 20.9. The output voltage as a function of time for a relatively low discharge rate is shown in Fig. 20.7 for a series of cell operating temperatures. The gradual drop in output voltage with an increasing discharge is a desirable characteristic. Not only does it give a simple measure of the state of the charge, but also it means that the charge regulating system can be set to supply a constant voltage equal to that at the fully charged condition so that overcharging will not occur. (Overcharging may damage the battery.)

The electrochemistry of the lead-acid cell is such that subtle changes occur during the life of the battery that affect its capacity. Figure 20.8 shows this effect for a typical case. These changes depend on the depth of discharge, as well as the discharge and charge rates, with the latter depending on the charging voltage imposed. Figure 20.9 indicates that these effects can be quite large.

TABLE 20.2 Summary of Characteristics of Typical Batteries
(1978 dollars)

Type	Electrode materials	Electrolyte	Operating tempera-ture, °C	Specific energy, Wh/kg		Specific power, W/kg	Cycle life	Energy efficiency, %	Cost, $/kWh	Ref.	Comments
				Ideal	Practical						
Lead-acid	Pb-PbO$_2$	H$_2$SO$_4$	−40 to 50	35	4–8	~14	50–5000	~65	20–100	14, 15	Cost increases with life rating
Nickel-iron	Fe-NiO$_2$	KOH		60	4–6	~45	3000		~100	14, 22	
Nickel-zinc	Zn-NiOOH	KOH		75	11–16	~45	~300		~100	14	
Zinc-air	Zn-Air	KOH	50–60	220	~27	~14	200–1000		~40	14	
Iron-air	Fe-Air (C + Ag)	KOH	40–50	230	~22	~14	~1000		~40	14, 17	Porous graphite cathode with Ag catalyst
Zinc-chlorine	Zn-Cl	ZnCl$_2$	10–50	170	~22	~30	~1000	~77	15–25	16, 22	Porous graphite cathode
Sodium-sulfur	Na-S	Al$_2$O$_3$	350–400	155	~140	~45	>100		~15	19	
Lithium-sulfur	Li-S	LiCl-KCl		225	~45	~45	~1000		15	14	Development dropped in favor of Li–FeS for improved life
Li-Al/FeS	Li-Al/FeS	LiCl-KCl	400–450		~60	60	>1000	~80		20	Melting pts. of Li-Al, Li-Si, and FeS are above operating temp.
Li-Si/FeS$_2$	Li-Si/FeS$_2$	LiCl-KCl	410–450		115	27	>300	~82		18	Melting point of LiCl-KCl = 352°C
Sodium--chloro-aluminate					~22	~22	~1000		~15	14	
Mercury	Zn-HgO	KOH-K$_2$Zn$_2$O$_2$	4–50		~22		1			22	
Nickel-cadmium	Cd-Ni(OH)$_2$	KOH	−16 to 50		3–6		100–200			23, 24	
Silver-cadmium	Cd-AgO	KOH	−40 to 40		5–18		300–1000			22	
Leclanché (common dry cell)	C/MnO$_2$-Zn	NH$_4$Cl-ZnCl$_2$	4–50		1–12		1			22	

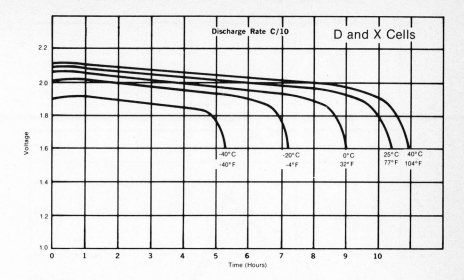

Figure 20.7 Effects of discharge time and operating temperature on the output voltage of an advanced type of lead-acid battery cell.[23] *(Courtesy Gates Energy Products, Inc.)*

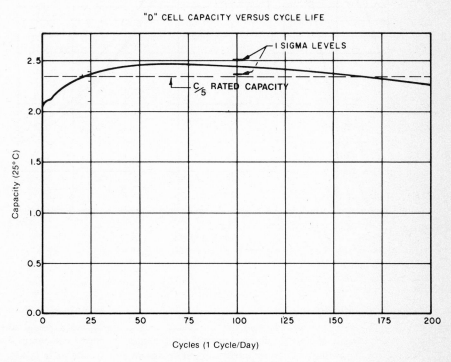

Figure 20.8 Effects of the number of power cycles on the capacity of an advanced type of lead-acid battery.[23] *(Courtesy Gates Energy Products, Inc.)*

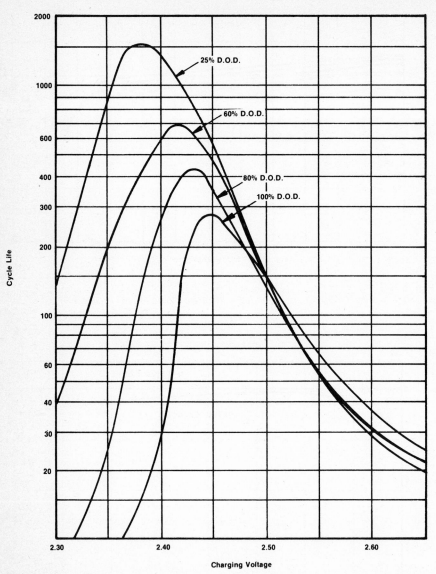

Figure 20.9 Effects of depth of discharge and charging voltage on the life of an advanced type of lead-acid battery for operation at 23°C (73°F) for a 16-h charge period. *(Courtesy of Gates Energy Products, Inc.)*

Cost and Specific Energy

The first step toward reducing the cost of a piece of equipment is to reduce the cost of the materials required, and this in turn implies either reducing the weight or choosing less expensive materials, or both. A highly significant point in this connection is made by Table 20.3, which shows the allowable material cost as a function of the energy storage capacity per unit of weight for three typical ratios of material cost to total cost. Table 20.4 shows the commodity prices for the basic electrode and electrolyte materials required for the batteries of Table 20.2.

Inspection of Tables 20.3 and 20.4 indicates the reason for putting the bulk of the developmental effort into the alkali-metal–sulfur and zinc-chlorine types of cell: They have an inherently high potential output per unit weight, and the basic materials are relatively low in cost. By the same token, cells based on silver, mercury, cadmium, or nickel electrodes are not promising candidates, because the potential output per unit weight is relatively low and the cost per unit weight is high.

One of the factors affecting the cost, specific energy, and specific power of some types of cell is the use of a catalyst. Small amounts of platinum or silver are often added to the electrodes of some types of battery, mainly to increase the current density obtainable. This is particularly important for vehicular applications where a critical requirement is a high output in kilowatts per unit weight for short periods under acceleration and hill-climbing conditions.

An indication of the ranges for both the specific power and the specific energy of representative types of battery is given in Fig. 20.6 with additional curves for fuel cells and internal combustion engines for comparison. For a utility application in which only a short period of, say, 2 h, of peaking power is expected, the wattage output capability should be half of the watthours of capacity, whereas, if the peak power requirement is for 5 h, the wattage output capability need be only 20 percent of the watthours of capacity. These figures of merit are expressed in terms of unit weight in Fig. 20.6, which was prepared with vehicular and space power applications in mind.

Cell Life

Probably the most crucial questions to be resolved in the development programs are the lives of both individual cells and complete batteries. As with fuel cells, deterioration of the electrodes and insulators tends to occur, reducing the capacity and possibly causing failure. The rate at which this occurs is likely to depend on both the depth of discharge and the charge and discharge rates; e.g., as shown in Fig. 20.9, the number of cycles to failure for lead-acid batteries may be over 1000 for a 25 percent discharge per cycle with the proper charging voltage, but only 50 for an 80 percent discharge per cycle with the same charging voltage. On the other hand, the zinc-chlorine battery operates best if fully discharged each cycle, because in a large cell the rate of metallic zinc buildup during the charging operation is not perfectly uniform over the surface of the electrode, and over a sufficient number of cycles this leads to an excessive buildup in some areas with possible shorting. Fully discharging the battery each cycle strips all the zinc from the electrode grid so that in effect there is a fresh start each cycle. The rates of charge and discharge are also important factors affecting the life of the cell, in part because the temperature rises accompanying these operations may lead to distortion. The effects vary widely with the type of battery.

TABLE 20.3   Allowable Material Cost as a Function of Specific Energy Storage Capacity for a Permissible Battery Cost of $30/kWh Assuming That the Material Cost May Be Half the Selling Price

Specific energy, Wh/kg	Allowable material cost	
	$/kg	$/lb
10	0.15	0.07
25	0.38	0.17
50	0.75	0.34
100	1.50	0.68
150	2.25	1.02
200	2.00	1.36

TABLE 20.4   Bulk Material Prices in Dollars per Pound (January 1979)

Aluminum	0.57	Lithium	11.60
Cadmium	2.25	Mercury	2.40
Copper	0.70	Nickel	2.11
Graphite	1.00	Platinum	3600.00
Iron	0.10	Sodium	0.72
Lead	0.40	Sulfur	0.08
Zinc	0.35	KCL	0.02
Al_2O_3	0.10	KOH	0.61
H_2SO_4	0.027	LiCl	1.65

Efficiency

Three different efficiencies are used in measuring the performance of storage batteries. The *coulombic efficiency* is the ratio of the ampere-hour output to the ampere-hour input, the *voltaic efficiency* is the ratio of the discharge voltage to the charge voltage, and the product of these two is the *energy efficiency*. As indicated in Fig. 20.10, the latter (the ratio of the energy output to the energy input) depends in large measure on the rates of charge and discharge. That is, as the current flow increases, the internal resistance losses increase and so the voltage output is reduced for the discharge mode and the voltage required for charging is increased. The values given for the energy efficiency in Table 20.2 are for representative conditions for utility service, i.e., about 6 h for charging and 4 h for discharging.

In addition to the losses in the battery itself, in a complete system other losses are introduced by the step-down

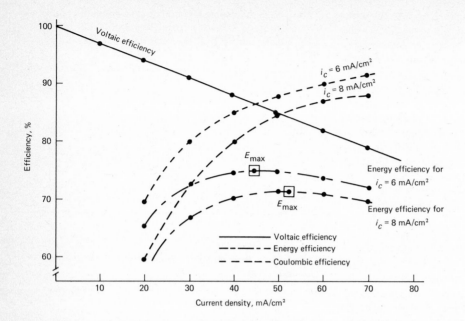

Figure 20.10 Cell efficiencies for a zinc-chlorine battery. The voltaic efficiency depends only on the applied current density. The coulombic and energy efficiencies depend on both the corrosion current and the applied current densities.[22]

transformers and rectifiers required during the charging operation and by the inverter and transformer during the discharge mode. These losses commonly total 15 to 20 percent. The extent of the losses in the transmission lines depends on the location of the battery station. If used near the end of a transmission line, the battery may actually reduce the transmission losses, because the charging operation will be carried out when the transmission lines are lightly loaded and the load in the lines will be reduced during the peak power period, thus cutting the I^2R losses.

Safety

Safety problems are difficult to quantify. The potential for unpleasant accidents with lead-acid batteries is obvious (not the least of which is the possibility of a hydrogen explosion set off by an electric spark). In sodium- or lithium-sulfur cells the fire hazards associated with both the alkali metal and with sulfur are causes for concern, as is the possible release of chlorine gas if it is used.

THERMAL ENERGY STORAGE

The need for the storage of heat energy in hot water for district heating and industrial use was cited in Chap. 17. In a related but different application, some utilities are employing various devices and incentives for off-peak use of electric hot-water heaters to reduce their peak loads and improve their load factors. Similarly, heat storage in hot water is employed in solar energy systems for building heating. For electric power production, storage of heat from high-temperature nuclear reactors or solar-energy collectors for subsequent use in a thermodynamic cycle offers interesting possibilities.

Low-Temperature Heat Storage for Building Heating

Heat has been stored in hot-water tanks for district heating systems since the 1960s to accommodate diurnal differences between the availability of waste heat from electric power plants and the load on the heating system at a capital cost of $3.40/kWht, or $1.00/1000 Btu (1976 dollars) for a 70°C (126°F) temperature drop in the stored water. Experiments with large-diameter (11 to 16 m) tanks excavated in the earth for seasonal storage of solar energy are interesting, but a critical analysis indicates that the heat losses over periods of many months are rather high. Thus the cost of heat from the system is high in relation to that from conventional systems at 1978 prices.[25, 26] (The estimated cost for a reservoir 60 m in diameter and 12 m deep was $0.25/kWh for a 45°C change in temperature.)

Another approach to energy storage for leveling electric utility loads is to offer residential consumers special night rates that induce them to employ home heating systems with substantial heat-storage capacities. The Hamburg Electric Works in Germany instituted this approach in 1969 and by 1973 had succeeded surprisingly well in flattening their daily load profile[27] (see Fig. 20.11). The heat-storage system employs thermally insulated cabinets containing bricks of a suitable refractory that are heated electrically to as much as 650°C (1200°F). A thermostatically controlled fan circulates room air through the bricks to maintain the desired room temperature. The principal problem that they encountered was that some extra expenses they had not foreseen developed so that their initial estimates of appropriate rates for off-peak use of electric power had to be revised upward somewhat.

A somewhat similar system that provides load leveling in the summer, as well as in the winter, employs a heat pump that makes ice in a large reservoir (30 m³ for a single-family residence) during the winter, when the outside air temperature is commonly lower than 0°C, so that the heat pump has a better performance factor than if the heat were drawn from the colder outside air. The ice can then be melted in the summer to provide air conditioning, thus cutting the overall energy consumption about in half.[28]

Energy Storage in Domestic Hot-Water Heaters

Detroit Edison was the first of a number of U.S. utilities to institute an arrangement involving special meters and automatic control units on residential electric hot-water heaters so that the operator in a central station can send a pulse through the lines that will turn on or cut off the power supply to the heaters so fitted in the system and thus shift that load from peak to slack periods. This arrangement involves extra costs for the equipment required and usually much greater than normal capacity in the hot-water tanks, but it appears to be economically attractive and technically sound. Special meters that make possible different rates for peak and off-peak residential power use are also being market-tested in an effort to get householders to shift as many loads as possible (e.g., those for automatic washers) to off-peak periods. In principle, these steps ought to be economically more attractive than either extra capacity or pumped-hydro storage systems.

High-Temperature Thermal-Energy Storage for Electric Utilities

Heat could be made available from nuclear reactors or coal-fired furnaces during off-peak periods at the incremental cost of production and stored in the form of superheated water, steam,

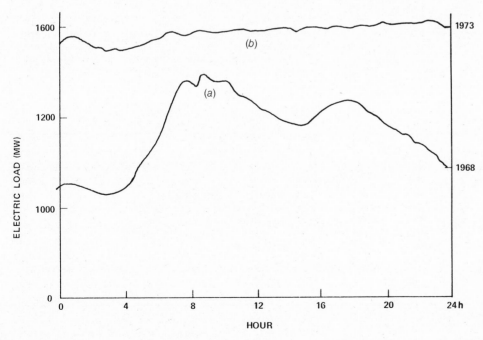

Figure 20.11 Hamburg electric works daily load variations on typical January days (a) in 1968 before initiation of thermal-energy storage program and (b) after development of a major off-peak electric load for building heating with thermal-energy storage.[27]

or hot oil for subsequent use either to generate steam for a separate peaking-power steam turbine or to provide feed-water heating under peak load conditions. Heat storage could reduce drastically the amount of steam bled from the main turbine, increasing its output.[29,30] (This requires a specially designed turbine that will perform well with large changes in bleed-off.) Heat storage becomes economically attractive if the capital cost of the heat-storage system is low as compared to that for the nuclear reactor or coal-fired furnace, and if the thermal energy stored can be used almost as efficiently as if it were put directly into steam turbines. Studies indicate that the cost of thermal-energy storage in superheated water or steam is high, because of the inherently high pressures, but that heat storage in hot oil or a molten salt appears to be in the range of interest.[29,30]

TABLE 20.5 Properties of Selected Salts[a] (Ref. 30.)

Comp	Salt[b]	Melting point, °F	ΔH_f Btu/lb	Heat capacity at melting point $C_p(s)$ Btu/(lb·°F)	$C_p(l)$	Thermal conductivity at melting point $k(s)$ Btu/(h·°F·ft)	$k(l)$	Viscosity, cp At melting point	Melting point + 50°F
1	LiOH	864	376	0.59	0.94				
2	LiBr	1017	88	0.26	0.18		0.23	1.8	1.7
3	NaOH	606	69	0.48	0.51		0.53	4.8	4.0
4	B₂O₃	842	142	0.41	0.44	0.90	0.58		
5	54KCl-46ZnCl₂	810	94	0.16	0.21		0.48	406	208
6	61KCl-39MgCl₂	815	151	0.19	0.23		0.47		
7	48NaCl-52MgCl₂	842	185	0.22	0.24		0.55		
8	36KCl-64MgCl₂	878	167	0.20	0.23		0.48		
9	33NaCl-67CaCl₂	932	121	0.20	0.24		0.59	30	14
10	37MgCl₂-63SrCl₂	995	103	0.16	0.19		0.61		
11	47Li₂CO₃-53K₂CO₃	910	147	0.26	0.32		1.15	21	17
12	44Li₂CO₃-56Na₂CO₃	925	159	0.43	0.50		1.21	23	19
13	28Li₂CO₃-72K₂CO₃	928	113	0.35	0.43		1.07	17	16
14	51K₂CO₃-49Na₂CO₃	1310	70	0.40	0.37		1.00	5.4	5.2
15	33LiF-67KF	918	266	0.32	0.39	2.4–4.8	2.30		
16	67NaF-33MgF₂	1530	265	0.34	0.33	2.4–4.8	2.69		
17	45NaBr-55MgBr	808	91	0.12	0.14		0.52		
18	20LiF-80LiOH	799	374	0.21	0.24				
19	25KCl-27CaCl₂-48MgCl₂	909	147	0.19	0.22		0.51		
20	5KCl-29NaCl-66CaCl₂	939	120	0.28	0.24		0.58	17	13
21	13KCl-19NaCl-68SrCl₂	939	96	0.16	0.20		0.61	198	22
22	28KCl-19NaCl-53BaCl₂	1008	95	0.15	0.19		0.50	107	71
23	24KCl-47BaCl₂-29CaCl₂	1024	94	0.16	0.20		0.55	96	68
24	32Li₂CO₃-35K₂CO₃-33Na₂CO₃	747	119	0.40	0.39		1.17	171	60
25	12NaF-59KF-29LiF	849	257	0.32	0.38	2.4–4.8	2.60		
26	40KCl-23KF-37K₂CO₃	982	122	0.24	0.30		0.69		
27	17NaF-21KCl-62K₂CO₃	968	118	0.28	0.33		0.87		
28	35Li₂CO₃-65K₂CO₃	941	148	0.32	0.42		1.09	18	15
29	20Li₂CO₃-60Na₂CO₃-20K₂CO₃	1022	122	0.38	0.45		1.05	4.1	3.7
30	22Li₂CO₃-16Na₂CO₃-62K₂CO₃	1022	124	0.43	0.50		1.13	3.9	3.5
31	68K₂CO₃-32MgCO₃	860	—	—	—		—	—	—
32	40NaNO₂-7NaNO₃-53KNO₃	288	—	—	0.37		0.35	—	7.5

[a]Where data were available, values given. Other values are estimates.

[b]Mixture composition on weight basis.

[c]Estimated from experimental values of single salts.

[d]Toxic hazard rating code: 0: None, 1: slight, 2: moderate, 3: high.

[e]When strongly heated, they emit highly toxic fumes.

[f]Will react with water or steam to produce heat and will attack living tissue.

[g]Suggested material if suitable corrosion inhibitors are used.

[h]Based on ΔH_f value only.

Heat Storage in Hot Oil

A highly refined, high-boiling-point petroleum oil costing $0.69/gal (in 1976 dollars) with good thermal stability at temperatures up to 270°C (520°F) was used as the basis for an Exxon design study of a system to supply feed-water heating under peak load conditions.[29] This method of heat storage permitted a maximum increase in the station power output of 24 percent and a total heat-storage capacity for 10 h of peaking power with an energy efficiency of 73.4 percent. The estimated capital costs were $183/kW for power-related equipment and $10.50/kWh for the thermal-energy storage. The cost of the oil inventory represented about 20 percent of the capital cost.

Density at 25°C, lb/ft³	Volume change on fusion at mp, %	Safety hazard[d] (inhalation)	Containment material	Salt cost, $/lb	Capacity cost,[h] $/10⁶ Btu	Specific capacity[h] (based on ρ at 25°C), Btu/ft³
91.1		2	Mild Steel[g]	1.97	5,240	34.255
216.0		2[e]	SS	3.00	34,090	19,010
133.0		2[f]	Mild Steel[f]	0.12	1,740	9,120
115.5		1	SS	0.19	1,340	16,400
150.5	17.5[c]	2[e]	SS	0.27	2,870	14,150
131.7	23.6[c]	2[e]	SS	0.06	395	19,890
139.1	27.6[c]	2	SS	0.08	430	25,735
137.3	26.2[c]	2[e]	SS	0.09	540	22,930
134.8	14.5[c]	1	SS	0.03	250	16,310
173.6	16.3[c]	2	SS	0.33	3,205	17,880
137.3	8.5	2	SS	0.47	3,200	20,185
144.8	13.5	2	SS	0.35	2,200	23,025
139.8	9.3	2	SS	0.36	3,185	15,800
149.8		2	Protected SS	0.11	1,570	10,485
157.9		3[e]	316 SS[g]	1.21	4,550	42,000
133.6	22.8	3[e]	316 SS[g]	0.57	2,150	35,405
217.9		2[e]	316 SS	1.62	17,800	19,830
99.9		3	300 Series SS	2.06	5,510	37,365
157.9	20.5[c]	2[e]	SS	0.08	545	23,210
134.2	12.2[c]	2[e]	SS	0.03	250	16,105
171.7	12.9[c]	2[e]	SS	1.37	14,270	16,485
189.2		2[e]	SS	0.09	945	17,975
182.9		2[e]	SS	0.09	960	17,195
143.6		2	SS	0.33	2,775	17,090
157.9		3[e]	316 SS[g]	1.11	4,320	40,580
142.5		3[e]	SS	0.23	1,885	17,385
148.9		3[e]	SS	0.18	1,525	17,570
141.5	9.6	2	SS	0.40	2,705	20,940
148.6		2	SS	0.21	1,720	18,130
146.3		2	SS	0.30	2,420	18,140
—	—	—	—	—	—	—
118	—	—	—	0.20	—	—

Thermal-Energy Storage in Molten Salts

For high-temperature nuclear reactors, for some high-temperature solar-energy collection systems, and for some vehicular applications, the higher thermal efficiency obtainable at high temperatures favors the use of a molten salt for heat storage. The heat of fusion of a molten salt can be employed to give a compact, constant-temperature, thermal-energy source at any of a set of temperatures ranging from 100 to 800 °C.[30,31] Table 20.5 shows the costs, melting points, heats of fusion, and other pertinent characteristics for a typical set of salts that appear to be suitable from the standpoints of chemical stability, corrosion, etc.[30] Studies and some experiments indicate that certain of these systems may prove economically attractive for some applications.[30,31] It has been suggested, for example, that electric energy from the utility could be converted to heat in off-peak hours and stored in LiF tanks in vehicles so that the heat could be used in Stirling engines to give a lighter, lower-cost system than lead-acid batteries. Such a system, however, would still fall far short of being competitive with the gasoline engine for vehicular use unless the cost of petroleum rises far more than seems likely for decades to come. From both the energy efficiency and cost standpoints, a more attractive possibility appears to be the use of the heat of fusion of a salt less expensive than LiF

to store thermal energy from a solar collector or from a high-temperature nuclear reactor. For such an application, data from Table 20.5 indicate that thermal energy could be stored at 551 °C (1024 °F) in a potassium-barium-calcium chloride melt for a capital investment in the salt of $3.27/kWht ($0.96/1000 Btu) in 1977 dollars. The cost of the tank would be roughly 50 percent of the cost of the salt.

The poor thermal conductivity of a frozen melt makes it necessary to employ a large amount of heat-transfer area in the storage tank. This, together with difficulties in designing the equipment to avoid damage from the volumetric changes in freezing and thawing cycles, has led to studies of systems employing a much lower melting point salt, such as HTS,* and utilizing the sensible heat of the salt over a substantial temperature range rather than the heat of fusion.[30,32] Such an approach entails some temperature loss because of the temperature drop in the salt during the heat-removal phase, but it greatly reduces the problems associated with freezing of the salt. Within the limits of the estimates the overall costs appear to be about the same, and they are roughly competitive with those for pumped-hydro power.

*Composition No. 32 of Table 20.5.

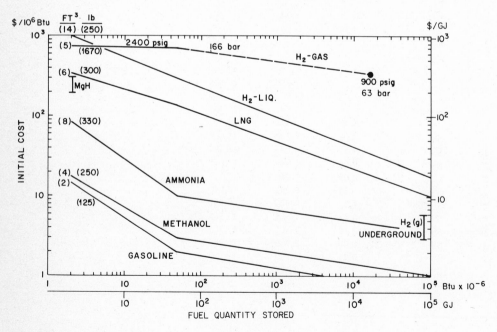

Figure 20.12 Fuel storage investment (1972 basis).[33] *(Courtesy Oak Ridge National Laboratory.)*

ENERGY STORAGE IN HYDROGEN

Extensive studies of energy storage by using off-peak electricity for the electrolysis of water to yield hydrogen and then storing it have produced some attractive pictures of possible systems of the future.[33-35] Hydrogen produced on site at nuclear power plants during off-peak hours, for example, might be distributed through piping systems in the same manner as natural gas, or it might be liquefied, stored, and used in aircraft or automotive vehicles. The basic problems and costs involved in these and related applications are outlined in this section.

Electrolysis Systems

The equipment for hydrogen production by electrolysis is well-developed and the costs and efficiency are well-defined.[33,34] For large-scale production, the capital cost of electrolytic cells in 1978 dollars is ~ \$160/kWe, giving a modest charge per kilowatt-hour if the load factor is high, but the capital charges become less attractive rapidly as the load factor is dropped. The efficiency of commercial electrolytic cells is in the 60 to 72 percent range; somewhat higher efficiencies have been obtained in experimental cells of advanced design, hence a 75 percent efficiency seems attainable.[34] The hydrogen might be burned in conventional peaking-power gas-turbine engines having a thermal efficiency of 28 percent to give an overall energy efficiency for the system of roughly $0.75 \times 0.28 = 21$ percent, if losses for auxiliaries are neglected. Of these latter losses, the most important is the power required to compress the hydrogen for storage. If stored in tanks at ~ 50 atm (a condition giving close to minimum costs for storage in gaseous form), the energy required for the compression process is roughly 2.4 kWh/kg of H_2, or ~ 7 percent of the heat of combustion. The capital cost of the storage system is also substantial: ~ \$7/kWhe. The hydrogen can be liquefied for storage at atmospheric pressure and 20 K with an energy input of ~ 13 kWh/kg, or ~ 39 percent of the heat of combustion, to cut the storage system cost to ~ \$2/kWhe. A still less expensive storage system can be obtained by converting the H_2 to NH_3 at relatively little additional capital and energy costs and storing it in liquid form, in which case the energy for liquefaction is reduced to 0.4 kWh/kg of contained hydrogen and the tankage costs would be reduced to \$0.40/kWhe. Note that the energy cost of liquefying methane is 2.4 kWh/kg of methane, or about 17 percent of the heat of combustion.[33] These costs were obtained from Refs. 33 and 34 and are summarized in Fig. 20.12 to facilitate comparison with other energy storage systems.

The above costs do not include a credit for the oxygen produced, because this factor depends on the size of the hydrogen production system in relation to that of the market for oxygen. An indication of an appropriate credit for the oxygen produced is given by the cost of gaseous oxygen for blast furnaces; this ran \$20 per ton in the 1978 market and would offset about 10 to 15 percent of the cost of the hydrogen production. Note that up to 1978 most of the hydrogen production was from natural gas or petroleum, and most of the oxygen from air liquefaction systems.

Market for Hydrogen

The market for hydrogen is indicated by Table 20.6, which gives a breakdown by usage classification. Also included is an indication of the amount of H_2 that might be produced if the capacity of 20 percent of the U.S. electric utility system (that with the lowest fuel costs) were more fully utilized by employing electrolysis systems to provide the requisite additional load during off-peak hours, assuming that the load factor otherwise would be 65 percent and the availability 75 percent. Note that abrupt swings in the grid load could be absorbed readily by varying the current density and switching banks of electrolysis cells on or off. It is apparent that the higher value uses of H_2 in industry constitute a more than adequate market to absorb the output of the hydrogen from electrolysis carried out with the off-peak electricity from utilities. Further, the market for H_2 would be greatly increased if there were a large-scale system of coal gasification plants that required H_2 for methanation of their output. As discussed in the next section on distribution system costs, such a market may develop, but the cost of synthetic

TABLE 20.6 **Principal Uses of Hydrogen in the United States in 1973**

Use	Energy equivalent	
	10^{15} J	10^9 kWh
Ammonia	379	105
Petroleum refining	492	137
Methanol and other chemicals	103	28.6
Hydrogenation of oils, etc.	5	1.4
Welding, cooling, etc.	69	19
Total	1048	291
Potential H_2 production from 20% of U.S. utility capacity during off-peak hours		80

methane from coal appears likely to be higher than that of natural gas from high-cost wells in the less accessible reserves.

Off-Peak Production of Aluminum and Magnesium

In principle, electrolytic cells for producing Al and Mg might use off-peak power, but they cannot be turned on and off readily as can the cells for H_2 production, and their capital costs are much higher, e.g., \$435/kWe for Al in 1969.[36] Further, the potential markets are much smaller. U.S. production of Al and Mg in 1975 required roughly 100×10^9 kWh and 4×10^9 kWh, respectively, which compares with the equivalent of 291×10^9 kWh given in Table 20.6 as the amount that could be utilized to supply the market for H_2.

COMPARISON OF PEAKING-POWER SYSTEMS

Comparisons of many of the various peaking-power systems are included in Refs. 1, 12, 14, 15, 29, and 32, and an excellent overall review of most of the systems treated above is presented in Ref. 35. A comprehensive summary of these comparisons is presented in Table 20.7. Inconsistencies were resolved by averaging or the use of other data, together with allowances for the rapid escalation of costs in the 1970s, to put all the costs in terms of 1978 dollars. The same values for the energy-storage capacity, capital charge rate, and off-peak electric power cost were used throughout.

A review of Table 20.7 indicates that the lead-acid battery and the hydrogen electrolysis systems seem to be clearly out of the running, while the thermal-energy storage systems are of only marginal interest. Although also marginal from the standpoint of the costs shown in Table 20.7, advanced storage batteries have several advantages: if they can be developed, they can probably be procured and installed more rapidly than any other system listed except possibly gas turbines, and they can be readily placed at end-of-the-line locations, where they would reduce transmission line losses under critical peak load conditions. The prime candidates appear to be pumped-storage hydro and underground storage of compressed air where salt domes make possible low-cost storage cavities, and possibly pumped-storage hydro with mined cavities where rock formations are favorable (Fig 20.2). Actual cost experience with installations will be required to reduce the uncertainties in mined storage cavity costs.

TABLE 20.7 Comparison of Costs and Efficiencies for Utility Peaking-Power Systems (Based on 1978 costs, 5 h/day, 250 days/year for peaking power and 15 mills/kWh for off-peak nuclear power)

	Installed capacity in U.S. (1978), MW	Power-related costs, \$/kWh	Storage-related costs, \$/kWh	Incremental cost of peak elec. power, mills/kWh	Overall energy efficiency, %	References
Gas turbines (oil-fired)*	48,000	100–120	0.01	50–60	25–28	34
Pumped hydro						
Aboveground storage	14,100	100–175	2–12	35–50	70–75	1, 35
Underground storage	0	100–175	10–40	41–73	70–75	1
Compressed air storage	0	110–230	5–35	28–72	~25*	11–13, 35
Batteries						
Lead-acid	0	75–85	70–120	82–115	60–70	14, 35
Advanced type	0	75–85	25–65	48–78	65–75	15–20, 22
Hydrogen electrolysis						
Stored as gas	0	260–300	7	~120	~20	33, 34
Stored as liquid	0	260–300	2	~135	~16	33, 34
Stored as NH_3	0	300–340	0.40	~135	~17	34
Thermal-energy storage						
Hot oil (300°C)	0	160–270†	11–16	50–70†	65–75	29, 30, 35
Molten salt (550°C)	0	220–300†	11–16	55–70†	85–95	30

*Assumed No. 2 fuel oil cost at \$3.30/10^6 Btu, a heat rate for the gas turbine of 5000 Btu/kWh plus 0.8 kWh to compressors per kWe output.

†Assumed heat available from a nuclear reactor at \$1.30/10^6 Btu.

ENERGY TRANSMISSION

The costs of transporting energy over long distances are likely to exceed the cost of production, whether the energy be in the form of raw coal or high-tension electricity. This section summarizes both the capital costs and the energy losses involved in transmitting energy over moderate to long distances.

Electric Power Transmission

The average capital and operating costs for producing, transmitting, and distributing electricity in the United States are summarized in Table 20.8.[37] The average cost of transmission is relatively low, because most power plants are sited close to load centers to keep these costs low. There was a total of 113,400 mi of 230- to 765-kV transmission lines in the United States in 1977 with an average capacity of about 500 MW. This amount was expected to be increased about 50 percent by 1985.

The cost of electric power transmission depends on the amount of power, the distance, and the voltage as shown in Fig. 20.13. The greater the power, the higher the voltage required to avoid excessive line losses.[38] Note that the curves for 230, 345, and 500 kV give minimum costs for powers of about 400, 750, and 1400 MW, respectively. The cost of the transformers and switchgear is a larger fraction of the total costs for the shorter distances. Also note that the voltage drop at full load must not be too large, because the increase in voltage at the end of the line associated with an abrupt drop in load must be kept relatively small.

For still greater distances (over 800 km, or 500 mi) and higher powers than included in Fig. 20.13, it appears worthwhile to go to high-voltage dc transmission to avoid losses to the induced currents in conducting material near an ac line. A number of these lines are in use in Sweden, and one is in service in the United States. A major cost factor, of course, is the conversion of alternating current to direct current and vice versa.

For aesthetic reasons it is desirable to employ underground rather than overhead transmission lines, but the costs are much higher. In urban areas the cost of putting in new lines is greatly increased by problems in avoiding existing water, gas, sewer and other underground lines, and hence the cost in urban areas is commonly 10 to 20 times that for overhead lines. However, for lines through relatively open country the cost differential may be much less. In one case, a 1977 study made by the Philadelphia Electric Company for a 10,000-MVA line and a representative 110-km (66-mi) route, it was found to be more nearly a factor of 2, i.e., 1340×10^6 versus 670×10^6 for overhead transmission.[37] Of course, I^2R losses in underground cables present heat-removal problems that limit the capacity of an uncooled cable to

TABLE 20.8 Breakdown of Average Power Costs for 1972, % (Ref. 37)

Component of cost	Residential and small commercial users	Large industrial users
Transmission structure	16.5	28.9
Transmission substation	3.9	5.0
Transmission operation and maintenance	1.8	0.9
Distribution pole miles	17.6	—
Distribution substation	2.7	0.2
Distribution transformers	4.6	1.4
Meters	1.3	3.1
Distribution operation and maintenance	8.6	0.9
General operation and maintenance	11.9	5.0
Cost of generation	31.1	54.6
Total	100	100

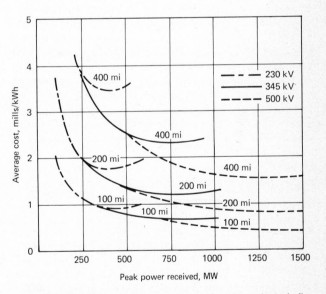

Figure 20.13 Cost of energy transmission for selected distances at 50 percent load factor. (*Courtesy Federal Energy Regulating Commission.*)

1000 MVA. Hence for higher capacities the cables must be cooled or a multiplicity of cables must be employed, and the trenching costs are therefore increased. The Philadelphia Electric Company study included consideration of a superconduct-

TABLE 20.9   Power Loss and Refrigeration Requirements for 30-mi Lengths of 3000-MVA Cable (Ref. 37)

Cable type	Electric power loss, MW	Refrigeration requirement, MW	Total, MW
Oil—paper, static (6 parallel cables)	28.5	None	28.5
Oil—paper, force-cooled (3 parallel cables)	56.0	28.0	84.0
Compressed gas (SF6) insulated	11.0	None	11.0
Liquid nitrogen cooled	2.6	36.0	38.6
Superconducting	0.003–0.03	7.2–20	7.2–20
Overhead	20–30	None	20–30

ing cable and found that its cost was a little lower than for the conventional types.[37] Another study summarized in Table 20.9 showed a marked reduction in energy losses for superconducting cable: the power for the helium cryogenic system designed to hold the cable temperature to 4.7 K was roughly half the amount of the I^2R losses in the conventional cables.[39] Note that the cable types listed in Table 20.9 include a liquid nitrogen-cooled cable designed to take advantage of the reduction in the resistance of copper by a factor of 20 obtainable by operating at a temperature of ~ 80 K. The power losses associated with this cable are much greater than those for the superconducting cable in spite of the far greater power required to pump out a kilowatt of heat from 4.7 K than from 80 K (500 kWe is required to remove 1 kWt from 4 K). This is because the heat leakage through the thermal insulation can be kept to a far lower value than the I^2R losses in copper at 80 K.

Gas Pipelines

The U.S. natural gas utility industry supplied 29 percent of the nation's energy in 1977 with a network of 472,000 km (290,000 mi) of transmission lines and 1,200,000 km (740,000 mi) of distribution mains, not to mention 130,000 km (80,000 mi) of lines to gather the gas from wells in the gas fields.[40] The total investment in capital equipment was estimated to have a book value of 34×10^9 in 1975.

The usual balance found between the pumping-power cost and the capital investments in the pipeline, including the pumping stations, commonly leads to an energy loss of 0.2 percent per 100 km (0.3 percent per 100 mi) for 75-cm-diameter (30-in-diameter) pipelines with the gas at a pressure of ~ 50 atm (750

psig).[33] (This is assuming that the pumps are driven with gas turbines fueled with gas bled from the pipeline.) The energy expenditure per unit of energy transmitted through pipelines is about one-eighth as much as for high-voltage overhead transmission lines. The capital investment for long lines (~ 1000 km) is reduced by a somewhat smaller factor, commonly running less than 20 percent as great for a given energy throughput as for overhead electric transmission lines. For example, a 75-cm-diameter, 50-atm (30-in, 750-psi) gas pipeline typically cost $150,000/km ($244,000/mi) in 1972 dollars.[33]

Oil Pipelines

For a given size of pipe the cost of oil pipelines is about the same as for gas if the gas-line design pressure is ~ 50 atm, and for minimum overall costs the pumping-station power requirements are about the same. However, the higher density of the oil gives a higher energy-transport capacity even though it is necessary to cut the fluid velocity from about 7.6 m/s (25 ft/s) for gas to about 1.8 m/s (6 ft/s) for crude oil (depending on its viscosity). The energy requirement for pumping runs 0.07 percent per 100 km (0.12 percent per 100 mi) of the energy in the liquid pumped after allowing for the thermal efficiency of the prime mover required to convert the chemical energy in the fuel into pumping power.

Note that for both gas and oil pipelines the pumping-power requirement is approximately inversely proportional to the size of the line, because the fluid velocity for minimum overall costs is nearly constant irrespective of pipe diameter, while the length-diameter ratio and hence the pressure drop are inversely proportional to the pipe diameter. The choice of pipe diameter for a given flow rate falls in a rather narrow range, because the pumping power varies inversely as the fifth power of the pipe diameter, whereas the pipeline cost is directly proportional to the pipe diameter.

Coal Slurry Pipelines

Probably the least expensive way to transport large quantities of coal long distances is to crush it and pump it as a slurry. This was recognized at an early date and the concept was tested in 1891.[41] The first commercial operation was a 20-cm (8-in) line that began carrying a coal slurry into London in 1914. The first commercial installation in the United States was a 25-cm (10-in) line 176 km (108 mi) long started up in 1956 by the Consolidation Coal Company in Ohio. Operation of this system was terminated after 6 years when a competing railroad, to avoid losing further business, cut its rates drastically to provide lower-cost transportation for coal in that area. Other pipelines for handling other minerals in slurry form were built,[42] but it was not until 1971 that a second coal slurry pipeline was built in the United States, this time in Arizona, where the topography and the location of the existing railroad lines made it possible to get a much shorter route with a pipeline.[43] The coal is transported 44 km (270 mi) from the Black Mesa mine in Arizona to the 1510-MW Mohave power plant of the Southern California Edison Company near Las Vegas, Nevada, at the rate of 660 tons per hour. The coal is crushed to −14 mesh, mixed with water in a 50:50 ratio by weight, and pumped at 1.77 m/s (5.8 ft/s) through a 46-cm-diameter (18-in-diameter) pipe. The pumping is carried out with Moyno pumps (which employ twin parallel rotors with meshing, high-pitch, helical lobes), the pumping load being greatly eased by a 1700-m (5600-ft) drop in elevation between the mine and the power plant. The slurry can be stored in tanks without serious settling for as much as a month and then dewatered at the power plant to 15 percent moisture by weight with centrifuges. Effluent from the centrifuges at the Mohave plant flows to settling tanks, where the fine coal particles are flocculated and settle out. The water leaving the settling tanks is further purified and used as boiler feed water, while the sludge is withdrawn from the bottom of the settling tanks and recycled through the centrifuges.[43] It is interesting to note than in one experiment in the 1950s Consolidation Coal pumped a slurry of 70 percent solids by weight directly to burners for a boiler with no serious difficulties. There was, of course, a heat loss for vaporizing the water of ~5 percent.

In comparing the coal slurry pipeline with an oil pipeline, one may note that the density of coal is typically ~1.35. Thus the density of the slurry is ~1.15 and the heat content of the fluid flowing through the line is typically 0.445 kWh/L (0.43 × 10⁶ Btu/ft³), as compared to 1.069 kWh/L (1.03 × 10⁶ Btu/ft³) for crude oil, or less than half as much. The density of the coal slurry is greater and the velocity for minimum costs is roughly the same, hence the energy required for pumping is roughly three times as great per unit of energy available from the flow through a given size of pipeline. Thus the energy requirement for transport would be roughly 0.2 percent per 100 km (0.3 percent per 100 mi) of the energy available from the fluid pumped for a 75-cm-diameter (30-in-diameter) pipeline. The capital cost for the pipeline would be about the same for the same pipe diameter but roughly twice as great for the larger pipe required for the same energy throughput.

The amount of water required for coal slurry pipelines used to deliver coal from the coal-rich but water-short areas of Montana and Wyoming raises ecological questions and is a politically sensitive matter. However, the water requirements are low in relation to the annual runoff even from these areas of relatively little rainfall. For example, Table A9.6 gives the annual runoff from the Missouri River basin as 75 km³ per year. If the entire U.S. coal consumption in 1980 had been conveyed from that area by coal slurry pipeline, the water requirement would have been ~1 percent of the annual runoff. From the cost-effectiveness standpoint, diverting water from irrigation would entail a loss in value of the crops produced of $0.05 per ton of coal shipped out in slurry form.

Coal Transport by Truck, Rail, Barge, and Ship

Virtually all coal shipments up to 1980 have been made by conventional bulk carriers, because they provide a great deal of flexibility in coal purchasing and routing from the mine to the plant, as well as opportunities for mixing different types of coal to get satisfactory combustion conditions in the furnace. A study of the energy requirements for coal transportation indicates that haulage by either mixed or unit trains entails an energy consumption of 0.3 percent per 100 km (0.5 percent per 100 mi), while haulage by river barge requires about half as much.[44] Haulage by heavy trucks is used only for short distances—generally less than 20 km for large quantities, although as much as 162 km (100 mi) for single truckloads—and the energy requirement per unit of distance is about twice as great. Because of higher labor charges, the dollar costs are substantially higher for truck than for train transport.

Haulage by ship, the oldest means of bulk commodity transport, is most efficient, but the energy requirements for long hauls are still substantial. A 30,000-ton tanker carrying oil from Kuwait to Philadelphia, for example, has a round-trip fuel consumption equivalent to about 12 percent of its payload. The energy efficiency is much improved in a supertanker; even with the longer run around the Cape of Good Hope because it can not pass through the Suez Canal, the fuel consumption runs about 3 percent of the payload.

Comparison of Dollar Costs for Energy Transmission

The relative costs of conveying various forms of energy depend on both the quantity of energy to be transported and the distance. For long hauls of over 800 km (500 mi) and amounts of energy of the order of 3000 MW, the lowest costs in both dollars and energy losses are obtained with pipelines. For a 75-cm-diameter (30-in-diameter) pipeline, the capital cost for the pipeline would be about the same irrespective of whether it was used for gas, oil, or a coal-water slurry, but the energy throughputs would be roughly in the ratio of 1, 2, and 5 for gas, coal, and oil, respectively. In terms of 1975 dollars the pipeline capital charges for gas transmission would be ~0.07 mill per thermal kilowatthour per 162 km (100 mi), or ~0.2 mill per electric kilowatthour per 162 km (100 mi). For an 800-km

(500-mi) line they would be ~ 1 mill per electric kilowatthour. The capital costs for coal-water slurry and oil transport would be, respectively, about 50 and 20 percent as much. Operating costs for fuel and maintenance would be less than 10 percent of the capital charges in each case, i.e., well within the differences in construction costs from one area to another. The corresponding costs for unit train transport in 1975 came to ~ 3 mills per kilowatthour for an 800-km (500-mi) route, while for 500-kV electric power transmission for 800 km (500 mi) the cost was ~ 2 mills per kilowatthour, again in 1975 dollars.

PROBLEMS

1. Estimate both the interest on invested capital of lead-acid batteries for utility energy storage for peak shaving and the prorated cost of the consumption of their useful life, using a typical currently advertised value for an automotive battery cost and life. (The usual storage capacity is 750 Wh for the battery in a 3000-lb gasoline automobile.) Take the battery life as a function of depth of discharge as being that given by Fig. 20.9 for a charging voltage of 2.4 V per cell, and calculate values for 20, 40, 60 and 80 percent discharge per cycle.

2. Estimate the capital cost of the transmission lines and the energy losses that would be entailed in the transmission if electricity were generated by a solar power plant in Arizona and transmitted 2460 mi to New York. Assume that, because of equipment availability limitations, about 60 percent of the energy that ideally could be generated an average of ~ 7 h per day (Figs. 21.14, 21.17, and 21.18) with a 1000-MWe-peak-capacity solar-energy conversion system could be fed to the transmission lines and transmitted with an energy loss of 1 percent per 100 mi. Estimate the transmission losses in percent, the transmission cost per kilowatthour, and the factor by which the capital cost of the solar-power plant should be increased if its capital cost were expressed in terms of the power delivered to New York rather than the power generated at the site.

3. If 10^8 tons of coal per year were mined in the Wyoming-Montana region and piped in slurry form to the Chicago area, what fraction of the mean annual runoff of the water in the Missouri Basin would be required? (See Fig. 19.2 and Table A9.6.) What would be the average pipeline water flow rate in cubic meters per second? Compare this with the average flow rate for a small river in your vicinity.

4. If 10,000 m^3 H_2O/ha is required for irrigation, and the gross annual crop yield from the irrigated cropland is $500/ha, how would the value of the crop that could be grown with a given amount of water compare with the value of the coal that could be transported?

REFERENCES

1. *An Assessment of Energy Storage Systems Suitable for Use by Electric Utilities,* prepared by Public Service Electric and Gas Co. for ERDA and EPRI, EPRI EM-264, July 1976.

2. Fairfield, J.J., et al.: "Underground Pumped Hydro Storage: An Evaluation of the Concept," *Proceedings of the Fourteenth IECEC*, Aug. 5–10, 1979, pp. 399–404.

3. Lee, W. S.: "7,750-MW Project Meshes Nuclear, Hydro, and Pumped Storage," *Electrical World,* Oct. 16, 1967, p. 89.

4. *Racoon Mt. Pumped-Storage Project, Final Environmental Statement,* Tennessee Valley Authority, 1976.

5. Johnson, G. D.: "Worldwide Pumped Storage Projects," *Power Engineering,* October 1968, p. 68.

6. Terry, R. V., and F. E. Jaski: "Test Characteristics of a Combined Pump-Turbine Model with Wicket Gates," *Trans. ASME,* vol. 64, 1942, p. 731.

7. McCormack, W. J.: "Performance of Reversible Francis Pump-Turbines," *Trans. ASME*, vol. 78, 1956, p. 417.

8. Deriaz, P., and J. G. Warnock: "Reversible Pump-Turbines for Sir Adam Beck-Niagara Pumping-Generating Station," *Journal of Basic Engineering, Trans. ASME,* vol. 81, 1959, p. 521.

9. Warnock, J. G., and D. C. Willet: "Pumped Storage Underground," presented at the Symposium on Hydro Electric Pumped Storage Systems conducted by the U.N. Commission for Europe, Athens, November 1972.

10. Carson, J., and S. Fogleman: "Considerations in Converting Conventional Power Plants to Pumped Storage Facilities," ASCE Engineering Foundation Conference Publication *Pumped Storage,* August 1974.

11. Bush, J. B., Jr., et al.: "Compressed Air Storage—A Near-Term Option for Utility Application," *Proceedings of the Eleventh IECEC,* Sept. 12–17, 1976, p. 578.

12. *Economic and Technical Feasibility Study of Compressed Air Storage,* ERDA-76-76 prepared by the General Electric Company, March 1976.

13. Stys, Z. S.: "Compressed Air Storage for Load Leveling of Nuclear Power Plants," *Proceedings of the Twelfth IECEC,* Aug. 28–Sept. 2, 1977, p. 1023.

14. Maskalick, N. J., et al.: "The Case for Lead-Acid Storage Battery Peaking Systems," *Proceedings of the Tenth IECEC,* Aug. 18–22, 1975, p. 1135.

15. Yao, N. P., and J. R. Birk: "Battery Energy Storage for Utility Load Leveling and Electric Vehicles: A Review of Advanced Secondary Batteries," *Proceedings of the Tenth IECEC,* Aug. 18–22, 1975, p. 1107.

16. Warde, C. J., et al.: "100 MWh Zinc-Chlorine Peak-Shaving Battery Plants," *Proceedings of the Thirteenth IECEC,* Aug. 20–25, 1978, p. 755.

17. Buzzelli, E. S., et al.: "Iron-Air Batteries for Electric Vehicles," *Proceedings of the Thirteenth IECEC,* Aug. 20–25, 1978, p. 745.

18. Zeitner, E. J., Jr., and J. S. Dunning: "High Performance Lithium/Iron Disulfide Cells," *Proceedings of the Thirteenth IECEC,* Aug. 20–25, 1978, p. 697.

19. Bird, J. M., et al.: "Sodium/Sulfur Cell Designed for Quantity Production," *Proceedings of the Thirteenth IECEC,* Aug. 20–25, 1978, p. 685.

20. Gay, E. C., et al.: "Review of Industrial Participation in the ANL Lithium/Iron Sulfide Battery Development Program," *Proceedings of the Thirteenth IECEC,* Aug. 20–25, 1978, p. 690.

21. Appleby, A. J., et al.: "Economic and Technical Aspects of the C.G.E. Zn-Air Vehicle Battery," *Proceedings of the Tenth IECEC,* Aug. 18–22, 1975, p. 811.

22. *Development of High Efficiency, Cost-Effective, Zinc Chloride Batteries for Utility Peak Shaving,* EPRI 711, prepared by Energy Development Associates, March 1978.

23. Hammel, R.: "Sealed Lead-Acid Batteries," *Advanced Battery Technology*, vol. 16, nos. 2/3, February–March 1980, pp. 47–49.

24. *Handbook of Tables for Applied Engineering Science,* Chemical Rubber Company, 1970, p. 459.

25. Margen, P.: "Central Solar Heat Stations and the Studsvik Demonstration Plant," *Proceedings of the Thirteenth IECEC,* Aug. 20–25, 1978, p. 1614.

26. Beard, J. T., et al.: "Energy Loss Analysis from Hot Water Storage Pool," *Proceedings of the Fourteenth Southeastern Seminar on Thermal Sciences, North Carolina State University,* Apr. 6–7, 1978, p. 226.

27. Swisher, J. H.: "Issues in the Near-Term Commercialization of Underground Pumped Hydro, Compressed Air, and Customer-Side-of-the-Meter Thermal Energy Storage," *Trans. ANS Annual Meeting,* June 3–7, 1979, pp. 3–7.

28. Fischer, H. C., and E. A. Nephew: "Application of the Ice-Maker Pump to an Annual Cycle Energy System," ASME Paper No. 76-WA/Ener-4, December 1976.

29. Nicholson, E. W., and R. P. Cahn: "Storage in Oil of Off-Peak Thermal Energy from Large Power Stations," *Proceedings of the Eleventh IECEC,* Sept. 12–17, 1976, p. 598.

30. Fox, E. C., et al.: "An Assignment of High Temp. Nuclear Energy Storage Systems for Production of Intermediate and Peak-Load Electric Power," Oak Ridge National Laboratory Report No. ORNL-TM-5821, May 1977.

31. Marianowski, L. G., and H.C. Maru: "Latent Heat Thermal Energy Storage Systems above 450°C," *Proceedings of the Twelfth IECEC,* Aug. 28–Sept. 2, 1977, p. 555.

32. Vrable, D. L., and R. N. Quade: "High Efficiency Thermal Energy Storage System for Utility Applications," *Proceedings of the Thirteenth IECEC,* Aug. 20–25, 1978, p. 917.

33. Gregory, D. P.: *A Hydrogen-Energy System,* American Gas Association, Catalog No. L21173, August 1972.

34. Michel, J. W.: "Hydrogen and Exotic Fuels," Oak Ridge National Laboratory Report No. ORNL-TM-4461, June 1973.

35. Surface, M. O.: "Exotic Power and Energy Storage," *Power Engineering,* December 1977, p. 36.

36. Goeller, H. E., and J. R. Mrochek: "Generalized Capital and Operating Costs for Power-Intensive and Allied Industries," Oak Ridge National Laboratory Report No. ORNL-4296, December 1969.

37. Forsyth, E. B.: "Overview of Electric Transmission (Technology and Economics)," *Proceedings of the Thirteenth IECEC,* Aug. 20–25, 1978, p. 1004.

38. Baughman, M. L., and D. J. Bottaro: "Electric Power Transmission and Distribution Systems: Costs and their Allocation," *IEEE Trans. on Power Apparatus and Systems,* vol. PAS-93, no. 3, May–June 1976, pp. 782–790.

39. Haid, D. A.: "Power Transmission Via the Superconducting Cable," *Mechanical Engineering,* vol. 98, no. 1, January 1976, p. 20.

40. Statistical Abstract of the U.S. Department of Commerce, Bureau of Census, 1970, p. 515.

41. Thompson, T. L., and E. J. Wasp: "Coal Pipelines—A Reappraisal," ASME Paper No. 68-PWR-8, September 1968.

42. Wasp, E. J., et al.: "Process Basis and Economics of Pipeline Systems," *Canadian Mining and Metallurgical Bulletin,* December 1970, p. 1373.

43. "270-Mile Pipeline to Bring Coal Slurry to Mohave Plant," *Electrical World,* Sept. 23, 1968.

44. *Energy Alternatives: A Computer Analysis, Science and Public Policy Program*, University of Oklahoma, May 1975, pp. 1–125.

21

GEOTHERMAL, SOLAR, WIND, WAVE, AND OCEAN THERMAL DIFFERENCE ENERGY SYSTEMS

People are always eager to get something for nothing, and hence are much intrigued by the prospect of getting free energy from the sun or the wind or ocean waves. However, all these are low-head systems comparable to low-head hydroelectric systems, and thus would entail high capital investments. Often, as is the case with energy in tides as discussed in Chap. 19, these forms of energy are available in the wrong place at the wrong time hence would require large investments both for storage and for transmission of the electricity produced to areas in which it could be used to advantage. There is no question but that electric energy can be produced from these various "natural" energy sources. The question is one of cost. If the cost of electricity is to run 2 to 20 times that of electricity from fossil fuel, it seems highly doubtful that most people will prefer expensive electricity from a source of free, low-grade energy. It is significant that efforts have been under way for many years to produce electricity from the "free" energy sources, but only two of these sources—geothermal and "solar biomass"—are currently yielding commercial electric power anywhere in the world. Only a few plants are presently utilizing this energy, and they are close to electric load centers.

Capital costs are the heart of the problem in each case. Although they are difficult to estimate in a definitive fashion, it is possible to get some significant insights by examining the proposed systems and estimating the costs of major components that are essentially similar to those in widespread commercial use, and this approach is followed here.

GEOTHERMAL ENERGY

There were five commercial power plants producing geothermal energy in 1980. The first of these—that at Larderello in Italy—began operation in 1904; that in New Zealand in 1957; the Geysers near San Francisco, California, in 1960; the 150-MWe Cerro Prieto plant in Baja California in 1972; and the most recent—the one in El Salvador—in 1975.[1-4] The fields in Italy

and New Zealand appear to be limited in scope, and so these plants have not been expanded in recent years. The Geysers plant in California, however, has been steadily increased in size to a capacity of 900 MW at the time of writing, and further expansion to around 1500 MW is under way.[5] The new fields in El Salvador and Baja California also appear capable of providing for further expansion. The principal characteristics of these plants are summarized in Table 21.1.

Types of Geothermal Heat Sources

All the geothermal heat sources of interest have been created by the intrusion of hot magma from deep in the earth up into rock strata close to the surface.[6] Geologically recent intrusions of this sort have been responsible for extensive volcanic activity in the region stretching up through California, Oregon, and Washington. Note that the most recent volcanic eruptions in this area occurred between 1914 and 1921 at Mount Lassen, about 200 km north of Sacramento, and at Mount St. Helens in Washington in 1980. (The Geysers power plant is about 75 km west-north-west of Sacramento.) The thermal conductivity of rock is low and the depth of these masses of magma is usually of the order of many kilometers, and so the time required for cooling is of the order of 1 million years. A good idea of the location of these potential sources of heat is provided by extensive measurement of the temperature gradient as a function of depth as measured in wells drilled in efforts to find oil and gas deposits. These temperature measurements yield the data from which the vertical thermal gradient as a function of depth can be computed. The U.S. Coast and Geodetic Service has organized these data to provide the map shown in Fig. 21.1, which gives geothermal temperature gradient contours as determined from the average temperature gradient to a depth of 6 km.[7] The regions of interest for power production are those in which the temperature gradient exceeds about 20°C/km; these regions have been cross-hatched in Fig. 21.1.

TABLE 21.1   Engineering Data for Commercial Geothermal Power Plants Operating in 1979
(Data from Refs. 1–4)

	Larderello	Wairakei	Geysers	Ahuachapan	Cerro Prieto
Country	Italy	New Zealand	U.S.A.	El Salvador	Mexico
Year of initial operation	1904	1957	1960	1975	1972
Normal peak power output, MWe	400	180	908	60	150
Total installed power capacity, MWe	440	240	908	60	150
Steam temperature, °C	140–190	260 max.	172–240	250	167
Steam pressure, bars	7–40	12, 4.6, 1.14	6.5–7.5	14.6	5
Type of geothermal source	Dry steam	Steam @ 20% quality	Dry steam	Dry steam	Hot water
Number of wells					40
Well depth, m	< 1000	171–1220	1200–3000		1100–1400
Well bore, cm	34	20			18 (liner)
Size of well field, km			4 × 11		
Weight fraction noncondensables, %	5–30	0.26	0.54		

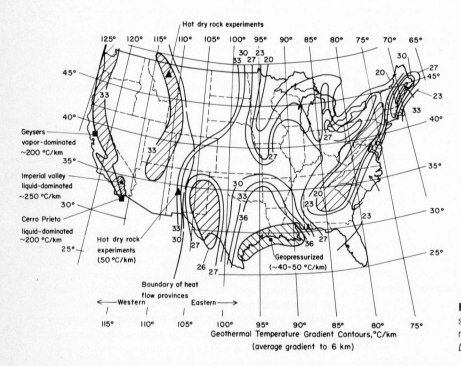

Figure 21.1 Map of the United States showing lines of constant average geothermal gradient.[6] (*Courtesy Oak Ridge National Laboratory.*)

The problems of extracting heat from a geothermal heat source vary widely with the particular type of rock strata in the area. In a few regions, porous rock is overlain by a low permeability stratum and above that an aquifer which allows water to trickle into the hot porous rock at a rate such that a steady flow of dry steam is generated. It is this type of formation that has been tapped at Larderello, the Geysers, and in El Salvador. A similar formation exists under the power plant in New Zealand except that there the aquifer feeds in an excess of water, so that only about 20 percent by weight of the steam-water mixture emerging from the wells is in the form of steam and the balance is in the form of superheated water. Favorable rock formations of this sort are rare, which is the reason that there are so few commercial geothermal power plants.

A second major geothermal source is the hot brines in porous strata. These brines sometimes contain as much as 30 percent salt by weight and hence are both highly corrosive and difficult to handle, because salt precipitates out rapidly as heat is removed from them. The geothermal heat sources in the Imperial Valley area of California are of this type. The hot water in the nearby Baja California field, on the other hand, has a low salt content. The geopressured hot-brine deposits found along the Texas-Louisiana gulf coast are not at as high a temperature as the brines in the Imperial Valley, but they contain a substantial amount of dissolved methane. In fact, the first well drilled into these brines to explore their potential as an energy resource disclosed that the amount of methane obtained from the hot brine emerging from the well was roughly double the amount that would have been obtained from a brine saturated with methane. Thus it appears that there are pockets of methane gas in these deposits, so that small bubbles of methane are entrained in the brine drawn from the strata. Analyses indicate that neither the thermal energy in the brine nor the contained methane would be sufficient by itself to make commercial exploitation of these deposits economically attractive, but by exploiting both the thermal energy and the methane, the economics may prove to be favorable. Unfortunately, the rock has been under such great pressure that its compressibility is low; hence only ~ 4 or 5 percent of the fluid is recoverable.[8]

A third type of geothermal heat source is the pools of molten magma beneath active volcanoes, such as those in Hawaii and Central America. No good way of exploiting this type of geothermal source has been proposed.

The fourth and largest type of geothermal heat source is hot, dry rock, and a substantial amount of effort was under way at the time of writing to develop means for exploiting it. The approach being followed is to drill into hot, dry rock in regions with a high thermal gradient and then hydrofracture the rock by pumping water in at pressures of the order of 200 atm.[6] This produces a system of vertical cracks radiating from the bore of the well. In principle, wells could be drilled in a checkerboard pattern and water injected into alternate wells to yield hot water and/or steam flowing out of the intermediate wells. Extensive analytical efforts indicate that percolation of water from the feed wells through the cracks in the rock could yield attractive rates of steam formation and collection from the steam wells. As the rock is cooled, further fracturing should occur as a consequence of thermal stresses, and so the yield of steam from a given set of wells may actually improve. On the other hand, the water flow through some cracks may tend to open them up to a greater degree than others, so that channeling will occur and the effective heat-transfer surface will be greatly reduced. Extensive and expensive tests will be required to determine what can actually be accomplished.

The extent of these various types of geothermal resources in the United States is indicated by Table 21.2,[5] which gives the amount of thermal energy available in terms of quads (10^{15} Btu) and in terms of units of 10^{18} J (virtually the same quantity of energy).[5] As shown in Table 1.1, the U.S. energy consumption in 1977 totalled about 76 quads per year of which about 10 percent was in the form of electricity. It must be remembered that the thermal efficiency in converting relatively low-temperature geothermal heat energy into electricity is likely to be around 10 percent, and the U.S. consumption of electricity in 1978, if derived from geothermal heat, would have required 75 quads of heat. Thus, the estimates of Table 21.2 indicate that, if economical ways to exploit them completely can be found, the geohydrothermal resources of the United States might provide our electric power needs at the 1978 level of consumption for roughly 160 years, and the geopressurized resources might serve for an equal period if one allows for the low recovery rate of ~ 5 percent that appears likely. However, the hot, dry rock resource represents an amount of energy roughly 1000 times that available from the geohydrothermal deposits.

Geothermal Wells

One hundred years of experience in drilling oil and gas wells have produced a mature technology for drilling and a good basis for estimating the costs of wells. The generally used drilling technique employs a drill bit consisting of three toothed rotors set 120 degrees apart, each tangent to the bore of the hole being drilled. These are made of very hard, wear-resistant material. The assembly is rotated at slow speed in a bath of "mud," which is a thick suspension of clay particles. The mud acts as a lubricant and coolant as well as the vehicle for removing the debris from the drilling. A heavy axial load of about 10,000 kg imposed on the drill bit acts to crush the rock under the points of contact

TABLE 21.2   Estimated U.S. Geothermal Resources[5]

Type of source	Thermal energy	
	Quads (10^{15} Btu)	10^{18} J
Geohydrothermal	12,000	12,600
Geopressurized	190,000	200,000
Magma	200,000	210,000
Hot, dry rock	13,000,000	13,700,000

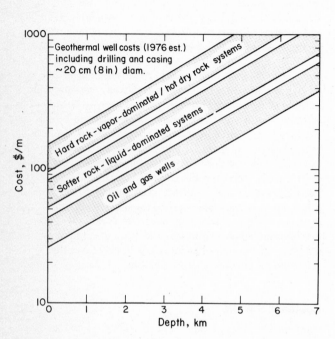

Figure 21.2 Estimated costs for drilling deep wells.[6] *(Courtesy Oak Ridge National Laboratory.)*

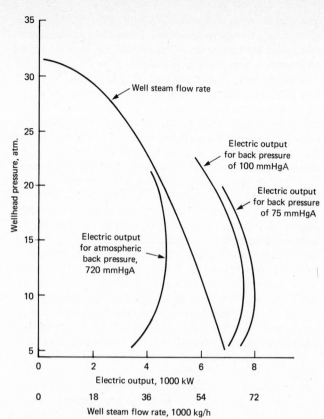

Figure 21.3 Relation of wellhead pressure to steam flow rate and electric output.[3]

between the teeth of the rotors and the rock. The penetration rate depends on the type of rock but is commonly of the order of 3 to 7 m/h in the sedimentary rock strata encountered in oil and gas well drilling. After being drilled, the well is commonly lined with a casing whose OD is 3.8 cm (1.5 in) less than the bore of the well to facilitate its insertion. For deep wells a larger diameter is used for the upper portion and the diameter is stepped down at depth.

Experience in drilling wells in the Geysers area has shown that the conventional technique for lubricating and cooling the drill bit with mud is not satisfactory for geothermal wells because the high temperature leads to both caking of the mud and clogging of the porous wall of the well so that the effectiveness of the well for producing steam is severely reduced.[9] This has necessitated the use of compressed air for blowing out the debris from the

drilling operation. When the drill begins to enter the steam stratum, the mass of high-temperature steam escaping up the bore of the hole is substantial so that relatively little cooling can be accomplished with this compressed air. The combination of high temperatures and lack of lubrication leads to a drastically shortened life for the drill bits, and this greatly increases the expense of drilling the wells.

The cost of drilling deep wells in the search for new deposits of oil and gas is so great that new techniques for drilling have been investigated. One of these makes use of tiny water jets employing pressures of the order of 1500 to 3000 atm.[10,11] These jets enter cleavage planes between the rock crystals and act to break crystallites loose from the matrix. Rapid cutting action in porous rock has been obtained, but penetration rates are low for the hard, dense rocks, such as basalt, encountered in drilling geothermal wells. To date there has been no commercial application of this drilling technique. Another technique under investigation is designed to penetrate hard rock with a piercing tool having a heated nose made of a material, such as tungsten or molybdenum, that can withstand a high temperature and melt its way through the rock.[12] While intriguing, the results of analyses and experiments have not served to demonstrate the value of the system for penetrating hard rock. The heated piercing tool has also been suggested for use in drilling through salt formations, but these pose another problem; namely, the plastic character of the salt tends to close the hole and crush the casing. This problem stems from the fact that, aside from the nominal compressive stresses, bending stresses are also induced in the casing as a consequence of deviations from perfect circularity. Bending stresses are likely to lead to the phenomenon of creep buckling of cylindrical shells under external pressure, as described in Chap. 6. A much thicker and hence more expensive casing is therefore required.

Cost and Output of Wells

A good indication of the cost of wells as a function of depth is given in Fig. 21.2, taken from Ref. 6. Note the much higher cost associated with drilling through hard, porous rock permeated by hot steam. It must be remembered that the permeability of the rock is not high, so that, in order to get a large yield from a well, it is necessary to drill into the hot-rock region quite some distance to get sufficient wall surface area to yield a high steam flow rate.

An important factor limiting the energy obtainable from a well is the pressure drop in the steam or hot water flowing to the surface from the region tapped. An indication of the importance of this effect is given by Fig. 21.3, which was obtained for a well in the Geysers area. In this instance the most economical operating condition is at a high steam flow rate that yields a wellhead pressure only about one-third of that for zero flow. This loss stems from the pressure drop within the porous rock along with the pressure drop within the casing bringing the steam to the surface. The cost data plotted in Fig. 21.2 were for wells having diameters in the range of 15 to 30 cm, in part because drill rigs are not equipped for drilling larger wells and in part because the limitations on the flow rate imposed by rock permeability ordinarily limit the output of the well to a value within the capacity of a 15- to 30-cm bore.

The limitations on well output imposed by the pressure drop for the flow to the surface are indicated by Fig. 21.4 for some

typical water and steam conditions for three sizes of casing. The electric output obtainable from the heat energy in the hot water or steam is indicated by Fig. 21.5. These charts can be used for the rough estimation of the useful output of a well if the pressure drops through the porous rock strata and the surface piping are small.

The value of a well is also a function of its operating life. This is determinable only from operation in each particular type of rock strata. In New Zealand, for example, the output of the wells fell off within the first year to a value not much more than half of that expected from the original test wells, but apparently the output has stabilized at that reduced value. It remains to be seen what the useful life of these wells will prove to be. In the Geysers area the average life span of a well appears to be about 8 years. Thus it is evident that the cost of the well must be written off over a period of the order of 10 years.[13] The relatively short life of the wells as compared to the operating life of the power plant indicates that the power plant should be situated within a reasonable radius of a number of wells. In any case, the steam obtained is not free: The capital charges are high and in the Geysers wells have led to a 1976 price for the steam at the wellhead equivalent to about 7 mills per kilowatthour.[13]

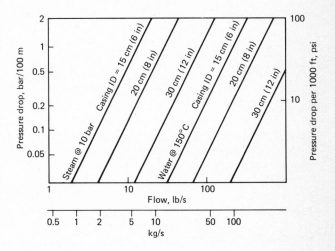

Figure 21.4 Chart for rough estimates of pressure gradients in geothermal wells. (For rough approximations, the pressure drop for a given flow varies inversely as the square root of the steam pressure, the electric output obtainable from steam at 10 bars is ~ 800 kW per 1.0 kg/s, and the electric output obtainable from 180°C water is ~ 120 kW per 1.0 kg/s. (See Fig. 21.5.)

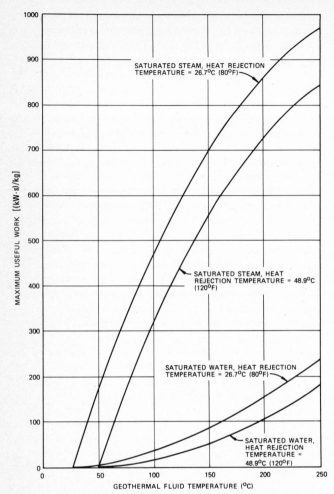

Figure 21.5 Maximum useful work or availability plotted as a function of geothermal fluid temperature for saturated-steam and saturated-water sources.[6] *(Courtesy Oak Ridge National Laboratory.)*

Thermodynamic Cycles

For wells such as those at the Geysers, the obvious course is to use the steam directly in the turbine after filtering it to remove particulates. In New Zealand, where 80 percent by weight of the fluid emerging from the wells is water, or in hot-brine fields such as that in the Imperial Valley, it is advantageous to consider more complex systems. Steam can be obtained from superheated water by flashing it in one or more stages, with some of the sensible heat in the water employed to vaporize a portion of

the water in each stage. In the New Zealand plant this has led to a system in which steam is fed to the turbine at three pressures: 12 bars for the steam that is drawn from the well, 4.6 bars for the steam from the first-stage flash boiler, and 1.14 bars for the steam from the second-stage flash boiler. The thermal efficiency with which the heat can be utilized from each of these three pressure levels, of course, drops with the pressure, thus making the performance of the plant particularly sensitive to the temperature of the cooling water available for the condenser.

The high mineral content of hot brine makes it advantageous to have a heat exchanger between the brine and the thermodynamic working fluid. Further, in order to reduce the size of the turbine for these low-temperature vapor conditions, it may be advantageous to employ a working fluid with a higher vapor pressure than water, e.g., a Freon or an organic fluid. Systems of this sort are discussed in Chap. 17. The principal additional problem for hot-brine systems is the tendency toward heavy scaling and severe corrosion in the boiler. It has been suggested that the scaling problem could be handled by passing the hot brine through a liquid fluidized bed containing the boiler heat-transfer surfaces so that the particles in the bed will scrub off any scale that forms. A preliminary test indicates that this approach can be effective.[14]

Special Problems

Problems peculiar to geothermal power plants include many different types of corrosion stemming from impurities in the hot fluid, air pollution by noncondensables such as H_2S and NH_3, and surface water pollution by materials such as bromine and boron that may be present in the condensate from a dry steam well. An indication of the seriousness of these problems is that 20 percent of the electric output of the Larderello plant in Italy must be used for scavenging noncondensable gases from the condensers, and about 7 percent of the cost for the steam in the Geysers field is for disposing of the condensate by reinjecting it into the field because of the substantial boron and bromine contents. These problems are all very much site-dependent so will not be discussed further here.

SOLAR-ENERGY UTILIZATION

An enormous amount of energy flows to the earth from the sun: a total of over 10^{11} MW around the clock, or 1.39 kW/m² of area normal to the sun's rays.[15] (Eccentricity in the earth's orbit causes the latter value to vary from 1.35 to 1.44 kW/m².) A substantial fraction of the energy incident on the earth is absorbed in the earth's atmosphere or reflected by clouds or particulates so that the energy reaching the earth's surface on a clear day when the sun is directly overhead is commonly $0.9\,kW/m^2$. This energy serves to heat the earth and keep its temperature high enough to make it habitable. A small fraction of the solar energy incident on foliage induces chemical reactions that yield stored chemical energy in biomass, which can be burned to produce electricity. This commonly represents ~2 percent (in a few plants as much as 5 percent) of the incident light energy. At first glance it appears that some sort of system devised by humans ought to yield a higher conversion efficiency for the production of electricity. There are, however, practical problems, which are treated in this section.

Although some people include wind and waves and even tides and hydroelectric power units under the heading of solar energy, for purposes of this section only three basic types of energy are classified as solar energy: biomass, direct thermal conversion of solar energy into electric energy by a thermodynamic cycle, and direct conversion of solar energy into electric energy by photovoltaic cells. Of these three, the only one that has been and is currently being used to produce electricity commercially is biomass; a substantial amount of wood is being employed as fuel in steam plants.

Biomass

Although less than 1 percent of the wood produced by the U.S. forest products industry is employed as fuel, nearly half of the forest products harvested in the world are employed as fuel, largely in the underdeveloped countries.[16] The reason for this disparity is that harvesting the timber is a labor-intensive operation, and so the cost of the product is too high for it to be compe-

titive with other fuels unless labor costs are very low. Even then, commercial operations on a substantial scale require tree farming, i.e., the planting of trees of highly productive species in uniform rows on relatively level land so that they can be harvested by cutting over large areas at a time. In traveling thousands of miles in rural areas in India and Brazil, the writer has seen numerous tree farms dedicated to the growth of firewood. Fast-growing evergreen species are employed: banyan trees in India and eucalyptus trees in Brazil. It is interesting to note that in Minas Gerais, Brazil, over 400,000 ha (10^6 acres) are producing ~ 10^7 tons of wood per year for fuel, much of it for making charcoal to smelt iron.

In the United States, wood has been getting some attention recently as a possible fuel for use by electric utilities. One example of a U.S. electric utility fueled by wood is a small, 10-MW, coal-burning plant of the Burlington Electric Department in Vermont.[17] In this instance the wood is harvested in the wake of timbering operations by removing the culls—stunted or otherwise undesirable trees not suitable for lumber. (As is evident in Fig. 21.6, the heating value of wood per unit weight is independent of the specie of tree; it depends only on the moisture content.[18] The density of wood varies widely; e.g., a cord of light wood such as poplar has a heating value only about half that for dense woods such as hickory or oak.) The trees harvested in this way are reduced to chips, hauled to the mill, and burned with oil, the proportions depending on the supply of wood chips. Inasmuch as wood chips are bulky, they are expensive to haul and store. Hence they ordinarily are brought in from a radius of not more than 80 km (50 mi), but usually less than 50 km (30 mi), and are burned within a day or two of their arrival at the plant so that the high cost of large storage bins can be avoided.[19]

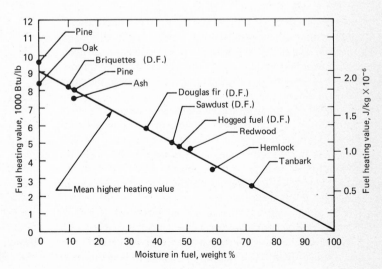

Figure 21.6   Higher heating value of wood fuel.[18]

TABLE 21.3   U.S. Commercial Forest Production and Its Energy Equivalent (Data from Refs. 16 and 20)

U.S. forestland area, ha	305×10^6 (750×10^6 acres)
U.S. forestland suitable for timber, ha	202×10^6 (500×10^6 acres)
U.S. forest industry wood production, tonnes/year (63% lumber, 35% fiber products)	226×10^6 (250×10^6 tons)
Wood waste fuel used by paper industry, J	0.96×10^{12} (0.91×10^{15} Btu)
Energy equivalent of U.S. 1977 forest products, J (assuming HHV = 5000 Btu/lb for green wood)	2.6×10^{18} (2.5×10^{15} Btu)
Total U.S. energy consumption in 1977, J	80×10^{18} (76×10^{15} Btu)

Forest Resources of the United States

In view of Burlington Electric's successful use of wood fuel obtained by following up timber clear-cutting operations, a survey of the U.S. forest products industry from the energy standpoint appears in order. Table 21.3 was compiled to indicate the amount of energy potentially available. If all the trees cut for timber and paper in the United States in 1977 had been employed instead as fuel, the energy available would have amounted to only about 3 percent of the total U.S. energy consumption. This utilization, of course, would not have been practical, because both lumber and paper represent higher value uses than fuel. Note that even the low-grade wood used for pulp was bringing ~ $28 per cord, or about ~ $20 per ton in Tennessee in 1980. This amounts to about $2/$10^9$ J ($2/$10^6$ Btu). It is interesting to note that although the paper industry is well integrated, with the paper mill facilities including huge tracts of land on which much of their own pulpwood is grown and harvested, the economics are such that only about 30 percent of the fuel used in the paper mills is in the form of wood waste; the balance is either coal or fuel oil. While it might be argued that a much larger fraction of the fuel could be obtained from the small branches and even the leaves of the trees that are cut, it is common practice to leave this material at the site to provide nutrients and organic matter for the soil. In fact, to improve the production rate from the forest areas, commercial fertilizer is commonly applied, usually by dispersal from an airplane.

It is often suggested that dead and wind- or lightning-damaged trees in forested areas could be culled and used for fuel. However, even when these areas are readily accessible by truck, it is difficult for a skilled lumberman to take out more than about one cord per day (128 ft^3 of firewood or 160 ft^3 of pulpwood). A cord of pulpwood weighs about 1.5 tons, hence even if labor were paid the minimum wage and there were no charges for overhead or transportation, the fuel cost per 10^6 Btu would be about double that for coal. For such operations it is not practicable to bring in heavy equipment because this not only damages the standing timber but also deeply ruts the forest floor and is likely to lead to serious gullying. Finally, it is evident from Table 21.3 that, at most, this approach would yield only a small fraction of the U.S. energy requirements.

Agricultural and Municipal Solid Wastes

Other forms of biomass have been suggested as fuel. A crop such as hay, for example, might be grown expressly for fuel. Again, a higher use of the land, i.e., for agricultural production, is clearly in order, particularly in view of the world food shortage. Agricultural wastes such as cornstalks might be employed, but again they have an important use as fertilizer and to renew the organic material in the soil. Even if all the agricultural and forest wastes together with all the municipal solid wastes were employed, estimates indicate that the total would at most be equivalent to about 7 percent of the 1976 U.S. energy consumption (Ref. 10 of Chap. 17).

Harvesting biomass from the sea offers interesting possibilities. To reduce harvesting costs, extensive sea farms have been suggested, possibly including some deliberate arrangements to bring nutrients from deep waters to the surface to promote the rapid growth of kelp. (See the last section of this chapter.) After harvesting, the kelp would be sun-dried and then could be used as fuel. A number of experiments to investigate the possibilities of this approach are under way at the time of writing; results of this work should give a good indication as to its commercial feasibility. The ecological effects of massive harvesting of kelp will be more difficult to determine.

Conversion to Methane or Alcohol Fuels

Conversion of organic wastes such as manure into methane by anaerobic fermentation has received some attention.[21] This is technically feasible but expensive, and the heating value of the methane obtained is only about 60 percent that of the dried cattle dung widely used as fuel in India and other underdeveloped countries.

Conversion of crops such as grain or sugar cane to ethyl alcohol has been a popular possibility for many years. The technology is well established and the concept is strongly supported by political pressure groups. However, the costs are high and the energy efficiency is poor.

The basic cost of the raw material—a food crop—is inherently high as compared to the cost of fossil fuel. Even more important, the energy conversion efficiency is poor. Extensive studies have shown that the lowest cost and the highest energy efficiency are obtained with sugar cane; grains, nuts, cassava, etc., give a less favorable ratio of the energy available in the alcohol produced to the energy required for their production (tractor fuel, insecticides, refining, etc.).[21-24] Inasmuch as virtually all the water must be removed if the alcohol is to be mixed with gasoline, it is necessary to follow the distillation process by treatment with a dessicant such as quicklime. Thus the energy required is greater than that for conventional alcohol distillation. (This step may not be necessary if the alcohol is emulsified in diesel fuel.) If sugar cane is used as the source material, the energy for refining can be obtained by burning the bagasse (the cane residue). In fact, more steam is available from burning the bagasse than is required for the refining operations, and it is possible to take advantage of this energy by establishing an energy balance for the system.

Table 21.4 summarizes data for the most energy-efficient system that could be used in the United States, i.e., sugar cane production in Louisiana.[24] Three cases that differ in the degree of utilization of the energy in the bagasse are covered. For case 1,

credit is taken for all the energy obtainable from burning the bagasse to generate steam, case 2 credits the bagasse just with the energy required for operation of the refinery, while case 3 makes no use of the bagasse as fuel. From the energy standpoint the best yield is for the case 1 conditions, but even for this most favorable case the energy equivalent of roughly half of the alcohol produced would have to go back into the agricultural operations.

The capital charge for the land use is high. Using the cane yield of 53 tons per hectare per year and the net energy production of 16.3×10^6 kcal per hectare per year of Table 21.4 and a value of $1000 per acre for the agricultural land gives a capital charge of $\sim \$6/10^6$ Btu for the land. In addition, there would be capital charges for the agricultural equipment, trucks, etc. The capital cost of the refinery is also high; e.g., according to Ref. 22, in 1978 the cost of a refinery for producing 10^5 L per day of ethanol from sugar cane was around $\$10 \times 10^6$. This represents a capital charge of ~ 5¢/L, or $\sim \$2.00/10^6$ Btu for the gross output. (Both of these figures should be roughly doubled if they are to be put in terms of the net useful output.) Although the cost of the refinery per unit output could be reduced by going to a greater capacity (petroleum refineries commonly have capacities ~ 100 times greater), this is not practicable, because the costs of

TABLE 21.4 Energy Balance of Alcohol Production[24]

Quantity	Sub-total	Total	Net energy balance	Output-input
Agricultural yield, [ton/(ha · year)]		53		
Alcohol production, L				
Per ton		66		
Per hectare per year		3498		
Energy expended [$\times 10^6$ kcal/(ha · year)]				
Agricultural	8.5			
Industrial structure	0.4			
Industrial fuel	10.4	19.3		
Energy produced [$\times 10^6$ kcal/(ha · year)]				
Case 1				
Converting all bagasse to steam	17.2			
Content of alcohol produced	18.4	35.6	+ 16.3	1.8:1
Case 2				
Converting enough bagasse to meet all industrial requirements	10.4			
Content of alcohol produced	18.4	28.8	+ 9.5	1.5:1
Case 3				
Burning fossil fuel to meet all industrial requirements	0			
Content of alcohol produced	18.4	18.4	− 0.9	0.9:1

hauling sugar cane from the field to the refinery become excessive if a larger cane field area is served.[22] Unfortunately, this limitation also affects the value of the excess steam that can be produced from the bagasse. If employed for generating electricity, as assumed in Refs. 22 and 24, the steam-electric plant will be small (~ 7 MWe), and hence both its operating costs and its efficiency will be much less favorable than would be the case for a large plant.

In the world as a whole, the production of alcohol for fuel has occurred principally in Brazil, where the addition of alcohol to gasoline has been required by law since the 1940s. More favorable growing conditions and a more labor-intensive economy have yielded a higher value than indicated by Table 21.4 for the energy output-input ratio,[23] i.e., ~ 2.4:1 as compared to the 1.8:1 of Table 21.4.

A recent Brazilian study by one of the authors of Ref. 22 indicates that, by burning the bagasse in a high-pressure, high-temperature steam generator and taking a more favorable credit for the steam generated, Brazil may be able to achieve an energy output-input ratio as high as 3.6:1 for sugar cane. The same study indicates that methanol might be produced from wood to give an even more favorable hydrocarbon fuel energy output-input ratio of ~ 10. No details were given in the paper, but this probably assumed that most of the harvesting would be done by

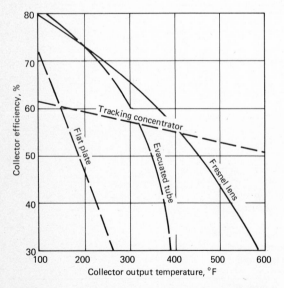

Figure 21.7 Estimated performance of several types of solar collector. Ambient temperature = 21°C (70°F). Solar intensity = 948 W/m² [300 Btu/(h · ft²)], all available as direct.[28] *(Courtesy Barber-Nichols Engineering Co.)*

manual labor and draft animals. It was stated that the capital cost of the refinery and the gross energy output per hectare should be about the same as for ethanol production from sugar cane. The yield of methanol from the destructive distillation of wood is not high (~ 25 kg/t of wood), but most methanol is synthesized by the reaction $CO + 2H_2 \rightarrow CH_3OH$, using a catalyst and operating at ~ 200 atm and ~ 350°C. (This process is under development for making methanol from coal.) If wood is used as the source material, it was claimed that a methanol yield of 300 kg/t of wood can be obtained.[25] Thus the chemical energy in the hydrocarbon fuel produced would be about half that in the wood supplied, a value close to the upper limit theoretically obtainable. (See the section on methanation in Chap. 5.)

A significant strategic consideration with respect to small-scale ethanol production in rural areas is that if U.S. farms were equipped to generate alcohol from agricultural wastes, they would not be vulnerable to loss of their fuel supply in the event of a major war, and this would be of enormous strategic value.

It can be argued that costs should be ignored, in which case the question becomes one of the land area required. Assuming the use of sugar cane and the data of Table 21.4 as the point of departure, a net production of 16.3×10^6 kcal per hectare per year in the form of ethanol and prime steam from combustion of the bagasse could be obtained for the most favorable combination of conditions for energy production from biomass in the United States. To meet a U.S. annual energy demand of 80 quads per year would require 12×10^6 km². This is about 10 times the total acreage under cultivation in the United States, or about 100 times the area of Louisiana. It has been suggested that plants might be developed to yield a higher efficiency for photosynthesis, but existing plants represent the culmination of over 3×10^9 years of evolution; hence substantial further improvements seem highly unlikely. Further, this energy would not be from a "renewable" resource: U.S. Department of Agriculture figures for soil erosion rates indicate that 7 kg of topsoil would be lost for each kilogram of alcohol produced. This represents a more serious disadvantage than the erosion associated with the unregulated strip mining of coal, and the world food shortage presents humanitarian problems that preclude large-scale diversion of agricultural land to fuel production.

Solar-Thermal-Energy Conversion Systems

Solar heating systems for domestic hot water were fairly common in Florida in the 1920s, and some of these systems are still in use. The installation of new systems was brought to a halt when natural gas became generally available, because this gave a much lower system cost. The use of solar energy for heating houses has been under investigation at M.I.T. since about 1930 with funds from a substantial endowment. The M.I.T. work has included the building and testing of a number of single-family houses.[26] Solar heating systems for swimming pools became fairly common in California in the 1970s, extending the swimming season by a month or two in both the spring and the fall. Note that in the domestic hot-water and swimming-pool systems it is easy and inexpensive to provide a large amount of heat capacity relative to the heat loss rate so that diurnal variations in the solar-energy flux do not represent a serious problem. It is for this reason that the most promising application of solar energy is generally considered to be the heating of domestic hot water and buildings. A huge federally funded program of subsidies was under way at the time of writing to encourage these applications of solar energy.

The use of solar energy to generate steam and produce power is also not new; the first power plant of record was built in Egypt in 1912 to pump irrigation water.[27] This plant employed parabolic trough reflectors to concentrate sunlight on boiler tubes. The steam generated was used to drive a 15-hp engine, the largest amount of power to be developed with a solar power plant for the next 65 years.

In the 1950s a substantial program to develop the thermoelectric solar power units for spacecraft was initiated and has been continued up to the time of writing. Much interesting work has been done, but no operational system has been installed in a spacecraft.

It is implicit in the record that production of electric power from a solar-thermal power plant is technically feasible but that the capital costs have been so high that it has not proved to be commercially attractive in spite of the zero cost of solar energy. This section outlines the principal problem areas.

Solar-Energy Collectors

Simple, relatively inexpensive flat-plate collectors are normally employed for domestic hot-water, swimming-pool, and building-heating applications. However, as the temperature at which the energy is collected is increased, the efficiency of the collector falls off rapidly because of increasing losses via thermal radiation and conduction. These losses can be greatly reduced by concentrating the sun's rays on a boiler tube placed at the focus of a parabolic trough reflector and even further by enclosing the boiler tube in an evacuated glass tube to reduce heat losses by thermal convection. Still higher receiver temperatures can be obtained by employing a cavity-type receiver which is well insulated on all sides except for an aperture on the side toward the concentrator.

The efficiencies with which typical collector systems convert the solar energy input into useful heat in the collector are indi-

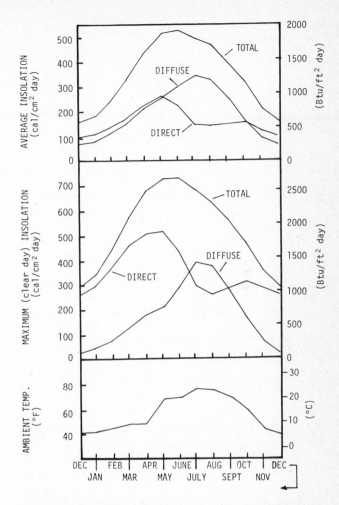

Figure 21.8   Monthly average and maximum insolation incident on a horizontal surface and ambient temperatures based on 16 years of data (1955 to 1971) for the Oak Ridge area (latitude 36.03 degrees).[29]

cated in Fig. 21.7.[28] These data are for operation on a clear day; a substantial amount of haze and/or upper-air turbidity make the comparison more favorable to simple flat plates because a much higher percentage of the incident energy at the first surface is in the form of diffuse rather than direct radiation, and the diffuse radiation is not concentrated by a lense or a parabolic mirror. The average fraction of diffuse radiation varies with the geographical location. Figure 21.8 shows data for the Oak Ridge area in Tennessee.[29] The upper curves show averages on a weekly basis and indicate that over half of the total incident solar radiation is diffuse, and for 4 months of the year the dif-

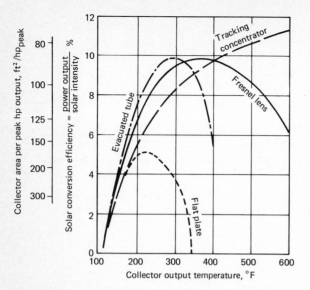

Figure 21.9 Estimated solar conversion system efficiency as a function of collector temperature. Assumptions: (1) total solar intensity = 948 W/m² [300 Btu/(h · ft²)] with 90% direct, (2) maximum cycle temperature = 95% of collector temperature, (3) collector efficiency from Fig. 21.7, (4) Rankine cycle with regeneration, (5) indirect component lost to Fresnel and concentrator.[28] *(Courtesy Barber-Nichols Engineering Co.)*

fuse radiation markedly exceeds the direct radiation. Note, too, that the middle set of curves indicates that even on a nominally clear day the diffuse radiation exceeds the direct radiation for about 3 months of the year. Partly for this reason and partly because of their inherently lower cost, the flat-plate collectors are usually preferred for building heating and domestic hot-water systems. However, the poor thermal efficiency associated with a low collector temperature makes it equally clear that it is best to employ some form of solar-energy concentrator if one wishes to produce electricity via a thermodynamic cycle. Even under the most favorable conditions at midday on a clear summer day, it is difficult to get useful power from a flat-plate collector. This effect is shown in Fig. 21.9, in which the useful power output as a percentage of the solar-energy input is plotted against collector outlet temperature for several types of solar-energy collectors.[28] An additional scale on the ordinate gives the collector area required per peak horsepower output at noon on a clear day.

The nature and magnitude of the losses in a solar-energy collector are illustrated by data for a typical parabolic trough collector designed by the author and shown in Fig. 21.10.[29] To reduce heat losses to the surrounding air, particularly in a strong

wind, the unit was covered with a glass plate. This also provided an easily cleaned surface and minimized deterioration in the reflectivity of the parabolic mirror. However, some of the solar energy is reflected from a cover glass surface, and some is absorbed in passing through the glass. In addition, the reflectivity of the mirror surface is less than perfect: commonly about 0.83 for a highly polished, Alzak aluminum surface. Further, depending on the absorptivity of its surface coating, the boiler tube will absorb at best around 90 percent of the incident light energy, the remaining 10 percent being reflected. Even with the glass cover and with thermal insulation lining the box containing the parabolic trough, a substantial amount of energy is lost from the boiler tube by thermal convection and conduction within the insulated enclosure. The cumulative effects of these losses are indicated in Fig. 21.11 for a representative parabolic trough reflector that was not enclosed.[30] Enclosing the parabolic trough and flat plate collectors of Fig. 21.10 reduced the conduction and convection losses sufficiently to much more than offset the losses introduced by the glass cover plate.[29]

The discussion up to this point has been concerned mainly with collector performance when the plane of the reflector is normal to the incident rays of the sun. If this condition is to be maintained, the reflector must be rotated about two axes. Rotation about the polar axis is required to track the sun throughout the course of the day. It is obvious that the elevation of the sun at noon varies with the day of the year, the total variation being 47 degrees through the course of the year. While less obvious, the pointing angle about the horizontal axis also varies with the time of the day except at the equinoxes, and the variation is a maximum at the summer and winter solstices.[29] This effect is shown in Fig. 21.12. A parabolic trough reflector can be mounted on an east-west axis so that it need not track the sun about a polar axis, and its elevation angle can be varied from day to day so that its inlet face will be perpendicular to the sun's rays at noon. However, either a low-concentration factor must be employed or the elevation must be varied throughout the day if a good collector efficiency is to be obtained during the months close to the summer and winter solstices. It was for this reason that vertical fins were placed on the tube of Fig. 21.10; they reduced the concentration factor so that the pointing angle of the collector would not have to be changed throughout the day.

It might be noted that, to keep the cost low, the parabolic mirror of Fig. 21.10 was formed by a simple bowstring tensioning of an initially flat Alzac aluminum sheet to yield a close approximation to a parabolic trough. This gave a concentration factor of approximately 40, an appropriate amount for the mirror aperture width of 47 cm. With the reflector at a fixed elevation angle, variations in pointing error throughout the course of the day near the summer or winter solstice caused the focused band

PARABOLIC FOCUSING COLLECTOR

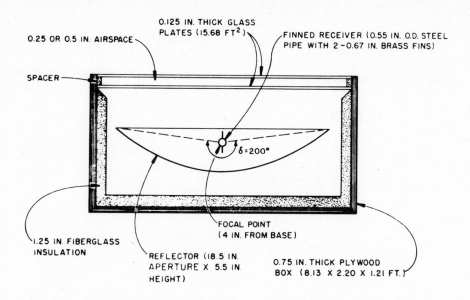

0.25 OR 0.5 IN. AIRSPACE

0.125 IN. THICK GLASS
PLATES (15.68 FT^2)

FINNED RECEIVER (0.55 IN. O.D. STEEL
PIPE WITH 2 –0.67 IN. BRASS FINS)

SPACER

$\delta = 200°$

1.25 IN. FIBERGLASS
INSULATION

FOCAL POINT
(4 IN. FROM BASE)

REFLECTOR (18.5 IN.
APERTURE X 5.5 IN.
HEIGHT)

0.75 IN. THICK PLYWOOD
BOX (8.13 X 2.20 X 1.21 FT.)

FLAT PLATE COLLECTOR

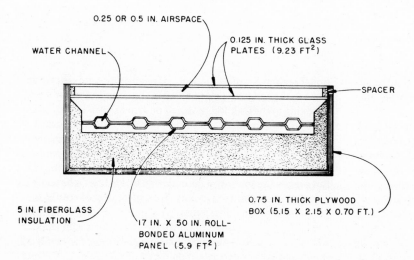

0.25 OR 0.5 IN. AIRSPACE

0.125 IN. THICK GLASS
PLATES (9.23 FT^2)

WATER CHANNEL

SPACER

5 IN. FIBERGLASS
INSULATION

17 IN. X 50 IN. ROLL-
BONDED ALUMINUM
PANEL (5.9 FT^2)

0.75 IN. THICK PLYWOOD
BOX (5.15 X 2.15 X 0.70 FT.)

Figure 21.10 Sections through two solar-energy
collectors tested.[29]

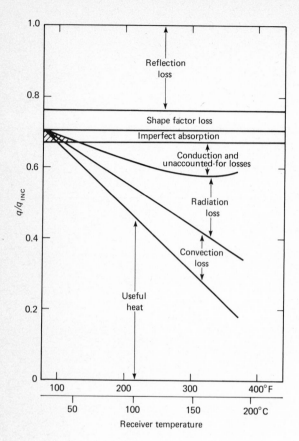

Figure 21.11 Distribution of the incident energy q_{inc} for a 6.19-ft-aperture reflector, 2.375-in-diameter receiver, as a function of receiver surface temperature. (Smoothed data from 15 runs, incident radiation 275 to 325 Btu/min, wind speed 65 to 140 fpm, ambient temperature 70 to 88°F.)[30]

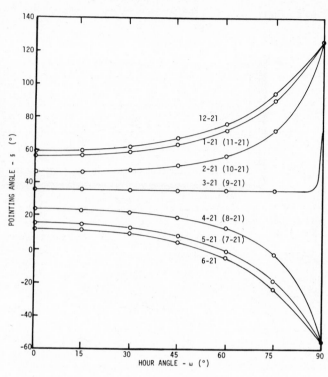

Figure 21.12 Pointing angle ϕ as a function of hour angle ω throughout the year, with $\omega = 0$ corresponding to solar noon. Month-day indicated.[29]

of light to move out onto the fins, whose thermal conduction was high so that the collector tube still functioned. A larger tube could have been employed, but it would have presented a greater area to produce heat losses and would have had a much higher heat capacity—which is undesirable during warm-up conditions early in the day and during transients when clouds pass over, common events throughout the year. The heat capacity is an important consideration, because often, even with the low thermal capacity of the system shown in Fig. 21.10, on partly cloudy days the system would warm up only sufficiently to reach the desired operating temperature before another cloud would come along and the system temperature would drop. In fact, the irregularities in the heat input to the collector caused by occa-

sional scattered clouds made it impossible to get a comprehensive set of good heat-balance data in the course of many weeks of testing in the Oak Ridge area. The system would rarely reach thermal equilibrium before another small cloud perturbed it and spoiled the heat balance. It is for this reason that most testing in the United States is carried out either with synthetic energy sources from sun lamps or in the American southwest. But even in the southwest, as indicated by Fig. 21.13, haze, light overcasts, and scattered clouds give a great deal of scatter that poses experimental difficulties.[31] For example, one of the best sets of data available was obtained by a team from Minneapolis which chose a test site near Phoenix, Arizona. Figure 21.14 shows one of their best sets of data as presented in Ref. 32. Note

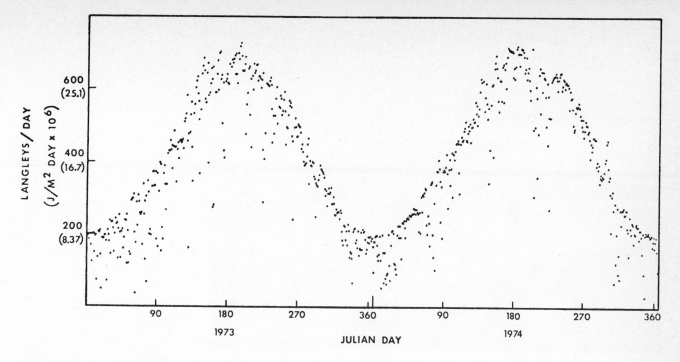

Figure 21.13 Daily total global irradiance for the 2-year period 1973 to 1974 at Las Vegas, Nevada (36 degrees N).[31]

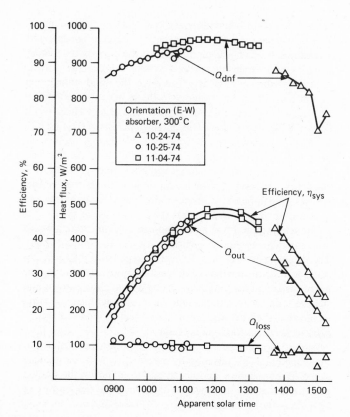

Figure 21.14 Test data from a site near Phoenix, Arizona for a cylindrical parabolic trough reflector with an east-west orientation and a selective coated absorber tube at 300°C. The total direct normal flux is given at the top, the collector efficiency in the middle, and the heat loss at the bottom.[32]

Figure 21.15 Schematic diagram of a solar power tower and heliostat field.

that in preparing Fig. 21.14, they found it necessary to take data obtained on three different days in order to get a complete picture of the collector performance as a function of the time of day. The extreme sensitivity of concentrating solar-energy collectors to variations in superficially clear sky conditions can only be appreciated if one attempts to get good heat-balance data over extended periods.

Solar Power Tower Systems

If one attempts to couple a large number of parabolic trough collectors, such as that of Fig. 21.10, to form a large heat source for a central station power plant, one finds that the extensive system of pipes required to collect the steam, hot water, or other high-temperature heat-transfer fluid leads to a large heat capacity, substantial thermal losses, and high costs. The problems are so serious that a more attractive approach is to employ a large field of two-axis tracking mirrors to concentrate the collected solar energy on a boiler located in a cavity mounted on a tower. A French system employs a field of flat mirrors that direct sunlight to a parabolic mirror that in turn focuses the rays on the boiler.[33] As indicated in Fig. 21.15, a similar system favored in the United States entails focusing the beams from parabolic mirror heliostats directly on the cavity of the boiler.[34] In either case the "power tower" approach has the advantages that the heat losses from the boiler can be kept to a low value—of the

order of 5 percent—and the heat capacity is kept to a minimum. This system has the disadvantage that it requires a high degree of accuracy in the two-axis mountings for the mirrors and a fairly complex and expensive tracking system to keep the mirrors focused on the boiler cavity. It also has the disadvantage that abrupt changes in energy input to the boiler tubes associated with a passing cloud induce severe thermal stresses. This effect is shown in Fig. 21.16, which was obtained in tests of the first unit of this type for which test data are available. In this instance the designers had given a great deal of attention to the thermal-stress problem under transient conditions, and test data indicated that by designing for moderate heat fluxes the thermal stresses were in fact kept to acceptable levels. Probably the key question of such an approach is the capital cost of the elegant equipment required for the double-axis tracking system which must maintain a pointing accuracy of less than 1 minute of arc. The mechanism must be sufficiently rugged and rigid to withstand a storm with winds of at least 161 km/h (100 mph). The mirror itself and the frame supporting it must also be strong and rigid to take severe aerodynamic forces without distortion that would spoil the high precision of the optical surfaces required to give the accurate focusing necessary.

A disadvantage of the power tower–heliostat approach is that the individual heliostats must be spaced fairly well apart so that one does not shadow another at any time during the day. In practice this means that the area of the heliostat field must be 6 to 10 times the area of the heliostats[34] (Table 21.5).

Average Solar-Energy Flux in the United States

The distribution of the solar-energy flux for the United States is shown in Figs. 21.17 and 21.18 for both direct radiation to surfaces continuously oriented so as to be roughly normal to the sun's rays (as for the heliostats of Fig. 21.15) and for the total energy flux—both normal and diffuse—to horizontal surfaces (such as simple flat plates for heating hot water). These maps indicate that, as one would expect, the solar-energy flux in December varies by as much as a factor of 3 between the southwest and the northwest, and for most regions by a factor of about 3 between June and December. The solar energy that can be collected and used varies by a much greater factor, because heat losses from the collection system commonly represent about 20 percent of the energy collected at noon and perhaps 50 percent of the energy collected at 9 A.M. or 3 P.M. on a clear summer day (Fig. 21.14). Unfortunately, up to 1980 no experimental data have been obtained to show a yearly summation of the net energy collected by a typical system at a high enough temperature to be useful for a thermodynamic cycle. It is partly because of this problem that efforts to develop solar-thermal power systems are concentrated in the hot, dry, desert areas of the U.S. southwest, where conditions are the most favorable.

Cost of Electricity

The costs of solar-thermal–electric systems are high. Even the enthusiasts cite capital costs several times those for conventional systems, and these costs are based on the peak system output at noon in the summer with no allowances for energy storage. The situation is particularly confusing because the high costs of the experimental systems that have been built (e.g., $300/m² for heliostats in 1979) are dismissed by advocates as not indicative of the cost of actual production systems, which they believe should be much lower. Thus, for purposes of this text it appears that the most meaningful approach to the cost problem is to estimate the minimum cost per pound of equipment that one might hope to get for heliostats on the basis of experience in manufacturing high production items such as trucks, and, from this, estimate the cost of electricity on an annual basis with due allowance for diurnal and seasonal variations in the energy collection and utilization. Accepted values for the cost of the other equipment in the plant, which would be essentially conventional, would be used in the estimate.

In attempting to estimate the cost of heliostats in large-scale production, one finds, on inspection of designs that have been proposed and experimental units that have been built, that a substantial amount of structure is required to provide the necessary rigidity for the mirror. In addition, the mounting for the two-axis tracking system also entails a substantial amount of

weight. Examination of several designs indicates that it is unlikely that two-axis tracking heliostats can be built for less than about 22 kg/m² (5 lb/ft²). Extensive experience in manufacturing roughly comparable equipment indicates that with large-scale production the unit cost would be at least $5/kg ($2.30/lb) in 1979 dollars, or about the same cost per pound as a pickup truck. Note that the cost of precision machinery such as

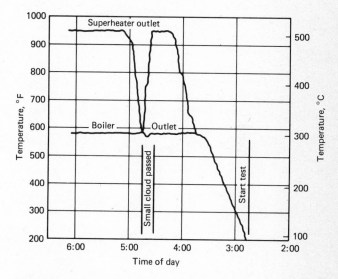

Figure 21.16 Test data from a typical run using a steam generator receiver with the heliostat field of the Centre National de la Recherche Scientifique (CNRS) facility in the Pyrenees. Curves from the strip chart recorder give the boiler drum and superheater outlet temperatures as functions of time following a startup at 2:48 P.M. It took until 4:15 P.M. to reach the design conditions of 8619 kPa/510°C (1250 psig/950°F). A cloud passed at 4:37 P.M.; equilibrium at design conditions was re-established by 5:10 P.M. The temperature rise rate for the boiler in the startup was 278°C (500°F) per hour, and the superheater outlet temperature rise rate after the cloud passed was 26°C (47°F) per minute.[34]

TABLE 21.5   Heliostat and Heliostat Field Areas Required for Typical System Peak Outputs at Noon on a Clear Summer Day (Based on area data in Ref. 20)

Power, MWe	No. of power towers required	Heliostat area, m²	Heliostat field area, km²
10	1	71,100	0.53
100	5	711,000	5.3
1000	50	7,110,000	53

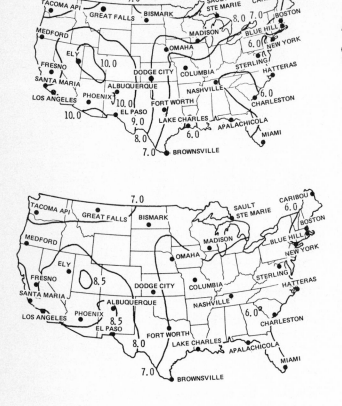

Figure 21.17 Mean daily direct-normal (top) and total-horizontal (bottom) solar radiation for June (kWh/m²) (Ref. 35).

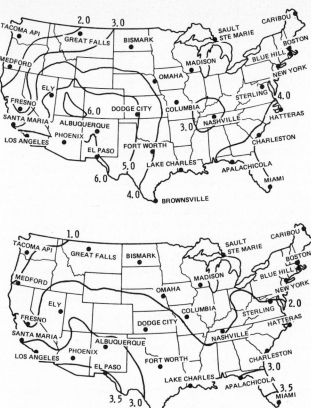

Figure 21.18 Mean daily direct-normal (top) and total-horizontal (bottom) solar radiation for December (kWh/m²) (Ref. 35).

lathes may be appropriate and is very much higher. On the "pickup truck" unit cost basis, a low estimate for the cost of heliostats in large-scale production would be $110/m², including the two-axis mounting and tracking system. In estimating the useful energy collected on an annual basis, one finds that the sun is high enough to give a useful electric output for an average of 6 h per day. In the American southwest the output is further reduced by the weather conditions to about 80 percent of the time, and on the average, even on nominally sunny days, the direct radiation will comprise no more than about 90 percent of the total, the balance being in the form of diffuse radiation which cannot be concentrated on the boiler cavity. Further, the reflectivity of the mirror surface will probably be about 80 percent. Thus the annual total thermal energy input to the boiler, if 0.9 kW/m² is assumed to be the insolation rate to the earth's surface, is given by the following equation:

Thermal energy collected

$$= 0.9 \times 365 \times 6 \times 0.9 \times 0.8 \times 0.8$$
$$= 1135 \text{ kWh/m}^2 \text{ per year}$$

Taking the capital charge for the system as being 16 percent, the cost of this energy would be $17.60, or $4.31/10⁹ J ($4.55/10⁶ Btu). At first glance this appears to be about double the 1979 cost of fuel oil and about four times the cost of coal. However, it does not include allowances for storage costs and energy losses associated with storage operations. If the energy collected were used immediately to generate electricity and the electricity employed in a pumped-hydro storage system, the efficiency of the storage system would be about 70 percent, and hence the effective cost of the thermal energy would be increased by about

40 percent. If the electric energy were transmitted to a load center 1600 km (1000 mi) away, the effective cost would be increased further by about \$3.33/10^9 J (\$3.50/10^6 Btu). The capital charges for the steam plant utilizing the solar heat collected would be higher than those for a conventional plant by roughly a factor of 5 because of the low utilization factor, and the capital charges for the pumped-hydro system would be much higher than for the usual peaking-power pumped-hydro system because of the much longer drawdown period and the shorter pumping period.

A thermal-energy storage system such as the molten-salt system discussed in the previous chapter may be preferable to a pumped-storage hydro system because, although its costs may be somewhat higher, the capital cost of the steam plant can be much reduced. On the basis of data given in the preceding chapter, and assuming a heat-storage capacity sufficient to carry through one cloudy day, the capital cost of the high-temperature thermal-energy storage system would be equivalent to about \$60/m^2 of collector. Allowing for heat losses, this implies about a 65 percent increase in the cost of the solar energy collected.

The cumulative effects of the above factors indicate an overall cost of electricity from a solar-thermal power system to be about 10 times that for electricity from a coal-fired power plant. These costs may seem high, but actually the estimates are probably low because they are based on a number of optimistic assumptions, such as a low cost for the heliostats required by the solar-thermal system.

In view of these unfavorably high costs, it appears that the 1135 kWh per year per square meter estimated above for a heliostat system should be checked by making the estimate on a quite different basis, e.g., the insolation maps of Figs. 21.17 and 21.18. If the solar-energy system is to be self-sufficient, its average output in the winter must be as great as in the summer, because the storage of energy over a period of many months would be prohibitively expensive. Thus the full capacity of the heliostats could not be used to good advantage in the summer because the system output must be based on the insolation rate in December. Using southern Arizona and Fig. 21.18 as the basis, one finds that the average daily energy input to the heliostats would be approximately 6 kWh/m^2. Allowing for an 80 percent mirror efficiency and a 90 percent boiler heat absorption efficiency gives an average net energy collection rate of 4.3 kWh/m^2 per day, or 1580 kWh/m^2 per year. The actual rate would be substantially lower because of heat losses during start-up in the morning and during the passage of intermittent clouds, together with the poor performance early in the morning and late in the afternoon. Thus, the value of 1135 kWh per year per square meter appears to be a reasonable estimate. In fact, nei-

ther of these estimates includes allowances for the overall plant availability; even if an unusually high degree of reliability were obtained in the mechanical and electrical equipment so that an availability as high as 80 percent could be obtained, both estimates would have to be reduced by 20 percent.

The high costs of energy storage suggest that it might be best to attempt only to save fuel during the day and provide 100 percent backup in the form of coal-fired plants. If this approach were followed, the capital cost of the solar plant would still have to be written off against a low operating time per year, giving capital charges at least six times, and more probably 10 times, those for a coal-fired plant. The cost of the solar-thermal energy would still be high, and the amount of fossil fuel saved would be only about 20 percent of the usual coal plant consumption. Also, the severe thermal cycling would increase maintenance costs for the coal-fired plant.

The above rough estimates do not include allowances for degradation in performance stemming from reductions in the specular reflectivity of the mirror surface as a consequence of accumulations of dirt or etching by sandstorms—which are fairly common in the southwest. Note that a single severe sandstorm could completely ruin the specular surfaces.

Solar Photovoltaic Systems

Solar cells have been a useful source of electricity since about 1960 and have been finding increasingly widespread use for small amounts of electric energy in remote locations. They have proved particularly well suited for use in spacecraft; in fact, much of the photovoltaic R&D has been funded by the space program. By 1977 the total annual sales amounted to about 750 kW, largely in the form of silicon solar cells. The efficiency of the individual cells is about 12 percent, but the voltage output is low so that, by the time allowances are made for losses in the connections and the area lost to gaps between individual cells, the overall efficiency of the array is commonly only 6 to 8 percent. In 1979 the cost amounted to about \$15,000/kW on the basis of the peak output on a clear day with the array normal to the incident light. Fortunately, the cells respond to both direct and diffuse sunlight so that, in general, the cell output is essentially proportional to the sine of the angle of incidence of the direct rays.

Principles of Operations

Light photons can give up their energy to electrons in certain semiconductors such as silicon, and this extra energy may be sufficient to dislodge the electrons from their positions in the crystal lattice.[36] To utilize this reaction in silicon solar cells to produce an electric current, a thin layer on the front face of the crystal lattice of the silicon is "doped" with a trace amount of foreign atoms, such as phosphorus, which take the places in the crystal lattice of some of the silicon atoms. Phosphorus has five valence electrons as compared to four in silicon. Thus there will be a small excess of electrons, which causes electrons energized by photons to migrate with their negative charge toward a positive electrode on the face of the crystal lattice. The balance of the crystal lattice is "doped" with a small percentage of atoms such as boron. Each boron atom has only three valence electrons. Hence the region "doped" with boron will be deficient in electrons, giving electron vacancies, or "electron holes," that have a positive charge and will tend to migrate through the crystal lattice toward the negatively charged opposite face. The electrons activated by photons can then flow from the conductor on the front face of the cell through an external circuit to the conductor on the rear face to fill the "holes" that migrate there, giving a driving voltage of ~ 1 electronvolt (eV). For good effi-

ciency the electron-rich region on the front face (the negative, or N-type crystal lattice) should be only ~ 0.5 μm thick so that incident photons dislodge electrons from their lattice locations in the region close to the interface between the two types of crystal lattice. Thus the resulting holes migrate through the positive, or P-type, crystal lattice to a layer of metallic conductor on the rear face of the composite crystal, while the electrons move to a fine grid of metallic conductors on the front face through which sunlight enters the crystal lattice.

Solar-Cell Construction

Up to the time of writing, most of the solar cells have been prepared by growing a silicon crystal from a pool of molten silicon that is very pure except for a small amount of boron to give a P-type crystal lattice. The crystal is then sliced into thin layers with a diamond saw, and one surface of each slice is exposed to a phosphorus atmosphere at high temperature for long enough to give the proper thickness for the N-type layer. Both surfaces are plated with a high-electric-conductivity metal such as silver, and then, by photographic techniques, the silver is etched away from most of the area of the front face to leave a fine grid of conductors (such as that in Fig. 21.19). Thus ~ 90 percent of the front face is open for light to enter the N-type layer. To protect the silver grid from gradual deterioration by corrosion, the conductor grid on the front face is ordinarily given a protective coating of a metal such as nickel, titanium, gold, or palladium.[37] To reduce the losses caused by light reflection from the surface of the silicon, the front face is given a final coating of antireflection material. The conductor network on the front face is then coupled to the intercell wiring system by soldering[37] (Fig. 21.19).

The above method of fabricating solar cells is inherently complex and expensive. Other materials combinations and fabrication techniques are under development in an effort to reduce costs. However, it is implicit in the high degree of purity required in the materials (≤ 1 ppm of impurities), the complexity of the fabrication processes, and the requirements for tight quality control in all phases of the fabrication that the cost of photovoltaic cells tends to be high.

Basic Factors Affecting the Efficiency of Solar Cells

The efficiency of conversion of light energy into electricity in solar cells is much less than 100 percent because of the cumulative effects of many factors.[36,38] In the first place, ideally the energy of the incident photon should be just that required to dislodge an electron from the crystal lattice so that it becomes mobile. Photons of lower energy will be ineffective, while photons of somewhat higher energy than that required will yield

Figure 21.19 Photograph of a 1-cm-diameter GaAlAs/GaAs concentrator cell after packaging in an aluminum holder/heat sink. The contact ring is connected to the electrode plate with 2-mil gold wire by means of multiple ultrasonic bonds. Visible in the left bottom area are the thermocouple leads used to measure the cell temperature.[37] *(Courtesy Rockwell International.)*

the excess energy as heat, and photons of too high an energy will be completely ineffective. The electron energy range in which the crystal becomes conducting is known as the *conduction band*. As a consequence, only a certain band width of the solar spectrum can be effective in generating electricity in any given photovoltaic material. The resulting *quantum efficiency* as a function of photon energy is shown in Fig. 21.20. The overall efficiencies ideally obtainable from sunlight outside the atmosphere are shown in Fig. 21.21 for some typical semiconductors. Note that AlGaAs is inherently more efficient than silicon and hence is receiving much attention. In practice, small defects in the crystals and other basic deviations from ideality reduce the efficiency by a factor of ~ 20 percent. Further losses stem from reflection of some of the incident light from the front surface, the obscuring effect of the conductor grid which blocks off ~ 10 percent of the area of the front face, and cell internal electrical resistance. (These reductions in cell efficiency are not included in Fig. 21.21.) Additional losses are introduced when cells are coupled in arrays, e.g., gaps between cells for the supporting framework, the resistance in intercell conductors, and losses in the power-conditioning equipment including conversion from direct to alternating current.

The efficiency of solar cells depends on operating conditions, one of the most important factors being the current density. Figure 21.22 shows a typical curve relating the voltage output to the current for an AlGaAs cell. A curve for the cell temperature is also shown, the minimum temperature indicating the region for minimum losses and the maximum efficiency. For some types of cell the efficiency can be increased by concentrating the sunlight with a Fresnel lens or a parabolic mirror; Fig. 21.23 shows this effect for the AlGaAs cell of Fig. 21.19. However, if a concentrator is used, it has the disadvantage that it utilizes only the direct rays of the sun, and so, as discussed above for thermal energy systems, about one-third of the total incident sunlight is lost on an annual basis.

If a concentrator is employed, not only can the cost of the cell per unit of electric output be greatly reduced, but also it is possible to split the concentrated beam with a filter into two spectral bands and focus each on a type of cell suited to that band.[39] Figure 21.24 shows the performance of such a filter used with silicon and AlGaAs cells to give an overall efficiency for the filter and cell subsystem of 28.5 percent in terms of the input in the form of the concentrated light beam. This is a most encouraging development, though its cost-effectiveness is reduced because two cells plus a beam-splitting filter are required instead of a single cell. Another approach to the use of a greater fraction of the spectrum is to employ a multiplicity of layers of very thin cells of different types to give multicolor cells—an excellent concept in principle but difficult to apply in practice.

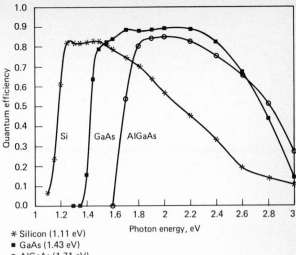

Figure 21.20 Spectral response of Si, GaAs, and AlGaAs solar cells. Quantum efficiency is not corrected for 10 percent contact obscuration. Area = 0.56 cm².[39] *(Courtesy Hewlett Packard.)*

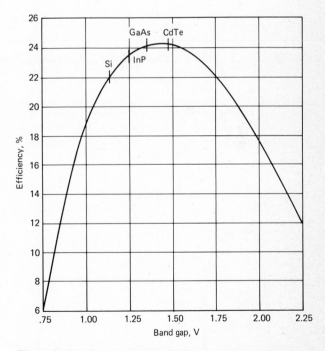

Figure 21.21 The theoretical efficiency of photovoltaic cells as a function of the band gap (conducting region) in electronvolts. The band gaps for four typical semiconductors are indicated on the theoretical curve.[36] *(Courtesy Scientific American.)*

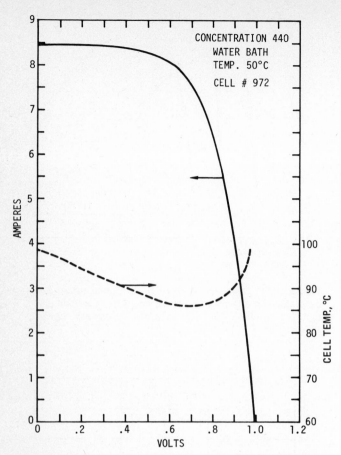

Figure 21.22 The cell temperature variation during *I-V* curve measurement. For this particular plot, the recirculating water temperature and, therefore, the heat-sink temperature, was maintained at 50°C. The cell temperature was measured by a thermocouple placed directly under the center of the cell.[37] (*Courtesy Rockwell International.*)

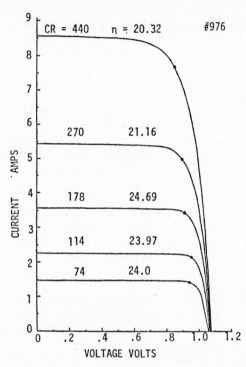

Figure 21.23 *I-V* curves for a concentrator cell No. 976 measured at Table Mountain, California. Each curve is also labeled with the concentration ratio (CR) and the efficiency (η) at the maximum power point (the dot just below the knee of each curve). The cell temperature at the maximum power point was maintained at 50°C.[37] (*Courtesy Rockwell International.*)

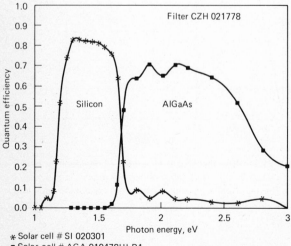

✱ Solar cell # SI 020301
■ Solar cell # AGA 010478HLR1

Figure 21.24 Spectral response of Si and AlGaAs concentrator cells mounted with a spectral splitting filter. Each cell is 0.56 cm² in area.[39] (*Courtesy Hewlett Packard.*)

Cell	I_{sc}	V_{oc}	P_m	Fill factor	Efficiency, %
AlGaAs (1.61 eV)	1.382	1.26	1.440	0.827	17.4
Si (1.1 eV)	1.711	0.738	0.915	0.725	11.1

Developmental Choices

The development engineers and physicists have a difficult choice. To increase the efficiency of simple flat-plate arrays seems likely to increase the cost of the cells. If concentrators are used, the concentrator cost is likely to be high. Cost estimates by advocates of concentrator systems are commonly low, but where reasonably detailed structural designs that include weight estimates are available, the weights run from 50 to 68 kg/m² (Refs. 40 and 41), or two to three times the 22 kg/m cited earlier as the minimum likely for the simpler heliostats of the solar-thermal system. Hence, even if the solar cells were inexpensive, following the same reasoning as used for heliostats but crediting the photovoltaics with both a 30 percent efficiency and a longer effective day because they are not sensitive to heat losses, one finds the costs of the concentrators to be at least $0.10/kWh in 1980 dollars. Allowance for other charges would probably more than double this amount, giving an overall cost of electricity that would be at least five times the average 1980 charge to consumers—too high to be acceptable to the public except in a few special applications.

Space Applications

Data on solar-cell efficiencies sometimes appear inconsistent because some of the cells are intended for space applications, others for terrestrial applications. The broader solar spectrum in space means that a smaller fraction of the incident light can be utilized by the cell and hence the cell efficiency is roughly 20 percent lower for space applications. Another problem with solar cells in space applications is that they are subject to degradation as a consequence of damage to the crystal lattice caused by pro-tons and electrons in the solar wind or the Van Allen radiation belt. This effect is shown in Fig. 21.25.[42]

In reviewing the more comprehensive record for space solar-cell development to assess the probable rate of future solar-cell developments, one finds that by 1960 after about 5 years of development the efficiency of silicon cells for space applications was as high as 15 percent, and extensive work was under way on other types of cell.[42,43] A 1978 status report listed the best efficiency achieved in the more promising types of solar cell for space applications and gave an efficiency for silicon cells of 15.6 percent.[44] These values compare with a theoretical upper limit of 22 percent. The most encouraging developments have been for gallium-arsenide cells, the efficiency of which for space applications was increased from around 6 percent in 1960 to around 17 percent in 1977. The latter compares with an upper theoretical limit of 25 percent efficiency. These values, of course, are for individual cells and are summarized in Table 21.6. When solar cells are integrated into arrays, additional losses are substantial. As an example, according to 1978 data for a typical array, a flat panel of silicon cells oriented normal to the incident solar radiation was used for the Orbiting Solar Observatory and gave an overall energy conversion efficiency of 7 percent during its first year of operation.[45] In short, the most vital problem is the cell efficiency. The solid-state physics involved is beyond the scope of this text. Hence it will simply be stated that according to expert opinion the highest efficiency ideally obtainable is around 40 percent,[32] and an efficiency of the order of 20 to 30 percent will be required with relatively inexpensive materials to yield an attractive system.[33-40] This is well above values obtained in actual solar-cell arrays. Thus many people believe that the most promising avenue to achieving costs suffi-

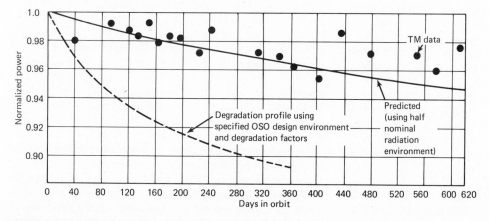

Figure 21.25 Orbiting satellite observatory OSO-8 in-orbit predicted performance and telemetry data versus time in orbit.[45]

TABLE 21.6   Maximum Efficiency in Percent of Incident Solar Energy for the Principal Types of Solar Cell Reported in 1961 and 1978

	Operation in space			Terrestrial operation	
	1961 (Ref. 42)	1978 (Ref. 45)	Theoretical limit	1980	Theoretical limit
Single crystal silicon	15	15.6	22	19	
Gallium arsenide	7	18.5	25	23	
Cadmium sulfide	6	8.5			
Cu_2O	0.5				
Selenium	1				
InPCdS		15			
InP/InSnO		15			
CuInSe		3	17		
Cd/CuInSe		6.2	17		
Two-color cell				28.5	~35
Multicolor cell					~60

ciently low to make solar cells commercially attractive for electric utility applications is to carry out basic research that would yield materials both low in cost and capable of utilizing a large fraction of the energy in the complete solar spectrum. Note that a satisfactory cell must also have a long life. Also, its characteristics in production should be quite uniform so that large numbers of cells can be coupled in series-parallel arrays to give a well-balanced system with a minimum of losses associated with local imbalances in electric current flows. Some notion of the complexity of the problem is given by Figs. 21.22 and 21.23, which indicate the effects of current flow and insolation rate, respectively, on cell output and efficiency.

One of the major advantages of solar cells is that their efficiency is at its best in small arrays. Hence, if the cost were reasonable, solar-cell arrays could be installed on the buildings in which the electric energy is to be used. This arrangement would avoid the losses and expenses associated with transmission systems from a central utility. It would also, however, introduce the requirement for a much larger electric energy storage capacity than would be required if the solar cells were in central stations because there would not be a network of transmission lines to supply power from a multiplicity of sources.

Satellite Power Systems

In the 1960s Peter Glaser proposed that a large array of solar cells be placed in a geosynchronous orbit and that the electric power generated be beamed to the earth in the form of microwave radiation. It seemed doubtful that a beam could be fo-

cused sharply enough for the energy to be transmitted efficiently from a satellite 35,000 km (22,000 mi) above the earth, but an experiment carried out at JPL in 1975 showed that 30 kW could be transmitted for a distance of over a mile with an overall efficiency of 82 percent.[46] Analyses have indicated that by scaling up the system to a very large size—at least 5000 MWe—an overall efficiency of around 60 percent might be obtained with a transmitting antenna diameter of about 1 km and a rectenna or receiver array having a diameter of around 10 km. A fairly detailed picture of what such a system might look like is given in Refs. 46 to 49, and data for a typical design are given in Table 21.7. The solar-cell array would be huge, e.g., 21.4 km long by 5.3 km wide and mounted on a triangular girder structure having a girder depth of about half a kilometer. At either end there would be an antenna having a diameter of 1 km, each antenna directing a beam at a receiver on the earth's surface with a microwave frequency of 2.45 GHz. Solar cells would generate 25,500 MW of electricity which would go through a collection system and be converted to beams of microwave energy totaling 17,000 MW. The output of the terrestrial receivers would be 10,000 MW. The energy distribution envisioned for the beams at the earth's surface is indicated by Fig. 21.26.

The problems in developing a system of this sort include roughly doubling the efficiency of solar-cell arrays, cutting the specific weight of the cells by a factor of 4, and increasing their life by a factor of roughly 6. At the same time the cost of the cells would have to be reduced by a factor of roughly 100 through large-scale production. For the system to be viable, all these objectives must be realized.The 1-km-diameter antenna and the

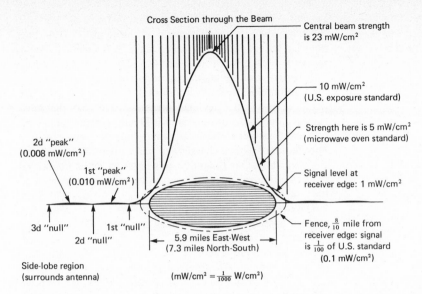

Figure 21.26 Energy distribution across the microwave beam from a satellite power plant in the vicinity of the rectenna at the receiving station.[49] *(Courtesy Boeing Aerospace Co.)*

~ 10-km-diameter rectenna must be built with a high degree of precision; the position of each conductor in the arrays must be held within ~ 1 cm. This will be very difficult to do in these large-diameter arrays, particularly with the light structure required for the satellite antenna. Perturbations to be expected include thermal distortion during terrestrial or lunar eclipses, the distorting forces of the gravity gradient to which the satellite will be subjected, and inertial forces associated with keeping the satellite solar array normal to the sun's rays while at the same time keeping the antennae directed toward the earth. There is also a question as to whether the beam can be directed with the required precision so that it lands squarely on the rectenna. For safety reasons the peak energy flux in the center of the beam is to be only 23 mW/cm², which is only a little more than twice the standard U.S. tolerance value for microwave radiation. Thus, if as a consequence of a pointing error the beam should stray from the receiver, it should not represent a serious hazard to the personnel in the surrounding area unless by some fluke it became more sharply focused and yielded more concentrated energy in a local region.

Another basic feasibility question is concerned with the cost of putting the equipment in orbit, assembling it, and maintaining it. Much of the projected cost for such a system involves the space shuttle system required to transport roughly 100,000 tons of material into orbit for each 10,000-MW power plant (i.e., ~ 10 percent of the weight of the Menkaura pyramid at Gizeh). This will require a reduction in the cost per unit of weight conveyed into orbit by a factor of 10 relative to the cost projected for the first NASA space shuttle system. To obtain this large reduction in cost, it will be necessary to have much larger shuttles

TABLE 21.7   Design Data for a Proposed Space Power System in a Geosynchronous Orbit (35,924 km altitude) (Ref. 49)

Design power output	
Solar cells, MWe	25,500
Energy to antennas, MWe	17,000
Terrestrial receiver, MWe	10,000
Satellite structure	
Length, km	24.6 (14.96 mi)
Width, km	5.3 (3.29 mi)
Depth, km	0.47 (0.292 mi)
Solar-cell array	
Area of array, km²	114.5 (44 mi²)
Solar-cell efficiency, %	17.3
No. of solar cells	21×10^9
Size of solar cells, cm	6.6 × 7.4 (2.6 × 2.9 in)
Cell thickness, η m	50 (0.002 in)
Cover thickness, η m	75 (0.003 in)
Substrate thickness, η m	50 (0.002 in)
Antennas	
No. required	2
Diameter, km	1.0 (0.6214 mi)
Beam frequency, GHz	2.45
Rectenna	
Number required	2
Shape	Elliptical
Size (each), km	9.5 × 11.8 (5.9 × 7.3 mi)
Microwave beam power density	
Peak, mW/cm²	23
Value at perimeter fence, mW/cm²	0.01
Total weight in orbit, t	100,000 (220,000,000 lb)

TABLE 21.8   Energy Cost of Materials*

	kWh/kg	Btu/ton × 10^{-6}
Metals		
Aluminum	78.6	244
Chromium	45.7	142
Copper	36.1	112
Lead	9.7	30
Magnesium	127	395
Manganese	16.5	55
Molybdenum	51.5	160
Nickel	140	436
Steel	8.1	25
Titanium	155	482
Silicon	72	224
Zinc	23.2	72
Ceramics		
Cement (Portland)	2.7	8.4
Concrete	0.35	1.1
Firebrick	1.3	4.2
Glass	8	25
Mica	5.8	18
Organics		
Lumber	0.0003	0.0009
Paper	7.1	22
Polyethylene	34	106
Polystyrene	21	64
Polyvinylchloride	15.8	49

*Data from Refs. 50 and 51 and from J. P. Albers et al., *Demand and Supply of Nonfuel Minerals and Materials for the U.S. Energy Industry, 1975–1990—A Preliminary Report,* Geological Survey Professional Paper 1006-A, B, U.S. Government Printing Office, Washington, 1976.

and operate them on a round-the-clock, year-round basis for many years to recuperate the initial investment in the shuttle vehicles. This in turn implies the construction of a complex of space power plant satellites that will provide a very large fraction of the U.S. electric power requirements.

Monetary and Energy Costs of Solar Power Plants

Estimates of the monetary costs of the various solar power plants outlined above generally run from $2000/kW to $20,000/kW of peak output, depending on the degree of optimism of the estimator. Advocates of particular systems generally seem to be able to make a superficially plausible case for a cost of $2000/kW. For terrestrial power plants the effective capital cost would be ~10 times higher because the year-round utilization factor would be low—certainly less than 20

percent—and there would be large additional capital charges. If the plant could operate 24 h per day, as described for satellite plants, a cost of $2000/kW would make such a plant competitive with a fossil fuel or nuclear plant if the cost of fossil fuel were to double or the capital cost of nuclear plants were to double as a consequence of environmental requirements. It may be noted, however, that the huge increases in fuel costs that have followed the Yom Kippur War in 1973 have been accompanied by roughly similar increases in capital charges; thus the relative importance of capital charges and fuel costs probably will not change greatly. One reason for this situation is that the energy costs of producing materials represent a substantial fraction of the total production costs.[50,51]

The energy costs of material, i.e., the amount of energy that must be invested in producing a given quantity of material, is substantial, as indicated by Table 21.8.[50] The energy investment in making silicon solar cells, for example, in the commonly used processes is equal to the total output obtainable from the solar cells during their first 10 years of operating life if they are used without concentrators.[51] This is particularly important if the operating life of the cells proves to be only 10 or 20 years. If an aluminum mirror concentrator is employed, the energy investment in a 1-mm-thick aluminum parabolic reflector for a favorable location in the U.S. southwest is equivalent to the total amount of electric energy that could be produced from the power plant in a period of 3 years. This is one reason why it will be necessary to minimize the quantity of material required by making use of thin-film solar cells if solar cells are to prove a practicable approach to the production of electric energy. For comparison, it should be mentioned that the energy investment in a conventional coal-fired fossil fuel plant is roughly equal to the energy produced by the plant during its first month of operation, and for a nuclear plant the energy investment in materials is about equal to the energy received from the plant during 2 months of operation. It should be noted also that labor involves an energy cost, because people are consumers of energy, and consequently each man-day should properly carry with it an energy charge. For the 1970s in the United States, this appears to have been about 3.5×10^9 J per man-day (3.5×10^6 Btu per man-day), or about 30 gal of fuel oil per man-day.

It is not easy to appraise the degree of difficulty associated with the many developmental problems of the various solar-thermal and photovoltaic energy systems that are under active development, as well as the probable capital cost of the completed power plants if built on a large scale in large numbers. This situation is further confused by the emotional arguments of zealous advocates. The best indication of the prospects is probably given by the results of studies by committees or panels of experts who do not have large personal stakes in the solar-energy program. One such panel organized by the American Physical Society at the request of the White House and the Department of Energy reported early in 1979 that "major scientific and technological advances" will be required before the direct conversion of sunlight into electricity can become a significant source of power.[52] The panel concluded that at best a 20 billion dollar investment in photovoltaic cells might serve to provide 1 percent of the estimated U.S. electric power capacity by the year 2000, and that even this would require some major developments which the panel was not certain could be achieved.

WINDMILLS

As mentioned in the first chapter, windmills have been in use in Europe since at least the eleventh century and were a major factor in initiating the industrial revolution.[53,54] They were introduced into the United States shortly after the initial settlement of New Amsterdam, and they became widely used in this country by the middle of the nineteenth century for pumping water on farms.[55] About 5 million were built, of which around 150,000 were still in use in 1979. The majority had 8-ft-diameter rotors with about 12 blades. In the early 1920s the need for electricity on farms led to the production of wind-powered electric generators, and over a million of these were produced in the 1920s and 1930s. These machines were expensive, however, and rural electrification provided a much less expensive and more dependable source of power.[55] In Europe, steam engines largely supplanted windmills by the early part of this century, and, at the time of writing, even in the Netherlands the very few windmills still being operated were supported with a strong government subsidy.

(It costs about $1000 per year for maintenance on one of the Dutch windmills.) Further, they cannot be allowed to operate unattended, because variations in wind velocity require continual trimming of the sails.

The first effort to build a wind-powered electric generator for utility service in the United States was constructed in Burlington, New Jersey, in 1933.[55] This plant depended on a set of vertical rotors mounted on railcars which operated on a circular railroad track. The system was destroyed in a storm before it could be tested and the concept was abandoned. The next serious effort to produce electric power for utilities was carried out with a 1.25-MW unit put into service by the Central Vermont Public Service Corporation in 1941 and operated through 1945, when the rotor failed.[55,56] No further operation was attempted because the cost of repairs appeared higher than could be justified economically. Note that this machine was designed with the help of the eminent aerodynamicist Theodore von Karman and by professors on the staff at M.I.T.

The abrupt change in the cost of energy in the 1970s led to a number of efforts to produce new systems that would exploit the latest technology to supply economical wind power. Through one of these efforts the Grumman Corporation developed a 20-kW generator with a 7.62-m-diameter (25-ft-diameter) three-bladed rotor. A much more ambitious NASA program has led to the construction of a series of machines ranging from a 100-kW unit at Plumbrook, Ohio, to a 2-MW unit on a mountain (elevation 1347 m, or 4420 ft) near Boone, North Carolina.[57,58] The principal design data for these machines, together with data on older windmills, are summarized in Table 21.9.

TABLE 21.9   Characteristics of Typical Windmills

| Type of windmill | Rotor diameter, m | Tower height, m | Wind velocity | | | Design power output, kW | Cost in 1977 dollars, $/kW |
			Min., m/s	Max., m/s	Design, m/s		
Nineteenth-century Dutch drainage mill	18				~8.	~25	
U.S. farm mill for pumping water	2.44	10				~0.5	~2400
U.S. farm mill for generating electricity	6					3	
Grumman Windstream	7.62	40	4	22.3	12.5	20	1093†
Smith-Putman generator, Grandpa's Knob, Vt.	55	34				1250	
NASA MOD-O	38	30	4.5	13.3	8.1	100	5500*
NASA MOD-OA	38	30	3.6	17.8	10	200	
NASA MOD-1	61	42.7	7	20	16	1500	2900
NASA MOD-2	91					2500	
Indian NAL 6-blade windmill	10	12	1.67		2.8	100	11,800

*NASA estimates production quantities will cost ~ $1500/kWe.

†Cost is for a kit; does not include assembly and installation.

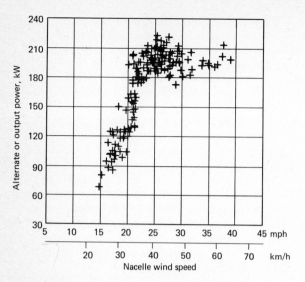

Figure 21.27 Test data for the NASA MOD-0 100-kW wind turbine. Each point represents the average for one revolution of the rotor (1.5 s).[57] *(Courtesy NASA.)*

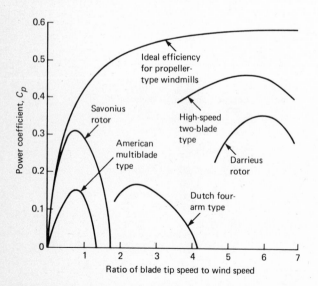

Figure 21.28 Typical performance curves for various types of wind machines.[55]

Characteristics and Problems

Probably the most difficult set of problems posed by windmill operation is the continually fluctuating velocity of the wind. Even if the wind appears to be steady, there are always minor fluctuations that tend to cause the windmill speed to vary. In the big old-fashioned windmills, small changes in wind speed were accommodated by turning the windmill so that the plane of the rotor was inclined away from the wind direction in order to reduce the output. Larger changes were accommodated by reefing the cloth sails. In the new Grumman and NASA machines, variations in wind velocity are accommodated by varying the pitch of the rotor blades. In practice, however, the fluctuations in wind velocity are rapid enough to make holding the rotor speed constant a major problem.[57,58,59] An indication of this is given by the width of the scatter band of points obtained in the test covered by Fig. 21.27. For large rotors these fluctuations are not uniform across the diameter of the disk swept by the rotor, and the resulting fluctuations in the forces on the blades lead to serious vibration problems with the blades commonly vibrating in many different modes. The resulting vibratory stress problems appear to be so severe that many experts consider the maximum practicable rotor diameter to be around 60 m (200 ft),[55] i.e., the size of the MOD-1 NASA machine in Table 21.9. Although blade vibration in that machine has proved to be a problem, the problems appear to be tractable, and a 91-m-diameter (300-ft-diameter) windmill is to be built for NASA in 1981. Even though windmills are designed so that the rotor is stopped if the average wind velocity exceeds a value somewhere between 15 and 25 m/s, severe gusts in storms are likely to produce failures such as that experienced with the Grandpa's Knob rotor in 1945.

A second major problem with windmills is that when the wind speed drops below the value for which the machine was designed to produce its rated output, the power output drops rapidly and normally falls to zero at a wind speed of around 4 m/s (9 mph) as a result of frictional and electrical losses. Thus, under the low-wind-velocity conditions that prevail through much of the summer and fall, the windmill will produce no power at all. If one looks at this quantitatively, the energy ideally available in an airstream with standard sea-level air density is given by $E = \frac{1}{2} \varrho v^3 = 0.0006124 \, v^3$, where v is in meters per second and the energy is in kilowatts. Because of the finite number of blades in the rotor and aerodynamic losses, the actual energy that can be removed from the airstream is substantially less and is commonly obtained by multiplying the ideal energy available by the power coefficient. Figure 21.28 gives typical values for the power coefficient for representative types of rotor as a function of the ratio of the rotor tip speed to the wind speed.[55] Thus, if the power coefficient for a given rotor is 0.4, and the design wind speed is 10 m/s, the energy obtainable would be 0.61

kW/m² of rotor disk area. However, as stated before, the useful power output of this rotor would drop to zero for a wind speed of about 4 m/s where the energy in the airstream would be 39 W/m².

Windmills also give problems with icing during winter storms, and large windmills have caused both a high noise level and electromagnetic interference with TV sets. The latter problem can be eased by using blades constructed of fiberglass rather than metal, or by installing cable TV systems for viewers in the area (a step that NASA found necessary for its windmill in Rhode Island). The only solution to the noise problem appears to be the location of windmills remote from inhabited areas. (The noise problem was a major factor when NASA decided to dismantle the MOD-1 windmill near Boone, North Carolina.)

The gusty character of the wind also gives frequency control problems even when the unit is coupled into a utility grid that can tolerate the resulting fluctuations in power output. The problem is particularly difficult in mountainous terrain in which the ridges inherently generate large-scale eddies, evidence of which is easily seen in low clouds rolling over mountain ridges. Both the control and blade vibration problems are aggravated by the boundary-layer effect, which gives a substantial gradient in velocity from the ground surface to the highest level swept by the blade tips.

In attempting to assess the applicability of windmills to electric-utility power generation, it is necessary to consider the local wind velocities available. Figure 21.29 shows the average energy available in the wind as a function of geographical position in the United States.[60] The solid contour lines represent a series of constant values for the energy ideally available, and the dashed lines indicate boundaries for mountainous terrain. Note that most of the metropolitan areas in the United States are in regions in which the energy in the wind during the summer would average less than 100 W/m²—barely enough to turn the rotor of a windmill. This is quite a serious matter because, as indicated in the previous chapter, the cost of storing energy for more than perhaps 6 h becomes very high indeed. Further, the availability of windmill energy is not much better in the autumn months. Thus, as was concluded regarding solar-energy systems in the previous section, the capital charges for the electricity produced by windmills would be much higher than implied by the nominal capital cost per kilowatt of design capacity. Of course, it would be possible to design windmills so that they would function at the lower wind velocities implied by Fig. 21.29. This, however, would reduce the peak output obtainable from a given size of rotor and would actually increase the cost of the electricity that could be produced, because the tower and rotor strength, weight, and cost are determined by gust forces in storms rather than by the wind velocity at which the windmill is designed to function.

A good insight into the cost problem of windmills is given by a proposal for their use in India as sources of power for irrigation.[61] The Indian National Aeronautical Laboratory (NAL) has developed a low-cost unit using cloth sails in an effort to give the 70 percent of Indian farms that do not have electricity a means for irrigating their fields during much or all of the 9-month dry season, thus doubling or tripling the crop output in the many areas where ample groundwater supplies are available. Figure 21.30 gives NAL estimates, which compare the costs of irrigation using various energy sources, including two types of windmill, bullocks, diesels, and electric motors. Two sets of curves are given: one for an area of 1 ha (2.5 acres) and one for 3 ha (7.5 acres). The former is for an average plot, while the latter is for the maximum size that can usually be served from a single well. Two cases are shown for electrically-driven pumps, one including the total capital charge for the electric-power transmission system and the other neglecting that charge if it is written off against other uses for the electricity. It seems highly significant that windmills are barely competitive with other systems even for this favorable application involving irrigation water pumping in areas where labor costs are low and the wind speed during the pumping season would be above the design value of 2.8 m/s for an average of 10 h per day.

WAVE POWER SYSTEMS

It has been recognized for a long time that there is an enormous amount of energy in ocean waves, which in a sense concentrate some of the energy in the wind flowing over the ocean surface. At the time of writing, small-scale efforts to tap this energy were under way in Norway, the United States, Great Britain, and Japan, with the latter two nations giving their projects greater emphasis. Five principal problems are inherent with wave power systems:[56,62,63,64]

1. The energy is diffuse, hence systems are inherently large.

2. The wave forces in storms can be enormous; hence the structure must be very rugged.

3. Depending on wind conditions, waves vary widely in size and wavelength and vary in direction both on short- and long-term scales.

4. The mean water surface level changes with the tide.

5. The energy available varies widely and sometimes is nearly zero.

Many types of mechanisms have appeared promising as a means of converting wave energy into electricity. One employed by the Japanese is a cylindrical buoy whose mass and buoyancy

SPRING-WIND POWER (WATTS/M^2) AT 50 M

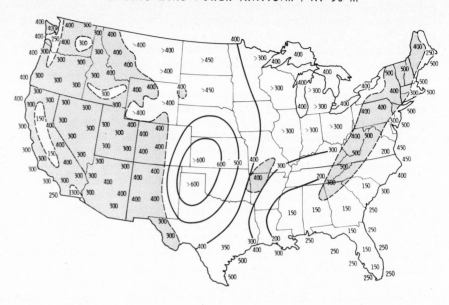

SUMMER-WIND POWER (WATTS/M^2) AT 50 M

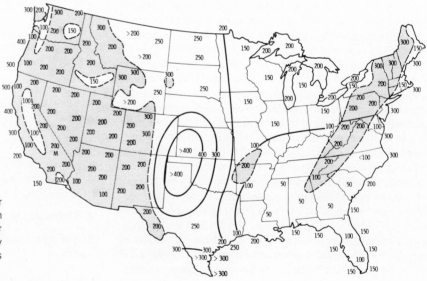

Figure 21.29 Available wind power in watts per square meter of rotor disk area at an elevation of 50 m above the ground. The solid contour lines are for constant values of wind energy while the dashed lines indicate the boundaries of mountainous terrain.[60]

FALL-WIND POWER (WATTS/M^2) AT 50 M

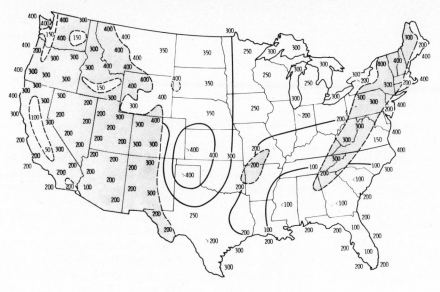

WINTER-WIND POWER (WATTS/M^2) AT 50 M

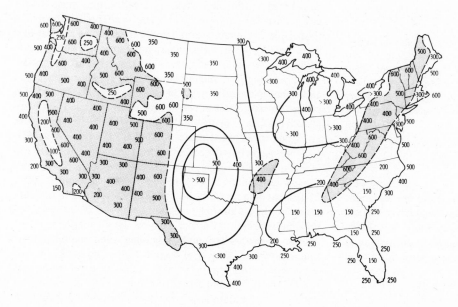

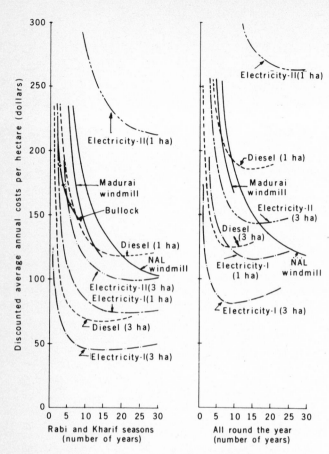

Figure 21.30 Discounted average annual costs in India for irrigating 1 ha (under wheat or equivalent water-consuming crops) from open wells with an average depth of 6.67 m for the water table. ("Electricity-II" includes transmission system costs, "Electricity-I" charges transmission system costs to other uses.) Windmills would prove to be economical if they were designed to operate effectively at low wind speeds and all year round.[61]

can be adjusted so that its natural frequency for bobbing with the waves is the same as the wave frequency, and thus its amplitude is perhaps three times that of the wave height. The vertical motion through the water drives a propeller turbine whose blade angle automatically changes when the buoy reverses direction so that the generator is always driven in the same direction by the turbine. Some of these units have been used to supply power to lights and horns on marker buoys, with a storage battery providing power when the sea is relatively quiet.

Other mechanisms of interest include one in which water acts

as a piston to compress air in a cylinder and toroidal floats that move up and down over a central cylinder. In all cases the problem is to obtain a simple, reliable device that will withstand severe storms and yet with its complete electric energy collection system be cost-effective. The meagre cost data available do not look promising.

OCEAN THERMAL ENERGY CONVERSION

The roughly 20°C temperature difference between surface and deep water in tropical regions has been considered a potential source of power for a century. In the 1920s a plant designed to exploit this source was built on the coast of Cuba, but it failed to produce useful power.[56,65] The concept was ardently promoted in the 1970s, and major funding has led to extensive studies. These indicate that the cold water must be brought up from a depth of 400 m or more and that, to keep the pumping power to an acceptable level, the pipe size must be large (to keep the L/D ratio low) and should extend vertically downward from a floating power plant.[65,66] This favors a minimum capacity of ~ 100 MWe in a plant located ~ 130 km (80 mi) off shore.

Size and Costs

The inherently low thermal efficiency of the ideal Carnot cycle (~ 4 percent) means that enormous quantities of water must be handled and the heat exchangers and turbine must be very large. There are two principal approaches. The first involves a flash boiler to obtain steam directly from the warm seawater and accepting the problems of designing for a very large turbine rotor diameter—45 m (150 ft). The second is to use a thermodynamic working fluid such as ammonia or Freon and accept an additional temperature loss in the heat-transfer matrix of a boiler. The large cost of the turbine casing for the former approach has caused many to favor the latter with special surfaces giving improved heat-transfer coefficients (see pages 479–483).

Some insight into the cost of the heat exchangers required can be gained by drawing a parallel with condensers for conventional steam plants, because these are essentially similar. Table 21.10 gives data for both the TVA Bull Run plant and what appeared to the author to be the most promising OTEC design at the time of writing.[65] Perhaps the most significant points are the amount of surface area and the cooling-water pumping power required per net kilowatt of output. Note from line 10 that the condenser surface area required is almost 200 times as great for the OTEC plant, and note from line 8 that the condenser cooling-water pumping power requirements are 75 times as great. The pumping-power requirements for circulating the warm seawater and for deaerating the condenser are also large—about 60 and 70 percent, respectively, of the pumping power required for the cold seawater. This large parasitic power—equal to about half of the net output of even a very large power plant—is a serious matter, because if these estimates prove to be optimistic and the actual parasitic pumping power proves to be half again as great, the net plant output would be halved and the capital cost doubled. Similarly, the estimated cost of the condenser, $370/kWe, represents about 25 percent of the cost of the plant, and if that cost were to be increased, it would have a major effect on the overall plant cost. Note that the actual cost of the condenser for a conventional nuclear steam plant built in 1978 ran $10/ft², or nearly twice the value of $6.10 used in the estimates of Ref. 65. The cost of the condenser would be higher if ammonia or Freon were used as the working fluid because they give lower heat-transfer coefficients than water. This, coupled with the extra temperature loss in the boiler heat-transfer matrix required for an intermediate fluid, led the design team of Ref. 65 to conclude that an "open" cycle with a flash boiler would be a much more promising approach if the problems presented by a 45-m-diameter (150-ft-diameter) turbine rotor could be solved. They evolved a clever and promising rotor design based on helicopter rotor technology. The problems of getting a reasonable cost for the structure of the turbine casing, which must withstand the enormous external pressure loads involved (the turbine must operate in a near vacuum—1 percent of atmospheric pressure), were solved by making the turbine casing of reinforced concrete and an integral element of the plant structure. Figure 21.31 shows a section through the complete plant in which most of the walls are surfaces of revolution.[65] Note that this design is particularly clean from the hydrodynamic standpoint, and the bulk of the structure (virtually all of which is reinforced concrete) serves the dual functions of providing fluid flow passages and hull structure.

Two interesting and difficult problems not always considered in OTEC plant designs but accommodated in the design of Fig. 21.31 stem from wave motion in high seas. One problem with an open cycle is that waves could cause large percentage variations in the pressure in the flash boiler or large fluctuations in the local flow rates unless the plant hull is designed so that its elevation stays constant in relation to the mean sea level and the warm seawater inlets and outlets are sufficiently far below the sea surface to keep the local pressures constant. A second problem is that rolling motions induce severe bending stresses of the joint between the cold seawater inlet pipe and the hull. These problems in the dynamics of the system of Ref. 65 were handled by providing much of the hull flotation force with the toroidal air cushion shown in Fig. 21.31, using a large diameter hull (107 m), and choosing a set of proportions and a mass such that the plant would heave with really large waves in very heavy seas but would

TABLE 21.10   Data Related to Condenser Size and Cost for Typical Coal-Fired and OTEC Plants

Line no.	Parameter	TVA Bull Run plant (Ref. 17 of Chap. 11)	Westinghouse OTEC plant (Ref. 65)
1.	Plant net output, MWe	850	100
2.	Plant gross output, MWe	914	148
3.	Steam flow, kg/s (lb/h)	478.2 (3.787×10^6)	1379 (10.9×10^6)
4.	Steam flow, m³/s (ft³/s)	1533 (53,870)	1692 (5.95×10^6)
5.	Steam flow, heat load, MWt (Btu/h)	1043 (3.561×10^9)	3296 (11.25×10^9)
6.	Condenser cooling water flow, kg/s (gpm)	25,070 (397,500)	419,936 (6.657×10^6)
7.	Condenser cooling water pumping power, kWe	2355	21,000
8.	Condenser cooling water pumping power, %	0.28	21.0
9.	Condenser surface area, m² (ft²)	29,730 (320,000)	606,700 (6.5×10^6)
10.	Condenser surface area, m²/kW (ft²/kW)	0.03498 (0.376)	6.07 (65)
11.	Condenser steam temperature, °C (°F)	30.5 (87)	7.78 (46)
12.	Cooling water inlet temperature, °C (°F)	12.8 (55)	4.44 (40)
13.	Cooling water exit temperature, °C (°F)	22.7 (72.9)	6.25 (43.2)
14.	LMTD, °C (°F)	12.2 (22)	2.33 (4.2)
15.	Heat flux, W/m² [Btu/(h · ft²)]	35,098 (11,128)	5441 (1725)
16.	Overall heat-transfer coefficient, W/(m² · °C) [Btu/(h · ft² · °F)]	89.1 (506)	72.4 (411)
17.	Cooling water pipe length, m (ft)		973 (3192)

not respond to short-wavelength sea surface motions.

The cost of the plant of Fig. 21.31 was estimated to be about $1500/kWe in 1977 dollars not including provisions for mooring or power transmission to shore, both of which would be substantial. A separate but related study yielded a cost of about $420/kWe in 1976 dollars for a dc transmission line to a plant 130 km (80 mi) off shore. The costs for ac transmission lines were higher.

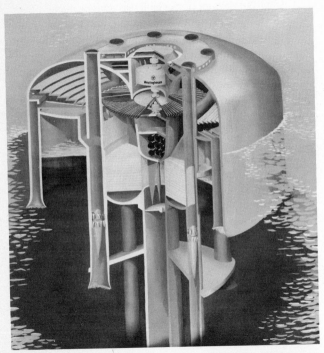

Figure 21.31 Conceptual design for a 100-MWe OTEC power plant having a diameter of 107 m (351 ft).[56] *(Courtesy Westinghouse Electric Corp.)*

Operating Experience

Although several systems have been built, the first to produce a net power output was an experimental OTEC unit built by the state of Hawaii and tested off its coast in 1979.[67] The turbine produced 50 kWe with a 22 °C (37 °F) difference in temperature between the surface water and water drawn up at a rate of 2700 gpm from a depth of 655 m (2150 ft) through a 56-cm-ID (22.1-in-ID) polyethylene pipe. By using a low velocity (0.68 m/s, or 2.2 ft/s) in the cold-water intake pipe, the pumping power required for the system was kept down to 40 kW so that there was a net output of ~ 10 kWe.[66] Testing was terminated after about six weeks when the cold-water intake pipe was broken off in a storm.

Combining an OTEC Plant with Mariculture

It has long been recognized that the plant nutrients in cold, deep ocean water can greatly increase the productivity of the surface water. In fact, the open ocean (90 percent of the total area) produces only ~ 0.7 percent of the fish because most of the nutrients in the surface water are extracted by plants and drift down to the ocean floor in the remains of plant or animal life. The waters in the coastal zones are continually supplied with fresh nutrients in the runoff from the adjacent land, and hence support a high level of plant life activity and produce 54 percent of the fish. Only 0.1 percent of the ocean area lies in the upwelling regions, where nutrient-laden water is brought up from the ocean depths, yet these regions produce 44 percent of the fish.[68] The reason for this spectacular difference can be seen in Table 21.11, which shows that the nitrate and phosphorus concentrations in deep seawater are ~ 150 and 5 times, respectively, their concentrations in surface water at a typical site.

There have been several efforts to produce an upwelling effect. At the urging of the oceanographer C. O. Iselin at Wood's Hole in 1956, the author examined the possibility of using a system of nuclear power plants on the sea floor to produce upwelling but concluded that the cost would be excessive. However, in

TABLE 21.11 **Concentrations of Nutrients in Both Surface Water and Water from a Depth of 870 m at St. Croix in the U.S. Virgin Islands**[68]

	Nutrients (μg-atoms/L)*				
	$(NO_3 + NO_2) - N$	$NO_2 - N$	$NH_3 - N$	$PO_4 - P$	$SiO_4 - Si$
Surface water (3 km offshore)	0.2	0.2	0.9	0.2	4.9
870-m-deep water	31.3	0.2	0.7	2.1	20.6

*Concentrations given for the key nutrient atoms, and do not include other atoms of O or H in the ion.

the 1970s, O. A. Roels studied the possibility of using a shore-based OTEC plant to supply nutrient-laden water to a mariculture system, and since 1972 he has supported his analyses with a series of experiments carried out at St. Croix in the U.S. Virgin Islands. At that site the ocean is 1000 m deep only 1.6 km offshore. Three polyethylene pipelines 6.9 cm (ID) and 1830 m long have brought ~250 L/min of bottom water into 5-m³ pools where diatoms from laboratory cultures are grown. The food-laden effluent flows through metered channels to pools where shellfish are raised. The resulting protein production rate has been excellent; 78 percent of the inorganic nitrogen in the deep seawater has been converted to phytoplankton protein-nitrogen, and 22 percent of that was converted to clam-meat–protein-nitrogen. This compares with plant-protein/animal-protein conversion ratios of 31 percent for cows' milk production and 6.5 percent for feedlot beef production.

Coupling an OTEC plant to a mariculture operation would require shore-based plants because in the open sea the cold water from the depths would sink back down with relatively little mixing. For a shore-based plant roughly 50 km² would be required for the mariculture tanks per 100 MWe produced. Suitable sites with deep water close to shore would be limited largely to volcanic islands, hence the bulk of the electric power produced would probably go into energy-intensive operations such

TABLE 21.12 Estimate of the Annual Output from a Combined OTEC-Mariculture System Using a Deep-Water Flow Rate of 3.75 m³/s (Ref. 68)

System	Product	Assumptions	Unit price	Gross sales value
OTEC	1MWe	90% of time on line	4¢/kWh bus bar	$ 315.360
Mariculture	1693 t shellfish	42% meat	$1/lb meat	1,564,200
	Whole wet weight	711 t meat	(= $2.20/kg)	

as the production of aluminum or nitrogen fertilizer. (See pages 545–546.) Chlorine probably could not be used to prevent biofouling of heat exchanger surfaces, hence the open cycle of the system of Fig. 21.31 would probably be preferable. These and related disadvantages of coupling OTEC and mariculture systems appear to be much more than offset by the value of the sea food produced which Roels estimates would run about 5 times the value of the electricity. (See Table 21.12.)

PROBLEMS

1. Estimate the cost of drilling two holes for 20-cm-ID casings in hot, dry rock in a region having a temperature gradient of 50°C/km (Fig. 21.1) to a depth of 3 km. Neglect the cost of fracturing the rock between the wells. Assume that the temperature difference between the rock and the water leaving it after circulating through the rock from one well to the other is 30°C irrespective of the water flow rate. Estimate the amount of heat obtained per hour and the capital charge for the wells at 18 percent plus a write-off, assuming a 10-year life. Express the cost in 1976 dollars per 10⁹ J of heat obtained, assuming a 70 percent load factor. Consider circulating the water under pressure down one well and up the other at flow velocities of 2, 4, 8, 16, and 32 m/s with a temperature rise in the circuit of 60°C.

2. Estimate the pumping power required in kilowatts per 10⁹ J as a function of the water flow rate for the above cases, assuming that the pressure drop through the fractured rock and the field collection lines is 25 percent that through the wells, and plot as a function of flow.

3. Estimate the net thermal efficiency for a steam plant (Fig. A2.9), assuming dry cooling towers giving a condenser temperature of 50°C. Use the heat yielded by the wells of Prob. 1, the pumping power of Prob. 2, and a capital cost for the steam power plant of $400/kWe (gross) plus that for the wells from Prob. 1 but with no allowances for well field piping and related auxiliaries. Estimate the net power output as a function of flow and the cost of electricity in cents per kilowatthour for these capital costs, using a capital charge for the steam plant of 18 percent per year.

4. The roof of a house that faces to the south with no trees shading it has an area of 100 m². The house is well insulated so that the heat loss is 1.5 MJ/(°C · h) and during the heating season averages a total of 75 GJ per year. Calculate the heat loss for an average day in December with an outside air temperature of −5°C and an inside temperature of 20°C. If the mean daily total solar radiation in December is given by Fig. 21.18 as 3.1 kWh/m², estimate the solar heat available as a fraction of the heat required to hold the temperature in the

house at 20°C (68°F). Assume simple flat plates collecting heat at 50°C with the collection efficiency given by Fig. 21.7. Assume also that the roof angle has been chosen so that it is nearly normal to the incident sunlight at noon in December and that ample heat storage capacity is available.

5. For the same house as above, consider the use of a parabolic trough collector similar to that of Fig. 21.9 with the collection efficiency given by Fig. 21.14 and an average direct insolation of 2.1 kWh per day per square meter (as given by Fig. 21.18). What is the average collector efficiency? What is the total heat collected? If the heat collected at 300°C is cascaded through an organic Rankine cycle with a heat rejection temperature of 50°C, estimate the electric energy that could be obtained and the fraction of the heat load for the house that could be obtained. Use an efficiency for the actual cycle equal to 50 percent of that for the ideal Carnot cycle. Consider that all the electric energy generated ends up as heat in the house.

6. If the local cost of gas for heating is 25 percent of the cost of electricity per unit of energy, compare the total value of the energy obtained for the systems of Probs. 4 and 5. Assume that the capital cost of the system of Prob. 5 is double that for the system of Prob. 4 but allowance is made for the useful electric output during the 50 percent of the year when heat is required only for domestic hot water (30 percent of the average December heat load) and the electric output will average

double the value obtainable in December. Compare the costs for the two systems.

7. A town having an average electric load of 30 MW is in an area having an average wind energy in the summer of 100 W/m² (Fig. 21.25). Assume that 25 percent of this energy is lost because half of the time the wind velocity is too low to generate power. Estimate the number of windmills and the size of the area required to supply an average power demand of 30 MWe. Assume a high-speed, two-blade type of windmill having a power coefficient as given by Fig. 21.24 designed for the summer wind conditions, assume that 38-m-diameter windmills will be spaced on 50-m centers in lines 200 m apart, and allow for energy storage with a 65 percent efficiency.

8. Estimate the capital cost of both the windmills and the lead-acid storage batteries required if the energy stored must be equal to one day's consumption, the peak windmill output is double the average, and the windmill output is zero one-half the time. (Use data from Table 21.8 for the NASA MOD-OA and express in 1977 dollars per kilowatt for the 30-MWe average power consumption of the town. Take the cost for production windmills. Use a battery cost of $70/kWh. See also Table 20.2.) Estimate the capital charges for the windmills and batteries, assuming a 10-year life and 18 percent on the initial investment. Neglect charges for land use, the electric-power collection system, instrumentation, controls, and maintenance. Compare with average U.S. power costs in 1977 (Fig. 10.7).

REFERENCES

1. Burnham, J. B., and D. H. Stewart: *Foreign and Domestic Discussions on Natural Geothermal Power and Potential Use of Plowshare to Stimulate These Natural Systems*, Battelle Northwest Laboratories Report No. BNWL-B-110, July 1971.

2. Lengquist, R., and F. Hirschfeld: "Geothermal Power, the 'Sleeper' in the Energy Race," *Mechanical Engineering*, vol. 98, no. 12, December 1976.

3. Kiewicz, J. Pietrusz: "Optimum Design Conditions for a Power Plant at a Vapor-Dominated Geothermal Resource, P.G. and E.'s Geysers Power Plant Unit No. 16," *Proceedings of the Thirteenth IECEC*, Aug. 20–25, 1978.

4. "Geothermal Energy–Central America," *Mechanical Engineering*, vol. 97, no. 9, September 1975, p. 57.

5. "Furnace Beneath Your Feet," *Mechanical Engineering*, vol. 100, no. 9, September 1978, p. 47.

6. Milora, S. L., and J. W. Tester: *Geothermal Energy as a Source of Electric Power*, The M.I.T. Press, Cambridge, Mass., 1976.

7. White, D. E., and D. L. Williams: *Assessment of Geothermal Resources of the United States—1975*, Geological Survey Circular 726, U.S. Geological Survey, Reston, Va., 1975.

8. Samuels, G.: "Geopressure Energy Resource Evaluation," Oak Ridge National Laboratory Report No. ORNL/PPA-79/2, May 1979.

9. Barker, L. M., et al.: "Geothermal Environmental Effects

on Drill Bit Life," *Proceedings of the Eleventh IECEC*, Sept. 12–17, 1976, p. 711.

10. Cristy, G. A., and W. C. McClain: "Examination of High Pressure Water Jets for Use in Rock Excavation," Oak Ridge National Laboratory Report No. ORNL/HUD-1, January 1970.

11. *The Second International Symposium on Jet Cutting Technology*, BHRA Fluid Engineering, Cambridge, England, April 1974.

12. Altseimer, J. H.: "The Subterrene Rock-Melting Concept Applied to the Production of Deep Geothermal Wells," *Proceedings of the Eleventh IECEC*, Sept. 12–17, 1976, p. 717.

13. Dutcher, J. L., and L. H. Moir: "Geothermal Steam Pricing at the Geysers, Lake and Sonoma Counties, Cal.," *Proceedings of the Eleventh IECEC*, Sept. 12–17, 1976, p. 786.

14. Allen, C. A., et al.: "Fluidized Bed Heat Exchangers for Geothermal Application," *Proceedings of the Eleventh IECEC*, Sept. 12–17, 1976, p. 761.

15. Johnson, F. S.: "The Solar Constant," *Journal of Meteorology*, November 1959, p. 431.

16. Spurr, S. H.: "Silviculture," *Scientific American*, vol. 240, no. 2, February 1979, p. 76.

17. "Utilities Put the Sun to Work," *EPRI Journal*, vol. 3, no. 2, March 1978, p. 26.

18. Hager, K. G., and C. A. Berg: "Wood Residue–Fired Gas Turbine," paper presented at the Forest Product Research Society Conference, Atlanta, Ga., November 1976.

19. *An Assessment of Solar Energy as a National Energy Resource*, NSF/NASA Solar Energy Panel, December 1972.

20. Reding, J. T., and B. P. Shepherd: *Energy Consumption: Fuel Utilization and Conservation in Industry*, Dow Chemical, U.S.A., Texas Division, prepared for U.S. Environmental Protection Agency, EPA-650/2-75-032-d, August 1975.

21. Parker, H. W.: "Disadvantages of Fermentation Ethanol as a Motor Fuel," paper presented at the Central Carolinas Section of the AIChE, Mar. 20, 1980.

22. Goldemberg, J.: "A Madeira como Fonte de Carburantes Liquidos, Anais, V Encontro," *Jornal dos Reflorestadores*, ano 1, no. 2, Brazil, April 1979, pp. 13–17.

23. da Silva, J. G., et al.: "Energy Balance for Ethyl Alcohol Production from Crops," *Science*, vol. 201, no. 4359, September 1978, pp. 903–906.

24. Hopkinson, C. S., Jr., and J. W. Day, Jr.: "Net Energy Analysis of Alcohol Production from Sugarcane," *Science*, vol. 207, no. 4428, Jan. 18, 1980, pp. 302–303.

25. Villar Ferrin, A. B.: "Sintese do Metanol Derivado dos Residuos da Madeira, Anais, V Encontro," *Jornal dos Reflorestadores*, ano 1, no. 2, Brazil, April 1979, p. 28.

26. Hottel, H. C.: "Residential Uses of Solar Energy," *Proceedings of the World Symposium on Applied Solar Energy, Phoenix, Ariz.,* 1955.

27. Yellott, J. I.: "Power from Solar Energy," *Trans. ASME*, vol. 79, 1957, pp. 1349–1359.

28. Barber, R. E.: "Current Costs of Solar-Powered Organic Rankine Cycle Engines," *Solar Energy*, vol. 20, no. 1, 1978, pp. 1–6.

29. Tester, J. W., et al.: "Comparative Performance Characteristics of Cylindrical Parabolic and Flat Plate Solar Energy Collectors," ASME paper, 74-WA/Ener-3, November 1974.

30. Lof, G. O. G., et al.: "Energy Balance on a Parabolic Cylinder Solar Collector," *Journal of Engineering for Power, Trans. ASME*, vol. 84, no. 1, January 1962, pp. 24–32.

31. Spight, L. D.: "Solar Insolation Measurements at Las Vegas, Nevada," *Solar Energy*, vol. 20, no. 2, 1978, pp. 197–203.

32. Ramsey, J. W., et al.: "Experimental Evaluation of a Cylindrical Parabolic Solar Collector," *Journal of Heat Transfer, Trans. ASME*, vol. 99, 1977, pp. 163–173.

33. Tracey, T. R., et al.: "1 MW(t) Solar Cavity Steam Generator Solar Test Program," *Proceedings of the Twelfth IECEC*, 1977, pp. 1224–1230.

34. Schweinberg, R. W., and J. N. Reeves: "Solar One Project—A 10-MW Solar Thermal Central Receiver Pilot Plant," *Proceedings of the Fourteenth IECEC*, Aug. 5–10, 1979, pp. 181–182.

35. Kreider, J. F., and F. Kreith: *Solar Energy Handbook*, McGraw-Hill, New York, 1981.

36. Chalmers, B.: "The Photovoltaic Generation of Electricity," *Scientific American*, vol. 235, no. 4, October 1976, pp. 34–43.

37. Sahai, R., et al.: "High Efficiency AlGaAs/GaAs Solar

Cell Development," *Proceedings of the Thirteenth IEEE Photovoltaic Specialists Conference*, 1978, pp. 946–952.

38. Queisser, H. J., and W. Shockley: "Some Theoretical Aspects of the Physics of Solar Cells," Energy Conversion for Space Power, vol. 3, *Progress in Astronautics and Rocketry*, Academic Press, New York, 1961, p. 317.

39. Moon, R. L., et al.: "Multigap Solar Cell Requirements and the Performance of AlGaAs and Si Cells in Concentrated Sunlight," *Proceedings of the Thirteenth IEEE Photovoltaic Specialists Conference*, 1978, pp. 859–867.

40. Donovan, R. L., and S. Broadbent: "10 kW Photovoltaic Concentrator Array, Martin-Marietta Aerospace Corp.," Sandia Laboratories Report No. SAND-78-702A, May 1978.

41. "Concentrating Array Production Process Design," Final Report, General Electric Company Report No. 78SDS4266, Sandia Report No. 78-7072, Dec. 15, 1978.

42. Wolf, M.: "Advances in Silicon Solar Cell Development," Energy Conversion for Space Power, vol. 3, *Progress in Astronautics and Rocketry*, Academic Press, New York, 1961, p. 231.

43. Middleton, A. E., et al.: "Evaporated C & S Film Photovoltaic Cells for Solar Energy Conversion," Energy Conversion for Space Power, vol. 3, *Progress in Astronautics and Rocketry*, Academic Press, New York, 1961, p. 275.

44. Donovan, R. L., and S. Broadbent: "Photovoltaic Concentrating Array," *Proceedings of the IECEC*, 1978, p. 1593.

45. Brooks, G. R., et al.: "Orbiting Solar Observatory (050-8) Solar Panel Design and In-Orbit Performance," *Proceedings of the IECEC*, 1978, pp. 105–109.

46. Denman, O. S.: "From Sunlight in Space to 60 Hz on Earth —the Losses along the Way," *Proceedings of the IECEC*, 1978, pp. 178–182.

47. Caputo, R.: "An Initial Comparative Assessment of Orbital and Terrestrial Central Power Systems," Final Report, Jet Propulsion Laboratory Technical Report No. 900-780, March 1977.

48. "Solar Power Satellites," *Proceedings of the IECEC*, 1978, pp. 178–204.

49. Woodcock, G. R., *Solar Power Satellite*, vol. 1, Boeing Aerospace Co. Report No. D180-22876-1, December 1977.

50. Payne, P. R., and D. W. Doyle: "The Fossil Fuel Cost of Solar Heating," *Proceedings of the IECEC*, 1978, p. 1650.

51. Hunt, L. P.: "Total Energy Use in the Production of Silicon Solar Cells from Raw Materials to Finished Product," *Proceedings of the Twelfth IEEE Photovoltaic Specialists Conference—1976*, Nov. 15-18, 1976, pp. 347–352.

52. Robinson, A. L.: "American Physical Society Gives a Long-Term Yes to Electricity from the Sun," *Science*, vol. 203, no. 16, February 1979, p. 629.

53. Freese, S.: *Windmills and Millwrighting*, David and Charles, Ltd., Devon, England, 1971.

54. Spier, P.: *Of Dikes and Windmills*, Doubleday & Company, Inc., Garden City, N.Y., 1969.

55. Hirschfeld, F.: "Wind Power—Pipe Dream or Reality?" *Mechanical Engineering*, vol. 99, no. 9, September 1977, p. 20.

56. "The Earth as a Solar Heat Engine," *EPRI Journal*, March 1978, p. 43.

57. Glasgow, J. C., and A. G. Birchenough: "Design and Operating Experience on the U.S. D.O.E. Experimental MOD-O 100 kW Wind Turbine," *Proceedings of the Thirteenth IECEC*, Aug. 20-25, 1978, pp. 2052–2055.

58. Richards, T.R., and H. E. Neustadter: "DOE/NASA MOD-OA Turbine Performance," *Proceedings of the Thirteenth IECEC*, Aug. 20-25, 1978, pp. 2060–2062.

59. Linscott, B. S., et al.: "Experimental Data and Theoretical Analysis of an Operating 100 kW Wind Turbine," *Proceedings of the Twelfth IECEC*, 1977, p. 1633.

60. Elliott, D. L.: "Synthesis of National Wind Energy Assessments," Pacific Northwest Laboratories Report No. BNWL-2220, WIND-5, July 1977.

61. Tewari, S. K.: "Economics of Wind Energy Use for Irrigation in India," *Science*, vol. 202, no. 4367, Nov. 3, 1978, p. 481.

62. "Will Japan Be First with Solar Wave Power?" *Mechanical Engineering*, vol. 100, no. 1, January 1978, p. 53.

63. Merriam, M. F.: "Wind, Waves, and Tides," *Annual Review Energy*, 1978, p. 29.

64. Swann, M.: "Power from the Sea," *Environment*, vol. 18, no. 4, May 1976, p. 25.

65. "100 MW(e) OTEC Alternate Power Systems," Final Report, Westinghouse Electric Corp., Power Generation Divisions, Nov. 29, 1978.

66. *A Conceptual Feasibility and Cost Study for a 100 MW(e) Sea Solar Power Plant*, United Engineers and Constructors, Inc., 1976.

67. Hartline, B. K.: "Tapping Sun-Warmed Ocean Water for Power," *Science*, vol. 209, no. 4458, Aug. 15, 1980, pp. 794–796.

68. Roels, O. A.: "From the Deep Sea; Food, Energy, and Fresh Water," *Mechanical Engineering*, vol. 102, no. 6, June 1980, pp. 37–43.

22

COMPARISON OF ADVANCED ENERGY CONVERSION SYSTEMS

A meaningful comparison of advanced energy systems is extremely difficult technically because of the wide range of disciplines involved—as is evident from the scope of the previous chapters. The situation has been greatly confused by enthusiasts for one system or another because they each promote their favorite system with optimistic claims while often failing to mention the crucial problems other than the need for large amounts of research and development funds. The situation is further confused by the utopian visions of social theorists and altruists who are so fearful of systems they don't like that they disregard the many problems of systems they favor, thus failing to put the whole complex of problems in perspective. Clearly, a critical, comprehensive evaluation and comparison of the various energy conversion systems has been greatly needed, and it should be a comparison based on facts rather than on wishful thinking. It is not surprising that in the face of the growing shortage of cheap energy, many people find it enticing and plausible to believe that the machinations of big corporations have maliciously prevented us from enjoying the benefits of cheap electricity obtained from the sun, the wind, the waves, or the tides. However great the temptation to join in this popular way of thinking, questions so vital to the welfare of the world deserve a rational approach, no matter how tedious and demanding this may be.

The objective of this chapter is to sort out and organize the salient facts presented in the previous chapters to give a good perspective on the principal characteristics of the various energy systems that appear to be of interest. This should then provide a rational basis for planning both R&D efforts and the construction of new facilities for electric-power production.

SUMMARY OF EFFORTS TO COMPARE ADVANCED ENERGY SYSTEMS IN THE 1970s

An appreciation for the serious implications of our limited energy supplies became sufficiently widespread in government offices and the electric utilities by 1970 that a series of efforts was initiated to examine the relative merits of advanced power conversion systems. The first such effort to arouse widespread interest was undertaken by the Electric Research Council in 1970. (The council was an association of private and public utilities plus a representative from the Department of the Interior.) A task force was set up in the fall of 1970 to consider R&D needs for the utility industry and to make recommendations. A comprehensive study was summarized in a report that included consideration of various advanced energy conversion systems, improved means for transmission and distribution of electric power, environmental problems, energy utilization, and the general problems of industry growth and system development.[1] The report included estimates of the funding required for the various R&D projects recommended, but it did not include specific estimates for the thermal efficiency, capital cost, or cost of electric power for the various advanced energy conversion systems considered. This report was widely circulated and led both to some further work[2] and to the establishment of the Electric Power Research Institute (EPRI).

The Office of Science and Technology (OST) launched a quite comprehensive study in January 1972 by organizing a set of task forces composed of people from the Bureau of Mines, National Aeronautics and Space Administration (NASA), the Department of the Interior, the National Bureau of Standards (NBS), the U.S. Atomic Energy Commission (AEC), etc., in an effort to appraise the feasibility of the various advanced systems, their thermal efficiencies, capital costs, costs of electricity, costs of the developmental programs, and their respective effects on the environment and the consumption of natural resources. Each task force (usually made up of advocates of the systems assigned) submitted its report to OST in the summer of 1972. These reports were reviewed by a panel of experts, and the results of the task force work, together with the evaluation by the review panel, were summarized in a report prepared at the Brookhaven National Laboratory.[3]

In the summer of 1972, the National Science Foundation

(NSF) initiated a series of workshops in which the various aspects of the national energy situation were examined by groups made up of representatives from industry, government, and universities. The series included a one-week workshop at Saxtons River, Vermont, which covered essentially all aspects of the energy situation.[4] A second one-week conference at Berwick, Maine, was devoted to problems of the electric utilities.[5] A third, held in June 1973, was concerned with the utilization of coal.[6] Other, smaller conferences in 1973 were devoted to more specialized energy sources, including fission reactors, solar power, geothermal heat, and ocean thermal differences. In each case the agenda consisted of presentations by advocates of particular advanced systems, pleas or demands made by members of environmentalist organizations, statements on legal or financial problems made by experts in those areas, estimates of mineral reserves made by university or government geologists and by people from the petroleum and mining industries, etc. The group as a whole discussed these presentations. Reports summarizing the first three workshops were published, but the reports covering the later, smaller workshops did not get beyond the draft stage.

In 1973, in response to a presidential request, Dixie Lee Ray, then Chairman of the AEC, organized a set of task forces similar to that set up by OST in 1972. In this set of studies, greater emphasis was given to means for energy conservation than in the OST study. The report, commonly referred to as the Dixie Lee Ray Report, was widely disseminated and was a major factor in the development of federal government plans that culminated in the establishment of the Energy Research and Development Administration (ERDA).[7]

Essentially concurrent with the Dixie Lee Ray Study, the Federal Power Commission (FPC) established a committee of representatives from government agencies, universities, and industry to appraise the potential of advanced systems for use by the electric utilities. Both the study and the resulting report[8] expressed the value of a proposed system in terms of the amount of fuel oil that might be saved as a consequence of the development of that system.

In 1974, in response to an express request from the President of the United States, the American Society of Mechanical Engineers (ASME) set up a task force to consider the relative merits of advanced energy conversion systems. A committee of about 40 ASME members from both industry and government was set up, a series of meetings was held at which brief presentations were made by advocates of advanced systems, and the relative merits of the systems were discussed by the experts on the committee. Efforts to arrive at a reasonable consensus proved difficult, but a report was prepared and approved by the ASME Policy Board and Council,[9] and a paper summarizing

this report was presented at the ASME Winter Annual Meeting in December 1976.

In 1974 an effort to obtain a more definitive study of fossil fuel systems than any made previously was initiated by representatives of NSF, the Office of Management and Budget (OMB), and the Office of Coal Research (OCR). This led to a well-funded study entitled Energy Conversion Alternatives Study (ECAS) that was carried out under the detailed supervision of NASA Lewis Laboratory, which awarded contracts to General Electric Company and Westinghouse. While NASA did some comparative work in-house, General Electric and Westinghouse did the bulk of the work following a set of ground rules arrived at through conferences with NSF, OCR, and NASA. They made an architect-engineer type of study and prepared quite detailed estimates of the thermal efficiency, capital cost, cost of electricity, and development program costs for advanced fossil energy systems for utility applications. No serious effort was made to assess the appropriateness of the component efficiencies or operating lifetimes assumed by the system advocates. This work was completed by the latter part of 1976. The role played by OCR when the project was launched initially was continued by ERDA–Fossil Energy Division (ERDA-FE), when that agency was formed. The resulting reports have been widely circulated.[10-16]

In 1975, ERDA-FE funded a study at MITRE which was similar to the ECAS study but was aimed primarily at program planning and hence was limited in scope. A report covering this study was prepared, and a few copies were printed for internal use at ERDA in the summer of 1976.[17]

A review of the above studies disclosed that in the cases where estimated values for the performance of many of the advanced energy conversion systems were included, there were substantial inconsistencies. The values for the estimated thermal efficiency, for example, have been extracted and tabulated in Table 22.1. A review of this table discloses that the estimates for any given energy conversion system differ substantially from one report to another. Some of these differences stem from quite different design conditions. For example, the OST values for the alkali-metal-vapor topping cycle were for a 955°C (1750°F) peak temperature with a refractory metal system and a thermonuclear reactor heat source (i.e., no heat losses to the stack gas), whereas the other studies were for fossil-fuel heat sources with stack losses and lower cycle temperature limits imposed by the use of iron-chromium-nickel alloys. Having been an active participant in the studies of Refs. 3 to 9 and involved in the review of Refs. 10 to 16, this author felt that the most important factors leading to disparities between the studies were the limited time available in each case and strong political pressures that restricted the ranges covered. In no instance was there a

TABLE 22.1 Summary of Efforts to Compare Overall Thermal Efficiency for Advanced Energy Systems in the 1970–1976 Period

Scale of study in $	10^4 to 10^5		10^5 to 10^6							Greater than 10^6	
Prime sponsor	Electric Research Council	Nat'l Air Pollution Control	OST	NSF	AEC	FPC	ASME	ERDA-FE	EPRI	ERDA	
Approximate total cost, $ $\times$ 10^{-3}	100	300	1000	100	1000	200	100		400	2500	
Study group	Utility industry	United Aircraft and Burns & Roe	Gov't agencies	Gov't agencies, universities and industry	Gov't agencies & industry	Gov't agencies, universities and industry	Industry and gov't	MITRE	G.E.	NASA, G.E., and Westinghouse	
Reference no. of report	1	2	3	4, 5, 6	7	8	9	17		10, 11, 12	13–16
Common term for study	ERC	United Aircraft	OST		Dixy Lee Ray Report	FPC				ECAS I	ECAS II
Period of study	1970–1971	1970	1972	1972–1973	1973	1973–1974	1974–1976	1976–1977	1975–1976	1975–1977	
Improved conventional fossil fuel plants											
Steam plants	—	38.6[a]	—	40.5	—	40.0[a]	32	36.0–38.0	39.0[g]		
Open-cycle gas turbines	—	35.0[b]	—	—	—	35.0	—	—	31	17.0–19.0[h]	—
Combined steam and gas turbines	—	47.6[c]	55.0	—	—	42.0–45.0	—	38.0[d]	38	36.0–42.0[j]	46.8
Bottoming cycles	—	41.0	—	—	—	—	—	—	—	—	—
Nuclear plants											
Water-cooled fission reactor	—	—	—	—	—	33.0	—	32.0[a]	33	—	—
Gas-cooled fission reactor	—	—	38.0	—	—	39.0	—	—	—	—	—
Molten-salt fission reactor	—	—	44.0	—	—	—	—	—	—	—	—
LMFBR	—	—	42.0	—	—	—	—	40.0[a]	—	—	—
Fusion reactors	—	—	—	—	—	—	—	—	—	—	—
Advanced fossil fuel systems											
Fuel cells	—	—	60.0–70.0	—	—	—	—	—	—	32.0–53.0	—
Open-cycle MHD	—	—	50.0–60.0	60.0	—	48.0–52.0	—	48.0[d]	48	48.0–49.0	48.3
Closed-cycle MHD	—	—	52.0–54.0	—	—	50.0	—	—	—	33.0–46.0	—
Closed-cycle gas turbine	—	41.4[c]	45.0	—	—	45.0	—	38.0[d,f]	40	31.0–32.0[j]	39.9[f]
Alkali-metal-vapor topping cycle	—	44.6	55.0	—	—	50.0	—	44.0[d,g]	44	35.0–43.0[f,g]	44.4[g]
Supercritical CO_2	—	38.2[c]	50.0	—	—	35.0	—	—	—	39.0–41.0[f,g]	—
Thermionic cells	—	—	—	—	—	—	—	—	—	—	—
Solar, wind, geothermal, etc., systems											
Solar cells	—	—	14.0	—	—	10.0–15.0	—				
Solar vapor cycles	—	—	—	—	—	—	—				
Biomass conversion	—	—	3.0	—	—	—	—				
Windmills	—	—	—	—	—	—	—				
Geothermal plants	—	—	—	—	—	—	—				
Ocean thermal differences	—	—	2.0	2.0	—	—	—				
Energy storage systems											
Batteries	—	—	—	—	—	—	—				
H_2 production	—	—	—	—	—	—	—				
Compressed-air storage	—	—	—	—	—	—	—		—	—	—

[a] 70% load.
[b] Converted coal fuel.
[c] Includes coal conversion efficiency, 78%.
[d] 60% load.
[e] 65% load.
[f] Atmospheric fluidized bed.
[g] Pressurized fluidized bed.
[h] Includes coal conversion efficiency, 50%.
[i] Integrated low-Btu gasifier.
[j] Pressurized furnace, low-Btu gasifier.

thoroughgoing review of the background of R&D experience for each system to provide a basis for judging the reasonableness of the assumptions implicit in the performance estimates.

The above shortcomings were recognized by many of the people involved, and, in an effort to remedy the situation, ERDA commissioned a study of the developmental background for each of the systems included in the ECAS work. This work (for which the author was responsible) was carried out at the Oak Ridge National Laboratory. The ensuing detailed reports (totaling over 600 pages) although reviewed extensively by leading experts were given only a limited circulation, but the key results were summarized in an ANS paper.[18] The ORNL study, which entailed about 5 man-years of effort, has been an important source of material for Chaps. 11 through 18 of this book, which are concerned with fossil fuel systems. This book as a whole as summarized in this chapter has been prepared to cover the entire range of systems for producing electric energy and thus to the author's knowledge is unique both in scope and in the depth to which the developmental background for each system has been examined.

THE NATURE OF RESEARCH AND DEVELOPMENT WORK

The widespread misconception that a whole set of advanced energy systems will soon be serving the public stems from difficulties most people have in appreciating two major problem areas: the serious uncertainties in both the research and the development phases and the time spans required for not only these two phases but also a third phase, the commercialization to the point where the system makes a significant contribution to the economy as a whole. These three phases must be conducted in series, and the time required for each depends in part on the difficulty of basic technical problems such as corrosion, in part on the size of the system required for commercial viability (and hence the time required for construction), and in part on the competition. Pocket-size solid-state electronic calculators represent a dramatic example of technical innovation that was developed and virtually took over the market from its predecessor (slide rules) in less than 10 years. This rapid growth was possible because the basic R&D on silicon chip microcircuits was well in hand, the units were small, inexpensive, and quickly manufactured, and they were much superior to the product they displaced. Advanced systems for producing electricity present vastly more time-consuming problems. Conventional steam power plants are highly developed so that it is difficult to compete with them. Moreover, the economics of scale strongly favor very large units each of which costs roughly a billion dollars, takes 5 to 10 years to build, and—because of its size—introduces environmental effects which in turn make for

political problems that often dominate technical considerations and introduce years of delays. Thus it is a formidable task indeed to carry a new concept for an advanced energy system through to commercialization, the time required is at least two decades, and there is no assurance of success. In fact, it is a bit like a horse race: only a few of the entrants can be expected to win, place, or show.

An excellent illustration of the nature of the situation is given by an historical review of the time required for the research, development, and commercialization phases for fission power reactors. The record is summarized in Fig. 22.1. Fifteen different reactor types were under development in the 1945–1970 period, but by 1980 only four types had reached the stage at which at least six units were in operation or under construction. One or a few units of each of four other types had produced substantial amounts of electricity, but by 1980 only one of these, the LMFBR, was still under active development, and all R&D work on 7 of the 15 types had been terminated.

In searching for the characteristics that distinguished the more successful from the less successful reactor concepts, it is instructive to examine Fig. 22.1 in detail. A number of points appear to be significant and have important implications with respect to other advanced energy systems.

Limitations Imposed by Materials Compatibility

Probably the first point is that all four of the commercial reactor types operate at relatively low temperatures—around 300°C (575°F)—and made use of fairly conventional materials that, from the standpoints of both extensive experience and basic chemical thermodynamics, one would expect to be compatible and relatively free of corrosion difficulties. No scientific or technological "breakthroughs" were required for their development, and—in particular—the materials problems (except for radiation damage) were largely in hand in 1945. Yet even in the most successful types, the pressurized and boiling-water reactor plants, the most serious troubles have been with corrosion of stainless-steel systems by nominally pure water. Most of this corrosion has occurred where leaks in steam condensers have permitted chloride salts to enter a system that was supposed to contain very pure water. Similarly, the CO_2-cooled-graphite-uranium-magnox reactors have been troubled by reactions between the CO_2 and the carbon steel of the pressure envelope. The aqueous homogeneous reactor employed an aqueous solution of uranyl sulfate and was attractive because it completely eliminated expensive fuel element fabrication operations. However, uranyl sulfate is far more corrosive than pure water, and the corrosion problems were sufficiently formidable that government funding was halted, even though two reactor experiments demonstrated the feasibility of generating electric

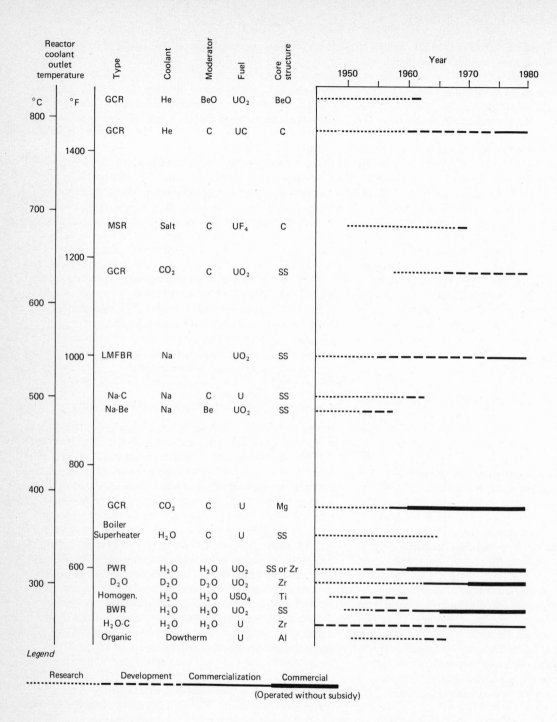

Figure 22.1   Research and developmental history of free world fission reactors arranged on a temperature scale.

power via this concept. Even the lowest-temperature reactor concept of Fig. 22.1 was dropped from development because of a materials problem: the organic coolant (whose low vapor pressure would have made it possible to avoid the need for the massive and expensive pressure vessels of the water reactors) was subject to polymerization under intense irradiation, and the resulting gummy solids fouled the vital heat-transfer surfaces in the reactor core, making the system completely unacceptable.

Capital Costs

A second major factor affecting system acceptability has been the capital investment for both the plant and the fuel. Of the reactor concepts requiring an expensive moderator, that is, Be, BeO, or D_2O, only the D_2O reactor has reached commercialization, although reactors of the other two types were built and operated. D_2O reactors entail a high capital investment for the D_2O, but they have the important advantage that the extremely low neutron absorption cross sections of D and O make the neutron economy of the D_2O reactors the best of any of the thermal solid fuel reactors. Not only does this excellent neutron economy permit them to employ relatively inexpensive and readily available natural uranium as the fuel, but also they can use it in the form of UO_2 (which permits a fuel element life many times that for metallic uranium) and they have sufficient reactivity to burn up a larger fraction of the uranium mined than is the case for light-water reactors. The CO_2-cooled, graphite-metallic uranium-magnox reactors have a good neutron economy so that, while they cannot do as well as the D_2O reactors, they are able to burn a larger fraction of the ^{235}U in the natural uranium than the water-cooled graphite reactors. (The limitations on the life of different forms of uranium fuel are discussed in Chap. 5.) Costs are similar; hence the CO_2-cooled graphite reactors have been commercialized (mainly in the United Kingdom), whereas only a few water-cooled graphite reactors have been used for commercial power production, mostly in the U.S.S.R.

System Sensitivity to Leaks

A third major factor affecting commercialization is the sensitivity of the system to small leaks and the consequent maintainability. However thorough the design work, and however stringent the quality control in the course of fabrication, occasional leaks are bound to occur. The PWR and BWR systems have shown themselves to be the least sensitive to leaks of the 15 types in Fig. 22.1. In the first place, the fuel is separated into thousands of discrete fuel element capsules so that a leak in any one capsule can release only a tiny fraction of the total fission product inventory. Further, the vast majority of the fission products are re-

tained within the UO_2 crystal lattice, further reducing the consequences of a fuel capsule leak. In regard to the water system, leakage of mildly radioactive water from the system, or of air into the system, does not present an immediately serious corrosion problem, though of course a concurrent fuel element leak can make the radioactivity in a water leak quite troublesome. A similar relatively low sensitivity to leaks holds for the gas-cooled reactors in which the fuel is contained in metallic capsules. The gas-cooled reactors in which the fuel is fabricated in ceramic form (both to permit operation at high temperatures and to minimize neutron losses) inherently have microcracks and other imperfections in the ceramic envelopes that give a much higher level of fission products in the coolant system. Thus these systems are quite sensitive to leaks.

The sodium-cooled reactors are very sensitive to leaks. This was particularly true for the sodium-cooled graphite reactor, because sodium reacts with graphite causing it to swell. To prevent this problem, the graphite blocks were canned in sheet stainless steel, which had to be thin in order to avoid excessive neutron losses. The coefficient of thermal expansion of the stainless steel was three times that of the graphite, hence the cans tended to expand away from the graphite. The external pressure of the sodium, however, acted to collapse them and thus wrinkled the can walls. As predicted by critics of the concept, when a reactor was built and operated, the cans cracked in the wrinkles upon thermal cycling, the sodium got into the graphite, and swelling occurred. The graphite swelling closed the gaps between graphite blocks, pinching the control plates that operated in these gaps and thus rendering first one and then another control plate inoperative. This forced termination of reactor operation after relatively little running, and it terminated the development program because no good solution to this fundamental problem was in sight.

The aqueous homogeneous and molten-salt reactors are also inherently sensitive to small leaks because all the fuel is in liquid form and there is no primary barrier to retain the fission products as there is with solid-fuel elements. Fear of leaks by government officials was a major factor in termination of support for these two programs, though experimental reactors of both types were operated for periods of a year or more. The operating experience with the molten-salt reactor was particularly good, and its breeding potential appears attractive.

System Stability and Control Characteristics

A fourth major factor is system stability. As discussed in Chap. 5, the steam bubbles in a boiling-water reactor are voids in the water moderator, and increasing their volume reduces the reactivity. The control problems that this posed in the initial design studies appeared formidable. But clever design followed by successful reactor operating experience demonstrated that these control problems were tractable, and in any event they did not hold up the development program. Similarly, the short neutron lifetime of fast reactors appeared to present such serious control problems that their development was delayed until extensive analyses supported by tests of small experimental reactors demonstrated that satisfactory control could be achieved. On the other hand, various attempts to design water reactors that would serve as both boilers and superheaters proved unsuccessful because local fuel burnout conditions could be induced by either potential boiling instabilities or shifts in the heat-load distribution between boiling and superheating caused by changes in load or system pressure. As a consequence, development of this type of reactor was abandoned at one stage or another in every country in which it was considered.

Implications of Reactor Developmental Experience

In reviewing the above, it is apparent that some fission reactor concepts presented relatively few technological problems. Their development proceeded fairly rapidly with only ~ 15 years required for R&D before commercialization could begin and another 15 years required to penetrate ~ 5 percent of the market. At the other extreme, many of the concepts presented such serious materials or basic stability and control problems that they were abandoned. A few concepts of intermediate difficulty have been pursued to the point where one or a small number of reactors produced commercial electric power, but because of slowness in development, costs, and/or maintenance problems, these types are not widely used and seem unlikely to take over a larger segment of the market. Only one of these concepts, the LMFBR, is under active development. In spite of high capital costs and a marked sensitivity to system leaks, there is a strong incentive to proceed with the breeder development because the price of natural uranium is expected to rise enough by the early twenty-first century so that the much lower fuel costs obtainable by breeding fissionable material will more than offset the higher capital costs.

In applying the lessons implicit in fission reactor development to an appraisal of concepts for advanced energy systems, it appears that a long period of development will be required for any advanced type. Serious materials or other problems requiring scientific or technological breakthroughs can be expected at least to increase greatly the time required for the R&D phases and in many cases to prevent concepts from ever becoming commercially useful. In fact, one must expect that the majority of the concepts for advanced energy systems will fall by the wayside

and that only a few will become commercially viable sources of electric power.

COMPARISON OF SYSTEM CHARACTERISTICS, R&D EFFORTS, AND DEVELOPMENTAL STATUS

The diversity of the advanced energy systems of interest is so great that, in comparing them, it is necessary to consider many different parameters. Even then, comparisons are complicated because of the different levels of development that have been achieved—a point illustrated by Fig. 22.2, which is similar to Fig. 22.1. Comparisons are further complicated by differences in the character of environmental effects, cycle efficiency, and costs. Several different attempts by the author at such comparisons in preparing for lectures on the subject in recent years have led to the format of Table 22.2, which summarizes a great deal of information. The principal systems of interest listed at the left are grouped on the basis of the energy source, i.e., fossil fuel, nuclear, or renewable (sometimes called "natural"). The first four columns of data are for parameters closely related to the thermal efficiency of the thermodynamic cycles—which of course are not pertinent to systems such as wind or hydro. (Note that parentheses are used to distinguish estimated values from those well validated by experiments.) The next five columns indicate the status of development of each system in terms of the number of years of experience, the summation of the shaft or electric energy output that has been obtained in the United States, both cumulative past and fiscal 1980 Federal R&D funding, and the demonstrated status of materials problems for the operating conditions given in the first set of four columns. The next three columns give figures of merit for commercial use where the status of the developmental effort is sufficient to define the projected range of values within a factor of 2 or less. The last column indicates briefly the most vital problems confronted in each case. Most of the material in the table was drawn from preceding chapters. This section presents the qualifications involved in the selection of the particular values chosen for Table 22.2, and points out some of the more important implications of the information.

Reference Designs

In choosing the reference designs on which Table 22.2 was based, it was necessary to narrow the field and take representative examples to keep the size of the table manageable. For example, the first item, the pulverized coal-fired steam system, was taken as a supercritical pressure system with one reheat and

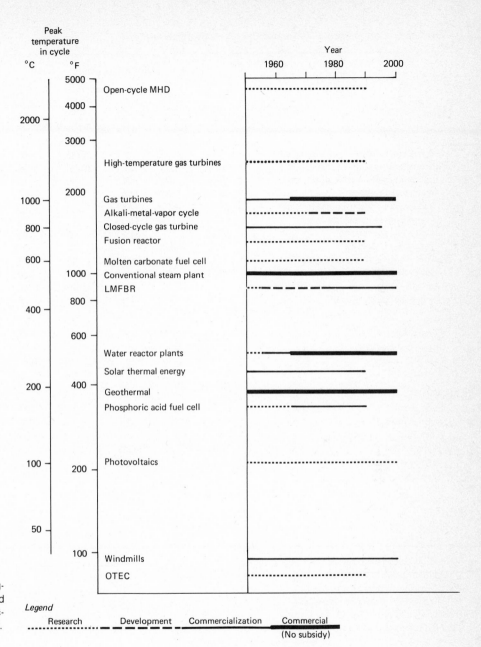

Figure 22.2 Past and projected progress in research, development, and commercialization of systems for electric power production in the free world.

SO$_2$ wet stack gas scrubbers, and it is considered here as representative (with some qualifications) of similar units with oil-firing or fluidized-bed coal combustion systems. Similarly, a typical utility gas turbine for peaking service was considered for generation of electricity, while for cogeneration conditions the example chosen was a base-load unit that the author visited in the Munich Statwerke which employs an 800 °C gas-turbine inlet temperature and a regenerator. This unit was chosen because of the demonstrated high performance and operating reliability of this plant. Use of the combined cycle for base or intermediate

TABLE 22.2 Comparison of Advanced Energy Conversion Systems

System	Data for reference designs				Years of hardware develop-ment	Total power output in U.S., kWh	R&D funding	
	Cycle T_{min}, °F	Cycle P_{max}, psia	Thermal eff., %				Total thru 1978, $	U.S. Govt. FY 80 10^6
			Electr. only, η_e	Cogen,[a] η_e/η_{e+t}				
Fossil fuel								
Conventional steam—pulverized coal	1050	3500	38	24/85	100	10^{14}		
Gas turbines—gas or distillate fuel	1800	200	28	30/80[b]	90	10^{12}	10^{10}	
Compound cycle—gas or distillate fuel	1700	200	42	32/90	50	10^{12}		
Direct coal-fired[c]	(1700)	(200)	(40)		35	10^7	5×10^7	
Coal gasifier—water-cooled turbine[c]	(2500)	(250)	(40)		80	10^5	2×10^7	
Coal gasifier—ceramic turbine[c]	(2500)	(250)	(46)		35	10^2	6×10^7	
Closed-cycle gas turbine[d]	1380		31.3	31/65	40	10^6	4×10^7	
Supercritical CO_2 cycle[c]	(1350)	(3780)			20	0	2×10^6	
Alkali-metal-vapor topping cycle[e]	1540	30	(46)	(35/80)	20	10^6	5×10^7	0
Open-cycle MHD[c]	4640	150			40	10^4	3×10^8	72
Fuel cells—phos. acid—gas or distillate[f]	280		32	32/40	50	10^6	4×10^8	
Molten carbonate—coal[c]	1110		(42)		25	10^4	5×10^7	20
Nuclear								
Water reactors[g]	540	975	33	16/90	35	10^{13}	7×10^8	37
Gas-cooled reactors	1000	2400	40	(26/90)	35	10^9	2×10^8	13
LMFBR[h]	900	1450	40	(24/90)	33	10^9	2×10^9	462
Fusion[i]	(1800)	(90)	(56)	(40/90)	30	0	10^9	360
Renewable								
Geothermal[j]	350	65	14		75	10^{10}	3×10^8	139
Solar—biomass[k]	900	900	35		280	10^{13}	10^8	58
Thermal (heat engine)					65	10^6	6×10^8	121
Photovoltaic	200		7 (17)		35	10^7	5×10^8	130
Satellite			7 (17)		20	10^6		
Wind					1000	10^8	10^8	67
Waves						0		
Tides[l]					60	0		
OTEC					50	0	10^8	35
Hydro					1500	10^{13}		

[a]Heat from cogeneration supplied as hot water at 160°C.
[b]Data from Munich Statwerke unit with regenerator & 800°C turbine inlet.
[c]Performance data from ECAS reports.
[d]Performance data from Oberhausen unit.
[e]Data from Ref. 54 of Chap. 15.
[f]Data from Ref. 43 of Chap. 18.

[g]Data from Ref. 3, Chap. 17.
[h]Data from CRBR.
[i]Data from Ref. 43 of Chap. 5.
[j]Data for typical Geysers unit.
[k]Data for typical paper mill.
[l]Data from Rance River plant in France — 10^9 kWh.

loads makes it necessary to employ a lower turbine inlet temperature than for peaking service. The data for the direct coal-fired, water-cooled, and ceramic turbines are given in parentheses for each of these because neither the practicality of the turbine nor that of the integrated system has been demonstrated anywhere in the world. For the closed-cycle gas turbine it seemed best to use the demonstrated performance of the latest European unit (at Oberhausen) rather than any of the various sets of data estimated for U.S. conceptual designs. For the alkali-metal-vapor cycle and phosphoric acid fuel cells, the data

Relative status of mat'ls probs.	Relative cost of electricity	Rel. envir. effect		Key problems
		Health	Land use	
Good	1.0	30	1	At limit of potential development
Good	1.9	0.1/36	~0.01	Clean fuel expensive
Good	1.5	0.1/25	~0.01	Clean fuel expensive
Poor		30	1	Turbine erosion and deposits
Poor		30	1	10^4 passages near burnout heat flux
Bad		30	1	Flameout thermal stress $20\times$ water quench
Good	~1.3	30	1	High cost of heat exchangers
Poor	(~1.9)	30		Very high heat exchanger costs
Good	(~1.0)	27	0.8	Sensitive to design and construction errors
Bad				Low gen. efficiency—can't even drive compressor
Fair		31	~0.01	Average life in latest tests only ~3000 h
Poor		28	1	Corrosion
Good	0.7	0.2	0.02	Emotional reaction of organized minority
Good	0.7	0.2	0.02	Emotional reaction of organized minority
Good		0.2	10^{-4}	Emotional reaction of organized minority
Fair			10^{-6}	Uncertainties in plasma physics
Good	~1		10^3	Few economic supplies, R&D may help
Good	~1		500	Limited supplies, high cost
Good	(~10)	9	25	Poor availability; high cost; most R&D diffuse
Fair	(~20)	10	25	Poor availability; high cost; low efficiency
Fair			10^3	400×10^9 required to establish costs
Good	~2	12	500	Large energy storage required
Good				Must withstand severe storms
Good	~4			Long transmission distances
Fair		0.4		Heat ex. area and pump hp $20\times$ steam plant
Good	~1	0.7	250	Few good sites remain

used were from the most recent and complete conceptual design studies available. The most complete studies found for MHD systems and for molten carbonate fuel cells were in ECAS reports. In view of the potentially great importance of cogeneration with district heating systems, data for light-water reactors coupled to such systems were obtained from Ref. 3 of Chap. 17. Data in that same reference were also used to obtain the estimated values for cogeneration with the other nuclear systems.

Typical representative examples for the systems using renewable energy sources were taken from references used in the preceding chapter with preference given to data from actual operating systems where such data were available. This was the case for the Geysers geothermal units, paper mills burning wood wastes, windmills, photovoltaic units in satellites, the La Rance tidal power project, and hydroelectric units. The most thorough design studies available were used for the solar-thermal energy conversion and OTEC cases.

Developmental Background

The first of the five columns on the developmental background for the systems of Table 22.2 indicates that all have been under active consideration for at least 20 years and two—wind and hydro—for over 1000 years. In looking at the column for the total power output of each system in service in the United States, it is interesting to note that there are six well-established, mature systems that have produced over 10^{12} kWh, namely, coal-steam, clean-fuel gas turbines, clean-fuel-gas-turbine–steam combined cycles, biomass (mainly wood waste–fueled steam plants in paper mills), water reactors, and hydroelectric plants. Note, too, that the power produced by water reactors in 1978 actually exceeded the total output of the entire U.S. central station complex in 1946. On the other hand, although windmills have been in use in the United States since the early days of New Amsterdam (the first two units were prefabricated in Holland and shipped over), their total output has been only the equivalent of about 10^8 kWh (mostly for pumping water on farms). It is also interesting to note from the next column (which was prepared from annual U.S. Federal Government budget data) that the R&D funding for either solar-thermal or solar-voltaic energy development up to 1979 approached that for commercial water reactors. All the numbers in this column were taken from U.S. Federal Government budget data, and hence are underestimates. For example, there has been a substantial amount of in-house work at NASA on solar and wind energy systems that was not itemized in the general budget figures, and the commercial water reactors have benefited greatly from fuel development and other work funded under the submarine reactor program. There has also been substantial funding of R&D work by industry; in fact most of the development of conventional steam systems has been privately funded. Efforts by the author to arrive at estimates of funding by private industry have yielded such nebulous data that it seemed best not to include any private funding. R&D funding by the Edison Electric Institute (EEI) and EPRI has been important, but except for conventional steam systems the amounts have generally been less that 10 percent of U.S. government funding. In attempting to analyze the budget data for gas-turbine development, a still different set of problems was encountered: virtually all the funding was for aircraft power plants, yet much of the development work was generally applicable to stationary power applications. In this case the total for the NASA and Air Force R&D budgets for turbine engine development was used. Note, too, that the data are for the actual dollars in the budget with no correction for the effects of inflation and no distinction between R&D funds and subsidies for commercial demonstrations. Thus, although the effectiveness of a dollar spent in 1950 was perhaps double that for a dollar spent in 1979, this fact has relatively little significance in the overall picture and hence was neglected for the purpose of Table 22.2. The next column, that for the status of the materials problems for each case, represents the author's interpretation of extensive discussions with materials experts deeply involved in work on these systems.

Costs and Environmental Effects

In attempting to compare the various systems from the standpoints of both the utilities and the general public, it appears that there are three major areas: overall costs, effects on health, and land use (with their relative importance probably in that order).

Costs

As discussed in Chap. 10, the principal cost elements are capital, fuel, and nonfuel operating costs. Not only do these differ in relative importance from one system to another, but also, over an extended period, differences in the rate of inflation cause shifts in their relative magnitude. However, even the dramatic change in the cost of crude oil in the 1970s caused relatively little shift in the relative importance of capital and fuel costs for utilities except for gas turbines, which had an exceptionally low capital and a high fuel cost even with cheap fuel oil. The relative capital and fuel costs of coal-fired steam, water reactor, and hydroelectric installations did not change greatly because inflation in capital and labor costs roughly paralleled that in fuel costs, and for all three systems construction was beset in one way or another by severe environmental restrictions. For simplicity therefore, only one column in Table 22.2 gives the cost of electricity (the total cost) rather than four columns (one each for capital, fuel, nonfuel operating, and total cost). To minimize the effects of inflation and simplify comparisons, the cost of electricity as stated in each case is relative to that for a typical coal-fired steam plant with wet stack gas scrubbers for SO_2 removal.

In reviewing the relative costs given in Table 22.2 it is striking to note that the high cost of clean fuel oil or gas yields a cost of electricity almost twice as great for peaking-power gas turbines as for coal-fired plants and yields 50 percent higher costs for combined-cycle units. This has led many utilities, such as TVA, to make drastic cutbacks in their use of peaking-power gas turbines; instead, when possible, they buy power from adjacent utilities having a surplus of less expensive power available. Water reactors, on the other hand, have consistently demonstrated the lowest costs of any of the systems available, generally entailing from 60 to 75 percent of the overall costs for coal-fired plants, depending on the plant location and hence the cost of coal transportation. Only one gas-cooled reactor is in operation in the United States at the time of writing, and so there is a lack of good operating data on costs. Both European experience and numerous studies, however, have indicated that gas-cooled reactors should yield electric power costs roughly competitive with water reactors while also offering the possibility of better fuel utilization. Operating data on fast breeder reactors are limited, but the data available indicate somewhat higher capital costs for the plant excluding the fuel inventory, while the fuel inventory capital charges are perhaps four times those for water reactors. In the long run, overall fuel costs for breeders should be lower, however, and the breeders should make our uranium reserves suffice for over 10,000 years instead of the less than 100 years obtainable with light-water reactors.

In the Geysers field, where geothermal energy has proved to be readily available, the local utility buys steam from the well-drilling companies at a price that yields electric power at the same costs as in the other utility plants burning fossil fuel, so that the cost of the electricity obtained is the same. Costs for steam from hot, dry rock or deep, geopressured aquifers have yet to be determined, but seem likely to be substantially higher, and surely highly site-dependent, so that these sources will be attractive for only a limited number of areas. Similarly, biomass has traditionally supplied a significant amount of energy where wood and other solid wastes have been available at a low cost, but the supplies of low-cost wood wastes are relatively small and highly site-dependent. Hydroelectric power has been and will continue to be an important contributor to the U.S. energy economy, but, as in the past, new hydro installations will be made only where sites are sufficiently favorable to give costs that are competitive with other types of plant. It is interesting to note that when TVA decided to build the Tellico Dam (of snail-darter fame) it was initially proposed that hydroelectric generators be installed. At that time (1965), however, the capital and operating costs for coal-fired plants were lower than just the incremental costs of installing turbine-generators in the dam that was being built for flood control and navigation; this was for a dam giving a substantial head (24 m) with existing transmission lines nearby and a need for peaking-power capacity in the system.

Health Effects

In attempting to compare the overall effects of the various systems on the health and life expectancy of mining, fabrication, construction, and operating personnel as well as of the general public, it seemed best to utilize the results of the study summarized in Ref. 42 of Chap 9. As discussed in Chap. 9, this study applied standard generic data on lost-time accidents, reductions in life span, and deaths in various types of mining, processing, fabrication, construction, operation, and maintenance work to obtain a measure of the occupational health effects for each of eleven energy sources including coal, oil, and gas-fired steam plants, together with windmills, solar thermal, solar photovoltaic, OTEC, and hydroelectric power units. The study included the losses associated with the backup power for such ephemeral energy sources as solar and wind energy plants. Using similar statistical data, the effects on public health of both stack emissions and catastrophic accidents such as dam failures and nuclear accidents were also estimated. The man-days lost to accidents were related to deaths by considering a death equivalent to 6000 man-days, i.e., an average of 26 man-years. As discussed in Chap. 9, the results give a very different picture than conveyed by enthusiasts for "natural" sources of energy, with coal yielding adverse health effects only about three times more serious than solar-electric systems, while the health effects of nuclear plants appear to be only one-fiftieth those of solar-electric systems and about one-third of the value for hydroelectric power. As Dr. F. C. Fobbins, chairman of a panel of the NSF, stated early in 1979 when releasing the panel's study on the effects of low levels of carcinogens, "We cannot live in a risk-free environment. It does not exist." Even common foods such as corn, milk, and peanuts contain small amounts of aflatoxins which in large amounts are carcinogenic. Thus in considering health effects, a meaningful comparison must be comprehensive.

Land Use

One factor in considering environmental effects is land use. Strip mines, whether for coal, copper, uranium, or iron, are obvious offenders, as are the piles of tailings from deep mines and the piles of coal ash from power plants. Electric transmission lines, oil and gas well fields, pipelines, clear-cutting of forests, fields of heliostats for solar power towers, and fields of microwave antennas for satellite solar power plants represent other environmentally objectionable uses of land. The degree to which these effects appear objectionable varies so much from one person to another that no attempt was made to apply factors to allow for these subjective differences in compiling the numbers given in Table 22.2. The amount of land area affected per year per unit of electric energy output was simply calculated for each system by using data given in the previous chapter for the "renewable" energy systems, together with mining statistics for the fossil- and nuclear-fueled plants. For the latter, it was assumed that strip-mined land would be restored and reseeded within 2 years.

In reviewing the numbers given in the land use column, it is perhaps surprising to see that nuclear plants entail a land requirement only a few percent of that for coal-fired plants, and it may be even more surprising to find that the "renewable" energy sources require from 25 to 1000 times as much land area as coal-fired plants. Note that the factor of 250 for hydroelectric units was obtained using the lake surface area and electric output of the TVA dams that produce power. On the one hand, the lakes are lovely when full; on the other hand, the mud bottoms are unsightly when the lake levels are dropped during the fall and winter months, and the lakes have certainly flooded a great deal of prime agricultural bottomlands, to the despair of local farmers.

KEY FEASIBILITY PROBLEMS

In considering the various systems of Table 22.2, one may divide them into a number of different categories. First are those systems that have proved to be, and can be expected to continue to be, economical and reliable sources of power, at least for some applications. Next are the systems that are under development and can be made to work to some degree, but the extent of their future use will depend on the success of their development programs in reducing costs. The third and last category includes those systems requiring one or more scientific breakthroughs to make them feasible.

Proven Systems

The mainstay of the U.S. electric utility system is the coal-fired steam plant. The technology is mature, so that after a century of development the system thermal efficiency has been increased by a factor of about 8, and little further improvement can be expected. Partly through the increased thermal efficiency, but mainly through the economics of increasing scale, the cost of electricity in cents per kilowatthour was reduced by a factor of 3 between 1930 and 1970, and, if allowance is made for inflation, the reduction in cost was by a factor of 5. The limitations imposed by environmentalists, coupled with the skyrocketing cost of petroleum, have reversed that trend in the 1970s, roughly doubling the nominal cost in cents per kilowatthour (without allowing for inflation, which would reduce the effect). The principal prospects for future improvements lie in reductions in the cost and performance penalties associated with emissions from

sulfur in the fuel. A supercharged fluidized-bed combustion system or a type of dry stack gas scrubber show the greatest promise (Chaps. 11 and 13).

Open-cycle gas turbines operating on clean fuel—gas or distillates—have a limited market, because the cost of these clean fuels has risen to the extent that other systems are clearly more attractive. For applications involving cogeneration, particularly for those in which the ratio of heat required to the electric output is less than 4 and the loads are variable, the coal-fired fluidized-bed, closed-cycle gas turbine appears promising. For these conditions the thermal efficiency is competitive with a steam plant, and, though not shown in Table 22.2, European experience indicates that the costs compare favorably with those for coal-fired steam plants.

The light-water reactor plants have proved to be the most economical and least polluting of the many types of electric power system of Table 22.2. They have suffered severely from the highly emotional attacks from those who fear radioactivity but do not recognize that there are risks associated with any system: There were lethal kicks from horses in the horse-drawn economy of the past, and there are, in our current world, hazards ranging from explosions caused by leaks of natural gas and failures of dams to the possibilities of wars that may be fought over the earth's dwindling supplies of petroleum.

The performance of gas-cooled reactors is roughly competitive with that of light-water reactors, but they have not been as well developed and seem unlikely to take over much of the market. In the future—probably in the first half of the twenty-first century—unless breeders are developed, light-water reactors will be confronted by steeply rising fuel costs because the richer reserves of uranium will have been exhausted. Thus, in spite of their higher costs, breeder reactors provide the only sure route currently available to long-term, economical electric power.

Hydroelectric power has been and will continue to be an important element in the electric utility field. Most of the economical sites in the United States have already been developed, however, and so its relative contribution to the nation's energy supply will probably continue the decline of recent decades.

Advanced Systems for Which Cost Problems Are Dominant

The technical feasibility of the majority of the systems in Table 22.2 has already been demonstrated in the sense that at least a small unit has been built and has produced a net power output. The key questions have to do with costs, the most vital factor in these costs in many cases is the life of key components and their maintenance costs. The direct coal-fired gas turbine, for example, yielded a useful output in the latter 1940s, but turbine erosion and deposits have so limited the life of every unit run since then that there seems little likelihood of obtaining an economically attractive system. Every type of particulate-removal system for hot gases that has been tested has been insufficiently effective except for turbine inlet temperatures below $\sim 600\,°C$. This is too low to produce economical electric power from the gas turbine, though sufficient to provide the power required to drive a compressor having a pressure ratio of 3:1 and hence satisfactory for operating a supercharged, fluidized-bed, boiler furnace.

Water-cooled turbines designed to operate with inlet gas temperatures of $1370\,°C$ ($2500\,°F$) or more require an extremely complex and expensive blade construction. The heat flux to the blade is exceedingly high—over 315 W/cm^2 [10^6 Btu/ (h · ft^2)]—hence the blades are subject to such severe thermal stresses that their life may be too limited to give an economically attractive system. Further, of the order of 20,000 cooling passages only about 2 mm in diameter will be required in each turbine. To keep the heat losses associated with blade cooling to an acceptable level, each passage must be orificed so that it will operate with two-phase boiling flow with a blade exit quality of ~ 70 percent, and anything causing a small loss in flow might throw it into a burnout heat flux condition. If this occurred, local melting would take place in ~ 2 s. Thus the development program must include a long endurance test of a full-scale engine to demonstrate feasibility. This should be done with the hot gas entering the turbine containing the level of particulates to be expected from a coal gasifier or a direct coal-fired combustion chamber because, with the high temperature differences prevailing between the hot gas and the cold blades, thermophoresis may lead to the rapid buildup of adherent deposits. (See the last section in Chap. 12.) Ceramic turbine blades are superficially intriguing, but their brittle nature makes them subject to thermal shock; e.g., a flameout at full power gives thermal stresses 20 times as severe as a water quench (Chap. 12). Thus it is not surprising that 40 years of effort have yielded less than 100 h of total running time with ceramic turbines.

Supercritical CO_2 and dissociating gas cycles show little or no performance advantage over conventional closed-cycle gas turbines, and clearly entail higher capital costs. Hence, although undoubtedly capable of producing power, they do not appear to be attractive prospects for development.

The alkali-metal-vapor topping cycle shows promise of giving a markedly higher thermal efficiency than any other system that has been proposed, but the maintenance difficulties experienced with sodium-cooled reactor systems have made many people doubt that such systems can be shown to be practical. Certainly an unusually high level of competence is required throughout every phase of the design, fabrication, construc-

tion, operation, and maintenance of such systems. Perhaps this is a strong argument for the development of the alkali-metal-vapor cycle, since it would provide a highly useful nonnuclear system that would serve for training personnel for LMFBR work.

At the time of writing, after over 20 years of intensive development, the phosphoric acid fuel cell has yet to show an adequate life to give an economically attractive system. As is the case with the gas turbine, it must be supplied with clean fuel, and its overall thermal efficiency is only a few points higher than that of the gas turbine, not enough to offset what appear to be much higher capital charges. As a consequence, it appears that its development potential is not sufficient to justify the remaining hundreds of millions of dollars in development costs estimated to be required by its advocates. The molten carbonate fuel cell can be fueled with coal and its thermal efficiency looks promising, but the corrosion and related materials compatibility problems are so formidable that years of further basic materials work will be required before its economic feasibility can be determined (Chap. 18).

The further development of geothermal power depends on developing low-cost deep wells for geopressured hot brines or evolving techniques for drilling into and fracturing hot, dry rock at depths of 3 km in such a way that water can be pumped through the fractured region to provide a sustained yield of large amounts of heat over a period of years. The principal problem for the latter case is to avoid channeling of the water flow—a problem analogous to the channeling difficulties that for 35 years have frustrated efforts to carry out efficient, in-situ, gasification of coal.

The use of biomass as a fuel will probably expand from the very low level prevailing at the time of writing, but supplies are inherently too limited and collection and transportation costs too high for biomass to contribute more than a few percent of the nation's electric power supply.

The various solar-thermal and photovoltaic systems for producing electric power have so far proved to be many times more expensive than conventional power systems, in part because for terrestrial plants they require additional expensive back-up or storage systems. As a consequence, the overall cost of solar-electric power for general applications appears to be at least 10 times as high as for conventional systems—a level unacceptable to consumers. A satellite solar power system would avoid the energy storage problem, but even the advocates make cost projections that are favorable only if a scientific breakthrough is achieved in reducing the weight and cost of solar cells while simultaneously increasing their efficiency. Such a breakthrough is conceivable but is by no means certain of achievement. (Note that a special panel convened by the American Physical Society at the request of the White House reported early in 1979 after an extensive study that economical electric power from photovoltaic systems will require "major scientific and technological advances" and that it is not certain that these advances can be made.) Further, to obtain favorable costs, the advocates assume an enormous complex of large systems with dozens of satellites each being 5×25 km.

The systems designed to generate electric power from the wind, waves, and tides are all handicapped by being low-head systems so that the capital investment is inherently high to begin with, and is further increased manyfold both by the requirement for extremely rugged construction to resist damage in severe storms and by the need for either back-up power or energy storage.

To obtain electric power from the $\sim 20\,°C$ temperature difference available from certain areas of the ocean entails an inherently low thermal efficiency (~ 2 percent) and a very large pumping power so that under any circumstances the capital costs will be high. If there are somewhat higher losses than assumed in the conceptual designs (e.g., biofouling of the heat exchangers), the net power output may actually be negative. Only by going to a large-scale system is there a possibility of obtaining a net power output that is a large fraction of the gross electric output, and this makes a definitive experiment very expensive.

Systems Whose Feasibility Requires Both Scientific and Technological Breakthroughs

As of 1979, two of the systems of Table 22.2—open-cycle MHD and fusion reactors—have not yet been shown to be scientifically feasible by an experiment that has shown a net electric

energy output, and a demonstration of that sort appears to be a decade or more in the future. There are additional extremely formidable practical engineering problems that must be solved—such as coping with high heat fluxes and severe thermal stresses, materials compatibility, and high capital costs—so that in both cases the demonstration of scientific feasibility is a necessary but not sufficient condition for their success in an economic sense. Note that there is a fundamental difference in their potential value, because at best the MHD system may yield an increase of perhaps 10 points in cycle efficiency, with an attendant reduction in fossil fuel consumption so that only a limited developmental cost could be justified. On the other hand, fusion gives promise of yielding low-cost electric power by burning deuterium, the cost of which is low and the supply so abundant that it would provide all human energy requirements for many times the life of the sun. Thus the stakes in fusion are so high that a high-cost developmental program is readily justified.

SUMMARY

The author is well aware that some people may not want to accept the conclusions explicit in the paragraphs above and implicit in Table 22.2 and may be inclined to attribute them to bias. It is hoped that the information presented in this text will provide sufficient background and insights so that the readers will not reject these conclusions out-of-hand but instead will formulate incisive questions on any of the above statements that they doubt, investigate these questions themselves, and then, as the author has done, draw their own conclusions on the basis of a reasonably complete complement of the facts.

PROBLEMS

1. When it becomes evident that the natural gas supply for a municipally-owned 200 MWe gas-fired steam boiler plant will no longer be available in 5 years, it is decided that the principal option available is the use of a 20 percent ash, 3 percent sulfur, and 8 percent moisture coal from a nearby mine. Three systems are considered: a new furnace designed for pulverized-coal firing with wet stack gas scrubbers, a new furnace employing a fluidized bed of limestone, and a gasifier utilizing an atmospheric fluidized bed of limestone to produce low-Btu gas of low sulfur content that would be fed directly without cooling to new burners installed in the existing furnace.

(a) Consider that all the hydrogen in the coal in the gasifier reacts to form H_2O, all the carbon forms CO, and all the sulfur forms CaS. If no heat-transfer surface is installed in the gasifier and air is supplied at $20\,°C$, estimate the temperature of the hot gas leaving the gasifier.

(b) List the principal new components required for each of the three systems under consideration and make a rough estimate of the capital cost of each in current dollars per kilowatt.

2. For the above case, prepare a response to a city council de-

mand for an estimate of the capital and operating costs if the plant were replaced with a system of windmills and lead-acid storage batteries. (See Probs. 7 and 8 of Chap. 21.)

3. Following the ground rules used in preparing Table 22.2 and consulting references in the library, make an independent estimate of 10 different items in the two right-hand columns of Table 22.2. Compare your results with those in the table and discuss possible reasons for the differences.

REFERENCES

1. *Electric Utilities Industry Research and Development Goals Through the Year 2000,* Report of the Research and Development Goals Task Force to the Electric Research Council, ERC Publication No. 1-71, June 1971.

2. Robson F. L., et al.: *Technological and Economic Feasibility of Advance Power Cycles and Methods of Producing Nonpolluting Fuels for Utility Power Stations,* UARL Report J-970855-13, December 1970.

3. *An Assessment of New Options in Energy Research and Development,* Energy Advisory Panel, Executive Office of the President, November 1973.

4. *Summary Report of Engineering Foundation Conference of Energy Technologies for the Future, Saxtons River, Vt.,* July 24–27, 1972.

5. *Energy Research Priorities Conference, Berwick Academy, South Berwick, Maine,* July 30–Aug. 4, 1972.

6. "A Program of Research, Development and Demonstration for Enhancing Coal Utilization to Meet National Energy Needs," *Results of the Carnegie-Mellon University Workshop on Advanced Coal Technology,* October 1973.

7. *The Nation's Energy Future,* WASH-1281, submitted by Dr. Dixy Lee Ray to President Richard M. Nixon, Dec. 1, 1973.

8. *Report on Technical Aspects of Conservation of Energy Related to Electrical Production and Utilization,* prepared by Task Force on Technical Aspects of the Technical Advisory Committee on Conservation of Energy of the Federal Power Commission, Nov. 1, 1973.

9. *Research Needs Report: Energy Conversion Research,* prepared by Task Force on Energy Conversion Research, The American Society of Mechanical Engineers, 1976.

10. Corman, J. C.: *Energy Conversion Alternatives Study,* ECAS, General Electric Phase I Final Report, NASA-CR-134948, vol. I–III, February 1976.

11. Isenberg, A. O., R. J. Ruka, and C. J. Warde: *Energy Conversion Alternatives Study,* Westinghouse Phase I Final Report, NASA-CR-134941, vol. I–XII, Feb. 12, 1976.

12. *Comparative Evaluation of Phase I Results from the Energy Conversion Alternatives Study,* NASA-TM X-71855, February 1976.

13. Corman, J. C., and G. R. Fox: *Energy Conversion Alternatives Study,* General Electric Phase II Final Report, NASA-CR-134949, vol. I–III, December 1976.

14. Beecher, D. T., et al.: *Energy Conversion Alternatives Study,* Westinghouse Phase II Final Report, NASA-CR-134942, vol. I–III, Nov. 1, 1976.

15. King, J. M.: *Energy Conversion Alternatives Study,* United Technologies Phase II Final Report, NASA-CR-134955, Oct. 19, 1976.

16. Fox, G. R., and J. C. Corman: "A Study of Advanced Energy Conversion Techniques for Utility Applications Using Coal or Coal-Derived Fuels," A Panel Presentation at the ASME Winter Annual Meeting, Houston, Texas, Dec. 3, 1975.

17. Bertman, L., et al.: *Benefit-Cost Evaluation of the ERDA Fossil Energy Combustion and Advanced Power Development Program,* The MITRE Corporation, Report No. MTR-7206, July 1976.

18. Fraas, A. P.: "Assessment of Advanced Power Cycles," *Transactions of ANS Annual Meeting,* June 3–7, 1979.

APPENDIXES

NOMENCLATURE, UNITS, CONVERSION FACTORS, AND CONSTANTS

TABLE A1.1   Nomenclature

A	Area, m²	T	Temperature, K
b	Thickness, m	δt	Temperature rise (or drop), °C
b'	Thickness, mm	Δt	Temperature difference, °C
C	Coefficient	Δt_m	Log mean temperature difference, °C
c_p	Specific heat at constant pressure, J/(kg·K)	U	Overall heat conductance between two fluids, W/(m²·°C)
c_v	Specific heat at constant volume, J/(kg·K)	u	Fluid velocity, m/s
D	Diameter, m	V	Specific volume, m³/kg
D_{eq}	Equivalent diameter of noncircular cross sections, m	W	Weight, kg
d	Diameter, mm	W'	Weight flow rate, kg/s
d_{eq}	Equivalent diameter of noncircular cross sections, mm	W'	Width (or height), m
E	Modulus of elasticity, Pa	w	Width (or height), mm
F	Form factor (for radiant heat transmission)	w'	Vapor quality (ratio of vapor weight to total weight of vapor plus moisture)
f_d	Flow friction factor (based on equivalent passage diameter)	α	Coefficient of thermal expansion, m/(m·°C)
f_r	Flow friction factor (based on hydraulic radius)	δ	Deflection, mm
G	Mass flow rate, kg/(m²·s)	ϵ	Emissivity
GTD	Greatest temperature difference, °C	η	Efficiency or heating (or cooling) effectiveness
g	Acceleration of gravity, m/s²	θ	Temperature correction factor
H	Enthalpy	ω	Angular velocity, rad/s
ΔH_v	Enthalpy of vaporization, J/kg	μ	Viscosity, Pa·s
ΔH_s	Enthalpy of superheat, J/kg	ϱ	Density, kg/m³
h	Heat-transfer coefficient, W/(m²·K)	σ	Ratio of gas density to that of air at standard conditions
ITD	Inlet temperature difference, °C		
k	Thermal conductivity, W/(m·K)		
k	Ratio of specific heats, c_p/c_v		
L	Length, m		

l	Length, mm		**Subscripts**
LMTD	Log mean temperature difference, °C	a	Air
LTD	Least temperature difference, °C	b	Bulk fluid conditions
M	Molecular weight	c	Compressor
Nu	Nusselt number hD/k	e	Effectiveness
N	rpm	eq	Equivalent
n	Number in a series	f	Film (refers to arithmetical mean temperature between the wall and the bulk-free stream)
P	Pressure, Pa		
Pr	Prandtl number		
ΔP	Pressure drop, Pa	w	Wall
Q	Heat flow rate, W	g	Gas
q	Dynamic head, Pa	i	Internal
R	Radius, m	id	Ideal
r	Radius, mm	l	Liquid
R_h	Hydraulic radius, m	m	Mean
r_h	Hydraulic radius, mm	o	Outer
Re	Reynolds number, $\varrho VD/\mu$	s	Saturated
R_g	Gas constant, J/(kg·°C)	t	Turbine
R_u	Universal gas constant, J/(kg·mol·°C)	th	Thermal
S	Stress, Pa	sh	Superheated
s	Spacing, mm	0	Initial, or base condition
		1	Inlet
		2	Outlet

TABLE A1.2 The International System of Units (SI)

Quantity	Unit	SI symbol	Formula
Base units			
length	meter	m	...
mass	kilogram	kg	...
time	second	s	...
electric current	ampere	A	...
thermodynamic temperature	kelvin	K	...
amount of substance	mole	mol	...
luminous intensity	candela	cd	...
Supplementary units			
plane angle	radian	rad	...
solid angle	steradian	sr	...
Derived units			
acceleration	meter per second squared	...	m/s^2
activity (of a radioactive source)	disintegration per second	...	(disintegrations)/s
angular acceleration	radian per second squared	...	rad/s^2
angular velocity	radian per second	...	rad/s
area	square meter	...	m^2
density	kilogram per cubic meter	...	kg/m^3
electric capacitance	farad	F	$(A \cdot s)/V$
electrical conductance	siemens	S	A/V
electric field strength	volt per meter	...	V/m
electric inductance	henry	H	$(V \cdot s)/A$
electric potential difference	volt	V	W/A
electric resistance	ohm	Ω	V/A
electromotive force	volt	V	W/A
energy	joule	J	$N \cdot m$
entropy	joule per kelvin	...	J/K
force	newton	N	$(kg \cdot m)/s^2$
frequency	hertz	Hz	(cycle)/s
illuminance	lux	lx	lm/m^2
luminance	candela per square meter	...	cd/m^2
luminous flux	lumen	lm	$cd \cdot sr$
magnetic field strength	ampere per meter	...	A/m

TABLE A1.2   The International System of Units (SI) (*continued*)

Derived units		SI symbol	Formula
Quantity	Unit		
magnetic flux	weber	Wb	$V \cdot s$
magnetic flux density	tesla	T	Wb/m^2
magnetomotive force	ampere	A	$\ldots$
power	watt	W	J/s
pressure	pascal	Pa	N/m^2
quantity of electricity	coulomb	C	$A \cdot s$
quantity of heat	joule	J	$N \cdot m$
radiant intensity	watt per steradian	$\ldots$	W/sr
specific heat	joule per kilogram-kelvin	$\ldots$	$J/(kg \cdot K)$
stress	pascal	Pa	N/m^2
thermal conductivity	watt per meter-kelvin	$\ldots$	$W/(m \cdot K)$
velocity	meter per second	$\ldots$	m/s
viscosity, dynamic	pascal-second	$\ldots$	$Pa \cdot s$
viscosity, kinematic	square meter per second	$\ldots$	m^2/s
voltage	volt	V	W/A
volume	cubic meter	$\ldots$	m^3
wavenumber	reciprocal meter	$\ldots$	$(wave)/m$
work	joule	J	$N \cdot m$

SI prefixes		
Mulitiplication factors	Prefix	SI symbol
$1\ 000\ 000\ 000\ 000 = 10^{12}$	tera	T
$1\ 000\ 000\ 000 = 10^{9}$	giga	G
$1\ 000\ 000 = 10^{6}$	mega	M
$1\ 000 = 10^{3}$	kilo	k
$100 = 10^{2}$	hecto*	h
$10 = 10^{1}$	deka*	da
$0.1 = 10^{-1}$	deci*	d
$0.01 = 10^{-2}$	centi*	c
$0.001 = 10^{-3}$	milli	m
$0.000\ 001 = 10^{-6}$	micro	m
$0.000\ 000\ 001 = 10^{-9}$	nano	n
$0.000\ 000\ 000\ 001 = 10^{-12}$	pico	p
$0.000\ 000\ 000\ 000\ 001 = 10^{-15}$	femto	f
$0.000\ 000\ 000\ 000\ 000\ 001 = 10^{-18}$	atto	a

*To be avoided.

TABLE A1.3 Factors for Converting Common Units to SI Units

Acceleration	
	m/s²
1 ft/s²	0.3048

Area	
	mm²
1 in²	645.16
1 circular mil	5.0671 × 15⁻⁴
	m²
1 ft²	0.092903
1 yd²	0.83613
1 acre	4046.9
1 ha	10,000

Calorific value-volume basis	
	kJ/m³
1 kJ/m³	1
1 kcal/m³	4.1868
1 Btu/ft³	37.260
1 therm/ft³	3,726,000

Density	
	mg/L
1 mg/L	1
1 grain/ft³	2288.4
	kg/L
1 kg/m³	0.001
1 lb/ft³	0.016018
1 lb/US gal	0.11983

Energy, heat, work	
	J
1 Mev	1.6 × 10⁻¹³
1 erg	10⁻⁷
	kJ
1 ft · pdl	0.000042139
1 ft · lbf	0.0013558
1 kgf · m	0.0098067
1 Btu	1.0551
1 kcal	4.1868
	MJ
1 hp-h	2.6845
1 kWh	3.6000
1 therm	105.51
1 quad (10¹⁵ Btu)	1.055 1 × 10¹²

Enthalpy, calorific value-mass basis	
	kJ/kg
1 J/kg	0.001
1 Btu/lb	2.3260
1 kcal/kg	4.1868

Entropy	
	kJ/(kg · K)
1 Btu/(lb · °F)	4.187

Force	
	N
1 pdl	0.13825
1 dyne	10⁻⁵
	kN
1 lbf	0.0044482
1 kgf (or kp)	0.0098067
1 tonf	9.9640

Heat flux	
	kW/m²
1 W/m²	0.001
1 kcal/(h · m²)	0.001163
1 Btu/(h · ft²)	0.0031546

Heat rate	
	kJ/kWh
1 Btu/kWh	1.0551
1 Btu/(hp · h) (British)	1.4149
1 Btu/(hp · h) (metric)	1.4346
1 kcal/kWh	4.1868
1 kcal/(hp · h) (British)	5.6146
1 kcal/(hp · h) (metric)	5.6926

Heat-transfer coefficient	
	W/(m² · °C)
1 W/(m² · °C) [or W/(m² · K)]	1
1 kcal/(h · m² · °C)	1.163
1 Btu/(h · ft² · °F	5.6784

TABLE A1.3   Factors for Converting Common Units to SI Units (*continued*)

Length	
	mm
1 cm	10
1 in	25.4
1 ft	304.8
1 Å	10^{-7}
	m
1 yd	0.9144
	km
1 mi	1.6093

Mass (weight)	
	mg
1 mg	1
1 grain	64.8
	kg
1 g	0.001
1 lb	0.45359
1 kg	1
	tonne (metric)
1 ton (short)	0.90714
1 ton (long)	1.016

Mass flow	
	kg/h
1 lb/h	0.45359
1 kg/h	1
1 ton/h (short)	907.2
1 t/h	1000
1 ton/h (long)	1106
1 kg/s	3600
1 lb/(s · ft²)	4.8917 kg/(s · m²)

Power, heat flow	
	kW
1 Btu/h	0.00029308
1 W (or J/s)	0.001
1 kcal/h	0.001163
1 (kgf · m)/s [or (kp · m)/s]	0.0098067
1 hp (metric)	0.73548
1 hp (British)	0.7457
1 Btu/s	1.0551
1 ton refrigeration	3.5169
1 kcal/s	4.1868
1 therm/h	29.308

Pressure	
	Pa
1 atm	101,325
1 bar	100,000
1 kgf/m²	9.807
1 lbf/ft²	47.88
1 mbar	100
1 torr (mmHg)	133.32
1 lbf/in² (psi)	6894.8
1 in H_2O	249.09
1 ft H_2O	2989.1
1 in Hg	3386.6
1 m H_2O	9806.7

Specific heat, entropy	
	kJ/(kg · °C)
1 J/(kg · °C)	0.001
1 kJ/(kg · °C)	1
1 kcal/(kg · °C)	4.1868
1 Btu/(lb · °F)	4.1868

Specific fuel consumption	
	kg/kWh
1 g/kWh	0.001
1 g/(hp · h) (British)	0.001341
1 g/(hp · h) (metric)	0.0013597
1 lb/(hp · h) (British)	0.60827
1 lb/(hp · h) (metric)	0.61673

Specific volume	
	m³/kg
1 L/kg	0.001
1 ft³/lb	0.062428

Thermal conductivity	
	W/(m · °C)
1 Btu/(h · ft²) (°F/in)	0.14423
1 kcal/(h · m · °C)	1.1630
1 Btu/(h · ft · °F)	1.7308

Velocity	
	m/s
1 ft/min	0.00508
1 ft/s	0.3048
1 mi/h	0.44704

TABLE A1.3 Factors for Converting Common Units to SI Units (*continued*)

Viscosity	
1 cP	0.001 Pa·s
1 cS	0.000001 m²/s
1 ft²/s	0.0929 m²/s
1 (lbf·s)/ft²	47.88 Pa·s
1 lb/(h·ft)	4.134×10^{-4} Pa·s
1 lb/(s·ft)	1488.2 Pa·s

Volume flow	
	L/h
1 L/h	1
1 US gal/h	3.7853
1 UK gal/h	4.546

Volume	
	L
1 in³	0.016387
1 US gal	3.7853
1 UK gal	4.546
1 US fluid oz	0.02957

	m³/h
1 ft³/h	0.028317
1 L/min	0.06
1 US gal/min	0.22712
1 UK gal/min	0.27276
1 m³/h	1
1 ft³/min (or cumin)	1.699
1 L/s	3.6
1 ft³/s (or cusec)	101.94

	m³
1 ft³	0.028317
1 US bbl (42 gal)	0.15898
1 yd³	0.76455
1 acre·ft	1233.5

TABLE A1.4 Useful Conversion Factors

Area

1 ft^2 = 929 cm^2
1 m^2 = 10.764 ft^2
1 mi^2 = 2.5898 km^2

Density

1 lb/in^3 = 27.680 g/cm^3
1 kg/m^3 = 0.06243 lb/ft^3

Energy, heat, power

1 Btu = 252 cal
1 Btu = 778.161 ft · lb
1 hp = 550 (ft · lb)/s
1 hp = 745.7 W
1 kWh = 3413 Btu
1 hph = 2545 Btu

Heat flux

1 W/m^2 = 0.31709 Btu/(h · ft^2)
1 Btu/(h · ft^2) = 0.3154 × 10^{-3} W/cm^2

Heat-transfer coefficient

1 Btu/(h · ft^2 · °F) = 5.677 × 10^{-4} W/(cm^2 · °C)
1 W/(m^2 · °C) = 0.1761 Btu/(h · ft^2 · °F)
1 Btu/(h · ft^2 · °F) = 4.882 kcal/(h · m^2 · °C)

Mass

1 oz = 28.35 g
1 kg = 2.2046 lb
1 g = 15.432 gr
1 slug = 32.1739 lb
1 ton (short) = 2000 lb
1 long ton = 2240 lb
1 t = 1.102 tons (short)
1 ton crude oil = 6.65 bbl
1 t crude oil = 7.33 bbl

Mass flux

1 lb/(ft^2 · h) = 1.3563 × 10^{-3} kg/(m^2 · s)
1 kg/(m^2 · s) = 737.3 lb/(ft^2 · h)

Pressure

1 atm = 14.696 psi
1 atm = 1.0132 bars
1 bar = 14.5045 psi
1 in Hg = 0.491154 psi

Specific heat

1 Btu/(lb · °F) = 1 kcal/(kg · °C) = 1 cal/(g · °C)
1 J/(g · °C) = 0.23885 Btu/(lb · °F)

Temperature

$T(°F) = 1.8 (K - 273) + 32$

$T(K) = \dfrac{1}{1.8} (°F - 32) + 273$

$T(°C) = \dfrac{1}{1.8} (°R - 492)$

$\Delta T(°C) = \dfrac{1}{1.8} (°F)$

Viscosity

1 P = 1 g/(cm · s)
1 P = 10^{-2} cP
1 cP = 241.9 lb/(ft · h)

Volume

1 in^3 = 16.387 cm^3
1 L = 61.024 in^3
1 ft^3 = 62.39 lb H$_2$O at 59°F
1 ft^3 = 7.4805 gal (U.S.)
1 gal (U.S.) = 8.338 lb H$_2$O at 59°F
1 gal (U.S.) = 0.8327 gal (imperial)
1 gal (U.S.) = 231 in^3
1 bbl = 42 gal
1 bbl = 159 L

TABLE A1.5 Constants

Acceleration of gravity = 32.1739 ft/s^2
= 980.67 cm/s^2

Avogadro's number = 6.022 × 10^{23}/mol

Gas constant (universal) = 1544 (ft · lb)/(lb · mol · °R)
= 0.730 (ft^3 · atm)/(lb · mol · °R)
= 0.08205 (m^3 · atm)/(kg · mol · K)
= 8314 J/(kg · mol · K)
= 1.987 cal/(g · mol · K)

Molar volume at STP = 22.415 L/(g · mol)

Stefan-Boltzmann constant = 0.1714 × 10^{-8} Btu/(h · ft^2 · °R^4)
= 0.56697 × 10^{-8} W/(m^2 · K^4)

Water density at STP = 0.9991 g/cm^3
= 62.39 lb/ft^3
= 8.338 lb/gal

TABLE A1.6 Temperature Conversion Table

Find the known temperature to be converted in the center column.
Then read the Celsius conversion to left or Fahrenheit to right.

Example: −3.9 25 77.0
therefore   25°C = 77.0°F
            25°F = −3.9°C

°C		°F	°C		°F	°C		°F	°C		°F	°C		°F	°C		°F
−73	−100	−148	62.8	145	293.0	249	480	896	493	920	1688	738	1360	2480	982	1800	3272
−68	−90	−130	65.6	150	302.0	254	490	914	499	930	1706	743	1370	2498	988	1810	3290
−62	−80	−112	68.3	155	311.0	260	500	932	504	940	1724	749	1380	2516	993	1820	3308
−57	−70	−94	71.1	160	320.0	266	510	950	510	950	1742	754	1390	2534	999	1830	3326
−51	−60	−76	73.9	165	329.0	271	520	968	516	960	1760	760	1400	2552	1004	1840	3344
−45.6	−50	−58.0	76.7	170	338.0	277	530	986	521	970	1778	766	1410	2570	1010	1850	3362
−42.8	−45	−49.0	79.4	175	347.0	282	540	1004	527	980	1796	771	1420	2588	1016	1860	3380
−40.0	−40	−40.0	82.2	180	356.0	288	550	1022	532	990	1814	777	1430	2606	1021	1870	3398
−37.2	−35	−31.0	85.0	185	365.0	293	560	1040	538	1000	1832	782	1440	2624	1027	1880	3416
−34.4	−30	−22.0	87.8	190	374.0	299	570	1058	543	1010	1850	788	1450	2642	1032	1890	3434
−31.7	−25	−13.0	90.6	195	383.0	304	580	1076	549	1020	1868	793	1460	2660	1038	1900	3452
−28.9	−20	−4.0	93.3	200	392.0	310	590	1094	554	1030	1886	799	1470	2678	1043	1910	3470
−26.1	−15	5.0	96.1	205	401.0	316	600	1112	560	1040	1904	804	1480	2696	1049	1920	3488
−23.3	−10	14.0	98.9	210	410.0	321	610	1130	566	1050	1922	810	1490	2714	1054	1930	3506
−20.6	−5	23.0	101.7	215	419.0	327	620	1148	571	1060	1940	816	1500	2732	1060	1940	3524
−17.8	0	32.0	104.0	220	428.0	332	630	1166	577	1070	1958	821	1510	2750	1066	1950	3542
−15.0	5	41.0	107.2	225	437.0	338	640	1184	582	1080	1976	827	1520	2768	1071	1960	3560
−12.2	10	50.0	110.0	230	446.0	343	650	1202	588	1090	1994	832	1530	2786	1077	1970	3578
−9.4	15	59.0	112.8	235	455.0	349	660	1220	593	1100	2012	838	1540	2804	1082	1980	3596
−6.7	20	68.0	115.6	240	464.0	354	670	1238	599	1110	2030	843	1550	2822	1088	1990	3614
−3.9	25	77.0	118.3	245	473.0	360	680	1256	604	1120	2048	849	1560	2840	1093	2000	3632
−1.1	30	86.0	121	250	482	366	690	1274	610	1130	2066	854	1570	2858	1099	2010	3650
1.7	35	95.0	127	260	500	371	700	1292	616	1140	2084	860	1580	2876	1104	2020	3668
4.4	40	104.0	132	270	518	377	710	1310	621	1150	2102	866	1590	2894	1110	2030	3686
7.2	45	113.0	138	280	536	382	720	1328	627	1160	2120	871	1600	2912	1116	2040	3704
10.0	50	122.0	143	290	554	388	730	1346	632	1170	2138	877	1610	2930	1121	2050	3722
12.8	55	131.0	149	300	572	393	740	1364	638	1180	2156	882	1620	2948	1127	2060	3740
15.6	60	140.0	154	310	590	399	750	1382	643	1190	2174	888	1630	2966	1132	2070	3758
18.3	65	149.0	160	320	608	404	760	1400	649	1200	2192	893	1640	2984	1138	2080	3776
21.1	70	158.0	166	330	626	410	770	1418	654	1210	2210	899	1650	3002	1143	2090	3794
23.9	75	167.0	171	340	644	416	780	1436	660	1220	2228	904	1660	3020	1149	2100	3812
26.7	80	176.0	177	350	662	421	790	1454	666	1230	2246	910	1670	3038	1154	2110	3830
29.4	85	185.0	182	360	680	427	800	1472	671	1240	2264	916	1680	3056	1160	2120	3848
32.2	90	194.0	188	370	698	432	810	1490	677	1250	2282	921	1690	3074	1166	2130	3866
35.0	95	203.0	193	380	716	438	820	1508	682	1260	2300	927	1700	3092	1171	2140	3884
37.8	100	212.0	199	390	734	443	830	1526	688	1270	2318	932	1710	3110	1177	2150	3902
40.6	105	221.0	204	400	752	449	840	1544	693	1280	2336	938	1720	3128	1182	2160	3920
43.3	110	230.0	210	410	770	454	850	1562	699	1290	2354	943	1730	3146	1188	2170	3938
46.1	115	239.0	216	420	788	460	860	1580	704	1300	2372	949	1740	3164	1193	2180	3956
48.9	120	248.0	221	430	806	466	870	1598	710	1310	2390	954	1750	3182	1199	2190	3974
51.7	125	257.0	227	440	824	471	880	1616	716	1320	2408	960	1760	3200	1204	2200	3992
54.4	130	266.0	232	450	842	477	890	1634	721	1330	2426	966	1770	3218			
57.2	135	275.0	238	460	860	482	900	1652	727	1340	2444	971	1780	3236			
60.0	140	284.0	243	470	878	488	910	1670	732	1350	2462	977	1790	3254			

Values of single degrees

°C		°F	°C		°F	°C		°F	°F		°C	°F		°C	°F		°C
1	=	1.8	4	=	7.2	7	=	12.6	1	=	0.56	4	=	2.22	7	=	3.89
2	=	3.6	5	=	9.0	8	=	14.4	2	=	1.11	5	=	2.78	8	=	4.44
3	=	5.4	6	=	10.8	9	=	16.2	3	=	1.67	6	=	3.33	9	=	5.0

THERMODYNAMIC DATA

TABLE A2.1 Standard Atmospheric Conditions (STP)

Pressure $= 101325$ N/m^2
$= 14.696$ psi
$= 29.92$ inHg
$= 760$ mmHg

Temperature $= 288.15$ K
$= 15°C$
$= 59°F$

Density $= 1.2250$ kg/m^3
$= 0.076474$ lb/ft^3

Gas constant $= 8.31432$ J/(kg $\cdot$ mol $\cdot$ K)
$= 1545.31$ ft lb/(lb $\cdot$ mol $\cdot$ °R)

Molecular weight $= 28.966$

	Composition	
Gas	**% by volume**	**% by weight**
N_2	78.09	75.55
O_2	20.95	23.13
A	0.93	1.27
CO_2	0.03	0.05

	ppm by volume	**ppm by weight**
Ne	18	12.9
He	5.2	0.74
CH_4	2.2	1.3
Kr	1	3
NO	1	1.6
H_2	0.5	0.03
Xe	0.08	0.37
O_3	0.01	0.02
Rn	0.06×10^{-12}	

TABLE A2.2 Values of the Y-Factor for Air at Low Pressure and for Perfect Diatomic Gases

$$Y = \left(\frac{P_2}{P_1}\right)^{0.283} - 1$$

P_2/P_1	0	1	2	3	4	5	6	7	8	9
1.0	0.0000	0028	0056	0084	0112	0139	0166	0193	0220	0247
1.1	0273	0300	0326	0352	0378	0404	0429	0454	0480	0505
1.2	0530	0554	0579	0603	0628	0652	0676	0700	0724	0747
1.3	0771	0794	0817	0841	0864	0886	0909	0932	0954	0977
1.4	0999	1021	1043	1065	1087	1109	1130	1152	1173	1195
1.5	0.1216	1237	1258	1279	1300	1321	1341	1362	1382	1402
1.6	1423	1443	1463	1483	1503	1523	1542	1562	1581	1601
1.7	1620	1640	1659	1678	1697	1716	1735	1754	1773	1791
1.8	1810	1828	1847	1865	1884	1902	1920	1938	1956	1974
1.9	1992	2010	2028	2045	2063	2080	2098	2115	2133	2150
2	0.2167	2336	2500	2658	2812	2960	3105	3246	3383	3516
3	3647	3774	3898	4020	4139	4255	4369	4481	4591	4698
4	4804	4908	5010	5110	5209	5306	5401	5495	5588	5679
5	5769	5858	5945	6031	6116	6200	6283	6365	6446	6525
6	6604	6682	6759	6835	6910	6985	7058	7131	7203	7274
7	0.7345	7414	7483	7552	7620	7687	7753	7819	7884	7949
8	8013	8076	8139	8201	8263	8324	8385	8445	8505	8564
9	8623	8681	8739	8797	8854	8910	8966	9022	9077	9132
10	9187	9241	9295	9348	9401	9453	9506	9558	9609	9660
11	9711	9762	9812	9862	9912	9961	1.0010	1.0058	1.0107	1.0155
12	1.0202	1.0250	1.0297	1.0344	1.0390	1.0437	1.0484	1.0529	1.0575	1.0620
16	1.1916	1.1955	1.1993	1.2032	1.2070	1.2108	1.2146	1.2183	1.2221	1.2258
20	1.3345	1.3378	1.3411	1.3443	1.3476	1.3509	1.3541	1.3573	1.3605	1.3637

TABLE A2.3 **Properties of Dry Air at Atmospheric Pressure—SI Units**

Symbols and units

K = absolute temperature, degrees Kelvin
°C = temperature, degrees Celsius
°F = temperature, degrees Fahrenheit
ϱ = density, kg/m³
c_p = specific heat capacity, kJ/(kg · K)
c_p/c_v = specific heat capacity ratio, dimensionless
μ = viscosity. For (N · s)/m² [= kg/(m · s)] multiply tabulated values by 10^{-6}
k = thermal conductivity, MW/(m · K)
P_r = relative pressure
Pr = Prandtl number, dimensionless
H = enthalpy, kJ/kg
V_s = sound velocity, m/s

Temperature			Properties								
K	°C	°F	ϱ	c_p	c_p/c_v	μ	k	Pr	H	V_s	P_r
100	− 173.15	− 280	3.598	1.028		6.929	9.248	0.770	98.42	198.4	0.0297
110	− 163.15	− 262	3.256	1.022	1.420 2	7.633	10.15	0.768	108.7	208.7	0.0415
120	− 153.15	− 244	2.975	1.017	1.416 6	8.319	11.05	0.766	118.8	218.4	0.0562
130	− 143.15	− 226	2.740	1.014	1.413 9	8.990	11.94	0.763	129.0	227.6	0.0744
140	− 133.15	− 208	2.540	1.012	1.411 9	9.646	12.84	0.761	139.1	236.4	0.0964
150	− 123.15	− 190	2.367	1.010	1.410 2	10.28	13.73	0.758	149.2	245.0	0.1227
160	− 113.15	− 172	2.217	1.009	1.408 9	10.91	14.61	0.754	159.4	253.2	0.1537
170	− 103.15	− 154	2.085	1.008	1.407 9	11.52	15.49	0.750	169.4	261.0	0.1900
180	− 93.15	− 136	1.968	1.007	1.407 1	12.12	16.37	0.746	179.5	268.7	0.2323
190	− 83.15	− 118	1.863	1.007	1.406 4	12.71	17.23	0.743	189.6	276.2	0.2802
200	− 73.15	− 100	1.769	1.006	1.405 7	13.28	18.09	0.739	199.7	283.4	0.3352
205	− 68.15	− 91	1.726	1.006	1.405 5	13.56	18.52	0.738	204.7	286.9	0.3654
210	− 63.15	− 82	1.684	1.006	1.405 3	13.85	18.94	0.736	209.7	290.5	0.3974
215	− 58.15	− 73	1.646	1.006	1.405 0	14.12	19.36	0.734	214.8	293.9	0.4315
220	− 53.15	− 64	1.607	1.006	1.404 8	14.40	19.78	0.732	219.8	297.4	0.4677
225	− 48.15	− 55	1.572	1.006	1.404 6	14.67	20.20	0.731	224.8	300.8	0.5059
230	− 43.15	− 46	1.537	1.006	1.404 4	14.94	20.62	0.729	229.8	304.1	0.5462
235	− 38.15	− 37	1.505	1.006	1.404 2	15.20	21.04	0.727	234.9	307.4	0.5889
240	− 33.15	− 28	1.473	1.005	1.404 0	15.47	21.45	0.725	239.9	310.6	0.6338
245	− 28.15	− 19	1.443	1.005	1.403 8	15.73	21.86	0.724	244.9	313.8	0.6808
250	− 23.15	− 10	1.413	1.005	1.403 6	15.99	22.27	0.722	250.0	317.1	0.7310
255	− 18.15	− 1	1.386	1.005	1.403 4	16.25	22.68	0.721	255.0	320.2	0.7833
260	− 13.15	8	1.359	1.005	1.403 2	16.50	23.08	0.719	260.0	323.4	0.8385
265	− 8.15	17	1.333	1.005	1.403 0	16.75	23.48	0.717	265.0	326.5	0.8962
270	− 3.15	26	1.308	1.006	1.402 9	17.00	23.88	0.716	270.1	329.6	0.9567
275	+ 1.85	35	1.285	1.006	1.402 6	17.26	24.28	0.715	275.1	332.6	1.0201
280	6.85	44	1.261	1.006	1.402 4	17.50	24.67	0.713	280.1	335.6	1.0864
285	11.85	53	1.240	1.006	1.402 2	17.74	25.06	0.711	285.1	338.5	1.1558
290	16.85	62	1.218	1.006	1.402 0	17.98	25.47	0.710	290.2	341.5	1.2288
295	21.85	71	1.197	1.006	1.401 8	18.22	25.85	0.709	295.2	344.4	1.3040

TABLE A2.3 Properties of Dry Air at Atmospheric Pressure—SI Units (*continued*)

Temperature			Properties								
K	°C	°F	p	c_p	c_p/c_v	μ	k	Pr	H	V_s	P_r
300	26.85	80	1.177	1.006	1.401 7	18.46	26.24	0.708	300.2	347.3	1.3830
305	31.85	89	1.158	1.006	1.401 5	18.70	26.63	0.707	305.3	350.2	1.4656
310	36.85	98	1.139	1.007	1.401 3	18.93	27.01	0.705	310.3	353.1	1.5513
315	41.85	107	1.121	1.007	1.401 0	19.15	27.40	0.704	315.3	355.8	1.6412
320	46.85	116	1.103	1.007	1.400 8	19.39	27.78	0.703	320.4	358.7	1.7344
325	51.85	125	1.086	1.008	1.400 6	19.63	28.15	0.702	325.4	361.4	1.8312
330	56.85	134	1.070	1.008	1.400 4	19.85	28.53	0.701	330.4	364.2	1.9319
335	61.85	143	1.054	1.008	1.400 1	20.08	28.90	0.700	335.5	366.9	2.037
340	66.85	152	1.038	1.008	1.399 9	20.30	29.28	0.699	340.5	369.6	2.145
345	71.85	161	1.023	1.009	1.399 6	20.52	29.64	0.698	345.6	372.3	2.258
350	76.85	170	1.008	1.009	1.399 3	20.75	30.03	0.697	350.6	375.0	2.375
355	81.85	179	0.994 5	1.010	1.399 0	20.97	30.39	0.696	355.7	377.6	2.496
360	86.85	188	0.980 5	1.010	1.398 7	21.18	30.78	0.695	360.7	380.2	2.622
365	91.85	197	0.967 2	1.010	1.398 4	21.38	31.14	0.694	365.8	382.8	2.752
370	96.85	206	0.953 9	1.011	1.398 1	21.60	31.50	0.693	370.8	385.4	2.887
375	101.85	215	0.941 3	1.011	1.397 8	21.81	31.86	0.692	375.9	388.0	3.027
380	106.85	224	0.928 8	1.012	1.397 5	22.02	32.23	0.691	380.9	390.5	3.171
385	111.85	233	0.916 9	1.012	1.397 1	22.24	32.59	0.690	386.0	393.0	3.321
390	116.85	242	9.905 0	1.013	1.396 8	22.44	32.95	0.690	391.0	395.5	3.475
395	121.85	251	0.893 6	1.014	1.396 4	22.65	33.31	0.689	396.1	398.0	3.635
400	126.85	260	0.882 2	1.014	1.396 1	22.86	33.65	0.689	401.2	400.4	3.800
410	136.85	278	0.860 8	1.015	1.395 3	23.27	34.35	0.688	411.3	405.3	4.147
420	146.85	296	0.840 2	1.017	1.394 6	23.66	35.05	0.687	421.5	410.2	4.515
430	156.85	314	0.820 7	1.018	1.393 8	24.06	35.75	0.686	431.7	414.9	4.908
440	166.85	332	0.802 1	1.020	1.392 9	24.45	36.43	0.684	441.9	419.6	5.325
450	176.85	350	0.784 2	1.021	1.392 0	24.85	37.10	0.684	452.1	424.2	5.767
460	186.85	368	0.767 7	1.023	1.391 1	25.22	37.78	0.683	462.3	428.7	6.236
470	196.85	386	0.750 9	1.024	1.390 1	25.58	38.46	0.682	472.5	433.2	6.733
480	206.85	404	0.735 1	1.026	1.389 2	25.96	39.11	0.681	482.8	437.6	7.258
490	216.85	422	0.720 1	1.028	1.388 1	26.32	39.76	0.680	493.0	442.0	7.814
500	226.85	440	0.705 7	1.030	1.387 1	26.70	40.41	0.680	503.3	446.4	8.401
510	236.85	458	0.691 9	1.032	1.386 1	27.06	41.06	0.680	513.6	450.6	9.020
520	246.85	476	0.678 6	1.034	1.385 1	27.42	41.69	0.680	524.0	454.9	9.672

TABLE A2.3   Properties of Dry Air at Atmospheric Pressure—SI Units (*continued*)

Temperature			Properties								
K	°C	°F	p	c_p	c_p/c_v	μ	k	Pr	H	V_s	P_r
530	256.85	494	0.665 8	1.036	1.384 0	27.78	42.32	0.680	534.3	459.0	10.360
540	266.85	512	0.653 5	1.038	1.382 9	28.14	42.94	0.680	544.7	463.2	11.083
550	276.85	530	0.641 6	1.040	1.381 8	28.48	43.57	0.680	555.1	467.3	11.844
560	286.85	548	0.630 1	1.042	1.380 6	28.83	44.20	0.680	565.5	471.3	12.644
570	296.85	566	0.619 0	1.044	1.379 5	29.17	44.80	0.680	575.9	475.3	13.484
580	306.85	584	0.608 4	1.047	1.378 3	29.52	45.41	0.680	586.4	479.2	14.366
590	316.85	602	0.598 0	1.049	1.377 2	29.84	46.01	0.680	596.9	483.2	15.292
600	326.85	620	0.588 1	1.051	1.376 0	30.17	46.61	0.680	607.4	486.9	16.261
620	346.85	656	0.569 1	1.056	1.373 7	30.82	47.80	0.681	628.4	494.5	18.340
640	366.85	692	0.551 4	1.061	1.371 4	31.47	48.96	0.682	649.6	502.1	20.62
660	386.85	728	0.534 7	1.065	1.369 1	32.09	50.12	0.682	670.9	509.4	23.11
680	406.85	764	0.518 9	1.070	1.366 8	32.71	51.25	0.683	692.2	516.7	25.83
700	426.85	800	0.504 0	1.075	1.364 6	33.32	52.36	0.684	713.7	523.7	28.78
720	446.85	836	0.490 1	1.080	1.362 3	33.92	53.45	0.685	735.2	531.0	32.00
740	466.85	872	0.476 9	1.085	1.360 1	34.52	54.53	0.686	756.9	537.6	35.47
760	486.85	908	0.464 3	1.089	1.358 0	35.11	55.62	0.687	778.6	544.6	39.24
780	506.85	944	0.452 4	1.094	1.355 9	35.69	56.68	0.688	800.5	551.2	43.31
800	526.85	980	0.441 0	1.099	1.354	36.24	57.75	0.689	822.4	557.8	47.71
850	576.85	1 070	0.415 2	1.110	1.349	37.63	60.30	0.693	877.5	574.1	60.24
900	626.85	1 160	0.392 0	1.121	1.345	38.97	62.76	0.696	933.4	589.6	75.23
950	676.85	1 250	0.371 4	1.132	1.340	40.26	65.20	0.699	989.7	604.9	93.01
1 000	726.85	1 340	0.352 9	1.142	1.336	41.53	67.54	0.702	1 046	619.5	113.95
1 100	826.85	1 520	0.320 8	1.161	1.329	43.96			1 162	648.0	166.96
1 200	926.85	1 700	0.294 1	1.179	1.322	46.26			1 279	675.2	237.9
1 300	1 026.85	1 880	0.271 4	1.197	1.316	48.46			1 398	701.0	330.7
1 400	1 126.85	2 060	0.252 1	1.214	1.310	50.57			1 518	725.9	450.3
1 500	1 220.85	2 240	0.235 3	1.231	1.304	52.61			1 640	749.4	601.6
1 600	1 326.85	2 420	0.220 6	1.249	1.299	54.57			1 764	772.6	790.9
1 800	1 526.85	2 780	0.196 0	1.288	1.288	58.29			2 018	815.7	1310
2 000	1 726.85	3 140	0.176 4	1.338	1.274				2 280	855.5	2068
2 400	2 126.85	3 860	0.146 7	1.574	1.238				2 853	924.4	4607
2 800	2 526.85	4 580	0.124 5	2.259	1.196				3 599	983.1	9159

Source: *Handbook of Engineering Science*, courtesy CRC Press, Inc., and Keenan and Kaye, *Gas Tables*, courtesy John Wiley and Sons, Inc.

TABLE A2.4 Properties of Saturated Steam, Water, and Ice—SI Units

Subscripts

f	refers to a property of liquid in equilibrium with vapor	fg refers to a change by evaporation
g	refers to a property of vapor in equilibrium with liquid	ig refers to a change by sublimation
i	refers to a property of solid in equilibrium with vapor	

Temperature		Pressure, MN/m²	Specific volume, m³/kg		Specific internal energy, kJ/kg		Specific enthalpy, kJ/kg			Specific entropy, kJ/(kg · K)	
C	K		v_i	v_g	u_i	u_g	h_i	h_{ig}	h_g	s_i	s_g
Solid-Vapor											
−40	233.15	0.0000129	0.0010841	83.54	−411.70	2319.6	−411.70	2838.9	2427.2	−1.532	10.64
−30	243.15	0.0000381	0.0010858	29.43	−393.23	2333.6	−393.23	2839.0	2445.8	−1.455	10.22
−20	253.15	0.0001035	0.0010874	11.286	−374.03	2347.5	−374.03	2838.4	2464.3	−1.377	9.83
−10	263.15	0.0002602	0.0010891	4.667	−354.09	2361.4	−354.09	2837.0	2482.9	−1.299	9.48
0	273.15	0.0006108	0.0010908	2.063	−333.43	2375.3	−333.43	2834.8	2501.3	−1.221	9.15
0.01	273.16	0.0006113	0.0010908	2.061	−333.40	2375.3	−333.40	2834.8	2501.4	−1.221	9.15

			v_f	v_g	u_f	u_g	h_f	h_{fg}	h_g	s_f	s_g
Liquid-Vapor											
0	273.15	0.0006109	0.0010002	206.278	−0.03	2375.3	−0.02	2501.4	2501.3	−0.0001	9.15
0.01	273.16	0.0006113	0.0010002	206.136	0	2375.3	+0.01	2501.3	2501.4	0	9.15
5.00	278.15	0.0008721	0.0010001	147.120	+20.97	2382.3	20.98	2489.6	2510.6	+0.0761	9.02
6.98	280.13	0.0010000	0.0010002	129.208	29.30	2385.0	29.30	2484.9	2514.2	0.1059	8.90
10.00	283.15	0.0012276	0.0010004	106.379	42.00	2389.2	42.01	2477.7	2519.8	0.1510	8.90
13.03	286.18	0.0015000	0.0010007	87.980	54.71	2393.3	54.71	2470.6	2525.3	0.1957	8.82
15.00	288.15	0.0017051	0.0010009	77.926	62.99	2396.1	62.99	2465.9	2528.9	0.2245	8.78
17.50	290.65	0.0020000	0.0010013	67.004	73.48	2399.5	73.48	2460.0	2533.5	0.2607	8.72
20.00	293.15	0.002339	0.0010018	57.791	83.95	2402.9	83.96	2454.1	2538.1	0.2966	8.66
24.08	297.23	0.0030000	0.0010027	45.665	101.04	2408.5	101.05	2444.5	2545.5	0.3545	8.57
25.00	298.15	0.003169	0.0010029	43.360	104.88	2409.8	104.89	2442.3	2547.2	0.3674	8.558
28.96	302.11	0.004000	0.0010040	34.800	121.45	2415.2	121.46	2432.9	2554.4	0.4226	8.477
30.00	303.15	0.004246	0.0010043	32.894	125.78	2416.6	125.79	2430.5	2556.3	0.4369	8.450
32.88	306.03	0.005000	0.0010053	28.192	137.81	2420.5	137.82	2423.7	2561.5	0.4764	8.39
35.00	308.15	0.005628	0.0010060	25.216	146.67	2423.4	146.68	2418.6	2565.3	0.5053	8.35
36.16	309.31	0.006000	0.0010064	23.739	151.53	2425.0	151.53	2415.9	2567.4	0.5210	8.330
39.00	312.15	0.007000	0.0010074	20.530	163.39	2428.8	163.40	2409.1	2572.5	0.5592	8.275
40.00	313.15	0.007384	0.0010078	19.523	167.56	2430.1	167.57	2406.7	2574.3	0.5725	8.257
41.51	314.66	0.00800	0.0010084	18.108	173.87	2432.2	173.88	2403.1	2577.0	0.5926	8.228
43.76	316.91	0.009000	0.0010094	16.203	183.27	2435.2	183.29	2397.7	2581.0	0.6224	8.187
45.00	318.15	0.009593	0.0010099	15.258	188.44	2436.8	188.45	2394.8	2583.2	0.6387	8.164
45.81	318.96	0.010000	0.0010102	14.674	191.82	2437.9	191.83	2392.8	2584.7	0.6493	8.150
50.00	323.15	0.012349	0.0010121	12.032	209.32	2443.5	209.33	2382.7	2592.1	0.7038	8.076
53.97	327.12	0.015000	0.0010141	10.022	225.92	2448.7	225.94	2373.1	2599.1	0.7549	8.008
55.00	328.15	0.015758	0.0010146	9.568	230.21	2450.1	230.23	2370.7	2600.9	0.7679	7.991

TABLE A2.4 Properties of Saturated Steam, Water, and Ice—SI Units (*continued*)

Temperature		Pressure, MN/m²	Specific volume, m³/kg		Specific internal energy, kJ/kg		Specific enthalpy, kJ/kg			Specific entropy, kJ/(kg · K)	
C	K		v_f	v_g	u_f	u_g	h_f	h_{fg}	h_g	s_f	s_g
		Liquid-Vapor									
60.00	333.15	0.019940	0.0010172	7.671	251.11	2456.6	251.13	2358.5	2609.6	0.8312	7.909
60.06	333.21	0.020000	0.0010172	7.649	251.38	2456.7	251.40	2358.3	2609.7	0.8320	7.908
65.00	338.15	0.025030	0.0010199	6.197	272.02	2463.1	272.06	2346.2	2618.3	0.8935	7.831
69.10	342.25	0.030000	0.0010223	5.229	289.20	2468.4	289.23	2336.1	2625.3	0.9439	7.768
70.00	343.15	0.031190	0.0010228	5.042	292.95	2469.6	292.98	2333.8	2626.8	0.9549	7.755
75.00	348.15	0.038580	0.0010259	4.131	313.90	2475.9	313.93	2321.4	2635.3	1.0155	7.682
75.87	349.02	0.040000	0.0010265	3.993	317.53	2477.0	317.58	2319.2	2636.8	1.0259	7.670
80.00	353.15	0.047390	0.0010291	3.407	334.86	2482.2	334.91	2308.8	2643.7	1.0753	7.612
81.33	354.48	0.050000	0.0010300	3.240	340.44	2483.9	340.49	2305.4	2645.9	1.0910	7.593
85.00	358.15	0.057830	0.0010325	2.828	355.84	2488.4	355.90	2296.0	2651.9	1.1343	7.544
85.94	359.09	0.060000	0.0010331	2.732	359.79	2489.6	359.86	2293.6	2653.5	1.1453	7.532
89.95	363.10	0.070000	0.0010360	2.365	376.63	2494.5	376.70	2283.3	2660.0	1.1919	7.479
90.00	363.15	0.070140	0.0010360	2.361	376.85	2494.5	376.92	2283.2	2660.1	1.1925	7.479
93.50	366.65	0.080000	0.0010386	2.087	391.58	2498.8	391.66	2274.1	2665.8	1.2329	7.434
95.00	368.15	0.084550	0.0010397	1.9819	397.88	2500.6	397.96	2270.2	2668.1	1.2500	7.415
96.71	369.86	0.090000	0.0010410	1.869	405.06	2502.6	405.15	2265.7	2670.9	1.2695	7.3949
99.63	372.78	0.100000	0.0010432	1.6940	417.36	2506.1	417.46	2258.0	2675.5	1.3026	7.3594
100.00	373.15	0.101350	0.0010435	1.6729	418.94	2506.5	419.04	2257.0	2676.1	1.3069	7.3549
110.00	383.15	0.143270	0.0010516	1.2102	461.14	2518.1	461.30	2230.2	2691.5	1.4185	7.2387
111.37	384.52	0.150000	0.0010528	1.1593	466.94	2519.7	467.11	2226.5	2693.6	1.4336	7.2233
120.00	393.15	0.198530	0.0010603	0.8919	503.50	2529.3	503.71	2202.6	2706.3	1.5276	7.1296
120.23	393.38	0.200000	0.0010605	0.8857	504.49	2529.5	504.70	2201.9	2706.7	1.5301	7.1271
130.00	403.15	0.270100	0.0010697	0.6685	546.02	2539.9	546.31	2174.2	2720.5	1.6344	7.0269
133.55	406.70	0.300000	0.0010732	0.6058	561.15	2543.6	561.47	2163.8	2725.3	1.6718	6.9919
140.00	413.15	0.361300	0.0010797	0.5089	588.74	2550.0	589.13	2144.7	2733.9	1.7391	6.9299
143.63	416.78	0.400000	0.0010836	0.4625	604.31	2553.6	604.74	2133.8	2738.6	1.7766	6.8959
150.00	423.15	0.475800	0.0010905	0.3928	613.68	2559.5	632.20	2114.3	2746.5	1.8418	6.8379
151.86	425.01	0.500000	0.0010926	0.3749	639.68	2561.2	640.23	2108.5	2748.7	1.8607	6.8213
160.00	433.15	0.617800	0.0011020	0.3701	674.87	2568.4	675.55	2082.6	2758.1	1.9427	6.7502
170.00	443.15	0.791700	0.0011143	0.2428	718.33	2576.5	719.21	2049.5	2768.7	2.0419	6.6663
179.91	453.06	1.000000	0.0011273	0.19444	761.68	2583.6	762.81	2015.3	2778.1	2.1387	6.5865
180.00	453.15	1.002100	0.0011274	0.19405	762.09	2583.7	763.22	2015.0	2778.2	2.1396	6.5857
190.00	463.15	1.254400	0.0011414	0.15654	806.19	2590.0	807.62	1978.8	2786.4	2.2359	6.5079
198.32	471.47	1.500000	0.0011539	0.13177	843.16	2594.5	844.89	1947.3	2792.2	2.3150	6.4448
200.00	473.15	1.553800	0.0011565	0.12736	850.65	2595.3	852.45	1940.7	2793.2	2.3309	6.4323

TABLE A2.4 Properties of Saturated Steam, Water, and Ice—SI Units (*continued*)

Temperature		Pressure, MN/m²	Specific volume, m³/kg		Specific internal energy, kJ/kg		Specific enthalpy, kJ/kg			Specific entropy, kJ/(kg · K)	
C	K	Liquid-Vapor	v_f	v_g	u_f	u_g	h_f	h_{fg}	h_g	s_f	s_g
210.00	483.15	1.906200	0.0011726	0.10441	895.53	2599.5	897.76	1900.7	2798.5	2.4248	6.3585
212.42	485.57	2.000000	0.0011767	0.09963	906.44	2600.3	908.79	1890.7	2799.5	2.4474	6.3409
220.00	493.15	2.318000	0.0011900	0.08619	940.87	2602.4	943.62	1858.5	2802.1	2.5178	6.2861
223.99	497.14	2.500000	0.0011973	0.07998	959.11	2603.1	962.11	1841.0	2803.1	2.5547	6.2575
230.00	503.15	2.795000	0.0012088	0.07158	986.74	2603.9	990.12	1813.8	2804.0	2.6099	6.2146
233.90	507.05	3.000000	0.0012165	0.06668	1004.78	2604.1	1008.42	1795.7	2804.2	2.6457	6.1869
240.00	513.15	3.344000	0.0012291	0.05976	1033.21	2604.0	1037.32	1766.5	2803.8	2.7015	6.1437
242.60	515.75	3.500000	0.0012347	0.05707	1045.43	2603.7	1049.75	1753.7	2803.4	2.7253	6.1253
250.00	523.15	3.973000	0.0012512	0.05013	1080.39	2602.4	1085.36	1716.2	2801.5	2.7927	6.0730
250.40	523.55	4.000000	0.0012522	0.04978	1082.31	2602.3	1087.31	1714.1	2801.4	2.7964	6.0701
260.00	533.15	4.688000	0.0012755	0.04221	1128.39	2599.0	1134.37	1662.5	2796.9	2.8838	6.0019
263.99	537.14	5.000000	0.0012859	0.03944	1147.81	2597.1	1154.23	1640.1	2794.3	2.9202	5.9734
270.00	543.15	5.499000	0.0013023	0.03564	1177.36	2593.7	1184.51	1605.2	2789.7	2.9751	5.9301
275.64	548.79	6.000000	0.0013187	0.03244	1205.44	2589.7	1213.35	1571.0	2784.3	3.0267	5.8892
280.00	553.15	6.412000	0.0013321	0.03017	1227.46	2586.1	1235.99	1543.6	2779.6	3.0668	5.8571
285.88	559.03	7.000000	0.0013513	0.02737	1257.55	2580.5	1267.00	1505.1	2772.1	3.1211	5.8133
290.00	563.15	7.436000	0.0013656	0.02557	1278.92	2576.0	1289.07	1477.1	2766.2	3.1594	5.7821
295.06	568.21	8.000000	0.0013842	0.02352	1305.57	2569.8	1316.64	1441.3	2758.0	3.2068	5.7432
300.00	573.15	8.581000	0.0014036	0.02167	1332.0	2563.0	1344.0	1404.9	2749.0	3.2534	5.7045
303.40	576.55	9.000000	0.0014178	0.02048	1350.51	2557.8	1363.26	1378.9	2742.1	3.2858	5.6772
310.00	583.15	9.856000	0.0014474	0.018350	1387.1	2546.4	1401.3	1326.0	2727.3	3.3493	5.6230
311.06	584.21	10.000000	0.0014524	0.018026	1393.04	2544.4	1407.56	1317.1	2724.7	3.3596	5.6141
320.00	593.15	11.274000	0.0014988	0.015488	1444.6	2525.5	1416.5	1238.6	2700.1	3.4480	5.5362
324.75	597.90	12.000000	0.0015267	0.014263	1473.0	2513.7	1491.3	1193.6	2684.9	3.4962	5.4924
330.00	603.15	12.845000	0.0015607	0.012966	1505.3	2498.9	1525.3	1140.6	2665.9	3.5507	5.4417
336.75	609.90	14.000000	0.0016107	0.011485	1548.6	2476.8	1571.1	1066.5	2637.6	3.6232	5.3717
340.00	613.15	14.586000	0.0016379	0.010797	1570.3	2464.6	1594.2	1027.9	2622.0	3.6594	5.3357
347.44	620.59	16.000000	0.0017107	0.009306	1622.7	2431.7	1650.1	930.6	2580.6	3.7461	5.2455
350.00	623.15	16.513000	0.0017403	0.008813	1641.9	2418.4	1670.6	893.4	2563.9	3.7777	5.2112
357.06	630.21	18.000000	0.0018397	0.007489	1698.9	2374.3	1732.0	777.1	2509.1	3.8715	5.1044
360.00	633.15	18.651000	0.0018925	0.006945	1725.2	2351.5	1760.5	720.5	2481.0	3.9147	5.0526
365.81	638.96	20.000000	0.002036	0.005834	1785.6	2293.0	1826.3	583.4	2409.7	4.0139	4.9269
370.00	643.15	21.030000	0.002213	0.004925	1844.0	2228.5	1890.5	441.6	2332.1	4.1106	4.7971
373.80	646.95	22.000000	0.002742	0.003568	1961.9	2087.1	2022.2	143.4	2164.5	4.3110	4.5327
374.136	647.286	22.090000	0.003155	0.003155	2029.6	2029.6	2099.3	0	2099.3	4.4298	4.4298

Source: Condensed from J.H. Keenan, E.G. Keyes, P.G. Hill, and J.G. Moore, *Steam Tables: Thermodynamic Properties of Water Including Vapor, Liquid and Solid Phases,* copyright © 1969, by John Wiley & Sons, Inc. Reprinted by permission of the authors and publisher and CRC Press, Inc.

TABLE A2.5   Properties of Gases at Atmospheric Pressure and 25°C (298 K, 77°F)

Gas	Air	Nitrogen N₂	Oxygen O₂	Carbon dioxide CO₂	Helium He
Chemical and physical properties					
Molecular weight	28.966	28.0134	31.9988	44.01	4.0026
Specific gravity, air = 1	1.00	0.967	1.105	1.52	0.138
Specific volume, ft³/lb	13.5	13.98	12.24	8.8	97.86
Specific volume, m³/kg	0.842	0.872	0.764	0.55	6.11
Density of liquid (at atm bp), lb/ft³	54.6	50.46	71.27	—	7.80
Density of liquid (at atm bp), kg/m³	879.	808.4	1,142.		125.
Vapor pressure at 25°C, psia				931.	
Vapor pressure at 25°C, MN/m²				6.42	
Viscosity (abs.), lbm/(ft · s)	12.1×10^{-6}	12.1×10^{-6}	13.4×10^{-6}	9.4×10^{-6}	13.4×10^{-6}
Viscosity (abs), centipoises	0.018	0.018	0.020	0.014	0.02
Sound velocity in gas, m/s	346.	353.	329.	270.	1,015.
Thermal and thermodynamic properties					
Specific heat, c_p, Btu/(lb · °F) or cal/(g · °C)	0.2403	0.249	0.220	0.205	1.24
Specific heat, c_p, J/(kg · K)	1,005.	1,040.	920.	876.	5,188.
Specific heat ratio, c_p/c_v	1.40	1.40	1.40	1.30	1.66
Gas constant R, ft-lb/(lb · °F)	53.3	55.2	48.3	35.1	386.
Gas constant R, J/(kg · °C)	286.8	297.	260.	189.	2,077.
Thermal conductivity, Btu/(h · ft · °F)	0.0151	0.015	0.015	0.01	0.086
Thermal conductivity, W/(m · °C)	0.026	0.026	0.026	0.017	0.149
Boiling point (sat 14.7 psia), °F	−320	−320.4	−297.3	−109.4[b]	−452.
Boiling point (sat 760 mm), °C	−195	−195.8	−182.97	−78.5	4.22 K
Latent heat of evap (at bp), Btu/lb	88.2	85.5	91.7	246.	10.0
Latent heat of evap (at bp), J/kg	205,000.	199,000.	213,000.	572,000.	23,300.
Freezing (melting) point, °F (1 atm)	−357.2	−346.	−361.1		[b]
Freezing (melting) point, °C (1 atm)	−216.2	−210.	−218.4		
Latent heat of fusion, Btu/lb	10.0	11.1	5.9	—	—
Latent heat of fusion, J/kg	32,200.	25,800.	13.700.	—	—
Critical temperature, °F	−220.5	−232.6	−181.5	88.	−450.3
Critical temperature, °C	−140.3	−147.	−118.6	31.	5.2 K
Critical pressure, psia	550.	493.	726.	1,072.	33.22
Critical pressure, MN/m²	3.8	3.40	5.01	7.4	
Critical volume, ft³/lb	0.050	0.051	0.040		0.231
Critical volume, m³/kg	0.003	0.00318	0.0025		0.0144
Flammable (yes or no)	No	No	No	No	No
Heat of combustion, Btu/ft³	—	—	—	—	—
Heat of combustion, Btu/lb	—	—	—	—	—
Heat of combusion, kJ/kg	—	—	—	—	—

TABLE A2.5 Properties of Gases at Atmospheric Pressure and 25°C (298 K, 77°F) (continued)

Gas	Neon Ne	Argon Ar	Krypton Kr	Xenon Xe	Hydrogen H_2
Chemical and physical properties					
Molecular weight	20.179	39.948	83.80	131.30	2.016
Specific gravity, air = 1	0.697	1.38	2.89	4.53	0.070
Specific volume, ft³/lb	19.41	9.80	4.67	2.98	194.
Specific volume, m³/kg	1.211	0.622	0.291		12.1
Density of liquid (at atm bp), lb/ft³	75.35	87.0	150.6	190.8	4.43
Density of liquid (at atm bp), kg/m³	1,207.	1,400.	2,413.	3,060.	71.0
Vapor pressure at 25°C, psia				931.	
Vapor pressure at 25°C, MN/m²				6.42	
Viscosity (abs.), lbm/(ft · s)	21.5×10^{-6}	13.4×10^{-6}	16.8×10^{-6}	15.5×10^{-6}	6.05×10^{-6}
Viscosity (abs), centipoises	0.032	0.02	0.025	0.023	0.009
Sound velocity in gas, m/s	454.	322.	223.	177.	1,315.
Thermal and thermodynamic properties					
Specific heat, c_p, Btu/(lb · °F) or cal/(g · °C)	0.246	0.125	0.059	0.115	3.42
Specific heat, c_p, J/(kg · K)	1,030.	523.	247.	481.	14,310.
Specific heat ratio, c_p/c_v	1.64	1.67	1.68	1.67	1.405
Gas constant R, ft-lb/(lb · °F)	76.6	38.7	18.4	11.8	767.
Gas constant R, J/(kg · °C)	412.	208.	99.0	63.5	4,126.
Thermal conductivity, Btu/(h · ft · °F)	0.028	0.0102	0.0054	0.003	0.105
Thermal conductivity, W/(m · °C)	0.048	0.0172	0.0093	0.0052	0.0182
Boiling point (sat 14.7 psia), °F	-410.9	$-303.$	$-244.$	-162.5	$-423.$
Boiling point (sat 760 mm), °C	$-246.$	$-186.$	$-153.$	$-108.$	20.4 K
Latent heat of evap (at bp), Btu/lb	37.	70.	46.4	41.4	192.
Latent heat of evap (at bp), J/kg	86,100.	163,000.	108,000.	96,000.	447,000.
Freezing (melting) point, °F (1 atm)	-415.6	-308.5	$-272.$	$-220.$	-434.6
Freezing (melting) point, °C (1 atm)	-248.7	-189.2	$-169.$	$-140.$	-259.1
Latent heat of fusion, Btu/lb	6.8		4.7	10.	25.0
Latent heat of fusion, J/kg	15,800.		10,900.	23.300.	58,000.
Critical temperature, °F	-379.8	-187.6		61.9	-399.8
Critical temperature, °C	-228.8	$-122.$	-63.8	16.6	-240.0
Critical pressure, psia	396.	707.	800.	852.	189.
Critical pressure, MN/m²	2.73	4.87	5.52	5.87	1.30
Critical volume, ft³/lb	0.033	0.0299	0.0177	0.0145	0.53
Critical volume, m³/kg	0.0020	0.00186	0.0014	0.00090	0.033
Flammable (yes or no)	No	No	No	No	Yes
Heat of combustion, Btu/ft³	—	—	—	—	320.
Heat of combustion, Btu/lb	—	—	—	—	62,050.
Heat of combusion, kJ/kg	—	—	—	—	144,000.

TABLE A2.5   Properties of Gases at Atmospheric Pressure and 25°C (298 K, 77°F) (*continued*)

Gas	Methane CH$_4$	Propane C$_3$H$_8$	Ammonia, anhyd. NH$_3$	Nitric oxide NO	Carbon monoxide CO
Chemical and physical properties					
Molecular weight	16.044	55.097	17.02	30.006	28.011
Specific gravity, air = 1	0.554	1.52	0.59	1.04	0.967
Specific volume, ft^3/lb	24.2	8.84	23.0	13.05	14.0
Specific volume, m^3/kg	1.51	0.552	1.43	0.814	0.874
Density of liquid (at atm bp), lb/ft^3	26.3	36.2	42.6		
Density of liquid (at atm bp), kg/m^3	421.	580.	686.		
Vapor pressure at 25°C, psia	135.7	145.4			
Vapor pressure at 25°C, MN/m^2	0.936	1.00			
Viscosity (abs.), lbm/(ft · s)	7.39×10^{-6}	53.8×10^{-6}	6.72×10^{-6}	12.8×10^{-6}	12.1×10^{-6}
Viscosity (abs), centipoises	0.011	0.080	0.010	0.019	0.018
Sound velocity in gas, m/s	446.	253.	415.	341.	352.
Thermal and thermodynamic properties					
Specific heat, c_p, Btu/(lb · °F) or cal/(g · °C)	0.54	0.39	0.52	0.235	0.25
Specific heat, c_p, J/(kg · K)	2,260.	1,630.	2,175.	983.	1,046.
Specific heat ratio, c_p/c_v	1.31	1.2	1.3	1.40	1.40
Gas constant R, ft-lb/(lb · °F)	96.	35.0	90.8	51.5	55.2
Gas constant R, J/(kg · °C)	518.	188.	488.	277.	297.
Thermal conductivity, Btu/(h · ft · °F)	0.02	0.010	0.015	0.015	0.014
Thermal conductivity, W/(m · °C)	0.035	0.017	0.026	0.026	0.024
Boiling point (sat 14.7 psia), °F	− 259.	− 44.	− 28.	− 240.	− 312.7
Boiling point (sat 760 mm), °C	− 434.2	− 42.2	− 33.3	− 151.5	− 191.5
Latent heat of evap (at bp), Btu/lb	219.2	184.	589.3		92.8
Latent heat of evap (at bp), J/kg	510,000.	428,000.	1,373,000.		216,000.
Freezing (melting) point, °F (1 atm)	− 296.6	− 309.8	− 107.9	− 258.	− 337.
Freezing (melting) point, °C (1 atm)	− 182.6	− 189.9	− 77.7	− 161.	− 205.
Latent heat of fusion, Btu/lb	14.	19.1	143.0	32.9	12.8
Latent heat of fusion, J/kg	32,600.	44,400.	332,300.	76,500.	
Critical temperature, °F	− 116.	205.	271.4	− 136.	− 220.
Critical temperature, °C	− 82.3	96.	132.5	− 93.3	− 140.
Critical pressure, psia	673.	618.	1,650.	945.	507.
Critical pressure, MN/m^2	4.64	4.26	11.4	6.52	3.49
Critical volume, ft^3/lb	0.099	0.073	0.068	0.0332	0.053
Critical volume, m^3/kg	0.0062	0.0045	0.00424	0.00207	0.0033
Flammable (yes or no)	Yes	Yes	No	No	Yes
Heat of combustion, Btu/ft^3	985.	2,450.	—	—	310.
Heat of combustion, Btu/lb	2,290.	21,660.	—	—	4,340.
Heat of combusion, kJ/kg		50,340.	—	—	10,100.

TABLE A2.5   Properties of Gases at Atmospheric Pressure and 25°C (298 K, 77°F) (continued)

Gas	Sulfur dioxide SO$_2$	Fluorocarbons CCl$_2$F$_2$ Freon 12	CClF$_3$ Freon 13
Chemical and physical properties			
Molecular weight	64.06	120.91	104.46
Specific gravity, air = 1	2.21	4.17	3.61
Specific volume, ft^3/lb	6.11	3.12	3.58
Specific volume, m^3/kg		0.195	0.224
Density of liquid (at atm bp), lb/ft^3	42.8	93.0	95.0
Density of liquid (at atm bp), kg/m^3	585.	1,490.	1,522.
Vapor pressure at 25°C, psia	56.6	94.51	516.
Vapor pressure at 25°C, MN/m^2	0.390	0.652	3.56
Viscosity (abs.), lbm/(ft · s)	8.74×10^{-6}	8.74×10^{-6}	
Viscosity (abs), centipoises	0.013	0.013	
Sound velocity in gas, m/s	220.		
Thermal and thermodynamic properties			
Specific heat, c_p, Btu/(lb · °F) or cal/(g · °C)	0.11	0.146	0.154
Specific heat, c_p, J/(kg · K)	460.	611.	644.
Specific heat ratio, c_p/c_v	1.29	1.14	1.145
Gas constant R, ft-lb/(lb · °F)	24.1		
Gas constant R, J/(kg · °C)	518.	130.	
Thermal conductivity, Btu/(h · ft · °F)	0.006	0.006	
Thermal conductivity, W/(m · °C)	0.010	0.0104	
Boiling point (sat 14.7 psia), °F	14.0	−21.8	−114.6
Boiling point (sat 760 mm), °C	−10.	−29.9	−81.4
Latent heat of evap (at bp), Btu/lb	155.5	74.1	63.0
Latent heat of evap (at bp), J/kg	362,000.	165,000.	147,000.
Freezing (melting) point, °F (1 atm)	−104.	−252.	−294.
Freezing (melting) point, °C (1 atm)	−75.5	−157.8	−181.1
Latent heat of fusion, Btu/lb	58.0		
Latent heat of fusion, J/kg	135,000.		
Critical temperature, °F	315.5	233.	83.9
Critical temperature, °C	157.6	111.7	28.8
Critical pressure, psia	1,141.	582.	559.
Critical pressure, MN/m^2	7.87	4.01	3.85
Critical volume, ft^3/lb	0.03	0.287	0.0277
Critical volume, m^3/kg	0.0019	0.018	0.00173
Flammable (yes or no)	No	No	No
Heat of combustion, Btu/ft^3	—	—	—
Heat of combustion, Btu/lb	—	—	—
Heat of combusion, kJ/kg	—	—	—

Source: Reprinted by permission of Iowa State University Press.

TABLE A2.6 Properties of Combustion Gases at High Temperatures (For Hydrocarbon Fuels of the Composition C_nH_{2n})

Symbols and Units

P = total pressure, atmospheres. For psia multiply by 14.696. For kg/cm² multiply by 1.033. For MN/m² multiply by 0.1013.

T = temperature, °R. For K, multiply by $\frac{5}{9}$.

H = enthalpy, Btu/lb$_m$. For cal/g multiply by 0.5555. For J/kg multiply by 2324.

s = entropy, Btu/(lb$_m$ · °R) = cal/(g · K)

k = specific heat ratio, c_p/c_v

a = sonic velocity, ft/s. For m/s multiply by 0.3048.

PURE AIR, N/O = 3.73, 0% EXCESS O_2 | PURE OXYGEN, N/O = 0, 0% EXCESS O_2

P, atm	T, °R	H, Btu/lb$_m$	s, Btu/lb$_m$·°R	Mol wt	k, c_p/c_v	Sonic vel, ft/s	P, atm	T, °R	H, Btu/lb$_m$	s, Btu/lb$_m$·°R	Mol wt	k, c_p/c_v	Sonic vel, ft/s
0.1	2700.	634.8	2.272	28.90	1.268	2427.	0.1	2700.	763.7	2.282	31.00	1.187	2268.
	3600.	992.9	2.381	28.74	1.256	2797.		3600.	1245.	2.432	30.61	1.178	2625.
	4500.	1674.	2.543	27.56	1.264	3204.		4500.	2516.	2.724	27.67	1.196	3109.
	5400.	3084.	2.818	24.36	1.303	3790.		5400.	5898.	3.365	20.53	1.278	4088.
1.0	2700.	634.2	2.113	28.90	1.268	2426.	1.0	2700.	762.3	2.134	31.01	1.187	2267.
	3600.	956.8	2.215	28.83	1.255	2792.		3600.	1187.	2.268	30.83	1.177	2614.
	4500.	1453.	2.334	28.25	1.255	3153.		4500.	1992.	2.456	29.38	1.181	2999.
	5400.	2350.	2.506	26.52	1.273	3590.		5400.	3832.	2.800	25.18	1.215	3600.
20.0	2700.	633.8	1.907	28.90	1.268	2426.	20.0	2700.	761.6	1.942	31.01	1.187	2267.
	3600.	942.2	2.005	28.88	1.255	2789.		3600.	1156.	2.067	30.95	1.176	2608.
	4500.	1326.	2.099	28.65	1.250	3125.		4500.	1702.	2.198	30.39	1.173	2939.
	5400.	1894.	2.209	27.90	1.255	3476.		5400.	2669.	2.382	28.52	1.183	3338.

PURE AIR, N/O = 3.73, 20% EXCESS O_2 | PURE OXYGEN, N/O = 0, 20% EXCESS O_2

P, atm	T, °R	H, Btu/lb$_m$	s, Btu/lb$_m$·°R	Mol wt	k, c_p/c_v	Sonic vel, ft/s	P, atm	T, °R	H, Btu/lb$_m$	s, Btu/lb$_m$·°R	Mol wt	k, c_p/c_v	Sonic vel, ft/s
0.1	2700.	625.5	2.269	28.91	1.274	2438.	0.1	2700.	733.4	2.278	31.14	1.197	2271.
	3600.	947.1	2.370	28.84	1.261	2798.		3600.	1145.	2.405	30.98	1.186	2617.
	4500.	1575.	2.517	27.84	1.267	3190.		4500.	2225.	2.649	28.60	1.198	3062.
	5400.	2911.	2.778	24.77	1.304	3759.		5400.	5425.	3.253	21.37	1.279	4009.
1.0	2700.	625.3	2.111	28.91	1.274	2432.	1.0	2700.	732.9	2.131	31.14	1.197	2271.
	3600.	932.8	2.208	28.89	1.261	2795.		3600.	1115.	2.250	31.08	1.185	2612.
	4500.	1378.	2.314	28.47	1.258	3145.		4500.	1775.	2.402	30.12	1.185	2967.
	5400.	2225.	2.476	26.85	1.275	3570.		5400.	3464.	2.715	26.11	1.217	3538.
20.0	2700.	625.1	1.905	28.91	1.274	2432.	20.0	2700.	732.5	1.940	31.14	1.197	2271.
	3600.	926.2	2.001	28.91	1.260	2794.		3600.	1101.	2.056	31.12	1.185	2610.
	4500.	1278.	2.086	28.79	1.255	3123.		4500.	1565.	2.166	30.85	1.180	2925.
	5400.	1802.	2.187	28.17	1.258	3463.		5400.	2396.	2.321	29.37	1.186	3293.

PURE AIR, N/O = 3.73, 100% EXCESS O_2 | PURE OXYGEN, N/O = 0.0, 100% EXCESS O_2

P, atm	T, °R	H, Btu/lb$_m$	s, Btu/lb$_m$·°R	Mol wt	k, c_p/c_v	Sonic vel, ft/s	P, atm	T, °R	H, Btu/lb$_m$	s, Btu/lb$_m$·°R	Mol wt	k, c_p/c_v	Sonic vel, ft/s
0.1	2700.	668.3	2.232	31.43	1.222	2284.	0.1	2700.	668.3	2.232	31.43	1.222	2284.
	3600.	1031.	2.343	31.32	1.210	2630.		3600.	1031.	2.343	31.32	1.210	2630.
	4500.	1850.	2.525	29.69	1.217	3028.		4500.	1850.	2.525	29.69	1.217	3028.
	5400.	4463.	3.016	23.17	1.289	3865.		5400.	4463.	3.016	23.17	1.289	3865.
1.0	2700.	667.8	2.086	31.44	1.222	2284.	1.0	2700.	667.8	2.086	31.44	1.222	2284.
	3600.	1009.	2.193	31.39	1.210	2627.		3600.	1009.	2.193	31.39	1.210	2627.
	4500.	1541.	2.313	30.76	1.207	2963.		4500.	1541.	2.313	30.76	1.207	2963.
	5400.	2846.	2.551	27.65	1.231	3457.		5400.	2846.	2.551	27.65	1.231	3457.
20.0	2700.	667.4	1.897	31.44	1.222	2284.	20.0	2700.	667.4	1.897	31.44	1.222	2284.
	3600.	997.1	2.001	31.42	1.210	2625.		3600.	997.1	2.001	31.42	1.210	2625.
	4500.	1400.	2.096	31.23	1.203	3936.		4500.	1400.	2.096	31.23	1.203	3936.
	5400.	2446.	2.406	28.94	1.218	3362.		5400.	2052.	2.215	30.25	1.206	3271.

Source: Reprinted from: *Handbook of Applied Engineering Science,* courtesy CRC Press, Inc.

TABLE A2.7   Properties of Common Liquids—SI Units [At 1.0 Atm Pressure (0.101 325 MN/m²), 300 K, Except as Noted]

Common name	Density, kg/m³	Specific heat, kJ/(kg·K)	Viscosity, (N·s)/m²	Thermal conductivity, W/(m·K)	Freezing point, K	Latent heat of fusion, kJ/kg	Boiling point, K	Latent heat of evaporation, kJ/kg	Coefficient of cubical expansion per K
Acetic acid	1049	2.18	.001155	0.171	290	181	391	402	0.0011
Acetone	784.6	2.15	.000316	0.161	179.0	98.3	329	518	0.0015
Alcohol, ethyl	785.1	2.44	.001095	0.171	158.6	108	351.46	846	0.0011
Alcohol, methyl	786.5	2.54	.00056	0.202	175.2	98.8	337.8	1100	0.0014
Alcohol, propyl	800.0	2.37	.00192	0.161	146	86.5	371	779	
Ammonia (aqua)	823.5	4.32		0.353					
Benzene	873.8	1.73	.000601	0.144	278.68	126	353.3	390	0.0013
Bromine		.473	.00095		245.84	66.7	331.6	193	0.0012
Carbon disulfide	1261	.992	.00036	0.161	161.2	57.6	319.40	351	0.0013
Carbon tetrachloride	1584	.866	.00091	0.104	250.35	174	349.6	194	0.0013
Castor oil	956.1	1.97	.650	0.180	263.2				
Chloroform	1465	1.05	.00053	0.118	209.6	77.0	334.4	247	0.0013
Decane	726.3	2.21	.000859	0.147	243.5	201	447.2	263	
Dodecane	754.6	2.21	.001374	0.140	247.18	216	489.4	256	
Ether	713.5	2.21	.000223	0.130	157	96.2	307.7	372	0.0016
Ethylene glycol	1097	2.36	.0162	0.258	260.2	181	470	800	
Fluorine refrigerant R-11	1476	.870*	.00042	0.093*	162		279.0	180†	
Fluorine refrigerant R-12	1311	.971*		0.071*	115	34.4	243.4	165‡	
Fluorine refrigerant R-22	1194	1.26*		0.086*	113	183	232.4	232†	
Glycerine	1259	2.62	.950	0.287	264.8	200	563.4	974	0.00054
Heptane	679.5	2.24	.000376	0.128	182.54	140	371.5	318	
Hexane	654.8	2.26	.000297	0.124	178.0	152	341.84	365	
Iodine		2.15			386.6	62.2	457.5	164	
Kerosene	820.1	2.09	.00164	0.145				251	
Linseed oil	929.1	1.84	.0331		253		560		
Mercury		.139	.00153		234.3	11.6	630	295	0.00018
Octane	698.6	2.15	.00051	0.131	216.4	181	398	298	0.00072
Phenol	1072	1.43	.0080	0.190	316.2	121	455		0.00090
Propane	493.5	2.41*	.00011		85.5	79.9	231.08	428†	
Propylene	514.4	2.85	.00009		87.9	71.4	225.45	342	
Propylene glycol	965.3	2.50	.042		213		460	914	
Seawater	1025	3.76 – 4.10			270.6				
Toluene	862.3	1.72	.000550	0.133	178	71.8	383.6	363	
Turpentine	868.2	1.78	.001375	0.121	214		433	293	0.00099
Water	997.1	4.18	.00089	0.609	273	333	373	2260	0.00020

*At 297 K, liquid.

†At 101, 325 MN saturation temperature.

Source: *Handbook of Applied Engineering Science,* courtesy CRC Press, Inc.

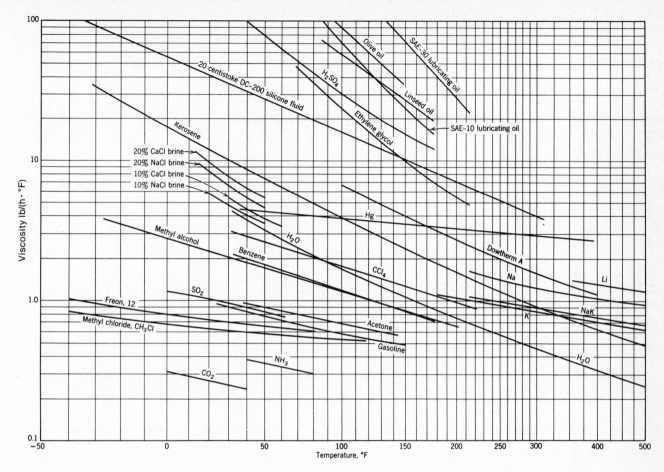

Figure A2.1 Effects of temperature on the viscosities of typical liquids (Ref. 3, Chap. 4). (*Courtesy John Wiley & Sons, Inc.*)

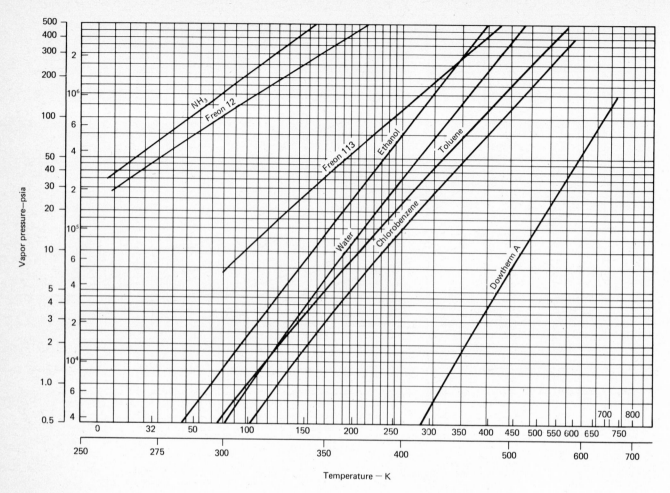

Figure A2.2 Vapor pressures of typical fluids of interest for special applications of Rankine cycles.

TABLE A2.8 **Thermodynamic Data for the Saturated Liquid and Vapor of Potassium** (Ref. 54, Chap. 15)

Temperature, °R	P, atm	Specific volume, ft³/lb		Enthalpy, Btu/lb			Entropy, Btu/(lb·°R)	
		V_f	V_g	H_f	H_{fg}	H_g	S_f	S_g
960	0.00012	0.02045	151939.	127.2	922.0	344.8	0.5505	1.5109
970	0.00014	0.02049	126236.	198.9	921.5	1120.5	0.5521	1.5022
980	0.00017	0.02052	105294.	200.5	921.1	1121.6	0.5538	1.4937
990	0.00021	0.02055	88162.	202.1	920.6	1122.7	0.5554	1.4854
1000	0.00025	0.02059	74090.	203.7	920.1	1123.9	0.5571	1.4772
1010	0.00030	0.02062	62487.	205.3	919.6	1125.0	0.5587	1.4692
1020	0.00036	0.02066	52884.	207.0	919.1	1126.1	0.5603	1.4614
1030	0.00043	0.02069	44909.	208.6	918.6	1127.2	0.5619	1.4537
1040	0.00051	0.02072	38260.	210.3	918.0	1128.3	0.5635	1.4462
1050	0.00060	0.02076	32700.	211.9	917.5	1129.4	0.5650	1.4389
1060	0.00070	0.02079	28034.	213.5	916.9	1130.5	0.5666	1.4317
1070	0.00082	0.02083	24106.	215.2	916.3	1131.6	0.5681	1.4246
1080	0.00096	0.02086	20788.	216.9	915.8	1132.7	0.5697	1.4177
1090	0.00112	0.02090	17978.	218.5	915.2	1133.7	0.5712	1.4109
1100	0.00131	0.02093	15591.	220.2	914.5	1134.8	0.5727	1.4042
1110	0.00152	0.02097	13556.	221.9	913.9	1135.8	0.5743	1.3977
1120	0.00176	0.02101	11818.	223.5	913.3	1136.9	0.5758	1.3912
1130	0.00203	0.02104	10329.	225.2	912.6	1137.9	0.5773	1.3849
1140	0.00233	0.02108	9050.	226.9	912.0	1138.9	0.5787	1.3788
1150	0.00268	0.02111	7948.	228.6	911.3	1139.9	0.5802	1.3727
1160	0.00307	0.02115	6996.	230.3	910.6	1141.0	0.5817	1.3667
1170	0.00350	0.02119	6173.	232.0	909.9	1141.9	0.5832	1.3604
1180	0.00399	0.02122	5458.	233.7	909.2	1142.9	0.5846	1.3551
1190	0.00454	0.02126	4837.	235.4	908.4	1143.9	0.5661	1.3495
1200	0.00516	0.02130	4295.	237.2	907.7	1144.9	0.5875	1.3439
1210	0.00584	0.02134	3822.	238.9	906.9	1145.8	0.5889	1.3385
1220	0.00660	0.02137	3407.	240.6	906.1	1146.8	0.5903	1.3331
1230	0.00745	0.02141	3044.	242.3	905.3	1147.7	0.5918	1.3278
1240	0.00838	0.02145	2724.	244.1	904.5	1148.7	0.5932	1.3227
1250	0.00942	0.02149	2443.	245.8	903.7	1149.6	0.5946	1.3176
1260	0.01056	0.02152	2104.	247.6	902.9	1150.5	0.5960	1.3126
1270	0.01182	0.02156	1974.	249.3	902.0	1151.4	0.5973	1.3076
1280	0.01321	0.02160	1780.	251.1	901.2	1152.3	0.5987	1.3028
1290	0.01473	0.02164	1607.	252.9	900.3	1153.2	0.6001	1.2980
1300	0.01641	0.02168	1453.	254.6	899.4	1154.0	0.6015	1.2933
1310	0.01824	0.02172	1316.	256.4	898.4	1154.9	0.6028	1.2887
1320	0.02024	0.02176	1194.	258.2	897.5	1155.8	0.6042	1.2842
1330	0.02243	0.02180	1085.	260.0	896.6	1156.6	0.6055	1.2797
1340	0.02481	0.02184	987.	261.8	895.6	1157.4	0.6069	1.2753
1350	0.02741	0.02188	900.	263.6	894.7	1158.3	0.6082	1.2709
1360	0.03023	0.02191	821.	265.4	893.7	1159.1	0.6095	1.2667
1370	0.03329	0.02196	750.	267.2	892.7	1159.9	0.6109	1.2625
1380	0.03661	0.02200	687.	269.0	891.6	1160.7	0.6122	1.2583
1390	0.04021	0.02204	629.	270.8	890.6	1161.5	0.6135	1.2543
1400	0.04410	0.02208	577.	272.6	889.6	1162.2	0.6148	1.2502
1410	0.04831	0.02212	530.	274.4	888.5	1163.0	0.6161	1.2463
1420	0.05284	0.02216	488.	276.3	887.4	1163.7	0.6174	1.2424
1430	0.05773	0.02220	449.	278.1	886.3	1164.5	0.6187	1.2385
1440	0.06298	0.02224	414.	280.0	885.2	1165.2	0.6200	1.2347
1450	0.06864	0.02228	382.	281.8	884.1	1166.0	0.6212	1.2310
1460	0.07471	0.02232	353.	283.7	883.0	1166.7	0.6225	1.2273
1470	0.08122	0.02237	327.	285.4	881.9	1167.4	0.6237	1.2237
1480	0.08820	0.02241	303.	287.3	880.7	1168.1	0.6250	1.2201

TABLE A2.8 Thermodynamic Data for the Saturated Liquid and Vapor of Potassium (continued)

Temper-ature, °R	P, atm	Specific volume, ft³/lb		Enthalpy, Btu/lb			Entropy, Btu/(lb·°R)	
		V_f	V_g	H_f	H_{fg}	H_g	S_f	S_g
1490	0.09567	0.02245	281.	289.2	879.5	1168.8	0.6263	1.2166
1500	0.10365	0.02249	260.	291.1	878.3	1169.4	0.6275	1.2131
1510	0.11218	0.02254	242.	292.9	877.1	1170.1	0.6288	1.2097
1520	0.12129	0.02258	225.	294.8	875.9	1170.8	0.6300	1.2063
1530	0.13099	0.02262	209.	296.7	874.7	1171.4	0.6313	1.2030
1540	0.14133	0.02267	195.	298.6	873.4	1172.1	0.6325	1.1997
1550	0.15233	0.02271	182.	300.3	872.3	1172.7	0.6336	1.1964
1560	0.16403	0.02275	170.	302.2	871.0	1173.3	0.6348	1.1932
1570	0.17646	0.02280	159.	304.1	869.7	1173.9	0.6361	1.1901
1580	0.18964	0.02284	148.	306.1	868.4	1174.6	0.6373	1.1870
1590	0.20363	0.02289	139.	308.0	867.1	1175.2	0.6385	1.1839
1600	0.21844	0.02293	130.	309.9	865.8	1175.7	0.6397	1.1808
1610	0.23413	0.02297	122.	312.1	864.1	1176.3	0.6411	1.1778
1620	0.25072	0.02302	114.	314.0	862.8	1176.9	0.6422	1.1749
1630	0.26825	0.02306	107.	315.9	861.5	1177.5	0.6434	1.1720
1640	0.28677	0.02311	101.	317.9	860.1	1178.0	0.6446	1.1691
1650	0.30631	0.02316	95.3	319.8	858.7	1178.6	0.6458	1.1662
1660	0.32692	0.02320	89.7	321.7	857.3	1179.1	0.6469	1.1634
1670	0.34864	0.02325	84.5	323.7	855.9	1179.7	0.6481	1.1607
1680	0.37151	0.02329	79.7	325.6	854.5	1180.2	0.6493	1.1579
1690	0.39557	0.02334	75.2	327.6	853.1	1180.7	0.6504	1.1552
1700	0.42088	0.02339	71.0	329.5	851.7	1181.2	0.6517	1.1526
1710	0.44747	0.02343	67.1	331.5	850.2	1181.7	0.6527	1.1499
1720	0.47539	0.02348	63.4	333.4	848.8	1182.2	0.6539	1.1473
1730	0.50469	0.02353	60.0	335.3	847.4	1182.8	0.6549	1.1448
1740	0.53542	0.02358	56.8	337.3	845.9	1183.2	0.6561	1.1423
1750	0.56763	0.02362	53.8	339.2	844.4	1183.7	0.6572	1.1398
1760	0.60136	0.02367	51.0	341.2	843.0	1184.2	0.6583	1.1373
1770	0.63668	0.02372	48.4	343.2	841.5	1184.7	0.6594	1.1349
1780	0.67362	0.02377	45.9	345.1	840.0	1185.2	0.6605	1.1325
1790	0.71225	0.02382	43.6	347.1	838.5	1185.6	0.6616	1.1301
1800	0.75261	0.02387	41.4	349.1	837.0	1186.1	0.6627	1.1277
1810	0.79476	0.02392	39.4	351.1	835.4	1186.5	0.6638	1.1254
1820	0.83876	0.02397	37.5	353.0	833.9	1187.0	0.6649	1.1231
1830	0.88466	0.02402	35.7	355.0	832.4	1187.5	0.6660	1.1209
1840	0.93252	0.02407	34.0	357.0	830.9	1187.9	0.6671	1.1187
1850	0.98239	0.02412	32.4	358.8	829.5	1188.3	0.6681	1.1165
1860	1.034	0.0241	30.9	360.8	827.9	1188.8	0.6691	1.1143
1870	1.088	0.0242	29.4	362.8	826.4	1189.2	0.6702	1.1121
1880	1.144	0.0242	28.1	364.8	824.8	1189.7	0.6713	1.1100
1890	1.203	0.0243	26.8	366.7	823.3	1190.1	0.6723	1.1079
1900	1.263	0.0243	25.6	368.7	821.7	1190.5	0.6733	1.1059
1910	1.327	0.0244	24.5	370.7	820.2	1190.9	0.6744	1.1038
1920	1.392	0.0244	23.4	372.7	818.6	1191.4	0.6754	1.1018
1930	1.460	0.0245	22.4	374.7	817.1	1191.8	0.6764	1.0998
1940	1.531	0.0245	21.5	376.6	815.5	1192.2	0.6775	1.0979
1950	1.604	0.0245	20.5	378.6	813.9	1192.6	0.6785	1.0959
1960	1.680	0.0246	19.7	380.6	812.4	1193.0	0.6795	1.0940
1970	1.759	0.0247	18.9	382.3	811.1	1193.5	0.6804	1.0921
1980	1.841	0.0247	18.1	384.3	809.5	1193.9	0.6814	1.0903
1990	1.925	0.0248	17.4	386.3	807.9	1194.3	0.6824	1.0884

TABLE A2.8 Thermodynamic Data for the Saturated Liquid and Vapor of Potassium (*continued*)

Temper-ature, °R	P, atm	Specific volume, ft³/lb		Enthalpy, Btu/lb			Entropy, Btu/(lb·°R)	
		V_f	V_g	H_f	H_{fg}	H_g	S_f	S_g
2000	2.012	0.0249	16.7	388.3	806.4	1194.7	0.6834	1.0866
2010	2.103	0.0249	16.0	390.3	804.8	1195.1	0.6844	1.0848
2020	2.196	0.0250	15.4	392.3	803.2	1195.6	0.6854	1.0830
2030	2.292	0.0250	14.8	394.2	801.7	1196.0	0.6863	1.0813
2040	2.392	0.0251	14.2	396.2	800.1	1196.4	0.6873	1.0795
2050	2.495	0.0251	13.7	398.2	798.6	1196.8	0.6883	1.0778
2060	2.602	0.0252	13.1	400.2	797.0	1197.2	0.6892	1.0761
2070	2.711	0.0252	12.7	402.2	795.4	1197.6	0.6902	1.0745
2080	2.825	0.0253	12.2	404.1	793.9	1198.1	0.6911	1.0728
2090	2.941	0.0254	11.7	406.1	792.3	1198.5	0.6921	1.0712
2100	3.062	0.0254	11.3	408.1	790.8	1198.9	0.6930	1.0696
2110	3.186	0.0255	10.9	409.7	789.7	1199.4	0.6938	1.0681
2120	3.313	0.0255	10.5	411.6	788.1	1199.8	0.6947	1.0665
2130	3.445	0.0256	10.1	413.6	786.6	1200.2	0.6956	1.0649
2140	3.581	0.0256	9.8	415.6	785.0	1200.7	0.6966	1.0634
2150	3.720	0.0257	9.5	417.6	783.5	1201.1	0.6975	1.0619
2160	3.863	0.0258	9.1	419.5	782.0	1201.6	0.6984	1.0604
2170	4.011	0.0258	8.8	421.5	780.4	1202.0	0.6993	1.0590
2180	4.163	0.0259	8.5	423.5	778.9	1202.4	0.7002	1.0575
2190	4.319	0.0259	8.2	425.5	777.4	1202.9	0.7011	1.0561
2200	4.479	0.0260	8.0	427.4	775.9	1203.3	0.7020	1.0547
2210	4.643	0.0261	7.7	429.4	774.4	1203.8	0.7029	1.0533
2220	4.812	0.0261	7.5	431.4	772.8	1204.3	0.7038	1.0519
2230	4.986	0.0262	7.2	433.3	771.3	1204.7	0.7047	1.0506
2240	5.164	0.0262	7.0	435.3	769.8	1205.2	0.7055	1.0492
2250	5.347	0.0263	6.8	436.7	769.0	1205.7	0.7062	1.0480
2260	5.534	0.0264	6.6	438.6	767.5	1206.2	0.7070	1.0467
2270	5.726	0.0264	6.4	440.6	766.0	1206.7	0.7079	1.0454
2280	5.923	0.0265	6.2	442.6	764.5	1207.2	0.7088	1.0441
2290	6.125	0.0266	6.0	444.6	763.1	1207.7	0.7096	1.0429
2300	6.332	0.0266	5.8	446.5	761.6	1208.1	0.7105	1.0416
2310	6.544	0.0267	5.6	448.5	760.1	1208.6	0.7113	1.0404
2320	6.761	0.0267	5.5	450.5	758.6	1209.1	0.7122	1.0392
2330	6.983	0.0268	5.3	454.5	757.1	1209.6	0.7130	1.0380
2340	7.210	0.0269	5.1	454.5	755.7	1210.2	0.7139	1.0368
2350	7.443	0.0269	5.0	456.4	754.2	1210.7	0.7147	1.0357
2360	7.681	0.0270	4.8	458.4	752.7	1211.2	0.7156	1.0345
2370	7.925	0.0271	4.7	460.4	751.2	1211.7	0.7164	1.0334
2380	8.174	0.0271	4.6	462.5	749.7	1212.2	0.7172	1.0323
2390	8.428	0.0272	4.4	464.5	748.3	1212.8	0.7181	1.0312
2400	8.689	0.0273	4.3	465.5	747.9	1213.4	0.7185	1.0302
2410	8.954	0.0273	4.2	467.4	746.5	1213.9	0.7193	1.0291
2420	9.226	0.0274	4.1	469.4	745.0	1214.5	0.7202	1.0280
2430	9.504	0.0275	4.0	471.4	743.5	1215.0	0.7210	1.0270
2440	9.787	0.0275	3.9	473.4	742.1	1215.6	0.7218	1.0260
2450	10.076	0.0276	3.8	475.5	740.6	1216.1	0.7226	1.0249
2460	10.372	0.0277	3.7	477.5	739.1	1216.6	0.7235	1.0239
2470	10.673	0.0277	3.6	479.6	737.5	1217.2	0.7243	1.0229
2480	10.980	0.0278	3.5	481.5	736.2	1217.8	0.7251	1.0219
2490	11.294	0.0279	3.4	483.7	734.6	1218.3	0.7259	1.0210
2500	11.614	0.0280	3.3	485.7	733.1	1218.9	0.7268	1.0200

TABLE A2.9   Thermodynamic Data for the Saturated Liquid and Vapor of Cesium (Ref. 54, Chap. 15)

Temperature, °R	P, atm	Specific volume, ft³/lb		Enthalpy, Btu/lb			Entropy, Btu/(lb·°R)	
		V_f	V_g	H_f	H_{fg}	H_g	S_f	S_g
960	0.00089	0.00932	5928.2	54.3	234.8	289.2	0.1987	0.4433
970	0.00105	0.00934	5056.2	54.7	234.5	289.3	0.1991	0.4410
980	0.00124	0.00935	4327.5	55.1	234.3	289.5	0.1996	0.4387
990	0.00146	0.00937	3716.3	55.6	234.0	289.7	0.2000	0.4364
1000	0.00171	0.00939	3201.7	56.0	233.8	289.9	0.2005	0.4343
1010	0.00200	0.00940	2766.9	56.5	233.5	290.1	0.2009	0.4322
1020	0.00233	0.00942	2398.5	57.0	233.3	290.3	0.2014	0.4302
1030	0.00270	0.00944	2085.1	57.4	233.1	290.6	0.2018	0.4282
1040	0.00313	0.00946	1817.9	57.9	232.9	290.8	0.2023	0.4262
1050	0.00361	0.00947	1589.3	58.4	232.6	291.1	0.2027	0.4244
1060	0.00416	0.00949	1393.1	58.9	232.4	291.3	0.2032	0.4225
1070	0.00477	0.00951	1224.3	59.4	232.2	291.6	0.2037	0.4207
1080	0.00546	0.00953	1078.7	59.8	232.0	291.9	0.2041	0.4190
1090	0.00624	0.00954	952.7	60.3	231.8	292.1	0.2046	0.4172
1100	0.00711	0.00956	843.4	60.8	231.5	292.4	0.2050	0.4156
1110	0.00808	0.00958	748.4	61.3	231.3	292.7	0.2055	0.4139
1120	0.00916	0.00960	665.6	61.8	231.1	293.0	0.2059	0.4123
1130	0.01036	0.00962	593.2	62.3	230.9	293.2	0.2064	0.4107
1140	0.01169	0.00964	529.8	62.8	230.6	293.5	0.2068	0.4092
1150	0.01317	0.00965	474.2	63.3	230.4	293.8	0.2072	0.4076
1160	0.01481	0.00967	425.2	63.8	230.2	294.1	0.2077	0.4061
1170	0.01661	0.00969	382.1	64.4	229.9	294.3	0.2081	0.4047
1180	0.01859	0.00971	343.9	64.9	229.7	294.6	0.2086	0.4032
1190	0.02077	0.00973	310.2	65.4	229.4	294.9	0.2090	0.4018
1200	0.02317	0.00975	280.3	65.9	229.2	295.1	0.2094	0.4005
1210	0.02579	0.00977	253.7	66.4	228.9	295.4	0.2099	0.3991
1220	0.02866	0.00979	230.6	66.9	228.7	295.7	0.2103	0.3978
1230	0.03179	0.00981	208.9	67.5	228.4	295.9	0.2107	0.3965
1240	0.03520	0.00982	190.0	68.0	228.1	296.2	0.2111	0.3952
1250	0.03892	0.00984	173.1	68.5	227.9	296.4	0.2116	0.3939
1260	0.04295	0.00986	158.0	69.0	227.6	296.7	0.2120	0.3927
1270	0.04733	0.00988	144.3	69.5	227.3	296.9	0.2124	0.3914
1280	0.05208	0.00990	132.1	70.1	227.1	297.2	0.2128	0.3902
1290	0.05721	0.00992	121.1	70.6	226.8	297.4	0.2132	0.3890
1300	0.06276	0.00994	111.1	71.2	226.5	297.7	0.2136	0.3879
1310	0.06875	0.00996	102.1	71.7	226.2	297.9	0.2141	0.3867
1320	0.07520	0.00998	94.0	72.2	225.9	298.2	0.2145	0.3856
1330	0.08214	0.01000	86.6	72.7	225.6	298.4	0.2148	0.3845
1340	0.08961	0.01002	79.9	73.3	225.3	298.6	0.2153	0.3834
1350	0.09762	0.01005	73.8	73.8	225.0	298.9	0.2157	0.3824
1360	0.10621	0.01007	68.3	74.4	224.7	299.1	0.2161	0.3813
1370	0.11542	0.01009	63.2	74.9	224.3	299.3	0.2165	0.3803
1380	0.12527	0.01011	58.6	75.5	224.0	299.5	0.2169	0.3792
1390	0.13579	0.01013	54.4	76.0	223.7	299.8	0.2173	0.3782
1400	0.14702	0.01015	50.5	76.6	223.3	300.0	0.2177	0.3772
1410	0.159	0.01017	46.06	77.1	223.1	300.2	0.2180	0.3763
1420	0.171	0.01019	43.82	77.6	222.7	300.4	0.2184	0.3753
1430	0.185	0.01021	40.84	78.2	222.4	300.6	0.2188	0.3744
1440	0.199	0.01024	38.11	78.8	222.0	300.8	0.2192	0.3734
1450	0.215	0.01026	35.59	79.3	221.7	301.1	0.2196	0.3725
1460	0.231	0.01028	33.28	79.9	221.3	301.3	0.2200	0.3716
1470	0.248	0.01030	31.14	80.5	220.9	301.5	0.2204	0.3707
1480	0.266	0.01032	29.17	81.0	220.6	301.7	0.2208	0.3698

TABLE A2.9   Thermodynamic Data for the Saturated Liquid and Vapor of Cesium (*continued*)

Temperature, °R	P, atm	Specific volume, ft³/lb		Enthalpy, Btu/lb			Entropy, Btu/(lb·°R)	
		V_f	V_g	H_f	H_{fg}	H_g	S_f	S_g
1490	0.286	0.01035	27.34	81.6	220.2	301.9	0.2212	0.3690
1500	0.306	0.01037	25.68	82.0	220.0	302.1	0.2215	0.3681
1510	0.328	0.01039	24.11	82.6	219.6	302.3	0.2218	0.3673
1520	0.351	0.01042	22.66	83.2	219.2	302.4	0.2222	0.3665
1530	0.375	0.01044	21.32	83.8	218.8	302.6	0.2226	0.3656
1540	0.400	0.01046	20.07	84.4	218.4	302.8	0.2230	0.3648
1550	0.427	0.01048	18.92	85.0	218.0	303.0	0.2234	0.3640
1560	0.455	0.01051	17.84	85.5	217.6	303.2	0.2237	0.3633
1570	0.484	0.01053	16.83	86.1	217.2	303.4	0.2241	0.3625
1580	0.515	0.01056	15.90	86.7	216.8	303.5	0.2245	0.3617
1590	0.548	0.01058	15.05	87.0	216.6	303.7	0.2247	0.3610
1600	0.582	0.01060	14.23	87.6	216.2	303.9	0.2251	0.3602
1610	0.618	0.01063	13.44	88.7	215.3	304.0	0.2257	0.3595
1620	0.655	0.01065	12.73	89.3	214.8	304.2	0.2261	0.3588
1630	0.694	0.01068	12.06	89.9	214.4	304.4	0.2265	0.3580
1640	0.735	0.01070	11.44	90.5	214.0	304.5	0.2268	0.3573
1650	0.778	0.01072	10.86	91.1	213.5	304.7	0.2272	0.3566
1660	0.823	0.01075	10.31	91.7	213.1	304.9	0.2276	0.3560
1670	0.870	0.01077	9.80	92.3	212.7	305.0	0.2279	0.3553
1680	0.919	0.01080	9.32	92.9	212.3	305.2	0.2283	0.3546
1690	0.970	0.01083	8.87	93.5	211.8	305.3	0.2286	0.3540
1700	1.023	0.01085	8.44	94.1	211.4	305.5	0.2290	0.3533
1710	1.078	0.01088	8.04	94.7	210.9	305.6	0.2293	0.3527
1720	1.136	0.01090	7.67	95.3	210.5	305.8	0.2297	0.3521
1730	1.195	0.01093	7.31	95.9	210.0	305.9	0.2300	0.3514
1740	1.258	0.01095	6.98	96.5	209.5	306.1	0.2304	0.3508
1750	1.323	0.01098	6.66	97.1	209.1	306.2	0.2307	0.3502
1760	1.390	0.01101	6.36	97.7	208.6	306.4	0.2311	0.3496
1770	1.460	0.01103	6.08	98.3	208.1	306.5	0.2314	0.3490
1780	1.532	0.01106	5.82	98.9	207.7	306.6	0.2317	0.3484
1790	1.607	0.01109	5.57	99.5	207.3	306.6	0.2320	0.3479
1800	1.685	0.01111	5.33	100.1	206.8	306.9	0.2324	0.3473
1810	1.766	0.01114	5.11	100.7	206.3	307.1	0.2327	0.3467
1820	1.850	0.01117	4.89	101.3	205.9	307.2	0.2331	0.3462
1830	1.936	0.01120	4.69	101.9	205.4	307.3	0.2334	0.3456
1840	2.026	0.01122	4.50	102.5	204.9	307.5	0.2337	0.3451
1850	2.119	0.01125	4.32	103.1	204.4	307.6	0.2340	0.3446
1860	2.21	0.01128	4.14	103.7	203.9	307.7	0.2344	0.3440
1870	2.31	0.01131	3.98	104.3	203.5	307.8	0.2347	0.3435
1880	2.41	0.01134	3.82	104.9	203.0	308.0	0.2350	0.3430
1890	2.52	0.01137	3.68	105.6	202.5	308.1	0.2354	0.3425
1900	2.63	0.01140	3.54	106.2	202.0	308.2	0.2357	0.3420
1910	2.74	0.01142	3.40	106.8	201.5	308.3	0.2360	0.3415
1920	2.85	0.01145	3.27	107.4	201.0	308.5	0.2363	0.3410
1930	2.97	0.01148	3.16	107.9	200.7	308.6	0.2366	0.3406
1940	3.10	0.01151	3.04	108.5	200.2	308.7	0.2369	0.3401
1950	3.22	0.01154	2.93	109.1	199.7	308.8	0.2372	0.3396
1960	3.36	0.01157	2.82	109.7	199.2	309.0	0.2375	0.3392
1970	3.49	0.01160	2.72	110.3	198.7	309.1	0.2378	0.3387
1980	3.63	0.01163	2.63	110.9	198.2	309.2	0.2381	0.3383
1990	3.77	0.01167	2.54	111.6	197.7	309.3	0.2384	0.3378
2000	3.92	0.01170	2.45	112.2	197.2	309.5	0.2387	0.3374
2010	4.07	0.01173	2.37	112.8	196.8	309.6	0.2390	0.3369

TABLE A2.9 Thermodynamic Data for the Saturated Liquid and Vapor of Cesium (*continued*)

Temperature, °R	P, atm	Specific volume, ft³/lb		Enthalpy, Btu/lb			Entropy, Btu/(lb·°R)	
		V_f	V_g	H_f	H_{fg}	H_g	S_f	S_g
2020	4.22	0.01176	2.29	113.4	196.3	309.7	0.2393	0.3365
2030	4.38	0.01179	2.21	114.0	195.8	309.8	0.2396	0.3361
2040	4.54	0.01182	2.14	114.6	195.3	309.9	0.2399	0.3357
2050	4.71	0.01185	2.07	115.2	194.8	310.1	0.2402	0.3353
2060	4.88	0.01189	2.00	115.8	194.3	310.2	0.2405	0.3349
2070	5.06	0.01192	1.94	116.4	193.8	310.3	0.2408	0.3345
2080	5.24	0.01195	1.88	116.8	193.5	310.4	0.2410	0.3341
2090	5.43	0.01198	1.82	117.5	193.1	310.6	0.2413	0.3337
2100	5.62	0.01202	1.76	118.1	192.6	310.7	0.2416	0.3333
2110	5.81	0.01205	1.71	118.7	192.1	310.8	0.2419	0.3329
2120	6.01	0.01208	1.65	119.3	191.6	310.9	0.2422	0.3326
2130	6.22	0.01212	1.60	119.9	191.1	311.1	0.2425	0.3322
2140	6.42	0.01215	1.56	120.5	190.7	311.2	0.2427	0.3318
2150	6.64	0.01219	1.51	121.1	190.2	311.3	0.2430	0.3315
2160	6.86	0.01222	1.47	121.7	189.7	311.4	0.2433	0.3311
2170	7.08	0.01226	1.42	122.3	189.2	311.5	0.2436	0.3308
2180	7.31	0.01229	1.38	122.9	188.7	311.7	0.2438	0.3304
2190	7.55	0.01233	1.34	123.5	188.3	311.8	0.2441	0.3301
2200	7.78	0.01236	1.30	124.1	187.8	311.9	0.2444	0.3298
2210	8.03	0.01240	1.27	124.7	187.3	312.1	0.2447	0.3294
2220	8.28	0.01243	1.23	125.1	187.1	312.2	0.2448	0.3291
2230	8.53	0.01247	1.20	125.6	186.6	312.3	0.2451	0.3288
2240	8.80	0.01251	1.17	126.2	186.2	312.5	0.2454	0.3285
2250	9.06	0.01254	1.14	126.9	185.7	312.6	0.2456	0.3282
2260	9.33	0.01258	1.11	127.5	185.2	312.7	0.2459	0.3279
2270	9.61	0.01262	1.08	128.1	184.7	312.9	0.2462	0.3276
2280	9.89	0.01266	1.05	128.7	184.3	313.0	0.2464	0.3273
2290	10.18	0.01270	1.02	129.3	183.7	313.1	0.2467	0.3270
2300	10.48	0.01273	1.00	129.9	183.3	313.2	0.2469	0.3267
2310	10.78	0.01277	0.974	130.5	182.8	313.4	0.2472	0.3264
2320	11.08	0.01281	0.950	131.2	182.3	313.5	0.2475	0.3261
2330	11.39	0.01285	0.926	131.8	181.8	313.6	0.2477	0.3258
2340	11.71	0.01289	0.903	132.4	181.3	313.8	0.2480	0.3255
2350	12.03	0.01293	0.881	133.0	180.8	313.9	0.2483	0.3252
2360	12.36	0.01297	0.859	133.6	180.4	314.0	0.2485	0.3250
2370	12.70	0.01301	0.839	134.2	179.9	314.2	0.2488	0.3247
2380	13.04	0.01305	0.820	134.5	179.8	314.3	0.2489	0.3245
2390	13.39	0.01309	0.799	135.4	178.9	314.4	0.2493	0.3242
2400	13.74	0.01313	0.782	135.7	178.8	314.6	0.2494	0.3239
2410	14.10	0.01317	0.764	136.3	178.4	314.7	0.2497	0.3237
2420	14.46	0.01322	0.746	136.9	177.9	314.9	0.2499	0.3234
2430	14.84	0.01326	0.729	137.6	177.4	315.0	0.2502	0.3232
2440	15.21	0.01330	0.713	138.2	176.9	315.1	0.2504	0.3229
2450	15.60	0.01334	0.696	138.9	176.3	315.2	0.2507	0.3227
2460	15.99	0.01339	0.681	139.5	175.8	315.4	0.2509	0.3224
2470	16.38	0.01343	0.666	140.2	175.3	315.5	0.2512	0.3222
2480	16.79	0.01348	0.651	140.8	174.7	315.6	0.2515	0.3219
2490	17.20	0.01352	0.636	141.5	174.2	315.7	0.2517	0.3217
2500	17.61	0.01356	0.623	142.1	173.7	315.8	0.2520	0.3215

TABLE A2.10 **Thermodynamic Data for Saturated Liquid and Vapor of Mercury** (Data from Ref. 23 of Chap. 15.)

Temperature, °R	Vapor pressure, psia	Liquid specific volume, ft³/lb	Vapor specific volume, ft³/lb	Frozen specific heat of vapor Btu/(lb·°R)	Specific heat of liquid Btu/(lb·°R)	Enthalpy of liquid $(H-H_o)_l$, Btu/lb	Enthalpy of vapor $(H-H_o)_v$, Btu/lb
900	0.766	1.228×10^{-3}	62.87	0.02476	0.0323	35.75	162.53
1000	3.231	1.241×10^{-3}	16.57	0.02476	0.0323	38.98	165.31
1100	10.41	1.254×10^{-3}	5.654	0.02476	0.0323	42.21	168.09
1200	27.45	1.267×10^{-3}	2.340	0.02476	0.0324	45.44	170.88
1300	62.00	1.281×10^{-3}	1.122	0.02476	0.0326	48.69	173.68
1400	124.1	1.294×10^{-3}	0.6038	0.02476	0.0329	51.96	176.50
1500	225.6	1.307×10^{-3}	0.3559	0.02476	0.0332	55.27	179.36
1600	379.4	1.320×10^{-3}	0.2258	0.02476	0.0336	58.61	182.25
1700	598.4	1.333×10^{-3}	0.1521	0.02476	0.0341	61.99	185.18
1800	894.8	1.347×10^{-3}	0.1077	0.02476	0.0347	65.43	188.17
1900	1279	1.360×10^{-3}	0.07950	0.02476	0.0353	68.93	191.22
2000	1761	1.373×10^{-3}	0.06078	0.02476	0.0361	72.50	194.35
2100	2348	1.386×10^{-3}	0.04788	0.02476	0.0369	76.15	197.54
2200	3044	1.399×10^{-3}	0.03869	0.02476	0.0378	79.88	200.83
2300	3852	1.413×10^{-3}	0.03196	0.02476	0.0387	83.71	204.20
2400	4773	1.426×10^{-3}	0.02692	0.02476	0.0398	87.64	207.68
2500	5805	1.439×10^{-3}	0.02305	0.02476	0.0409	91.67	211.26
2600	6947	1.452×10^{-3}	0.02003	0.02476	0.0421	95.82	214.96
2700	6195	1.465×10^{-3}	0.01764	0.02476	0.0434	100.1	218.79

TABLE A2.11 Saturation Properties of Dowtherm A (Metric Units)

| Temperature | | Vapor pressure, kg/cm² | | Enthalpy, kcal/kg | | | Specific heat Liquid, cal/(g·°C) | Density | | Specific gravity |
°F	°C	Absolute	Gauge	Liquid	Latent	Vapor		Liquid, g/cm³	Vapor, kg/m³	Liquid, t/25°C
53.6	12.0	0.0000		0.0	97.3	97.3	0.371	1.066	0.000	1.069
60	15.6	0.0000		1.3	96.9	98.2	0.374	1.063	0.000	1.066
70	21.1	0.0000		3.4	96.2	99.6	0.377	1.059	0.000	1.062
80	26.7	0.0000		5.5	95.6	101.1	0.381	1.054	0.000	1.057
90	32.2	0.0001		7.7	94.8	102.5	0.385	1.050	0.000	1.053
100	37.8	0.0001		9.8	94.2	104.0	0.388	1.046	0.000	1.049
110	43.3	0.0001		11.9	93.6	105.5	0.392	1.041	0.002	1.044
120	48.9	0.0002		14.2	92.9	107.1	0.396	1.037	0.002	1.040
130	54.4	0.0003		16.4	92.2	108.6	0.400	1.032	0.002	1.035
140	60.0	0.0005		18.6	91.6	110.2	0.403	1.028	0.003	1.031
150	65.6	0.0007		20.8	91.0	111.8	0.407	1.023	0.005	1.026
160	71.1	0.0010		23.1	90.4	113.5	0.411	1.019	0.006	1.022
170	76.7	0.0014		25.4	89.8	115.2	0.414	1.014	0.008	1.017
180	82.2	0.0019		27.7	89.1	116.8	0.418	1.010	0.011	1.013
190	87.8	0.0026		30.1	88.6	118.6	0.422	1.005	0.014	1.008
200	93.3	0.0036		32.4	87.9	120.3	0.426	1.001	0.019	1.003
210	98.9	0.0048		34.8	87.3	122.1	0.429	0.996	0.026	0.999
220	104.4	0.0064		37.2	86.8	123.9	0.433	0.991	0.034	0.994
230	110.0	0.0084		39.6	86.2	125.8	0.437	0.987	0.043	0.990
240	115.6	0.0112		42.1	85.6	127.6	0.440	0.982	0.054	0.985
250	121.1	0.0141		44.5	85.0	129.5	0.444	0.977	0.070	0.980
260	126.7	0.0183		46.9	84.4	131.4	0.448	0.972	0.088	0.975
270	132.2	0.0232		49.4	83.9	133.3	0.451	0.968	0.111	0.971
280	137.8	0.0288		52.0	83.3	135.3	0.455	0.963	0.138	0.966
290	143.3	0.0359		54.5	82.7	137.2	0.459	0.958	0.170	0.961
300	148.9	0.0443		57.1	82.2	139.2	0.463	0.953	0.207	0.956
310	154.4	0.0548		59.7	81.6	141.2	0.466	0.948	0.251	0.951
320	160.0	0.0675		62.3	81.0	143.3	0.470	0.943	0.306	0.946
330	165.5	0.0823		64.9	80.4	145.3	0.474	0.939	0.367	0.941
340	171.1	0.0991		67.5	79.9	147.4	0.477	0.934	0.439	0.936
350	176.7	0.1195		70.2	79.3	149.5	0.481	0.928	0.522	0.931
360	182.2	0.1427		72.9	78.7	151.6	0.485	0.923	0.617	0.926
370	187.8	0.170		75.6	78.2	153.8	0.488	0.918	0.727	0.921
380	193.3	0.201		78.3	77.7	155.9	0.492	0.913	0.852	0.916
390	198.9	0.236		81.1	77.1	158.1	0.496	0.908	0.993	0.911
400	204.4	0.278		83.8	76.4	160.3	0.500	0.903	1.153	0.906
410	210.0	0.324		86.6	75.9	162.5	0.503	0.898	1.334	0.901
420	215.6	0.337		89.4	75.3	164.7	0.507	0.893	1.536	0.895
430	221.1	0.437		92.2	74.7	166.9	0.511	0.887	1.762	0.890
440	226.7	0.504		95.1	74.2	169.2	0.514	0.882	2.015	0.885

Source: Reprinted from *The Dowtherm Heat Transfer Fluids,* courtesy Dow Chemical Co.

TABLE A2.11   Saturation Properties of Dowtherm A (Metric Units) (*continued*)

Temperature		Vapor pressure, kg/cm²		Enthalpy, kcal/kg			Specific heat	Density		Specific gravity
°F	°C	Absolute	Gauge	Liquid	Latent	Vapor	Liquid, cal/(g · °C)	Liquid, g/cm³	Vapor, kg/m³	Liquid, t/25°C
450	232.2	0.580		97.9	73.6	171.5	0.518	0.877	2.294	0.879
460	237.8	0.663		100.8	72.9	173.8	0.522	0.871	2.605	0.874
470	243.3	0.757		103.7	72.4	176.1	0.526	0.866	2.947	0.868
480	248.9	0.860		106.7	71.7	178.4	0.529	0.860	3.325	0.863
490	254.4	0.974		109.6	71.1	180.7	0.533	0.855	3.740	0.857
494.8	**257.1**	**1.033**	**0.000**	**111.1**	**70.8**	**181.8**	**0.535**	**0.852**	**3.957**	**0.854**
500	260.0	1.100	0.067	112.6	70.5	183.1	0.537	0.849	4.194	0.851
510	265.6	1.238	0.205	115.6	69.8	185.4	0.541	0.834	4.692	0.846
520	271.1	1.390	0.357	118.6	69.2	187.8	0.545	0.838	5.233	0.840
530	276.7	1.556	0.523	121.7	68.5	190.2	0.549	0.832	5.824	0.834
540	282.2	1.737	0.704	124.7	67.8	192.6	0.554	0.826	6.467	0.828
550	287.8	1.934	0.901	127.8	67.1	194.9	0.558	0.820	7.163	0.823
560	293.3	2.147	1.114	130.9	66.4	197.4	0.562	0.814	7.918	0.816
570	298.9	2.378	1.345	134.1	65.7	199.8	0.567	0.808	8.733	0.811
580	304.4	2.628	1.595	137.3	64.9	202.2	0.571	0.802	9.616	0.804
590	310.0	2.897	1.864	140.4	64.2	204.6	0.575	0.796	10.56	0.798
600	315.6	3.187	2.154	143.7	63.4	207.1	0.579	0.790	11.59	0.792
610	321.1	3.498	2.465	146.9	62.6	209.5	0.582	0.783	12.69	0.785
620	326.7	3.832	2.799	150.1	61.8	211.9	0.586	0.777	13.88	0.779
630	332.2	4.189	3.156	153.4	61.0	214.4	0.589	0.770	15.16	0.772
640	337.8	4.572	3.539	156.7	60.2	216.9	0.593	0.764	16.53	0.766
650	343.3	4.979	3.946	160.0	59.3	219.3	0.596	0.757	18.00	0.759
660	348.9	5.413	4.380	163.3	58.4	221.8	0.599	0.750	19.59	0.752
670	354.4	5.875	4.842	166.7	57.6	224.3	0.602	0.743	21.27	0.745
680	360.0	6.367	5.334	170.1	56.7	226.7	0.605	0.736	23.09	0.738
690	365.6	6.887	5.854	173.4	55.7	229.2	0.608	0.729	25.05	0.731
700	371.1	7.438	6.405	176.8	54.8	231.6	0.611	0.721	27.15	0.723
710	376.7	8.029	6.996	180.3	53.8	234.1	0.615	0.714	29.41	0.716
720	382.2	8.647	7.614	183.7	52.8	236.6	0.619	0.706	31.84	0.708
730	387.8	9.301	8.268	187.2	51.7	239.0	0.623	0.698	34.45	0.700
740	393.3	9.990	8.957	190.8	50.7	241.4	0.628	0.690	37.27	0.692
750	398.9	10.72	9.687	194.3	49.6	243.8	0.633	0.682	40.31	0.684
760	404.4	11.48	10.447	197.9	48.4	246.3	0.640	0.673	43.61	0.675
770	410.0	12.29	11.257	201.5	47.2	248.7	0.647	0.665	47.19	0.667
780	415.6	13.15	12.117	205.2	45.9	251.0	0.655	0.656	51.09	0.658
790	421.1	14.04	13.007	208.8	44.5	253.3	0.664	0.646	55.36	0.648
800	426.7	14.99	13.957	212.6	43.1	255.7	0.675	0.637	60.05	0.638

TABLE A2.12 Thermodynamic Data for NH₃ and Freon 12

SATURATED AMMONIA

(All values are Btu per lb of refrigerant)

Tempera-ture, °F	psia	psig	Liquid density, lb/ft³	Vapor sp vol, ft³/lb	Sat. liquid	Enthalpy Evap	Sat. vapor
−40	10.41	8.7*	43.08	24.86	0.0	597.6	597.6
−30	13.90	1.6*	42.65	18.97	10.7	590.7	601.4
−20	18.30	3.6	42.23	14.68	21.4	583.6	605.0
−10	23.74	9.0	41.78	11.50	32.1	576.4	608.5
0	30.47	15.7	41.34	9.116	42.9	568.9	611.8
5	34.27	19.6	41.11	8.150	48.3	565.0	613.3
10	38.51	23.8	40.89	7.304	53.8	561.1	614.9
20	48.21	33.5	40.43	5.910	64.7	553.1	617.8
30	59.74	45.0	39.96	4.825	75.7	544.8	620.5
40	73.32	58.6	39.49	3.971	86.8	536.2	623.0
50	89.19	74.5	39.00	3.294	97.9	527.3	625.2
60	107.6	92.9	38.50	2.751	109.2	518.1	627.3
70	128.8	114.1	38.00	2.312	120.5	508.6	629.1
80	153.0	138.3	37.48	1.955	132.0	498.7	630.7
86	169.2	154.5	37.16	1.772	138.9	492.6	631.5
90	180.6	165.9	36.95	1.661	143.5	485.5	632.0
100	211.9	197.2	36.40	1.419	155.2	477.8	633.0
110	247.0	232.3	35.84	1.217	167.0	466.7	633.7
120	286.4	271.7	35.26	1.047	179.0	455.0	634.0

*Inches of mercury below one standard atmosphere (29.92 in.)

SUPERHEATED AMMONIA

(V—specific volume; H—enthalpy)

Total tempera-ture, °F	20 (−16.64) V	H	30 (−0.57) V	H	40 (11.66) V	H	140 (74.79) V	H	170 (86.29) V	H	200 (96.34) V	H
−10	13.74	610.0	—	—	—	—	—	—	—	—	—	—
0	14.09	615.5	9.250	611.9	—	—	—	—	—	—	—	—
20	14.78	626.4	9.731	623.5	7.203	620.4	—	—	—	—	—	—
40	15.45	637.0	10.20	634.6	7.568	632.4	—	—	—	—	—	—
60	16.12	647.5	10.65	645.5	7.922	643.4	—	—	—	—	—	—
80	16.78	658.0	11.10	656.2	8.268	654.4	2.168	613.8	—	—	—	—
100	17.43	668.5	11.55	666.9	8.609	665.3	2.285	647.8	1.837	641.9	1.570	635.6
150	19.05	694.7	12.65	693.5	9.444	692.2	2.569	679.9	2.081	675.9	1.740	671.8
200	20.66	721.2	13.73	720.3	10.27	715.4	2.830	709.9	2.303	706.9	1.935	703.9
250	—	—	14.31	747.5	11.08	746.4	3.080	719.2	2.514	736.8	2.118	734.5

Source: Reprinted from *Power Handbook*, issued by Power, a McGraw-Hill publication.

TABLE A2.12   Thermodynamic Data for NH₃ and Freon 12 (*continued*)

(All values are Btu per lb of refrigerant)   **SATURATED FREON-12**

Tempera-ture, °F	psia	psig	Liquid density, lb/ft³	Vapor sp vol ft³/lb	Sat. liquid	Enthalpy	
						Evap	Sat. vapor
−40	9.317	10.96*	0.01057	3.911	0.00	73.50	73.50
−30	12.02	5.45*	0.0107	3.088	2.03	72.67	74.70
−20	15.28	0.58	0.0108	2.474	4.07	71.80	75.37
−10	19.20	4.50	0.0109	2.003	6.14	70.91	77.05
0	23.87	9.17	0.0110	1.637	8.25	69.95	78.21
5	26.51	11.81	0.0111	1.485	9.32	69.47	78.79
10	29.35	14.65	0.0112	1.351	10.33	68.97	79.36
20	35.75	21.05	0.0113	1.121	12.55	67.54	80.49
30	43.16	28.46	0.0115	0.939	14.75	66.85	81.61
40	51.68	36.98	0.0116	0.792	17.00	65.71	82.71
50	61.39	46.69	0.0118	0.673	19.27	64.51	83.78
60	72.41	57.71	0.0119	0.575	21.57	63.25	84.82
70	84.82	70.12	0.0121	0.493	23.90	61.32	85.82
80	98.76	84.06	0.0123	0.425	26.28	60.52	86.80
86	109.9	93.2	0.0124	0.383	27.72	59.65	87.37
90	114.3	99.6	0.0125	0.368	28.70	59.04	87.74
100	131.6	116.9	0.0127	0.319	31.05	57.45	88.62
110	150.7	136.0	0.0129	0.277	33.65	55.78	89.43
120	171.8	157.1	0.0132	0.240	36.15	53.99	90.15

(*V*—specific volume; *H*—enthalpy)   **SUPERHEATED FREON-12**

Total tempera-ture, °F	Absolute pressure, psia (Saturation temperature, °F)											
	20 (−8.2)		40 (25.9)		60 (48.7)		120 (93.4)		140 (104.5)		160 (114.5)	
	V	H	V	H	V	H	V	H	V	H	V	H
20	2.060	81.14	—	—	—	—	—	—	—	—	—	—
30	2.107	82.55	1.019	81.76	—	—	—	—	—	—	—	—
40	2.155	83.97	1.044	83.20	—	—	—	—	—	—	—	—
50	2.203	85.40	1.070	84.65	0.690	83.83	—	—	—	—	—	—
60	2.250	86.85	1.095	86.11	0.708	85.33	—	—	—	—	—	—
80	2.343	89.78	1.144	89.09	0.743	88.35	—	—	—	—	—	—
100	2.437	92.75	1.194	92.09	0.778	91.41	0.357	89.13	—	—	—	—
120	2.510	95.78	1.242	95.15	0.812	94.51	0.377	92.38	0.114	91.60	0.264	90.68
140	2.671	98.85	1.291	98.25	0.846	97.65	0.375	95.65	0.332	94.56	0.282	94.12
160	2.716	101.97	1.340	101.47	0.880	100.84	0.417	98.96	0.350	98.34	0.298	97.57

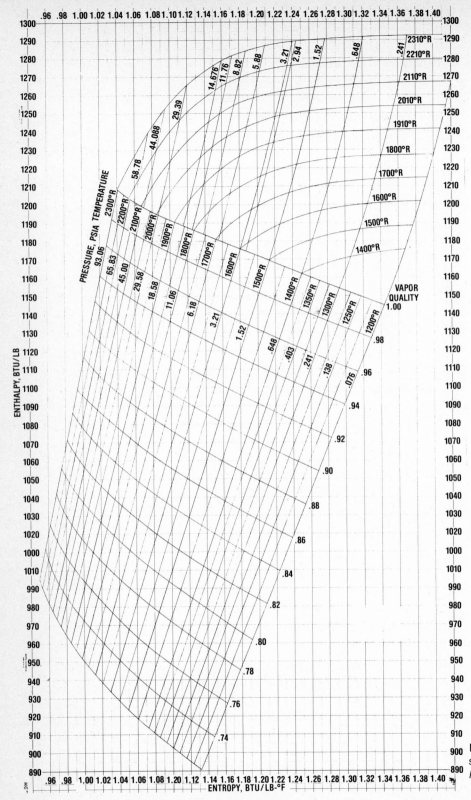

Figure A2.3 Mollier diagram for potassium. (Ref. 54, Chap. 15.) (*Courtesy Oak Ridge National Laboratory.*)

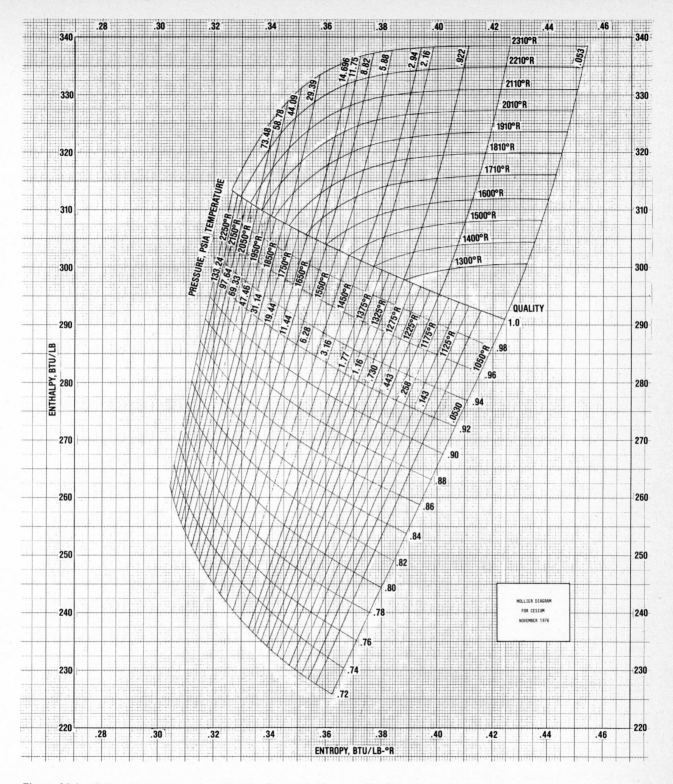

Figure A2.4 Mollier diagram for cesium. (Ref. 54, Chap. 15.) *(Courtesy Oak Ridge National Laboratory.)*

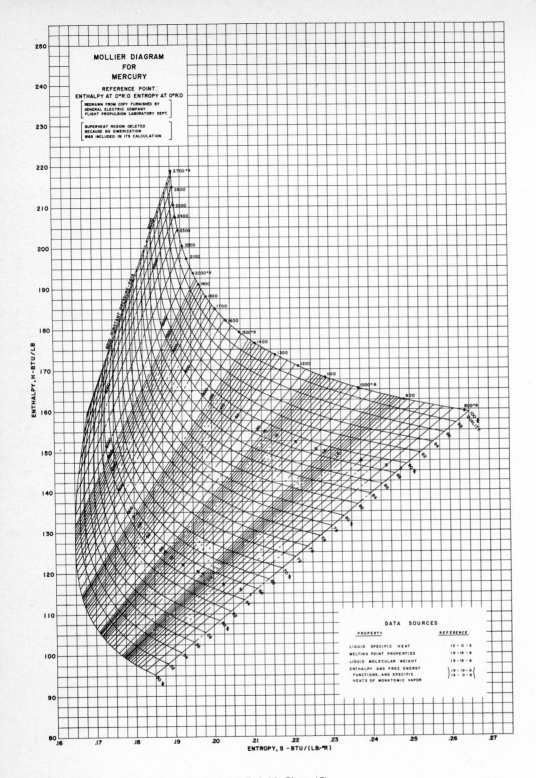

Figure A2.5 Mollier diagram for mercury (Ref. 23, Chap. 15).

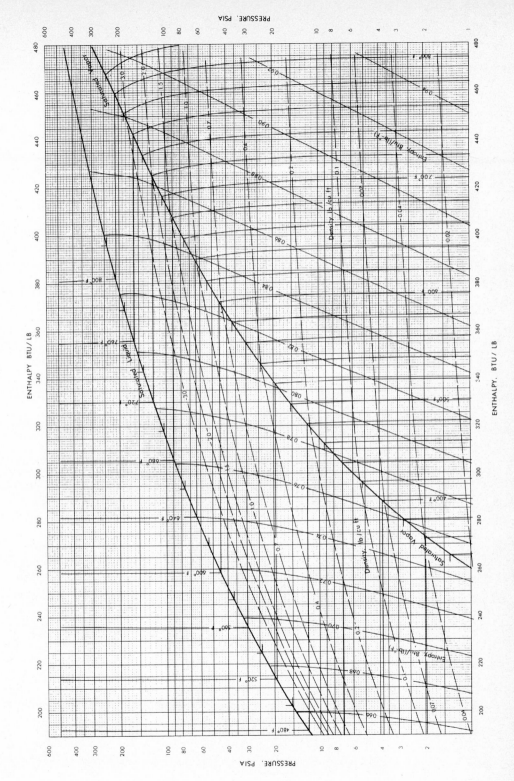

Figure A2.6 Thermodynamic chart for Dowtherm A. (*Courtesy Dow Chemical Co.*)

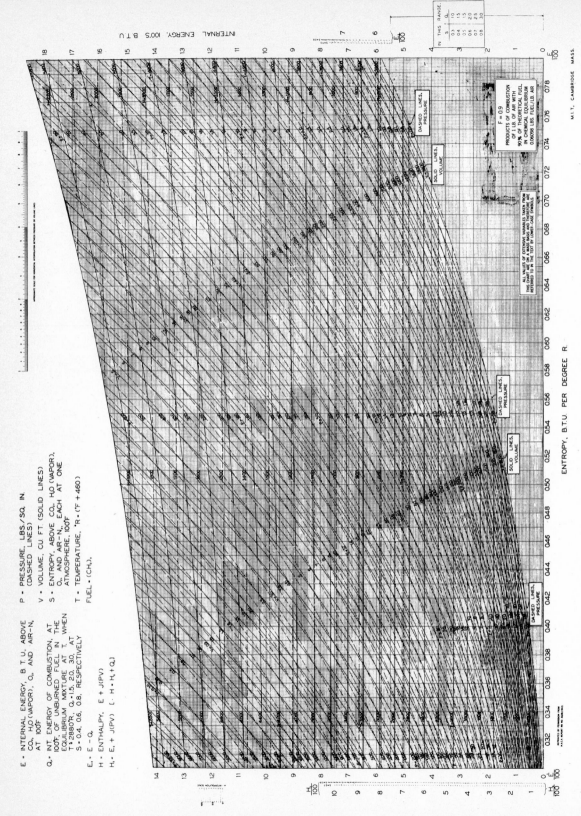

Figure A2.7 Thermodynamic chart for the products of combustion of a stoichiometric mixture of octane and air. (*Courtesy NASA.*)

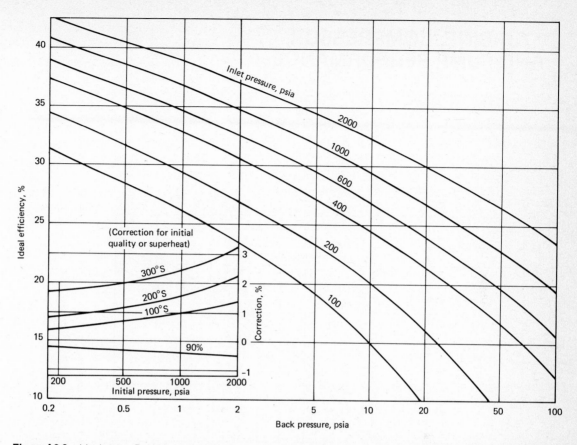

Figure A2.8   Ideal steam Rankine cycle efficiency as a function of turbine inlet pressure and back pressure. (The actual efficiency will be reduced by the turbine efficiency, auxiliary power requirements, stack losses, etc.) (*Courtesy Oak Ridge National Laboratory.*)

A3

TURBINE, COMPRESSOR, AND PUMP PERFORMANCE

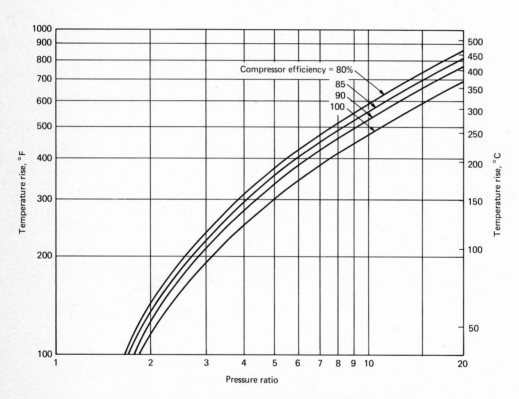

Figure A3.1 Compressor air temperature rise versus pressure ratio for a 38°C (60°F) compressor inlet temperature. (The temperature rise is directly proportional to the inlet temperature, and the work input per unit weight flow is the temperature rise times the average specific heat.)

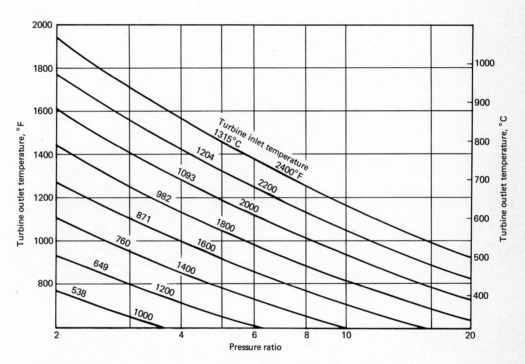

Figure A3.2   Gas-turbine outlet temperature versus pressure ratio for a 90 percent turbine efficiency. (The work output per unit weight flow is the temperature drop times the specific heat.)

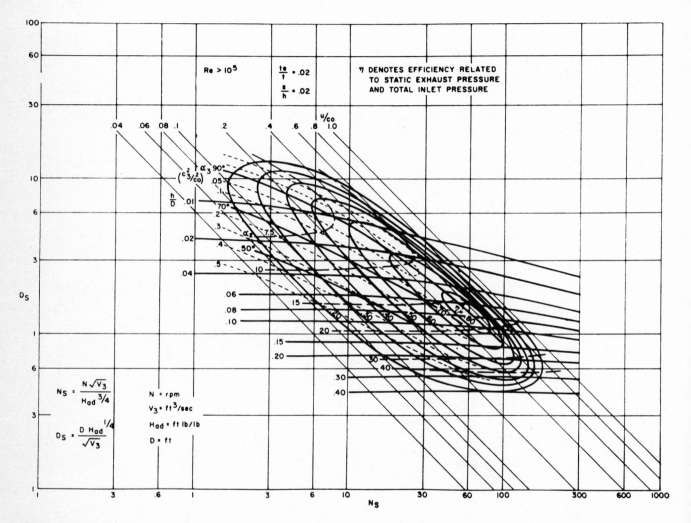

Figure A3.3 N_S-D_S-diagrams for single-stage, full-admission, axial-impulse turbines. (Ref. 10, Chap. 3)

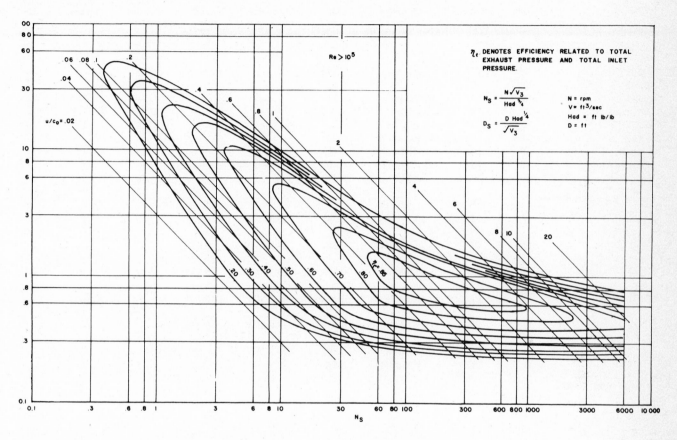

Figure A3.4 N_s-D_s-diagram for single-disk reaction turbines showing total efficiency. (Ref. 10, Chap. 3)

TABLE A3.1 Data for Typical Gas Turbines for Utility Service

Model	Use	Status	Normal hp, MW	sfc	Maximum hp, MW	sfc	Compr. stages	Turbine stages	Turb. inlet, °F
GENERAL ELECTRIC COMPANY, GAS TURBINE DIVISION, SCHENECTADY, NEW YORK 12345									
Mechanical drive gas turbines									
M1502 (B)	SP	PD	5,050 hp	0.574	—	—	15	1	—
M3142	SP	PD	14,600 hp	0.524	—	—	15	1	—
M5251	SP	PD	25,000 hp	0.524	—	—	—	—	—
M5262(A)	SP	PD	26,250 hp	0.532	—	—	15	1	—
M5332(B)	SP	PD	33,550 hp	0.484	—	—	16	1	—
M7652	SP	PD	65,400 hp	0.503	—	—	15	1	—
M3132R	SP	PD	13,750 hp	0.403	—	—	15	1	—
M5252R(A)	SP	PD	25,200 hp	0.402	—	—	15	1	—
M5322R(B)	SP	PD	32,000 hp	0.390	—	—	16	1	—
Gas turbine—generator sets									
G3142	G	PD	10.15 MW	0.742	10.55 MW	0.739	15	1	—
G5341	G	PD	24.05 MW	0.669	25.95 MW	0.657	—	—	—
G5261	G	PD	18.9 MW	0.728	20.35 MW	0.724	—	—	—
G7821	G	PD	60.0 MW	0.596	66.3 MW	0.590	—	—	—
G7981	G	PD	73.2 MW	0.572	79.9 MW	0.570	—	—	—
G9111	G	PD	85.2 MW	0.597	94.2 MW	0.591	—	—	—
G3132R	G	PD	9.35 MW	0.573	—	—	15	1	—
Package power plants									
PG5341	PG	PD	23.45 MW	0.678	25.3 MW	0.672	—	—	—
PG7821	PG	PD	58.5 MW	0.605	64.6 MW	0.599	—	—	—
PG7981	PG	PD	71.6 MW	0.579	77.0 MW	0.578	—	—	—
PG9111	PG	PD	83.3 MW	0.608	92.3 MW	0.601	—	—	—
PG7791R	PG	PD	55.9 MW	0.523	61.2 MW	0.508	—	—	—
Combined steam turbine and gas turbine generator with heat recovery steam generator									
STAG-100	PG	PD	98.0 MW	0.443	—	—	—	—	—
STAG-400	PG	PD	396.0 MW	0.439	—	—	—	—	—
STAG-600	PG	PD	595.0 MW	0.438	—	—	—	—	—

Power shaft		Press. ratio	No. comb.	Exhaust		Heat exch.	Fuel type	Weight dry lb	Dimensions, in		
Turb. stages	rpm			Flow lb/s	°F				D	W	H
1	10,290	7.0	1	45	967	—	—	70,000	325	100	130
1	6,500	6.0	6	113	990	—	—	120,000	420	130	144
2	4,860	8.0	10	202	975	—	—	165,000	458	130	150
1	4,670	6.9	12	215	975	—	—	253,000	600	130	150
	4,670	8.2	12	257	930	—	—	257,000	600	130	150
1	3,020	8.2	12	526	932	—	—	510,000	612	300	216
1	6,500	7.2	6	113	660	RE	—	120,000	420	130	144
1	4,670	7.0	12	215	638	RE	—	253,000	600	130	150
1	4,670	8.3	12	250	679	RE	—	257,000	600	130	150
		7.1			990						
1	6,500	7.2	6	113	1,040	—	—	242,000	765	130	150
		10.0			904						
2	5,105	10.2	10	261	954	—	—	398,000	930	130	150
		8.0			955						
2	5,100	8.1	10	213	1,010	—	—	318,000	816	126	150
		9.5			947						
3	3,600	9.7	10	529	1,016	—	—	587,000	838	348	158
		11.3			974						
3	3,600	11.5	10	590	1,036	—	—	587,000	838	348	158
		9.4			945						
3	3,000	9.6	14	760	1,015	—	—	650,000	996	180	228
1	6,500	7.2	6	113	655	RE	—	242,000	765	130	150
		10.0			908						
2	5,105	10.2	10	259	958	—	—	570,000	1,385	228	424
		9.5			954						
3	3,600	9.7	10	527	1,023	—	—	1,095,000	1,394	736	384
		11.3			979						
3	3,600	11.5	10	587	1,041	—	—	1,070,000	1,590	850	370
		9.4			950						
3	3,000	9.6	14	756	1,020	—	—	1,425,000	1,680	576	405
		9.5			744						
3	3,600	9.7	10	527	769	RE	—	1,560,000	1,390	1,180	384
3	3,600	10.0	10	563[11]	1,000	B	—	—	2,250	1,428	840
3	3,600	10.0	10	563[11]	1,000	B	—	—	—	—	—
3	3,600	10.0	10	563[11]	1,000	B	—	—	—	—	—
3	3,600	10.0	10	563[11]	1,000	B	—	—	—	—	—

TABLE A3.1   Data for Typical Gas Turbines for Utility Service (*continued*)

Model	Use	Status[d]	Power rating Normal hp, MW	Power rating Normal sfc	Power rating Maximum hp, MW	Power rating Maximum sfc	Compressor shaft Compr. stages	Compressor shaft Turbine stages	Turb. inlet, °F
UNITED TECHNOLOGIES CORPORATION, POWER SYSTEMS DIVISION, 1690 NEW BRITAIN AVENUE, FARMINGTON, CONNECTICUT 06032									
GG4C 1D,GF, LF&DF[b]	I/M	—	43,500 Ghp	—	47,500 Ghp 55,400 Ghp[c] 50,000 Ghp	—	16	3	—
FT4C 1D,GF, LF&DF	I/M	—	36,900 Shp	8,550[f]	40,400 Shp 46,000 Shp[c] 42,000 Shp	8,450 8,500 8,450	16	3	—
FT4C 3F,GF,LF&DF	I/M	—	39,000 Shp	7,960[f]	48,850 Shp[c] 44,250 Shp	7,750[f] 7,825	16	3	—
TP4 2 (C1D1)	I	—	54,000 MW	11,700[e]	59.0 MW 67,300 MW[c] 61,400 MW	11,550 11,600[e] 11,550	32	6	—
TP2-2 (C3F)	I	—	57.0 MW	10.900[e]	71.4 MW[c] 64.7 MW	10,600[e] 10,700	32	6	—

[a]Customer should consult turbine manufacturer for other fuels.
[b]GF = Gas fuel, LF = Liquid fuel, DF = Dual fuel, GG = Gas generator, FT = Gas turbine.
[c]MAX. CAPABILITY or MAX. INTERMITTENT for Marine rating, other rating below is PEAKING or MAX. CONTINUOUS for Marine rating.
[d]Customer should consult turbine manufacturer for information desired.
[e]Heat rate Btu/kWh.
[f]Heat rate Btu/Shp-h.

Model	Use	Status[d]	Power rating Normal hp, MW	Power rating Normal sfc	Power rating Maximum hp, MW	Power rating Maximum sfc	Compressor shaft Compr. stages	Compressor shaft Turbine stages	Turb. inlet, °F
WESTINGHOUSE ELECTRIC CORPORATION, GENERATION SYSTEMS, DIVISION, P.O. BOX 9175, LESTER BRANCH, PHILADELPHIA, PENNSYLVANIA 19113									
W251M	MC	PD	45.4 MW[a, b]	0.67[c]	—	—	18	—	—
W251G	G	PD	32.5 MW[a, b]	0.67[c]	—	—	18	—	—
W501G	G	PD	86.9 MW[a, b]	0.61[c]	—	—	19	—	—
PACE 320[e]	G	PD	282.3 MW[a, b]	0.45[c]	—	—	—	—	—
W1101G[d]	G	PD	97.9 MW[a, b]	0.60[c]	—	—	—	—	—

[a]ISO-BASE-OIL.
[b]For other fuels refer to factory.
[c]LB/KWH.
[d]50 Hz only built in Belgium.
[e]Combined cycle system.

Note: Ratings are at International Standards Organization (I.S.O.) Conditions: Sea level and 15C, 59F. SFC rating is based on LVH of 18,400 Btu/lb unless otherwise specified in footnotes.

Source: Sawyer's Gas Turbine Catalog 1977–1978, Gas Turbine Publications, Inc., Stamford, Conn.

Power shaft		Press. ratio	No. comb.	Exhaust		Heat exch.	Fuel type	Weight dry lb	Dimensions, in		
Turb. stages	rpm			Flow lb/s	°F				D	W	H
—	—	—	8	—	—	—	Gng,LF-1,2[a]	8,000[d]	154	45	53
2	3,600	—	8	—	—	—	Gng,LF-1,2[a]	17,600[d]	327	77	86
3	3,600	—	8	—	—	—	Gng,LF-1,2[a]	30,600[d]	327	124	106
4	3,600	—	16	—	—	—	Gng,LF-1,2[a]	—	—	—	—
6	3,600	—	16	—	—	—	Gng,LF-1,2[a]	—	—	—	—
	5,000	10.6	8	359	920	—	Gng GT2	220,000	458	122	126
	4,894	10.6	8	355	884	—	Gng GT2	220,000	1,368	360	300
	3,600	12.1	16	797	994	—	Gng GT2	270,000	1,884	456	480
							Gng GT2	—	3,040	2,996	
						BB	Gng GT2	—	3,384	2,916	636

HEAT EXCHANGERS

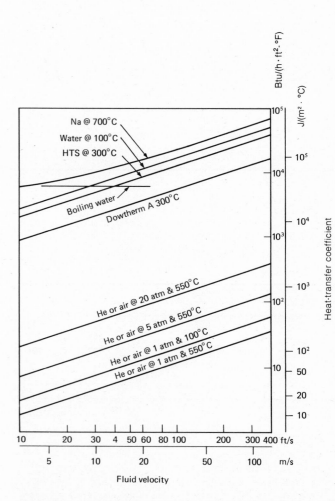

Figure A4.1 Heat-transfer coefficients for typical fluids flowing inside 1.0-in-ID (2.54-cm-ID) tubes.

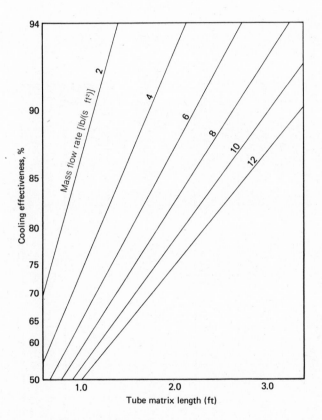

Figure A4.2 Cooling effectiveness as a function of tube matrix length in the airflow direction for air flowing across 1.0-in-OD finned tubes of a waste-heat boiler. Al fin OD = 1.737 in, fin pitch = 8.8/in, fin thickness = 0.012 in, tube staggered spacing = 1.959 in (transverse) by 2.063 in (axial), free flow area fraction = 0.439, air-side heat-transfer area = 91.2 ft²/ft³. (*Courtesy Oak Ridge National Laboratory.*)

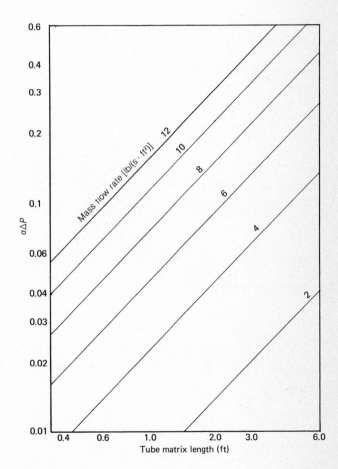

Figure A4.3 The pressure drop parameter $\sigma \Delta P$ as a function of tube matrix length in the direction of airflow for 1.0-in-OD finned tubes in the waste-heat boiler of Fig. A4.2. (*Courtesy Oak Ridge National Laboratory.*)

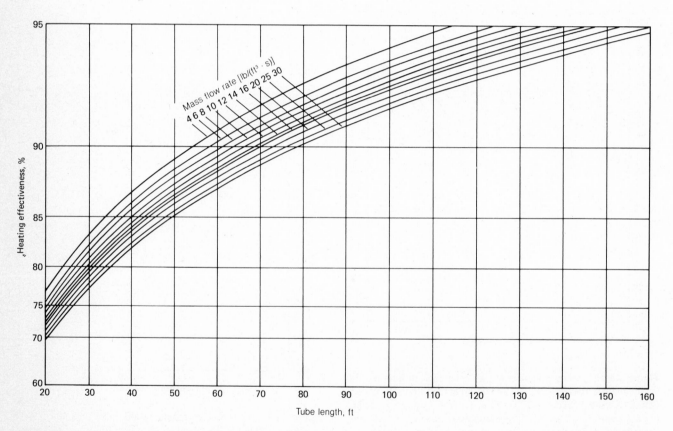

Figure A4.4 Heating effectiveness as a function of tube length in the recuperator for flow through the inside of 0.50-in-OD and 0.028-in-thick-wall tubes with a shell-side mass flow rate equal to 40 percent of that inside the tubes and a shell-side flow passage area 2.5 times the tube-side flow-passage area. *(Courtesy Oak Ridge National Laboratory.)*

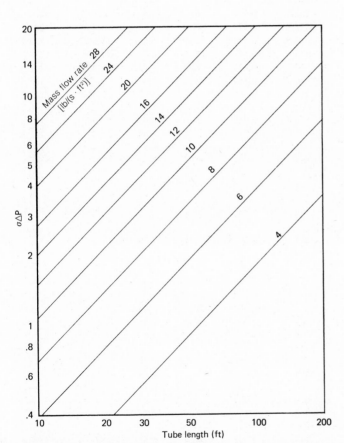

Figure A4.5 The pressure drop parameter $\sigma\Delta P$ as a function of tube length for the turbine airflow through the 0.50-in-OD and 0.028-in-thick-wall tubes of the unit of Fig. A4.4. (*Courtesy Oak Ridge National Laboratory.*)

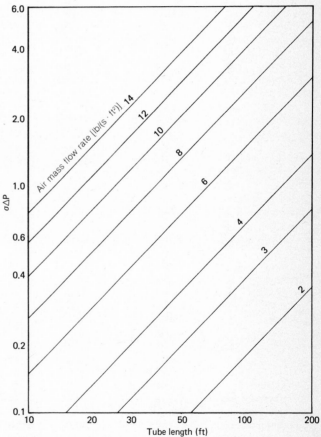

Figure A4.6 The pressure drop parameter $\sigma\Delta P$ as a function of tube length for the turbine exhaust airflow on the shell-side of the unit of Fig. A4.4 for 0.50-in-OD tubes with the shell-side air mass flow rate equal to 40 percent of the tube-side mass flow rate. (*Courtesy Oak Ridge National Laboratory.*)

HEAT SOURCES AND FUELS

TABLE A5.1   Properties of Representative Compounds in Hydrocarbon Fuels

	Mol formula	Mol wt	Melting point at 14.7 psia, °F	Boiling point at 14.7 psia, °F	Density of liquid			Specific gravity (air = 1.00)	Ratio gas vol per liquid vol
					Specific gravity at 60/60, °F	°API	lb/gal at 60°F		
Hydrogen	H_2	2.016	—	−423	—	—	—	.0695	—
Carbon	C	12.010	—	—	—	—	—	—	—
Methane	CH_4	16.042	−296.5	−258.5	0.3	340	2.5	0.554	443
Ethane	C_2H_6	30.068	−297.8	−128.2	0.374	247	3.11	1.038	294.5
Propane	C_3H_8	44.094	−305.9	− 43.8	0.508	147.0	4.23	1.522	272.7
n-Butane	C_4H_{10}	58.120	−216.9	+ 31.1	0.584	110.8	4.86	2.006	237.8
n-Pentane	C_5H_{12}	72.146	−201.5	97.0	0.631	92.7	5.25	2.491	207.0
n-Hexane	C_6H_{14}	86.172	−139.7	155.7	0.664	81.6	5.53	2.975	182.4
n-Heptane	C_7H_{16}	100.198	−131.1	209.1	0.688	74.2	5.73	3.459	162.5
n-Octane	C_8H_{18}	114.224	− 70.0	258.1	0.706	68.5	5.89	3.943	146.5
n-Nonane	C_9H_{20}	128.250	− 64.7	303.3	0.722	64.5	6.01	4.428	133.2
n-Decane	$C_{10}H_{22}$	142.276	− 21.5	345.2	0.734	61.5	6.11	4.912	122.1
n-Cetane	$C_{16}H_{34}$	226.4	65	536	0.774	51.5	6.43	—	—
Pentatriacontane	$C_{35}H_{72}$	492.3	176	628	0.781	49.5	6.51	—	—
Isopentane, or 2-methyl-butane	C_5H_{12}	72.1	− 256	82	0.613	99.0	5.17	—	—
Triptane, or 2,2,3-tri-methylbutane	C_7H_{16}	100.2	− 13	178	0.690	72.9	5.76	—	—
Isooctane, or 2,2,4-tri-methylpentane	C_8H_{18}	114.2	− 162	211	0.692	73.5	5.77	—	—
n-Pentene	C_5H_{10}	70.1	− 230	86	0.642	87.0	5.39	—	—
n-Octene	C_8H_{16}	112.2	− 152	251	0.751	—	—	3.87	—
Cyclopentane	C_5H_{10}	70.1	− 138	121	0.746	56.7	6.22	—	—
Cyclohexane	C_6H_{12}	84.1	44	177	0.778	51.6	6.50	—	—
Benzene	C_6H_6	78.1	42	176	0.88	29.0	7.34	—	—
Toluene	C_7H_8	92.1	− 139	231	0.87	31.0	7.23	—	—
Naphthalene	$C_{10}H_8$	128.2	177	424	0.975	—	—	4.42	—
Cumene, or isopropyl benzene	C_9H_{12}	120.2	− 141	306	0.862	32.6	7.19	—	—
Aniline	$C_6H_5NH_2$	93.1	+ 21	364	1.022	—	8.53	—	—
2,4-Xylidine	$C_6H_3NH_2(CH_3)_2$	121.2	—	420	0.974	13.8	8.13	—	—
α-methyl naphthalene	$C_{11}H_{10}$	142.2	− 7.6	470	1.025	—	—	—	—
Tetraethyl lead	$PB(C_2H_5)4$	323.4	− 213	360	1.653	—	13.8	—	—
Ethylene dibromide	$C_2H_4Br_2$	187.8	+ 50	269	2.181	—	18.2	—	—
Methanol	CH_3OH	32.0	− 144	149	0.792	46.4	6.62	—	—
Ethanol	C_2H_5OH	46.0	− 179	172	0.785	47.1	6.58	—	—
Ether	$(C_2H_5)_2O$	74.08	− 177	94	0.714	—	—	—	—
Diethyl-diglycol ether	—	—	—	340	0.91	—	—	—	—
Acetaldehyde	CH_3CHO	44.0	− 190	68	0.781	—	—	—	—

Source: Much of the data in this table came in the form of isolated fragments from dozens of widely scattered papers, most of them in the *SAE Transactions* or the *Petroleum Refiner.*

Lower heating value		Combustibility limits		ft³ air to burn 1 ft³ gas	Ignition tempera- ture, °F	Heat of va- porization 14.7 psia, Btu/lb at boiling point	Specific heat 14.7 psia			Octane rating	Cetane rating
Btu/lb	Btu/gal liquid at 60°F	Lower F/A	Upper F/A				c_p vapor, Btu/(lb) (°F)	c_v vapor, Btu/(lb) (°F)	c_p liquid, Btu/(lb) (°F)		
51,608	—	0.029	0.172	2.38	1076	192	—	—	—	—	—
14,600	—	—	—	—	—	—	—	—	—	—	—
21,597	—	0.026	0.088	9.55	1346	245	0.526	0.402	—	110	—
20,597	—	0.031	0.131	16.71	1050	211	0.415	0.347	—	104	—
20,015	—	0.031	0.131	23.87	995	183	0.390	0.346	—	100	—
19,795	—	0.030	0.155	31.03	961	166	0.396	0.363	0.55 at 32	92	—
19,745	101,000	0.008	0.172	38.19	933	153	0.402	0.376	0.55	61	—
19,432	—	0.035	—	45.35	909	146	0.406	0.384	0.536	95	—
19,408	—	0.033	0.165	52.06	893	138	0.415	0.397	0.522	0	57
19,329	—	0.038	—	59.67	880	131	0.420	0.404	0.519	− 17	—
19,292	—	0.037	—	66.85	871	125	0.429	0.415	0.518	− 45	—
19,205	—	0.033	—	73.99	866	120	0.436	0.424	0.517	—	—
19,091	—	—	—	—	390	—	—	—	—	—	100
19,032	—	—	—	—	—	—	—	—	—	—	—
19,597	100,000	0.034	—	—	—	146	—	—	—	90	—
19,100	110,000	—	—	—	—	125	—	—	—	140	—
19,159	110,000	—	—	—	1350	117	—	—	—	100	—
18,034	—	—	—	—	—	—	—	—	—	80	—
18,180	—	—	—	—	—	132	—	—	0.466	34	—
17,494	117,000	—	—	—	—	—	—	—	—	82	—
18,800	122,000	—	—	—	—	156	—	—	—	77	—
17,190	126,000	0.038	0.234	35.7	1364	169	—	—	0.40	97	− 10
17,460	126,000	0.054	0.230	42.9	1490	156	—	—	0.40	104	− 21
16,730	—	—	—	57.1	—	—	—	—	0.305	—	—
17,700	128,000	—	—	—	—	134	—	—	—	78	—
15,000	128,000	—	—	—	—	187	—	—	—	—	—
15,700	128,000	—	—	—	—	150	—	—	—	—	—
—	—	—	—	—	—	—	—	—	—	—	0
—	—	—	—	—	—	73	—	—	—	—	—
—	—	—	—	—	—	82	—	—	—	—	—
8,644	56,000	0.064	0.310	—	—	502	—	—	—	98	—
11,604	76,000	0.058	0.350	—	—	396	—	—	—	99	—
—	—	0.052	0.72	—	—	—	—	—	—	—	185
12,420	—	—	—	—	—	—	—	—	—	—	—
—	—	0.062	2.00	—	752	—	—	—	—	—	—

TABLE A5.2 Comparison of Fuels Used to Produce Electric Energy

The following table of statistics is for the entire fuel cycle—*from fuel extraction to power generation and including waste disposal* —of a 1000-MWe plant at 75% capacity generating 6.6 billion kWhe. (Courtesy Atomic Industrial Forum, Inc.)

	Coal	Residual fuel oil	Gas	Nuclear (LWR)
Annual fuel consumption[1]	2.3 million tons	10 million bbl[2]	64 billion ft^3	30 tons[3]
Daily fuel production requirements	7000 tons	enough crude to yield 35,000 bbl	x[4]	225 tons of ore
Portion of known domestic fuel reserves consumed annually	0.000006	0.0001[5]	0.0004	0.0002
kWhe costs (mills in 1980 dollars):[6]				
Capital	6.80	6.20	5.50	8.50
Operation and maintenance	0.53	0.45	0.45	0.73
Fuel[7]	4.40	6.70	9.10	2.10
Selected abatement costs such as land reclamation, sulfur dioxide removal	2.50	1.50	0.40	0.60
Total[7]	14.23	14.85	15.45	11.93
Fraction of power cost assigned to fuel[7]	31%	45%	59%	18%
Acres required by entire plant and supporting fuel cycle plus acreage consumed in fuel extraction and waste disposal over lifetime of plant	22,400	1,600	3,600	1,000[8]
(Acres/year consumed by mining, waste disposal)	740	Small	Small	12
(Required fuel storage space)	45-acre pile for 2-month reserve	20 acres for 18 million-ft^3 tanks	Small	Small
Fuel transportation requirements	27,000 railroad cars/year	100 typical U.S. tankers/year	0	5 truckloads/year
Sulfur dioxide releases (tons/year)				
Without abatement	120,000	38,600	20	3,000
With abatement	24,000	21,000	0	720
Oxides of nitrogen releases (tons/year)				
Without abatement	27,000	26,000	13,400	810

[1]These amounts vary considerably among individual plants because of fuel quantity and type difference. Coal consumption, for instance, can run up to more than 3 million tons and oil from 8 to 12 million barrels. 50,000 acres of redwoods, 25,000 acres of solar cells, and 7 million tons of garbage are other energy equivalents of the consumption estimates.

[2]One barrel equals 42 gallons.

[3]This amount is fabricated from approximately 130 tons of U_3O_x (uranium oxide). Of the 30 tons of fuel, 97% can be recycled and reused.

[4]x-Unknown, unavailable or unevaluated

[5]Estimated reserves of U.S., Africa, and Venezuela included in base

[6]These are average costs and vary with plant locations.

[7]The figures are conservative because coal and oil prices have been increasing sharply.

[8]Including required exclusion areas

TABLE A5.2 Comparison of Fuels Used to Produce Electric Energy (continued)

The following table of statistics is for the entire fuel cycle—from fuel extraction to power generation and including waste disposal—of a 1000-MWe plant at 75% capacity generating 6.6 billion kWhe. (Courtesy Atomic Industrial Forum, Inc.)

	Coal	Residual fuel oil	Gas	Nuclear (LWR)
Particulate releases (tons/year)				
Without abatement	270,000	26,000	518	8,000
With abatement	2,000	150	4	60
Thermal discharges from stack (billion kWht/year)	1.64	1.71	2.20	0
Annual wastes disposed (partial)	100,000 tons of fly ash, 60,000 tons of sulfur (based on 3.5% sulfur coal with 80% stack efficiency)	x	x	90 ft³ of solidified high-level radioactive waste concentrates, 175 ft³ of low-level
Cooling water flow (billion gallons/year)	263	263	263	424
Process water use (billion gallons/year)	1.46	1.75	1.42	0.095
Other water impacts (billion gallons/year)	16.80	7.90	0	Small
Power plant thermal discharge (billion kWht/year)	9	9	9	14
Power plant thermal efficiencies	38%	39%	38%	32%
Fuel transportation injuries to the public				
Deaths/year	0.55	x	x	0.009
Nonfatal injuries/year	1.20	x	x	0.080
Person-days lost/year	3500	x	x	60
On-job injuries to workers[9]				
Deaths/year	1.10	0.17	0.08	0.10
Nonfatal injuries/year	46.80	13.10	5.30	6.50
Person-days lost/year	9250	1725	780	950
Effects on health[10] of workers				
Person-days lost/year	600	x	x	480

[9]In coal and LWR, largely attributable to mining

[10]Mostly due to "black lung" among underground coal miners, lung cancers among uranium miners and cancers from exposures at the reactor and reprocessing plants. Estimates are that one case of "black lung" can be attributed to mining requirements of the coal-fired power plant per year and one malignancy in the fuel cycle of the LWR per 30 years, approximately, or the life of the plant.

MATERIALS

TABLE A6.1 Selected Properties of the Elements

	1	2	3	4	5	6	7	8	9	10	11	12	13
Period	Atomic No. Z	Symbol	Average thermal neutron absorption cross section barns*	Density at 20°C except as noted g/cm³	Modulus of elasticity (tension) × 10⁶ psi	Electrical resistivity μΩ/cm	Latent heat of fusion cal/g	Specific heat cal/g/°C, 20°C	Coefficient of linear thermal expansivity × 10⁻⁶ 20°C	Thermal conductivity cal/cm²/cm/°C/s, 20°C	Melting point, °C	Boiling point, °C	Atomic No. Z
1	1	H	0.33	0.00008988			15.0	3.45		4.061×10^{-4}	−259.4	−252.5	1
	2	He	0.007	0.0001785				1.25		3.32×10^{-4}	>−272.2	−268.6	2
2	3	Li	71.	0.534		8.55(0°)	104.2	0.79	56.	0.17	179.	1317.	3
	4	Be	0.010	1.848	40.-44.	4.0(20°)	260.	0.45	11.6	0.35	1278. ± 5	2970.	4
	5	B	755.	2.34	64.	1.8×10^{12}(0°)		0.309	8.3		2300.	2550.	5
	6	C	0.00037	1.9-2.3	0.7	1375.(0°)		0.165	0.6-4.3	0.057	3550.	4827.	6
	7	N	1.9	0.0012506			6.2	0.247		0.600×10^{-4}	−209.86	−195.8	7
	8	O	<0.0002	0.001429			3.3	0.218		0.590×10^{-4}	−218.4	−183.07	8
	9	F	0.009	0.001696			10.1	0.18			−219.62	−188.14	9
	10	Ne	<1.	0.00089990						1.1×10^{-4}	−248.67	−245.92	10
3	11	Na	0.53	0.971^{20}		4.2(0°)	27.5	0.295	71.	0.32	97.81 ± 03	883.	11
	12	Mg	0.069	1.738	5.77	4.45(20°)	89.	0.245	26.	0.38	651.	1107.	12
	13	Al	0.24	2.702	9.0	26548(20°)	94.6	0.215	23.9	0.53	659.7	2057.	13
	14	Si	0.16	2.33^{25}	16.	10^5(0°)	432.	0.162	2.8-7.3	0.20	1410.	2355.	14
	15	P	0.20	$2.07\text{-}1.957^{20}$		10^{17}(11°)	5.0	0.177	125.		44.1 ± 01	280.	15
	16	S	0.52	2.07		2×10^{23}(20°)	9.3	0.175	64.	6.31×10^{-4}	112.8—119.0	444.6	16
	17	Cl	34.	0.003214			21.6	0.116		$0.172 \times^- \times$	−100.98	−34.6	17
	18	A	0.66	0.001784			6.7	0.125		0.406×10^{-4}	−189.2	−185.7	18
4	19	K	2.1	0.862		6.15(0°)	14.5	0.177	83.	0.24	63.65	774.	19
	20	Ca	0.44	1.55	3.2-3.8	3.91(0°)	52.0	0.149	22.	0.3	845. ± 3	1487.	20
	21	Sc	24.	2.992		61.0(22°)	84.5	0.134			1539.	2727.	21
	22	Ti	5.8	4.507	16.8	80.(0°)	104.(est)	0.124	8.41		1675.	3260.	22
	23	V	5.00	$6.01^{18.7}$	18.-20.	24.8-26.0(20°)		0.120	8.3	0.074	1890. ± 10	± 3000.	23
	24	Cr	3.1	7.18-7.20	36.	12.9(0°)	96.	0.11	6.2	0.16	1890.	2482.	24
	25	Mn	13.2	7.24-7.44	23.	185.(23°)	64.	0.115	22.		1244. ± 3	2097.	25
	26	Fe	2.6	7.874	28.5	9.71(20°)	65.	0.11	11.7	0.18	1555.	3500.	26
	27	Co	38.	8.85	30.	6.24(20°)	85.4	0.099	13.8	0.165	1495.	2900.	27
	28	Ni	4.6	8.902^{25}	30	6.84(20°)	74.	0.105	13.3	0.22	1453.	2732.	28
	29	Cu	3.8	8.96	16.	1.673(20°)	50.6	0.092	16.5	0.94	1083.0 ± 0.1	2595.	29
	30	Zn	1.1	7.133^{25}	12.	5.916(20°)	24.09	0.0915	39.7	0.27	419.4	907.	30
	31	Ga	2.8	5.907		17.4(20°)	19.2	0.079	18.	0.07-0.09	29.78	2403.	31
	32	Ge	2.5	5.323^{25}		46×10^6(22°)		0.073	5.75	0.14	937.4	2830.	32

TABLE A6.1 Selected Properties of the Elements (*continued*)

	1	2	3	4	5	6	7	8	9	10	11	12	13
Period	Atomic No. Z	Symbol	Average thermal neutron absorption cross section barns*	Density at 20°C except as noted g/cm³	Modulus of elasticity (tension) × 10⁶ psi	Electrical resistivity μΩ/cm	Latent heat of fusion cal/g	Specific heat cal/g/°C, 20°C	Coefficient of linear thermal expansivity × 10⁻⁶ 20°C	Thermal conductivity cal/cm²/cm/°C/s, 20°C	Melting point, °C	Boiling point, °C	Atomic No. Z
	33	As	4.3	5.727^{14}		33.3(20°)	88.5	0.082	4.7		814(at 36 atm)	615.	33
	34	Se	12.	4.79	8.4	12.0(0°)	16.4	0.084	37.	$7\text{-}18 \times 10^{-4}$	217.	684.9 ± 1	34
	35	Br	6.7	3.12			16.2	0.070			− 7.2	58.78	35
	36	Kr	31.	0.003743						0.21×10^{-4}	− 156.6	− 152.30	36
5	37	Rb	0.73	1.532		12.5(20°)	6.5	0.080	90.		38.89	688.	37
	38	Sr	1.2	2.60		23.(20°)	25.	0.176			769	1384.	38
	39	Y	1.3	4.45	17.0	29.0(25°)	46.0	0.071		0.035	1495 ± 5	2927.	39
	40	Zr	0.18	6.53	13.7	40.0(20°)	60(est)	0.066	5.	0.211	1852 ± 2	3578.	40
	41	Nb	1.2	8.57	15.0^{25}	12.5(°)	69.0	0.065	7.1	0.125	2468 ± 10	4927.	41
	42	Mo	2.7	10.22	50.	5.2(0°)	70(est)	0.066	4.9	0.35	2610	5560.	42
	43	Tc	22.	11.50							2200 ± 50		43
	44	Ru	2.6	12.41	60.	7.6(0°)		0.057	9.1		2250.	3900.	44
	45	Rh	150.	12.41	42.	4.51(20°)		0.059	8.3	0.21	1960 ± 3	3727 ± 100	45
	46	Pd	8.	12.02	17.	10.8(20°)	34.2	0.058	11.8	0.17	1552.	2927.	46
	47	Ag	63.	10.50^{20}	11.	1.59(20°)	25.	0.056	19.7	0.975	960.8	2212.	47
	48	Cd	2450.	8.65	8.	6.83(0°)	13.2	0.055	29.8	0.22	320.9	765.	48
	49	In	190.	7.31		8.37(20°)	6.8	0.057	33.	0.057	156.61	2000 ± 10	49
	50	Sn	0.62	5.75 (1)	6.	11.0(0°)	14.5	0.054	23.	0.016	231.89	2270.	50
	51	Sb	5.7	6.684^{25}	11.3	39.0(0°)	38.3	0.049	8.5-10.8	0.045	630.5	1380.	51
	52	Te	4.7	6.24	6.	$4.36 \times 10^{5}(25°)$	32.0	0.047	16.8	0.014	449.5 ± 0.3	989.8 ± 3.8	52
	53	I	7.0	4.93		$1.3 \times 10^{15}(20°)$	14.2	0.052	93.	10.4×10^{-4}	113.5	184.35	53
	54	Xe	35.	0.005887						1.25×10^{-4}	111.9	− 107.1 ± 3	54
6	55	Cs	28.	1.873		20(20°)	3.8	0.048	97.		28.52	690.	55
	56	Ba	1.2	3.51				0.068			725.	1140.	56
	57	La	8.9	5.98-6.186		5.70(25°)	17.3	0.048	5.0	0.033	920.	3469.	57
	58	Ce	0.73	6.67-8.23		75.0(25°)	8.6	0.045	8.	0.026	795.	3468.	58
	59	Pr	11.3	6.782		68.(25°)	11.8	0.046	4.	0.028	935.	3127.	59
	60	Nd	46.	6.80-7.004		64.0(25°)	11.8	0.045	6.	0.031	1024.	3027.	60
	61	Pm									1035.	2730.	61
	62	Sm	5600	7.536	8.0	88.(25°)		0.043			1072.	1900.	62
	63	Eu	4300	5.259		90.0(25°)	16.9	0.040	26.		826.	1439.	63
	64	Gd	46,000	7.9895	8-14	140.5(25°)	16.4	0.071	4	0.021	1312.	≐ 3000.	64
	65	Tb	46	8.272			23.6	0.044	7.0		1356. ± 50	2800.	65
	66	Dy	950	8.537	10-14	57.0(25°)	26.4	0.041	9.0	0.024	1407.	2600.	66
	67	Ho	65	8.803	11.0	87.0(25°)		0.039			146	2600.	67
	68	Er	173	9.051		107.0(25°)	24.6	0.040	9.0	0.023	1497.	2900.	68
	69	Tm	127	9.332		79.0(25°)	26.0	0.038			1545.	1727.	69
	70	Yb	37	6.977		29.0(25°)	12.7	0.035	25.		824. ± 5	1427.	70

TABLE A6.1 Selected Properties of the Elements (*continued*)

1		2	3	4	5	6	7	8	9	10	11	12	13
Pe-riod	Atom-ic No. Z	Sym-bol	Average thermal neutron absorption cross section barns*	Density at 20°C except as noted g/cm³	Modulus of elasticity (tension) × 10⁶ psi	Electrical resistivity μΩ/cm	Latent heat of fusion cal/g	Specific heat cal/g/°C, 20°C	Coefficient of linear thermal expansivity × 10⁻⁶ 20°C	Thermal conductivity cal/cm²/cm/°C/s, 20°C	Melting point, °C	Boiling point, °C	Atom-ic No. Z
	71	Lu	112	9.7872		79.0(25°)	26.4	0.037			1652.	3327.	71
	72	Hf	105	13.29	20.	35.1(25°)		0.035	5.9		2150.	5400.	72
	73	Ta	21	16.6	27.	12.45(25°)	38.0	0.034	6.5	0.13	2996.	5425 ± 100	73
	74	W	19.2	19.3	50.	5.76(27°)	44.	0.032	4.6	0.397	3410 ± 20	5927.	74
	75	Re	86.	21.02	66.7	19.3(20°)		0.033	6.7	0.17	3180.	5627.	75
	76	Os	15.3	22.57	80.	9.5(20°)		0.031	4.6		3000. ± 10	5000.	76
	77	Ir	440.	22.42¹⁷	75.	5.3(20°)		0.031	6.8	0.14	2410.	4527 ± 100	77
	78	Pt	8.8	21.45	21.	10.6(20°)	27.	0.032	8.9	0.17	1769.	3827 ± 100	78
	79	Au	98.8	19.32	12.	2.35(20°)	16.1	0.031	14.2	0.71	1063.	2966.	79
	80	Hg	380.	13.546		98.4(50°)	2.7	0.033		0.0201	− 38.87	356.58	80
	81	Ti	3.4	11.85		18.0(0°)	5.04	0.031	28.	0.093	303.5	1457 ± 10	81
	82	Pb	0.17	11.35	2.6	20.648(20°)	6.3	0.031	29.3	0.083	327.5	1744.	82
	83	Bi	0.034	9.747	4.6	106.8(0°)	12.5	0.029	13.3	0.020	271.3 ± 0.1	1560 ± 5	83
	84	Po		9.32		42.					254.	0962.	84
	85	At											85
	86	Rn		0.000973							− 71.	− 61.8	86
7	87	Fr											87
	88	Ra	20.	± 5.							700.	1737.	88
	89	Ac	510.								1050. ± 50	± 3000.	89
	90	Th	7.56	11.66	7.-10.	13.0(0°)	19.82	0.034	12.5	0.090	1700.	± 4000.	90
	91	Pa	260.	15.37									91
	92	U	7.68	18.95	24.0	30.	11.3	0.028		0.064	1132.3 ± 0.8	3818.	92
	93	Np		18.0-20.45							640 ± 1		93
	94	Pu		19.84²⁵	14.0	145.4(107°)	3.3	0.033	48.4	0.020	630.5 ± 2	3235 ± 19	94
	95	Am		11.7		143.					> 800.	2600.	95
	96	Cm		± 7.									96
	97	Bk											97
	98	Cf											98
	99	Es											99
	100	Fm											100
	101	Md											101
	102	No											102
	103	Lw											103
	104												104
	105												105
	106												106

*Thermal neutron absorption cross section (2200 m/s) barns BNL-325, Second Edition and Supplement No. 1.

Source: Courtesy NUMEC

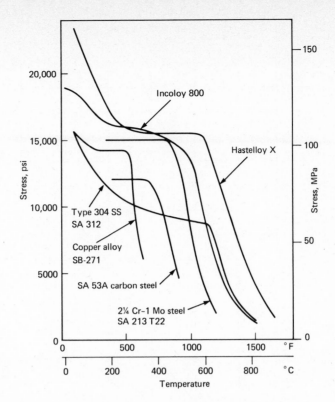

Figure A6.1 ASME pressure vessel code stresses as functions of temperature (values for a given alloy vary somewhat with heat treatment, amount of cold work, etc.).

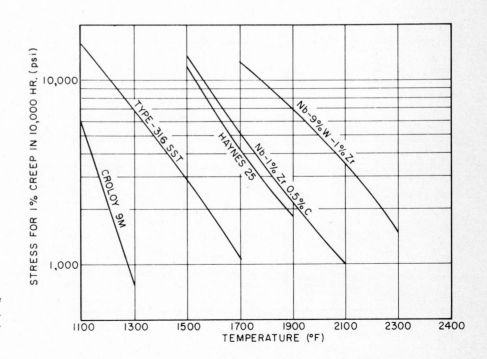

Figure A6.2 Effects of temperature on the creep strength of typical alloys. (*Courtesy Oak Ridge National Laboratory.*)

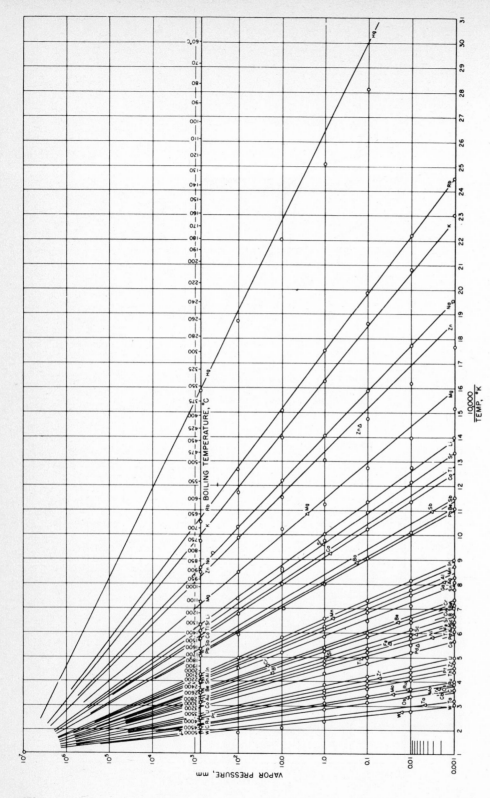

Figure A6.3 A vapor pressure chart for metals. *(Reprinted from "A Vapor Pressure Chart for Metals," R. L. Loftness, NAA-SR-132, July 10, 1952.)*

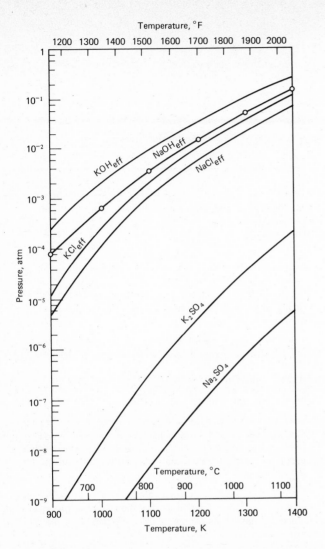

Figure A6.4 Vapor pressures of NaOH, KOH, NaCl, Na_2SO_4 and K_2SO_4. $NaOH_{eff}$, KOH_{eff}, etc., are effective equilibrium pressures which represent the sums of the pressures of the monomer and twice the dimer species (Ref. 38, Chap. 16).

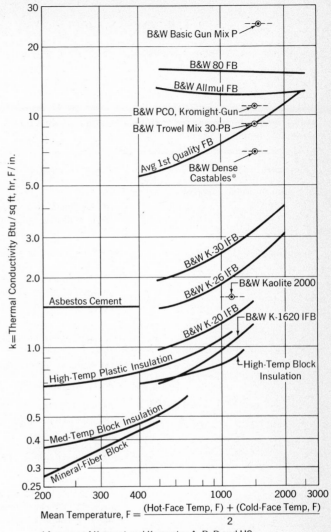

$$\text{Mean Temperature, } F = \frac{(\text{Hot-Face Temp, F}) + (\text{Cold-Face Temp, F})}{2}$$

*Average of Kaocast and Kaocretes A, B, D and HS.

Figure A6.5 Effects of temperature on the thermal conductivities of thermal insulating materials. (*Courtesy Babcock and Wilcox Co.*)

A7

I & C

TABLE A7.1 Recommended Performance Requirements for the Control Systems of Turbine-Generator Units
(From Ref. 3 of Chap. 7.)

Performance need	Response required
Frequency governing in normal operations	(a) Respond $+1.3\%$, -0.7% of unit nameplate megawatt rating in 2 s in prompt, stable fashion. (b) Maximum dead band of 0.06% frequency. (c) Overall steady-state regulation of 5%. (d) Linearity—per standards.
Normal daily load following	Able to go from 100 to 50% of nameplate megawatt rating at rates up to 2% per minute over a 2-h period, stay at 50% load for 4–6 h and return to 100% in 2 h at rates up to 2% per minute. Periods of zero response rate are permissible during loading and unloading while plant components are being added or removed. In addition, peaking units should be able to load or unload over 70% of nameplate load in 10 to 20 min.
Normal startup and shutdown	(a) Startup of base-load units following a brief shutdown in 2–4 h; 6–10 h following a more extended shutdown. Intermediate units at full load in 1 to 2 h from startup. Peaking units at full load in 30 min. (b) Shutdown rates same as startup are acceptable.
Tie-line backup	Rates of response as for daily load following, over spinning reserve range.
System emergency-off-nominal frequency	Ability of steam supply and auxiliaries to maintain operation at full load with off-normal frequencies for maximum permissible times as specified by turbine and generator manufacturers. A typical limit of 1% change from rated frequency for sustained operation has been cited. The permissible time of operation at greater frequency deviations decreases, until immediate trip is required for deviations of approximately 5% or 3 Hz. Rapid response under governor control from 100% to some lower value and return to 100% in 20 min. The larger the total possible excursion, the better, but unit controls should be coordinated in a fashion so as to keep the unit on the line. A minimum generation of 70% under these abnormal conditions would be a desirable objective.
System emergency-off-nominal voltage	Capability for continuous operation at rated load at any terminal voltage within ± 5 percent of rated. Capable of maintaining auxiliaries supporting load with auxiliary bus voltages in the range of 80 to 110% of normal voltage. Generator operation below 95% of rated voltage is possible with suitable reduction of load as defined by the generator manufacturer. Generator voltage may be restricted by volts per Hertz considerations.
Unit emergency-total load rejection	Response is desired in a manner which would permit reloading to 100% power in 20–30 min. after prompt resynchronizing, particularly of peaking and cycling plants.

RELIABILITY

TABLE A8.1 Failure Rate Data on Selected Electronic, Electrical, and Mechanical Equipment Used in the Nuclear Power Plant and Aerospace Fields*

Source	AHSB(S) R 117	High	Mean	Low	Others	Source	AHSB(S) R 117	High	Mean	Low	Others
		Failure rate, failures/10^6 h						Failure rate, failures/10^6 h			
		MI-60-54-(Rev 1)						MI-60-54-(Rev 1)			
Accumulators		19.3	7.2	0.4		Filters		0.8	0.3	0.045	
Actuators		13.7	5.1	0.35		Blockage	1				
Alternators	7	2.94	0.7	0.033		Leakage	1				
Baffles		1.3	1.0	0.12		Fuse	5	0.82	0.5	0.30	0.1
Batteries	1.0					Gaskets	0.5				
Rechargeable		14.29	1.4	0.5		D ring	0.2	0.03	0.02	0.01	
Bearings						Phenolic		0.07	0.05	0.01	
Ball						Rubber		0.03	0.02	0.011	
Heavy duty	2.0	3.53	1.8	0.072		Generators		2.41	0.9	0.40	
Light duty	1.0	1.72	0.875	0.035		d.c.	9	6.27	0.9	0.30	
Roller	5.0	1.0	0.5	0.02		Diesel, battery start					89
Sleeve	5.0	1.0	0.5	0.02		Heaters, electrical elements		0.04	0.02	0.01	
Bellows	5.0	4.38	2.237	0.040		Heat exchangers		18.6	15.0	2.21	
Blowers		3.57	2.4	0.89		Hose		3.22	2.0	0.05	
Buzzer		1.30	0.60	0.05		Heavily stressed	40				
Circuit breakers	2.0	0.40	0.1375	0.045		Lightly stressed	4				
Thermal		0.50	0.3	0.25	1.0	Instruments					
Magnetic					0.5	Electric		5.77	1.375	1.35	
Coils		0.088	0.050	0.033		Pressure					
Connectors, electrical						Gage	10	7.8	4.0	0.135	
General, each pin	0.2	0.47	0.2	0.03		Sensor		6.6	3.5	1.7	
Contactors	3	0.4/c	0.25/c	0.1/c		Temperature					
Covers						Bulb		3.30	1.0	0.05	
Dust		0.01	0.006	0.002		Sensor		6.4	3.3	1.5	
Protective		0.061	0.038	0.015		Meters (moving coil)	3				
Cylinders	0.1	0.81	0.007	0.005		Recorders	25				
Hydraulic		0.12	0.008	0.005		Lamps		35.0	8.625	3.45	
Pneumatic		0.013	0.004	0.002		Fluorescent	10				
Diaphragms		9.0	6.00	0.10		Incandescent		32.0	8.0	5.20	1.0
Metal	5					Indicator	5				
Rubber	8					Neon	2	18.8	10.25	4.50	0.2
Ducts	1	1.3	0.5125	0.21		Motors	10	7.5	0.625	0.15	
Fans						Blower		5.5	0.2	0.05	
Exhaust	90	9.0	0.225	0.21		Electric		.58	0.3	0.11	
Fasteners						Hydraulic		7.15	4.3	1.45	
Bolts	0.02					Servo		0.35	0.23	0.11	
Nuts	0.02					Stepper	5	0.71	0.37	0.22	
Screws	5										

*Data from B. J. Garrick et al., "Reliability Analysis of Nuclear Power Plant Protective Systems," USAEC Report HN–190, Holmes and Narver, May 1967.

TABLE A8.1   Failure Rate Data on Selected Electronic, Electrical, and Mechanical Equipment Used in the Nuclear Power Plant and Aerospace Fields* (continued)

Source	AHSB(S) R 117	High	Mean	Low	Others	Source	AHSB(S) R 117	High	Mean	Low	Others
		Failure rate, failures/10^6 h						Failure rate, failures/10^6 h			
		MI-60-54-(Rev 1)						MI-60-54-(Rev 1)			
Mechanism, power transmittal						Restrictors	5	0.983	0.59	0.197	
Belts	40	15.0	3.875	0.142		Seals					
Clutches		1.1	0.4	0.06		Rotating	7	1.12	0.7	0.25	
Friction	3					Sliding	3	0.92	0.3	0.11	
Magnetic	6	0.93	0.6	0.45		Solenoids		0.55	0.05	0.036	
Slip		0.94	0.3	0.07							
Coupling	5					Springs		0.221	0.1125	0.004	
Flexible		1.348	0.6875	0.027		Heavily stressed	1				
Rigid		0.049	0.025	0.001		Lightly stressed	0.2				
Gear		0.20	0.12	0.0118		Hair	1				
Helical	10	0.098	0.05	0.002		Calibration		0.42	0.22	0.009	
Spur	1	4.3	2.175	0.087		Creep	2				
Shafts		0.62	0.35	0.15		Breakage	0.2				
Heavily stressed	0.2					Switches		0.14/c	0.5/c	0.009/c	
Lightly stressed	0.02					General, each contact	0.2				
Rack and pinion	2					Micro	2	0.50/c	0.25/c	0.09/c	
Mounts, resilient	9	1.60	0.875	0.20		Push button	0.5	0.11/c	0.063/c	0.043/c	
Orifices						Rotary	2	0.660/c	0.175/c	0.118/c	
Fixed		2.11	0.15	0.01		Thermal		0.261/c	0.161/c	0.114/c	
Variable		3.71	0.55	0.045		heater	1				
Pumps		24.3	13.5	2.7		contacts	1				
Electric drive		27.4	13.5	2.9		Toggle		0.123/c	0.06/c	0.015/c	
Piping						General	1				
Pipes	0.2					Each pair contact	0.2				
Pipe joints	0.5					Synchros	8	0.61	0.35	0.09	
Union and junctions	0.4										
Pressure vessels						Tanks		0.27	0.15	0.083	
General	3					Pressure, small		0.324	0.018	0.10	
High standard	0.3					High pressure, small		0.144	0.08	0.044	
Regulators		5.54	2.14	0.70							
Flow and pressure		5.54	2.14	0.70		Tachometers	5	0.55	0.3	0.25	
Pneumatic		6.21	2.40	0.77							
Relays						Transducers		45.0	30.0	20.0	
General		0.48/c	0.25/c	0.10/c		Liquid level		3.73	2.6	1.47	
Each coil	0.3					Light		6.66	4.7	2.70	
Each contact pair	0.2					Photoelectric cells	15				
High speed	5					Pressure	15	52.2	35.0	23.2	
Heavy duty	5	0.81/c	0.5/c	0.30/c		Radioactivity					
Hermatically sealed	0.5	0.19/c	0.04/c	0.02/c		Beta ray		21.3	14.00	6.70	
Miniature		0.25/c	0.06/c	0.03/c		Ion chamber and leads	5				
High speed		1.13/c	0.7/c	0.42/c		Strain					
Power		4.10/c	0.3/c	0.15/c		Gage	25	20.0	12.0	7.0	
P.O. type						Temperature		6.4	3.3	1.5	
General	2					Thermistors		28.00	15.0	10.0	0.30
Fully tropicalized	1					Thermocouples	10				

TABLE A8.1   Failure Rate Data on Selected Electronic, Electrical, and Mechanical Equipment Used in the Nuclear Power Plant and Aerospace Fields* (*continued*)

Source	AHSB(S) R 117	MI-60-54-(Rev 1) High	Mean	Low	Others
		Failure rate, failures/10^6 h			
Transformers		2.0	0.2	0.07	
General, each winding	0.3				
Mains					
Encapsulated	5				
Oil filled	1				
Power		2.08	1.04	0.46	
Low voltage		0.60	0.3	0.13	
High voltage		1.88	0.94	0.407	
Pulse	1.5				
Low voltage		0.235	0.15	0.065	
High voltage					
Variable	1	0.31	0.1	0.035	
A. F.	0.3	0.04	0.02	0.01	
L. F.	1	0.31	0.1	0.035	
AIEE Class					
O					0.21–10.00
A					0.20–19.50
B					0.22–12.00
H					0.21–18.50
C					0.20– 1.00
Valves		8.0	5.1	2.00	
Ball	0.5	7.7	4.6	1.11	
Butterfly		5.33	3.4	1.33	
Check		8.10	5.0	2.02	
Control	30	19.8	8.5	1.68	
Relief		14.1	5.7	3.27	
Leakage	2				
Blockage	0.5				
Shutoff	15	10.2	6.5	1.98	
Solenoid	30	19.7	11.0	2.27	
Selector		19.7	16.0	3.70	
3-Way		7.41	4.6	1.87	
4-Way		7.22	4.6	1.81	
Vibrators	5	0.80	0.5	0.4	
Wiring					
Joints					
Soldered	0.02	0.005	0.004	0.0002	
Wrapped	0.01				
Terminals	0.5	0.27	0.05	0.041	
Wires	0.1	0.12	0.015	0.008	

A9

ENVIRONMENTAL EFFECTS

TABLE A9.1   U.S. EPA Standards for Stack Gas Emissions (Adopted May 25, 1979)

Heat release in furnace	Steam Boilers (Emission limit in lb/10^6 Btu in fuel.)					
	50–100 10^6 Btu/h		100–250 10^6 Btu/h		> 250 10^6 Btu/h	
	Coal	Oil	Coal	Oil	Coal	Oil
SO_2 emissions	0.35–2.0	0.2–0.8	0.35–1.2	0.2–0.8	0.35–1.2 S 90% reduction if > 0.6 S and 70% reduction if < 0.6 S	0.2–0.8 S 90% reduction if > 0.2 S
NO_x emissions	0.7	0.3	0.5[a] 0.6[b]	0.3	0.5* 0.6†	0.3
Particulates	0.03–0.10	0.03–0.10	0.03–0.10	0.03–0.10	0.03–0.10	0.03–0.10

	Gas Turbines (Emission limit in ppm of stack gas.)
	(Adopted Sept. 10, 1979)
NO_x emissions	75 ppm (plus special allowances for amount of N_2 in fuel and gas-turbine efficiency, which may total as much as 65 ppm)
	150 ppm for oil field and pipeline service
SO_2 emissions	150 ppm (or 0.8% max. S in fuel)
CO emissions	0
Opacity	0

*Sub-bituminous coal.
†Bituminous coal.

TABLE A9.2   EPA Standards on Aqueous Emissions from Steam Electric Power Plants

Pollutant	NSPS effluent limitations, mg/L	
	Maximum for any 1 day	Average of daily values for 30 days
Total suspended solids	100	30
Oil and grease	20	15
Copper	1	1
Iron	1	1

Source: Effluent Limitations Guidelines, Pretreatment Standards, and New Source Performance Standards Under Clean Water Act; Steam Electric Power Generating Source Category, Federal Register, vol. 45, no. 200, Tuesday, Oct. 14, 1980, "Proposed Rules," pp. 68353–68356.

TABLE A9.3 National Ambient Air Quality Standards

Pollutant	Average period	Allowable excesses	Maximum concentration, mg/m³
Sulfur dioxide			
Primary standards	12 months (arithmetic mean) 24 h	None	80
Secondary standards	3 h	Once a year	1300
Nitrogen dioxide			
Primary and secondary standards	12 months (arithmetic mean)	None	100
Suspended particulate matter			
Primary standards	12 months (geometric mean)	None	75
	24 h	Once a year	260
Secondary standards	12 months (geometric mean)	Guideline only	60
	24 h	Once a year	150
Carbon monoxide			
Primary and secondary standards	8 h	Once a year	10
	1 h	Once a year	40
Ozone			
Primary and secondary standards	1 h	Once a year	235
Hydrocarbons			
Primary and secondary standards	3 h	Once a year	160
Lead			
Primary and secondary standards	3 months (arithmetic mean)	None	1.5

TABLE A9.4 U.S. Nuclear Regulatory Commission Standards for Exposure to Radiation

Occupational		Public	
Maximum permissible doses for personnel in restricted areas, rems per calendar quarter		Whole body dose	6 mrem/year above natural background
Whole body; head and trunk; active blood-forming organs; lens of eyes; or gonads	1.25		
Hands and forearms; feet and ankles	18.75		
Skin of whole body	7.5		

In special cases a whole body dose of 3 rem per calendar quarter may be permitted provided that the total dose does not exceed 5 rem/year and that the total accumulated dose to the person involved does not exceed $5(N - 18)$ rem where N is the person's age in years.

TABLE A9.5 Maximum Permissible Concentrations of Typical Radionuclides in Air and Water as Specified by the Nuclear Regulatory Commission for Both Occupational Exposure (40 h/wk) and the Public (168 h/wk) (From Ref. 20 of Chap. 9)
The values are specified in mCi/mL, that is, Ci/m^3.

Isotope			Table I Occupational Air	Table I Occupational Water	Table II Public Air	Table II Public Water
Argon	A 37	Sub[b]	6×10^{-3}	—	1×10^{-4}	—
	A 41	Sub	2×10^{-6}	—	4×10^{-8}	—
Carbon	C 14	S[a]	4×10^{-6}	2×10^{-2}	1×10^{-7}	8×10^{-4}
	CO_2)	Sub	5×10^{-5}	—	1×10^{-6}	—
Cesium (55)	Cs 131	S	1×10^{-5}	7×10^{-2}	4×10^{-7}	2×10^{-3}
		I	3×10^{-6}	3×10^{-2}	1×10^{-7}	9×10^{-4}
	CS 134m	S	4×10^{-5}	2×10^{-1}	1×10^{-6}	6×10^{-3}
		I	6×10^{-6}	3×10^{-2}	2×10^{-7}	1×10^{-3}
	Cs 134	S	4×10^{-8}	3×10^{-4}	1×10^{-9}	9×10^{-8}
		I	1×10^{-8}	1×10^{-3}	4×10^{-10}	4×10^{-5}
	Cs 135	S	5×10^{-7}	3×10^{-3}	2×10^{-8}	1×10^{-4}
		I	9×10^{-8}	7×10^{-3}	3×10^{-9}	2×10^{-4}
	Cs 136	S	4×10^{-7}	2×10^{-3}	1×10^{-8}	9×10^{-5}
		I	2×10^{-7}	2×10^{-3}	6×10^{-9}	6×10^{-5}
	Cs 137	S	6×10^{-8}	4×10^{-4}	2×10^{-9}	2×10^{-5}
		I	1×10^{-8}	1×10^{-3}	5×10^{-10}	4×10^{-5}
Chlorine (17)	Cl 36	S	4×10^{-7}	2×10^{-3}	1×10^{-8}	8×10^{-5}
		I	2×10^{-8}	2×10^{-3}	8×10^{-10}	6×10^{-5}
	Cl 38	S	3×10^{-6}	1×10^{-2}	9×10^{-8}	4×10^{-4}
		I	2×10^{-6}	1×10^{-2}	7×10^{-8}	4×10^{-4}
Chromium (24)	Cr 51	S	1×10^{-5}	5×10^{-2}	4×10^{-7}	2×10^{-3}
		I	2×10^{-6}	5×10^{-2}	8×10^{-8}	2×10^{-3}
Cobalt (27)	Co 57	S	3×10^{-6}	2×10^{-2}	1×10^{-7}	5×10^{-4}
		I	2×10^{-7}	1×10^{-2}	6×10^{-9}	4×10^{-4}
	Co 58m	S	2×10^{-5}	8×10^{-2}	6×10^{-7}	3×10^{-3}
		I	9×10^{-6}	6×10^{-2}	3×10^{-7}	2×10^{-3}
	Co 58	S	8×10^{-7}	4×10^{-2}	3×10^{-8}	1×10^{-4}
		I	5×10^{-8}	3×10^{-3}	2×10^{-9}	9×10^{-5}
	Co 60	S	3×10^{-7}	1×10^{-3}	1×10^{-8}	5×10^{-5}
		I	9×10^{-9}	1×10^{-3}	3×10^{-10}	3×10^{-5}
Copper (29)	Cu 64	S	2×10^{-6}	1×10^{-2}	7×10^{-8}	3×10^{-4}
		I	1×10^{-6}	6×10^{-3}	4×10^{-8}	2×10^{-4}
Hydrogen (1)	H3	S	5×10^{-6}	1×10^{-1}	2×10^{-7}	3×10^{-3}
		I	5×10^{-6}	1×10^{-1}	2×10^{-7}	3×10^{-3}
		Sub	2×10^{-3}	—	4×10^{-5}	—
Indium (49)	In 113m	S	8×10^{-6}	4×10^{-2}	3×10^{-7}	1×10^{-3}
		I	7×10^{-6}	4×10^{-2}	2×10^{-7}	1×10^{-3}
	In 114m	S	1×10^{-7}	5×10^{-4}	4×10^{-9}	2×10^{-5}
		I	2×10^{-8}	5×10^{-4}	7×10^{-10}	2×10^{-5}
	In 115m	S	2×10^{-6}	1×10^{-2}	8×10^{-8}	4×10^{-4}
		I	2×10^{-6}	1×10^{-2}	6×10^{-8}	4×10^{-4}
	In 115	S	2×10^{-7}	3×10^{-3}	9×10^{-9}	9×10^{-5}
		I	3×10^{-8}	3×10^{-3}	1×10^{-9}	9×10^{-5}
Iodine (53)	I 125	S	5×10^{-9}	4×10^{-5}	8×10^{-11}	2×10^{-7}
		I	2×10^{-7}	6×10^{-3}	6×10^{-9}	2×10^{-4}
	I 126	S	8×10^{-9}	5×10^{-5}	9×10^{-11}	3×10^{-7}
		I	3×10^{-7}	3×10^{-3}	1×10^{-8}	9×10^{-5}
	I 129	S	2×10^{-9}	1×10^{-5}	2×10^{-11}	6×10^{-8}
		I	7×10^{-8}	6×10^{-3}	2×10^{-9}	2×10^{-4}
	I 131	S	9×10^{-9}	6×10^{-5}	1×10^{-10}	3×10^{-7}
		I	3×10^{-7}	2×10^{-3}	1×10^{-8}	6×10^{-5}
	I 132	S	2×10^{-7}	2×10^{-3}	3×10^{-9}	8×10^{-6}
		I	9×10^{-7}	5×10^{-3}	3×10^{-8}	2×10^{-4}
	I 133	S	3×10^{-8}	2×10^{-4}	4×10^{-10}	1×10^{-6}
		I	2×10^{-7}	1×10^{-3}	7×10^{-9}	4×10^{-5}
	I 134	S	5×10^{-7}	4×10^{-3}	6×10^{-9}	2×10^{-5}
		I	3×10^{-6}	2×10^{-2}	1×10^{-7}	6×10^{-4}
	I 135	S	1×10^{-7}	7×10^{-4}	1×10^{-9}	4×10^{-6}
		I	4×10^{-7}	2×10^{-3}	1×10^{-8}	7×10^{-5}
Krypton (36)	Kr 85m	Sub	6×10^{-6}	—	1×10^{-7}	
	Kr 85	Sub	1×10^{-5}	—	3×10^{-7}	
Plutonium (94)	Pu 238	S	2×10^{-12}	1×10^{-4}	7×10^{-14}	5×10^{-6}
		I	3×10^{-11}	8×10^{-4}	1×10^{-12}	3×10^{-5}
	Pu 239	S	2×10^{-12}	1×10^{-4}	6×10^{-14}	5×10^{-6}
		I	4×10^{-11}	8×10^{-4}	1×10^{-12}	3×10^{-5}
	Pu 240	S	2×10^{-12}	1×10^{-4}	6×10^{-14}	5×10^{-6}
		I	4×10^{-11}	8×10^{-4}	1×10^{-12}	3×10^{-5}
	Pu 241	S	9×10^{-11}	7×10^{-3}	3×10^{-12}	2×10^{-4}
		I	4×10^{-8}	4×10^{-2}	1×10^{-9}	1×10^{-3}
	Pu 242	S	2×10^{-12}	1×10^{-4}	6×10^{-14}	5×10^{-6}
		I	4×10^{-11}	9×10^{-4}	1×10^{-12}	3×10^{-5}
	Pu 243	S	2×10^{-6}	1×10^{-2}	6×10^{-8}	3×10^{-4}
		I	2×10^{-6}	1×10^{-2}	8×10^{-8}	3×10^{-4}
	Pu 244	S	2×10^{-12}	1×10^{-4}	6×10^{-14}	4×10^{-6}
		I	3×10^{-11}	8×10^{-4}	7×10^{-12}	3×10^{-5}
Polonium (84)	Po 210	S	5×10^{-10}	2×10^{-4}	2×10^{-11}	7×10^{-7}
		I	2×10^{-10}	8×10^{-4}	7×10^{-12}	3×10^{-5}
Potassium (19)	K 42	S	2×10^{-6}	9×10^{-3}	7×10^{-8}	3×10^{-4}
		I	1×10^{-7}	6×10^{-4}	4×10^{-9}	2×10^{-5}
Radium (88)	Ra 223	S	2×10^{-9}	2×10^{-5}	6×10^{-11}	7×10^{-7}
		I	2×10^{-10}	1×10^{-4}	8×10^{-12}	4×10^{-6}

[a]Soluble (S); Insoluble (I).

[b]"Sub" means that values given are for submersion in a semispherical infinite cloud of airborne material.

[c]For soluble mixtures of U-238, U-234 and U-235 in air chemical toxicity may be the limiting factor. If the percent by weight (enrichment) of U-235 is less than 5, the concentration value for a 40-hour workweek, Table I, is 0.2 milligram uranium per cubic meter of air average. For any enrichment, the product of the average concentration and time of exposure during a 40-hour workweek shall not exceed 8×10^{-3} SA mCi-h/mL, where SA is the specific activity of the uranium inhaled. The concentration value for Table II is 0.007 milligram uranium per cubic meter of air. The specific activity for natural uranium is 6.77×10^{-7} curies per gram U. The specific activity for other mixtures of U-238, U-235 and U-234, if not known, shall be:

$$SA = 3.6 \times 10^{-7} \text{ curies/gram U} \qquad \text{U-depleted}$$
$$SA = (0.4 + 0.38 E + 0.0034 E^2) 10^{-6} \qquad E \geq 0.72$$

where E is the percentage by weight of U-235, expressed as percent.

NOTE: In any case where there is a mixture in air or water of more than one radionuclide, the limiting values for purposes of this Appendix should be determined as follows:

1. If the identity and concentration of each radionuclide in the mixture are known, the limiting values should be derived as follows: Determine, for each radionuclide in the mixture, the ratio between the quantity present in the mixture and the limit otherwise established in Apendix B for the specific radionuclide when not in a mixture. The sum of such ratios for all the radionuclides in the mixture may not exceed "1" (i.e., "unity")

TABLE A9.5 Maximum Permissible Concentrations of Typical Radionuclides in Air and Water as Specified by the Nuclear Regulatory Commission for Both Occupational Exposure (40 h/wk) and the Public (168 h/wk) (From Ref. 20 of Chap. 9)
The values are specified in mCi/mL, that is, Ci/m^3. (*continued*)

Isotope			Table I Occupational Air	Table I Occupational Water	Table II Public Air	Table II Public Water
	Ra 224	S	5×10^{-9}	7×10^{-5}	2×10^{-10}	2×10^{-6}
		I	7×10^{-10}	2×10^{-4}	2×10^{-11}	5×10^{-6}
	Ra 226	S	3×10^{-11}	4×10^{-7}	3×10^{-12}	3×10^{-8}
		I	5×10^{-11}	9×10^{-4}	2×10^{-12}	3×10^{-5}
	Ra 228	S	7×10^{-11}	8×10^{-7}	2×10^{-12}	3×10^{-8}
		I	4×10^{-11}	7×10^{-9}	1×10^{-12}	3×10^{-5}
Radon (86)	Rn 222[3]	S	3×10^{-8}	—	1×10^{-9}	—
		I	—	—	—	—
Sodium (11)	Na 22	S	2×10^{-7}	1×10^{-3}	6×10^{-9}	4×10^{-5}
		I	9×10^{-9}	9×10^{-4}	3×10^{-10}	3×10^{-5}
	Na 24	S	1×10^{-6}	6×10^{-3}	4×10^{-8}	2×10^{-4}
		I	1×10^{-7}	8×10^{-4}	5×10^{-9}	3×10^{-5}
Strontium (38)	Sr 85m	S	4×10^{-5}	2×10^{-1}	1×10^{-6}	7×10^{-3}
		I	3×10^{-5}	2×10^{-1}	1×10^{-6}	7×10^{-3}
	Sr 85	S	2×10^{-7}	3×10^{-2}	8×10^{-9}	1×10^{-4}
		I	1×10^{-7}	5×10^{-3}	4×10^{-9}	2×10^{-4}
	Sr 89	S	3×10^{-8}	3×10^{-4}	3×10^{-10}	3×10^{-6}
		I	4×10^{-8}	8×10^{-4}	1×10^{-9}	3×10^{-5}
	Sr 90	S	1×10^{-9}	1×10^{-5}	3×10^{-11}	3×10^{-7}
		I	5×10^{-9}	1×10^{-3}	2×10^{-10}	4×10^{-5}
	Sr 91	S	4×10^{-7}	2×10^{-3}	2×10^{-8}	7×10^{-5}
		I	3×10^{-7}	1×10^{-3}	9×10^{-9}	5×10^{-5}
	Sr 92	S	4×10^{-7}	2×10^{-3}	2×10^{-8}	7×10^{-5}
		I	3×10^{-7}	2×10^{-3}	1×10^{-8}	6×10^{-5}
Sulfur (16)	S 35	S	3×10^{-7}	2×10^{-3}	9×10^{-9}	6×10^{-5}
		I	3×10^{-7}	8×10^{-3}	9×10^{-9}	3×10^{-4}
Thorium (90)	Th 227	S	3×10^{-10}	5×10^{-4}	1×10^{-11}	2×10^{-5}
		I	2×10^{-10}	5×10^{-4}	6×10^{-12}	2×10^{-5}
	Th 228	S	9×10^{-12}	2×10^{-4}	3×10^{-13}	7×10^{-6}
		I	6×10^{-12}	4×10^{-4}	2×10^{-13}	10^{-5}
	Th 230	S	2×10^{-12}	5×10^{-5}	8×10^{-14}	2×10^{-6}
		I	10^{-11}	9×10^{-4}	3×10^{-13}	3×10^{-5}
	Th 231	S	1×10^{-6}	7×10^{-3}	5×10^{-8}	2×10^{-4}
		I	1×10^{-6}	7×10^{-3}	4×10^{-8}	2×10^{-4}
	Th 232	S	3×10^{-11}	5×10^{-5}	10^{-12}	2×10^{-6}
		I	3×10^{-11}	10^{-3}	10^{-12}	4×10^{-5}
	Th natural	S	6×10^{-11}	6×10^{-5}	2×10^{-12}	2×10^{-6}
		I	6×10^{-11}	6×10^{-4}	2×10^{-12}	2×10^{-5}
	Th 234	S	6×10^{-8}	5×10^{-4}	2×10^{-9}	2×10^{-5}
		I	3×10^{-8}	5×10^{-4}	10^{-9}	2×10^{-5}

Isotope			Table I Occupational Air	Table I Occupational Water	Table II Public Air	Table II Public Water
Uranium (92)	U230	S	3×10^{-10}	1×10^{-4}	1×10^{-11}	5×10^{-6}
		I	1×10^{-10}	1×10^{-4}	4×10^{-12}	5×10^{-6}
	U232	S	1×10^{-10}	8×10^{-4}	3×10^{-12}	3×10^{-6}
		I	3×10^{-11}	8×10^{-4}	9×10^{-13}	3×10^{-5}
	U233	S	5×10^{-10}	9×10^{-4}	2×10^{-11}	3×10^{-5}
		I	1×10^{-10}	9×10^{-4}	4×10^{-12}	3×10^{-5}
	U 234[c]	S	6×10^{-10}	9×10^{-4}	2×10^{-11}	3×10^{-5}
		I	1×10^{-10}	9×10^{-4}	4×10^{-12}	3×10^{-5}
	U 235[c]	S	5×10^{-10}	8×10^{-4}	2×10^{-11}	3×10^{-5}
		I	1×10^{-10}	8×10^{-4}	4×10^{-12}	3×10^{-5}
	U 236	S	6×10^{-10}	1×10^{-3}	2×10^{-11}	3×10^{-5}
		I	1×10^{-10}	1×10^{-3}	4×10^{-12}	3×10^{-5}
	U 236[c]	S	7×10^{-11}	1×10^{-3}	3×10^{-12}	4×10^{-5}
		I	1×10^{-10}	1×10^{-3}	5×10^{-12}	4×10^{-5}
	U 240	S	2×10^{-7}	1×10^{-3}	8×10^{-9}	3×10^{-5}
		I	2×10^{-7}	1×10^{-3}	6×10^{-9}	3×10^{-5}
	U-natural[c]	S	1×10^{-10}	1×10^{-3}	5×10^{-12}	3×10^{-5}
		I	1×10^{-10}	1×10^{-3}	5×10^{-12}	3×10^{-5}
Vanadium (23)	V48	S	2×10^{-7}	9×10^{-4}	6×10^{-9}	3×10^{-5}
		I	6×10^{-8}	8×10^{-4}	2×10^{-9}	3×10^{-5}
Xenon (54)	Xe131m	Sub	2×10^{-5}	—	4×10^{-7}	—
	Xe133	Sub	1×10^{-5}	—	3×10^{-7}	—
	Xe133m	Sub	1×10^{-5}	—	3×10^{-7}	—
	Xe135	Sub	4×10^{-6}	—	1×10^{-7}	—
Zinc (30)	Zn 65	S	1×10^{-7}	3×10^{-3}	4×10^{-9}	1×10^{-4}
		I	6×10^{-8}	5×10^{-3}	2×10^{-9}	2×10^{-4}
	Zn69m	S	4×10^{-7}	2×10^{-3}	1×10^{-8}	7×10^{-5}
		I	3×10^{-7}	2×10^{-3}	1×10^{-8}	6×10^{-5}
	Zn 69	S	7×10^{-6}	5×10^{-2}	2×10^{-7}	2×10^{-3}
		I	9×10^{-6}	5×10^{-2}	3×10^{-7}	2×10^{-3}
Zirconium (40)	Zr 93	S	1×10^{-7}	2×10^{-2}	4×10^{-9}	8×10^{-4}
		I	3×10^{-7}	2×10^{-2}	1×10^{-8}	8×10^{-4}
	Zr 95	S	1×10^{-7}	2×10^{-3}	4×10^{-9}	6×10^{-5}
		I	3×10^{-8}	2×10^{-3}	1×10^{-9}	6×10^{-5}
	Zr 97	S	1×10^{-7}	5×10^{-4}	4×10^{-9}	2×10^{-5}
		I	9×10^{-8}	5×10^{-4}	3×10^{-9}	2×10^{-5}

EXAMPLE: If radionuclides A, B, and C are present in concentrations C_A, C_B, and C_C, and if the applicable MPC's, are MPC_A, MPC_B, and MPC_C respectively, then the concentrations shall be limited so that the following relationship exists:

$$(C_A/MPC_A) + (C_B/MPC_B) + (C_C/MPC_C) \leq 1$$

2. If either the identity or the concentration of any radionuclide in the mixture is not known, the limiting values for purposes of Appendix B shall be:
a. For purposes of Table I, Col. 1—6×10^{-13}
b. For purposes of Table I, Col. 2—4×10^{-7}
c. For purposes of Table II, Col. 1—2×10^{-14}
d. For purposes of Table II, Col. 2—3×10^{-8}

3. If any of the conditions specified below are met, the corresponding values specified below may be used in lieu of those specified in paragraph 2 above.
a. If the identity of each radionuclide in the mixture is known but the concentration of one or more of the radionuclides in the mixture is not known the concentration limit for the mixture is the limit specified in Appendix "B" for the radionuclide in the mixture having the lowest concentration limit; or
b. If the identity of each radionuclide in the mixture is not known, but it is known that certain radionuclides specified in Appendix "B" are not present in the mixture, the concentration limit for the mixture is the lowest concentration limit specified in Appendix "B" for any radionuclide which is not known to be absent from the mixture.

TABLE A9.6   Regional Runoff, 1975 Consumption, per Capita Runoff, and Consumption per Unit Runoff
(A map showing the boundaries of the drainage basins is shown in Fig. 19.2)

Drainage basin	Mean annual runoff, km³/year	Data for 1975		
		Consumption, km³/year	Per capita runoff, 10³m³/ person/year	Consumption/ mean annual runoff
New England	93.0	0.61	7.9	0.0066
Mid-Atlantic	120.0	2.2	3.0	0.018
South Atlantic Gulf	270.0	5.1	10.2	0.019
Great Lakes	100.0	1.5	4.5	0.015
Ohio River	170.0	1.7	8.0	0.01
Tennessee River	57.0	0.39	17.0	0.0068
Upper Mississippi	90.0	1.3	4.6	0.014
Lower Mississippi	100.0	7.6	17.0	0.069
Souris-Red-Rainy	8.6	0.17	12.0	0.016
Missouri	75.0	24.0	8.4	0.32
Arkansas	100.0	16.0	16.0	0.16
Texas Gulf	44.0	13.0	4.2	0.30
Rio Grande	6.9	6.0	3.5	0.87
Upper Colorado	18.0	3.4	40.0	0.19
Lower Colorado	4.4	10.0	1.7	2.3
Great Basin	10.0	5.5	7.0	0.55
Columbia–North Pacific	290.0	18.0	44.0	0.067
California	86.0	34.0	4.1	0.40
Alaska	800.0	0.0077	2000.0	9.6×10^{-6}
Hawaii	18.0	0.77	22.0	0.043
United States	2471.0	151.0	11.0	0.060
United States excluding Alaska and Hawaii	1653.0	150.0	7.8	0.091

Source: Reprinted from "Energy and Science" by John Harte, *Science*, Feb. 10, 1980.

TABLE A9.7   Freshwater Use in the United States, 1975, Expressed as Cubic Kilometers per Year

Use category	Withdrawal	Consumption
Municipal use including domestic and commercial	40.0	9.2
Industrial mining and manufacturing	52.0	5.8
Thermal electric power plant cooling	180.0	2.6
Irrigation, livestock, and rural use	200.0	115.0
Evaporation from constructed reservoirs	18.0	18.0
Total	490.0	151.0

Source: Reprinted from "Energy and Science" by John Harte, *Science*, Feb. 10, 1980.

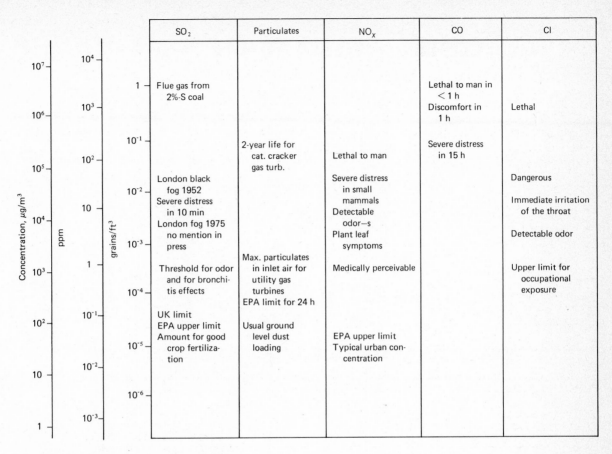

Figure A9.1 Scale of effects of air pollutants.

A10

COSTS

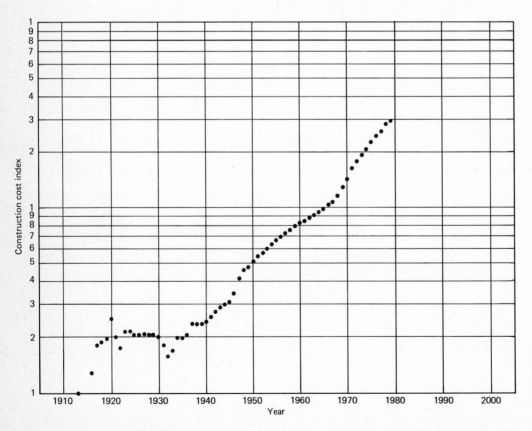

Figure A10.1 U.S. construction cost index as a function of year. (*Data plotted from Engineering News Record, Mar. 22, 1979, p. 73.*)

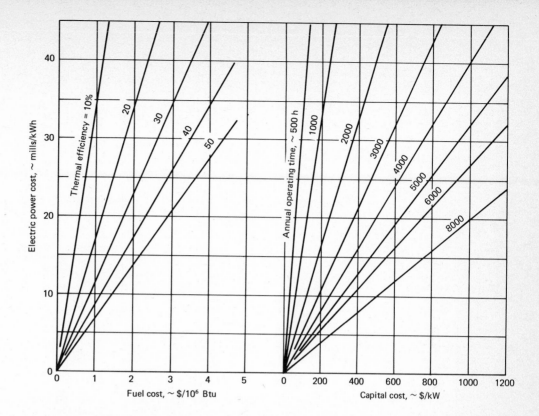

Figure A10.2 Electric power costs as functions of fuel cost and plant capital cost assuming a 16 percent annual capital charge.

Electric power cost, ~ mills/kWh

Thermal efficiency = 10%   20   30   40   50

Annual operating time, ~ 500 h   1000   2000   3000   4000   5000   6000   8000

Fuel cost, ~ $/10^6 Btu

Capital cost, ~ $/kW

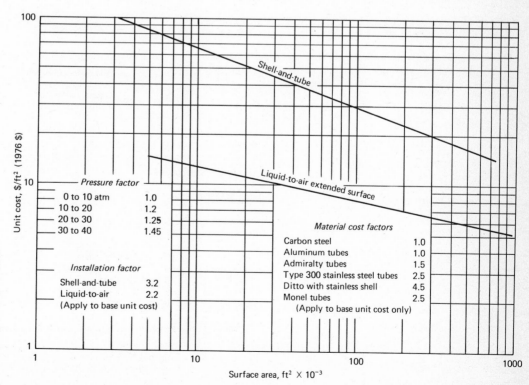

Figure A10.3 Unit costs of conventional shell- and tube- and liquid-to-air extended surface heat exchangers of carbon steel with aluminum fins for extended surfaces. (*Based on data in Chemical Engineering Progress, July 1976, p. 73.*)

Unit cost, $/ft² (1976 $)

Shell-and-tube

Liquid-to-air extended surface

Pressure factor

0 to 10 atm	1.0
10 to 20	1.2
20 to 30	1.25
30 to 40	1.45

Installation factor

| Shell-and-tube | 3.2 |
| Liquid-to-air | 2.2 |

(Apply to base unit cost)

Material cost factors

Carbon steel	1.0
Aluminum tubes	1.0
Admiralty tubes	1.5
Type 300 stainless steel tubes	2.5
Ditto with stainless shell	4.5
Monel tubes	2.5

(Apply to base unit cost only)

Surface area, ft² × 10⁻³

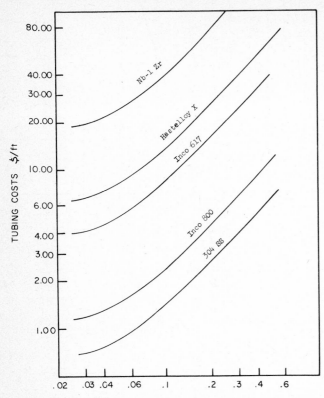

Figure A10.4 Effects of choice of material and tube-wall thickness on the unit cost of 0.50-in-OD alloy tubing in 10,000-ft lots in 1975 dollars. (*Courtesy Oak Ridge National Laboratory.*)

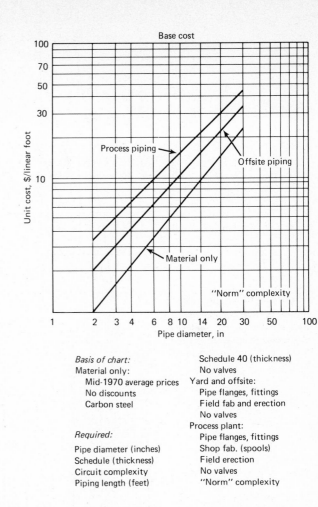

Basis of chart:
Material only:
    Mid-1970 average prices
    No discounts
    Carbon steel

Required:
Pipe diameter (inches)
Schedule (thickness)
Circuit complexity
Piping length (feet)

Schedule 40 (thickness)
No valves
Yard and offsite:
    Pipe flanges, fittings
    Field fab and erection
    No valves
Process plant:
    Pipe flanges, fittings
    Shop fab. (spools)
    Field erection
    No valves
    "Norm" complexity

Piping system cost (M & L) $ = [base cost $\times$ ($F_c + F_m + F_t + F_p$] $\times$ piping length $\times$ Escalation index

ADJUSTMENT FACTORS:

Circuit complexity	F_c
Tight	1.08
"Norm"	1.00
Loose	0.85

Material	F_M*	Wall thickness	F_t*	Pressure rating	F_p
Carbon steel	0.00	Schedule 40	0.00	150 lb	0.00
Chrome/moly	1.58	Extra strong	0.30	300 lb	0.59
Stainless (ave)	3.22	Double extra strong	1.19	600 lb	0.68
Monel	3.45				

Note: If these factors are used separately, add 1.00 to the above values.

Figure A10.5 Piping system unit costs (Ref. 17, Chap. 10). (*Courtesy Craftsman Book Company of America.*)

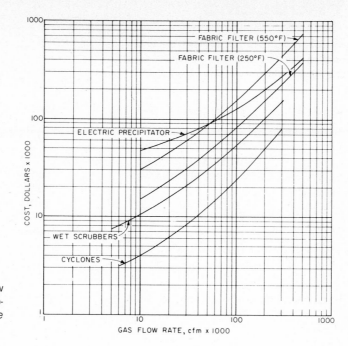

Figure A10.6 Effects of design gas flow rate on the costs of particle removal equipment in 1975 dollars. (*Courtesy Oak Ridge National Laboratory.*)

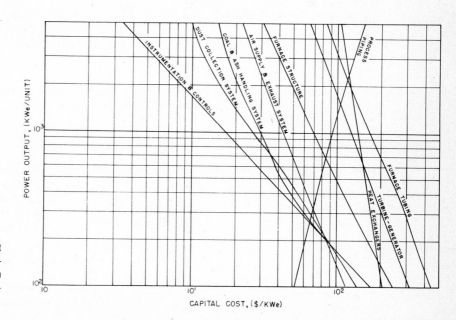

Figure A10.7 Relations between the unit costs of major steam power plant components and the plant design power output in 1975. (*Courtesy Oak Ridge National Laboratory.*)

ANSWERS
TO PROBLEMS

(Values should be expected to differ by ± 5 percent and in a few cases much more, depending on detailed assumptions, choice of physical property data, calculational method, and degree of approximation.)

Chapter 1. (1) 130 years; (2) 22%; (3) 78% (4) 24 years, 2000 years.

Chapter 2. (1a) 62.5%; (b) 43.2%; (c) 10%, 4%, 1.3%, 1%, 1%, 1%; (2a) 77%; (b) 73 MW, 173 MW, 46%; (c) 0.0192; (d) 5%, 8%, 4%, 54%, 2%; (3) 280 kW, 560 kW; (4) 6%.

Chapter 3. (1) 835 m/s, 100°C, 98%; (2a) 282 m/s, (b) 564 N, (c) 159 kW, (d) 88.3%.

Chapter 4. (1) 2.1 kW/cm²; (2) 16 W/cm²; (3) 114°C, 136 W/cm² − °C, 257°C; (4) 78°C, 0.136 W/cm² − °C, 800°C; (6) 7.75 m, 20.1 m.

Chapter 5. (1) 3.4%; (2) 0.6%, 1.6%; (3) 53%−bed, 47%−gas; (4a) 57.1%, 36.7%, 12.5%; (4b) 300/700, 690/1090, 1650/2050°F; (5) 37.5% savings in steam generator vs. 29% increase for air preheater; (6) 69 bar; (7) 2.5 points.

Chapter 6. (1) 2 mm; (2) 2.4 mm; (3) 222°C, 200 h; (4) 2.2 mm, $131/kWe, 10% (2 points), $9.46 per year versus $21 per year.

Chapter 7. (2) 9.5 s, 193.5 s; (3) 1.4 s; (4) 2750 kW/m², 1.5°C/s; (5) 33°C; (6) $18,000.

Chapter 8. (1) 2 weeks; (2) 7000 cycles; (3) 3800 h; (4) 1400 h.

Chapter 9. (1) 20 min, 3.4 h; (2) 1235 (automobile accidents), 2330 (cancer); (3) ~ 1000; (4) 2 × 10¹¹ m³

Chapter 10. (2) 16.8 mills/kWh (nuclear), 21 mills/kWh (coal); (3) 9.5 mills/kWh (nuclear), 16.6 mills/kWh (coal); (6) $735,000

Chapter 11. (1a) 60%, (b) 22.8%, (c), 75%; (2) 0.27 points drop in thermal efficiency; (3) 2.2, 5.0, 90, and 230%; (4) 2 points; (5) 63,000 kg/h

Chapter 12. (2a) 20,000 kg/h and 690,000 kg/h, (b) 1.74, 0.67, 0.80, and 2.4 m; (3a) 60 m³, (b) 3.6 m diameter and 11 m long, (c) 35,000 kg and $165,000, (d) 2 (i.e., 2 s lag).

Chapter 13. (1) 30%; (2) 39.4%; (3) 0.3 points increase in thermal efficiency.

Chapter 14. (1) Regenerator—14.3 m long, 0.23 m² cross-sectional area, 318 m² inside surface area; waste-heat boiler—0.50 m long, 0.142 m² cross-sectional area, 13 m² surface area; (2) 190 kg and $32,000 for regenerator, 77 kg and $1300 for boiler.

Chapter 15. (1) 105 MWe and 26.5% gas turbine, 134 MWe and 12.3% K vapor, 360 MWe and 50% steam, and 600 MWe and 51% overall; (2) 105 MWe and 26.5% gas turbine, 91.5 MWe and 8.4% K vapor, 374 MWe and 50% steam, 571.5 MWe and 48.5% overall; (3) 43.5%; (4) 19% Cs, 41.3% steam, 53.6% overall.

Chapter 16. (1) 1222 kJ/kg, 45.4%, 384 kJ/kg; (2) 5000 V, 50 kPa, 1.02 m²; (3) 454 V, 9.1%, 1000 V, 20%; (4) 125%, 61%; (5) 180 kJ/kg, 99 kJ/kg.

Chapter 17. (1) 23.3% electricity, 66.7% heat, 10% stack; (2) 27% electricity, 63% heat, 10% hot water; (3) 6%, $1.41/10⁶ Btu; (4) 2.43 m, $83 × 10⁶, 13,000 kW; (5) $1250 per residential unit, $2.50, $1.25, and $3.75/10⁶ Btu (1969 costs); (6) 19.5% for electricity

Chapter 18. (1) 400 cells, 0.267 m², 2 m high; (2) 385 L/s, 38.5 L/s, 385 cm², 38.5 cm²; (3) 500 h, ~ 1500 h.

Chapter 19. (1) 77.3%, 9.3%, 85.2%; (3) 10 mills/kWh.

Chapter 20. (2) 24.6%, $0.01/kWh, 24.6%; (3) ~ 1%, 3.2 m³/s; (4) crops—$0.05/m³ of H_2O, coal—$30/m³ of H_2O.

Chapter 21. (1) $3.3 × 10⁶; 15, 30, 60, 120, 240 MWt; $2.80, $1.40, $0.70, $0.35, and $0.18/10⁶ Btu; (2) 100, 400, 1600, 6400, and 25,600 kW; (3) 22.7%, 21%, 20.3%, 17.7%, 12.6%; 3400 kWe, 6300 kWe; 12,200 kWe; 21,200 kWe, 30,400 kWe; $0.057/kWh, $0.038/kWh, $0.026/kWh, $0.023/kWh, $0.026/kWh; (4) 250 kWh/day, 83%; (5) 35%, 73.5 kWh/day; 13.7 kWh/day; 29% (6) Value of energy from system of Prob. 5 is ~ 60% that for Prob. 4; capital cost per dollar savings in gas and electricity is ~ 56% as great for system of Prob. 4 as for system of Prob. 5; (7) 1120 mills, 1120 km²; (8) $168 × 10⁶ for windmills, $50 × 10⁶ for batteries, $0.237/kWe.

Chapter 22. (1) 970°C.

INDEX